POWER QUALITY PROBLEMS AND MITIGATION TECHNIQUES

POWER QUALITY PROBLEMS AND MITIGATION TECHNIQUES

Bhim Singh
Indian Institute of Technology Delhi, India

Ambrish Chandra
École de Technologie Supérieure, Canada

Kamal Al-Haddad
École de Technologie Supérieure, Canada

This edition first published 2015

Registered office
John Wiley & Sons Ltd, The Atrium, Southern Gate, Chichester, West Sussex, PO19 8SQ, United Kingdom

For details of our global editorial offices, for customer services and for information about how to apply for permission to reuse the copyright material in this book please see our website at www.wiley.com.

Library of Congress Cataloging-in-Publication Data

Singh, Bhim (Electrical engineer)
Power quality problems and mitigation techniques / Professor Bhim Singh, Professor Ambrish Chandra, Professor Kamal Al-Haddad.
pages cm
Includes bibliographical references and index.
ISBN 978-1-118-92205-7 (cloth)
1. Electric power systems–Quality control. 2. Electric power systems—Management. 3. Electric power system stability. I. Chandra, Ambrish. II. Al-Haddad, Kamal. III. Title.
TK1010.S56 2015
621.37'45—dc23

2014031901

A catalogue record for this book is available from the British Library.

ISBN: 9781118922057

Set in 9/11 pt TimesLTStd-Roman by Thomson Digital, Noida, India

1 2015

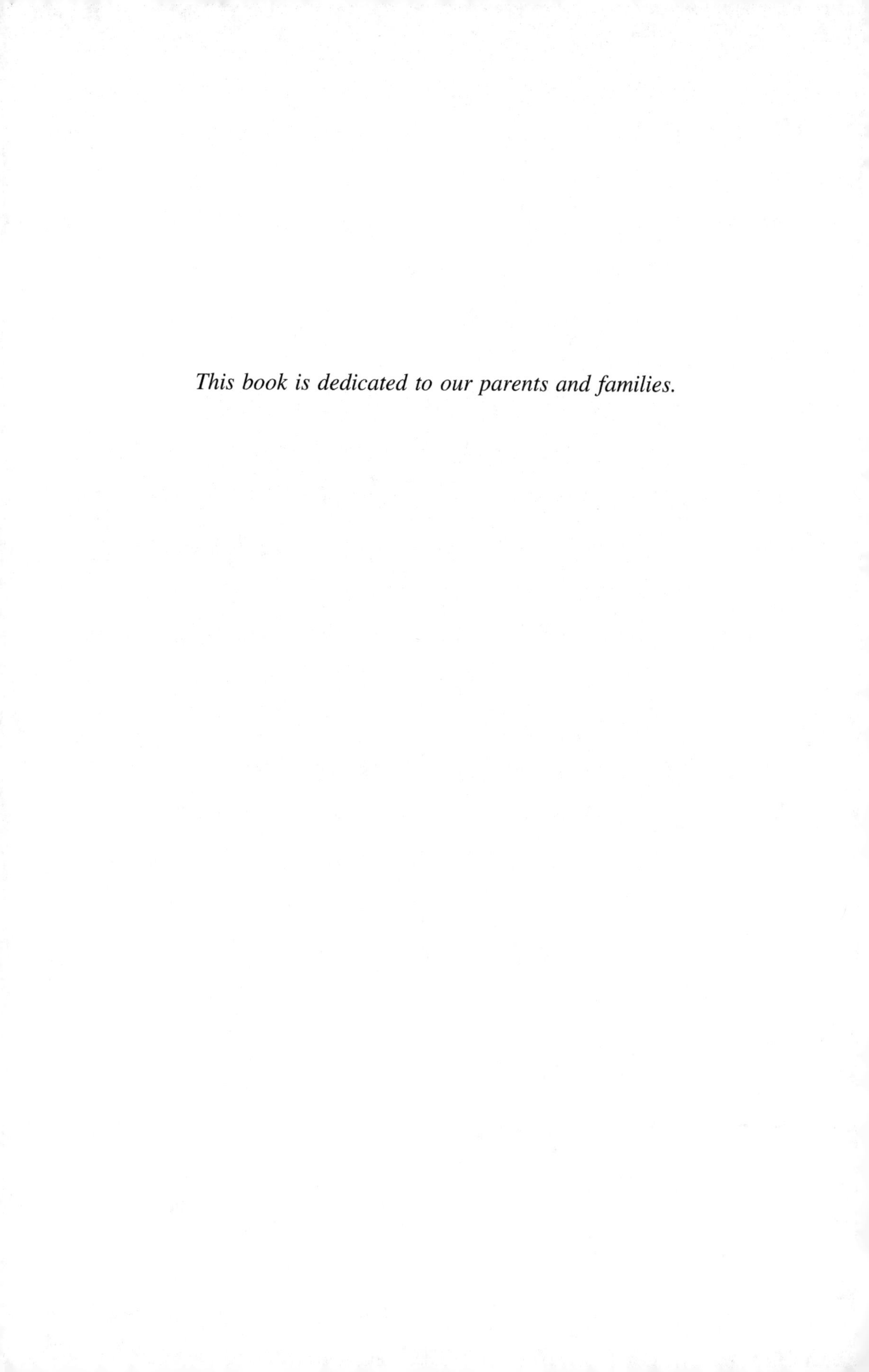

This book is dedicated to our parents and families.

Contents

Preface

Due to the increased use of power electronic converters in domestic, commercial, and industrial sectors, the quality of power in distribution networks is deteriorating at an alarming rate. This is causing a number of problems such as increased losses, poor utilization of distribution systems, mal-operation of sensitive equipment, and disturbance to nearby consumers, protective devices, and communication systems. These problems are also aggravated by the direct injection of non-steady power from renewable energy sources in the distribution system. It is expected that in the next few years, more than 80% of AC power is to be processed through power converters owing to their benefits of energy conservation, flexibility, network interconnection, and weight and volume reduction in a number of equipment such as lighting, HVAC, computers, fans, and so on. In view of these facts, it is considered timely to write this book to identify, classify, analyze, simulate, and quantify the associated power quality problems and thereby provide mitigation techniques to these power quality problems that will help practicing engineers and scientist to design better energy supply systems and mitigate existing ones.

Motivation

This book is aimed at both undergraduate and postgraduate students in the field of energy conversion and power quality in more than 10,000 institutions around the word. The book aims to achieve the following:

- Easy explanation of the subject matter through illustrations, waveforms, and phasor diagrams using minimum texts, which is one of the most efficient methods of understanding complex phenomenon.
- Simple learning of the subject through numerical examples and problems, which is one of the most favorite techniques of learning by engineering graduates.
- To gain an in-depth knowledge of the subject through computer simulation-based problems, which is the most favored skill of today's young engineers.
- To get the confidence to find the solutions of latest practical problems, which are encountered in the field of power quality.
- To develop enthusiasm for logical thinking in students and instructors.
- To gain an in-depth understanding of latest topics on power quality in minimum time and with less efforts.

Focus and Target

This book is planned in a unique and different manner compared with existing books on the subject. It consists of rare material for easy learning of the subject matter and a large number of simple derivations are included in a simplified mathematical form for solving most of the power quality problems in analytical form and designing their mitigation devices. Aside from this, the book provides essential theory supported by a reasonable number of solved numerical examples with illustrations, waveforms and phasor diagrams, small review questions, unsolved numerical problems, computer simulation-based problems, and references.

In addition to undergraduate and postgraduate students in the field of power quality, this book will also prove useful for researchers, instructors, and practicing engineers in the field.

This book facilitates simplified mathematical formulations in closed form solution through calculation, computation, and modeling of power quality problems and designing their mitigation devices.

Structure

This book consists of 11 chapters. Chapter 1 gives an introduction on power quality (PQ), causes and effects of PQ problems, requirement of PQ improvements, and mitigation aspects of PQ problems. Chapter 2 deals with PQ definitions, terminologies, standards, benchmarks, monitoring requirements, financial loss, and analytical quantification through numerical problems.

In Chapters 3–6, passive shunt and series compensation using lossless passive LC components, active shunt compensation using DSTATCOM (distribution static compensators), active series compensation using DVR (dynamic voltage restorer), and combined compensation using UPQC (unified power quality compensator) are covered for mitigation of current-based PQ problems such as reactive power compensation to achieve power factor correction (PFC) or voltage regulation (VR), load balancing, and neutral current reduction and mitigation of voltage-based PQ problems such as compensation of voltage drop, sag, swell, unbalance, and so on in the single-phase and three-phase three-wire and four-wire loads and supply systems.

In Chapter 7, various types of nonlinear loads, which cause these power quality problems, are illustrated, classified, modeled, quantified, and analyzed for associated power quality problems.

Chapters 8–11 deal with different kinds of power filters such as passive filters, active shunt filters, active series filters, and hybrid filters to meet the requirements of various kinds of power quality problems such as current and voltage harmonic elimination, reactive power compensation, and so on caused by harmonics-producing single-phase and three-phase nonlinear loads. Moreover, these power filters are also used for elimination of voltage harmonics present in the supply systems.

The major strength of this book is its 175 numerical examples, 250 review questions, 175 numerical problems, 250 computer simulation-based problems, and 600 references in different chapters.

Acknowledgments

The authors would like to thank faculty colleagues for their support and encouragement in writing this book. Professor Singh gratefully acknowledges the support from the Indian Institute of Technology Delhi, and École de technologie supérieure, Montréal, Canada (ÉTS).

Professor Singh would like to thank his research students for their contributions to some of the chapters in terms of making diagrams, figures, simulations, numerical verifications, proofreading, and constructive suggestions. Many of them deserve mention, namely, Chinmay Jain, Sabharj Arya, Ikhlaq Hussain, Rajan Kumar, Raj Kumar Garg, N. Krishna Swami Naidu, M. Sandeep, Vashist Bist, Shailendra Sharma, P. Jayaprakash, V. Rajagopal, Yash Pal, Raja Sekhara Reddy Chilipi, Stuti Shukla, M. Rajesh, Miloud Rezkallah, Sanjeev Singh, Vishal Verma, Ram Niwas, Vinod Khadkikar, Jitendra Solanki, Parag Knajiya, Sunil Kumar, Sai Pranith, Sunil Kumar Dubey, Sagar Goel, Subrat Kumar, Arun Kumar Verma, Krishan Kant Bhalla, Geeta Pathak, Nidhi Mishra, Chandu Valluri, Mohit Joshi, and Mishal Chauhan. Some of the work presented in this book primarily comes from the research conducted by them under the supervision of Professor Singh at IIT Delhi. Professor Singh expresses his gratitude to IIT Delhi for sponsoring this research. Professor Singh also expresses his gratitude to his wife Sushila Singh, children Mohit Singh, Ankita, Tripti Singh, Mohit Gupta, Gyanit Singh, and Pallavi Singh, grand children Advika and Ishan, parents, other family members, relatives, and friends for providing support during writing of this book. Professor Singh also expresses his gratitude to Professor Anjali Agarwal, Niharika, Radhika, and their relative and friends who provided excellent hospitality and family environment during his stay in Canada for writing this book.

Professor Chandra gratefully acknowledges the support from Natural Sciences and Engineering Research Council of Canada, Ecole de Technologie Supérieure (ÉTS), and Groupe de Recherche en Électronique de Puissance et Commande Industrielle GRÉPCI. Professor Chandra will like to thank

Professor Bhim Singh with whom he has personal and professional relationship for the past more than 40 years. Professor Chandra would like to thank all his past and present students, especially Brij N. Singh, Vinod Khadkikar, Mukhtiar Singh, Slaven Kincic, Etienne Tremblay, Jorge Lara Cardoso, and Miloud Rezkallah. There is no doubt that a professor learns a lot from his students. Some of the work presented in this book comes from their research work done under the supervision of Professor Chandra at ÉTS. Professor Chandra has learnt a lot from his students doing graduation on power quality, the topic that he has been teaching for the past 15 years. He is thankful to all of them. Professor Chandra is also grateful to his wife Anjali and children Niharika, Radhika, and Avneesh to provide the support and the pleasant family environment at home.

Professor Al-Haddad would like to express his gratitude and high appreciation to his colleague and friend Professor Bhim Singh who is collaborating with him since 1995 on a continuous basis. This long time collaboration with Canada Research Chair in Electric Energy Conversion and Power Electronics (CRC-EECPE) led by professor Al-Haddad started with Dr Bhim Singh first visit to Montreal in 1995 as a visiting researcher, followed by invited professor position, adjunct professor member of the CRC, and since then it is a continuous work between the team from IIT Delhi and the CRC-EECPE researchers. Professor Al-Haddad also would like to express his gratitude to the following agencies who strongly supported this work: Canada Research Chairs (CRC), Natural Sciences and Engineering Research Council of Canada (NSERC), Fond de Recherche Nature et Technologies du gouvernement du Québec (FRQNT), ÉTS, and Groupe de Recherche en Électronique de Puissance et Commande Industrielle GRÉPCI. He also expresses his gratitude to his numerous hard working researchers, his brilliant dedicated graduate and undergraduate students, a long list of more than 150 names, his colleagues in the electrical engineering department as well as those from IEEE Industrial Electronics Society and equally important to his wife Fadia, and daughters Isabel and Joelle for their support.

This book would not have been possible without the excellent cooperation among the authors of this book and a number of students. Our sincere thanks go to all contributors, proofreaders, the publisher, and our families for making this book possible.

Bhim Singh
Indian Institute of Technology Delhi,
New Delhi, India, India

Ambrish Chandra
Kamal Al-Haddad
École de technologie supérieure (ÉTS),
Montréal, Canada

About the Companion Website

This book has a companion website:
www.wiley.com/go/singh/power

The website includes:
* Solutions to numerical problems

1

Power Quality: An Introduction

1.1 Introduction

The term electric power quality (PQ) is generally used to assess and to maintain the good quality of power at the level of generation, transmission, distribution, and utilization of AC electrical power. Since the pollution of electric power supply systems is much severe at the utilization level, it is important to study at the terminals of end users in distribution systems. There are a number of reasons for the pollution of the AC supply systems, including natural ones such as lightening, flashover, equipment failure, and faults (around 60%) and forced ones such as voltage distortions and notches (about 40%). A number of customer's equipment also pollute the supply system as they draw nonsinusoidal current and behave as nonlinear loads. Therefore, power quality is quantified in terms of voltage, current, or frequency deviation of the supply system, which may result in failure or mal-operation of customer's equipment. Typically, some power quality problems related to the voltage at the point of common coupling (PCC) where various loads are connected are the presence of voltage harmonics, surge, spikes, notches, sag/dip, swell, unbalance, fluctuations, glitches, flickers, outages, and so on. These problems are present in the supply system due to various disturbances in the system or due to the presence of various nonlinear loads such as furnaces, uninterruptible power supplies (UPSs), and adjustable speed drives (ASDs). However, some power quality problems related to the current drawn from the AC mains are poor power factor, reactive power burden, harmonic currents, unbalanced currents, and an excessive neutral current in polyphase systems due to unbalancing and harmonic currents generated by some nonlinear loads.

These power quality problems cause failure of capacitor banks, increased losses in the distribution system and electric machines, noise, vibrations, overvoltages and excessive current due to resonance, negative-sequence currents in generators and motors, especially rotor heating, derating of cables, dielectric breakdown, interference with communication systems, signal interference and relay and breaker malfunctions, false metering, interferences to the motor controllers and digital controllers, and so on.

These power quality problems have become much more serious with the use of solid-state controllers, which cannot be dispensed due to benefits of the cost and size reduction, energy conservation, ease of control, low wear and tear, and other reduced maintenance requirements in the modern electric equipment. Unfortunately, the electronically controlled energy-efficient industrial and commercial electrical loads are most sensitive to power quality problems and they themselves generate power quality problems due to the use of solid-state controllers in them.

Because of these problems, power quality has become an important area of study in electrical engineering, especially in electric distribution and utilization systems. It has created a great challenge to both the electric utilities and the manufacturers. Utilities must supply consumers with good quality power for operating their equipment satisfactorily, and manufacturers must develop their electric equipment either to be immune to such disturbances or to override them. A number of techniques

Power Quality Problems and Mitigation Techniques, First Edition.
Bhim Singh, Ambrish Chandra and Kamal Al-Haddad.

have evolved for the mitigation of these problems either in existing systems or in equipment to be developed in the near future. It has resulted in a new direction of research and development (R&D) activities for the design and development engineers working in the fields of power electronics, power systems, electric drives, digital signal processing, and sensors. It has changed the scenario of power electronics as most of the equipment using power converters at the front end need modifications in view of these newly visualized requirements. Moreover, some of the well-developed converters are becoming obsolete and better substitutes are required. It has created the need for evolving a large number of circuit configurations of front-end converters for very specific and particular applications. Apart from these issues, a number of standards and benchmarks are developed by various organizations such as IEEE (Institute of Electrical and Electronics Engineers) and IEC (International Electrotechnical Commission), which are enforced on the customers, utilities, and manufacturers to minimize or to eliminate the power quality problems.

The techniques employed for power quality improvements in exiting systems facing power quality problems are classified in a different manner from those used in newly designed and developed equipment. These mitigation techniques are further subclassified for the electrical loads and supply systems, since both of them have somewhat different kinds of power quality problems. In existing nonlinear loads, having the power quality problems of poor power factor, harmonic currents, unbalanced currents, and an excessive neutral current, a series of power filters of various types such as passive, active, and hybrid in shunt, series, or a combination of both configurations are used externally depending upon the nature of loads such as voltage-fed loads, current-fed loads, or a combination of both to mitigate these problems. However, in many situations, the power quality problems may be other than those of harmonics such as in distribution systems, and the custom power devices such as distribution static compensators (DSTATCOMs), dynamic voltage restorers (DVRs), and unified power quality conditioners (UPQCs) are used for mitigating the current, voltage, or both types of power quality problems. Power quality improvement techniques used in newly designed and developed systems are based on the modification of the input stage of these systems with power factor corrected (PFC) converters, also known as improved power quality AC–DC converters (IPQCs), multipulse AC–DC converters, matrix converters for AC–DC or AC–AC conversion, and so on, which inherently mitigate some of the power quality problems in them and in the supply system by drawing clean power from the utility. This book is aimed at providing an awareness of the power quality problems, their causes and adverse effects, and an exhaustive exposure of the mitigation techniques to the customers, designers, manufacturers, application engineers, and researchers dealing with the power quality problems.

1.2 State of the Art on Power Quality

The power quality problems have been present since the inception of electric power. There have been several conventional techniques for mitigating the power quality problems and in many cases even the equipment are designed and developed to operate satisfactorily under some of the power quality problems. However, recently the awareness of the customers toward the power quality problems has increased tremendously because of the following reasons:

- The customer's equipment have become much more sensitive to power quality problems than these have been earlier due to the use of digital control and power electronic converters, which are highly sensitive to the supply and other disturbances. Moreover, the industries have also become more conscious for loss of production.
- The increased use of solid-state controllers in a number of equipment with other benefits such as decreasing the losses, increasing overall efficiency, and reducing the cost of production has resulted in the increased harmonic levels, distortion, notches, and other power quality problems. It is achieved, of course, with much more sophisticated control and increased sensitivity of the equipment toward power quality problems. Typical examples are ASDs and energy-saving electronic ballasts, which have substantial energy savings and some other benefits; however, they are the sources of waveform distortion and much more sensitive to the number of power quality disturbances.

- The awareness of power quality problems has increased in the customers due to direct and indirect penalties enforced on them, which are caused by interruptions, loss of production, equipment failure, standards, and so on.
- The disturbances to other important appliances such as telecommunication network, TVs, computers, metering, and protection systems have forced the end users to either reduce or eliminate power quality problems or dispense the use of power polluting devices and equipment.
- The deregulation of the power systems has increased the importance of power quality as consumers are using power quality as performance indicators and it has become difficult to maintain good power quality in the world of liberalization and privatization due to heavy competition at the financial level.
- Distributed generation using renewable energy and other local energy sources has increased power quality problems as it needs, in many situations, solid-state conversion and variations in input power add new problems of voltage quality such as in solar PV generation and wind energy conversion systems.
- Similar to other kinds of pollution such as air, the pollution of power networks with power quality problems has become an environmental issue with other consequences in addition to financial issues.
- Several standards and guidelines are developed and enforced on the customers, manufacturers, and utilities as the law and discipline of the land.

In view of these issues and other benefits of improving power quality, an increased emphasis has been given on quantifying, monitoring, awareness, impacts, and evolving the mitigation techniques for power quality problems. A substantial growth is observed in developing the customer's equipment with improved power quality and improving the utilities' premises. Starting from conventional techniques used for mitigating power quality problems in the utilities, distribution systems, and customers' equipment, a substantial literature has appeared in research publications, texts, patents, and manufacturers' manuals for the new techniques of mitigating power quality problems. Most of the technical institutions have even introduced courses on the power quality for teaching and training the forthcoming generation of engineers in this field.

A remarkable growth in research and development work on evolving the mitigation techniques for power quality problems has been observed in the past quarter century. A substantial research on power filters of various types such as passive, active, and hybrid in shunt, series, or a combination of both configurations for single-phase two-wire, three-phase three-wire, and three-phase four-wire systems has appeared for mitigating not only the problems of harmonics but also additional problems of reactive power, excessive neutral current, and balancing of the linear and nonlinear loads. Similar evolution has been seen in custom power devices such as DSTATCOMs for power factor correction, voltage regulation, compensation of excessive neutral current, and load balancing; DVRs and series static synchronous compensators (SSSCs) for mitigating voltage quality problems in transient and steady-state conditions; and UPQCs as a combination of DSTATCOM and DVR for mitigating current and voltage quality problems in a number of applications. These mitigation techniques for power quality problems are considered either for retrofit applications in existing equipment or for the utilities' premises. An exponential growth is also made in devising a number of circuit configurations of input front-end converters providing inherent power quality improvements in the equipment from fraction of watts to MW ratings. The use of various AC–DC and AC–AC converters of buck, boost, buck–boost, multilevel, and multipulse types with unidirectional and bidirectional power flow capability in the input stage of these equipment and providing suitable circuits for specific applications have changed the scenario of power quality improvement techniques and the features of these systems.

1.3 Classification of Power Quality Problems

There are a number of power quality problems in the present-day fast-changing electrical systems. These may be classified on the basis of events such as transient and steady state, the quantity such as current, voltage, and frequency, or the load and supply systems.

The transient types of power quality problems include most of the phenomena occurring in transient nature (e.g., impulsive or oscillatory in nature), such as sag (dip), swell, short-duration voltage variations, power frequency variations, and voltage fluctuations. The steady-state types of power quality problems include long-duration voltage variations, waveform distortions, unbalanced voltages, notches, DC offset, flicker, poor power factor, unbalanced load currents, load harmonic currents, and excessive neutral current

The second classification can be made on the basis of quantity such as voltage, current, and frequency. For the voltage, these include voltage distortions, flicker, notches, noise, sag, swell, unbalance, undervoltage, and overvoltage; similarly for the current, these include reactive power component of current, harmonic currents, unbalanced currents, and excessive neutral current.

The third classification of power quality problems is based on the load or the supply system. Normally, power quality problems due to nature of the load (e.g., fluctuating loads such as furnaces) are load current consisting of harmonics, reactive power component of current, unbalanced currents, neutral current, DC offset, and so on. The power quality problems due to the supply system consist of voltage- and frequency-related issues such as notches, voltage distortion, unbalance, sag, swell, flicker, and noise. These may also consist of a combination of both voltage- and current-based power quality problems in the system. The frequency-related power quality problems are frequency variation above or below the desired base value. These affect the performance of a number of loads and other equipment such as transformers in the distribution system.

1.4 Causes of Power Quality Problems

There are a number of power quality problems in the present-day fast-changing electrical systems. The main causes of these power quality problems can be classified into natural and man-made in terms of current, voltage, frequency, and so on. The natural causes of poor power quality are mainly faults, lightening, weather conditions such as storms, equipment failure, and so on. However, the man-made causes are mainly related to loads or system operations. The causes related to the loads are nonlinear loads such as saturating transformers and other electrical machines, or loads with solid-state controllers such as vapor lamp-based lighting systems, ASDs, UPSs, arc furnaces, computer power supplies, and TVs. The causes of power quality problems related to system operations are switching of transformers, capacitors, feeders, and heavy loads.

The natural causes result in power quality problems that are generally transient in nature, such as voltage sag (dip), voltage distortion, swell, and impulsive and oscillatory transients. However, the man-made causes result in both transient and steady-state types of power quality problems. Table 1.1 lists some of the power quality problems and their causes.

However, one of the important power quality problems is the presence of harmonics, which may be because of several loads that behave in a nonlinear manner, ranging from classical ones such as transformers, electrical machines, and furnaces to new ones such as power converters in vapor lamps, switched-mode power supplies (SMPS), ASDs using AC–DC converters, cycloconverters, AC voltage controllers, HVDC transmission, static VAR compensators, and so on.

1.5 Effects of Power Quality Problems on Users

The power quality problems affect all concerned utilities, customers, and manufacturers directly or indirectly in terms of major financial losses due to interruption of process, equipment damage, production loss, wastage of raw material, loss of important data, and so on. There are many instances and applications such as automated industrial processes, namely, semiconductor manufacturing, pharmaceutical industries, and banking, where even a small voltage dip/sag causes interruption of process for several hours, wastage of raw material, and so on.

Some power quality problems affect the protection systems and result in mal-operation of protective devices. These interrupt many operations and processes in the industries and other establishments. These also affect many types of measuring instruments and metering of the various quantities such as voltage, current, power, and energy. Moreover, these problems affect the monitoring systems in much critical, important, emergency, vital, and costly equipment.

Harmonic currents increase losses in a number of electrical equipment and distribution systems and cause wastage of energy, poor utilization of utilities' assets such as transformers and feeders, overloading of power capacitors, noise and vibrations in electrical machines, and disturbance and interference to electronics appliances and communication networks.

Table 1.1 Power quality problems: causes and effects [15,23]

Problems	Category	Categorization	Causes	Effects
Transients	Impulsive	Peak, rise time, and duration	Lightning strikes, transformer energization, capacitor switching	Power system resonance
	Oscillatory	Peak magnitude and frequency components	Line, capacitor, or load switching	System resonance
Short-duration voltage variation	Sag	Magnitude, duration	Motor starting, single line to ground faults	Protection malfunction, loss of production
	Swell	Magnitude, duration	Capacitor switching, large load switching, faults	Protection malfunction, stress on computers and home appliances
	Interruption	Duration	Temporary faults	Loss of production, malfunction of fire alarms
Long-duration voltage variation	Sustained interruption	Duration	Faults	Loss of production
	Undervoltage	Magnitude, duration	Switching on loads, capacitor de-energization	Increased losses, heating
	Overvoltage	Magnitude, duration	Switching off loads, capacitor energization	Damage to household appliances
Voltage imbalance		Symmetrical components	Single-phase load, single-phasing	Heating of motors
Waveform distortion	DC offset	Volts, amperes	Geomagnetic disturbance, rectification	Saturation in transformers
	Harmonics	THD, harmonic spectrum	ASDs, nonlinear loads	Increased losses, poor power factor
	Interharmonics	THD, harmonic spectrum	ASDs, nonlinear loads	Acoustic noise in power equipment
	Notching	THD, harmonic spectrum	Power electronic converters	Damage to capacitive components
	Noise	THD, harmonic spectrum	Arc furnaces, arc lamps, power converters	Capacitor overloading, disturbances to appliances
Voltage flicker		Frequency of occurrence, modulating frequency	Arc furnaces, arc lamps	Human health, irritation, headache, migraine
Voltage fluctuations		Intermittent	Load changes	Protection malfunction, light intensity changes
Power frequency variations			Faults, disturbances in isolated customer-owned systems and islanding operations	Damage to generator and turbine shafts

1.6 Classification of Mitigation Techniques for Power Quality Problems

In view of increased problems due to power quality in terms of financial loss, loss of production, wastage of raw material, and so on, a wide variety of mitigation techniques for improving the power quality have evolved in the past quarter century. These include passive components such as capacitors, reactors, custom power devices, a series of power filters, improved power quality AC–DC converters, and matrix converters.

However, the power quality problems may not be because of harmonics in many situations such as in distribution systems where problems of poor voltage regulation, low power factor, load unbalancing, excessive neutral current, and so on are observed. Some of these power quality problems such as poor power factor because of reactive power requirements may be mitigated using lossless passive elements such as capacitors and reactors. Moreover, the custom power devices such as DSTATCOMs, DVRs, and UPQCs are extensively used for mitigating the current, voltage, or both types of power quality problems.

In the presence of harmonics in addition to other power quality problems, a series of power filters of various types such as active, passive, and hybrid in shunt, series, or a combination of both configurations in single-phase two-wire, three-phase three-wire, and three-phase four-wire systems are used externally as retrofit solutions for mitigating power quality problems through compensation of nonlinear loads or voltage-based power quality problems in the AC mains. Since there are a large number of circuits of filters, the best configuration of the filter is decided depending upon the nature of loads such voltage-fed loads, current-fed loads, or a combination of both to mitigate their problems.

Power quality improvement techniques used in newly designed and developed equipment are based on the modification of the input stage of these systems with PFC converters, also known as IPQCs, multipulse AC–DC converters, matrix converters for AC–DC or AC–AC conversion, and so on, which inherently mitigate some of the power quality problems in them and in the supply system by drawing clean power from the utility. There are a large number of circuits of the converters of boost, buck, buck–boost, multilevel, and multipulse types for unidirectional and bidirectional power flow with and without isolation in single-phase and three-phase supply systems to suit very specific applications. These are used as front-end converters in the input stage as a part of the total equipment and in many situations they make these equipment immune to power quality problems in the supply system.

1.7 Literature and Resource Material on Power Quality

Power quality has become an important area of specialization in engineering. Many technical institutions, industries, and R&D organizations are offering regular and short-term courses on power quality and many of them have developed laboratories for research and teaching the power quality. There are a number of texts, standards, and patents relating to power quality and many journals, magazines, and conferences, among others, are publishing a number of research publications and case studies on power quality. Some of the journals, magazines, and conferences dealing with power quality are as follows:

IEEE Transactions on Aerospace and Systems
IEEE Transactions on Energy Conversion
IEEE Transactions on Industrial Electronics
IEEE Transactions on Industry Applications
IEEE Transactions on Industrial Informatics
IEEE Transactions on Magnetics
IEEE Transactions on Power Delivery
IEEE Transactions on Power Electronics
IEEE Transactions on Power Systems
IEEE Transactions on Smart Grid
IEEE Transactions on Sustainable Energy
IEEE Industry Applications Magazine
IEE/IET Proceedings on Electric Power Applications (EPA)

IEE/IET Proceedings on Generation, Transmission and Distribution (GTD)
IET Power Electronics (PE)
IET Renewable Power Generation (RPG)
Electrical Engineering in Japan
Electric Power Systems Research
International Journal of Electrical Engineering Education
International Journal of Electric Power Components and Systems (EPCS)
International Journal of Electrical Power & Energy Systems (EPES)
European Transactions on Electrical Power Engineering (ETEP)
European Journal of Power Electronics (EPE)
International Journal of Emerging Electric Power Systems
Journal of Power Electronics (JPE)
International Journal of Power Electronics
International Journal of Power Electronics and Drive Systems (IJPEDS)
International Journal of Power Electronics and Systems
International Journal of Energy Technology and Policy
International Journal of Global Energy Issues (IJGEI)
International Journal of Power System and Power Electronics Engineering (IJPSPEE)
International Journal of Power Electronics and Energy (IJPEE)
IEEE Applied Power Electronics Conference (APEC)
IEEE Energy Conversion Congress and Exposition (IEEE-ECCE)
IEEE International Telecommunications Energy Conference (IEEE-INTELEC)
European Power Electronics Conference (EPEC)
IEEE Industrial Electronics Conference (IECON)
IEEE International Symposium on Industrial Electronics (ISIE)
IEEE Industry Applications Society (IAS) Annual Meeting
IEEE International Conference on Power Electronics and Electric Drives (PEDS)
IEEE International Conference on Power Electronics, Drives and Energy Systems for Industrial Growth (PEDES)
IEEE Intersociety Energy Conversion Engineering Conference (IECEC)
IEEE International Power Electronics Specialist Conference (PESC)
IEEE Canadian Electrical and Computer Engineering Conference (CECEC)
IEEE International Electric Machines and Drives Conference (IEMDC)
International Power Electronics Conference (IPEC)
International Power Electronics Congress (CIEP)
IEEE Power Quality Conference
International Conference on Electrical Machines (ICEM)
Power Conversion Intelligent Motion (PCIM)
IEEE India International Power Electronics Conference (IICPE)
National Power Electronics Conference (NPEC)

1.8 Summary

Recently, power quality has become an important subject and area of research because of its increasing awareness and impacts on the consumers, manufacturers, and utilities. There are a number of economic and reliability issues for satisfactory operation of electrical equipment. As power quality problems are increasing manifold due to the use of solid-state controllers, which cannot be dispensed due to many financial benefits, energy conservation, and other production benefits, the research and development in mitigation techniques for power quality problems is also becoming relevant and important to limit the pollution of the supply system. In such a situation, it is quite important to study the causes, effects, and mitigation techniques for power quality problems.

1.9 Review Questions

1. What is power quality?
2. What are the power quality problems in AC systems?
3. Why is power quality important?
4. What are the causes of power quality problems?
5. What are the effects of power quality problems?
6. What is a nonlinear load?
7. What is voltage sag (dip)?
8. What is voltage swell?
9. What are the harmonics?
10. What are the interharmonics?
11. What are the subharmonics?
12. What is the role of a shunt passive power filter?
13. What is the role of a series passive power filter?
14. What is an active power filter?
15. What is the role of a shunt active power filter?
16. What is the role of a series active power filter?
17. What is the role of a DSTATCOM?
18. What is the role of a DVR?
19. What is the role of a UPQC?
20. What is a PFC?
21. What is an IPQC?
22. Why is the excessive neutral current present in a three-phase four-wire system?
23. How can the excessive neutral current be eliminated?
24. Which are the standards for harmonic current limits?
25. What are the permissible limits on harmonic current?

References

1. ABB Power Systems (1988) Harmonic Currents, Static VAR Systems. Information NR500-015E.
2. Mitsubishi Electric Corporation (1989) Active Filters: Technical Document, 2100/1100 Series.
3. Kikuchi, A.H. (1992) Active Power Filter. Toshiba GTR Module (IGBT) Application Notes.
4. Schaefer, J. (1965) *Rectifier Circuits: Theory and Design*, John Wiley & Sons, Inc., New York.
5. Miller, T.J.E. (ed.) (1982) *Reactive Power Control in Electric Systems*, Wiley-Interscience, Toronto, Canada.
6. Mathur, R.M. (1984) *Static Compensators for Reactive Power Control*, Contexts Publications, Winnipeg, Canada.
7. Seguier, G. (1986) *Power Electronic Converters – AC/DC Conversion*, McGraw-Hill.
8. IEEE Standard 1030-1987 (1987) *IEEE Guide for Specification of High-Voltage Direct Current Systems. Part I. Steady State Performance*, IEEE.
9. Griffith, D.C. (1989) *Uninterruptible Power Supplies*, Marcel Dekker, New York.

10. Clark, J.W. (1990) *AC Power Conditioners – Design Applications*, Academic Press, San Diego, CA.
11. IEEE Standard 519-1992 (1992) *IEEE Guide for Harmonic Control and Reactive Compensation of Static Power Converters*, IEEE.
12. Kazibwe, W.E. and Sendaula, M.H. (1993) *Electrical Power Quality Control Techniques*, Van Nostrand Reinhold Company.
13. Heydt, G.T. (1994) *Electric Power Quality*, 2nd edn, Stars in a Circle Publications, West Lafayette, IN.
14. IEC 61000-3-2 (1995) *Electromagnetic Compatibility (EMC) – Part 3: Limits – Section 2: Limits for Harmonic Current Emissions (Equipment Input Current <16 A Per Phase)*, IEC.
15. IEEE Standard 1159-1995 (1995) *IEEE Recommended Practice for Monitoring Electric Power Quality*, IEEE.
16. Paice, D.A. (1996) *Power Electronic Converter Harmonics*, IEEE Press, New York.
17. Arrillaga, J., Smith, B.C., Watson, N.R., and Wood, A.R. (1997) *Power System Harmonic Analysis*, John Wiley & Sons, Ltd, Chichester, UK.
18. Porter, G.J. and Sciver, J.A.V. (eds) (1999) *Power Quality Solutions: Case Studies for Troubleshooters*, The Fairmont Press, Inc., Lilburn, GA.
19. Arrillaga, J., Watson, N.R., and Chen, S. (2000) *Power System Quality Assessment*, John Wiley & Sons, Inc., New York.
20. Bollen, H.J. (2001) *Understanding Power Quality Problems*, 1st edn, Standard Publishers Distributors, New Delhi.
21. Schlabbach, J., Blume, D., and Stephanblome, T. (2001) *Voltage Quality in Electrical Power Systems*, Power Engineering and Energy Series, IEEE Press.
22. Ghosh, A. and Ledwich, G. (2002) *Power Quality Enhancement Using Custom Power Devices*, Kluwer Academic Publishers, London.
23. Sankaran, C. (2002) *Power Quality*, CRC Press, New York.
24. Das, J.C. (2002) *Power System Analysis – Short-Circuit Load Flow and Harmonics*, Marcel Dekker, New York.
25. IEEE Standard 1573-2003 (2003) *IEEE Guide for Application and Specification of Harmonic Filters*, IEEE.
26. Emadi, A., Nasiri, A., and Bekiarov, S.B. (2005) *Uninterruptible Power Supplies and Active Filters*, CRC Press, New York.
27. Wu, B. (2006) *High-Power Converters and AC Drives*, IEEE Press, Hoboken, NJ.
28. Bollen, M.H.J. and Gu, I.Y.-H. (2006) *Signal Processing of Power Quality Disturbances*, IEEE Press, Hoboken, NJ.
29. Dugan, R.C., McGranaghan, M.F., and Beaty, H.W. (2006) *Electric Power Systems Quality*, 2nd edn, McGraw-Hill, New York.
30. Short, T.A. (2006) *Distribution Reliability and Power Quality*, CRC Press, New York.
31. Moreno-Munoz, A. (2007) *Power Quality: Mitigation Technologies in a Distributed Environment*, Springer, London.
32. Akagi, H., Watanabe, E.H., and Aredes, M. (2007) *Instantaneous Power Theory and Applications to Power Conditioning*, John Wiley & Sons, Inc., Hoboken, NJ.
33. Padiyar, K.R. (2007) *FACTS Controllers in Transmission and Distribution*, New Age International, New Delhi.
34. Kusko, A. and Thompson, M.T. (2007) *Power Quality in Electrical Systems*, McGraw-Hill.
35. Fuchs, E.F. and Mausoum, M.A.S. (2008) *Power Quality in Power Systems and Electrical Machines*, Elsevier/Academic Press, London.
36. Sastry Vedam, R. and Sarma, M.S. (2008) *Power Quality VAR Compensation in Power Systems*, CRC Press, New York.
37. Baggini, A. (2008) *Handbook on Power Quality*, John Wiley & Sons, Inc., Hoboken, NJ.
38. Benysek, G. and Pasko, M. (eds) (2012) *Power Theories for Improved Power Quality*, Springer, London.
39. Grady, W.M., Samotyj, M.J., and Noyola, A.H. (1990) Survey of active power line conditioning methodologies. *IEEE Transactions on Power Delivery*, **5**, 1536–1542.
40. Van Wyk, J.D. (1993) Power quality, power electronics and control. Proceedings of EPE'93, pp. 17–32.
41. Akagi, H. (1996) New trends in active filters for power conditioning. *IEEE Transactions on Industry Applications*, **32**, 1312–1322.
42. Singh, B., Al-Haddad, K., and Chandra, A. (1999) A review of active filters for power quality improvement. *IEEE Transactions on Industrial Electronics*, **46**, 960–971.
43. El-Habrouk, M., Darwish, M.K., and Mehta, P. (2000) Active power filters: a review. *IEE Proceedings – Electric Power Applications*, **147**, 493–413.
44. Peng, F.Z. (2001) Harmonic sources and filtering approaches. *IEEE Industry Applications Magazine*, **7**(4), 18–25.
45. Singh, B., Verma, V., Chandra, A., and Al-Haddad, K. (2005) Hybrid filters for power quality improvement. *IEE Proceedings – Generation, Transmission and Distribution*, **152**(3), 365–378.

46. Boroyevich, D. and Hiti, S. (1996) Three-phase PWM converter: modeling and control design. Seminar 9, IEEE-APEC'96.
47. Enjeti, P. and Pitel, I. (1999) Design of three-phase rectifier systems with clean power characteristics. *Tutorial, PESC'99.*
48. Kolar, J.W. and Sun, J. (2001) Three-phase power factor correction technology. *Seminar 1&4, PESC'01.*
49. Singh, B., Singh, B.N., Chandra, A. *et al.* (2003) A review of single-phase improved power quality AC–DC converters. *IEEE Transactions on Industrial Electronics*, **50**(5), 962–981.
50. Singh, B., Singh, B.N., Chandra, A. *et al.* (2004) A review of three-phase improved power quality AC–DC converters. *IEEE Transactions on Industrial Electronics*, **51**(3), 641–660.
51. Wheeler, P.W., Rodriguez, J., Clare, J.C. *et al.* (2002) Matrix converters: a technology review. *IEEE Transactions on Industrial Electronics*, **49**(2), 276–288.

2

Power Quality Standards and Monitoring

2.1 Introduction

There has been exponentially growing interest in power quality (PQ) in the past quarter century, which may be witnessed by the published literature in terms of research publications, texts, standards, patents, and so on. Some of the main reasons for this have been enhanced sensitivity of equipment, awareness of consumers, increased cost of electricity globally, increased use of solid-state controllers in energy-intensive equipment with the aim of energy conservation, power loss reduction, better utilization of utility assets, environmental pollution such as interference to telecommunication systems, malfunction of protection systems, and so on.

The power quality problems affect the customers in a number of ways such as economic penalty in terms of power loss, equipment failure, mal-operation, interruption in the process, and loss of production. In view of these facts, various terms and definitions are used to quantify the power quality problems in terms of different performance indices. Moreover, a number of standards have been developed by various organizations and institutes that are enforced on the customers, manufacturers, and utilities to maintain an acceptable level of power quality. Apart from these factors, various techniques and instruments are developed to study and monitor the level of power quality pollution and their causes. Many industries are developing a number of instruments, recorders, and analyzers to measure, record, and analyze the data at the site or in the research laboratories. In view of these increasing issues of power quality and awareness of power quality, it is considered relevant to introduce various terminologies, definitions, standards, and monitoring systems to quantify and assess the threshold level of power quality.

This chapter deals with the state of the art on power quality standards and monitoring, power quality terminologies, power quality definitions, power quality standards, power quality monitoring, monitoring equipment, summary, numerical examples, review questions, numerical and computer simulation-based problems, and references.

2.2 State of the Art on Power Quality Standards and Monitoring

There have been power quality problems and issues since the inception of electric power. However, the terminology of power quality does not date back to the early days and it has been identified by various other names. In the past few decades, it has become a very common terminology and widely known as power quality. Similarly, several standards have been developed, modified, recommended, and enforced depending upon the evolution of technology to maintain and quantify the level of power quality. At present, there is a long list of standards on various aspects of power quality, such as permissible level of

Power Quality Problems and Mitigation Techniques, First Edition.
Bhim Singh, Ambrish Chandra and Kamal Al-Haddad.

Table 2.1 List of some standards on various issues of power quality

Standards	Description
IEEE Standard 519-1992	Recommended Practices and Requirements for Harmonic Control in Electrical Power Systems
IEEE Standard 1159-1995	Recommended Practice for Monitoring Electric Power Quality
IEEE Standard 1100-1999	Recommended Practice for Powering and Grounding Sensitive Electronic Equipment
IEEE Standard 1250-1995	Guide for Service to Equipment Sensitive to Momentary Voltage Disturbances
IEEE Standard 1366-2012	Electric Power Distribution Reliability Indices
IEC 61000-2-2	Compatibility Levels for Low-Frequency Conducted Disturbances and Signaling in Public Supply Systems
IEC 61000-2-4	Compatibility Levels in Industrial Plants for Low-Frequency Conducted Disturbances
IEC 61000-3-2	Limits for Harmonic Current Emissions (Equipment Input Current Up to and Including 16 A Per Phase)
IEC 61000-4-15	Flicker Meter – Functional and Design Specifications
EN 50160	Voltage Characteristics of Public Distribution Systems

deviations, mitigation, and monitoring. Some of them are given here; however, new standards are continuously being developed, with modifications in the existing ones on various aspects such as limits, monitoring, and mitigation devices.

Several standards such as IEEE 519-1992, IEC 61000, and many others in different countries have been developed on the permissible limits in the levels of deviations and distortions in various electrical quantities such as voltage, current, and power factor. Moreover, there are several standards on the level of power quality in specific equipment such as lighting and variable-frequency drives in many countries. Table 2.1 shows a list of some currently available standards on various aspects of power quality.

2.3 Power Quality Terminologies

Since the power quality issues, awareness, and mitigation techniques are reported to a level of concern, various terminologies are defined to quantify power quality problems.

For reference, see the following terms and definitions, which are defined in detail in IEEE Standards [24]:

- *Flicker*: Impression of unsteadiness of visual sensation induced by a light stimulus whose luminance or spectral distribution fluctuates with time.
- *Fundamental (component)*: The component of order 1 (e.g., 50 Hz, 60 Hz) of the Fourier series of a periodic quantity.
- *Imbalance (voltage or current)*: The ratio of the negative-sequence component to the positive-sequence component, usually expressed as a percentage. *Syn*: unbalance (voltage or current).
- *Impulsive transient*: A sudden non-power frequency change in the steady-state condition of voltage or current that is unidirectional in polarity (primarily either positive or negative).
- *Instantaneous*: When used to quantify the duration of a short-duration root-mean-square (rms) variation as a modifier, it refers to a time range from 0.5 to 30 cycles of the power frequency.
- *Interharmonic (component)*: A frequency component of a periodic quantity that is not an integer multiple of the frequency at which the supply system is operating (e.g., 50 Hz, 60 Hz).
- *Long-duration rms variation*: A variation of the rms value of the voltage or current from the nominal value for a time greater than 1 min. The term is usually further described using a modifier indicating the magnitude of a voltage variation (e.g., undervoltage, overvoltage, and voltage interruption).
- *Momentary interruption*: A type of short-duration rms voltage variation where a complete loss of voltage (<0.1 pu) on one or more phase conductors is for a time period between 0.5 cycle and 3 s.

- *Root-mean-square variation*: A term often used to express a variation in the rms value of a voltage or current measurement from the nominal value. See sag, swell, momentary interruption, temporary interruption, sustained interruption, undervoltage, and overvoltage.
- *Short-duration rms variation*: A variation of the rms value of the voltage or current from the nominal value for a time greater than 0.5 cycle of the power frequency but less than or equal to 1 min. When the rms variation is voltage, it can be further described using a modifier indicating the magnitude of a voltage variation (e.g., sag, swell, and interruption) and possibly a modifier indicating the duration of the variation (e.g., instantaneous, momentary, and temporary).
- *Sustained interruption*: A type of long-duration rms voltage variation where the complete loss of voltage (<0.1 pu) on one or more phase conductors is for a time greater than 1 min.
- *Temporary interruption*: A type of short-duration rms variation where the complete loss of voltage (<0.1 pu) on one or more phase conductors is for a time period between 3 s and 1 min.
- *Voltage change*: A variation of the rms or peak value of a voltage between two consecutive levels sustained for definite but unspecified durations.
- *Voltage fluctuation*: A series of voltage changes or a cyclic variation of the voltage envelope.
- *Voltage interruption*: The disappearance of the supply voltage on one or more phases. It is usually qualified by an additional term indicating the duration of the interruption (e.g., momentary, temporary, and sustained).
- *Waveform distortion*: A steady-state deviation from an ideal sine wave of power frequency principally characterized by the spectral content of the deviation.

For the purposes of standardization, the following additional terms and definitions are also used [24]:

- *Accuracy*: The quality of freedom from mistake or error, that is, of conformity to truth or to a rule (as in instrumentation and measurement). The accuracy of an indicated or recorded value is expressed by the ratio of the error of the indicated value to the true value. It is usually expressed in percent. See accuracy rating of an instrument (as indicated or recorded value).
- *Calibration*: The adjustment of a device to have the designed operating characteristics, and the subsequent marking of the positions of the adjusting means, or the making of adjustments necessary to bring operating characteristics into substantial agreement with standardized scales or marking. Comparison of the indication of the instrument under test, or registration of the meter under test, with an appropriate standard (as in metering).
- *Common-mode voltage*: The voltage that, at a given location, appears equally and in phase from each signal conductor to ground.
- *Coupling*: The association of two or more circuits or systems in such a way that power or signal information may be transferred from one system or circuit to another.
- *Current transformer (CT)*: An instrument transformer designed for use in the measurement or control of current (as in metering).
- *Dropout*: A loss of equipment operation (discrete data signals) due to noise, voltage sags, or interruption.
- *Electromagnetic compatibility (EMC)*: A measure of equipment tolerance to external electromagnetic fields. The ability of a device, equipment, or system to function satisfactorily in its electromagnetic environment without introducing intolerable electromagnetic disturbances to anything in that environment.
- *Electromagnetic disturbance*: An electromagnetic phenomenon that may be superimposed on a wanted signal. Any electromagnetic phenomenon that may degrade the performance of a device, a piece of equipment, or a system.
- *Equipment grounding conductor*: The conductor used to connect the noncurrent-carrying parts of conduits, raceways, and equipment enclosures to the grounding electrode at the service equipment (main panel) or secondary of a separately derived system.
- *Failure mode*: The manner in which failure occurs; generally categorized as electrical, mechanical, thermal, and contamination.

- *Frequency deviation*: An increase or decrease in the power frequency from the nominal value. The duration of a frequency deviation can be from several cycles to several hours.
- *Ground*: A conducting connection, whether intentional or accidental, by which an electric circuit or equipment is connected to the earth, or to some conducting body of relatively large extent that serves in place of the earth. Grounds are used for establishing and maintaining the potential of the earth (or of the conducting body) or approximately that potential, on conductors connected to it, and for conducting ground currents to and from earth (or the conducting body).
- *Ground loop*: A potentially detrimental loop formed when two or more points in an electrical system that are nominally at ground potential are connected by a conducting path such that either or both points are not at the same ground potential.
- *Harmonic*: A sinusoidal component of a periodic wave or quantity having a frequency that is an integral multiple of the fundamental frequency. For example, a component having a frequency twice the fundamental frequency is called a second harmonic.
- *Harmonic components*: The components of the harmonic content expressed in terms of the order and rms values of the Fourier series terms describing the periodic function.
- *Harmonic content*: The function obtained by subtracting the DC and fundamental components from a nonsinusoidal periodic function. The deviation from the sinusoidal form, expressed in terms of the order and magnitude of the Fourier series terms describing the wave. Distortion of a sinusoidal waveform characterized by indication of the magnitude and order of the Fourier series terms describing the wave.
- *Immunity (to a disturbance)*: The ability of a device, equipment, or system to perform without degradation in the presence of an electromagnetic disturbance.
- *Impulse*: A pulse that begins and ends within a time so short that it may be regarded mathematically as infinitesimal, although the area remains finite. An impulse is a surge of unidirectional polarity.
- *Isolated equipment ground*: An isolated equipment grounding conductor run in the same conduit or raceway as the supply conductors. This conductor may be insulated from the metallic raceway and all ground points throughout its length. It originates at an isolated ground-type receptacle or equipment input terminal block and terminates at the point where neutral and ground are bonded at the power source.
- *Isolation*: Separation of one section of a system from undesired influences of other sections.
- *Maximum demand*: The largest of a particular type of demand occurring within a specified period.
- *Momentary*: When used as a modifier to quantify the duration of a *short-duration variation*, it refers to a time range from 30 cycles to 3 s.
- *Momentary interruption*: A type of short-duration variation. The complete loss of voltage (<0.1 pu) on one or more phase conductors for a time period between 0.5 cycle and 3 s.
- *Noise*: Electrical noise is unwanted electrical signals that produce undesirable effects in the circuits of the control systems in which they occur.
- *Nominal voltage*: A nominal value assigned to a circuit or system for the purpose of conveniently designating its voltage class (as 208 V/120 V, 480 V/277 V, 600 V).
- *Nonlinear load*: A load that draws a nonsinusoidal current wave when supplied by a sinusoidal voltage source.
- *Normal-mode voltage*: The voltage that appears differentially between two signal wires and that acts on the circuit in the same manner as the desired signal.
- *Notch*: A switching (or other) disturbance of the normal power voltage waveform, lasting less than a half cycle, which is initially of opposite polarity to the waveform and is thus subtracted from the normal waveform in terms of the peak value of the disturbance voltage. This includes complete loss of voltage for up to a half cycle.
- *Oscillatory transient*: A sudden, non-power frequency change in the steady-state condition of voltage or current that includes either positive or negative polarity value.
- *Overvoltage*: When used to describe a specific type of *long-duration variation*, it refers to a measured voltage having a value greater than the *nominal voltage* for a time greater than 1 min. The typical values are 1.1–1.2 pu.

- *Phase shift*: The displacement in time of one waveform relative to another of the same frequency and harmonic content.
- *Point of common coupling (PCC)*: The point at which the electric utility and the customer interface occurs. Typically, this point is the customer side of the utility revenue meter.
- *Potential transformer (PT)*: An instrument transformer that is intended to have its primary winding connected in shunt with a power supply circuit, the voltage of which is to be measured or controlled.
- *Power disturbance*: Any deviation from the nominal value (or from some selected thresholds based on load tolerance) of the input AC power characteristics.
- *Power quality*: The concept of powering and grounding electronic equipment in a manner that is suitable to the operation of that equipment and compatible with the premise wiring system and other connected equipment.
- *Pulse*: A wave that departs from an initial level for a limited duration of time and ultimately returns to the original level.
- *Sag*: A decrease in rms voltage or current for durations of 0.5 cycle to 1 min. The typical values are 0.1–0.9 pu.
- *Shield*: A metallic sheath, usually copper or aluminum, applied over the insulation of a conductor(s) for the purpose of providing means for reducing electrostatic coupling between the conductor(s) so shielded and others that may be susceptible to or that may be generating unwanted (noise) electrostatic fields.
- *Shielding*: The process of applying a conductive barrier between a potentially disturbing noise source and electronic circuitry. Shields are used to protect cables (data and power) and electronic circuits. Shielding may be accomplished by the use of metal barriers, enclosures, or wrappings around source circuits and receiving circuits.
- *Sustained*: When used to quantify the duration of a voltage interruption, it refers to the time frame associated with a long-duration variation (i.e., greater than 1 min).
- *Sustained interruption*: A type of long-duration variation. The complete loss of voltage (<0.1 pu) on one or more phase conductors for a time greater than 1 min.
- *Swell*: An increase in rms voltage or current for durations from 0.5 cycle to 1 min. The typical values are 1.1–1.8 pu.
- *Temporary interruption*: A type of short-duration variation. The complete loss of voltage (<0.1 pu) on one or more phase conductors for a time period between 3 s and 1 min.
- *Total demand distortion (TDD)*: The total rms harmonic current distortion, in percent of the maximum demand load current (15 or 30 min demand).
- *Total harmonic distortion (THD) (HF: harmonic factor)*: The ratio of the rms value of the harmonic content to the rms value of the fundamental quantity, expressed as a percent of the fundamental.
- *Transient*: Pertaining to or designating a phenomenon or a quantity that varies between two consecutive steady states during a time interval that is short compared to the timescale of interest. A transient can be a unidirectional impulse of either polarity or a damped oscillatory wave with the first peak occurring in either polarity.
- *Undervoltage*: When used to describe a specific type of *long-duration variation*, it refers to a measured voltage having a value less than the *nominal voltage* for a time greater than 1 min. The typical values are 0.8–0.9 pu.
- *Voltage distortion*: Any deviation from the nominal sine wave form of the AC line voltage.
- *Voltage imbalance (unbalance)*: The ratio of the negative- or zero-sequence component to the positive-sequence component, usually expressed as a percentage in polyphase systems.
- *Voltage regulation*: The degree of control or stability of the rms voltage at the load. Often specified in relation to other parameters, such as input voltage changes, load changes, or temperature changes.

2.4 Power Quality Definitions

Power quality is defined in many sources, which give different meaning to different people and sometimes conflicting statements of power quality due to a lot of confusion on the meaning of the term power quality.

Therefore, its definition has not been universally agreed upon. It is used synonymously with supply reliability, service quality, voltage quality, quality of supply, and quality of consumption. The definitions given by IEEE (Institute of Electrical and Electronics Engineers) and IEC (International Electrotechnical Commission) are provided here.

The definition of power quality given in the IEEE dictionary states that "Power quality is the concept of powering and grounding sensitive equipment in a manner that is suitable to the operation of that equipment."

The IEC definition of power quality, given in IEC 61000-4-30, states "Characteristics of the electricity at a given point on an electrical system, evaluated against a set of reference technical parameters."

Electromagnetic compatibility is a term related to power quality used in IEC 61000-1-1, which states that "Electromagnetic compatibility is the ability of an equipment or system to function satisfactorily in its electromagnetic environment without introducing intolerable electromagnetic disturbances to anything in that environment."

Recently, power quality is referred to "the ability of the electric utilities to supply electric power without interruption."

Power quality is considered as a combination of current and voltage quality as per many published literature. Voltage quality is concerned with the deviation of actual voltage from the ideal voltage and an equivalent definition exists for the current quality. Any deviation in the voltage or current from the ideal value is a power quality disturbance. However, in the power system, it is difficult to distinguish between the voltage and current disturbances because an event leads to different disturbances for different customers. Therefore, in general, power quality is related to disturbances in voltage, current, frequency, and power factor.

2.5 Power Quality Standards

If power quality problems increase to a level that these start affecting not only those who are creating them, but also other consumers, then it becomes a matter of concern. In view of these power pollution problems, a number of organizations such as IEC, IEEE, American National Standards Institute (ANSI), British Standards (BS), European Norms (EN), Computer Business Equipment Manufacturers Association (CBEMA), and Information Technology Industry Council (ITIC) have developed different standards to specify the permissible limits of various performance indices to maintain the level of power quality to an acceptable benchmark and to provide guidelines to the customers, manufactures, and utilities on curbing the various events causing the power quality problems. Tables 2.2–2.8 show some important limits on voltages and currents in these standards.

Table 2.2 IEEE Standard 519-1992: current distortion limits for general distribution systems (120–69 000 V) [21]

	Maximum harmonic current distortion (in percent of I_L)					
	Individual harmonic order (odd harmonics)					
I_{SC}/I_L	$h<11$	$11 \le h<17$	$17 \le h<23$	$23 \le h<35$	$35 \le h$	TDD (%)
<20*	4.0	2.0	1.5	0.6	0.3	5.0
20 to <50	7.0	3.5	2.5	1.0	0.5	8.0
50 to <100	10.0	4.5	4.0	1.5	0.7	12.0
100 to <1000	12.0	5.5	5.0	2.0	1.0	15.0
>1000	15.0	7.0	6.0	2.5	1.4	20.0

- Even harmonics are limited to 25% of the odd harmonic limits above.
- Current distortions that result in a DC offset, for example, half-wave converters, are not allowed.
- *All power generation equipment is limited to these values of current distortion, regardless of actual I_{SC}/I_L, where I_{SC} = maximum short-circuit current at PCC and I_L = maximum demand load current (fundamental frequency component) at PCC.

Table 2.3 IEEE Standard 519-1992: current distortion limits for general distribution systems (>161 kV), dispersed generation and cogeneration [21]

Maximum harmonic current distortion (in percent of I_L)						
Individual harmonic order (odd harmonics)						
I_{SC}/I_L	$h<11$	$11 \leq h < 17$	$17 \leq h < 23$	$23 \leq h < 35$	$35 \leq h$	TDD (%)
<50	2.0	1.0	0.75	0.3	0.15	2.5
≥50	3.0	1.5	1.15	0.45	0.22	3.75

- Even harmonics are limited to 25% of the odd harmonic limits above.
- Current distortions that result in a DC offset, for example, half-wave converters, are not allowed.

Table 2.4 IEC 61000-3-2: maximum permissible harmonic current for class D equipment (current limited to less than or equal to 16 A per phase) (class D: PC, PC monitors, radio, or TV receivers) (input power $P \leq 600$ W) [43]

Harmonic order, h	Maximum permissible harmonic current per watt (mA/W)	Maximum permissible harmonic current (A)
3	3.4	2.30
5	1.9	1.14
7	1.0	0.77
9	0.5	0.40
11	0.35	0.33
$13 \leq h \leq 39$ (odd harmonics only)	$3.85/h$	$0.15 - 0.15/h$

Table 2.5 IEEE Standard 519-1992: voltage distortion limits [21]

Bus voltage at PCC	Individual voltage distortion (%)	Total voltage distortion (%)
69 kV and below	3.0	5.0
69.001–161 kV	1.5	2.5
161.001 kV and above	1.0	1.5

Note: High-voltage systems can have up to 2.0% THD, where the cause is an HVDC terminal that will attenuate by the time it is tapped for a user.

Table 2.6 IEC 61000-2-2: voltage distortion limits in public low-voltage network (class 1) [43]

Odd harmonics		Even harmonics		Triplen harmonics	
H	V_h (pu)	h	V_h (pu)	h	V_h (pu)
5	6	2	2	3	5
7	5	4	1	9	1.5
11	3.5	6	0.5	15	0.3
13	3	8	0.5	≥21	0.2
17	2	10	0.5		
19	1.5	≥12	0.2		
23	1				
25	1.5				
≥29	$0.2 + 12.5/h$				

Table 2.7 IEC 61000-2-4: voltage distortion limits in industrial plants (class 2) [43]

Odd harmonics		Even harmonics		Triplen harmonics	
H	V_h (pu)	h	V_h (pu)	h	V_h (pu)
5	6	2	2	3	5
7	5	4	1	9	1.5
11	3.5	6	0.5	15	0.3
13	3	8	0.5	≥21	0.2
17	2	10	0.5		
19	1.5	≥12	0.2		
23	1.5				
25	1.5				
≥29	$0.2 + 12.5/h$				

Table 2.8 IEC 61000-2-4: voltage distortion limits in industrial plants (class 3) [43]

Odd harmonics		Even harmonics		Triplen harmonics	
H	V_h (pu)	h	V_h (pu)	h	V_h (pu)
5	6	2	3	3	6
7	5	4	1.5	9	2.5
11	3.5	≥6	1	15	2
13	3			21	1.75
17	2			≥27	1
19	1.5				
23	1				
25	1.5				
≥29	$5\sqrt{(11/h)}$				

2.6 Power Quality Monitoring

PQ events are random in nature, which occur arbitrarily. Therefore, monitoring of the PQ phenomena becomes almost essential for critical and sensitive equipment in which a huge loss of revenue is expected by PQ problems. The monitoring system used for assessing PQ events may provide enough data to decide for curing and mitigating the power quality problems provided these recording/measuring instruments are selected properly to record PQ events. There are many standards [24] and texts, which are fully devoted to PQ monitoring. Here only a brief introduction is given to justify and awareness of the PQ monitoring.

2.6.1 Objectives of PQ Monitoring

PQ monitoring is required to quantify PQ phenomena at a particular location on electric power equipment. In some situations, the objective of the monitoring may be to diagnose incompatibilities between the supply and the consumer loads. In other cases, it is used to evaluate the electrical environment at a particular location for the required machinery or equipment. In some cases, monitoring may be used to predict performance of the load equipment and to select power quality mitigating systems. PQ monitoring requires the right selection of monitoring equipment, the method of collecting data, and so on. The objective may be as simple as verifying voltage variations at PCC or analyzing the harmonic level within a distribution system. The recorded information needs to meet only the monitoring objectives in order for the monitoring to be successful. The methodology for quantifying monitoring objectives may differ in nature. For example, when PQ monitoring is required to find out shutdown problems in critical equipment, the aim may be to record tolerance events of a few types. Preventive and predictive

Table 2.9 IEEE-519: parameters that can be determined from acquired voltage and current data [21]

ANSI transformer derating factor	Interharmonic rms current	True power factor
Arithmetic sum power factor	Interharmonic rms voltage	Unsigned harmonic power
Arithmetic sum displacement power factor	Current–time product	Vector sum displacement factor
Arithmetic sum volt-amperes	Negative-sequence current	Vector sum power factor
Current crest factor	Negative-sequence voltage	Vector sum volt-amperes
Current THD	Net current	Voltage crest factor
Current THD (rms)	Positive-sequence current	Voltage THD
Current total interharmonic distortion (TID)	Positive-sequence voltage	Voltage THD (rms)
Current TID (rms)	Residual current	Voltage TID
Current imbalance	rms current	Voltage TID (rms)
Displacement power factor	rms current individual harmonics	Voltage telephone interference factor (TIF)
Frequency	rms harmonic current (total)	Voltage TIF (rms)
Fund frequency arithmetic sum volt-amperes	rms voltage	Voltage imbalance
Fund frequency vector sum volt-amperes	rms voltage individual harmonics	Watt-hours
Harmonic power (sum)	Total fund frequency reactive power	Zero-sequence current
IEEE 519 current TDD	Transformer K-factor	Zero-sequence voltage

monitoring may require recorded voltages and currents to quantify the existing level of power quality. Measurement of PQ includes both time- and frequency-domain variables, which may be in the form of overvoltages and undervoltages, interruptions, sags and swells, surges, spikes, notches, transients, phase imbalance, frequency deviations, and harmonic distortion. PQ monitoring may be provided by the utility, the customers, or any other personnel such as energy auditors.

Table 2.9 shows some important parameters that can be determined using suitable algorithms from the voltage and current waveforms, which are acquired, digitized, and stored in the monitors' memory [21].

2.6.2 Justifications for PQ Monitoring

There are many reasons and requirements of power quality monitoring. The major reason for monitoring PQ is the financial damages caused by PQ events in critical and sensitive equipment. PQ problems and events may cause malfunctions, damages, process interruptions, and other anomalies in the equipment and their operations. PQ monitoring needs resources in terms of equipment, training, education, and, of course, time. There are benefits of PQ monitoring, but industry management and plant and production engineers must agree with the investment. The PQ monitoring may be used as a tool for ensuring the availability of power to the customers. Some of the following aspects may be used to convince users for PQ monitoring:

- To find out the need for mitigation of PQ problems
- To schedule preventive and predictive maintenance
- To ensure the performance of equipment
- To assess the sensitivity of equipment to PQ disturbances
- To identify power quality events and problems
- To reduce the power losses in the process and distribution system
- To reduce the loss in production and to improve equipment availability

These are a few points; however, PQ monitoring may also be used for upgrading, modernizing, removal of obsolescence, and renovation process.

Power quality problems caused by various events and disturbances are specified in terms of different performance indices, which are monitored by various instruments.

2.7 Numerical Examples

Example 2.1

For a square wave of current with an amplitude I of 100 A (shown in Figure E2.1), calculate (a)crest factor (CF), (b)distortion factor (DF), and (c) total harmonic distortion (THD).

Solution: Given a square wave that has an amplitude $I = 100$ A.

The rms value of the fundamental component of a square wave is $I_1 = (2\sqrt{2}/\pi)I = 0.9$ times its amplitude $= 0.9I$.

The rms value of a square wave is $I_{rms} =$ amplitude of a square wave $= I$.

a. CF of a square wave = peak value/rms value of a square wave $= I/I = 1$.
b. DF = fundamental component of a square wave/rms value of a square wave $= I_1/I = 0.9$.
c. THD of a square wave $= \sqrt{(I_{rms}^2 - I_1^2)}/I_1 = \sqrt{\{I^2 - (0.9I)^2\}}/0.9I = 0.4843 = 48.34\%$.

Example 2.2

For a quasi-square wave (120° pulse width) of current with an amplitude I of 100 A (shown in Figure E2.2), calculate (a) crest factor (CF), (b) distortion factor (DF), and (c) total harmonic distortion (THD).

Solution: Given a quasi-square wave (120° pulse width) that has an amplitude $I = 100$ A.

The rms value of the fundamental component of a quasi-square wave is $I_1 = (2\sqrt{2}/\pi)\sin(120°/2)$ $I = (\sqrt{6}/\pi)I$ times its amplitude $= 0.7797I$.

The rms value of a quasi-square wave is $I_{rms} = \sqrt{\{(2/3)I\}} = 0.8165I$.

a. CF of a quasi-square wave = peak value/rms value of a quasi-square wave $= I/0.8165I = 1.225$.
b. DF = fundamental component of a quasi-square wave/rms value of a quasi-square wave $= I_1/I = 0.9549$.
c. THD of a quasi-square wave $= \sqrt{(I_{rms}^2 - I_1^2)}/I_1 = 0.3108 = 31.08\%$.

Example 2.3

For a triangular wave of current with an amplitude I of 100 A (shown in Figure E2.3), calculate (a) crest factor (CF), (b) distortion factor (DF), and (c) total harmonic distortion (THD).

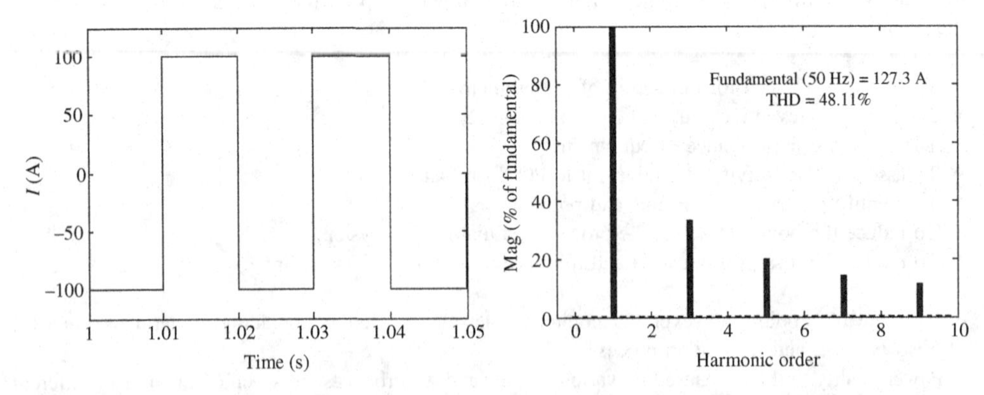

Figure E2.1 Waveform and harmonic spectrum of a square wave of current with an amplitude I of 100 A

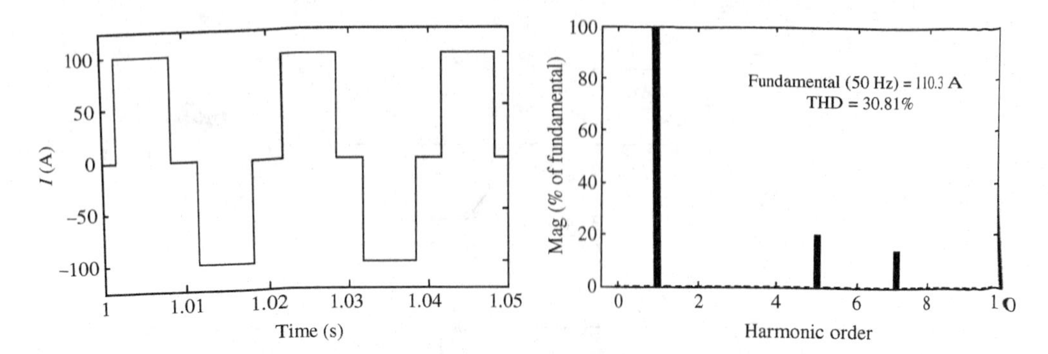

Figure E2.2 Waveform and harmonic spectrum of a quasi-square wave of current with an amplitude I of 100 A

Solution: Given a triangular wave that has an amplitude $I = 100$ A. From Fourier analysis of a triangular wave, the amplitude of the hth harmonic of a triangular wave is

$$I_h = \sum_{h=1}^{\infty}[\{8/(h\pi)^2\}(-1)^{(h-1)/2}\sin(h\theta)]/\sqrt{2} = 8I/\{\sqrt{2}(\pi h)^2\}.$$

The rms value of the fundamental component of a triangular wave is $I_1 = (4\sqrt{2}/\pi^2)I = 0.5732I$. The rms value of a triangular wave is $I_{\mathrm{rms}} = \sqrt{(1/3)}I = 0.5774I$.

a. CF of a triangular wave = peak value/rms value of a triangular wave = $I/0.5774I = \sqrt{3} = 1.732$.
b. DF = fundamental component of a triangular wave/rms value of a triangular wave = $I_1/I_{\mathrm{rms}} = 0.573/0.577\,35 = 0.9927$.
c. THD of a triangular wave = $\sqrt{(I_{\mathrm{rms}}^2 - I_1^2)}/I_1 = 0.121\,15 = 12.12\%$.

Example 2.4

For a trapezoidal wave (90° flat portion) of current with an amplitude I of 100 A (shown in Figure E2.4), calculate (a) crest factor (CF), (b) distortion factor (DF), and (c) total harmonic distortion (THD).

Solution: Given a trapezoidal wave (90° flat portion) that has an amplitude $I = 100$ A. From Fourier analysis of a trapezoidal wave (90° flat portion), the amplitude of the hth harmonic is

$$I_h = \frac{8}{\pi^2}\left[\frac{\sin(h\pi/4) + \sin(3h\pi/4)}{h^2}\right].$$

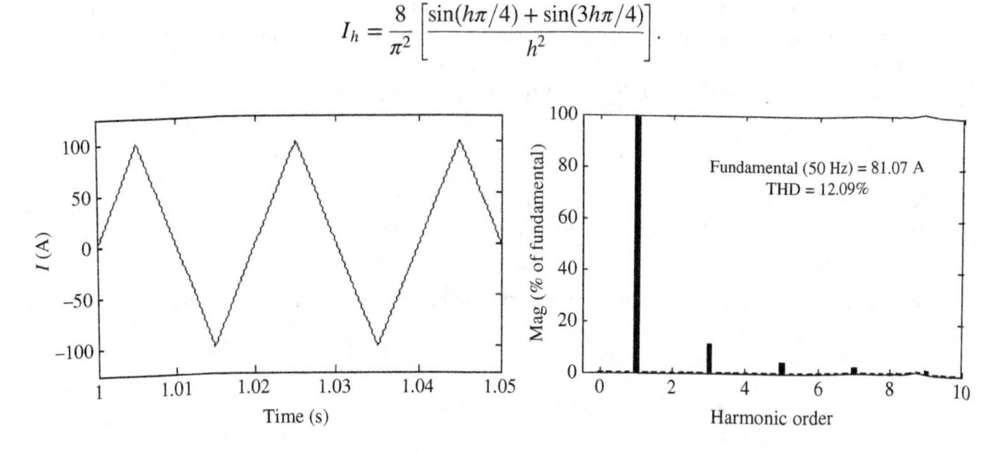

Figure E2.3 Waveform and harmonic spectrum of a triangular wave of current with an amplitude I of 100 A

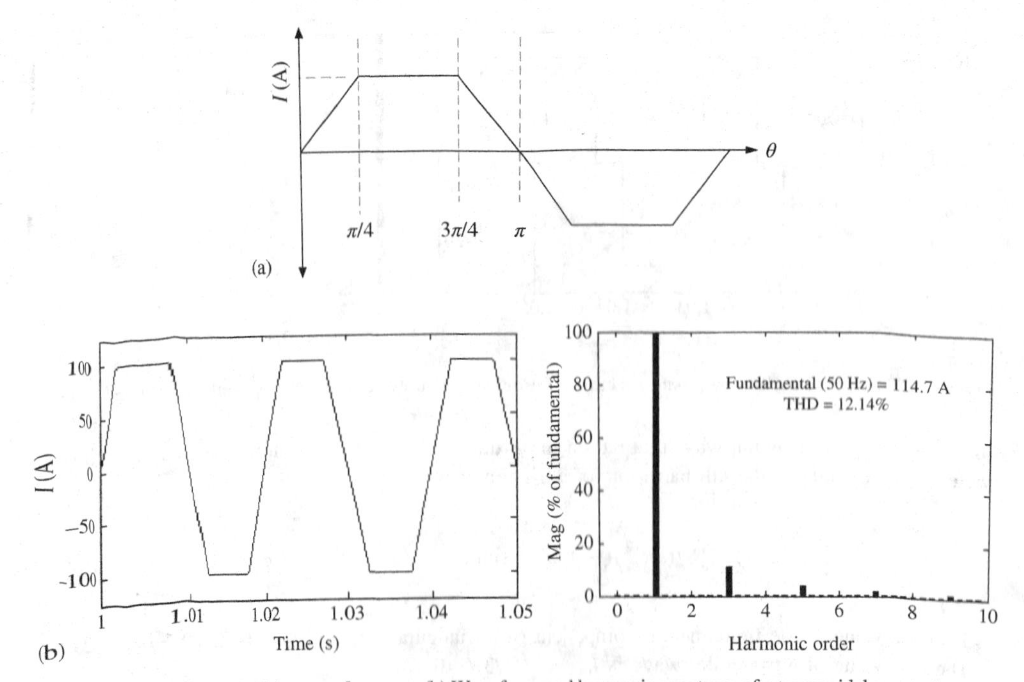

Figure E2.4 (a) Trapezoidal wave of current. (b) Waveform and harmonic spectrum of a trapezoidal wave of current with an amplitude I of 100 A

The rms value of the fundamental component of a trapezoidal wave (90° flat portion) is $I_1=(8/\pi^2)I=0.8106I$.

The rms value of a trapezoidal wave (90° flat portion) is

$$I_{\text{rms}}=\sqrt{\frac{1}{\pi}\left[2\int_0^{\pi/4}\left(\frac{4I}{\pi}\theta\right)^2 \mathrm{d}\theta+\int_{\pi/4}^{3\pi/4} I^2\,\mathrm{d}\theta\right]}=\sqrt{\frac{2}{3}}\times I=0.8165I.$$

a. CF of a trapezoidal wave = peak value/rms value of a trapezoidal wave = $I/0.8165I=1.2247$.
b. DF = fundamental component of a trapezoidal wave/rms value of a trapezoidal wave = $I_1/I_{\text{rms}}=0.8106I/0.8165I=0.9927$.
c. THD of a trapezoidal wave = $\sqrt{(I_{\text{rms}}^2-I_1^2)}/I_1=0.1211=12.12\%$.

Example 2.5

A three-phase unbalanced supply system has the following phase voltages: $V_a=0.9\angle 0°$ pu, $V_b=1.1\angle 240°$ pu, and $V_c=0.95\angle 120°$ pu. Find the positive-, negative-, and zero-sequence components of supply voltages.

Solution: Given a three-phase unbalanced supply system having the following voltages: $V_a=0.9\angle 0°$ pu, $V_b=1.1\angle 240°$ pu, and $V_c=0.95\angle 120°$ pu.

The zero-sequence component for phase a is

$$V_{a0}=(V_a+V_b+V_c)/3=(0.9-0.55-j0.953-0.475+j0.823)=-0.0417-j0.043$$
$$=0.0601\angle 226.12°\text{ pu}.$$

The zero-sequence components for phases b and c are

$$V_{a0} = V_{b0} = V_{c0} = 0.0601\angle 226.12^\circ \text{ pu.}$$

The positive-sequence component for phase a is

$$\begin{aligned} V_{a1} &= (V_a + aV_b + a^2V_c)/3 \\ &= (0.9\angle 0^\circ + 1\angle 120^\circ \times 1.1\angle 240^\circ + 1\angle 240^\circ \times 0.95\angle 120^\circ)/3 \\ &= (0.9 + 1.1 + 0.95)/3 = 0.9835\angle 0^\circ \text{ pu.} \end{aligned}$$

The positive-sequence components for phases b and c are

$$V_{b1} = a^2V_{a1} = 1\angle 240^\circ \times 0.983\angle 0^\circ = 0.9835\angle 240^\circ \text{ pu,}$$

$$V_{c1} = aV_{a1} = 1\angle 120^\circ \times 0.983\angle 0^\circ = 0.9835\angle 120^\circ \text{ pu.}$$

The negative-sequence component for phase a is

$$\begin{aligned} V_{a2} &= (V_a + a^2V_b + aV_c)/3 \\ &= (0.9\angle 0^\circ + 1\angle 240^\circ \times 1.1\angle 240^\circ + 1\angle 120^\circ \times 0.95\angle 120^\circ)/3 \\ &= (0.9\angle 0^\circ + 1.1\angle 480^\circ + 0.95\angle 240^\circ)/3 = -0.0418 + j0.0433 = 0.0602\angle 134^\circ \text{ pu.} \end{aligned}$$

The negative-sequence components for phases b and c are

$$V_{b2} = aV_{a2} = 1\angle 120^\circ \times 0.0602\angle 134^\circ = 0.0602\angle 254^\circ \text{ pu,}$$

$$V_{c2} = a^2V_{a2} = 1\angle 240^\circ \times 0.0602\angle 134^\circ = 0.0602\angle 14^\circ \text{ pu.}$$

Example 2.6

A three-phase balanced supply system has phase voltages $V_a = 1.0\angle 0^\circ$ pu, $V_b = 1.0\angle 240^\circ$ pu, and $V_c = 1.0\angle 120^\circ$ pu and unbalanced load currents $I_a = 0.75\angle -20^\circ$ pu, $I_b = 0.65\angle 270^\circ$ pu, and $I_c = 0.35\angle 90^\circ$ pu. Find (a) the total complex power, (b) the positive-sequence component of power, (c) the negative-sequence component of power, and (d) the zero-sequence component of power.

Solution: Given a three-phase balanced supply system having phase voltages $V_a = 1.0\angle 0^\circ$ pu, $V_b = 1.0\angle 240^\circ$ pu, and $V_c = 1.0\angle 120^\circ$ pu and unbalanced load currents $I_a = 0.75\angle -20^\circ$ pu, $I_b = 0.65\angle 270^\circ$ pu, and $I_c = 0.35\angle 90^\circ$ pu, with $a = 1.0\angle 120^\circ$ and $a^2 = 1.0\angle 240^\circ$.

a. The total complex power is

$$\begin{aligned} P_{abc} + jQ_{abc} &= V_aI_a^* + V_bI_b^* + V_cI_c^* \\ &= 1.0\angle 0^\circ \times 0.75\angle 20^\circ + 1.0\angle 240^\circ \times 0.65\angle -270^\circ + 1.0\angle 120^\circ \times 0.35\angle -90^\circ \\ &= 0.75\angle 20^\circ + 0.65\angle -30^\circ + 0.35\angle 30^\circ = (1.5707 + j0.1065) \text{ pu.} \end{aligned}$$

b. The positive-sequence component of power is

$$\begin{aligned} P_1 + jQ_1 &= V_{a1}I_{a1}^* = (1/3)\{(V_a + aV_b + a^2V_c)(I_a + aI_b + a^2I_c)^*\} \\ &= (1/3)\{(1.0\angle 0^\circ + 1.0\angle 120^\circ \times 1.0\angle 240^\circ + 1.0\angle 240^\circ \times 1.0\angle 120^\circ) \\ &\quad (0.75\angle -20^\circ + 1.0\angle 120^\circ \times 0.65\angle 270^\circ + 1.0\angle 240^\circ \times 0.35\angle 90^\circ)^*\} \\ &= (1/3)\{(1.0 + 1.0 + 1.0)(1.571 + j0.1065) = (1.5707 + j0.1065) \text{ pu.} \end{aligned}$$

c. The negative-sequence component of power is

$$\begin{aligned}P_2 + jQ_2 &= V_{a2}I_{a2}^* = (1/3)\{(V_a + a^2V_b + aV_c)(I_a + a^2I_b + aI_c)^*\}\\ &= (1/3)\{(1.0\angle 0° + 1.0\angle 240° \times 1.0\angle 240° + 1.0\angle 120° \times 1.0\angle 120°)(0.75\angle -20°\\ &\quad + 1.0\angle 240° \times 0.65\angle 270° + 1.0\angle 120° \times 0.35\angle 90°)^*\}\\ &= (1/3)\{(1.0 + 1.0\angle 120° + 1.0\angle 240°)(0.75\angle -20° + 1.0\angle 240° \times 0.65\angle 270°\\ &\quad + 1.0\angle 120° \times 0.35\angle 90°)^*\}\\ &= (1/3)\{(0.0)(0.75\angle - 20° + 1.0\angle 240° \times 0.65\angle 270° + 1.0\angle 120° \times 0.35\angle 90°)^*\} = 0.0\,\text{pu}.\end{aligned}$$

d. The zero-sequence component of power is

$$\begin{aligned}P_0 + jQ_0 &= V_{a0}I_{a0}^* = (1/3)\{(V_a + V_b + V_c)(I_a + I_b + I_c)^*\}\\ &= (1/3)\{(1.0\angle 0° + 1.0\angle 240° + 1.0\angle 120°)(0.75\angle -20° + 0.65\angle 270° + 0.35\angle 90°)^*\}\\ &= (1/3)\{(0.0)(0.75\angle -20° + 0.65\angle 270° + 0.35\angle 90°)^*\} = 0.0\,\text{pu}.\end{aligned}$$

It means that the total complex power is equal to the sum of all three components of power:

$$P_{abc} + Q_{abc} = P_0 + jQ_0 + P_1 + jQ_1 + P_2 + jQ_2 = V_{a0}I_{a0}^* + V_{a1}I_{a1}^* + V_{a2}I_{a2}^* = (1.5706 + j0.1065)\,\text{pu}.$$

It can be observed that the powers of the zero-sequence component and negative-sequence component are zero if AC mains three-phase voltages are balanced even if three-phase currents are unbalanced. Similarly, the powers of the zero-sequence component and negative-sequence component will be zero if AC mains three-phase voltages are unbalanced when three-phase currents are balanced.

Example 2.7

A three-phase unbalanced supply system has phase voltages $V_a = 1.1\angle 0°$ pu, $V_b = 1.0\angle 230°$ pu, and $V_c = 0.9\angle 120°$ pu and unbalanced load currents $I_a = 0.75\angle -20°$ pu, $I_b = 0.65\angle 270°$ pu, and $I_c = 0.35\angle 90°$ pu. Find (a) the total complex power, (b) the positive-sequence component of power, (c) the negative-sequence component of power, and (d) the zero-sequence component of power.

Solution: Given a three-phase unbalanced supply system having phase voltages $V_a = 1.1\angle 0°$ pu, $V_b = 1.0\angle 230°$ pu, and $V_c = 0.9\angle 120°$ pu and unbalanced load currents $I_a = 0.75\angle -20°$ pu, $I_b = 0.65\angle 270°$ pu, and $I_c = 0.35\angle 90°$ pu, with $a = 1.0\angle 120°$ and $a^2 = 1.0\angle 240°$.

a. The total complex power is

$$\begin{aligned}P_{abc} + jQ_{abc} &= V_aI_a^* + V_bI_b^* + V_cI_c^*\\ &= 1.1\angle 0° \times 0.75\angle 20° + 1.0\angle 230° \times 0.65\angle -270° + 0.9\angle 120° \times 0.35\angle -90°\\ &= 0.825\angle 20° + 0.65\angle -40° + 0.315\angle 30° = (1.5459 + j0.0219)\,\text{pu}.\end{aligned}$$

b. The positive-sequence component of power is

$$\begin{aligned}P_1 + jQ_1 &= V_{a1}I_{a1}^* = (1/3)\{(V_a + aV_b + a_2V_c)(I_a + aI_b + a^2I_c)^*\}\\ &= (1/3)\{(1.1\angle 0° + 1.0\angle 120° \times 1.0\angle 230° + 1.0\angle 240° \times 0.9\angle 120°)(0.75\angle -20°\\ &\quad +1.0\angle 120° \times 0.65\angle 270° + 1.0\angle 240° \times 0.35\angle 90°)^*\}\\ &= (1/3)\{(2.9846 - j0.1758)(1.5708 + j0.1065)^*\} = (1.5688 + j0.0151)\,\text{pu}.\end{aligned}$$

c. The negative-sequence component of power is

$$\begin{aligned}P_2 + jQ_2 &= V_{a2}I_{a2}^* = (1/3)\{(V_a + a^2V_b + aV_c)(I_a + a^2I_b + aI_c)^*\}\\ &= (1/3)\{(1.1\angle 0° + 1.0\angle 240° \times 1.0\angle 230° + 1.0\angle 120° \times 0.9\angle 120°)(0.75\angle -20°\\ &\quad +1.0\angle 240° \times 0.65\angle 270° + 1.0\angle 120° \times 0.35\angle 90°)^*\}\\ &= (1/3)\{(0.3088 + j0.1612)(-0.1612 + j0.1065)^*\} = (-0.0222 + j0.0023)\,\text{pu}.\end{aligned}$$

d. The zero-sequence component of power is

$$
\begin{aligned}
P_0 + jQ_0 &= V_{a0}I_{a0}^* = (1/3)\{(V_a + V_b + V_c)(I_a + I_b + I_c)^*\} \\
&= (1/3)\{(1.1\angle 0° + 1.0\angle 230° + 0.9\angle 120°)(0.75\angle -20° + 0.65\angle 270° + 0.35\angle 90°)^*\} \\
&= (1/3)\{(0.007 + j0.013)(0.7048 - 0.5565i)^*\} = (1/3)(-0.0023 + j0.0135) \\
&= (-0.0008 + j0.0045)\ \text{pu}.
\end{aligned}
$$

It means that the total complex power is equal to the sum of all three components of power: $P_{abc} + jQ_{abc} = P_0 + jQ_0 + P_1 + jQ_1 + P_2 + jQ_2 = V_{a0}I_{a0}^* + V_{a1}I_{a1}^* + V_{a2}I_{a2}^* = (1.5458 + j0.0219)$ pu.

Example 2.8

In a three-phase AC mains, there is a voltage sag at PCC of 10, 20, and 30% on three phases for 10, 15, and 20 cycles, respectively. Calculate (a) Detroit Edison sag score (SS) and (b)voltage sag lost energy index (VSLEI) of this sag event.

Solution: Given a three-phase AC mains having a voltage sag at PCC. It results in $V_1 = 0.9$ pu, $V_2 = 0.8$ pu, $V_3 = 0.7$ pu, $t_1 = 200$ ms, $t_2 = 300$ ms, and $t_3 = 400$ ms.

Qualifying sag for Detroit Edison sag score has at least one phase equal to or below 0.75 pu.

a. Detroit Edison sag score is $SS = (V_a + V_b + V_c)/3 = (0.9 + 0.8 + 0.7)/3 = 0.8$.
b. VSLEI of this sag event is

$$
\begin{aligned}
\text{VLSEI} &= (1 - V_a/V_{nom})^{3.14}t_a + (1 - V_b/V_{nom})^{3.14}t_b + (1 - V_c/V_{nom})^{3.14}t_c \\
&= 0.1^{3.14} \times 200 + 0.2^{3.14} \times 300 + 0.3^{3.14} \times 400 \\
&= 0.145 + 1.916 + 9.125 = 11.1855.
\end{aligned}
$$

Example 2.9

Estimate the K-factor rating of a single-phase transformer used to feed a single-phase diode bridge rectifier with constant DC load current.

Solution: Given that the input current (transformer secondary current) to the single-phase diode bridge rectifier with constant DC load current is a square wave, which can be expressed as

$$i(t) = (4I_a/\pi)\{\sin \omega t + (1/3)\sin 3\omega t + (1/5)\sin 5\omega t + (1/7)\sin 5\omega t + (1/9)\sin 9\omega t + (1/11)\sin 11\omega t + \cdots\}.$$

K-factor $= \sum I_h^2 h^2$, where I_h is the fraction of the total rms load current at harmonic h.
$I_{Trms} = I_a$, as total rms current of a square wave is equal to its amplitude.

$$I_1 = (4I_a/\pi)/\sqrt{2} = 0.9I_a = 0.9I_{Trms}.$$

Let us consider only first six harmonics as the magnitude of 11th-order harmonic is less than 1%. Moreover, such small currents of high order do not cause much eddy current losses due to reduced depth of penetration and, in actual practice, magnitude of higher order harmonics further reduces drastically because of source impedance and other nonlinearity.

$$I_h/I_1 = I_1/h = 1, 1/3, 1/5, 1/7, 1/9, 1/11 = 1, 0.333, 0.2, 0.143, 0.111, 0.091 \quad \text{for } h = 1, 3, 5, 7, 9, 11.$$

$$I_h/I_{Trms} = (I_h/I_1) \times (I_1/I_{Trms}) = (I_h/I_1) \times 0.9 = 0.9, 0.9/3, 0.9/5, 0.9/7, 0.9/9, 0.9/11 \quad \text{for } h = 1, 3, 5, 7, 9, 11.$$

I_h (pu) with a base $I_{\text{Trms}} = I_h/I_{\text{Trms}} = 0.9, 0.3, 0.18, 0.1285, 0.1, 0.0819$ for $h = 1, 3, 5, 7, 9, 11$.

$$K\text{-factor} = \sum I_h^2 h^2 = 0.9^2 \times 1^2 + 0.3^2 \times 3^2 + 0.18^2 \times 5^2 + 0.1285^2 \times 7^2 + 0.1^2 \times 9^2 + 0.08197^2 \times 11^2 = 4.86.$$

Example 2.10

Estimate the K-factor rating of a three-phase transformer used to feed a three-phase diode bridge rectifier with constant DC load current.

Solution: The harmonic components of input current of a three-phase diode bridge rectifier with constant DC load current are

$$i_a(t) = (2\sqrt{3}I/\pi)\{\cos\omega t - (1/5)\cos 5\omega t + (1/7)\cos 7\omega t - (1/13)\cos 13\omega t + (1/17)\cos 17\omega t - \cdots\}.$$

$$K\text{-factor} = \sum I_h^2 h^2 \quad \text{for } h = 1, \ldots, n.$$

$$I_h/I_1 = 1, 0.2, 0.1429, 0.0909, 0.0769, 0.0588 \quad \text{for } h = 1, 5, 7, 11, 13, 17.$$

$$(I_h/I_1)^2 = 1, 0.04, 0.0204, 0.0083, 0.0059, 0.0035.$$

$$I_{\text{Trms}}/I_1 = \sqrt{\sum (I_h/I_1)^2} = 1.0383$$

$$I_h\ (\text{pu}) = (I_1/I_{\text{Trms}}) \times (I_h/I_1) = 0.9631, 0.1926, 0.1376, 0.0876, 0.0741, 0.0567.$$

$$K\text{-factor} = \sum I_h^2 h^2 = 5.56.$$

Example 2.11

A single-phase transformer is used to feed a single-phase diode bridge rectifier with constant DC load current of 100 A. The transformer has been rated for a winding eddy current loss density of 10% (0.1 pu). Calculate its derating factor.

Solution: In a single-phase diode bridge rectifier, supply rms current is $I_s = I_0 = 100$ A.
The harmonics present in the current are 3, 5, 7, 9, 11, . . .
The maximum rated eddy current loss density is $P_{\text{EC-R(pu)}} = 0.1$.
The maximum load current loss density is $P_{\text{LL-R(pu)}} = 1 + 0.1 = 1.1$.

$$\sum (I_h/I_1)^2 = 1^2 + (1/3)^2 + (1/5)^2 + (1/7)^2 + (1/9)^2 + (1/11)^2 = 1.1921.$$

$$\sum (I_h/I_1)^2 h^2 = 1^2 + (1/3)^2 3^2 + (1/5)^2 5^2 + (1/7)^2 7^2 + (1/9)^2 9^2 + (1/11)^2 11^2 = 6.$$

The harmonic loss factor is $F_{\text{HL}} = \sum (I_h/I_1)^2 h^2 / \sum (I_h/I_1)^2 = 6/1.1921 = 5.033$ for $h = 1, 3, 5, 7, 9, 11$.

$$I_{\max(\text{pu})} = \sqrt{\{P_{\text{LL-R(pu)}}/(1 + F_{\text{HL}} P_{\text{EC-R(pu)}})\}} = \sqrt{1.1/(1 + 5.033 \times 0.1)} = 0.8554.$$

So, the derating factor is estimated to be on the order of 85.5% (0.855 pu).

Example 2.12

A single-phase fully controlled bridge converter (shown in Figure E2.12) is fed from a supply of 220 V at 50 Hz at a thyristor firing angle of $\alpha = 60°$. Consider continuous load current of 20 A. Compute (a) total harmonic distortion (THD) of AC mains current, (b)distortion index (DIN) of AC mains current, (c) total

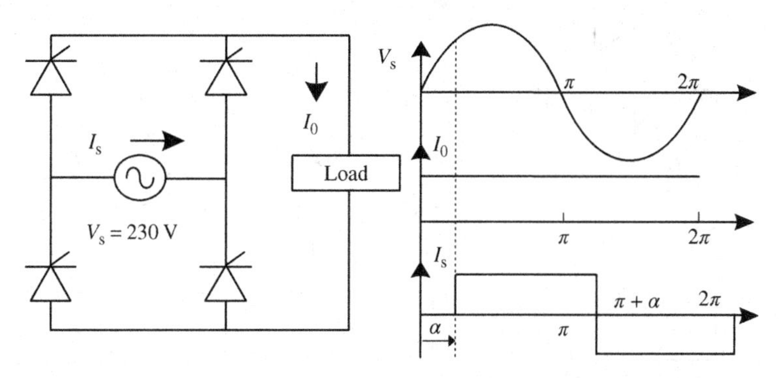

Figure E2.12 Single-phase converter-based current-fed type of a nonlinear load

demand distortion (TDD) of AC mains current, (d) distortion factor (DF), (e)displacement factor (DPF), and (f)power factor (PF).

Solution: Given supply rms voltage $V_s = 220$ V, frequency of the supply $(f) = 50$ Hz, $I_O = 20$ A, and $\alpha = 60°$.

In a single-phase thyristor bridge converter, the waveform of the supply current (I_s) is a square wave with the amplitude of DC link current (I_0). Moreover, the rms value of the fundamental component of a square wave is 0.9 times its amplitude. Therefore, $I_s = I_0 = 20$ A and $I_{s1} = 0.9I_0 = 18$ A.

a. THD of AC current $= \sqrt{(I_s^2 - I_{s1}^2)}/I_{s1} = 0.4843 = 48.43\%$.
b. DIN of AC mains current $= \text{THD}/\sqrt{(1 + \text{THD}^2)} = 0.435\ 889 = 43.59\%$.
c. TDD of AC mains current = total current demand distortion = calculated harmonic current distortion against the full load (demand) level of the electrical system.
   At full load, $\text{TDD}(I) = \text{THD}(I) = 0.4843 = 48.43\%$.
   Therefore, TDD gives us better insight into the impact of harmonic distortion in the system. For example, one could have very high THD, but the load of the system is low. In this case, the impact on the system is also low.
d. $\text{DF} = 1/\sqrt{(1 + \text{THD}^2)} = I_{s1}/I_s = 0.90$.
e. $\text{DPF} = \cos\theta_1 = \cos\alpha = \cos 60° = 0.5$.
f. $\text{PF} = \text{DPF} \times \text{DF} = 0.9 \times 0.5 = 0.45$.

Example 2.13

A single-phase AC voltage controller (shown in Figure E2.13) has a heating load (resistive load) of 20 Ω. The input voltage is 220 V (rms) at 50 Hz. The delay angle of thyristors is $\alpha = 120°$. Feeder conductors have the resistance (R_s) of 0.20 Ω each. Calculate (a) AC source rms current (I_s) and (b) losses in the distribution system. If an ideal shunt compensator is used to compensate power factor to unity of this load, then calculate (c) AC source rms current (I_{sc}), (d) losses in the distribution system, and (e) ratio of losses in the distribution system without and with a compensator.

Solution: Given supply rms voltage $V_s = 220$ V, frequency of the supply $(f) = 50$ Hz, $R = 20\,\Omega$, $R_s = 0.20\,\Omega$, and $\alpha = 120°$. The total resistance of the circuit is $R_T = R + 2R_s = 20.4\,\Omega$.

In a single-phase, phase-controlled AC controller, the waveform of the supply current (I_s) has a value of V_s/R_T from angle α to π. $V_{sm} = 220\sqrt{2} = 311.13$ V.

a. The supply rms current is $I_s = V_{sm}[\{1/(2\pi)\}\{(\pi - \alpha) + \sin 2\alpha/2\}]^{1/2}/R_T = 4.7683$ A.
   The active power of the load is $P_L = I_s^2 R = 4.768^2 \times 20 = 454.7411$ W.
b. Losses in the distribution system are $P_{Loss} = 2I_s^2 R_s = 2 \times 4.768^2 \times 0.20 = 9.0948$ W.

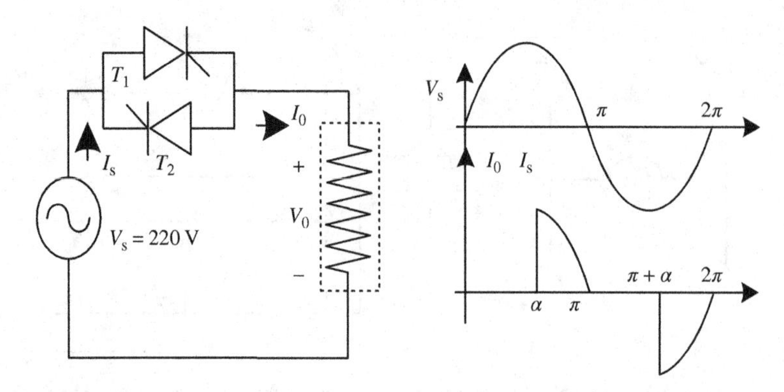

Figure E2.13 A single-phase AC voltage controller feeding a heating load with phase control

c. After the compensation, the power factor is corrected to unity of the AC mains by a shunt compensator. The current in the AC mains becomes sinusoidal in phase with the phase voltage.
The new supply current is the rms value of the fundamental active power component of load current, $I_{sc} = I_{s1a} = P_L/V_s = 454.75/220 = 2.067$ A.
d. Losses in the distribution system are $P_{Lossc} = 2I_{sc}^2 R_s = 2 \times 2.067^2 \times 0.2 = 1.709$ W.
e. Ratio of losses without and with a compensator is $P_{Loss}/P_{Lossc} = 9.1/1.71 = 5.3217$.
It means that such nonlinear loads cause up to 5.3217 times increased losses in the distribution system.

Example 2.14

A single-phase AC voltage controller (shown in Figure E2.14) is used to control the heating of packing element in a food vending machine at a power of 200 W at 20 V fed from a single-phase AC mains of 230 V at 50 Hz. Feeder conductors have the resistance of 0.25 Ω each. Calculate (a) AC source rms current (I_s) and (b) losses in the distribution system. If an ideal shunt compensator is used to compensate power factor to unity, then calculate (c) AC source rms current (I_{sc}), (d) losses in the distribution system, and (e) ratio of losses in the distribution system without and with a compensator.

Solution: Given supply rms voltage $V_s = 230$ V, frequency of the supply $(f) = 50$ Hz, $P = 200$ W, and $R_s = 0.25\ \Omega$.

The load resistance is $R_L = V_{Ls}^2/P = 20^2/200 = 2.0\ \Omega$.
The rms voltage across the load is $V_{Ls} = I_s R_L = 20$ V.

a. The supply rms current is $I_s = \sqrt{(P/R_L)} = V_{Ls}/R_L = 10$ A.
b. Losses in the distribution system are $P_{Loss} = 2I_s^2 R_s = 2 \times 10^2 \times 0.25 = 50$ W.

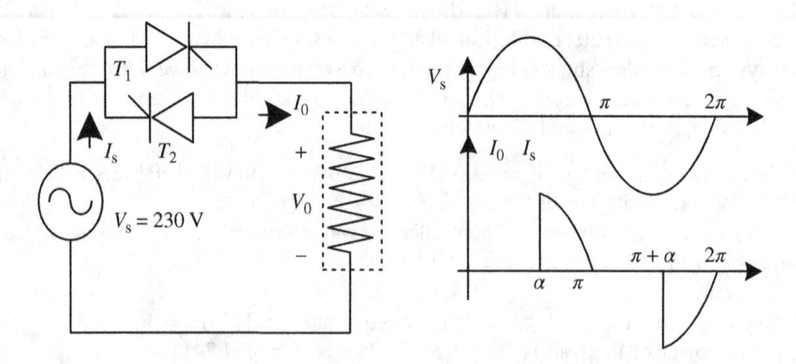

Figure E2.14 A single-phase AC voltage controller feeding a heating load with phase control

c. After the compensation, the power factor is corrected to unity of the AC mains by a shunt compensator.
   The current in the AC mains becomes sinusoidal in phase with the phase voltage.
   The new supply current is the rms value of the fundamental active power component of load current, $I_{sc} = I_{s1a} = P/V_s = 200/230 = 0.8696$ A.
d. Losses in the distribution system are $P_{Lossc} = 2I_{sc}^2R_s = 2 \times 0.8696^2 \times 0.25 = 0.3781$ W.
e. Ratio of losses without and with a compensator is $P_{Loss}/P_{Lossc} = 50/0.378 = 132.25$.
   It means that such nonlinear loads cause up to 132.25 times increased losses in the distribution system.

Example 2.15

A single-phase uncontrolled bridge converter (shown in Figure E2.15) has a *RE* load with $R = 1.0\,\Omega$ and $E = 275$ V. The input AC voltage is $V_s = 220$ V at 50 Hz. Feeder conductors have the resistance of $0.05\,\Omega$ each. Calculate (a) AC source rms current (I_s), (b) losses in the distribution system, (c) total harmonic distortion (THD) in current, and (d) crest factor (CF) of supply current. If an ideal shunt compensator is used to compensate power factor to unity, then calculate (e) AC source rms current (I_{sc}), (f) losses in the distribution system, and (g) ratio of losses in the distribution system without and with a compensator.

Solution: Given supply voltage $V_s = 220$ V, $V_{sm} = 311.13$ V, frequency of the supply (f) = 50 Hz, $R = 1\,\Omega$, and $E = 275$ V.

In a single-phase diode bridge converter, with *RE* load, the current flows from angle α when AC voltage is equal to E to angle β at which AC voltage reduces to E.

The total resistance of the circuit is $R_T = 2R_s + R = 2 \times 0.05 + 1.0 = 1.1\,\Omega$.

$\alpha = \sin^{-1}(E/V_{sm}) = \sin^{-1}(275/311.13) = 62.11°$, $\beta = \pi - \alpha = 117.89°$, and the conduction angle is $\beta - \alpha = 55.78°$.

The active power drawn from the AC mains is $P = I_s^2R_T + EI_0 = 194.99 + 1858.39 = 2053.39$ W.

Fundamental rms current from the AC mains is $I_{s1} = P/V_s = 2053.39/220 = 9.33$ A.

Supply AC peak current is $I_{peak} = (V_{sm} - E)/R_T = 32.84$ A.

Load average current (I_0) is $I_0 = \{1/(\pi R_T)\}(2V_{sm}\cos\alpha + 2E\alpha - \pi E) = 6.76$ A.

a. rms supply current (I_s) is the rms value of discontinuous current in the AC mains: $I_s = [\{1/(\pi R_T^2)\}\{(0.5V_{sm}^2 + E^2)(\pi - 2\alpha) + 0.5V_{sm}^2 \sin 2\alpha - 4V_{sm}E\cos\alpha\}]^{1/2} = 13.31$ A.
b. Losses in the distribution system are $P_{Loss} = 2I_s^2R_s = 2 \times 13.31^2 \times 0.05 = 17.72$ W.
c. THD of AC current $= \sqrt{(I_s^2 - I_{s1}^2)}/I_{s1} = \sqrt{(13.31^2 - 9.33^2)}/9.33 = 1.0173 = 101.73\%$.
d. CF of supply current = peak value/rms value = 32.85/9.33 = 3.52.
e. After the compensation, the power factor is corrected to unity of the AC mains by a shunt compensator.
   The current in the AC mains becomes sinusoidal in phase with the phase voltage.
   The new supply current is the rms value of the fundamental active power component of load current, $I_{sc} = I_{s1a} = P/V_s = I_{s1} = P/V_s = 2053.39/220 = 9.33$ A.

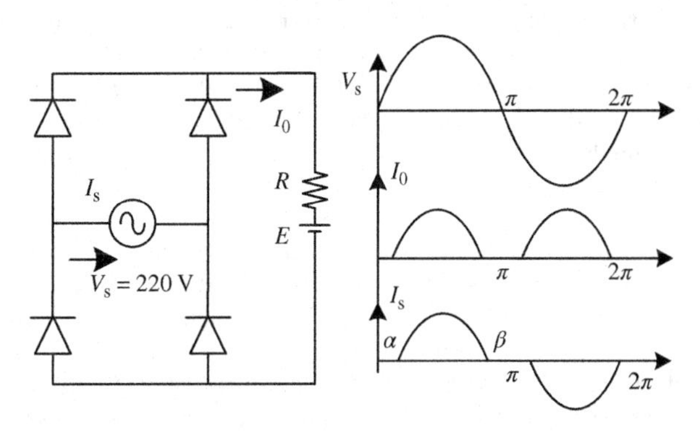

Figure E2.15 A single-phase uncontrolled bridge converter with a *RE* load

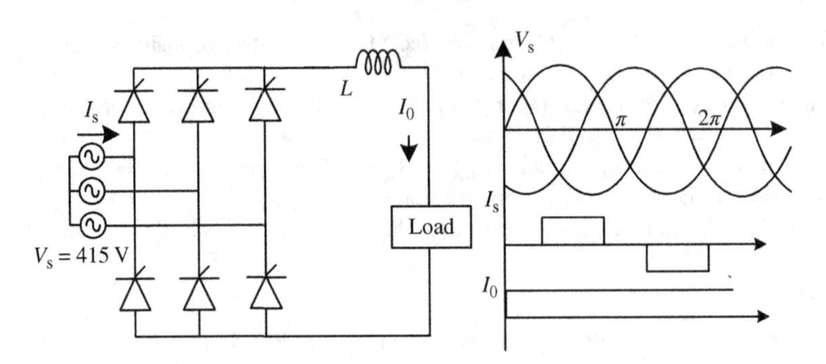

Figure E2.16 A three-phase fully controlled bridge converter feeding *RL* load

f. Losses in the distribution system are $P_{Lossc} = 2I_s^2R_s = 2 \times 9.33^2 \times 0.05 = 8.709$ W.
g. Ratio of losses without and with a compensator is $P_{Loss}/P_{Lossc} = 17.72/8.709 = 2.04$.
   It means that such nonlinear loads cause up to 2.04 times increased losses in the distribution system.

Example 2.16

A three-phase fully controlled bridge converter (shown in Figure E2.16) feeds power to a load having a resistance (R) of 10 Ω and very large inductance to result in continuous current with an input from a three-phase supply of 415 V at 50 Hz. Feeder conductors have the resistance of 0.1 Ω each. For firing angles of 60°, calculate (a) AC source rms current (I_s) and (b) losses in the distribution system. If an ideal shunt compensator is used to compensate power factor to unity, then calculate (c) AC source rms current (I_{sc}), (d) losses in the distribution system, and (e) ratio of losses in the distribution system without and with a compensator.

Solution: Given supply rms voltage $V_s = 415/\sqrt{3} = 239.6$ V, frequency of the supply (f) = 50 Hz, $R_{DC} = 10\,\Omega$, and firing angle $\alpha = 60°$.

In a three-phase thyristor bridge converter, the waveform of the supply current (I_s) is a quasi-square wave with the amplitude of DC link current (I_0).

Average output DC voltage is $V_{DC} = (3\sqrt{3}\sqrt{2}V_s/\pi)\cos\alpha = 280.223$ V.

The DC link current is $I_0 = V_{DC}/R_{DC} = 28.02$ A.

a. The AC source rms current is $I_s = \sqrt{(2/3)}I_0 = 0.816\,49 I_0 = 22.88$ A.
b. Losses in the distribution system are $P_{Loss} = 3I_s^2R_s = 3 \times 22.88^2 \times 0.1 = 157.05$ W.
c. After the compensation, the power factor is corrected to unity of the AC mains by a shunt compensator.
   The three-phase currents in the AC mains become sinusoidal in phase with the phase voltage.
   The new supply current is the rms value of the fundamental active power component of load current, $I_{sc} = I_{s1a} = I_{s1}\cos\alpha = (\sqrt{6})/\pi) I_{DC}\cos\alpha = 28.01 \times 0.78 \times \cos 60° = 10.92$ A.
d. Losses in the distribution system are $P_{Lossc} = 3I_s^2R_s = 3 \times 10.92^2 \times 0.1 = 35.77$ W.
e. Ratio of losses without and with a compensator is $P_{Loss}/P_{Lossc} = 157.05/35.77 = 4.39$.
   It means that such nonlinear loads cause up to 4.39 times increased losses in the distribution system.

Example 2.17

In a three-phase four-wire distribution system with a line voltage of 415 V at 50 Hz, three single-phase loads (connected between phases and neutral terminal) have a single-phase uncontrolled bridge converter having a *RE* load with $R = 1.0\,\Omega$ and $E = 300$ V (shown in Figure E2.17). Feeder and neutral conductors have the resistance of 0.05 Ω each. Calculate (a) AC source rms current (I_s), (b) neutral current (I_{sn}), and (c) losses in the distribution system. If an ideal four-wire shunt compensator is used to compensate power factor to unity in each phase, then calculate (d) AC source rms current (I_{sc}), (e) neutral current (I_{snc}),

(f) losses in the distribution system, and (g) ratio of losses in the distribution system without and with a compensator.

Solution: Given supply voltage $V_s = 415/\sqrt{3}$ V = 239.6 V, frequency of the supply (f) = 50 Hz, $R_s = 0.05\,\Omega$, and a single-phase uncontrolled bridge converter (shown in Figure E2.17) having a *RE* load with $R = 1.0\,\Omega$ and $E = 300$ V.

In a single-phase diode bridge converter, with *RE* load, the current flows from angle α when AC voltage is equal to E to angle β at which AC voltage reduces to E.

The total resistance of the circuit is $R_T = 2R_s + R = 2 \times 0.05 + 1.0 = 1.1\,\Omega$.

$\alpha = \sin^{-1}(E/V_{sm}) = \sin^{-1}(300/338.85) = 62.29°$, $\beta = \pi - \alpha = 117.70°$, and the conduction angle is $\beta - \alpha = 55.41°$.

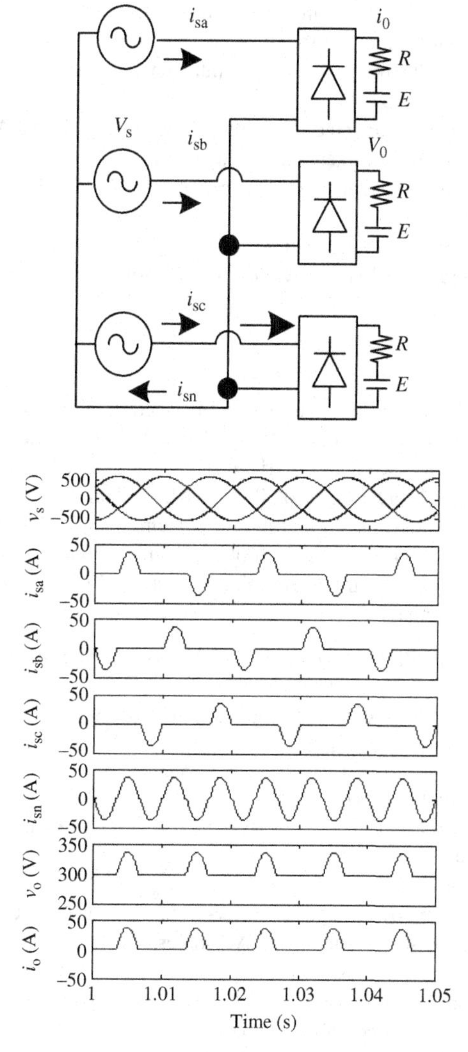

Figure E2.17 A three-phase four-wire distribution system with three single-phase loads (connected between phases and neutral terminal) having a single-phase uncontrolled bridge converter having a *RE* load with $R = 1.0\,\Omega$ and $E = 300$ V

The active power drawn from the AC mains is $P = I_s^2 R_T + EI_0 = 223.95 + 2165.6 = 2389.55$ W.
Fundamental rms current from the AC mains is $I_{s1} = P/V_s = 2389.56/239.6 = 9.97$ A.
Supply AC peak current is $I_{peak} = (V_{sm} - E)/R_T = 35.31$ A.
Load average current is $I_0 = \{1/(\pi R_T)\}(2V_{sm} \cos\alpha + 2E\alpha - \pi E) = 7.22$ A.

a. rms supply current (I_s) is the rms value of discontinuous current in the AC mains: $I_s = [\{1/(\pi R_T^2)\}\{(0.5V_{sm}^2 + E^2)(\pi - 2\alpha) + 0.5V_{sm}^2 \sin 2\alpha - 4V_{sm}E \cos\alpha\}]^{1/2} = 14.27$ A.
b. The neutral current is $I_{sn} = [\{3/(\pi R_T^2)\}\{(0.5V_{sm}^2 + E^2)(\pi - 2\alpha) + 0.5V_{sm}^2 \sin 2\alpha - 4V_{sm}E \cos\alpha\}]^{1/2} =$ 24.71 A (since it is a segment of sine wave with three times the fundamental frequency phase current).
   This neutral current is 1.7321 times the phase current.
c. Losses in the distribution system are $P_{Loss} = 3I_s^2 R_s + I_{sn}^2 R_s = 3 \times 14.272 \times 0.05 + 24.712 \times 0.05 =$ 61.08 W.
d. After the compensation, the power factor is corrected to unity of the AC mains by a four-wire shunt compensator. The three-phase currents in the AC mains become sinusoidal in phase with the phase voltage and neutral current becomes zero.
   The new supply current is the rms value of the fundamental active power component of load current, $I_{sc} = I_{s1a} = 9.97$ A.
e. In this case after the compensation, since the three-phase currents in the AC mains are balanced and sinusoidal, neutral current becomes zero: $I_{snc} = 0.0$ A.
f. Losses in the distribution system are $P_{Lossc} = 3I_{sc}^2 R_s + I_{sn}^2 R_s = 3 \times 9.97^2 \times 0.05 + 0^2 \times 0.05 =$ 14.92 W.
g. Ratio of losses without and with a compensator is $P_{Loss}/P_{Lossc} = 61.08/14.92 = 4.09$.
   It means that such nonlinear loads cause up to 4.09 times increased losses in the distribution system.

Example 2.18

In a three-phase four-wire distribution system with a line voltage of 415 V at 50 Hz, three single-phase loads (connected between phases and neutral terminal) have a single-phase thyristor bridge converter drawing equal 15 A constant DC current at a thyristor firing angle of 60° (shown in Figure E2.18). Feeder and neutral conductors have the resistance of 0.1 Ω each. Calculate (a) AC source rms current (I_s), (b) neutral current (I_{sn}), and (c) losses in the distribution system. If an ideal four-wire shunt compensator is used to compensate power factor to unity in each phase, then calculate (d) AC supply rms current (I_{sc}), (e) neutral current (I_{snc}), (f) losses in the distribution system, and (g) ratio of losses in the distribution system without and with a compensator.

Solution: Given supply voltage $V_s = 415/\sqrt{3}$ V $= 239.6$ V, frequency of the supply (f) = 50 Hz, $R_s = 0.1\ \Omega$, DC link current $I_0 = 15$ A, and firing angle $\alpha = 60°$.

In a single-phase thyristor bridge converter, the waveform of the supply current (I_s) is a square wave with the amplitude of DC link current (I_0).

a. The AC source rms current is $I_s = I_0 = 15$ A.
b. The neutral current is $I_{sn} = 15$ A (since it will also be a square wave with three times the fundamental frequency phase current).
c. Losses in the distribution system are $P_{Loss} = 3I_s^2 R_s + I_{sn}^2 R_s = 3 \times 15^2 \times 0.1 + 15^2 \times 0.1 = 90$ W.
d. After the compensation, the power factor is corrected to unity of the AC mains by a four-wire shunt compensator. The three-phase currents in the AC mains become sinusoidal in phase with the phase voltage and neutral current becomes zero.
   The new supply current is the rms value of the fundamental active power component of load current, $I_{sc} = I_{s1a} = I_{s1} \cos\alpha = 15 \times 0.9 \times \cos 60° = 6.75$ A.
e. In this case after the compensation, since the three-phase currents in the AC mains are balanced and sinusoidal, neutral current becomes zero: $I_{snc} = 0.0$ A.
f. Losses in the distribution system are $P_{Lossc} = 3I_{sc}^2 R_s + I_{sn}^2 R_s = 3 \times 6.75^2 \times 0.1 + 0^2 \times 0.1 = 13.67$ W.

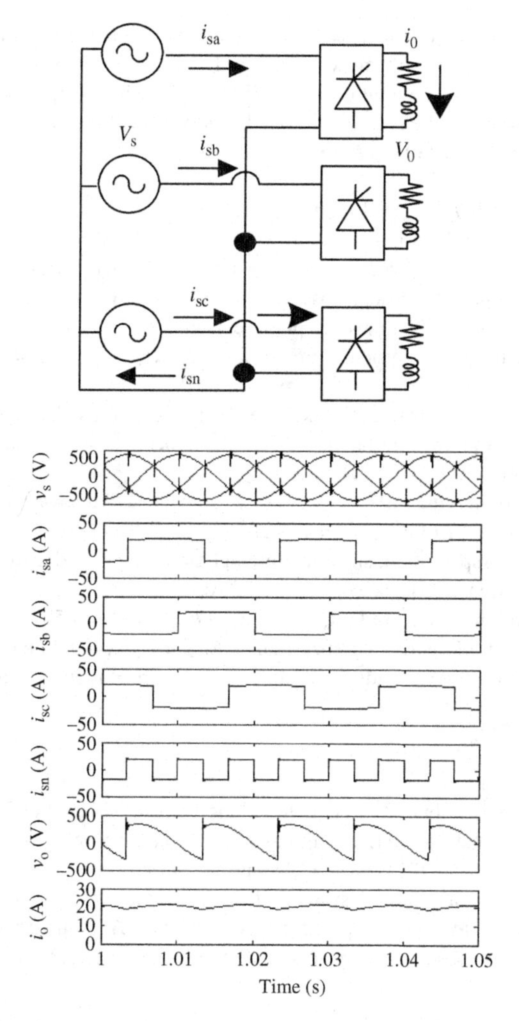

Figure E2.18 A three-phase four-wire distribution system with three single-phase loads (connected between phases and neutral terminal) having a single-phase thyristor bridge converter drawing equal 15 A constant DC current at a thyristor firing angle of 60°

g. Ratio of losses without and with a compensator is $P_{Loss}/P_{Lossc} = 90/13.67 = 6.584$.
 It means that such nonlinear loads cause up to 6.584 times increased losses in the distribution system.

Example 2.19

In a three-phase four-wire distribution system with a line voltage of 415 V at 50 Hz, three single-phase loads (connected between phases and neutral terminal) have a single-phase AC voltage controller for heating loads (resistive load) of 10 Ω. The delay angle of thyristors is $\alpha = 120°$. Feeder and neutral conductors have the resistance of 0.1 Ω each. Calculate (a) AC supply rms current (I_s), (b) neutral current (I_{sn}), and (c) losses in the distribution system. If an ideal four-wire shunt compensator is used to compensate power factor to unity in each phase, then calculate (d) AC supply rms current (I_{sc}), (e) neutral current (I_{snc}), (f) losses in the distribution system, and (g) ratio of losses in the distribution system without and with a compensator.

Solution: Given supply phase voltage $V_s = 415/\sqrt{3}\,V = 239.6\,V$, frequency of the supply $(f) = 50\,Hz$, $R_s = 0.1\,\Omega$, and a single-phase AC voltage controller for heating loads (resistive load) of $10\,\Omega$. The delay angle of thyristors is $\alpha = 120°$. The total circuit resistance of each phase is $R_T = R + 2R_s = 10.2\,\Omega$.

In a single-phase, phase-controlled AC controller, the waveform of the supply current (I_s) has a value of V_s/R_T from angle α to π. $V_{sm} = 239.6\sqrt{2} = 338.85\,V$.

a. AC supply rms current is $I_s = V_{sm}[\{1/(2\pi)\}\{(\pi - \alpha) + \sin 2\alpha/2\}]^{1/2}/R_T = 10.386\,A$.
   The active power of the load is $P_L = 3I_s^2R = 3 \times 10.386^2 \times 10 = 3236.27\,W$.
b. The neutral current is $I_{sn} = V_{sm}[\{3/(2\pi)\}\{(\pi - \alpha) + \sin 2\alpha/2\}]^{1/2}/R_T = 17.99\,A$ (since it is a segment of sine wave with three times the fundamental frequency phase current).
   This neutral current is 1.7321 times the phase current.
c. Losses in the distribution system are $P_{Loss} = 3I_s^2R_s + I_{sn}^2R_s = 3 \times 10.386^2 \times 0.1 + 17.99^2 \times 0.1 = 64.72\,W$.
d. After the compensation, the power factor is corrected to unity of the AC mains by a four-wire shunt compensator. The three-phase currents in the AC mains become sinusoidal in phase with the phase voltage and neutral current becomes zero.
   The new supply current is the rms value of the fundamental active power component of load current, $I_{sc} = I_{s1a} = P_L/3V_s = 3236.27/(3 \times 239.6) = 4.50\,A$.
e. In this case after the compensation, since the three-phase currents in the AC mains are balanced and sinusoidal, neutral current becomes zero: $I_{snc} = 0.0\,A$.
f. Losses in the distribution system are $P_{Lossc} = 3I_{sc}^2R_s + I_{snc}^2R_s = 3 \times 4.50^2 \times 0.1 + 0^2 \times 0.1 = 6.08\,W$.
g. Ratio of losses without and with a compensator is $P_{Loss}/P_{Lossc} = 64.72/6.08 = 10.643$.
   It means that such nonlinear loads cause up to 10.643 times increased losses in the distribution system.

Example 2.20

In a three-phase four-wire distribution system with a line voltage of 415 V at 50 Hz, a single-phase load (connected between phase and neutral terminal) has a 125 A, 0.8 lagging power factor. Feeder and neutral conductors have the resistance of $0.1\,\Omega$ each. Calculate (a) AC supply rms current (I_s), (b) neutral current (I_{sn}), and (c) losses in the distribution system. If an ideal four-wire shunt compensator is used to compensate power factor to unity in each phase, then calculate (d) AC source rms current (I_{sc}), (e) neutral current (I_{snc}), (f) losses in the distribution system, and (g) ratio of losses in the distribution system without and with a compensator.

Solution: Given supply voltage $V_s = 415/\sqrt{3}\,V = 239.6\,V$, frequency of the supply $(f) = 50\,Hz$, $R_s = 0.1\,\Omega$, and a single-phase load (connected between phase and neutral) having a 125 A, 0.8 lagging power factor.

a. The AC source rms current is $I_s = 125\,A$.
b. The neutral current is $I_{sn} = 125\,A$ (same as phase current).
c. Losses in the distribution system are $P_{Loss} = 2I_s^2R_s = 2 \times 125^2 \times 0.1 = 3125\,W$.
d. After the compensation, the power factor is corrected to unity of the AC mains by a four-wire shunt compensator. The three-phase currents in the AC mains become sinusoidal in phase with the phase voltage and neutral current becomes zero. In this case, because of loading, this current is divided in all three phases of the AC mains.
   The new supply current is the rms value of the fundamental active power component of load current, $I_{sc} = I_{s1a} = I_{s1}\cos\theta = 125 \times 0.8/3 = 100/3 = 33.33\,A$.
e. In this case after the compensation, since the three-phase currents in the AC mains are balanced and sinusoidal, neutral current becomes zero: $I_{snc} = 0.0\,A$.
f. Losses in the distribution system are $P_{Lossc} = 3I_{sc}^2R_s = 3 \times 33.33^2 \times 0.1 = 333.33\,W$.
g. Ratio of losses without and with a compensator is $P_{Loss}/P_{Lossc} = 3125/333.33 = 9.375$.
   It means that such unbalanced lagging power factor loads cause up to 9.375 times increased losses in the distribution system.

Example 2.21

A three-phase 22 kW, 415 V, 50 Hz, four-pole delta connected squirrel cage induction motor is used to drive a compressor load of constant torque. It runs at 4% slip at full load and rated voltage and frequency. If terminal voltage reduces to 360 V, calculate its (a) slip, (b) shaft speed, (c) output power, and (d) rotor winding loss as a ratio of rated rotor winding loss at rated voltage. Consider small slip approximation.

Solution: Given that a three-phase 22 kW, 415 V, 50 Hz, four-pole delta connected squirrel cage induction motor is used to drive a compressor load of constant torque. It runs at 4% slip at full load and rated voltage and frequency. Its terminal voltage is 360 V.

a. For a small slip approximation, $S \propto 1/V^2$, the new slip at reduced voltage is $S_n = 0.04(415/360)^2 = 0.053115 = 5.3155\%$.

   The synchronous speed is $N_s = 120f/p = 120 \times 50/4 = 1500$ rpm.
b. The shaft speed at reduced voltage is $N_m = N_s(1 - S_n) = 1500 \times (1 - 0.053\,115) = 1420.27$ rpm.
c. The output power at reduced voltage (at constant torque load) is $P_{on} = \omega_m T_m = \{(1 - S_n)/(1 - S)\}P_O = \{(1 - 0.053\,115)/(1 - 0.04)\} \times 22\,000 = 21\,698.51\text{ W} = 21.698\text{ kW}$.
d. Because of constant torque load, $T = P_g/\omega_{ms}$ = (air gap power/synchronous speed); therefore, P_g is constant. So, rotor winding loss at reduced voltage is $P_{rwn} = S_n P_g = (S_n/S)P_{rwr} = (0.053\,115/0.04)P_{rwr} = 1.328 P_{rwr}$.

   It can be concluded that the decrease in terminal voltage results in an increase in rotor winding loss. However, a decrease in terminal voltage at constant frequency decreases its core loss and magnetizing current, which partly offset its effect.

Example 2.22

A three-phase 22 kW, 415 V, 50 Hz, four-pole delta connected squirrel cage induction motor is used to drive a compressor load of constant torque. It runs at 4% slip at full load and rated voltage and frequency. If terminal voltage increases to 440 V, calculate its (a) slip, (b) shaft speed, (c) output power, and (d) rotor winding loss as a ratio of rated rotor winding loss at rated voltage. Consider small slip approximation.

Solution: Given that a three-phase 22 kW, 415 V, 50 Hz, four-pole delta connected squirrel cage induction motor is used to drive a compressor load of constant torque. It runs at 4% slip at full load and rated voltage and frequency. Its terminal voltage is 440 V.

a. For a small slip approximation, $S \propto 1/V^2$, the new slip at increased voltage is $S_n = 0.04(415/440)^2 = 0.035\,58 = 3.558\%$.

   The synchronous speed is $N_s = 120f/p = 120 \times 50/4 = 1500$ rpm.
b. The shaft speed at increased voltage is $N_m = N_s(1 - S_n) = 1500 \times (1 - 0.035\,58) = 1446.62$ rpm.
c. The output power at increased voltage (at constant torque load) is $P_{on} = \omega_m T_m = \{(1 - S_n)/(1 - S)\}P_O = \{(1 - 0.035\,58)/(1 - 0.04)\} \times 22\,000 = 22\,101.21\text{ W} = 22.101\text{ kW}$.
d. Because of constant torque load, $T = P_g/\omega_{ms}$ = (air gap power/synchronous speed); therefore, P_g is constant. So, rotor winding loss at increased voltage is $P_{rwn} = S_n P_g = (S_n/S)P_{rwr} = (0.035\,58/0.04)P_{rwr} = 0.8895 P_{rwr}$.

   It can be concluded that the increase in terminal voltage results in a decrease in rotor winding loss. However, an increase in terminal voltage at constant frequency increases its core loss and magnetizing current, which partly offset its effect.

Example 2.23

A three-phase 3.7 kW/5 hp, 1420 rpm, delta connected squirrel cage induction motor is used to drive a load. It has its per-phase equivalent circuit parameters referred to stator: $R_1 = 5.39\,\Omega$, $R_2 = 5.72\,\Omega$, $X_1 = X_2 = 8.22\,\Omega$, $R_m = 2000\,\Omega$, and $X_m = 192\,\Omega$ at rated voltage and frequency of 415 V, 50 Hz. Calculate its (a) supply current, (b) losses, and (c) input power at rated speed and balanced rated voltage. If it has applied line voltages of 440 V ∠0°, 390 V ∠−120°, and 415 V ∠120° in three phases, calculate (d) positive-sequence voltage, (e) negative-sequence voltage, (f) positive-sequence supply current, (g) negative-sequence supply current, (h) losses, (i) input power at rated speed, (j) output power, (k) supply voltage unbalance factor, and (l) supply current unbalance factor.

Solution: Given that a three-phase 3.7 kW, 1420 rpm, delta connected squirrel cage induction motor is used to drive a compressor load. It has its per-phase equivalent circuit parameters referred to stator: $R_1 = 5.39\,\Omega$, $R_2 = 5.72\,\Omega$, $X_1 = X_2 = 8.22\,\Omega$, $R_m = 2000\,\Omega$, and $X_m = 192\,\Omega$ at rated voltage and frequency of 415 V, 50 Hz.

For rated speed and balanced rated voltage, its performance is slip $(S) = (N_s - N_r)/N_s = \{(120 \times 50/4) - 1420\}/(120 \times 50/4) = 0.0533$.

From an equivalent circuit of a three-phase induction motor, its impedances are as follows:

- The stator impedance is $Z_1 = R_1 + jX_1 = (5.39 + j8.22)\Omega$.
- The rotor impedance is $Z_2 = (R_2/S) + jX_2 = (107.25 + j8.22)\Omega$.
- The magnetizing impedance is $Z_M = 1.0/\{(1/R_m) + (1/jX_m)\} = (18.264 + j190.25)\Omega$.
- The total equivalent impedance is $Z_T = Z_1 + Z_2Z_M/(Z_2 + Z_M) = R_T + jX_T = (80.2661 + j53.5829)\Omega$.

The stator phase current is $I_1 = V/Z_T = 4.3002$ A.
The rotor current is $I_2 = I_1Z_M/(Z_2 + Z_M) = 3.4999$ A.
The air-gap voltage is $E_M = V - I_1Z_1 = 376.4604$ V.
The power factor is $\mathrm{PF} = \cos(R_T/Z_T) = 0.8317$.

a. The supply current is $I_{Line} = \sqrt{3}V/Z_T = 7.4481$ A.
b. The total losses are $P_L = 3(I_1^2R_1 + I_2^2R_2 + E_M^2/R_m) = 721.7826$ W.
c. The input power is $P_{in} = 3VI_1 \times \mathrm{PF} = 4452.70$ W.
The output power is $P_o = P_{in} - P_L = 4452.70 - 721.7826 = 3730.92$ W.

For unbalanced line voltages of $V_a = 440\,\mathrm{V} \angle 0°$, $V_b = 390\,\mathrm{V} \angle -120°$, and $V_c = 415\,\mathrm{V} \angle 120°$, the results are as follows:

d. The positive-sequence voltage is $V_P = (V_a + aV_b + a^2V_c)/3 = 415$ V.
e. The negative-sequence voltage is $V_N = (V_a + a^2V_b + aV_c)/3 = 14.4389$ V.

The positive-sequence quantities are as follows:

- The stator impedance is $Z_{1P} = R_1 + jX_1 = (5.39 + j8.22)\Omega$.
- The rotor impedance is $Z_{2P} = (R_2/S) + jX_2 = (107.25 + j8.22)\Omega$.
- The magnetizing impedance is $Z_{MP} = 1.0/\{(1/R_m) + (1/jX_m)\} = (18.264 + j190.25)\Omega$.
- The total equivalent impedance is $Z_{TP} = Z_{1P} + Z_{2P}Z_{MP}/(Z_{2P} + Z_{MP}) = R_{TP} + jX_{TP} = (80.2661 + j53.5829)\Omega$.

f. Positive-sequence supply current is $I_{1P} = V/Z_{TP} = 4.3001$ A.
The rotor current is $I_{2P} = I_{1P}Z_{MP}/(Z_{2P} + Z_{MP}) = 3.4998$ A.
The air-gap voltage is $E_{MP} = V - I_{1P}Z_{1P} = 376.461$ V.
The total losses for positive sequence are $P_{LP} = 3(I_{1P}^2R_1 + I_{2P}^2R_2 + E_{MP}^2/R_m) = 721.7684$ W.

The negative-sequence quantities are as follows:

- The stator impedance is $Z_{1N} = R_1 + jX_1 = (5.39 + j8.22)\Omega$.
- The rotor impedance is $Z_{2N} = \{R_2/(2 - S)\} + jX_2 = (2.9384 + j8.22)\Omega$.

- The magnetizing impedance is $Z_{MN} = 1.0/\{(1/R_m) + (1/jX_m)\} = (18.264 + j190.25)\Omega$.
- The total equivalent impedance is $Z_{TN} = Z_1 + Z_{2N}Z_M/(Z_2 + Z_M) = R_{TN} + jX_{TN} = (8.1191 + j16.1207)\Omega$.

g. The negative-sequence supply current is $I_{1N} = V_N/Z_{TN} = 0.7999$ A.
   The rotor current is $I_{2N} = I_{1N}Z_{MN}/(Z_{2N} + Z_{MN}) = 0.766$ A.
   The air-gap voltage is $E_{MN} = V_N - I_{1N}Z_{1N} = 6.6865$ V.
   The total losses for negative sequence are $P_{LN} = 3(I_{1N}^2 R_1 + I_{2N}^2 R_2 + E_{MN}^2/R_m) = 20.4827$ W.
h. The total losses are $P_{LT} = P_{LP} + P_{LN} = 742.2511$ W.
i. The total input power is $P_{in} = 4452.7$ W.
j. The total output power is $P_o = [3(1 - \text{slip})\{I_{2P}^2(R_2/\text{slip})\} - \{I_{2N}^2 R_2/(2.0 - \text{slip})\}] = 3729.09$ W.
k. The supply voltage unbalance factor is $U_V = V_N/V_P = 14.4389/415 = 0.0348 = 3.48\%$.
l. The supply current unbalance factor is $U_I = I_{1N}/I_{1P} = 0.7999/4.3001 = 0.1860 = 18.60\%$.
   It is evident that a voltage unbalance factor of 3.48% causes a current unbalance factor of 18.60% in the winding of the motor. It reduces output power and increases input power.

Example 2.24

A three-phase 3.7 kW/5 hp, 1420 rpm, delta connected squirrel cage induction motor is used to drive a compressor load. It has its per-phase equivalent circuit parameters referred to stator: $R_1 = 5.39\,\Omega$, $R_2 = 5.72\,\Omega$, $X_1 = X_2 = 8.22\,\Omega$, $R_m = 2000\,\Omega$, and $X_m = 192\,\Omega$ at rated voltage and frequency of 415 V, 50 Hz. At 440 V, 50 Hz, it has $R_m = 1900\,\Omega$ and $X_m = 172\,\Omega$. Calculate its (a) supply current, (b) losses, and (c) input power at rated speed and voltage. Calculate its (d) supply current, (e) losses, and (f) input power at rated speed and 440 V, 50 Hz. Consider small slip approximation.

Solution: Given that a three-phase 3.7 kW, 1420 rpm, delta connected squirrel cage induction motor is used to drive a compressor load. It has its per-phase equivalent circuit parameters referred to stator: $R_1 = 5.39\,\Omega$, $R_2 = 5.72\,\Omega$, $X_1 = X_2 = 8.22\,\Omega$, $R_m = 2000\,\Omega$, and $X_m = 192\,\Omega$ at rated voltage and frequency of 415 V, 50 Hz.

At 440 V, 50 Hz, it has $R_m = 1900\,\Omega$ and $X_m = 172\,\Omega$.

At rated speed and voltage, its performance is $S = (N_s - N_r)/N_s = \{(120 \times 50/4) - 1420\}/(120 \times 50/4) = 0.0533$.

From an equivalent circuit of a three-phase induction motor, its impedances are as follows:

- The stator impedance is $Z_1 = R_1 + jX_1 = (5.39 + j8.22)\Omega$.
- The rotor impedance is $Z_2 = (R_2/S) + jX_2 = (107.25 + j8.22)\Omega$.
- The magnetizing impedance is $Z_M = 1.0/\{(1/R_m) + (1/jX_m)\} = (18.264 + j190.25)\Omega$.
- The total equivalent impedance is $Z_T = Z_1 + Z_2Z_M/(Z_2 + Z_M) = R_T + jX_T = (80.2661 + j53.5829)\Omega$.

The stator phase current is $I_1 = V/Z_T = 4.3002$ A.
The rotor current is $I_2 = I_1Z_M/(Z_2 + Z_M) = 3.4999$ A.
The air-gap voltage is $E_M = V - I_1Z_1 = 376.4604$ V.
The power factor is $\text{PF} = \cos(R_T/Z_T) = 0.8317$.

a. The supply current is $I_{Line} = \sqrt{3}V/Z_T = 7.4481$ A.
b. The total losses are $P_L = 3(I_1^2 R_1 + I_2^2 R_2 + E_M^2/R_m) = 721.7826$ W.
c. The input power is $P_{in} = 3VI_1 \times \text{PF} = 4452.70$ W.
   At rated speed and 440 V, 50 Hz, $R_m = 1900\,\Omega$, and $X_m = 172\,\Omega$, the results are given below.
   For a small slip approximation, $S \propto 1/V^2$, the new slip at increased voltage is $S_n = 0.0533 \times (415/440)^2 = 0.0474 = 4.74\%$.
   The synchronous speed is $N_s = 120f/p = 120 \times 50/4 = 1500$ rpm.
   The shaft speed at increased voltage is $N_m = N_s(1 - S_n) = 1500 \times (1 - 0.0474) = 1428.9$ rpm.

From an equivalent circuit of a three-phase induction motor, its impedances are as follows:

- The stator impedance is $Z_1 = R_1 + jX_1 = (5.39 + j8.22)\Omega$.
- The rotor impedance is $Z_2 = (R_2/S_n) + jX_2 = (120.56 + j8.22)\Omega$.
- The magnetizing impedance is $Z_M = 1.0/\{(1/R_m) + (1/jX_m)\} = (15.444 + j170.60)\Omega$.
- The total equivalent impedance is $Z_T = Z_1 + Z_2Z_M/(Z_2 + Z_M) = R_T + jX_T = (79.94 + j62.35)\Omega$.

The stator phase current is $I_1 = V/Z_T = 4.3398$ A.
The rotor current is $I_2 = I_1Z_M/(Z_2 + Z_M) = 3.3089$ A.
The air-gap voltage is $E_M = V - I_1Z_1 = 399.85$ V.
The power factor is $\text{PF} = \cos(R_T/Z_T) = 0.7885$.

d. The supply current is $I_{\text{Line}} = \sqrt{3}V/Z_T = 7.5167$ A.
e. The total losses are $P_L = 3(I_1^2R_1 + I_2^2R_2 + E_M^2/R_m) = 744.8685$ W.
f. The input power is $P_{in} = 3VI_1 \times \text{PF} = 4517$ W.
It means that at 6% higher voltage, for same output power, it has higher losses, input power, and input current.

Example 2.25

A three-phase 3.7 kW/5 hp, 1420 rpm, delta connected squirrel cage induction motor is used to drive a constant power load. It has its per-phase equivalent circuit parameters referred to stator: $R_1 = 5.39\,\Omega$, $R_2 = 5.72\,\Omega$, $X_1 = X_2 = 8.22\,\Omega$, $R_m = 2000\,\Omega$, and $X_m = 192\,\Omega$ at rated voltage and frequency of 415 V, 50 Hz. At 360 V, 50 Hz, it has $R_m = 2100\,\Omega$ and $X_m = 240\,\Omega$. Calculate its (a) supply current, (b) losses, and (c) input power at rated speed and voltage. Calculate its (d) supply current, (e) losses, and (f) input power at rated speed and 360 V, 50 Hz. For constant output power, consider new slip = rated slip × $(V_{\text{rated}}/V_{\text{New}})^{2.4}$ approximation.

Solution: Given that a three-phase 3.7 kW, 1420 rpm, delta connected squirrel cage induction motor is used to drive a compressor load. It has its per-phase equivalent circuit parameters referred to stator: $R_1 = 5.39\,\Omega$, $R_2 = 5.72\,\Omega$, $X_1 = X_2 = 8.22\,\Omega$, $R_m = 2000\,\Omega$, and $X_m = 192\,\Omega$ at rated voltage and frequency of 415 V, 50 Hz.

At 360 V, 50 Hz, it has $R_m = 2100\,\Omega$ and $X_m = 240\,\Omega$.

At rated speed and voltage, its performance is $S = (N_s - N_r)/N_s = \{(120 \times 50/4) - 1420\}/(120 \times 50/4) = 0.0533$.

From an equivalent circuit of a three-phase induction motor, its impedances are as follows:

- The stator impedance is $Z_1 = R_1 + jX_1 = (5.39 + j8.22)\Omega$.
- The rotor impedance is $Z_2 = (R_2/S) + jX_2 = (107.25 + j8.22)\Omega$.
- The magnetizing impedance is $Z_M = 1.0/\{(1/R_m) + (1/jX_m)\} = (18.264 + j190.25)\Omega$.
- The total equivalent impedance is $Z_T = Z_1 + Z_2Z_M/(Z_2 + Z_M) = R_T + jX_T = (80.2661 + j53.5829)\Omega$.

The stator phase current is $I_1 = V/Z_T = 4.3002$ A.
The rotor current is $I_2 = I_1Z_M/(Z_2 + Z_M) = 3.4999$ A.
The air-gap voltage is $E_M = V - I_1Z_1 = 376.4604$ V.
The power factor is $\text{PF} = \cos(R_T/Z_T) = 0.8317$.

a. The supply current is $I_{\text{Line}} = \sqrt{3}V/Z_T = 7.4481$ A.
b. The total losses are $P_L = 3(I_1^2R_1 + I_2^2R_2 + E_M^2/R_m) = 721.7826$ W.
c. The input power is $P_{in} = 3VI_1 \times \text{PF} = 4452.70$ W.
At rated speed and 360 V, 50 Hz, $R_m = 2100\,\Omega$, and $X_m = 240\,\Omega$, the results are given below.
The slip is $S = (N_s - N_r)/N_s = [\{(120 \times 50/4) - 1420\}/(120 \times 50/4)] \times (415/360)^{2.4} = 0.0750$.

From an equivalent circuit of a three-phase induction motor, its impedances are as follows:

- The stator impedance is $Z_1 = R_1 + jX_1 = (5.39 + j8.22)\Omega$.
- The rotor impedance is $Z_2 = (R_2/S) + jX_2 = (76.2444 + j8.22)\Omega$.
- The magnetizing impedance is $Z_M = 1.0/\{(1/R_m) + (1/jX_m)\} = (27.075 + j236.91)\Omega$.
- The total equivalent impedance is $Z_T = Z_1 + Z_2Z_M/(Z_2 + Z_M) = R_T + jX_T = (68.9030 + j34.5133)\Omega$.

The stator phase current is $I_1 = V/Z_T = 4.6715$ A.
The rotor current is $I_2 = I_1Z_M/(Z_2 + Z_M) = 4.1874$ A.
The air-gap voltage is $E_M = V - I_1Z_1 = 321.1185$ V.
The power factor is $PF = \cos(R_T/Z_T) = 0.8941$.
The synchronous speed is $w_s = 2 \times 3.14 \times N_s/60.0 = 157.10$ rad/s.
The developed torque is $T = 3I_2^2(R_2/S)/w_s = 25.53$ N m.
The output power is $P_o = Tw_s(1 - S) = 3709.90$ W.

d. The supply current is $I_{Line} = \sqrt{3}V/Z_T = 8.0912$ A.
e. The total losses are $P_L = 3(I_1^2R_1 + I_2^2R_2 + E_M^2/R_m) = 801.0755$ W.
f. The input power is $P_{in} = 3VI_1 \times PF = 4510.90$ W.
It means that at 13.25% lower voltage (at 360 V), for same output power, it has higher losses, input power, and input current.

2.8 Summary

Because of increased awareness of power quality and its associated problems, standardization, assessment, monitoring, and mitigation have become almost essential for manufacturers, customers, utilities, and researchers. In view of these power quality issues, a number of organizations such as IEC and IEEE have published different standards to specify the permissible limits of various power quality indices to limit the level of power quality to an acceptable benchmark and to provide guidelines to the customers, manufactures, and utilities on curbing the various events causing the power quality problems. Similarly, a number of instruments for monitoring and assessing the power quality indices are developed by many manufactures with different names such as power quality analyzers, monitors, and meters for assessing at customer sites, utility premises, and manufacturing stages of various electrical equipment. Because of direct or indirect penalty and loss of revenue, not only power quality mitigation is used at the retrofit level, but also many manufactures have started introducing it in their equipment. This exhaustive exposure of these standards, definitions, monitoring, and assessment of power quality will be beneficial to the designers, users, manufacturers, and research engineers dealing with power quality improvement.

2.9 Review Questions

1. What is the power quality monitoring?
2. What are the power quality standards?
3. What is the total harmonic distortion factor?
4. What is the total demand distortion factor?
5. What is the point of common coupling?
6. What is the distortion factor?
7. What is the displacement factor?
8. What is the power factor in the presence of harmonics in AC voltage and AC current?
9. What is the crest factor?

10. What is the quality factor?
11. What is the active power in the presence of harmonics in AC voltage and AC current?
12. What is the fundamental reactive power in the presence of harmonics in AC voltage and AC current?
13. What is the unbalance factor?
14. What is the telephone influence factor?
15. What is the C-message weight?
16. What is the distortion index?
17. What is the flicker factor?
18. What is the harmonic spectrum?
19. What is the power analyzer?
20. What is the power monitor?
21. What is the power scope?
22. What is the spectrum analyzer?
23. What is the IEC flicker meter?
24. What is the negative-sequence voltage and how does it affect the different loads?
25. In a three-phase unbalanced load, how the instantaneous negative- and positive-sequence currents can be extracted?

2.10 Numerical Problems

1. For a quasi-square wave (150° pulse width), calculate (a) crest factor, (b) distortion factor, and (c) total harmonic distortion.
2. For a trapezoidal wave (60° flat portion), calculate (a) crest factor, (b) distortion factor, and (c) total harmonic distortion.
3. For a trapezoidal wave (120° flat portion), calculate (a) crest factor, (b) distortion factor, and (c) total harmonic distortion.
4. For a rising sawtooth wave, calculate (a) crest factor, (b) distortion factor, and (c) total harmonic distortion.
5. A three-phase unbalanced supply system has the following phase voltages: $V_a = 1.0 \angle 20°$ pu, $V_b = 1.0 \angle 220°$ pu, and $V_c = 1.0 \angle 110°$ pu. Find the positive-, negative-, and zero-sequence components of supply voltages.
6. A three-phase unbalanced supply system has the following phase voltages: $V_a = 0.92 \angle 0°$ pu, $V_b = 1.05 \angle 240°$ pu, and $V_c = 0.96 \angle 120°$ pu. Find the positive-, negative-, and zero-sequence components of supply voltages.
7. A three-phase balanced supply system has phase voltages $V_a = 0.90 \angle 0°$ pu, $V_b = 0.90 \angle 240°$ pu, and $V_c = 0.90 \angle 120°$ pu and unbalanced load currents $I_a = 0.75 \angle -20°$ pu, $I_b = 0.65 \angle 270°$ pu, and $I_c = 0.45 \angle 90°$ pu. Find (a) the total complex power, (b) the positive-sequence component of power, (c) the negative-sequence component of power, and (d) the zero-sequence component of power.
8. A three-phase balanced supply system has phase voltages $V_a = 0.90 \angle 0°$ pu, $V_b = 1.1 \angle 240°$ pu, and $V_c = 0.95 \angle 120°$ pu and unbalanced load currents $I_a = 0.75 \angle -20°$ pu, $I_b = 0.65 \angle 270°$ pu, and

$I_c = 0.35\angle 90°$ pu. Find (a) the total complex power, (b) the positive-sequence component of power, (c) the negative-sequence component of power, and (d) the zero-sequence component of power.

9. A three-phase unbalanced supply system has phase voltages $V_a = 1.1\angle 0°$ pu, $V_b = 1.0\angle 230°$ pu, and $V_c = 0.9\angle 120°$ pu and unbalanced load currents $I_a = 0.75\angle -20°$ pu, $I_b = 0.75\angle 260°$ pu, and $I_c = 0.75\angle 140°$ pu. Find (a) the total complex power, (b) the positive-sequence component of power, (c) the negative-sequence component of power, and (d) the zero-sequence component of power.

10. In a three-phase AC mains, there is a voltage sag at PCC of 15, 25, and 20% on three phases for 15, 25, and 30 cycles, respectively. Calculate (a) Detroit Edison sag score and (b) voltage sag lost energy index of this sag event.

11. Estimate the *K*-factor rating of a single-phase transformer used to feed a single-phase AC voltage controller with resistive load at a thyristor firing angle of 90°.

12. Estimate the *K*-factor rating of a three-phase transformer used to feed a three-phase 12-pulse diode bridge rectifier with constant DC load current.

13. A single-phase transformer is used to feed a single-phase diode bridge rectifier with constant DC load current of 50 A. The transformer has been rated for a winding eddy current loss density of 15% (0.15 pu). Calculate its derating factor.

14. A single-phase AC voltage controller is used to control the heating load in a food vending machine at a power of 500 W at 50 V fed from a single-phase AC mains of 230 V at 50 Hz. Feeder conductors have the resistance of 0.2 Ω each. Calculate (a) AC source rms current (I_s) and (b) losses in the distribution system. If an ideal shunt compensator is used to compensate power factor to unity, then calculate (c) AC source rms current (I_{sc}), (d) losses in the distribution system, and (e) ratio of losses in the distribution system without and with a compensator.

15. A single-phase uncontrolled bridge converter has a *RE* load with $R = 1.0\,\Omega$ and $E = 285$ V. The input AC voltage is $V_s = 230$ V at 50 Hz. Feeder conductors have the resistance of 0.05 Ω each. Calculate (a) AC source rms current (I_s), (b) losses in the distribution system, (c) total harmonic distortion in current, and (d) crest factor of supply current. If an ideal shunt compensator is used to compensate power factor to unity, then calculate (e) AC source rms current (I_{sc}), (f) losses in the distribution system, and (g) ratio of losses in the distribution system without and with a compensator.

16. A three-phase fully controlled bridge converter feeds power to a load having a resistance (R) of 5 Ω and very large inductance to result in continuous current with an input from a three-phase supply of 415 V at 50 Hz. Feeder conductors have the resistance of 0.125 Ω each. For firing angles of 30°, calculate (a) AC source rms current (I_s) and (b) losses in the distribution system. If an ideal shunt compensator is used to compensate power factor to unity, then calculate (c) AC source rms current (I_{sc}), (d) losses in the distribution system, and (e) ratio of losses in the distribution system without and with a compensator.

17. In a three-phase four-wire distribution system with a line voltage of 415 V at 50 Hz, three single-phase loads (connected between phases and neutral terminal) have a single-phase uncontrolled bridge converter (shown in Figure E2.17) having a *RE* load with $R = 0.9\,\Omega$ and $E = 290$ V. Feeder and neutral conductors have the resistance of 0.05 Ω each. Calculate (a) AC source rms current (I_s), (b) neutral current (I_{sn}), and (c) losses in the distribution system. If an ideal four-wire shunt compensator is used to compensate power factor to unity in each phase, then calculate (d) AC source rms current (I_{sc}), (e) neutral current (I_{snc}), (f) losses in the distribution system, and (g) ratio of losses in the distribution system without and with a compensator.

18. In a three-phase four-wire distribution system with a line voltage of 415 V at 50 Hz, three single-phase loads (connected between phases and neutral terminal) have a single-phase thyristor bridge converter drawing equal 20 A constant DC current at a thyristor firing angle of 30°. Feeder and neutral

conductors have the resistance of 0.1 Ω each. Calculate (a) AC source rms current (I_s), (b) neutral current (I_{sn}), and (c) losses in the distribution system. If an ideal four-wire shunt compensator is used to compensate power factor to unity in each phase, then calculate (d) AC source rms current (I_{sc}), (e) neutral current (I_{snc}), (f) losses in the distribution system, and (g) ratio of losses in the distribution system without and with a compensator.

19. In a three-phase four-wire distribution system with a line voltage of 415 V at 50 Hz, three single-phase loads (connected between phases and neutral terminal) have a single-phase AC voltage controller for heating loads (resistive load) of 5 Ω. The delay angle of thyristors is $\alpha = 130°$. Feeder and neutral conductors have the resistance of 0.1 Ω each. Calculate (a) AC source rms current (I_s), (b) neutral current (I_{sn}), and (c) losses in the distribution system. If an ideal four-wire shunt compensator is used to compensate power factor to unity in each phase, then calculate (d) AC source rms current (I_{sc}), (e) neutral current (I_{snc}), (f) losses in the distribution system, and (g) ratio of losses in the distribution system without and with a compensator.

20. In a three-phase four-wire distribution system with a line voltage of 415 V at 50 Hz, a single-phase load (connected between phase and neutral terminal) has a 75 A, 0.85 lagging power factor. Feeder and neutral conductors have the resistance of 0.1 Ω each. Calculate (a) AC source rms current (I_s), (b) neutral current (I_{sn}), and (c) losses in the distribution system. If an ideal four-wire shunt compensator is used to compensate power factor to unity in each phase, then calculate (d) AC source rms current (I_{sc}), (e) neutral current (I_{snc}), (f) losses in the distribution system, and (g) ratio of losses in the distribution system without and with a compensator.

21. A three-phase 37 kW, 415 V, 50 Hz, four-pole delta connected squirrel cage induction motor is used to drive a compressor load of constant torque. It runs at 3.5% slip at full load and rated voltage and frequency. If terminal voltage reduces to 360 V, calculate its (a) slip, (b) shaft speed, (c) output power, and (d) rotor winding loss as a ratio of rated rotor winding loss at rated voltage. Consider small slip approximation.

22. A three-phase 37 kW, 415 V, 50 Hz, four-pole delta connected squirrel cage induction motor is used to drive a compressor load of constant torque. It runs at 3.5% slip at full load and rated voltage and frequency. If terminal voltage increases to 440 V, calculate its (a) slip, (b) shaft speed, (c) output power, and (d) rotor winding loss as a ratio of rated rotor winding loss at rated voltage. Consider small slip approximation.

23. A three-phase 10 kW, 1440 rpm, delta connected squirrel cage induction motor is used to drive a compressor load. It has its per-phase equivalent circuit parameters referred to stator: $R_1 = 3.55\,\Omega$, $R_2 = 3.72\,\Omega$, $X_1 = X_2 = 5.22\,\Omega$, $R_m = 1500\,\Omega$, and $X_m = 120\,\Omega$ at rated voltage and frequency of 415 V, 50 Hz. Calculate its (a) supply current, (b) losses, and (c) input power at rated speed and balanced rated voltage. If it has applied line voltages of $440\,V \angle 0°$, $415\,V \angle 120°$, and $400\,V \angle -120°$ in three phases, calculate (d) positive-sequence voltage, (e) negative-sequence voltage, (f) positive-sequence supply current, (g) negative-sequence supply current, (h) losses, and (i) input power at rated speed.

24. A three-phase 10 kW, 1440 rpm, delta connected squirrel cage induction motor is used to drive a compressor load. It has its per-phase equivalent circuit parameters referred to stator: $R_1 = 3.55\,\Omega$, $R_2 = 3.72\,\Omega$, $X_1 = X_2 = 5.22\,\Omega$, $R_m = 1500\,\Omega$, and $X_m = 120\,\Omega$ at rated voltage and frequency of 415 V, 50 Hz. At 440 V, 50 Hz, it has $R_m = 1450\,\Omega$ and $X_m = 100\,\Omega$. Calculate its (a) supply current, (b) losses, and (c) input power at rated speed and 440 V, 50 Hz.

25. A three-phase 10 kW, 1440 rpm, delta connected squirrel cage induction motor is used to drive a power load. It has its per-phase equivalent circuit parameters referred to stator: $R_1 = 3.55\,\Omega$, $R_2 = 3.72\,\Omega$, $X_1 = X_2 = 5.22\,\Omega$, $R_m = 1500\,\Omega$, and $X_m = 120\,\Omega$ at rated voltage and frequency of 415 V, 50 Hz. At 360 V, 50 Hz, it has $R_m = 1600\,\Omega$ and $X_m = 150\,\Omega$. Calculate its (a) supply current, (b) losses, and (c) input power at rated speed and 360 V, 50 Hz.

2.11 Computer Simulation-Based Problems

1. Compute (a) fundamental rms voltage (V_{1rms}), (b) total rms voltage (V_{Trms}), (c) crest factor, (d) distortion factor, and (e) total harmonic distortion for a trapezoidal AC voltage wave (varying flat-top portion from 0° to 180°) with an amplitude of 100 V.

2. Compute (a) fundamental rms voltage (V_{1rms}), (b) total rms voltage (V_{Trms}), (c) crest factor, (d) distortion factor, and (e) total harmonic distortion for a rectangular AC voltage wave (varying flat-top portion from 1° to 180°) with an amplitude of 100 V.

3. A three-phase unbalanced supply system has the following phase voltages: $V_a = 1.0\angle\alpha°$ pu, $V_b = 1.0\angle 240°$ pu, and $V_c = 1.0\angle 120°$ pu. Compute (a) positive-, (b) negative-, and (c) zero-sequence components of supply voltages if angle α varies from 0° to 60° with a variation of 1°.

4. A three-phase unbalanced supply system has the following phase voltages: $V_a = K\angle 0°$ pu, $V_b = 1.0\angle 240°$ pu, and $V_c = 1.0\angle 120°$ pu. Compute (a) positive-, (b) negative-, and (c) zero-sequence components of supply voltages if K varies from 0.8 to 1.2 with a variation of 0.01.

5. A three-phase balanced supply system has phase voltages $V_a = 1.0\angle 0°$ pu, $V_b = 1.0\angle 240°$ pu, and $V_c = 1.0\angle 120°$ pu and unbalanced load currents $I_a = 1\angle\alpha°$ pu, $I_b = 1.0\angle 240°$ pu, and $I_c = 1.0\angle 90°$ pu. If angle α varies from 0° to 60° with a variation of 1°, compute and plot (a) the total complex power, (b) the positive-sequence component of power, (c) the negative-sequence component of power, and (d) the zero-sequence component of power.

6. A three-phase balanced supply system has phase voltages $V_a = 1.0\angle 0°$ pu, $V_b = 1.0\angle 240°$ pu, and $V_c = 1.0\angle 120°$ pu and unbalanced load currents $I_a = K\angle 0°$ pu, $I_b = 1.0\angle 240°$ pu, and $I_c = 1.0\angle 90°$ pu. If K varies from 0.8 to 1.2 with a variation of 0.01, compute and plot (a) the total complex power, (b) the positive-sequence component of power, (c) the negative-sequence component of power, and (d) the zero-sequence component of power.

7. Estimate the K-factor rating of a single-phase transformer used to feed a single-phase AC voltage controller with resistive load at a thyristor firing angle varying from 0° to 160° with a variation of 1°.

8. Estimate the K-factor rating of a single-phase transformer used to feed a single-phase semicontrolled bridge rectifier with constant current DC load at a thyristor firing angle varying from 0° to 150° with a variation of 1°.

9. Estimate the K-factor rating of a three-phase transformer used to feed a three-phase semicontrolled bridge rectifier with constant current DC load at a thyristor firing angle varying from 0° to 150° with a variation of 1°.

10. Estimate the K-factor rating of a three-phase transformer used to feed a three-phase controlled bridge rectifier with pure resistive DC load at a thyristor firing angle varying from 0° to 150° with a variation of 1°.

11. A single-phase AC voltage controller has a heating load (resistive load) of 10 Ω. The input voltage is 230 V (rms) at 50 Hz. The delay angle of thyristors is varying from $\alpha = 0°$ to $\alpha = 180°$. Feeder and neutral conductors have the resistance of $R_s = 0.25\ \Omega$ each. Compute (a) AC source rms current (I_s) and (b) losses in the distribution system. If an ideal shunt compensator is used to compensate the power factor to unity of this load, then compute (c) load active power, (d) AC source rms current (I_{sc}), (e) current rating of the shunt compensator, (f) VA rating of the shunt compensator, (g) losses in the distribution system, and (h) ratio of losses in the distribution system without and with a compensator.

12. A single-phase 230 V, 50 Hz, uncontrolled bridge converter has a parallel capacitive DC filter of 1500 μF and an equivalent resistive load varying from 250 to 50 Ω with a variation of 10 Ω. Feeder and neutral conductors have the resistance of 0.25 Ω each. Compute (a) AC supply rms current (I_s), (b) losses in the distribution system, (c) total harmonic distortion in supply current, and (d) crest factor of supply current. If an ideal shunt compensator is used to compensate power factor to unity, then

compute (e) load active power, (f) AC supply rms current (I_{sc}), (g) current rating of the compensator, (h) VA rating of the compensator, (i) losses in the distribution system, and (j) ratio of losses in the distribution system without and with a compensator.

13. A single-phase uncontrolled bridge converter (shown in Figure E2.15) has a *RE* load with $R = 1.50\,\Omega$ and E varying from 250 to 300 V with a variation of 1 V. The input AC voltage is $V_s = 230$ V at 50 Hz. Feeder conductors have the resistance of 0.1 Ω each. Compute and plot (a) AC source rms current (I_s), (b) losses in the distribution system, (c) total harmonic distortion in current, and (d) crest factor of supply current. If an ideal shunt compensator is used to compensate its power factor to unity, then compute and plot (e) its active power, (f) AC source rms current (I_{sc}), (g) losses in the distribution system, and (h) ratio of losses in the distribution system without and with a compensator.

14. A three-phase 415 V, 50 Hz, uncontrolled bridge converter has a *RE* load with $R = 1.50\,\Omega$ and E varying from 500 to 580 V with a variation of 2 V. The input AC line voltage is $V_s = 415$ V at 50 Hz. Feeder conductors have the resistance of 0.15 Ω each. Compute and plot (a) AC source rms current (I_s), (b) losses in the distribution system, (c) total harmonic distortion in current, and (d) crest factor of supply current. If an ideal shunt compensator is used to compensate its power factor to unity, then compute and plot (e) its active power, (f) AC source rms current (I_{sc}), (g) losses in the distribution system, and (h) ratio of losses in the distribution system without and with a compensator.

15. A three-phase 415 V, 50 Hz, uncontrolled bridge converter has a parallel capacitive DC filter of 3000 μF and a resistive load varying from 200 to 10 Ω with a variation of 5 Ω. Feeder conductors have the resistance of 0.2 Ω each. Compute (a) AC supply rms current (I_s), (b) losses in the distribution system, (c) total harmonic distortion in supply current, and (d) crest factor of supply current. If an ideal shunt compensator is used to compensate power factor to unity, then compute (e) load active power, (f) AC supply rms current (I_{sc}), (g) current rating of the compensator, (h) VA rating of the compensator, (i) losses in the distribution system, and (j) ratio of losses in the distribution system without and with a compensator.

16. A three-phase fully controlled bridge converter feeds power to a load having a resistance (R) of 20 Ω and very large inductance to result in continuous current with an input from a three-phase supply of 415 V at 50 Hz. Feeder conductors have the resistance of 0.15 Ω each. For firing angles varying from $\alpha = 0°$ to $\alpha = 80°$ with a variation of 1°, compute (a) AC supply rms current (I_s), (b) losses in the distribution system, (c) total harmonic distortion in supply current, and (d) crest factor of supply current. If an ideal shunt compensator is used to compensate power factor to unity, then compute (e) load active power, (f) AC supply rms current (I_{sc}), (g) current rating of the compensator, (h) VA rating of the compensator, (i) losses in the distribution system, and (j) ratio of losses in the distribution system without and with a compensator.

17. A three-phase nonlinear load is fed from a three-phase supply of 415 V at 50 Hz, having a 12-pulse thyristor bridge converter with a series connected inductive load of 50 mH and an equivalent resistive load of 5 Ω. It has a source impedance of 0.2 Ω resistive element and 2.0 Ω inductive element and the firing angle of its thyristors is varying from $\alpha = 0°$ to $\alpha = 80°$ with a variation of 1°. It consists of an ideal transformer with single primary star connected winding and two secondary windings connected in star and delta with same line voltages as the input supply voltage to provide 30° phase shift between two sets of three-phase output voltages. Two 6-pulse thyristor bridges are connected in series to provide a 12-pulse AC–DC converter. Compute and plot (a) AC supply rms current (I_s), (b) losses in the distribution system, (c) total harmonic distortion in supply current, and (d) crest factor of supply current. If an ideal shunt compensator is used to compensate power factor to unity at the AC mains, then compute and plot (e) load active power, (f) AC supply rms current (I_{sc}), (g) current rating of the compensator, (h) VA rating of the compensator, (i) losses in the distribution system, and (j) ratio of losses in the distribution system without and with a compensator.

18. In a three-phase four-wire distribution system with a line voltage of 415 V at 50 Hz, three single-phase loads (connected between phases and neutral) have a single-phase AC voltage controller for

heating loads (resistive load) of 10 Ω. The delay angle of thyristors is varying from $\alpha = 0°$ to $\alpha = 180°$ with a variation of 1°. Feeder and neutral conductors have the resistance of 0.15 Ω each. Compute (a) AC source rms current (I_s), (b) neutral current (I_{sn}), and (c) losses in the distribution system. If an ideal four-wire shunt compensator is used to compensate power factor to unity in each phase, then compute (d) load active power, (e) AC source rms current (I_{sc}), (f) current rating of the shunt compensator, (g) VA rating of the shunt compensator, (h) losses in the distribution system, and (i) ratio of losses in the distribution system without and with a compensator.

19. In a three-phase four-wire distribution system with a line voltage of 380 V at 50 Hz, a three single-phase uncontrolled bridge converter has a parallel capacitive DC filter of 1500 μF and an equivalent resistive load varying from 250 to 50 Ω with a variation of 10 Ω. Feeder and neutral conductors have the resistance of 0.25 Ω each. Compute (a) AC source rms current (I_s), (b) neutral current (I_{sn}), and (c) losses in the distribution system. If an ideal four-wire shunt compensator is used to compensate power factor to unity in each phase, then compute (d) load active power, (e) AC source rms current (I_{sc}), (f) current rating of the shunt compensator, (g) VA rating of the shunt compensator, (h) losses in the distribution system, and (i) ratio of losses in the distribution system without and with a compensator.

20. In a three-phase four-wire distribution system with a line voltage of 415 V, 50 Hz, a single-phase load (connected between phase and neutral) has a current varying from 15 to 150 A with a variation of 5 A, at 0.8 lagging power factor. Feeder and neutral conductors have the resistance of 0.15 Ω each. Calculate (a) AC source rms current (I_s), (b) neutral current (I_{sn}), and (c) losses in the distribution system. If an ideal four-wire shunt compensator is used to compensate power factor to unity and balance the load in each phase, then compute (d) load active power, (e) AC source rms current (I_{sc}), (f) current rating of the shunt compensator, (g) VA rating of the shunt compensator, (h) losses in the distribution system, and (i) ratio of losses in the distribution system without and with a compensator.

21. A three-phase 3.7 kW, four-pole, delta connected squirrel cage induction motor is used to drive a constant power load. It has its per-phase equivalent circuit parameters referred to stator: $R_1 = 5.39\,\Omega$, $R_2 = 5.72\,\Omega$, $X_1 = X_2 = 8.22\,\Omega$, $R_m = 2100\,\Omega$, and $X_m = 192\,\Omega$ at rated voltage and frequency of 415 V, 50 Hz. Compute its (a) supply current, (b) losses, (c) input power, (d) rotor speed, (e) output torque, (f) efficiency, and (g) rotor current at rated output power if it has applied line voltages with voltage unbalance varying from 1 to 15%.

22. A three-phase 3.7 kW, four-pole, delta connected squirrel cage induction motor is used to drive a constant power load. It has its per-phase equivalent circuit parameters referred to stator: $R_1 = 5.39\,\Omega$, $R_2 = 5.72\,\Omega$, $X_1 = X_2 = 8.22\,\Omega$, $R_m = 2100\,\Omega$, and $X_m = 192\,\Omega$ at rated voltage and frequency of 415 V, 50 Hz. The voltage across the motor is reducing from 100 to 80% with a variation of 0.1% at constant frequency of 50 Hz. Its R_m remains constant and X_m is increasing inversely proportional to the supply voltage. Compute its (a) supply current, (b) losses, (c) input power, (d) efficiency, (e) power factor, (f) slip, and (g) rotor speed at rated output power.

23. A three-phase 3.7 kW, four-pole, delta connected squirrel cage induction motor is used to drive a rated power load. It has its per-phase equivalent circuit parameters referred to stator: $R_1 = 5.39\,\Omega$, $R_2 = 5.72\,\Omega$, $X_1 = X_2 = 8.22\,\Omega$, $R_m = 2100\,\Omega$, and $X_m = 192\,\Omega$ at rated voltage and frequency of 415 V, 50 Hz. The voltage across the motor is reducing from 100 to 80% with a variation of 0.1% at constant frequency of 50 Hz. Its R_m remains constant and X_m is increasing inversely proportional to the supply voltage. Compute its (a) supply current, (b) output power, (c) input power, (d) efficiency, (e) power factor, (f) slip, and (g) derating factor at rated power losses.

24. A three-phase 3.7 kW, four-pole, delta connected squirrel cage induction motor is used to drive a rated power load. It has its per-phase equivalent circuit parameters referred to stator: $R_1 = 5.39\,\Omega$, $R_2 = 5.72\,\Omega$, $X_1 = X_2 = 8.22\,\Omega$, $R_m = 2100\,\Omega$, and $X_m = 192\,\Omega$ at rated voltage and frequency of 415 V, 50 Hz. The voltage across the motor is increasing from 100 to 120% with a variation of 0.1% at constant frequency of 50 Hz. Its R_m remains constant and X_m is decreasing inversely proportional to the supply voltage. Compute its (a) supply current, (b) output power, (c) input power, (d) efficiency, (e) power factor, (f) slip, and (g) derating factor at rated power losses.

25. A three-phase 3.7 kW, four-pole, delta connected squirrel cage induction motor is used to drive a rated power load. It has its per-phase equivalent circuit parameters referred to stator: $R_1 = 5.39\,\Omega$, $R_2 = 5.72\,\Omega$, $X_1 = X_2 = 8.22\,\Omega$, $R_m = 2100\,\Omega$, and $X_m = 192\,\Omega$ at rated voltage and frequency of 415 V, 50 Hz. Compute its (a) supply current, (b) rotor losses, (c) input power, (d) rotor speed, (e) output power, (f) efficiency, and (g) rotor current at rated power losses if it has applied line voltages with voltage unbalance varying from 1 to 15%.

References

1. IEEE Working Group on Power System Harmonics (1983) Power system harmonics: an overview. *IEEE Transactions on Power Apparatus and Systems*, **102**(8), 2455–2460.
2. Shuter, T.C., Vollkommer, H.T., Jr., and Kirkpatrick, J.L. (1989) Survey of harmonic levels on the American electric power distribution system. *IEEE Transactions on Power Delivery*, **4**(4), 2204–2213.
3. Liew, A.C. (1989) Excessive neutral currents in three-phase fluorescent lighting circuits. *IEEE Transactions on Industry Applications*, **25**(4), 776–782.
4. Gruzs, T.M. (1990) A survey of neutral currents in three-phase computer power systems. *IEEE Transactions on Industry Applications*, **26**(4), 719–725.
5. Subjak, J.S., Jr. and McQuilkin, J.S. (1990) Harmonics – causes, effects, measurements, and analysis: an update. *IEEE Transactions on Industry Applications*, **26**(6), 1034–1042.
6. Amoli, M.E. and Florence, T. (1990) Voltage and current harmonic control of a utility system – a summary of 1120 test measurements. *IEEE Transactions on Power Delivery*, **5**(3), 1552–1557.
7. Grady, W.M., Samotyj, M.J., and Noyola, A.H. (1990) Survey of active power line conditioning methodologies. *IEEE Transactions on Power Delivery*, **5**, 1536–1542.
8. Beides, H.M. and Heydt, G.T. (1992) Power system harmonics estimation and monitoring. *Electric Machines & Power Systems*, **20**, 93–102.
9. Emanuel, A.E., Orr, J.A., Cyganski, D., and Gulchenski, E.M. (1993) A survey of harmonics voltages and currents at the customer's bus. *IEEE Transactions on Power Delivery*, **8**(1), 411–421.
10. Ling, P.J.A. and Eldridge, C.J. (1994) Designing modern electrical systems with transformers that inherently reduce harmonic distortion in a PC-rich environment. Proceedings of the Power Quality Conference, pp. 166–178.
11. Packebush, P. and Enjeti, P. (1994) A survey of neutral current harmonics in campus buildings, suggested remedies. Proceedings of the Power Quality Conference, pp. 194–205.
12. Mansoor, A., Grady, W.M., Staats, P.T. *et al.* (1994) Predicting the net harmonic currents produced by large numbers of distributed single-phase computer loads. *IEEE Transactions on Power Delivery*, **10**(4), 2001–2006.
13. IEEE Working Group on Nonsinusoidal Situations (1996) A survey of North American electric utility concerns regarding nonsinusoidal waveforms. *IEEE Transactions on Power Delivery*, **11**(1), 73–78.
14. Domijan, A., Jr., Santander, E.E., Gilani, A. *et al.* (1996) Watthour meter accuracy under controlled unbalanced harmonic voltage and current conditions. *IEEE Transactions on Power Delivery*, **11**(1), 64–72.
15. IEEE Working Group on Nonsinusoidal Situations (1996) Practical definitions for powers in systems with nonsinusoidal waveforms and unbalanced loads: a discussion. *IEEE Transactions on Power Delivery*, **11**(1), 79–101.
16. Duffey, C.K. and Stratford, R.P. (1989) Update of harmonic standard IEEE-519: IEEE recommended practices and requirements for harmonic control in electric power systems. *IEEE Transactions on Industry Applications*, **25**(6), 1025–1034.
17. Miller, T.J.E. (1982) *Reactive Power Control in Electric Systems*, John Wiley & Sons, Inc., Toronto, Canada, pp. 32–48.
18. Clark, J.W. (1990) *AC Power Conditioners – Design and Applications*, Academic Press, San Diego, CA.
19. IEC SC 77A (1990) *Draft Revision of Publication IEC 555-2: Harmonics, Equipment for Connection to the Public Low Voltage Supply System*, IEC.
20. Heydt, G.T. (1991) *Electric Power Quality*, Stars in a Circle Publications, West Lafayette, IN.
21. IEEE Standard 519-1992 (1992) *IEEE Recommended Practices and Requirements for Harmonic Control in Electric Power Systems*, IEEE.
22. Kazibwe, W.E. and Sendaula, M.H. (1993) *Electrical Power Quality Control Techniques*, Van Nostrand Reinhold Company.

23. Heydt, G.T. (1994) *Electric Power Quality*, 2nd edn, Stars in a Circle Publications, West Lafayette, IN.
24. IEEE Standard 1159-2009 (2009) *IEEE Recommended Practice for Monitoring Electric Power Quality*, IEEE.
25. Paice, D.A. (1996) *Power Electronic Converter Harmonics – Multipulse Methods for Clean Power*, IEEE Press, New York.
26. Duagan, R.C., McGranaghan, M.F., and Beaty, H.W. (1996) *Electric Power System Quality*, McGraw-Hill, New York.
27. Arrillaga, J., Smith, B.C., Watson, N.R., and Wood, A.R. (1997) *Power System Harmonic Analysis*, John Wiley & Sons, Ltd, Chichester, UK.
28. Porter, G.J. and Sciver, J.A.V. (eds) (1999) *Power Quality Solutions: Case Studies for Troubleshooters*, The Fairmont Press, Inc., Lilburn, GA.
29. Kennedy, B.W. (2000) *Power Quality Primer*, McGraw-Hill.
30. Bollen, M.H.J. (2000) *Understanding Power Quality Problems: Voltage Sags and Interruptions*, Series on Power Engineering, IEEE Press, New York.
31. Arrillaga, J., Watson, N.R., and Chen, S. (2000) *Power System Quality Assessment*, John Wiley & Sons, Inc., New York.
32. Wakileh, M.G.J. (2001) *Power Systems Harmonics*, Springer, New York.
33. Bollen, H.J. (2001) *Understanding Power Quality Problems*, 1st edn, Standard Publishers Distributors, New Delhi.
34. Schlabbach, J., Blume, D., and Stephanblome, T. (2001) *Voltage Quality in Electrical Power Systems*, Power Engineering and Energy Series, IEEE Press.
35. Sankaran, C. (2002) *Power Quality*, CRC Press, New York.
36. Ghosh, A. and Ledwich, G. (2002) *Power Quality Enhancement Using Custom Power Devices*, Kluwer Academic Publishers, London.
37. Das, J.C. (2002) *Power System Analysis – Short-Circuit Load Flow and Harmonics*, Marcel Dekker, New York.
38. IEEE Standard 1573-2003 (2003) *IEEE Guide for Application and Specification of Harmonic Filters*, IEEE.
39. Emadi, A., Nasiri, A., and Bekiarov, S.B. (2005) *Uninterruptible Power Supplies and Active Filters*, CRC Press, New York.
40. Wu, B. (2006) *High-Power Converters and AC Drives*, IEEE Press, Hoboken, NJ.
41. Bollen, M.H.J. and Gu, I.Y.-H. (2006) *Signal Processing of Power Quality Disturbances*, IEEE Press, Hoboken, NJ.
42. Dugan, R.C., McGranaghan, M.F., and Beaty, H.W. (2006) *Electric Power Systems Quality*, 2nd edn, McGraw-Hill, New York.
43. IEC 61000-3-2 (1995) *Electromagnetic Compatibility (EMC) – Part 3: Limits – Section 2: Limits for Harmonic Current Emissions (Equipment Input Current <16 A Per Phase)*, IEC.
44. Moreno-Munoz, A. (2007) *Power Quality: Mitigation Technologies in a Distributed Environment*, Springer, London.
45. Akagi, H., Watanabe, E.H., and Aredes, M. (2007) *Instantaneous Power Theory and Applications to Power Conditioning*, John Wiley & Sons, Inc., Hoboken, NJ.
46. Padiyar, K.R. (2007) *FACTS Controllers in Transmission and Distribution*, New Age International, New Delhi.
47. Kusko, A. and Thompson, M.T. (2007) *Power Quality in Electrical Systems*, McGraw-Hill.
48. Fuchs, E.F. and Mausoum, M.A.S. (2008) *Power Quality in Power Systems and Electrical Machines*, Elsevier/Academic Press, London.
49. Sastry Vedam, R. and Sarma, M.S. (2008) *Power Quality: VAR Compensation in Power Systems*, CRC Press, New York.
50. Baggini, A. (2008) *Handbook on Power Quality*, John Wiley & Sons, Inc., Hoboken, NJ.

3

Passive Shunt and Series Compensation

3.1 Introduction

Passive shunt and series compensators have been in the service since the inception of the AC supply system to improve the power quality of the power system by enhancing the efficiency and utilization of equipment in transmission and distribution networks. The passive compensators normally consist of lossless reactive elements such as capacitors and inductors with and without switching devices. The passive compensators are used for improving transient, steady state, and dynamic, voltage and angle stabilities. Moreover, these also help in reducing losses, enhancing the loadability, improving transmission capacity, damping power system oscillations, and mitigating subsynchronous resonance (SSR) and other contingency problems in transmission systems. The passive shunt and series compensators are also extensively used in distribution systems for improving the voltage profile at the point of common coupling (PCC), reducing losses, power factor correction (PFC), load balancing, neutral current compensation, and for better utilization of distribution equipment. Ideally, the passive compensators can supply or absorb variable or fixed reactive power locally to mitigate the power quality problems. This chapter focuses on the concepts and methodologies of passive lossless compensation in distribution systems, especially on load compensation. It includes power factor correction, voltage regulation (VR), load balancing, and neutral current compensation.

3.2 State of the Art on Passive Shunt and Series Compensators

Passive compensation is now a mature technology for providing reactive power compensation for power factor correction and/or voltage regulation, load balancing, and reduction of neutral current in AC networks. It has evolved during the past century with development in terms of varying configurations and requirements. Passive compensators are used for regulating the terminal voltage, suppressing voltage flicker, improving voltage balance, power factor correction, load balancing, and neutral current mitigation in three-phase distribution systems. These objectives are achieved either individually or in combination depending upon the requirements and configurations that need to be selected appropriately. This section describes the history of development and the current status of the passive compensation technology.

The reactive power compensation employing lossless passive components in distribution systems has been used in practice for a long time for improving the voltage profile at the load end by the utilities and enhancing the power factor in the industries for avoiding the penalty by the utilities. In the early twentieth century, Steinmetz had investigated that an unbalanced single-phase resistive load may be realized as a

Power Quality Problems and Mitigation Techniques, First Edition.
Bhim Singh, Ambrish Chandra and Kamal Al-Haddad.

balanced load using lossless passive elements in a three-phase supply system. This concept was later on extended in many directions such as balancing of three-phase unbalanced loads, power factor correction at the supply system, compensation of negative-sequence and zero-sequence currents, and voltage regulation. It has become quite important and relevant because in practice there are many single-phase and unbalanced loads such as traction, metros, furnaces, residential, and commercial loads. There are many methods to implement these compensators in practice for improving power quality, especially voltage quality, for the consumers nearby the fluctuating loads such as arc furnaces. Since these compensators are simple, cost effective, and easily realizable in practice, they are still used in large power rating. The chronological development of the passive compensation technology, popularly known as classical load compensation, has led to many concepts, theories, and design formulations in the past. Some of these derived mathematical formulations have become reasonably important as the basic framework for the design of these compensators. This chapter illustrates these concepts of load compensation with suitable formulations and numerical examples that are expected to meet the requirements of the design and practice engineers.

3.3 Classification of Passive Shunt and Series Compensators

The passive compensators can be classified based on the topology and the number of phases. The topology can be shunt, series, or a combination of both. The other classification is based on the number of phases, such as two-wire (single-phase) and three- or four-wire (three-phase) systems.

3.3.1 Topology-Based Classification

The passive compensators can be classified based on the topology, for example, series, shunt, or hybrid compensators. Figure 3.1 shows the examples of basic series, shunt, and hybrid compensators. Passive series compensators have limited applications in distribution systems as they affect the performance of the loads to a great extent and have resonance problems. The passive series compensators are used in transmission systems to improve power transfer capability, of course, with restricted capacity to avoid series resonance. The passive series compensators are also used in stand-alone self-excited induction generators for improving the voltage profile and enhancing the stability. In majority of the cases, mainly shunt compensators are used in practice as they are connected in parallel to the loads and do not disturb the operation of the loads. These are mainly used at the load end. So, current-based compensation is used at the load end. These inject equal compensating currents, opposite in phase, to cancel reactive power components of the load current for power factor correction at the point of connection. The passive shunt compensators are also used for voltage regulation and load balancing at the load end. These are also used as static VAR generators in the power system network for stabilizing and improving the voltage profile. The passive hybrid compensators shown in Figure 3.1c and d as combinations of passive series and shunt elements in both short-shunt and long-shunt configurations are used in stand-alone self-excited induction generators for improving the voltage profile and enhancing the stability.

3.3.2 Supply System-Based Classification

As mentioned earlier, mainly passive shunt compensators are used in the distribution system for reactive power compensation and load balancing, so these are studied in detail here. This classification of passive compensators is based on the supply and/or the load systems having single-phase (two-wire) and three-phase (three-wire and four-wire) systems. There are many varying loads such as domestic appliances connected to single-phase supply systems. Some three-phase unbalanced loads are without neutral terminal, such as AC motors, traction, metros, and furnaces fed from three-phase three-wire supply systems. There are many other single-phase loads distributed on three-phase four-wire supply systems, such as heating and lighting systems, among others. Hence, passive compensators may also be classified as single-phase two-wire, three-phase three-wire, and three-phase four-wire passive shunt compensators.

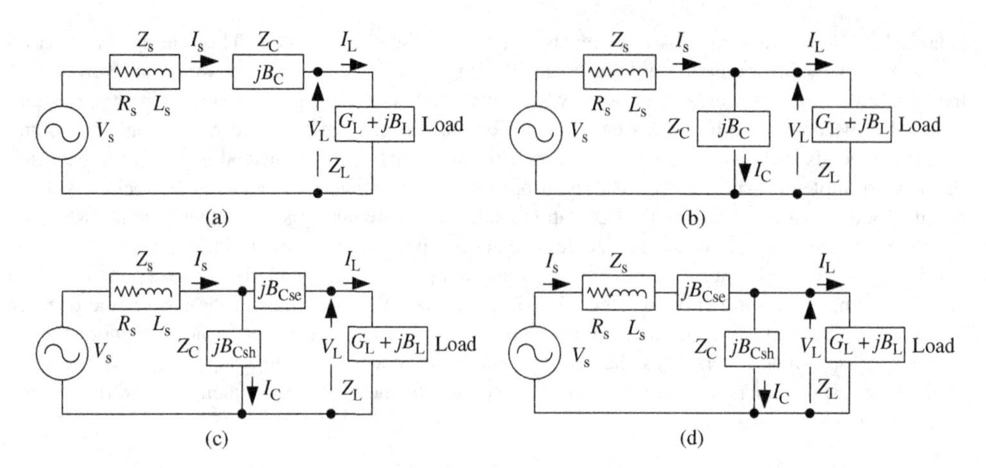

Figure 3.1 Load compensation using (a) a series compensator, (b) a shunt compensator, (c) a short-shunt hybrid compensator, and (d) a long-shunt hybrid compensator

3.3.2.1 Two-Wire Passive Compensators

Single-phase two-wire passive compensators are used in all three modes, that is, series, shunt, and a combination of both. Figure 3.1a–d shows four possible configurations of passive series, passive shunt, and a combination of both as short-shunt and long-shunt configurations. Passive series compensators are normally used for reducing voltage sags, swell, fluctuations, and so on, while shunt compensators are used for voltage regulation or power factor correction using reactive power compensation. Therefore, shunt compensators are commonly used in the distribution systems. Figure 3.2a–d shows a typical configuration of a passive shunt compensator along with its phasor diagrams for power factor correction and zero voltage regulation (ZVR) at the load end.

3.3.2.2 Three-Wire Passive Compensators

Three-phase three-wire loads such as AC motors are one of the major applications. In addition, there are many unbalanced loads on a three-wire supply system such as traction, metros, and furnaces, which are

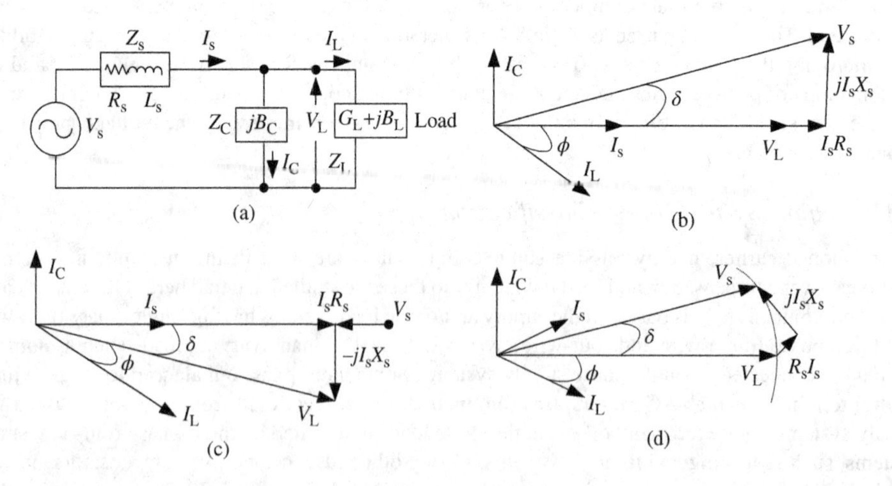

Figure 3.2 (a) A shunt compensator, and phasor diagrams for (b) PFC at load terminals, (c) PFC at substation, and (d) ZVR at load terminals

fed from a three-wire supply system. Passive shunt compensators are also designed sometimes with isolation transformers for proper voltage matching, independent phase control, and reliable compensation in unbalanced systems. Figure 3.3a–f shows typical configurations of a passive shunt compensator for power factor correction and zero voltage regulation at the load end.

3.3.2.3 Four-Wire Passive Shunt Compensators

A large number of single-phase loads may be supplied from a three-phase AC distribution system with the neutral conductor. They cause neutral current and reactive power burden and unbalanced currents. To reduce these problems, four-wire passive compensators have been used in practice. Figure 3.4a and b shows typical configurations of a passive shunt compensator with delta (D) and star (Y) connections of lossless passive elements for power factor correction and zero voltage regulation with neutral current mitigation at the load end.

3.4 Principle of Operation of Passive Shunt and Series Compensators

The main objectives of passive shunt compensators are to provide reactive power compensation for linear AC loads for improving the voltage profile (even for zero voltage regulation or power factor correction) at the AC mains in single-phase and three-phase circuits using lossless passive elements such as capacitors and inductors. In three-phase three-wire circuits, the passive shunt compensators using lossless passive elements also provide load balancing at the AC mains in addition to ZVR or PFC. Moreover, in three-phase four-wire circuits, the passive shunt compensators using lossless passive elements also provide neutral current mitigation at the AC mains in addition to load balancing, ZVR, or PFC. This aspect of passive shunt compensators has been perceived long back and used in practice for a long time, even before the introduction of solid-state control. However, with the introduction of solid-state control, their performance is further improved in terms of response, flexibility, reliability, and so on. It is mainly known as classical load compensation and used in many applications such as furnaces, traction, metros, industries, and distribution systems. Nowadays, the passive shunt compensators are also used in distributed, stand-alone, and renewable power generating systems.

The passive compensators are also used in a series configuration and a combination of shunt and series configurations depending upon application and their effectiveness. The passive series compensators are used for voltage regulation and enhancing power flow control in transmission systems. The passive series compensators are more effective in large power transmission systems. However, they have much severe resonance problems than passive shunt compensators; therefore, they are used cautiously and up to a certain part of compensation to avoid such divesting resonance problems. In a hybrid configuration, the series elements are used with shunt elements in some applications such as stand-alone self-excited induction generators. However, the series compensators are connected in series with the loads and affect the voltage across the loads; thus, they are not very popular in distribution systems.

3.5 Analysis and Design of Passive Shunt Compensators

In recent years, there has been an increased demand for the compensators to compensate large rating loads such as arc furnaces, traction, metros, commercial lighting, and air conditioning. If these loads are not compensated, then these create system unbalance and lead to fluctuations in the supply voltages. Therefore, such a supply system cannot be used to feed sensitive loads such as computers and electronic equipment. However, the importance of balanced load on the supply system has already been felt long back. The unbalanced loads cause neutral current and reactive power burden, which in turn result in low system efficiency, poor power factor, and disturbance to other consumers.

The passive shunt and series compensators are used for reactive power compensation for power factor correction or voltage regulation in single-phase systems. In addition, these are used for load balancing in three-phase three-wire systems. In three-phase four-wire systems, the passive compensators are also used for neutral current compensation along with load balancing and reactive power compensation for power factor correction or voltage regulation.

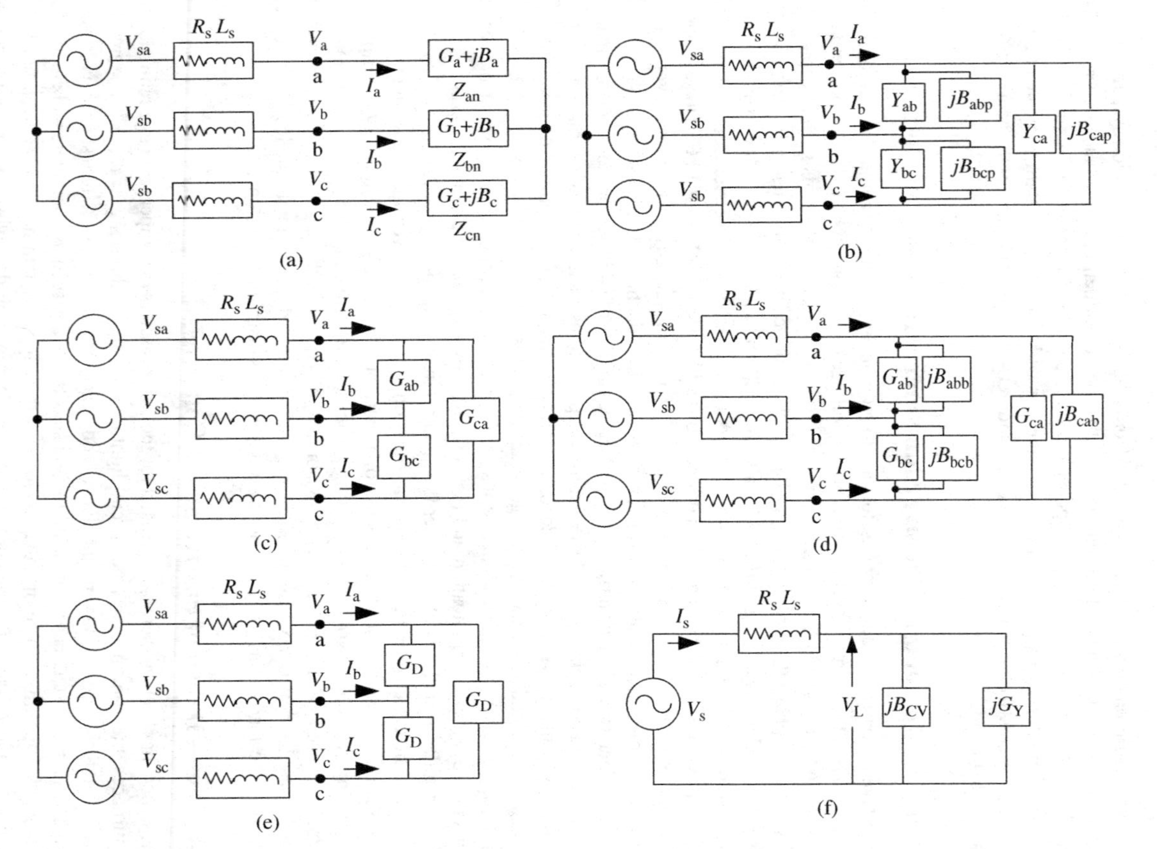

Figure 3.3 (a) A three-phase three-wire star connected load with isolated neutral terminal. (b) Compensation for PFC of a three-phase three-wire delta connected load as an equivalent of (a) after star–delta transformation. (c) An unbalanced delta connected unity power load after PFC at each phase load as an equivalent of (b). (d) Load balancing of a delta connected unbalanced unity power load of (c). (e) A balanced delta connected unity power load after compensation of load of (d). (f) Compensation for ZVR of a per-phase basis balanced star connected unity power load as an equivalent of (e)

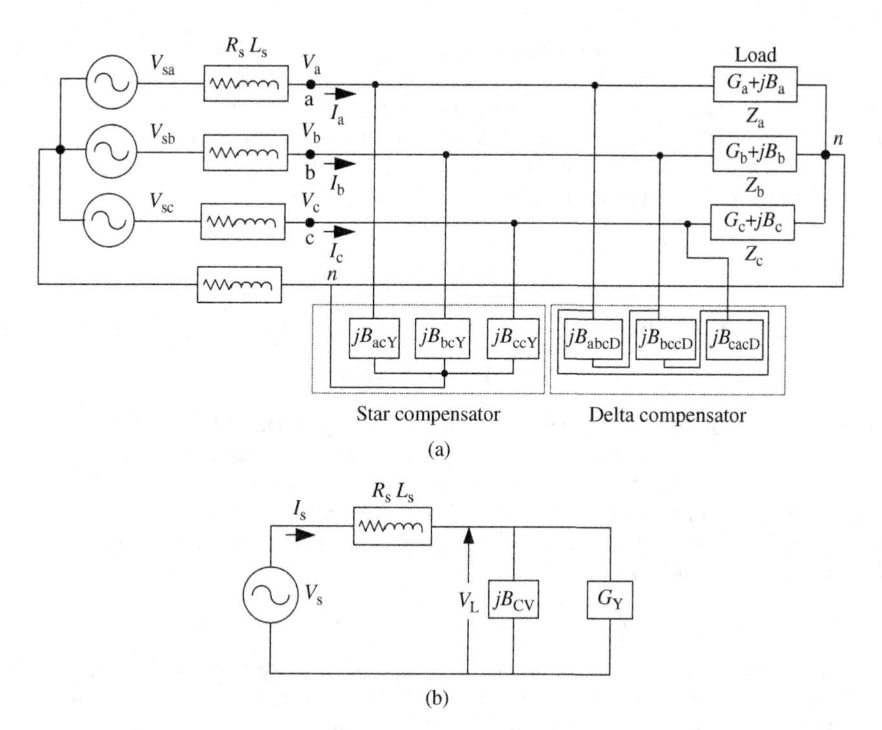

Figure 3.4 (a) Compensation for PFC, load balancing, and neutral current of a three-phase four-wire unbalanced load. (b) Compensation for ZVR of a per-phase basis balanced star connected load as an equivalent of (a)

3.5.1 Analysis and Design of Single-Phase Passive Shunt Compensators

Single-phase passive shunt compensators are used for power factor correction or zero voltage regulation across the loads. Figure 3.2a–d shows the circuit of a shunt compensator along with its phasor diagrams for these two cases. The rating of the compensator may be estimated using the system data and given load data, for which compensation is to be made.

3.5.1.1 Analysis and Design of Shunt Compensators for Power Factor Correction

Normally for the power factor correction of the load at the AC mains, a passive shunt compensator is used as it is connected directly across the load to be compensated. This shunt compensator does not affect the voltage across the loads to a great extent. Passive series compensators can also improve/correct the power factor, but they may affect the voltage across the load depending upon the load power factor and its current magnitude; therefore, they are not much preferred in the distribution system.

Figure 3.2a shows a single-phase load having impedance $Z_L = R_L + jX_L$ and admittance $Y_L = 1/Z_L = G_L + jB_L$ fed from an AC voltage of V_L. The current drawn by the load (I_L) is

$$I_L = V_L/Z_L = V_L Y_L = V_L G_L + jV_L B_L = I_R + jI_X. \tag{3.1}$$

In this equation, I_R is the active power component of the load current that is in phase with the load voltage V_L. The other component is the reactive power component of the load current that is in quadrature with the load voltage V_L. If I_X is negative, then I_L is lagging the load voltage V_L. The angle ϕ between V_L and I_L is known as the power factor angle.

The apparent power of the load is

$$S_L = V_L I_L^* = V_L^2 G_L - jV_L^2 B_L = P_L + jQ_L. \tag{3.2}$$

The apparent power S_L has an active power component P_L (real power responsible for doing actual work) and a reactive power component Q_L (imaginary power that is not used for doing any work, but is responsible for loading the system components). Typical examples of reactive powers are power corresponding to the magnetizing currents of an induction motor and a transformer. For lagging reactive power loads, B_L and I_X are negative and Q_L is positive, as shown in the phasor diagram of Figure 3.2b.

The load current drawn from the supply is higher than the current required for doing actual work (I_R), by a factor

$$I_R/I_L = \cos\phi. \tag{3.3}$$

Here $\cos\phi$ is known as the power factor and it can be expressed in terms of power triangle quantities as

$$\text{power factor (PF)} = \cos\phi. \tag{3.4}$$

$\cos\phi$ is the fraction of the apparent power that is used for doing actual work. Power loss in the feeder or cable is proportional to I_L^2, and is increased by a factor $1/(\cos^2\phi)$. Therefore, feeder rating has to be increased from I_R to I_L and this loss has to be borne by the consumers.

The basic objective of power factor correction or improvement is to compensate for this reactive power locally by connecting a compensator with admittance $Y_C = -jB_L$ in parallel to the consumer loads and to reduce the supply current to I_R:

$$I_s = I_R = I_L + I_C = V_L(G_L + jB_L) - V_L Y_C = V_L G_L. \tag{3.5}$$

After compensation, the supply current is minimum and in phase with the load/supply voltage. It corrects the power factor of the supply to unity; however, the power factor of the load remains the same as $\cos\phi$. The reactive power of the load is supplied locally by the compensator and the load is fully compensated. After this load compensation, the supply can feed additional load without exceeding the current limit of the feeder.

The compensator current is

$$I_C = V_L Y_C = -jB_L V_L. \tag{3.6}$$

The apparent power rating of the compensator is

$$S_C = P_C + jQ_C = V_L I_C^* = jV_L^2 B_L. \tag{3.7}$$

It means that the compensator does not need any active power, as $P_C = 0$ and $S_C = Q_C = V_L^2 B_L = -Q_L$. Majority of the loads are inductive in nature and thus the compensator has to supply capacitive current. Figure 3.2b shows the phasor diagram demonstrating full compensation.

With this type of compensation, the power factor at the load bus is corrected to unity. However, sometimes it is required to maintain unity power factor (UPF) at the substation feeders for high energy efficiency. In that case, the distribution feeders and the loads are viewed from the substation as pure resistive loads. Figure 3.2 shows a circuitry that consists of an impedance of the distribution feeder ($Z_s = R_s + jX_s$), an equivalent impedance of the load ($Z_L = R_L + jX_L$), an impedance of the compensator ($Z_C = jX_C = 1/Y_C = -j/B_C$) connected at the load end after the feeder, and AC mains voltage V_s.

This additional susceptance of the compensator (B_C) is estimated as follows. The imaginary part of the impedance viewed at the substation must be zero:

$$\text{Imag}\{R_s + jX_s + 1/(Y_L + jB_C)\} = \text{Imag}\{R_s + jX_s + 1/(G_L + jB_T)\} = 0. \tag{3.8}$$

Solving the above equation, the total susceptance of the compensator is

$$B_T = B_C + B_L = -[\{1 - \sqrt{(1 - 4X_s^2 G_L^2)}\}/2X_s]. \tag{3.9}$$

By connecting this total susceptance of the compensator, a unity power factor may be maintained at the substation.

3.5.1.2 Analysis and Design of Shunt Compensators for Zero Voltage Regulation

In many situations, it is considered relevant to maintain the load terminal voltage equal to the AC mains voltage (for zero voltage regulation) by using a compensator connected at the load end. It means to recover the voltage drop in the distribution feeder. It has the following advantages:

- Avoids the voltage swells caused by capacitor switching.
- Reduces the voltage sags due to common feeder faults.
- Controls the voltage fluctuations caused by customer load variations.
- Reduces the frequency of mechanical switching operations in load tap changing (LTC) transformers and mechanically switched capacitors for drastic reduction in their maintenance.
- Enhances the loadability of the system, especially for improving the stability of the load such as an induction motor under major disturbances.

A compensator with its additional susceptance B_{CV} is used across the load for ZVR as shown in the circuitry in Figure 3.2a along with the phasor diagram in Figure 3.2d, which consists of an impedance of the distribution feeder ($Z_s = R_s + jX_s$), an equivalent impedance of the load ($Z_L = R_L + jX_L$), an impedance of the compensator ($Z_{CV} = jX_{CV} = 1/Y_{CV} = -j/B_{CV}$) connected at the load end after the feeder, and AC mains voltage V_s.

This additional susceptance of the compensator (B_{CV}) for ZVR at the load terminals is estimated as follows. The load terminal voltage V_L must be the same as the AC mains voltage V_s:

$$\begin{aligned} |V_s| &= |V_L| \\ &= |V_s/\{R_s + jX_s + 1/(Y_L + jB_{CV})\}| \times |1/(Y_L + jB_{CV})| \\ &= |V_s/\{R_s + jX_s + 1/(G_L + jB_T)\}| \times |1/(G_L + jB_T)|. \end{aligned} \tag{3.10}$$

Solving the above equation, the total susceptance of the compensator is

$$B_T = B_{CV} + B_L = [X_s \pm \sqrt{\{X_s^2 - A(2R_sG_L - AG_L^2)\}}]/A, \tag{3.11}$$

where $A = R_s^2 + X_s^2$.

By connecting this total susceptance of the compensator at the load terminals, its load terminal voltage V_L is to be maintained equal to the AC mains voltage V_s.

3.5.2 Analysis and Design of Three-Phase Three-Wire Passive Shunt Compensators

Three-phase passive compensators may be used for power factor correction or zero voltage regulation along with load balancing by connecting lossless passive elements across the unbalanced three-phase three-wire loads. The rating of the lossless passive elements of the comparator may be estimated using the system data and given load data, for which compensation is to be made as given in the following section.

3.5.2.1 Analysis and Design of Shunt Compensators for Power Factor Correction

Any three-phase unbalanced ungrounded star connected load, which is shown in Figure 3.3a, may be transformed to a three-phase unbalanced delta connected load (see Figure 3.3b) by star–delta transformation as follows:

$$Y_{ab} = 1/Z_{ab} = Z_{cn}/(Z_{an}Z_{bn} + Z_{bn}Z_{cn} + Z_{cn}Z_{an}), \tag{3.12}$$

$$Y_{ba} = 1/Z_{ba} = Z_{an}/(Z_{an}Z_{bn} + Z_{bn}Z_{cn} + Z_{cn}Z_{an}), \tag{3.13}$$

$$Y_{ca} = 1/Z_{ca} = Z_{bn}/(Z_{an}Z_{bn} + Z_{bn}Z_{cn} + Z_{cn}Z_{an}), \tag{3.14}$$

where Z_{an}, Z_{bn}, and Z_{cn} are three-phase load impedances of any three-phase unbalanced ungrounded star connected load.

Therefore, any three-phase unbalanced ungrounded star connected load, shown in Figure 3.3a, may be converted to an equivalent three-phase delta connected unbalanced reactive load shown in Figure 3.3b.

Moreover, any three-phase delta connected unbalanced reactive load may also be represented as an equivalent load shown in Figure 3.3b. This three-phase delta connected unbalanced reactive load can be defined in terms of admittances, impedances, conductances, and susceptances as follows:

$$Y_{ab} = 1/Z_{ab} = G_{ab} - jB_{ab}, \quad Y_{bc} = 1/Z_{bc} = G_{bc} - jB_{bc}, \quad Y_{ca} = 1/Z_{ca} = G_{ca} - jB_{ca}. \tag{3.15}$$

Imaginary parts of these admittances (susceptances) of the three-phase delta connected unbalanced reactive load are estimated to compensate their reactive part for power factor correction as follows:

$$B_{abp} = -(\text{imaginary part of } Y_{ab}) = -B_{ab}, \tag{3.16}$$

$$B_{bcp} = -(\text{imaginary part of } Y_{bc}) = -B_{bc}, \tag{3.17}$$

$$B_{cap} = -(\text{imaginary part of } Y_{ca}) = -B_{ca}. \tag{3.18}$$

After connecting these compensating susceptances (B_{abp}, B_{bcp}, B_{cap}) in parallel to the three-phase delta connected load, the resulting delta connected load shown in Figure 3.3c becomes an unbalanced resistive load with three unequal conductances (G_{ab}, G_{bc}, G_{ca}). This unbalanced delta connected resistive load (Figure 3.3c) may be realized as a balanced delta connected resistive load (Figure 3.3e) by connecting three lossless passive elements (susceptances B_{abb}, B_{bcb}, B_{cab}) across them (Figure 3.3d). The values of these susceptances (B_{abb}, B_{bcb}, B_{cab}) may be estimated using symmetrical component theory given below.

The unbalanced delta connected resistive load shown in Figure 3.3c is fed from balanced three-phase per-phase voltages

$$V_a = V_p, \quad V_b = a^2 V_p, \quad V_c = aV_p, \quad \text{where } a = e^{j2\pi/3} = -(1/2) + j(\sqrt{3}/2). \tag{3.19}$$

The line voltages are

$$V_{ab} = (V_a - V_b) = (1 - a^2)V_p = V\angle 30°, \quad V_{bc} = (V_b - V_c) = (a^2 - a)V_p = V\angle -90°,$$
$$V_{ca} = (V_c - V_a) = (a - 1)V_p = V\angle -210°. \tag{3.20}$$

The line voltage is $V=\sqrt{3}V_p$ and the rms phase voltage is V_p.

Three-phase load currents in three branches of the delta loop are

$$I_{ab} = G_{ab}V_{ab} = G_{ab}(1 - a^2)V_p, \quad I_{bc} = G_{bc}V_{bc} = G_{bc}(a^2 - a)V_p, \quad I_{ca} = G_{ca}V_{ca} = G_{ca}(a - 1)V_p. \tag{3.21}$$

The line currents of the unbalanced delta connected resistive load are

$$I_a = I_{ab} - I_{ca} = \{G_{ab}(1 - a^2) - G_{ca}(a - 1)\}V_p, \tag{3.22}$$

$$I_b = I_{bc} - I_{ab} = \{G_{bc}(a^2 - a) - G_{ab}(1 - a^2)\}V_p, \tag{3.23}$$

$$I_c = I_{ca} - I_{bc} = \{G_{ca}(a - 1) - G_{bc}(a^2 - a)\}V_p. \tag{3.24}$$

Symmetrical components of the line currents of the unbalanced delta connected resistive load are

$$I_0 = (I_a + I_b + I_c)/\sqrt{3}, \quad I_1 = (I_a + aI_b + a^2 I_c)/\sqrt{3}, \quad I_2 = (I_a + a^2 I_b + aI_c)/\sqrt{3}. \tag{3.25}$$

The factor of $1/\sqrt{3}$ is to realize power invariance symmetrical component transformation.

I_0, I_1, and I_2 are the zero, positive, and negative symmetrical components of the line currents of the unbalanced delta connected resistive load. These are expressed in terms of voltages and conductances using Equations 3.22–3.24:

$$I_0 = 0, \quad I_1 = (G_{ab} + G_{bc} + G_{ca})\sqrt{3}V_p, \quad I_2 = -(a^2 G_{ab} + G_{bc} + aG_{ca})\sqrt{3}V_p. \tag{3.26}$$

Similarly, symmetrical components of the line currents of the delta connected lossless elements (reactive)-based compensator used for load balancing with susceptances (B_{abb}, B_{bcb}, B_{cab}) are

$$I_{0c} = 0, \quad I_{1c} = j(B_{abb} + B_{bcb} + B_{cab})\sqrt{3}V_p, \quad I_{2c} = -j(a^2 B_{abb} + B_{bcb} + aB_{cab})\sqrt{3}V_p. \tag{3.27}$$

The two negative-sequence currents of the symmetrical components of the compensator and the load must cancel each other for loading balancing at the AC mains. It results in the following relation:

$$I_2 + I_{2c} = 0, \quad \text{or} \quad -(a^2 G_{ab} + G_{bc} + aG_{ca})\sqrt{3}V_p + \{-j(a^2 B_{abb} + B_{bcb} + aB_{cab})\sqrt{3}V_p\} = 0. \tag{3.28}$$

It results in the following relation among conductances of the load and susceptances (B_{abb}, B_{bcb}, B_{cab}) of the compensator:

$$a^2 G_{ab} + G_{bc} + aG_{ca} = -j(a^2 B_{abb} + B_{bcb} + aB_{cab}). \tag{3.29}$$

Moreover, the positive-sequence current of the symmetrical components of the compensator currents I_{1c} must be zero for power factor correction and load balancing:

$$I_{1c} = 0 = j(B_{abb} + B_{bcb} + B_{cab})\sqrt{3}V_p. \tag{3.30}$$

It results in the following relation among susceptances (B_{abb}, B_{bcb}, B_{cab}) of the compensator:

$$B_{abb} + B_{bcb} + B_{cab} = 0. \tag{3.31}$$

Solving Equations 3.29 and 3.31, a set of susceptances may be estimated for load balancing and power factor correction at the three-phase AC mains:

$$B_{abb} = (-G_{bc} + G_{ca})/\sqrt{3}, \tag{3.32}$$

$$B_{bcb} = (G_{ab} - G_{ca})/\sqrt{3}, \tag{3.33}$$

$$B_{cab} = (-G_{ab} + G_{bc})/\sqrt{3}. \tag{3.34}$$

The rms balanced line current I is

$$I = |V_p|(G_{ab} + G_{bc} + G_{ca}). \tag{3.35}$$

Total three-phase susceptances (B_{abc}, B_{cac}, B_{bcc}) to be connected across the three lines of the AC mains for power factor correction and load balancing of three-phase delta connected unbalanced reactive loads are

$$B_{abc} = B_{abp} + B_{abb}, \tag{3.36}$$

$$B_{cac} = B_{cap} + B_{cab}, \tag{3.37}$$

$$B_{bcc} = B_{bcp} + B_{bcb}. \tag{3.38}$$

Therefore, by connecting the three-phase susceptances (B_{abc}, B_{cac}, B_{bcc}) across three lines of the AC mains, any unbalanced delta connected load may be realized as an equivalent balanced delta connected unity power factor load.

The parameters of an equivalent balanced delta connected unity power factor load are

$$G_D = 1/R_D = (G_{ab} + G_{ca} + G_{bc})/3. \tag{3.39}$$

The parameters of an equivalent balanced star connected unity power factor load are

$$G_Y = 1/R_Y = 3G_D = G_{ab} + G_{ca} + G_{bc}. \tag{3.40}$$

The total active power consumed by the load is

$$P = V^2 G_D = 3V_p^2 G_Y. \tag{3.41}$$

3.5.2.2 Analysis and Design of Shunt Compensators for Zero Voltage Regulation

After load balancing and power factor correction to unity, a balanced star connected load with $G = G_Y$ may be considered as an equivalent per-phase circuit as shown in Figure 3.3f. To maintain the load terminal voltage equal to the AC mains voltage (for zero voltage regulation), another balanced star connected compensator may be used at the load end.

A compensator with an additional susceptance B_{CV} is used across the load for ZVR as shown in the circuitry in Figure 3.3f, which consists of an impedance of the distribution feeder ($Z_s = R_s + jX_s$), an equivalent balanced conductance of the load (G_Y), an impedance of the compensator ($Z_{CV} = jX_{CV} = 1/Y_{CV} = -j/B_{CV}$) connected at the load end after the feeder, and AC mains voltage V_s.

The additional susceptance of the compensator (B_{CV}) for ZVR at the load terminals is estimated as follows. The load terminal voltage V_L must be the same as the AC mains voltage V_s:

$$|V_s| = |V_L| = |V_s/\{R_s + jX_s + 1/(G_Y + jB_{CV})\}| \times |1/(G_Y + B_{CV})|. \tag{3.42}$$

Solving the above equation, the additional susceptance of the compensator is

$$B_{CV} = [X_s \pm \sqrt{\{X_s^2 - A(2R_s G_Y - AG_Y^2)\}}]/A, \tag{3.43}$$

where $A = R_s^2 + X_s^2$.

By connecting three equal-valued susceptances of the compensator each with a value of B_{CV} in star connection across the load terminals, its load terminal voltage V_L is maintained equal to the AC mains voltage V_s, resulting in zero voltage regulation.

3.5.3 *Analysis and Design of Three-Phase Four-Wire Passive Shunt Compensators*

Three-phase four-wire passive shunt compensators may be used for power factor correction or zero voltage regulation along with load balancing by connecting lossless passive elements across the unbalanced three-phase four-wire loads. The rating of the lossless passive elements of the compensator may be estimated using the system data and given load data, for which compensation is to be made as given in the following section.

3.5.3.1 Analysis and Design of Shunt Compensators for Power Factor Correction

Any three-phase four-wire unbalanced load (grounded star connected load) may be compensated by using six lossless passive elements (inductors and capacitors) as shown in Figure 3.4a. Out of the six lossless passive elements, three passive elements are connected in star configuration with their neutral terminal connected to load neutral terminal to mitigate AC mains neutral current and to feed load neutral current, resulting in an equivalent three-phase unbalanced ungrounded star connected load (an equivalent three-phase three-wire load). However, out of the three passive elements, one element may be eliminated and two elements are sufficient to be connected across any two phases and common neutral terminal for mitigating AC mains neutral current. Here, all three lossless passive elements are considered for generality.

Other three lossless passive elements are connected in delta configuration across three-phase lines for load balancing and power factor correction.

In order to balance the line currents and improve the power factor, star connected three lossless passive elements are connected and other delta connected three lossless passive elements are placed at the load bus to provide a different amount of reactive power compensation to each phase.

The unbalanced four-wire load shown in Figure 3.4 is fed from balanced three-phase per-phase voltages

$$V_a = V_p, \quad V_b = a^2 V_p, \quad V_c = aV_p, \quad \text{where } a = e^{j2\pi/3} = -(1/2) + j(\sqrt{3}/2). \tag{3.44}$$

The load currents in terms of load admittances are

$$I_a = (G_a + jB_a)V_p, \quad I_b = (G_b + jB_b)a^2 V_p, \quad I_c = (G_c + jB_c)aV_p. \tag{3.45}$$

Symmetrical components of the line currents of the unbalanced four-wire load are

$$I_0 = (I_a + I_b + I_c)/\sqrt{3}, \quad I_1 = (I_a + aI_b + a^2 I_c)/\sqrt{3}, \quad I_2 = (I_a + a^2 I_b + aI_c)/\sqrt{3}. \tag{3.46}$$

The factor of $1/\sqrt{3}$ is to realize power invariance symmetrical component transformation.

I_0, I_1, and I_2 are the zero, positive, and negative symmetrical components of the line currents of the unbalanced four-wire load. These are expressed in terms of voltages and conductances using Equations 3.44–3.45:

$$I_0 = (I_a + I_b + I_c)/\sqrt{3} = \{(G_a + a^2 G_b + aG_c) + j(B_a + a^2 B_b + aB_c)\}V_p/\sqrt{3}, \tag{3.47}$$

$$I_1 = (I_a + aI_b + a^2 I_c)/\sqrt{3} = \{(G_a + G_b + G_c) + j(B_a + B_b + B_c)\}\sqrt{3}V_p, \tag{3.48}$$

$$I_2 = (I_a + aI_b + a^2 I_c)/\sqrt{3} = \{(G_a + a^2 G_b + aG_c) + j(B_a + aB_b + a^2 B_c)\}V_p/\sqrt{3}. \tag{3.49}$$

Similarly, zero, positive, and negative symmetrical components of the line currents of the star connected compensator are

$$I_{0cY} = (I_{acY} + I_{bcY} + I_{ccY})/\sqrt{3} = \{j(B_{acY} + a^2 B_{bcY} + aB_{ccY})\}V_p/\sqrt{3}, \tag{3.50}$$

$$I_{1cY} = (I_{acY} + aI_{bcY} + a^2 I_{ccY})/\sqrt{3} = \{j(B_{acY} + B_{bcY} + B_{ccY})\}\sqrt{3}V_p, \tag{3.51}$$

$$I_{2cY} = (I_{acY} + aI_{bcY} + a^2 I_{ccY})/\sqrt{3} = \{j(B_{acY} + aB_{bcY} + a^2 B_{ccY})\}V_p/\sqrt{3}. \tag{3.52}$$

The zero, positive, and negative symmetrical components of the line currents of the delta connected compensator are

$$I_{0cD} = (I_{acD} + I_{bcD} + I_{ccD})/\sqrt{3} = 0, \tag{3.53}$$

$$I_{1cD} = (I_{acD} + aI_{bcD} + a^2 I_{ccD})/\sqrt{3} = \{j(B_{abcD} + B_{bccD} + B_{cacD})\}\sqrt{3}V_p, \tag{3.54}$$

$$I_{2cD} = (I_{acD} + aI_{bcD} + a^2 I_{ccD})/\sqrt{3} = \{j(a^2 B_{abcD} + B_{bccD} + aB_{cacD})\}V_p/\sqrt{3}. \tag{3.55}$$

With the use of six lossless elements of the compensator (three elements connected in star configuration and other three connected in delta configuration) across the load, the negative- and zero-sequence components of the load currents are to be eliminated and the power factor at the load bus is to be improved to unity. It results in balanced sinusoidal unity power supply currents.

For this purpose, it must satisfy the following conditions:

1. Real parts of negative-sequence components of load currents and compensator currents must cancel each other. It means that the sum of real parts of negative-sequence components of load currents and compensator currents must be zero:

$$\mathrm{Real}(I_2) + \mathrm{Real}(I_{2\mathrm{cY}}) + \mathrm{Real}(I_{2\mathrm{cD}}) = 0, \tag{3.56}$$

$$\begin{aligned}&\mathrm{Real}[\{(G_\mathrm{a} + a^2G_\mathrm{b} + aG_\mathrm{c}) + j(B_\mathrm{a} + aB_\mathrm{b} + a^2B_\mathrm{c})\}V_\mathrm{p}/\sqrt{3}]\\&+ \mathrm{Real}[\{j(B_\mathrm{acY} + aB_\mathrm{bcY} + a^2B_\mathrm{ccY})\}V_\mathrm{p}/\sqrt{3}] + \mathrm{Real}[\{j(a^2B_\mathrm{abcD} + B_\mathrm{bccD} + aB_\mathrm{cacD})\}V_\mathrm{p}/\sqrt{3}] = 0.\end{aligned} \tag{3.57}$$

2. Imaginary parts of negative-sequence components of load currents and compensator currents must cancel each other. It means that the sum of imaginary parts of negative-sequence components of load currents and compensator currents must be zero:

$$\mathrm{Imag}(I_2) + \mathrm{Imag}(I_{2\mathrm{cY}}) + \mathrm{Imag}(I_{2\mathrm{cD}}) = 0, \tag{3.58}$$

$$\begin{aligned}&\mathrm{Imag}[\{(G_\mathrm{a} + a^2G_\mathrm{b} + aG_\mathrm{c}) + j(B_\mathrm{a} + aB_\mathrm{b} + a^2B_\mathrm{c})\}V_\mathrm{p}/\sqrt{3}]\\&+ \mathrm{Imag}[\{j(B_\mathrm{acY} + aB_\mathrm{bcY} + a^2B_\mathrm{ccY})\}V_\mathrm{p}/\sqrt{3}] + \mathrm{Imag}[\{j(a^2B_\mathrm{abcD} + B_\mathrm{bccD} + aB_\mathrm{cacD})\}V_\mathrm{p}/\sqrt{3}] = 0.\end{aligned} \tag{3.59}$$

3. Real parts of zero-sequence components of load currents and compensator currents must cancel each other. It means that the sum of real parts of zero-sequence components of load currents and compensator currents must be zero. In this case, delta connected elements of the compensator do not contribute as there is no flow of zero-sequence current in the delta connected network. This condition results in the following equation:

$$\mathrm{Real}(I_0) + \mathrm{Real}(I_{0\mathrm{cY}}) = 0, \tag{3.60}$$

$$\begin{aligned}&\mathrm{Real}[\{(G_\mathrm{a} + a^2G_\mathrm{b} + aG_\mathrm{c}) + j(B_\mathrm{a} + a^2B_\mathrm{b} + aB_\mathrm{c})\}V_\mathrm{p}/\sqrt{3}]\\&+ \mathrm{Real}[\{j(B_\mathrm{acY} + a^2B_\mathrm{bcY} + aB_\mathrm{ccY})\}V_\mathrm{p}/\sqrt{3}] = 0.\end{aligned} \tag{3.61}$$

4. Imaginary parts of zero-sequence components of load currents and compensator currents must cancel each other. It means that the sum of imaginary parts of zero-sequence components of load currents and compensator currents must be zero. In this case also, delta connected elements of the compensator do not contribute as there is no flow of zero-sequence current in the delta connected network. This condition results in the following equation:

$$\mathrm{Imag}(I_0) + \mathrm{Imag}(I_{0\mathrm{cY}}) = 0, \tag{3.62}$$

$$\begin{aligned}&\mathrm{Imag}[\{(G_\mathrm{a} + a^2G_\mathrm{b} + aG_\mathrm{c}) + j(B_\mathrm{a} + a^2B_\mathrm{b} + aB_\mathrm{c})\}V_\mathrm{p}/\sqrt{3}]\\&+ \mathrm{Imag}[\{j(B_\mathrm{acY} + a^2B_\mathrm{bcY} + aB_\mathrm{ccY})\}V_\mathrm{p}/\sqrt{3}] = 0.\end{aligned} \tag{3.63}$$

5. For unity power factor at the AC mains, imaginary parts of positive-sequence components of load currents and compensator currents must cancel each other. It means that the sum of imaginary parts of positive-sequence components of load currents and compensator currents must be zero:

$$\mathrm{Imag}(I_1) + \mathrm{Imag}(I_{1\mathrm{cY}}) + \mathrm{Imag}(I_{1\mathrm{cD}}) = 0, \tag{3.64}$$

$$\begin{aligned}&\mathrm{Imag}[\{(G_\mathrm{a} + G_\mathrm{b} + G_\mathrm{ca}) + j(B_\mathrm{a} + B_\mathrm{b} + B_\mathrm{c})\}\sqrt{3}V_\mathrm{p}] + \mathrm{Imag}[\{j(B_\mathrm{acY} + B_\mathrm{bcY} + B_\mathrm{ccY})\}\sqrt{3}V_\mathrm{p}]\\&+ \mathrm{Imag}[\{j(B_\mathrm{abcD} + B_\mathrm{bccD} + B_\mathrm{cacD})\}\sqrt{3}V_\mathrm{p}] = 0.\end{aligned} \tag{3.65}$$

These five conditions have infinite solutions because there are six unknowns (susceptances of the Y-connected and the D-connected parts of the compensator) with five equations. An additional condition together with the above five equations may offer a unique solution.

An additional condition may be that the imaginary part of the positive-sequence component of load currents is eliminated by the Y-connected part of the compensator alone; that is, the D-connected part of the compensator does not generate the imaginary part of positive-sequence currents. Conversely, the imaginary part of the positive-sequence component of load currents is eliminated by the D-connected part of the compensator alone; that is, the Y-connected part of the compensator does not generate the imaginary part of positive-sequence currents.

Here, the first condition is considered that the imaginary part of the positive-sequence component of load currents is eliminated by the Y-connected part of the compensator alone; that is, the D-connected part of the compensator does not generate the imaginary part of positive-sequence currents. In this case, the sixth condition is

$$\text{Imag}[I_{1\text{cD}}] = \text{Imag}[\{j(B_{\text{abcD}} + B_{\text{bccD}} + B_{\text{cacD}})\}\sqrt{3}V_{\text{p}}] = 0 \quad \text{or} \quad B_{\text{abcD}} + B_{\text{bccD}} + B_{\text{cacD}} = 0. \tag{3.66}$$

The solution of Equations 3.57, 3.59, 3.61, 3.63, 3.65, and 3.66 gives susceptances of the Y-connected part and the D-connected part of the compensator in terms of conductances and susceptances of the four-wire load:

$$B_{\text{acY}} = -B_{\text{a}} + (G_{\text{b}} - G_{\text{c}})/\sqrt{3}, \tag{3.67}$$

$$B_{\text{bcY}} = -B_{\text{b}} + (G_{\text{c}} - G_{\text{a}})/\sqrt{3}, \tag{3.68}$$

$$B_{\text{ccY}} = -B_{\text{c}} + (G_{\text{a}} - G_{\text{b}})/\sqrt{3}, \tag{3.69}$$

$$B_{\text{abcD}} = (2/3)(G_{\text{a}} - G_{\text{b}})/\sqrt{3}, \tag{3.70}$$

$$B_{\text{bccD}} = (2/3)(G_{\text{b}} - G_{\text{c}})/\sqrt{3}, \tag{3.71}$$

$$B_{\text{cacD}} = (2/3)(G_{\text{c}} - G_{\text{a}})/\sqrt{3}. \tag{3.72}$$

These six lossless passive elements (L and/or C) connected in star and delta configurations across the four-wire load are expected to result in unity power factor balanced sinusoidal supply currents.

Similarly, another solution may be achieved by considering that the imaginary part of the positive-sequence component of load currents is eliminated by the D-connected part of the compensator alone; that is, the Y-connected part of the compensator does not generate the imaginary part of positive-sequence currents.

It is also interesting to note as reported in the literature that such compensation of the four-wire load for unity power factor balanced sinusoidal supply currents may be achieved with only any five lossless passive elements of the six elements using five equations (Equations 3.57, 3.59, 3.61, 3.63, and 3.65) given in the first five conditions. It means that one of the six elements may be taken out of the service for maintenance, faults, or outages. However, it is possible only if the values of these elements are variable in nature. Usually, the loads are varying in practice, so these elements are also continuously variable in nature. It results in six different solutions as any one element out of the six may be zero. Here, only one such case is given, in which the first element (B_{acY}) is considered zero. It gives the following design of the compensator:

$$B_{\text{bcY}} = B_{\text{a}} - B_{\text{b}} + (G_{\text{a}} - G_{\text{b}} + 2G_{\text{c}})/\sqrt{3}, \tag{3.73}$$

$$B_{\text{ccY}} = B_{\text{a}} - B_{\text{c}} + (G_{\text{a}} - 2G_{\text{b}} + G_{\text{c}})/\sqrt{3}, \tag{3.74}$$

$$B_{\text{abcD}} = -B_{\text{a}} + (2G_{\text{a}} - G_{\text{b}} - G_{\text{c}})/\sqrt{3}, \tag{3.75}$$

$$B_{bccD} = -B_a + (G_b - G_c)/\sqrt{3}, \tag{3.76}$$

$$B_{cacD} = -B_a + (1/3)(-2G_a + G_c + G_a)/\sqrt{3}. \tag{3.77}$$

Similarly, other five solutions may be achieved by considering any one of the five elements to be zero.

The parameters of an equivalent balanced star connected unity power factor load are as follows:

$$G_Y = 1/R_Y = (G_a + G_b + G_c)/3. \tag{3.78}$$

The total active power consumed by the load is

$$P = 3V_p^2 G_Y = 3V_p^2/R_Y. \tag{3.79}$$

3.5.3.2 Analysis and Design of Shunt Compensators for Zero Voltage Regulation

After load balancing and power factor correction to unity, a balanced star connected load with $G = G_Y$ may be considered as an equivalent per-phase circuit as shown in Figure 3.4b. To maintain the load terminal voltage equal to the AC mains voltage (for zero voltage regulation), another balanced star connected compensator may be used at the load end.

A compensator with an additional susceptance B_{CV} is used across the load for ZVR as shown in the circuitry in Figure 3.4b, which consists of an impedance of the distribution feeder ($Z_s = R_s + jX_s$), an equivalent balanced conductance of the load (G_Y), an impedance of the compensator ($Z_{CV} = jX_{CV} = 1/Y_{CV} = -j/B_{CV}$) connected at the load end after the feeder, and AC mains voltage V_s.

The additional susceptance of the compensator (B_{CV}) for ZVR at the load terminals is estimated as follows. The load terminal voltage V_L must be the same as the AC mains voltage V_s:

$$|V_s| = |V_L| = |V_s/\{R_s + jX_s + 1/(G_Y + jB_{CV})\}| \times |1/(G_Y + B_{CV})|. \tag{3.80}$$

Solving the above equation, the additional susceptance of the compensator is

$$B_{CV} = [X_s \pm \sqrt{\{X_s^2 - A(2R_s G_Y - AG_Y^2)\}}]/A, \tag{3.81}$$

where $A = R_s^2 + X_s^2$.

By connecting three equal-valued susceptances of the compensator each with a value of B_{CV} in star connection across the equivalent star connected load terminals, its load terminal voltage V_L is maintained equal to the AC mains voltage V_s, resulting in zero voltage regulation.

3.6 Modeling, Simulation, and Performance of Passive Shunt and Series Compensators

Modeling and simulation of passive shunt and series compensators are carried out to demonstrate their performance for their effectiveness and basic understanding of load compensation through voltage and current waveforms. After design of the passive compensators, these are connected in the system configuration and waveform analysis is done through simulation to study their effect on the system and to observe their interactions with the system and occurrence of any phenomena such as subsynchronous resonance and parallel resonance considering all the practical conditions, which are not considered in the design of the passive compensators. Earlier, the simulation study of these compensators with the system has been quite cumbersome; however, with various available simulation packages such as MATLAB, PSCAD, EMTP, PSPICE, SABER, PSIM, ETEPP, and desilent, the simulation of the performance of these compensators has become quite simple and straightforward. Nowadays, for a particular application, after the design of these compensators, their performance is studied in simulation before these are implemented in practice. In view of these requirements, a few examples of the simulation studies done in MATLAB are presented here to provide an exposure to the designers and application engineers.

3.7 Numerical Examples

Example 3.1

A single-phase load having $Z_L = (4.0 + j1.0)$ pu is fed from an AC supply with an input AC voltage of 230 V at 50 Hz and a base impedance of 4.15 Ω. It is to be realized as a unity power factor load on the AC supply system using a shunt connected lossless passive element (L or C) as shown in Figure E3.1. Calculate (a) the value of the compensator element (in farads or henries) and (b) equivalent resistance (in ohms) of the compensated load.

Solution: Given supply voltage $V_s = 230$ V, frequency of the supply (f) = 50 Hz, and a single-phase load having $Z = (4.0 + j1.0)$ pu with a base impedance of 4.15 Ω per phase.

The load resistance is $R_L = 4.15 \times 4.0\,\Omega = 16.6\,\Omega$. The load reactance is $X_L = 4.15 \times 1\,\Omega = 4.15\,\Omega$.

The load impedance is $Z_L = (16.6 + j4.15)\,\Omega = 17.11\angle 14.024°\,\Omega$.

The load current before compensation is $I_{sold} = V/Z_L = (230/17.11)\angle -14.024°\text{A} = 13.44\angle -14.024°\text{A}$.

The reactive current is $I_r = I_{sold} \sin\theta = I_{sold}X_L/Z_L = 3.2575$ A, where θ is the power factor angle of the load.

The compensating capacitor should supply the same reactive current as the load; hence, $I_C = I_r = 3.2575$ A.

a. The value of the capacitor for power factor correction is $C = I_C/(V\omega) = 3.2575/(314 \times 230) = 45.104\,\mu\text{F}$.
   The supply current after compensation is $I_{snew} = I_{sold} \cos\theta = I_{sold}R_L/Z_L = I_a = 13.0394$ A.
b. The equivalent resistance of the compensated load is $R_{eq} = V/I_{snew} = 230/13.0394 = 17.639\,\Omega$.

Example 3.2

A single-phase AC supply has rms voltage of 230 V at 50 Hz and a feeder (source) impedance of 1.0 Ω resistance and 4.0 Ω inductive reactance after which a single-phase load having $Z_L = (16 + j12)\,\Omega$ is connected. Calculate (a) the voltage drop across the source impedance and (b) the voltage across the load. If a shunt compensator consisting of a lossless passive element (L or C) is used to raise the voltage to the input voltage (230 V) as shown in Figure E3.2, calculate (c) the value of the compensator element (in farads or henries), (d) its kVA rating, and (e) the voltage drop across the source impedance after compensation.

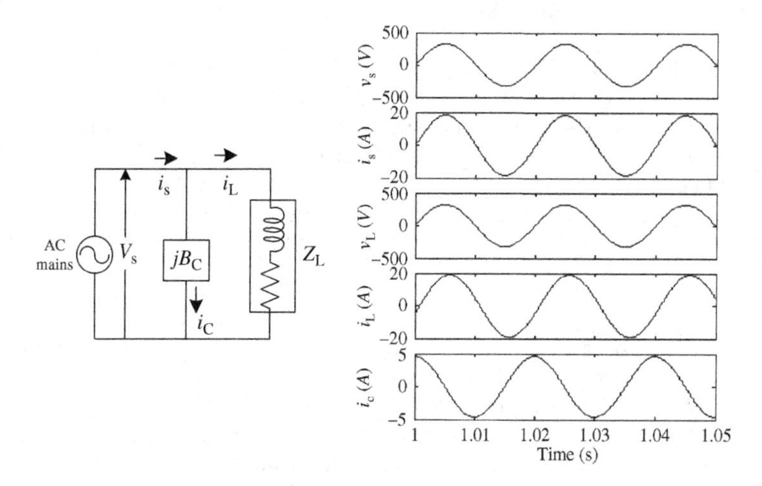

Figure E3.1 Schematic diagram and performance waveforms

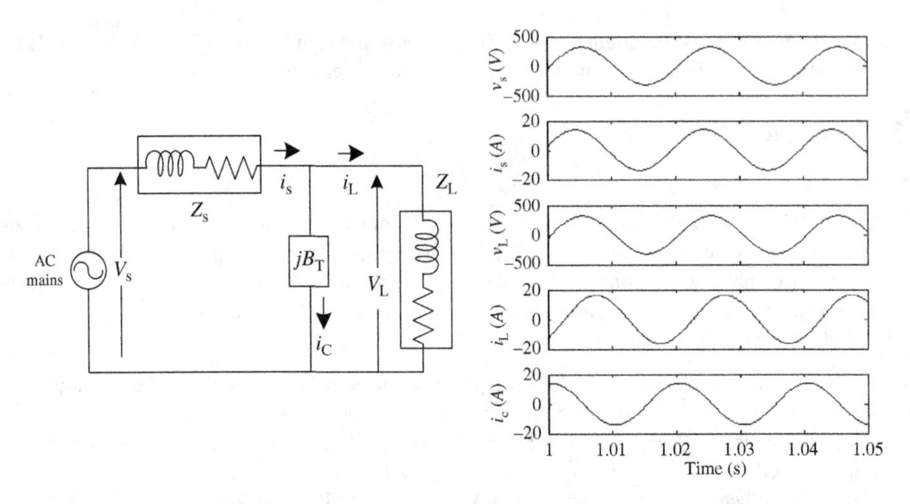

Figure E3.2 Schematic diagram and performance waveforms

Solution: Given supply voltage $V_s = 230$ V, frequency of the supply (f) = 50 Hz, and a single-phase load having $Z_L = (16 + j12)$ Ω. The source impedance is $Z_s = R_s + jX_s = (1 + j4)$ Ω.

The load impedance is $Z_L = (16 + j12)\ \Omega = 20\angle 36.86°\ \Omega$.

Total impedance is $Z_T = Z_s + Z_L = (17 + j16)\ \Omega = 23.345\angle 43.26°\ \Omega$.

The load current before compensation is $I_{sold} = V_s/|Z_T| = 230/23.345\ \text{A} = 9.856\ \text{A}$.

a. The voltage drop across the source impedance is $V_{Zs} = I_{sold}Z_s = 40.637$ V.
b. The voltage across the load is $V_{ZL} = I_{sold}Z_L = 197.1$ V.
c. The load admittance is $Y_L = G_L + jB_L = 1/Z_L = (0.04 - j0.03)$ mhos.

   The load power factor can be made unity by connecting a susceptance of $B_{PF} = -B_L = 0.03$ mhos.

   Let the susceptance of value B_T mhos is connected in parallel to the load to raise the load voltage to the input voltage ($V_L = V_s$). B_T has two components, one for power factor correction and other for raising the load voltage to the input voltage. Hence, $B_T = B_{PF} + B_{CV}$.

   The basic equation to maintain the load voltage equal to the input voltage is $|V_s/[\{(R_s + jX_s) + 1/(G_L + jB_{CV})\}]| \times |1/(G_L + jB_{CV})| = |V_s|$.

   Solving this equation, the value of B_{CV} is $B_{CV} = [X_s \pm \{\sqrt{X_s^2 - (X_s^2 + R_s^2)(2R_sG_L + R_s^2G_L^2 + X_s^2G_L^2)}\}]/(X_s^2 + R_s^2)$.

   Substituting the values of $X_s = 4\ \Omega$, $R_s = 1\ \Omega$, and $G_L = 0.04$ mhos and considering "–" sign in "±", for lower value of the compensator, the value of B_{CV} is 0.013 804 976 mhos.

   Total susceptance for voltage regulation is $B_T = B_{CV} + B_{PF} = 0.043\,804\,976$ mhos.

   The total value of the capacitor for voltage regulation is $C_T = B_T/\omega = 139.51\ \mu\text{F}$.
d. The kVA rating of the compensator is $Q_{eq} = V^2B_T = 230^2 \times 0.043\,804\,976 = 2.317$ kVA.
e. The voltage drop across the source impedance after compensation is $V_{Zs} = I_{snew}Z_s = V_L(G_L + jB_{CV})Z_s = 230 \times (0.04 + j0.013\,804\,976) \times (1 + j4) = (9.2 + j3.175) \times 4.123 = (37.9316 + j13.09) = 40.12$ V.

Example 3.3

A single-phase AC supply has rms voltage of 230 V at 50 Hz and a feeder (source) impedance of 4.0 Ω inductive reactance after which a single-phase load having $Z_L = (16 + j12)$ Ω is connected. If it is to be realized as a unity power factor load on the AC supply system using a series compensator consisting of lossless passive elements (L and/or C) as shown in Figure E3.3, calculate (a) the voltage drop across the source impedance and (b) the voltage across the load without and with a series compensator. Moreover, calculate (c) the values of compensator elements (in farads or henries) and (d) its kVA rating. If this series

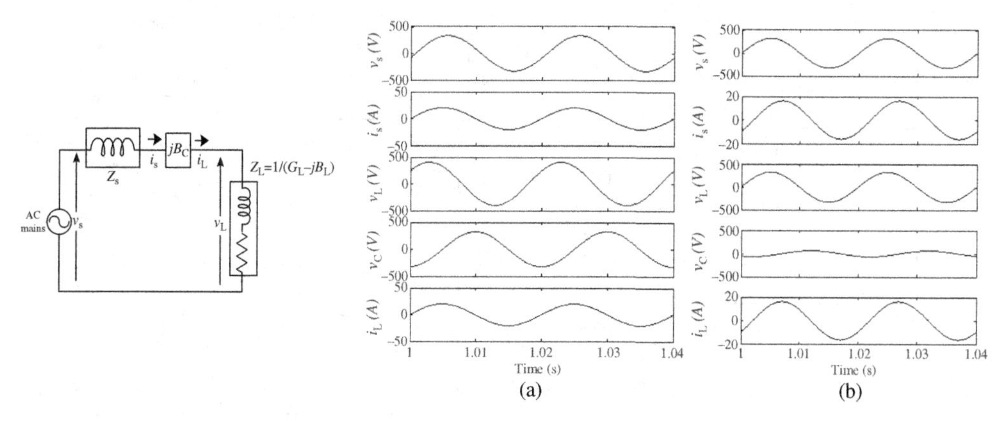

Figure E3.3 Schematic diagram and performance waveforms for (a) power factor correction and (b) voltage regulation

compensator is used to raise the load voltage to the input voltage (230 V), calculate (e) the values of compensator elements (in farads or henries) and (f) its kVA rating.

Solution: Given supply voltage $V_s = 230$ V, frequency of the supply (f) = 50 Hz, and a single-phase load having $Z_L = (16 + j12)\ \Omega$. The source impedance is $Z_s = jX_s = j4.0\ \Omega$.

Total impedance is $Z_T = Z_s + Z_L = (16 + j16)\ \Omega = 22.627\angle 45°\Omega$.

The load current before compensation is $I_{sold} = V_s/|Z_T| = 230/22.627\ \text{A} = 10.16$ A.

a. The voltage drop across the source impedance before compensation is $V_{Zsold} = I_{sold}Z_s = 40.64$ V.
b. The voltage across the load before compensation is $V_{Lold} = I_{sold}Z_L = 203.2$ V.
c. Since after compensation, it is desired to maintain unity power factor at the AC mains, the load current is $I_s = V_s/R_L = 14.375$ A.

   The voltage drop across the source impedance is $V_{Zs} = I_{snew}Z_s = 57.5$ V.

   The voltage across the load is $V_L = I_sZ_L = 20 \times 14.375 = 287.5$ V.

   Since circuit is an inductive circuit, the compensator has to be capacitive in nature. The net reactance of the circuit is same as the compensator (capacitive) reactance.

   Therefore, the net capacitive reactance is $X_C = X_s + X_L = 4 + 12 = 16\ \Omega$.

   The value of the capacitor for power factor correction is $C = 1/X_C\omega = 199.04\ \mu$F.
d. The kVA rating of the compensator is $Q_{eq} = I_s^2X_C = 14.375^2 \times 16 = 3.306\ 25$ kVA.

   Note: Since it has increased the load voltage higher than the supply voltage for which normally the loads are not designed, such passive series compensation is not used in practice.
e. This series compensator is also used to raise the load voltage to the input voltage (230 V).

   The value of the capacitor for voltage regulation is calculated as follows: $X_{CV} = -X_s$. Thus, $C = 1/X_{CV}\omega = 795.77\ \mu$F.
f. The kVA rating of the compensator for voltage regulation is $Q_{eq} = I^2X_{CV} = (230/20)^2 \times 4 = 0.529$ kVA.

Example 3.4

A three-phase three-wire shunt compensator consisting of lossless passive elements (L and/or C) is employed at a 415 V, 50 Hz distribution system to provide load balancing and power factor correction of a single-phase 25 kW unity power factor load connected between two lines as shown in Figure E3.4. Calculate (a) supply line currents, (b) compensator currents, (c) the values of compensator elements (in farads or henries), and (d) its kVA rating.

Solution: Given supply line voltage $V_s = 415$ V, frequency of the supply (f) = 50 Hz, and a single-phase 25 kW unity power factor load connected between two lines that is to be compensated as a balanced load on the three-phase distribution system.

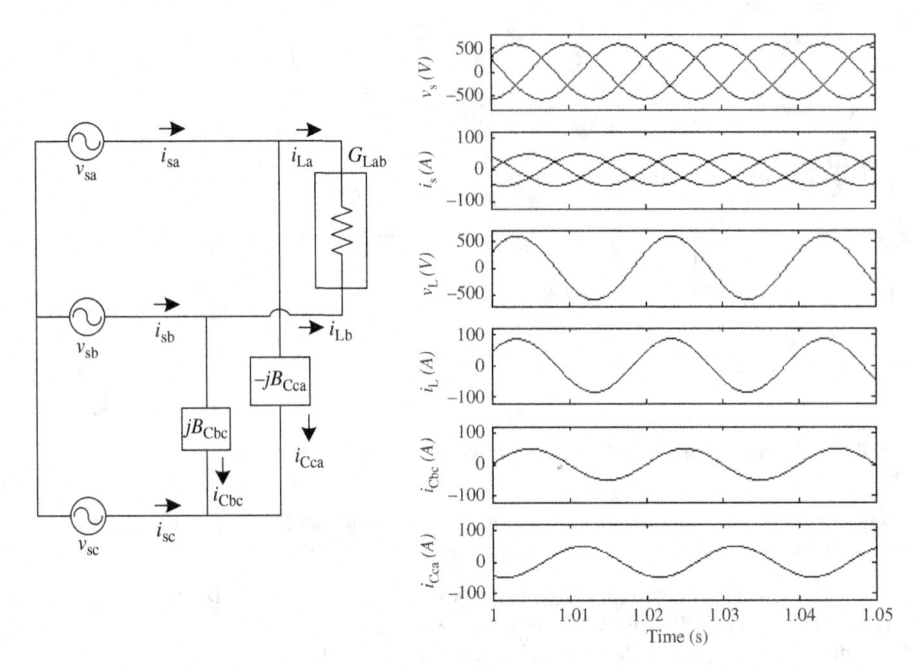

Figure E3.4 Schematic diagram and performance waveforms

a. After compensation, this load is realized as a balanced three-phase unity power factor load; therefore, the supply line currents are $I_{sa} = 25\,000/(\sqrt{3} \times 415) = 34.78$ A, $I_{sb} = 34.78$ A, and $I_{sc} = 34.78$ A.
b. For such a load compensation, the real admittance $G_{Lab} = 25\,000/415^2 = 0.145\,158\,949$ mhos has to be complemented with a reactive admittance network across other two lines to obtain a resultant balanced load on the AC supply. This problem had been investigated originally by Steinmetz at the beginning of the twentieth century. It can easily be shown that the load on the AC supply becomes balanced (and remains real) if a capacitive susceptance $B_{Cbc} = G_{Lab}/\sqrt{3}$ is connected between phases b and c, and an inductive susceptance $B_{Cca} = -G_{Lab}/\sqrt{3}$ is connected between phases c and a. Therefore, compensator currents are $I_{Cbc} = I_{Cca} = V_{LL} G_{Lab}/\sqrt{3} = 415 \times 0.083\,807\,558 = 34.78$ A.
c. Therefore, $B_{Cbc} = G_{Lab}/\sqrt{3} = 0.083\,807\,558$ mhos, $C_{Cbc} = B_{Cbc}/\omega = 266.90\,\mu$F, and $B_{Cca} = -G_{Lab}/\sqrt{3} = 0.083\,807\,558$ mhos, $L_{Cca} = 1/\omega B_{Cca} = 38.00$ mH.
d. The kVA rating of the compensator is $Q_{Cbc} = Q_{Cca} = Q = V^2 G_{Lab}/\sqrt{3} = 14.433$ kVA.

Example 3.5

A three-phase three-wire shunt compensator consisting of lossless passive elements (L and/or C) is employed at a 415 V, 50 Hz distribution system to provide load balancing and power factor correction of a single-phase 50 kVA, 0.8 lagging power factor load connected between two lines as shown in Figure E3.5. Calculate (a) supply line currents, (b) compensator currents, (c) the values of compensator elements (in farads or henries), and (d) its kVA rating.

Solution: Given supply line voltage $V_s = 415$ V, frequency of the supply (f) = 50 Hz, and a single-phase 50 kVA, 0.8 lagging power factor load connected between two lines that is to be compensated as a balanced load on the three-phase distribution system.

a. After compensation, this load is realized as a balanced three-phase unity power factor load; therefore, the supply line currents are $I_{sa} = 50\,000 \times 0.8/(\sqrt{3} \times 415) = 55.65$ A, $I_{sb} = 55.65$ A, and $I_{sc} = 55.65$ A.

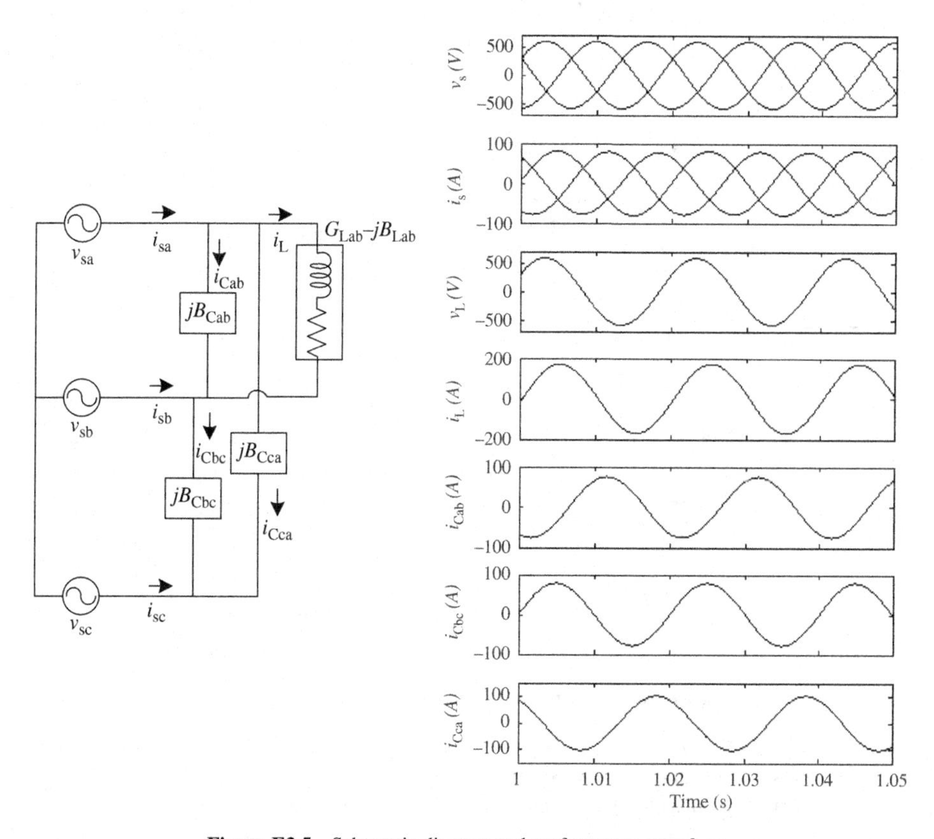

Figure E3.5 Schematic diagram and performance waveforms

b. For such a load compensation, the real admittance $G_{Lab} = 50\,000 \times 0.8/415^2 = 0.232\,254\,318$ mhos has to be complemented with a reactive admittance network across other two lines to obtain a resultant balanced load on the AC supply. It may be simply derived that the load on the AC supply becomes balanced (and remains real) if a capacitive susceptance $B_{Cbc} = G_{Lab}/\sqrt{3}$ is connected between phases b and c, and an inductive susceptance $B_{Cca} = -G_{Lab}/\sqrt{3}$ is connected between phases c and a. Therefore, the compensator currents are $I_{Cab} = Q_{Lab}/V = 50\,000 \times 0.6/415 = 30\,000/415 = 72.29$ A and $I_{Cbc} = I_{Cca} = V_s G_{Lab}/\sqrt{3} = 415 \times 0.134\,092\,093 = 55.65$ A.
c. Therefore, $B_{Cbc} = G_{Lab}/\sqrt{3} = 0.134\,092\,093$ mhos, $C_{Cbc} = B_{Cbc}/\omega = 427\,\mu$F, and $B_{Cca} = -G_{Lab}/\sqrt{3} = 0.134\,092\,093$ mhos, $L_{Cca} = 1/\omega B_{Cca} = 23.75$ mH. Moreover, for power factor correction at line ab, $Q_{Lab} = Q_{Cab} = 50\,000 \times 0.6 = 30\,000$ VA. Thus, $C_{Cab} = Q_{Cab}/V_s^2\omega = 554.75\,\mu$F.
d. The kVA rating of the compensator elements is $Q_{Cab} = 50\,000 \times 0.6 = 30$ kVA and $Q_{Cbc} = Q_{Cca} = Q = V_s^2 G_{Lab}/\sqrt{3} = 23.094$ kVA.

Example 3.6

A three-phase three-wire delta connected balanced load having $Z_L = (5.0 + j2.0)$ pu is fed from an AC supply with an input AC line voltage of 415 V at 50 Hz and a base impedance of 4.15 Ω per phase. It is to be realized as a unity power factor load on the AC supply system using shunt connected lossless passive elements (L and/or C) as shown in Figure E3.6. Calculate (a) supply line currents, (b) compensator currents, (c) the values of compensator elements (in farads or henries), (d) its kVA rating, and (e) equivalent per-phase resistance (in ohms) of the compensated load.

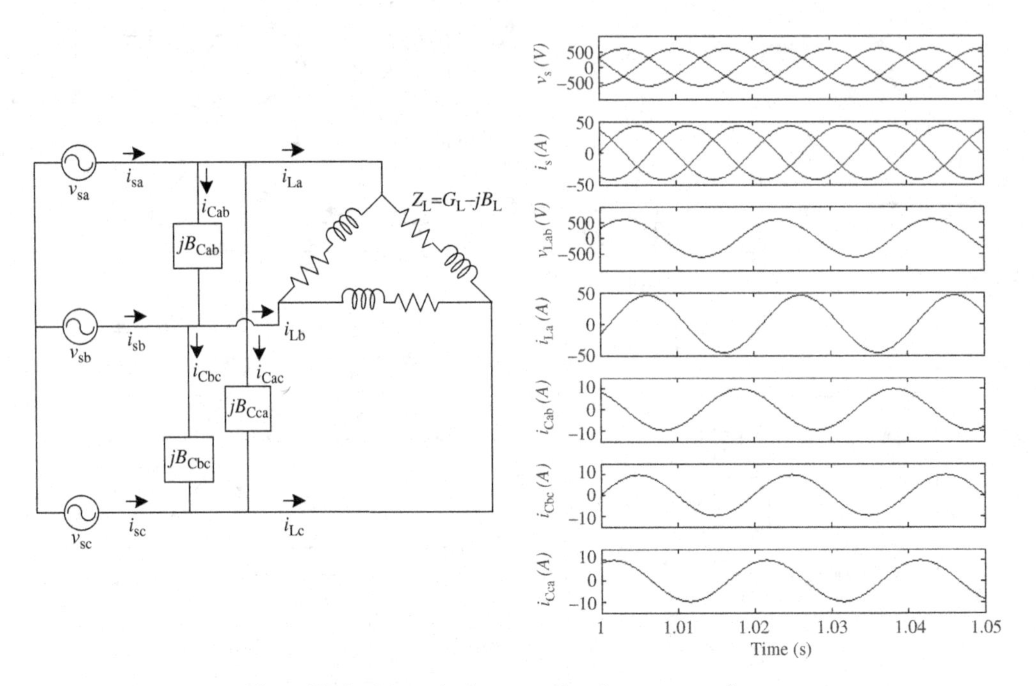

Figure E3.6 Schematic diagram and performance waveforms

Solution: Given supply voltage $V_s = 415$ V, frequency of supply (f) = 50 Hz, and a three-phase load having $Z_L = (5.0 + j2.0)$ pu with a base impedance of 4.15 Ω per phase.

The load resistance is $R_L = 4.15 \times 5.0\,\Omega = 20.75\,\Omega$. The load reactance is $X_L = 4.15 \times 2\,\Omega = 8.30\,\Omega$.

The load admittance is $Y_L = 1/Z_L = G_L - jB_L = (0.041\,5455 - j0.016\,6182)$ mhos.

a. The supply current after compensation is $I_s = \sqrt{3}V_sG_L = 29.863$ A.
   Because of balanced and symmetrical operation, $I_{sa} = I_{sb} = I_{sc} = I_s = 29.863$ A.
   For unity power factor operation, shunt capacitors are connected in parallel to all load branches. So, $B_L = -B_C = 0.016\,6182$.
b. The compensator phase current is $I_C = V_sB_C = 415 \times 0.016\,6182 = 6.897$ A.
   All compensator branches share equal compensator currents.
c. The value of the capacitor for power factor correction is $C = B_C/\omega = 52.924\,\mu$F.
d. The compensator kVA rating is $Q_C = 3VI_C = 3 \times 415 \times 6.897 = 8.586$ kVA.
e. The equivalent resistance of the compensated load is $R_{eq} = 1/G_L = 24.07\,\Omega$.

Example 3.7

A three-phase three-wire AC distribution system has a line voltage of 415 V at 50 Hz and a feeder (source) impedance of 1.0 Ω resistance and 3.0 Ω inductive reactance per phase after which a balanced delta connected load having $Z_L = (24 + j18)\,\Omega$ per phase is connected. Calculate (a) the voltage drop across the source impedance and (b) the voltage across the load. A shunt compensator consisting of lossless passive elements (L and/or C) is used to raise the voltage to the input voltage (415 V) as shown in Figure E3.7. Calculate (c) supply line currents, (d) compensator currents, (e) the values of compensator elements (in farads or henries), and (f) its kVA rating.

Solution: Given the supply phase voltage $V_s = 415/\sqrt{3}$ V = 239.6 V, frequency of the supply (f) = 50 Hz, and a delta connected load having $Z_L = (24 + j18)\,\Omega$ per phase. The source impedance is $Z_s = R_s + jX_s = (1 + j3)\,\Omega$.

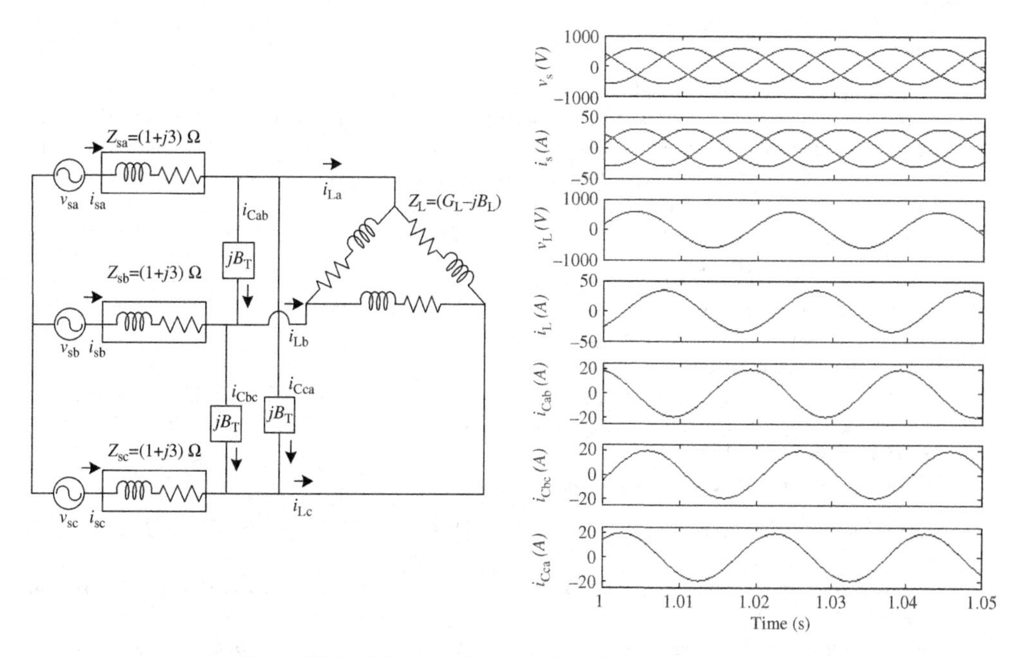

Figure E3.7 Schematic diagram and performance waveforms

The star equivalent of the load is $Z_{LY} = Z_L/3 = (24 + j18)/3\ \Omega$ per phase $= (8 + j6)\ \Omega$ per phase.

Total impedance per phase is $Z_T = Z_s + Z_{LY} = (9 + j9)\ \Omega = 12.73\ \Omega$.

The load line current before compensation is $I_{sold} = V_s/Z_T = 239.6/12.73\ \text{A} = 18.82\ \text{A}$.

a. The voltage drop across the source impedance before compensation is $V_{Zsold} = I_{sold}Z_s = 59.53\ \text{V}$.
b. The voltage across the load before compensation is $V_{ZLold} = I_{sold}Z_{LY} \times \sqrt{3} = 326.05\ \text{V}$.

   The load admittance is $Y_{LY} = G_{LY} + jB_{LY} = 1/Z_{LY} = (0.08 - j0.06)$ mhos.

   The load power factor can be made unity by connecting a susceptance of $B_{PF} = -B_{LY} = 0.06$ mhos.

   Let the susceptance of value B_{CV} mhos is connected in parallel to the load to raise the load voltage to the input voltage ($V_L = V_s$). The basic equation to maintain the load voltage equal to the input voltage is $|V_L| = |V_s|/[\{(R_s + jX_s) + 1/(G_{LY} + jB_{CV})\}]| \times |1/(G_{LY} + jB_{CV})| = |V_s|$.

   Solving this equation, the value of B_{CV} is $B_{CV} = [X_s \pm \{\sqrt{X_s^2 - (X_s^2 + R_s^2)}\ (2R_sG_{LY} + R_s^2G_{LY}^2 + X_s^2G_{LY}^2)\}]/(X_s^2 + R_s^2)$.

   Substituting the values of $X_s = 3\ \Omega$, $R_s = 1\ \Omega$, and $G_{LY} = 0.08$ mhos and considering "–" sign in "±", for lower value of the compensator, the value of B_{CV} is 0.04 mhos.

   The total susceptance for power factor correction and voltage regulation is $B_T = B_{CV} + B_{PF} = 0.10$ mhos.
c. The supply line current is $|V_s|/|[\{(R_s + jX_s) + 1/(G_{LY} + jB_{CV})\}]| = 239.6/|[\{(1 + j3) + 1/(0.08 + j.04)\}]| = 21.43\ \text{A}$.

   The susceptance to be connected line to line for power factor correction is $B_D = B_T/3 = 0.0333$.

   After compensation, the load line–line voltage is $V_{LL} = 415\ \text{V}$.
d. The compensator currents are $V_{LL}B_D = 415 \times 0.0333 = 13.83\ \text{A}$.
e. The value of the capacitor for power factor correction and voltage regulation is $C_D = B_T/3\omega = 106.15\ \mu\text{F}$.
f. The kVA rating of the compensator is $Q_{eq} = 3V_{LL}^2B_D = 3 \times 415^2 \times 0.0333 = 17.2225\ \text{kVA}$.

Example 3.8

A three-phase three-wire unbalanced delta connected load having $Z_{Lab} = (6.0 + j3.0)$ pu, $Z_{Lbc} = (3.0 + j1.5)$ pu, and $Z_{Lca} = (7.5 + j1.5)$ pu) is fed from an AC supply with an input line voltage of 415 V at 50 Hz and a base impedance of 4.15 Ω per phase. It is to be realized as a balanced unity power factor load on the three-phase supply system using a shunt compensator consisting of lossless passive elements (L and/or C) as shown in Figure E3.8. Calculate (a) supply line currents, (b) the values of compensator elements (in farads or henries), (c) compensator currents, (d) its kVA rating, and (e) equivalent per-phase resistance (in ohms) of the compensated load.

Solution: Given supply voltage $V_s = 415$ V, frequency of the supply $(f) = 50$ Hz, and a three-phase three-wire unbalanced delta connected load having $Z_{ab} = (6.0 + j3.0)$ pu, $Z_{bc} = (3.0 + j1.5)$ pu, and $Z_{ca} = (7.5 + j1.5)$ pu with a base impedance of 4.15 Ω per phase.

Load impedances are $Z_{Lab} = (24.9 + j12.45)\ \Omega$, $Z_{Lbc} = (12.45 + j6.225)\ \Omega$, and $Z_{Lca} = (31.125 + j6.225)\ \Omega$. Load admittances are $Y_{Lab} = (0.032\ 129 - j0.016\ 0643)$ mhos, $Y_{Lbc} = (0.064\ 257 - j0.032\ 129)$ mhos, and $Y_{Lca} = (0.030\ 893 - j0.006\ 1786)$ mhos.

The total load conductance is $G_{LT} = G_{Lab} + G_{Lbc} + G_{Lca} = 0.127\,279$ mhos.

a. Supply line currents after compensation are $I_{sa} = (V_s G_{LT}/\sqrt{3})\angle 0°$ A $= I\angle 0°$ A, $I_{sb} = I\angle 120°$ A, and $I_{sc} = I\angle 240°$ A.

   Substituting the values, $I_{sa} = (V_s G_{LT}/\sqrt{3})\angle 0°$ A $= 30.496\angle 0°$ A, $I_{sb} = 30.496\angle 120°$ A, and $I_{sc} = 30.496\angle 240°$ A.

b. Compensator susceptances are $B_{Cab} = -B_{Lab} + (G_{Lca} - G_{Lbc})/\sqrt{3}$, $B_{Cbc} = -B_{Lbc} + (G_{Lab} - G_{Lca})/\sqrt{3}$, and $B_{Cca} = -B_{Lca} + (G_{Lbc} - G_{Lab})/\sqrt{3}$.

   Substituting the values, $B_{Cab} = -0.003\,198\,41$ mhos, $B_{Cbc} = 0.032\,8426$ mhos, and $B_{Cca} = 0.024\,7277$ mhos.

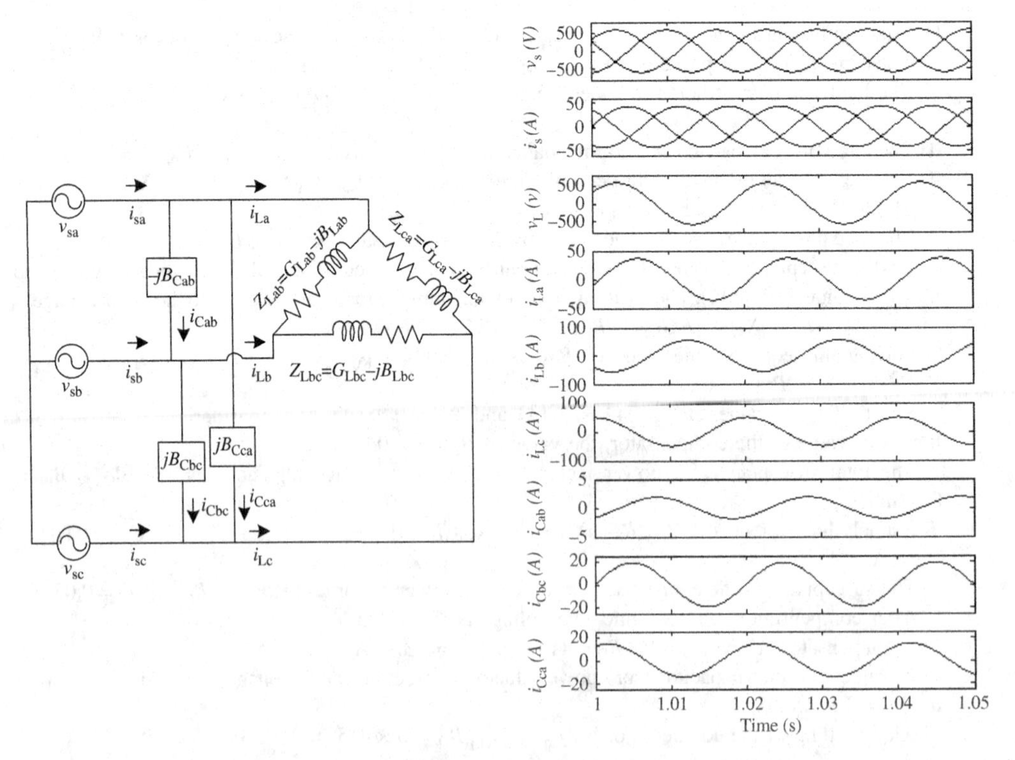

Figure E3.8 Schematic diagram and performance waveforms

The values of compensator elements are $L_{Cab} = 1/B_{Cab}\ \omega = 0.9957$ H, $C_{Cbc} = B_{Cbc}/\omega = 104.59\ \mu$F, and $C_{Cca} = B_{Cca}/\omega = 78.75\ \mu$F.

c. The compensator phase currents are $I_{Cab} = V_s B_{Cab} = 1.327$ A, $I_{Cbc} = V_s B_{Cbc} = 13.629$ A, and $I_{Cca} = V_s B_{Cca} = 10.262$ A.
d. The compensator kVA rating is $Q_C = V(|I_{Cab}| + |I_{Cbc}| + |I_{Cca}|) = 10.465\,47$ kVA.
e. The equivalent delta connected resistance of the compensated load is $R_{eq} = 3/G_{LT} = 23.57\ \Omega$.

Example 3.9

A three-phase AC supply system has a line voltage of 415 V at 50 Hz and a feeder (source) impedance of 0.25 Ω resistance and 3.50 Ω inductive reactance per phase after which an unbalanced delta connected load having $Z_{Lab} = 24\ \Omega$, $Z_{Lbc} = 30\ \Omega$, and $Z_{Lca} = j18\ \Omega$ is connected. Calculate (a) the voltage drop across the source impedance and (b) the voltage across the load if a shunt compensator consisting of lossless passive elements (L and/or C) is used to balance this load and to realize it as a unity power factor load as shown in Figure E3.9. Moreover, calculate (c) supply line currents, (d) the values of compensator elements (in farads or henries), (e) compensator currents, (f) its kVA rating, and (g) equivalent per-phase resistance (in ohms) of the compensated load.

Solution: Given supply voltage $V_s = 415$ V, frequency of the supply $(f) = 50$ Hz, and a three-phase three-wire unbalanced delta connected load having $Z_{Lab} = 24\ \Omega$, $Z_{Lbc} = 30\ \Omega$, and $Z_{Lca} = j18\ \Omega$ with a feeder (source) impedance of 0.25 Ω resistance and 3.50 Ω inductive reactance per phase. $Z_s = (0.25 + j3.50)\ \Omega$.

Load admittances are $Y_{Lab} = 0.041\,666$ mhos, $Y_{Lbc} = 0.033\,333$ mhos, and $Y_{Lca} = -j0.055\,5555$ mhos.

Equivalent star connected conductance and resistance after compensation are $G_{LT} = G_{Lab} + G_{Lbc} + G_{Lca} = 0.0750$ mhos and $R_{LT} = 13.333\ \Omega$.

Total impedance per phase in star connection is $Z_{TY} = Z_s + R_{LT} = (13.5833 + j3.5)\ \Omega = 14.026\ \angle 14.45°\ \Omega$.

The source current after compensation is $I_s = V_s/\sqrt{3}|Z_{TY}| = 17.08$ A.

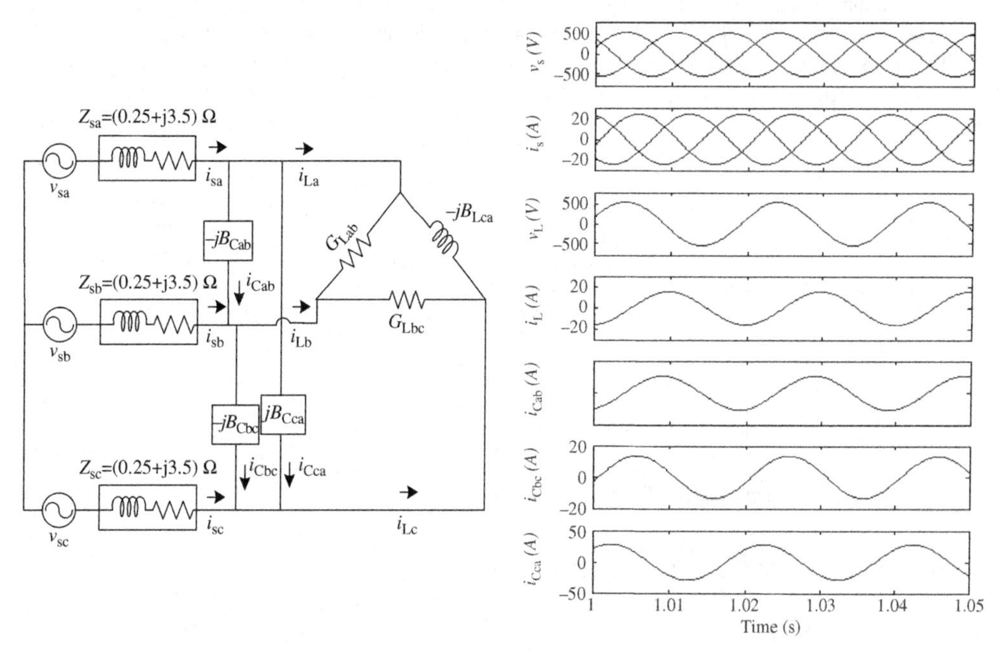

Figure E3.9 Schematic diagram and performance waveforms

a. The voltage drop across the source impedance is $V_{Zs} = I_s Z_s = 59.9371$ V.
b. The voltage across the load is $V_{ZL} = I_s R_{LT} = 227.9$ V per phase = 394.477 V per line.
c. Supply line currents after compensation are $I_{sa} = (V_s/\sqrt{3}Z_{TY})\angle -14.45°$ A = $I\angle -14.45°$ A, $I_{sb} = I\angle -134.45°$ A, and $I_{sc} = I\angle -254.45°$ A.
   Substituting the values, $I_{sa} = (V_s/\sqrt{3}Z_{TY})\angle -14.45°$ A = $17.028\angle -14.45°$ A, $I_{sb} = 17.028\angle -134.45°$ A, and $I_{sc} = 17.028\angle -254.45°$ A.
d. Compensator susceptances are $B_{Cab} = -B_{Lab} + (G_{Lca} - G_{Lbc})/\sqrt{3}$, $B_{Cbc} = -B_{Lbc} + (G_{Lab} - G_{Lca})/\sqrt{3}$, and $B_{Cca} = -B_{Lca} + (G_{Lbc} - G_{Lab})/\sqrt{3}$.
   Substituting the values, $B_{Cbc} = 0.0241$ mhos, $B_{Cab} = -0.0192$ mhos, and $B_{Cca} = 0.0507$ mhos.
   The values of compensator elements are $L_{Cbc} = 1/B_{Cbc}\,\omega = 0.1323$ H, $C_{Cab} = B_{Cab}/\omega = 61$ µF, and $C_{Cca} = B_{Cca}/\omega = 161.52$ µF.
e. The compensator phase currents are $I_{Cab} = V_s B_{Cab} = 7.9866$ A, $I_{Cbc} = V_s B_{Cbc} = 9.9833$ A, and $I_{Cca} = V_s B_{Cca} = 21.0589$ A.
f. The compensator kVA rating is $Q_C = V(|I_{Cab}| + |I_{Cbc}| + |I_{Cca}|) = 16.197$ kVA.
g. The equivalent delta connected resistance of the compensated load is $R_{eq} = 3/G_{LT} = 40\,\Omega$.

Example 3.10

A three-phase AC supply has a line voltage of 415 V at 50 Hz and a feeder (source) impedance of 1.0 Ω resistance and 3.0 Ω inductive reactance per phase after which an unbalanced delta connected load having $Z_{Lab} = 20\,\Omega$, $Z_{Lbc} = 25\,\Omega$, and $Z_{Lca} = 30\,\Omega$ is connected. Calculate (a) the voltage drop across the source impedance and (b) the voltage across the load if a shunt compensator consisting of lossless passive elements (L and/or C) is used to balance these loads and to raise the voltage to the input voltage (415 V) as shown in Figure E3.10. Moreover, calculate (c) supply line currents, (d) the values of compensator elements (in farads or henries), (e) compensator currents, (f) its kVA rating, and (g) equivalent per-phase resistance (in ohms) of the compensated load.

Solution: Given supply voltage $V_s = 415$ V, frequency of the supply $(f) = 50$ Hz, and a three-phase three-wire unbalanced delta connected load having $Z_{Lab} = 20\,\Omega$, $Z_{Lbc} = 25\,\Omega$, and $Z_{Lca} = 30\,\Omega$ with a feeder (source) impedance of 1.0 Ω resistance and 3.0 Ω inductive reactance per phase.

Load admittances are $Y_{Lab} = 0.05$ mhos, $Y_{Lbc} = 0.04$ mhos, and $Y_{Lca} = 0.033\,333$ mhos.

Compensator susceptances for load balancing and power factor correction are $B_{CabL} = -B_{Lab} + (G_{Lca} - G_{Lbc})/\sqrt{3}$, $B_{CbcL} = -B_{Lbc} + (G_{Lab} - G_{Lca})/\sqrt{3}$, and $B_{CcaL} = -B_{Lca} + (G_{Lbc} - G_{Lab})/\sqrt{3}$.

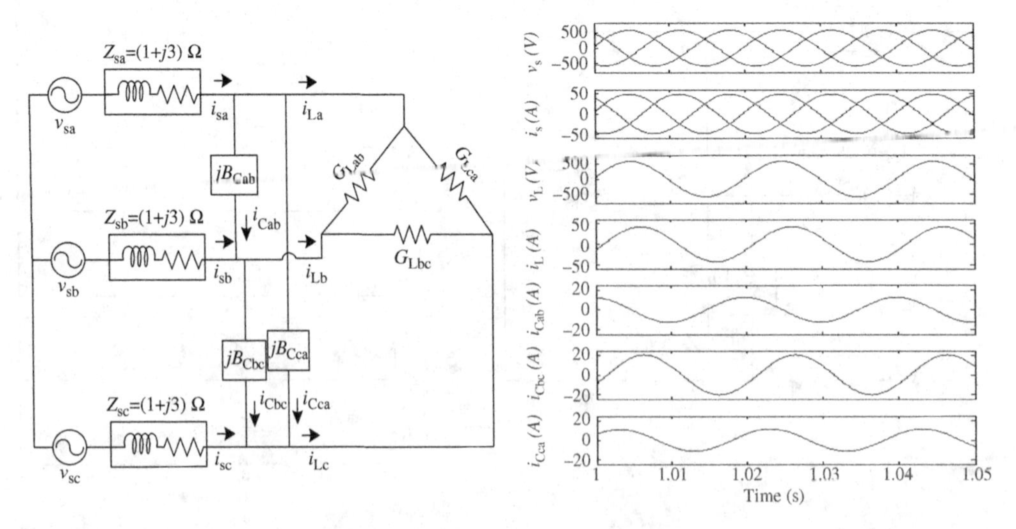

Figure E3.10 Schematic diagram and performance waveforms

Substituting the values, $B_{CabL} = -0.003\,849$ mhos, $B_{CbcL} = 0.009\,6225$ mhos, and $B_{CcaL} = -0.005\,7735$ mhos.

The star equivalent conductance after load balancing and power factor correction is $G_{LT} = G_{ab} + G_{bc} + G_{ca} = 0.123\,3333$ mhos.

The star equivalent resistance after load balancing and power factor correction is $R_{LT} = 1/G_{LT} = 8.108\,\Omega$.

Let the susceptance of value B_{CV} mhos is connected in parallel to the star equivalent load $G_{LY} = G_{LT}$ (as per phase representation) to raise the load voltage to the input voltage ($V_L = V_s$). The basic equation to maintain the load voltage equal to the input voltage is $|V_L| = |V_s/[\{(R_s + jX_s) + 1/(G_{LY} + jB_{CV})\}]| \times |1/(G_{LY} + jB_{CV})| = |V_s|$.

Solving this equation, the value of B_{CV} is $B_{CV} = [X_s \pm \{\sqrt{X_s^2 - (X_s^2 + R_s^2)(2R_sG_{LY} + R_s^2G_{LY}^2 + X_s^2G_{LY}^2)}\}]/(X_s^2 + R_s^2)$.

Substituting the values of $X_s = 3.0\,\Omega$, $R_s = 1.0\,\Omega$, and $G_{LY} = 0.123\,3333$ mhos and considering "–" sign in "±", for lower value of the compensator, the value of B_{CV} is 0.076 120 071 mhos.

$$jX_{CV} = 1/jB_{CV} = -j13.137\,\Omega.$$

$$Z_{LT} = R_{LT}jX_{CV}/(R_{LT} + jX_{CV}) = 5.8715 - j3.623 = 6.899\angle -31.68^\circ\,\Omega.$$

Supply line currents after compensation are $I_s = V_{sa}/(Z_{LT}\sqrt{3})\angle 0\,\text{A} = 239.6\angle 0/(6.899\angle -31.67^\circ) = 34.725\angle 31.68^\circ$ A.

a. The voltage drop across the source impedance is $V_{Zs} = I_sZ_s = 109.81$ V.
b. The voltage across the load is $V_{ZL} = I_sZ_{LT} = 239.6$ V per phase.
c. Supply line currents after compensation are $I_{sa} = V_s/(Z_{LT}\sqrt{3})\angle 31.68^\circ\,\text{A} = 34.725\angle 31.68^\circ$ A, $I_{sb} = 34.725\angle -88.32^\circ$ A, and $I_{sc} = 34.725\angle -208.32^\circ$ A.

   Susceptance for voltage regulation is $B_{V\Delta} = B_{CV}/3 = 0.025\,373\,357$ mhos.

   Total susceptances for load balancing and voltage regulation are $B_{Cab} = B_{V\Delta} + B_{CabL} = 0.021\,524\,357$ mhos, $B_{Cbc} = B_{V\Delta} + B_{CbcL} = 0.034\,995\,857$ mhos, and $B_{Cca} = B_{V\Delta} + B_{CcaL} = 0.019\,599\,857$ mhos.
d. The values of compensator elements are $C_{Cab} = B_{Cab}/\omega = 68.55\,\mu$F, $C_{Cbc} = B_{Cab}/\omega = 111.45\,\mu$F, and $C_{Cca} = B_{Cca}/\omega = 62.42\,\mu$F.
e. The compensator phase currents are $I_{Cab} = V_sB_{Cab} = 8.9326$ A, $I_{Cbc} = V_sB_{Cbc} = 14.2328$ A, and $I_{Cca} = V_sB_{Cca} = 8.133\,94$ A.
f. The compensator kVA rating is $Q_C = V(|I_{Cab}| + |I_{Cbc}| + |I_{Cca}|) = 13.11$ kVA.
g. The equivalent delta connected resistance of the compensated load is $R_{eq} = 3/G_{LT} = 24.324\,\Omega$.

Example 3.11

A three-phase three-wire unbalanced isolated neutral star connected load having $Z_{La} = (6.0 + j3.0)$ pu, $Z_{Lb} = (3.0 + j1.5)$ pu, and $Z_{Lc} = (7.5 + j1.5)$ pu is fed from an AC supply with an input line voltage of 415 V at 50 Hz and a base impedance of 4.15 Ω per phase. It is to be realized as a balanced unity power factor load on the three-phase supply system using a shunt compensator consisting of lossless passive elements (L and/or C) as shown in Figure E3.11. Calculate (a) supply line currents, (b) compensator currents, (c) the values of compensator elements (in farads or henries), (d) its kVA rating, and (e) equivalent per-phase resistance (in ohms) of the compensated load.

Solution: Given supply voltage $V_s = 415$ V, frequency of the supply (f) = 50 Hz, and a three-phase three-wire unbalanced delta connected load having $Z_{La} = (6.0 + j3.0)$ pu, $Z_{Lb} = (3.0 + j1.5)$ pu, and $Z_{Lc} = (7.5 + j1.5)$ pu with a base impedance of 4.15 Ω per phase.

Load impedances in star connection are $Z_a = (24.9 + j12.45)\,\Omega$, $Z_b = (12.45 + j6.225)\,\Omega$, and $Z_c = (31.125 + j6.225)\,\Omega$.

Load admittances in star connection are $Y_a = (0.032\,129 - j0.016\,0643) = 0.035\,921\,224\angle -30.68^\circ$ mhos, $Y_b = (0.064\,257 - j0.032\,129) = 0.071\,841\,733\angle -26.57^\circ$ mhos, and $Y_c = (0.030\,893 - j0.006\,1786) = 0.031\,504\,801\angle -11.31^\circ$ mhos.

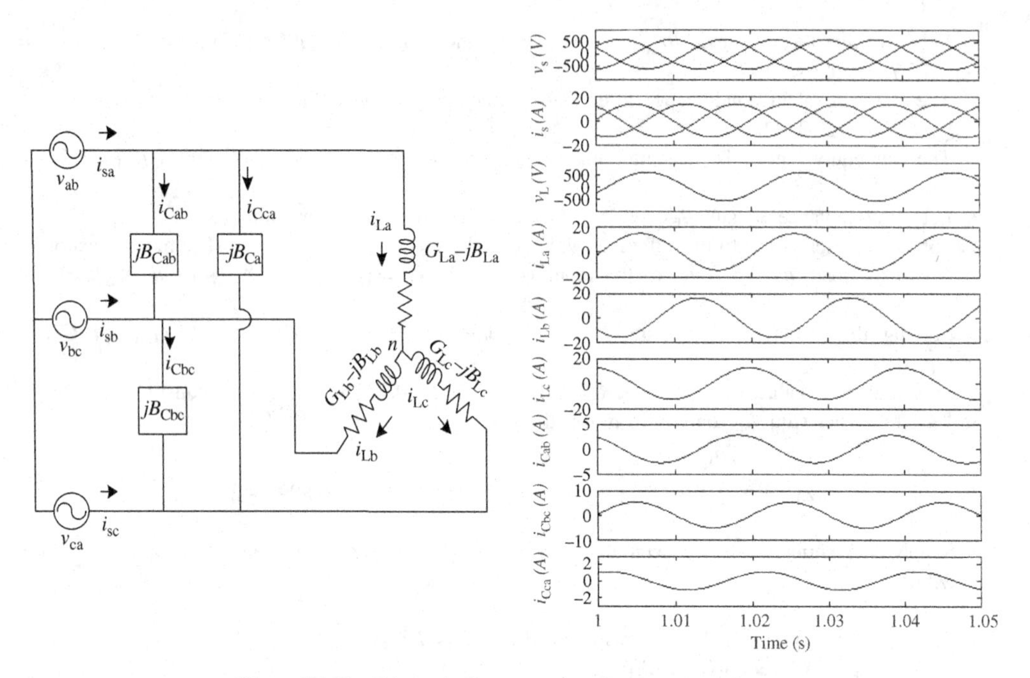

Figure E3.11 Schematic diagram and performance waveforms

Load admittances in equivalent delta connection are $Y_{ab} = Y_a Y_b/(Y_a + Y_b + Y_c)$, $Y_{bc} = Y_b Y_c/(Y_a + Y_b + Y_c)$, and $Y_{ca} = Y_c Y_a/(Y_a + Y_b + Y_c)$.

Substituting the values, $Y_{ab} = 0.018\,645\,443\angle -30.01° = (0.016\,147\,427 - j0.009\,3227)$ mhos, $Y_{bc} = 0.016\,353\,033\angle -14.75° = (0.015\,814\,243 - j0.004\,163\,542\,757)$ mhos, and $Y_{ca} = 0.008\,176\,598103\angle -14.74° = (0.007\,907\,509 - j0.002\,080\,3977)$ mhos.

$$G_{LT} = G_{ab} + G_{bc} + G_{ca} = 0.039\,869\,179 \text{ mhos}.$$

a. Supply line currents after compensation are $I_{sa} = (V_s G_{LT}/\sqrt{3})\angle 0°$ A $= I\angle 0°$ A, $I_{sb} = I\angle -120°$ A, and $I_{sc} = I\angle -240°$ A.

   Substituting the values, $I_{sa} = (V_s G_{LT}/\sqrt{3})\angle 0°$ A $= 9.552\,6697\angle 0°$ A, $I_{sb} = 9.552\,6697\angle -120°$ A, and $I_{sc} = 9.552\,6697\angle -240°$ A.
b. Compensator susceptances are $B_{Cab} = -B_{ab} + (G_{ca} - G_{bc})/\sqrt{3}$, $B_{Cbc} = -B_{bc} + (G_{ab} - G_{ca})/\sqrt{3}$, and $B_{Cca} = -B_{ca} + (G_{bc} - G_{ab})/\sqrt{3}$.

   Substituting the values, $B_{Cab} = 0.004\,757\,744\,997$ mhos, $B_{Cbc} = 0.008\,920\,861\,632$ mhos, and $B_{Cca} = 0.0018$ mhos.

   The values of compensator elements are $C_{Cab} = B_{Cab}/\omega = 15.152\ \mu$F, $C_{Cbc} = B_{Cbc}/\omega = 28.41\ \mu$F, and $C_{Cca} = B_{Cca}/\omega = 6.0117\ \mu$F.
c. The compensator phase currents are $I_{Cab} = V_s B_{Cab} = 1.974\,46$ A, $I_{Cbc} = V_s B_{Cbc} = 3.7021$ A, and $I_{Cca} = V_s B_{Cca} = 0.7838$ A.
d. The compensator kVA rating is $Q_C = V(|I_{Cab}| + |I_{Cbc}| + |I_{Cca}|) = 2.6807$ kVA.
e. The equivalent delta connected resistance of the compensated load is $R_{eq} = 3/G_{LT} = 75.246\ \Omega$.

Example 3.12

A three-phase four-wire 415 V (line), 50 Hz AC supply system has a single-phase 10 kW unity power factor load connected across line and neutral terminal. If it is required to eliminate the neutral current using a shunt compensator consisting of lossless passive elements (L and/or C) as shown in Figure E3.12,

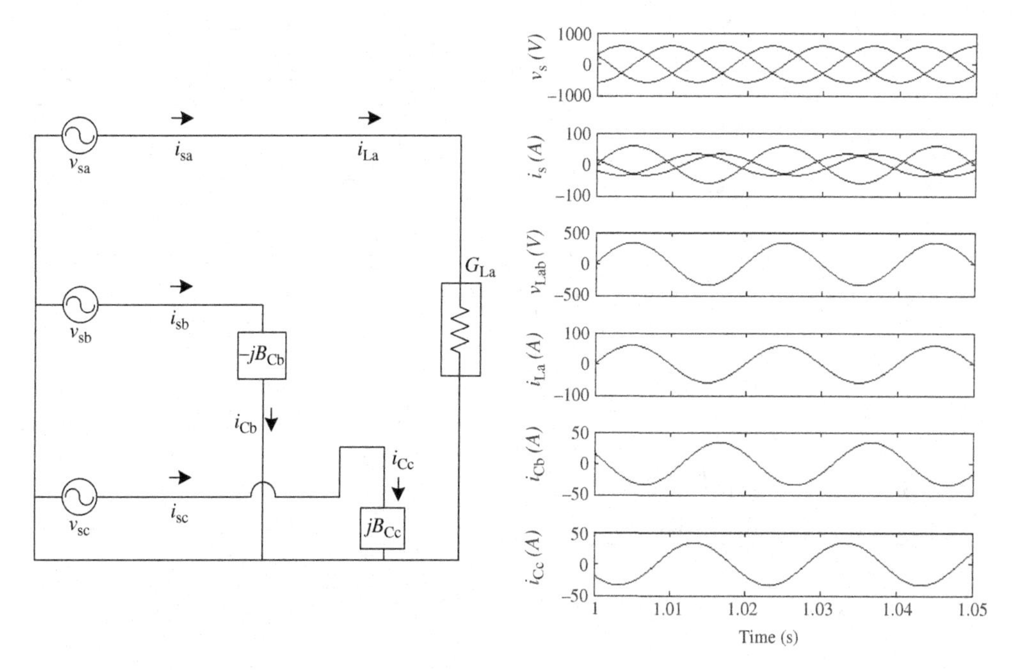

Figure E3.12 Schematic diagram and performance waveforms

calculate (a) supply line currents, (b) the values of compensator elements (in farads or henries), (c) compensator currents, and (d) its kVA rating.

Solution: Given supply phase voltage $V_{sp} = 415/\sqrt{3}$ V = 239.6 V, frequency of the supply (f) = 50 Hz, and a single-phase 10 kW unity power factor load connected across line and neutral terminal.

The load impedance per phase is $Z_L = V_{sp}^2/P = 239.6^2/10\,000 = 5.740\,816\,\Omega$.

The load admittance per phase is $Y_L = 0.174\,191\,264$ mhos.

It requires two susceptances across other two phases and neutral terminal to eliminate the neutral current and a susceptance across the load to correct the power factor to unity of the single-phase load. However, the load is a unity power factor load; hence, only two susceptances for neutral current compensation are required.

a. Supply line currents after compensation are $I_{sa} = (V_{sp}G_{LT}/\sqrt{3})\angle 0°$ A, $I_{sb} = (V_{sp}G_{LT}/3)\angle 150°$ A, and $I_{sc} = (V_{sp}G_{LT}/3)\angle 210°$ A.

   Substituting the values, $I_{sa} = (V_{sp}G_{LT}/\sqrt{3})\angle 0°$ A $= 41.74\angle 0°$ A, $I_{sb} = 24.096\angle 150°$ A, and $I_{sc} = 24.096\angle 210°$ A.

b. Compensator susceptances are $B_{Ca} = -B_{La} + (G_{Lb} - G_{Lc})/\sqrt{3}$, $B_{Cb} = -B_{Lb} + (G_{Lc} - G_{La})/\sqrt{3}$, and $B_{Cc} = -B_{Lc} + (G_{La} - G_{Lb})/\sqrt{3}$.

   Substituting the values, $B_{Ca} = 0.0$ mhos, $B_{Cb} = -0.100\,569\,373$ mhos, and $B_{Cc} = 0.100\,569\,373$ mhos.

   The values of compensator elements are $L_{Cb} = 1/B_{Cb}\omega = 31.67$ mH and $C_{Cc} = B_{Cc}/\omega = 320.28$ μF.

c. The compensator phase currents are $I_{Ca} = V_{sp}B_{Ca} = 0.0$ A, $I_{Cb} = V_{sp}B_{Cb} = 24.096\angle 150°$ A, and $I_{Cc} = V_{sp}B_{Cc} = 24.096\angle 210°$ A.

d. The compensator kVA rating is $Q_C = V(|I_{Ca}| + |I_{Cb}| + |I_{Cc}|) = 11.5468$ kVA.

Example 3.13

A three-phase four-wire 415 V (line–line), 50 Hz AC supply system has a single-phase 15 kVA, 0.8 lagging power factor load connected across line and neutral terminal. If it is required to eliminate neutral current using a shunt compensator consisting of lossless passive elements (L and/or C) as shown in

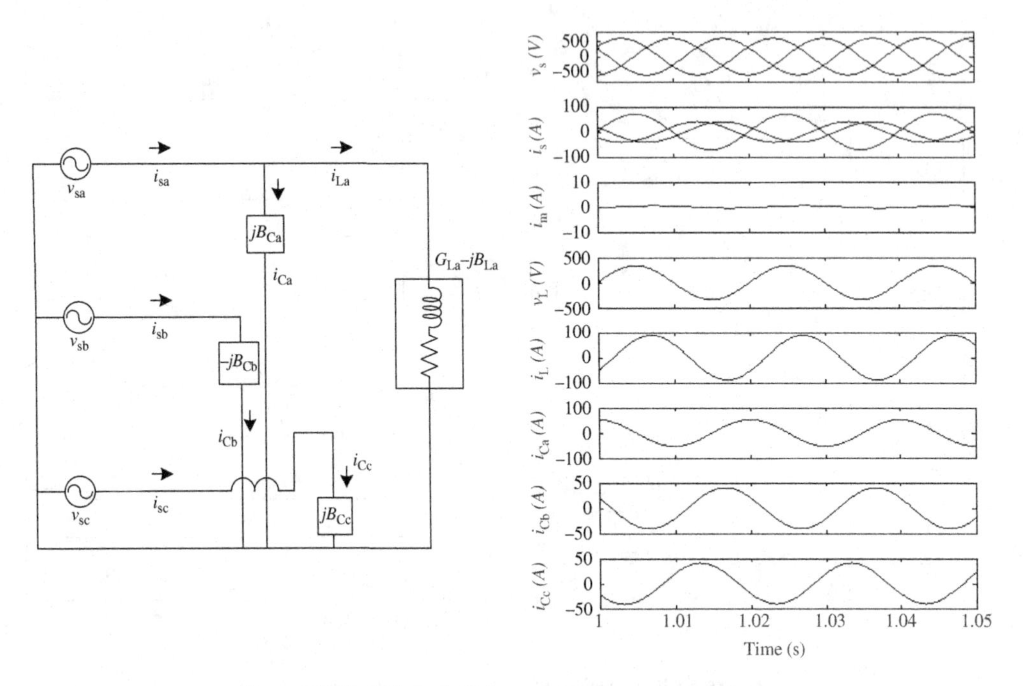

Figure E3.13 Schematic diagram and performance waveforms

Figure E3.13, calculate (a) supply line currents, (b) the values of compensator elements (in farads or henries), (c) compensator currents, and (d) its kVA rating.

Solution: Given supply phase voltage $V_{sp}=415/\sqrt{3}$ V $=239.6$ V, frequency of the supply $(f)=50$ Hz, and a single-phase 15 kVA, 0.8 lagging power factor load connected across line and neutral terminal.

The load impedance per phase is $Z_{La}=3.827\,21\angle 36.87°\ \Omega$.

The load admittance per phase is $Y_{La}=0.261\,286\,897\angle{-36.87°}=(0.209\,029\,517-j0.156\,772\,138)$ mhos.

It requires two susceptances across other two phases and neutral terminal to eliminate neutral current and a susceptance across the load to correct the power factor to unity of the single-phase load.

a. Supply line currents after compensation are $I_{sa}=(V_{sp}G_{LT}/\sqrt{3})\angle 0°$ A, $I_{sb}=(V_{sp}G_{LT}/3)\angle 150°$ A, and $I_{sc}=(V_{sp}G_{LT}/3)\angle 210°$ A.

   Substituting the values, $I_{sa}=(V_{sp}G_{LT}/\sqrt{3})\angle 0°$ A $=41.74\angle 0°$ A, $I_{sb}=28.92\angle 150°$ A, and $I_{sc}=28.92\angle 210°$ A.

b. Compensator susceptances are $B_{Ca}=-B_{La}+(G_{Lb}-G_{Lc})/\sqrt{3}$, $B_{Cb}=-B_{Lb}+(G_{Lc}-G_{La})/\sqrt{3}$, and $B_{Cc}=-B_{Lc}+(G_{La}-G_{Lb})/\sqrt{3}$.

   Substituting the values, $B_{Ca}=0.156\,772\,138$ mhos, $B_{Cb}=-0.0.120\,683\,43$ mhos, and $B_{Cc}=0.120\,683\,43$ mhos.

   The values of compensator elements are $C_{Ca}=B_{Ca}/\omega=499.27\ \mu F$, $L_{Cb}=1/B_{Cb}\omega=26.39$ mH, and $C_{Cc}=B_{Cc}/\omega=384.34\ \mu F$.

c. The compensator phase currents are $I_{Ca}=V_sB_{Ca}=37.56\angle 90°$ A, $I_{Cb}=V_sB_{Cb}=28.92\angle 150°$ A, and $I_{Cc}=V_sB_{Cc}=28.92\angle 210°$ A.

d. The compensator kVA rating is $Q_C=V(|I_{Ca}|+|I_{Cb}|+|I_{Cc}|)=22.872$ kVA.

Example 3.14

A three-phase four-wire shunt compensator consisting of lossless passive elements (L and/or C) is employed at a 415 V, 50 Hz supply system to provide load balancing and unity power factor of a

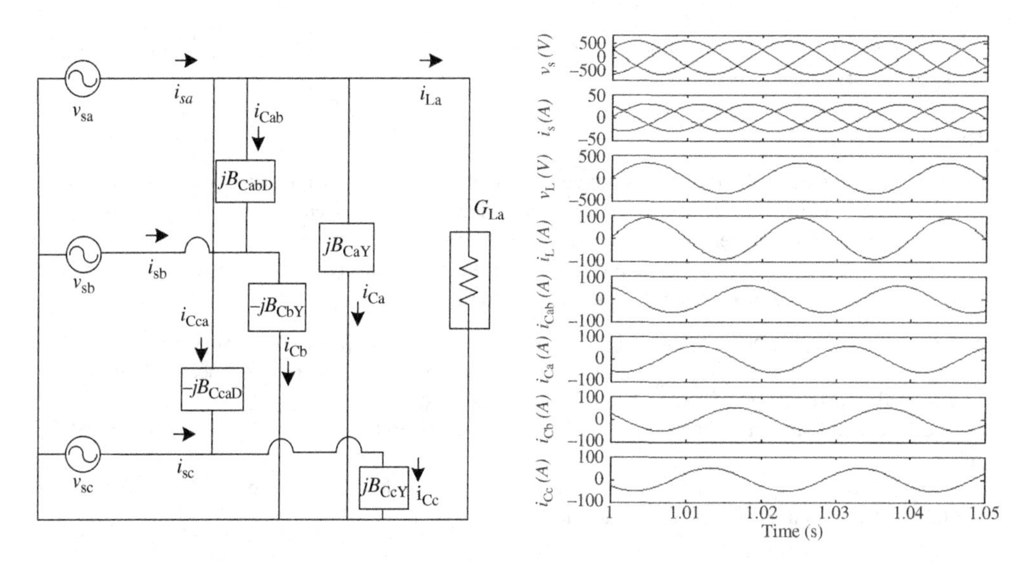

Figure E3.14 Schematic diagram and performance waveforms

single-phase 15 kW unity power factor load connected across line and neutral terminal as shown in Figure E3.14. Calculate (a) the values of compensator elements (in farads or henries), (b) compensator currents, (c) supply line currents, and (d) its kVA rating.

Solution: Given supply phase voltage $V_{sp} = 415/\sqrt{3}$ V = 239.6 V, frequency of the supply $(f) = 50$ Hz, and a single-phase 15 kW unity power factor load connected across line and neutral terminal.

The load impedance per phase is $Z_{La} = 3.82721\,\Omega$. The load admittance per phase is $Y_{La} = 0.261286897$ mhos.

It requires two set of susceptances, one star connected and other delta connected, across the lines for neutral current elimination and load balancing.

a. Compensator star connected susceptances for neutral current and power factor correction are $B_{CaY} = -B_{La} + (G_{Lb} - G_{Lc})/\sqrt{3}$, $B_{CbY} = -B_{Lb} + (G_{Lc} - G_{La})/\sqrt{3}$, and $B_{CcY} = -B_{Lc} + (G_{La} - G_{Lb})/\sqrt{3}$.

   Substituting the values, $B_{CaY} = 0.0$ mhos, $B_{CbY} = -0.15085406$ mhos, and $B_{CcY} = 0.15085406$ mhos.

   The values of star connected compensator elements are $C_{CaY} = B_{CaY}/\omega = 0.0\ \mu$F, $L_{CbY} = 1/B_{Cb}\omega = 21.11$ mH, and $C_{CcY} = B_{CcY}/\omega = 480.43\ \mu$F.

   Compensator delta connected susceptances are $B_{CabD} = 2(G_{La} - G_{Lb})/3\sqrt{3}$, $B_{CbcD} = 2(G_{Lb} - G_{Lc})/3\sqrt{3}$, and $B_{CcaD} = 2(G_{Lc} - G_{La})/3\sqrt{3}$.

   Substituting the values, $B_{CabD} = 0.100569373$ mhos, $B_{CbcD} = 0.0$ mhos, and $B_{CcaD} = -0.100569373$ mhos.

   The values of delta connected compensator elements are $C_{CabD} = B_{CabD}/\omega = 320.28\ \mu$F and $L_{CcaD} = 1/B_{CcaD}\omega = 31.67$ mH.

b. The compensator star connected phase currents are $I_{Ca} = V_{sp}B_{CaY} = 0.0$ A, $I_{Cb} = V_{sp}B_{CbY} = 36.14$ A, and $I_{Cc} = V_{sp}B_{CcY} = 36.14$ A.

   The compensator delta connected phase currents are $I_{Cab} = V_{sp}B_{CabD} = 41.74$ A, $I_{Cbc} = V_{sp}B_{CbcD} = 0.0$ A, and $I_{Cca} = V_{sp}B_{CcaD} = 41.74$ A.

c. Supply line currents after compensation are $I_{sa} = (V_{sp}G_{LT}/3)\angle 0°$ A, $I_{sb} = (V_{sp}G_{LT}/3)\angle -120°$ A, and $I_{sc} = (V_{sp}G_{LT}/3)\angle -240°$ A.

   Substituting the values, $I_{sa} = (V_{sp}G_{LT}/3)\angle 0°$ A $= 20.87\angle 0°$ A, $I_{sb} = 20.87\angle -120°$ A, and $I_{sc} = 20.87\angle -240°$ A.

d. The compensator kVA rating is $Q_C = V_s(|I_{Ca}| + |I_{Cb}| + |I_{Cc}|) + \sqrt{3}V_s(|I_{Cab}| + |I_{Cbc}| + |I_{Cca}|) = 17\,318.28 + 34\,644.2\,\text{VA} = 51.962\,\text{kVA}$.

Example 3.15

A three-phase four-wire shunt compensator consisting of lossless passive elements (L and/or C) is used at a 415 V, 50 Hz supply system to provide load balancing and power factor correction of a single-phase 30 kVA, 0.8 lagging power factor load connected across line and neutral terminal as shown in Figure E3.15. Calculate (a) supply line currents, (b) the values of compensator elements (in farads or henries), (c) compensator currents, and (d) its kVA rating.

Solution: Given supply phase voltage $V_{sp} = 415/\sqrt{3}\,\text{V} = 239.6\,\text{V}$, frequency of the supply $(f) = 50\,\text{Hz}$, and a single-phase 30 kVA, 0.8 lagging power factor load connected across line and neutral terminal.

The load impedance per phase is $Z_{La} = 1.913\,605\,333\angle 36.87°\,\Omega$.

The load admittance per phase is $Y_{La} = 0.522\,573\,794\angle{-36.87°} = (0.418\,059\,035 - j0.313\,544\,276)$ mhos.

It requires two set of susceptances, one star connected and other delta connected, across the lines for neutral current elimination and load balancing.

a. Compensator star connected susceptances for neutral current and power factor correction are $B_{CaY} = -B_{La} + (G_{Lb} - G_{Lc})/\sqrt{3}$, $B_{CbY} = -B_{Lb} + (G_{Lc} - G_{La})/\sqrt{3}$, and $B_{CcY} = -B_{Lc} + (G_{La} - G_{Lb})/\sqrt{3}$.

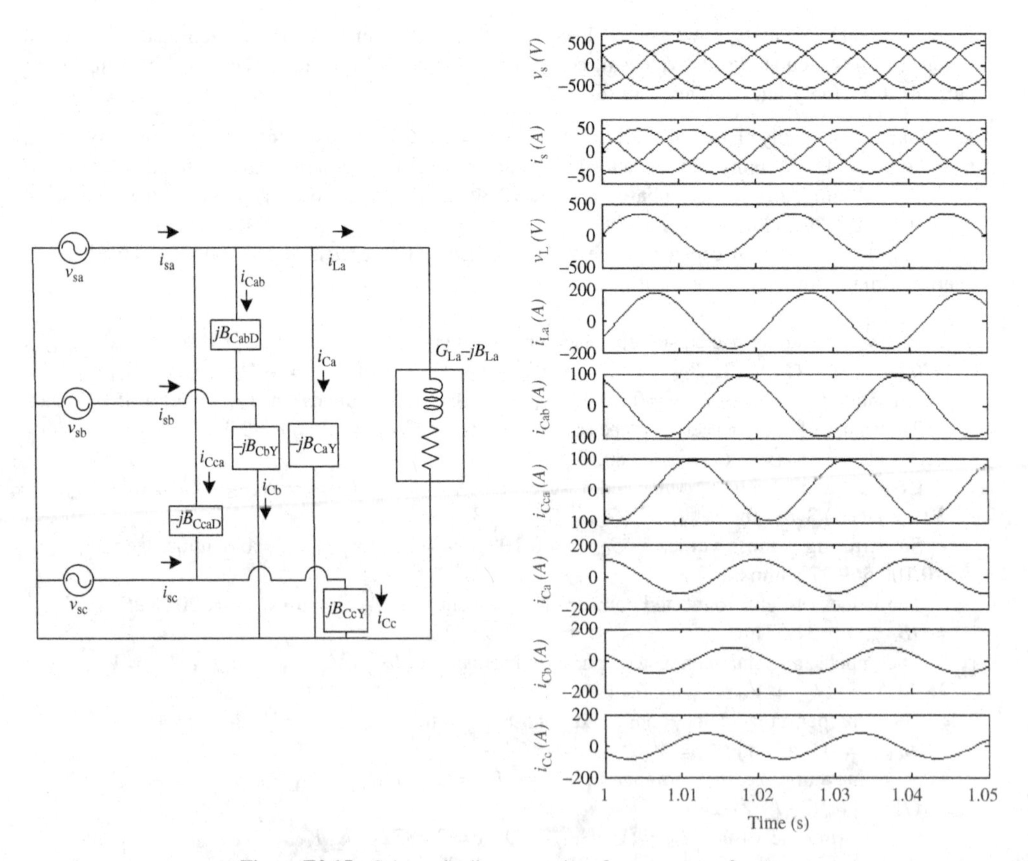

Figure E3.15 Schematic diagram and performance waveforms

Substituting the values, $B_{CaY} = 0.313\,544$ mhos, $B_{CbY} = -0.241\,366\,496$ mhos, and $B_{CcY} = 0.241\,366\,496$ mhos.

The values of star connected compensator elements are $C_{CaY} = B_{CaY}/\omega = 998.04\ \mu F$, $L_{CbY} = 1/B_{CbY}\omega = 13.2$ mH, and $C_{CcY} = B_{CcY}/\omega = 768.29\ \mu F$.

Compensator delta connected susceptances are $B_{CabD} = 2(G_{La} - G_{Lb})/3\sqrt{3}$, $B_{CbcD} = 2(G_{Lb} - G_{Lc})/3\sqrt{3}$, and $B_{CcaD} = 2(G_{Lc} - G_{La})/3\sqrt{3}$.

Substituting the values, $B_{CabD} = 0.160\,910\,997$ mhos, $B_{CbcD} = 0.0$ mhos, and $B_{CcaD} = -0.160\,910\,997$ mhos.

The values of delta connected compensator elements are $C_{CabD} = B_{CabD}/\omega = 512.46\ \mu F$ and $L_{CcaD} = 1/B_{CcaD}\omega = 19.79$ mH.

b. The compensator star connected phase currents are $I_{Ca} = V_{sp}B_{CaY} = 75.1253$ A, $I_{Cb} = V_{sp}B_{CbY} = 57.83$ A, and $I_{Cc} = V_{sp}B_{CcY} = 57.83$ A.

The compensator delta connected phase currents are $I_{Cab} = V_sB_{CabD} = 66.78$ A, $I_{Cbc} = V_sB_{CbcD} = 0.0$ A, and $I_{Cca} = V_sB_{CcaD} = 66.78$ A.

c. Supply line currents after compensation are $I_{sa} = (V_{sp}G_{LT}/3\sqrt{3})\angle 0°$ A, $I_{sb} = (V_{sp}G_{LT}/3\sqrt{3})\angle -120°$ A, and $I_{sc} = (V_{sp}G_{LT}/3\sqrt{3})\angle -240°$ A.

Substituting the values, $I_{sa} = (V_{sp}G_{LT}/3\sqrt{3})\angle 0°$ A $= 33.3889\angle 0°$ A, $I_{sb} = 33.3889\angle -120°$ A, and $I_{sc} = 33.3889\angle -240°$ A.

d. The compensator kVA rating is $Q_C = V_s(|I_{Ca}| + |I_{Cb}| + |I_{Cc}|) + \sqrt{3}V_s(|I_{Cab}| + |I_{Cbc}| + |I_{Cca}|) = 45\,713 + 55\,426$ VA $= 101.14$ kVA.

Example 3.16

A three-phase four-wire unbalanced load having $Z_{La} = (6.0 + j3.0)$ pu connected between phase a and neutral terminal is fed from an AC supply with an input line voltage of 415 V at 50 Hz and a base impedance of 4.15 Ω per phase. It is to be realized as a three-phase balanced unity power factor load on the three-phase supply system using a shunt compensator consisting of lossless passive elements (L and/or C) as shown in Figure E3.16. Calculate (a) the values of compensator elements (in farads or henries), (b) supply line currents, (c) compensator currents, (d) its kVA rating, and (e) equivalent per-phase resistance (in ohms) of the compensated load.

Solution: Given supply phase voltage $V_{sp} = 415/\sqrt{3}$ V $= 239.6$ V, frequency of the supply $(f) = 50$ Hz, and a single-phase load having $Z_{La} = (6.0 + j3.0)$ pu connected between phase a and neutral terminal with a base impedance of 4.15 Ω per phase.

The load impedance per phase is $Z_{La} = 27.839\angle 26.57°\ \Omega = (24.9 + j12.45)\ \Omega$.

The load admittance per phase is $Y_{La} = 0.035\,920\,77\angle -26.57° = 0.032\,128\,513 - j0.016\,067\,725$ mhos.

It requires two set of susceptances, one star connected and other delta connected, across the lines for neutral current elimination and load balancing.

a. Compensator star connected susceptances for neutral current and power factor correction are $B_{CaY} = -B_{La} + (G_{Lb} - G_{Lc})/\sqrt{3}$, $B_{CbY} = -B_{Lb} + (G_{Lc} - G_{La})/\sqrt{3}$, and $B_{CcY} = -B_{Lc} + (G_{La} - G_{Lb})/\sqrt{3}$.

Substituting the values, $B_{CaY} = 0.016\,067\,725$ mhos, $B_{CbY} = -0.018\,549\,405$ mhos, and $B_{CcY} = 0.018\,549\,405$ mhos.

The values of star connected compensator elements are $C_{CaY} = B_{CaY}/\omega = 51.17\ \mu F$, $L_{CbY} = 1/B_{CbY}\omega = 171.69$ mH, and $C_{CcY} = B_{CcY}/\omega = 59.075\ \mu F$.

Compensator delta connected susceptances are $B_{CabD} = 2(G_{La} - G_{Lb})/3\sqrt{3}$, $B_{CbcD} = 2(G_{Lb} - G_{Lc})/3\sqrt{3}$, and $B_{CcaD} = 2(G_{Lc} - G_{La})/3\sqrt{3}$.

Substituting the values, $B_{CabD} = 0.012\,366\,27$ mhos, $B_{CbcD} = 0.0$ mhos, and $B_{CcaD} = -0.012\,366\,27$ mhos.

The values of delta connected compensator elements are $C_{CabD} = B_{CabD}/\omega = 39.38\ \mu F$ and $L_{CcaD} = 1/B_{CcaD}\omega = 257.53$ mH.

b. The compensator star connected phase currents are $I_{Ca} = V_{sp}B_{CaY} = 3.85$ A, $I_{Cb} = V_{sp}B_{CbY} = 4.44$ A, and $I_{Cc} = V_{sp}B_{CcY} = 4.44$ A.

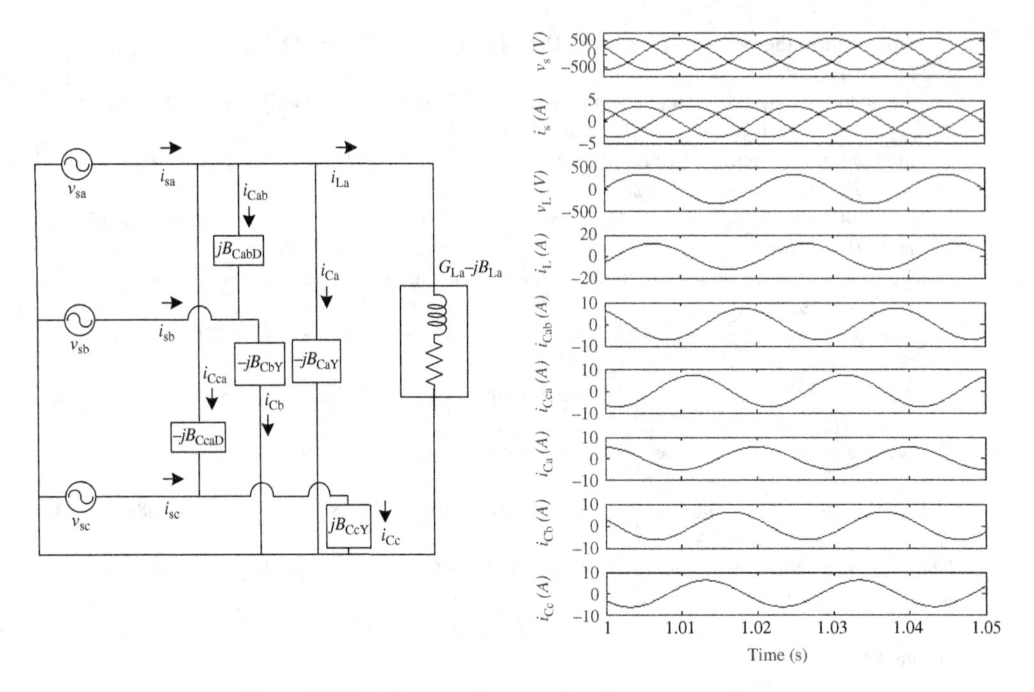

Figure E3.16 Schematic diagram and performance waveforms

The compensator delta connected phase currents are $I_{Cab} = V_{sp}B_{CabD} = 5.132$ A, $I_{Cbc} = V_{sp}B_{CbcD} = 0.0$ A, and $I_{Cca} = V_{sp}B_{CcaD} = 5.132$ A.

c. Supply line currents after compensation are $I_{sa} = (V_{sp}G_{LT}/3\sqrt{3})\angle 0°$ A, $I_{sb} = (V_{sp}G_{LT}/3\sqrt{3})\angle -120°$ A, and $I_{sc} = (V_{sp}G_{LT}/3\sqrt{3})\angle -240°$ A.

   Substituting the values, $I_{sa} = (V_{sp}G_{LT}/3\sqrt{3})\angle 0°$ A $= 2.566\angle 0°$ A, $I_{sb} = 2.566\angle -120°$ A, and $I_{sc} = 2.566\angle -240°$ A.

d. The compensator kVA rating is $Q_C = V_s(|I_{Ca}| + |I_{Cb}| + |I_{Cc}|) + \sqrt{3}V_s(|I_{Cab}| + |I_{Cbc}| + |I_{Cca}|) =$ 3052.23+ 4259.56 VA = 7.312 kVA.

e. Equivalent per-phase resistance (in star connection) of the compensated load is calculated as follows: $G_{eq} = (G_{La} + G_{Lb} + G_{Lc})/3 = 0.010\,7095$ ℧. Thus, $R_{eq} = 1/G_{eq} = 93.377\ \Omega$.

Example 3.17

A three-phase four-wire unbalanced load having $Z_{La} = (5.0 + j2.0)$ pu, $Z_{Lb} = (3.0 + j1.5)$ pu, and $Z_{Lc} = (7.5 + j1.5)$ pu is fed from an AC supply with an input line voltage of 415 V at 50 Hz and a base impedance of 4.15 Ω per phase. It is to be realized as a three-phase balanced unity power factor load on the three-phase supply system using a shunt compensator consisting of lossless passive elements (L and/or C) as shown in Figure E3.17. Calculate (a) the values of compensator elements (in farads and henries) and (b) equivalent per-phase resistance (in ohms) of the compensated load.

Solution: Given supply voltage $V_s = 415/\sqrt{3}$ V $= 239.6$ V, frequency of the supply $(f) = 50$ Hz, and an unbalanced load having $Z_{La} = (5.0 + j2.0)$ pu, $Z_{Lb} = (3.0 + j1.5)$ pu, and $Z_{Lc} = (7.5 + j1.5)$ pu fed from an AC supply with an input line–line voltage of 415 V at 50 Hz and a base impedance of 4.15 Ω per phase.

The load impedances per phase are $Z_{La} = (20.75 + j8.3)\ \Omega$, $Z_{Lb} = (12.45 + j6.225)\ \Omega$, and $Z_{Lc} = (31.125 + j6.225)\ \Omega$.

The load admittances per phase are $Y_{La} = (0.041\,545\,492 - j0.016\,618\,196)$ mhos, $Y_{Lb} = (0.063\,845\,949 - j0.031\,922\,974)$ mhos, and $Y_{Lc} = (0.030\,8928 - j0.006\,178\,560\,395)$ mhos.

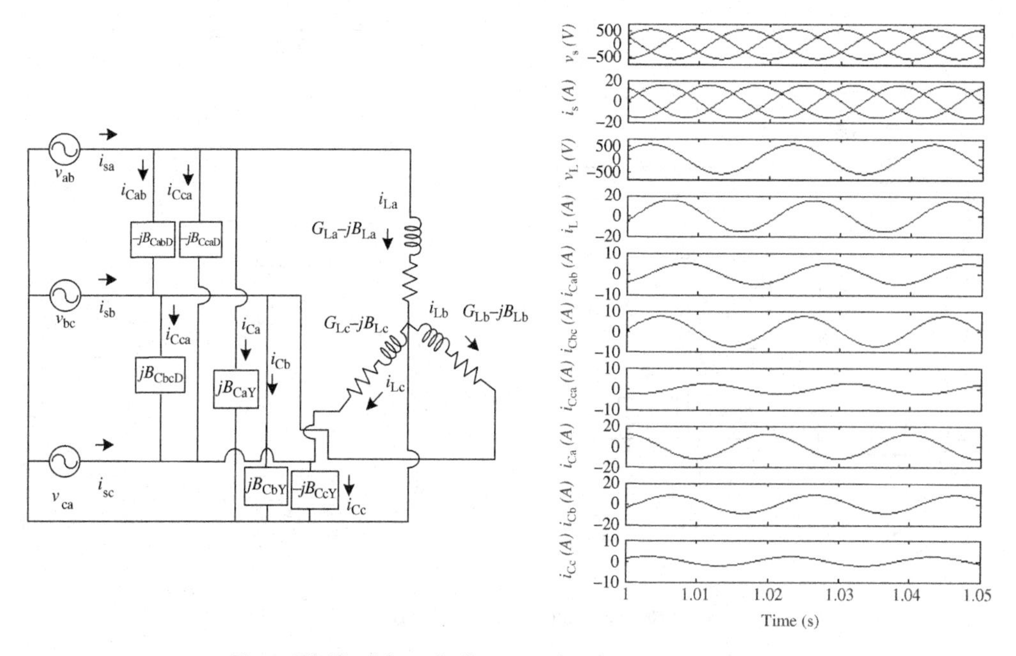

Figure E3.17 Schematic diagram and performance waveforms

It requires two set of susceptances, one star connected and other delta connected, across the lines for neutral current elimination and load balancing.

a. Compensator star connected susceptances for neutral current and power factor correction are $B_{CaY} = -B_{La} + (G_{Lb} - G_{Lc})/\sqrt{3}$, $B_{CbY} = -B_{Lb} + (G_{Lc} - G_{La})/\sqrt{3}$, and $B_{CcY} = -B_{Lc} + (G_{La} - G_{Lb})/\sqrt{3}$.
   Substituting the values, $B_{CaY} = 0.035\,643\,705$ mhos, $B_{CbY} = 0.025\,772\,639$ mhos, and $B_{CcY} = -0.0069$ mhos.
   The values of star connected compensator elements are $C_{CaY} = B_{CaY}/\omega = 113.52\ \mu F$, $C_{CbY} = B_{CbY}/\omega = 82.078\ \mu F$, and $L_{CcY} = 1/B_{CcY}\omega = 459.1$ mH.
   Compensator delta connected susceptances are $B_{CabD} = 2(G_{La} - G_{Lb})/3\sqrt{3}$, $B_{CbcD} = 2(G_{Lb} - G_{Lc})/3\sqrt{3}$, and $B_{CcaD} = 2(G_{Lc} - G_{La})/3\sqrt{3}$.
   Substituting the values, $B_{CabD} = -0.0087$ mhos, $B_{CbcD} = 0.0128$ mhos, and $B_{CcaD} = -0.004\,100\,223\,063$ mhos.
   The values of delta connected compensator elements are $L_{CabD} = 1/B_{CabD}\omega = 364.1$ mH, $C_{CbcD} = B_{CbcD}/\omega = 40.87\ \mu F$, and $L_{CcaD} = 1/B_{CcaD}\omega = 776.3$ mH.
b. Equivalent per-phase resistance (in star connection) of the compensated load is calculated as follows: $G_{eq} = (G_{La} + G_{Lb} + G_{Lc})/3 = 0.004\,542\,808$ ℧. Thus, $R_{eq} = 1/G_{eq} = 21.9466\ \Omega$.

Example 3.18

A three-phase AC supply system has a line voltage of 415 V at 50 Hz and a feeder (source) impedance of 1.0 Ω resistance and 3.0 Ω inductive reactance per phase after which an unbalanced isolated neutral and star connected load having $Z_{La} = 10\ \Omega$, $Z_{Lb} = 20\ \Omega$, and $Z_{Lc} = 30\ \Omega$ is connected. Calculate (a) the voltage drop across the source impedance and (b) the voltage across the load if a shunt compensator consisting of lossless passive elements (L and/or C) is used (c) to balance at UPF and then (d) to raise the voltage to the input voltage (415 V), in both the cases as shown in Figure E3.18.

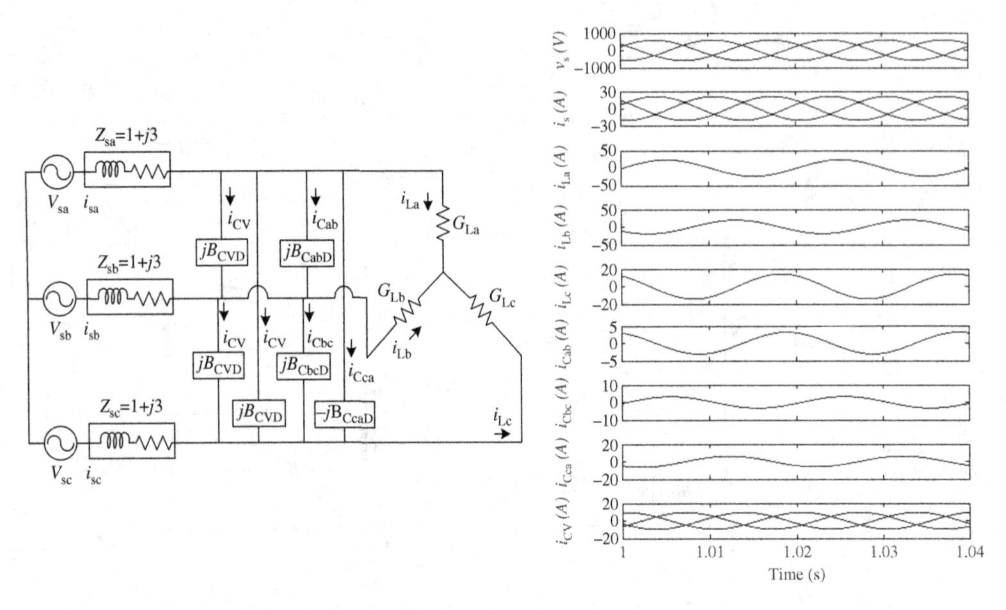

Figure E3.18 Schematic diagram and performance waveforms

Solution: Given supply phase voltage $V_{sp} = 415/\sqrt{3}$ V = 239.6 V, frequency of the supply (f) = 50 Hz, and an unbalanced isolated neutral and star connected load having $Z_{La} = 10\,\Omega$, $Z_{Lb} = 20\,\Omega$, and $Z_{Lc} = 30\,\Omega$ with a feeder (source) impedance of 1.0 Ω resistance and 3.0 Ω inductive reactance per phase.

Load admittances per phase are $Y_{La} = 0.1$ mhos, $Y_{Lb} = 0.05$ mhos, and $Y_{Lc} = 0.033\,333$ mhos.

a. For power factor correction, load balancing, and neutral current elimination, the equivalent load conductance in star configuration is $G_{LY} = (G_{La} + G_{Lb} + G_{Lc})/3 = 0.061\,111$ mhos and $R_{LY} = 16.3636\,\Omega$.

   Total impedance across the supply is $Z_{LT} = (17.3636 + j3)\,\Omega = 17.62\,\Omega$.

   The supply current is $I_s = V_{sp}/Z_{LT} = 239.6/17.62 = 13.5975$ A.

   The voltage across the source impedance is $V_{Zs} = I_s Z_s = 43$ V.

   The voltage across the load impedance is $V_{ZL} = Z_{LY} I_s = R_{LY} I_s = 222.50$ V.

   The impedance for equivalent delta connected load is $Z_{Lab} = 36.66\,\Omega$, $Z_{Lbc} = 110\,\Omega$, $Z_{Lca} = 55\,\Omega$.

   The load conductance is, $G_{Lab} = 0.02727768$ mhos, $G_{Lbc} = 0.00909091$ mhos, $G_{Lca} = 0.01818181$ mhos.

   Compensator susceptances, $B_{CabD} = -B_{Lab} + (G_{Lca} - G_{Lbc})/\sqrt{3}$, $B_{CbcD} = -B_{Lbc} + (G_{Lab} - G_{Lca})/\sqrt{3}$, $B_{CcaD} = -B_{Lca} + (G_{Lbc} - G_{Lab})/\sqrt{3}$.

   Substituting the values, $B_{CbcD} = 0.0052486$ mhos, $B_{CabD} = 0.00554023$ mhos, $B_{CcaD} = -0.01050014$ mhos.

   Values of elements of compensator, $C_{CabD} = B_{CabD}/\omega = 16.71\,\mu$F, $C_{CbcD} = B_{CabD}/\omega = 17.64\,\mu$F, $L_{CcaD} = 1/B_{CcaD}\omega = 0.3033$ H.

b. Let the susceptance of value B_{CV} mhos is connected in parallel to the load to raise the load voltage to the input voltage ($V_L = V_{sp}$). The basic equation to maintain the load voltage equal to the input voltage is $|V_L| = |V_{sp}/[\{(R_s + jX_s) + 1/(G_L + jB_{CV})\}]| \times |1/(G_L + jB_{CV})| = |V_{sp}|$.

   Solving this equation, the value of B_{CV} is $B_{CV} = [X_s \pm \{\sqrt{X_s^2 - (X_s^2 + R_s^2)(2R_sG_L + R_s^2G_L^2 + X_s^2G_L^2)}\}]/(X_s^2 + R_s^2)$.

Substituting the values of $X_s = 3\,\Omega$, $R_s = 1\,\Omega$, and $G_L = G_{LY} = 0.061111$ mhos and considering "–" sign in "±", for lower value of the compensator, the value of B_{CV} is 0.027 890 746 mhos.

The value of the capacitor is $C_{CV} = B_{CV}/\omega = 0.027\,890\,746/314.15 = 88.82\,\mu F$. The value of equivalent delta connected capacitor for voltage regulation is $C_{CVD} = C_{CV}/3 = 88.82\,\mu F/3 = 29.60\,\mu F$.

The equivalent per phase admittance of load and compensator as seen from PCC is

$$Y_{LV} = (0.061\,111 + j0.027\,890\,746)\text{ mhos} = 0.067\,174\,757\angle 24.5317°\text{ mhos}.$$

$$Z_{LV} = (14.886\,5442\angle -24.5317°)\,\Omega = (13.542\,75 - j6.180\,8430)\,\Omega.$$

$$Z_{LT} = (14.542\,75 - j3.180\,843\,027)\,\Omega = 14.886\,368\,85\,\Omega.$$

$$I_s = V_{sp}/|Z_{LT}| = 239.6/14.886\,368\,85 = 16.095\,261\,54\text{ A}.$$

c. The voltage across the source impedance is $V_{Zs} = I_s Z_s = 50.897\,686$ V.
d. The voltage across the load impedance is $V_{ZL} = Z_{LV} I_s = 239.60$ V.

Example 3.19

A three-phase AC supply has a line voltage of 415 V at 50 Hz and a feeder (source) impedance of 1.0 Ω resistance and 3.0 Ω inductive reactance per phase after which a balanced star connected load having $Z = (32 + j24)\,\Omega$ per phase is connected. Calculate (a) the voltage drop across the source impedance and (b) the voltage across the load. If a shunt compensator consisting of lossless passive elements (L and/or C) is used to raise the voltage to the input voltage (415 V) as shown in Figure E3.19, calculate (c) the values of compensator elements (in farads or henries), (d) compensator currents, (e) supply line currents, and (f) its kVA rating.

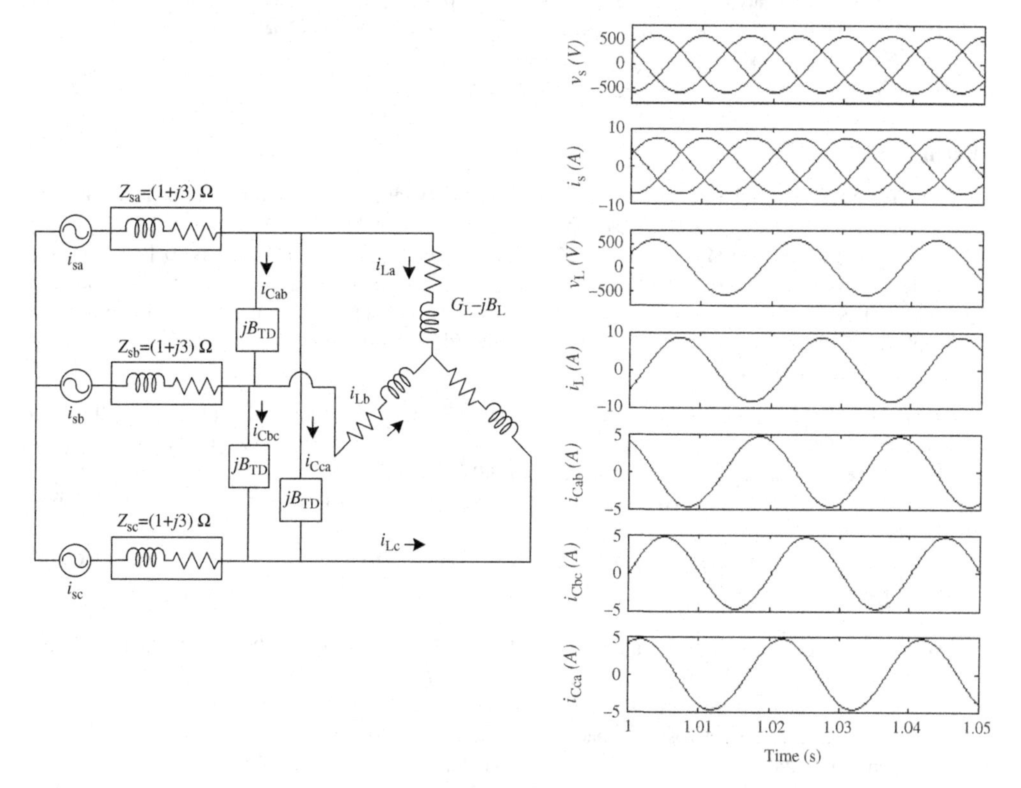

Figure E3.19 Schematic diagram and performance waveforms

Solution: Given supply line–line voltage $V_{LL}=415$ V, hence phase to neutral voltage $V_{sp}=415/\sqrt{3}$ V $=239.6$ V, frequency of the supply (f) = 50 Hz, and an unbalanced isolated neutral and star connected load having $Z_{La}=(32+j24)\ \Omega$, $Z_{Lb}=(32+j24)\ \Omega$, and $Z_{Lc}=(32+j24)\ \Omega$ with a feeder (source) impedance of 1.0 Ω resistance and 3.0 Ω inductive reactance per phase.

Load admittances per phase are $Y_{La}=(0.02-j0.015)$ mhos, $Y_{La}=(0.02-j0.015)$ mhos, and $Y_{La}=(0.02-j0.015)$ mhos.

$$G_{LY}=(G_{La}+G_{Lb}+G_{Lc})/3=0.02\text{ mhos and }R_{LY}=50\ \Omega.$$

A set of star connected susceptances in parallel to the load is required for power factor correction of the value $B_{PF}=-B_L=0.015$ mhos.

a. The voltage across the source impedance is $V_{Zs}=I_sZ_s=14.83$ V.
b. The voltage across the load impedance is $V_{ZL}=Z_{LY}I_s=234.49$ V.
   Let the susceptance of value B_{CV} mhos is connected in parallel to the load to raise the load voltage to the input voltage ($V_L=V_s$). The basic equation to maintain the load voltage equal to the input voltage is $|V_{sp}|=|V_{sp}/[\{(R_s+jX_s)+1/(G_L+jB_{CV})\}]|\times|1/(G_L+jB_{CV})|=|V_{sp}|$.
   Solving this equation, the value of B_{CV} is $B_{CV}=[X_s\pm\{\sqrt{X_s^2-(X_s^2+R_s^2)(2R_sG_L+R_s^2G_L^2+X_s^2G_L^2)}\}]/(X_s^2+R_s^2)$.
c. Substituting the values of $X_s=3\ \Omega$, $R_s=1\ \Omega$, and $G_L=0.02$ mhos and considering "−" sign in "±", for lower value of the compensator, the value of B_{CV} is 0.007 425 2232 mhos, $B_T=B_{PF}+B_{CV}=0.022\,425\,223$, $B_{TD}=B_T/3=0.007\,4750$, and $C_{BD}=B_T/3\omega=23.80\ \mu$F (per phase for the delta connected compensator).
d. Compensator star connected phase currents are $I_{Cab}=V_{ab}(jB_{TD})=415\angle30°\times0.007\,4750\angle90°=3.1\angle120°$ A, $I_{Cbc}=3.1\angle0°$ A, and $I_{Cca}=3.1\angle-120°$ A.
e. Supply line currents are calculated as follows: $Y_{LV}=(0.02+j0.007\,425\,2232)$ mhos, $Z_{LV}=(43.94-j16.31)\ \Omega$, and $Z_{LT}=Z_{LV}+Z_s=44.94-j13.31=46.87\angle-16.50°\ \Omega$. Thus, $I_s=V_{sp}/Z_{LT}=239.6/46.87=5.11$ A.
f. Its kVA rating is $Q_C=3|V_{LL}|\times|I_{Cab}|=3.8595$ kVA.

Example 3.20

A three-phase AC supply has a line voltage of 415 V at 50 Hz and a feeder (source) impedance of 1.0 Ω resistance and 3.0 Ω inductive reactance per phase after which an unbalanced delta connected load having $Z_{ab}=(24+j18)\ \Omega$, $Z_{bc}=30\ \Omega$, and $Z_{ca}=(24-j18)\ \Omega$ is connected. If this load is to be realized as a balanced unity power factor load using a shunt compensator consisting of lossless passive elements (L and/or C) as shown in Figure E3.20, calculate (a) supply line currents, (b) the values of compensator elements (in farads or henries), (c) compensator currents, (d) its kVA rating, and (e) power losses in source impedance after compensation.

Solution: Given supply phase voltage $V_{sp}=415/\sqrt{3}$ V $=239.6$ V, frequency of the supply (f) = 50 Hz, and an unbalanced delta connected load having $Z_{ab}=(24+j18)\ \Omega$, $Z_{bc}=30\ \Omega$, and $Z_{ca}=(24-j18)\ \Omega$ with a feeder (source) impedance of 1.0 Ω resistance and 3.0 Ω inductive reactance per phase.

Load admittances are $Y_{Lab}=(0.026\,666-j0.02)$ mhos, $Y_{Lbc}=0.033\,333$ mhos, and $Y_{Lca}=(0.026\,666+j0.02)$ mhos.

Equivalent star connected conductance after compensation is $G_{LY}=G_{Lab}+G_{Lbc}+G_{Lca}=0.086\,666$ mhos and $R_{LY}=1/G_{LY}=11.538\,48\ \Omega$.

a. Supply line currents are calculated as follows: $Z_{LT}=R_s+R_{LY}+jX_s=(12.538\,48+j3.0)=12.8924\ \Omega$. Thus, $I_s=V_s/Z_{LT}=239.6/12.8924=18.584\,62$ A.
   The phase voltage across the load and compensator is $V_{Lp}=R_{LY}I_s=214.44$ V.
   The line–line voltage across the load is $V_{LL}=\sqrt{3}V_{Lp}=371.4$ V.
b. Compensator delta connected susceptances for load balancing and power factor correction are $B_{Cab}=-B_{Lab}+(G_{Lca}-G_{Lbc})/\sqrt{3}$, $B_{Cbc}=-B_{Lbc}+(G_{Lab}-G_{Lca})/\sqrt{3}$, and $B_{Cca}=-B_{Lca}+(G_{Lbc}-G_{Lab})/\sqrt{3}$.

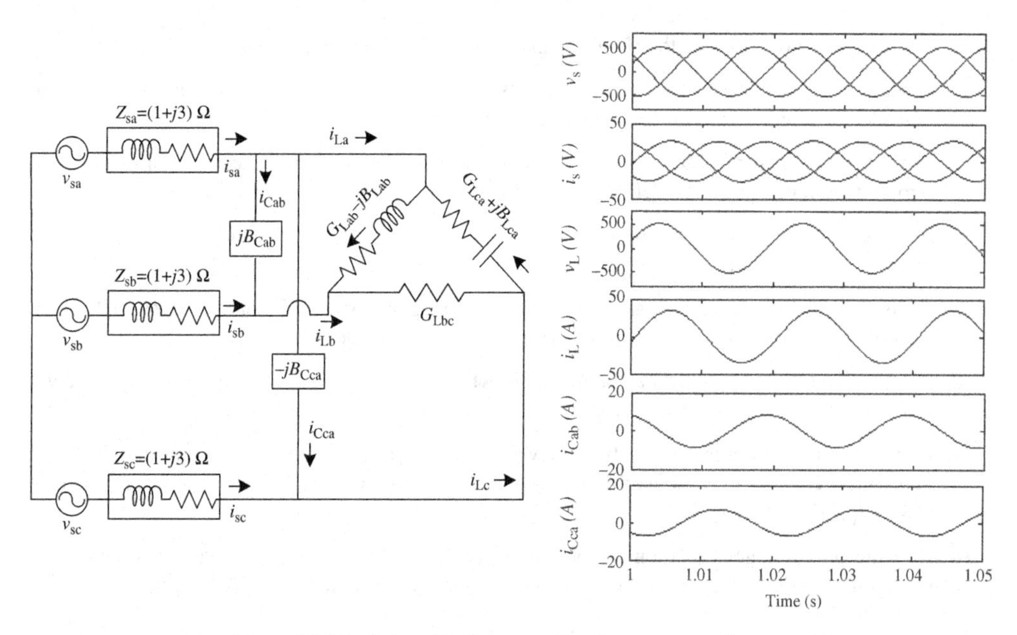

Figure E3.20 Schematic diagram and performance waveforms

Substituting the values, $B_{Cab} = 0.016150998$ mhos, $B_{Cbc} = 0.0$ mhos, and $B_{Cca} = -0.016509 98$ mhos.

The values of delta connected compensator elements are $C_{Cab} = B_{Cab}/\omega = 51.436\,\mu$F, $L_{Ccb} = 1/B_{Ccb}\omega = 0.0$ mH, and $L_{Cca} = 1/B_{Cca}\omega = 197.18$ mH.

c. The compensator currents are $I_{Cab} = V_{LL}B_{Cab} = 6$ A, $I_{Cbc} = V_{LL}B_{Cbc} = 0.0$ A, and $I_{Cca} = V_{LL}B_{Cca} = 6$ A.

d. Its kVA rating is $Q_C = V_{LL}(|I_{Cab}| + |I_{Cbc}| + |I_{Cbc}|) = 4.456$ kVA.

e. Power losses in source impedance after compensation are $P_{Ls} = 3I_s^2R_s = 1036.16$ W.

3.8 Summary

An exhaustive study of passive shunt and series compensation has been made to provide a wide exposure on various issues of power quality that can easily be mitigated using lossless passive components such as capacitors and inductors for enhancing the efficiency and utilization of equipment in distribution systems. The use of passive shunt and series compensators has been demonstrated through several numerical examples in distribution systems for improving the voltage profile at the point of common coupling, reducing losses, power factor correction, load balancing, neutral current compensation, and for better utilization of distribution equipment. The passive shunt and series compensators are considered as a better alternative for power quality improvement due to voltage- and current-based power quality mitigation, simple design, and high reliability compared with other options of power quality improvement, especially in the absence of harmonic voltages and currents. These are considered to be beneficial to the designers, users, manufacturers, and research engineers dealing with power quality improvement in the distribution systems such as furnaces, traction systems, and rural supply systems to balance consumer loads, to reduce negative-sequence voltages at the point of common coupling, to improve power factor, and to improve voltage regulation.

3.9 Review Questions

1. What are the lossless passive elements used for shunt and series compensation in an AC supply distribution system?

2. What are the limitations of the series compensation using lossless passive components?

3. What are the limitations of the shunt compensation using lossless passive components?
4. In which types of linear loads, a series compensation using lossless passive components can provide zero voltage regulation with definite source inductance (L_s)?
5. In which types of linear loads, a shunt compensation using lossless passive components can provide zero voltage regulation with definite source impedance (L_s, R_s)?
6. How the unity power factor can be achieved at the single-phase AC mains with a lagging power factor linear load using lossless passive elements?
7. Is it possible to get unity power at the AC mains while regulating the AC voltage across the load with some definite source impedance using lossless passive shunt devices?
8. Is it possible to get unity power at the AC mains while regulating the AC voltage across the load with some definite source impedance using lossless passive series devices?
9. How the unity power factor can be achieved at the single-phase AC mains with definite source impedance (L_s, R_s) and without affecting the voltage across a lagging power factor linear load using lossless passive elements?
10. How the zero voltage regulation can be achieved across a single-phase lagging power factor linear load at the single-phase AC mains with definite source impedance (L_s, R_s) using lossless passive elements?
11. How the unity power factor can be achieved at the single-phase AC mains with definite source impedance while regulating the AC terminal voltage across a single-phase lagging power factor linear load by using lossless passive elements?
12. What are the different ways to get zero neutral current in the AC mains with a single-phase lagging power factor linear load across line and neutral, in a three-phase four-wire supply system using lossless passive elements?
13. Which scheme has the lowest power rating of the lossless passive compensator to get zero neutral current in the AC mains with a single-phase lagging power factor linear load across line and neutral, in a three-phase four-wire supply system?
14. How a single-phase lagging power factor linear load can be realized as a balanced unity power factor load when it is connected across two lines of a three-phase three-wire supply system by using lossless passive elements?
15. How a single-phase lagging power factor linear load can be realized as a balanced unity power factor load when it is connected across line and neutral of a three-phase four-wire supply system by using lossless passive elements?
16. What is the % loss reduction in the feeders of a three-phase four-wire supply system having 1.0 Ω resistance and a single-phase 125 A, 0.8 lagging power single-phase load connected across line and neutral after realizing it as a balanced unity power load?
17. What are the power factors of two equal single-phase loads connected between lines and neutral in a three-phase four-wire supply system, which cause maximum neutral current?
18. What maximum value of neutral current (as a percentage of phase current) can flow in the AC mains with two equal single-phase varying power factor linear loads across two lines and neutral, in a three-phase four-wire supply system?
19. What is the % reduction in total rms currents in all three phases and neutral after realizing the balanced unity power factor loads on the supply system through shunt compensation using lossless passive elements of the load system having varying power factors of two equal single-phase loads connected between lines and neutral in a three-phase four-wire supply system, which cause maximum neutral current?

20. How a zero neutral current in the AC mains can be achieved with the lagging power factor linear loads across two lines and neutral, in a three-phase four-wire supply system by using lossless passive elements?

21. Which compensation out of shunt, series, and their combination using lossless passive components has the lowest power rating for zero voltage regulation across a lagging power factor load with definite source impedance (L_s, R_s) in a single-phase supply system?

22. In which case out of power factor correction to unity of a lagging power linear load and maintaining its AC terminal voltage regulation to zero, a lossless passive shunt compensator has lower power rating?

23. What are the economic factors for the justification for using lossless passive shunt compensators in an AC system?

24. What are the factors that decide the rating of lossless passive shunt compensators?

25. What are the benefits of using a lossless passive shunt compensator in a three-phase isolated gasoline engine-driven squirrel cage induction generator set (DG set) for economic justification?

3.10 Numerical Problems

1. A single-phase load having $Z = (3.0 + j0.75)$ pu is fed from an AC supply with an input AC voltage of 220 V at 50 Hz and a base impedance of 6.25 Ω. It is to be realized as a unity power factor load on the AC supply system using shunt connected lossless passive elements (L and/or C). Calculate (a) the values of compensator elements (in farads and henries) and (b) equivalent resistance (in ohms) of the compensated load.

2. A single-phase AC supply has AC mains voltage of 220 V at 50 Hz and a feeder (source) impedance of 3.0 Ω inductive reactance after which a single-phase load having $Z = (12 + j9)$ Ω is connected. Calculate (a) the voltage drop across the source impedance and (b) the voltage across the load. If a shunt compensator consisting of lossless passive elements (L and/or C) is used to raise the voltage to the input voltage (220 V), calculate (c) the values of compensator elements (in farads or henries), (d) its kVA rating, and (e) the voltage drop across the source impedance after compensation.

3. A single-phase AC supply has AC mains voltage of 220 V at 50 Hz and a feeder (source) impedance of 3.0 Ω inductive reactance after which a single-phase load having $Z_L = (12 + j9)$ Ω is connected. Calculate (a) the voltage drop across the source impedance and (b) the voltage across the load without and with a series compensator. If a series compensator consisting of lossless passive elements (L and/ or C) is used to raise the voltage across the load to the input voltage (220 V), calculate (c) the values of compensator elements (in farads or henries) and (d) its kVA rating.

4. A three-phase three-wire shunt compensator consisting of lossless passive elements (L and/or C) is employed at a 415 V, 50 Hz system to provide load balancing and power factor correction of a single-phase 50 kW unity power factor load connected between two lines. Calculate (a) supply line currents, (b) compensator currents, (c) the values of compensator elements (in farads or henries), and (d) its kVA rating.

5. A three-phase three-wire shunt compensator consisting of lossless passive elements (L and/or C) is employed at a 415 V, 50 Hz system to provide load balancing and power factor correction of a single-phase 25 kVA, 0.8 lagging power factor load connected between two lines. Calculate (a) supply line currents, (b) compensator currents, (c) the values of compensator elements (in farads or henries), and (d) its kVA rating.

6. A three-phase three-wire delta connected balanced load having $Z_L = (3.0 + j1.0)$ pu is fed from an AC supply with an input AC line voltage of 415 V at 50 Hz and a base impedance of 8.25 Ω per phase. It is to be realized as a unity power factor load on the AC supply system using shunt connected lossless

passive elements (L and/or C). Calculate (a) supply line currents, (b) compensator currents, (c) the values of compensator elements (in farads or henries), (d) its kVA rating, and (e) equivalent per-phase resistance (in ohms) of the compensated load.

7. A three-phase three-wire AC supply system has a line voltage of 415 V at 50 Hz and a feeder (source) impedance of 0.5 Ω resistance and 2.5 Ω inductive reactance per phase after which a balanced delta connected load having $Z = (16 + j12)\ \Omega$ per phase is connected. Calculate (a) the voltage drop across the source impedance and (b) the voltage across the load. If a shunt compensator consisting of lossless passive elements (L and/or C) is used to raise the voltage to the input voltage (415 V), calculate (c) supply line currents, (d) compensator currents, (e) the values of compensator elements (in farads or henries), and (f) its kVA rating.

8. A three-phase three-wire unbalanced delta connected load having $Z_{ab} = (5.0 + j2.0)$ pu, $Z_{bc} = (4.0 + j1.2)$ pu, and $Z_{ca} = (7 + j2)$ pu is fed from an AC supply with an input line–line voltage of 415 V at 50 Hz and a base impedance of 5.25 Ω per phase. It is to be realized as a balanced unity power factor load on the three-phase supply system using a shunt compensator consisting of lossless passive elements (L and/or C). Calculate (a) supply line currents, (b) the values of compensator elements (in farads or henries), (c) compensator currents, (d) its kVA rating, and (e) equivalent per-phase resistance (in ohms) of the compensated load.

9. A three-phase AC supply system has a line–line voltage of 415 V at 50 Hz and a feeder (source) impedance of 0.5 Ω resistance and 3.0 Ω inductive reactance per phase after which an unbalanced delta connected load having $Z_{ab} = 15\ \Omega$, $Z_{bc} = 20\ \Omega$, and $Z_{ca} = j25\ \Omega$ is connected. Calculate (a) the voltage drop across the source impedance and (b) the voltage across the load if a shunt compensator consisting of lossless passive elements (L and/or C) is used to balance this load and to realize it as UPF load. Also, calculate (c) supply line currents, (d) the values of compensator elements (in farads or henries), (e) compensator currents, (f) its kVA rating, and (g) equivalent per-phase resistance (in ohms) of the compensated load.

10. A three-phase AC supply has a line–line voltage of 415 V at 50 Hz and a feeder (source) impedance of 0.5 Ω resistance and 2.5 Ω inductive reactance per phase after which an unbalanced delta connected load having $Z_{ab} = 15\ \Omega$, $Z_{bc} = 20\ \Omega$, and $Z_{ca} = 25\ \Omega$ is connected. Calculate (a) the voltage drop across the source impedance and (b) the voltage across the load if a shunt compensator consisting of lossless passive elements (L and/or C) is used to balance these loads and to raise the voltage to the input voltage (415 V). Also, calculate (c) supply line currents, (d) the values of compensator elements (in farads or henries), (e) compensator currents, (f) its kVA rating, and (g) equivalent per-phase resistance (in ohms) of the compensated load.

11. A three-phase three-wire unbalanced isolated neutral star connected load having $Z_a = (5.0 + j2.0)$ pu, $Z_b = (4.0 + j1.2)$ pu, and $Z_c = (6 + j2)$ pu is fed from an AC supply with an input line–line voltage of 415 V at 50 Hz and a base impedance of 5.30 Ω per phase. It is to be realized as a balanced unity power factor load on the three-phase supply system using a shunt compensator consisting of lossless passive elements (L and/or C). Calculate (a) supply line currents, (b) compensator currents, (c) the values of compensator elements (in farads or henries), (d) its kVA rating, and (e) equivalent per-phase resistance (in ohms) of the compensated load.

12. A three-phase four-wire 415 V (line–line), 50 Hz AC supply system has a single-phase 7.5 kW unity power factor load connected across line and neutral. If it is required to eliminate neutral current using a shunt compensator consisting of lossless passive elements (L and/or C), calculate (a) supply line currents, (b) the values of compensator elements (in farads or henries), (c) compensator currents, and (d) its kVA rating.

13. A three-phase four-wire 415 V (line–line), 50 Hz AC supply system has a single-phase 25 kVA, 0.8 lagging power factor load connected across line and neutral terminal. If it is required to eliminate neutral current using a shunt compensator consisting of lossless passive elements (L and/or C),

calculate (a) supply line currents, (b) the values of compensator elements (in farads or henries), (c) compensator currents, and (d) its kVA rating.

14. A three-phase four-wire shunt compensator consisting of lossless passive elements (L and/or C) is employed at a 415 V, 50 Hz supply system to provide load balancing and unity power factor of a single-phase 25 kW unity power factor load connected across line and neutral terminal. Calculate (a) the values of compensator elements (in farads or henries), (b) compensator currents, (c) supply line currents, and (d) its kVA rating.

15. A three-phase four-wire shunt compensator consisting of lossless passive elements (L and/or C) is used at a 415 V, 50 Hz supply system to provide load balancing and power factor correction of a single-phase 25 kVA, 0.8 lagging power factor load connected across line and neutral terminal. Calculate (a) supply line currents, (b) the values of compensator elements (in farads or henries), (c) compensator currents, and (d) its kVA rating.

16. A three-phase four-wire unbalanced load having $Z_a = (5.0 + j2.0)$ pu connected between phase a and neutral terminal is fed from an AC supply with an input line–line voltage of 415 V at 50 Hz and a base impedance of 5.3 Ω per phase. It is to be realized as a three-phase balanced unity power factor load on the three-phase supply system using a shunt compensator consisting of lossless passive elements (L and/or C). Calculate (a) the values of compensator elements (in farads or henries), (b) supply line currents, (c) compensator currents, (d) its kVA rating, and (e) equivalent per-phase resistance (in ohms) of the compensated load.

17. A three-phase four-wire unbalanced load having $Z_a = (3.0 + j1.0)$ pu, $Z_b = (4.0 + j2)$ pu, and $Z_c = (6 + j1.2)$ pu is fed from an AC supply with an input line–line voltage of 415 V at 50 Hz and a base impedance of 5.3 Ω per phase. It is to be realized as a three-phase balanced unity power factor load on the three-phase supply system using a shunt compensator consisting of lossless passive elements (L and/or C). Calculate (a) the values of compensator elements (in farads and henries) and (b) equivalent per-phase resistance (in ohms) of the compensated load.

18. A three-phase AC supply system has a line voltage of 415 V at 50 Hz and a feeder (source) impedance of 0.5 Ω resistance and 2.5 Ω inductive reactance per phase after which an unbalanced isolated neutral and star connected load having $Z_a = 12\,\Omega$, $Z_b = 24\,\Omega$, and $Z_c = 36\,\Omega$ is connected. Calculate (a) the voltage drop across the source impedance and (b) the voltage across the load if a shunt compensator consisting of lossless passive elements (L and/or C) is used (c) to balance at UPF and then (d) to raise the voltage to the input voltage (415 V), in both the cases.

19. A three-phase AC supply has a line–line voltage of 415 V at 50 Hz and a feeder (source) impedance of 0.5 Ω resistance and 2.5 Ω inductive reactance per phase after which a balanced star connected load having $Z = (24 + j16)\,\Omega$ per phase is connected. Calculate (a) the voltage drop across the source impedance and (b) the voltage across the load if a shunt compensator consisting of lossless passive elements (L and/or C) is used to raise the voltage to the input voltage (415 V). Also, calculate (c) the values of compensator elements (in farads or henries), (d) compensator currents, (e) supply line currents, and (f) its kVA rating.

20. A three-phase AC supply has a line voltage of 415 V at 50 Hz and a feeder (source) impedance of 0.5 Ω resistance and 2.5 Ω inductive reactance per phase after which an unbalanced delta connected load having $Z_{ab} = (20 + j15)\,\Omega$, $Z_{bc} = 25\,\Omega$, and $Z_{ca} = (20 - j15)\,\Omega$ is connected. If this load is to be realized as a balanced unity power factor load using a shunt compensator consisting of lossless passive elements (L and/or C), calculate (a) supply line currents, (b) the values of compensator elements (in farads or henries), (c) compensator currents, (d) its kVA rating, and (e) power losses in source impedance after compensation.

3.11 Computer Simulation-Based Problems

1. Design a passive shunt compensator (a combination of a capacitor and/or an inductor) system for a single-phase AC supply of 230 V at 50 Hz with a feeder (source) impedance of 0.75 Ω resistance and

3.75 Ω inductive reactance after which (a) a single-phase 5 kVA load (at rated voltage) of varying power factor from 0.5 lagging to 0.5 leading is connected at PCC and (b) a single-phase 0.8 lagging power factor load of varying kVA from 0.5 to 5 kVA (at rated voltage) is connected at PCC. Calculate the current in the shunt compensator, the voltage drop across the source impedance, and the voltage across the load. If a shunt compensator consisting of lossless passive elements (L and/or C) is used to raise the voltage to the input voltage (230 V), calculate the values of compensator elements (in farads or henries), its kVA rating, and losses in source impedance under all load conditions.

2. Design a passive series compensator (a combination of a capacitor and/or an inductor) system for a single-phase AC supply having AC mains voltage of 220 V at 50 Hz and a feeder (source) impedance of 4.0 Ω inductive reactance after which (a) a single-phase 5 kVA load (at rated voltage) of varying power factor from 0.5 lagging to 0.5 leading is connected at PCC and (b) a single-phase 0.8 lagging power factor load of varying kVA from 0.5 to 5 kVA (at rated voltage) is connected at PCC. Calculate the voltage drop across the source impedance, the voltage drop across the compensator, and the voltage across the load. If a series compensator consisting of lossless passive elements (L and/or C) is used to raise the voltage to the input voltage (220 V), calculate the values of compensator elements (in farads or henries), its kVA rating, and losses in source impedance under all load conditions.

3. Design a passive shunt compensator (a combination of a capacitor and/or an inductor) system for a single-phase AC supply having AC mains voltage of 230 V at 50 Hz and a feeder (source) impedance of 0.75 Ω resistance and 3.75 Ω inductive reactance after which (a) a single-phase 5 kVA load (at rated voltage) of varying power factor from 0.5 lagging to 0.5 leading is connected at PCC and (b) a single-phase 0.8 lagging power factor load of varying kVA from 0.5 to 5 kVA (at rated voltage) is connected at PCC. Calculate the current in and voltage across the shunt compensator, the voltage drop across the source impedance, and the voltage across the load. If a shunt compensator consisting of lossless passive elements (L and/or C) is used to improve the power factor to unity at the supply, calculate the values of compensator elements (in farads or henries), its kVA rating, and losses in source impedance under all load conditions.

4. Design a passive series compensator (a combination of a capacitor and/or an inductor) system for a single-phase AC supply having AC mains voltage of 230 V at 50 Hz and a feeder (source) impedance of 3.75 Ω inductive reactance after which (a) a single-phase 5 kVA load (at rated voltage) of varying power factor from 0.5 lagging to 0.5 leading is connected at PCC and (b) a single-phase 0.8 lagging power factor load of varying kVA from 0.5 to 5 kVA (at rated voltage) is connected at PCC. Calculate the current in and voltage across the series compensator, the voltage drop across the source impedance, and the voltage across the load. If a series compensator consisting of lossless passive elements (L and/or C) is used to improve the power factor to unity at the supply, calculate the values of compensator elements (in farads or henries), its kVA rating, and losses in source impedance under all load conditions.

5. Design a three-phase three-wire shunt compensator consisting of lossless passive elements (L and/or C) for a 415 V, 50 Hz system to provide load balancing and power factor correction to unity of a single-phase load varying from 1.0 to 30 kVA (with a step of 1 kVA) at 0.8 lagging power factor connected between two supply lines. Calculate (a) supply line currents, (b) compensator currents, (c) the values of compensator elements (in farads or henries), and (d) its kVA rating.

6. Design a three-phase three-wire shunt compensator consisting of lossless passive elements (L and/or C) for a three-phase three-wire AC supply system having a line voltage of 415 V at 50 Hz and a feeder (source) impedance of 1.0 Ω resistance and 5.0 Ω inductive reactance per phase to provide load balancing and voltage regulation to rated voltage of a single-phase load varying from 1.0 to 30 kVA (with a step of 1 kVA) at 0.8 lagging power factor connected between two supply lines. Calculate the voltages drop across source impedance and the voltage across the load. If a shunt compensator consisting of lossless passive elements (L and/or C) is used to raise the voltage to the input voltage (415 V), calculate the values of compensator elements (in farads or henries) and its kVA rating.

7. Design a three-phase three-wire shunt compensator consisting of lossless passive elements (L and/or C) for a three-phase three-wire AC supply system having a line voltage of 415 V at 50 Hz and a feeder (source) impedance of 0.5 Ω resistance and 3.0 Ω inductive reactance per phase to regulate the PCC voltage to rated voltage (415 V) after which a balanced delta connected load varying from 1.0 to 50 kVA (with a step of 1 kVA) at 0.8 lagging power factor is connected. Calculate the voltage drop across source impedance and the voltage across the load in each case. If a shunt compensator consisting of lossless passive elements (L and/or C) is used to raise the voltage to the input voltage (415 V), calculate (a) supply line currents, (b) compensator currents, (c) the values of compensator elements (in farads or henries), and (d) its kVA rating.

8. Design a three-phase three-wire series compensator consisting of lossless passive elements (L and/or C) for a three-phase AC supply system having a line–line voltage of 415 V at 50 Hz and a feeder (source) impedance of 3.50 Ω inductive reactance per phase to regulate the PCC voltage to rated voltage (415 V) after which a balanced delta connected load varying from 1.0 to 50 kVA (with a step of 1 kVA) at 0.8 lagging power factor is connected. Calculate the voltage drop across source impedance and the voltage across the load in each case. If a series compensator consisting of lossless passive elements (L and/or C) is used to raise the voltage to the input voltage (415 V), calculate (a) supply line currents, (b) compensator currents, (c) the values of compensator elements (in farads or henries), (d) its kVA rating, and (e) equivalent per-phase resistance (in ohms) of the compensated load.

9. Design a three-phase three-wire series compensator consisting of lossless passive elements (L and/or C) to improve the power factor to unity at the AC mains for a three-phase AC supply system having a line–line voltage of 415 V at 50 Hz and a feeder (source) impedance of 3.0 Ω inductive reactance per phase after which a balanced delta connected load varying from 1.0 to 50 kVA (with a step of 1 kVA) at 0.8 lagging power factor is connected. Calculate the voltage drop across source impedance and the voltage across the load in each case. If a series compensator consisting of lossless passive elements (L and/or C) is used to improve the power factor to unity at the AC mains, calculate (a) supply line currents, (b) compensator currents, (c) the values of compensator elements (in farads or henries), (d) its kVA rating, and (e) equivalent per-phase resistance (in ohms) of the compensated load.

10. Design a three-phase three-wire shunt compensator consisting of lossless passive elements (L and/or C) to provide load balancing and to improve the power factor to unity at the AC mains for a three-phase AC supply system having a line–line voltage of 415 V at 50 Hz and three-phase three-wire unbalanced delta connected loads (1–50 kVA at 0.8 lagging power factor across line ab, 51–100 kVA at 0.8 leading power factor across line bc, and 101–150 kVA at 0.8 lagging power factor across line ca). It is to be realized as a balanced unity power factor load on the three-phase supply system using a shunt compensator consisting of lossless passive elements (L and/or C). Compute (a) supply line currents, (b) compensator currents, (c) the values of compensator elements (in farads or henries), (d) its kVA rating, and (e) equivalent per-phase resistance (in ohms) of the compensated loads.

11. Design a three-phase three-wire shunt compensator consisting of lossless passive elements (L and/or C) to provide load balancing and to regulate the rated voltage for a three-phase AC supply system. It has a line–line voltage of 415 V at 50 Hz and a feeder (source) impedance of 0.5 Ω resistance and 3.5 Ω inductive reactance per phase after which it has three-phase three-wire unbalanced delta connected loads (1–50 kVA at 0.8 lagging power factor across line ab, 51–100 kVA at 0.8 leading power factor across line bc, and 101–150 kVA at 0.8 lagging power factor across line ca). It is to be realized as a balanced load on the three-phase supply system with voltage regulation using a shunt compensator consisting of lossless passive elements (L and/or C). Compute (a) supply line currents, (b) compensator currents, (c) the values of compensator elements (in farads or henries), (d) its kVA rating, and (e) equivalent per-phase resistance (in ohms) of the compensated loads.

12. Design a three-phase four-wire shunt compensator consisting of lossless passive elements (L and/or C) to eliminate neutral current of a three-phase four-wire 415 V (line–line), 50 Hz AC supply system, which has a single-phase load varying from 1 to 25 kW at unity power factor connected across line and neutral.

Calculate (a) supply line currents, (b) compensator currents, (c) the values of compensator elements (in farads or henries), and (d) its kVA rating.

13. Design a three-phase four-wire shunt compensator consisting of lossless passive elements (L and/or C) for a 415 V, 50 Hz supply system to provide load balancing of a single-phase load varying from 1 to 15 kW at unity power factor connected across line and neutral. Calculate (a) supply line currents, (b) compensator currents, (c) the values of compensator elements (in farads or henries), and (d) its kVA rating.

14. Design a three-phase four-wire shunt compensator consisting of lossless passive elements (L and/or C) for a 415 V, 50 Hz supply system to provide load balancing of two single-phase loads varying from 1 to 15 kW at unity power factor each connected across two lines and neutral. Calculate (a) supply line currents, (b) compensator currents, (c) the values of compensator elements (in farads or henries), (d) its kVA rating, and (e) neutral currents of the load, AC mains, and the compensator.

15. Design a three-phase four-wire shunt compensator consisting of lossless passive elements (L and/or C) for a 415 V, 50 Hz supply system to provide load balancing and power factor correction of two single-phase loads varying from 1 to 15 kVA at 0.8 lagging power factor each connected across two lines and neutral. Calculate (a) supply line currents, (b) compensator currents, (c) the values of compensator elements (in farads or henries), (d) its kVA rating, and (e) neutral currents of the load, AC mains, and the compensator.

16. Design a three-phase four-wire shunt compensator consisting of lossless passive elements (L and/or C) for a 415 V, 50 Hz supply system to provide load balancing and reactive power compensation of a single-phase load varying from 1 to 30 kVA at 0.8 lagging power factor connected across line and neutral. Calculate (a) supply line currents, (b) compensator currents, (c) the values of compensator elements (in farads or henries), and (d) its kVA rating.

17. Design a three-phase four-wire shunt compensator consisting of lossless passive elements (L and/or C) for a 415 V, 50 Hz system to provide load balancing and power factor correction to unity of a single-phase load varying from 1.0 to 30 kVA (with a step of 1 kVA) at 0.8 lagging power factor connected between line and neutral terminal. Calculate (a) supply line currents, (b) compensator currents, (c) the values of compensator elements (in farads or henries), (d) its kVA rating, and (e) equivalent per-phase resistance (in ohms) of the compensated loads.

18. A three-phase four-wire unbalanced load varying from 1.0 to 30 kVA (with a step of 1 kVA) at 0.8 lagging power factor is connected between phase a and neutral with an input AC line–line voltage of 415 V at 50 Hz. It is to be realized as a three-phase balanced unity power factor load on the three-phase supply system using a shunt compensator consisting of lossless passive elements (L and/or C). Calculate (a) supply line currents, (b) compensator currents, (c) the values of compensator elements (in farads or henries), (d) its kVA rating, and (e) equivalent per-phase resistance (in ohms) of the compensated load.

19. Design a three-phase four-wire shunt compensator consisting of lossless passive elements (L and/or C) to provide load balancing and to improve the power factor to unity at the AC mains for a three-phase AC supply system having a line voltage of 415 V at 50 Hz and three-phase four-wire unbalanced star connected loads (1–50 kVA at 0.8 lagging power factor across line an, 51–100 kVA at 0.8 leading power factor across line bn, and 101–150 kVA at 0.8 lagging power factor across line cn, with a step of 1 kVA). It is to be realized as a balanced unity power factor load on the three-phase supply system using a shunt compensator consisting of lossless passive elements (L and/or C). Compute (a) supply line currents, (b) compensator currents, (c) the values of compensator elements (in farads or henries), (d) its kVA rating, (e) equivalent per-phase resistance (in ohms) of the compensated loads, and (f) load neutral currents.

20. Design a three-phase four-wire shunt compensator consisting of lossless passive elements (L and/or C) to provide load balancing and voltage regulation for a three-phase AC supply system having a line voltage of 415 V at 50 Hz and a feeder (source) impedance of 0.5 Ω resistance and 3.50 Ω inductive reactance per phase and three-phase four-wire unbalanced star connected loads (1–50 kVA at 0.8

lagging power factor across line an, 51–100 kVA at 0.8 leading power factor across line bn, and 101–150 kVA at 0.8 lagging power factor across line cn, with a step of 1 kVA). It is to be realized as a balanced load with rated voltage across the loads using a shunt compensator consisting of lossless passive elements (L and/or C). Compute (a) supply line currents, (b) compensator currents, (c) the values of compensator elements (in farads or henries), (d) its kVA rating, (e) equivalent per-phase resistance (in ohms) of the compensated loads, and (f) load neutral currents.

21. A three-phase AC supply has a line voltage of 415 V at 50 Hz and a feeder (source) impedance of 1.0 Ω resistance and 3.0 Ω inductive reactance per phase. It has three-phase four-wire unbalanced star connected loads (1–25 kVA at 0.8 lagging power factor across line an, 1–25 kVA at 0.8 leading power factor across line bn, and 26–50 kVA at 0.8 lagging power factor across line cn, with a step of 1 kVA). Calculate (a) the voltage drop across the source impedance and (b) the voltage across the load. If a shunt compensator consisting of lossless passive elements (L and/or C) is used (c) to balance at UPF and then (d) to raise the voltage to the input voltage (415 V), calculate the compensator line currents and its kVA rating in both cases.

22. A three-phase AC supply has a line–line voltage of 415 V at 50 Hz and a feeder (source) impedance of 0.75 Ω resistance and 3.50 Ω inductive reactance per phase after which a balanced star connected load varying from 25 to 50 kVA (with a step of 1 kVA) at 0.8 lagging power factor is connected. Calculate the voltage drop across the source impedance and the voltage across the load. If a shunt compensator consisting of lossless passive elements (L and/or C) is used to raise the voltage to the input voltage (415 V), calculate supply currents, load currents, compensator line currents, and its kVA rating.

23. A three-phase AC supply has a line voltage of 415 V at 50 Hz and a feeder (source) impedance of 1.0 Ω resistance and 4.0 Ω inductive reactance per phase after which a balanced delta connected load varying from 50 to 100 kVA (with a step of 1 kVA) at 0.8 lagging power factor is connected. Calculate the voltage drop across the source impedance and the voltage across the load. If a shunt compensator consisting of lossless passive elements (L and/or C) is used to raise the voltage to the input voltage (415 V), calculate the values of compensator elements (in farads or henries), supply currents, load currents, compensator currents, and its kVA rating.

24. A four-wire unbalanced star connected load (1–25 kVA at 0.8 lagging power factor across line an, 1–25 kVA at 0.8 leading power factor across line bn, and 26–50 kVA at 0.8 leading power factor across line cn, with a step of 1 kVA) is fed from an AC supply with an input line–line voltage of 415 V at 50 Hz. It is to be realized as a three-phase balanced unity power factor load on the three-phase supply system using a shunt compensator consisting of lossless passive elements (L and/or C). Calculate (a) the values of compensator elements (in farads and henries), (b) equivalent per-phase resistance (in ohms) of the compensated load, (c) supply currents, and (d) load currents.

25. A four-wire unbalanced star connected load (1–25 kVA at 0.8 lagging power factor across line an, 1–25 kVA at 0.8 leading power factor across line bn, and 26–50 kVA at 0.8 lagging power factor across line cn, with a step of 1 kVA) is fed from an AC supply with an input line–line voltage of 415 V at 50 Hz and a feeder (source) impedance of 1.0 Ω resistance and 4.0 Ω inductive reactance per phase. It is to be realized as a three-phase balanced load with rated voltage (415 V) across the load using a shunt compensator consisting of lossless passive elements (L and/or C). Calculate (a) the values of compensator elements (in farads and henries), (b) equivalent per-phase resistance (in ohms) of the compensated load, (c) supply currents, (d) load currents, (e) drop in feeder impedance, and (f) load voltages.

References

1. Miller, T.J.E. (1982) *Reactive Power Control in Electric Systems*, John Wiley & Sons, Inc., Toronto, Canada, pp. 32–48.
2. Kazibwe, W.E. and Sendaula, M.H. (1993) *Electrical Power Quality Control Techniques*, Van Nostrand Reinhold Company.

3. Ghosh, A. and Ledwich, G. (2002) *Power Quality Enhancement Using Custom Power Devices*, Kluwer Academic Publishers, London.
4. Acha, E., Ageligis, V.G., Anaya-Lara, O., and Miller, T.J.E. (2002) *Power Electronic Control in Electrical Systems*, Newnes Power Engineering Series, Newnes Publishers, Oxford, UK.
5. Sastry Vedam, R. and Sarma, M.S. (2008) *Power Quality: VAR Compensation in Power Systems*, CRC Press, New York.
6. Gyugyi, L., Otto, R.A., and Putman, T.H. (1978) Principles and applications of static, thyristor-controlled shunt compensators. *IEEE Transactions on Power Apparatus and Systems*, **97**(5), 1935–1945.
7. Gyugyi, L. (1979) Reactive power generation and control by thyristor circuits. *IEEE Transactions on Industry Applications*, **15**(5), 521–532.
8. Tremayne, J.F. (1983) Impedance and phase balancing of mains-frequency induction furnaces. *IEEE Transactions on Signal Processing*, **130**, (3), 161–170.
9. Vasu, E., Rao, V.V.B., and Sankaran, P. (1985) An optimization criterion for three-phase reactive power compensation. *IEEE Transactions on Power Apparatus and Systems*, **104**(11), 3216–3220.
10. Kneschke, T.A. (1985) Control of utility system unbalance caused by single-phase electric traction. *IEEE Transactions on Industry Applications*, **21**(6), 1559–1570.
11. Cox, M.D. and Mirbod, A. (1986) A new static VAR compensator for an arc furnace. *IEEE Transactions on Power Systems*, **1**(3), 110–119.
12. Thukaram, D., Ramakrishna Iyengar, B.S., and Parthasarathy, K. (1986) An algorithm for optimum control of static VAR compensators to meet the phase-wise unbalanced reactive power demands. *Journal of Electrical Power System Research*, **11**, 129–137.
13. El-Sadek, M.Z. (1987) Balancing of unbalanced loads using static VAR compensators. *Journal of Electric Power System Research*, **12**, 137–148.
14. Le, T.N. (1987) Flicker reduction performance of static VAR compensators with arc furnaces. Second EPE Conference, pp. 1259–1263.
15. Gueth, G., Enstedt, P., Rey, A., and Menzies, R.W. (1987) Individual phase control of a static compensator for load compensation and voltage balancing and regulation. *IEEE Transactions on Power Systems*, **2**(4), 898–904.
16. Gyugyi, L. (1988) Power electronics in electric utilities: static VAR compensators. *Proceedings of the IEEE*, **76**(4), 483–494.
17. Lin, C.E., Chen, T.C., and Huang, C.L. (1988) Optimal control of a static VAR compensator for minimization of line loss. *Electric Power Systems Research*, **15**, 51–61.
18. Lin, C.E., Chen, T.C., and Huang, C.L. (1989) A real-time calculation method for optimal reactive power compensator. *IEEE Transactions on Power Systems*, **4**(2), 643–652.
19. Czarnecki, L.S. (1989) Reactive and unbalanced current compensation in three-phase asymmetrical circuits under nonsinusoidal conditions. *IEEE Transactions on Instrumentation and Measurement*, **38**(3), 754–759.
20. Baghzouz, Y. and Cox, M.D. (1991) Optimal shunt compensation for unbalanced linear loads with nonsinusoidal supply voltages. *Electric Machines and Power Systems*, **19**, 171–183.
21. Kearly, J., Chikhani, A.Y., Hackam, R. *et al.* (1991) Microprocessor controlled reactive power compensator for loss reduction in radial distribution feeders. *IEEE Transactions on Power Delivery*, **6**(4), 1848–1855.
22. Chen, D.K.C., Lee, Y.S., and Wu, C.J. (1991) Design of an on-line microprocessor based individual phase control of a static VAR compensator. Proceedings of the IEEE International Conference on Advances in Power System Control, Operation, and Management, APSCOM-91, November, Hong Kong, pp. 119–122.
23. Lin, C.E. and Chen, T.C. (1991) Real-time optimal reactive power control of static VAR compensators. *Electrical Power & Energy Systems*, **13**, 103–110.
24. Czarnecki, L.S. (1992) Minimisation of unbalanced and reactive currents in three-phase asymmetrical circuits with nonsinusoidal voltage. *IEE Proceedings B*, **139**(4), 347–354.
25. Willems, J.L. (1993) Current compensation in three-phase power systems. *European Transactions on Electrical Power Engineering*, **3**(1), 61–66.
26. Czarnecki, L.S. (1993) Power factor improvement of three-phase unbalanced loads with nonsinusoidal supply voltage. *European Transactions on Electrical Power Engineering*, **3**(1), 67–74.
27. Lee, S.-Y. and Wu, C.-J. (1993) On-line reactive power compensation schemes for unbalanced three-phase four-wire distribution feeders. *IEEE Transactions on Power Delivery*, **8**(4), 1958–1965.
28. Wu, C.-J., Liaw, C.-M., and Lee, S.-Y. (1994) Microprocessor-based static reactive power compensators for unbalanced loads. *Electric Power Systems Research*, **31**, 51–59.
29. Kern, A., Lux, K.-J., and Teichmann, K. (1994) A load balancing and control scheme to eliminate feedback problems in industrial appliances. Proceedings of the Conference on Electronics and Variable Speed Drives, October 26–28, IEE, Publication No. 399.

30. Kern, A. and Schrtider, G. (1994) A novel approach to power factor control and balancing problems. Proceedings of the 20th International Conference on Industrial Electronics, Control and Instrumentation, IECON '94, vol. **I**, pp. 428–433.
31. Depenbrock, M., Marshall, D.A., and Wyk, J.D.V. (1994) Formulating requirements for a universally applicable power theory as control algorithm in power compensators. *European Transactions on Electrical Power Engineering*, **4**(6), 445–455.
32. Czarnecki, L.S. and Hsu, S.M. (1994) Thyristor controlled susceptances for balancing compensators operated under nonsinusoidal conditions. *IEE Proceedings – Electric Power Applications*, **141**(4), 177–185.
33. Lee, S.-Y., Chang, W.-N., and Wu, C.-J. (1995) A compact algorithm for three-phase three-wire system reactive power compensation and load balancing. Proceedings of the International Conference on Energy Management and Power Delivery, EMPD '95, vol. **1**, pp. 358–363.
34. Czarnecki, L.S., Hsu, S.M., and Chen, G. (1995) Adaptive balancing compensator. *IEEE Transactions on Power Delivery*, **10**, 1663–1669.
35. Anuradha, Singh, B., and Kothari, D.P. (1996) Generalized concepts for balancing three-phase load fed from three-phase supply. Proceedings of the Ninth National Power System Conference, December 19–21, IIT Kanpur.
36. El-Sadek, M.Z. (1998) Static VAR compensation for phase balancing and power factor improvement of single-phase train loads. *Electric Machines and Power Systems*, **26**, 347 361.
37. Singh, B., Saxena, A., and Kothari, D.P. (1998) Power factor correction and load balancing in three-phase distribution systems. Proceedings of the IEEE Region 10 International Conference on Global Connectivity in Energy, Computer, Communication and Control, vol. 2, pp. 479–488.
38. Nikolaenko, V.G. (1998) Optimal balancing of large unbalanced loads using shunt compensators. Proceedings of the Eighth International Conference on Harmonics and Quality of Power, ICHQP '98, October 14–16, Athens, Greece, pp. 537–542.
39. Chen, J.H., Lee, W.J., and Chen, M.S. (1999) Using a static VAR compensator to balance a distribution system. *IEEE Transactions on Industry Applications*, **35**(2), 298–304.
40. Anuradha, Singh, B., and Kothari, D.P. (1999) Generalized concepts for balancing single-phase load fed from three-phase supply. *Electric Machines and Power Systems*, **27**(1), 63–78.
41. Lee, S.Y. and Wu, C.J. (2000) Reactive power compensation and load balancing for unbalanced three-phase four-wire system by an SVC. *IEE Proceedings – Electric Power*, **147**(6), 563–570.
42. Ghosh, A. and Joshi, A. (2000) A new approach to load balancing and power factor correction in power distribution system. *IEEE Transactions on Power Delivery*, **15**(1), 417–422.
43. Dehnavi, G.R., Shayanfar, H.A., Mahdavi, J., and Marami Saran, M. (2001) Some new aspects of design and implementation of TCR for load balancing and power factor correction in distribution system. Proceedings of the IEEE Porto Power Tech Conference, September 10–13, Porto, Portugal.
44. Lee, S.-Y., Wu, C.-J., and Chang, W.-N. (2001) A compact control algorithm for reactive power compensation and load balancing with static VAR compensator. *Electric Power Systems Research*, **58**, 63–70.
45. Talebi, M.A., Kazemi, A., Gholami, A., and Rajabi, M. (2005) Optimal placement of static VAR compensators in distribution feeders for load balancing by genetic algorithm. Proceedings of the 18th International Conference on Electricity Distribution (CIRED), June 6–9, Session 5.
46. Sainz, L., Pedra, J., and Caro, M. (2005) Steinmetz circuit influence on the electric system harmonic response. *IEEE Transactions on Power Delivery*, **20**(2), 1143–1150.
47. Vigneau, P., Destombes, J.-M., and Grünbaum, R. (2006) SVC for load balancing and maintaining of power quality in an island grid feeding a nickel smelter. Proceedings of the 32nd Annual Conference of IEEE Industrial Electronics, IECON 2006, pp. 1981–1986.
48. Pană, A., Băloi, A., and Molnar-Matei, F. (2014) Experimental validation of power mechanism for load balancing using variable susceptances in three-phase four-wire distribution networks. Proceedings of EUROCON 2007 – The International Conference on "Computer as a Tool", September 9–12, Warsaw, pp. 1567–1572.
49. Said, I.K. and Pirouti, M. (2009) Neural network-based load balancing and reactive power control by static VAR compensator. *International Journal of Computer and Electrical Engineering*, **1**(1), 25–31.
50. Jing, B. and Ning-Qiang, J. (2010) Optimization of dynamic VAR compensation for asymmetric loads considering harmonic suppression. Proceedings of the IEEE Asia-Pacific Power and Energy Engineering Conference (APPEEC), pp. 1–5.

4

Active Shunt Compensation

4.1 Introduction

Present-day AC distribution systems are facing a number of power quality problems, especially due to the use of sensitive equipment in most of the industrial, residential, commercial, and traction applications. These power quality problems are classified as voltage and current quality problems in distribution systems. The custom power devices (CPDs), namely, DSTATCOMs (distribution static compensators), DVRs (dynamic voltage restorers), and UPQCs (unified power quality conditioners), are used to mitigate some of the problems depending upon the requirements. Out of these CPDs, DSTATCOMs are extensively used for mitigating the current-based power quality problems. There are a number of current-based power quality problems such as poor power factor, or poor voltage regulation, unbalanced currents, and increased neutral current. Therefore, depending upon the problems, the configuration of the DSTATCOM is selected in the practice. With the objective of mitigating the current-based power quality problems especially in distribution systems, this chapter focuses on the configurations, design, control algorithms, modeling, and illustrative examples of DSTATCOMs.

These problems further aggravate in the presence of harmonics either in the voltage or in the currents. The shunt active compensators are also reported with some modifications as cost-effective shunt active power filters to eliminate harmonic currents in nonlinear loads. Of course, the main objective of shunt active power filters has been to eliminate harmonic currents at the PCC (point of common coupling) voltage normally created by nonlinear loads. In view of the additional applications of the shunt compensators as shunt active power filters, a separate chapter (Chapter 9) is devoted to the shunt active power filters to deal with the elimination of harmonics in currents along with some specific applications and case studies.

4.2 State of the Art on DSTATCOMs

The DSTATCOM technology is now a mature technology for providing reactive power compensation, load balancing, and/or neutral current and harmonic current compensation (if required) in AC distribution networks. It has evolved in the past quarter century with development in terms of varying configurations, control strategies, and solid-state devices. These compensating devices are also used to regulate the terminal voltage, suppress voltage flicker, and improve voltage balance in three-phase systems. These objectives are achieved either individually or in combination depending upon the requirements and the control strategy and configuration that need to be selected appropriately. This section describes the history of development and the current status of the DSTATCOM technology.

In AC distribution systems, current-based power quality problems have been faced for a long time in terms of poor power factor, poor voltage regulation, load unbalancing, and enhanced neutral current.

Power Quality Problems and Mitigation Techniques, First Edition.
Bhim Singh, Ambrish Chandra and Kamal Al-Haddad.

Classical technology of using power capacitors and static VAR compensators using TCRs (thyristor-controlled reactors) and TSCs (thyristor-switched capacitors) has been used to mitigate some of these power quality problems. However, DSTATCOM technology is considered the best technology to mitigate all the current-based power quality problems.

DSTATCOMs are basically categorized into three types, namely, single-phase two-wire, three-phase three-wire, and three-phase four-wire configurations, to meet the requirements of three types of consumer loads on supply systems. Single-phase loads such as domestic lights and ovens, TVs, computer power supplies, air conditioners, laser printers, and Xerox machines cause power quality problems. Single-phase two-wire DSTATCOMs have been investigated in varying configurations and control strategies to meet the needs of single-phase systems. Starting from 1984, many configurations have been developed and commercialized for many applications. Both current source converters (CSCs) with inductive energy storage and voltage source converters (VSCs) with capacitive energy storage are used to develop single-phase DSTATCOMs.

A major amount of AC power is consumed by three-phase loads. A substantial work has been reported on three-phase three-wire DSTATCOMs, starting from 1984. Many configurations and control strategies such as instantaneous reactive power theory, synchronous frame d–q theory, and synchronous detection method are used in the development of three-phase DSTATCOMs.

The problem of increased neutral current in addition to poor power factor and load unbalancing is observed in three-phase four-wire systems mainly due to unbalanced loads such as computer power supplies and fluorescent lighting. The problems of neutral current and unbalanced load currents can be resolved by using four-wire DSTATCOMs in four-wire distribution systems, which cause reduction of neutral current, load balancing, reactive power compensation, and/or harmonic compensation (if required).

The problems of reactive power and load unbalancing have been recognized long ago and they have got aggravated in the presence of nonlinear loads. Many publications are reported on solid-state compensators for voltage flicker, reactive power, and balancing the reactive loads such as arc furnaces and traction loads. Many more terminologies such as static VAR compensators, static flicker compensators, and static VAR generators have been used in the literature.

One of the major factors in advancing the DSTATCOM technology is the advent of fast, self-commutating solid-state devices. In the initial stages, BJTs (bipolar junction transistors) and power MOSFETs (metal-oxide semiconductor field-effect transistors) have been used to develop DSTATCOMs; later, SITs (static induction thyristors) and GTOs (gate turn-off thyristors) have been employed to develop DSTATCOMs. With the introduction of IGBTs (insulated gate bipolar transistors), the DSTATCOM technology has got a real boost and at present it is considered as an ideal solid-state device for DSTATCOMs. The improved sensor technology, especially Hall effect current and voltage sensors, has also contributed to the enhanced performance of DSTATCOMs. The availability of Hall effect sensors and isolation amplifiers at reasonable cost and with adequate ratings has improved the performance of DSTATCOMs.

The next breakthrough in DSTATCOM development has resulted from the microelectronics revolution. Starting from the use of discrete analog and digital components, the progression has been to microprocessors, microcontrollers, and DSPs (digital signal processors). Now it is possible to implement complex algorithms online for the control of DSTATCOMs at a reasonable cost. This development in DSPs has made it possible to use different control algorithms such as PI (proportional–integral) controller, variable structure control, fuzzy logic control, and neural network control for improving the dynamic and steady-state performance of DSTATCOMs. With these improvements, the DSTATCOMs are capable of providing fast corrective action even with dynamically changing loads such as furnaces and traction.

4.3 Classification of DSTATCOMs

DSTATCOMs can be classified based on the type of converter used, topology, and the number of phases. The converter used in the DSTATCOM can be either a current source converter or a voltage source converter. Different topologies of DSTATCOMs can be realized by using transformers and various

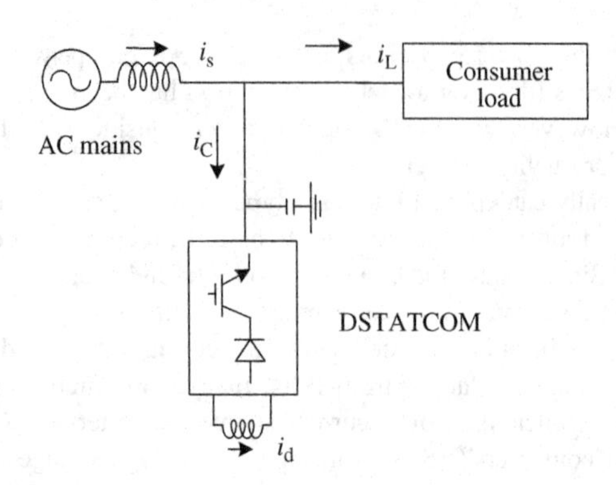

Figure 4.1 A CSC-based DSTATCOM

circuits of VSCs. The third classification is based on the number of phases, namely, single-phase two-wire, three-phase three-wire, and three-phase four-wire systems.

4.3.1 Converter-Based Classification

Two types of converters are used to develop DSTATCOMs. Figure 4.1 shows a DSTATCOM using a CSC bridge. A diode is used in series with the self-commutating device (IGBT) for reverse voltage blocking. However, GTO-based DSTATCOM configurations do not need the series diode, but they have restricted frequency of switching. They are considered sufficiently reliable, but have high losses and require high values of parallel AC power capacitors. Moreover, they cannot be used in multilevel or multistep modes to improve the performance of DSTATCOMs in higher power ratings.

The other converter used in a DSTATCOM is a voltage source converter shown in Figure 4.2. It has a self-supporting DC voltage bus with a large DC capacitor. It is more widely used because it is light, cheap, and expandable to multilevel and multistep versions, to enhance the performance with lower switching frequencies.

4.3.2 Topology-Based Classification

DSTATCOMs can also be classified based on the topology, for example, VSCs without transformers, VSCs with non-isolated transformers, and VSCs with isolated transformers. DSTATCOMs are also used as advanced static VAR generators (STATCOMs) in the power system network for stabilizing and improving the voltage profile. Therefore, a large number of circuits of DSTATCOMs with and without transformers are evolved for meeting the specific requirements of the applications.

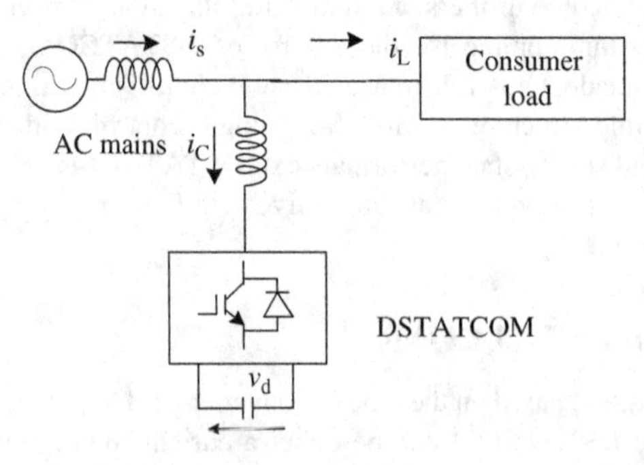

Figure 4.2 A VSC-based DSTATCOM

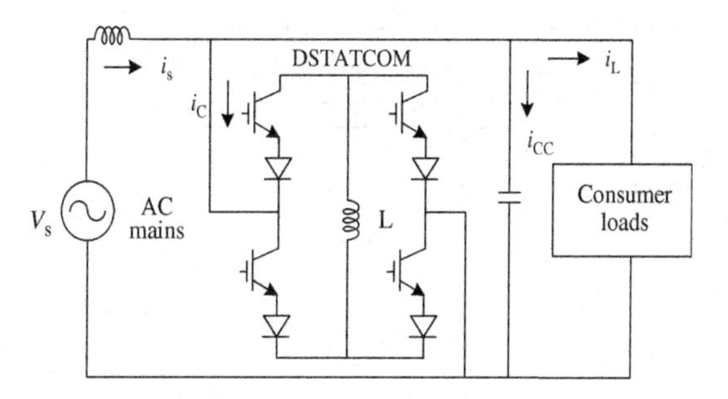

Figure 4.3 A two-wire DSTATCOM with a CSC

4.3.3 Supply System-Based Classification

This classification of DSTATCOMs is based on the supply and/or the load system, for example, single-phase two-wire, three-phase three-wire, and three-phase four-wire systems. There are many varying loads such as domestic appliances connected to single-phase supply systems. Some three-phase loads are without neutral terminals, such as traction, furnaces, and ASDs (adjustable speed drives) fed from three-wire supply systems. There are many single-phase loads distributed on three-phase four-wire supply systems, such as computers and commercial lighting. Hence, DSTATCOMs may also be classified accordingly as two-wire, three-wire, and four-wire DSTATCOMs.

4.3.3.1 Two-Wire DSTATCOMs

Two-wire (single-phase) DSTATCOMs are used in both converter configurations, a CSC bridge with inductive energy storage elements and a VSC bridge with capacitive DC bus energy storage elements, to form two-wire DSTATCOM circuits.

Figure 4.3 shows a configuration of a DSTATCOM with a CSC bridge using inductive energy storage elements. A similar configuration based on a VSC bridge with capacitive energy storage at its DC bus is obtained by considering only two wires (phase and neutral terminals) as shown in Figure 4.4.

4.3.3.2 Three-Wire DSTATCOMs

There are various configurations of capacitor-supported DSTATCOMs based on the type of VSC used and auxiliary circuits. The classification of three-phase three-wire DSTATCOMs is shown in Figure 4.5, consisting of isolated and non-isolated VSC-based topologies of DSTATCOMs. The non-isolated configurations include three-leg VSC-based DSTATCOMs and two-leg VSC-based DSTATCOMs; these circuit configurations are shown in Figures 4.6 and 4.7, respectively. The two-leg VSC-based DSTATCOM has the advantage that it requires only four switching devices, but there are two capacitors

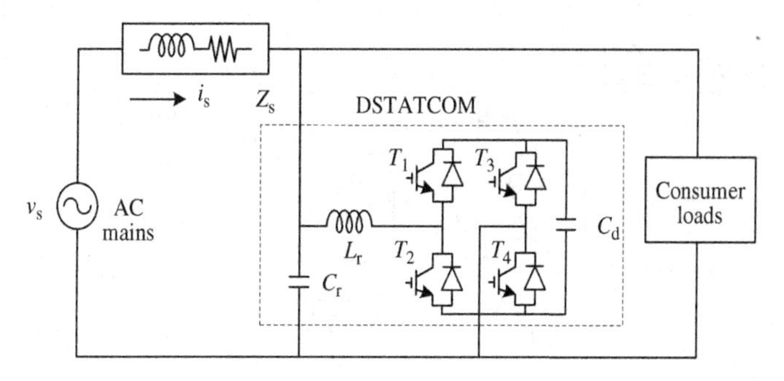

Figure 4.4 A two-wire DSTATCOM with a VSC

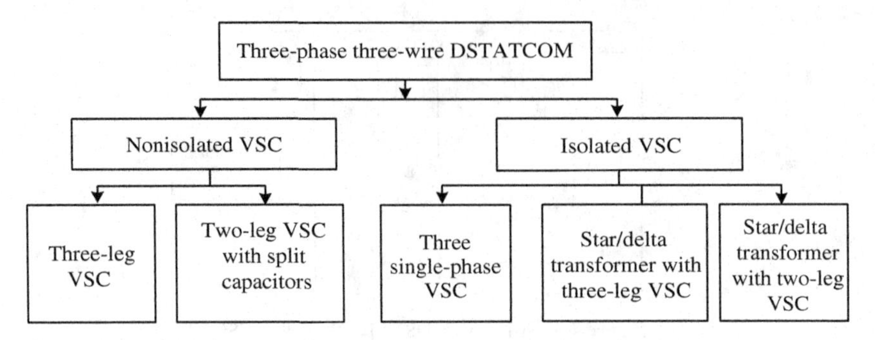

Figure 4.5 Topology classification of three-phase three-wire DSTATCOMs

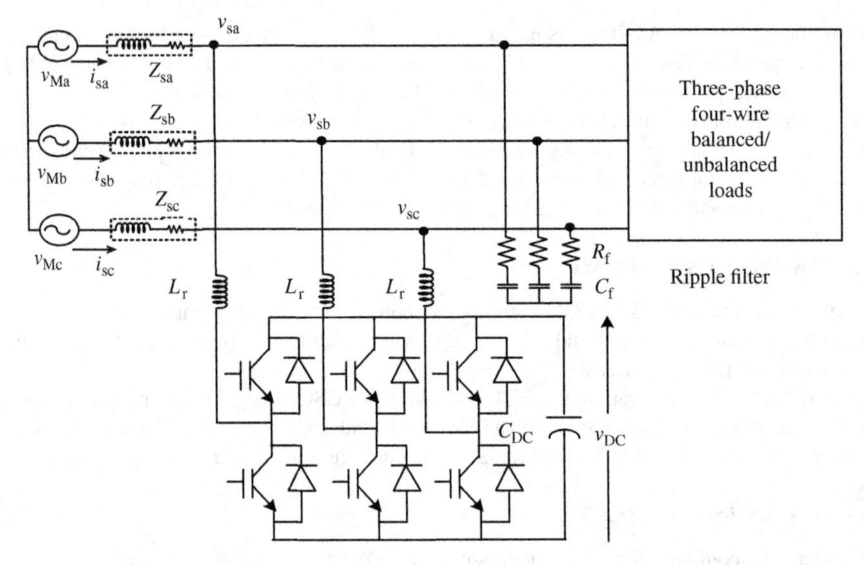

Figure 4.6 A three-leg VSC-based three-phase three-wire DSTATCOM

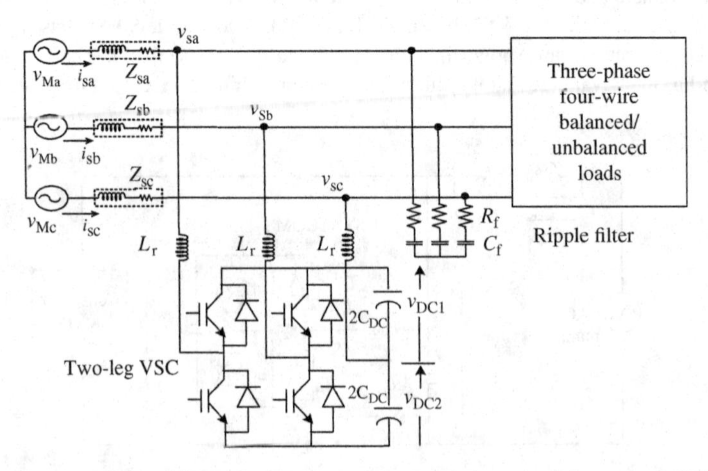

Figure 4.7 An H-bridge VSC and midpoint capacitor-based three-phase three-wire DSTATCOM

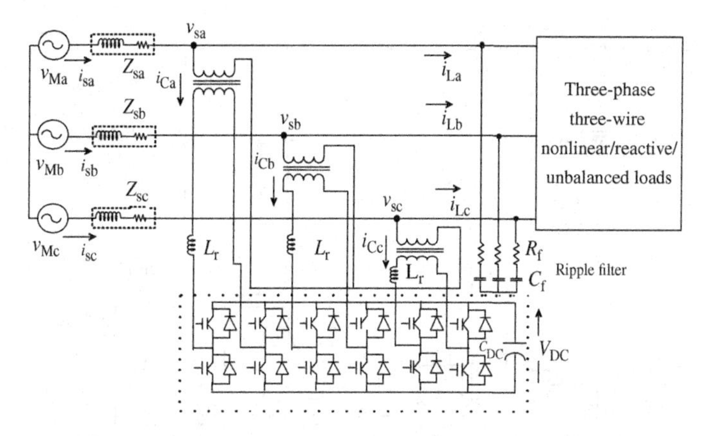

Figure 4.8 A three single-phase VSC-based three-phase three-wire DSTATCOM

connected in series and the total DC capacitor voltage is twice the DC bus voltage of the three-leg VSC topology. The isolated configurations include three single-phase VSC-based DSTATCOMs, three-leg VSC-based DSTATCOMs, and two-leg VSC-based DSTATCOMs; these configurations are shown in Figures 4.8–4.10, respectively. The advantage of the isolated VSC-based DSTATCOM topology is that the voltage rating of the VSC can be optimally designed as there is an interfacing transformer. Three single-phase VSC-based DSTATCOMs require 12 semiconductor switches, whereas in three-leg VSC-based DSTATCOMs there are only 6 switches. However, two-leg VSC-based DSTATCOMs require only four switches.

4.3.3.3 Four-Wire DSTATCOMs

In a three-phase four-wire distribution system, there are three-phase loads and single-phase loads depending upon the consumers' demands. This results in severe burden of unbalanced currents along with the neutral current on the distribution feeder. To prevent the unbalanced currents from being drawn from the distribution bus, a shunt compensator, also called DSTATCOM, can be used. It ensures that the

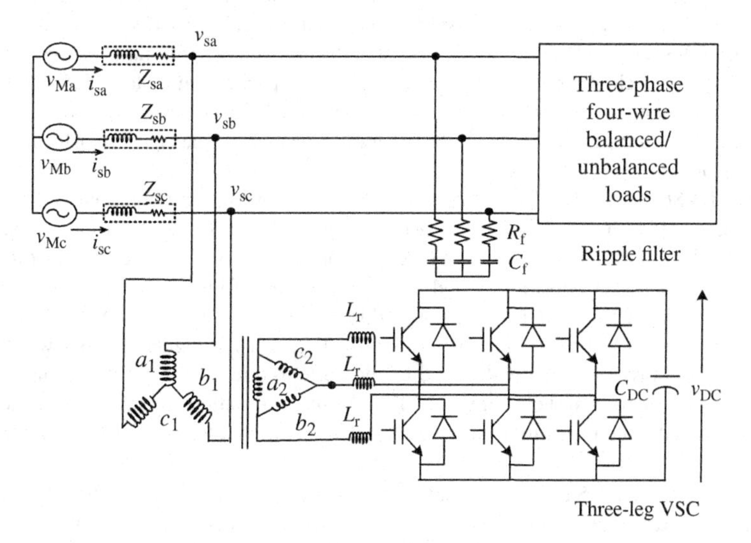

Figure 4.9 An isolated three-leg VSC-based three-phase three-wire DSTATCOM

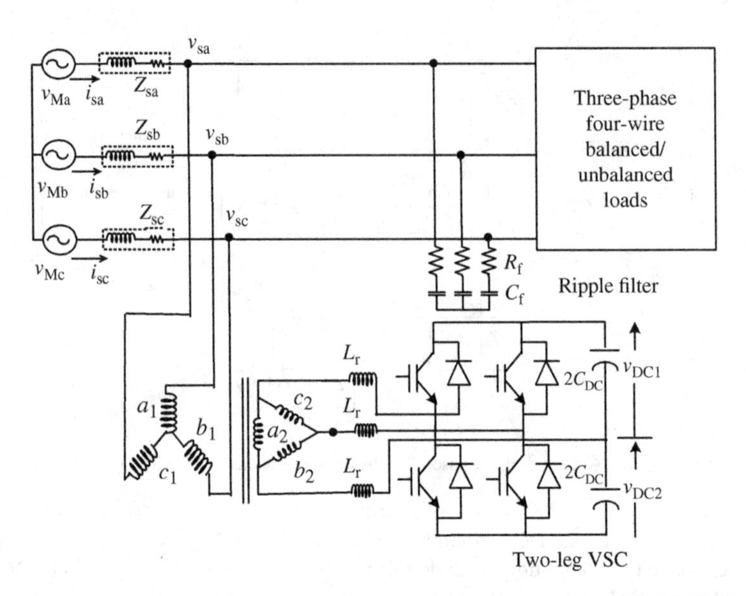

Figure 4.10 An isolated H-bridge VSC and midpoint capacitor-based DSTATCOM

currents drawn from the distribution bus are balanced and sinusoidal and, moreover, the neutral current is compensated.

A DSTATCOM is a fast-response, solid-state power controller that provides power quality improvements at the point of connection to the utility distribution feeder. It is the most important controller for distribution networks. It has been widely used to precisely regulate the system voltage and/or for load compensation. It can exchange both active and reactive powers with the distribution system by varying the amplitude and phase angle of the voltage of the VSC with respect to the PCC voltage, if an energy storage system (ESS) is included into the DC bus. However, a capacitor-supported DSTATCOM is preferred for power quality improvement in the currents, such as reactive power compensation for unity power factor or voltage regulation at PCC, load balancing, and neutral current compensation.

The classification of three-phase four-wire DSTATCOM topologies is shown in Figure 4.11, based on the type of VSC used. They are mainly classified as non-isolated and isolated VSC-based DSTATCOMs. The non-isolated VSC-based DSTATCOMs consist of the following configurations: four-leg VSC, three-leg VSC with split capacitors, three-leg VSC with three DC capacitors, three-leg VSC with transformers, and two-leg VSC with transformers. The transformers used are a zigzag transformer, a star/delta transformer, a Scott transformer, a T-connected transformer, a star/hexagon transformer, and a star/polygon transformer.

The isolated VSC-based DSTATCOMs consist of the following configurations: three single-phase VSCs, three-leg VSC with transformers, and two-leg VSC with transformers. Various transformers used for isolation are a zigzag transformer, a star/delta transformer, a T-connected transformer, a Scott transformer, a star/hexagon transformer, and a star/polygon transformer.

The schematic diagram of a four-leg VSC-based three-phase four-wire DSTATCOM connected to a three-phase four-wire distribution system is shown in Figure 4.12. Figure 4.13 shows the schematic diagram of a three single-phase VSC-based three-phase four-wire DSTATCOM connected to a three-phase four-wire distribution system. Figure 4.14 shows the schematic diagram of a three-leg VSC with split capacitor-based three-phase four-wire DSTATCOM connected to a three-phase four-wire distribution system.

Three-phase four-wire DSTATCOM configurations based on non-isolated three-leg VSCs with a zigzag transformer, a star/delta transformer, a T-connected transformer, a star/hexagon transformer, a star/polygon transformer, and a Scott transformer are shown in Figures 4.15–4.20, respectively. Similarly,

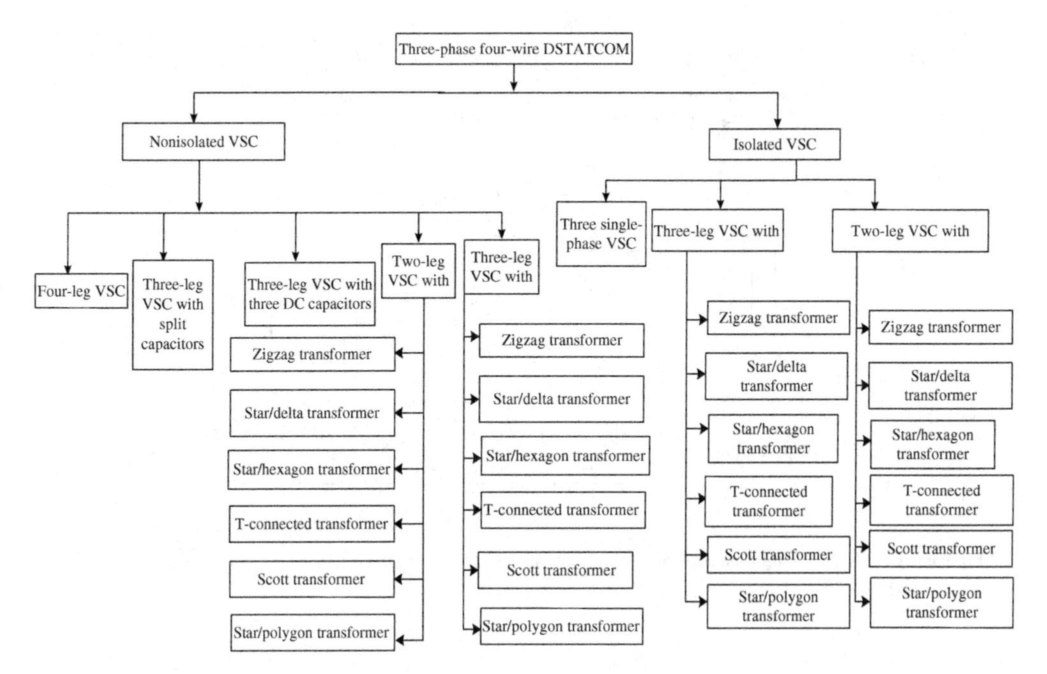

Figure 4.11 Topology classification of three-phase four-wire DSTATCOMs

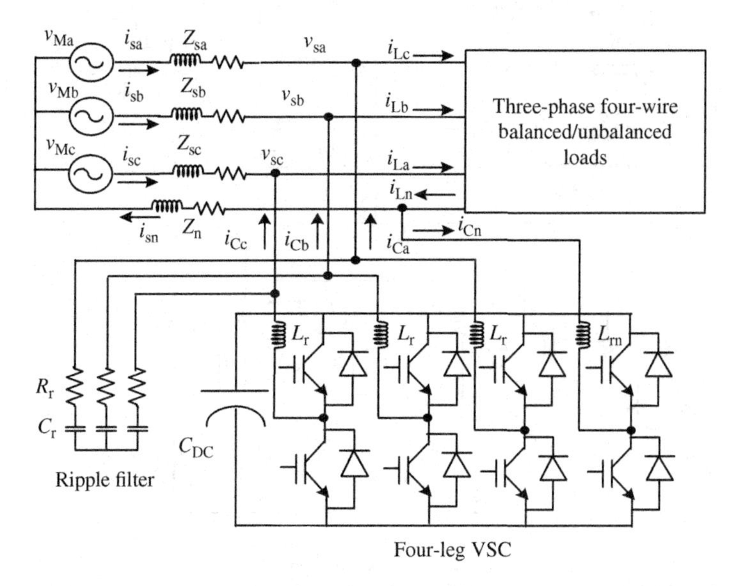

Figure 4.12 A four-leg VSC-based three-phase four-wire DSTATCOM connected to a three-phase four-wire system

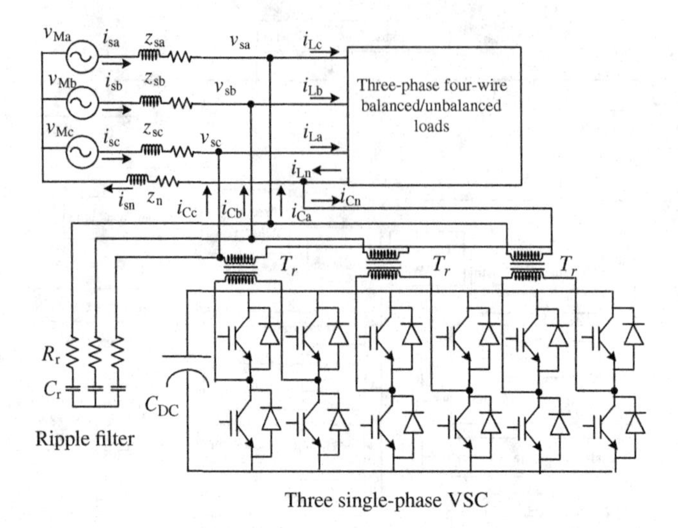

Figure 4.13 A three single-phase VSC-based three-phase four-wire DSTATCOM connected to a three-phase four-wire system

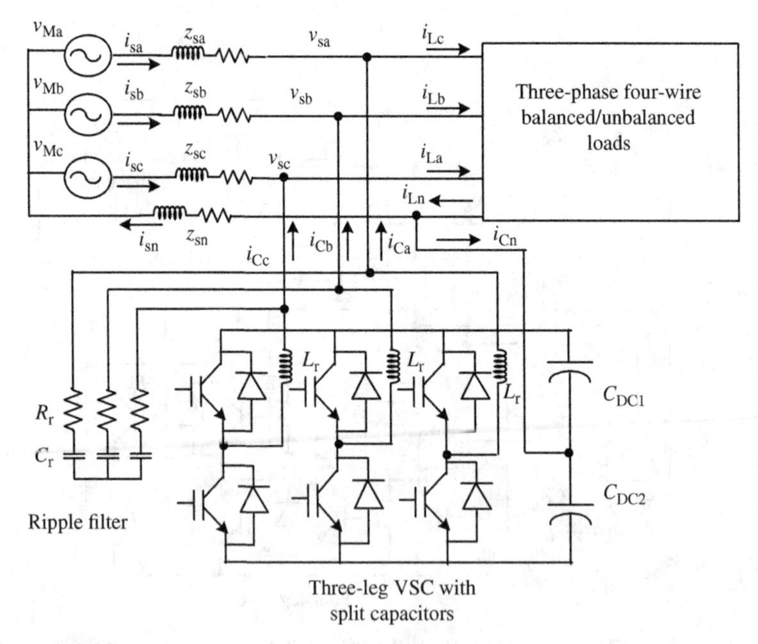

Figure 4.14 A three-leg VSC and split capacitor-based three-phase four-wire DSTATCOM connected to a three-phase four-wire system

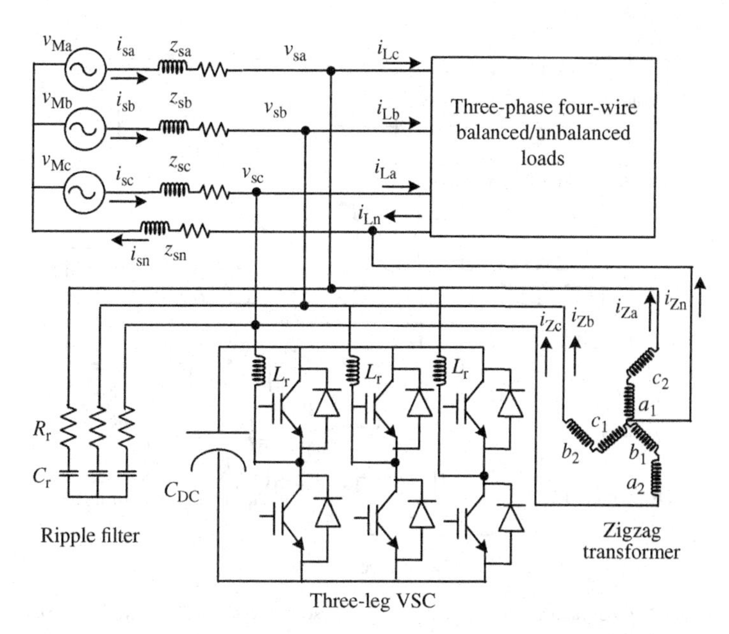

Figure 4.15 A three-leg VSC and zigzag transformer-based three-phase four-wire DSTATCOM connected to a three-phase four-wire system

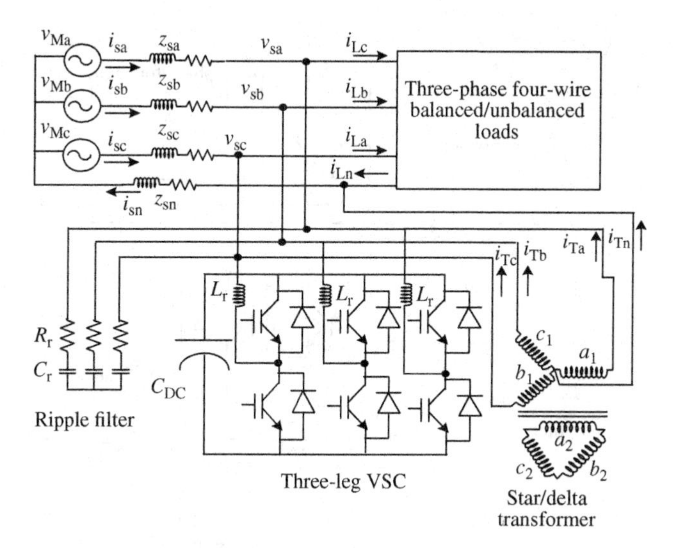

Figure 4.16 A three-leg VSC and star/delta transformer-based three-phase four-wire DSTATCOM connected to a three-phase four-wire system

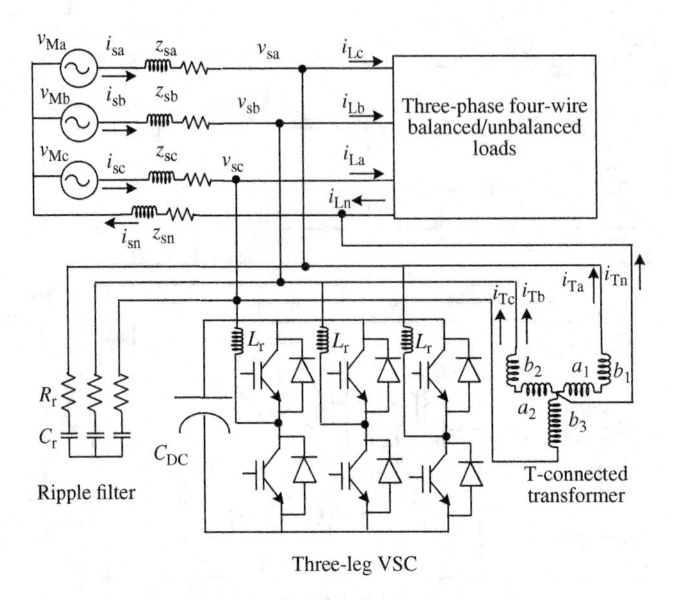

Figure 4.17 A three-leg VSC and T-connected transformer-based three-phase four-wire DSTATCOM connected to a three-phase four-wire system

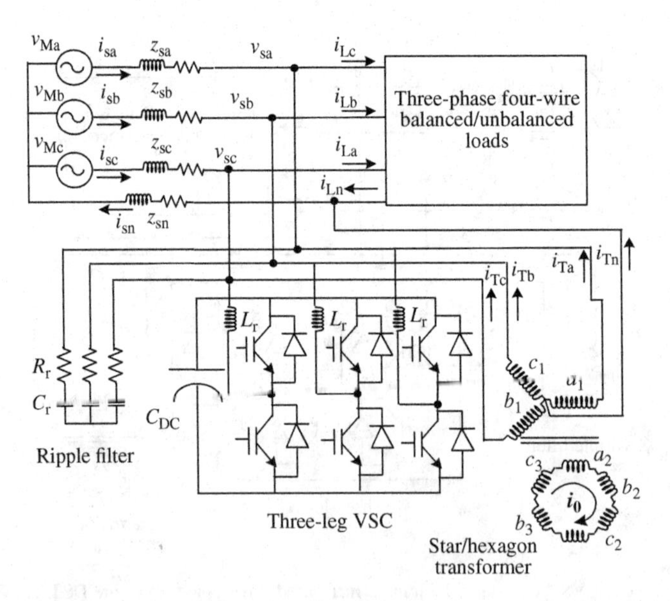

Figure 4.18 A three-leg VSC and star/hexagon transformer-based three-phase four-wire DSTATCOM connected to a three-phase four-wire system

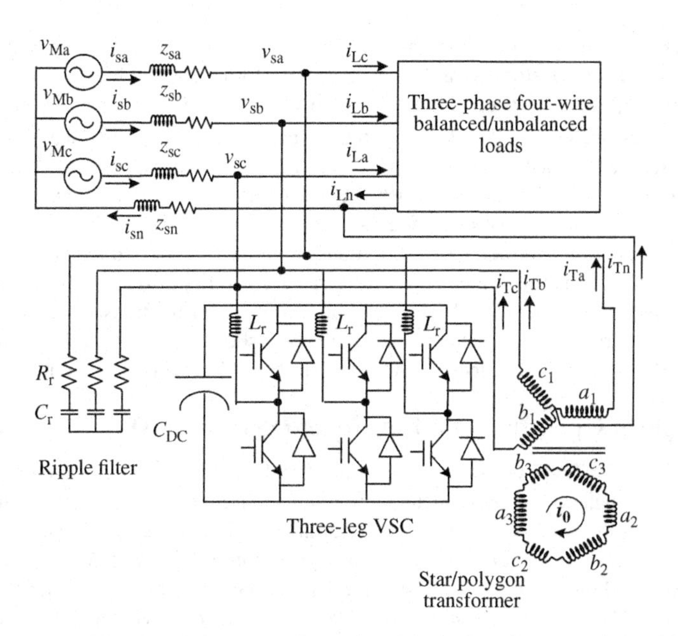

Figure 4.19 A three-leg VSC and star/polygon transformer-based three-phase four-wire DSTATCOM connected to a three-phase four-wire system

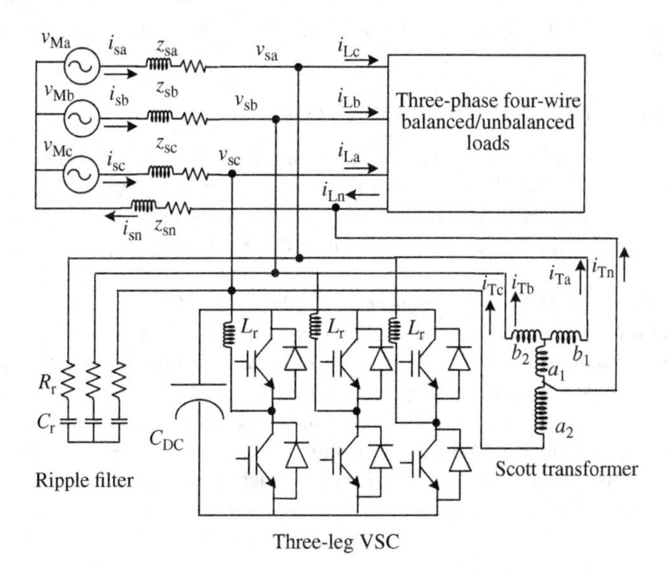

Figure 4.20 A three-leg VSC and Scott transformer-based three-phase four-wire DSTATCOM connected to a three-phase four-wire system

three-phase four-wire DSTATCOM configurations based on non-isolated two-leg VSCs with a zigzag transformer, a star/delta transformer, a T-connected transformer, a star/hexagon transformer, a star/polygon transformer, and a Scott transformer may be realized for load compensation.

Three-phase four-wire DSTATCOM configurations based on isolated three-leg VSCs with a zigzag transformer, a star/delta transformer, a T-connected transformer, a star/hexagon transformer, a star/polygon transformer, and a Scott transformer may be realized in a similar manner to the three-wire DSTATCOM configurations. Three-phase four-wire DSTATCOM configurations based on isolated two-leg VSCs with a zigzag transformer, a star/delta transformer, a T-connected transformer, a star/hexagon transformer, a star/polygon transformer, and a Scott transformer may also be realized in a similar manner to non-isolated three-wire DSTATCOMs using neutral terminal of the transformers of the supply side.

In total, 28 configurations of four-wire DSTATCOMs as shown in Figure 4.11 may be realized using two-leg VSCs, three-leg VSCs, four-leg VSCs, and six-leg VSCs with or without transformers.

4.4 Principle of Operation and Control of DSTATCOMs

The basic function of DSTATCOMs is to mitigate most of the current-based power quality problems such as reactive power, unbalanced currents, neutral current, and harmonics (if any) and to provide sinusoidal balanced currents in the supply with the self-supporting DC bus of the VSC used as a DSTATCOM.

A fundamental circuit of the DSTATCOM for a three-phase three-wire AC system with balanced/unbalanced loads is shown in Figure 4.6. An IGBT-based current-controlled voltage source converter (CC-VSC) with a DC bus capacitor is used as the DSTATCOM. Using a control algorithm, the reference DSTATCOM currents are directly controlled by estimating the reference DSTATCOM currents. However, in place of DSTATCOM currents, the reference supply currents may be estimated for an indirect current control of the VSC. The gating pulses to the DSTATCOM are generated by employing hysteresis (carrierless PWM (pulse-width modulation)) or PWM (fixed frequency) current control over reference and sensed supply currents resulting in an indirect current control. Using the DSTATCOM, the reactive power compensation and unbalanced current compensation are achieved in all the control algorithms. In addition, zero voltage regulation (ZVR) at PCC is also achieved by modifying the control algorithm suitably.

4.4.1 *Principle of Operation of DSTATCOMs*

The main objective of DSTATCOMs is to mitigate the current-based power quality problems in a distribution system. A DSTATCOM mitigates most of the current quality problems, such as reactive power, unbalance, neutral current, harmonics (if any), and fluctuations, present in the consumer loads or otherwise in the system and provides sinusoidal balanced currents in the supply with its DC bus voltage regulation.

In general, a DSTATCOM has a VSC connected to a DC bus and its AC sides are connected in shunt normally across the consumer loads or across the PCC as shown in Figures 4.6–4.8. The VSC uses PWM control; therefore, it requires small ripple filters to mitigate switching ripples. It requires Hall effect voltage and current sensors for feedback signals and normally a DSP is used to implement the required control algorithm to generate gating signals for the solid-state devices of the VSC of the DSTATCOM. The VSC is normally controlled in PWM current control mode to inject appropriate currents in the system. The DSTATCOM also needs many passive elements such as a DC bus capacitor, AC interacting inductors, injection and isolation transformers, and small passive filters.

4.4.2 *Control of DSTATCOMs*

The main objective of a control algorithm of DSTATCOMs is to estimate the reference currents using feedback signals. These reference currents along with corresponding sensed currents are used in PWM current controllers to derive PWM gating signals for switching devices (IGBTs) of the VSC used as a DSTATCOM. Reference currents for the control of DSTATCOMs have to be derived accordingly and these signals may be estimated using a number of control algorithms. There are many control algorithms reported in the literature for the control of DSTATCOMs, which are classified as time-domain and frequency-domain control algorithms. There are more than a dozen of time-domain control algorithms that are used for the control of DSTATCOMs. A few of these control algorithms are as follows:

- Unit template technique or PI controller-based theory
- Power balance theory (BPT)
- $I \cos \varphi$ control algorithm
- Current synchronous detection (CSD) method
- Instantaneous reactive power theory (IRPT), also known as PQ theory or α–β theory
- Synchronous reference frame (SRF) theory, also known as d–q theory
- Instantaneous symmetrical component theory (ISCT)
- Singe-phase PQ theory
- Singe-phase DQ theory
- Neural network theory (Widrow's LMS-based Adaline algorithm)
- Enhanced phase locked loop (EPLL)-based control algorithm
- Conductance-based control algorithm
- Adaptive detecting control algorithm, also known as adaptive interference canceling theory

These control algorithms are time-domain control algorithms. Most of them have been used for the control of DSTATCOMs and other compensating devices.

Similarly, there are around the same number of frequency-domain control algorithms. Some of them are as follows:

- Fourier series theory
- Discrete Fourier transform theory
- Fast Fourier transform theory
- Recursive discrete Fourier transform theory
- Kalman filter-based control algorithm
- Wavelet transformation theory
- Stockwell transformation (S-transform) theory
- Empirical decomposition (EMD) transformation theory
- Hilbert–Huang transformation theory

These control algorithms are frequency-domain control algorithms. Most of them are used for power quality monitoring for a number of purposes in the power analyzers, PQ instruments, and so on. Some of these algorithms have been used for the control of DSTATCOMs. However, these algorithms are sluggish and slow, requiring heavy computation burden; therefore, these control methods are not too much preferred for real-time control of DSTATCOMs compared with time-domain control algorithms.

All these control algorithms may be used for the control of DSTATCOMs. However, because of space limitation and to give a basic understanding, only some of them are explained here.

4.4.2.1 Unit template- or PI Controller-Based Control Algorithm of DSTATCOMs

The unit template- or PI controller-based control algorithm is a simple control algorithm for active compensating devices such as DSTATCOMs for AC voltage regulation at load terminals (at PCC) and load balancing of unbalanced loads. This control algorithm of the DSTATCOM is made flexible and it can be modified either for power factor correction (unity power factor (UPF) at PCC) or for voltage control (zero voltage regulation at PCC) through reactive power compensation along with load balancing of unbalanced loads. This control algorithm inherently provides a self-supporting DC bus of the VSC used as a DSTATCOM. It can be used for the direct current control of VSC currents of the DSTATCOM and provides an estimation of reference compensator currents. However, an indirect current control of supply currents is preferred to obtain PWM switching signals for the devices used in the CC-VSC working as a

DSTATCOM. This indirect current control of the DSTATCOM offers the advantages of fast control, reduced burden on the processor (DSP used for implementation), inherent elimination of sharp notches in currents, and so on. For this purpose, three-phase reference supply currents are derived using sensed AC voltages (at PCC) and DC bus voltage of the DSTATCOM as feedback signals. Two PI voltage controllers, one to regulate the DC bus of the VSC used as a DSTATCOM and other to regulate amplitude of the PCC voltages, are used to estimate the amplitudes of in-phase and quadrature components of reference supply currents.

4.4.2.1.1 Control of DSTATCOMs in UPF Mode of Operation

Figure 4.21 shows the unit template-based control algorithm of DSTATCOMs for power factor correction (PFC) at PCC. In this control algorithm, three-phase voltages at PCC along with the DC bus voltage of the DSTATCOM are used for implementing this control algorithm. In real-time implementation of DSTATCOMs, a band-pass filter (BPF) plays an important role. Three-phase voltages are sensed at PCC and are conditioned in a band-pass filter to filter out any distortion. The three-phase load voltages are inputs and three-phase filtered voltages (v_{sa}, v_{sb}, v_{sc}) are outputs of band-pass filters.

For the control of the DSTATCOM, the self-supporting DC bus is realized using a PI voltage controller over the sensed (v_{DC}) and reference (v^*_{DC}) values of the DC bus voltage of the DSTATCOM. The PI voltage controller on the DC bus voltage of the DSTATCOM provides the amplitude (I^*_{spp}) of in-phase components (i^*_{sa}, i^*_{sb}, i^*_{sc}) of reference supply currents. The three-phase unit current vectors (u_{sa}, u_{sb}, u_{sc}) are derived in phase with the filtered supply voltages (v_{sa}, v_{sb}, v_{sc}). The multiplication of the in-phase amplitude with in-phase unit vectors results in the in-phase components of three-phase reference supply currents (i^*_{sa}, i^*_{sb}, i^*_{sc}). Hence, for fundamental unity power factor supply currents, the in-phase reference supply currents, which are estimated in the above-described procedure, become the reference supply currents.

The amplitude of reference supply currents is computed using a PI voltage controller over the average value of the DC bus voltage (v_{DCa}) of the DSTATCOM and its reference value (v^*_{DC}). A comparison of average and reference values of the DC bus voltage of the DSTATCOM results in a voltage error, which is fed to a PI voltage controller:

$$v_{DCe}(n) = v^*_{DC}(n) - v_{DCa}(n). \tag{4.1}$$

This voltage error is fed to a PI voltage controller, and the output of the PI voltage controller is

$$I_{spp}(n) = I_{spp}(n-1) + K_{pd}\{v_{DCe}(n) - v_{DCe}(n-1)\} + K_{id}v_{DCe}(n), \tag{4.2}$$

where $v_{DCe}(n) = v^*_{DC}(n) - v_{DC}(n)$ is the error between the reference (v^*_{DC}) and the sensed (v_{DC}) DC voltage at the nth sampling instant, and K_{pd} and K_{id} are the proportional and integral gain constants of the DC bus voltage PI controller, respectively.

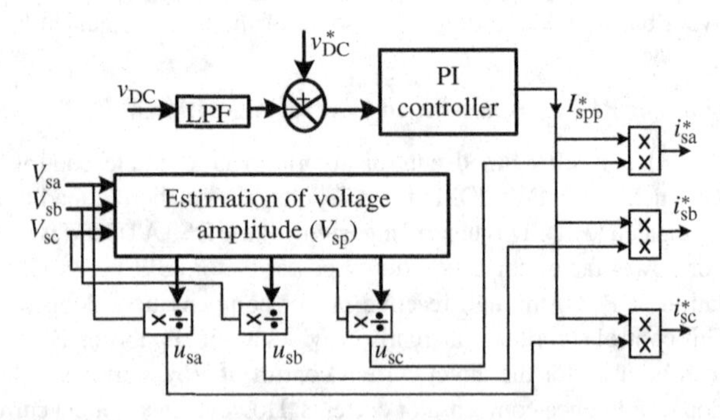

Figure 4.21 Unit template-based control algorithm of DSTATCOMs for power factor correction mode of operation

Here, proportional (K_{pd}) and integral (K_{id}) gain constants are chosen such that a desired DC bus voltage response is achieved. The output of the PI controller is taken as the amplitude (I^*_{spp}) of the reference supply currents. Now, in-phase components of the three-phase reference supply currents are computed using their amplitude and in-phase unit vectors derived in phase with the PCC voltages, and are given as

$$i^*_{sa} = I^*_{spp}u_{sa}, \quad i^*_{sb} = I^*_{spp}u_{sb}, \quad i^*_{sc} = I^*_{spp}u_{sc}, \tag{4.3}$$

where u_{sa}, u_{sb}, and u_{sc} are in-phase unit templates and are derived as

$$u_{sa} = v_{sa}/V_{sp}, \quad u_{sb} = v_{sb}/V_{sp}, \quad u_{sc} = v_{sc}/V_{sp}, \tag{4.4}$$

where V_{sp} is the amplitude of the PCC voltage and is computed as

$$V_{sp} = \left\{2/3(v^2_{sa} + v^2_{sb} + v^2_{sc})\right\}^{1/2}. \tag{4.5}$$

4.4.2.1.2 Control of DSTATCOMs in ZVR Mode of Operation

The DSTATCOM can compensate the reactive power and negative-sequence currents of the loads. However, because of finite (nonzero) internal impedance of the utility, which is represented by Z_s (L_s, R_s), the voltage waveforms at PCC to other loads are not regulated and result in a voltage drop. The DSTATCOM should regulate the PCC voltages so that other loads connected at PCC are not affected by this voltage drop. The voltage drops are caused by many loads such as inrush currents by the direct-on line starting of motors. Thus, it is necessary to switch the operating mode of the DSTATCOM to a voltage regulator.

As mentioned earlier, in addition to load balancing, the DSTATCOM can also be operated to maintain constant voltage at PCC. For this purpose, the DSTATCOM takes normally a leading current component (in general) due to lagging power factor loads and it is explained using phasor diagrams shown in Figure 4.22. When the system is operating without a DSTATCOM, the voltage at PCC (V_s) is less than the supply voltage (V_M) due to the drop in the supply impedance Z_s (L_s, R_s) as shown in Figure 4.22a. Now with a DSTATCOM connected in the system and drawing a leading current component, the supply current and hence the drop across the supply impedance can be controlled so that the magnitudes of the PCC voltage and supply voltage become equal ($|V_s| = |V_M|$) as shown in Figure 4.22b. By controlling the DSTATCOM current, the amplitude and phase of the supply current may be changed to maintain the desired load voltage. Hence, at the same time, both UPF and ZVR functions cannot be achieved.

The control algorithm to maintain the desired PCC voltage, the DSTATCOM for ZVR operation at PCC, is shown in Figure 4.23. Using this algorithm, one can achieve AC voltage regulation at load terminals (at PCC) and load balancing of unbalanced loads. For regulation of voltage at PCC, the three-phase reference supply currents (i^*_{sa}, i^*_{sb}, i^*_{sc}) have two components. The first component (i^*_{sad}, i^*_{sbd}, i^*_{scd}) is in phase with the voltages at PCC to feed active power to the loads and the losses of the DSTATCOM. The second component (i^*_{saq}, i^*_{sbq}, i^*_{scq}) is in quadrature with the voltages at PCC to feed reactive power to the loads and to compensate the line voltage drop by reactive power injection at PCC. For power factor correction to unity and balancing of unbalanced loads, the quadrature component of reference supply

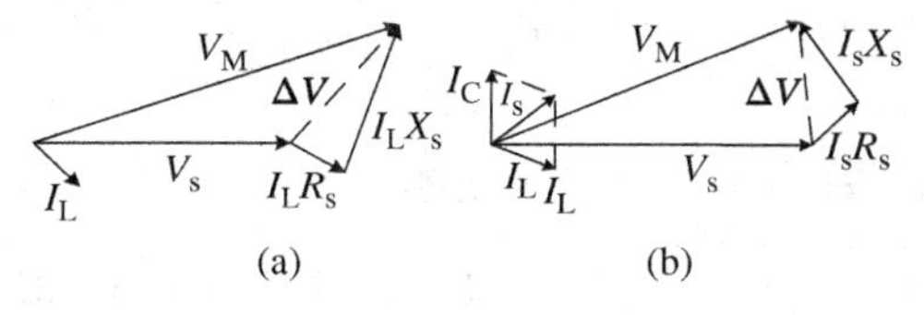

Figure 4.22 Phasor diagrams for ZVR mode of operation: (a) without a DSTATCOM and (b) with a DSTATCOM

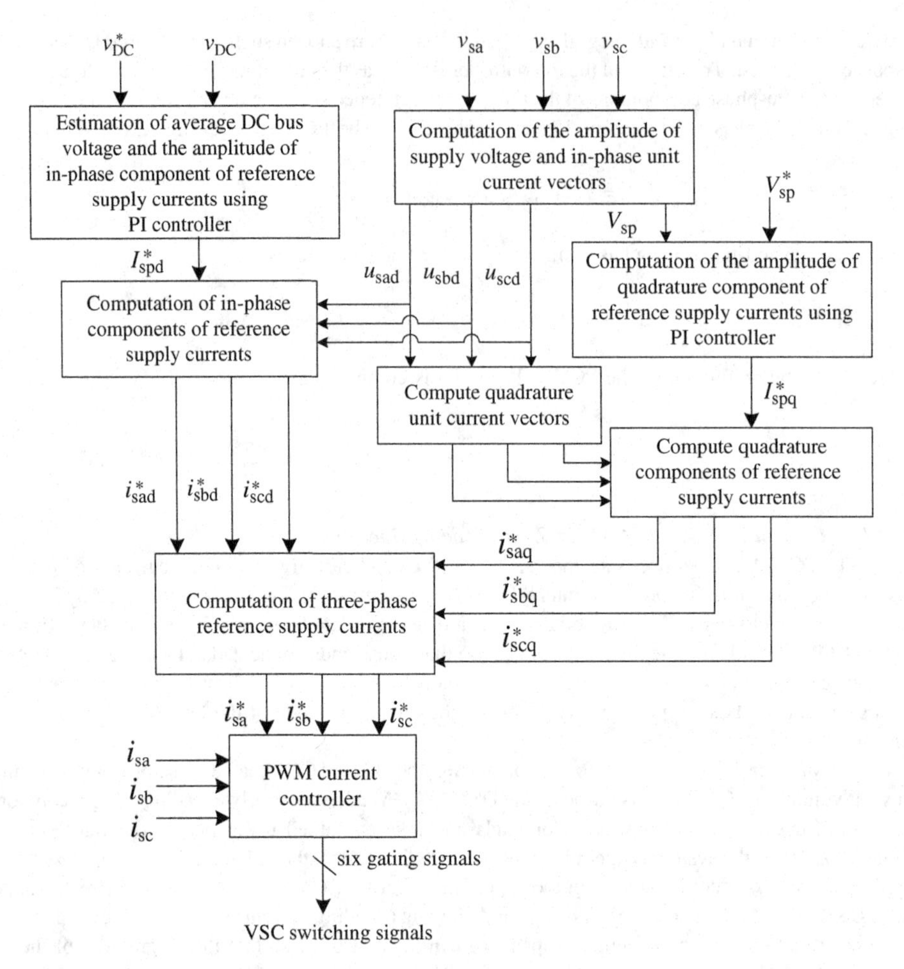

Figure 4.23 Unit template-based control algorithm of DSTATCOMs for ZVR mode of operation

currents is set to zero. For voltage regulation at PCC, the supply currents should lead the supply voltages for lagging PF loads, while for the power factor control to unity, the supply currents should be in phase with the supply voltages. These two conditions, namely, voltage regulation at PCC and power factor control to unity, cannot be achieved simultaneously. Therefore, the control algorithm of the DSTATCOM is made flexible to achieve either voltage regulation or power factor correction to unity and load balancing. The operation of DSTATCOMs in UPF mode is already explained in the previous section. Therefore, the three in-phase components of supply currents are the quantities estimated in Equation 4.3.

The amplitude (I_{spq}^*) of the quadrature component of reference supply currents is estimated using another PI controller over the filtered amplitude (V_{sp}) of the PCC voltage and its reference value (V_{sp}^*). A comparison of the reference value with the amplitude of the PCC voltage results in a voltage error (V_{spe}). This voltage error signal is processed in a PI controller. The output signal of the PI voltage controller $I_{spq}^*(n)$ for maintaining the PCC terminal voltage at a constant value at the nth sampling instant is expressed as

$$I_{spq}^*(n) = I_{spq}^*(n-1) + K_{pt}\{v_{spe}(n) - v_{spe}(n-1)\} + K_{it}v_{spe}(n), \tag{4.6}$$

where K_{pt} and K_{it} are the proportional and integral gain constants of the AC bus voltage PI controller, respectively, $v_{spe}(n)$ and $v_{spe}(n-1)$ are the voltage errors at the nth and $(n-1)$th instants, respectively, and

$I^*_{spq}(n-1)$ is the required reactive power at the $(n-1)$th instant. The term $I^*_{spq}(n)$ is considered as the amplitude (I^*_{spq}) of the quadrature component of reference supply currents. Three-phase quadrature components of reference supply currents are estimated using their amplitude and quadrature unit current templates as

$$i^*_{saq} = I^*_{spq} u_{saq}, \quad i^*_{sbq} = I^*_{spq} u_{sbq}, \quad i^*_{scq} = I^*_{spq} u_{scq}, \tag{4.7}$$

where u_{saq}, u_{sbq}, and u_{scq} are quadrature unit current templates and are derived as

$$\begin{aligned} u_{saq} &= (-u_{sbd} + u_{scd})/\sqrt{3}, \\ u_{sbq} &= (3u_{sad} + u_{sbd} - u_{scd})/2\sqrt{3}, \\ u_{scq} &= (-3u_{sad} + u_{sbd} - u_{scd})/2\sqrt{3}, \end{aligned} \tag{4.8}$$

where $u_{sad} = u_{sa}$, $u_{sbd} = u_{sb}$, $u_{scd} = u_{sc}$ are in-phase unit templates of phase voltages.

Three-phase instantaneous reference supply currents are estimated by adding in-phase and quadrature components expressed in Equations 4.3 and 4.7. For power factor correction and load balancing, the amplitude of quadrature components is set to zero and in this condition the in-phase components of reference supply currents become the total reference supply currents. These estimated three-phase reference supply currents and sensed supply currents are given to the hysteresis/PWM current controller to generate the switching signals for switches of the VSC of the DSTATCOM.

4.4.2.2 PBT-Based Control Algorithm of DSTATCOMs

Figure 4.24 shows the PBT-based control algorithm of DSTATCOMs. This control algorithm is based on extraction of fundamental components of load currents from instantaneous power consumed by the loads. This algorithm needs the sensing of PCC line voltages, supply currents, load currents, and DC bus voltage of the VSC of the DSTATCOM. The fundamental active power component of load currents is added to the output of the DC link PI voltage controller in order to generate the fundamental active power component of reference supply currents. The fundamental reactive power component of load currents is subtracted

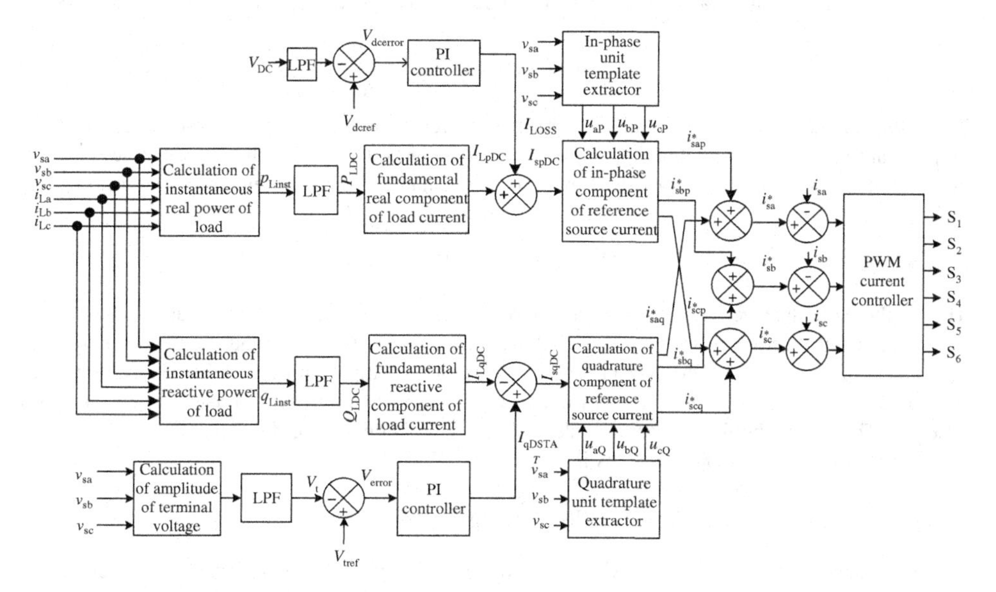

Figure 4.24 The PBT-based control algorithm for extracting reference supply currents

from the output of the AC terminal voltage PI controller to estimate the fundamental reactive power component of reference supply currents. These active and reactive power reference supply currents are DC quantities corresponding to fundamental currents. The instantaneous active power components of reference supply currents are estimated by multiplying the amplitude of active power component of reference supply currents by in-phase unit templates and the reactive power components of reference supply currents are obtained by multiplying the amplitude of reactive reference supply currents by quadrature unit templates. Finally, reference supply currents of each phase are generated by adding in-phase and quadrature reference supply currents of the corresponding phases.

The instantaneous active and reactive powers of the load are calculated as

$$p_L = v_{sa}i_{La} + v_{sb}i_{Lb} + v_{sc}i_{Lc} = P_{LDC} + P_{LAC} \tag{4.9}$$

$$q_L = (1/\sqrt{3})\{(v_{sa} - v_{sb})i_{Lc} + (v_{sb} - v_{sc})i_{La} + (v_{sc} - v_{sa})i_{Lb}\} = Q_{LDC} + Q_{LAC} \tag{4.10}$$

From the instantaneous active power (p_L) and reactive power (q_L) of the loads, the fundamental components of active power (P_{LDC}) and reactive power (Q_{LDC}) of the loads are extracted using LPFs (low-pass filters).

The amplitude of the fundamental active power component of load currents is obtained from average load power (P_{LDC}) as

$$I_{LpDC} = (2/3)(P_{LDC}/V_{sp}). \tag{4.11}$$

Similarly, the amplitude of the fundamental reactive power component of load currents is obtained from the DC component of reactive power (Q_{LDC}) as

$$I_{LqDC} = (2/3)(Q_{LDC}/V_{sp}). \tag{4.12}$$

The amplitude of the active power component of reference supply currents is estimated by adding the output of the DC voltage PI controller of the DSTATCOM to the fundamental active power component of the load currents. The output of the DC voltage PI controller (I_{loss}) is considered as the loss component of the DSTATCOM and estimated similarly to Equation 4.2. Therefore, the instantaneous value of the amplitude of the fundamental active power component of reference supply currents is

$$I_{spDC} = I_{loss} + I_{LpDC}. \tag{4.13}$$

The instantaneous values of the fundamental active power component of reference three-phase supply currents are estimated by multiplying in-phase unit templates and the amplitude of the active power component of reference supply currents. Three-phase instantaneous in-phase reference supply currents are

$$i^*_{sap} = u_{sa}I_{spDC}, \quad i^*_{sbp} = u_{sb}I_{spDC}, \quad i^*_{scp} = u_{sc}I_{spDC}. \tag{4.14}$$

The amplitude of the reactive power component of reference supply currents is estimated by subtracting the fundamental reactive power component of the load currents from the fundamental reactive power component of the DSTATCOM current. The reactive power component of the DSTATCOM current is estimated by using a terminal voltage PI controller similarly to Equation 4.6 and it is considered as I_{qDSTAT}. Therefore, the instantaneous value of the amplitude of the fundamental reactive power component of reference supply currents is

$$I_{sqDC} = I_{qDSTAT} - I_{LqDC}. \tag{4.15}$$

The instantaneous values of the fundamental reactive power component of reference three-phase supply currents are estimated by multiplying quadrature unit templates by the amplitude of the reactive component of reference supply currents given in Equation 4.15. Three-phase instantaneous quadrature reference supply currents are estimated as

$$i^*_{saq} = u_{saq} I_{sqDC}, \quad i^*_{sbq} = u_{sbq} I_{sqDC}, \quad i^*_{scq} = u_{scq} I_{sqDC}. \tag{4.16}$$

Instantaneous fundamental reference supply currents are estimated by adding in-phase and quadrature reference supply currents:

$$i^*_{sa} = i^*_{sap} + i^*_{saq}, \quad i^*_{sb} = i^*_{sbp} + i^*_{sbq}, \quad i^*_{sc} = i^*_{scp} + i^*_{scq}. \tag{4.17}$$

These reference three-phase supply currents (i^*_{sa}, i^*_{sb}, i^*_{sc}) are compared with respective sensed three-phase supply currents (i_{sa}, i_{sb}, i_{sc}) to estimate current errors. These current errors are amplified and amplified current errors are compared with the triangular carrier wave to generate the switching pulses of the VSC of the DSTATCOM.

4.4.2.3 $I\cos\varphi$-Based Control Algorithm of DSTATCOMs

Figure 4.25 shows a block diagram of the $I\cos\varphi$-based control algorithm used for the estimation of reference supply currents. The load currents (i_{La}, i_{Lb}, i_{Lc}), PCC voltages (v_{sa}, v_{sb}, v_{sc}), supply currents, and DC bus voltage (v_{DC}) of the VSC of the DSTATCOM are sensed as feedback control signals.

The amplitude of the active power component of current (I_{Lpa}) of phase a is extracted from the fundamental component of load current (i_{Lfa}) (achieved after filtering) at the zero crossing of the in-phase unit template ($\cos\varphi_{pa} = u_{sa}$) of three-phase PCC voltages. A zero-crossing detector and "sample and hold" logic are used to extract the I_{Lpa} (amplitude of filtered load current at zero crossing of the corresponding in-phase unit template). Similarly, amplitudes of the active power component of currents of phase b (I_{Lpb}) and phase c (I_{Lpc}) are extracted. In the case of a balanced system, the amplitude of the active power

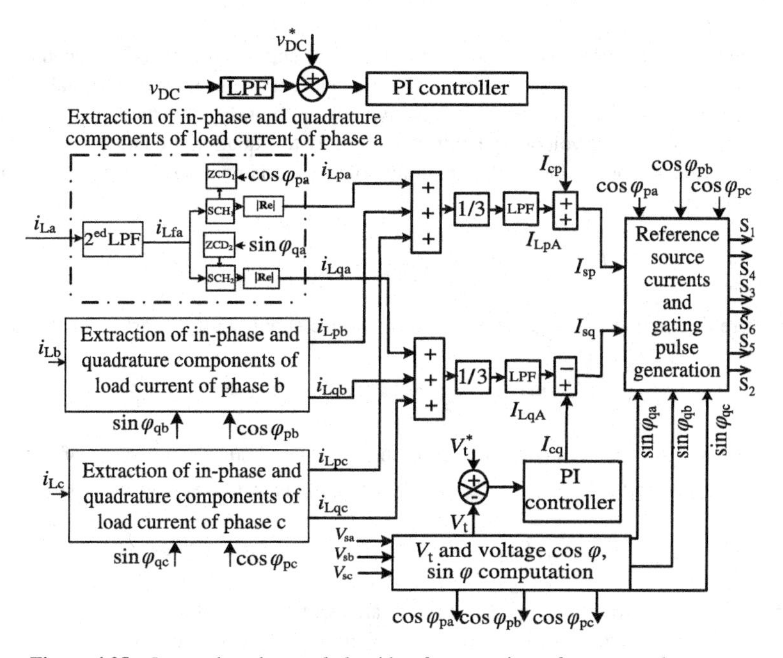

Figure 4.25 $I\cos\varphi$-based control algorithm for extracting reference supply currents

component of load currents can be expressed as

$$I_{LpA} = (I_{Lpa} + I_{Lpb} + I_{Lpc})/3, \tag{4.18}$$

where I_{Lpa}, I_{Lpb}, and I_{Lpc} are the amplitudes of the active power component of three-phase load currents.

For a self-supporting DC bus of the VSC, a PI controller is used for regulating the DC bus voltage of the VSC using Equation 4.2. The output of the PI controller (I_{Cd}) is considered as part of the active power component of reference supply currents.

The amplitude of the active power component (I_{sp}) of reference supply currents is estimated as

$$I_{sp} = I_{LpA} + I_{Cp}. \tag{4.19}$$

In-phase components of reference supply currents are estimated as

$$i^*_{sap} = I_{sp} \cos \varphi_{pa}, \quad i^*_{sbp} = I_{sp} \cos \varphi_{pb}, \quad i^*_{scp} = I_{sp} \cos \varphi_{pc}, \tag{4.20}$$

where in-phase templates are $\cos \varphi_{pa} = u_{sa}$, $\cos \varphi_{pb} = u_{sb}$, and $\cos \varphi_{pc} = u_{sc}$.

The unit amplitude templates ($\sin \varphi_{qa} = u_{saq}$, $\sin \varphi_{qb} = u_{sbq}$, $\sin \varphi_{qc} = u_{scq}$) in quadrature with three-phase PCC voltages (v_{sa}, v_{sb}, v_{sc}) are derived using a quadrature transformation of the in-phase unit vectors $\cos \varphi_{pa}$, $\cos \varphi_{pb}$, and $\cos \varphi_{pc}$ as in Equation 4.8.

The amplitude of the reactive power component of load current (I_{Lqa}) of phase a is extracted from the fundamental component of load current (i_{Lfa}) (achieved after filtering) at the zero crossing of the quadrature unit template of the PCC voltage ($\sin \varphi_{qa}$). A zero-crossing detector and "sample and hold" logic are used to extract the I_{Lqa} (amplitude of filtered load current at zero crossing of the corresponding quadrature unit template). Similarly, the amplitudes of the reactive power component of currents of phase b (i_{Lqb}) and phase c (i_{Lqc}) are estimated. The amplitude of the reactive power component of load currents can be expressed as

$$I_{LqA} = (i_{Lqa} + i_{Lqb} + i_{Lqc})/3. \tag{4.21}$$

The amplitude of the reactive power component of reference supply currents is estimated by subtracting the fundamental reactive power component of the load currents from the fundamental reactive power component of the DSTATCOM current. The reactive power component of the DSTATCOM current is estimated by using a terminal voltage PI controller similarly to Equation 4.6 and it is considered as I_{Cq}. Therefore, the instantaneous value of the amplitude of the fundamental reactive power component of reference supply currents is

$$I_{sq} = I_{Cq} - I_{LqA}. \tag{4.22}$$

The quadrature or reactive power components of reference supply currents are estimated as

$$i^*_{saq} = I_{sq} \sin \varphi_{qa}, \quad i^*_{sbq} = I_{sq} \sin \varphi_{qb}, \quad i^*_{scq} = I_{sq} \sin \varphi_{qc}. \tag{4.23}$$

Total reference supply currents are the sum of in-phase and quadrature components of reference supply currents, which are computed as

$$i^*_{sa} = i^*_{sap} + i^*_{saq}, \quad i^*_{sb} = i^*_{sbp} + i^*_{sbq}, \quad i^*_{sc} = i^*_{scp} + i^*_{scq}. \tag{4.24}$$

These reference supply currents (i^*_{sa}, i^*_{sb}, i^*_{sc}) are compared with the sensed supply currents (i_{sa}, i_{sb}, i_{sc}) in PI current controllers. The outputs of PI current controllers for all the phases are used to generate gating signals of IGBTs of the VSC using a PWM current controller.

4.4.2.4 CSD Control Algorithm of DSTATCOMs

Figure 4.26 shows a block diagram of the CSD control algorithm used for the estimation of reference supply currents. Sensed instantaneous PCC line voltages v_{ab} and v_{bc} are converted to phase voltages v_{sa}, v_{sb}, and v_{sc} as

$$v_{sa} = \frac{2}{3}v_{ab} + \frac{1}{3}v_{bc}, \quad v_{sb} = -\frac{1}{3}v_{ab} + \frac{1}{3}v_{bc}, \quad v_{sc} = \frac{1}{3}v_{ab} + \frac{2}{3}v_{bc}. \tag{4.25}$$

Since these PCC phase voltages may be distorted, a set of band-pass filters (30–70 Hz) is used to filter the voltage distortion. These filtered three-phase PCC voltages (v_{saf}, v_{sbf}, v_{scf}) are used for the estimation of balanced positive-sequence components (v_{sa1}, v_{sb1}, v_{sc1})

$$v_{sa1} = \frac{1}{3}(v_{saf} + av_{sbf} + a^2v_{scf}), \quad v_{sb1} = \frac{1}{3}(v_{sbf} + av_{scf} + a^2v_{saf}), \quad v_{sc1} = \frac{1}{3}(v_{scf} + av_{saf} + a^2v_{sbf}), \tag{4.26}$$

where $a = e^{j120}$ and $a^2 = e^{j240}$.

The amplitude of positive-sequence PCC voltages (V_t) after passing through a low-pass filter is

$$V_t = \sqrt{\{(2/3)(v_{sa1}^2 + v_{sb1}^2 + v_{sc1}^2)\}}. \tag{4.27}$$

These extracted reference currents are used for generating switching pulses of the voltage source converter. In a single-phase system, for average load power (P) an active current (I_{sp}) can be calculated as

$$I_{sp} = \{2P/V_{sm1}^2\}v_{s1}, \tag{4.28}$$

where P is the average load power, V_{sm1} is the amplitude of the ideal phase voltage, and v_{s1} is the instantaneous ideal phase voltage.

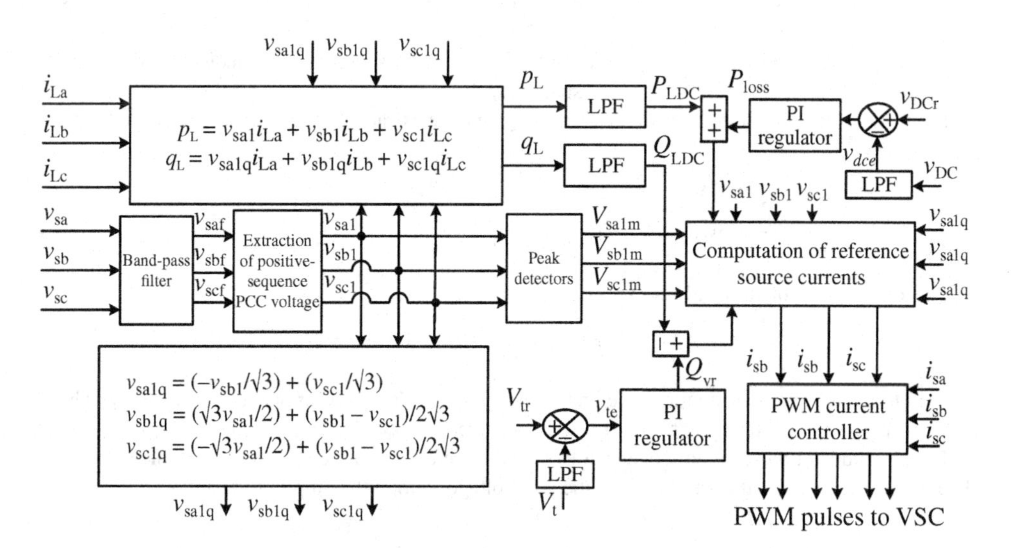

Figure 4.26 CSD control algorithm for extracting reference supply currents

Based on the single-phase concept, the three-phase active power of connected loads can be computed as

$$p_L = v_{sa1} i_{La} + v_{sb1} i_{Lb} + v_{sc1} i_{Lc} = P_{LDC} + P_{LAC}, \tag{4.29}$$

where P_{LDC} and P_{LAC} can be considered as the average and oscillating components, respectively. The average component of power (P_{LDC}) can be extracted using a low-pass filter.

For a self-supporting DC bus of the VSC, the losses of the VSC as an active power component must be fed by the supply. This component of active power is achieved using a PI regulator over the DC bus voltage of the VSC. For this purpose, an error in the DC bus voltage of the VSC ($v_{DCe}(n)$) of the DSTATCOM at the nth sampling instant is

$$v_{DCe}(n) = v_{DCr}(n) - v_{DC}(n), \tag{4.30}$$

where $v_{DCr}(n)$ is the reference DC voltage and $v_{DC}(n)$ is the sensed DC bus voltage of the VSC.

The output of the PI regulator for maintaining the DC bus voltage of the VSC of the DSTATCOM at the nth sampling instant is expressed as

$$P_{loss}(n) = P_{loss}(n-1) + k_{pDC}\{v_{DCe}(n) - v_{DCe}(n-1)\} + k_{iDC} v_{DCe}(n), \tag{4.31}$$

where $P_{loss}(n)$ is considered as part of active power of supply currents, k_{pDC} and k_{iDC} are the proportional and integral gain constants of the PI regulator, respectively, and v_{DCe} is the error in the DC bus voltage of the VSC.

Total active power component (P_{tA}) of reference supply currents can be written as

$$P_{tA} = P_{LDC} + P_{loss}. \tag{4.32}$$

The active components of reference supply currents are estimated using this estimated power as

$$i^*_{sap} = \{2(P_{tA})v_{sa1}\}/\{V_{sa1m}(V_{sa1m} + V_{sb1m} + V_{sc1m})\}, \tag{4.33}$$

$$i^*_{sbp} = \{2(P_{tA})v_{sb1}\}/\{V_{sb1m}(V_{sa1m} + V_{sb1m} + V_{sc1m})\}, \tag{4.34}$$

$$i^*_{scp} = \{2(P_{tA})v_{sc1}\}/\{V_{sc1m}(V_{sa1m} + V_{sb1m} + V_{sc1m})\}, \tag{4.35}$$

where V_{sa1m}, V_{sb1m}, V_{sc1m} and v_{sa1}, v_{sb1}, v_{sc1} are amplitudes and instantaneous positive-sequence quantities of three-phase PCC voltages, respectively.

The voltages in quadrature with v_{sa1}, v_{sb1}, and v_{sc1} may be extracted using a quadrature transformation:

$$v_{sa1q} = -v_{sb1}/\sqrt{3} + v_{sc1}/\sqrt{3}, \quad v_{sb1q} = \sqrt{3}v_{sa1}/2 + (v_{sb1} - v_{sc1})/2\sqrt{3},$$
$$v_{sc1q} = -\sqrt{3}v_{sa1}/2 + (v_{sb1} - v_{sc1})/2\sqrt{3}. \tag{4.36}$$

The reactive power of the loads is expressed as

$$q_L = v_{sa1q} i_{La} + v_{sb1q} i_{Lb} + v_{sc1q} i_{Lc} = Q_{LDC} + Q_{LAC}, \tag{4.37}$$

where Q_{LDC} and Q_{LAC} are the average and non-DC or oscillating components, respectively. The DC component of this reactive power is extracted using a low-pass filter and is expressed as Q_{LDC}.

The amplitude of the terminal voltage at PCC and its reference value are fed to a PI voltage regulator. The voltage error v_{te} is the amplitude of the AC voltage at the nth sampling instant:

$$v_{te}(n) = V_{tr}(n) - V_t(n), \tag{4.38}$$

where $v_{tr}(n)$ is the reference amplitude of the load terminal voltage and $v_t(n)$ is the amplitude of the sensed voltage at the nth sampling instant.

The output of the PI regulator $Q_{vr}(n)$ for maintaining the amplitude of the AC terminal voltage to a constant value at the nth sampling instant is

$$Q_{vr}(n) = Q_{vr}(n-1) + k_{pAC}\{v_{te}(n) - v_{te}(n-1)\} + k_{iAC}v_{te}(n), \tag{4.39}$$

where k_{pAC} and k_{iAC} are the proportional and integral gain constants of the PI regulator, respectively, and $v_{te}(n)$ and $v_{te}(n-1)$ are the voltage errors at the nth and $(n-1)$th sampling instants, respectively.

Total reactive power (Q_{tR}) to be supplied by reference supply currents can be written as

$$Q_{tR} = Q_{vr} - Q_{LDC}. \tag{4.40}$$

The reactive power components of reference supply currents are estimated as

$$i^*_{saq} = \{2(Q_{tR})v_{sa1q}\}/\{V_{sa1m}(V_{sa1m} + V_{sb1m} + V_{sc1m})\}, \tag{4.41}$$

$$i^*_{sbq} = \{2(Q_{tR})v_{sb1q}\}/\{V_{sa1m}(V_{sa1m} + V_{sb1m} + V_{sc1m})\}, \tag{4.42}$$

$$i^*_{scq} = \{2(Q_{tR})v_{sc1q}\}/\{V_{sa1m}(V_{sa1m} + V_{sb1m} + V_{sc1m})\}, \tag{4.43}$$

where v_{sa1q}, v_{sb1q}, and v_{sc1q} are the 90° shifted instantaneous positive-sequence quantities of three-phase PCC voltages.

Total reference supply currents are the sum of active and reactive power components of reference supply currents, which are computed as

$$i^*_{sa} = i^*_{saq} + i^*_{sap}, \quad i^*_{sb} = i^*_{sbq} + i^*_{sbp}, \quad i^*_{sc} = i^*_{scq} + i^*_{scp}. \tag{4.44}$$

These reference supply currents are compared with the sensed supply currents (i_{sa}, i_{sb}, i_{sc}) to estimate the three-phase current error components (i_{ea}, i_{eb}, i_{ec}). Amplified current errors are fed to a PWM controller to generate gating signals of IGBTs of the VSC used as a DSTATCOM.

4.4.2.5 IRPT-Based Control Algorithm of DSTATCOMs

The IRPT-based control algorithm of DSTATCOMs is shown in Figure 4.27. Three-phase load currents and PCC voltages are sensed and used to calculate the instantaneous active and reactive powers. Three-phase PCC voltages are sensed and processed through BPFs before their transformation to eliminate their ripple contents and are denoted as (v_{sa}, v_{sb}, v_{sc}). A first-order Butterworth filter is used as a BPF.

These three-phase filtered load voltages are transformed into two-phase α–β orthogonal coordinates (v_α, v_β) as

$$\begin{pmatrix} v_\alpha \\ v_\beta \end{pmatrix} = \sqrt{\frac{2}{3}} \begin{pmatrix} 1 & -\frac{1}{2} & -\frac{1}{2} \\ 0 & \frac{\sqrt{3}}{2} & -\frac{\sqrt{3}}{2} \end{pmatrix} \begin{pmatrix} v_{sa} \\ v_{sb} \\ v_{sc} \end{pmatrix}. \tag{4.45}$$

Similarly, three-phase load currents (i_{La}, i_{Lb}, i_{Lc}) are transformed into two-phase α–β orthogonal coordinates ($i_{L\alpha}$, $i_{L\beta}$) as

$$\begin{pmatrix} i_{L\alpha} \\ i_{L\beta} \end{pmatrix} = \sqrt{\frac{2}{3}} \begin{pmatrix} 1 & -\frac{1}{2} & -\frac{1}{2} \\ 0 & \frac{\sqrt{3}}{2} & -\frac{\sqrt{3}}{2} \end{pmatrix} \begin{pmatrix} i_{La} \\ i_{Lb} \\ i_{Lc} \end{pmatrix}. \tag{4.46}$$

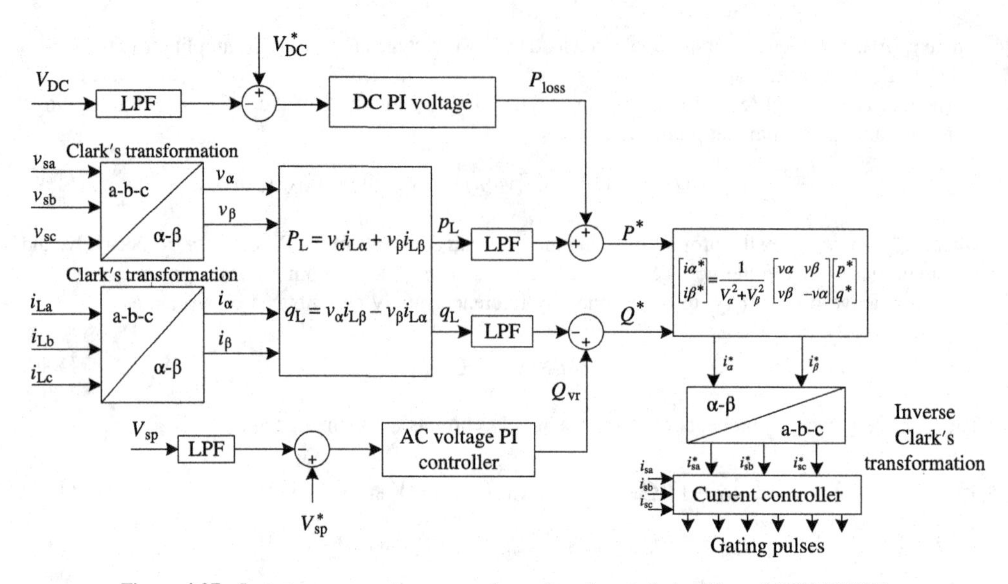

Figure 4.27 Instantaneous reactive power theory-based control algorithm of DSTATCOMs

From these two sets of expressions, the instantaneous active power p_L and the instantaneous reactive power q_L flowing into the load side are computed as

$$\begin{pmatrix} p_L \\ q_L \end{pmatrix} = \begin{pmatrix} v_\alpha & v_\beta \\ v_\beta & -v_\alpha \end{pmatrix} \begin{pmatrix} i_{L\alpha} \\ i_{L\beta} \end{pmatrix}. \tag{4.47}$$

Let $\bar{p}_L$ and $\tilde{p}_L$ are the DC component and the AC component of p_L, respectively, and $\bar{q}_L$ and $\tilde{q}_L$ are the DC component and the AC component of q_L, respectively, Therefore, these may expressed as

$$p_L = \bar{p}_L + \tilde{p}_L, \quad q_L = \bar{q}_L + \tilde{q}_L. \tag{4.48}$$

In these expressions, the fundamental load power is transformed to DC components $\bar{p}_L$ and $\bar{q}_L$, and the distortion or negative sequence is transformed to AC components $\tilde{p}_L$ and $\tilde{q}_L$. The DC components of active and reactive powers are extracted by using two LPFs.

The reference three-phase supply currents i^*_{sa}, i^*_{sb}, and i^*_{sc} are estimated as

$$\begin{pmatrix} i^*_{sa} \\ i^*_{sb} \\ i^*_{sc} \end{pmatrix} = \sqrt{\frac{2}{3}} \begin{pmatrix} 1 & 0 \\ -\frac{1}{2} & \frac{\sqrt{3}}{2} \\ -\frac{1}{2} & -\frac{\sqrt{3}}{2} \end{pmatrix} \begin{pmatrix} v_\alpha & v_\beta \\ -v_\beta & v_\alpha \end{pmatrix}^{-1} \begin{pmatrix} p^* \\ q^* \end{pmatrix}. \tag{4.49}$$

This IRPT-based control algorithm may easily be modified for the control on supply currents for indirect current control. In this case, for power factor correction mode of operation of the DSTATCOM, $p^* = \bar{p}_L + p_{loss}$ and $q^* = \bar{q}_L - q_{vr} = 0$ in Equation 4.49 and after the transformation from the α–β frame to the abc frame, three-phase transformed currents are reference supply currents and these must be compared with sensed supply currents in the PWM current controllers as shown in Figure 4.27 for indirect current control of the DSTATCOM. The term P_{loss} is an instantaneous active power necessary to adjust the voltage of the DC capacitor of the VSC used as a DSTATCOM to its reference value. In addition, q_{vr} is instantaneous reactive power necessary to adjust the PCC voltage to its reference value (these are achieved using a PI controller similarly to the above algorithms as shown in Figure 4.27), and $\overline{p}_L$ and $\overline{q}_L$ are the extracted load fundamental active and reactive power components, respectively.

In the case of ZVR at PCC (voltage regulation mode of operation of the DSTATCOM), a PI voltage controller over the PCC voltage is used similarly to the above algorithms and its output is used to estimate p^* and q^* as $p^* = \overline{p}_L + p_{loss}$ and $q^* = q_{vq} - \overline{q}_L$. After the transformation, three-phase transformed currents are reference supply currents and these are compared with sensed supply currents as shown in Figure 4.27 for indirect current control of the DSTATCOM.

4.4.2.6 SRF Theory-Based Control Algorithm of DSTATCOMs

The synchronous reference frame theory is reported in the literature for the control of DSTATCOMs. A block diagram of the control algorithm is shown in Figure 4.28. The load currents (i_{La}, i_{Lb}, i_{Lc}), PCC voltages (v_{sa}, v_{sb}, v_{sc}), and DC bus voltage (v_{DC}) of the DSTATCOM are sensed as feedback signals. The load currents in the three phases are converted into the dq0 frame using the Park's transformation as follows:

$$\begin{bmatrix} i_{Ld} \\ i_{Lq} \\ i_{L0} \end{bmatrix} = \frac{2}{3} \begin{bmatrix} \cos\theta & -\sin\theta & \frac{1}{2} \\ \cos\left(\theta - \frac{2\pi}{3}\right) & -\sin\left(\theta - \frac{2\pi}{3}\right) & \frac{1}{2} \\ \cos\left(\theta + \frac{2\pi}{3}\right) & \sin\left(\theta + \frac{2\pi}{3}\right) & \frac{1}{2} \end{bmatrix} \begin{bmatrix} i_{La} \\ i_{Lb} \\ i_{Lc} \end{bmatrix}. \quad (4.50)$$

A three-phase PLL (phase locked loop) is used to synchronize these signals with the PCC voltages. These d–q current components are then passed through a LPF to extract the DC components of i_{Ld} and i_{Lq}. The d-axis and q-axis currents consist of fundamental and harmonic components as

$$i_{Ld} = i_{dDC} + i_{dAC}, \quad (4.51)$$

$$i_{Lq} = i_{qDC} + i_{qAC}. \quad (4.52)$$

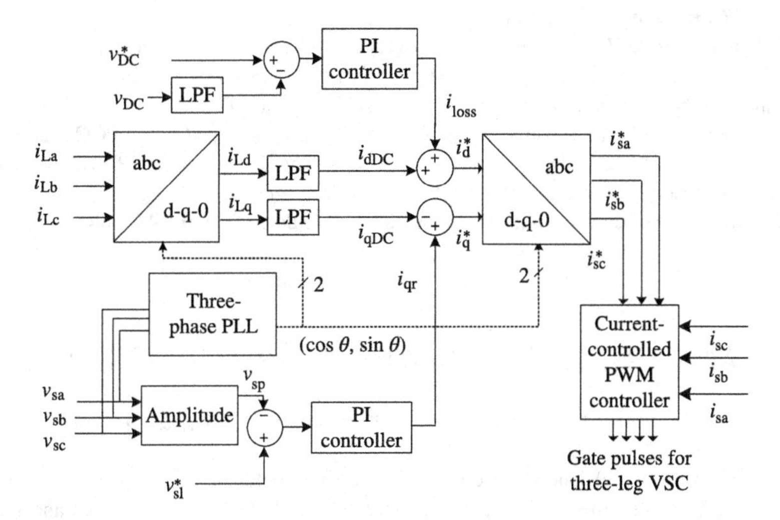

Figure 4.28 Block diagram of SRF theory-based control algorithm of DSTATCOMs

A SRF controller extracts DC quantities by a LPF and hence the non-DC quantities are separated from the reference signals. It can be operated in UPF and ZVR modes as shown below.

4.4.2.6.1 UPF Operation of DSTATCOMs

The control strategy for reactive power compensation for UPF operation considers that the supply must deliver the DC component of the direct-axis component of the load current (i_{dDC}) along with the active power component for maintaining the DC bus and meeting the losses (i_{loss}) in the DSTATCOM. The output of the PI controller at the DC bus voltage of the DSTATCOM is considered as the current (i_{loss}) for meeting its losses:

$$i_{loss}(n) = i_{loss}(n-1) + K_{pd}\{v_{de}(n) - v_{de}(n-1)\} + K_{id}v_{de}(n), \tag{4.53}$$

where $v_{de}(n) = v^*_{DC}(n) - v_{DC}(n)$ is the error between the reference (v^*_{DC}) and the sensed (v_{DC}) DC voltage at the nth sampling instant, and K_{pd} and K_{id} are the proportional and integral gain constants of the DC bus voltage PI controller, respectively.

Therefore, the reference direct-axis supply current is

$$i^*_d = i_{dDC} + i_{loss}. \tag{4.54}$$

The reference supply current must be in phase with the voltage at PCC but with no zero-sequence component. It is therefore obtained by the following reverse Park's transformation with i^*_d as in Equation 4.54 and i^*_q and i^*_0 as zero:

$$\begin{bmatrix} i^*_{sa} \\ i^*_{sb} \\ i^*_{sc} \end{bmatrix} = \begin{bmatrix} \cos\theta & \sin\theta & 1 \\ \cos\left(\theta - \dfrac{2\pi}{3}\right) & \sin\left(\theta - \dfrac{2\pi}{3}\right) & 1 \\ \cos\left(\theta + \dfrac{2\pi}{3}\right) & \sin\left(\theta + \dfrac{2\pi}{3}\right) & 1 \end{bmatrix} \begin{bmatrix} i^*_d \\ i^*_q \\ i^*_0 \end{bmatrix}. \tag{4.55}$$

4.4.2.6.2 ZVR Operation of DSTATCOMs

The control strategy for ZVR operation of the DSTATCOM considers that the supply must deliver the same direct-axis component i^*_d as mentioned in Equation 4.54 along with the difference of quadrature-axis current (i_{qDC}) of the load and the component obtained from the PI voltage controller (i_{qr}) used for regulating the voltage at PCC. The amplitude of the AC terminal voltage (V_{sp}) at PCC is controlled to its reference voltage (V^*_{sp}) using the PI voltage controller. The output of the PI voltage controller is considered as the reactive power component of current (i_{qr}) for zero voltage regulation of the AC voltage at PCC. The amplitude of the AC voltage (V_{sp}) at PCC is calculated from the AC voltages (v_{sa}, v_{sb}, v_{sc}) as

$$V_{sp} = (2/3)^{1/2}(v^2_{sa} + v^2_{sb} + v^2_{sc})^{1/2}. \tag{4.56}$$

Then, a PCC voltage PI controller is used to regulate this voltage to a reference value:

$$i_{qr}(n) = i_{qr}(n-1) + K_{pq}\{v_{te}(n) - v_{te}(n-1)\} + K_{iq}v_{te}(n), \tag{4.57}$$

where $v_{te}(n) = v^*_{sp}(n) - v_{sp}(n)$ denotes the error voltage between reference $v^*_{sp}(n)$ and actual $v_{sp}(n)$ terminal voltage amplitudes at the nth sampling instant, and K_{pq} and K_{iq} are the proportional and integral gain constants of the AC voltage PI controller, respectively.

The reference quadrature-axis supply current is

$$i_q^* = i_{qr} - i_{qDC}. \tag{4.58}$$

Three-phase reference supply currents are obtained by reverse Park's transformation using Equation 4.55 with i_d^* as in Equation 4.54, i_q^* as in Equation 4.58, and i_0^* as zero. These reference supply currents (i_{sa}^*, i_{sb}^*, i_{sc}^*) with the respective sensed supply currents (i_{sa}, i_{sb}, i_{sc}) are fed to a PWM current controller to generate the switching signals of the VSC used as a DSTATCOM.

4.4.2.7 ISCT-Based Control Algorithm of DSTATCOMs

Figure 4.29 shows a block diagram of the ISCT-based control algorithm used for the estimation of reference supply currents. This algorithm uses PCC phase voltages, average load power, and power factor angle for the estimation of reference supply currents (i_{sa}^*, i_{sb}^*, i_{sc}^*). The average power is estimated by using instantaneous load currents and PCC phase voltages. The product of the instantaneous PCC phase voltages (v_{sa}, v_{sb}, v_{sc}) and load currents (i_{La}, i_{Lb}, i_{Lc}) gives instantaneous load power (p_{Linst}), which is composed of an AC component (P_{LAC}) and a DC component (P_{LDC}). The average load power (P_{LDC}) is extracted from instantaneous power by passing it through a LPF. The reference supply currents estimated by the ISCT-based controller are given as

$$i_{sa}^* = \frac{v_{sa} + (v_{sb} - v_{sc})\beta}{|A|} \times P_{LDC}, \tag{4.59}$$

$$i_{sb}^* = \frac{v_{sb} + (v_{sc} - v_{sa})\beta}{|A|} \times P_{LDC}, \tag{4.60}$$

$$i_{sc}^* = \frac{v_{sc} + (v_{sa} - v_{sb})\beta}{|A|} \times P_{LDC}. \tag{4.61}$$

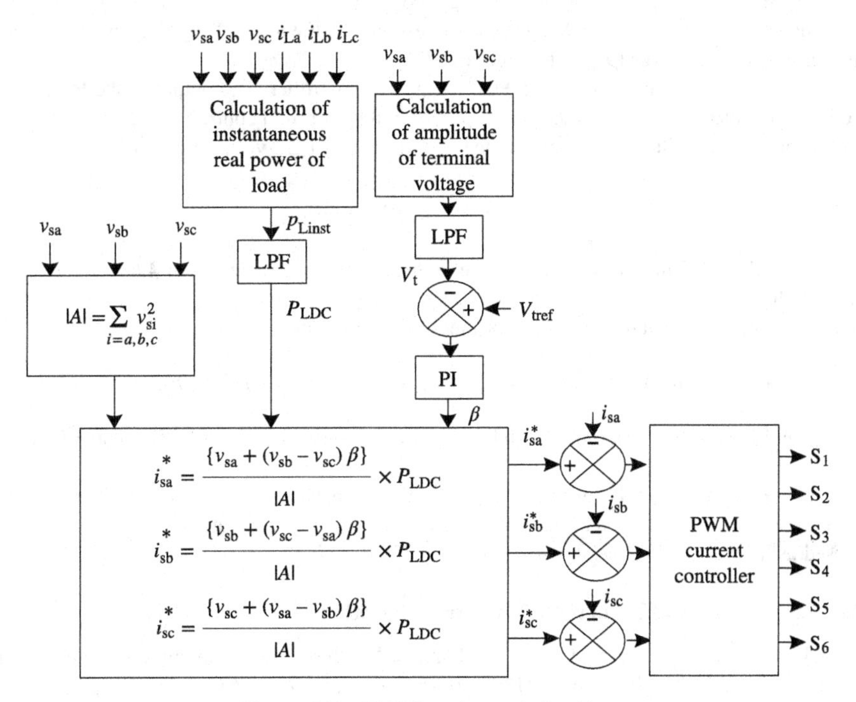

Figure 4.29 ISCT-based control algorithm

The term $|A|$ is defined as

$$|A| = \sum_{i=\mathrm{a,b,c}} v_{si}^2 = 3(\text{rms phase voltage})^2. \tag{4.62}$$

The angle β is defined as

$$\beta = \frac{\tan\phi}{\sqrt{3}}, \tag{4.63}$$

where ϕ is the phase difference between PCC phase voltage and fundamental supply current.

The PCC phase voltages are estimated using two sensed line voltages (v_{sab}, v_{sbc}) as

$$\begin{bmatrix} v_{sa} \\ v_{sb} \\ v_{sc} \end{bmatrix} = \frac{1}{3}\begin{bmatrix} 2 & 1 \\ -1 & 1 \\ -1 & -2 \end{bmatrix}\begin{bmatrix} v_{sab} \\ v_{sbc} \end{bmatrix}. \tag{4.64}$$

The amplitude of PCC phase voltages is estimated using instantaneous PCC phase voltages as

$$V_t = \sqrt{\frac{2}{3}\left(v_{sa}^2 + v_{sb}^2 + v_{sc}^2\right)}. \tag{4.65}$$

The average load power is estimated from instantaneous active power (p_{Linst}), which is obtained from phase voltages and sensed load currents as

$$p_{Linst} = v_{sa}i_{La} + v_{sb}i_{Lb} + v_{sc}i_{Lc} = P_{LDC} + P_{LAC}. \tag{4.66}$$

The instantaneous active power consists of a DC component (P_{LDC}) and a pulsating AC component (P_{LAC}). The DC component is the average power consumed by the loads. This component is extracted by passing instantaneous power (p_{Linst}) through a LPF.

Here a function β is computed using a PCC voltage PI controller. The output of the PCC voltage PI controller is considered as function β. The function β in the proposed controller is used to keep the PCC voltage at constant magnitude. The input of the PI controller is a PCC voltage error, which can be given as

$$V_{err}(n) = V_{tref}(n) - V_t(n), \tag{4.67}$$

where $V_t(n)$ is the amplitude of the PCC phase voltage, estimated in Equation 4.65, and $V_{tref}(n)$ is the reference voltage.

At the nth sampling instant, the output of the voltage controller is given as

$$\beta(n) = \beta(n-1) + k_{pv}\{V_{err}(n) - V_{err}(n-1)\} + k_{iv}V_{err}(n), \tag{4.68}$$

where k_{pv} and k_{iv} are the proportional and integral gain constants of the PCC voltage PI controller, respectively.

Three-phase reference supply currents (i_{sa}^*, i_{sb}^*, i_{sc}^*) are estimated using Equations 4.59–4.61 and sensed supply currents (i_{sa}, i_{sb}, i_{sc}) are fed to a PWM current controller to generate the switching signals of the VSC used as a DSTATCOM as shown in Figure 4.29.

4.4.2.8 Single-Phase PQ Theory-Based Control Algorithm of DSTATCOMs

Figure 4.30 shows a block diagram of the single-phase PQ theory-based control algorithm used for the estimation of reference supply currents. The load currents (i_{La}, i_{Lb}, i_{Lc}), PCC voltages (v_{sa}, v_{sb}, v_{sc}), supply currents, and DC bus voltage (v_{DC}) of the VSC of the DSTATCOM are sensed as feedback control signals.

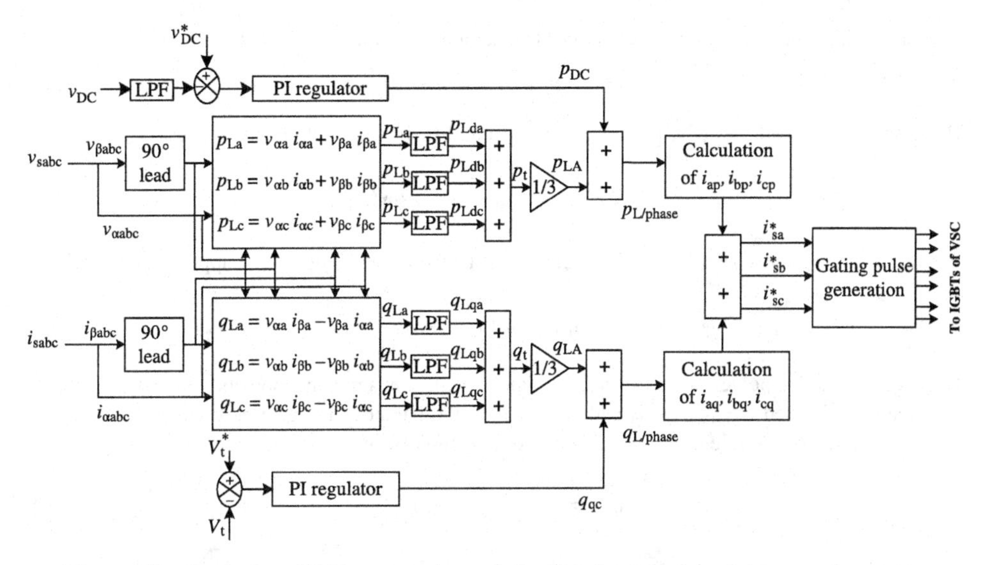

Figure 4.30 Single-phase PQ theory-based control algorithm for extracting reference supply currents

Any single-phase system can be realized as a two-phase system by giving a $\pi/2$ lag or lead to the PCC voltages and currents. The PCC voltages and currents are considered as quantities in the α-axis, whereas $\pi/2$ lead voltages and currents are considered as β-axis quantities.

The α–β quantities of the PCC voltages and currents are given as

$$v_{\alpha a} = v_{sa}(\omega t), \quad v_{\beta a} = v_{sa}(\omega t + \pi/2), \tag{4.69}$$

$$i_{\alpha a} = i_{sa}(\omega t + \varphi_L), \quad i_{\beta a} = i_{sa}[(\omega t + \varphi_L) + \pi/2)], \tag{4.70}$$

$$v_{\alpha b} = v_{sb}(\omega t), \quad v_{\beta b} = v_{sb}(\omega t + \pi/2), \tag{4.71}$$

$$i_{\alpha b} = i_{sb}(\omega t + \varphi_L), \quad i_{\beta b} = i_{sb}[(\omega t + \varphi_L) + \pi/2)], \tag{4.72}$$

$$v_{\alpha c} = v_{sc}(\omega t), \quad v_{\beta c} = v_{sc}(\omega t + \pi/2), \tag{4.73}$$

$$i_{\alpha c} = i_{sc}(\omega t + \varphi_L), \quad i_{\beta c} = i_{sc}[(\omega t + \varphi_L) + \pi/2)]. \tag{4.74}$$

The instantaneous active powers, using the single-phase approach, can be derived as

$$p_{La} = v_{\alpha a} i_{\alpha a} + v_{\beta a} i_{\beta a}, \quad p_{Lb} = v_{\alpha b} i_{\alpha b} + v_{\beta b} i_{\beta b}, \quad p_{Lc} = v_{\alpha c} i_{\alpha c} + v_{\beta c} i_{\beta c}. \tag{4.75}$$

The instantaneous active powers (p_{La}, p_{Lb}, p_{Lc}) of three phases consist of average, DC and oscillating, AC components. The average active powers are extracted from these instantaneous active powers using a low-pass filter. The average active power components can be represented as p_{Lda}, p_{Ldb}, and p_{Ldc}.

To achieve the balanced active power demand in all three phases, the average single-phase active power is estimated as

$$P_{LA} = (p_{Lda} + p_{Ldb} + p_{Ldc})/3. \tag{4.76}$$

The filtered values of the sensed DC bus voltage and the reference DC bus voltage of the VSC used as a DSTATCOM are compared to extract the error signal. The error in the DC bus voltage of the VSC is fed to the DC bus PI regulator and its output is used for maintaining the DC bus voltage. It is represented as p_{DC}.

Total fundamental active power component of reference supply currents is

$$p_t = p_{LA} + p_{DC}. \tag{4.77}$$

Three-phase PCC voltages (v_{sa}, v_{sb}, v_{sc}) are sensed and their amplitude is computed as

$$V_t = \sqrt{\{(2/3)(v_{sa}^2 + v_{sb}^2 + v_{sc}^2)\}}. \tag{4.78}$$

The instantaneous reactive powers of the load currents, using the single-phase approach, are derived as

$$q_{La} = v_{\alpha a} i_{\beta a} - v_{\beta a} i_{\alpha a}, \quad q_{Lb} = v_{\alpha b} i_{\beta b} - v_{\beta b} i_{\alpha b}, \quad q_{Lc} = v_{\alpha c} i_{\beta c} - v_{\beta c} i_{\alpha c}. \tag{4.79}$$

The average reactive powers are extracted from the instantaneous reactive powers using a low-pass filter. These are represented as q_{Lqa}, q_{Lqb}, and q_{Lqc}. To get the balanced reactive power demand in all three phases, the balanced single-phase reactive power is estimated as

$$q_{LA} = (q_{Lqa} + q_{Lqb} + q_{Lqc})/3. \tag{4.80}$$

Total fundamental reactive power component of the supply is given as

$$q_t = q_{qc} - q_{LA}, \tag{4.81}$$

where q_{qc} is the output of the PCC voltage PI controller. It is required for regulating the PCC voltage in ZVR mode of the DSTATCOM.

Three-phase reference supply currents (i_{sa}^*, i_{sb}^*, i_{sc}^*) are estimated as

$$i_{sa}^* = \left(\frac{v_{\alpha a}}{v_{\alpha a}^2 + v_{\beta a}^2}\right) p_t + \left(\frac{v_{\beta a}}{v_{\alpha a}^2 + v_{\beta a}^2}\right) q_t, \tag{4.82}$$

$$i_{sb}^* = \left(\frac{v_{\alpha b}}{v_{\alpha b}^2 + v_{\beta b}^2}\right) p_t + \left(\frac{v_{\beta b}}{v_{\alpha b}^2 + v_{\beta b}^2}\right) q_t, \tag{4.83}$$

$$i_{sc}^* = \left(\frac{v_{\alpha c}}{v_{\alpha c}^2 + v_{\beta c}^2}\right) p_t + \left(\frac{v_{\beta c}}{v_{\alpha c}^2 + v_{\beta c}^2}\right) q_t. \tag{4.84}$$

These estimated reference supply currents (i_{sa}^*, i_{sb}^*, i_{sc}^*) are compared with the sensed supply currents (i_{sa}, i_{sb}, i_{sc}) to generate the current errors. Amplified values of current errors through the PI current controllers are compared with fixed-frequency carrier signals to generate the gating signals of the IGBTs of the VSC used as a DSTATCOM.

4.4.2.9 Single-Phase DQ Theory-Based Control Algorithm of DSTATCOMs

Figure 4.31 shows a block diagram of the single-phase synchronous DQ frame theory-based control algorithm of DSTATCOMs. The reference supply currents are estimated using a single-phase synchronous DQ frame theory applied to the three-phase system.

4.4.2.9.1 Single-Phase Synchronously Rotating DQ Frame Theory

It is simple to design controllers for three-phase systems in a synchronously rotating DQ frame because all time-varying signals of the system become DC quantities and time-invariant. In case of a three-phase system, initially three-phase voltage or current signals (in the abc frame) are transformed to a stationary

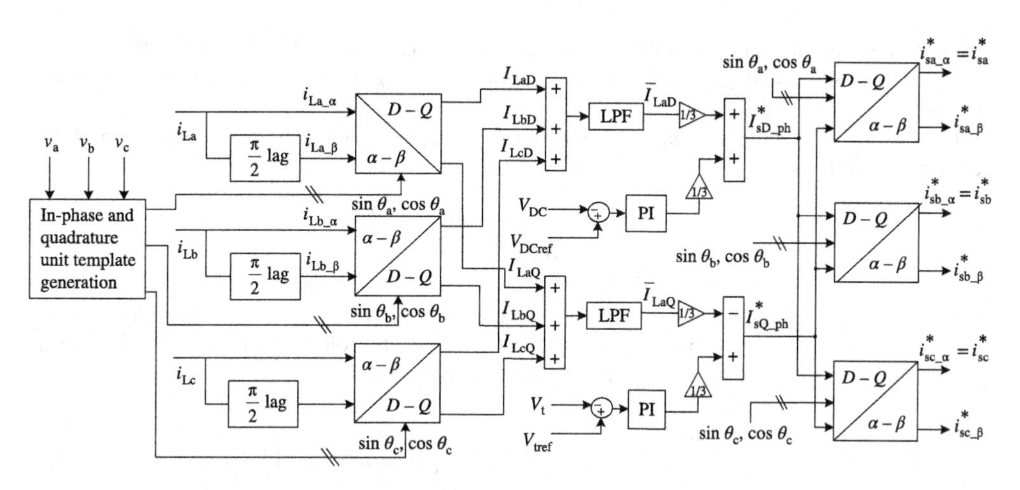

Figure 4.31 Single-phase synchronous DQ theory-based control algorithm for estimating reference supply currents

frame (α–β) and then to a synchronously rotating DQ frame. Similarly, to transform an arbitrary signal "$x(t)$" of a single-phase system into a synchronously rotating DQ frame, initially that variable is transformed into a stationary α–β frame and then into a synchronously rotating DQ frame. Therefore, to transform a signal into a stationary α–β frame, at least two phases are needed. Hence, a pseudo-second phase for the arbitrary signal $x(t)$ is created by giving a 90° lag to the original signal. The original signal represents the component of the α-axis and the 90° lag signal is the β-axis component of the stationary reference frame.

Therefore, an arbitrary periodic signal $x(t)$ with a time period of T can be represented in a stationary α–β frame as

$$x_\alpha(t) = x(t), \quad x_\beta(t) = x\left(t - \frac{T}{4}\right). \tag{4.85}$$

The signal $x(t)$ is represented as vector $\vec{x}$ and the vector $\vec{x}$ can be decomposed into two components x_α and x_β. Then as $\vec{x}$ vector rotates around the center, its components x_α and x_β that are the projections on the α–β axes vary in time accordingly. Now considering that there is a synchronously rotating DQ coordinate that rotates with the same angular frequency and direction as $\vec{x}$, the position of $\vec{x}$ with respect to its components x_D and x_Q is the same regardless of time. Therefore, it is clear that the x_D and x_Q do not vary with time and only depend on the magnitude of $\vec{x}$ and its relative phase with respect to the DQ rotating frame. The angle θ is the rotating angle of the DQ frame and is defined as

$$\theta = \int_0^t \omega \mathrm{d}t, \tag{4.86}$$

where ω is the angular frequency of the arbitrary variable $\vec{x}$.

The components of an arbitrary single-phase variable $x(t)$ in a stationary reference frame are transformed into a synchronously rotating DQ frame using the transformation matrix as

$$\begin{bmatrix} x_D \\ x_Q \end{bmatrix} = \begin{bmatrix} \cos\theta & \sin\theta \\ -\sin\theta & \cos\theta \end{bmatrix} \begin{bmatrix} x_\alpha \\ x_\beta \end{bmatrix}. \tag{4.87}$$

4.4.2.9.2 Reference Current Estimation Using Single-Phase Synchronously Rotating DQ Frame Theory

The control algorithm of the DSTATCOM employs an AC voltage PI controller to regulate the PCC voltage and a DC link voltage PI controller to maintain the DC bus capacitor voltage fixed and greater than the peak value of line voltage of PCC for successful operation of the DSTATCOM. The PCC voltages (v_a, v_b, v_c), supply currents (i_{sa}, i_{sb}, i_{sc}), load currents (i_{La}, i_{Lb}, i_{Lc}), and DC link voltage (V_{DC}) are sensed and used as feedback signals.

Considering PCC voltages as balanced and sinusoidal, the amplitude of the terminal voltage at PCC is estimated as follows:

$$V_t = \sqrt{\frac{2}{3}(v_a^2 + v_b^2 + v_c^2)}. \tag{4.88}$$

Consider one of the three phases at a time and then transform the voltages and currents of that particular phase into a stationary α–β frame using Equation 4.87. Then, the voltages and load currents of all the phases in a stationary α–β frame are represented as

$$v_{a\beta}(t) = v_a\left(t - \frac{T}{4}\right), \quad v_{b\beta}(t) = v_b\left(t - \frac{T}{4}\right), \quad v_{c\beta}(t) = v_c\left(t - \frac{T}{4}\right), \tag{4.89}$$

$$i_{La\alpha}(t) = i_{La}(t), \quad i_{Lb\alpha}(t) = i_{Lb}(t), \quad i_{Lc\alpha}(t) = i_{Lc}(t), \tag{4.90}$$

$$i_{La\beta}(t) = i_{La}\left(t - \frac{T}{4}\right), \quad i_{Lb\beta}(t) = i_{Lb}\left(t - \frac{T}{4}\right), \quad i_{Lc\beta}(t) = i_{Lc}\left(t - \frac{T}{4}\right). \tag{4.91}$$

Considering a synchronously rotating DQ frame for phase a that is rotating in the same direction as $v_a(t)$ and the projections of load current $i_{La}(t)$ to the D–Q axes gives the D and Q components of load currents. Therefore, the D and Q components of load currents in phase a are estimated as

$$\begin{bmatrix} I_{LaD} \\ I_{LaQ} \end{bmatrix} = \begin{bmatrix} \cos\theta_a & \sin\theta_a \\ -\sin\theta_a & \cos\theta_a \end{bmatrix} \cdot \begin{bmatrix} i_{La\alpha} \\ i_{La\beta} \end{bmatrix}, \tag{4.92}$$

where $\cos\theta_a$ and $\sin\theta_a$ are estimated using $v_{a\alpha}$ and $v_{a\beta}$ as follows:

$$\begin{bmatrix} \cos\theta_a \\ \sin\theta_a \end{bmatrix} = \frac{1}{\sqrt{v_{a\alpha}^2 + v_{a\beta}^2}} \begin{bmatrix} v_{a\alpha} \\ v_{a\beta} \end{bmatrix}. \tag{4.93}$$

Similarly, the D and Q components for the load currents in phases b and c are estimated as

$$\begin{bmatrix} I_{LbD} \\ I_{LbQ} \end{bmatrix} = \begin{bmatrix} \cos\theta_b & \sin\theta_b \\ -\sin\theta_b & \cos\theta_b \end{bmatrix} \cdot \begin{bmatrix} i_{Lb\alpha} \\ i_{Lb\beta} \end{bmatrix}, \tag{4.94}$$

$$\begin{bmatrix} I_{LcD} \\ I_{LcQ} \end{bmatrix} = \begin{bmatrix} \cos\theta_c & \sin\theta_c \\ -\sin\theta_c & \cos\theta_c \end{bmatrix} \cdot \begin{bmatrix} i_{Lc\alpha} \\ i_{Lc\beta} \end{bmatrix}. \tag{4.95}$$

The D-axis components of load currents of all the phases are added together to obtain an equivalent D-axis component current that represents the load as

$$I_{LD} = I_{LaD} + I_{LbD} + I_{LcD}. \tag{4.96}$$

Similarly, the equivalent Q-axis component current of the load is estimated as

$$I_{LQ} = I_{LaQ} + I_{LbQ} + I_{LcQ}. \tag{4.97}$$

These equivalent D-axis and Q-axis component currents of the total load are decomposed into two parts, namely, average and oscillatory parts, as

$$I_{LD} = \bar{I}_{LD} + \tilde{I}_{LD}, \quad (4.98)$$

$$I_{LQ} = \bar{I}_{LQ} + \tilde{I}_{LQ}. \quad (4.99)$$

The reason for the existence of the oscillatory AC part is the unbalanced loads in the system. The reference D-axis and Q-axis components of supply currents must be free from these oscillatory components. Hence, these signals are passed through a LPF to extract the fundamental (or DC) components.

To maintain the DC bus capacitor voltage of the DSTATCOM to a constant value, it is sensed and compared with the reference value and then the voltage error is processed in a PI controller.

The DC bus voltage error V_{DCer} at the kth sampling instant is expressed as

$$V_{DCer}(k) = V_{DCref}(k) - V_{DC}(k), \quad (4.100)$$

where $V_{DCref}(k)$ is the reference DC voltage and $V_{DC}(k)$ is the sensed DC link voltage of the DSTATCOM at the kth sampling instant. The output of the PI controller for maintaining the DC bus voltage of the DSTATCOM at the kth sampling instant is expressed as

$$I_{loss}(k) = I_{loss}(k-1) + K_{pd}\{V_{DCer}(k) - V_{DCer}(k-1)\} + K_{id}V_{DCer}(k), \quad (4.101)$$

where $I_{loss}(k)$ is considered as the power loss component of supply currents that must be supplied to meet the losses in the DSTATCOM, and K_{pd} and K_{id} are the proportional and integral gain constants of the DC bus voltage PI controller, respectively.

In order to ensure balanced and sinusoidal supply currents, the D-axis component of supply currents must be equal for all the phases and it should not contain any ripple. Now, the D-axis component of reference supply currents for each phase can be expressed as

$$I^*_{sDph} = \frac{\bar{I}_{LD} + I_{loss}}{3}. \quad (4.102)$$

For regulating the system voltage, the amplitude of the PCC voltage (V_t) is calculated using Equation 4.88 and compared with the reference voltage (V_{tref}).

The AC voltage error V_{er} at the kth sampling instant is

$$V_{er}(k) = V_{tref}(k) - V_t(k), \quad (4.103)$$

where V_{tref} is the amplitude of the reference AC terminal voltage and $V_t(k)$ is the amplitude of the sensed three-phase AC voltages at PCC at the kth instant.

The output of the PI controller for maintaining the PCC voltage constant at the kth sampling instant is expressed as

$$I_Q(k) = I_Q(k-1) + K_{pa}\{V_{er}(k) - V_{er}(k-1)\} + K_{ia}V_{er}(k), \quad (4.104)$$

where K_{pa} and K_{ia} are the proportional and integral gain constants of the PI controller, respectively, $V_{er}(k)$ and $V_{er}(k-1)$ are the voltage errors at the kth and $(k-1)$th instants, respectively, and I_Q is the equivalent Q-axis current required to meet the reactive power requirements of both the load and supply.

Therefore, the per-phase Q-axis component required for reference supply currents in order to regulate the system voltage is defined as

$$I^*_{sQph} = \frac{I_Q - \bar{I}_{LQ}}{3}. \quad (4.105)$$

For phase a, α-axis and β-axis components of reference supply currents can be estimated from the following matrix:

$$\begin{bmatrix} i^*_{sa\alpha} \\ i^*_{sa\beta} \end{bmatrix} = \begin{bmatrix} \cos\theta_a & \sin\theta_a \\ -\sin\theta_a & \cos\theta_a \end{bmatrix}^{-1} \cdot \begin{bmatrix} I^*_{sDph} \\ I^*_{sQph} \end{bmatrix}. \tag{4.106}$$

In the above matrix, the α-axis current represents the original reference supply current for phase a and the β-axis current represents the current that is at $\pi/2$ phase lag with respect to the original current.

Therefore, the reference supply current for phase a is

$$i^*_{sa} = I^*_{sDph}\cos\theta_a - I^*_{sQph}\sin\theta_a. \tag{4.107}$$

Similarly, the reference supply currents for phases b and c can be estimated as

$$i^*_{sb} = I^*_{sDph}\cos\theta_b - I^*_{sQph}\sin\theta_b, \tag{4.108}$$

$$i^*_{sc} = I^*_{sDph}\cos\theta_c - I^*_{sQph}\sin\theta_c. \tag{4.109}$$

Three-phase reference supply currents (i^*_{sa}, i^*_{sb}, i^*_{sc}) are compared with the sensed supply currents (i_{sa}, i_{sb}, i_{sc}). These current error signals are fed to a PWM current controller for switching of the IGBTs of the DSTATCOM.

4.4.2.10 Neural Network LMS-Adaline-Based Control Algorithm of DSTATCOMs

The modified Adaline (adaptive linear) NN (neural network)-based control algorithm for the estimation of reference supply currents is shown in Figure 4.32. The basic equations of this control strategy are given below.

4.4.2.10.1 Active Power Component of Reference Supply Currents

The three-phase PCC voltages (v_a, v_b, v_c) are sensed and filtered. Their amplitude is computed as

$$V_t = \{(2/3)(v_a^2 + v_b^2 + v_c^2)\}^{1/2}. \tag{4.110}$$

The in-phase unit vectors of v_a, v_b, and v_c are computed as

$$u_{pa} = v_a/V_t, \quad u_{pb} = v_b/V_t, \quad u_{pc} = v_c/V_t. \tag{4.111}$$

This error in the DC bus voltage of the VSC ($V_{DCe}(k)$) between the reference DC bus voltage ($V_{DCref}(k)$) and the sensed DC bus voltage ($V_{DC}(k)$) of the VSC of the DSTATCOM at the kth sampling instant is given as

$$V_{DCe}(k) = V_{DCref}(k) - V_{DC}(k). \tag{4.112}$$

The error in the DC bus voltage is given to a DC voltage PI controller. The output of the DC voltage PI controller is used for maintaining the DC bus voltage of the DSTATCOM at the kth sampling instant:

$$W_L(k) = W_L(k-1) + K_{pd}\{V_{DCe}(k) - V_{DCe}(k-1)\} + K_{id}V_{DCe}(k), \tag{4.113}$$

where $W_L(k)$ is part of the d-axis component of supply currents, and K_{pd} and K_{id} are the proportional and integral gain constants of the DC bus voltage PI controller, respectively.

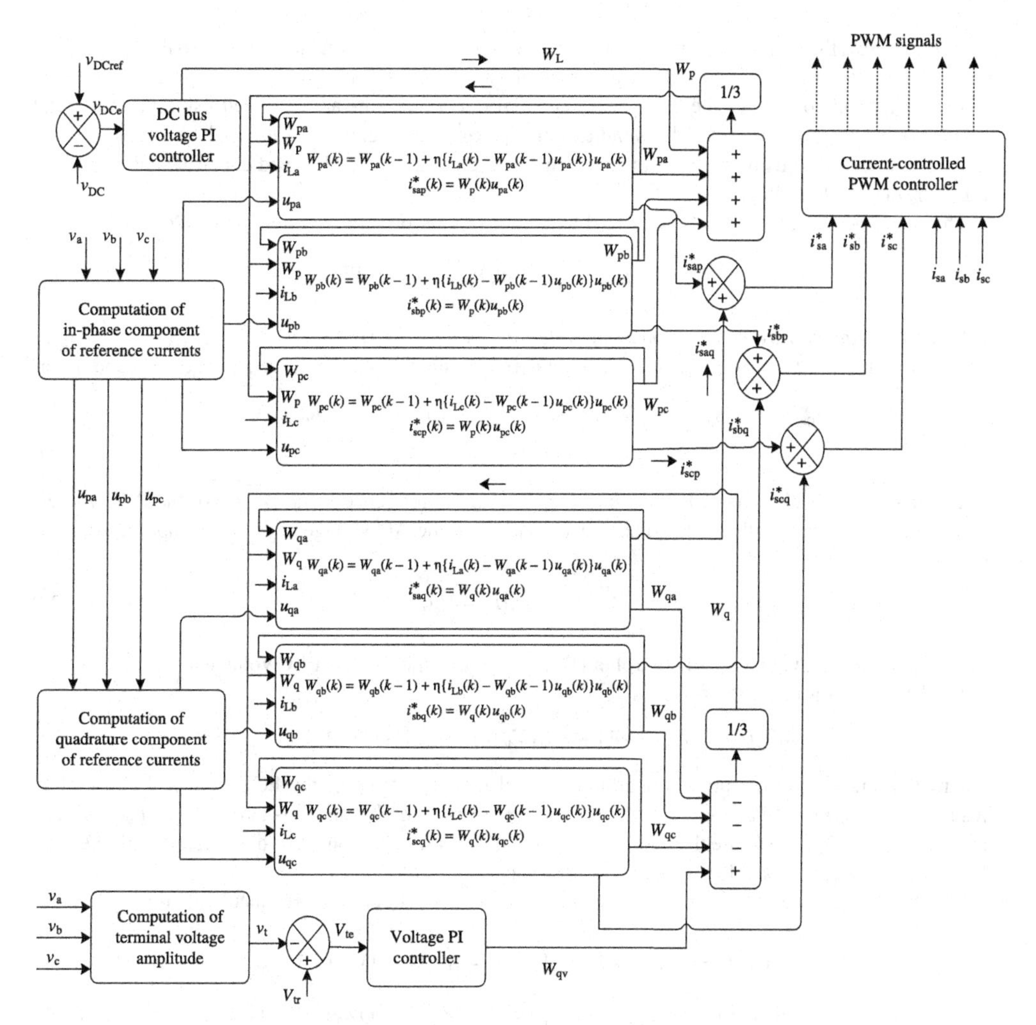

Figure 4.32 Neural network LMS-Adaline-based control algorithm for estimating reference supply currents

The average weight of the fundamental d-axis component of reference supply currents is given as

$$W_{\mathrm{p}}(k) = \{W_{\mathrm{L}}(k) + W_{\mathrm{pa}}(k) + W_{\mathrm{pb}}(k) + W_{\mathrm{pc}}(k)\}/3. \tag{4.114}$$

The extraction of weights of the fundamental d-axis components of the load currents is based on the least mean square (LMS) algorithm and its training through the Adaline NN control algorithm. The weights of the d-axis components of three-phase consumer load currents are estimated as

$$W_{\mathrm{pa}}(k) = W_{\mathrm{pa}}(k-1) + \eta\{i_{\mathrm{La}}(k) - W_{\mathrm{pa}}(k-1)u_{\mathrm{pa}}(k)\}u_{\mathrm{pa}}(k), \tag{4.115}$$

$$W_{\mathrm{pb}}(k) = W_{\mathrm{pb}}(k-1) + \eta\{i_{\mathrm{Lb}}(k) - W_{\mathrm{pb}}(k-1)u_{\mathrm{pb}}(k)\}u_{\mathrm{pb}}(k), \tag{4.116}$$

$$W_{\mathrm{pc}}(k) = W_{\mathrm{pc}}(k-1) + \eta\{i_{\mathrm{Lc}}(k) - W_{\mathrm{pc}}(k-1)u_{\mathrm{pc}}(k)\}u_{\mathrm{pc}}(k). \tag{4.117}$$

The accuracy of estimation and the rate of convergence are based on the value of convergence factor (η). When the value of η is high, the rate of convergence is fast, but the accuracy of estimation is less; when the value of η is low, the rate of convergence is slow, but the accuracy of estimation is high. So the value of η is selected to make a trade-off between the accuracy of estimation and the rate of convergence. The weights of the d-axis components of the three-phase consumer load currents are extracted using an individual Adaline in each phase. Similarly, the weights of the q-axis components of the three-phase consumer load currents are extracted using an individual Adaline in each phase. The observed experimental value of η varies from 0.01 to 1.0 for best results.

The fundamental three-phase reference d-axis components of supply currents are calculated as

$$i^*_{\mathrm{sap}} = W_{\mathrm{p}} u_{\mathrm{pa}}, \quad i^*_{\mathrm{sbp}} = W_{\mathrm{p}} u_{\mathrm{pb}}, \quad i^*_{\mathrm{scp}} = W_{\mathrm{p}} u_{\mathrm{pc}}. \tag{4.118}$$

4.4.2.10.2 Reactive Power Component of Reference Supply Currents

The quadrature unit vectors of v_{a}, v_{b}, and v_{c} are derived using in-phase unit vectors u_{pa}, u_{pb}, and u_{pc} as

$$\begin{aligned} u_{\mathrm{qa}} &= -u_{\mathrm{pb}}/\sqrt{3} + u_{\mathrm{pc}}/\sqrt{3}, \quad u_{\mathrm{qb}} = \sqrt{3}u_{\mathrm{pa}}/2 + (u_{\mathrm{pb}} - u_{\mathrm{pc}})/2\sqrt{3}, \\ u_{\mathrm{qc}} &= -\sqrt{3}u_{\mathrm{pa}}/2 + (u_{\mathrm{pb}} - u_{\mathrm{pc}})/2\sqrt{3}. \end{aligned} \tag{4.119}$$

The voltage error of the sensed PCC voltage ($V_{\mathrm{t}}(k)$) and the reference value ($V_{\mathrm{tr}}(k)$) of the PCC voltage is fed to an AC PI controller. The voltage error ($V_{\mathrm{te}}(k)$) of the AC voltage at the kth sampling instant is given as

$$V_{\mathrm{te}}(k) = V_{\mathrm{tr}}(k) - V_{\mathrm{t}}(k). \tag{4.120}$$

The output of the PCC voltage PI controller ($W_{\mathrm{qv}}(k)$) for maintaining the PCC voltage to a constant value at the kth sampling instant is given as

$$W_{\mathrm{qv}}(k) = W_{\mathrm{qv}}(k-1) + K_{\mathrm{pa}}\{V_{\mathrm{te}}(k) - V_{\mathrm{te}}(k-1)\} + K_{\mathrm{ia}} V_{\mathrm{te}}(k), \tag{4.121}$$

where K_{pa} and K_{ia} are the proportional and integral gain constants of the PCC voltage PI controller, respectively, $V_{\mathrm{te}}(k)$ and $V_{\mathrm{te}}(k-1)$ are the AC voltage errors at the kth and $(k-1)$th instants, respectively, and $W_{\mathrm{qv}}(k)$ and $W_{\mathrm{qv}}(k-1)$ are the amplitudes of the quadrature component of the fundamental reference current at the kth and $(k-1)$th instants, respectively.

The three-phase weights of the reactive components of load currents are calculated as

$$W_{\mathrm{qa}}(k) = W_{\mathrm{qa}}(k-1) + \eta\{i_{\mathrm{La}}(k) - W_{\mathrm{qa}}(k-1)u_{\mathrm{qa}}(k)\}u_{\mathrm{qa}}(k), \tag{4.122}$$

$$W_{\mathrm{qb}}(k) = W_{\mathrm{qb}}(k-1) + \eta\{i_{\mathrm{Lb}}(k) - W_{\mathrm{qb}}(k-1)u_{\mathrm{qb}}(k)\}u_{\mathrm{qb}}(k), \tag{4.123}$$

$$W_{\mathrm{qc}}(k) = W_{\mathrm{qc}}(k-1) + \eta\{i_{\mathrm{Lc}}(k) - W_{\mathrm{qc}}(k-1)u_{\mathrm{qc}}(k)\}u_{\mathrm{qc}}(k). \tag{4.124}$$

The average weight of the reference reactive power components of supply currents is given as

$$W_{\mathrm{q}}(k) = [W_{\mathrm{qv}}(k) - \{W_{\mathrm{qa}}(k) + W_{\mathrm{qb}}(k) + W_{\mathrm{qc}}(k)\}]/3. \tag{4.125}$$

The reference three-phase reactive power components of supply currents are given as

$$i^*_{\mathrm{saq}} = W_{\mathrm{q}} u_{\mathrm{qa}}, \quad i^*_{\mathrm{sbq}} = W_{\mathrm{q}} u_{\mathrm{qb}}, \quad i^*_{\mathrm{scq}} = W_{\mathrm{q}} u_{\mathrm{qc}}. \tag{4.126}$$

Total reference supply currents are the sum of active and reactive power components of the reference supply currents, which are computed as

$$i^*_{\mathrm{sa}} = i^*_{\mathrm{saq}} + i^*_{\mathrm{sap}}, \quad i^*_{\mathrm{sb}} = i^*_{\mathrm{sbq}} + i^*_{\mathrm{sbp}}, \quad i^*_{\mathrm{sc}} = i^*_{\mathrm{scq}} + i^*_{\mathrm{scp}}. \tag{4.127}$$

These estimated reference supply currents (i^*_{sa}, i^*_{sb} i^*_{sc}) are compared with the sensed supply currents (i_{sa}, i_{sb}, i_{sc}) to generate the current errors. Amplified values of current errors through the PI current controllers are compared with carrier signals to generate the gating signals of the IGBTs of the VSC used as a DSTATCOM.

4.5 Analysis and Design of DSTATCOMs

The analysis and design of DSTATCOMs include the detailed analysis for deriving the design equations for calculating the values of different components used in their circuit configurations. As already discussed in the previous section, there are a large number of topologies of DSTATCOMs; therefore, it is not practically possible to include here the design of all circuit configurations due to space constraints. In view of these facts, the design of selected three topologies of DSTATCOMs, one for three-phase three-wire DSTATCOMs and two for three-phase four-wire DSTATCOMs, is given here through a step-by-step design procedure.

The design of a three-phase three-wire DSTATCOM includes the design of the VSC and its other passive components. The DSTATCOM includes a VSC, interfacing inductors, and a ripple filter. The design of the VSC includes the DC bus voltage level, the DC capacitance, and the rating of IGBTs.

A three-phase three-wire DSTATCOM topology is considered for detailed analysis. Figure 4.6 shows a schematic diagram of one of the DSTATCOMs for a three-phase three-wire distribution system. It uses a three-leg VSC-based DSTATCOM. The design of the DSTATCOM is discussed in the following sections through the example of a 50 kVA, 415 V DSTATCOM.

4.5.1 Design of a Three-Phase Three-Wire DSTATCOM

The design of a DSTATCOM involves the estimation and selection of various components of the VSC of the DSTATCOM such as DC capacitor value, DC bus voltage, interfacing AC inductor, and a ripple filter. A ripple filter is used to filter the switching ripples from the voltage at PCC. The design of the interfacing inductors and a ripple filter is carried out to limit the ripple in the currents and voltages. The design of a DC bus capacitor depends on the energy storage capacity needed during transient conditions. The rating of the DSTATCOM depends on the required reactive power compensation and degree of unbalance in the load. Hence, the current rating of the DSTATCOM is affected by the load power rating and its voltage rating depends on the DC bus voltage. The design equations for the estimation and selection of the components of the VSC of the DSTATCOM and the voltage levels are given in the following sections.

4.5.1.1 Design of a Three-Phase Three-Leg VSC-Based DSTATCOM

A three-leg VSC is used as a distribution static compensator as shown in Figure 4.6 and this topology has six IGBTs, three AC inductors, and a DC capacitor. The required compensation to be provided by the DSTATCOM decides the rating of the VSC components. The VSC is designed for compensating a reactive power of 50 kVA (with a safety factor of 0.1) in a 415 V, 50 Hz, three-phase distribution system. The selection of an interfacing inductor, a DC capacitor, and a ripple filter is given in the following sections.

4.5.1.1.1 Selection of the DC Bus Voltage

The minimum DC bus voltage of the VSC of the DSTATCOM should be greater than twice of the peak of the phase voltage of the distribution system. The DC bus voltage is calculated as

$$V_{DC} = 2\sqrt{2}V_{LL}/(\sqrt{3}m), \tag{4.128}$$

where m is the modulation index and is considered as 1 and V_{LL} is the AC line output voltage of the DSTATCOM. Thus, V_{DC} is obtained as 677.69 V for a V_{LL} of 415 V and it is selected as 700 V.

4.5.1.1.2 Selection of a DC Bus Capacitor

The value of the DC capacitor (C_{DC}) of the VSC of the DSTATCOM depends on the instantaneous energy available to the DSTATCOM during transients. The principle of energy conservation is applied as

$$\frac{1}{2}C_{DC}\left(V_{DC}^2 - V_{DC1}^2\right) = k_1 3VaIt, \tag{4.129}$$

where V_{DC} is the nominal DC voltage equal to the reference DC voltage and V_{DC1} is the minimum voltage level of the DC bus, a is the overloading factor, V is the phase voltage, I is the phase current, and t is the time by which the DC bus voltage is to be recovered.

Considering the minimum voltage level of the DC bus (V_{DC1}) = 677.69 V, V_{DC} = 700 V, V = 239.60 V, I = 76.51 A, t = 30 ms, a = 1.2, and variation of energy during dynamics = 10% (k_1 = 0.1), the calculated value of C_{DC} is 12 882.75 μF and it is selected as 13 000 μF.

4.5.1.1.3 Selection of an AC Inductor

The selection of the AC inductance (L_r) of a VSC depends on the current ripple $i_{cr,pp}$, switching frequency f_s, and DC bus voltage (V_{DC}), and it is given as

$$L_r = \sqrt{3}mV_{DC}/(12af_s I_{cr,pp}), \tag{4.130}$$

where m is the modulation index and a is the overloading factor. Considering $I_{cr,pp}$ = 15%, f_s = 1.8 kHz, m = 1, V_{DC} = 700 V, and a = 1.2, the value of L_r is calculated to be 4 mH. The round-off value of 4 mH is selected in this investigation.

4.5.1.1.4 Selection of a Ripple Filter

A high-pass first-order filter tuned at half the switching frequency is used to filter out the noise from the voltage at PCC. The time constant of the filter should be very small compared with the fundamental time period (T), $R_f C_f \leq T_f$, considering $R_f C_f = T_s/10$, where R_f, C_f, and T_s are the ripple filter resistance, ripple filter capacitance, and switching time, respectively. Considering switching frequency equal to 1.8 kHz, the ripple filter parameters are selected as R_f = 10 Ω and C_f = 5.5 μF. The impedance offered for switching frequency is 18.93 Ω and impedance offered to fundamental frequency is 579.12 Ω, which is sufficiently large and hence the ripple filter draws negligible fundamental frequency current.

4.5.1.1.5 Voltage and Current Ratings of the Solid-State Switches

The voltage rating (V_{sw}) of the device can be calculated under dynamic conditions as

$$V_{sw} = V_{DC} + V_d, \tag{4.131}$$

where V_d is the 10% overshoot in the DC link voltage under dynamic conditions. The voltage rating of the switch is calculated as 770 V. With an appropriate safety factor, 1200 V, IGBTs are selected for the VSC used in the DSTATCOM.

The current rating (I_{sw}) of the device can be calculated under dynamic conditions as

$$I_{sw} = 1.25(I_{cr,pp} + I_{peak}). \tag{4.132}$$

From these equations, the voltage and current ratings of the IGBT switches can be estimated. The current rating of the switch is calculated as 149.59 A. Thus, a solid-state switch (IGBT) for the VSC is selected with the next available higher rating of 1200 V and 300 A.

4.5.2 Design of a Three-Phase Four-Wire DSTATCOM

The design of a DSTATCOM involves the estimation of their values and selection of components of the VSC and transformers of the DSTATCOM. This includes the selection of a DC capacitor, a DC capacitor bus voltage, an AC inductor, and a ripple filter. The rating of the DSTATCOM depends on the compensation required for the reactive power and unbalanced currents of the loads. Hence, the current rating of the DSTATCOM is affected by the load power rating and the voltage rating depends on the bus voltage. Depending on the switching frequency, the design of the inductor and ripple filter is carried out to reduce the effect of switching ripples. The design of the DC bus capacitor depends on the energy storage capacity needed during transient conditions. The details of the design of various configurations of three-phase four-wire DSTATCOMs are discussed in the following sections. These DSTATCOMs are designed for compensating a reactive power of 50 kVA.

4.5.2.1 Design of a Four-Leg VSC-Based DSTATCOM

The schematic diagram of a three-phase four-wire DSTATCOM based on a four-leg VSC is shown in Figure 4.12. It is clear that the eight IGBT-based switches are controlled for the required compensation using the DSTATCOM. The design of the DSTATCOM includes estimation and selection of DC bus voltage, AC inductors, and the VSC. The design of the DC bus voltage and DC capacitor is similar to a three-leg VSC given in Section 4.5.1.

4.5.2.1.1 Selection of the DC Bus Voltage

The value of the DC bus voltage (V_{DC}) depends on the minimum voltage level required for getting the desired AC output during PWM operation of the VSC of the DSTATCOM. For a three-leg VSC with split capacitors, the DC bus voltage across one capacitor is defined as

$$V_{DC} = 2\sqrt{2}V_{LL}/\sqrt{3}m, \quad (4.133)$$

where m is the modulation index and is considered as 1 and V_{LL} is the line–line PCC voltage of the DSTATCOM.

Thus, one may obtain the value of V_{DC} as 677.6 V for a V_{LL} of 415 V and it is selected as 700 V.

4.5.2.1.2 Selection of a DC Bus Capacitor

The design of the DC capacitor is governed by the reduction in the DC bus voltage of the DSTATCOM upon the application of the load and rise in the DC bus voltage on removal of the load. However, second harmonic appears in the DC bus voltage under unbalanced load currents or unbalanced voltages. By considering the second harmonic ripple voltage across the capacitor, the capacitance is calculated as

$$C_{DC} = I_0/(2\omega v_{DC,pp}), \quad (4.134)$$

where I_0 is the capacitor current, ω is the angular frequency, and $v_{DC,pp}$ is the ripple in capacitor voltage.

Considering the ripple as 1.5%, $v_{DC,pp} = 0.015 \times 700 = 10.5$ V, and $I_0 = 50\,000/700 = 71.42 \times 1.1 = 78.57$ A, C_{DC} is obtained as 11 909.337 μF. Thus, the capacitance is chosen to be 12 000 μF.

4.5.2.1.3 Selection of an AC Inductor

The ripple filter inductance for the neutral leg is

$$L_{rn} = mV_{DC}/(3\sqrt{3}af_s I_{cr,pp}), \quad (4.135)$$

Considering a 15% current ripple, $f_s = 1.8$ kHz, $m = 1$, $V_{DC} = 700$ V, and $a = 1.2$, the neutral leg ripple inductance (L_{rn}) is calculated to be 5.43 mH. The round-off value of 5.5 mH is selected in this investigation.

The selection of the AC inductance for phase depends on the current ripple $I_{cr,pp}$. The AC filter inductance is calculated as

$$L_r = \sqrt{3}mV_{DC}/(12af_s I_{cr,pp}). \quad (4.136)$$

Considering a 15% current ripple, $f_s = 1.8$ kHz, $m = 1$, $V_{DC} = 700$ V, and $a = 1.2$, L_r is calculated to be 4 mH. The value of 4 mH is selected in this investigation.

The rating of solid-state devices can be calculated as in Equations 4.131 and 4.132.

4.5.2.2 Design of a Three Single-Phase VSC-Based DSTATCOM

The three-phase four-wire DSTATCOM with three single-phase VSCs is shown in Figure 4.13. The design of the DC bus voltage and DC capacitor is similar to a three-leg VSC given in Section 4.5.1. At switching frequency f_s, the AC inductance (L_r) is determined using Equation 4.130. The selection of voltage and current ratings of solid-state switches is also given in Section 4.5.1.

4.5.2.2.1 Selection of the DC Bus Voltage

The value of the DC bus voltage (V_{DC}) depends on the minimum voltage level required for getting the desired AC output during PWM operation of the VSC of the DSTATCOM. For a six-leg VSC, the DC bus voltage across one capacitor is defined as

$$V_{DC} = \sqrt{2}V/m, \tag{4.137}$$

where m is the modulation index and is considered as 1 and V is the AC output phase voltage of the DSTATCOM.

Thus, one may obtain the value of V_{DC} as 338.8 V for a V of 239.6 V and it is selected as 360 V.

4.5.2.2.2 Selection of a DC Bus Capacitor

The design of the DC capacitor is governed by the reduction in the DC bus voltage of the DSTATCOM upon the application of the load and rise in the DC bus voltage on removal of the load. However, second harmonic appears in the DC bus voltage under unbalanced load currents or unbalanced voltages. By considering the second harmonic ripple voltage across the capacitor, the capacitance is calculated as

$$C_{DC} = I_0/(2\omega v_{DC,pp}), \tag{4.138}$$

where I_0 is the capacitor current, ω is the angular frequency, and $v_{DC,pp}$ is the ripple in capacitor voltage.

Considering the ripple as 5%, $v_{DC,pp} = 0.05 \times 360 = 18$ V, and $I_0 = 50\,000/360 = 138.88$ A, C_{DC} is obtained as 12 286.7 μF. Thus, each capacitance is chosen to be 12 500 μF.

4.5.2.2.3 Selection of an AC Inductor

The selection of the AC inductance depends on the current ripple $I_{cr,pp}$. The AC inductance is given as

$$L_r = mV_{DC}/(4af_sI_{cr,pp}). \tag{4.139}$$

Considering a 15% current ripple, $f_s = 1.8$ kHz, $m = 1$, $V_{DC} = 360$ V, and $a = 1.2$, L_r is calculated to be 2 mH. The value of 2 mH is selected in this investigation.

4.6 Modeling, Simulation, and Performance of DSTATCOMs

The MATLAB models of different topologies of DSTATCOMs are developed using Simulink and SimPowerSystems (SPS) toolboxes to simulate the performance of these DSTATCOMs in single-phase and three-phase distribution systems. A large number of cases of these topologies of DSTATCOMs are given in solved examples. Here, the performances of three topologies of three-phase DSTATCOMs with selected control algorithms are demonstrated for power factor correction and zero voltage regulation along with load balancing and neutral current compensation. The performance of DSTATCOMs is analyzed under balanced/unbalanced load conditions.

4.6.1 Performance of a SRF-Based Three-Leg VSC-Based DSTATCOM

The performance of a SRF-based three-leg VSC-based three-phase three-wire DSTATCOM is demonstrated for PFC and ZVR modes along with load balancing. The performance of the DSTATCOM is analyzed under varying loads.

4.6.1.1 Performance of a Three-Phase Three-Wire DSTATCOM for Load Balancing and ZVR Mode of Operation

The dynamic performance of the DSTATCOM under linear lagging power factor unbalanced load condition is shown in Figure 4.33. At 1.2 s, the load is changed to a two-phase load. The load is again applied at 1.4 s. The PCC voltages (v_s), supply currents (i_s), load currents (i_L), DC bus voltage (v_{DC}), and amplitude of voltage (V_s) at PCC are depicted in this figure. It is also observed that the DC bus voltage of the DSTATCOM is maintained close to the reference value under all disturbances. The amplitude of the PCC voltage is maintained at the reference value under various load disturbances, which shows the ZVR mode of operation of the DSTATCOM.

4.6.1.2 Performance of a Three-Phase Three-Wire DSTATCOM for Load Balancing and PFC Mode of Operation

The dynamic performance of the DSTATCOM under linear lagging power factor unbalanced load condition is shown in Figure 4.34. At 1.2 s, the load is changed to a two-phase load. The load is again applied at 1.4 s. The PCC voltages (v_s), source currents (i_s), load currents (i_L), DC bus voltage (v_{DC}), and amplitude of voltage (V_s) at PCC are depicted in this figure. It is also observed that the DC bus voltage of the DSTATCOM is maintained close to the reference value under all disturbances. The amplitude of the PCC voltage is maintained at the reference value under various load disturbances, which shows the PFC mode of operation of the DSTATCOM.

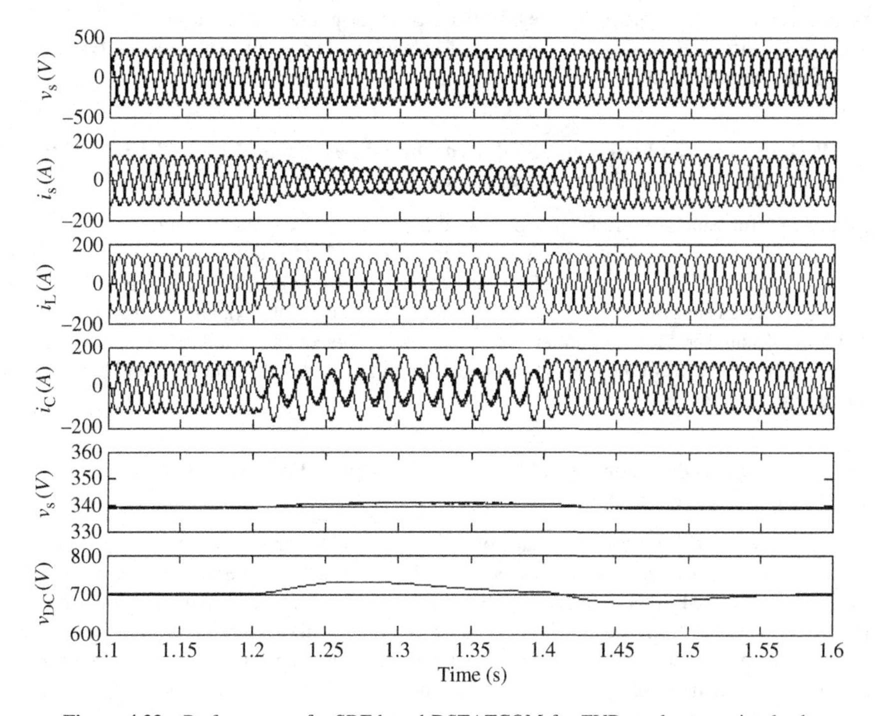

Figure 4.33 Performance of a SRF-based DSTATCOM for ZVR mode at varying loads

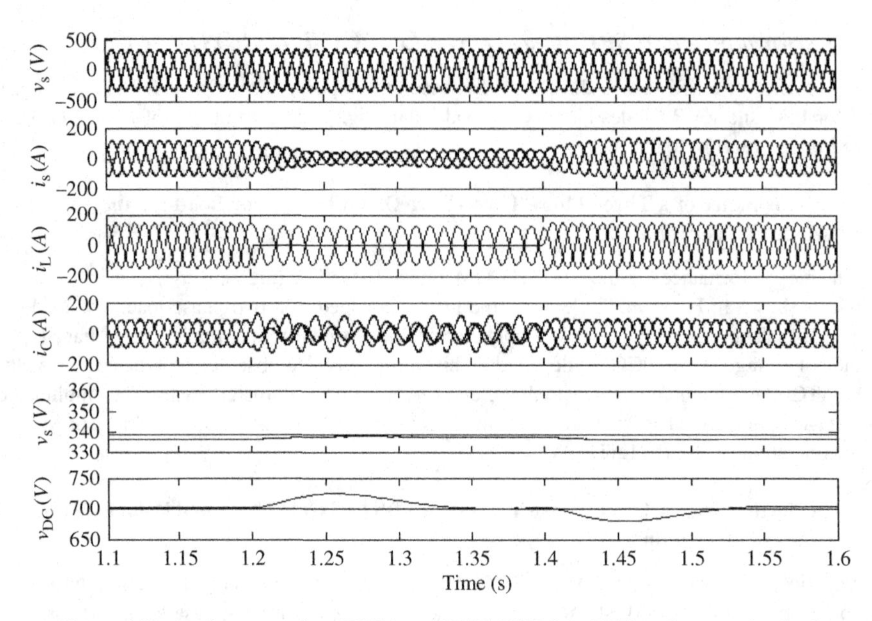

Figure 4.34 Performance of an IRPT-based DSTATCOM for UPF mode at varying loads

4.6.2 Performance of a Four-Leg VSC-Based Three-Phase Four-Wire DSTATCOM

The performance of a SRF-based four-leg VSC-based three-phase four-wire DSTATCOM is demonstrated for PFC and ZVR modes along with load balancing. The performance of the DSTATCOM is analyzed under varying loads.

4.6.2.1 Performance of a Four-Leg VSC-Based Three-Phase Four-Wire DSTATCOM for Load Balancing and PFC Mode of Operation

The dynamic performance of a four-leg VSC-based DSTATCOM for the UPF mode of operation along with neutral current compensation and load balancing at linear load is depicted in Figure 4.35. The PCC voltages (v_s), balanced source currents (i_s), load currents (i_L), compensator currents (i_C), load neutral current (i_{Ln}), compensator neutral current (i_{Cn}), supply neutral current (i_{sn}), and DC bus voltage (v_{DC}) of the DSTATCOM are shown under unbalanced load. The zero-sequence fundamental current of the load neutral current resulting from the unbalanced load current is circulated in the VSC and hence the supply neutral current is maintained at nearly zero.

4.6.2.2 Performance of a Four-Leg VSC-Based Three-Phase Four-Wire DSTATCOM for Load Balancing and ZVR Operation

Figure 4.36 shows the dynamic performance of the DSTATCOM for zero voltage regulation along with neutral current compensation and load balancing at linear load. It is observed that the amplitude of the PCC voltage is regulated to the reference amplitude. The zero-sequence fundamental current of the load neutral current resulting from the unbalanced load current is circulated in the VSC and hence the source neutral current is maintained at nearly zero. The PI controller regulates the DC bus voltage of the VSC of the DSTATCOM and the amplitude of voltage (v_s) is maintained near the reference voltage under different load disturbances.

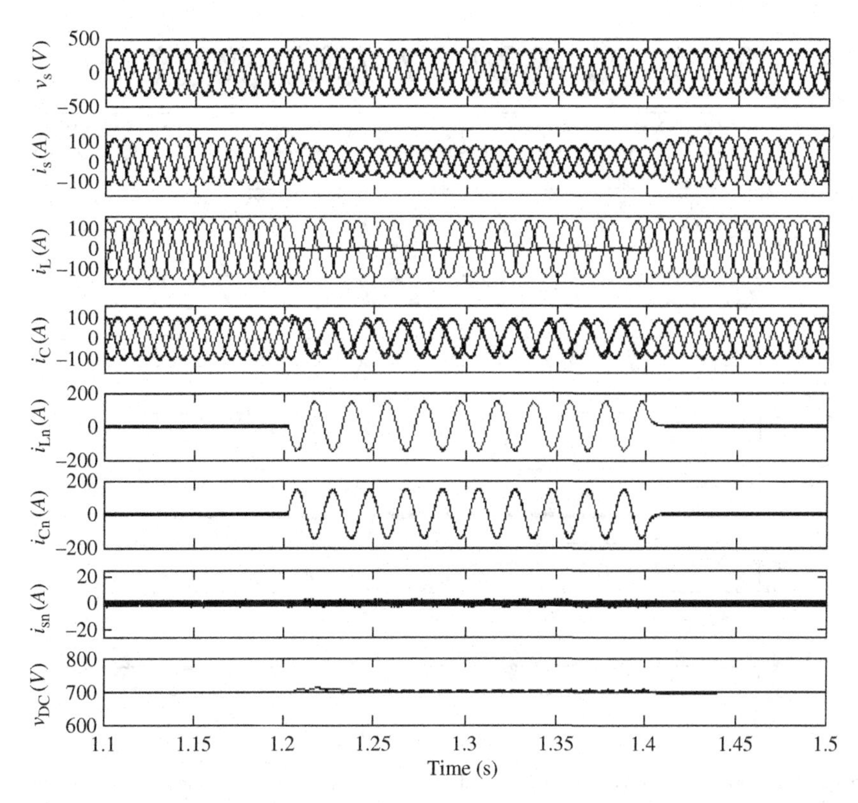

Figure 4.35 Performance of a four-leg VSC-based DSTATCOM for power factor correction along with neutral current compensation and load balancing

4.6.3 Performance of a Three Single-Phase VSC-Based Three-Phase Four-Wire DSTATCOM

The performance of a SRF-based three single-phase VSC-based three-phase four-wire DSTATCOM is demonstrated for PFC and ZVR modes along with load balancing. The performance of the DSTATCOM is analyzed under varying loads.

4.6.3.1 Performance of a Three Single-Phase VSC-Based Three-Phase Four-Wire DSTATCOM for Load Balancing and PFC Mode of Operation

The dynamic performance of the DSTATCOM for PFC and load balancing and neutral current compensation at linear load is depicted in Figure 4.37. The PCC voltages (v_s), balanced source currents (i_s), load currents (i_{La}, i_{Lb}, i_{Lc}), compensator currents (i_C), source neutral current (i_{sn}), load neutral current (i_{Ln}), compensator neutral current (i_{Cn}), DC bus voltage (v_{DC}), and amplitude of the PCC voltage (V_s) are shown for PFC at PCC. The zero-sequence fundamental current of the load neutral current resulting from the unbalanced load current is circulated in the VSC and hence the source neutral current is maintained at nearly zero. The PI controller regulates the DC bus voltage of the VSC of the DSTATCOM under different load disturbances.

4.6.3.2 Performance of a Three Single-Phase VSC-Based Three-Phase Four-Wire DSTATCOM for Load Balancing and ZVR Mode of Operation

Figure 4.38 shows the dynamic performance of the DSTATCOM for zero voltage regulation along with neutral current compensation and load balancing at linear load. The PCC voltages (v_s), balanced source

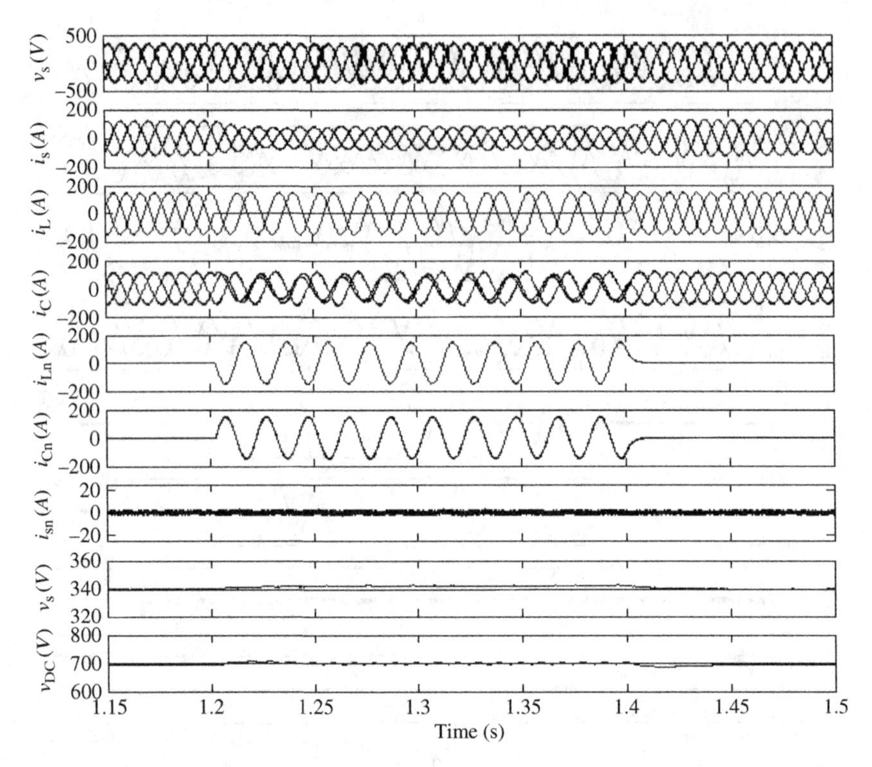

Figure 4.36 Performance of a four-leg VSC-based DSTATCOM for voltage regulation along with neutral current compensation and load balancing

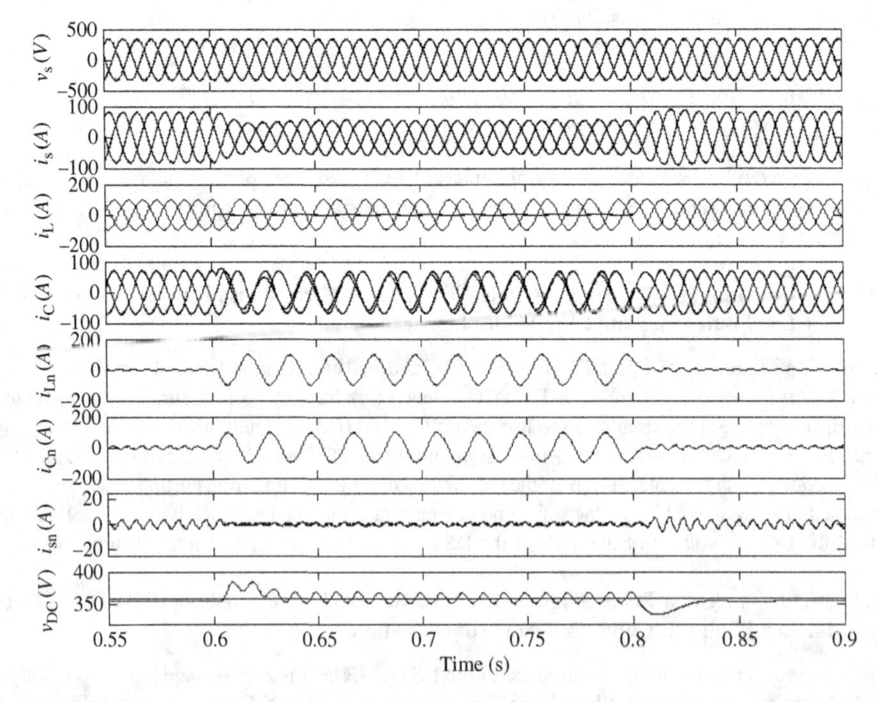

Figure 4.37 Performance of a three single-phase VSC-based DSTATCOM for UPF operation, load balancing, and neutral current compensation at linear load

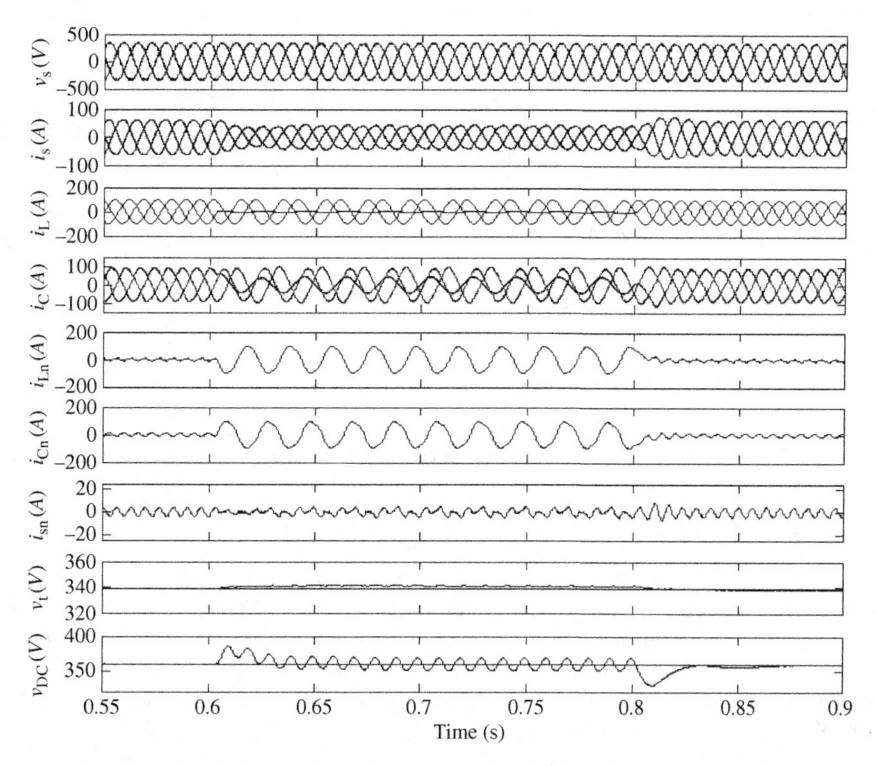

Figure 4.38 Performance of a three single-phase VSC-based DSTATCOM for voltage regulation operation, load balancing, and neutral current compensation

currents (i_s), load currents (i_{La}, i_{Lb}, i_{Lc}), source neutral current (i_{sn}), load neutral current (i_{Ln}), compensator neutral current (i_{Cn}), DC bus voltage (v_{DC}), and amplitude of the PCC voltage (V_s) are demonstrated under change of load conditions. It is observed that the amplitude of the PCC voltage is regulated to the reference amplitude.

4.7 Numerical Examples

Example 4.1

A single-phase load having $Z = (9.0 + j3.0)$ pu is fed from an AC supply with an input AC voltage of 230 V at 50 Hz and a base impedance of 9.15 Ω per phase. It is to be realized as a unity power factor load on the AC supply system using a PWM-based DSTATCOM, as shown in Figure E4.1. Calculate (a) the value of compensator current, (b) its VA rating, (c) supply current, and (d) equivalent resistance (in ohms) of the compensated load.

Solution: Given supply voltage $V_s = 230$ V, frequency of the supply $(f) = 50$ Hz, and a single-phase load having $Z = (9.0 + j3.0)$ pu with a base impedance of 9.15 Ω per phase.

The load resistance is $R = 9.15 \times 9.0\,\Omega = 82.35\,\Omega$. The load reactance is $X = 9.15 \times 3\,\Omega = 27.45\,\Omega$.

The load impedance is $Z = \sqrt{(82.35^2 + 27.45^2)} = 86.8\,\Omega$.

The load current before compensation is $I_{sold} = V/Z = 230/86.8$ A $= 2.65$ A.

a. The compensator current is $I_C = I_{sold} \sin\theta = I_{sold}X/Z = 2.65 \times 27.45/86.8$ A $= 0.84$ A.
b. The VA rating of the compensator is $S = VI_C = 230 \times 0.84 = 192.7$ VA.
c. The supply current after compensation is $I_{snew} = I_{sold} \cos\theta = I_{sold}R/Z = 2.65 \times 82.35/86.8$ A $= 2.51$ A.
d. The equivalent resistance of the compensated load is $R_{eq} = V/I_{snew} = 230/2.51 = 91.63\,\Omega$.

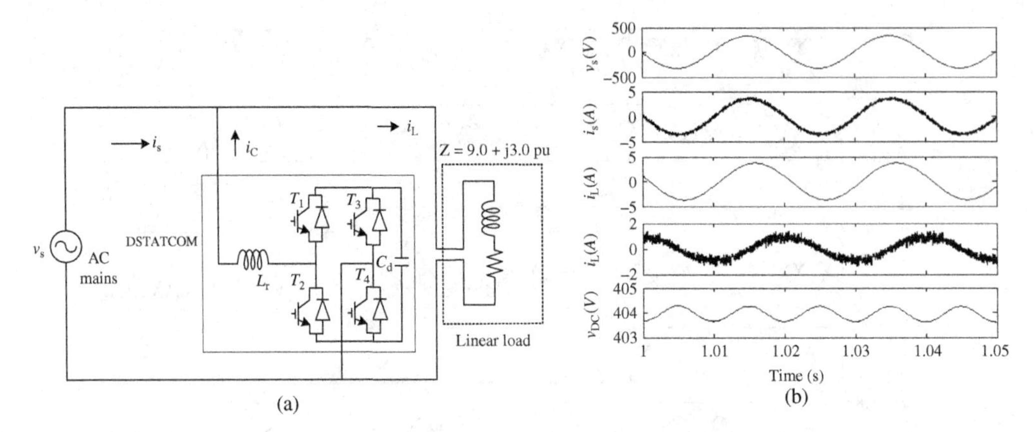

Figure E4.1 (a) A single-phase DSTATCOM for compensation of a linear load and (b) its waveforms

Example 4.2

A single-phase AC supply has AC mains voltage of 230 V at 50 Hz and a feeder (source) impedance of 1.0 Ω resistance and 3.0 Ω inductive reactance after which a load having $Z=(16+j12)$ Ω is connected, as shown in Figure E4.2. Calculate (a) the voltage drop across the source impedance and (b) the voltage across the load. If a PWM-based DSTATCOM is used to raise the voltage to the input voltage (230 V), calculate (c) the voltage rating of the DSTATCOM, (d) the VA rating of the compensator, and (e) the current rating of the compensator.

Solution: Given supply voltage $V_s=230$ V, frequency of the supply (f) = 50 Hz, and a single-phase load having $Z=(16+j12)$ Ω with a feeder (source) impedance of 1.0 Ω resistance and 3.0 Ω inductive reactance.

The load resistance is $R=16\,\Omega$. The load reactance is $X=12\,\Omega$.

The load admittance is $Y_L=(0.04-j0.03)$ mhos.

The active power consumed by the load after compensation is $P_L = V_s^2 G_L = 2116$ W.

The reactive power consumed by the load after compensation is $Q_L = V_s^2 B_L = -1587$ VAR.

The total impedance is $Z_{LT} = Z_s + Z_L = 17 + j15 = 22.672\angle 41.42°\,\Omega$.

The load current before compensation is $I_{sold} = V_s/Z_{Lt} = 230/22.672 = 10.144$ A.

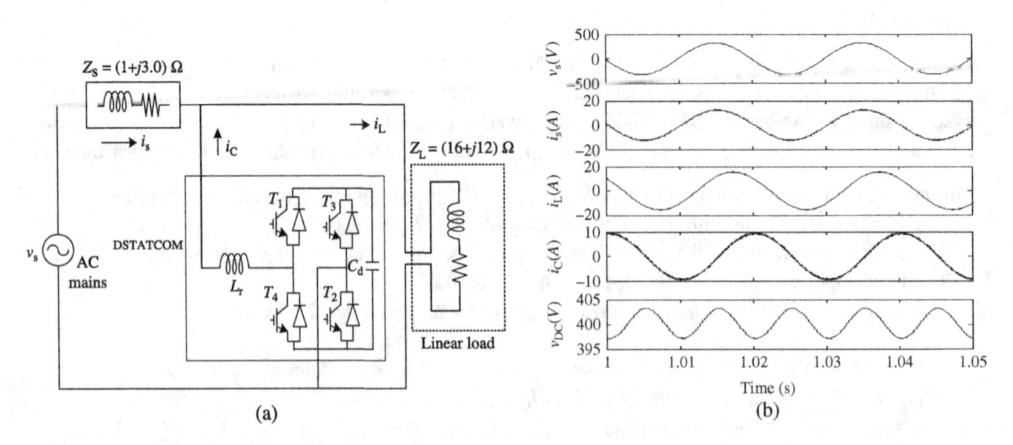

Figure E4.2 (a) A single-phase DSTATCOM for compensation of a linear load and (b) its waveforms

a. The voltage drop across the source impedance before compensation is $V_{Zs} = I_{sold}Z_s = 32.08$ V.
b. The voltage across the load before compensation is $V_{ZL} = I_{old}Z_L = 202.88$ V.
c. The voltage rating of the compensator is $V_C = V_s = 230$ V.
d. The VA rating of the compensator (Q_C) for $V_L = V_s$ is $Q_C = Q - Q_L$. The Q can be calculated from the quadratic equation $aQ^2 + bQ + c = 0$ as $Q = [-b \pm \{b^2 - 4ac\}^{1/2}]/2a$, where $a = R_s^2 + X_s^2$, $b = 2V_s^2X_s$, and $c = (V_L^2 + R_sP_L)^2 + X_s^2P_L^2 - V_s^2V_L^2$.
   Substituting the values, $a = (R_s^2 + X_s^2) = 1^2 + 3^2 = 10$, $b = 2V_L^2X_s = 317\,400$, $c = (V_L^2 + R_sP)^2 + X_s^2P^2 - V_s^2V_L^2 = 268\,647\,360$, and $Q = -870$ VA. Thus, $Q_C = Q + Q_L = -870 - 1587 = -2457$ VAR $= -2.457$ kVAR.
e. The current rating of the compensator is $I_C = Q_C/V_s = 2457/230 = 10.683$ A.

Example 4.3

A three-phase three-wire DSTATCOM is employed at a 415 V, 50 Hz system to provide reactive power compensation for unity power factor of a three-phase delta connected 25 kW induction motor operating at 0.8 lagging power factor, as shown in Figure E4.3. Calculate (a) supply line currents, (b) equivalent per-phase resistance (in ohms) of the compensated load, (c) DSTATCOM currents, (d) its kVA rating, (e) interfacing inductance, and (f) DC bus capacitance for the DSTATCOM, where the switching frequency is 2.5 kHz and DC bus voltage is 700 V and it has to be controlled within 8% range and ripple current in the inductor is 15%.

Solution: Given three-phase supply voltage $V_s = 415/\sqrt{3}$ V $= 239.6$ V, frequency of the supply $(f) = 50$ Hz, and a three-phase delta connected 25 kW induction motor operating at 0.8 lagging power factor. The switching frequency is 2.5 kHz and DC bus voltage is 700 V with 8% range and ripple current in the inductor is 15%.

The load line current is $I_{sold} = P/3V_s = 25\,000/(3 \times 239.6 \times 0.8) = 43.47$ A. (It is also supply current without compensation.)

a. The supply current after compensation is $I_{snew} = I_{sold} \cos\theta = I_{sold} \times PF = 43.47 \times 0.8 = 34.78$ A.
b. The equivalent resistance of the compensated load is $R_{eq} = V/I_{snew} = 239.6/34.78 = 6.88\ \Omega$.
c. The DSTATCOM current is $I_{DST} = I_{sold} \sin\theta = I_{sold}\{\sqrt{(1 - PF^2)}\} = 43.47 \times 0.6$ A $= 26.082$ A.
d. The VA rating of the DSTATCOM is $S = 3V_sI_{DST} = 3 \times 239.6 \times 26.082 = 18.75$ kVA.

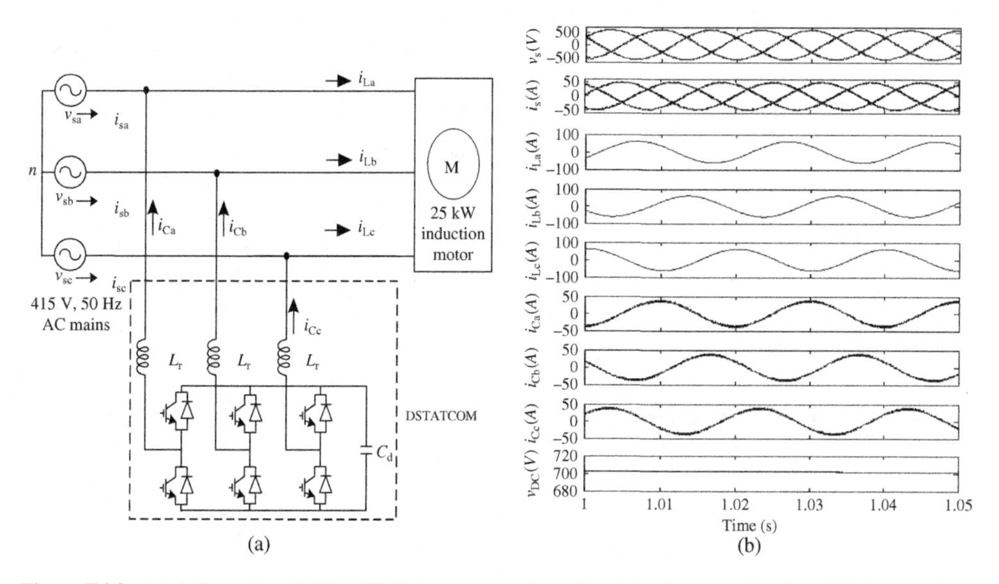

Figure E4.3 (a) A three-phase DSTATCOM for compensation of an induction motor load and (b) its waveforms

e. The interfacing inductance of the DSTATCOM is $L_{ST} = (\sqrt{3}/2) m_a V_{DCST}/6af_s\Delta I_{ST}$, where m_a (modulation index) = 1, $V_{DCST} = 700$ V, a (overloading factor) = 1.2, $f_s = 2.5$ kHz, $\Delta I_{ST} = 15\% = 0.15 \times 26.082 = 3.91$ A, and $V_{DCminST} = (1 - 0.05) \times 700 = 644$ V.

Therefore, $L_{ST} = (\sqrt{3}/2) \times 1.0 \times 700/(6 \times 1.2 \times 2500 \times 3.91) = 8.61$ mH.

f. The DC bus capacitance of the DSTATCOM is computed from change in stored energy during dynamics as follows.

The change in stored energy during dynamics is $\Delta E = (1/2)C_{DC}(V_{DCST}^2 - V_{DCminST}^2) = \sqrt{3} \times V_{ST} \times I_{ST} \times \Delta t$.

Substituting the values, $\Delta E = (1/2)C_{DC}(700^2 - 644^2) = \sqrt{3} \times 415 \times 26.082 \times (10/1000)$ (considering $\Delta t = 10$ ms).

$$C_{DC} = 4983.7\ \mu F.$$

Example 4.4

A three-phase three-wire DSTATCOM is employed at a 415 V, 50 Hz system to provide the load balancing and power factor correction of a single-phase 50 kW, unity power factor load across two lines, as shown in Figure E4.4. Calculate (a) supply line currents, (b) DSTATCOM currents, and (c) its kVA rating.

Solution: Given rms voltage $V_s = 415/\sqrt{3} = 239.6$ V, frequency of the supply (f) = 50 Hz, and the active power of a single-phase load (P) = 50 kW at UPF.

Load currents in lines a and b are equal before compensation: $I_{Lab} = -I_{Lba} = P/V_{ab} \times \text{PF} = 50\ 000/415 \times 1 = 120.48$ A.

This single-phase load connected across line to line is to be realized as a balanced load on the three-phase supply. A three-phase three-wire DSTATCOM is used for this purpose, which balances this load by circulating the currents on its AC sides. Let the load is connected across line ab in a three-phase abc system.

Load impedance is given by $|Z_L| = V_{ab}^2/S_L = 415^2/50\,000 = 3.44\ \Omega$.

$$Z_L = 3.44\angle 0°\ \Omega.$$

$$I_{La} = 415\angle 30°/3.44\angle 0° = 120.48\angle 30°\ \text{A}.$$

$$I_{Lb} = -I_{La} = -120.48\angle 30°\ \text{A}.$$

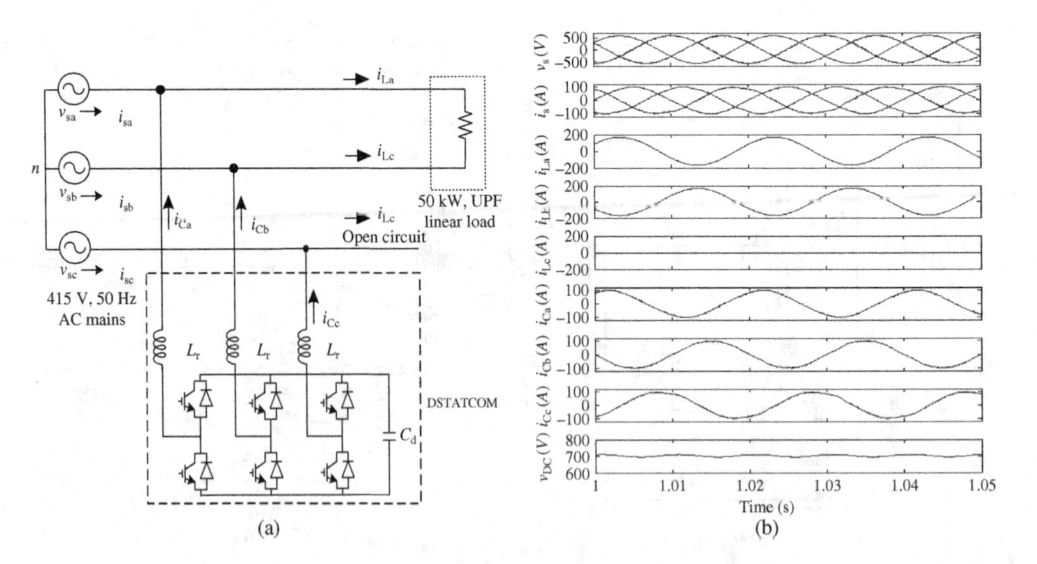

Figure E4.4 (a) A three-phase DSTATCOM for compensation of a load and (b) its waveforms

a. The supply line currents are $I_{sa} = I_{sb} = I_{sc} = I_s = P/3V_s = 50\,000/(3 \times 239.6) = 69.56$ A. (All three-phase lines have the same value of current as it is realized as a three-phase balanced UPF load.)
b. Three-phase DSTATCOM currents are $I_{Ca} = I_{La} - I_{sa} = 120.48\angle 30° - 69.56\angle 0° = 69.55\angle 60°$ A, $I_{Cb} = I_{Lb} - I_{sb} = -120.48\angle 30° - 69.56\angle -120° = 69.55\angle 180°$ A, and $I_{Cc} = I_{Lc} - I_{sc} = -I_{sc} = -69.56\angle -240° = 69.55\angle -60°$ A.
c. The VA rating of the DSTATCOM is $S = V_a I_{Ca} + V_b I_{Cb} + V_c I_{Cc} = 239.6(69.55 + 69.55 + 69.56) = 50$ kVA.

Example 4.5

A three-phase three-wire DSTATCOM is employed at a 415 V, 50 Hz system to provide load balancing and reactive power compensation for UPF of a single-phase 75 kVA, 0.8 lagging power factor load across two lines, as shown in Figure E4.5. Calculate (a) supply line currents, (b) DSTATCOM currents, and (c) its kVA rating.

Solution: Given rms voltage $V_s = 415/\sqrt{3} = 239.6$ V, frequency of the supply $(f) = 50$ Hz, and the active power of a single-phase load $(S_L) = 75$ kVA at 0.8 lagging power factor.

Load impedance is given by $|Z_L| = V_{ab}^2/S_L = 415^2/75\,000 = 2.296\ \Omega$.

$$Z_L = 2.296\angle 36.87°\ \Omega.$$

$$I_{La} = 415\angle 30°/2.296\angle 36.87° = 180.722\angle -6.87°\text{ A}.$$

$$I_{Lb} = -I_{La} = -180.722\angle -6.87°\text{ A}.$$

This single-phase load connected across line to line is to be realized as a balanced load on the three-phase supply. A three-phase three-wire DSTATCOM is used for this purpose, which balances this load by circulating the currents on its AC sides. Let the load is connected across line ab in a three-phase abc system.

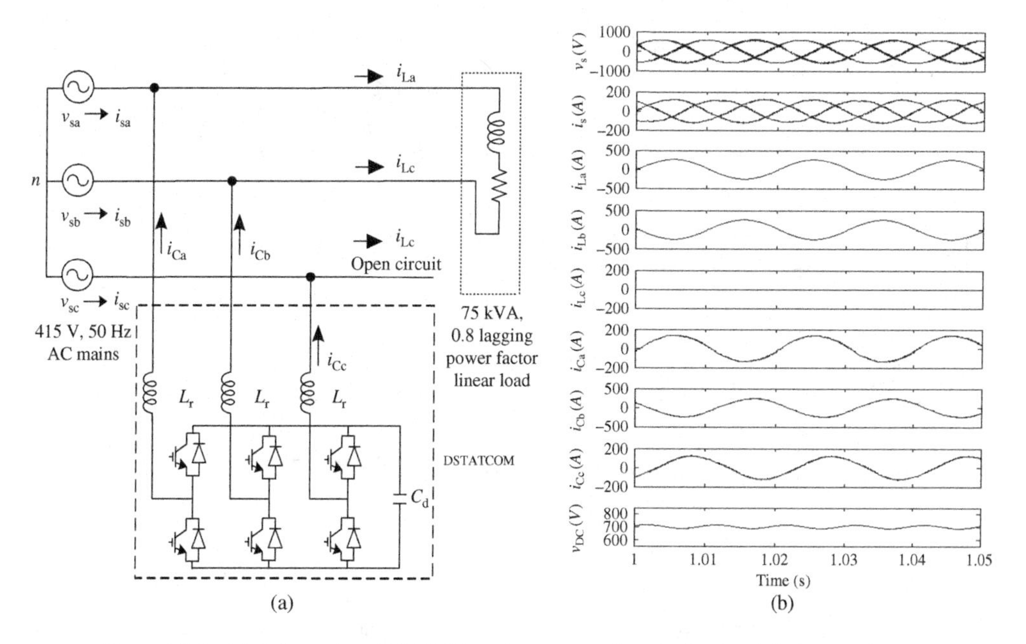

Figure E4.5 (a) A three-phase DSTATCOM for compensation of a load and (b) its waveforms

a. The supply line currents are $I_{sa} = I_{sb} = I_{sc} = I_s = S_L/3V_s = (75\ 000 \times 0.8)/(3 \times 239.6) = 83.47$ A. (All three-phase lines have the same value of current as it is realized as a three-phase balanced UPF load.)
b. Three-phase DSTATCOM currents are $I_{Ca} = I_{La} - I_{sa} = 181.22\angle{-6.87°} - 83.47\angle 0° = 98.35\angle{-12.66°}$ A, $I_{Cb} = I_{Lb} - I_{sb} = -181.22\angle{-6.87°} - 83.47\angle{-120°} = 166.65\angle 145.78°$ A, and $I_{Cc} = I_{Lc} - I_{sc} = -I_{sc} = -83.47\angle{-240°} = 83.47\angle{-60°}$ A (since there is no load current in this phase, $I_{Lc} = 0$).
c. The VA rating of the DSTATCOM is $S = V_a I_{Ca} + V_b I_{Cb} + V_c I_{Cc} = 239.6(98.85 + 167.10 + 83.47) = 83.72$ kVA.

Example 4.6

A three-phase unbalanced delta connected load having $Z_{ab} = (5.0 + j2.0)$ pu, $Z_{bc} = (2.5 + j1.75)$ pu, and $Z_{ca} = (9.5 + j2.5)$ pu is fed from an AC supply with an input line voltage of 415 V at 50 Hz and a base impedance of 9.15 Ω per phase, as shown in Figure E4.6. It is to be realized as a balanced unity power factor load on the three-phase supply system using a PWM-based DSTATCOM. Calculate (a) the supply line currents (in amperes), (b) DSTATCOM line currents (in amperes), (c) the kVA rating of the DSTATCOM, and (d) equivalent per-phase resistance (in ohms) of the compensated load.

Solution: Given supply voltage $V_s = 415$ V, frequency of the supply $(f) = 50$ Hz, and a three-phase three-wire unbalanced delta connected load having $Z_{ab} = (5.0 + j2.0)$ pu, $Z_{bc} = (2.5 + j1.75)$ pu, and $Z_{ca} = (9.5 + j2.5)$ pu with a base impedance of 9.15 Ω per phase.

The per-phase voltage is $V_{sp} = 415/\sqrt{3} = 239.6$ V.

Load impedances are $Z_{ab} = (45.75 + j18.3)$ Ω, $Z_{bc} = (22.87 + j16.01)$ Ω, and $Z_{ca} = (86.92 + j22.87)$ Ω. Load admittances are $Y_{ab} = (0.0188 - j0.007\ 53)$ mhos, $Y_{bc} = (0.0293 - j0.0205)$ mhos, and $Y_{ca} = (0.0108 - j0.002\ 83)$ mhos.

Load phase currents are $I_{Lab} = V_{ab}Y_{ab} = 415\angle 30° \times (0.0188 - j0.007\ 53) = 8.4\angle 8.17°$ A, $I_{Lbc} = V_{bc}Y_{bc} = 415\angle{-90°} \times (0.0293 - j0.0205) = 14.84\angle{-124.97°}$ A, and $I_{Lca} = V_{ca}Y_{ca} = 415\angle{-210°} \times (0.0107 - j0.002\ 83) = 4.59\angle 135.18°$ A.

The total active power of the load is $P_T = V_{sab}^2 G_{Lab} + V_{sbc}^2 G_{Lbc} + V_{sca}^2 G_{Lca} = 3237.83 + 5046.19 + 1853.81 = 10\ 126.83$ W.

The per-phase active power is $P = P_T/3 = 3375.61$ W.

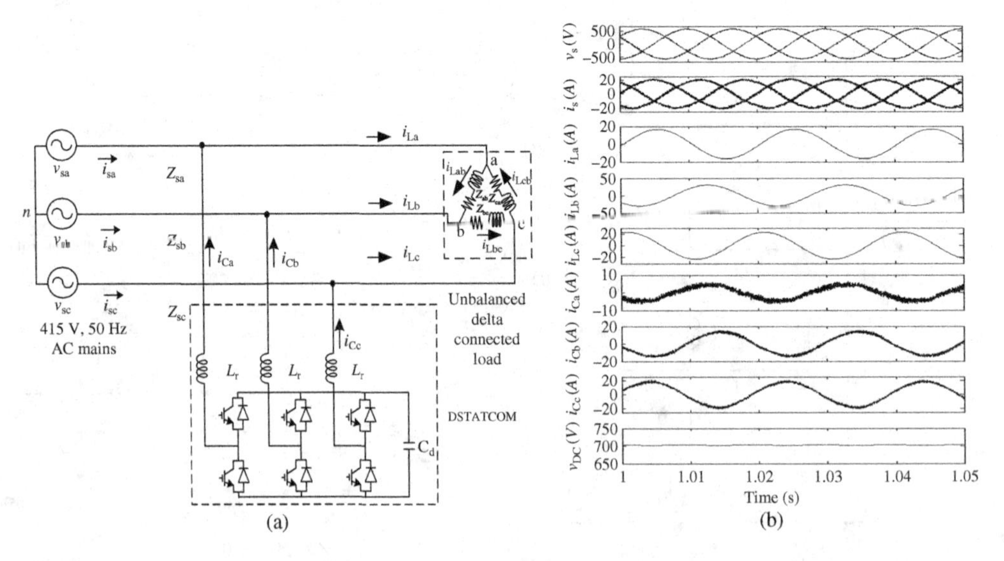

Figure E4.6 (a) A three-phase DSTATCOM for compensation of an unbalanced delta connected load and (b) its waveforms

a. Supply line currents after compensation are $I_{sa} = P/V_{sp}\angle 0°$ A, $I_{sb} = P_T/V_{sp}\angle -120°$ A, and $I_{sc} = P_T/V_{sp}\angle -120°$ A.
   Substituting the values, $I_{sa} = 14.09\angle 0°$ A, $I_{sb} = 14.09\angle -120°$ A, and $I_{sc} = 14.09\angle -240°$ A.
b. Three-phase DSTATCOM currents are $I_{Ca} = I_{La} - I_{sa} = I_{Lab} - I_{Lca} - I_{sa} = 8.4\angle 8.17° - 4.59\angle 135.18° - 14.09\angle 0° = 3.2378\angle -140.73°$ A, $I_{Cb} = I_{Lb} - I_{sb} = I_{Lbc} - I_{Lab} - I_{sb} = 14.84\angle -124.97° - 8.4\angle 8.17° - 14.09\angle -120° = 9.87\angle -173.2°$ A, and $I_{Cc} = I_{Lc} - I_{sc} = I_{Lca} - I_{Lbc} - I_{sc} = 4.59\angle 135.18° - 14.84\angle -124.97° - 14.09\angle -240° = 12.71\angle 14.56°$ A
c. The VA rating of the DSTATCOM is $Q_{DST} = V_s(|I_{Ca}| + |I_{Cb}| + |I_{Cc}|) = 6.186$ kVA.
d. The equivalent delta connected resistance of the compensated load is $R_{eq} = 3V_s^2/P_T = 3 \times 415^2/10\,126.83 = 50.89\ \Omega$.

Example 4.7

A three-phase four-wire unbalanced load having $Z_a = (3.0 + j2.0)$ pu connected between phase a and neutral terminal is fed from an AC supply with an input line voltage of 415 V at 50 Hz and a base impedance of 9.15 Ω per phase, as shown in Figure E4.7. It is to be realized as a balanced unity power factor load on the three-phase supply system using a four-leg PWM-based DSTATCOM. Calculate (a) the supply line currents (in amperes), (b) the DSTATCOM line currents (in amperes), (c) its neutral current (in amperes), (d) its kVA rating, and (e) equivalent per-phase resistance (in ohms) of the compensated load.

Solution: Given supply phase voltage $V_s = 415/\sqrt{3}$ V $= 239.6$ V, frequency of the supply $(f) = 50$ Hz, and a single-phase load having $Z_{La} = (3.0 + j2.0)$ pu connected between phase a and neutral terminal with a base impedance of 9.15 Ω per phase.

The load impedance per phase is $Z_{La} = 32.99\angle 33.69°\ \Omega = (27.45 + j18.3)\ \Omega$.

The load current before compensation is $I_{La} = 239.6/(32.99\angle 33.69°) = 7.26\angle -33.69°$ A.

The total load active power is $P_{La} = I_{La}^2 R_{La} = 7.26^2 \times 32.99 = 1446.8$ W.

a. Supply line currents after compensation are $I_{sa} = (P_{La}/3V_s)\angle 0°$ A, $I_{sb} = (P_{La}/3V_s)\angle -120°$ A, and $I_{sc} = (P_{La}/3V_s)\angle -240°$ A.

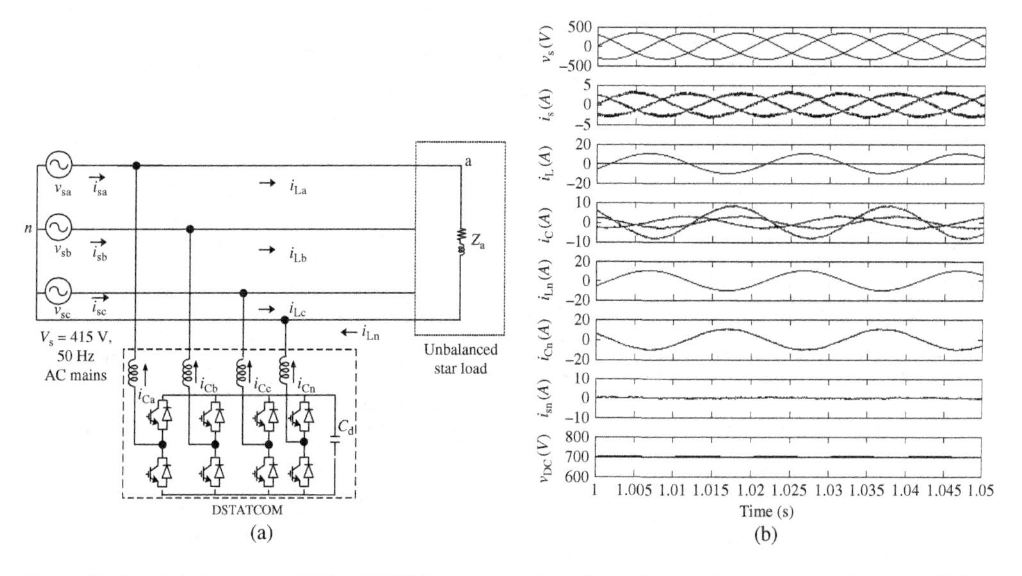

Figure E4.7 (a) A three-phase DSTATCOM for compensation of an unbalanced star connected load and (b) its waveforms

Substituting the values, $I_{sa} = \{1446.8/(3 \times 239.6)\}\angle 0°\text{ A} = 2.01\angle 0°\text{ A}$, $I_{sb} = 2.01\angle -120°\text{ A}$, and $I_{sc} = 2.01\angle -240°\text{ A}$.

b. Three-phase DSTATCOM currents are $I_{Ca} = I_{La} - I_{sa} = 7.26\angle -33.69° - 2.01\angle 0° = 5.69\angle -45°\text{ A}$, $I_{Cb} = I_{Lb} - I_{sb} = -2.01\angle -120° = 2.01\angle 60°\text{ A}$, and $I_{Cc} = I_{Lc} - I_{sc} = -2.01\angle -240° = 2.01\angle -60°\text{ A}$.

c. The load neutral current or DSTATCOM neutral current is $I_{Ln} = I_{Cn} = -(I_{Ca} + I_{Cb} + I_{Cc}) = 7.26\angle 146.3°\text{ A}$.

d. The VA rating of the DSTATCOM is $Q_{DST} = V_s(|I_{Ca}| + |I_{Cb}| + |I_{Cc}|) + V_s|I_{Cn}| = 239.6(5.69 + 2.01 + 2.01 + 7.26) = 4.066\text{ kVA}$.

e. Equivalent per-phase resistance (in star connection) of the compensated load is $R_{eq} = 3V_s^2/P_{La} = 3 \times 239.6^2/1446.8 = 119.03\ \Omega$.

Example 4.8

A three-phase four-wire unbalanced non-isolated star connected load having $Z_a = (9.0 + j3.0)$ pu, $Z_b = (3.0 + j1.5)$ pu, and $Z_c = (7.5 + j1.5)$ pu is fed from an AC supply with an input line voltage of 415 V at 50 Hz and a base impedance of 9.15 Ω per phase, as shown in Figure E4.8. It is to be realized as a balanced unity power factor load on the three-phase supply system using a four-leg PWM-based DSTATCOM. Calculate (a) the supply line currents (in amperes), (b) the DSTATCOM line currents (in amperes), (c) its neutral current (in amperes), (d) its kVA rating, and (e) equivalent per-phase resistance (in ohms) of the compensated load.

Solution: Given supply voltage $V_s = 415$ V, frequency of the supply $(f) = 50$ Hz, and a three-phase four-wire unbalanced non-isolated star connected load having $Z_a = (9.0 + j3.0)$ pu, $Z_b = (3.0 + j1.5)$ pu, and $Z_c = (7.5 + j1.5)$ pu with a base impedance of 9.15 Ω per phase.

Load impedances are $Z_a = (82.35 + j27.45)\ \Omega$, $Z_b = (27.45 + j13.72)\ \Omega$, and $Z_c = (68.62 + j13.72)\ \Omega$.

Three-phase load currents are $I_{La} = V_a/Z_a = (415/\sqrt{3})\angle 0°/(82.35 + j27.45) = 2.76\angle -18.43°\text{ A}$, $I_{Lb} = V_b/Z_b = (415/\sqrt{3})\angle -120°/(27.45 + j13.72) = 7.81\angle -146.56°\text{ A}$, and $I_{Lc} = V_c/Z_c = (415/\sqrt{3})\angle 120°/(68.62 + j13.72) = 3.42\angle 108.69°\text{ A}$.

Total active power of the load is $P_T = (2.76^2 \times 82.35 + 7.81^2 \times 27.45 + 3.42^2 \times 68.62)\text{ W} = 3104.26\text{ W}$.

The active power of the load per phase $= 3104.26/3 = 1034.75$ W.

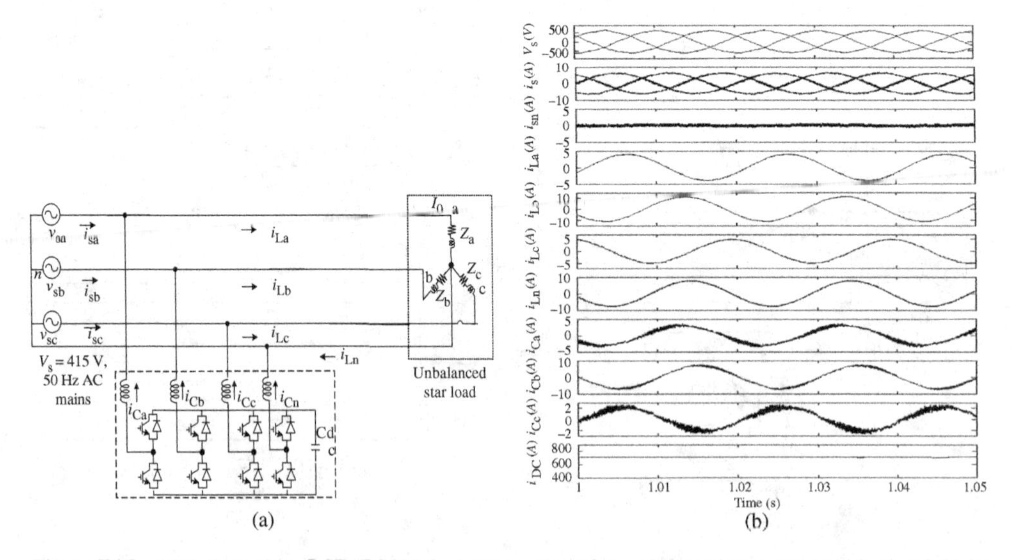

Figure E4.8 (a) A three-phase DSTATCOM for compensation of an unbalanced star connected load and (b) its waveforms

a. Three-phase source currents are $I_{sa} = P/V = 1034.75/(415/\sqrt{3}) = 4.32\angle 0°$ A, $I_{sb} = 4.32\angle -120°$ A, and $I_{sc} = 4.32\angle 120°$ A.
b. Three-phase compensator currents are $I_{Ca} = I_{La} - I_{sa} = 2.76\angle -18.43° - 4.32\angle 0° = 1.91\angle -152.8°$ A, $I_{Cb} = I_{Lb} - I_{sb} = 7.81\angle -146.56° - 4.32\angle -120° = 4.39\angle -172.64°$ A, and $I_{Cc} = I_{Lc} - I_{sc} = 3.42\ \angle 108.69° - 4.32\angle 120° = 1.17\angle -25°$ A.
c. The load neutral current is $I_{Ln} = -I_{Cn} = -(I_{La} + I_{Lb} + I_{Lc}) = -5.31\ \angle -158.76°$ A $= 5.31\angle 21.14°$ A.
d. The kVA rating is $Q_{DST} = V_s(|I_{Ca}| + |I_{Cb}| + |I_{Cc}| + |I_{Cn}|) = 3.062$ kVA.
e. The equivalent load resistance per phase (star equivalent) $= V_{ph}/I_s = 239.6/4.32 = 55.46\ \Omega$.

Example 4.9

A three-phase AC supply has a line voltage of 415 V at 50 Hz and a feeder (source) impedance of 1.0 Ω resistance and 3.0 Ω inductive reactance per phase after which a balanced delta connected load having $Z_L = 21\ \Omega$ per phase is connected, as shown in Figure E4.9. Calculate (a) the voltage drop across the source impedance and (b) the voltage across the load in each case. If a PWM-based DSTATCOM is used to maintain the voltage equal to the input voltage (415 V), calculate (c) the kVA rating of the DSTATCOM and (d) its line currents.

Solution: Given supply line voltage of 415 V at 50 Hz and a feeder (source) impedance of 1.0 Ω resistance and 3.0 Ω inductive reactance per phase after which a balanced delta connected load having $Z_{LD} = 21\ \Omega$ per phase is connected.

The equivalent load impedance in star connection is $Z_{LY} = Z_{LD}/3 = 21/3 = 7\ \Omega$ per phase.

The total impedance in star connection is $Z_{LT} = Z_s + Z_{LY} = (8 + j3) = 8.544\angle 20.56°\ \Omega$.

The load current before compensation is $I_{sold} = V/Z = 239.6/8.544 = 28.04$ A.

a. The voltage drop across the source impedance is $V_{Zs} = I_{old}Z_s = 88.68$ V.
b. The voltage across the load is $V_{ZL} = I_{old}Z_{LY} = 196.28$ V per phase (star) $= 339.96$ V (in delta connection).

   The active power consumed by the load after compensation is $P_T = 3V_s^2/Z_{LY} = 24\,603.57$ W.

   The active power consumed by the load per phase after compensation is $P = V_s^2/R_{LD} = 8201.19$ W.

   The reactive power consumed by the load after compensation is $Q_L = V_s^2 B_{LD} = 0.0$ VAR.

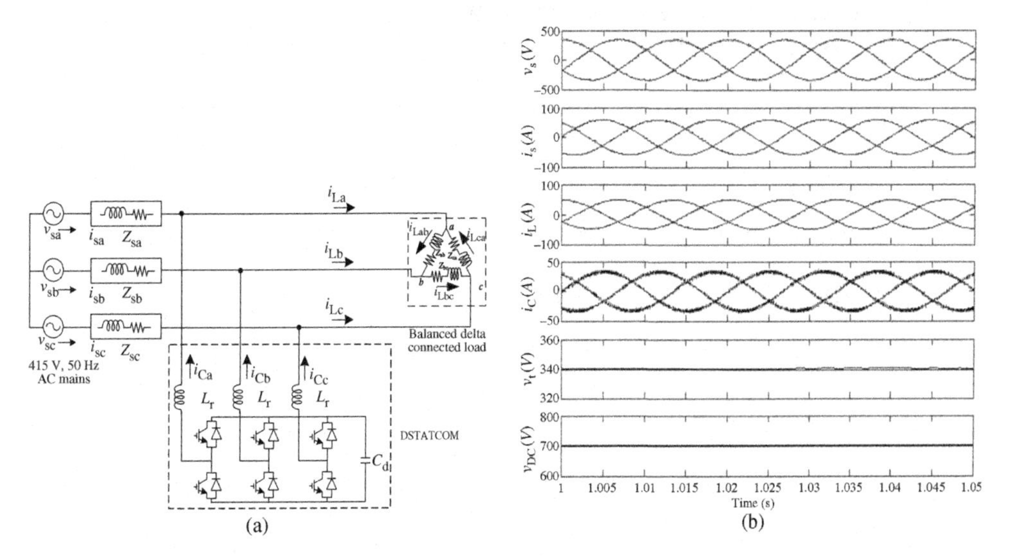

Figure E4.9 (a) A three-phase DSTATCOM for compensation of a balanced delta connected load and (b) its waveforms

c. The voltage rating of the DSTATCOM is $V_{DST} = V_s = V_L = 239.6$ V.

The VA rating of the DSTATCOM (Q_{DST}) for $V_L = V_s$ is $Q_{DST} = Q - Q_L$. The Q can be calculated from the quadratic equation $aQ^2 + bQ + c = 0$ as $Q = [-b \pm \{b^2 - 4ac\}^{1/2}]/2a$, where $a = R_s^2 + X_s^2$, $b = 2V_s^2 X_s$, and $c = (V_L^2 + R_s P)^2 + X_s^2 P^2 - V_s^2 V_L^2$.

Substituting the values, $a = R_s^2 + X_s^2 = 1^2 + 3^2 = 10$, $b = 2V_L^2 X_s = 344\,448.96$, $c = (V_L^2 + R_s P)^2 + X_s^2 P^2 - V_s^2 V_L^2 = 1614\,225\,629$, and $Q = -5595.32$ VA. Thus, $Q_{DST} = Q - Q_L = -5595.32 - 0.0 = -5595.32$ VAR $= -5595.32$ VAR per phase.

Total VA rating of the DSTATCOM $= 3Q_{DST} = 16.78$ kVA.

d. The current rating of the DSTATCOM is $I_{DST} = Q_{DST}/3V_s = 23.353$ A.

Example 4.10

A three-phase AC supply has a line voltage of 415 V at 50 Hz and a feeder (source) impedance of 1.0 Ω resistance and 3.0 Ω inductive reactance per phase after which a single-phase load having $Z_{ab} = (20 + j10)$ Ω is connected between two lines, as shown in Figure E4.10. If a three-leg PWM-based DSTATCOM is used to balance and maintain the voltage equal to the input voltage (415 V), calculate (a) the supply line currents (in amperes), (b) the DSTATCOM line currents (in amperes), (c) its kVA rating, (d) the voltage drop across the source impedance, and (e) equivalent per-phase resistance (in ohms) of the compensated load.

Solution: Given supply line voltage $V_{sab} = 415\ \text{V}\angle 30°$ V, frequency of the supply (f) = 50 Hz, and a feeder (source) impedance of 1.0 Ω resistance and 3.0 Ω inductive reactance per phase after which a single-phase load having $Z_{Lab} = (20 + j10)$ Ω is connected between two lines.

Load admittances are $Y_{Lab} = 1/Z_{Lab} = 1/(20 + j10) = (0.04 - j0.02)$ mhos, $Y_{Lbc} = 1/Z_{Lbc} = 0.0$ mhos, $Y_{Lca} = 1/Z_{Lca} = 0.0$ mhos.

Load currents in lines a and b are equal before compensation: $I_{Lab} = -I_{Lba} = V/(Z_{Lab} + 2Z_s) = 415\ \text{V}\angle 30°/(22^2 + 16^2)^{1/2} = 15.25\angle -6.02°$ A.

Load currents after compensation are $I_{Lab} = -I_{Lba} = V/Z_{Lab} = 415\ \text{V}\angle 30°/(20 + j10) = 18.55\angle 3.43°$ A.

All these loads have rated voltage after compensation as the DSTATCOM is connected in parallel to the loads.

$$P_T = V_{sab}^2 G_{Lab} + V_{sbc}^2 G_{Lbc} + V_{sca}^2 G_{Lca} = 6889\ \text{W}.$$

The per-phase active power is $P = P_T/3 = 2296.33$ W.

Reactive power required by the load $= I_{Lab}^2 X_L = 15.25^2 \times 10 = 2325.6$ VAR.

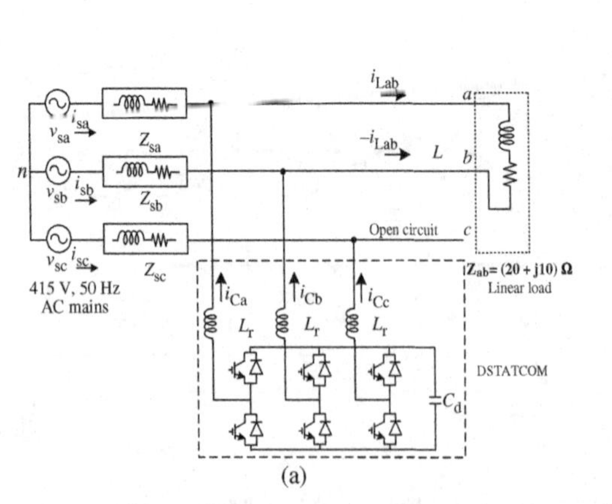

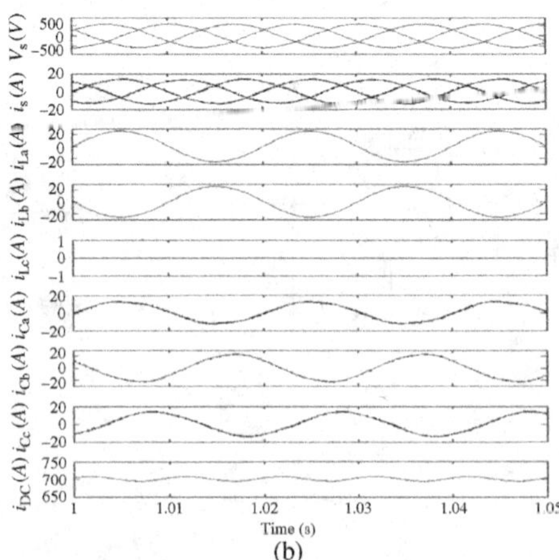

(a) (b)

Figure E4.10 (a) A three-phase DSTATCOM for compensation of a load and (b) its waveforms

The per-phase required reactive power Q_V at this UPF load of $P = 2296.33$ W per phase (in single line diagram of a star connected equivalent balanced load) for $V_L = V_s$ is $Q_V = Q - Q_L$. The Q can be calculated from the quadratic equation $aQ^2 + bQ + c = 0$ as $Q = [-b \pm \{b^2 - 4ac\}^{1/2}]/2a$, where $a = R_s^2 + X_s^2$, $b = 2V_s^2X_s$, and $c = (V_L^2 + R_sP)^2 + X_s^2P^2 - V_s^2V_L^2$.

Substituting the values $R_s = 1\,\Omega$, $X_s = 3\,\Omega$, $P = 2296.33$ W, and $V_s = V_L = 239.6$ V, $a = R_s^2 + X_s^2 = 1^2 + 3^2 = 10$, $b = 2V_L^2X_s = 344\,448.96$, and $c = (V_L^2 + R_sP)^2 + X_s^2P^2 - V_s^2V_L^2 = 316\,387\,474.8$, and $Q = -944.43$ VA per phase.

Total reactive power is $Q_T = 3 \times (-944.43) = -2833.29$ VAR. Thus, $Q_V = Q - Q_L = -2833.29 - 2325.6 = -5158.89$ VAR.

Per-phase reactive (capacitive) current for voltage regulation into the source is $I_V = Q_V/V_L = 944.43/239.6 = 3.94$ A.

a. The supply line currents after compensation have a resistive component because of the load active power and a reactive current I_V for the voltage regulation to compensate the drop in the source impedance: $I_{sa} = [(P/V_s)\angle 0° + jQ_V/V_L]$ A, $I_{sb} = [(P/V_s) + jQ_V/V_L]\angle -120°$ A, and $I_{sc} = [(P/V_s) + jQ_V/V_L]\angle -240°$ A.

   Substituting the values, $I_{sa} = [(2296.33/239.6)\angle 0° + j3.94]$ A $= 10.36\angle 22.35°$ A, I sb $= 10.36\angle -97.65°$ A, and $I_{sc} = 10.36\angle -217.65°$ A.

b. Three-phase DSTATCOM currents are $I_{Ca} = I_{Lab} - I_{Lca} - I_{sa} = 18.55\angle 3.43° - 10.36\angle 22.35° = 9.38\angle -17.54°$ A, $I_{Cb} = I_{Lbc} - I_{Lab} - I_{sb} = -18.56\angle -3.43° - 10.36\angle -97.65° = 19.43\angle 151.34°$ A, and $I_{Cc} = I_{Lca} - I_{Lbc} - I_{sc} = -10.36\angle -217.65°$ A.

c. The VA rating of the DSTATCOM is $Q_{DST} = V_s(|I_{Ca}| + |I_{Cb}| + |I_{Cc}|) = 239.6(9.38 + 19.43 + 10.36) = 9.385$ kVA.

d. The voltage drop across the source impedance is $V_{Zs} = (I_{LActive}\angle 0° + jQ_T/3V_s)Z_s = [(2296.33/239.6) + j3.94] \times 3.162 = 32.76$ V.

e. Equivalent per-phase resistance (in delta connection) of the UPF and balanced compensated load is $R_{eq} = 3V_{sab}^2/P_{LT} = 3 \times 415^2/6889 = 75\,\Omega$.

Example 4.11

A three-phase AC supply has a line voltage of 415 V at 50 Hz and a feeder (source) impedance of 1.0 Ω resistance and 3.0 Ω inductive reactance per phase after which an unbalanced isolated star configured load having $Z_{La} = 10\,\Omega$, $Z_{Lb} = 20\,\Omega$, and $Z_{Lc} = 40\,\Omega$ is connected, as shown in Figure E4.11. If a three-leg PWM-based DSTATCOM is used to balance and maintain the voltage equal to the input voltage (415 V), calculate (a) the DSTATCOM line currents, (b) the kVA rating of the DSTATCOM, (c) the voltage drop across the source impedance, and (d) equivalent per-phase resistance (in ohms) of the compensated load.

Solution: Given supply voltage $V_s = 415$ V, frequency of the supply $(f) = 50$ Hz, and a feeder (source) impedance of 1.0 Ω resistance and 3.0 Ω inductive reactance per phase after which a three-phase three-wire unbalanced isolated star connected load having $Z_{La} = 10\,\Omega$, $Z_{Lb} = 20\,\Omega$, and $Z_{Lc} = 40\,\Omega$ is connected.

Load impedances in star connection are $Z_{La} = 10\,\Omega$, $Z_{Lb} = 20\,\Omega$, and $Z_{Lc} = 40\,\Omega$.

Load admittances in star connection are $Y_a = 0.1\angle 0°$ mhos, $Y_b = 0.05\angle 0°$ mhos, and $Y_c = 0.025\angle 0°$ mhos.

Load admittances in delta connection are $Y_{ab} = Y_aY_b/(Y_a + Y_b + Y_c)$, $Y_{bc} = Y_bY_c/(Y_a + Y_b + Y_c)$, and $Y_{ca} = Y_cY_a/(Y_a + Y_b + Y_c)$.

Substituting the values, $Y_{ab} = 0.028\,757\,1428\angle 0°$ mhos, $Y_{bc} = 0.007\,142\,857\,143\angle 0°$ mhos, and $Y_{ca} = 0.014\,285\,714\angle 0°$ mhos.

Load phase currents are $I_{Lab} = V_{ab}Y_{ab} = 415\angle 30° \times 0.028\,757\,1428\angle 0° = 11.934\angle 30°$ A, $I_{Lbc} = V_{bc}Y_{bc} = 415\angle -90° \times 0.007\,142\,857\,143\angle 0° = 2.964\,285\,714\angle -90°$ A, and $I_{Lca} = V_{ca}Y_{ca} = 415\angle -210° \times 0.014\,285\,714\angle 0° = 5.928\,571\,429\angle -210°$ A.

The total active power of the load is $P_T = V_{sab}^2G_{Lab} + V_{sbc}^2G_{Lbc} + V_{sca}^2G_{Lca} = 8611.25$ W.

The per-phase active power is $P = P_T/3 = 2870.42$ W.

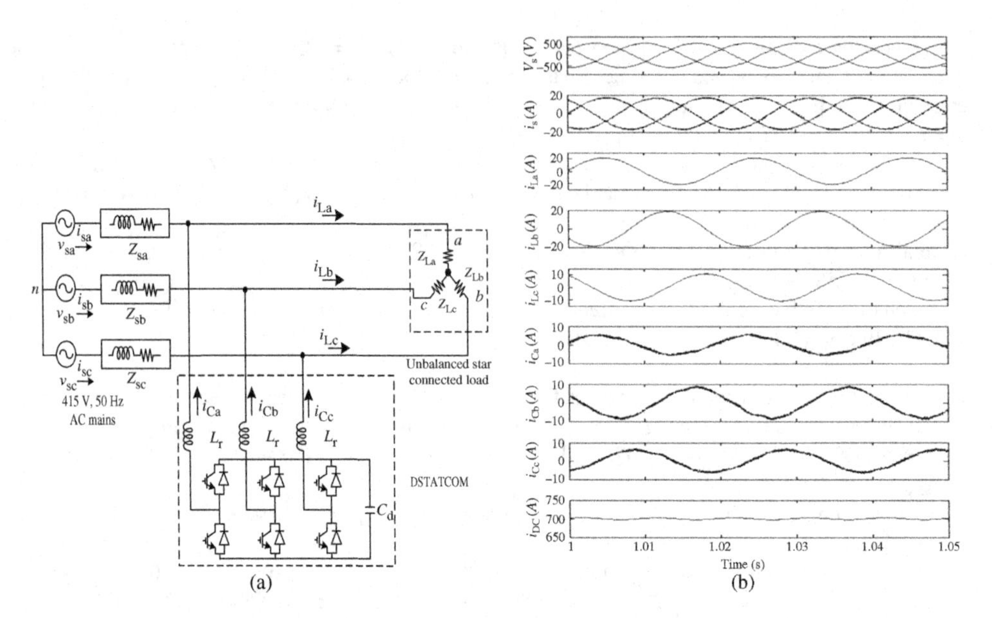

Figure E4.11 (a) A three-phase DSTATCOM for compensation of an unbalanced star connected load and (b) its waveforms

The per-phase required reactive power Q_V at this UPF load of $P = 2870.42$ W per phase (in single line diagram of a star connected equivalent balanced load) for $V_L = V_s$ is $Q_V = Q - Q_L$. The Q can be calculated from the quadratic equation $aQ^2 + bQ + c = 0$ as $Q = [-b \pm \{b^2 - 4ac\}^{1/2}]/2a$, where $a = R_s^2 + X_s^2$, $b = 2V_s^2 X_s$, and $c = (V_L^2 + R_s P)^2 + X_s^2 P^2 - V_s^2 V_L^2$.

Substituting the values $R_s = 1\,\Omega$, $X_s = 3\,\Omega$, $P = 2870.42$ W, and $V_s = V_L = 239.6$ V, $a = R_s^2 + X_s^2 = 1^2 + 3^2 = 10$, $b = 2V_L^2 X_s = 344\,448.96$, and $c = (V_L^2 + R_s P)^2 + X_s^2 P^2 - V_s^2 V_L^2 = 411\,964\,171.2$, and $Q = -1240.699$ VA. Thus, $Q_V = Q - Q_L = -1240.699 - 0 = -1240.699$ VAR.

Per-phase reactive (capacitive) current for voltage regulation into the source is $I_V = Q_V/V_L = 1240.699/239.6 = 5.18$ A.

a. The supply line currents after compensation have a resistive component because of the load active power and a reactive current I_V for the voltage regulation to compensate the drop in the source impedance: $I_{sa} = [(P/V_s)\angle 0° + jQ_V/V_L]$ A, $I_{sb} = [(P/V_s) + jQ_V/V_L]\angle -120°$ A, and $I_{sc} = [(P/V_s) + jQ_V/V_L]\angle -240°$ A.

   Substituting the values, $I_{sa} = [(2870.42/239.6)\angle 0° + j5.18]$ A $= 13.05\angle 23.38°$ A, $I_{sb} = 13.05\angle -99.62°$ A, and $I_{sc} = 13.05\angle -219.62°$ A.

b. Three-phase DSTATCOM currents are $I_{Ca} = I_{Lab} - I_{Lca} - I_{sa} = 11.934\angle 30° - 5.93\angle -210° - 13.05\angle 23.38° = (10.34 + j5.97) - (-5.14 + j2.965) - (11.98 + j5.18) = 3.50 - j2.17 = 4.076\angle -32.89°$ A, $I_{Cb} = I_{Lbc} - I_{Lab} - I_{sb} = 2.96\angle -90° - 11.93\angle 30° - 13.05\angle -99.62° = (0 - j2.96) - (10.34 + j5.97) - (-1.5 - j12.96) = -8.84 + j9.03 = 9.72\angle 155.49°$ A, and $I_{Cc} = I_{Lca} - I_{Lbc} - I_{sc} = 5.93\angle -210° - 2.96\angle -90° - 13.05\angle -219.62° = (-5.14 + j2.965) - (0 - j2.96) - (-10.47 + j7.78) = 5.33 - j1.86 = 5.65\angle -19.24°$ A.

c. The VA rating of the DSTATCOM is $Q_{DST} = V_s(|I_{Ca}| + |I_{Cb}| + |I_{Cc}|) = 239.6 \times (9.12 + 9.72 + 5.65) = 4669.80 = 4.6698$ kVA.

   The voltage drop across the source impedance is $V_{Zs} = (I_{LActive}\angle 0° + jQ_T/3V_s)Z_s = [(2870.42/239.6) + j5.18] \times 3.162 = 41.27$ V.

d. The equivalent delta connected resistance of the compensated load is $R_{eq} = 3V_s^2/P_{LT} = 3 \times 415^2/8611.25 = 60\,\Omega$.

Example 4.12

A three-phase AC supply has a line voltage of 415 V at 50 Hz and a feeder (source) impedance of 1.0 Ω resistance and 3.0 Ω inductive reactance per phase after which a single-phase load having $Z_a = (10 + j5)\ \Omega$ is connected between line and neutral terminals, as shown in Figure E4.12. A four-leg PWM-based DSTATCOM is used to balance and maintain the voltage equal to the input voltage (415 V). Calculate (a) the supply line currents (in amperes), (b) the line currents of the DSTATCOM, (c) the kVA rating of the DSTATCOM, and (d) the voltage drop across the source impedance.

Solution: Given supply phase voltage $V_s = 415/\sqrt{3}\ \text{V} = 239.6\ \text{V}$, frequency of the supply $(f) = 50\ \text{Hz}$, and an unbalanced isolated neutral and star connected load having $Z_{La} = (10 + j5)\Omega$, $Z_{Lb} = 0\ \Omega$, and $Z_{Lc} = 0\ \Omega$ with a feeder (source) impedance of 1.0 Ω resistance and 3.0 Ω inductive reactance per phase.

Load admittances per phase are $Y_{La} = (0.08 - j0.04)$ mhos, $Y_{La} = 0.00$ mhos, and $Y_{La} = 0.0$ mhos.

Load currents are $I_{La} = V_{La}Y_{La} = 239.6 \times (0.08 - j0.04) = 19.168 - j9.584 = 21.43\angle{-29.57°}\text{A}$, $I_{Lb} = 0.0$ A, and $I_{Lc} = 0.0$ A.

Since the DSTATCOM is connected in shunt with the load terminals, the total load active power is $P_T = I_{La}^2 R_{La} = 4592.6528$ W and $Q_L = I_{La}^2 X_{La} = 2296.3$ VAR.

The per-phase active power is $P = P_T/3 = 1530.88$ W.

The per-phase required reactive power Q_V at this UPF load of $P = 1530.88$ W per phase for $V_L = V_s$ is $Q_V = Q - Q_L$. The Q can be calculated from the quadratic equation $aQ^2 + bQ + c = 0$ as $Q = [-b \pm \{b^2 - 4ac\}^{1/2}]/2a$, where $a = R_s^2 + X_s^2$, $b = 2V_s^2 X_s$, and $c = (V_L^2 + R_s P)^2 + X_s^2 P^2 - V_s^2 V_L^2$.

Substituting the values $R_s = 1\ \Omega$, $X_s = 3\ \Omega$, $P = 1530.88$ W, and $V_s = V_L = 239.6$ V, $a = R_s^2 + X_s^2 = 1^2 + 3^2 = 10$, $b = 2V_L^2 X_s = 344\,448.96$, $c = (V_L^2 + R_s P)^2 + X_s^2 P^2 \ \ -V_s^2 V_L^2 = 199\,206\,061$, and $Q = -588.39$ VAR per phase. Thus, $Q_V = Q - Q_L = -588.39 - (2296.3/3) = -1353.82$ VAR.

Per-phase reactive (capacitive) current for voltage regulation into the source is $I_V = Q_V/V_L = j1353.82/239.6 = 5.65$ A.

a. The supply line currents after compensation have a resistive component because of the load active power and a reactive current I_V for the voltage regulation to compensate the drop in the source impedance: $I_{sa} = [(P/V_s)\angle 0° + jQ_V/V_L]$ A, $I_{sb} = [(P/V_s) + jQ_V/V_L]\angle -120°$ A, and $I_{sc} = [(P/V_s) + jQ_V/V_L]\ \angle -240°$ A.

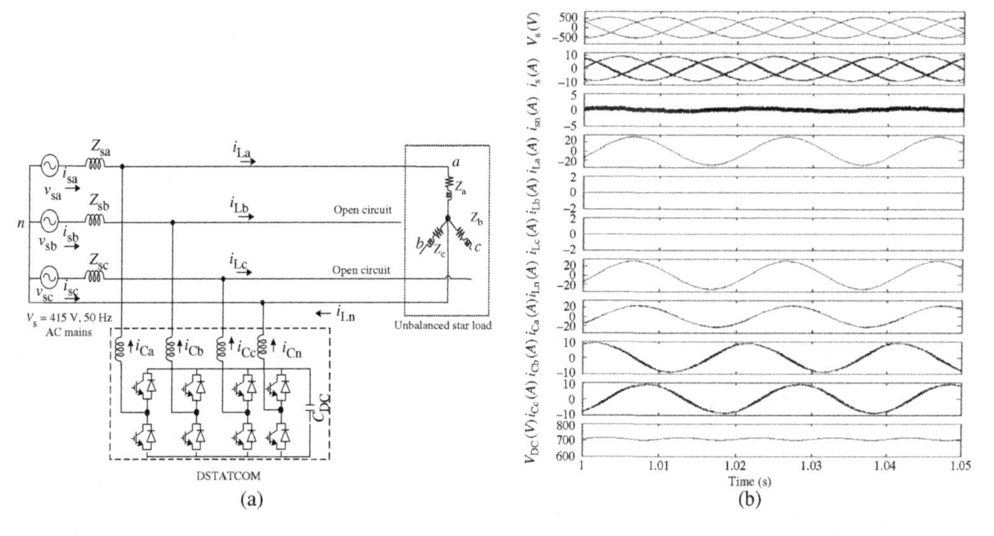

Figure E4.12 (a) A three-phase DSTATCOM for compensation of an unbalanced star connected load and (b) its waveforms

Substituting the values, $I_{sa} = [(1530.88/239.6)\angle 0° + j5.65]$ A = $8.529\angle 41.48°$ A, $I_{sb} = 8.529\angle -78.51°$ A, and $I_{sc} = 8.529\angle 161.48°$ A.

b. Three-phase DSTATCOM currents are $I_{Ca} = I_{La} - I_{sa} = 21.43\angle -29.57° - 8.529\angle 41.48°$ A = $20.32\angle -52.94°$ A, $I_{Cb} = I_{Lb} - I_{sb} = -8.529\angle -78.51°$ A $= 8.529\angle -101.49°$ A, and $I_{Cc} = I_{Lc} - I_{sc} = -8.529\angle 161.48°$ A $= 8.529\angle -18.51°$ A.

   The load neutral current or DSTATCOM neutral current is $I_{Ln} = -I_{Cn} = -(I_{La} + I_{Lb} + I_{Lc}) = -21.43\angle -29.57°$ A $= 21.43\angle -150.43°$ A.

c. The VA rating of the DSTATCOM is $Q_{DST} = V_s(|I_{Ca}| + |I_{Cb}| + |I_{Cc}|) + V_s|I_{Cn}| = 239.6(20.32 + 8.529 + 8.529 + 21.43) = 14.09$ kVA.

d. The voltage drop across the source impedance is $V_{Zs} = (I_{LActive}\angle 0° + jQ_T/3V_s)Z_s = [(1530.88/239.6) + j5.65] \times 3.162 = 26.96$ V.

Example 4.13

A three-phase AC supply has a line voltage of 415 V at 50 Hz and a feeder (source) impedance of 1.0 Ω resistance and 3.0 Ω inductive reactance per phase after which an unbalanced non-isolated star connected load having $Z_a = 10$ Ω, $Z_b = 20$ Ω, and $Z_c = 30$ Ω is connected between line and neutral terminals, as shown in Figure E4.13. If a four-leg PWM-based DSTATCOM is used to balance and maintain the load voltage equal to the input voltage (415 V), then calculate (a) supply line currents, (b) the line currents of the DSTATCOM, (c) its kVA rating, and (d) the voltage drop across the source impedance.

Solution: Given supply phase voltage $V_s = 415/\sqrt{3}$ V = 239.6 V, frequency of the supply (f) = 50 Hz, and an unbalanced non-isolated neutral and star connected load having $Z_{La} = 10$ Ω, $Z_{Lb} = 20$ Ω, and $Z_{Lc} = 30$ Ω with a feeder (source) impedance of 1.0 Ω resistance and 3.0 Ω inductive reactance per phase.

Load admittances per phase are $Y_{La} = 0.1$ mhos, $Y_{Lb} = 0.05$ mhos, and $Y_{Lc} = 0.033\,333$ mhos.

The load currents are $I_{La} = V_sY_{La}\angle 0° = 23.9\angle 0°$ A, $I_{Lb} = V_sY_{Lb}\angle -120° = 11.98\angle -120°$ A, and $I_{Lc} = V_sY_{Lc}\angle -240° = 7.98\angle -240°$ A.

Since the DSTATCOM is connected in shunt with the load terminals, the total load active power is $P_T = V_{sa}^2G_{La} + V_{sb}^2G_{Lb} + V_{sc}^2G_{Lc} = 10\,529.81$ W.

The per-phase active power is $P = P_T/3 = 3508.27$ W.

The per-phase required reactive power Q_V at this UPF load of $P = 3508.27$ W per phase for $V_L = V_s$ is $Q_V = Q - Q_L$. The Q can be calculated from the quadratic equation $aQ^2 + bQ + c = 0$ as $Q = [-b \pm \{b^2 - 4ac\}^{1/2}]/2a$, where $a = R_s^2 + X_s^2$, $b = 2V_s^2X_s$, and $c = (V_L^2 + R_sP)^2 + X_s^2P^2 - V_s^2V_L^2$.

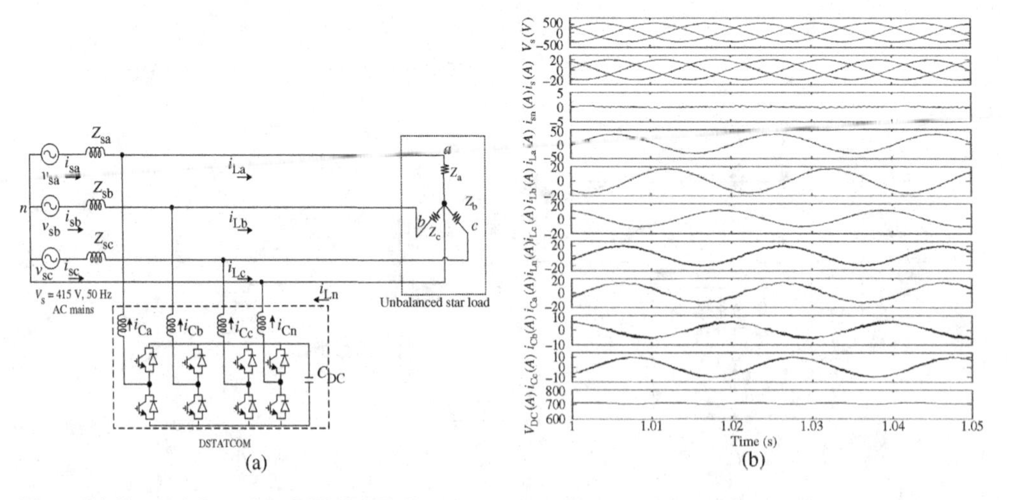

Figure E4.13 (a) A three-phase DSTATCOM for compensation of a current-fed type of an unbalanced star connected load and (b) its waveforms

Substituting the values $R_s = 1\,\Omega$, $X_s = 3\,\Omega$, $P = 3508.27$ W, and $V_s = V_L = 239.6$ V, $a = R_s^2 + X_s^2 = 1^2 + 3^2 = 10$, $b = 2V_L^2 X_s = 344\,448.96$, $c = (V_L^2 + R_s P)^2 + X_s^2 P^2 - V_s^2 V_L^2 = 525\,886\,235$, and $Q = -1601.2$ VA. Thus, $Q_V = Q - Q_L = -1601.2 - 0 = -1601.2$ VAR.

Per-phase reactive (capacitive) current for voltage regulation into the source is $I_V = Q_V/V_L = 1601.2/239.6 = 6.6828$ A.

a. The supply line currents after compensation have a resistive component because of the load active power and a reactive current I_V for the voltage regulation to compensate the drop in the source impedance: $I_{sa} = [(P/V_s)\angle 0° + jQ_V/V_L]$ A, $I_{sb} = [(P/V_s) + jQ_V/V_L]\angle -120°$ A, and $I_{sc} = [(P/V_s) + jQ_V/V_L]\angle -240°$ A.

   Substituting the values, $I_{sa} = [(3508.27/239.6)\angle 0° + j6.6828]$ A $= 16.09\angle 24.53°$ A, $I_{sb} = 16.09\angle -95.47°$ A, and $I_{sc} = 16.09\angle -215.47°$ A.

b. Three-phase DSTATCOM currents are $I_{Ca} = I_{La} - I_{sa} = 23.96\angle 0.0° - 16.09\angle 24.53° = 11.46\angle -35.64°$ A, $I_{Cb} = I_{Lb} - I_{sb} = 11.98\angle -120° - 16.09\angle -95.46° = 7.1935\angle 128.27°$ A, and $I_{Cc} = I_{Lc} - I_{sc} = 7.987\angle -240° - 16.09\angle 144.53° = 9.43\angle -14.87°$ A.

   The load neutral current or DSTATCOM neutral current is $I_{Ln} = -I_{Cn} = -(I_{La} + I_{Lb} + I_{Lc}) = -(23.96\angle 0.0° + 11.98\angle -120° + 7.987\angle -240°) = -(13.98 - j3.455) = -14.398\angle -13.89°$.

c. The VA rating of the DSTATCOM is $Q_{DST} = V_s(|I_{Ca}| + |I_{Cb}| + |I_{Cc}|) + V_s|I_{Cn}| = 239.6(11.89 + 7.79 + 9.89 + 14.40) = 10.181$ kVA.

d. The voltage drop across the source impedance is $V_{Zs} = (I_{LActive}\angle 0° + jQ_T/3V_s)Z_s = [(3508.27/239.6) + j6.68] \times 3.162 = 50.89$ V.

Example 4.14

A DSTATCOM with a 660 V battery at the DC side of the VSC is connected to an isolated three-phase three-wire 415 V, 75 kVA diesel engine generating (DG) system, as shown in Figure E4.14. This DG system is controlled such that the load on the system remains always between 80 and 100% to improve the efficiency of the DG system. There is a load shedding provision if load increases above 125%. If load is below 80%, the battery charging takes place; if the load is beyond 100%, discharging of the battery takes place; and in between 80 and 100% load, the battery remains in floating condition. Calculate (a) the watt-hour rating of the battery and (b) the current of the battery if at full load a backup of 4 h is desired. Moreover, calculate (c) the kVA rating of the DSTATCOM, (d) the voltage rating of the DSTATCOM, and (e) the current rating of the DSTATCOM if the maximum load is 125% of the active power (125% of 75 kW) at 0.8 power factor.

Solution: Given that a DSTATCOM with a 660 V battery at the DC side of the VSC is connected to an isolated three-phase three-wire 415 V, 75 kVA diesel engine generating (DG) system.

a. The watt-hour rating of the battery if at full load a backup of 4 h is desired is $E = \%$ of the full load $\times PH = 0.25 \times 75 \times 4 = 75$ kWh (since the battery is required to feed 25% of the full load active power).

b. The battery current is $I_b = E/V_b H = 75\ 000/(660 \times 4) = 28.41$ A.

   The active power through the DSTATCOM is $P_{DST} = 25\%$ of full load $= 18.75$ kW and $P_T = 75 \times 1.25 = 93.75$ kW.

   The reactive power of the DSTATCOM is $Q_{DST} = (P_T/\text{PF})\{\sqrt{(1 - \text{PF}^2)}\} = (93.75/0.8)\{\sqrt{(1 - 0.8^2)}\} = 70.3125$ kVAR.

c. The kVA rating of the DSTATCOM is $S_{DST} = \sqrt{(P_{DST}^2 + Q_{DST}^2)} = 72.7696$ kVA.

d. The voltage rating of the DSTATCOM is $V_{DST} = 415$ V (line) $= 239.6$ V (phase) (since it is connected at PCC).

e. The current rating of the DSTATCOM is $I_{DST} = S_{DST}/3V_{DSTph} = 101.24$ A.

Example 4.15

A three-phase self-excited 400 V, 22 kW, 50 Hz, six-pole delta connected squirrel cage induction generator needs the excitation of 9.38 kVAR at no load and 15.57 kVAR at a unity power factor full

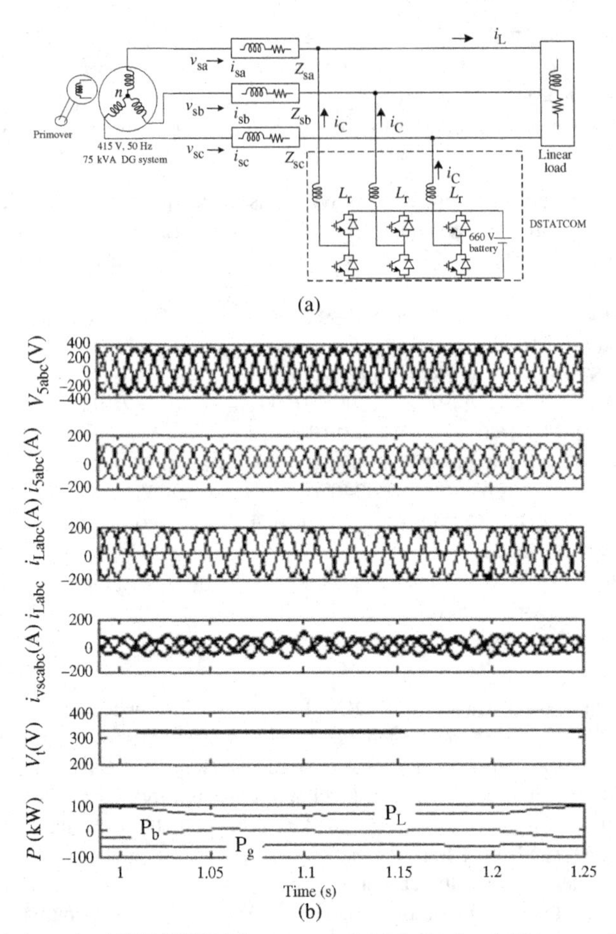

Figure E4.14 (a) A three-phase DSTATCOM for compensation of a load and (b) its waveforms

load, as shown in Figure E4.15. A PWM-based DSTATCOM is used for meeting the reactive power requirements of this induction generator. Considering fixed capacitor rating equal to no load excitation kVAR, calculate (a) the kVAR rating of the DSTATCOM and (b) its line current at (i) no load, (ii) unity power factor full load (22 kW), (iii) 0.8 lagging power factor full load (22 kW), and (iv) 0.8 leading power factor full load (22 kW). If a fixed capacitor bank is used for reducing the rating of the DSTATCOM and the DSTATCOM is used only for smooth control of voltage at rated value, then calculate (c) the reduced rating of the DSTATCOM and (d) its line currents at different load conditions.

Solution: Given that a three-phase self-excited 400 V, 22 kW, 50 Hz, six-pole delta connected squirrel cage induction generator needs the excitation of 9.38 kVAR at no load and 15.57 kVAR at a unity power factor full load. For the first case considering fixed capacitor kVAR rating equal to no load excitation, Q_0 (no load excitation requirement) $= Q_C = 9.38$ kVAR.

(i) At no load:

a. The kVAR rating of the DSTATCOM is $Q_{DST} = Q_0 - Q_C = 0$ kVAR.
b. The line current of the DSTATCOM is $I_C = Q_{DST}/\sqrt{3}V_s = 0/(\sqrt{3} \times 400) = 0$ A.

(ii) For a unity power factor full load (22 kW):

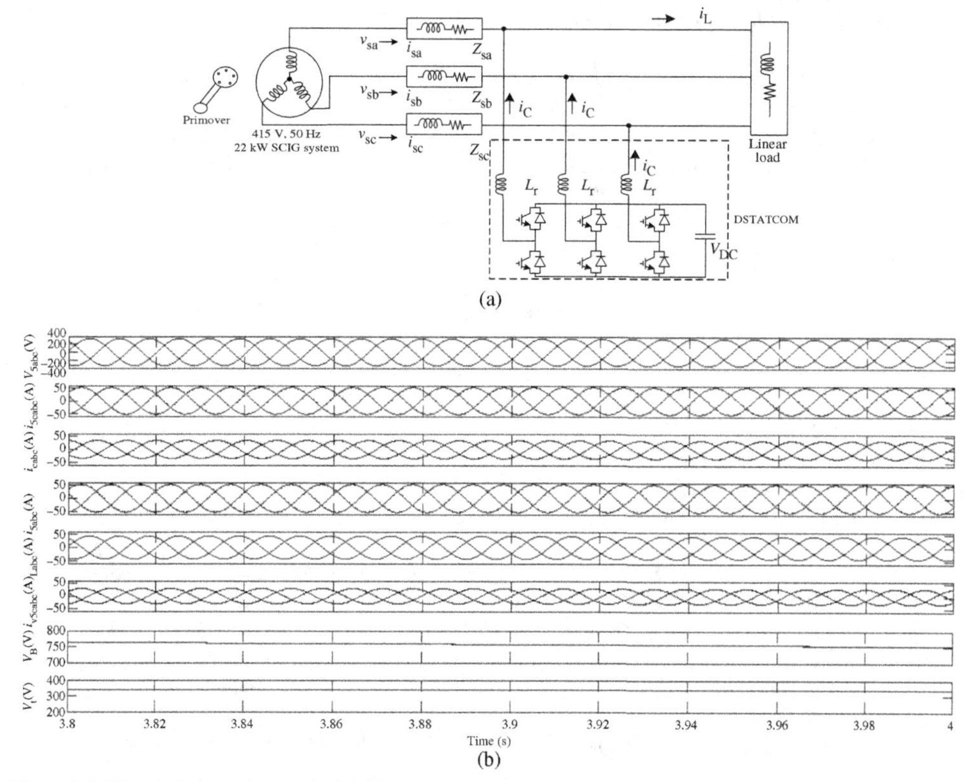

Figure E4.15 (a) A three-phase DSTATCOM for compensation of a load and (b) its waveforms

a. The kVAR rating of the DSTATCOM is $Q_{DST} = Q_{fl} - Q_C = (15.57 - 9.38)$ kVAR = 6.19 kVAR.
b. The line current of the DSTATCOM is $I_C = Q_{DST}/\sqrt{3}V_s = 6.19/(\sqrt{3} \times 400) = 8.93$ A.

(iii) For a 0.8 lagging power factor full load (22 kW):

a. The kVAR rating of the DSTATCOM is $Q_{DST} = Q_{fl} + Q_{load} - Q_C = (15.57 + 16.5 - 9.38)$ kVAR = 22.69 kVAR.
b. The line current of the DSTATCOM is $I_C = Q_{DST}/\sqrt{3}V_s = 22\,690/(\sqrt{3} \times 400) = 32.75$ A.

(iv) For a 0.8 leading power factor full load (22 kW):

a. The kVAR rating of the DSTATCOM is $Q_{DST} = Q_{fl} + Q_{load} - Q_C = (16.5 - 15.57 - 9.38)$ kVAR = −8.55 kVAR.
   The negative sign denotes that the DSTATCOM has to provide lagging reactive power to maintain unity power factor.
b. The line current of the DSTATCOM is $I_C = Q_{DST}/\sqrt{3}V_s = 8550/(\sqrt{3} \times 400) = 12.19$ A.

In a reduced rating DSTATCOM-based voltage regulator, as the DSTATCOM can work in leading and lagging reactive power modes depending on the extent of load, its rating can be reduced by using an appropriate rating AC capacitor at the induction generator terminals to cover total range of the operation. In most of the cases, its half the required rating (other than no load) can be selected as the rating of the DSTATCOM and its full load reactive power rating is selected as the rating of the AC capacitor. At no load, the DSTATCOM works as an inductor and at full load it works as a capacitor.

For a unity power factor full load (22 kW):

c. The kVA rating of the DSTATCOM is $Q_{DST} = (Q_{fl} - Q_0)/2 = (15.57 - 9.38)/2 = 3.095$ kVA. Since the capacitor rating is $Q_C = Q_0 + (Q_{fl} - Q_0)/2 = 9.38 + (15.57 - 9.38)/2 = 9.38 + 3.095 = 12.475$ kVAR, at no load the DSTATCOM is to provide 3.095 kVAR lagging reactive power and at full load the DSTATCOM is to provide 3.095 kVAR leading reactive power to the generator system.
d. The line current of the DSTATCOM is $I_{DST} = Q_{DST}/\sqrt{3}V_s = 3095/(\sqrt{3} \times 400) = 4.467$ A.

For a 0.8 lagging power factor full load (22 kW):

c. The kVA rating of the DSTATCOM is $Q_{DST} = (Q_{fl} + Q_{load} - Q_0)/2 = (15.57 + 16.5 - 9.38)/2 =$ 11.345 kVAR. Since the capacitor rating is $Q_C = Q_0 + (Q_{fl} + Q_{load} - Q_0)/2 = 9.38 + (15.57 + 16.5 - 9.38)/2 = 9.38 + 11.345 = 20.725$ kVAR, at no load the DSTATCOM is to provide 11.345 kVAR lagging reactive power and at full load the DSTATCOM is to provide 11.345 kVAR leading reactive power to the generator system.
d. The line current of the DSTATCOM is $I_{DST} = Q_{ST}/\sqrt{3}V_s = 11\ 345/(\sqrt{3} \times 400) = 16.375$ A.

For a 0.8 leading power factor full load (22 kW):

c. The kVA rating of the DSTATCOM is $Q_{DST} = (Q_{fl} + Q_{load} - Q_0)/2 = (15.57 - 16.5 - 9.38)/2 =$ -5.155 kVAR. Hence, a DSTATCOM rating of 5.155 kVAR is required with leading load. The capacitor rating is $Q_C = Q_0 + (Q_{fl} + Q_{load} - Q_0)/2 = 9.38 + (15.57 - 16.5 - 9.38)/2 = 9.38 - 5.155 =$ 4.225 kVAR; therefore, at no load the DSTATCOM is to provide 5.155 kVAR leading reactive power and at full load the DSTATCOM is to provide 5.155 kVAR lagging reactive power to the generator system.
d. The line current of the DSTATCOM is $I_{DST} = Q_{DST}/\sqrt{3}V_s = 5155/(\sqrt{3} \times 400) = 7.44$ A.

4.8 Summary

DSTATCOMs are used to compensate the power quality problems of the reactive power component of load current, fluctuating currents, unbalanced currents, excessive neutral current, and harmonic currents (if required) produced by a variety of consumer loads. DSTATCOMs offer the best solution for all current-based power quality problems. PWM-based VSCs are preferred to develop DSTATCOMs because of low cost, reduced size, light weight, and reduced losses compared with CSCs. A small passive ripple filter is also used at PCC where DSTATCOMs are connected to improve the voltage profile and to eliminate higher order harmonics. This passive ripple filter does not have much losses even being sometimes damped type because energy associated with higher order harmonics is quite small. However, these higher order harmonics produced by DSTATCOMs due to switching of VSCs cause disturbance to communication systems and other electronic appliances. An analytical study of various performance indices of DSTATCOMs for the compensation of a variety of consumer loads is made in detail with several numerical examples to study the rating of DSTATCOMs and how it is affected by various kinds of consumer loads. DSTATCOMs are observed as one of the best retrofit solutions for mitigating power quality problems of the reactive power component of load current, fluctuating currents, unbalanced currents, and excessive neutral current caused due to diverse kinds of consumer loads, for reducing losses and currents of the AC mains, and for improving the utilization of distribution system assets.

4.9 Review Questions

1. What is a synchronous shunt distribution static compensator?
2. What are the power quality problems that a DSTATCOM can mitigate?

3. In which case out of power factor correction to unity of a lagging power linear load and maintaining its AC terminal voltage regulation to zero, a DSTATCOM has lower power rating?
4. A DSTACTOM is used to balance an unbalanced three-phase three-wire linear load. How the degree of unbalance in the load affects the power rating of the DSTATCOM?
5. What are the benefits of using a DSTATCOM?
6. What are the economic factors for the justification for using a DSTATCOM in an AC system?
7. Can a current source inverter be used as a DSTATCOM?
8. Why voltage source inverters (VSIs) are preferred in a DSTATCOM?
9. What are the factors that decide the rating of a DSTATCOM?
10. How the value of an interfacing inductor is computed for a VSI used in a single-phase DSTATCOM?
11. How the value of a DC bus capacitor is computed for a VSI used in a single-phase DSTATCOM?
12. How the balancing of unbalanced loads affects the value of a DC bus capacitor in a three-phase DSTATCOM?
13. How the value of an interfacing inductor is computed for a VSI used in a three-phase three-wire DSTATCOM?
14. How the value of a DC bus capacitor is computed for a VSI used in a three-phase three-wire DSTATCOM?
15. How the self-supporting DC bus of a VSI is achieved in a DSTATCOM?
16. Why an indirect current control is considered superior than a direct current control scheme of a DSTATCOM?
17. Why an H-bridge VSI-based DSTATCOM is preferred by the industries?
18. Which configuration of a DSTATCOM is most effective to eliminate the neutral current?
19. How the profile of the voltage at PCC can be improved in a DSTATCOM?
20. Up to what highest order of harmonic can be eliminated by a DSTATCOM if the switching frequency is 3 kHz?
21. How the harmonics of higher order than the switching frequency can be eliminated in a DSTATCOM?
22. What are the benefits of using a DSTATCOM in a three-phase isolated diesel engine alternator set (DG set) for economic justification?
23. What are the benefits of using a DSTATCOM in a three-phase isolated gasoline engine-driven squirrel cage induction generator set (DG set) for economic justification?
24. Is it possible to get unity power at the AC mains while regulating the AC terminal voltage across the load by using a DSTATCOM?
25. How the power rating of a DSTATCOM may be reduced when it is used for AC voltage regulation?

4.10 Numerical Problems

1. A single-phase load having $Z = (2.5 + j1.25)$ pu is fed from an AC supply with an input AC voltage of 220 V at 50 Hz and a base impedance of 9.15 Ω per phase. It is to be realized as a unity power factor load on the AC supply system using a PWM-based DSTATCOM. Calculate (a) the value of

compensator current, (b) its VA rating, (c) supply current, and (d) equivalent resistance (in ohms) of the compensated load.

2. A single-phase AC supply has AC mains voltage of 230 V at 50 Hz and a feeder (source) impedance of 1.0 Ω resistance and 3.0 Ω inductive reactance after which a load of 4 kVA, 0.8 lagging power factor is connected. Calculate (a) the voltage drop across the source impedance and (b) the voltage across the load. If a PWM-based DSTATCOM is used to raise the voltage to the input voltage (230 V), calculate (c) the voltage rating of the compensator, (d) the current rating of the compensator, and (e) the VA rating of the compensator.

3. A three-phase AC supply has a line voltage of 415 V at 50 Hz and a feeder (source) impedance of 1.0 Ω resistance and 3.0 Ω inductive reactance per phase after which a balanced delta connected load having $Z_L = (30 + j24)\ \Omega$ per phase is connected. Calculate (a) the voltage drop across the source impedance and (b) the voltage across the load in each case. If a PWM-based DSTATCOM is used to maintain the voltage equal to the input voltage (415 V), calculate (c) the kVA rating of the DSTATCOM and (d) its line currents.

4. A three-phase three-wire DSTATCOM is employed at a 415 V, 50 Hz system to provide load balancing and power factor correction of a single-phase 30 kW, unity power factor load across two lines. Calculate (a) supply line currents, (b) DSTATCOM currents, and (c) its kVA rating.

5. A three-phase unbalanced delta connected load having $Z_{ab} = (9.0 + j1.0)$ pu, $Z_{bc} = (5.5 + j2.75)$ pu, and $Z_{ca} = (3.5 + j1.5)$ pu is fed from an AC supply with an input line voltage of 415 V at 50 Hz and a base impedance of 9.15 Ω per phase. It is to be realized as a balanced unity power factor load on the three-phase supply system using a PWM-based DSTATCOM. Calculate (a) the supply line currents (in amperes), (b) DSTATCOM line currents (in amperes), (c) the kVA rating of the DSTATCOM, and (d) equivalent per-phase resistance (in ohms) of the compensated load.

6. A three-phase four-wire unbalanced non-isolated star connected load having $Z_a = (7.0 + j9.0)$ pu, $Z_b = (9.0 + j2.5)$ pu, and $Z_c = (9.5 + j3.5)$ pu is fed from an AC supply with an input line–line voltage of 415 V at 50 Hz and a base impedance of 9.15 Ω per phase. It is to be realized as a balanced unity power factor load on the three-phase supply system using a four-leg PWM-based DSTATCOM. Calculate (a) the supply line currents (in amperes), (b) the compensator line currents (in amperes), (c) its neutral current (in amperes), (d) its kVA rating, and (e) equivalent per-phase resistance (in ohms) of the compensated load.

7. A three-phase four-wire unbalanced non-isolated star connected load having $Z_a = (9.0 + j2.0)$ pu, $Z_b = (3.0 + j1.5)$ pu, and $Z_c = (9.5 - j2.5)$ pu is fed from an AC supply with an input line voltage of 415 V at 50 Hz and a base impedance of 9.15 Ω per phase. It is to be realized as a balanced unity power factor load on the three-phase supply system using a four-leg PWM-based DSTATCOM. Calculate (a) the supply line currents (in amperes), (b) the DSTATCOM line currents (in amperes), (c) its neutral current (in amperes), (d) its kVA rating, and (e) equivalent per-phase resistance (in ohms) of the compensated load.

8. A three-phase AC supply has a line voltage of 415 V at 50 Hz and a feeder (source) impedance of 1.0 Ω resistance and 3.0 Ω inductive reactance per phase after which a balanced star connected load having $Z = (10 + j7)\ \Omega$ per phase is connected. Calculate (a) the voltage drop across the source impedance and (b) the voltage across the load. If a three-leg PWM-based DSTATCOM is used to maintain the voltage equal to the input voltage (415 V), then calculate (c) its kVA rating, (d) the DSTATCOM line currents, (e) the voltage drop across the source impedance, and (f) the voltage across the load.

9. A three-phase AC supply has a line voltage of 415 V at 50 Hz and a feeder (source) impedance of 1.0 Ω resistance and 9.0 Ω inductive reactance per phase after which a single-phase load having $Z_{ab} = (15 + j12)\ \Omega$ is connected between two lines. If a three-leg PWM-based DSTATCOM is used to

balance and maintain the voltage equal to the input voltage (415 V), calculate (a) the supply line currents (in amperes), (b) the DSTATCOM line currents (in amperes), (c) its kVA rating, (d) the voltage drop across the source impedance, and (e) equivalent per-phase resistance (in ohms) of the compensated load.

10. A three-phase AC supply has a line voltage of 415 V at 50 Hz and a feeder (source) impedance of 0.4 Ω resistance and 2.0 Ω inductive reactance per phase after which a single-phase load having 150 kVA at 0.8 lagging power factor is connected between two lines. If a three-leg PWM-based DSTATCOM is used to balance and maintain the voltage equal to the input voltage (415 V), calculate (a) the supply line currents (in amperes), (b) the DSTATCOM line currents (in amperes), (c) its kVA rating, (d) the voltage drop across the source impedance, and (e) equivalent per-phase resistance (in ohms) of the compensated load.

11. A three-phase AC supply has a line voltage of 415 V at 50 Hz and a feeder (source) impedance of 0.5 Ω resistance and 2.0 Ω inductive reactance per phase after which an unbalanced isolated star configured load having $Z_{La} = 15\,\Omega$, $Z_{Lb} = 30\,\Omega$, and $Z_{Lc} = 45\,\Omega$ is connected. If a three-leg PWM-based DSTATCOM is used to balance and maintain the voltage equal to the input voltage (415 V), calculate (a) the DSTATCOM line currents, (b) the kVA rating of the DSTATCOM, (c) the voltage drop across the source impedance, and (d) equivalent per-phase resistance (in ohms) of the compensated load.

12. A three-phase AC supply has a line voltage of 415 V at 50 Hz and a feeder (source) impedance of 0.25 Ω resistance and 1.0 Ω inductive reactance per phase after which a single-phase load having $Z_a = (5 + j4)\,\Omega$ is connected between line and neutral terminals. A four-leg PWM-based DSTATCOM is used to balance and maintain the voltage equal to the input voltage (415 V). Calculate (a) the supply line currents (in amperes), (b) the DSTATCOM line currents, (c) the kVA rating of the DSTATCOM, and (d) the voltage drop across the source impedance.

13. A three-phase AC supply has a line voltage of 415 V at 50 Hz and a feeder (source) impedance of 0.5 Ω resistance and 2.0 Ω inductive reactance per phase after which an unbalanced non-isolated star connected load having $Z_a = 5\,\Omega$, $Z_b = 10\,\Omega$, and $Z_c = 20\,\Omega$ is connected between line and neutral terminals. If a four-leg PWM-based DSTATCOM is used to balance and maintain the load voltage equal to the input voltage (415 V), then calculate (a) supply line currents, (b) the DSTATCOM line currents, (c) the kVA rating of the DSTATCOM, and (d) the voltage drop across the source impedance.

14. A DSTATCOM with a 720 V battery at the DC side of the VSC is connected to an isolated three-phase three-wire 415 V, 50 kVA diesel engine generating (DG) system. This DG system is controlled such that the load on the system remains always between 85 and 100% to improve the efficiency of the DG system. There is a load shedding provision if load increases above 125%. If load is below 85%, the battery charging takes place; if the load is beyond 100%, discharging of the battery takes place; and in between 85 and 100% load, the battery remains in floating condition. Calculate (a) the watt-hour rating of the battery, (b) the current of the battery if at full load a backup of 4 h is desired. Moreover, calculate (c) the kVA rating of the DSTATCOM, (d) the voltage rating of the DSTATCOM, and (e) the current rating of the DSTATCOM if the maximum load is 125% of the active power (125% of 50 kW) at 0.8 power factor.

15. A three-phase self-excited 230 V, 15 kW, 50 Hz, four-pole delta connected squirrel cage induction generator needs the excitation of 9.31 kVAR at no load and 12 kVAR at a unity power factor full load. If a PWM-based DSTATCOM is used for meeting the reactive power requirements of this induction generator, then calculate (a) the kVAR rating of the DSTATCOM and (b) its line current at (i) no load, (ii) unity power factor full load (15 kW), (iii) 0.8 lagging power factor full load (15 kW), and (iv) 0.8 leading power factor full load (15 kW). If a fixed capacitor bank is used for reducing the rating of the DSTATCOM and the DSTATCOM is used only for smooth control of voltage at rated value, then calculate (c) the reduced rating of the DSTATCOM and (d) its line currents at different load conditions.

4.11 Computer Simulation-Based Problems

1. Design a single-phase DSTATCOM and simulate its behavior to maintain almost UPF of a single-phase 230 V, 50 Hz, 4 kVA at 0.85 lagging PF at PCC. Plot the supply voltage, supply current, load current, DSTATCOM current and dc bus voltage of the DSTATCOM with time. Demonstrate the effect of change of the load from 4 kVA to 1 kVA and then to 4 kVA all at 0.85 lagging PF. Compute (a) rms value of supply current, (b) rms value of ac load current, (c) supply active, reactive powers and load active, reactive powers, (d) rms voltage rating of DSTATCOM, (e) rms current rating of DSTATCOM, (f) VA rating of DSTATCOM.

2. Design a single-phase DSTATCOM and simulate its behavior to maintain almost UPF of a single-phase 230 V, 50 Hz, 5 kVA at 0.8 lagging PF at PCC. It has source impedance of 0.5 Ω resistive element and 2.5 Ω inductive element. Plot the supply voltage, PCC voltage, supply current, load current, DSTATCOM current, and dc bus voltage of the DSTATCOM with time. Demonstrate the effect of change of the load from 5 kVA to 2 kVA and then to 5 kVA all at 0.8 lagging PF. Compute (a) rms value of supply current, (b) rms value of ac load current, (c) supply active, reactive powers and load active, reactive powers, (d) losses in the ac mains with and without DSTATCOM, (e) rms voltage rating of DSTATCOM, (f) rms current rating of DSTATCOM, (g) VA rating of DSTATCOM.

3. Design a single-phase DSTATCOM and simulate its behavior to regulate the PCC voltage of a single-phase 230 V, 50 Hz, 5 kVA at 0.8 lagging PF at PCC. It has source impedance of 0.5 Ω resistive element and 2.5 Ω inductive element. Plot the supply voltage, PCC voltage, and supply current, load current, DSTATCOM current, and dc bus voltage of the DSTATCOM with time. Demonstrate the effect of change of the load from 5 kVA to 2 kVA and then to 5 kVA all at 0.8 lagging PF. Compute (a) the voltage drop across the source impedance, (b) the voltage across the load, (c) rms value of ac mains current, (d) rms value of ac load current, (e) supply active, reactive powers and load active, reactive powers, (f) losses in the ac mains with and without DSTATCOM, (g) rms voltage rating of DSTATCOM, (h) rms current rating of DSTATCOM, (i) VA rating of DSTATCOM.

4. Design a single–phase VSC based DSTATCOM for an isolated single-phase, 230 V, 7.5 kVA, 50 Hz, 4-pole diesel engine driven permanent magnet synchronous generator (PMSG) system to regulate its voltage across loads. This PMSG system has the winding resistance of 1.5 ohms and the reactance of 4.5 ohms. If this PMSG is loaded with a single-phase load of 10 kVA at 0.75 lagging. Plot the PCC voltage, PMSG current, load current, DSTATCOM current, and dc bus voltage of the DSTATCOM with time. Demonstrate the effect of change of the load from 10 kVA to 4 kVA and then to 10 kVA all at 0.75 lagging PF. Compute (a) rms value of PMSG current, (b) rms value of ac load current, (c) PMSG active, reactive powers and load active, reactive powers, (d) rms voltage rating of DSTATCOM, (e) rms current rating of DSTATCOM, (f) VA rating of DSTATCOM.

5. Design a single-phase VSC based DSTATCOM and simulate its behavior with a 420 V battery at the dc side of VSC, which is connected to a single-phase, 230 V, 15 kVA distribution system. It has a load of single-phase, 230 V, 20 kVA, at 0.8 lagging, out of which half of the load is thrown off. Each such condition occurs for 5 hours. The DSTATCOM is controlled such that average power drawn from ac mains remains constant in all these conditions. Plot the supply voltage, supply current, DSTATCOM current, and dc bus voltage of the DSTATCOM with time. Compute (a) the maximum battery current and (b) the watt-hour rating of the battery. Moreover, compute (c) the kVA rating of the DSTATCOM, (d) the rms voltage rating of the DSTATCOM and (e) the rms current rating of DSTATCOM.

6. Design a single-phase VSC based DSTATCOM and simulate its behavior with a 360 V battery at the dc side of VSC, for an isolated single-phase, 220 V, 15 kVA, 50 Hz, 4-pole diesel engine driven permanent magnet synchronous generator (PMSG) system to regulate its voltage. This PMSG system has the winding resistance of 0.25 ohms and the reactance of 0.75 ohms. This PMSG is loaded with a single-phase load of 25 kVA at 0.8 lagging pf for two hours followed by single-phase load of 10 kVA at 0.8 lagging pf again for two hours. The DSTATCOM is controlled such that average power drawn from DG set remains constant in all these conditions. Plot the PCC voltage, PMSG current, load current, DSTATCOM current, dc bus voltage of the DSTATCOM and battery current with time. Compute (a) rms value of PMSG current, (b) rms value of ac load current, (c) PMSG active, reactive powers and load active, reactive powers, (d) voltage rating of DSTATCOM, (e) current rating of DSTATCOM, (f) VA rating of DSTATCOM, (g) the maximum battery current and (h) the watt-hour rating of the battery.
7. Design a 3–phase, 3-leg VSC based DSTATCOM and simulate its behavior for providing reactive power compensation for unity power factor of a three-phase star connected balanced load of 75 kVA, 0.8 lagging power-factor. Plot the supply voltages, supply currents, load currents, DSTATCOM currents, and dc bus voltage of the DSTATCOM with time. Demonstrate the effect of change of the load from 75 kVA to 25 kVA and then to 75 kVA all at 0.8 lagging PF. Compute (a) rms value of supply current, (b) rms value of ac load current, (c) supply active, reactive powers and load active, reactive powers, (d) voltage rating of DSTATCOM, (e) current rating of DSTATCOM, (f) VA rating of DSTATCOM.
8. Design a 3–phase, 3-leg VSC based DSTATCOM and simulate its behavior for providing reactive power compensation for unity power factor of a single-phase load of 25 kVA, 0.8 lagging power factor connected between two lines of ac mains. Plot the supply voltages, supply currents, load currents, DSTATCOM currents, and dc bus voltage of the DSTATCOM with time. Demonstrate the effect of change of the load from single-phase, 25 kVA to 10 kVA and then to 25 kVA all at 0.8 lagging PF. Compute (a) rms value of supply current, (b) rms value of ac load current, (c) supply active, reactive powers and load active, reactive powers, (d) voltage rating of DSTATCOM, (e) current rating of DSTATCOM, (f) VA rating of DSTATCOM.
9. Design a 3–phase, 3-leg VSC based DSTATCOM and simulate its behavior to regulate the PCC voltage through reactive power compensation of a three-phase AC supply, which has a line voltage of 415 V at 50 Hz and feeder (source) impedance of 0.5 ohms resistance and 2.0 ohms inductive reactance/phase after which a unbalanced isolated star configured load having $Z_{La} = 15$ ohms, $Z_{Lb} = 30$ ohms and $Z_{Lc} = 45$ ohms, is connected. If this three-leg PWM based DSTATCOM is used to balance and to regulate the voltage to same as input voltage (415 V), compute (a) the DSTATCOM line currents, (b) the kVA rating of the DSTATCOM, (c) the voltage drop across the source impedance and (d) equivalent per phase resistance (in ohms) of compensated load. Plot the supply voltages, PCC voltages, the voltage drop across the source impedance, supply currents, load currents, DSTATCOM currents, and dc bus voltage of the DSTATCOM with time.
10. Design a 3–phase, 3-leg VSC based DSTATCOM and simulate its behavior to correct the power factor to unity at the PCC voltage through reactive power compensation of a three-phase star connected balanced load of 45 kVA, 0.8 lagging power factor. It has source impedance of 0.5 Ω resistive element and 2.5 Ω inductive element. Plot the supply voltages, PCC voltages, and supply currents, load currents, DSTATCOM currents, dc bus voltage of the DSTATCOM with time. Demonstrate the effect of change of the load from 45 kVA to 15 kVA and then to 45 kVA all at 0.8 lagging power factor. Compute (a) the voltage drop across the source impedance, (b) the voltage across the load, (c) rms value of ac mains current, (d) rms value of ac load current, (e) supply active, reactive powers and load active, reactive powers, (e) losses in the ac mains with and without DSTATCOM, (f) rms voltage rating of DSTATCOM, (g) rms current rating of DSTATCOM, (h) VA rating of DSTATCOM.

11. Design a 3–phase, 3-leg VSC based DSTATCOM and simulate its behavior to regulate the PCC voltage through reactive power compensation of a three-phase star connected balanced load of 75 kVA, 0.8 lagging power factor. It has source impedance of 0.5 Ω resistive element and 2.5 Ω inductive element. Plot the supply voltages, PCC voltages, and supply currents, load currents, DSTATCOM currents, dc bus voltage of the DSTATCOM with time. Demonstrate the effect of change of the load from 75 kVA to 25 kVA and then to 75 kVA all at 0.8 lagging PF. Compute (a) the voltage drop across the source impedance, (b) the voltage across the load, (c) rms value of ac mains current, (d) rms value of ac load current, (e) supply active, reactive powers and load active, reactive powers, (e) losses in the ac mains with and without DSTATCOM, (f) rms voltage rating of DSTATCOM, (g) rms current rating of DSTATCOM, (h) VA rating of DSTATCOM.

12. Design a 3–phase, 3-leg VSC based DSTATCOM and simulate its behavior to regulate the PCC voltage through reactive power compensation of a single-phase load of 60 kVA, 0.8 lagging power factor connected between two lines of ac mains. It has source impedance of 0.5 Ω resistive element and 2.5 Ω inductive element. Plot the supply voltages, PCC voltages, and supply currents, load currents, DSTATCOM currents, dc bus voltage of the DSTATCOM with time. Demonstrate the effect of change of the load from 60 kVA to 20 kVA and then to 60 kVA all at 0.8 lagging PF. Compute (a) the voltage drop across the source impedance, (b) the voltage across the load, (c) rms value of ac mains current, (d) rms value of ac load current, (e) supply active, reactive powers and load active, reactive powers, (f) losses in the ac mains with and without DSTATCOM, (g) rms voltage rating of DSTATCOM, (h) rms current rating of DSTATCOM, (i) VA rating of DSTATCOM.

13. Design a three-phase, 3-leg VSC based DSTATCOM and simulate its behavior with a 660 V battery at the dc side of VSC, which is connected to a three-phase, 3-wire, 415 V, 250 kVA distribution system. It has a load of three-phase, 3-wire, 415 V, 300 kVA, at 0.8 lagging, out of which one phase load is thrown off. Each such condition occurs for three hours. The DSTATCOM is controlled such that average power drawn from ac mains remains constant in all these conditions. Plot the supply voltages, supply currents, DSTATCOM currents, and dc bus voltage of the DSTATCOM with time. Compute (a) the maximum battery current and (b) the watt-hour rating of the battery. Moreover, compute (c) the kVA rating of the DSTATCOM, (d) the rms voltage rating of the DSTATCOM and (e) the rms current rating of DSTATCOM.

14. Design a 3–phase, 3-leg VSC based DSTATCOM and simulate its behavior with a 760 V battery at the dc side of VSC, which is connected to a three-phase, 3-wire, 415 V, 150 kVA distribution system. This system is controlled such that the load on the supply side always remains between 80–100%. There is a cut-off provision if the load reduces below 40% or increases above 120%. If the load is in between 40–80% the battery charging takes place and if the load is beyond 100%, the discharging of the battery takes place, in between 80–100% of the load, the battery remains in floating condition. Plot the supply voltages, PCC voltages, supply currents, DSTATCOM currents, and dc bus voltage of the DSTATCOM with time. Compute (a) the current and (b) the watt-hour rating of the battery if at full load a backup of 5 hours is desirable. Moreover, compute (c) the kVA rating of the DSTATCOM, (d) the rms voltage rating of the DSTATCOM and (e) the rms current rating of DSTATCOM if the maximum load is 120% of the active power at 0.8 pf lagging.

15. Design a 3–phase, 3-leg VSC based DSTATCOM for an isolated three-phase, 3-wire, 415 V, 50 kVA, 50 Hz, 4-pole diesel engine driven permanent magnet synchronous generator (PMSG) system to regulate its voltage and to balance its loads. This PMSG system has the winding resistance of 0.25 ohms/phase and the reactance of 0.75 ohms/phase. If this PMSG is loaded with a single-phase line-line connected load of 60 kVA at 0.8 lagging power factor. Plot the PCC voltages, PMSG currents, load currents, DSTATCOM currents, and dc bus voltage of the DSTATCOM with time. Demonstrate the effect of change of the load from 60 kVA to 20 kVA and then to 60 kVA all at 0.8

lagging PF. Compute (a) rms value of PMSG current, (b) rms value of ac load current, (c) PMSG active, reactive powers and load active, reactive powers, (d) rms voltage rating of DSTATCOM, (e) rms current rating of DSTATCOM, (f) VA rating of DSTATCOM.

16. Design a three-phase, four-wire, three-leg, mid-point capacitors VSC based DSTATCOM and simulate its behavior for maintaining almost UPF and load balancing of a single-phase 230 V, 50 Hz load of 125 A at 0.85 lagging PF connected between line and neutral terminals. The supply system has its per-phase impedance of 0.25 Ω resistive element and 1.25 Ω inductive element. Plot the supply voltages, PCC voltages, supply currents, DSTATCOM currents, dc bus voltage of the DSTATCOM, supply and load neutral currents with time. Demonstrate the effect of change of the load from 125 A to 25 A and then to 125 A all at 0.85 lagging PF. Compute (a) rms value of supply current, (b) rms value of ac load current, (c) supply active, reactive powers and load active, reactive powers, (d) losses in the ac mains with and without DSTATCOM, (e) rms voltage rating of DSTATCOM, (f) rms current rating of DSTATCOM, (g) VA rating of DSTATCOM.

17. Design a three-phase, four-wire, three-leg, mid-point capacitors VSC based DSTATCOM and simulate its behavior for regulating the PCC voltages and load balancing of a single-phase 230 V, 50 Hz load of 125 A at 0.85 lagging PF connected between line and neutral terminals. Plot the supply voltages and supply currents, DSTATCOM currents, dc bus voltage of the DSTATCOM, supply and load neutral currents, output voltage and current with time. Demonstrate the effect of change of the load from 125 A to 25 A and then to 125 A all at 0.85 lagging PF. Compute (a) rms value of supply current, (b) rms value of ac load current, (c) supply active, reactive powers and load active, reactive powers, (d) rms voltage rating of DSTATCOM, (e) rms current rating of DSTATCOM, (f) VA rating of DSTATCOM.

18. Design a three-phase, four-wire, three-leg, mid-point capacitors VSC based DSTATCOM and simulate its behavior with a 760 V battery at the dc side of VSC, which is connected to a three-phase, 4-wire, 415 V, 75 kVA distribution system. It has a load of three-phase, 4-wire, 415 V, 100 kVA, at 0.8 lagging, out of which first one phase load is removed and then second-phase load is thrown off. Each such condition occurs for two hours. The DSTATCOM is controlled such that average power drawn from ac mains remains constant in all these conditions. Plot the supply voltages, supply currents, DSTATCOM currents, and dc bus voltage of the DSTATCOM with time. Compute (a) the maximum battery current and (b) the watt-hour rating of the battery. Moreover, compute (c) the kVA rating of the DSTATCOM, (d) the rms voltage rating of the DSTATCOM and (e) the rms current rating of DSTATCOM.

19. Design a three-phase, four-wire, four-leg VSC based DSTATCOM and simulate its behavior for maintaining UPF and load balancing of a single-phase 230 V, 50 Hz load of 125 A at 0.8 lagging PF connected between line and neutral terminals. The supply system has its per-phase impedance of 0.25 Ω resistive element and 1.25 Ω inductive element. Plot the supply voltages, PCC voltages, supply currents, DSTATCOM currents, dc bus voltage of the DSTATCOM, supply and load neutral currents with time. Demonstrate the effect of change of the load from 125 A to 25 A and then to 125 A all at 0.8 lagging PF. Compute (a) rms value of supply current, (b) rms value of ac load current, (c) supply active, reactive powers and load active, reactive powers, (d) losses in the ac mains with and without DSTATCOM, (e) rms voltage rating of DSTATCOM, (f) rms current rating of DSTATCOM, (g) VA rating of DSTATCOM.

20. Design a three-phase, four-wire, four-leg VSC based DSTATCOM and simulate its behavior with a 720 V battery at the dc side of VSC, which is connected to a three-phase, 4-wire, 415 V, 150 kVA distribution system. It has a load of three-phase, 4-wire, 415 V, 200 kVA, at 0.8 lagging power factor, out of which first one phase load is removed and then second-phase load is thrown off. Each such condition occurs for two hours. The DSTATCOM is controlled such that average power drawn from ac mains remains constant in all these conditions. Plot the supply voltages, supply currents, DSTATCOM currents, and dc bus voltage of the DSTATCOM with time. Compute (a) the maximum battery current

and (b) the watt-hour rating of the battery. Moreover, compute (c) the kVA rating of the DSTATCOM, (d) the rms voltage rating of the DSTATCOM and (e) the rms current rating of DSTATCOM.

21. Design a three-phase, four-wire, three single-phase VSC based DSTATCOM and simulate its behavior for regulating the PCC voltages and load balancing of a single-phase 230 V, 50 Hz load of 125 A at 0.8 lagging PF connected between line and neutral terminals. Plot the supply voltages and supply currents, DSTATCOM currents, dc bus voltage of the DSTATCOM, supply and load neutral currents, output voltage and current with time. Demonstrate the effect of change of the load from 125 A to 25 A and then to 125 A all at 0.8 lagging PF. Compute (a) rms value of supply current, (b) rms value of ac load current, (c) supply active, reactive powers and load active, reactive powers, (d) rms voltage rating of DSTATCOM, (e) rms current rating of DSTATCOM, (f) VA rating of DSTATCOM.

22. Design a three-phase, four-wire, three single-phase VSC based DSTATCOM and simulate its behavior for regulating the PCC voltages and load balancing of a single-phase 230 V, 50 Hz load of 125 A at 0.8 lagging PF connected between line and neutral terminals. The supply system has its per-phase impedance of 0.25 Ω resistive element and 1.25 Ω inductive element. Plot the PCC voltages, supply currents, load currents, DSTATCOM currents, and the dc bus voltage of the DSTATCOM with time. Demonstrate the effect of change of the load from 125 A to 25 A and then to 125 A all at 0.8 lagging PF. Compute (a) rms value of supply current, (b) rms value of ac load current, (c) supply active, reactive powers and load active, reactive powers, (d) rms voltage rating of DSTATCOM, (e) rms current rating of DSTATCOM, (f) VA rating of DSTATCOM.

23. Design a three-phase, four-wire, three single-phase VSC based DSTATCOM and simulate its behavior with a 480 V battery at the dc side of VSC, which is connected to a three-phase, 4-wire, 415 V, 300 kVA distribution system. It has a load of three-phase, 4-wire, 415 V, 400 kVA, at 0.8 lagging, out of which first one phase load is removed and then second-phase load is thrown off. Each such condition occurs for two hours. The DSTATCOM is controlled such that average power drawn from ac mains remains constant in all these conditions. Plot the supply voltages, supply currents, DSTATCOM currents, and dc bus voltage of the DSTATCOM with time. Compute (a) the maximum battery current and (b) the watt-hour rating of the battery. Moreover, compute (c) the kVA rating of the DSTATCOM, (d) the rms voltage rating of the DSTATCOM and (e) the rms current rating of DSTATCOM.

24. Design a 3–phase, 4-leg VSC based DSTATCOM for an isolated three-phase, 4-wire, 415 V, 25 kVA, 50 Hz, 4-pole diesel engine driven permanent magnet synchronous generator (PMSG) system to regulate its voltage, and to balance its loads and compensate its neutral current. This PMSG system has the winding resistance of 0.25 ohms/phase and the reactance of 0.75 ohms/phase. This PMSG is loaded with a single-phase line-neutral connected load of 30 kVA at 0.8 lagging. Plot the PCC voltages, PMSG currents, load currents, DSTATCOM currents, and dc bus voltage of the DSTATCOM with time. Demonstrate the effect of change of the load from 30 kVA to 5 kVA and then to 30 kVA all at 0.8 lagging PF. Compute (a) rms value of PMSG current, (b) rms value of ac load current, (c) PMSG active, reactive powers and load active, reactive powers, (d) voltage rating of DSTATCOM, (e) current rating of DSTATCOM, (f) VA rating of DSTATCOM.

25. Design a three-phase, four-wire, four-leg VSC based DSTATCOM and simulate its behavior with a 720 V battery at the dc side of VSC, for an isolated three-phase, 4-wire, 415 V, 15 kVA, 50 Hz, 4-pole diesel engine driven permanent magnet synchronous generator (PMSG) system to regulate its voltage, and to balance its loads and compensate its neutral current. This PMSG system has the winding resistance of 0.25 ohms/phase and the reactance of 0.75 ohms/phase. This PMSG is loaded with a single-phase line-neutral connected load of 25 kVA at 0.8 lagging pf for two hours followed by two-phase load of 10 kVA at 0.8 lagging pf again for two hours. The DSTATCOM is controlled such that average power drawn from DG set remains constant in all these conditions. Plot the PCC

voltages, PMSG currents, load currents, DSTATCOM currents, dc bus voltage of the DSTATCOM and battery current with time. Compute (a) rms value of PMSG current, (b) rms value of ac load current, (c) PMSG active, reactive powers and load active, reactive powers, (d) voltage rating of DSTATCOM, (e) current rating of DSTATCOM, (f) VA rating of DSTATCOM, (g) the maximum battery current and (h) the watt-hour rating of the battery.

References

1. Akagi, H., Kanazawa, Y., and Nabae, A. (1989) Instantaneous reactive power compensators comprising switching devices without energy storage components. *IEEE Transactions on Industry Applications*, **20**(3), 625–630.
2. Gyugyi, L. (1988) Power electronics in electric utilities: static VAR compensators. *Proceedings of the IEEE*, **76**(4), 483–494.
3. Moran, L.T., Ziogas, P.D., and Joos, G. (1989) Analysis and design of a three-phase synchronous solid-state VAR compensator. *IEEE Transactions on Industry Applications*, **25**(4), 598–608.
4. Kearly, J., Chikhani, A.Y., Hackam, R. *et al.* (1991) Microprocessor controlled reactive power compensator for loss reduction in radial distribution feeders. *IEEE Transactions on Power Delivery*, **6**(4), 1848–1855.
5. Kojori, H.A., Dewan, S.B., and Lavers, J.D. (1992) A large-scale PWM solid-state synchronous condenser. *IEEE Transactions on Industry Applications*, **28**(1), 41–49.
6. Schauder, C. and Mehta, H. (1993) Vector analysis and control of advanced static VAR compensators. *IEE Proceedings C*, **140**(4), 299–306.
7. Osborne, M.M., Kitchin, R.H., and Ryan, H.M. (1995) Custom power technology in distribution systems: an overview. Proceedings of the IEE North Eastern Centre Power Section Symposium on the Reliability, Security and Power Quality of Distribution Systems, April 5, pp. 10/1–10/11.
8. Hingorani, N.G. (1995) Introducing custom power. *IEEE Spectrum*, **32**(6), 41–48.
9. Stump, M.D., Keane, G.J., and Leong, F.K.S. (1998) The role of custom power products in enhancing power quality at industrial facilities. Proceedings of the International Conference on Energy Management and Power Delivery, March, vol. 2, pp. 507–517.
10. Singh, B., Chandra, A., Al-Haddad, K., Anuradha, and Kothari, D.P. (1998) Reactive power compensation and load balancing in electric power distribution systems. *International Journal of Electrical Power & Energy Systems*, **20**(6), 375–381.
11. Heydt, G.T. (1998) Electric power quality: a tutorial introduction. *IEEE Computer Applications in Power*, **11**(1), 15–19.
12. Sabin, D.D. (1999) Application of custom power devices for enhanced power quality. Proceedings of the IEEE Power Engineering Society Winter Meeting, February, vol. 2, pp. 1121–1129.
13. Nilsson, S. (1999) Special application considerations for custom power systems. Proceedings of the IEEE Power Engineering Society Winter Meeting, February, vol. 2, pp. 1127–1131.
14. Clouston, J.R. and Gurney, J.H. (1999) Field demonstration of a distribution static compensator used to mitigate voltage flicker. Proceedings of the IEEE Power Engineering Society Winter Meeting, vol. 2, pp. 1138–1141.
15. Chatterjee, K., Fernandes, B.G., and Dubey, G.K. (1999) An instantaneous reactive volt-ampere compensator and harmonic suppressor system. *IEEE Transactions on Power Electronics*, **14**(2), 381–392.
16. Singh, B.N., Chandra, A., and Al-Haddad, K. (2000) DSP-based indirect-current-controlled STATCOM. I. Evaluation of current control techniques. *IEE Proceedings – Electric Power Applications*, **147**(2), 107–112.
17. Singh, B.N., Chandra, A., and Al-Haddad, K. (2000) DSP-based indirect-current-controlled STATCOM. II. Multifunctional capabilities. *IEE Proceedings – Electric Power Applications*, **147**(2), 113–118.
18. Chandra, A., Singh, B., Singh, B.N., and Al-Haddad, K. (2000) An improved control algorithm of shunt active filter for voltage regulation, harmonic elimination, power-factor correction, and balancing of nonlinear loads. *IEEE Transactions on Power Electronics*, **15**(3), 495–507.
19. Madrigal, M., Anaya, O., Acha, E. *et al.* (2000) Single-phase PWM converters array for three-phase reactive power compensation. I. Time domain studies. Proceedings of the International Conference on Harmonics and Quality of Power, October, vol. 2, pp. 541–547.
20. Madrigal, M., Acha, E., Mayordomo, J.G. *et al.* (2000) Single-phase PWM converters array for three-phase reactive power compensation. II. Frequency domain studies. Proceedings of the International Conference on Harmonics and Quality of Power, October, vol. 2, pp. 645–651.
21. Xu, L., Anaya-Lara, O., Agelidis, V.G., and Acha, E. (2000) Development of prototype custom power devices for power quality enhancement. Proceedings of the International Conference on Harmonics and Quality of Power, October, vol. 3, pp. 775–783.

22. Papic, I. (2000) Power quality improvement using distribution static compensator with energy storage system. Proceedings of the International Conference on Harmonics and Quality of Power, October, vol. 3, pp. 916–920.
23. Sensarma, P.S., Padiyar, K.R., and Ramanarayanan, V. (2001) Analysis and performance evaluation of a distribution STATCOM for compensating voltage fluctuations. *IEEE Transactions on Power Delivery*, **16**(2), 259–264.
24. Lee, S.-Y., Wu, C.-J., and Chang, W.-N. (2001) A compact control algorithm for reactive power compensation and load balancing with static VAR compensator. *Electric Power Systems Research*, **58**(2), 63–70.
25. Xu, L., Agelidis, V.G., and Acha, E. (2001) Development considerations of DSP-controlled PWM VSC-based STATCOM. *IEE Proceedings – Electric Power Applications*, **148**(5), 449–455.
26. Chen, S. and Joos, G. (2001) Direct power control of DSTATCOMs for voltage flicker mitigation. Proceedings of the IEEE Industry Applications Society (IAS) Annual Meeting, October, vol. 4, 2683–2690.
27. Woo, S.-M., Kang, D.-W., Lee, W.-C., and Hyun, D.-S. (2001) The distribution STATCOM for reducing the effect of voltage sag and swell. IEEE Conference of the Industrial Electronics Society, December, vol. 2, pp. 1132–1137.
28. Funabashi, T., Koyanagi, K., and Yokoyama, R. (2002) Digital simulation examples of custom power controllers. Proceedings of the IEEE Power Engineering Society Winter Meeting, January, vol. 1, pp. 499–501.
29. Choma, K.N. and Etezadi-Amoli, M. (2002) The application of a DSTATCOM to an industrial facility. Proceedings of the IEEE Power Engineering Society Winter Meeting, January, vol. 2, pp. 725–728.
30. Jung, S.-Y., Kim, T.-H., Moon, S.-I., and Han, B.-M. (2002) Analysis and control of DSTATCOM for a line voltage regulation. Proceedings of the IEEE Power Engineering Society Winter Meeting, January, vol. 2, pp. 729–730.
31. Mishra, M.K., Ghosh, A., and Joshi, A. (2003) Operation of a DSTATCOM in voltage control mode. *IEEE Transactions on Power Delivery*, **18**(1), 258–264.
32. Ghosh, A. and Ledwich, G. (2003) Load compensating DSTATCOM in weak AC systems. *IEEE Transactions on Power Delivery*, **18**(4), 1302–1309.
33. McGranaghan, M.F. and Roettger, W.C. (2003) The economics of custom power. Proceedings of the IEEE Transmission and Distribution Conference and Exposition, September, vol. 3, pp. 944–948.
34. Wood, T. (2003) Experience with custom power applications for critical manufacturing facilities. Proceedings of the IEEE Transmission and Distribution Conference and Exposition, September, vol. 3, pp. 940–943.
35. Dinavahi, V.R., Iravani, M.R., and Bonert, R. (2009) Real-time digital simulation and experimental verification of a D-STATCOM interfaced with a digital controller. *International Journal of Electrical Power & Energy Systems*, **26**(9), 703–713.
36. Benhabib, M.C. and Saadate, S. (2005) New control approach for four-wire active power filter based on the use of synchronous reference frame. *Electric Power Systems Research*, **73**(3), 353–362.
37. Freitas, W., Morelato, A., Xu, W., and Sato, F. (2005) Impacts of AC generators and DSTATCOM devices on the dynamic performance of distribution systems. *IEEE Transactions on Power Delivery*, **20**(2), 1493–1501.
38. Domijan, A., Jr., Montenegro, A., Keri, A.J.F., and Mattern, K.E. (2005) Custom power devices: an interaction study. *IEEE Transactions on Power Systems*, **20**(2), 1111–1118.
39. Tavakoli Binaa, M., Eskandarib, M.D., and Panahloub, M. (2005) Design and installation of a ±250 kVAr DSTATCOM for a distribution substation. *Electric Power Systems Research*, **73**, 383–391.
40. Singh, B., Adya, A., Mittal, A.P., and Gupta, J.R.P. (2005) Modeling and control of DSTATCOM for three-phase, four-wire distribution systems, Proceedings of the IEEE Industry Applications Society (IAS) Annual Meeting, October, vol. 4, pp. 2428–2439.
41. Lin, B.-R. and Yang, T.-Y. (2009) Analysis and implementation of three-phase power quality compensator under the balanced and unbalanced load conditions. *Electric Power Systems Research*, **76**, 271–282.
42. Molina, M.G. and Mercado, P.E. (2006) Control design and simulation of DSTATCOM with energy storage for power quality improvements. Proceedings of TDC '06, August, pp. 1–7.
43. Milanés, M. I., Cadaval, E. R., and González, F. B. (2007) Comparison of control strategies for shunt active power filters in three-phase four-wire systems. *IEEE Transactions on Power Electronics*, **22**(1), 229–236.
44. Singh, B. and Solanki, J. (2008) An improved control approach for DSTATCOM with distorted and unbalanced AC mains. *Journal of Power Electronics*, **8**(2), 131–140.
45. Lohia, P., Mishra, M.K., Karthikeya, K., and Vasudevan, K. (2008) A minimally switched control algorithm for three-phase four-leg VSI topology to compensate unbalanced and nonlinear load. *IEEE Transactions on Power Electronics*, **23**(4), 1935–1944.
46. Singh, B., Jayaprakash, P., and Kothari, D.P. (2008) A three-phase four-wire DSTATCOM for power quality improvement. *Journal of Power Electronics*, **8**(3), 259–267.
47. Singh, B., Jayaprakash, P., and Kothari, D.P. (2008) A T-connected transformer and three-leg VSC based DSTATCOM for power quality improvement. *IEEE Transactions on Power Electronics*, **23**(6), 2710–2718.

48. Singh, B., Jayaprakash, P., and Kothari, D.P. (2008) Star/hexagon transformer based three-phase four-wire DSTATCOM for power quality improvement. *International Journal of Emerging Electric Power Systems*, **9**(6), 1–16.
49. Singh, B. and Solanki, J. (2009) A comparison of control algorithms for DSTATCOM. *IEEE Transactions on Industrial Electronics*, **56**(7), 2738–2745.
50. Singh, B., Jayaprakash, P., and Kothari, D.P. (2009) Three-leg VSC integrated with T connected transformer as three-phase four-wire DSTATCOM for power quality improvement. *International Journal on Electric Power Components and Systems*, **37**(8), 817–831.
51. Singh, B., Jayaprakash, P., Kothari, D.P. *et al.* (2014) Comprehensive study of DSTATCOM configurations. *IEEE Transactions of Industrial Informatics*, **10**(2), 854–870.

5

Active Series Compensation

5.1 Introduction

In modern distribution system, there are a number of voltage-based power quality (PQ) problems caused by substantial pollution and abnormal operating conditions. These power quality problems at point of common coupling (PCC) occur due to the voltage drop in feeders and transformers, various kinds of disturbances, faults, use of unbalanced lagging power factor consumer loads, and so on. Some of these voltage-related power quality problems are voltage spikes, surges, flickers, sags, swells, notches, fluctuations, voltage imbalance, waveform distortion, and so on. The active series compensators are extensively used to both inject the voltage of required magnitude and frequency and restore the voltage across the loads to protect the sensitive loads from these voltage quality problems. These compensators are known as solid-state synchronous series compensators (SSCs) and dynamic voltage restorers (DVRs). They use insulated gate bipolar transistor (IGBT)-based and metal–oxide–semiconductor field-effect transistor (MOSFET)-based PWM (pulse-width modulated) voltage source converters (VSCs) and current source converters (CSCs) to inject the equal and opposite voltages of disturbances in series synchronism with AC mains to protect and provide the clean regulated voltage waveform across the critical loads. The waveform of injected voltage is variable and it may consist of fundamental positive sequence, negative sequence, or even zero sequence, harmonic voltages, and so on. For generation of such varying voltage waveforms, PWM power converters would require instantaneous exchange of reactive and active powers. These PWM converters can generate reactive power locally itself, but they need an exchange of active power through its DC bus. This exchange of instantaneous active power in a series compensator is made possible through energy storage elements such as a large capacitor at DC bus of the VSC or a large inductor in case of CSCs with a self-supporting DC bus for short durations in most of the applications. The continuous and long duration exchange of the active power in these series compensators is achieved by installing a battery or another converter of similar nature or a simple rectifier with proper control. In many cases, a rectifier-supported well-regulated DC bus is used in these series compensators to both meet the need of exchange of the active power with suitable control to avoid the over- and undervoltages and reduce the cost of the system. Although the use of low-cost rectifier to support the DC bus of these series compensators can emulate the negative resistance in the series of line voltage to avoid the voltage drop, it may however cause current-based power quality problems. In present-day distribution systems, the need for these types of custom power devices, namely, SSCs and DVRs, is increasing substantially so as to provide the required voltage waveforms for critical and sensitive loads. Accordingly, the analysis, design, and control of these series compensators for the compensation of voltage-based power quality problems have become one of the most important research areas.

These active series compensators are recently reported with some modifications as cost-effective filters with series active power filters to eliminate harmonic currents in voltage-fed nonlinear loads and with

Power Quality Problems and Mitigation Techniques, First Edition.
Bhim Singh, Ambrish Chandra and Kamal Al-Haddad.

shunt passive filters to eliminate harmonic currents in current-fed nonlinear loads. Of course, the main objective of series active power filters has been to eliminate harmonic voltages either present in the supply system or created in the PCC voltage by the flow of highly distorted current due to supply impedance. In view of the new and additional application of these series compensators as series active power filters, Chapter 10 has been exclusively devoted to the study of series active power filters as well as harmonics elimination in currents and voltages along with some specific applications and case studies.

The active series compensators are classified on the basis of the converter used such as current source converter and voltage source converter, topology used such as half bridge and full bridge in the converter, supply systems such as single phase or three phase, and types of DC bus support of the converter used as the compensator. The SSCs are classified on the basis of supply system into three major categories: single-phase two-wire, three-phase three-wire, and three-phase four-wire SSCs. Similarly, another classification is also made on the basis of DC bus support of the converter used as the compensator: self-supported or capacitor-supported VSC-based series compensators, battery-supported SSCs, and rectifier-supported SSCs. This chapter deals with the state of the art on these SSCs, their classification, principle of operation and control, analysis and design, and modeling and simulation of performance and provides numerical examples, summary, review questions, numerical problems, computer simulation-based problems, and references.

5.2 State of the Art on Active Series Compensators

The custom power device technology is now mature enough for providing compensation for voltage-based power quality problems in AC distribution systems. It has evolved in the last decade of development with varying configurations, control strategies, and solid-state devices. Active series compensators are used to eliminate voltage spikes, sags, swells, notches, and harmonics, to regulate terminal voltage, to suppress voltage flicker, and to mitigate voltage unbalance in the three-phase systems. These wide range of objectives are achieved either individually or in combination depending upon the requirements, control strategy, and configuration, which have to be selected appropriately. This section describes a chronological development and present status of the active series compensator technology.

Since 1971, many configurations of active series compensators have been reported in the literature such as DVRs and static synchronous series compensators (SSSCs) for transmission systems, series compensation for furnaces, and many other fluctuating loads. The concepts based on both the CSC with inductive energy storage and the VSC with capacitive energy storage are used to develop single-phase SSSCs.

One of the major factors in the advancement of SSC technology is the advent of fast, self-commutating solid-state devices. In the initial stages, BJTs and power MOSFETs were used for DVR development; later, SITs and GTOs were employed to develop DVRs. With the introduction of IGBTs, the SSSC technology has got a real boost and at present it is considered an ideal solid-state device for SSCs. The improved sensor technology has also contributed to the enhanced performance of the SSC. The availability of Hall effect sensors and isolation amplifiers at a reasonable cost and with adequate ratings has improved the SSC performance substantially.

Another breakthrough in the development of SSC has resulted from the microelectronics revolution. From the initial use of discrete analog and digital components, the SSSCs are now equipped with microprocessors, microcontrollers, and DSPs. Now it is possible to implement complex algorithms online for the control of the SSC at a reasonable cost. This development has made it possible to use different control algorithms such as proportional–integral (PI) control, variable structure control, fuzzy logic control, and neural nets-based control for improving the dynamic and steady-state performance of the SSC. With these improvements, the SSCs are capable of providing fast corrective action even under dynamically changing loads.

5.3 Classification of Active Series Compensators

Active series compensators can be classified based on the power converter type, topology, and the number of phases. The type of power converter can be either CSC or VSC. The topology can be half-bridge VSC, full-bridge VSC, and so on. The third classification is based on the number of phases such as single-phase two-wire and three-phase three- or four-wire systems.

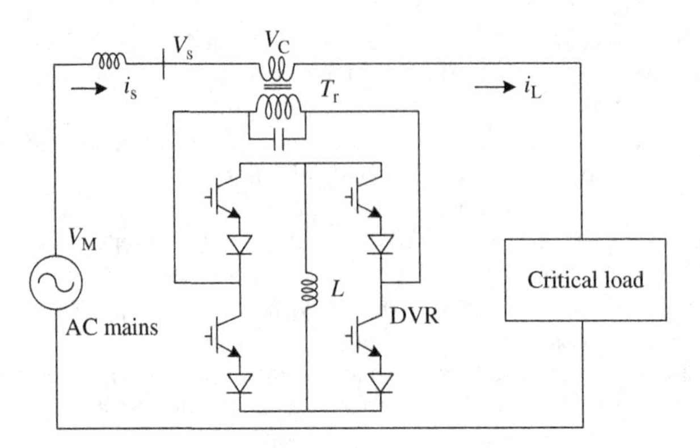

Figure 5.1 CSC-based single-phase DVR

5.3.1 Converter-Based Classification

Two types of power converters are used in the development of active series compensators. Figure 5.1 shows single-phase SSC (DVR) based on current source converter. In this CSC-based DVR, a diode is used in series with the self-commutating device (IGBT) for reverse-voltage blocking. However, GTO-based DVR configurations do not need the series diode, but they have restricted switching frequency. Although CSCs are considered sufficiently reliable, they cause high losses and require high-voltage parallel AC power capacitors. Moreover, they cannot be used in multilevel or multistep modes to improve performance in higher ratings.

The other power converter used in SSC is a VSC, as shown in Figure 5.2. It has a self-supporting DC voltage bus with a large DC capacitor. It has become more dominant since it is lighter, cheaper, and expandable to multilevel and multistep versions, to enhance the performance with lower switching frequencies. It is more popular in UPS-based applications because in the presence of AC mains, the same power converter can be used as an active series compensator for series compensation of critical and sensitive loads.

5.3.2 Topology-Based Classification

Active series compensators can be classified based on the topology used as half-bridge, full-bridge, and transformerless configurations. Figures 5.3–5.5 show the basic block of active series compensators. It is connected before the load in series with AC mains, using a matching transformer, to balance and regulate the terminal voltage of the load or line. It has been used to reduce negative sequence voltage and to regulate the voltage in three-phase systems. It can be installed by electric utilities to damp out harmonic propagation caused by resonance with line impedances and passive shunt compensators.

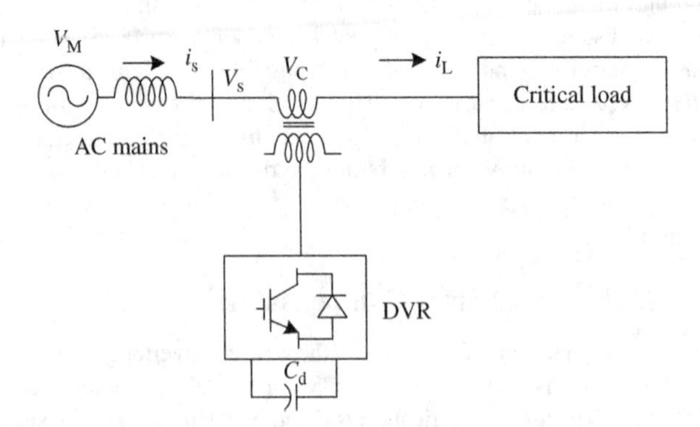

Figure 5.2 VSC-based single-phase DVR

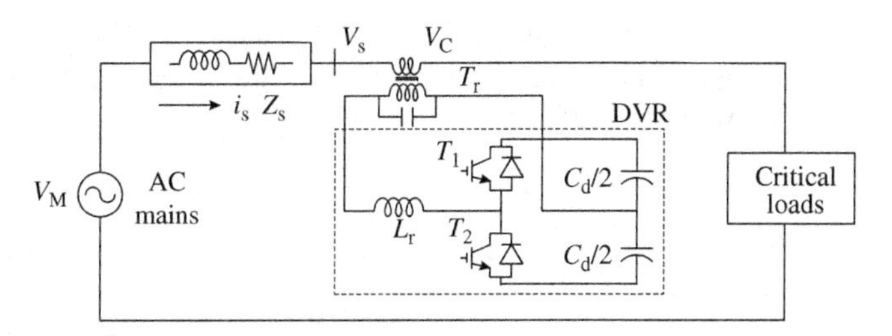

Figure 5.3 Half-bridge topology of VSC-based single-phase DVR

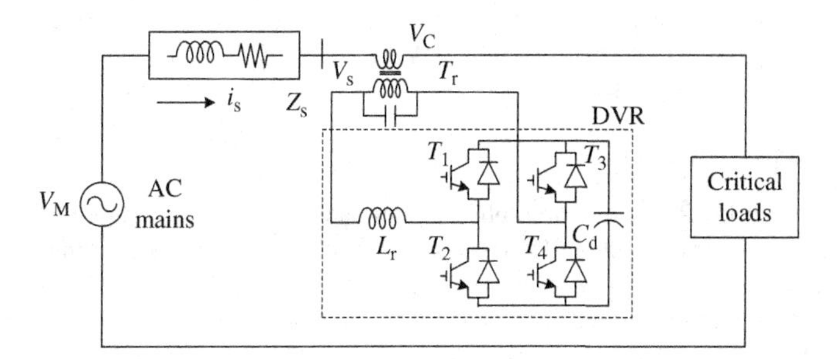

Figure 5.4 Full-bridge topology of VSC-based single-phase DVR

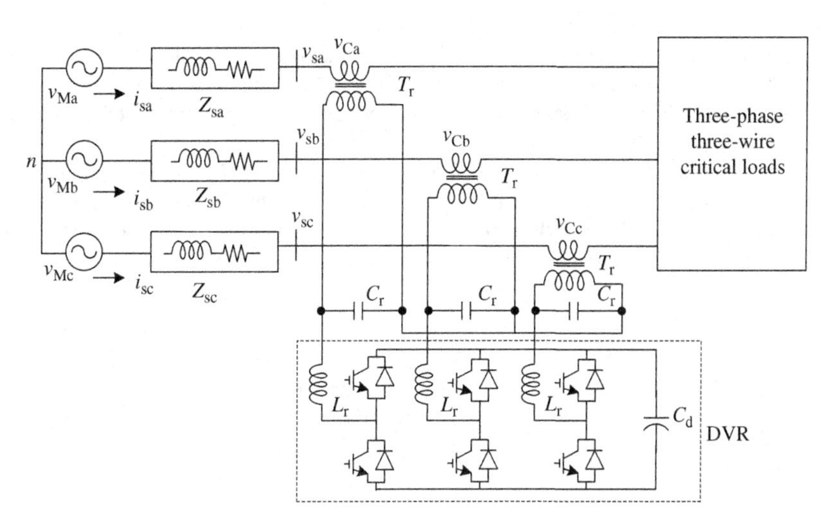

Figure 5.5 Three-phase three-wire DVR

5.3.3 Supply System-Based Classification

This classification of SSCs is based on the supply and/or the load system having single-phase (two-wire) and three-phase (three-wire or four-wire) systems. There are many sensitive critical loads such as domestic appliances connected to single-phase supply systems. Some three-phase consumer loads are without neutral terminal, such as ASDs (adjustable speed drives), fed from three-wire supply systems.

There are many single-phase loads distributed on four-wire, three-phase supply systems, such as computers, commercial lighting, and so on. Hence, SSCs may also be classified accordingly as two-wire, three-wire, and four-wire configurations.

5.3.3.1 Two-Wire DVRs

Single-phase two-wire DVRs are used in all three modes (capacitor-supported, rectifier-supported, and battery-supported DC bus) as active series compensators. Both power converter configurations, CSC with inductive energy storage elements and VSC with capacitive DC bus energy storage elements, are used to form two-wire DVR circuits. In some cases, active compensation is included in the power conversion stage to improve input properties at the supply end.

Figure 5.1 shows an active series compensator with CSC using an inductive storage element. Similar configurations based on VSC may be obtained by considering only two wires (phase and neutral) at each stage of Figures 5.3 and 5.4. In case of DVR with the VSC, sometimes the transformer is removed and the load is shunted with passive L–C components. A DVR is normally used to eliminate voltage spikes, sags, swells, notches, harmonics, and so on.

5.3.3.2 Three-Wire DVRs

The configurations of DVRs for the three-phase system are classified in Figure 5.6. This classification is based on the DC link of the VSC of DVRs and the number of required switching devices. Basically, there are rectifier-supported, battery-supported, and capacitor-supported DC links for the VSC of DVRs. The rectifier support may be achieved from the left of the injection transformer, namely, AC mains side, or from the right of the injection transformer, namely, load side. Similarly, a VSC is operated with three-leg VSC, two-leg VSC, and three single-phase VSCs. Figure 5.7a–c shows the left shunt rectifier-supported DVR with three-leg VSC, two-leg VSC, and three single-phase VSCs, respectively.

Figure 5.8a–c shows the right shunt rectifier-supported DVR with three-leg VSC, two-leg VSC, and three single-phase VSCs, respectively. The BESS (battery energy storage system)-supported DVRs with three-leg VSC, two-leg VSC, and three single-phase VSCs are shown in Figure 5.9a–c, respectively. The capacitor-supported DVR with three-leg VSC, two-leg VSC, and three single-phase VSCs are shown in Figure 5.10a–c, respectively.

Figure 5.11 shows a detailed schematic diagram of a three-phase DVR with BESS at its DC side and connected to three-phase critical load to restore the voltage. The three-phase supply is connected to a critical and sensitive load through three-phase series injection transformers. The supply terminal of a-phase is connected to PCC through short-circuit impedance. The voltage injected by the DVR in a-phase

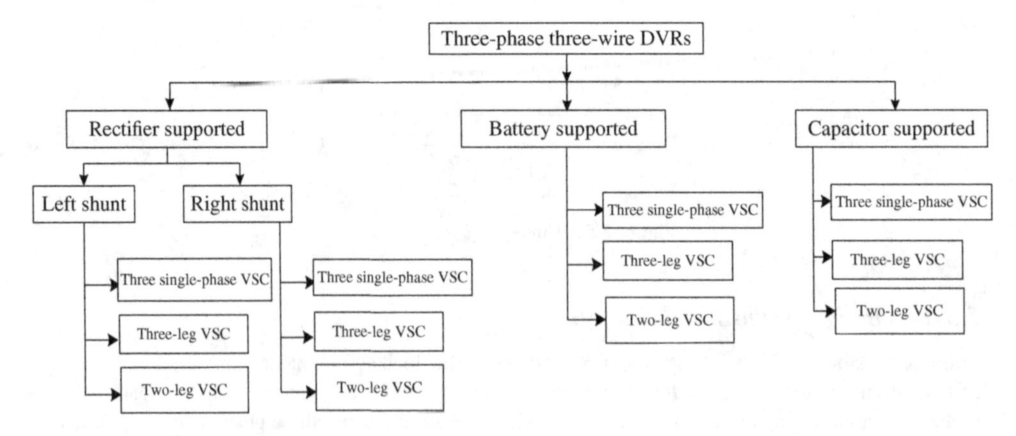

Figure 5.6 Classification of configurations of DVRs for three-phase, three-wire distribution system

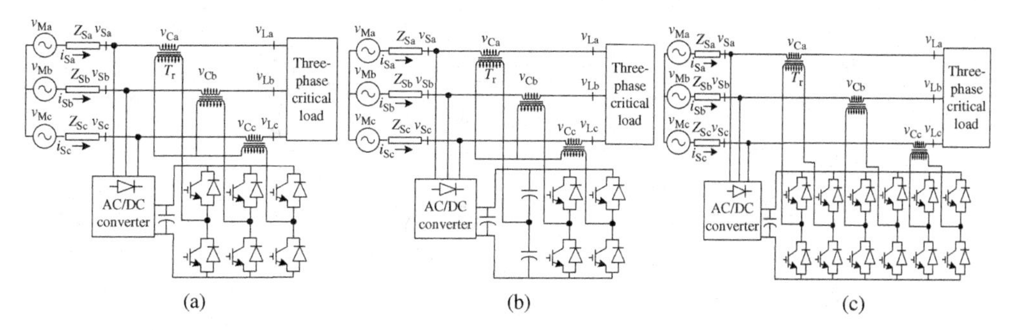

Figure 5.7 Schematic diagram of the left shunt rectifier-supported DVR connected system. (a) Three-leg VSC-based DVR. (b) Two-leg VSC-based DVR. (c) Three single-phase VSC-based DVRs

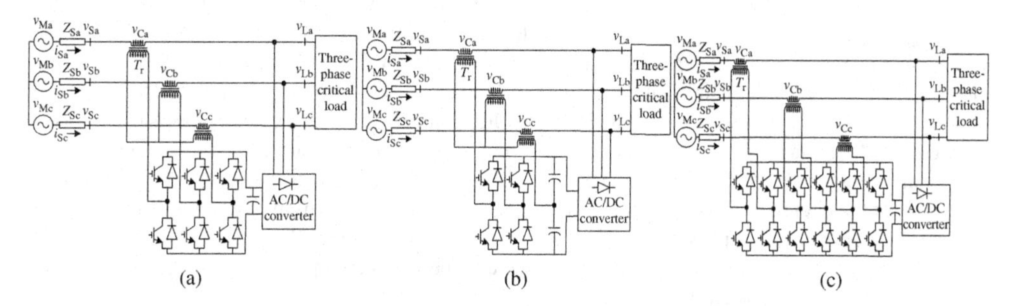

Figure 5.8 Schematic diagram of the right shunt rectifier-supported DVR connected system. (a) Three-leg VSC-based DVR. (b) Two-leg VSC-based DVR. (c) Three single-phase VSC-based DVRs

(v_{Ca}) is such that the load voltage (v_{La}) is undistorted and is achieved with rated magnitude. The three-phase DVR is connected to the line to inject a voltage in series using three single-phase transformers T_r. L_r and C_r represent the filter components used to filter the ripple in the injected voltage and R_r is the small resistance used in series with C_r. A three-leg voltage source converter with IGBTs is used as a DVR and the BESS is connected to the DC bus.

Figure 5.12 shows the detailed schematic diagram of a capacitor-supported DVR connected to three-phase critical loads. In a capacitor-supported DVR, the power absorbed/supplied is zero under steady-state condition and the voltage injected by the DVR is in quadrature with the feeder current. Figure 5.13 shows the phasor diagram of the DVR operation for the compensation of the sag, the swell, and the

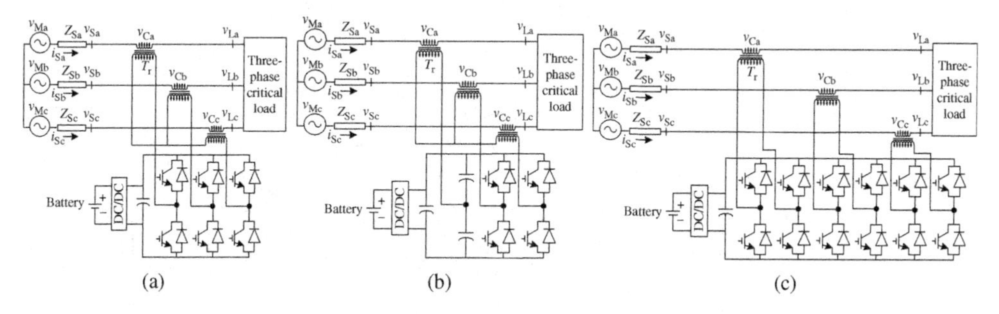

Figure 5.9 Schematic diagram of the BESS-supported DVR connected system. (a) Three-leg VSC-based DVR. (b) Two-leg VSC-based DVR. (c) Three single-phase VSC-based DVRs

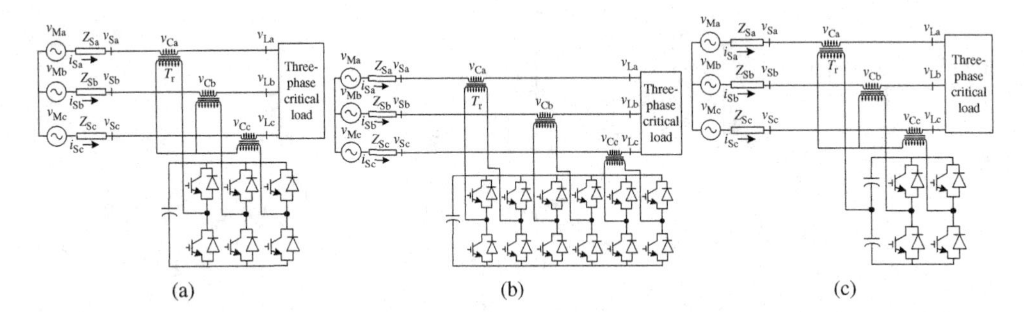

Figure 5.10 Capacitor-supported DVR using (a) three-leg VSC, (b) two-leg VSC, and (c) three single-phase VSCs

unbalance in the supply voltages. The load terminal voltage and the current during pre-sag are represented as $V_{L(presag)}$ and $I_{Sa'}$, as shown in Figure 5.13a. After the sag event, the terminal voltage (V_{Sa}) is lower in magnitude than that in pre-sag condition. The voltage injected by the DVR (V_{Ca}) is used to maintain the load voltage (V_{La}) at the rated magnitude and this has two components, V_{cad} and V_{caq}. The voltage in-phase with current (V_{cad}) is to regulate the DC bus voltage and also to meet the power losses in the VSC of the DVR. The voltage in quadrature with the current (V_{caq}) is to regulate the load voltage (V_{La}) at a constant magnitude. During swell in voltage, the voltage injection (V_{Ca}) is such that the load voltage lies in the locus of the circle, as shown in Figure 5.13b. The unbalanced compensation when voltage sag occurs in two phases is shown in Figure 5.13c. The unbalanced terminal voltages (V_{Sa}, V_{Sb}, V_{Sc}) and the injection voltages in each phase (V_{Ca}, V_{Cb}, V_{Cc}) are such that the line voltages ($V_{La\text{-}Lb}$, $V_{Lb\text{-}Lc}$, $V_{Lc\text{-}La}$) are equal in magnitude and are displaced by 120°.

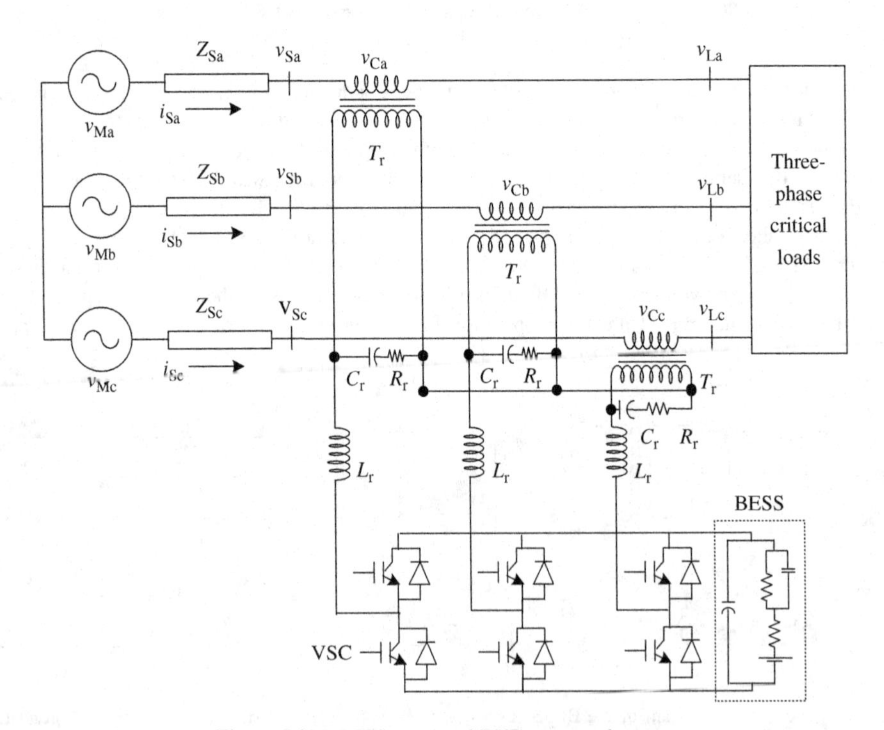

Figure 5.11 BESS-supported DVR connected system

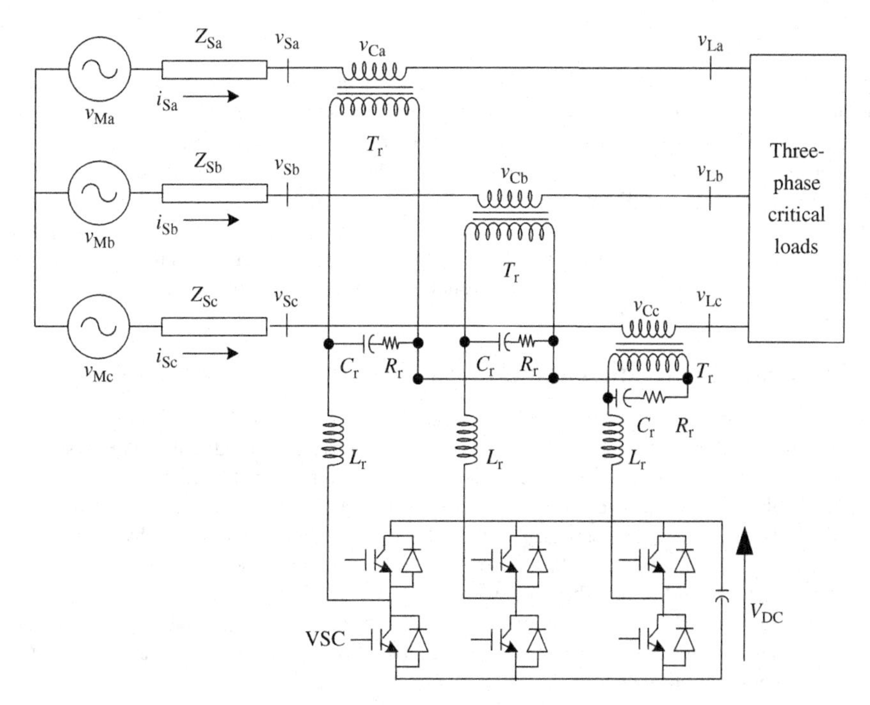

Figure 5.12 Capacitor-supported DVR connected system

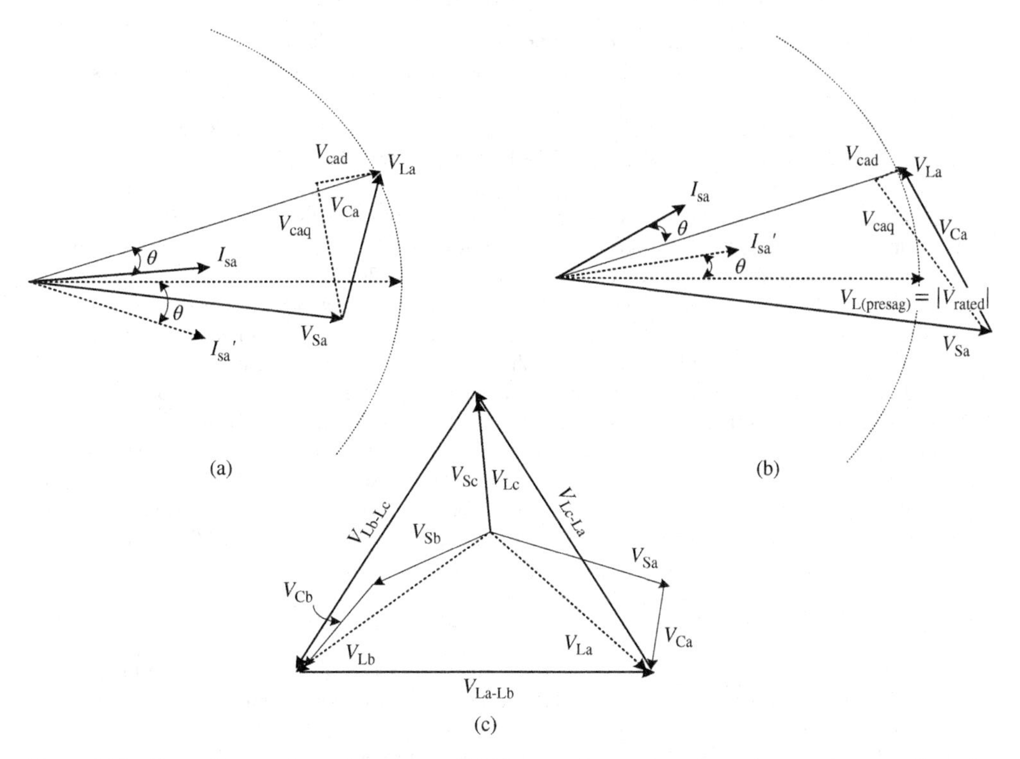

Figure 5.13 Phasor diagram of capacitor-supported DVR system for compensation under (a) voltage sag, (b) swell, and (c) unbalanced sag

5.3.3.3 Four-Wire DVRs

The basic configuration of a active series compensator involves inserting a voltage in series with the supply terminal voltage and load terminal voltage. An active series compensator is connected in series with the supply terminal and the load using an injection transformer. Therefore, a transformer is required to connect the VSC of the active series compensator. The VSC is controlled according to the control strategy and the application as DVR. In a DVR, a voltage is inserted in series with the supply voltage such that the load voltage is constant in magnitude and undistorted, although the supply voltage is not constant in magnitude or distorted. Similarly, in a DVR, a voltage is inserted in series with the supply voltage such that the AC mains current is free from power quality problems. The configurations for modeling and simulation are given in the following sections.

5.4 Principle of Operation and Control of Active Series Compensators

The fundamental circuit of the active series compensators for a three-phase, three-wire AC system is shown in Figure 5.12. An IGBT-based VSC with a DC bus capacitor is used as the DVR. Using a control algorithm, the injected voltages are directly controlled by estimating the reference injected voltages. However, in place of injected voltages, the reference load voltages may be estimated for an indirect voltage control of its VSC. The gate pulses for the DVR are generated by employing hysteresis (carrierless PWM) or PWM (fixed frequency) voltage control over reference and sensed load voltages, which result in an indirect voltage control. Using the DVR with a proper control algorithm, the voltage spikes, surges, flickers, sags, swells, notches, fluctuations, waveform distortion, voltage imbalance, and harmonics compensation are achieved. The detailed principle of operation and control of active series compensators are given in the following sections.

5.4.1 Principle of Operation of Active Series Compensators

The active series compensators are based on the principle of injecting a voltage in series with the supply and this is implemented in two ways. This compensator inserts a voltage of required waveform so that it can protect sensitive consumer loads from supply disturbances such as sag, swell, surges, spikes, notches, unbalance, harmonics, and so on in supply voltage and is known as dynamic voltage restorer.

Figure 5.14a shows a single line diagram of the DVR for power quality improvement in a distribution system. A voltage (V_C) is injected such that the load voltage (V_L) is constant in magnitude and undistorted, although the supply voltage (V_S) is not constant in magnitude or may be distorted. Figure 5.14b shows the phasor diagram of different voltage injection schemes of the DVR. A $V_{L(pre\text{-}sag)}$ is the voltage across the critical load prior to voltage sag condition. During the sag, the voltage is reduced to V_S with a phase lag angle of θ. Now the DVR injects a voltage so that the load voltage magnitude is maintained at the pre-sag condition. According to the phase angle of the load voltage, the injection of voltages can be realized in four ways. V_{C1} represents the voltage injected in-phase with the supply voltage. With the injection of V_{C2}, the load voltage magnitude remains the same, but it leads V_S by a small angle. In V_{C3}, the load voltage retains the same phase as that of the pre-sag condition, which may be an optimum angle considering the energy source at DC bus of the VSC used as a DVR. V_{C4} is the condition where the injected voltage is in quadrature with the current and this case is suitable for a capacitor-supported DVR as this injection involves no active power except small losses in the DVR. However, the minimum possible rating of the power converter is achieved by V_{C1} where the voltage injected is in-phase with the supply voltage. The DVR is operated in this scheme with a BESS, as shown in Figure 5.14a.

5.4.2 Control of Active Series Compensators

The control algorithms for the DVR are based on the estimation of either injected voltages or the reference load voltages for power quality improvement in a distribution system. The terminal voltages, source currents, load voltages, and the DC bus voltage are generally used as feedback signals and the reference load voltages are estimated using the control algorithms. There are many control algorithms reported in the literature for the control of DVR similar to other custom power devices such as

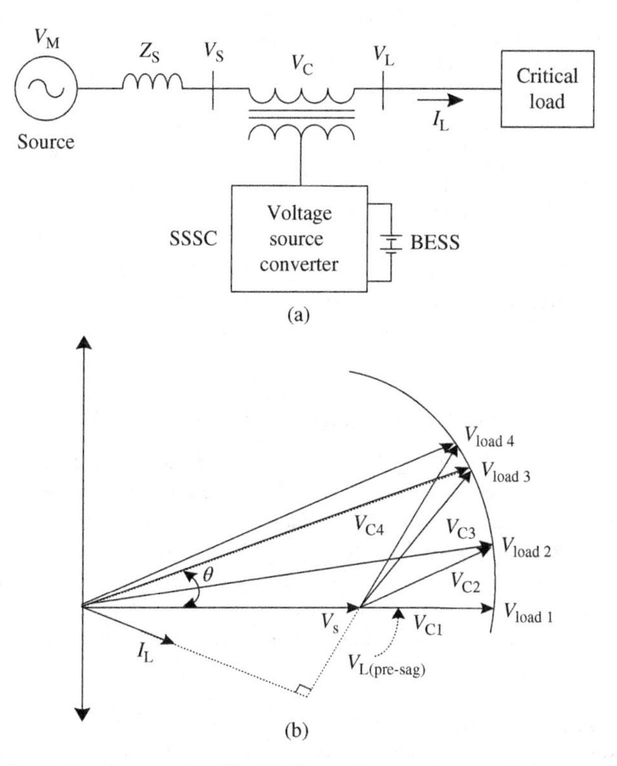

Figure 5.14 (a) Single line diagram of DVR. (b) Phasor diagram of the DVR voltage injection schemes

DSTATCOMs. These are classified as time-domain and frequency-domain control algorithms. There are more than dozen of time-domain control algorithms that may be used for the control of DVR. Few of these control algorithms are as follows:

- Synchronous reference frame (SRF) theory, also known as d–q theory
- Unit template technique or proportional-integral controller-based theory
- Instantaneous reactive power theory (IRPT), also known as PQ theory or α–β theory
- Instantaneous symmetrical component (ISC) theory
- Neural network theory (Widrow LMS-based Adaline algorithm)
- Singe-phase PQ theory
- Singe-phase DQ theory
- Enhanced phase locked loop (EPLL)-based control algorithm
- Adaptive detecting algorithm also known as adaptive interference canceling theory

Most of these time-domain control algorithms have been used for the control of DVRs.

Similarly, there are almost the same number of frequency-domain control algorithms. Some of them are as follows:

- Fourier series theory
- Discrete Fourier transform theory
- Fast Fourier transform theory
- Recursive discrete Fourier transform theory
- Kalman filter-based control algorithm
- Wavelet transformation theory
- Stockwell transformation (S-transform) theory

- Empirical decomposition (EMD) transformation theory
- Hilbert–Huang transformation theory

Most of these frequency-domain control algorithms are used for power quality monitoring for a number of purposes in the power analyzers, PQ instruments, and so on. Some of these algorithms have been used for the control of DVRs. However, these algorithms are sluggish and slow, requiring heavy computational burden; therefore these control methods are not much preferred for real-time control of the DVR compared with time-domain control algorithms.

All these control algorithms may be used in the control of DVR. However, because of space limitation and to just give a basic understanding, only the SRF theory, also known as d–q theory, is explained here. The basic principle of control algorithm with necessary mathematical expressions is discussed here.

5.4.2.1 Synchronous Reference Frame Theory-Based Control of DVRs

The compensation of voltage sags using a DVR can be performed by injecting/absorbing the reactive power or a real power, or the combination of both. When the injected voltage is in quadrature with the current at the fundamental frequency, the compensation is carried out by injecting a reactive power and the DVR may be realized with the self-supported DC bus. However, if the injected voltage is in-phase with the current, DVR may also need a real power and hence a battery or a rectifier is required at the DC side of the VSC. The control technique adopted should consider the limitations such as the voltage injection capability (power converter and transformer rating) and an optimization of the size of energy storage element.

5.4.2.1.1 Control of DVRs with BESS

Figure 5.15 represents the control algorithm of the DVR in which a synchronous reference frame theory is used for reference signals estimation. The voltages at PCC (v_S) and the load terminal voltages (v_L) are sensed to derive the IGBT gate signals. The reference load voltages (v_{La}^*, v_{Lb}^*, v_{Lc}^*) are extracted using the derived unit vectors. The amplitude of the load voltage (v_L) at PCC is calculated as

$$V_L = (2/3)^{1/2}(v_{La}^2 + v_{Lb}^2 + v_{Lc}^2)^{1/2} \tag{5.1}$$

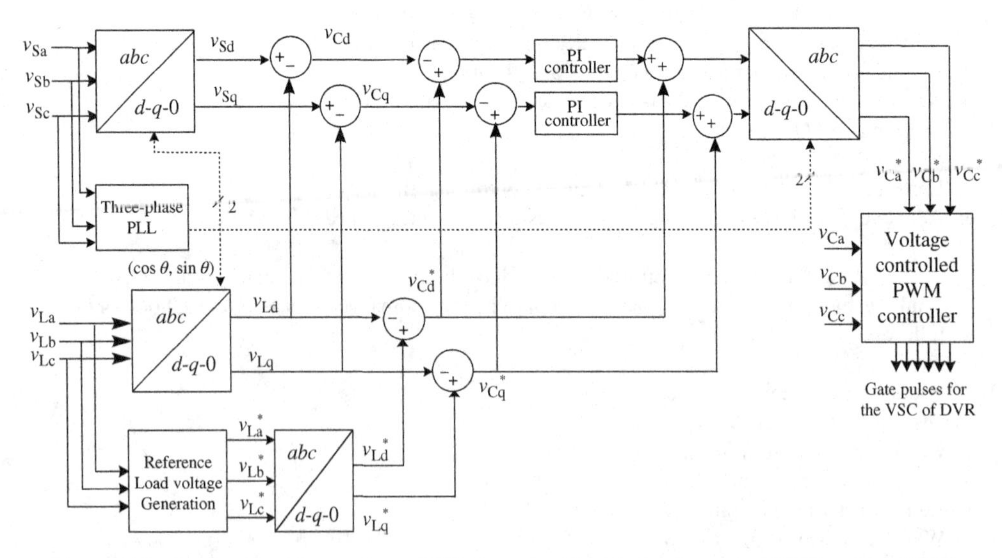

Figure 5.15 Synchronous reference frame (SRF) theory-based method for control of BESS-supported DVR

and the unit vectors are calculated as

$$\begin{bmatrix} u_a \\ u_b \\ u_c \end{bmatrix} = \frac{1}{V_L}\begin{bmatrix} v_{La} \\ v_{Lb} \\ v_{Lc} \end{bmatrix}. \tag{5.2}$$

Hence, the reference load voltages are estimated as

$$\begin{bmatrix} v_{La}^* \\ v_{Lb}^* \\ v_{Lc}^* \end{bmatrix} = V_L^*\begin{bmatrix} u_a \\ u_b \\ u_c \end{bmatrix}, \tag{5.3}$$

where V_L^* is the reference value of amplitude of load terminal voltage.

The load voltages (v_{La}, v_{Lb}, v_{Lc}) are converted into the rotating reference frame using the *abc-dq*$_0$ conversion using the Park's transformation with unit vectors (sin θ, cos θ) derived using a PLL (phase-locked loop):

$$\begin{bmatrix} v_{Ld} \\ v_{Lq} \\ v_{L0} \end{bmatrix} = \frac{2}{3}\begin{bmatrix} \cos\theta & -\sin\theta & \frac{1}{2} \\ \cos\left(\theta-\frac{2\pi}{3}\right) & -\sin\left(\theta-\frac{2\pi}{3}\right) & \frac{1}{2} \\ \cos\left(\theta+\frac{2\pi}{3}\right) & \sin\left(\theta+\frac{2\pi}{3}\right) & \frac{1}{2} \end{bmatrix}\begin{bmatrix} v_{La} \\ v_{Lb} \\ v_{Lc} \end{bmatrix}. \tag{5.4}$$

Similarly, the reference load voltages (v_{La}^*, v_{Lb}^*, v_{Lc}^*) and the voltages at PCC (v_S) are also converted into the rotating reference frame. Then, the DVR voltages are obtained in the rotating reference frame as

$$v_{Cd} = v_{Sd} - v_{Ld}, \tag{5.5}$$

$$v_{Cq} = v_{Sq} - v_{Lq}. \tag{5.6}$$

The reference DVR voltages are obtained in the rotating reference frame as

$$v_{Cd}^* = v_{Ld}^* - v_{Ld}, \tag{5.7}$$

$$v_{Cq}^* = v_{Lq}^* - v_{Lq}. \tag{5.8}$$

These voltage errors between the reference and actual DVR voltages in the rotating reference frame are regulated using two PI controllers.

The reference DVR voltage in abc frame is obtained from the reverse Park's transformation taking V_{Cd}^* from Equation 5.7, V_{Cq}^* from Equation 5.8, and V_{C0}^* as zero:

$$\begin{bmatrix} v_{Ca}^* \\ v_{Cb}^* \\ v_{Cc}^* \end{bmatrix} = \begin{bmatrix} \cos\theta & \sin\theta & 1 \\ \cos\left(\theta-\frac{2\pi}{3}\right) & \sin\left(\theta-\frac{2\pi}{3}\right) & 1 \\ \cos\left(\theta+\frac{2\pi}{3}\right) & \sin\left(\theta+\frac{2\pi}{3}\right) & 1 \end{bmatrix}\begin{bmatrix} v_{Cd}^* \\ v_{Cq}^* \\ v_{C0}^* \end{bmatrix}. \tag{5.9}$$

The reference DVR voltages (v^*_{Ca}, v^*_{Cb}, v^*_{Cc}) and the sensed DVR voltages (v_{Ca}, v_{Cb}, v_{Cc}) are used in a PWM controller unit to generate gate pulses for the VSC of DVRs. The PWM controller is operated with a constant switching frequency f_s on the order of 10 kHz.

5.4.2.1.2 *Control of Self-Supported DVRs*

Figure 5.16 shows the control algorithm of the DVR in which the SRF theory is used for the control of a self-supported DVR. The voltages at PCC (v_S) are converted to the rotating reference frame using the abc–dq0 conversion using the Park's transformation. The harmonics and the oscillatory components of the voltages are eliminated using low-pass filters (LPFs). The components of voltages in d- and q-axes are

$$v_{Sd} = v_{dDC} + v_{dAC}, \tag{5.10}$$

$$v_{Sq} = v_{qDC} + v_{qAC}. \tag{5.11}$$

The compensating strategy for compensation of voltage quality problems considers that the load terminal voltage should be of rated magnitude and undistorted in nature.

In order to maintain the DC bus voltage of the self-supported capacitor, a PI controller is used at the DC bus voltage of the DVR and the output is considered as the voltage loss (v_{loss}) for meeting its losses:

$$v_{loss\,(n)} = v_{loss(n-1)} + K_{p1}\left(v_{de(n)} - v_{de(n-1)}\right) + K_{i1}v_{de(n)}, \tag{5.12}$$

where $v_{de(n)} = v^*_{DC} - v_{DC(n)}$ is the error between the reference DC voltage (v^*_{DC}) and sensed DC voltage (v_{DC}) at the nth sampling instant. K_{p1} and K_{i1} are the proportional and the integral gains of the DC bus voltage PI controller.

Therefore, the reference d-axis load voltage is

$$v^*_d = v_{dDC} - v_{loss}. \tag{5.13}$$

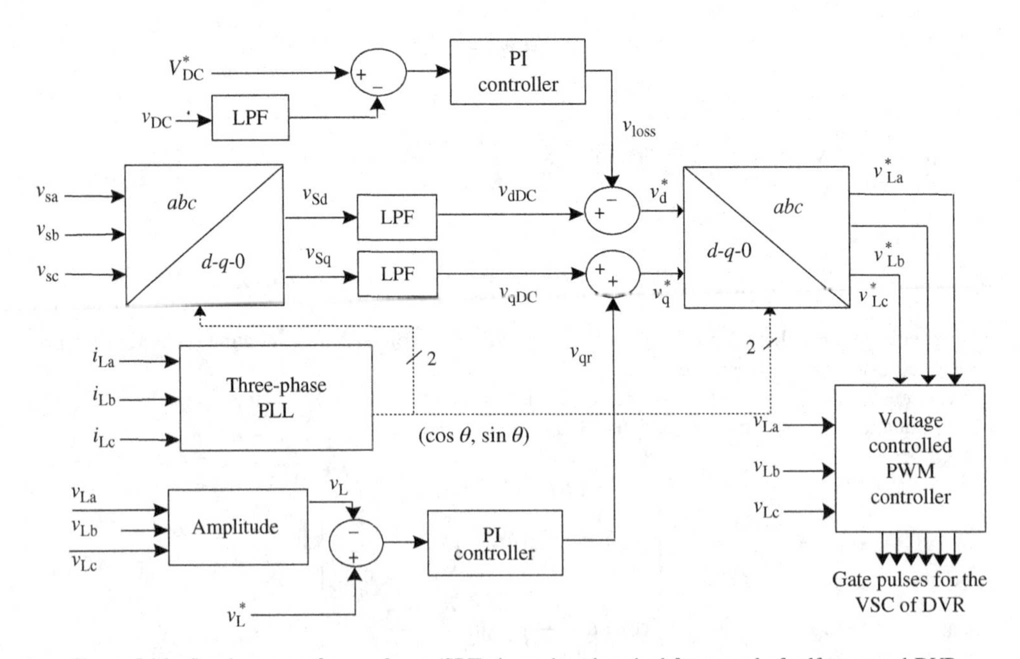

Figure 5.16 Synchronous reference frame (SRF) theory-based method for control of self-supported DVR

The amplitude of the load terminal voltage (v_L) is controlled to its reference voltage (v_L^*) using another PI controller. The output of PI controller is considered as the reactive component of voltage (v_{qr}) for voltage regulation of load terminal voltage. The amplitude of the load voltage (v_L) at PCC is calculated from the AC voltages (v_{La}, v_{Lb}, v_{Lc}) as in Equation 5.1.

Then, a PI voltage controller is used to regulate this to a reference value as

$$v_{qr(n)}=v_{qr(n-1)}+K_{p2}(v_{te(n)}-v_{te(n-1)})+K_{i2}v_{te(n)}, \tag{5.14}$$

where $v_{te(n)} = v_L^* - v_{L(n)}$ denotes the error between the reference load terminal voltage (v_L^*) and actual load terminal voltage ($v_{L(n)}$) amplitudes at the nth sampling instant. K_{p2} and K_{i2} are the proportional and the integral gains of the DC bus voltage PI controller.

The reference load quadrature axis voltage is

$$v_q^* = v_{qDC} + v_{qr}. \tag{5.15}$$

The reference load voltages (v_{La}^*, v_{Lb}, v_{Lc}^*) in abc frame are obtained from the reverse Park's transformation as in Equation 5.9. The errors between the sensed load voltages (v_{La}, v_{Lb}, v_{Lc}) and reference load voltages are used in the PWM controller to generate gate pulses for the VSC of the DVR.

Similarly, other control algorithms may be modified for the control of SSSCs for different operating conditions.

5.5 Analysis and Design of Active Series Compensators

Figure 5.12 shows a schematic diagram of a capacitor-supported DVR for power quality improvement in a distribution system. Three source voltages (v_{Ma}, v_{Mb}, v_{Mc}) represent a three-phase supply system and the series source impedance are shown as Z_{Sa} (R_s, L_s), Z_{Sb} (R_s, L_s), and Z_{Sc} (R_s, L_s). The PCC voltages (v_{Sa}, v_{Sb}, v_{Sc}) have power quality problems and the DVR uses injection transformers (T_r) to inject compensating voltages (v_{Ca}, v_{Cb}, v_{Cc}) to get undistorted load voltages (v_{La}, v_{Lb}, v_{Lc}). A VSC along with a DC capacitor (C_{DC}) is used as a DVR. The switching ripple in the injected voltage is filtered using a series inductor (L_r) and a parallel capacitor (C_r).

The design of a DVR includes the voltage rating of the VSC of the DVR, current rating of the VSC of the DVR, kVA rating of the VSC of the DVR, injection transformer rating, DC bus voltage, DC bus capacitance, AC interfacing inductance, and ripple filter. The design of DVR is illustrated through the following example.

5.5.1 *Voltage Rating of the VSC of DVRs*

Consider a voltage fluctuation between +20 and −30% and a 6% voltage unbalance in the terminal voltage in a three-phase 415 V, 20 kVA, 50 Hz critical load. The voltage rating of the VSC of a DVR depends on the maximum voltage to be injected in this condition of the critical load. Therefore, the voltage injection per phase is 30% of the phase voltage in case of a BESS-based DVR. However, in case of a self-supported VSC-based DVR, the injected voltage has to be in quadrature with the supply/load current and hence the VSC voltage rating is calculated accordingly as per the requirement of the load. The maximum sag in the source terminal voltage is obtained as $239.6 \times 0.7 = 167.72$ V.

Therefore, considering a unity power factor load, the injected voltage (V_C) is estimated as

$$V_C = \sqrt{(V_S^2 - V_L^2)} = \sqrt{(239.6^2 - 167.72^2)} = 171.1 \text{ V}. \tag{5.16}$$

This voltage has to be injected in quadrature with the supply current in case of a self-supported VSC-based DVR.

5.5.2 *Current Rating of the VSC of DVRs*

The current rating of the DVR depends on the load connected to the downstream of a DVR. For a 20 kVA load, the current is calculated as

$$\sqrt{3}V_S I_S = 20\,000, \tag{5.17}$$

where I_S is the current and V_S is the line voltage. Therefore, the current rating of the DVR is $I_S = 27.82$ A.

5.5.3 *kVA Rating of the VSC of DVRs*

The kVA rating of the VSC of a DVR is calculated as

$$S = 3V_C I_S/1000 = (3 \times 171.1\text{ V} \times 27.82\text{ A})/1000 = 14.28\text{ kVA}. \tag{5.18}$$

5.5.4 *Rating of an Injection Transformer of DVRs*

The injection transformer is designed considering the optimum voltage level of the VSC. For a step-down in the power converter voltage, the transformer is selected such that the voltage of the primary winding (supply side) is 171.10 V and the voltage of the secondary winding (VSC side) is 50 V. The kVA rating of the injection transformer is the same as that of the VSC rating and is calculated as

$$\text{kVA} = 3V_C I_S/1000 = 14.28\text{ kVA}. \tag{5.19}$$

Hence, the rating of the injection transformer is 14.27 kVA, $n = 50$ V/171.1 V.

5.5.5 *DC Capacitor Voltage of the VSC of DVRs*

The DC capacitor voltage is selected based on the following relation:

$$V_{DC} > 2\sqrt{2}V_{VSC}, \tag{5.20}$$

where the VSC voltage (V_{VSC}) is 50 V. The value of V_{DC} is obtained as 141.4 V and a V_{DC} of 150 V is selected for a DVR.

5.5.6 *DC Bus Capacitance of the VSC of DVRs*

The DC bus capacitance is selected based on the transient energy required during change in the load. Considering the energy stored in the DC bus capacitor for meeting the energy demand of the load for a fraction of power cycle, the relation can be expressed as

$$(1/2)C_d\left(V_{DC}^2 - V_{DC1}^2\right) = 3V_C I_s \Delta t, \tag{5.21}$$

where V_{DC} is the rated DC bus voltage, V_{DC1} is the drop in the DC bus voltage allowed during transients, and Δt is the time for which support is required.

Considering $\Delta t = 200\,\mu$s, $V_{DC} = 150$ V, $V_{DC1} = 150 - 5\%$ of 150 V = 142.5 V, the DC bus capacitance C_d is calculated as

$$1/2 \times C_d\left(150^2 - 142.5^2\right) = 3 \times 171.1 \times 27.82 \times 0.0002.$$

It gives $C_d = 2603.76\,\mu$F. Hence, a DC bus capacitor of 3000 μF, 200 V is selected for the DVR.

5.5.7 Interfacing Inductor for the VSC of DVRs

The interfacing inductance (L_r) is selected based on the current ripple in the current of the DVR (ΔI_s). Considering ripple current in the inductor is 2%, modulation index m is 1, and overloading factor $a = 1.2$, the inductor is calculated as

$$L_r = n \times (\sqrt{3}/2) m V_{DC}/(6 a f_s \Delta I_s). \tag{5.22}$$

Hence, the interfacing inductance (L_r) is estimated as

$$L_r = (50/171.1) \times (\sqrt{3}/2) \times 1 \times 150/\{6 \times 1.2 \times 10\,\text{k} \times (0.02 \times 27.82)\} = 0.946\,\text{mH}.$$

Hence, an interfacing inductor (L_r) of 1.0 mH and 50 A current carrying capacity is selected for the DVR.

5.5.8 Ripple Filter

A ripple filter is designed for eliminating the switching frequency ripples from the injected voltage of SSSC. The ripple filter, consisting of a capacitor C_r and a resistor R_r connected in series, is generally tuned at half of the switching frequency f_r, which is calculated as

$$f_r = 1/(2 \times \pi \times R_r \times C_r), \tag{5.23}$$

$$f_s/2 = 5000 = 1/(2 \times \pi \times R_r \times C_r). \tag{5.24}$$

Considering $R_r = 5\,\Omega$, it gives the value of the capacitor as $C_r = 6.37\,\mu\text{F}$.
Hence, $R_r = 5\,\Omega$ and $C_r = 10\,\mu\text{F}$ are selected to design a ripple filter.
Similarly, the design of other topologies of SSSCs may be achieved.

5.6 Modeling, Simulation, and Performance of Active Series Compensators

The performance of various topologies of three-phase DVR is simulated using MATLAB software using Sim-Power Systems (SPS) toolboxes. However, because of space limitation and to give just basic understanding, only a BESS-supported DVR and a capacitor-supported DVR are considered for the compensation of sag, swell, harmonics, and an unbalance in the terminal voltage for various injection schemes using SRF theory control algorithm. The simulated results are discussed in the following sections.

5.6.1 Performance of Synchronous Reference Frame Theory-Based BESS-Supported DVRs

The performance of the DVR is demonstrated under different supply voltage disturbances, such as sags and swells in the supply voltages. Figure 5.17 shows the transient performance of the DVR system under voltage sag and swells conditions. At 0.2 s, a sag in the supply voltage is created for 5 cycles; at 0.4 s, a swell in supply is created for 5 cycles. It is observed that the load voltage is regulated to constant amplitude under both sag and swell conditions. The PCC voltages (v_S), load voltages (v_L), DVR voltages (v_C), the amplitude of the load voltage (V_L) and the PCC voltage (V_S), source currents (i_S), reference load voltages (v_L^*), and the DC bus voltage (v_{DC}) are also depicted in Figure 5.17.

The compensation of harmonics in supply voltage is demonstrated in Figure 5.18. At 0.2 s, the supply voltage is distorted and continued for 5 cycles. The PCC voltages (v_S), load voltages (v_L), DVR voltages (v_C), the amplitude of the PCC voltage (V_S) and the load voltage (V_L), source currents (i_S), reference load voltages (v_L^*), and the DC bus voltage (v_{DC}) are also depicted in Figure 5.18. The load voltage is maintained sinusoidal by injecting proper compensation voltage by the DVR. The total harmonic

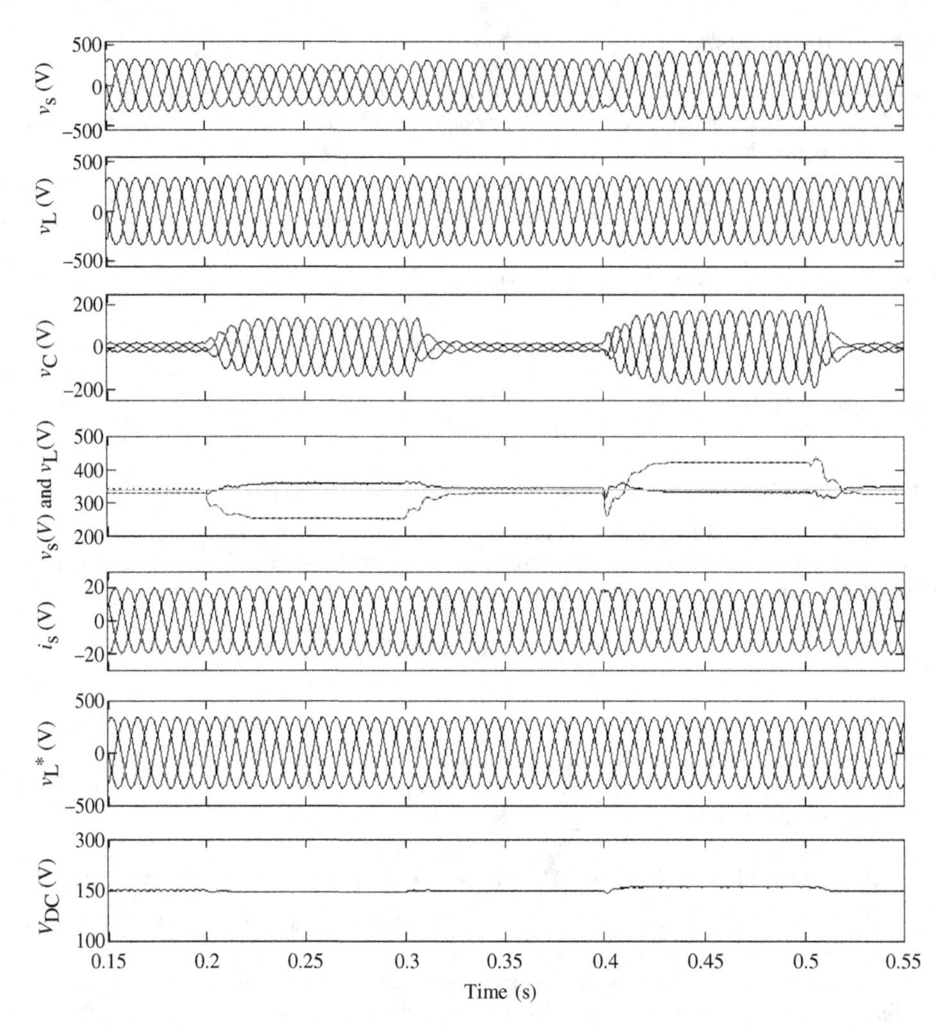

Figure 5.17 Dynamic performance of DVR with in-phase injection during voltage sag and swell

distortion (THD) of the voltage at PCC, the source current, and the load voltage are shown in Figure 5.19a–c. It is observed that THD of the load voltage is reduced to a level of 0.66% from the PCC voltage of 6.34%.

The magnitudes of the voltage injected by the DVR for mitigating the same kinds of sag in the supply with different angles of injection are observed. The injected voltage, supply current, and kVA ratings of the DVR for the four injection schemes are given in Table 5.1. Scheme I in Table 5.1 is the in-phase injection. Scheme II is the voltage injection at a small angle of 30° and scheme III is the voltage injection at

Table 5.1 Comparison of DVR rating for SAG mitigation

	Scheme I	Scheme II	Scheme III	Scheme IV
Injected phase voltage (V)	90	100	121	135
Phase current (A)	13	13	13	13
VA per phase	1170	1300	1573	1755
kVA (% of load)	37.5%	41.67%	50.42%	56.25%

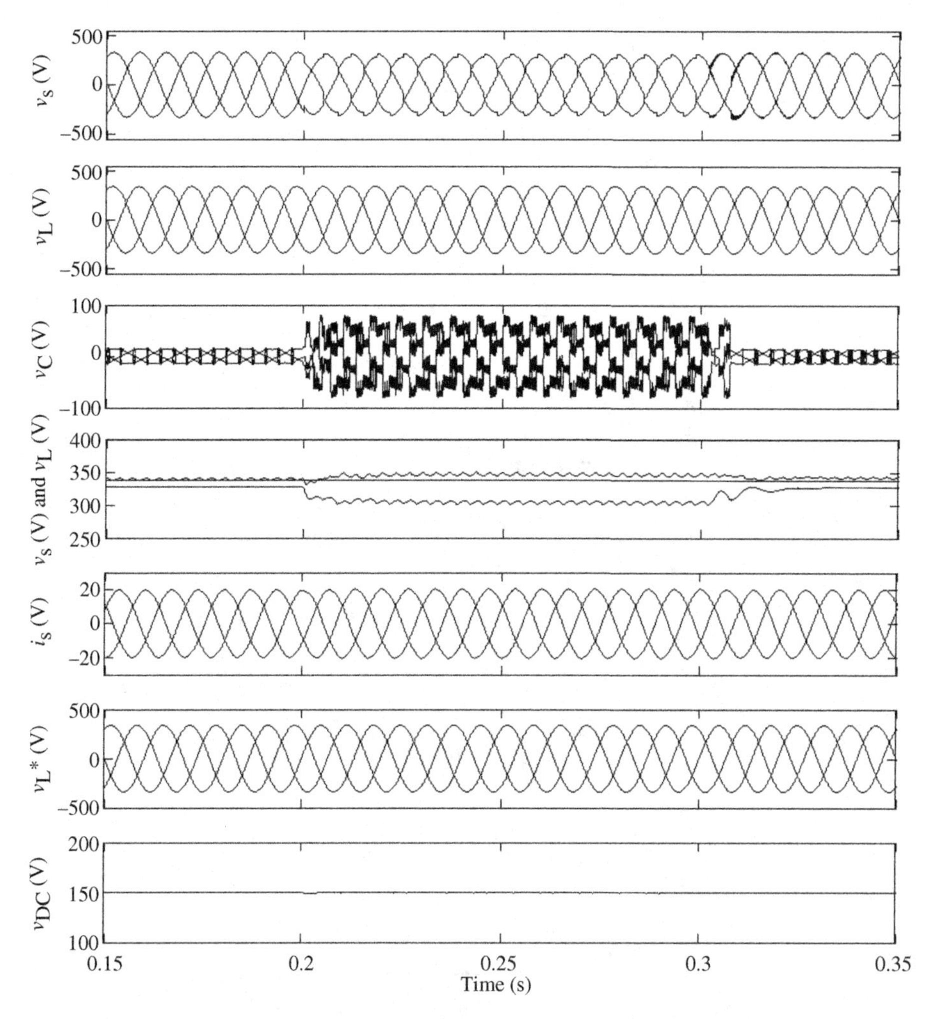

Figure 5.18 Dynamic performance of DVR during harmonics in supply voltage applied to critical load

an angle of 45°. Scheme IV is the voltage injection in the quadrature with the line current. The required rating of the compensator for the same sag using scheme I is much less than that of scheme IV.

5.6.2 Performance of SRF-Controlled Capacitor-Supported DVRs

The performance of SRF-controlled capacitor-supported DVR for different supply disturbances is tested under various operating conditions. The proposed control algorithm is tested for different power quality events such as voltage sag (Figure 5.20), voltage swell (Figure 5.21), unbalance in supply voltages (Figure 5.22), and harmonics in supply voltage (Figure 5.23). Figure 5.20 shows balanced sag of 30% in supply voltage at 0.15 s that occurs for 5 cycles of AC mains. The DVR injects fundamental voltage (v_C) in series with the supply voltage (v_s). The load terminal voltage (v_L) is regulated at the rated value. The source current (i_s), the amplitude of the supply voltage (V_S), the amplitude of the load terminal voltage (V_L), and the DC bus voltage (v_{DC}) are also shown in Figure 5.20.

The dynamic performance of the DVR for a swell in supply voltage is shown in Figure 5.21. This figure shows balanced swell of 30% in supply voltage at 0.12 s, which occurs for 5 cycles of AC mains. The load

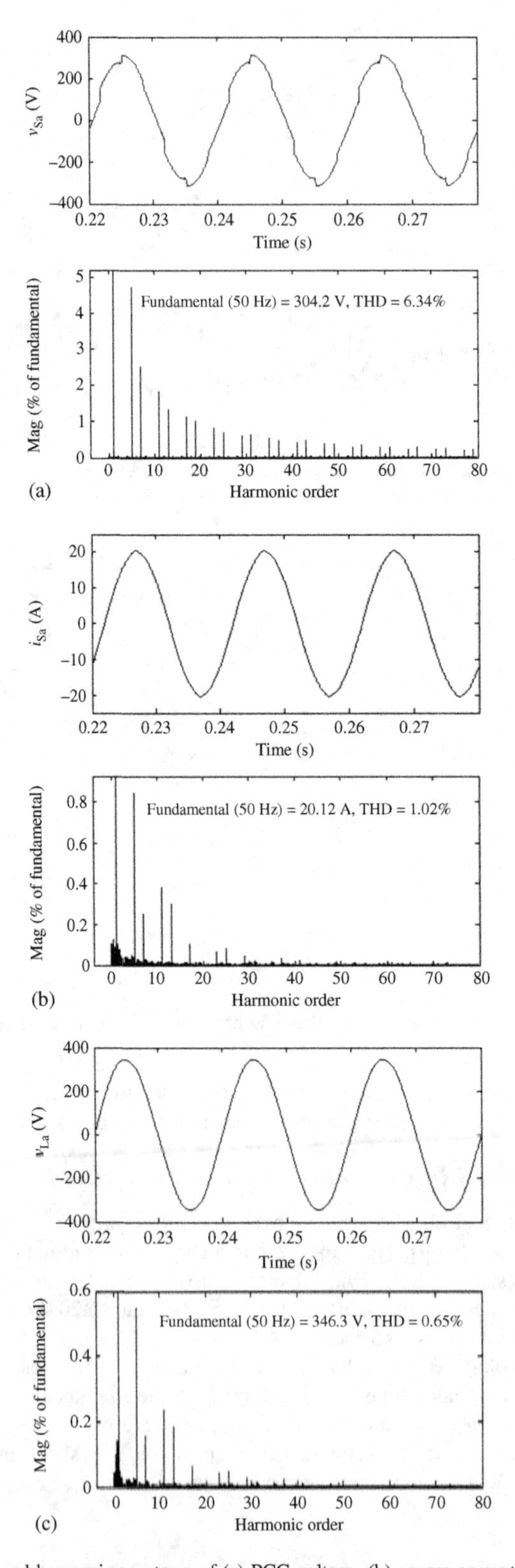

Figure 5.19 THD and harmonic spectrum of (a) PCC voltage, (b) source current and (c) Load voltage

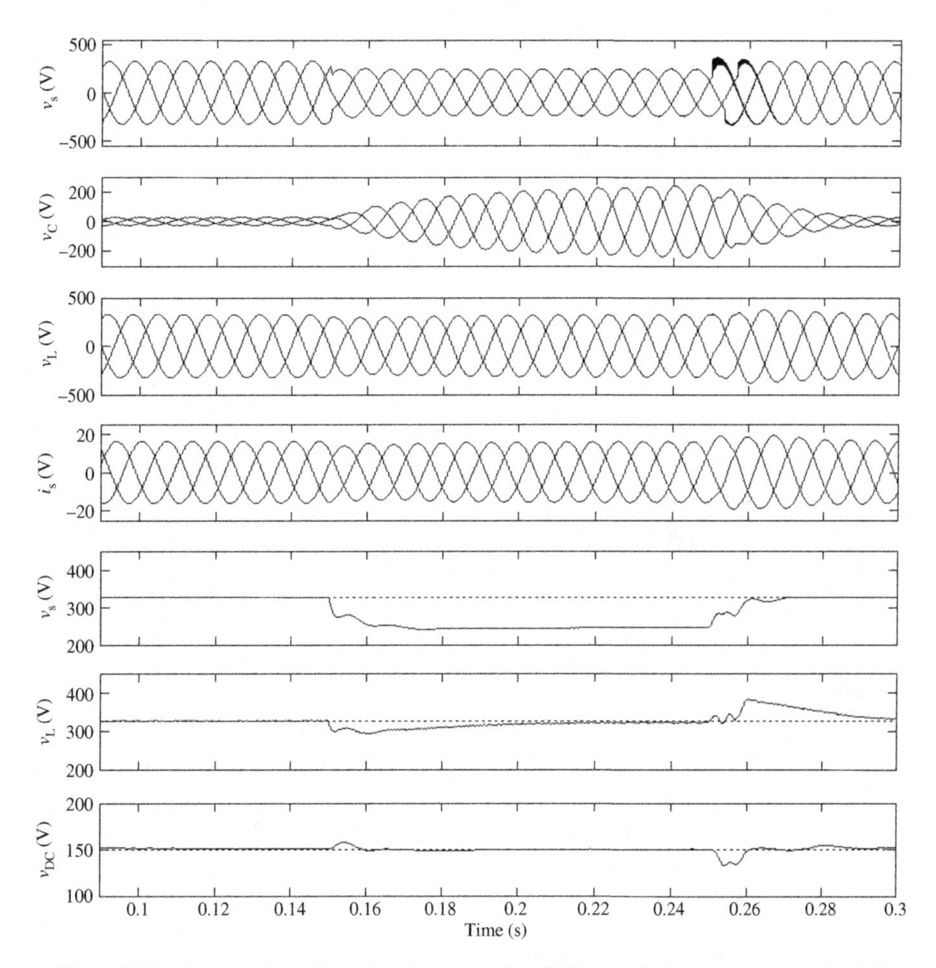

Figure 5.20 Compensation of supply voltage sag using SRF controlled capacitor supported DVR

voltage (v_L) is regulated at rated value, which shows the satisfactory performance of the DVR. The source current (i_s), the amplitude of the load terminal voltage (V_L), the amplitude of the supply voltage (V_S), and the DC bus voltage (v_{DC}) are also shown in Figure 5.21. The DC bus voltage is regulated at the reference value, although small fluctuations occur during transients.

The performance of DVR for an unbalance in supply voltages is shown in Figure 5.22. The three-phase voltages at PCC are different in magnitude at 0.7 s, as shown in the voltage at PCC (v_s). Now the DVR injects unequal fundamental voltages (v_C) so that the load terminal voltage (v_L) is regulated to constant magnitude. The fundamental positive sequence of the PCC voltages (v_{S1}) extracted using SRF theory, the source current (i_s), the amplitude of the load terminal voltage (V_L), the amplitude of the supply voltage (V_S), and the DC bus voltage (v_{DC}) are also shown in Figure 5.22 to demonstrate the satisfactory performance of DVR.

The harmonics compensation in supply voltage is tested and depicted in Figure 5.23. The voltage at PCC is disturbed by switching on and off a load in parallel at PCC. The load terminal voltage (v_L) is undistorted and constant in magnitude due to the injection of harmonic voltage (v_C) by the DVR. The load terminal voltage (v_L) has a THD of 1.2% (Figure 5.24a) at the time of disturbance and the voltage at PCC has a THD of 7.33% (Figure 5.24b). The source current is also sinusoidal with a THD of 0.14% (Figure 5.24c).

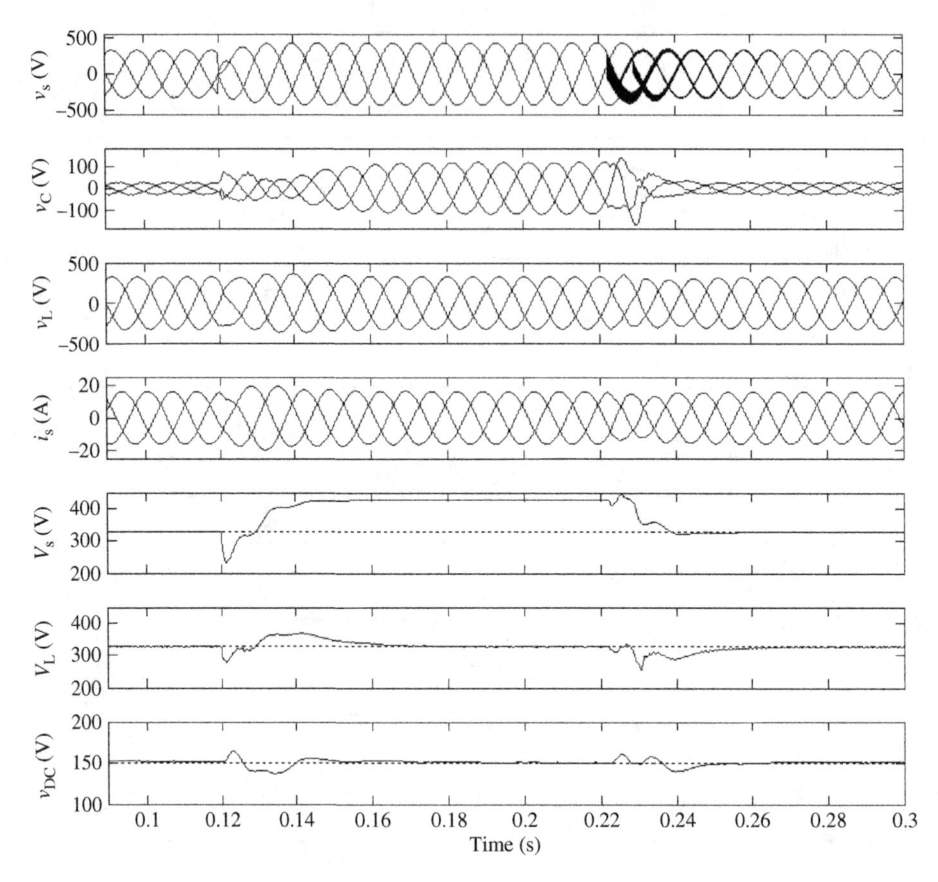

Figure 5.21 Compensation of supply voltage swell using SRF-controlled capacitor-supported DVR

5.7 Numerical Examples

Example 5.1

A single-phase load having $Z_L = (5.0 - j2.0)$ pu has an input AC voltage of 230 V at 50 Hz with a base impedance of 6.9 Ω. It is to be realized as unity power factor load on the AC supply system while maintaining the same rated load terminal voltage using PWM-based SSSC (as shown in Figure E5.1a). Calculate (a) the voltage rating of the compensator, (b) the current rating of the compensator, and (c) the VA rating of the compensator.

Solution: Given supply voltage $V_s = 230$ V (rms), frequency of the supply (f) = 50 Hz, and a critical load having $Z_L = 5.0 - j2.0$ pu with a base impedance of 6.9 Ω per phase.

The load resistance is $R_L = 5 \times 6.9\,\Omega = 34.5\,\Omega$. The load reactance is $X_C = 2 \times 6.9\,\Omega = 13.8\,\Omega$.

The total load impedance is $Z_L = R_L - jX_C = |Z_L| \angle \theta = 34.5 - j13.8 = 37.15 \angle -21.8°\,\Omega$.

The load current before compensation is $|I_L| = |V_L|/|Z_L| = 230/37.15 = 6.19$ A.

The magnitude of supply current under compensation is equal to that of the load current: $I_s = I_L = 6.19$ A.

Since under compensation an extra active power is pumped from the single-phase supply, a battery energy storage system is required at the DC link of SSSC to maintain active power balance.

The series-connected PWM SSSC is used to improve the power factor of the load to unity while maintaining the same rated voltage of 230 V across the load. It means that under compensation the load

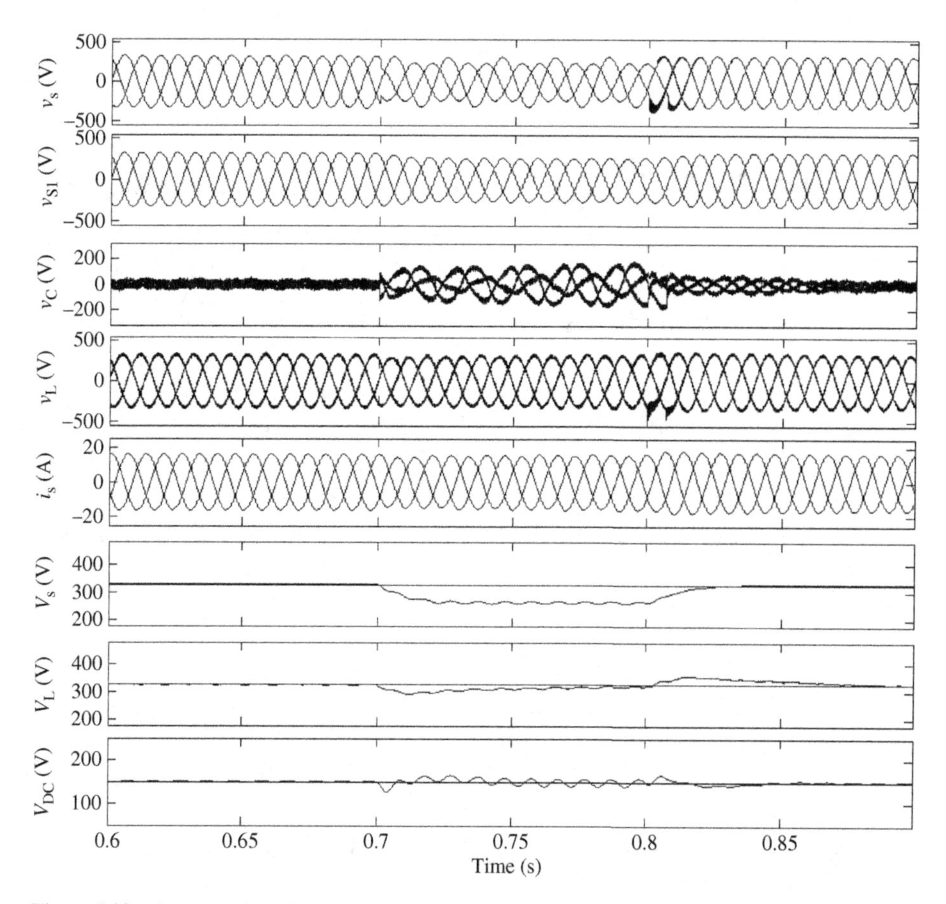

Figure 5.22 Compensation of supply voltage unbalance using SRF-controlled capacitor-supported DVR

voltage lags the supply voltage by the load power factor angle. Hence, the load current remains of the same magnitude as earlier, but the load current is in the phase of the supply voltage. The SSSC must inject a voltage, which is the difference between the supply voltage and load voltage.

The load voltage under compensation is $V_L = |V_L| \angle \theta = 230 \angle 26.56°$ V.

a. The voltage rating of the compensator is $V_{SSSC} = V_L - V_s = 230\angle 21.8° - 230 = 86.98\angle 100.9°$ V.
b. Since the load voltage magnitude and the load impedance are the same, the load current remains the same without the power factor correction. Hence, the current rating of the compensator is $|I_{SSSC}| = |I_L| = |I_s| = 6.19$ A.
c. The VA rating of the compensator is $S_{SSSC} = |V_{SSSC}||I_{SSSC}| = 86.98 \times 6.19 = 538.4$ VA.

Example 5.2

A single-phase AC mains has voltage of 230 V (rms) at 50 Hz and a feeder (source) impedance of 0.5 Ω resistance and 2.5 Ω inductive reactance. It feeds a single-phase load having $Z_L = (24 + j18)$ Ω. Calculate (a) the voltage drop across the source impedance and (b) the voltage across the load. If a self-supported DC bus-based PWM SSSC (as shown in Figure E5.2a) is used to raise the voltage to the input voltage (230 V), calculate (c) the voltage rating of the compensator, (d) the current rating of the compensator, and (e) the VA rating of the compensator.

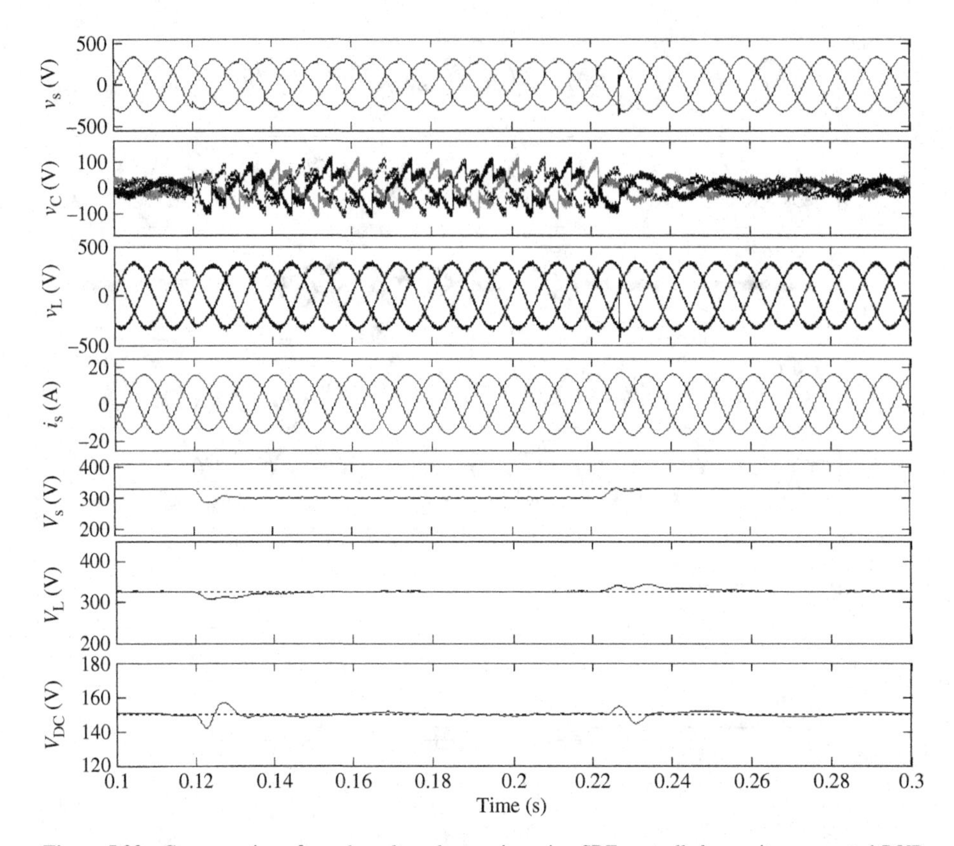

Figure 5.23 Compensation of supply voltage harmonics using SRF-controlled capacitor-supported DVR

Solution: Given supply voltage $V_s = 230$ V, frequency of the supply (f) = 50 Hz, and a single-phase load having $Z_L = (24 + j18)\ \Omega = 30 \angle 36.87°\ \Omega$ fed from an AC supply with an input AC voltage of 230 V at 50 Hz and a source impedance of $Z_s = (0.5 + j2.5)\ \Omega = 2.55 \angle 78.69°\ \Omega$.

Various circuit calculations before compensation are as follows:

The total impedance is $Z_s + Z_L = 24 + j18 + 0.5 + j2.5 = 24.5 + j20.5 = 31.945 \angle 39.92°\ \Omega$.

The load current before compensation is $I_L = V_s/(Z_s + Z_L) = 230\angle 0°/31.945 \angle 39.92° = 7.19 \angle -39.92°$ A.

The supply current is the same as the load current: $I_s = I_L = 7.19 \angle -39.92°$ A.

a. The voltage drop across the source impedance is $V_{Zs} = Z_s \times I_s = 2.55\angle 78.69° \times 7.19\angle -39.92° = 18.33\angle 38.77°$ V.
b. The voltage across the load terminals is $V_L = Z_L \times I_L = 30\angle 36.87° \times 7.19\angle -39.92° = 215.70\angle -3.05°$ V.

The series-connected PWM SSSC is used to raise the load terminal voltage to the input voltage (230 V). The SSSC consists of a VSC with self-supported DC link and hence the SSSC can inject only reactive power into the system. The SSSC injects a voltage in quadrature to load current. The phasor diagram under voltage drop compensation is shown in Figure E5.2b.

The load current after compensation is $|I'_L| = |V'_L|/|Z_L| = 230/30 = 7.67$ A.

The active power balance is expressed as $|V_s||I'_L|\cos\beta = |I'_L|^2 R + |I'_L|^2 R$.

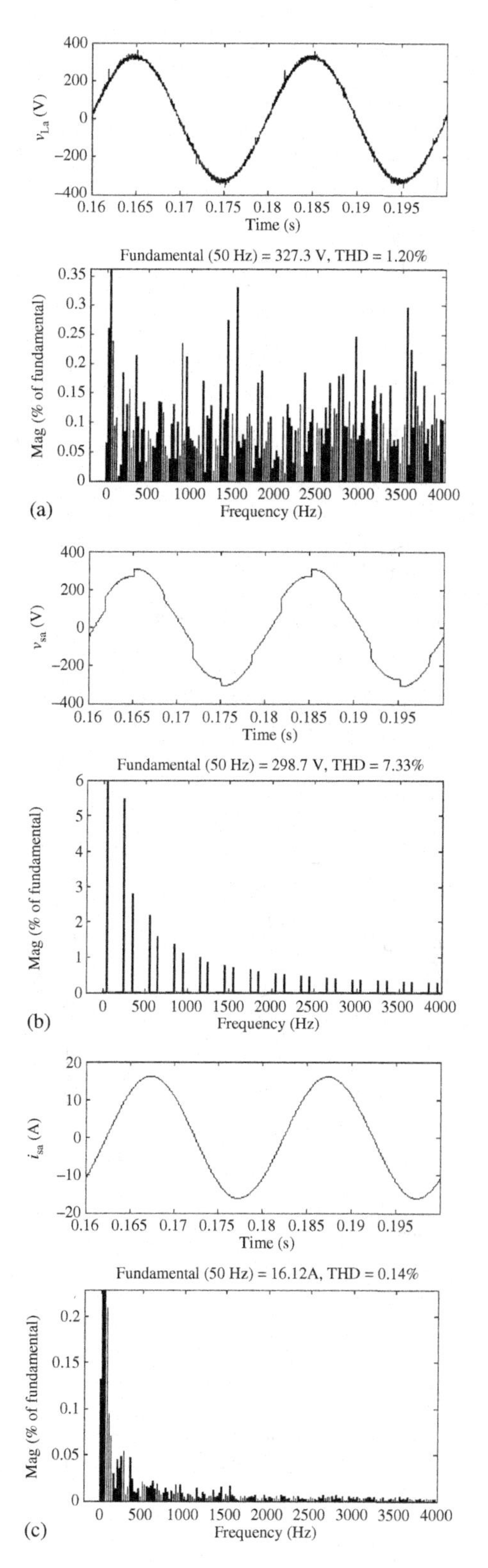

Figure 5.24 THD and harmonic spectra of the (a) load voltage, (b) PCC voltage, and (c) source current

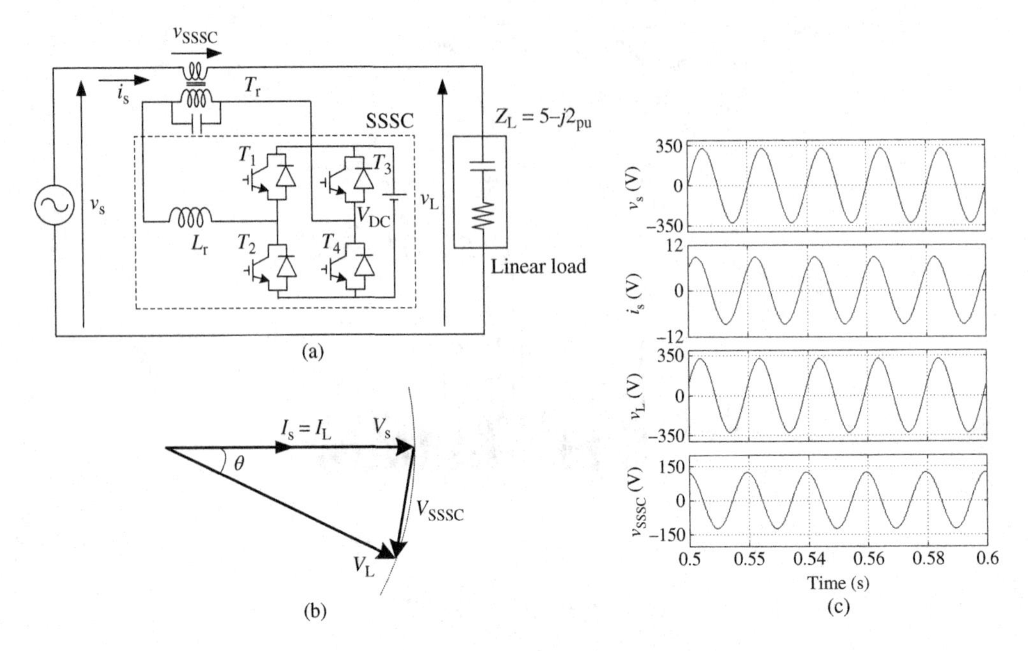

Figure E5.1 Single-phase SSSC for power factor correction of leading linear load. (a) Circuit configuration. (b) Phasor diagram under compensation. (c) Waveforms under compensation

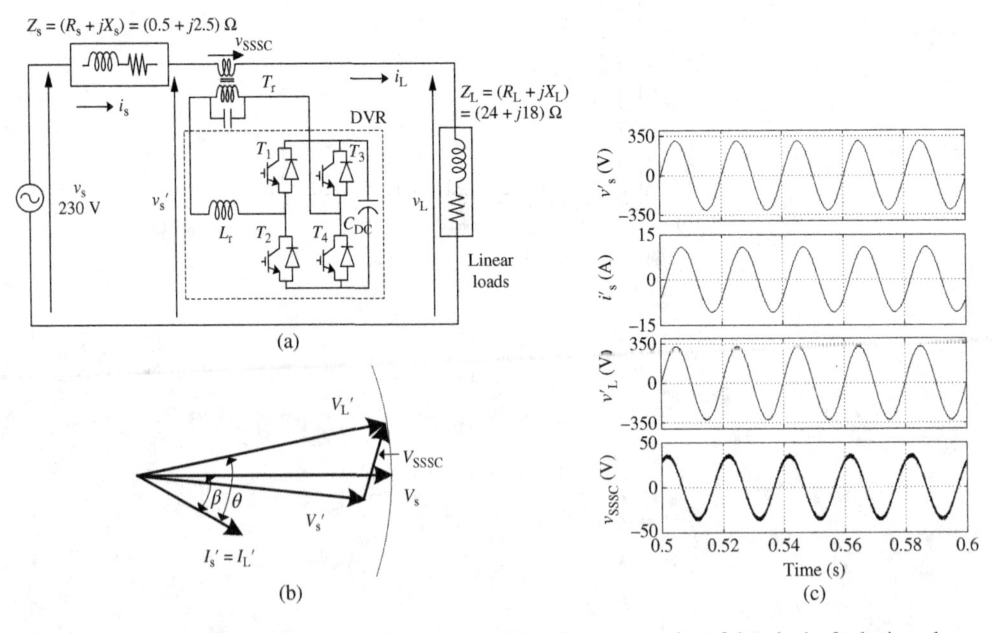

Figure E5.2 Single-phase SSSC with a self-supported DC bus for compensation of drop in the feeder impedance under lagging linear load. (a) Circuit configuration. (b) Phasor diagram under compensation. (c) Waveforms under compensation

Solving the above equation for angle β results in $230 \times 7.67 \times \cos\beta = 7.67^2 \times 0.5 + 7.67^2 \times 24$,

$$\beta = 35.27°.$$

The load current lags the supply voltage by an angle β: $I'_L = |I'_L|\angle -\beta = 7.67\angle -35.21°$ A.

The PCC voltage is estimated as $V'_s = V_s - Z_s I'_L = 230\angle 0° - 2.55\angle 78.69° \times 7.67\angle -35.21° = 216.23\angle -3.57°$ V.

The SSSC supplies leading reactive power into the system to increase the load voltage to 230 V. Therefore, the reactive power balance is $|V_s||I'_L|\sin\beta = |I'_L|^2 X + |I'_L|^2 X - |I'_L||V_{SSSC}|$.

Substituting the values of circuit parameters results in $230 \times 7.67 \sin 35.21° = 7.67^2 \times 2.5 + 7.67^2 \times 18 - 7.67 \times |V_{SSSC}|$.

On solving the above equation, $V_{SSSC} = 24.62$ V.

a. The voltage rating of the compensator is $|V_{SSSC}| = 24.62$ V.
b. The current rating of the compensator is $|I_{SSSC}| = |I'_L| = 7.67$ A (since it is connected in series with the load).
c. The VA rating of the compensator is $S = |V_{SSSC}||I_{SSSC}| = 188.83$ VA.

Example 5.3

A single-phase AC supply has a voltage of 230 V at 50 Hz and a feeder (source) impedance of 1.0 Ω resistance and 4.0 Ω inductive reactance. It feeds a single-phase load having $Z_L = 30\,\Omega$. Calculate (a) the voltage drop across the source impedance and (b) the voltage across the load. If a PWM-based SSSC (as shown in Figure E5.3a) is used to raise the voltage to the input voltage (230 V) with minimum rating, then calculate (c) the voltage rating of the compensator, (d) the current rating of the compensator, and (e) the VA rating of the compensator.

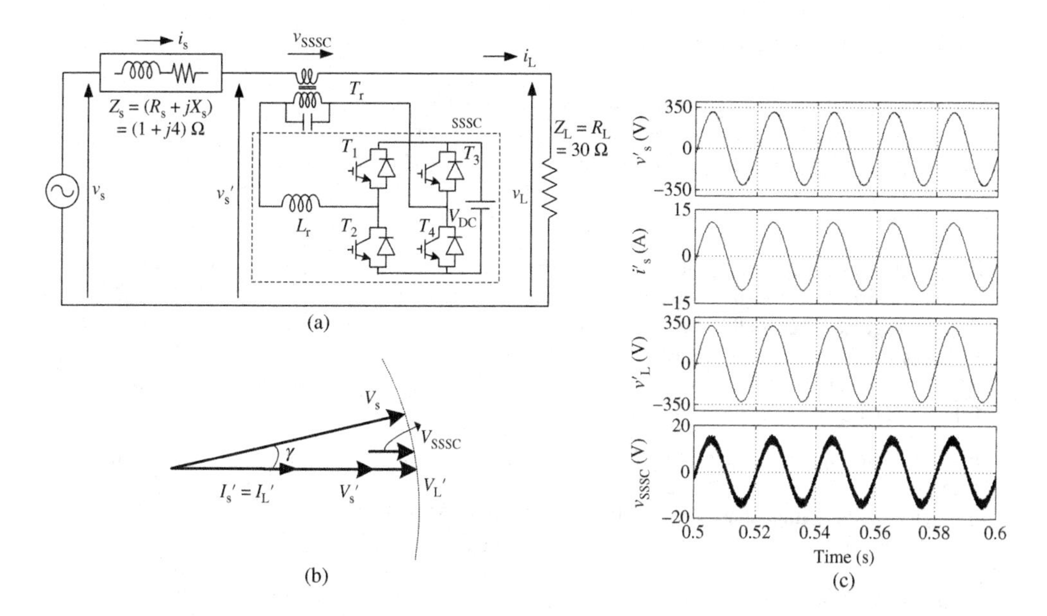

Figure E5.3 Single-phase SSSC with a battery energy storage system for compensation of drop in the feeder impedance under purely resistive load. (a) Circuit configuration. (b) Phasor diagram under compensation. (c) Waveforms under compensation

Solution: Given supply voltage $V_s = 230$ V (rms), frequency of the supply $(f) = 50$ Hz, and a single-phase load having $Z_L = 30\ \Omega = 30\angle 0°\ \Omega$ fed from an AC supply with an input AC voltage of 230 V at 50 Hz and a source impedance of $Z_s = (1 + j4)\ \Omega = 4.12\angle 75.96°\ \Omega$.

Various circuit calculations before compensation are as follows:

The total impedance is $Z_s + Z_L = 30 + 1 + j4 = 31 + j4 = 31.25\angle 7.35°\ \Omega$.
The load current before compensation is $I_L = V_s/(Z_s + Z_L) = 230\angle 0°/31.25\angle 7.35° = 7.36\angle -7.35°$ A.
The supply current is the same as the load current: $I_s = I_L = 7.36\angle -7.35°$ A.

a. The voltage drop across the source impedance is $V_{Zs} = Z_s \times I_s = 4.12\angle 75.96° \times 7.36\angle -7.35° = 30.32\angle 68.61°$ V.
b. The voltage across the load terminals is $V_L = Z_L \times I_L = 30\angle 0° \times 7.36\angle -7.35° = 220.8\angle -7.35°$ V.

The series-connected PWM SSSC is used to raise the load terminal voltage to the input voltage (230 V). The SSSC must inject voltage in phase with the PCC (V_s') voltage to attain minimum voltage rating. The SSSC voltage is injected in phase with the SSSC current. However, to inject this voltage, an active power is required from the DC link of SSSC and hence a battery energy storage is required. The load current after compensation is $|I_L'| = |V_L'|/|Z_L| = 230/30 = 7.67$ A.

The supply current is the same as the load current: $I_s' = I_L'$.

Considering the PCC voltage (V_s') at reference, the phasor diagram under voltage drop compensation is shown in Figure E5.3b.

Therefore, the reactive power balance can be expressed as $|V_s||I_L'|\sin\gamma = |I_L'|^2 X_s$.

Substituting the values of circuit parameters results in $230 \times 7.67 \sin\gamma = 7.67^2 \times 4$.

On solving this equation, $\gamma = 7.65°$.

The SSSC supplies active power into the network to increase the load voltage; hence, the active power balance can be expressed as $|V_s||I_L'|\cos\gamma = |I_L'|^2 R + |I_L'|^2 R - |V_{SSSC}||I_L'|$.

Substituting the values of circuit parameters results in $230 \times 7.67 \cos 7.65° = 7.67^2 \times 1 + 7.67^2 \times 30 - |V_{SSSC}| \times 7.67$.

On solving the above equation, $|V_{SSSC}| = 9.51$ V.

a. The voltage rating of the compensator is $|V_{SSSC}| = 9.51$ V.
b. The current rating of the compensator is $|I_{SSSC}| = |I_L'| = 7.67$ A.
c. The VA rating of the compensator is $S = |V_{SSSC}||I_{SSSC}| = 72.94$ VA.

Example 5.4

A single-phase AC supply has voltage of 230 V (rms) at 50 Hz and a feeder (source) impedance of 0.25 Ω resistance and a 2.5 Ω inductive reactance. It feeds a single-phase load having $Z_L = (24 - j18)\ \Omega$. Calculate (a) the voltage drop across the source impedance and (b) the voltage across the load. If a self-supported DC bus-based PWM SSSC (as shown in Figure E5.4a) is used to regulate load voltage to 220 V, calculate (c) the voltage rating of the compensator, (d) the current rating of the compensator, and (e) the VA rating of the compensator.

Solution: Given supply voltage $V_s = 230$ V, frequency of the supply $(f) = 50$ Hz, and a single-phase load having $Z_L = (24 - j18)\ \Omega = 30\angle -36.87°\ \Omega$ fed from an AC supply with an input voltage of 230 V at 50 Hz and a source impedance of $Z_s = (0.25 + j2.5)\ \Omega = 2.51\angle 84.28°\ \Omega$.

Various circuit calculations before compensation are as follows:

The total impedance is $Z_s + Z_L = 24 - j18 + 0.25 + j2.5 = 24.25 - j15.5 = 28.78\angle -32.59°\ \Omega$.
The load current before compensation is $I_L = V_s/(Z_s + Z_L) = 220\angle 0°/28.78\angle -32.59° = 7.64\angle 32.59°$ A.
The supply current is the same as the load current: $I_s = I_L = 7.64\angle 32.59°$ A.

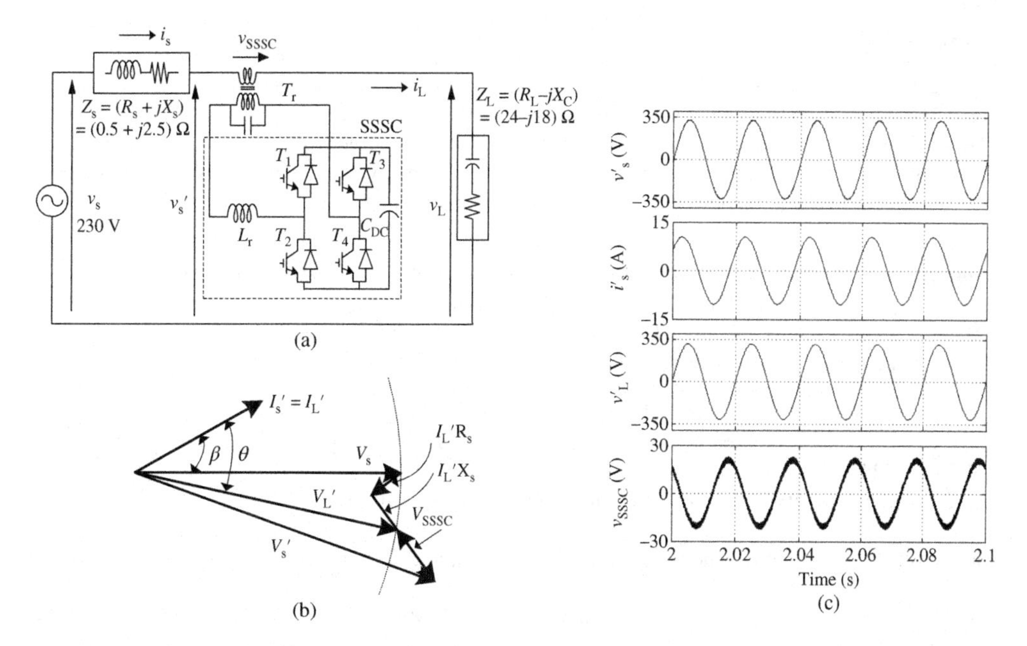

Figure E5.4 Single-phase SSSC with a self-supported DC bus for compensation of drop in the feeder impedance under lagging linear load. (a) Circuit configuration. (b) Phasor diagram under compensation. (c) Waveforms under compensation

a. The voltage drop across the source impedance is $V_{Zs} = Z_s \times I_s = 2.51\angle 84.28° \times 7.64\angle 32.59° = 19.17\angle 116.87°$ V.
b. The voltage across the load terminals is $V_{ZL} = Z_L \times I_L = 30\angle -36.87° \times 7.64\angle 32.59° = 229.2\angle -4.28°$ V.

The series-connected PWM SSSC is used to decrease the load terminal voltage to a voltage (220 V). The SSSC consists of a VSC with self-supported DC link and hence the SSSC can inject only reactive power into the system. The SSSC injects a voltage in quadrature to load current. The phasor diagram under voltage drop compensation is shown in Figure E5.4b.

The load current after compensation is $|I'_L| = |V'_L|/|Z_L| = 220/30 = 7.33$ A.

The active power balance may be expressed as $|V_s||I'_L|\cos\beta = |I'_L|^2 R + |I'_L|^2 R$.

Solving the above equation for angle β results in $220 \times 7.33 \times \cos\beta = 7.33^2 \times 0.25 + 7.33^2 \times 24$,

$$\beta = 36.10°.$$

The load current leads the supply voltage by an angle $\beta: I'_L = |I'_L|\angle\beta = 7.33\angle 36.10°$ A.

The PCC voltage is estimated as $V'_s = V_s - Z_s I'_L = 220\angle 0° - 2.51\angle 84.28° \times 7.33\angle 36.10° = 229.86\angle -3.96°$ V.

The SSSC absorbs reactive power from the system to decrease the load voltage to 220 V. Therefore, the reactive power balance is $-|V_s||I'_L|\sin\beta = |I'_L|^2 X + |I'_L|^2 X - |I'_L||V_{SSSC}|$.

Substituting the values of circuit parameters results in $-220 \times 7.33 \sin 36.10° = 7.33^2 \times 2.5 - 7.33^2 \times 18 - 7.33 \times |V_{SSSC}|$.

Solving the above equation, $V_{SSSC} = 16.00$ V.

c. The voltage rating of the compensator is $|V_{SSSC}| = 16$ V.
d. The current rating of the compensator is $|I_{SSSC}| = |I_{L'}| = 7.33$ A (since it is connected in series with the load).
e. The VA rating of the compensator is $S = |V_{SSSC}||I_{SSSC}| = 117.34$ VA.

Example 5.5

A single-phase AC supply has AC mains voltage of 230 V (rms) at 50 Hz and a feeder (source) impedance of 0.5 Ω resistance and 3.0 Ω inductive reactance. It feeds a load having $Z_L = (16 + j12)$ Ω. Calculate (a) the voltage drop across the source impedance and (b) the voltage across the load. If a PWM SSSC (as shown in Figure E5.5a) is used to regulate the voltage to the input voltage (230 V) with minimum rating, calculate (c) the voltage rating of the compensator, (d) the current rating of the compensator, and (e) the VA rating of the compensator.

Solution: Given supply voltage $V_s = 230$ V, frequency of the supply $(f) = 50$ Hz, and a single-phase load having $Z_L = (16 + j12)\ \Omega = |Z_L| \angle\theta = 20\angle 36.87°\ \Omega$ fed from an AC supply with an input AC voltage of 230 V at 50 Hz and a source impedance of $Z_s = (0.5 + j3)\ \Omega = 3.04\angle 80.5°\ \Omega$.

Various circuit calculations before compensation are as follows:

The total impedance is $Z_s + Z_L = 16 + j12 + 0.5 + j3 = 16.5 + j15 = 22.29\angle 42.27°\ \Omega$.

The load current before compensation is $I_L = V_s/(Z_s + Z_L) = 230\angle 0°/22.29\angle 42.27° = 10.31\angle -42.27°$ A.

The supply current is the same as the load current: $I_s = I_L = 10.31\angle -42.27°$ A.

a. The voltage drop across the source impedance is $V_{Zs} = Z_s \times I_s = 3.04\angle 80.5° \times 10.31\angle -42.27° = 31.36\angle 38.23°$ V.
b. The voltage across the load terminals is $V_L = Z_L \times I_L = 20\angle 36.87° \times 10.31\angle -42.27° = 206.2\angle -5.4°$ V.

The series-connected PWM SSSC is used to regulate the load terminal voltage to the input voltage (230 V). The SSSC must inject voltage in phase with the PCC (V_s') voltage to attain minimum rating. In order to inject voltage in phase with V_s', both real and reactive powers are required and hence a battery storage support is required at the DC link of SSSC. Figure E5.5b shows the phasor diagram under voltage drop compensation by SSSC.

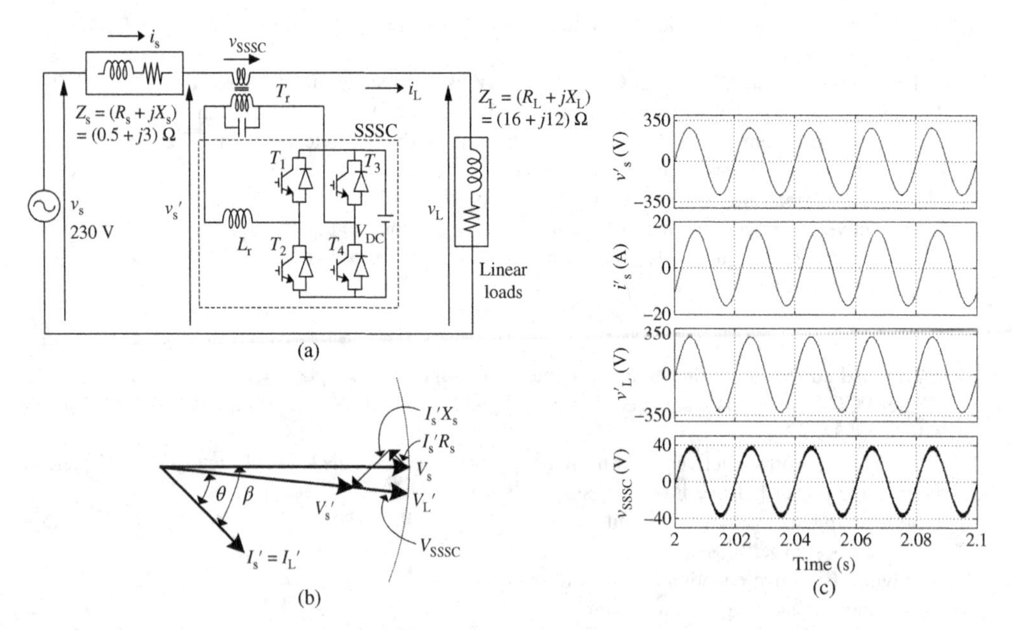

Figure E5.5 Single-phase SSSC with a battery energy storage system for compensation of drop in the feeder impedance under lagging linear load. (a) Circuit configuration. (b) Phasor diagram under compensation. (c) Waveforms under compensation

The magnitude of load current under feeder drop compensation is $|I'_L| = |V'_L|/|Z_L| = 230/20 = 11.5$ A.

The supply current and the load current are same for the given network: $I'_L = I'_s$.

From phasor diagram it can be observed that the SSSC supplies both active and reactive powers into the system. The power balance equations are used to calculate effective power factor angle β and the magnitude of injected voltage.

The active power balance may be expressed as $|V_s||I'_L|\cos\beta = |I'_L|^2R + |I'_L|^2R - |V_{SSSC}||I'_L|\cos\theta$.

Similarly, the reactive power balance may be expressed as $|V_s||I'_L|\sin\beta = |I'_L|^2X + |I'_L|^2X - |V_{SSSC}||I'_L|\sin\theta$.

Solving these two equations results in a quadratic equation. The two roots of the quadratic equation are $|V_{SSSC1}| = 26.57$ V and $|V_{SSSC2}| = 483$ V. By inspection the smaller value of $|V_{SSSC}|$ is selected.

c. The voltage rating of the compensator is $|V_{SSSC}| = 26.57$ V.
d. The current rating of the compensator is $|I_{SSSC}| = |I'_L| = 11.5$ A (since it is connected in series with the load).
e. The VA rating of the compensator is $S = |V_{SSSC}||I_{SSSC}| = 26.57 \times 11.5 = 305.55$ VA.

Example 5.6

A single-phase AC supply has AC mains voltage of 230 V (rms) at 50 Hz and a feeder (source) impedance of 0.25 Ω resistance and 3.5 Ω inductive reactance. It feeds a load having $Z_L = (25 - j15)$ Ω. Calculate (a) the voltage drop across the source impedance and (b) the voltage across the load. If a PWM SSSC (as shown in Figure E5.6a) is used to regulate the voltage to the input voltage (230 V) with minimum rating, calculate (c) the voltage rating of the compensator, (d) the current rating of the compensator, and (e) the VA rating of the compensator.

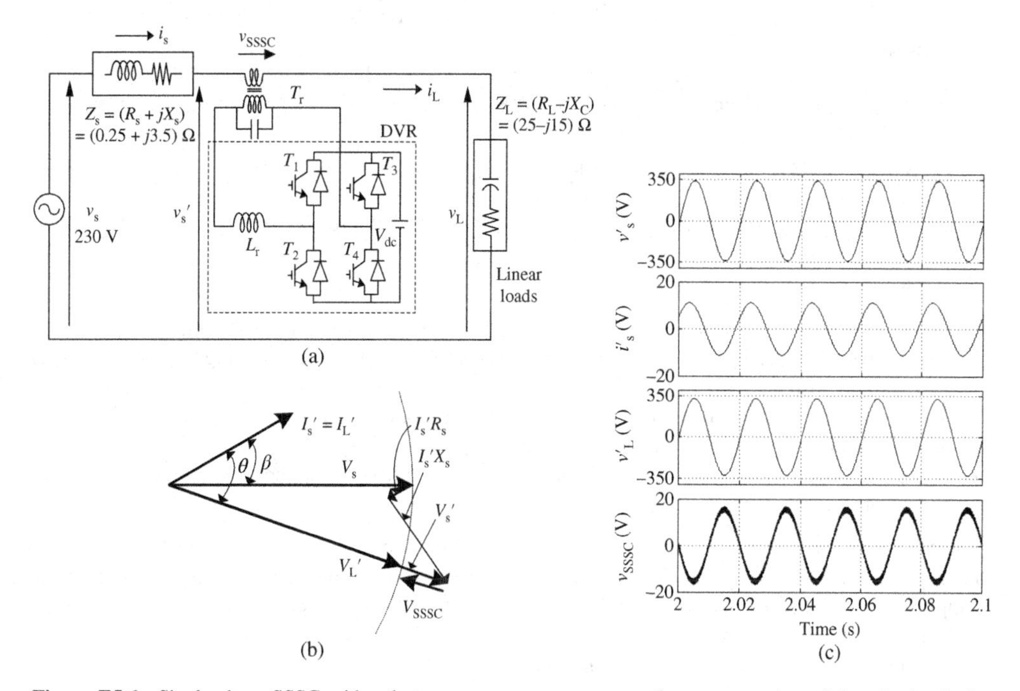

Figure E5.6 Single-phase SSSC with a battery energy storage system for compensation of drop in the feeder impedance under leading linear load. (a) Circuit configuration. (b) Phasor diagram under compensation. (c) Waveforms under compensation

Solution: Given supply voltage $V_s = 230$ V, frequency of the supply $(f) = 50$ Hz, and a single-phase load having $Z_L = (25 - j15)\ \Omega = |Z_L| \angle \theta = 29.15 \angle -30.96°\ \Omega$ fed from an AC supply with an input AC voltage of 230 V at 50 Hz and a source impedance of $Z_s = (0.25 + j3.5)\ \Omega = 3.51 \angle 85.91°\ \Omega$.

Various circuit calculations before compensation are as follows:

The total impedance is $Z_s + Z_L = 25 - j15 + 0.25 + j3.5 = 25.25 - j11.5 = 27.74 \angle -24.48°\ \Omega$.

The load current before compensation is $I_L = V_s/(Z_s + Z_L) = 230 \angle 0°/24.48 \angle -24.48° = 9.39 \angle 24.48°$ A.

The supply current is the same as the load current: $I_s = I_L = 9.39 \angle 24.48°$ A.

a. The voltage drop across the source impedance is $V_{Zs} = Z_s \times I_s = 3.51\angle 85.91° \times 9.39\angle 24.48° = 32.95\angle 110.39°$ V.
b. The voltage across the load terminals is $V_L = Z_L \times I_L = 29.15\angle -30.96° \times 9.39\angle -24.48° = 273.7\angle -6.48°$ V.

   The series-connected PWM SSSC is used to control the load terminal voltage to the input voltage (230 V). The SSSC must inject voltage in phase with the PCC (V'_s) voltage to attain minimum rating. In order to inject voltage in phase with V'_s, both real and reactive powers are required and hence a battery storage support is required at the DC link of SSSC. Figure E5.6b shows the phasor diagram under voltage drop compensation by SSSC.

   The magnitude of load current under feeder drop compensation is $|I'_L| = |V'_L|/|Z_L| = 230/29.15 = 7.89$ A.

   The supply current and the load current are same for the given network: $I'_L = I'_s$.

   From phasor diagram it can be observed that the SSSC absorbs both active and reactive powers from the system. The power balance equations are used to calculate effective power factor angle β and magnitude of the injected voltage.

   The active power balance is expressed as $|V_s||I'_L|\cos\beta = |I'_L|^2 R + |I'_L|^2 R + |V_{SSSC}||I'_L|\cos\theta$.

   Similarly, the reactive power balance is expressed as $|V_s||I'_L|\sin\beta = -|I'_L|^2 X + |I'_L|^2 X + |V_{SSSC}||I'_L|\sin\theta$.

   Solving these two equations results in a quadratic equation. The two roots of the quadratic equation are $|V_{SSSC1}| = 11.56$ V and $|V_{SSSC2}| = -449.5$ V. By inspection the smaller value of $|V_{SSSC}|$ is selected.
c. The voltage rating of the compensator is $|V_{SSSC}| = 11.56$ V.
d. The current rating of the compensator is $|I_{SSSC}| = |I'_L| = 7.89$ A (since it is connected in series with the load).
e. The VA rating of the compensator is $S = |V_{SSSC}||I_{SSSC}| = 91.2$ VA.

Example 5.7

A single-phase AC supply has AC voltage of 230 V (rms) at 50 Hz and a feeder (source) impedance of 0.25 Ω resistance and 1.25 Ω inductive reactance per phase. It is feeding a load having $Z_{L1} = (20 + j12)\ \Omega$. There is another load having $Z_{L2} = (5 - j4)\ \Omega$ connected in parallel for 20 cycles of AC mains. Calculate the voltage sag/swell across the first load under steady-state condition. If a self-supported DC bus-based PWM VSC is employed as SSSC (as shown in Figure E5.7a) in series of the load to maintain its terminal voltage to the input voltage (230 V), calculate the line current, voltage, and kVA rating of the series compensator in both cases.

Solution: Given supply voltage $V_s = 230$ V, frequency of the supply $(f) = 50$ Hz, two single-phase loads having $Z_{L1} = (20 + j12)\ \Omega = 23.32\angle 30.96°\ \Omega$ and $Z_{L2} = (5 - j4)\ \Omega = 6.4 \angle -38.65°\ \Omega$ fed from an AC supply with an input AC voltage of 230 V at 50 Hz and a source impedance of $Z_s = (0.25 + j1.25)\ \Omega = 1.27 \angle 78.69°$. The circuit diagram from the system is shown in Figure E5.7a.

Various circuit calculations before compensation are as follows:

The total impedance is $Z_T = Z_s + Z_{L1}||Z_{L2} = (0.25 + j1.25) + (20 + j12)||(5 - j4) = 5.38 - j1.19 = 5.51 \angle -12.49°\ \Omega$.

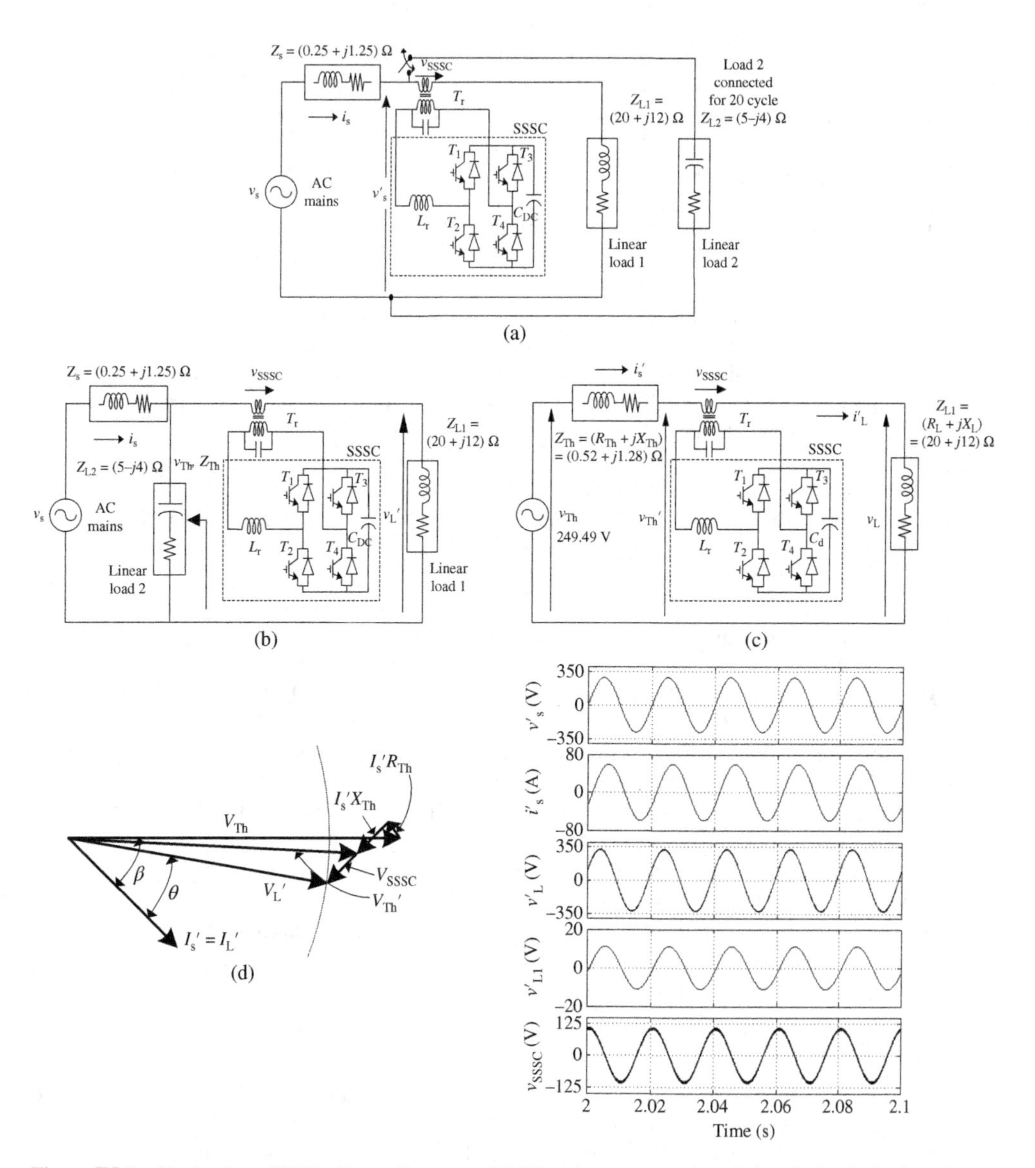

Figure E5.7 Single-phase SSSC with a self-supported DC bus for compensation of drop in the feeder impedance under linear load. (a) Circuit configuration. (b) Circuit diagram when second load is connected. (c) Thévenin's equivalent circuit. (d) Phasor diagram under compensation. (e) Waveforms under compensation

The load current before compensation is $I_L = V_s/Z_T = 230\angle 0°/5.51\angle -12.49° = 41.74\angle 12.49°$ A. The supply current is the same as the load current: $I_s = I_L = 41.74\angle 12.49°$ A.

a. The voltage drop across the source impedance is $V_{Zs} = Z_s \times I_s = 1.27\angle 78.69° \times 41.74\angle 12.49° = 53\angle 91.18°$ V.
b. The voltage across the load terminals is $V_{ZL} = Z_L \times I_L = (Z_{L1} \| Z_{L2}) \times I_L = 5.68\angle -25.43 \times 41.74\angle 12.49° = 237.11\angle -12.94°$ V.

The given system configuration can be simplified using *Thévenin's* equivalent circuit at PCC, as shown in Figure E5.7c. The value of Thévenin's voltage and impedence is estimated as $V_{Th} = V_s^* Z_{L2}/(Z_s + Z_{L2}) = (230\angle 0° \times (5 - j4))/(0.25 + j1.25 + 5 - j4) = 248.49\angle -11.01°$ V.

$$Z_{Th} = Z_s \| Z_{L2} = (0.25 + j1.25) \| (5 - j4) = 1.38\angle 67.90° = 0.52 + j1.28\ \Omega.$$

For simplicity of the solution, considering V_{Th} at reference, the various calculations are as follows:

The equivalent circuit and its phasor diagram for compensation are shown in Figure E5.7c and d. The series-connected PWM SSSC is used to decrease the load terminal voltage to the input voltage (230 V). The SSSC consists of a VSC with self-supported DC link and hence the SSSC can inject only reactive power into the system. The SSSC injects a voltage in quadrature to load current. The phasor diagram under voltage drop compensation is shown in Figure E5.7d.

The load current after compensation is $|I'_L| = |V'_L|/|Z_{L1}| = 230/23.32 = 9.86$ A.

The active power balance is $|V_{Th}||I'_L|\cos\beta = |I'_L|^2 R_{Th} + |I'_L|^2 R_{L1}$.

Solving the above equation for angle β results in $248.49 \times 9.86 \times \cos\beta = 9.86^2 \times 0.52 + 9.86^2 \times 20$,

$$\beta = 35.48°.$$

The load current lags the Thévenin's equivalent voltage by an angle β: $I'_L = |I'_L|\angle -\beta = 9.86\angle -35.48°$ A.

The PCC voltage is estimated as $V'_{Th} = V_{Th} - Z_{Th} I'_L = 248.49\angle 0° - 1.38\angle 67.90° \times\ 9.86\angle -35.48° = 237.1\angle -1.76°$ V.

The SSSC absorbs reactive power from the system to decrease the load voltage to 230 V. Therefore, the reactive power balance is $|V_{Th}||I'_L|\sin\beta = |I'_L|^2 X_{Th} + |I'_L|^2 X_{L1} + |I'_L||V_{SSSC}|$.

Substituting the values of circuit parameters results in $248.49 \times 9.86 \sin 35.48° = 9.86^2 \times 1.28 + 9.86^2 \times 12 + 9.86 \times |V_{SSSC}|$.

On solving the above equation, $|V_{SSSC}| = 13.27$ V.

c. The voltage rating of the compensator is $|V_{SSSC}| = 13.27$ V.
d. The current rating of the compensator is $|I_{SSSC}| = |I'_L| = 9.86$ A(since it is connected in series with the load).
e. The VA rating of the compensator is $S = |V_{SSSC}||I_{SSSC}| = 130.92$ VA.

Example 5.8

A single-phase AC supply has AC voltage of 230 V (rms) at 50 Hz and a feeder (source) impedance of 0.5 Ω resistance and 0.5 Ω inductive reactance. It is feeding a load having $Z_{L1} = (25 + j15)\ \Omega$. There is another load having $Z_{L2} = (5 + j3)\ \Omega$ connected in parallel for 20 cycles of AC mains. Calculate (a) the voltage across the source impedance and (b) the voltage across the first load under transient and steady-state conditions. If a self-supported DC bus-based PWM VSC is employed as SSSC (as shown in Figure E5.8a) in series of the first load to maintain its terminal voltage to the input voltage (230 V), calculate (c) the current rating of the SSSC, (d) voltage rating of the SSSC, and (e) VA rating of the SSSC in both cases.

Solution: Given supply voltage $V_s = 230$ V, frequency of the supply $(f) = 50$ Hz, two single-phase loads having $Z_{L1} = (25 + j15)\ \Omega = 29.15\angle 30.96°\ \Omega$ and $Z_{L2} = (5 + j3)\ \Omega = 5.83\angle 30.96°\ \Omega$ fed from an AC supply with an input AC voltage of 230 V at 50 Hz and a source impedance of $Z_s = (0.5 + j0.5)\ \Omega = 0.707\angle 45°\ \Omega$. The circuit diagram from the system is shown in Figure E5.8b.

Various circuit calculations before compensation are as follows:

The total impedance is $Z_T = Z_s + Z_{L1} || Z_{L2} = (0.5 + j0.5) + (25 + j15) || (5 + j3) = 4.66 + j3 = 5.54\angle 32.77°\ \Omega$.

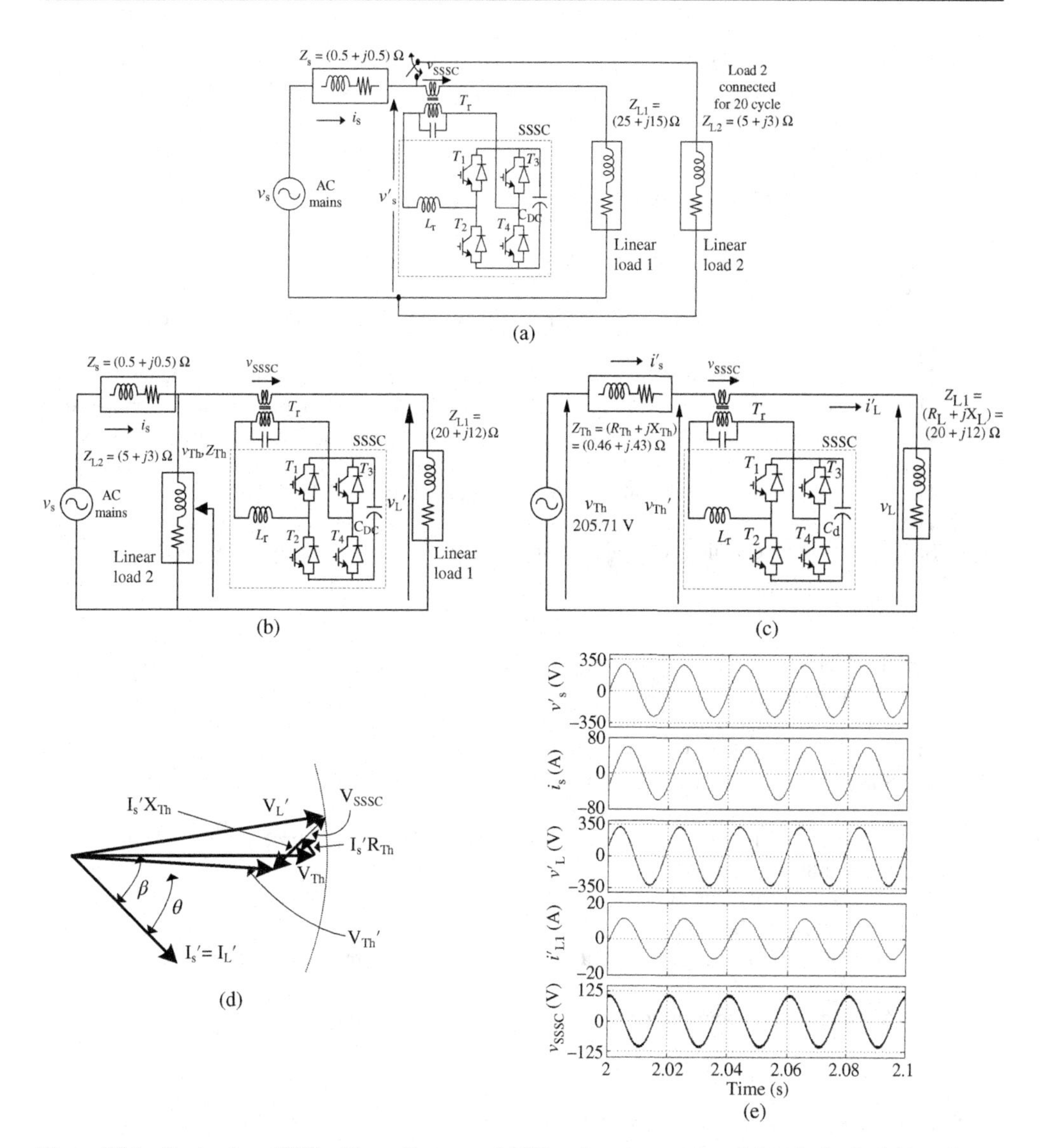

Figure E5.8 Single-phase SSSC with a self-supported DC bus for compensation of drop in the feeder impedance under linear load. (a) Circuit configuration. (b) Circuit diagram when second load is connected. (c) Thévenin's equivalent circuit. (d) Phasor diagram under compensation. (e) Waveforms under compensation

The load current before compensation is $I_L = V_s/Z_T = 230\angle 0°/5.54\angle 32.77° = 41.51\angle{-32.77}°$ A. The supply current is the same as the load current: $I_s = I_L = 41.51\angle{-32.77}°$ A.

a. The voltage drop across the source impedance is $V_{Zs} = Z_s \times I_s = 0.707\angle 45° \times 41.51\angle -32.77° = 29.31\angle 12.27°$ V.
b. The voltage across the load terminals is $V_{ZL} = Z_L \times I_L = (Z_{L1} \| Z_{L2}) \times I_L = 4.85\angle 31° \times 41.51\angle -32.77° = 201.40\angle -1.77°$ V.

The given system configuration can be simplified using *Thévenin's* equivalent circuit at PCC, as shown in Figure E5.8c. The value of Thévenin's voltage and impedance is estimated as

$$V_{Th} = V_s \times Z_{L2}/(Z_s + Z_{L2}) = (230\angle 0° \times (5 + j3))/(0.5 + j0.5 + 5 + j3) = 205.71\angle -1.51° \text{ V}.$$

$$Z_{Th} = Z_s \| Z_{L2} = (0.5 + j0.5) \| (5 + j3) = 0.63\angle 43.51° = 0.46 + j0.43\ \Omega.$$

For simplicity of the solution, considering V_{Th} at reference, the various calculations are as follows:

The equivalent circuit and its phasor diagram for compensation are shown in Figure E5.8c and d, respectively.

The series-connected PWM SSSC is used to increase the load terminal voltage to the input voltage (230 V). The SSSC consists of a VSC with self-supported DC link and hence the SSSC supplies only reactive power into the system. The SSSC injects a voltage in quadrature to load current. The phasor diagram under voltage drop compensation is shown in Figure E5.8d.

The load current after compensation is $|I'_L| = |V'_L|/|Z_{L1}| = 230/29.15 = 7.89$ A.

The active power balance is expressed as $|V_{Th}||I'_L|\cos\beta = |I'_L|^2 R_{Th} + |I'_L|^2 R_{L1}$

Solving the above equation for angle β results in $205.71 \times 7.89 \times \cos\beta = 7.89^2 \times 0.46 + 7.89^2 \times 25$,

$$\beta = 12.44°.$$

The load current lags the Thévenin's equivalent voltage by an angle β: $I'_L = |I'_L|\angle -\beta = 7.89\angle -12.44°$ A.

The PCC voltage is estimated as $V'_{Th} = V_{Th} - Z_{Th} I'_L = 205.71\angle 0° - 0.63\angle 43.51° \times 7.89\angle -12.44° = 200.85\angle -0.73°$ V.

The SSSC absorbs reactive power from the system to decrease the load voltage to 230 V. Therefore, the reactive power balance is $|V_{Th}||I'_L|\sin\beta = |I'_L|^2 X_{Th} + |I'_L|^2 X_{L1} + |I'_L||V_{SSSC}|$.

Substituting the values of circuit parameters results in $205.71 \times 7.89 \sin 12.44° = 7.89^2 \times 0.43 + 7.89^2 \times 15 - 7.89 \times |V_{SSSC}|$.

On solving the above equation, $|V_{SSSC}| = 77.42$ V.

c. The voltage rating of the compensator is $|V_{SSSC}| = 77.42$ V.
d. The current rating of the compensator is $|I_{SSSC}| = |I'_L| = 7.89$ A| (since it is connected in series with the load).
e. The VA rating of the compensator is $S = |V_{SSSC}||I_{SSSC}| = 610.91$ VA.

Example 5.9

A three-phase delta connected load having $Z_{LD} = (7.0 + j3.0)$ pu per phase is fed from an AC supply with an input AC line voltage of 415 V at 50 Hz and a base impedance of 4.15 Ω. It is to be realized as unity power factor load on the AC supply system while maintaining the same rated load terminal voltage using PWM-based SSSC (as shown in Figure E5.9a). Calculate (a) the voltage rating of the compensator, (b) the current rating of the compensator, and (c) the VA rating of the compensator.

Solution: Given supply magnitude of voltage per phase $|V_s| = 415/\sqrt{3} = 239.6$ V, frequency of the supply $(f) = 50$ Hz, and a three-phase delta connected load having $Z_{LD} = 7.0 + j3.0$ pu fed from an AC supply with an input AC voltage of 239.6 V at 50 Hz and a base impedance of 4.15 Ω.

The load resistance is $R_{LD} = 7 \times 4.15\ \Omega = 29.05\ \Omega$. The load reactance is $X_{LD} = 3 \times 4.15\ \Omega = 12.45\ \Omega$. This load is delta connected to the supply system. For per-phase calculations, the equivalent load per phase is $Z_{LY} = Z_{LD}/3 = (29.05 + j12.45)/3 = (9.68 + j4.15)\ \Omega = 10.53\angle 23.2°\ \Omega$.

The load current before compensation is $|I_L| = |I_s| = |V_L|/|Z_{LY}| = 239.6/10.53 = 22.74$ A.

The series-connected PWM SSSC is used to improve the power factor of the load to unity while maintaining the same rated voltage of 239.6 V per phase across the load. It means that now the load voltage must lead the source voltage by the load power factor angle. The magnitude of load current

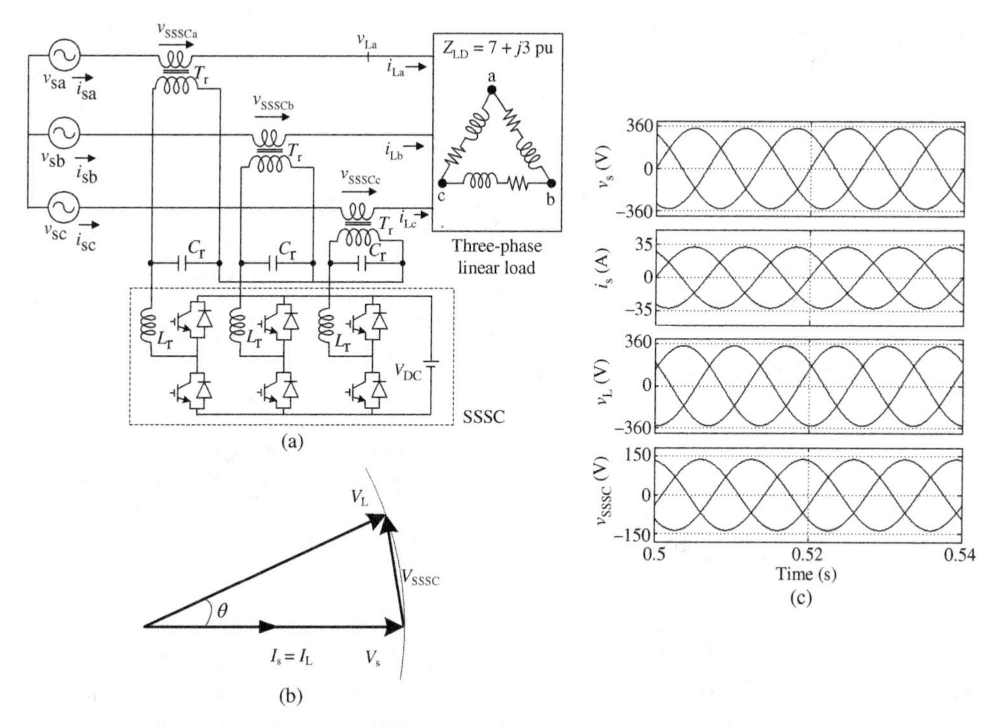

Figure E5.9 Three-phase SSSC for power factor correction of delta connected lagging linear load. (a) Circuit configuration. (b) Phasor diagram under compensation. (c) Waveforms under compensation

remains the same as earlier, whereas the phase angle of the load current is the same as the supply voltage. Under power factor correction mode, the SSSC injects the voltage difference between the supply voltage and load voltage. The phasor diagram under power factor correction mode is shown in Figure E5.9b.

Since under compensation an extra active power is pumped from the single-phase supply, a battery energy storage system is required at the DC link of SSSC to maintain active power balance.

The load voltage under compensation is $V_{La} = |V_{La}| \angle\theta = 239.6\angle 23.2°$ V.

a. The voltage rating of the compensator is $V_{SSSCa} = V_{La} - V_{sa} = 239.6\angle 23.2° - 239.6\angle 0° = 96.35\angle 101.6°$ V.

   As the system is a symmetrical system, the magnitude of injected voltage remains the same in all three phases.
b. Since the load voltage magnitude and the load impedance are same, the load current remains the same as without power factor correction. Hence, the current rating of the compensator is $|I_{SSSC}| = |I_L| = |I_s| = 22.74$ A.
c. The VA rating of the compensator is $S_{SSSC} = 3|V_{SSSC}||I_{SSSC}| = 3 \times 96.35 \times 22.74 = 6573$ VA.

Example 5.10

A three-phase, three-wire AC supply has AC mains voltage of 440 V (rms) at 50 Hz and a feeder (source) impedance of 0.25 Ω resistance and 2.5 Ω inductive reactance per phase. It feeds a three-phase star connected load having $Z_L = (24 + j18)$ Ω per phase. Calculate (a) the voltage drop across the source impedance and (b) the voltage across the load. If a PWM SSSC (as shown in Figure E5.10a) is used to raise the voltage to the input voltage (440 V) with minimum rating, calculate (c) the voltage rating of the compensator, (d) the current rating of the compensator, and (e) the VA rating of the compensator.

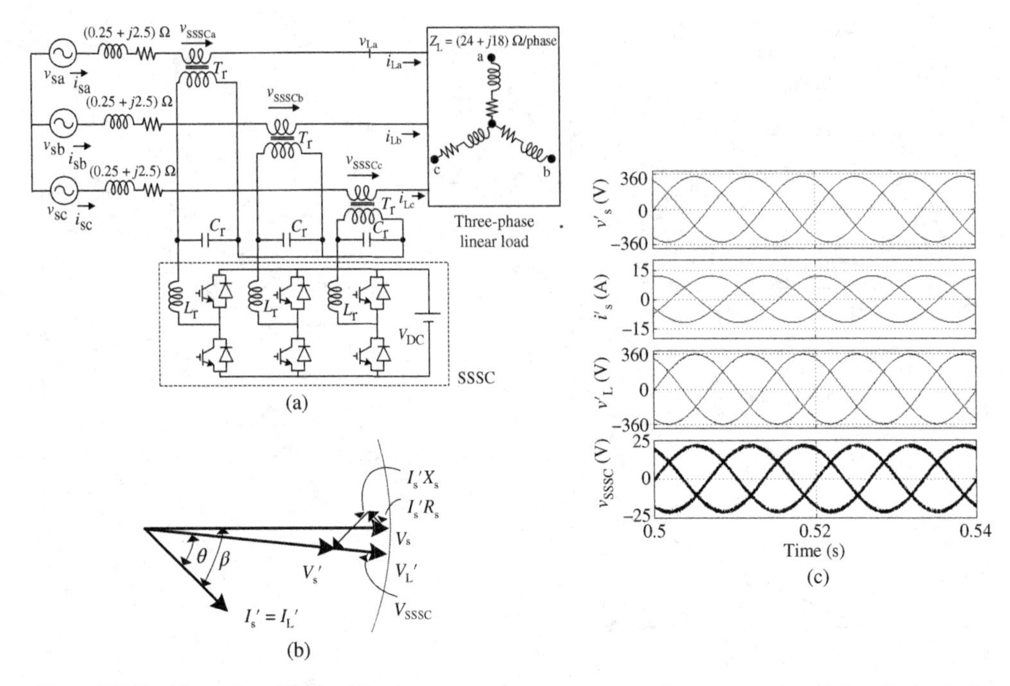

Figure E5.10 Three-phase SSSC with a battery energy storage system for compensation of drop in the feeder impedance under star connected lagging linear load. (a) Circuit configuration. (b) Phasor diagram under compensation. (c) Waveforms under compensation

Solution: Given supply voltage $V_s = 440/\sqrt{3} = 254$ V, frequency of the supply (f) = 50 Hz, and a single-phase load having $Z_L = (24 + j18)\ \Omega = |Z_L| \angle\theta = 30 \angle 36.87°\ \Omega$ fed from an AC supply with an input AC voltage of 254 V at 50 Hz and a source impedance of $Z_s = (0.25 + j2.5)\ \Omega = 2.51 \angle 84.29°\ \Omega$.

Various circuit calculations before compensation are as follows:

The total impedance $Z_T = Z_s + Z_L = 24 + j18 + 0.25 + j2.5 = 24.25 + j20.5 = 31.75 \angle 40.20°\ \Omega$.
The load current before compensation is $I_L = V_s/(Z_s + Z_L) = 254 \angle 0°/31.75\angle 40.20° = 8 \angle -40.20°$ A.
The supply current is the same as the load current: $I_s = I_L = 8 \angle -40.20°$ A.

a. The voltage drop across the source impedance is $V_{Zs} = Z_s \times I_s = 2.51\angle 84.29° \times 8\angle -40.20° = 20.08\angle 44.09°$ V.
b. The voltage across the load terminals is $V_L = Z_L \times I_L = 30\angle 36.87° \times 8\angle -40.20° = 240\angle -3.32°$ V.

The series-connected PWM SSSC is used to raise the load terminal voltage to the input voltage (254 V). The SSSC must inject voltage in phase with the PCC (V'_s) voltage to attain minimum rating. In order to inject voltage in phase with V'_s, both real and reactive powers are required and hence a battery storage support is required at the DC link of SSSC. Figure E5.10b shows the phasor diagram under voltage drop compensation by SSSC.

The magnitude of load current under feeder drop compensation is $|I'_L| = |V'_L|/|Z_L| = 254/30 = 8.46$ A.

The supply current and the load current are same for the given network: $I'_L = I'_s$.

From phasor diagram it can be observed that the SSSC supplies both active and reactive powers into the system. The power balance equations are used to calculate the effective power factor angle β and magnitude of the injected voltage.

The active power balance may be expressed as $|V_s||I'_L|\cos\beta = |I'_L|^2R + |I'_L|^2R - |V_{SSSC}||I'_L|\cos\theta$.

Similarly, the reactive power balance may be expressed as $|V_s||I'_L|\sin\beta = |I'_L|^2X + |I'_L|^2X - |V_{SSSC}||I'_L|\sin\theta$.

Solving these two equations results in a quadratic equation. The two roots of the quadratic equation are $|V_{SSSC1}| = 14.76$ V and $|V_{SSSC2}| = 519.81$ V. By inspection the smaller value of $|V_{SSSC}|$ is selected.

c. The voltage rating of the compensator is $|V_{SSSC}| = 14.76$ V.
d. The current rating of the compensator is $|I_{SSSC}| = |I'_L| = 8.46$ A (since it is connected in series with the load).
e. The VA rating of the compensator is $S = 3\times|V_{SSSC}||I_{SSSC}| = 374.60$ VA.

Example 5.11

A three-phase AC supply has line voltage of 415 V (rms) at 50 Hz and a feeder (source) impedance of 1.0 Ω resistance and 4.0 Ω inductive reactance per phase after which a balanced delta connected load having $Z_{LD} = (24 + j18)$ Ω per phase is connected. Calculate the voltage drop across the source impedance and the voltage across the load. If a self-supported DC bus-based PWM SSSC (as shown in Figure E5.11a) is used to raise the voltage to the input voltage (415 V), calculate the voltage, current, and kVA rating of the compensator.

Solution: Given supply voltage $V_s = 415/\sqrt{3} = 239.6$ V, frequency of the supply $(f) = 50$ Hz, and a three-phase delta connected load having $Z_{LD} = (24 + j18)\ \Omega = 30\angle 36.87°\ \Omega$ fed from an AC supply with an input AC voltage of 239.6 V at 50 Hz and a source impedance of $Z_s = (1 + j4)\ \Omega = 4.12\angle 75.96°\ \Omega$.

The equivalent load impedance for star connected load is $Z_{LY} = Z_{LD}/3 = (24 + j18)/3 = 8 + j6 = 10\angle 36.87°\ \Omega$.

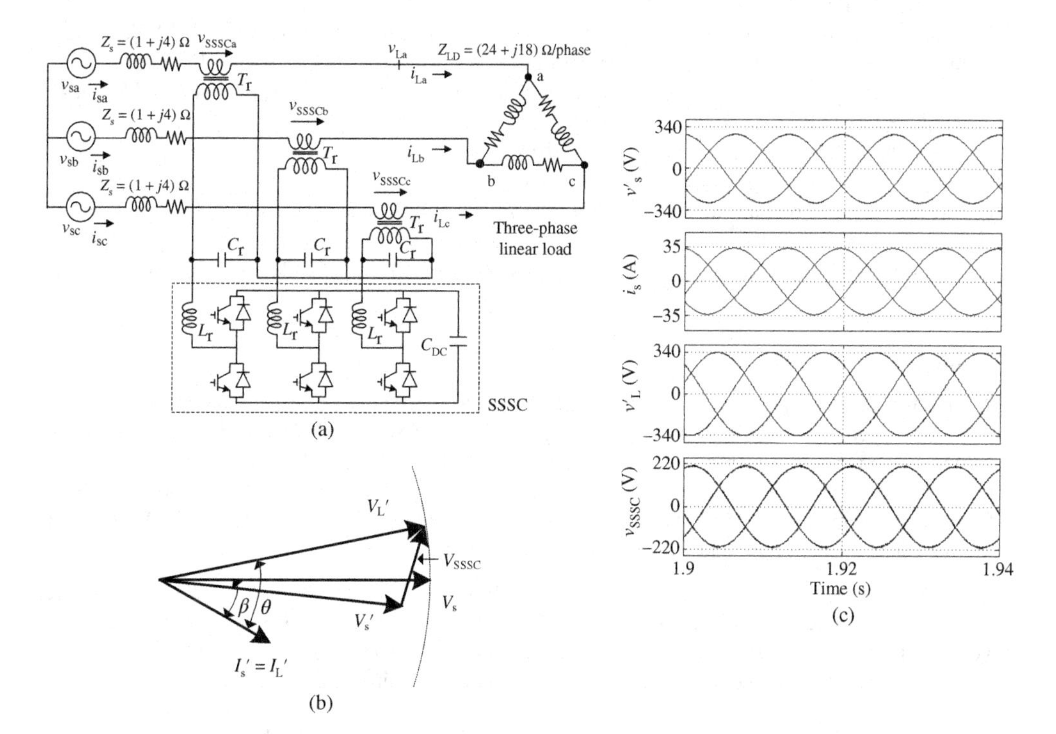

Figure E5.11 Three-phase SSSC with a self-supported DC bus for compensation of drop in the feeder impedance under delta connected lagging linear load. (a) Circuit configuration. (b) Phasor diagram under compensation. (c) Waveforms under compensation

Various circuit calculations before compensation are as follows:

The total impedance is $Z_T = Z_s + Z_{LY} = 1 + j4 + 8 + j6 = 9 + j10 = 13.45\angle 48.01°\,\Omega$.

The load current before compensation is $I_{La} = V_{sa}/(Z_s + Z_L) = 239.6\angle 0°/13.45\angle 48.01° = 17.81\angle -48.01°$ A.

The supply current is the same as the load current: $I_s = I_L = 17.81\angle -48.01°$ A.

a. The voltage drop across the source impedance is $V_{Zs} = Z_s \times I_s = 4.12\angle 75.96° \times 17.81\angle -48.01° = 73.37\angle 27.95°$ V.
b. The voltage across the load terminals is $V_{ZL} = Z_{LY} \times I_L = 10\angle 36.87° \times 17.81\angle -48.01° = 178.1\angle -11.14°$ V.

   The series-connected PWM SSSC is used to increase the load terminal voltage to the input voltage (239.6 V). The SSSC consists of a VSC with self-supported DC link and hence the SSSC can inject or absorb only reactive power. The SSSC injects a voltage in quadrature to load current. The phasor diagram under voltage drop compensation is shown in Figure E5.11b. The SSSC acts as an equivalent inductor to regulate the load voltage to the rated value.

   The load current after compensation is $|I'_L| = |V'_L|/|Z_L| = 239.6/10 = 23.96$ A.

   The active power balance can be expressed as $|V_s||I'_L|\cos\beta = |I'_L|^2 R + |I'_L|^2 R_{LY}$.

   Solving the above equation for angle β results in $239.6 \times 23.96 \times \cos\beta = 23.96^2 \times 1 + 23.96^2 \times 8$,

$$\beta = 25.84°.$$

   The load current lags the supply voltage by an angle β: $I'_L = |I'_L|\angle -\beta = 23.96\angle -25.84°$ A.

   The PCC voltage is estimated as $V'_s = V_s - Z_s I'_L = 239.6\angle 0° - 4.12\angle 75.96° \times 23.96\angle -25.84° = 191.89\angle -23.25°$ V.

   The SSSC injects reactive power into the system to increase the load voltage to 239.6 V. Therefore, the reactive power balance can be expressed as $|V_s||I'_L|\sin\beta = |I'_L|^2 X + |I'_L|^2 X_{LY} - |I'_L||V_{SSSC}|$.

   Substituting the values of circuit parameters results in $239.6 \times 23.96 \sin 25.84° = 23.96^2 \times 4 + 23.96^2 \times 6 - 23.96 \times |V_{SSSC}|$.

   Solving the above equation, $V_{SSSC} = 135.16$ V.
c. The per-phase voltage rating of the compensator is $|V_{SSSC}| = 135.16$ V.
d. The current rating of the compensator is $|I_{SSSC}| = |I'_L| = 23.96$ A (since it is connected in series with the load).
e. The VA rating of the compensator is $S = 3|V_{SSSC}||I_{SSSC}| = 3 \times 135.16 \times 23.96 = 9715.30$ VA.

Example 5.12

A three-phase AC supply has a line–line voltage of 415 V at 50 Hz and a feeder (source) impedance of 0.5 Ω resistance and 3.0 Ω inductive reactance per phase after which a balanced delta connected load having $Z_{LD} = (15 - j12)$ Ω per phase is connected. Calculate (a) the voltage drop across the source impedance and (b) the voltage across the load in each case. If a self-supported DC link- and PWM-based SSSC (as shown in Figure E5.12a) is used to maintain the voltage to the input voltage (415 V), calculate the (c) voltage, (d) current, and (e) kVA rating of the series compensator.

Solution: Given supply voltage $V_s = 415/\sqrt{3} = 239.6$ V, frequency of the supply (f) = 50 Hz, and a three-phase delta connected load having $Z_{LD} = (15 - j12)$ Ω per phase fed from an AC supply with an input AC voltage of 239.6 V at 50 Hz. The source impedance is $Z_s = (0.5 + j3) = 3.04\angle 80.53°\,\Omega$ per phase. This load is delta connected to the supply system. For per-phase single line diagram calculation, it is converted to equivalent star connected load.

The per phase impedance with this equivalent star connected load is one third of that with the balanced delta connected load: $Z_{LY} = R_{LY} + jX_{LY} = (\{15 - j12\}/3)\,\Omega = (5 - j4) = 6.40\angle -38.65\,\Omega$.

The total impedance per phase is $Z_T = Z_s + Z_{LY} = 0.5 + j3 + 5 - j4 = 5.5 - j1 = 5.59\angle -10.30\,\Omega$.

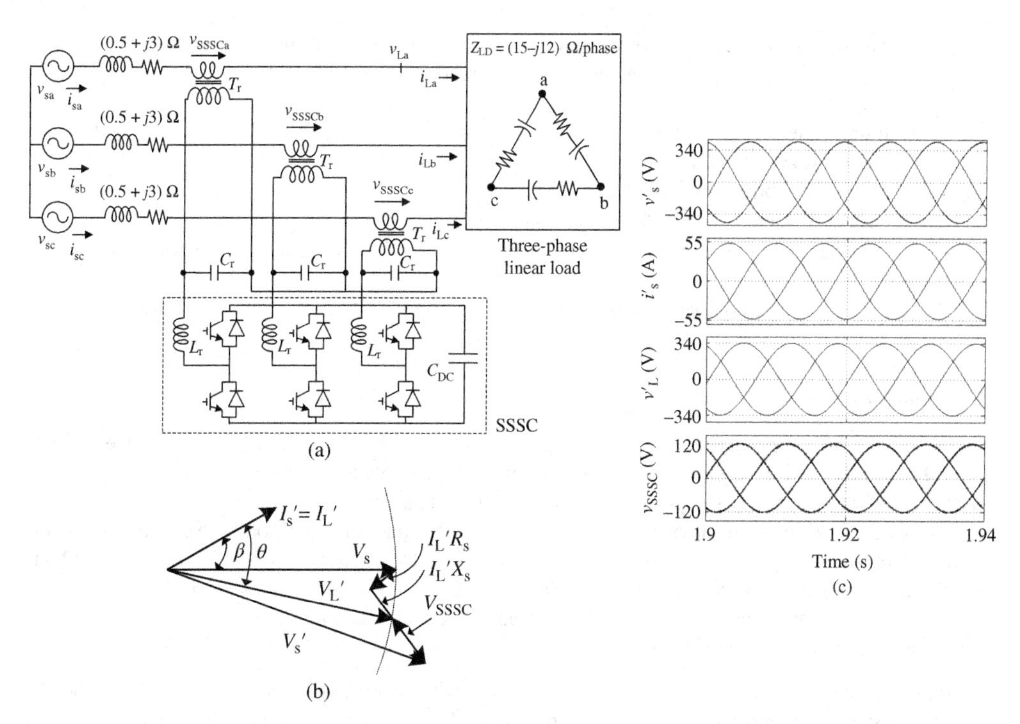

Figure E5.12 Three-phase SSSC with a self-supported DC bus for compensation of drop in the feeder impedance under delta connected leading linear load. (a) Circuit configuration. (b) Phasor diagram under compensation. (c) Waveforms under compensation

The load current before compensation is $I_{La} = V_{sa}/(Z_s + Z_L) = 239.6\angle 0°/5.59\angle -10.30° = 42.86\angle 10.30°$ A.

The supply current is the same as the load current: $I_s = I_L = 42.86\angle 10.30°$ A.

a. The voltage drop across the source impedance is $V_{Zs} = Z_s \times I_s = 3.04\angle 80.53° \times 42.86\angle 10.30° = 130.29\angle 90.53°$ V.
b. The voltage across the load terminals is $V_{ZL} = Z_{LY} \times I_L = 6.40\angle -38.65 \times 42.86\angle 10.30° = 274.30\angle -28.35°$ V.

The series-connected PWM SSSC is used to regulate the load terminal voltage to the input voltage (239.6 V). The SSSC consists of a VSC with a self-supported DC link and hence the SSSC can inject or absorb only reactive power. The SSSC injects a voltage in quadrature to load current. The phasor diagram under voltage drop compensation is shown in Figure E5.12b. The SSSC acts as an equivalent inductor to decrease the load voltage to the rated value.

The load current after compensation is $|I'_L| = |V'_L|/|Z_{LY}| = 239.6/6.4 = 37.43$ A.

The active power balance can be expressed as $|V_s||I'_L|\cos\beta = |I'_L|^2 R + |I'_L|^2 R_{LY}$.

Solving the above equation for angle β results in $239.6 \times 37.43 \times \cos\beta = 37.43^2 \times 0.5 + 37.43^2 \times 5$,

$$\beta = 30.77°.$$

The load current leads the supply voltage by an angle β: $I'_L = |I'_L|\angle\beta = 37.43\angle 30.77°$ A.

The PCC voltage is estimated as $V'_s = V_s - Z_s I'_L = 239.6\angle 0° - 3.04\angle 80.53° \times 37.43\angle 30.77° = 300.27\angle -20.67°$ V.

The SSSC absorbs reactive power from the system to decrease the load voltage to 239.6 V. Therefore, the reactive power balance can be expressed as $-|V_s||I'_L|\sin\beta = |I'_L|^2X + |I'_L|^2X_{LY} - |I'_L||V_{SSSC}|$.

Substituting the values of circuit parameters results in $-239.6 \times 37.43 \sin 30.77° = 37.43^2 \times 3 - 37.43^2 \times 4 - 37.43 \times |V_{SSSC}|$.

On solving the above equation, $|V_{SSSC}| = 85.15$ V.

c. The per-phase voltage rating of the compensator is $|V_{SSSC}| = 85.15$ V.
d. The current rating of the compensator is $|I_{SSSC}| = |I'_L| = 37.43$ A (since it is connected in series with the load).
e. The VA rating of the compensator is $S = 3|V_{SSSC}||I_{SSSC}| = 3 \times 85.15 \times 37.43 = 9561.49$ VA.

Example 5.13

A three-phase AC supply has a line–line voltage of 415 V at 50 Hz and a feeder (source) impedance of 0.25 Ω resistance and 0.5 Ω inductive reactance per phase after which a balanced delta connected load having $Z_{LD1} = (30 + j20)$ Ω per phase is connected. There is another load connected in parallel, which draws exactly five times the current at the same power factor than this load for 20 cycles of AC mains. Calculate (a) the voltage across the source impedance and (b) the voltage across the first load. If a PWM-IGBT-based VSC is employed as SSSC in series of the first load to maintain its terminal voltage to the input voltage (415 V) with minimum rating in steady state, calculate the (c) current rating, (d) voltage rating, and (e) VA rating of the SSSC.

Solution: Given the supply phase voltage $V_s = 415/\sqrt{3} = 239.6$ V, frequency of the supply $(f) = 50$ Hz, and the first delta connected three-phase load having $Z_{LD1} = (30 + j20)$ Ω per phase and the second delta connected load drawing five times the current drawn by load 1 at the same power factor and having $Z_{LD2} = \{(30 + j20)/5\}$ Ω per phase. These loads are fed from an AC supply with an input AC line voltage of 415 V at 50 Hz and a source impedance of $Z_s = (0.25 + j0.5) = 0.56\angle 63.43°$ Ω.

As the given network is a balanced three-phase network, it can be analyzed on per-phase basis. The equivalent star connected impedance for both the loads are given as

$$Z_{LY1} = Z_{LD1}/3 = (30 + j20)/3 = (10 + j6.66) = 12.02\angle 33.69°\ \Omega,$$

$$Z_{LY2} = Z_{LD2}/3 = (30 + j20)/(5 \times 3) = (2 + j1.33) = 2.40\angle 33.69°\ \Omega.$$

The total impedance is $Z_T = Z_s + Z_{LY1}||Z_{LY2} = (0.25 + j0.5) + (10 + j6.66)||(2 + j1.33) = 1.91 + j1.61 = 2.50\angle 40.12°$ Ω.

The load current before compensation is $I_{La} = V_{sa}/Z_T = 239.6\angle 0°/2.50\angle 40.12° = 95.84\angle -40.12°$ A. The supply current is the same as the load current: $I_{sa} = I_{La} = 95.84\angle -40.12°$ A.

a. The voltage drop across the source impedance is $V_{Zsa} = Z_{sa} \times I_{sa} = 0.56\angle 63.43° \times 95.84\angle -40.12° = 53.67\angle 23.31°$ V.
b. The voltage across the load terminals is $V_{ZL} = Z_L \times I_{La} = (Z_{LY1}||Z_{LY2}) \times I_{La} = \{(10 + j6.66)||(2 + j1.33)\} \times 95.84\angle -40.12° = 191.67\angle -6.36°$ V.

The given system configuration can be simplified using Thévenin's equivalent circuit at PCC, as shown in Figure E5.13c. The value of per-phase equivalent Thévenin's voltage and impedance is estimated as

$$V_{Tha} = V_{sa} \times Z_{L2}/(Z_s + Z_{L2}) = (239.6\angle 0° \times (2 + j1.33))/(0.25 + j0.5 + 2 + j1.33)$$
$$= 198.43\angle -5.48°\ \text{V},$$

$$Z_{Tha} = Z_s||Z_{L2} = R_{Th} + jX_{Th} = (0.25 + j0.5)||(2 + j1.33) = 0.46\angle 57.97° = 0.24 + j0.39\ \Omega.$$

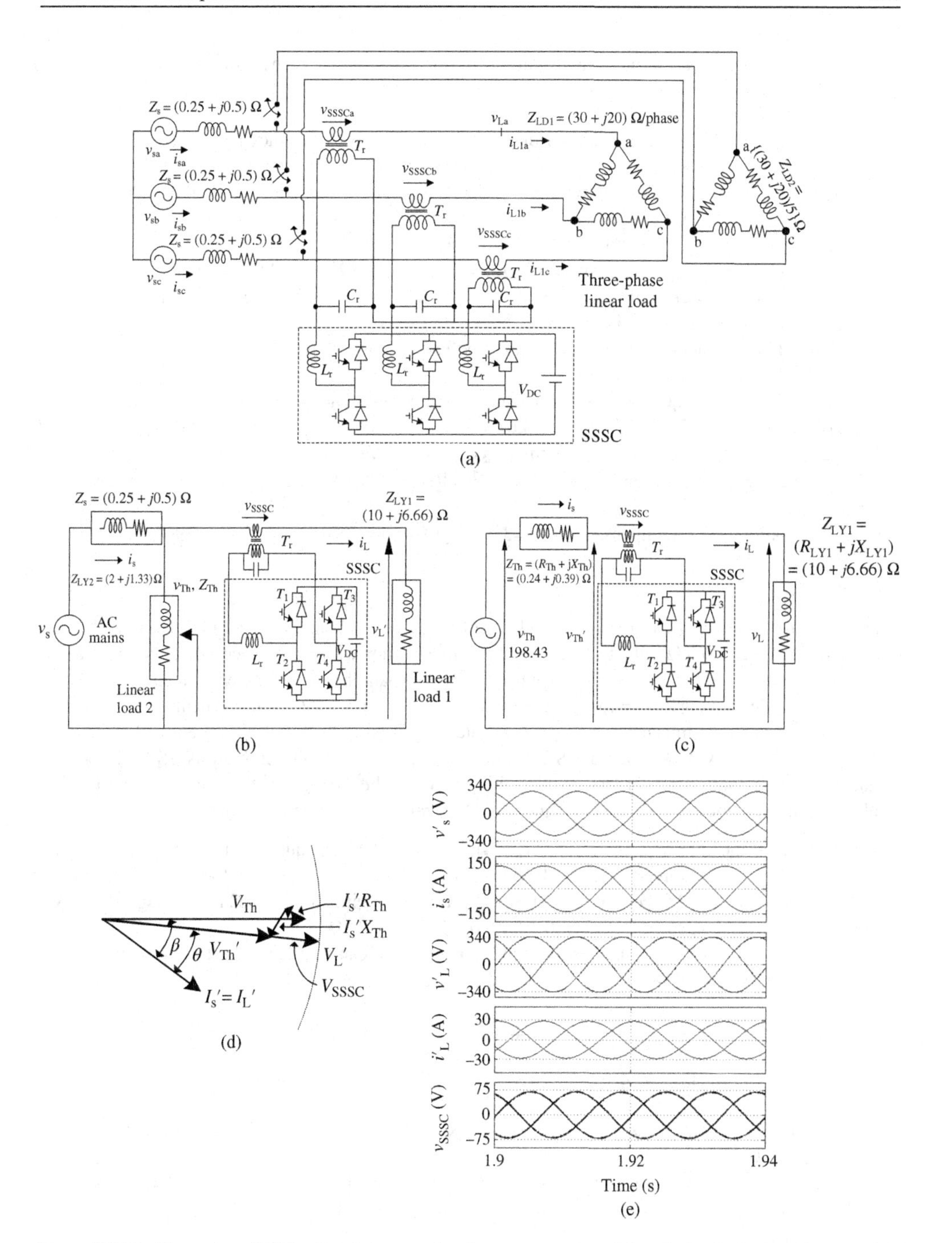

Figure E5.13 Three-phase SSSC with minimum rating for compensation of drop in the feeder impedance under star connected lagging linear load. (a) Circuit configuration. (b) Per-phase equivalent when second load is connected. (c) Per-phase Thévenin's equivalent circuit. (d) Phasor diagram under compensation. (e) Waveforms under compensation

For simplicity of the solution, considering V_{Tha} at new reference, the various calculations are as follows:

The equivalent circuit and its phasor diagram for compensation are shown in Figure E5.13c and d.

The series-connected PWM SSSC is used to increase the load terminal voltage to the input voltage (239.6 V). To attain minimum rating, SSSC injects a voltage in phase with the PCC voltage. However, to inject this voltage, an active power is required from the DC link of SSSC and hence a battery energy storage is required, as shown in Figure E5.13a.

The load current after compensation is $|I'_L| = |V'_L|/|Z_{LY1}| = 239.6/12.02 = 19.93$ A.

The active power balance may be expressed as $|V_{Th}||I'_L|\cos\beta = |I'_L|^2 R_{Th} + |I'_L|^2 R_{LY1} - |V_{SSSC}||I'_L|\cos\theta$, where $\cos\theta$ is the power factor of load 1.

Similarly, the reactive power balance may be expressed as $|V_{Th}||I'_L|\sin\beta = |I'_L|^2 X_{Th} + |I'_L|^2 X_{LY1} - |V_{SSSC}||I'_L|\sin\theta$.

Solving these two equations results in a quadratic equation. The two roots of the quadratic equation are $|V_{SSSC1}| = 49.51$ V and $|V_{SSSC2}| = 444.61$ V. By inspection the smaller value of $|V_{SSSC}|$ is selected.

c. The voltage rating of the compensator is $|V_{SSSC}| = 49.51$ V.
d. The current rating of the compensator is $|I_{SSSC}| = |I'_L| = 19.93$ A (since it is connected in series with the load).
e. The VA rating of the compensator is $S = 3\,|V_{SSSC}||I_{SSSC}| = 3 \times 49.51 \times 19.93 = 2960.20$ VA.

Example 5.14

A three-phase AC supply has a line–line voltage of 415 V at 50 Hz and a feeder (source) impedance of 0.1 Ω resistance and 0.5 Ω inductive reactance per phase after which a balanced delta connected load having $Z_{LD1} = (21 + j15)$ Ω per phase is connected. There is another delta connected load at the point of common coupling having $Z_{LD2} = (6 + j3)$ Ω per phase for 10 cycles of AC mains. Calculate (a) the voltage drop across the source impedance and (b) the voltage across the load under transient and steady-state conditions. If a PWM-IGBT-based VSC with a self-supported DC bus is used as SSSC for steady-state voltage regulation and transient conditions to raise the voltage to the input voltage (415 V), calculate the (c) voltage, (d) current, and (e) kVA rating of the series compensator.

Solution: Given the supply phase voltage $V_s = 415/\sqrt{3} = 239.6$ V, frequency of the supply $(f) = 50$ Hz, and first delta connected three-phase load having $Z_{LD1} = 21 + j15$ Ω per phase and second delta connected load having $Z_{LD2} = 6 + j3$ Ω per phase fed from an AC supply with an input AC line voltage of 415 V at 50 Hz and a source impedance of $Z_s = (0.1 + j0.5) = 0.51\angle 78.6°$ Ω.

As the given network is a balanced three-phase network, it can be analyzed on per-phase basis. The equivalent star connected impedance for both the loads are given as

$$Z_{LY1} = Z_{LD1}/3 = (21 + j15)/3 = (7 + j5) = 8.60\angle 35.53°\ \Omega,$$

$$Z_{LY2} = Z_{LD2}/3 = (6 + j3)/3 = (2 + j1) = 2.23\angle 26.56°\ \Omega.$$

The total impedance is $Z_T = Z_s + Z_{LY1}||Z_{LY2} = (0.1 + j0.5) + (7 + j5)||(2 + j1) = 1.66 + j1.34 = 2.13\angle 38.91°$ Ω.

The load current before compensation is $I_{La} = V_{sa}/Z_T = 239.6\angle 0°/2.13\angle 38.91° = 112.49\angle -38.91°$ A.

The supply current is the same as the load current: $I_{sa} = I_{La} = 112.49\angle -38.91°$ A.

a. The voltage drop across the source impedance is $V_{Zsa} = Z_{sa} \times I_{sa} = 0.51\angle 78.6° \times 112.49\angle -38.91° = 57.07\angle 39.73°$ V.
b. The voltage across the load terminals is $V_{ZL} = Z_L \times I_{La} = (Z_{LY1}||Z_{LY2}) \times I_{La} = \{(7 + j5)||(2 + j1)\} \times 112.49\angle -38.91° = 199.06\angle -10.56°$ V.

The given system configuration can be simplified using *Thévenin's* equivalent circuit at PCC, as shown in Figure E5.14c. The value of per-phase equivalent Thévenin's voltage and impedance is

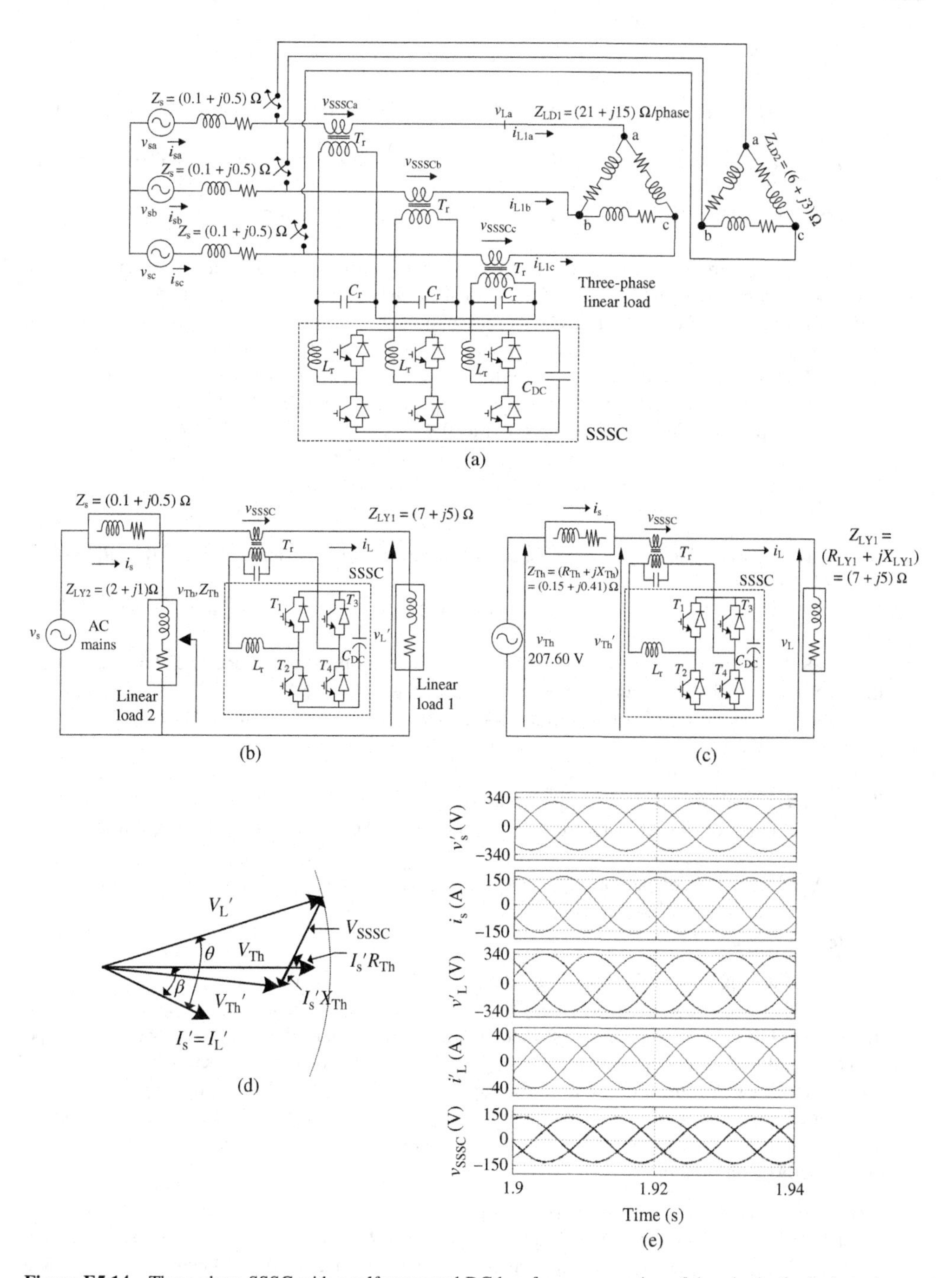

Figure E5.14 Three-phase SSSC with a self-supported DC bus for compensation of drop in the feeder impedance under star connected lagging linear load. (a) Circuit configuration. (b) Per-phase equivalent when second load is connected. (c) Per-phase Thévenin's equivalent circuit. (d) Phasor diagram under compensation. (e) Waveforms under compensation

estimated as follows:

$$V_{Tha} = V_{sa} \cdot Z_{LY2}/(Z_s + Z_{LY2}) = (239.6\angle 0° \cdot (2 + j1))/(0.1 + j0.5 + 2 + j1) = 207.6\angle -8.97° \text{ V}.$$

$$Z_{Tha} = Z_s \| Z_{LY2} = R_{Th} + jX_{Th} = (0.1 + j0.5)\|(2 + j1) = 0.44\angle 69.75° = 0.15 + j0.41\ \Omega.$$

For simplicity of the solution, considering V_{Th} at new reference, the various calculations are as follows:

The equivalent circuit and its phasor diagram for compensation are shown in Figure E5.14c and d. The series-connected PWM SSSC is used to increase the load terminal voltage to the input voltage (239.6 V). The SSSC consists of a VSC with self-supported DC link and hence the SSSC can inject or absorb only the reactive power from the system. The SSSC injects a voltage in quadrature to load current. The phasor diagram under voltage drop compensation is shown in Figure E5.14d.

The load current after compensation is $|I'_L| = |V'_L|/|Z_{LY1}| = 239.6/8.60 = 27.86$ A.

The active power balance can be expressed as $|V_{Th}||I'_L|\cos\beta = |I'_L|^2 R_{Th} + |I'_L|^2 R_{LY1}$.

Solving the above equation for angle β results in $207.6 \times 27.86 \times \cos\beta = 27.86^2 \times 0.15 + 27.86^2 \times 7$,

$$\beta = 16.35°.$$

The load current lags the Thévenin's equivalent voltage by an angle β: $I'_L = |I'_L|\angle -\beta = 27.86\angle -16.35°$ A.

The PCC voltage is estimated as $V'_{Th} = V_{Th} - Z_{Th} I'_{La} = 207.6\angle 0° - 0.44\angle 69.75° \times 27.86\angle -16.35° = 199.93\angle -2.82°$ V.

The SSSC injects reactive power into the system to increase the load voltage to 239.6 V. Therefore, the reactive power balance can be expressed as $|V_{Th}||I'_L|\sin\beta = |I'_L|^2 X_{Th} + |I'_L|^2 X_{LY1} - |I'_L||V_{SSSC}|$.

Substituting the values of circuit parameters results in $207.60 \times 27.86 \sin 6.35° = 27.86^2 \times 0.41 + 27.86^2 \times 5 - 27.86 \times |V_{SSSC}|$.

Solving the above equation, $|V_{SSSC}| = 92.28$ V.

c. The per-phase voltage rating of the compensator is $|V_{SSSC}| = 92.28$ V.
d. The current rating of the compensator is $|I_{SSSC}| = |I'_L| = 27.86$ A (since it is connected in series with the load).
e. The VA rating of the compensator is $S = 3|V_{SSSC}||I_{SSSC}| = 3 \times 92.28 \times 27.86 = 7712.76$ VA.

Example 5.15

A three-phase AC supply has AC line voltage of 415 V at 50 Hz and a feeder (source) impedance of 0.25 Ω resistance and 1.5 Ω inductive reactance per phase. It is feeding a delta connected load having $Z_{LD1} = (33 + j21)$ Ω per phase. There is another delta connected load having $Z_{LD2} = (9 - j6)$ Ω per phase connected at the point of common coupling for 20 cycles of AC mains. Calculate the voltage sag/swell across the first load under transient and steady-state conditions. If a PWM-IGBT-based VSC with a self-supported DC bus is used in the series compensator as SSSC for steady-state voltage regulation and transient conditions to control the voltage to the input voltage (415 V), calculate the voltage and kVA rating of the DVR in both cases.

Solution: Given the supply phase voltage $V_s = 415/\sqrt{3} = 239.6$ V, frequency of the supply $(f) = 50$ Hz, and first delta connected three-phase load having $Z_{LD1} = 33 + j21$ Ω per phase and second delta connected load having $Z_{LD2} = 9 - j6$ Ω per phase fed from an AC supply with an input AC line voltage of 415 V at 50 Hz and a source impedance of $Z_s = (0.25 + j1.5) = 1.52\angle 80.53°$ Ω.

As the given network is a balanced three-phase network, it can be analyzed on per-phase basis. The equivalent star connected impedance for both the loads are given as follows:

$$Z_{LY1} = Z_{LD1}/3 = (33 + j21)/3 = (11 + j7) = 13.04\angle 32.47°\ \Omega.$$

$$Z_{LY2} = Z_{LD2}/3 = (9 - j6)/3 = (3 - j2) = 3.60\angle -33.69°\ \Omega.$$

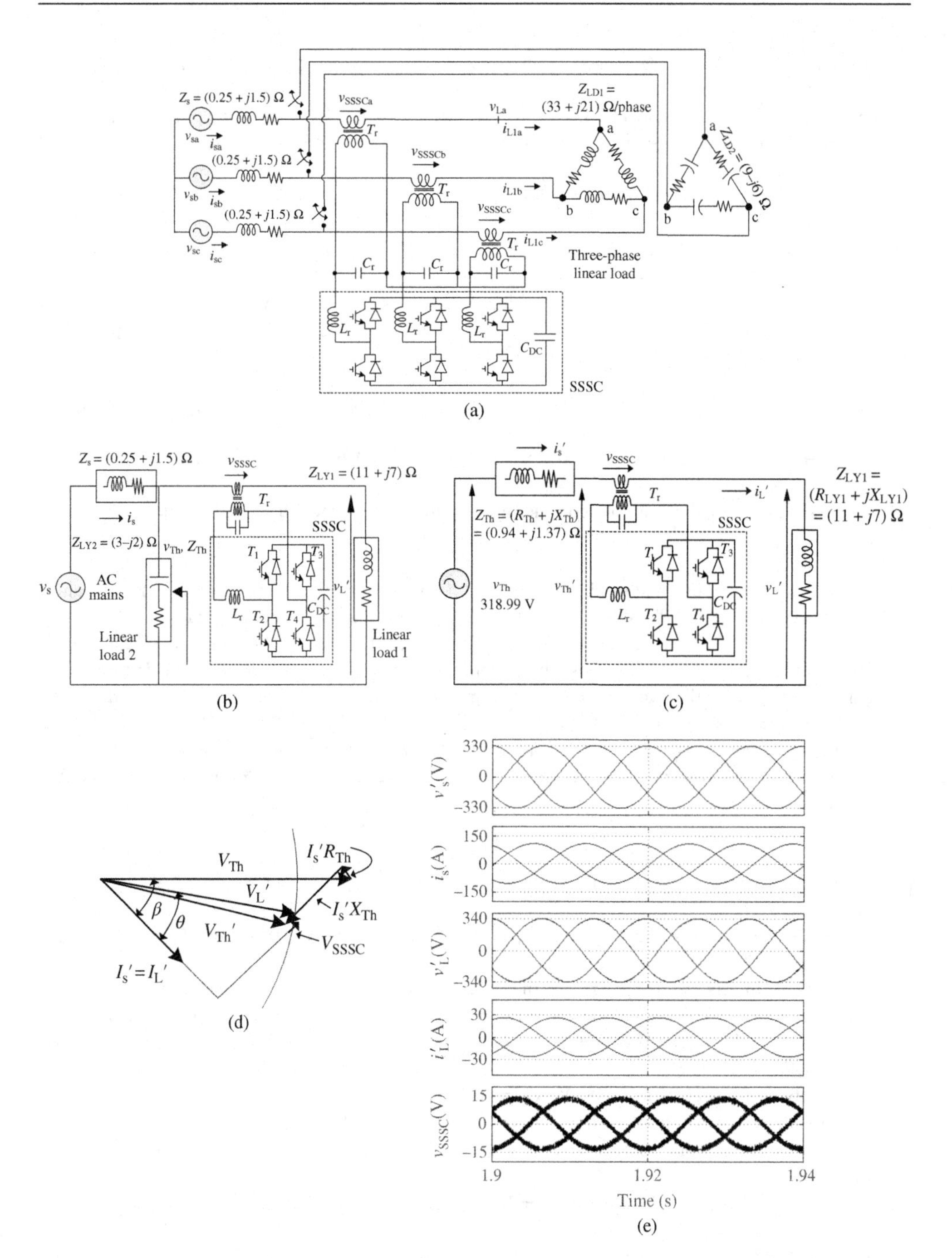

Figure E5.15 Three-phase SSSC with a self-supported DC bus for compensation of drop in the feeder impedance under delta connected linear load. (a) Circuit configuration. (b) Per-phase equivalent when second load is connected. (c) Per-phase Thévenin's equivalent circuit. (d) Phasor diagram under compensation. (e) Waveforms under compensation

The total impedance is $Z_T = Z_s + Z_{LY1}||Z_{LY2} = (0.25 + j1.5) + (11 + j7)||(3 - j_2) = 3.20 + j0.37 = 3.22\angle 6.59°\ \Omega$.

The load current before compensation is $I_{La} = V_{sa}/Z_T = 239.6\angle 0°/3.22\angle 6.59° = 74.40\angle{-6.59°}$ A.

The supply current is the same as the load current: $I_{sa} = I_{La} = 74.40\angle{-6.59°}$ A.

a. The voltage drop across the source impedance is $V_{Zsa} = Z_{sa} \times I_{sa} = 1.52\angle 80.53° \times 74.40\angle{-6.59°} = 112.92\angle 73.93°$ V.
b. The voltage across the load terminals is $V_{ZL} = Z_L \times I_{La} = (Z_{LY1}||Z_{LY2}) \times I_{La} = \{(11 + j7)||(3 - j2)\} \times 74.40\angle{-6.59°} = 234.83\angle{-27.53°}$ V.

   The given system configuration can be simplified using *Thévenin's* equivalent circuit at PCC, as shown in Figure E5.15c. The value of per-phase equivalent Thévenin's voltage and impedance is estimated as follows:

$$V_{Tha} = V_{sa} \times Z_{L2}/(Z_s + Z_{L2}) = (239.6\angle 0° \times (3 - j2))/(0.25 + j1.5 + 3 - j2) = 262.72\angle{-24.95°}\ \text{V}.$$

$$Z_{Tha} = Z_s||Z_{L2} = R_{Th} + jX_{Th} = (0.25 + j1.5)||(3 - j2) = 1.66\angle 55.62° = 0.94 + j1.37\ \Omega.$$

   For simplicity of the solution, considering V_{Th} at new reference, the various calculations are as follows:

   The equivalent circuit and its phasor diagram for compensation are shown in Figure E5.15c and d.

   The series-connected PWM SSSC is used to regulate the load terminal voltage to the input voltage (239.6 V). The SSSC consists of a VSC with self-supported DC link and hence the SSSC can inject or absorb only the reactive power from the system. The SSSC injects a voltage in quadrature to load current. The phasor diagram under voltage drop compensation is shown in Figure E5.15d.

   The load current after compensation is $|I'_L| = |V'_L|/|Z_{LY1}| = 239.6/13.04 = 18.37$ A.

   The active power balance can be expressed as $|V_{Th}||I'_L|\cos\beta = |I'_L|^2 R_{Th} + |I'_L|^2 R_{LY1}$.

   Solving the above equation for angle β results in $262.72 \times 18.37 \times \cos\beta = 18.37^2 \times 0.94 + 18.37^2 \times 11$,

$$\beta = 33.39°.$$

   The load current lags the Thévenin's equivalent voltage by an angle β: $I'_L = |I'_L|\angle{-\beta} = 18.37\angle{-33.39°}$ A.

   The PCC voltage is estimated as $V'_{Th} = V_{Th} - Z_{Th}I'_{La} = 262.72\angle 0° - 1.66\angle 55.62° \times 18.37\angle{-33.39°} = 234.77\angle{-2.81°}$ V.

   The SSSC injects reactive power into the system to increase the load voltage to 239.6 V. Therefore, the reactive power balance can be expressed as $|V_{Th}||I'_L|\sin\beta = |I'_L|^2 X_{Th} + |I'_L|^2 X_{LY1} - |I'_L||V_{SSSC}|$.

   Substituting the values of circuit parameters results in $262.72 \times 18.37 \sin 3.39° = 18.37^2 \times 1.37 + 18.37^2 \times 7 - 18.37 \times |V_{SSSC}|$.

   On solving the above equation, $|V_{SSSC}| = 9.17$ V.
c. The per-phase voltage rating of the compensator is $|V_{SSSC}| = 9.17$ V.
d. The current rating of the compensator is $|I_{SSSC}| = |I'_L| = 18.37$ A (since it is connected in series with the load).
e. The VA rating of the compensator is $S = 3|V_{SSSC}||I_{SSSC}| = 3 \times 9.17 \times 18.37 = 505.35$ VA.

5.8 Summary

Active solid-state SSCs are used to compensate the voltage quality problems of the supply system such as sag, dip, flicker, notch, swell, fluctuations, imbalance, regulation, and so on, and they protect sensitive loads from interruptions, which cause loss of production and mal-operation of other critical equipment such as medical and healthcare systems. These SSCs are also known as DVRs when they are used for dynamic compensation for short periods. The PWM-based VSCs are preferred to realize DVRs with three kinds of DC bus support, namely, DC capacitor, battery, and rectifier. The VSC of a SSC is used in voltage control mode to inject the required voltage in series with the supply to protect sensitive loads. An analytical study of various performance indices of SSCs for the compensation of voltage-based power quality problems of different types of loads has been made in detail with several numerical examples to study the rating of the DVR and how it is affected by the nature of loads. SSCs are observed as one of the best retrofit solution for mitigating the voltage quality problems across various types of loads and to protect critical and sensitive loads.

5.9 Review Questions

1. What is an active solid-state synchronous series compensator?
2. What is a dynamic voltage restorer?
3. What are the power quality problems that a SSC can mitigate?
4. What are the benefits of using a DVR?
5. What are the economic factors for the justification for using a DVR in an AC system?
6. Can a current source inverter be used as a SSC?
7. Why voltage source inverters (VSIs) are preferred in a SSC?
8. What are the factors that decide the rating of a SSC?
9. How the value of an interfacing inductor is computed for a SSC?
10. How the value of DC bus capacitor is computed for a VSI used in a SSC?
11. How the self-supporting DC bus of a VSI is achieved in a SSC?
12. What are the limitations of a SSC with a self-supported DC bus?
13. What are the limitations of a DVR with a rectifier-supported DC bus?
14. Which type of DC bus support for VSI (out of DC capacitor, battery, and rectifier) used as SSC have minimum power rating for voltage regulation?
15. Is it possible to get unity power at the AC mains while regulating the AC voltage across the load?
16. For balancing the unbalanced voltage across the load in a three-phase three-wire system, in how many phases is the injection of the external voltage required?
17. For regulation of the voltage across the load in a three-phase three-wire system, in how many phases is the injection of the external voltage required?
18. How is the rating of a SSC for the AC load voltage regulation affected with the power factor of the load?
19. Why the real power in a DVR change from the DC to AC side of it?
20. Which of the two has the lower power rating for voltage regulation across the load: DSTATCOM or DVR?
21. How DVRs protect sensitive loads from distortion in supply voltages?
22. What are the factors that must be considered in designing the injection transformers for DVR?
23. Under which conditions DVRs do not need injection transformers?
24. How can the power rating of the DVR be reduced when it is used for AC voltage regulation?
25. What are the variables required for sensing in the control of a three-phase DVR?

5.10 Numerical Problems

1. A single-phase load ($Z_L = 6.0 - j3.0$ pu) has an input AC voltage of 220 V at 50 Hz and a base impedance of 5 Ω. It is to be realized as a unity power factor load on the AC supply system while maintaining the same rated load terminal voltage using PWM-based SSSC (shown in Figure E5.1a). Calculate (a) the voltage rating of the compensator, (b) the current rating of the compensator, and (c) the VA rating of the compensator.

2. A single-phase AC mains has a voltage of 220 V at 50 Hz and a feeder (source) impedance of 0.25 Ω resistance and 1.5 Ω inductive reactance. It feeds a single-phase load having $Z_L = (30 + j20)$ Ω. Calculate (a) the voltage drop across the source impedance and (b) the voltage across the load. If a self-supported DC bus-based PWM SSSC (as shown in Figure E5.2a) is used to raise the voltage to the input voltage (220 V), calculate (c) the voltage rating of the compensator, (d) the current rating of the compensator, and (e) the VA rating of the compensator.

3. A single-phase AC supply has a voltage of 220 V at 50 Hz and a feeder (source) impedance of 1.5 Ω resistance and 3.0 Ω inductive reactance. It feeds a single-phase load having $Z_L = 40$ Ω. Calculate (a) the voltage drop across the source impedance and (b) the voltage across the load. If a PWM-based SSSC (as shown in Figure E5.3a) is used to raise the voltage to the input voltage (220 V) with minimum rating, then calculate (c) the voltage rating of the compensator, (d) the current rating of the compensator, and (e) the VA rating of the compensator.

4. A single-phase AC supply has a voltage of 230 V at 50 Hz and a feeder (source) impedance of 0.25 Ω resistance and 2.5 Ω inductive reactance. It feeds a single-phase load having $Z_L = (22 - j16)$ Ω. Calculate (a) the voltage drop across the source impedance and (b) the voltage across the load. If a self-supported DC bus-based PWM SSSC (as shown in Figure E5.4a) is used to raise the voltage to the input voltage (230 V), calculate (c) the voltage rating of the compensator, (d) the current rating of the compensator, and (e) the VA rating of the compensator.

5. A single-phase AC supply has AC mains voltage of 230 V at 50 Hz and a feeder (source) impedance of 0.25 Ω resistance and 3.5 Ω inductive reactance. It feeds a load having $Z_L = (21 + j15)$ Ω. Calculate (a) the voltage drop across the source impedance and (b) the voltage across the load. If a PWM SSSC (as shown in Figure E5.5a) is used to raise the voltage to the input voltage (230 V) with minimum rating, calculate (c) the voltage rating of the compensator, (d) the current rating of the compensator, and (e) the VA rating of the compensator.

6. A single-phase AC supply has AC mains voltage of 230 V at 50 Hz and a feeder (source) impedance of 0.25 Ω resistance and 4.0 Ω inductive reactance. It feeds a load having $Z_L = (20 - j15)$ Ω. Calculate (a) the voltage drop across the source impedance and (b) the voltage across the load. If a PWM SSSC (as shown in Figure E5.6a) is used to raise the voltage to the input voltage (230 V) with minimum rating, calculate (c) the voltage rating of the compensator, (d) the current rating of the compensator, and (e) the VA rating of the compensator.

7. A single-phase AC supply has AC voltage of 230 V at 50 Hz and a feeder (source) impedance of 0.25 Ω resistance and 1.25 Ω inductive reactance per phase. It is feeding a load having $Z_{L1} = (18 + j10)$ Ω. There is another load having $Z_{L2} = (6 - j5)$ Ω connected in parallel for 20 cycles of AC mains. Calculate the voltage sag/swell across the first load under steady-state condition. If a self-supported DC bus-based PWM VSC is employed as SSSC (as shown in Figure E5.7a) in series of the load to maintain its terminal voltage to input voltage (230 V), calculate the line current, voltage, and kVA rating of the series compensator in both cases.

8. A single-phase AC supply has AC voltage of 230 V at 50 Hz and a feeder (source) impedance of 0.25 Ω resistance and 0.4 Ω inductive reactance. It is feeding a load having $Z_{L1} = (20 + j12)$ Ω. There is another load having $Z_{L2} = (6 + j4)$ Ω connected in parallel for 20 cycles of AC mains. Calculate (a) the voltage across the source impedance and (b) the voltage across the first load under transient

and steady-state conditions. If a self-supported DC bus-based PWM VSC is employed as SSSC (as shown in Figure E5.8a) in series of the first load to maintain its terminal voltage to input voltage (230 V), calculate (c) the current rating of the SSSC, (d) the voltage rating of the SSSC, and (e) the VA rating of the SSSC in both cases.

9. A three-phase delta connected load having $Z_{LD} = (8.0 + j4.0)$ pu per phase is fed from an AC supply with an input AC line voltage of 415 V at 50 Hz and a base impedance of 6.15 Ω. It is to be realized as a unity power factor load on the AC supply system while maintaining the same rated load terminal voltage using PWM-based SSSC (as shown in Figure E5.9a). Calculate (a) the voltage rating of the compensator, (b) the current rating of the compensator, and (c) the VA rating of the compensator.

10. A three-phase, three-wire AC supply has AC mains voltage of 440 V at 50 Hz and a feeder (source) impedance of 0.5 Ω per phase resistance and 3.5 Ω per phase inductive reactance. It feeds a three-phase star connected load having $Z_L = (20 + j15)$ Ω per phase. Calculate (a) the voltage drop across the source impedance and (b) the voltage across the load. If a PWM SSSC (as shown in Figure E5.10a) is used to raise the voltage to the input voltage (440 V) with minimum rating, calculate (c) the voltage rating of the compensator, (d) the current rating of the compensator, and (e) the VA rating of the compensator.

11. A three-phase AC supply has a line voltage of 440 V at 50 Hz and a feeder (source) impedance of 0.5 Ω resistance and 3.0 Ω inductive reactance per phase after which a balanced delta connected load having $Z_{LD} = (20 + j15)$ Ω per phase is connected. Calculate the voltage drop across the source impedance and the voltage across the load. If a self-supported DC bus-based PWM SSSC (as shown in Figure E5.11a) is used to raise the voltage to the input voltage (440 V), calculate the voltage, current, and kVA rating of the compensator.

12. A three-phase AC supply has a line–line voltage of 415 V at 50 Hz and a feeder (source) impedance of 0.25 Ω resistance and 3.5 Ω inductive reactance per phase after which a balanced delta connected load having $Z_{LD} = (16 - j12)$ Ω per phase is connected. Calculate (a) the voltage drop across the source impedance and (b) the voltage across the load in each case. If a self-supported DC link- and PWM-based SSSC (as shown in Figure E5.12a) is used to maintain the voltage to the input voltage (415 V), calculate the (c) voltage, (d) current, and (e) kVA rating of the series compensator.

13. A three-phase AC supply has a line–line voltage of 415 V at 50 Hz and a feeder (source) impedance of 0.2 Ω resistance and 0.4 Ω inductive reactance per phase after which a balanced delta connected load having $Z_{LD1} = (35 + j21)$ Ω per phase is connected. There is another load connected in parallel, which draws exactly five times the current at the same power factor than this load for 20 cycles of the AC mains. Calculate (a) the voltage across the source impedance and (b) the voltage across the first load. If a PWM-IGBT-based VSC is employed as SSSC in series of the first load to maintain its terminal voltage to the input voltage (415 V) with minimum rating under steady state conditions, calculate (c) the current rating of the SSSC, (d) the voltage rating of the SSSC, and (e) the VA rating of the SSSC.

14. A three-phase AC supply has a line–line voltage of 440 V at 50 Hz and a feeder (source) impedance of 0.12 Ω resistance and 0.25 Ω inductive reactance per phase after which a balanced delta connected load having $Z_{LD1} = (25 + j18)$ Ω per phase is connected. There is another delta connected load having $Z_{LD2} = (8 + j4)$ Ω per phase at the point of common coupling for 10 cycles of the AC mains. Calculate (a) the voltage drop across the source impedance and (b) the voltage across the load under transient and steady-state conditions. If a PWM-IGBT-based VSC with a self-supported DC bus is used as SSSC for voltage regulation under steady-state and transient conditions to raise the voltage to the input voltage (440 V), calculate the (c) voltage, (d) current, and (e) kVA rating of the series compensator.

15. A three-phase AC supply has AC line voltage of 415 V at 50 Hz and a feeder (source) impedance of 0.5 Ω resistance and 2 Ω inductive reactance per phase. It is feeding a delta connected load having

$Z_{LD1} = (30 + j20)\ \Omega$ per phase. There is another delta connected load having $Z_{LD2} = (10 - j5)\ \Omega$ per phase at the point of common coupling for 20 cycles of the AC mains. Calculate the voltage sag/swell across the first load under transient and steady-state conditions. If a PWM-IGBT-based VSC with a self-supported DC bus is used as SSSC for voltage regulation under steady-state and transient conditions to control the voltage to the input voltage (415 V) of the series compensator, calculate the voltage and kVA rating of the DVR in both cases.

5.11 Computer Simulation-Based Problems

1. Design and develop a dynamic model of a DVR (consisting of a VSC, an injection transformer with an interfacing inductor, and a ripple filter on the AC side and a capacitor or a battery or a rectifier on the DC bus) and simulate its behavior for a single-phase AC supply having a voltage of 220 V at 50 Hz and a feeder (source) impedance of 0.5 Ω resistance and 3.5 Ω inductive reactance. It feeds a single-phase load having $Z = (24 - j18)\ \Omega$. It has another load having $Z = (4 + j3)\ \Omega$ connected at PCC for 20 cycles. Plot the supply voltage and supply current, load AC voltage, load AC current, DVR voltage and its current with time, and the harmonic spectra of supply voltage, supply current and load AC voltage, load AC current. Compute the kVA rating of the DVR, THD, crest factor, rms value of AC mains current, displacement factor, distortion factor, power factor, and input active and reactive powers. The DVR is controlled in the (a) capacitor-supported DC bus mode, (b) battery-supported DC bus mode, and (c) rectifier-supported DC bus of the DVR for regulating the load voltage.

2. Design and develop a dynamic model of a DVR (consisting of a VSC, an injection transformer with an interfacing inductor, and a ripple filter on the AC side and a capacitor or a battery or a rectifier on the DC bus) and simulate its behavior for a single-phase VSI with quasi-square wave AC output of 230 V (rms) at 50 Hz feeding a critical linear load of 4 kVA, 220 V, 50 Hz at 0.8 lagging PF to eliminate voltage harmonics, regulate fundamental 220 V (rms) across this load, and maintain UPF at the source. Plot the supply voltage and supply current, load AC voltage, load AC current, DVR voltage and its current with time, and the harmonic spectra of supply voltage, supply current and load AC voltage, load AC current. Compute the kVA rating of the DVR, THD, crest factor, rms value of AC mains current, displacement factor, distortion factor, power factor, and input active and reactive powers. The DVR is controlled in the (a) capacitor-supported DC bus mode, (b) battery-supported DC bus mode, and (c) rectifier-supported DC bus of the DVR for regulating the load voltage.

3. Design and develop a dynamic model of a DVR (consisting of a VSC, an injection transformer with an interfacing inductor, and a ripple filter on the AC side and a capacitor or a battery or a rectifier on the DC bus) and simulate its behavior for a single-phase 230 V, 50 Hz uncontrolled bridge converter used for charging a battery of 216 V. It has another linear load of 4 kVA at 0.8 lagging power factor connected in parallel at PCC. It has a source impedance of 1.5 Ω resistive element and 3.5 Ω inductive element. Plot the supply voltage and supply current, load AC voltage, load AC current, DVR voltage and its current with time, and the harmonic spectra of supply voltage, supply current and load AC voltage, load AC current. Compute the kVA rating of the DVR, THD, crest factor, rms value of AC mains current, displacement factor, distortion factor, power factor, and input active and reactive powers. The DVR is controlled in the (a) capacitor-supported DC bus mode, (b) battery-supported DC bus mode, and (c) rectifier-supported DC bus of the DVR for regulating the load voltage.

4. Design and develop a dynamic model of a DVR (consisting of a VSC, an injection transformer with an interfacing inductor, and a ripple filter on the AC side and a capacitor or a battery or a rectifier on the DC bus) and simulate its behavior for a single-phase 230 V, 50 Hz uncontrolled bridge converter used for charging a battery of 180 V. It has a source impedance of 0.75 Ω resistive element and 3.5 Ω inductive element. It has another load having $Z = (10 + j5)\ \Omega$ connected at PCC for 20 cycles. Plot the supply voltage and supply current, load AC voltage, load AC current, DVR voltage and its current

with time, and the harmonic spectra of supply voltage, supply current and load AC voltage, load AC current. Compute the kVA rating of the DVR, THD, crest factor, rms value of AC mains current, displacement factor, distortion factor, power factor, and input active and reactive powers. The DVR is controlled in the (a) capacitor-supported DC bus mode, (b) battery-supported DC bus mode, and (c) rectifier-supported DC bus of the DVR for regulating the load voltage.

5. Design and develop a dynamic model of a DVR (consisting of a VSC, an injection transformer with an interfacing inductor, and a ripple filter on the AC side and a capacitor or a battery or a rectifier on the DC bus) and simulate its behavior for a single-phase 230 V, 50 Hz controlled bridge converter with series-connected inductive load having 100 mH and an equivalent resistive load having 5 Ω. It has a source impedance of 0.5 Ω resistive element and 3.25 Ω inductive element. The delay angle of its thyristors is $\alpha = 60°$. It has another load having $Z = (5 - j3)$ Ω connected at PCC for 20 cycles. Plot the supply voltage and supply current, load AC voltage, load AC current, DVR voltage and its current with time, and the harmonic spectra of supply voltage, supply current and load AC voltage, load AC current. Compute the kVA rating of the DVR, THD, crest factor, rms value of AC mains current, displacement factor, distortion factor, power factor, and input active and reactive powers. The DVR is controlled in the (a) capacitor-supported DC bus mode, (b) battery-supported DC bus mode, and (c) rectifier-supported DC bus of the DVR for regulating the load voltage.

6. Design and develop a dynamic model of a DVR (consisting of a VSC, an injection transformer with an interfacing inductor, and a ripple filter on the AC side and a capacitor or a battery or a rectifier on the DC bus) and simulate its behavior for a single-phase AC supply system having AC mains voltage of 230 V at 50 Hz and a feeder (source) impedance of 0.5 Ω resistance and 3.0 Ω inductive reactance. It is feeding a load having $Z = (16 - j12)$ Ω. It has another load having $Z = (4 + j3)$ Ω connected at PCC for 20 cycles. Plot the supply voltage and supply current, load AC voltage, load AC current, DVR voltage and its current with time, and the harmonic spectra of supply voltage, supply current and load AC voltage, load AC current. Compute the kVA rating of the DVR, THD, crest factor, rms value of AC mains current, displacement factor, distortion factor, power factor, and input active and reactive powers. The DVR is controlled in the (a) capacitor-supported DC bus mode, (b) battery-supported DC bus mode, and (c) rectifier-supported DC bus of the DVR for regulating the load voltage.

7. Design and develop a dynamic model of a DVR (consisting of a VSC, an injection transformer with an interfacing inductor, and a ripple filter on the AC side and a capacitor or a battery or a rectifier on the DC bus) and simulate its behavior for a single-phase AC supply system having AC mains voltage of 220 V at 50 Hz and a feeder (source) impedance of 0.25 Ω resistance and 3.0 Ω inductive reactance. It is feeding a load having $Z = (15 + j10)$ Ω. It has another load having $Z = (3 - j3)$ Ω connected at PCC for 20 cycles. Plot the supply voltage and supply current, load AC voltage, load AC current, DVR voltage and its current with time, and the harmonic spectra of supply voltage, supply current and load AC voltage, load AC current. Compute the kVA rating of the DVR, THD, crest factor, rms value of AC mains current, displacement factor, distortion factor, power factor, and input active and reactive powers. The DVR is controlled in the (a) capacitor-supported DC bus mode, (b) battery-supported DC bus mode, and (c) rectifier-supported DC bus of the DVR for regulating the load voltage.

8. Design and develop a dynamic model of a DVR (consisting of a VSC, an injection transformer with an interfacing inductor and a ripple filter on the AC side and a capacitor or a battery or a rectifier on the DC bus) and simulate its behavior for a single-phase 220 V, 50 Hz diode bridge rectifier with a DC capacitor filter feeding at variable frequency a 4 kW three-phase VSI-fed induction motor drive in an air conditioner to reduce the harmonics in AC mains current to almost maintain UPF and regulate the DC bus voltage of a rectifier to 400 V. It has a source impedance of 0.25 Ω resistive element and 2.5 Ω inductive element. It has another load having $Z = (3 + j3)$ Ω connected at PCC for 20 cycles. Plot the supply voltage and supply current, load AC voltage, load AC current, DVR voltage and its current with time, and the harmonic spectra of supply voltage, supply current and load AC voltage, load AC

current. Compute the kVA rating of the DVR, THD, crest factor, rms value of AC mains current, displacement factor, distortion factor, power factor, and input active and reactive powers. The DVR is controlled in the (a) capacitor-supported DC bus mode, (b) battery-supported DC bus mode, and (c) rectifier-supported DC bus of the DVR for regulating the load voltage.

9. Design and develop a dynamic model of a DVR (consisting of a VSC, an injection transformer with an interfacing inductor and a ripple filter on the AC side and a capacitor or a battery or a rectifier on the DC bus) and simulate its behavior for filtering the single-phase square wave supply of 230 V (rms) to be converted to sine wave of 220 V (rms) to feed a critical linear lagging power load with $R = 20\,\Omega$ and $X = 15\,\Omega$. It has a source impedance of 0.5 Ω resistive element and 3.5 Ω inductive element. It has another load having $Z = (4 + j3)\,\Omega$ connected at PCC for 20 cycles. Plot the supply voltage and supply current, load AC voltage, load AC current, DVR voltage and its current with time, and the harmonic spectra of supply voltage, supply current and load AC voltage, load AC current. Compute the kVA rating of the DVR, THD, crest factor, rms value of AC mains current, displacement factor, distortion factor, power factor, input active and reactive powers. The DVR is controlled in the (a) capacitor-supported DC bus mode, (b) battery-supported DC bus mode, and (c) rectifier-supported DC bus of the DVR for regulating the load voltage.

10. Design and develop a dynamic model of a DVR (consisting of a VSC, an injection transformer with an interfacing inductor, and a ripple filter on the AC side and a capacitor or a battery or a rectifier on the DC bus) and simulate its behavior for filtering the single-phase quasi-square wave supply of 240 V (rms) to be converted to sine wave of 230 V (rms) to feed a critical linear lagging power load with $R = 25\,\Omega$ and $X = 10\,\Omega$. It has a source impedance of 0.5 Ω resistive element and 3.5 Ω inductive element. It has another load having $Z = (5 - j2)\,\Omega$ connected at PCC for 20 cycles. Plot the supply voltage and supply current, load AC voltage, load AC current, DVR voltage and its current with time, and the harmonic spectra of supply voltage, supply current and load AC voltage, load AC current. Compute the kVA rating of the DVR, THD, crest factor, rms value of AC mains current, displacement factor, distortion factor, power factor, and input active and reactive powers. The DVR is controlled in the (a) capacitor-supported DC bus mode, (b) battery-supported DC bus mode, and (c) rectifier-supported DC bus of the DVR for regulating the load voltage.

11. Design and develop a dynamic model of a DVR (consisting of a VSC, an injection transformer with an interfacing inductor, and a ripple filter on the AC side and a capacitor or a battery or a rectifier on the DC bus) and simulate its behavior to reduce the harmonic current in AC mains connected in the series and to maintain almost UPF of a single-phase AC supply of 230 V at 50 Hz feeding a three-phase diode rectifier with a capacitive filter of 470 μF and a resistive load having 10 Ω. It has a source impedance of 0.25 Ω resistive element and 3.0 Ω inductive element. It has another load having $Z = (10 + j6)\,\Omega$ connected at PCC for 20 cycles. Plot the supply voltage and supply current, load AC voltage, load AC current, DVR voltage and its current with time, and the harmonic spectra of supply voltage, supply current and load AC voltage, load AC current. Compute the kVA rating of the DVR, THD, crest factor, rms value of AC mains current, displacement factor, distortion factor, power factor, and input active and reactive powers. The DVR is controlled in the (a) capacitor-supported DC bus mode, (b) battery-supported DC bus mode, and (c) rectifier-supported DC bus of the DVR for regulating the load voltage.

12. Design and develop a dynamic model of a DVR (consisting of a VSC, an injection transformer with an interfacing inductor, and a ripple filter on the AC side and a capacitor or a battery or a rectifier on the DC bus) and simulate its behavior for a three-phase 415 V, 50 Hz, uncontrolled six-pulse bridge converter used for charging a battery of 560 V. It has source impedance of 0.5 Ω resistive element and 3.5 Ω inductive element. It has another load having $Z = (10 + j5)\,\Omega$ per phase connected in star at PCC for 20 cycles. Plot the supply voltage and supply current, load AC voltage, load AC current, DVR voltage and its current with time, and the harmonic spectra of supply voltage, supply current and load AC voltage, load AC current. Compute the kVA rating of the DVR, THD, crest factor, rms value of AC mains current, displacement factor, distortion factor, power factor, and input active and reactive powers. The DVR is controlled in the (a) capacitor-supported DC bus mode, (b) battery-supported DC bus mode, and (c) rectifier-supported DC bus of the DVR for regulating the load voltage.

13. Design and develop a dynamic model of a DVR (consisting of a VSC, an injection transformer with an interfacing inductor, and a ripple filter on the AC side and a capacitor or a battery or a rectifier on the DC bus) and simulate its behavior for a three-phase, three-wire AC supply system having AC mains voltage of 415 V at 50 Hz and a feeder (source) impedance of 0.75 Ω per phase resistance and 3.5 Ω per phase inductive reactance. It feeds a balanced three-phase star connected load having $Z = (20 + j12)\ \Omega$ per phase. It has another three-phase load having $Z = (3 - j5)\ \Omega$ connected between lines at PCC for 20 cycles. Plot the supply voltage and supply current, load AC voltage, load AC current, DVR voltage and its current with time, and the harmonic spectra of supply voltage, supply current and load AC voltage, load AC current. Compute the kVA rating of the DVR, THD, crest factor, rms value of AC mains current, displacement factor, distortion factor, power factor, and input active and reactive powers. The DVR is controlled in the (a) capacitor-supported DC bus mode, (b) battery-supported DC bus mode, and (c) rectifier-supported DC bus of the DVR for regulating the load voltage.

14. Design and develop a dynamic model of a DVR (consisting of a VSC, an injection transformer with an interfacing inductor, and a ripple filter on the AC side and a capacitor or a battery or a rectifier on the DC bus) and simulate its behavior for a three-phase AC mains of 400 V, 50 Hz with a voltage fluctuation between +10 and −15% and 5% voltage unbalance due to rural feeder and a nearby factory. A hospital needs a three-phase 400 V, 50 kVA, 0.8 lagging PF, 50 Hz supply for critical equipment. It has a source impedance of 0.5 Ω resistive element and 2.5 Ω inductive element. It has another single-phase load having $Z = (10 + j15)\ \Omega$ connected between two lines at PCC for 20 cycles. Plot the supply voltage and supply current, load AC voltage, load AC current, DVR voltage and its current with time, and the harmonic spectra of supply voltage, supply current and load AC voltage, load AC current. Compute the kVA rating of the DVR, THD, crest factor, rms value of AC mains current, displacement factor, distortion factor, power factor, and input active and reactive powers. The DVR is controlled in the (a) capacitor-supported DC bus mode, (b) battery-supported DC bus mode, and (c) rectifier-supported DC bus of the DVR for regulating the load voltage.

15. Design and develop a dynamic model of a DVR (consisting of a VSC, an injection transformer with an interfacing inductor, and a ripple filter on the AC side and a capacitor or a battery or a rectifier on the DC bus) and simulate its behavior for filtering the three-phase quasi-square wave supply of 415 V (rms) at 50 Hz to be converted to sine wave 400 V (rms) to feed a critical linear lagging power load of 50 kVA at 0.8 lagging power factor. It has a source impedance of 0.75 Ω resistive element and 3.5 Ω inductive element. It has another single-phase load having $Z = (10 - j15)\ \Omega$ connected between two lines at PCC for 20 cycles. Plot the supply voltage and supply current, load AC voltage, load AC current, DVR voltage and its current with time, and the harmonic spectra of supply voltage, supply current and load AC voltage, load AC current. Compute the kVA rating of the DVR, THD, crest factor, rms value of AC mains current, displacement factor, distortion factor, power factor, and input active and reactive powers. The DVR is controlled in the (a) capacitor-supported DC bus mode, (b) battery-supported DC bus mode, and (c) rectifier-supported DC bus of the DVR for regulating the load voltage.

16. Design and develop a dynamic model of a DVR (consisting of a VSC, an injection transformer with an interfacing inductor, and a ripple filter on the AC side and a capacitor or a battery or a rectifier on the DC bus) and simulate its behavior for a three-phase 415 V, 50 Hz, six-pulse diode bridge rectifier with a DC capacitor filter feeding a VSI-fed induction motor drive with DC of 600 V at 150 A average current at variable frequency. It has a source impedance of 0.5 Ω resistive element and 3.5 Ω inductive element. It has another single-phase load having $Z = (15 - j10)\ \Omega$ connected between two lines at PCC for 20 cycles. Plot the supply voltage and supply current, load AC voltage, load AC current, DVR voltage and its current with time, and the harmonic spectra of supply voltage, supply

current and load AC voltage, load AC current. Compute the kVA rating of the DVR, THD, crest factor, rms value of AC mains current, displacement factor, distortion factor, power factor, and input active and reactive powers. The DVR is controlled in the (a) capacitor-supported DC bus mode, (b) battery-supported DC bus mode, and (c) rectifier-supported DC bus of the DVR for regulating the load voltage.

17. Design and develop a dynamic model of a DVR (consisting of a VSC, an injection transformer with an interfacing inductor, and a ripple filter on the AC side and a capacitor or a battery or a rectifier on the DC bus) and simulate its behavior for a three-phase, 380 V line voltage, 50 Hz, four-wire distribution system with three single-phase uncontrolled bridge converters used for charging a battery of 172 V. It has a source impedance of 0.5 Ω resistive element and 3.5 Ω inductive element. It has another single-phase load having $Z=(10+j15)\ \Omega$ connected between line and neutral in one phase at PCC for 20 cycles. Plot the supply voltage and supply current, load AC voltage, load AC current, DVR voltage and its current with time, and the harmonic spectra of supply voltage, supply current and load AC voltage, load AC current. Compute the kVA rating of the DVR, THD, crest factor, rms value of AC mains current, displacement factor, distortion factor, power factor, and input active and reactive powers. The DVR is controlled in the (a) capacitor-supported DC bus mode, (b) battery-supported DC bus mode, and (c) rectifier-supported DC bus of the DVR for regulating the load voltage.

18. Design and develop a dynamic model of a DVR (consisting of a VSC, an injection transformer with an interfacing inductor, and a ripple filter on the AC side and a capacitor or a battery or a rectifier on the DC bus) and simulate its behavior for a single-phase AC supply system having AC voltage of 230 V at 50 Hz and a feeder (source) impedance of 0.5 Ω resistance and 3.5 Ω inductive reactance. It is feeding a load having $Z=(25+j15)\ \Omega$. There is another load connected in parallel to it, which has a load impedance of $Z=(5+j3)\ \Omega$ for 20 cycles of AC mains. Plot the supply voltage and supply current, load AC voltage, load AC current, DVR voltage and its current with time, and the harmonic spectra of supply voltage, supply current and load AC voltage, load AC current. Compute the kVA rating of the DVR, THD, crest factor, rms value of AC mains current, displacement factor, distortion factor, power factor, and input active and reactive powers. The DVR is controlled in the (a) capacitor-supported DC bus mode, (b) battery-supported DC bus mode, and (c) rectifier-supported DC bus of the DVR for regulating the load voltage.

19. Design and develop a dynamic model of a DVR (consisting of a VSC, an injection transformer with an interfacing inductor, and a ripple filter on the AC side and a capacitor or a battery or a rectifier on the DC bus) and simulate its behavior for a single-phase AC supply system having an AC voltage of 230 V at 50 Hz and a feeder (source) impedance of 0.5 Ω resistance and 3.0 Ω inductive reactance. It is feeding a load having $Z=(20+j10)\ \Omega$. There is another load connected in parallel to it, which has a load impedance of $Z=(5-j3)\ \Omega$ for 20 cycles of AC mains. Plot the supply voltage and supply current, load AC voltage, load AC current, DVR voltage and its current with time, and the harmonic spectra of supply voltage, supply current and load AC voltage, load AC current. Compute the kVA rating of the DVR, THD, crest factor, rms value of AC mains current, displacement factor, distortion factor, power factor, and input active and reactive powers. The DVR is controlled in the (a) capacitor-supported DC bus mode, (b) battery-supported DC bus mode, and (c) rectifier-supported DC bus of the DVR for regulating the load voltage.

20. Design and develop a dynamic model of a DVR (consisting of a VSC, an injection transformer with an interfacing inductor, and a ripple filter on the AC side and a capacitor or a battery or a rectifier on the DC bus) and simulate its behavior for a three-phase three-wire AC supply having a line voltage of 415 V at 50 Hz and a feeder (source) impedance of 0.5 Ω resistance and 2.5 Ω inductive reactance per phase. It is feeding an unbalanced delta connected load having $Z_{ab}=6.0+j3.0$ pu, $Z_{bc}=3.0+j1.5$ pu, and $Z_{ca}=7.5+j1.5$ pu with a base impedance of 4.15 Ω per phase. It has another load having $Z=(4-j3)\ \Omega$ connected across two lines at PCC for 20 cycles. Plot the supply voltage and supply current, load AC voltage, load AC current, DVR voltage and its current with time, and the harmonic spectra of supply voltage, supply current and load AC voltage, load AC current. Compute the kVA rating of the DVR, THD, crest factor, rms value of AC mains current, displacement factor, distortion factor, power factor, and input active and reactive powers. The DVR is controlled in the (a) capacitor-supported DC bus mode, (b) battery-supported DC bus mode, and (c) rectifier-supported DC bus of the DVR for regulating the load voltage.

21. Design and develop a dynamic model of a DVR (consisting of a VSC, an injection transformer with an interfacing inductor, and a ripple filter on the AC side and a capacitor or a battery or a rectifier on the DC bus) and simulate its behavior for a three-phase, three-wire AC supply system having an AC mains voltage of 415 V at 50 Hz and a feeder (source) impedance of 0.5 Ω resistance and 2.5 Ω inductive reactance per phase. It feeds a balanced three-phase star connected load having $Z = (25 + j15)\ \Omega$ per phase. It has another single-phase load having $Z = (4 - j5)\ \Omega$ connected between two lines at PCC for 20 cycles. Plot the supply voltage and supply current, load AC voltage, load AC current, DVR voltage and its current with time, and the harmonic spectra of supply voltage, supply current and load AC voltage, load AC current. Compute the kVA rating of the DVR, THD, crest factor, rms value of AC mains current, displacement factor, distortion factor, power factor, and input active and reactive powers. The DVR is controlled in the (a) capacitor-supported DC bus mode, (b) battery-supported DC bus mode, and (c) rectifier-supported DC bus of the DVR for regulating the load voltage.

22. Design and develop a dynamic model of a DVR (consisting of a VSC, an injection transformer with an interfacing inductor, and a ripple filter on the AC side and a capacitor or a battery or a rectifier on the DC bus) and simulate its behavior for a three-phase, 380 V line voltage, 50 Hz, four-wire distribution system with three single-phase uncontrolled bridge converters used for charging a battery of 172 V. It has a source impedance of 0.5 Ω resistive element and 3.5 Ω inductive element. It has another three-phase load having $Z = (3 - j3)\ \Omega$ connected at PCC between line and neutral for 20 cycles. Plot the supply voltage and supply current, load AC voltage, load AC current, DVR voltage and its current with time, and the harmonic spectra of supply voltage, supply current and load AC voltage, load AC current. Compute the kVA rating of the DVR, THD, crest factor, rms value of AC mains current, displacement factor, distortion factor, power factor, and input active and reactive powers. The DVR is controlled in the (a) capacitor-supported DC bus mode, (b) battery-supported DC bus mode, and (c) rectifier-supported DC bus of the DVR for regulating the load voltage.

23. Design and develop a dynamic model of a DVR (consisting of a VSC, an injection transformer with an interfacing inductor, and a ripple filter on the AC side and a capacitor or a battery or a rectifier on the DC bus) and simulate its behavior for a three-phase, 415 V line voltage, 50 Hz, four-wire distribution system with three single-phase loads (connected between phases and neutral) of a single-phase uncontrolled bridge converter having a parallel capacitive DC filter of 680 μF with a 15 Ω equivalent resistive load. It has a source impedance of 0.5 Ω resistive element and 2.5 Ω inductive element. It has another three-phase load having $Z = (10 + j10)\ \Omega$ connected at PCC between line and neutral for 20 cycles. Plot the supply voltage and supply current, load AC voltage, load AC current, DVR voltage and its current with time, and the harmonic spectra of supply voltage, supply current and load AC voltage, load AC current. Compute the kVA rating of the DVR, THD, crest factor, rms value of AC mains current, displacement factor, distortion factor, power factor, and input active and reactive powers. The DVR is controlled in the (a) capacitor-supported DC bus mode, (b) battery-supported DC bus mode, and (c) rectifier-supported DC bus of the DVR for regulating the load voltage.

24. Design and develop a dynamic model of a DVR (consisting of a VSC, an injection transformer with an interfacing inductor, and a ripple filter on the AC side and a capacitor or a battery or a rectifier on the DC bus) and simulate its behavior for a three-phase AC supply having an AC line voltage of 415 V at 50 Hz and a feeder (source) impedance of 0.25 Ω resistance and 3.5 Ω inductive reactance per phase. It is feeding a delta connected load having $Z = (20 + j12)\ \Omega$ per phase. There is another delta connected load having $Z = (5 - j4)\ \Omega$ per phase connected in parallel with the first load for 20 cycles of AC mains. Plot the supply voltage and supply current, load AC voltage, load AC current, DVR voltage and its current with time, and the harmonic spectra of supply voltage, supply current and load AC voltage, load AC current. Compute the kVA rating of the DVR, THD, crest factor, rms value of AC mains current, displacement factor, distortion factor, power factor, and input active and reactive powers. The DVR is controlled in the (a) capacitor-supported DC bus mode, (b) battery-supported DC bus mode, and (c) rectifier-supported DC bus of the DVR for regulating the load voltage.

25. Design and develop a dynamic model of a DVR (consisting of a VSC, an injection transformer with an interfacing inductor, and a ripple filter on the AC side and a capacitor or a battery or a rectifier on the DC bus) and simulate its behavior for a three-phase AC supply having an AC voltage of 415 V at 50 Hz and a feeder (source) impedance of 0.25 Ω resistance and 2.5 Ω inductive reactance per phase. It is feeding a balanced delta connected load having $Z = (25 + j15)\ \Omega$ per phase. There is another delta connected load having $Z = (5 + j3)\ \Omega$ per phase connected in parallel for 20 cycles of AC mains. Plot the supply voltage and supply current, load AC voltage, load AC current, DVR voltage and its current with time, and the harmonic spectra of supply voltage, supply current and load AC voltage, load AC current. Compute the kVA rating of the DVR, THD, crest factor, rms value of AC mains current, displacement factor, distortion factor, power factor, and input active and reactive powers. The DVR is controlled in the (a) capacitor-supported DC bus mode, (b) battery-supported DC bus mode, and (c) rectifier-supported DC bus of the DVR for regulating the load voltage.

References

1. Bollen, M.H.J. (2000) *Understanding Power Quality Problems: Voltage Sags and Interruptions*, IEEE Press Series on Power Engineering, John Wiley & Sons, Inc., New York.
2. Ghosh, A. and Ledwich, G. (2002) *Power Quality Enhancement Using Custom Power Devices*, Kluwer Academic Publishers, London.
3. Akagi, H., Watanabe, E.H., and Aredes, M. (2007) *Instantaneous Power Theory and Applications to Power Conditioning*, John Wiley & Sons, Inc., New York.
4. Padiyar, K.R. (2007) *FACTS Controllers in Transmission and Distribution*, New Age International, New Delhi.
5. Bhavraju, V.B. and Enjeti, P.N. (1994) A fast active power filter to correct voltage sags. *IEEE Transactions on Industrial Electronics*, **41**(3), 333–338.
6. Campos, A., Joos, G., Ziogas, P.D., and Lindsay, J.F. (1994) Analysis, design of a series voltage unbalance compensator based on a three-phase VSI operating with unbalanced switching functions. *IEEE Transactions on Power Electronics*, **9**(3), 269–274.
7. Vincenti, D., Jin, H., and Ziogas, P. (1994) Design and implementation of a 25-kVA three-phase PWM AC line conditioner. *IEEE Transactions on Power Electronics*, **9**(4), 384–389.
8. Nabae, A. and Yamaguchi, M. (1995) Suppression of flicker in an arc-furnace supply system by an active capacitance: a novel voltage stabilizer in power systems. *IEEE Transactions on Industry Applications*, **31**(1), 107–111.
9. Bhavraju, V.B. and Enjeti, P.N. (1996) An active line conditioner to balance voltages in a three phase system. *IEEE Transactions on Industry Applications*, **32**(2), 287–292.
10. Nelson, R.J. and Ramey, D.G. (1997) Dynamic power and voltage regulator for an AC transmission line. US Patent 5610501 (March 11).
11. Kara, A., Dahler, P., Amhof, D., and Gruning, H. (1998) Power supply quality improvement with a dynamic voltage restorer (DVR). Proceedings of the IEEE APEC'98, vol. 2, pp. 986–993.
12. Fang, M., Gardiner, M.I., MacDougall, A., and Mathieson, G.A. (1998) A novel series dynamic voltage restorer for distribution systems. Proceedings of the International Conference on Power System Technology (POWERCON'98), August 18–21, vol. 1, pp. 38–42.
13. Cheng, P.T., Lasseter, R., and Divan, D. (1910) Dynamic series voltage restoration for sensitive loads in unbalanced power systems. US Patent 5883796 (March 16).
14. Hochgraf, C.G. (1910) Power inverter apparatus using a transformer with its primary winding connected to the source end and a secondary winding connected to the load end of an AC power line to insert series compensation. US Patent 5905367 (May 18).

15. Woodley, N.H., Morgan, L., and Sundaram, A. (1910) Experience with an inverter-based dynamic voltage restorer. *IEEE Transactions on Power Delivery*, **14**(3), 1181–1186.
16. Vilathgamuwa, M., Perera, R., Choi, S., and Tseng, K. (1999) Control of energy optimized dynamic voltage restorer. Proceedings of the 25 th Annual Conference of the IEEE Industrial Electronics Society (IECON'99), vol. 2, pp. 873–8710.
17. Choi, S.S., Li, B.H., and Vilathgamuwa, D.M. (2000) Dynamic voltage restoration with minimum energy injection. Proceedings of the IEEE Power Engineering Society, Winter Meeting, January 23–27, vol. 2, pp. 1156–1161.
18. Daehler, P. and Affolter, R. (2000) Requirements and solutions for dynamic voltage restorer, a case study. Proceedings of the IEEE Power Engineering Society, Winter Meeting, vol. 4, pp. 2881–2885.
19. Brumsickle, W.E., Schneider, R.S., Luckjiff, G.A. *et al.* (2001) Dynamic sag correctors: cost-effective industrial power line conditioning. *IEEE Transactions on Industry Applications*, **37**, 212–217.
20. Nielsen, J.G., Blaabjerg, F., and Mohan, N. (2001) Control strategies for dynamic voltage restorer compensating voltage sags with phase jump. Proceedings of the 16th AnnualIEEE Applied Power Electronics Conference and Exposition (APEC'01), vol. 2, pp. 1267–1273.
21. Xu, L., Acha, E., and Agelidis, V.G. (2001) A new synchronous frame based control strategy for a series voltage and harmonic compensator. Proceedings of the 16th Annual IEEE Applied Power Electronics Conference and Exposition (APEC'01), vol. 2, pp. 1274–1280.
22. Ghosh, A. and Joshi, A. (2002) A new algorithm for the generation of reference voltages of a DVR using the method of instantaneous symmetrical components. *IEEE Power Engineering Review*, **22**(1), 63–65.
23. Jung, H.-J., Suh, I.-Y., Kim, B.-S. *et al.* (2002) A study on DVR control for unbalanced voltage compensation. Proceedings of the 17th Annual IEEE Applied Power Electronics Conference and Exposition (APEC '02), vol. 2, pp. 1068–1073.
24. Ghosh, A. and Ledwich, G. (2002) Compensation of distribution system voltage using DVR. *IEEE Transactions on Power Delivery*, **17**(4), 1030–1036.
25. Ding, H., Shuangyan, S., Xianzhong, D., and Jun, G. (2002) A novel dynamic voltage restorer and its unbalanced control strategy based on space vector PWM. *International Journal of Electrical Power & Energy Systems*, **24**(9), 693–699.
26. Awad, H., Svensson, J., and Bollen, M.H.J. (2003) Static series compensator for voltage dips mitigation. Proceedings of the IEEE Bologna Power Tech Conference, June 23–26, vol. 3, pp. 23–26.
27. Aeloíza, E.C., Enjeti, P.N., Morán, L.A. *et al.* (2003) Analysis and design of a new voltage sag compensator for critical loads in electrical power distribution systems. *IEEE Transactions on Industry Applications*, **39**(4), 1143–1150.
28. Affolter, R. and Connell, B. (2003) Experience with a dynamic voltage restorer for a critical manufacturing facility. *Proceedings of the IEEE Transmission and Distribution Conference and Exposition*, **3**, 937–9310.
29. Chung, I.-Y., Won, D.-J., Park, S.-Y. *et al.* (2003) The DC link energy control method in dynamic voltage restorer system. *International Journal of Electrical Power & Energy Systems*, **25**(7), 525–531.
30. Liu, J.W., Choi, S.S., and Chen, S. (2003) Design of step dynamic voltage regulator for power quality enhancement. *IEEE Transactions on Power Delivery*, **18**(4), 1403–1409.
31. Dahler, P. and Knap, G. (2003) Protection of a dynamic voltage restorer. US Patent 6633092 (October 14).
32. Ghosh, A., Jindal, A.K., and Joshi, A. (2004) Design of a capacitor-supported dynamic voltage restorer (DVR) for unbalanced and distorted loads. *IEEE Transactions on Power Delivery*, **19**(1), 405–413.
33. Singh, B.N. and Simina, M. (2004) Intelligent solid-state voltage restorer for voltage swell/sag and harmonics. *IEE Proceedings: Electric Power Applications*, **151**(1), 98–106.
34. Kim, H., Kim, J.-H., and Sul, S.-K. (2004) A design consideration of output filters for dynamic voltage restorers. Proceedings of the 35th Annual IEEE Power Electronics Specialists Conference (PESC '04), June 20–25, pp. 4268–4272.
35. Lee, S.-J., Kim, H., Sul, S.-K., and Blaabjerg, F. (2004) A novel control algorithm for static series compensators by use of PQR instantaneous power theory. *IEEE Transactions on Power Electronics*, **19**(3), 814–827.
36. Lee, S.-J., Kim, H., and Sul, S.-K. (2004) A novel control method for the compensation voltages in dynamic voltage restorers. Proceedings of the 19th Annual IEEE Applied Power Electronics Conference and Exposition (APEC'04) vol. 1, pp. 614–620.
37. Nielsen, J. G. and Blaabjerg, F. (2005) A detailed comparison of system topologies for dynamic voltage restorers. *IEEE Transactions on Industry Applications*, **41**(5), 1272–1280.
38. Newman, M.J., Holmes, D.G., Nielsen, J.G., and Blaabjerg, F. (2005) A dynamic voltage restorer (DVR) with selective harmonic compensation at medium voltage level. *IEEE Transactions on Industrial Electronics*, **51**(6), 1744–1753.

39. Banaei, M.R., Hosseini, S.H., Khanmohamadi, S., and Gharehpetian, G.B. (2006) Verification of a new energy control strategy for dynamic voltage restorer by simulation. *Simulation Modelling Practice and Theory*, **14**(2), 112–125.
40. Libano, F.B., Muller, S.L., Marques, A.M. *et al.* (2006) Simplified control of the series active power filter for voltage conditioning. Proceedings of the IEEE International Symposium on Industrial Economics, July 9–12, pp. 1706–1712.
41. Banaei, M.R., Hosseini, S.H., and Gharehpetian, G.B. (2006) Inter-line dynamic voltage restorer control using a novel optimum energy consumption strategy. *Simulation Modelling Practice and Theory*, **14**(7), 989–9910.
42. Vilathgamuwa, D. M., Wijekoon, H.M., and Choi, S.S. (2006) A novel technique to compensate voltage sags in multiline distribution system: the interline dynamic voltage restorer. *IEEE Transactions on Industrial Electronics*, **53**(5), 1603–1611.
43. Jindal, A.K., Ghosh, A., and Joshi, A. (2010) Critical load bus voltage control using DVR under system frequency variation. *Electric Power Systems Research*, **78**(2), 255–263.
44. Singh, B., Jayaprakash, P., Kothari, D.P. *et al.* (2008) Indirect control of capacitor supported DVR for power quality improvement in distribution system. Proceedings of the Power and Energy Society General Meeting, Pittsburgh, PA, July 20–24.
45. Jayaprakash, P., Singh, B., and Kothari, D.P. (2010) Control of reduced-rating dynamic voltage restorer with a battery energy storage system. IEEE Transactions on Industry Applications, vol. 50, pp. 1295–1303.
46. Singh, B., Jayaprakash, P., and Kothari, D.P. (2010) Current mode control of dynamic voltage restorer (DVR) for power quality improvement in distribution system. Proceedings of the 2nd IEEE International Conference on Power and Energy (PECon'08), Johor Bahru, Malaysia, December 1–3, pp. 301–306.
47. Singh, B., Jayaprakash, P., and Kothari, D.P. (2010) Adaline based control of capacitor supported DVR for distribution system. *Journal of Power Electronics*, **9**(3), 386–395.
48. Anil Kumar, R., Siva Kumar, G., Kalyan Kumar, B., and Mishra, M.K. (2009) Compensation of voltage sags with phase-jumps through DVR with minimum VA rating using particle swarm optimization. Proceedings of the World Congress on Nature & Biologically Inspired Computing, December 9–11, pp. 1361–1366.
49. Teke, A., Bayindir, K., and Tümay, M. (2010) Fast sag/swell detection method for fuzzy logic controlled dynamic voltage restorer. *IET Generation, Transmission & Distribution*, **4**(1), 1–12.

6

Unified Power Quality Compensators

6.1 Introduction

The main objective of electric utilities is to supply their customers an uninterrupted sinusoidal voltage of constant magnitude and frequency with sinusoidal balanced currents at the AC mains. However, present-day AC distribution systems are facing severe power quality (PQ) problems such as high reactive power burden, unbalanced loads, harmonic-rich load currents, and an excessive neutral current. In addition, these utilities are not able to avoid the voltage sag, swell, surges, notches, spikes, flicker, unbalance, and harmonics in the supply voltages across the consumers' load end. There are many critical and sensitive loads that require uninterrupted sinusoidal balanced voltages of constant magnitude and frequency, otherwise their protection systems operate due to power quality disturbances. Moreover, these critical loads use solid-state controllers and precision devices such as computers, processors, and other sensitive electronic components and they draw reactive power and harmonic currents that cause load unbalance and result in excessive neutral current. Some examples of these critical and sensitive loads are hospital equipment (life support systems, operation theaters, patient database, etc.), banking systems using computers with UPS (uninterruptable power supplies), semiconductor manufacturing industries, pharmaceutical industries, textile industries, food processing plants, and so on. Even small interruption in the operation of these sensitive and critical loads because of voltage disturbances may cause substantial loss of money due to loss of production, time, product quality, and services. A custom power device known as a unified power quality compensator (UPQC) is considered the right option for such critical and sensitive loads to compensate both voltage- and current-based power quality problems.

The UPQC, a combination of shunt and series compensators shown in Figure 6.1, is recommended in the literature [1–50] as a single solution for mitigating these multiple PQ problems of voltages and currents. The power circuit of a UPQC consists of two voltage source converters (VSCs) or current source converters (CSCs) joined back to back by a common DC link capacitor or an inductor at the DC bus, respectively. The shunt device of the UPQC, also known as a DSTATCOM (distribution compensator), provides reactive power compensation, load balancing, neutral current compensation, and elimination of harmonics (if required), and it is connected in parallel to the consumer load or AC mains depending upon the configuration as a right shunt or left shunt UPQC, respectively. The series device of the UPQC, also known as a DVR (dynamic voltage restorer), keeps the consumer load end voltages insensitive to the supply voltage quality problems such as sag/swell, surges, spikes, notches, fluctuations, depression, and unbalance. The DVR injects a compensating voltage between the supply and the consumer load, and restores the load voltage to its reference value.

Power Quality Problems and Mitigation Techniques, First Edition.
Bhim Singh, Ambrish Chandra and Kamal Al-Haddad.

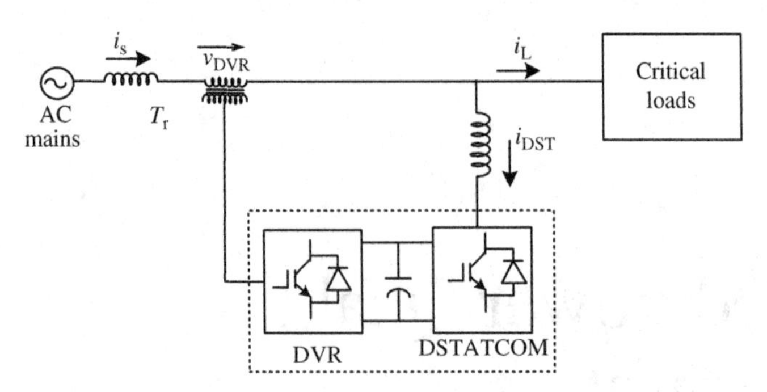

Figure 6.1 A VSC-based unified power quality compensator

This chapter deals with an exhaustive analysis and design of UPQCs. The UPQCs can be classified in many ways, such as supply-based classification (e.g., two-wire, three-wire, and four-wire UPQCs), converter-based classification (e.g., VSC- and CSC-based UPQCs), topology-based classification (e.g., right shunt UPQCs and left shunt UPQCs), classification based on the method of control (e.g., UPQC-Q: a DVR is used for series voltage injection in quadrature with supply current with almost zero active power injection; UPQC-P: a DVR is used for series voltage injection in phase with supply current with only an active power injection; UPQC-S: a DVR is used for series voltage injection at optimum phase angle with minimum kVA rating, S, or any other criterion decided by the requirement), and so on. There are many control algorithms for the control of UPQCs, such as IRPT (instantaneous reactive power theory), SRFT (synchronous reference frame theory), PBT (power balance theory), and ISCT (instantaneous symmetrical component theory). Starting with the introduction, the other sections include a state of the art on the UPQCs, their classification, principle of operation and control, analysis and design, modeling and simulation of performance, numerical examples, summary, review questions, numerical problems, computer simulation-based problems, and references.

6.2 State of the Art on Unified Power Quality Compensators

A UPQC, which is a combination of shunt and series compensators, is proposed as a single solution for mitigating multiple PQ problems. The power circuit of a UPQC consists of two VSCs joined back to back by a common DC link. The shunt device known as the DSTATCOM provides reactive power compensation along with load balancing, neutral current compensation, and elimination of harmonics (if required) and is positioned parallel to the consumer load. The series device known as the DVR keeps the load end voltage insensitive to the supply voltage quality problems such as sag/swell, surges, spikes, notches, or unbalance. The DVR injects a compensating voltage between the supply and the consumer load, and restores the load voltage to its reference value. The cost of PQ to manufacturing and emergency services together with the requirement of improved power quality in the current waveform justifies the cost and complex control required for UPQCs. There are many control techniques and topologies reported for the control of UPQCs.

Theory, modeling, control, and rating issues of UPQCs for load bus voltage control in distribution systems are reported in the literature [1–50]. An energy storage element at the DC bus is also used for the operation of an alternative unified power flow controller [6]. The DSTATCOM and DVR are controlled separately for power quality enhancement in the current and voltage, respectively. Most of the control algorithms reported for the DSTATCOM and DVR are applied to the UPQC. The instantaneous reactive power theory [6], synchronous reference frame theory [10], fuzzy control algorithm, instantaneous symmetrical component theory [50], and neural network theory, among others, are some control approaches reported in the literature.

The three-phase four-wire systems require a neutral current compensator along with other UPQC functions. A novel structure for a three-phase four-wire distribution system utilizing a unified power quality compensator is reported in the literature [45,46] and the control approaches of UPQCs for the three-phase four-wire system are also reported in the literature [48].

6.3 Classification of Unified Power Quality Compensators

UPQCs may be classified based on the type of converter used, topology configuration, supply system, and method of control, which affect their ratings. The converter can be either a CSC or a VSC. The topology depends on how the shunt device (DSTATCOM) and the series device (DVR) are connected to form a UPQC. For example, in a right shunt UPQC, the DSTATCOM is connected on the right-hand side of the DVR (connected across the consumer loads), and in a left shunt UPQC, the DSTATCOM is connected on the left-hand side of the DVR (connected across the PCC (point of common coupling)/AC mains). The topology of the UPQC may also differ depending on the internal configuration of the DSTATCOM and DVR. The third classification is based on the supply system, such as single-phase two-wire, three-phase three-wire, or three-phase four-wire UPQC systems. The fourth classification is based on the method of control, such as UPQC-Q (a DVR is used for series voltage injection in quadrature with supply current with almost zero active power injection), UPQC-P (a DVR is used for series voltage injection in phase with supply current with only an active power injection), and UPQC-S (a DVR is used for series voltage injection at optimum phase angle with minimum kVA rating, S, or any other criterion).

6.3.1 Converter-Based Classification of UPQCs

Two types of converters are used in the development of UPQCs. Figure 6.1 shows a UPQC using VSCs. VSC-based UPQCs have many advantages over CSC-based UPQCs. Figure 6.2 shows a UPQC using CSCs. A diode is used in series with the self-commutating device (IGBT: insulated gate bipolar transistor) for reverse voltage blocking. However, GTO (gate turn-off thyristor)-based CSC configurations of UPQCs do not need the series diode, but they have restricted frequency of switching. They are considered sufficiently reliable, but have higher losses and require higher values of parallel AC power capacitors or inductive energy storage at the DC bus, which is bulky, noisy, and costly and has high level of losses. Moreover, they cannot be used in multilevel or multistep modes to improve performance in higher ratings. Because of these reasons, VSC-based UPQCs have taken a lead in most of the applications.

6.3.2 Topology-Based Classification of UPQCs

UPQCs can also be classified based on the topology used, such as right shunt UPQCs and left shunt UPQCs. Figure 6.3 shows the basic configuration of a right shunt UPQC. Its DVR is connected before the

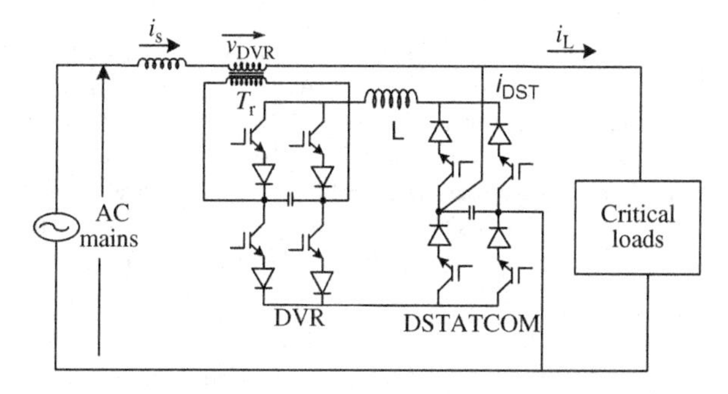

Figure 6.2 A two-wire CSC-based unified power quality compensator

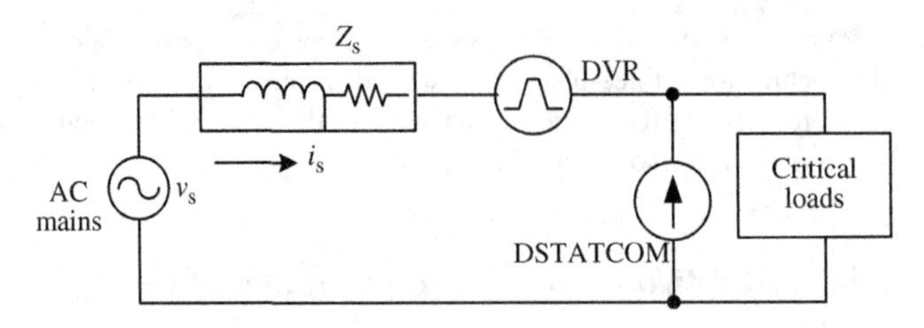

Figure 6.3 A right shunt UPQC as a combination of DSTATCOM and DVR

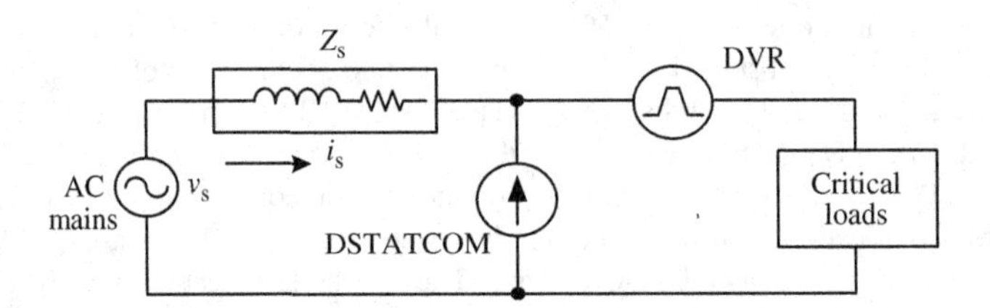

Figure 6.4 A left shunt UPQC as a combination of DSTATCOM and DVR

load in series with the AC mains, using a matching transformer, to mitigate sag, swell, spikes, and notches, to balance and regulate the terminal voltage across the consumer loads, and to eliminate voltage harmonics (if required). It has been used to eliminate negative-sequence voltage and to regulate the load voltage in three-phase systems. It can be installed by electric utilities to compensate voltage harmonics and to damp out harmonic propagation caused by resonance with line impedances and passive shunt compensators. It is considered a superior configuration as it has reduced ratings of both converters and requires simple control. Figure 6.4 shows a left shunt UPQC. The DC link storage element (either an inductor or a DC bus capacitor) is shared between two CSCs or VSCs operating as the DVR and DSTATCOM. It is considered an ideal compensator that mitigates voltage- and current-based power quality problems and is capable of giving clean power to critical and sensitive loads such as computers and medical equipment. It can balance and regulate the terminal voltage and eliminate negative-sequence currents. Its main drawbacks are high cost and control complexity. Therefore, a right shunt UPQC is considered a better option and dealt with in more detail.

6.3.3 Supply System-Based Classification of UPQCs

There are many consumer loads such as domestic appliances connected to single-phase supply systems. Some three-phase consumer loads are without neutral terminal, such as ASDs (adjustable speed drives) fed from three-wire supply systems. There are many single-phase consumer loads distributed on three-phase four-wire supply systems, such as computers and commercial lighting. Hence, UPQCs may also be classified according to supply systems as single-phase two-wire UPQCs, three-phase three-wire UPQCs, and three-phase four-wire UPQCs.

This classification of UPQCs is based on the supply and/or the load system having single-phase (two-wire) and three-phase (three-wire or four-wire) systems. A number of configurations of single-phase two-wire, three-phase three-wire, and three-phase four-wire UPQCs are given for enhancement of power quality in the currents as well as in the voltages. In three-phase four-wire UPQCs, various transformers may also be used for either isolating or deriving the fourth leg for neutral current compensation in the shunt connected VSC of the DSTATCOM, which may be a zigzag transformer, a T-connected transformer, a star/delta transformer, a star/hexagon transformer, and so on.

Two-Wire UPQCs

Two-wire (single-phase) UPQCs are used in both right shunt and left shunt UPQCs. Both converter configurations, a current source converter with PWM (pulse-width modulation) control having

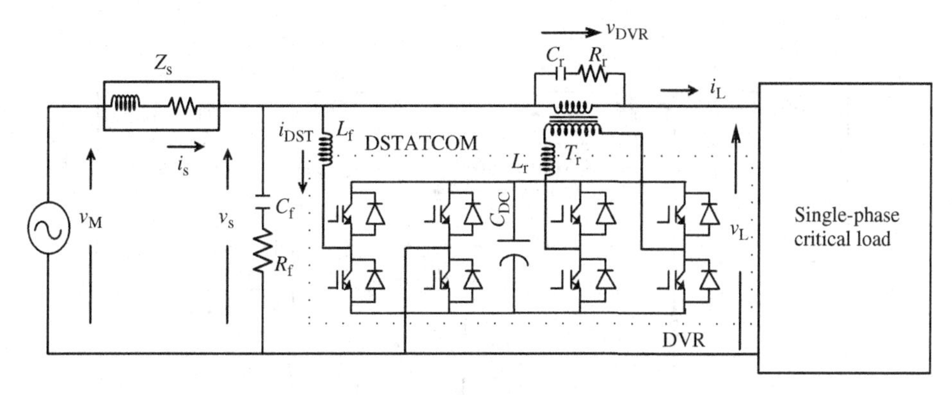

Figure 6.5 A single-phase left shunt UPQC

inductive energy storage elements and a voltage source converter with PWM control having capacitive DC bus energy storage elements, are used to form two-wire UPQC circuits. Figure 6.2 shows a configuration of a UPQC using current source converters with inductive energy storage elements. Similar configurations of UPQCs based on voltage source converters may be obtained by considering only two wires (phase and neutral) at each stage of Figures 6.5 and 6.6. In the case of DVR with a voltage source converter, sometimes the transformer is removed and the consumer load is shunted with passive L–C components.

Three-Wire UPQCs

Three-phase three-wire consumer loads such as ASDs are one of the major constituents of three-phase loads. All the configurations shown in Figures 6.1–6.6 are also developed, in three-wire UPQCs, with three wires on the AC side and two wires on the DC side. Figure 6.7 shows a classical VSC-based configuration of a three-phase three-wire UPQC. This most versatile configuration of UPQC is used in a number of applications. The use of ripple filters across both VSCs reduces switching ripples in the outputs of VSCs. Sometimes an isolation transformer is also used to isolate the shunt device, DSTATCOM, of the UPQC in high voltage rating applications to design power converters at optimum voltage and current ratings. However, there may be many configurations of three-phase three-wire UPQCs such as those using

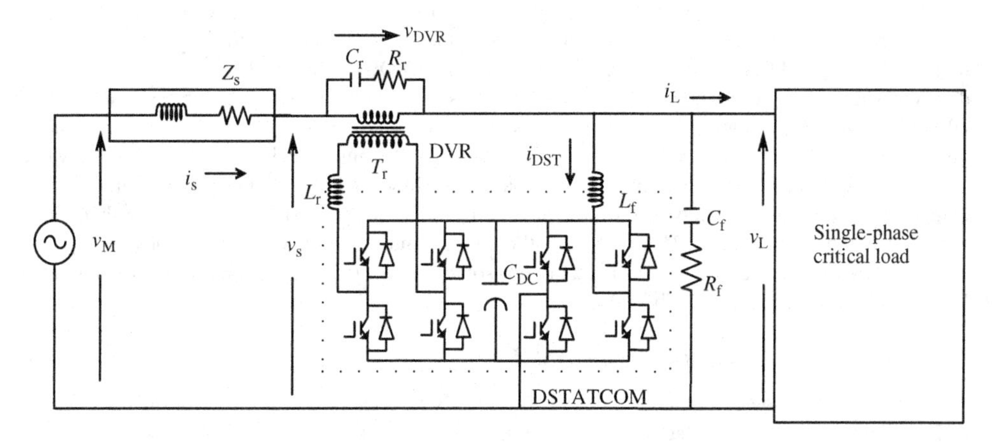

Figure 6.6 A single-phase right shunt UPQC

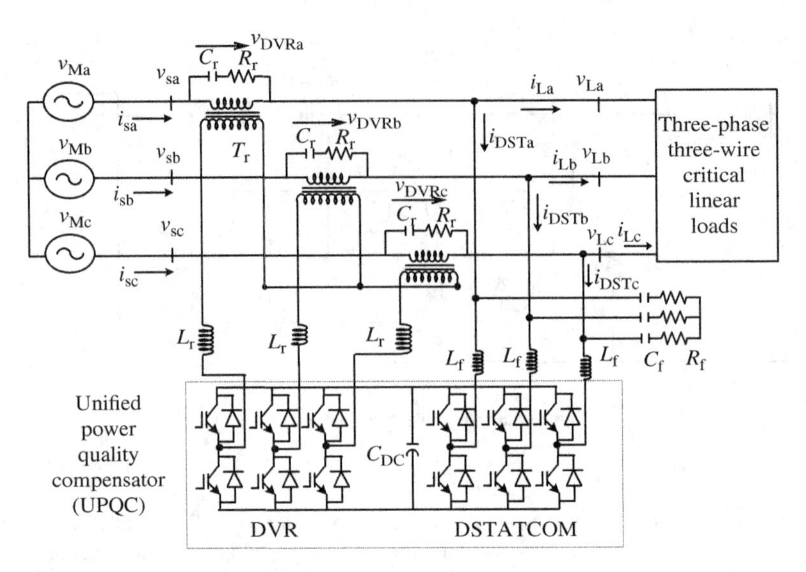

Figure 6.7 Three-phase three-wire right shunt UPQC topology with a three-leg VSC-based DSTATCOM and DVR

two-leg VSCs with midpoint DC bus capacitors to reduce the number of devices and the cost in low power rating UPQCs.

In a UPQC configuration, a three-leg VSC of the DSTATCOM is connected in shunt with the load as shown in Figure 6.7. The DC capacitor (C_{DC}) is connected at the DC link of the VSC of the DSTATCOM and its voltage is maintained at the desired reference DC bus voltage (v_{DC}) by controlling the VSC of the DSTATCOM. The VSC is connected using interfacing inductors (L_f). There is a ripple filter (C_f, R_f) connected at the load terminals of the UPQC in order to filter high-frequency components of PWM voltages from the VSC of the DSTATCOM. The VSC of the DVR is connected at the same DC link of the VSC of the DSTATCOM and is connected in series with the AC mains using three injection transformers (T_r). The ripple filter (L_r, C_r) is used to filter high-frequency components of PWM voltages from the VSC of the DVR. The DSTATCOM and DVR are controlled for improving the power quality in the PCC voltages as well as load currents.

Four-Wire UPQCs

A large number of single-phase loads may be supplied from three-phase AC mains with the neutral conductor. They cause excessive neutral current, harmonics, reactive power burden, and unbalance. To mitigate these problems, a set of three-phase four-wire UPQCs is used. These UPQCs have been developed in many configurations. Figure 6.8 shows the simplest configuration of a three-phase four-wire UPQC. In this configuration, the fourth wire may be replaced by a midpoint capacitor, used in smaller ratings. Here, the entire load neutral current flows through DC bus capacitors that are of large value. Another configuration may be a three single-phase VSC-based UPQC configuration, which is quite common, and this version allows the proper voltage matching for solid-state devices and enhances the reliability of the UPQC system. There are many three-phase four-wire UPQC configurations, in which the DSTATCOM may be realized with isolated or non-isolated transformers as the fourth leg by using three-leg VSCs and two-leg VSCs with midpoint capacitors and isolation transformers.

6.3.4 *Rating-Based Classification of UPQCs*

The fourth classification of UPQCs is based on the recently reported methods of control such as UPQC-Q (a DVR is used for series voltage injection in quadrature with supply current with almost zero active power injection), UPQC-P (a DVR is used for series voltage injection in phase with supply current with

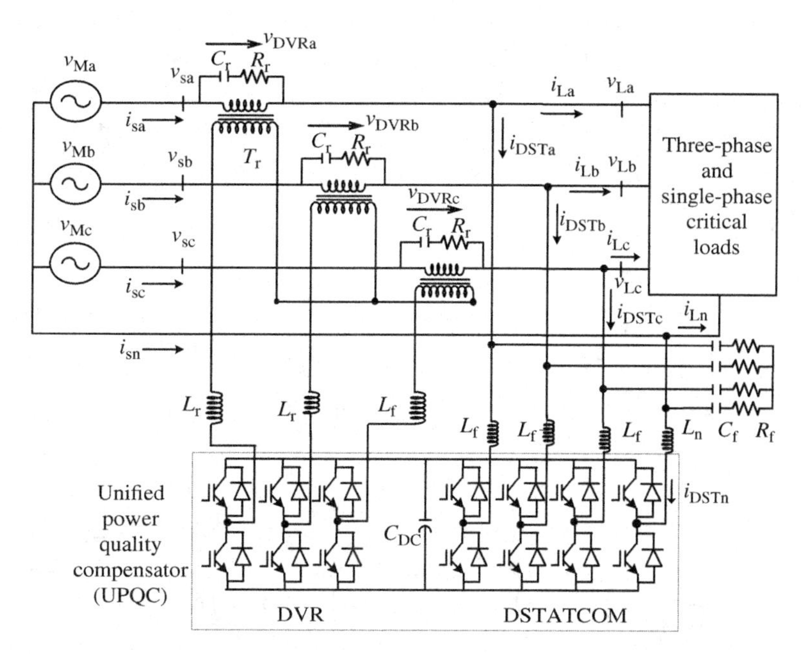

Figure 6.8 Three-phase four-wire right shunt UPQC topology with a four-leg VSC-based DSTATCOM and DVR

only an active power injection), and UPQC-S (a DVR is used for series voltage injection at optimum phase angle with minimum kVA rating, S, or any other criterion). These different types of control methods affect the ratings of the converters of UPQCs. The voltage sag on a system can be compensated through active power control and reactive power control methods. The phasor representation for voltage sag compensation using reactive power control in a UPQC-Q is shown in Figure 6.9a. For voltage swell compensation using a UPQC-Q (Figure 6.9b), the quadrature component injected by the series connected voltage source converter (DVR) does not intersect with the rated voltage locus. Therefore, a right-hand UPQC-Q cannot be used for swell compensation. Figure 6.9c and d depicts the compensation capability of a UPQC-P to compensate sag and swell in the system, respectively. Thus, the right shunt UPQC-Q approach is limited to compensate the sag in the system. However, the UPQC-P approach can effectively compensate both voltage sag and swell in the system. Furthermore, to compensate an equal percentage of sag, the UPQC-Q requires a larger magnitude of series injection voltage than the UPQC-P. Recently, the concept of UPQC-S with minimum VA rating of VSCs or phase angle control (PAC) of series voltage injection of the DVR has been introduced for different objectives. Figure 6.9e and f shows the voltage sag and swell compensation with series voltage injections at a suitable phase angle with required objectives, one of which may be used to reduce the overall VA rating of the UPQC.

UPQC-Q

In a UPQC-Q, the shunt compensator, DSTATCOM, is normally connected across consumer's load end as the right shunt UPQC. This DSTATCOM is used for all current-based compensation at the load end. In such a situation, the series compensator, DVR, injects the voltage in series between the AC mains and the load end in quadrature with the supply current and needs only reactive power from the series VSC (DVR) without any active power injection through the DVR with adequate voltage injection as shown in Figure 6.9a. This type of operation of UPQC-Q is quite satisfactory for voltage sag compensation. However, it cannot be used for voltage swell or voltage buck compensation as shown in Figure 6.9b. Since the DVR part of the UPQC injects only reactive power, this type of UPQC is known as the UPQC-Q.

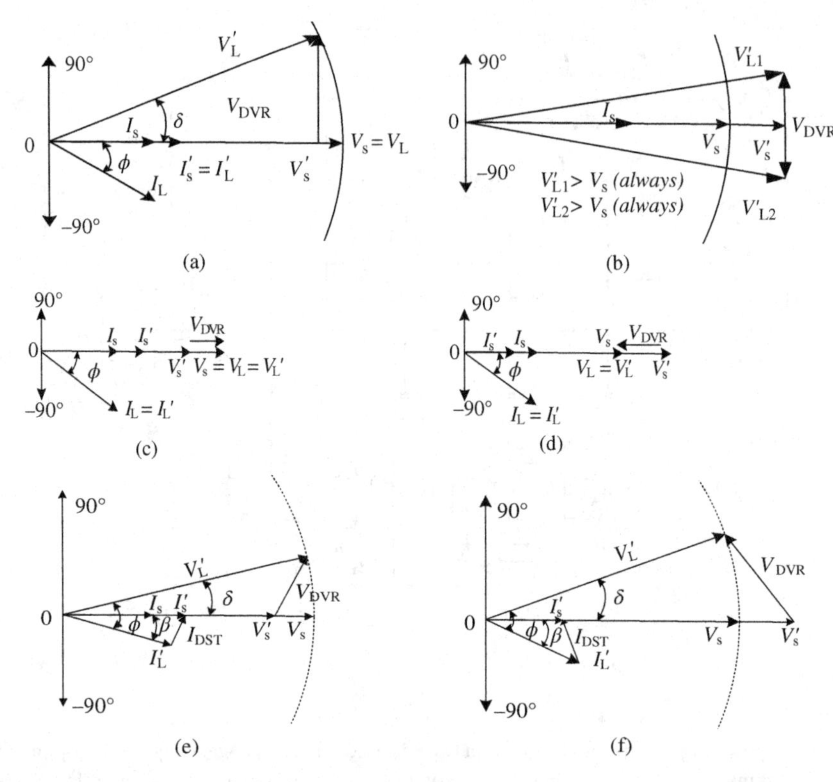

Figure 6.9 (a) Phasor diagram of a UPQC-Q for voltage sag compensation. (b) Phasor diagram of a UPQC-Q for voltage swell compensation. (c) Phasor diagram of a UPQC-P for voltage sag compensation. (d) Phasor diagram of a UPQC-P for voltage swell compensation. (e) Phasor diagram of a UPQC-S for voltage sag compensation. (f) Phasor diagram of a UPQC-S for voltage swell compensation

UPQC-P

In a UPQC-P, the shunt compensator, DSTATCOM, is used for all current-based compensation with unity power factor at the load end. In such a situation, the series compensator, DVR, injects a voltage in series between the AC mains and the load end in phase with the supply current/supply voltage and needs active power from the series VSC (DVR) without any reactive power injection through the DVR with minimum voltage injection as shown in Figure 6.9c and d. This type of operation of UPQC-P is quite satisfactory for both voltage sag and swell compensation. In this type of UPQC-P, the rating of the DVR is also quite low as it requires minimum series voltage injection. Since the DVR part of the UPQC injects active power, this type of UPQC is known as the UPQC-P.

UPQC-S

In a UPQC-S, the shunt compensator, DSTATCOM, is used for all current-based compensation other than full reactive power of the load/source at the load end. In such a situation, the series compensator, DVR, injects a voltage in series between the AC mains and the load end at a predetermined phase angle with the PCC voltage and needs both active power and reactive power through the series VSC (DVR) with the minimum VA rating of both VSCs or other criterion decided by the customer as shown in Figure 6.9e and f. This type of operation of UPQC-S is quite satisfactory for both voltage sag and swell compensation. In this type of UPQC-S, the ratings of both DVR and DSATCOM are minimized or utilized in proper coordination to supply reactive power of the system with proper sharing. Since the DVR part of the UPQC injects active and reactive powers with minimum VA rating, this type of UPQC is known as the UPQC-S. Its S denotes apparent power, which is the VA rating. This concept of UPQC-S with minimum VA rating of VSCs or phase angle control of the series voltage injection of the DVR has been introduced recently for different objectives.

6.4 Principle of Operation and Control of Unified Power Quality Compensators

Many configurations of UPQCs have been discussed in the previous section; for example, a CSC-based UPQC shown in Figure 6.2 and a VSC-based UPQC shown in Figures 6.5–6.8. Out of these two, VSC-based UPQCs are preferred due to a number of benefits of VSCs such as low passive filter requirements, low losses, and high switching frequency. Similarly, out of right shunt UPQCs and left shunt UPQCs, the former are preferred due to a number of benefits such as low losses, less circulation of power, easy and simple control, and better performance. Therefore, the principle of operation and control of UPQCs will be limited to VSC-based right shunt UPQCs, shown in Figures 6.6–6.8 for single-phase two-wire, three-phase three-wire, and three-phase four-wire configurations of UPQCs. Here, most of the concepts are given for three-phase UPQCs, which can also be applied to single-phase UPQCs.

6.4.1 Principle of Operation of UPQCs

The main objective of UPQCs is to mitigate multiple power quality problems in a distribution system. A UPQC mitigates most of the voltage quality problems such as sag, swell, surges, noise, spikes, notches, flicker, unbalance, fluctuations, regulation, and harmonics present in the supply/PCC system and a series compensator, DVR, provides clean, ideal, sinusoidal balanced voltages of constant magnitude at the consumer load end for satisfactory operation of the consumer equipment. At the same time, the shunt compensator of the UPQC, DSTATCOM, mitigates most of the current quality problems such as reactive power, unbalanced currents, neutral current, harmonics, and fluctuations present in the consumer loads or otherwise in the system and provides sinusoidal balanced currents in the supply, with its DC bus voltage regulation in proper coordination with the DVR.

In general, a UPQC has two VSCs connected to a common DC bus, one VSC is connected in series (known as the DVR or series compensator) of AC lines through an injection transformer and another VSC is connected in shunt (known as the DSTATCOM or shunt compensator) normally connected across the consumer loads or across the PCC as shown in Figures 6.6–6.8. Both the VSCs use PWM control; therefore, they require small ripple filters to mitigate switching ripples. They require Hall effect voltage and current sensors for feedback signals and normally a digital signal processor (DSP) is used to implement the required control algorithm to generate gating signals for the solid-state devices of both VSCs of the UPQC. The series VSC used as the DVR is normally controlled in PWM voltage control mode to inject appropriate voltages in series with the AC mains and the shunt VSC used as the DSTATCOM is normally controlled in PWM current control mode to inject appropriate currents in parallel with the load in the system. The UPQC also needs many passive elements such as a DC bus capacitor, AC interacting inductors, injection and isolation transformers, and small passive filters.

6.4.2 Control of UPQCs

As mentioned earlier, the criteria for the control of UPQCs are divided into three categories: UPQC-Q, UPQC-P, or UPQC-S. Reference signals for the control of both components of the UPQC, namely, DSTATCOM and DVR, have to be derived accordingly using a number of control algorithms normally used for the control of the DSTATCOM and DVR. There are more than a dozen of control algorithms that are used for the control of the DSTATCOM and DVR. A few of these control algorithms are as follows:

- Synchronous reference frame theory, also known as d–q theory
- Instantaneous reactive power theory, also known as PQ theory or α–β theory

- Instantaneous symmetrical component theory
- Power balance theory (BPT)
- Neural network theory (Widrow's LMS-based Adaline algorithm)
- PI controller-based algorithm
- Current synchronous detection (CSD) method
- I cos φ algorithm
- Single-phase PQ theory
- Enhanced phase locked loop (EPLL)-based control algorithm
- Conductance-based control algorithm
- Adaptive detecting algorithm, also known as adaptive interference canceling theory

These control algorithms are time-domain control algorithms. Most of them have been used for the control of the DSTATCOM and DVR.

Similarly, there are around the same number of frequency-domain control algorithms. Some of them are as follows:

- Fourier series theory
- Discrete Fourier transform theory
- Fast Fourier transform theory
- Recursive discrete Fourier transform theory
- Kalman filter-based control algorithm
- Wavelet transformation theory
- Stockwell transformation (S-transform) theory
- Empirical decomposition (EMD) transformation theory
- Hilbert–Huang transformation theory

These control algorithms are frequency-domain control algorithms. Most of them are used for power quality monitoring for a number of purposes in the power analyzers, PQ instruments, and so on. Some of these algorithms have been used for the control of the DSTATCOM and DVR. However, these algorithms are sluggish and slow, requiring heavy computation burden; therefore, these control methods are not too much preferred for real-time control of these compensators as time-domain control algorithms.

All these control algorithms may be used in the control of UPQCs. Some of these control algorithms are explained in earlier chapters in detail. However, because of space limitation and to give just a basic understanding, only one control algorithm, namely, synchronous reference frame theory, is explained here. However, more relevant are the three control criteria of UPQC-Q, UPQC-P, and UPQC-S used in the control of UPQCs for the coordination of the functions of two elements of UPQCs, namely, DSTATCOM and DVR.

6.4.2.1 Control Criteria of UPQCs

There may be a number of criteria for the control of UPQCs similarly to a UPFC (unified power flow controller) in transmission systems, which is mainly used to control the active power flow in both the directions with an injection of the series voltage with active and reactive powers. The UPFC is a very versatile power controller used for the control of active and reactive powers in four quadrants. Similarly, the UPQC may also be controlled to inject a series voltage between the AC mains and consumer loads through a series compensator (DVR) at any angle from 0° to 360° from any reference angle but at the cost of both active and reactive power injections. However, the main functions of the UPQC are load compensation (load balancing, harmonic current elimination, neutral current mitigation, and reactive power compensation), using a shunt device (DSTATCOM), and mitigation of voltage-based power quality problems present at PCC, using a series device (DVR), to provide clean and sinusoidal constant voltage across the consumer loads. In addition, the DVR may also be used for part of reactive power compensation of the system in coordination with the DSTATCOM. Moreover, if the DVR needs some active power for series injection, then the DSTATCOM has to provide it through a combined DC bus of both compensators. Therefore, there are many ways of coordination of these devices of UPQCs. So far, three criteria have been mainly considered, which are used here and illustrated in the following sections.

Control of UPQC as UPQC-Q

This control methodology of UPQCs has been perceived first because of similarity with the control of DVRs. In this mode, a DVR is used mainly for reactive power injection for fundamental voltage compensation. In this case, series voltage is injected in quadrature with the AC mains current or the current flowing through the DVR so that no active power is needed for series voltage injection.

As evident from Figure 6.9b, the DVR cannot compensate for voltage swell in the UPQC-Q mode of operation. In this case, the DSTATCOM compensates the reactive power of the consumer load; of course, a unity power factor is realized at PCC and total reactive power of the system is compensated by the DSTATCOM. Moreover, the DVR injects a voltage in quadrature with the AC mains current. In the case of voltage sag compensation also, there is a limitation of sag compensation depending upon the voltage rating of the DVR. From Figure 6.9a, the voltage sag can be expressed as

$$X = \{V_{Lc} - \sqrt{(V_{Lc}^2 - V_{DVR}^2)}\}/V_{Lc}, \tag{6.1}$$

where X is the percentage of voltage sag for compensation, V_{DVR} is the injected voltage by the DVR, and V_{Lc} is the load voltage after compensation (normally it is the rated load voltage or rated PCC voltage).

In this case, if the maximum or rated injected voltage (V_{DVRR}) is known, then from Equation 6.1 one can calculate the maximum level of voltage sag that can be compensated in this mode of operation. Of course, in this case now the DSTATCOM has to supply some amount of active power to meet the losses of the UPQC, but it does not supply active power to the DVR as the DVR does not inject any active power. Therefore, this mode of UPQC is considered most conservative and provides minimum level of compensation.

The rating calculation of both DVR and DSTATCOM may be done from the phasor diagram shown in Figure 6.9a as follows.

The injected voltage by the DVR is

$$V_{DVR} = \sqrt{(V_{Lc}^2 - V_s^2)}. \tag{6.2}$$

The VA rating of the DVR is

$$S_{DVR} = V_{DVR} I_s = V_s I_L \cos\phi \tan\delta, \tag{6.3}$$

where I_s and I_L are the supply and load currents, respectively, $\cos\phi$ is the load power factor, and δ is the angle between the load voltage and the PCC voltage after compensation. $\tan\delta = \sqrt{(V_{Lc}^2 - V_s^2)}/V_s$.

The current rating of the DSTATCOM is

$$I_{DST} = I_L[\sqrt{\{(1-X)^2 + \cos^2\phi - 2\cos\phi\cos(\phi-\delta)(1-X)\}}]/(1-X). \tag{6.4}$$

The VA rating of the DSTATCOM is

$$S_{DST} = V_{Lc} I_{DST} = V_{Lc} I_L[\sqrt{\{(1-X)^2 + \cos^2\phi - 2\cos\phi\cos(\phi-\delta)(1-X)\}}]/(1-X). \tag{6.5}$$

Therefore, total VA rating of the UPQC is

$$S_{UPQC\text{-}Q} = S_{DVR} + S_{DST}. \tag{6.6}$$

For real-time control, implementation of this mode of UPQC is now quite evident. For the control of the DSTATCOM part of the UPQC, sensing of PCC voltages, load currents, supply currents, and DC bus voltage is required to derive reference supply currents for the PWM current controller in SRF theory. The supply currents have to supply only active power, which has two parts: one as load active power and another to meet losses of the UPQC. The active power component of load currents is estimated by SRF transformation of load currents and then filtered for its DC component, which represents the required active power component of load currents. For this SRF transformation, the transformation angle is obtained from PLL (phase locked loop) over positive-sequence PCC voltages. The other part of the active power component of supply currents is estimated using a PI (proportional–integral) controller over the DC bus of the UPQC. The sum of these active power components of supply currents is considered as the amplitude of reference supply currents. This amplitude of reference supply currents and the transformation angle obtained from PLL over positive-sequence PCC voltages are used to compute three-phase instantaneous reference supply currents. These three-phase reference supply currents and sensed supply currents are used in PWM current controllers for generating gating signals of IGBTs of the DSTATCOM.

For the control of the DVR, reference injected voltages are computed from the difference of reference load voltages and sensed PCC voltages. Three-phase reference load voltages are computed from a desired/reference amplitude of load voltages and the transformation angle obtained from PLL over sensed DVR/supply currents. These three-phase reference injected voltages are given to a PWM voltage controller for generating gating signals of IGBTs of the DVR.

Control of UPQC as UPQC-P

This control algorithm of UPQCs has been conceptualized for the minimum voltage injection by the DVR. As evident from phasor diagrams given in Figure 6.9c and d, the DVR injects minimum voltages for both conditions of voltage sag and swell. In this mode, the DVR needs totally active power for voltage injection of series compensation. In this case, the series voltage is injected in phase with the AC mains current or PCC voltages, thus requiring only active power that has to be fed or received by the DSTATCOM through the DC bus. It increases the current and thus the kVA rating of the DSTATCOM. However, the kVA rating of the DVR is minimal in this operating mode of UPQC as a UPQC-P. Moreover, as evident from Figure 6.9c and d, the DVR can compensate for voltage sag and swell in the UPQC-P mode of operation up to rated voltage rating of the DVR. In this case also, the DSTATCOM compensates the reactive power of the consumer load; of course, a unity power factor is realized at PCC and total reactive power of the system is compensated by the DSTATCOM. From Figure 6.9c and d, the voltage sag can be expressed as

$$X = |V_{\mathrm{Lc}} - V_{\mathrm{s}}|/V_{\mathrm{Lc}} = |V_{\mathrm{DVR}}|/V_{\mathrm{Lc}}. \tag{6.7}$$

Therefore, the voltage rating of the compensator may be achieved from the required maximum value of sag or swell compensation. The UPQC-P mode of operation requires minimum injection voltage and thus minimum VA rating of the DVR but at the expense of the DSTATCOM rating, which increases due to large active power flow through the DSTATCOM. Therefore, this type of UPQC does not have minimum overall VA rating consisting of both elements of the UPQC.

The rating calculations of both DVR and DSTATCOM may be done from the phasor diagram shown in Figure 6.9c as follows.

The injected voltage by the DVR is

$$V_{\mathrm{DVR}} = V_{\mathrm{Lc}} - V_{\mathrm{s}} = V_{\mathrm{Lc}}X. \tag{6.8}$$

The VA rating of the DVR is totally an active power and is expressed as

$$S_{\mathrm{DVR}} = V_{\mathrm{DVR}}I_{\mathrm{s}} = |V_{\mathrm{Lc}} - V_{\mathrm{s}}|I_{\mathrm{L}} \cos\phi = V_{\mathrm{Lc}}XI_{\mathrm{L}} \cos\phi, \tag{6.9}$$

where I_{s} and I_{L} are supply and load currents, respectively, and $\cos\phi$ is the load power factor.

The current rating of the DSTATCOM is

$$I_{\mathrm{DST}} = \sqrt{[\{\sqrt{(I_{\mathrm{L}}^2 - I_{\mathrm{s}}^2)}\}^2 + (XI_{\mathrm{L}} \cos\phi)^2]}. \tag{6.10}$$

The VA rating of the DSTATCOM is

$$S_{\mathrm{DST}} = V_{\mathrm{Lc}} I_{\mathrm{DST}} = V_{\mathrm{Lc}} \sqrt{[\{\sqrt{(I_{\mathrm{L}}^2 - I_{\mathrm{s}}^2)}\}^2 + (XI_{\mathrm{L}} \cos\phi)^2]}. \tag{6.11}$$

Therefore, total VA rating of the UPQC is

$$S_{\mathrm{UPQC\text{-}Q}} = S_{\mathrm{DVR}} + S_{\mathrm{DST}}. \tag{6.12}$$

For real-time control, implementation of this mode of UPQC is also straightforward similarly to the DSTATCOM controlled in the UPQC-Q mode of operation. Therefore, it need not be discussed here again.

For the control of the DVR, reference injected voltages are computed from the difference of reference load voltages and sensed PCC voltages. Three-phase reference load voltages are computed from a desired/reference amplitude of load voltages and the transformation angle obtained from PLL over sensed PCC voltages as these three-phase reference load voltages are to be in phase with PCC voltages. These three-phase reference injected voltages are given to a PWM voltage controller for generating gating signals of IGBTs of the DVR.

Control of UPQC as UPQC-S

This control approach of UPQCs has been featured as the most generalized one with a number of objectives such as for the overall minimum VA of both DSTATCOM and DVR, full utilization of DVR rating with reduced burden on the DSTATCOM or optimum sharing of rating between the DVR and DSTATCOM. As evident from the phasor diagrams shown in Figure 6.9e and f, the DVR may inject voltages for both conditions of voltage sag and swell compensation. In the UPQC-S mode of operation, the DSTATCOM is normally used for all current-based compensation other than full reactive power of the system. In this case, the DVR injects a voltage in series between the AC mains and the load end at a predetermined phase angle with the PCC voltage and it needs both active power and reactive power through the DVR. This concept of UPQC-S with the phase angle control of the series voltage injection of the DVR has been perceived very recently for different objectives. Some of the objectives of UPQC-S operation are given here.

The first objective of this mode of control of UPQCs is based on the full utilization of DVR rating. Usually, VSCs of same VA rating are used for both DVR and DSTATCOM to reduce the inventory by the manufacturers and customers as identical VSCs are selected for both components of the UPQC. In general, the DVR is used with an injection transformer, but a transformer is not mandatory for the DSTATCOM unless the customer needs it. The control of this mode of UPQC is as follows.

The voltage injected by the DVR is its rated voltage ($V_{\mathrm{DVR}} = V_{\mathrm{R}}$) and the angle ψ_{DVR} at which it is to be injected with respect to the PCC voltage (shown in Figure 6.10) is

$$\psi_{\mathrm{DVR}} = \pi - \cos^{-1}\{(V_{\mathrm{s}}^2 + V_{\mathrm{DVR}}^2 - V_{\mathrm{Lc}}^2)/2V_{\mathrm{s}}V_{\mathrm{DVR}}\}, \tag{6.13}$$

where V_{s} is the PCC voltage under voltage sag condition and V_{Lc} is the load voltage after compensation.

The active power of the DVR is

$$P_{\mathrm{DVR}} = V_{\mathrm{DVR}} I_{\mathrm{s}} \cos\psi_{\mathrm{DVR}}. \tag{6.14}$$

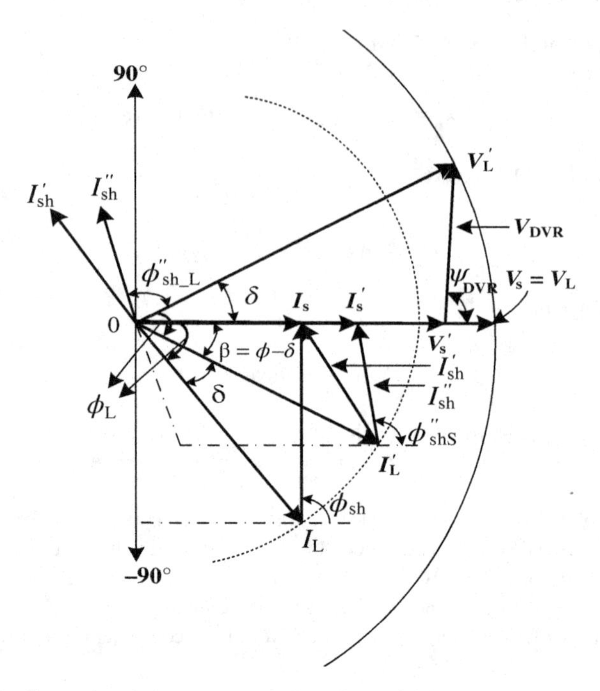

Figure 6.10 Current-based phasor representation of a UPQC-S under voltage sag condition

The reactive power of the DVR is

$$Q_{DVR} = V_{DVR} I_s \sin \psi_{DVR}. \tag{6.15}$$

This condition of DVR is considered for the compensation by the DVR with minimum active power and maximum reactive power with full utilization of its rating as $V_{DVR} = V_R$.

The VA rating of the DVR is

$$S_{DVR} = V_{DVR} I_s. \tag{6.16}$$

Let δ is the angle between the load voltage V_L and PCC voltage V_s after compensation. It can be calculated from the phasor diagram of Figure 6.10 as follows:

$$\delta = \sin^{-1}\{(V_{DVR} \sin \psi_{DVR})/V_L\}. \tag{6.17}$$

The angle β between the PCC voltage and the consumer load current after compensation is

$$\beta = \phi - \delta. \tag{6.18}$$

The current rating of the DSTATCOM is

$$I_{DST} = \sqrt{\{(I_L \sin \beta)^2 + [I_L\{(\cos \beta) - (\cos \phi)/(1 - X)\}]^2\}}. \tag{6.19}$$

The angle ψ_{DST} between the DSTATCOM current and the PCC voltage is

$$\psi_{DST} = (\pi/2) + \tan^{-1}[\{(\cos \beta) - (\cos \phi)/(1 - X)\}/\sin \beta]. \tag{6.20}$$

The active power of the DSTATCOM is

$$P_{\mathrm{DST}} = V_{\mathrm{s}} I_{\mathrm{DST}} \cos \psi_{\mathrm{DST}}. \tag{6.21}$$

The reactive power of the DSTATCOM is

$$Q_{\mathrm{DST}} = V_{\mathrm{s}} I_{\mathrm{DST}} \sin \psi_{\mathrm{DST}}. \tag{6.22}$$

The VA rating of the DSTATCOM is

$$S_{\mathrm{DST}} = V_{\mathrm{s}} I_{\mathrm{DST}}. \tag{6.23}$$

Therefore, total VA rating of the UPQC is

$$S_{\mathrm{UPQC\text{-}S}} = S_{\mathrm{DVR}} + S_{\mathrm{DST}}. \tag{6.24}$$

This condition of UPQC-S reduces the burden on the DSTATCOM rating as some part of reactive power of the load is supplied by the DVR.

This is one case of phase angle control of the UPQC-S. Similarly, there may be many criteria for UPQC-S control as minimum VA rating, which does not have any closed-form solution, and an optimum angle of the injection voltage are to be computed either iteratively or through some method of optimization.

6.4.2.2 Synchronous Reference Frame Theory for the Control of UPQCs

There are a large number of control algorithms that are used for the control of the DSTATCOM and DVR. Among the above-mentioned real-time control techniques for the generation of reference signals for the control of VSCs of the UPQC for a three-phase four-wire system, here only the SRFT is given with the indirect control for the operation of both the DSTATCOM and the DVR. A three-phase four-wire system covers the concept for both three-wire and four-wire systems; hence, the control algorithm for a three-phase four-wire system is discussed here. The SRFT control algorithm for the operation of the DSTATCOM and DVR is discussed in the following sections.

Control of the DSTATCOM of the UPQC

The control algorithm for the three-leg VSC-based DSTATCOM of the UPQC based on the synchronous reference frame theory is used for the control of the DSTATCOM. The objective of the DSTATCOM is to enhance the power quality of the supply current as well as to support the common DC bus of the DSTATCOM and the DVR by absorbing active power. A block diagram of the control scheme of the DSTATCOM is shown in Figure 6.11. The load currents (i_{La}, i_{Lb}, i_{Lc}), PCC voltages (v_{sa}, v_{sb}, v_{sc}), and DC bus voltage (v_{DC}) of the UPQC are sensed as feedback signals. The load currents from the abc frame are first converted to the dq0 frame as follows:

$$\begin{bmatrix} i_{\mathrm{Lq}} \\ i_{\mathrm{Ld}} \\ i_{\mathrm{L0}} \end{bmatrix} = \frac{2}{3} \begin{bmatrix} \cos\theta & \cos\left(\theta - \dfrac{2\pi}{3}\right) & \cos\left(\theta + \dfrac{2\pi}{3}\right) \\ \sin\theta & \sin\left(\theta - \dfrac{2\pi}{3}\right) & \sin\left(\theta + \dfrac{2\pi}{3}\right) \\ \dfrac{1}{2} & \dfrac{1}{2} & \dfrac{1}{2} \end{bmatrix} \begin{bmatrix} i_{\mathrm{La}} \\ i_{\mathrm{Lb}} \\ i_{\mathrm{Lc}} \end{bmatrix}, \tag{6.25}$$

where $\cos\theta$ and $\sin\theta$ are obtained using a three-phase PLL over PCC voltages. A PLL signal is obtained from PCC voltages for generating fundamental unit vectors for the conversion of sensed currents to the dq0 reference frame. The SRF controller extracts DC quantities by low-pass filters (LPFs) and hence

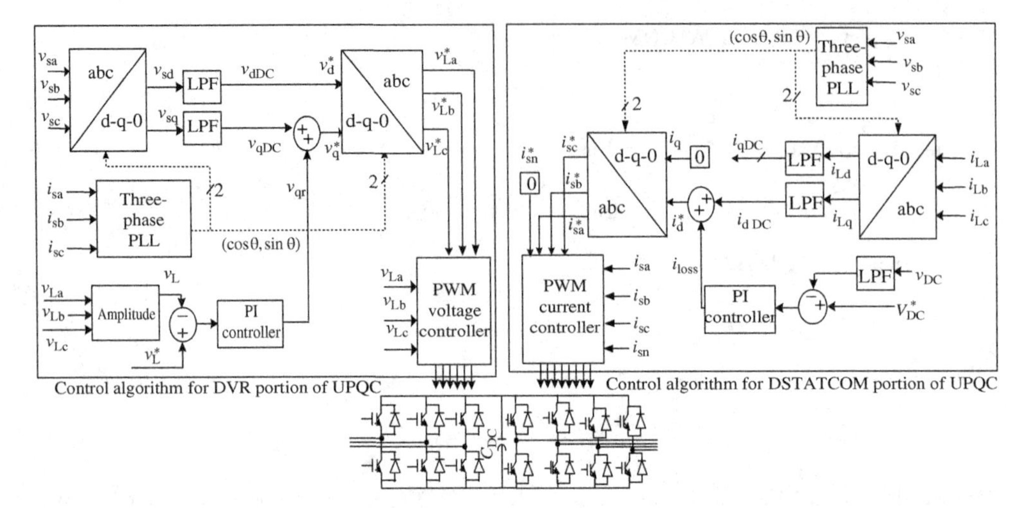

Figure 6.11 Control algorithm for a three-phase four-wire UPQC topology with a four-leg VSC-based DSTATCOM and a three-leg VSC-based DVR

the non-DC quantities (ripple) are separated from the reference signal. The d-axis and q-axis currents consist of DC (fundamental) and ripple (negative sequence and harmonics) components:

$$i_{Ld} = i_{dDC} + i_{dAC}, \tag{6.26}$$

$$i_{Lq} = i_{qDC} + i_{qAC}. \tag{6.27}$$

This control strategy considers that the AC mains must deliver the mean value of the direct-axis component of the load current along with the active power component of the current for maintaining the DC bus voltage and meeting the losses (i_{loss}) in the UPQC. The output of the PI controller at the DC bus voltage of the UPQC is considered as the current (i_{loss}) for meeting its losses:

$$i_{loss}(n) = i_{loss}(n-1) + K_{pd}\{v_{de}(n) - v_{de}(n-1)\} + K_{id}v_{de}(n), \tag{6.28}$$

where $v_{de}(n) = v_{DC}^* - v_{DC}(n)$ is the error between the reference (v_{DC}^*) and the sensed (v_{DC}) DC voltage at the nth sampling instant, and K_{pd} and K_{id} are the proportional and integral gain constants of the DC bus voltage PI controller, respectively.

Therefore, the amplitude of reference supply current is

$$i_d^* = i_{dDC} + i_{loss}. \tag{6.29}$$

The reference supply current must be in phase with the voltage at PCC but with no zero-sequence component. It is therefore obtained by the following reverse Park's transformation with i_q^* and i_0^* as zero. The resultant dq0 currents are again converted into the reference supply currents using the reverse Park's transformation.

$$\begin{bmatrix} i_{sa}^* \\ i_{sb}^* \\ i_{sc}^* \end{bmatrix} = \begin{bmatrix} \cos\theta & \sin\theta & 1 \\ \cos\left(\theta - \frac{2\pi}{3}\right) & \sin\left(\theta - \frac{2\pi}{3}\right) & 1 \\ \cos\left(\theta + \frac{2\pi}{3}\right) & \sin\left(\theta + \frac{2\pi}{3}\right) & 1 \end{bmatrix} \begin{bmatrix} i_q^* \\ i_d^* \\ i_0^* \end{bmatrix}. \tag{6.30}$$

The reference supply neutral current (i_{sn}^*) is set to zero for neutral current compensation.

Control of the DVR of the UPQC

The control block of the DVR of the UPQC is also shown in Figure 6.11, in which the synchronous reference frame theory is used for reference signal generation. The PCC voltages (v_s), supply currents (i_s), and load terminal voltages (v_L) are sensed for deriving the gate signals of IGBTs. The PCC voltages are converted to the rotating reference frame using the abc–dq0 conversion using the Park's transformation with unit vectors (sin θ, cos θ) derived from the supply currents using a PLL as follows:

$$\begin{bmatrix} v_{sq} \\ v_{Sd} \\ v_{s0} \end{bmatrix} = \frac{2}{3}\begin{bmatrix} \cos\theta & \cos\left(\theta-\frac{2\pi}{3}\right) & \cos\left(\theta+\frac{2\pi}{3}\right) \\ \sin\theta & \sin\left(\theta-\frac{2\pi}{3}\right) & \sin\left(\theta+\frac{2\pi}{3}\right) \\ \frac{1}{2} & \frac{1}{2} & \frac{1}{2} \end{bmatrix}\begin{bmatrix} v_{sa} \\ v_{sb} \\ v_{sc} \end{bmatrix}. \tag{6.31}$$

Then, d-axis and q-axis voltages consist of DC (fundamental) and ripple (harmonic) components:

$$v_{sd} = v_{dDC} + v_{LdAC}, \tag{6.32}$$

$$v_{sq} = v_{qDC} + v_{LqAC}. \tag{6.33}$$

The amplitude of AC load terminal voltage (V_L) is controlled to its reference voltage (V_L^*) using a PI controller and the output of the PI controller is considered as the voltage (v_{qr}) to be injected by the DVR. The amplitude of AC load terminal voltage (V_L) is calculated from the load terminal voltages (v_{La}, v_{Lb}, v_{Lc}) as follows:

$$V_L = (2/3)^{1/2}(v_{La}^2 + v_{Lb}^2 + v_{Lc}^2)^{1/2}. \tag{6.34}$$

Then, a PI controller is used to regulate this voltage to a reference value as follows:

$$v_{qr}(n) = v_{qr}(n-1) + K_{pq}\{v_{te}(n) - v_{te}(n-1)\} + K_{iq}v_{te}(n), \tag{6.35}$$

where $v_{te}(n) = V_L^*(n) - V_L(n)$ denotes the error between reference ($V_L^*(n)$) and sensed ($V_L(n)$) terminal voltage amplitudes at the nth sampling instant, and K_{pq} and K_{iq} are the proportional and integral gain constants of the DC bus voltage PI controller, respectively.

The reference quadrature-axis load voltage is

$$v_{Lq}^* = v_{qDC} + v_{qr}. \tag{6.36}$$

The reference direct-axis load voltage is

$$v_{Ld}^* = v_{dDC}. \tag{6.37}$$

The reference load voltage in the abc frame is obtained by reverse Park's transformation with v_{Ld}^* as in Equation 6.37, v_{Lq}^* as in Equation 6.36, and v_{L0}^* as zero:

$$\begin{bmatrix} v_{La}^* \\ v_{Lb}^* \\ v_{Lc}^* \end{bmatrix} = \begin{bmatrix} \cos\theta & \sin\theta & 1 \\ \cos\left(\theta-\frac{2\pi}{3}\right) & \sin\left(\theta-\frac{2\pi}{3}\right) & 1 \\ \cos\left(\theta+\frac{2\pi}{3}\right) & \sin\left(\theta+\frac{2\pi}{3}\right) & 1 \end{bmatrix}\begin{bmatrix} v_{Lq}^* \\ v_{Ld}^* \\ v_{L0}^* \end{bmatrix}. \tag{6.38}$$

The reference load voltages (v_{La}^*, v_{Lb}^*, v_{Lc}^*) and the sensed load voltages (v_{La}, v_{Lb}, v_{Lc}) are used in a PWM controller to generate gating pulses to the VSC of the DVR.

6.5 Analysis and Design of Unified Power Quality Compensators

The design of a three-phase four-wire UPQC includes the design of the DSTATCOM and DVR. The DSTATCOM includes a VSC, interfacing inductors, and a ripple filter. The design of the VSC includes the DC bus voltage level, the DC capacitance, and the rating of IGBTs used in VSCs. Similarly, the design of DVR includes design of the VSC, interfacing inductors, ripple filters, and injection transformers.

A three-phase four-wire UPQC topology is considered for detailed analysis. Figure 6.8 shows the schematic diagram of one of the UPQCs for a three-phase four-wire distribution system. It uses a four-leg VSC-based DSTATCOM. The designs of the DSTATCOM and DVR of the UPQC are discussed in the following sections.

6.5.1 Design of a DSTATCOM

Figure 6.8 shows the schematic diagram of one of the UPQCs for a three-phase four-wire distribution system. It uses a four-leg VSC-based DSTATCOM. The VSC, interfacing inductors, and the ripple filter are designed here. The design is carried out considering a right shunt UPQC-P. The supply system and the load considered are as follows.

Data for the three-phase UPQC are as follows: AC line voltage: 415 V, 50 Hz. Line impedance: $R_s = 0.1\,\Omega$, $L_s = 1.0\,\text{mH}$.

Load: linear balanced three-phase four-wire load: 30 kW, 0.80 lagging power factor.

The DC capacitor connected at the common DC bus acts as an energy buffer and establishes a DC voltage for the normal operation of the DSTATCOM as well as the DVR. The DC bus voltage (v_{DC}), DC bus capacitance (C_{DC}), interfacing inductors (L_f), and voltage and current ratings of switches are calculated as per the procedure given below.

A four-leg VSC is used as a DSTATCOM and this topology has eight IGBTs, four AC inductors, and a DC capacitor. The required compensation to be provided by the DSTATCOM decides the rating of the VSC components. The VSC is designed for compensating a reactive power of 22.5 kVAR. The selection of an interfacing inductor, a DC capacitor, and a ripple filter is given in the following sections.

Selection of the DC Bus Voltage

The minimum DC bus voltage of the VSC of the DSTATCOM should be greater than twice of the peak of the phase voltage of the distribution system. The DC bus voltage is calculated as

$$V_{DC} = 2\sqrt{2}V_{LL}/\sqrt{3}m, \tag{6.39}$$

where m is the modulation index and is considered as 1 and V_{LL} is the AC line output voltage of the DSTATCOM. Thus, V_{DC} is obtained as 677.69 V for a V_{LL} of 415 V and it is selected as 700 V.

Selection of a DC Bus Capacitor

The value of the DC capacitor (C_{DC}) of the VSC of the DSTATCOM depends on the instantaneous energy available to the DSTATCOM during transients. The principle of energy conservation is applied as

$$\frac{1}{2}C_{DC}(V_{DC}^2 - V_{DC1}^2) = k_1 3VaIt, \tag{6.40}$$

where V_{DC} is the nominal DC voltage equal to the reference DC voltage and V_{DC1} is the minimum voltage level of the DC bus, a is the overloading factor, V is the phase voltage, I is the phase current, and t is the time by which the DC bus voltage is to be recovered.

Considering the minimum voltage level of the DC bus (V_{DC1}) = 677.69 V, V_{DC} = 700 V, V = 239.60 V, I = 31.3 A, t = 30 ms, a = 1.2, and variation of energy during dynamics = 10% (k_1 = 0.1), the calculated value of C_{DC} is 5270.28 μF and it is selected as 5500 μF.

Selection of an AC Inductor for the Phase Leg of the VSC

The selection of the AC inductance (L_f) of a VSC depends on the current ripple, $I_{cr,pp}$, switching frequency f_s, and DC bus voltage (V_{DC}), and it is given as

$$L_f = \sqrt{3}mV_{DC}/12af_sI_{cr,pp}, \tag{6.41}$$

where m is the modulation index and a is the overloading factor. Considering $I_{cr,pp} = 15\%$, f_s = 10 kHz, m = 1, V_{DC} = 700 V, and a = 1.2, the value of L_r is calculated to be 1.79 mH. The round-off value of 2 mH is selected in this investigation.

Selection of an AC Inductor for the Neutral Leg

In case of unbalance in load currents on the three phases, a neural current is observed. The fourth leg in the VSC of the DSTATCOM is used to compensate this load neutral current. The ripple filter inductor for the neutral leg is

$$L_{fn} = mV_{DC}/3\sqrt{3}af_sI_{cr,pp}. \tag{6.42}$$

Considering a 15% current ripple, f_s = 10 kHz, m = 1, V_{DC} = 700 V, and a = 1.2, the neutral leg ripple inductance (L_{fn}) is calculated to be 2.37 mH. The round-off value of 2.5 mH is selected in this investigation.

Selection of a Ripple Filter

A high-pass first-order filter tuned at half the switching frequency is used to filter out the noise from the voltage at PCC. The time constant of the filter should be very small compared with the fundamental time period (T), $R_fC_f \ll T/10$, where R_f and C_f are the ripple filter resistance and its capacitance, respectively. Considering a low impedance of 8.1 Ω for the harmonic voltage at a frequency of 5 kHz, the ripple filter capacitor is designed as C_f = 5 μF. A series resistance (R_f) of 5 Ω is included in series with the capacitor (C_f). The impedance is found to be 637 Ω at fundamental frequency, which is sufficiently large, and hence the ripple filter draws negligible fundamental frequency current.

Selection of Voltage and Current Ratings of the Solid-State Switches

The voltage rating (V_{sw}) of the device can be calculated under dynamic conditions as

$$V_{sw} = V_{DC} + V_d, \tag{6.43}$$

where V_d is the 10% overshoot in the DC link voltage under dynamic conditions. The voltage rating of the switch is calculated as 770 V. The voltage rating is considered as 1200 V with adequate safety factor and considering availability of the practical device.

The current rating of the VSC is estimated as

$$I_{sw} = 1.25(I_{cr,pp} + I_{peak}). \tag{6.44}$$

From these equations, the voltage and current ratings of the IGBT switches can be estimated. The current rating of the switch is calculated as 63.96 A. Thus, a solid-state switch (IGBT) for the VSC is selected with next available higher rating of 1200 V and 100 A.

6.5.2 *Design of a DVR*

Figure 6.8 shows the schematic diagram of one of the UPQCs for a three-phase four-wire distribution system, which uses a DVR based on a three-leg VSC isolated with an injection transformer. The VSC, interfacing inductors, injection transformers, and the ripple filter are designed as follows.

Selection of an Injection Transformer of the DVR

The injection transformer is selected for connecting the VSC of a DVR in series with the supply. The voltage rating of the transformer depends on the voltage to be injected and the DC bus voltage of the VSC. For compensating a voltage variation of ±30%, the voltage to be injected is calculated as

$$V_{\mathrm{DVR}} = XV_{\mathrm{s}} = 239.6 \times 0.3 = 71.88\ \mathrm{V}. \tag{6.45}$$

The turns ratio of the injection transformer for the DVR is computed as follows.

The DC bus voltage of 700 V (decided for 415 V for the DSTATCOM) can be used to obtain 415 V across the line at the output of the VSC using a PWM controller. However, the DVR requires only 71.88 V per phase. Therefore, the maximum value of the turns ratio of the injection transformer is

$$K_{\mathrm{DVR}} = V_{\mathrm{VSC}}/V_{\mathrm{DVR}} = 415/\left(\sqrt{3} \times 71.88\right) = 3.33 \approx 3. \tag{6.46}$$

The VA rating of the injection transformer is

$$\begin{aligned} S_{\mathrm{DVR}} &= 3V_{\mathrm{DVR}}I_{\mathrm{DVR(under\ sag)}} = 3 \times 71.88 \times (30\,000/(3 \times (239.6 - 0.3 \times 239.6))) \\ &= 12.8\ \mathrm{kVA}. \end{aligned} \tag{6.47}$$

Selection of Interfacing Inductors of the DVR

The current through the secondary side of the injection transformer is decided by real power of the load. The minimum supply current occurs during voltage swell. The minimum value of supply current is

$$\begin{aligned} I_{\mathrm{DVR(under\ swell)}} &= P_{\mathrm{L}}/\{3(1+X)V_{\mathrm{s}}\} = 30\,000/\{3 \times (239.6 + 0.3 \times 239.6)\} \\ &= 32.1\ \mathrm{A}. \end{aligned} \tag{6.48}$$

Considering a 10% ripple in supply current, the value of filter inductance of the DVR is given as

$$\begin{aligned} L_{\mathrm{DVR}} &= \left(\sqrt{3}/2\right)m_{\mathrm{a}}V_{\mathrm{DC}}K_{\mathrm{DVR}}/6af_{\mathrm{s}}\Delta I_{\mathrm{DVR}} \\ &= \left(\sqrt{3}/2\right) \times 1 \times 700 \times 3/(6 \times 1.2 \times 10\,000 \times 0.1 \times 32.1) \\ &= 7.865 \approx 8\ \mathrm{mH}. \end{aligned} \tag{6.49}$$

Selection of a Ripple Filter of the DVR

A high-pass first-order filter tuned at half the switching frequency is used to filter the high-frequency noise from the voltage. Hence, the ripple filter is designed considering the cutoff frequency of 5 kHz. The time constant of the filter should be very small compared with the fundamental time period (T), $R_{\mathrm{f}}C_{\mathrm{f}} \ll T/10$, where R_{f} and C_{f} are the ripple filter resistance and its capacitance, respectively. Considering a low impedance of 8.1 Ω for the harmonic voltage at a frequency of 5 kHz, the ripple filter capacitor is designed as $C_{\mathrm{f}} = 5\ \mu\mathrm{F}$. A series resistance ($R_{\mathrm{f}}$) of 5 Ω is included in series with the capacitor (C_{f}). The impedance is found to be 637 Ω at fundamental frequency, which is sufficiently large, and hence the ripple filter draws negligible fundamental frequency current.

Selection of Device Ratings of a Three-Leg VSC of the DVR

The voltage rating of DVR devices is same as the DSTATCOM rating as both share the common DC link. The maximum current in the DVR device is

$$I_{DVR} = I_{DVR(\text{under sag})}/N_{DVR} = (30\,000/(3 \times (239.6 - 0.3 \times 239.6)))/N_{DVR}$$
$$= 19.87\ \text{A}. \quad (6.50)$$

With adequate safety factor, a current rating of 32 A can be selected for the device.

6.6 Modeling, Simulation, and Performance of UPQCs

The UPQCs are modeled in MATLAB platform for different configurations and operating conditions. Performance simulation is carried out in detail for a large number of cases, which are given in numerical examples. Here the performance of only three-phase four-wire UPQC is illustrated in detail.

6.6.1 Modeling and Simulation of UPQC

The modeling of UPQCs using MATLAB software with its Simulink and SimPowerSystems (SPS) toolboxes is given in the following sections. A three-phase four-wire UPQC system consists of a four-wire supply system, a three-phase four-wire load, a shunt connected four-leg VSC, a three-leg VSC connected in series via an injection transformer, and small passive filters to eliminate switching ripples. The design of all these elements has been discussed in detail in the previous section. The available models of three-phase supply with series impedance, three single-phase linear loads, IGBTs with antiparallel diodes, interfacing inductors, capacitors, and so on are used to model the VSC of the DSTATCOM and DVR along with their auxiliary circuits. Moreover, the available model of linear transformers, which includes losses, is used for modeling the series injection transformers of the DVR. Simulink and SPS toolboxes are used to implement control algorithm of the UPQC system. The modeling of individual VSCs and their control objective is described in the following discussion.

Modeling and Control Algorithm for a Four-Leg VSC-Based DSTATCOM of the UPQC

The UPQC system consists of a three-phase supply and three-phase four-wire critical loads. A shunt connected four-leg VSC (eight IGBTs) is interfaced to the supply system via interfacing inductors as shown in Figure 6.8. The DSTATCOM is modeled using the MATLAB/Simulink environment along with the SPS toolbox. The control objective for the shunt portion of the UPQC is described in the following discussion. The control algorithm for the DSTATCOM portion of the UPQC is shown in Figure 6.11.

The objective of the DSTATCOM is to estimate the reference supply currents and then the gating pulses for the IGBTs of the VSC of the DSTATCOM. The reference supply currents are derived from the sensed PCC voltages (v_{sa}, v_{sb}, v_{sc}), load currents (i_{La}, i_{Lb}, i_{Lc}), and the common DC bus voltage (v_{DC}). The load currents from the abc frame are first converted to the dq0 frame. A PLL signal is obtained from terminal voltages for generating fundamental unit vectors for the conversion of sensed currents to the dq0 reference frame. The SRF controller extracts DC quantities by LPFs and hence the non-DC quantities (ripples) are separated from the reference signal. The control strategy considers that the supply must deliver the mean value of the direct-axis component of the load current along with the active power component current for maintaining the DC bus and meeting the losses (i_{loss}) in the UPQC. The output of the PI controller at the DC bus voltage of the UPQC is considered as the current (i_{loss}) for meeting its losses.

The reference supply current must be in phase with the voltage at PCC but with no zero-sequence component. It is therefore obtained by the reverse Park's transformation with i_q^* and i_0^* as zero. The resultant dq0 currents are again converted into the reference supply currents using the reverse Park's transformation. A PWM current controller is used over the reference and sensed supply currents to generate the gating signals for the IGBTs of the VSC of the DSTATCOM.

Modeling and Control Algorithm for a DVR of the UPQC

The available model (in MATLAB SPS block set) of IGBTs with an antiparallel diode is used to realize the VSC of the DVR. The series *RLC* component of SPS block set is used to realize the filter inductor and ripple filter of the DVR. The linear transformers, which include losses, are used for modeling the series injection transformers of the DVR. The control algorithm is implemented using Simulink blocks. The control algorithm for the DVR of the UPQC is also shown in Figure 6.11, in which the synchronous reference frame theory is used for reference signal generation. The PCC voltages (v_s), supply currents (i_s), and load terminal voltages (v_L) are sensed for deriving the IGBT gate signals. The PCC voltages are converted to the rotating reference frame using the abc—dq0 conversion using the Park's transformation with unit vectors (sin θ, cos θ) derived from the supply currents using a PLL. The SRF controller extracts DC quantities by LPFs and hence the non-DC quantities (ripple) are separated from the reference signal. The control strategy considers that the reference load voltage consists of the DC component of d-axis PCC voltage and the DC component of q-axis PCC voltage along with the quadrature component estimated by the PI controller over the load terminal voltage. The amplitude of AC load terminal voltage (V_L) is controlled to its reference voltage (V_L^*) using a PI controller and the output of the PI controller is considered as the voltage (v_{qr}) to be injected by the DVR. The reference load voltage in the abc frame is obtained by reverse Park's transformation. The reference load voltages ($v_{La}^*, v_{Lb}^*, v_{Lc}^*$) and the sensed load voltages (v_{La}, v_{Lb}, v_{Lc}) are used in a PWM voltage controller unit to generate gating pulses to the VSC of the DVR. The PWM voltage controller is operated with a switching frequency of 10 kHz.

6.6.2 *Performance of UPQCs*

The performance of a three-phase four-wire UPQC is simulated using the developed model in MATLAB. A three-leg VSC is used to simulate the DVR portion of the UPQC, whereas a four-leg VSC is used to simulate the DSTATCOM portion of the UPQC as shown in Figure 6.8. The nominal grid voltage is 415 V at 50 Hz and a 30 kW, 0.8 lagging load is considered for simulation studies. The objective of this UPQC system is to regulate load voltage to rated value and to maintain balanced supply currents with unity power factor at the AC mains.

6.6.2.1 Performance of UPQCs at Critical Load for Neutral Current Compensation, Load Balancing, and Regulation of Load Voltage during Voltage Sag

The dynamic performance of the UPQC under critical linear lagging power factor balanced/unbalanced load condition is shown in Figure 6.12a. It shows the transient performance of the system under voltage sag condition, when a critical linear load is connected. At 0.3 s, a sag in supply voltage (0.3 pu) is created for 10 cycles and it is observed that the load voltage is regulated to constant amplitude under voltage sag and nominal operating conditions. It can be observed that only during sag, a voltage is injected by the DVR portion of the UPQC to regulate the load voltage. Due to sag in supply voltage, an increment in supply current is observed. The load is changed to a two-phase load at 0.4 s and to a single-phase load at 0.6 s. These loads are applied again at 0.7 s. It can be observed that supply currents are balanced and are in phase with supply voltage even under an unbalanced load at supply. The load neural current is supplied by the fourth leg of the DSTATCOM and hence the supply neutral current is approximately zero under all operating conditions.

6.6.2.2 Performance of UPQCs at Critical Load for Neutral Current Compensation, Load Balancing, and Regulation of Load Voltage during Voltage Swell

The dynamic performance of the UPQC under critical linear lagging power factor balanced/unbalanced load condition is shown in Figure 6.12b. It shows the transient performance of the system under voltage swell condition, when a critical linear load is connected. At 0.3 s, a swell in supply voltage (0.3 pu) is created for 10 cycles and it is observed that the load voltage is regulated to constant amplitude under voltage swell and nominal operating conditions. It can be observed that only during swell, a voltage is injected by the DVR portion of the UPQC to regulate the load voltage. Due to swell in supply voltage, a

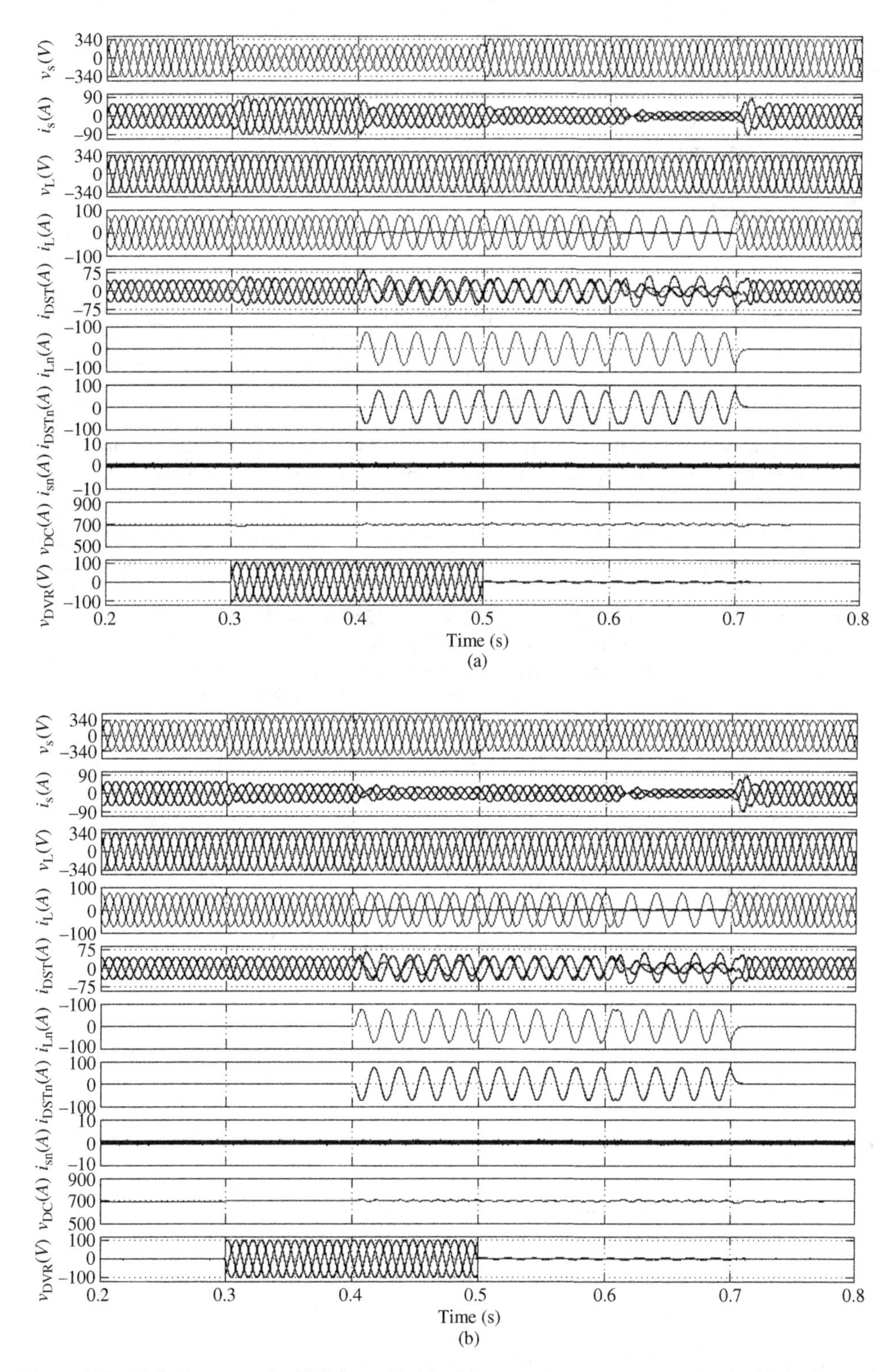

Figure 6.12 (a) Performance of a UPQC at critical load for neutral current compensation, load balancing, and regulation of load voltage during voltage sag. (b) Performance of a UPQC at critical load for neutral current compensation, load balancing, and regulation of load voltage during voltage swell

decrease in supply current is observed. The load is changed to a two-phase load at 0.4 s and to a single-phase load at 0.6 s. These loads are applied again at 0.7 s. It can be observed that supply currents are balanced and are in phase with supply voltage even under an unbalanced load at supply. The load neural current is supplied by the fourth leg of the DSTATCOM and hence the supply neutral current is approximately zero under all operating conditions.

6.7 Numerical Examples

Example 6.1

A single-phase unified power quality compensator (consisting of a DSTATCOM and a DVR using two VSCs with a common DC bus capacitor) is to be designed for a load of 230 V, 50 Hz, 25 A, 0.8 lagging power factor as a right-hand UPQC-Q (shown in Figure E6.1). There is a voltage sag of −20% in the supply system with a base value of 230 V. Calculate (a) the voltage rating of the DVR of the UPQC-Q, (b) the current rating of the DVR of the UPQC-Q, (c) the VA rating of the DVR of the UPQC-Q, (d) the voltage rating of the DSTATCOM of the UPQC-Q, (e) the current rating of the DSTATCOM of the UPQC-Q, (f) the VA rating of the DSTATCOM of the UPQC-Q, and (g) total VA rating of the UPQC-Q to provide reactive power compensation of the load for unity power factor at the AC mains with a constant regulated voltage of 230 V at 50 Hz across the load.

Solution: Given supply voltage $V_s = 230$ V, frequency of the supply $(f) = 50$ Hz, and a load of 230 V, 50 Hz, 25 A, 0.8 lagging power factor. There is a voltage sag of −20% in the supply system with a base value of 230 V.

The active power of the load is $P_L = V_L I_L \times PF = 230 \times 25 \times 0.8 = 4600$ W.

The supply current during nominal operating condition is $I_{sN} = P_L/V_s = 4600/230 = 20$ A.

During nominal grid operation, $V_{DSTN} = 230$ V and $I_{DSTN} = I_L \sin \phi = 25 \times \sqrt{(1 - 0.8^2)} = 15$ A. Hence, during nominal grid operation, the rating of the DSTATCOM is $S_{DSTN} = V_{DSTN} I_{DSTN} = 230 \times 15 = 3450$ VA. During nominal grid operation, $V_{DVRN} = 0$. Hence, during nominal grid operation, $S_{DVRN} = 0$.

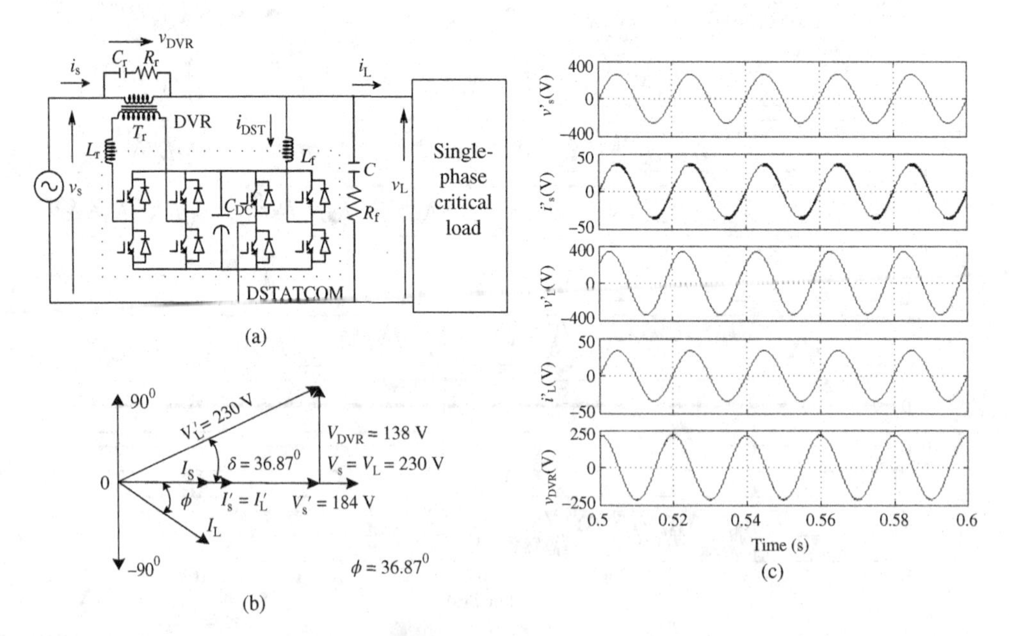

Figure E6.1 (a) A single-phase right-hand UPQC-Q. (b) Phasor diagram under voltage sag. (c) Waveforms of a single-phase right-hand UPQC-Q under voltage sag

Voltage Sag Compensation by a Right-Hand UPQC-Q

a. The voltage rating of the DVR of the UPQC-Q is computed as follows.
   There is a voltage sag of −20% in the supply system with a base value of 230 V. Therefore, the DVR of the UPQC-Q must inject a voltage in quadrature with supply current/voltage to provide the required voltage at the load end.
   The supply voltage reduces to $V'_s = 230 \times (1 - 0.20) = 184$ V and $X = 0.2$.
   The supply current during voltage sag compensation is $I'_s = P_L/V'_s = 4600/184 = 25$ A.
   The voltage rating of the DVR is $V_{DVR} = \sqrt{(V'^2_L - V'^2_s)} = \sqrt{(230^2 - 184^2)} = 138$ V.
b. The current rating of the DVR of the UPQC-Q must be same as supply current after shunt compensation: I_{DVR} = the fundamental active component of load current $(I'_s) = P_L/V'_s = 4600/184 = 25$ A.
c. The VA rating of the DVR of the UPQC-Q is calculated as follows.

$$Q_{DVR} = V_{DVR} I_{DVR} = 138 \times 25 \text{ VA} = 3450 \text{ VAR}.$$

$P_{DVR} = 0.0$, as the voltage of the DVR is injected in quadrature with its current.

$$S_{DVR} = \sqrt{(P^2_{DVR} + Q^2_{DVR})} = Q_{DVR} = 138 \times 25 \text{ VA} = 3450 \text{ VA}.$$

d. The voltage rating of the DSTATCOM of the UPQC-Q is equal to AC load voltage ($V_{sh} = 230$ V), since it is connected across the load of 230 V sine waveform.
e. The current rating of the DSTATCOM of the UPQC-Q is computed as follows.
   The DSTATCOM of the UPQC-Q needs to correct the power factor of the supply to unity. Hence, the required reactive power by the DSTATCOM is now lower than the load reactive power as the angle between the supply voltage and the load current is reduced to $\beta = \phi - \delta$, where $\cos\phi$ is the load power factor and δ is the angle between the load voltage and the PCC voltage after compensation. The angle δ is computed as follows: $\tan\delta = \sqrt{(V^2_{Lc} - V'^2_s)}/V'_s = 138/184 = 0.75$, $\delta = 36.87°$, and power factor angle is $\phi = \cos^{-1}(0.08) = 36.87°$.
   The current rating of the DSTATCOM is $I_{DST} = I_L[\sqrt{\{(1-X)^2 + \cos^2\phi - 2\cos\phi\cos(\phi - \delta)(1-X)\}}]/(1-X) = 25[\sqrt{\{0.8^2 + 0.8^2 - 2 \times 0.8\cos(36.87° - 36.87°) \times 0.8\}}]/0.8 = 0$ A.
f. The VA rating of the DSTATCOM is $S_{DST} = V_{DST} I_{DST} = V_{DST} I_L[\sqrt{\{(1-X)^2 + \cos^2\phi - 2\cos\phi\cos(\phi - \delta)(1-X)\}}]/(1-X) = 230 \times 0 = 0$ VA.
   Hence, the VA rating of the UPQC-Q under voltage sag is $S_{UPQC\text{-}Q(under\ sag)} = S_{DST} + S_{DVR} = 0 + 3450 = 3450$ VA.
   It means that the required reactive power for correcting the power factor to unity at supply is supplied by the DVR and it does not need any reactive power from the DSTATCOM during voltage sag.
   However, during nominal grid operation, all reactive power of the load is supplied by the DSTATCOM; hence, the VA rating of the DSTATCOM is 3450 VA.
   Hence, considering an overall rating (both normal and under voltage sag), the ratings of both the compensators are

$$V_{DST} = 230 \text{ V}, \quad I_{DST} = 15 \text{ A}, \quad S_{DST} = 230 \times 15 = 3450 \text{ VA}.$$

$$V_{DVR} = 138 \text{ V}, \quad I_{DVR} = 25 \text{ A}, \quad S_{DVR} = 3450 \text{ VA}.$$

g. The VA rating of the UPQC-Q is $S_{UPQC\text{-}Q} = S_{DST} + S_{DVR} = 3450 + 3450 = 6900$ VA.

Example 6.2

A single-phase unified power quality compensator (consisting of a DSTATCOM and a DVR using two VSCs with a common DC bus capacitor) is to be designed for a load of 230 V, 50 Hz, 25 A, 0.8 lagging

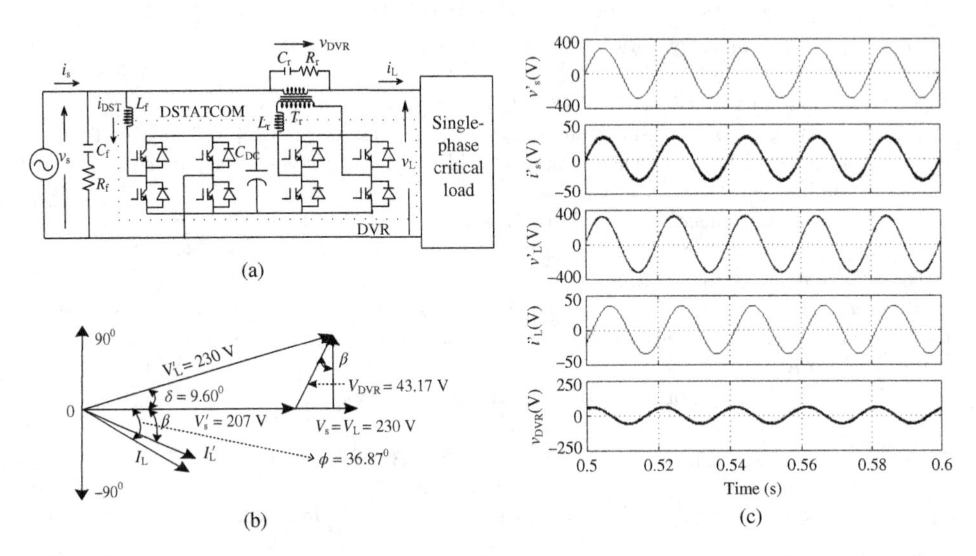

Figure E6.2 (a) A single-phase left-hand UPQC-Q. (b) Phasor diagram under voltage sag. (c) Waveforms of a single-phase left-hand UPQC-Q under voltage sag

power factor as a left-hand UPQC-Q (shown in Figure E6.2). There is a voltage sag of −10% in the supply system with a base value of 230 V. Calculate (a) the voltage rating of the DVR of the UPQC-Q, (b) the current rating of the DVR of the UPQC-Q, (c) the VA rating of the DVR of the UPQC-Q, (d) the voltage rating of the DSTATCOM of the UPQC-Q, (e) the current rating of the DSTATCOM of the UPQC-Q, (f) the VA rating of the DSTATCOM of the UPQC-Q, and (g) total VA rating of the UPQC-Q to provide reactive power compensation of the load for unity power factor at the AC mains with a constant regulated voltage of 230 V at 50 Hz across the load.

Solution: Given supply voltage $V_s = 230$ V, frequency of the supply $(f) = 50$ Hz, and a load of 230 V, 50 Hz, 25 A, 0.8 lagging power factor. There is a voltage sag of −10% in the supply system with a base value of 230 V.

It means that after the voltage sag, the supply voltage reduces to $V'_s = 230 \times (1 - 0.10) = 207$ V and $X = 0.1$.

The active power of the load is $P_L = V_L I_L \times \text{PF} = 230 \times 25 \times 0.8 = 4600$ W.

The supply current during voltage sag compensation is $I'_s = P_L/V'_s = 4600/207 = 22.22$ A.

During nominal grid operation, $V_{DSTN} = 230$ V and $I_{DSTN} = I_L \sin\phi = 25 \times \sqrt{(1 - 0.8^2)} = 15$ A. Hence, during nominal grid operation, the rating of the DSTATCOM is $S_{DSTN} = V_{DSTN} I_{DSTN} = 230 \times 15 = 3450$ VA. Since $V_{DVRN} = 0$, $S_{DVRN} = 0$.

Voltage Sag Compensation by a Left-Hand UPQC-Q

a. The voltage rating of the DVR of the UPQC-Q is computed as follows.

   There is a voltage sag of −10% in the supply system with a base value of 230 V. Therefore, the DVR of the left-hand UPQC-Q must inject in quadrature with the load current to provide the required voltage at the load end.

$$V'_L \cos\phi = V'_s \cos\beta, \quad \cos\beta = V_L \cos\phi / V'_s = 230 \times 0.8/(230 \times 0.9) = 8/9, \quad \beta = 27.27°.$$

$$\delta = \phi - \beta = 36.87° - 27.27° = 9.60°.$$

$$V_L \sin\delta = V_{DVR} \sin(90° - \beta), \quad V_{DVR} = V'_L \sin\delta / \sin(90° - \beta) = 43.17 \text{ V}.$$

   The voltage rating of the DVR is $V_{DVR} = V_L \sin\delta/\sin(90° - \beta) = 43.17$ V.

b. The current rating of the DVR of the UPQC-Q must be same as load current: $I_{DVR} = I'_L = 25$ A.
c. The VA rating of the DVR of the UPQC-Q is calculated as follows.

$$Q_{DVR} = V_{DVR} I_{DVR} = 43.17 \times 25 \text{ VA} = 1079.25 \text{ VAR}.$$

$P_{DVR} = 0.0$, as the voltage of the DVR is injected in quadrature with its current.

$$S_{DVR} = \sqrt{(P^2_{DVR} + Q^2_{DVR})} = Q_{DVR} = 1079.25 \text{ VA}.$$

d. The voltage rating of the DSTATCOM of the UPQC-Q is equal to AC supply voltage ($V_{DST} = 207$ V), since it is connected across PCC of 207 V sine waveform.
e. The current rating of the DSTATCOM of the UPQC-Q is computed as follows.
   The DSTATCOM of the UPQC-Q needs to correct the power factor of the source to unity. Hence, the required reactive power by the DSTATCOM is now lower than the load reactive power as the angle between the supply voltage and the load current is reduced to $\beta = \phi - \delta$.
   The current rating of the DSTATCOM is $I_{DST} = I'_L \sin\beta = 25 \sin 27.27° = 11.45$ A.

$$Q_{DST} = V'_s I'_L \sin\beta = 207 \times 11.45 = 2370.15 \text{ VA}.$$

$P_{DST} = 0.0$, as there is no active power consumed by either DSTATCOM or DVR.
f. The VA rating of the DSTATCOM is $S_{DST(\text{under sag})} = \sqrt{(P^2_{DST} + Q^2_{DST})} = Q_{DST} = V'_s I_{DST} = V'_s I'_L \sin\beta = 207 \times 11.45 = 2370.15$ VA.
   Hence, the VA rating of the UPQC-Q under voltage sag is $S_{UPQC\text{-}Q(\text{under sag})} = S_{DST} + S_{DVR} = 2370.15 + 1079.25 = 3449.4$ VA.
   However, during nominal grid condition, the DSTATCOM has to supply all reactive power of the load; hence, its rating needs to be 3450 VA.

$$S_{DST} = 3450 \text{ VA}.$$

Hence, considering an overall rating (both normal and under voltage sag), the ratings of both the compensators are

$$V_{DST} = 230 \text{ V}, \quad I_{DST} = 15 \text{ A}, \quad S_{DST} = V_{DST} I_{DST} = 230 \times 15 = 3450 \text{ VA}.$$

$$V_{DVR} = 43.17 \text{ V}, \quad I_{DVR} = 25 \text{ A}, \quad S_{DVR} = V_{DVR} I_{DVR} = 1019.25 \text{ VA}.$$

g. The VA rating of the UPQC-Q is $S_{UPQC\text{-}Q} = S_{DST} + S_{DVR} = 3450 + 1079.25 = 4529.25$ VA.

Example 6.3

A single-phase unified power quality compensator (consisting of a DSTATCOM and a DVR using two VSCs with a common DC bus capacitor) is to be designed for a load of 230 V, 50 Hz, 25 A, 0.8 lagging power factor as a left-hand UPQC-Q (shown in Figure E6.3). There is a voltage swell of +10% in the supply system with a base value of 230 V. Calculate (a) the voltage rating of the DVR of the UPQC-Q, (b) the current rating of the DVR of the UPQC-Q, (c) the VA rating of the DVR of the UPQC-Q, (d) the voltage rating of the DSTATCOM of the UPQC-Q, (e) the current rating of the DSTATCOM of the UPQC-Q, (f) the VA rating of the DSTATCOM of the UPQC-Q, and (g) total VA rating of the UPQC-Q to provide reactive power compensation of the load for unity power factor at the AC mains with a constant regulated voltage of 230 V at 50 Hz across the load.

Solution: Given supply voltage $V_s = 230$ V, frequency of the supply $(f) = 50$ Hz, and a load of 230 V, 50 Hz, 25 A, 0.8 lagging power factor.

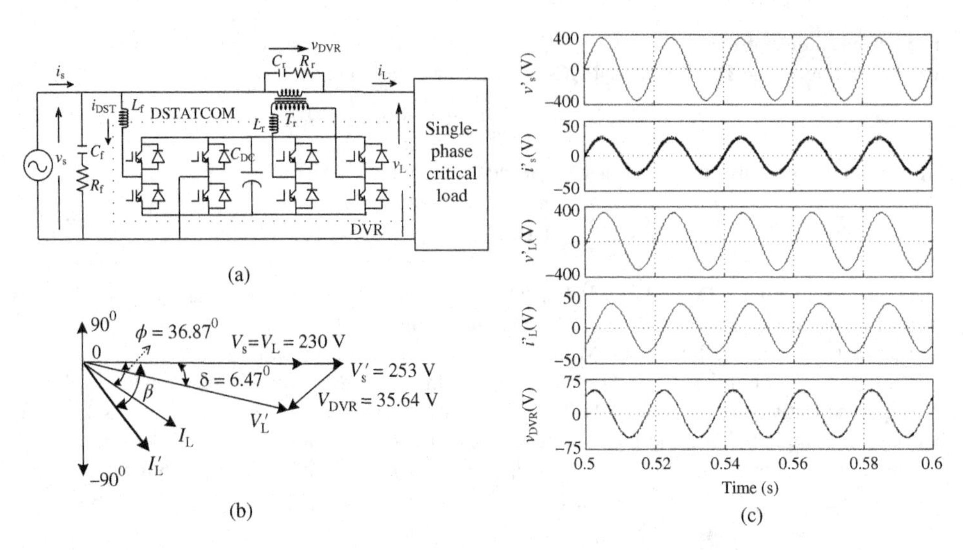

Figure E6.3 (a) A single-phase left-hand UPQC-Q. (b) Phasor diagram under voltage swell. (c) Waveforms of a single-phase left-hand UPQC-Q under voltage swell

During nominal grid operation, $V_{DSTN} = 230$ V and $I_{DSTN} = I_L \sin\phi = 25 \times \sqrt{(1-0.8^2)} = 15$ A. Hence, the rating of the DSTATCOM is $S_{DSTN} = V_{DSTN} I_{DSTN} = 230 \times 15 = 3450$ VA. The rating of the DVR is $V_{DVRN} = 0$; hence, $S_{DVRN} = 0$.

The active power of the load is $P_L = V_L I_L \times \text{PF} = 230 \times 25 \times 0.8 = 4600$ W.

Voltage Swell Compensation by a Left-Hand UPQC-Q

There is a voltage swell of +10% in the supply system with a base value of 230 V.

It means that during the voltage swell, the supply voltage increases to $V_s' = 230 \times (1+0.10) = 253$ V.

The supply current during voltage swell compensation is $I_s' = P_L/V_s' = 4600/253 = 18.18$ A.

a. The voltage rating of the DVR of the UPQC-Q is computed as follows.

   There is a voltage swell of +10% in the supply system with a base value of 230 V. Therefore, the DVR of the left-hand UPQC-Q must inject in quadrature with the load current to provide the required voltage at the load end.

$$V_L \cos\phi = V_s' \cos\beta, \quad \cos\beta = V_L \cos\phi / V_s' = 230 \times 0.8/(230 \times 1.1) = 0.8/1.1, \quad \beta = 43.34^\circ.$$

$$\delta = \beta - \phi = 43.34^\circ - 36.87^\circ = 6.47^\circ.$$

$$V_L' \sin\delta = V_{DVR} \sin(90^\circ - \beta), \quad V_{DVR} = V_L' \sin\delta / \sin(90^\circ - \beta) = 35.64 \text{ V}.$$

   The voltage rating of the DVR is $V_{DVR} = V_L' \sin\delta / \sin(90^\circ - \beta) = 35.64$ V.

b. The current rating of the DVR of the UPQC-Q must be same as load current: $I_{DVR} = I_L = 25$ A.
c. The VA rating of the DVR of the UPQC-Q is calculated as follows.

$$Q_{DVR} = V_{DVR} I_{DVR} = 35.64 \times 25 \text{ VA} = 891.15 \text{ VAR}.$$

$P_{DVR} = 0.0$, as the voltage of the DVR is injected in quadrature with its current.

$$S_{DVR(\text{during swell})} = \sqrt{(P_{DVR}^2 + Q_{DVR}^2)} = Q_{DVR} = 891.15 \text{ VA}.$$

d. The voltage rating of the DSTATCOM of the UPQC-Q is equal to AC source voltage ($V_{DST} = V'_s = 253$ V), since it is connected across PCC of 253 V sine waveform.
e. The current rating of the DSTATCOM of the UPQC-Q is computed as follows.
   The DSTATCOM of the UPQC-Q needs to correct the power factor of the source to unity. Hence, the required reactive power by the DSTATCOM is now higher than the load reactive power as the angle between the supply voltage and the load current is increased to $\beta = \phi + \delta$.
   The current rating of the DSTATCOM is $I_{DST} = I'_L \sin\beta = 25 \sin 43.34° = 17.16$ A.

$$Q_{DST} = V'_s I'_L \sin\beta = 253 \times 17.16 = 4341.01 \text{ VAR}.$$

$P_{DST} = 0.0$, as there is no active power consumed by either DSTATCOM or DVR.
f. The VA rating of the DSTATCOM is $S_{DST} = \sqrt{(P_{DST}^2 + Q_{DST}^2)} = Q_{DST} = V_s I_{DST} = V'_s I'_L \sin\beta = 253 \times 17.16 = 4341.01$ VA.
   Hence, the VA rating of the UPQC-Q under voltage swell is $S_{UPQC\text{-}Q(under\ swell)} = S_{DST} + S_{DVR} = 4341.01 + 891.15 = 5233.13$ VA.
   Hence, considering an overall rating (both normal and under voltage swell), ratings of both the compensators are

$$V_{DST} = 253 \text{ V}, \quad I_{DST} = 17.16 \text{ A}, \quad S_{DST} = V_{DST} I_{DST} = 253 \times 15 = 4341 \text{ VA}.$$

$$V_{DVR} = 35.64 \text{ V}, \quad I_{DVR} = 25 \text{ A}, \quad S_{DVR} = V_{DVR} I_{DVR} = 891.15 \text{ VA}.$$

g. The VA rating of the UPQC-Q is $S_{UPQC\text{-}Q} = S_{DST} + S_{DVR} = 4341.01 + 891.15 = 5233.13$ VA.

Example 6.4

A single-phase unified power quality compensator (consisting of a DSTATCOM and a DVR using two VSCs with a common DC bus capacitor) is to be designed for a load of 230 V, 50 Hz, 25 A, 0.8 lagging power factor as a right-hand UPQC-P (shown in Figure E6.4). There is a voltage sag of −20% in the

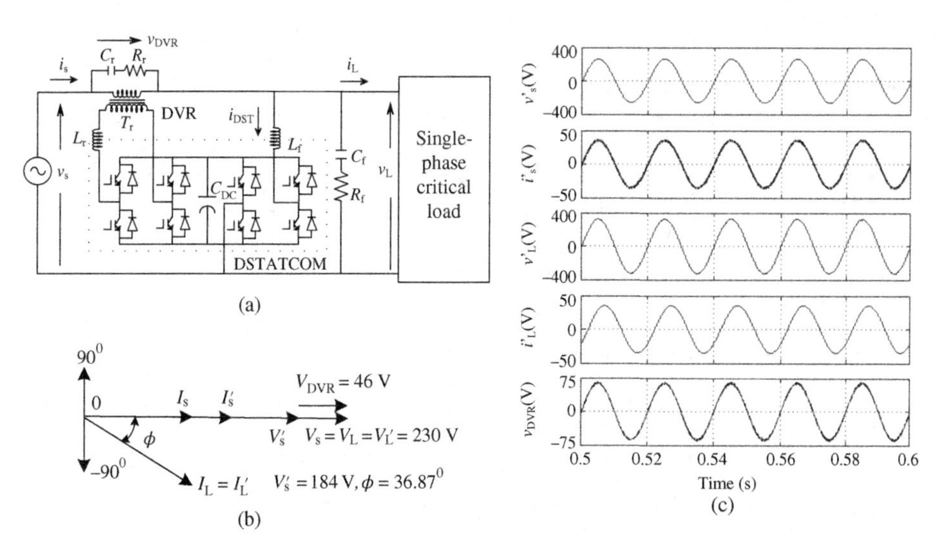

Figure E6.4 (a) A single-phase right-hand UPQC-P. (b) Phasor diagram under voltage sag. (c) Waveforms of a single-phase right-hand UPQC-P under voltage sag

supply system with a base value of 230 V. Calculate (a) the voltage rating of the DSTATCOM of the UPQC-P, (b) the current rating of the DSTATCOM of the UPQC-P, (c) the VA rating of the DSTATCOM of the UPQC-P, (d) the voltage rating of the DVR of the UPQC-P, (e) the current rating of the DVR of the UPQC-P, (f) the VA rating of the DVR of the UPQC-P, and (g) total VA rating of the UPQC-P to provide reactive power compensation for unity power factor at the AC mains with a constant regulated voltage of 230 V at 50 Hz across the load.

Solution: Given supply voltage $V_s = 230$ V, frequency of the supply $(f) = 50$ Hz, and a load of 230 V, 50 Hz, 25 A, 0.8 lagging power factor. There is a voltage sag of −20% ($X = 0.2$ pu) in the supply system with a base value of 230 V.

The active power of the load is $P_L = V_L I_L \times PF = 230 \times 25 \times 0.8 = 4600$ W.

During nominal grid operation, $V_{DSTN} = 230$ V and $I_{DSTN} = I_L \sin\phi = 25 \times \sqrt{(1 - 0.8^2)} = 15$ A. Hence, the rating of the DSTATCOM is $S_{DSTN} = V_{DSTN} I_{DSTN} = 230 \times 15 = 3450$ VA. The rating of the DVR is $V_{DVRN} = 0$; hence, $S_{DVRN} = 0$.

Voltage Sag Compensation by a Right-Hand UPQC-P

It means that during the voltage sag, the supply voltage reduces to $V_s' = V_s(1 - X) = 230 \times (1 - 0.20) = 184$ V.

The supply current during voltage sag compensation is $I_s' = P_L / V_s' = 4600/184 = 25$ A.

a. The voltage rating of the DSTATCOM of the UPQC-P is equal to AC load voltage ($V_{DST} = 230$ V), since it is connected across the load of 230 V sine waveform.
b. The current rating of the DSTATCOM of the UPQC-P is computed as follows.

   The DSTATCOM of the UPQC-P needs to correct the power factor load to unity. Hence, the required reactive power of the load it must supply is I_{DSTR} = reactive current of the load = $I_L\sqrt{(1 - PF^2)} = 25 \times 0.6 = 15$ A.

   The active power component of the DSTATCOM current is $I_{DSTA} = (V_s X) I_s / V_s = (46 \times 25)/230 = 5$ A.

   Total current rating of the DSTATCOM is $I_{DST} = \sqrt{(I_{DSTA}^2 + I_{DSTR}^2)} = 15.81$ A.
c. The VA rating of the DSTATCOM of the UPQC-P is $S_{DST} = V_{DST} I_{DST} = 230 \times 15.81$ VA = 3636.62 VA.
d. The voltage rating of the DVR of the UPQC-P is computed as follows.

   There is a voltage sag of −20% in the supply system with a base value of 230 V. Therefore, the DVR of the UPQC-P must inject the difference of these two voltages to provide the required voltage at the load end.

   The voltage rating of the DVR is $V_{DVR} = V_s X = 230 \times 0.20 = 46$ V.
e. The current rating of the DVR of the UPQC-P is same as the supply current after shunt compensation: I_{DVR} = the fundamental active component of load current $(I_s') = P_L / V_s' = 4600/184 = 25$ A.
f. The VA rating of the DVR of the UPQC-P is $S_{DVR} = V_{DVR} I_{DVR} = 46 \times 25$ VA = 1150 VA.

   The VA rating of the UPQC-P during voltage sag is $S_{UPQC\text{-}P(under\ sag)} = S_{DST} + S_{DVR} = 3636.62 + 1150 = 4786.62$ VA.

   Hence, considering an overall rating (both normal and under voltage sag), ratings of both the compensators are

$$V_{DST} = 230\text{ V},\quad I_{DST} = 15.81\text{ A},\quad S_{DST} = V_{DST} I_{DST} = 230 \times 15.81 = 3636.62\text{ VA}.$$

$$V_{DVR} = 46\text{ V},\quad I_{DVR} = 25\text{ A},\quad S_{DVR} = V_{DVR} I_{DVR} = 1150\text{ VA}.$$

g. The overall VA rating of the UPQC-P is $S_{UPQC\text{-}P} = S_{DST} + S_{DVR} = 3636.62 + 1150 = 4786.62$ VA.

Example 6.5

A single-phase unified power quality compensator (consisting of a DSTATCOM and a DVR using two VSCs with a common DC bus capacitor) is to be designed for a load of 230 V, 50 Hz, 25 A, 0.8 lagging

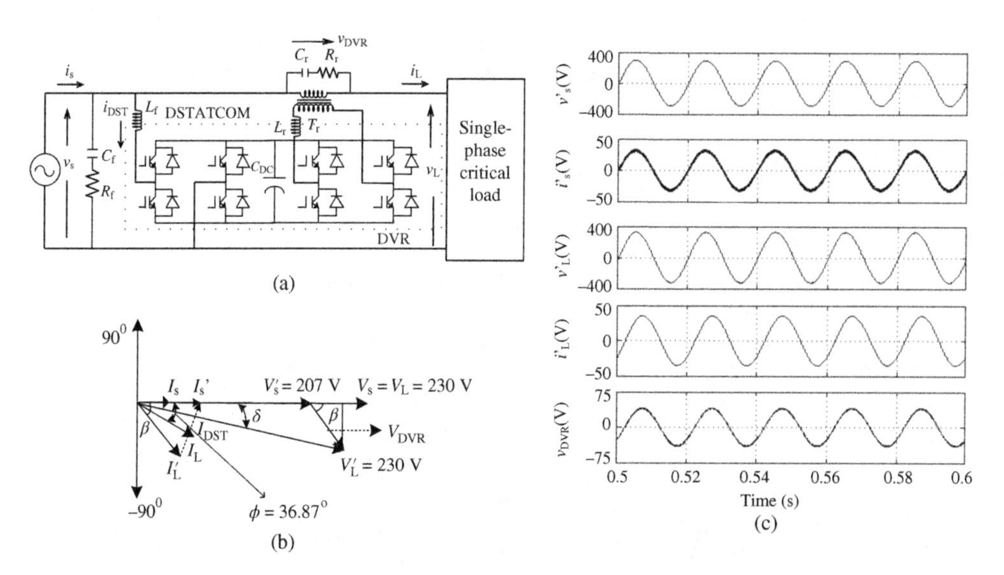

Figure E6.5 (a) A single-phase left-hand UPQC-P. (b) Phasor diagram under voltage sag. (c) Waveforms of a single-phase left-hand UPQC-P under voltage sag

power factor as a left-hand UPQC-P (shown in Figure E6.5). There is a voltage sag of −10% in the supply system with a base value of 230 V. Calculate (a) the voltage rating of the DVR of the UPQC-P, (b) the current rating of the DVR of the UPQC-P, (c) the VA rating of the DVR of the UPQC-P, (d) the voltage rating of the DSTATCOM of the UPQC-P, (e) the current rating of the DSTATCOM of the UPQC-P, (f) the VA rating of the DSTATCOM of the UPQC-P, and (g) total VA rating of the UPQC-P to provide reactive power compensation for unity power factor at the AC mains with a constant regulated voltage of 230 V at 50 Hz across the load.

Solution: Given supply voltage $V_s = 230$ V, frequency of the supply $(f) = 50$ Hz, and a load of 230 V, 50 Hz, 25 A, 0.8 lagging power factor. There is a voltage sag of −10% ($X = 0.1$ pu) in the supply system with a base value of 230 V.

During nominal grid operation, $V_{DSTN} = 230$ V and $I_{DSTN} = I_L \sin\phi = 25 \times \sqrt{(1-0.8^2)} = 15$ A. Hence, the rating of the DSTATCOM is $S_{DSTN} = V_{DSTN} I_{DSTN} = 230 \times 15 = 3450$ VA. The rating of the DVR is $V_{DVRN} = 0$; hence, $S_{DVRN} = 0$.

Voltage Sag Compensation by a Left-Hand UPQC-P

It means that during the voltage sag, the supply voltage reduces to $V_s' = V_s(1 - X) = 230 \times (1 - 0.10) = 207$ V.

The active power of the load is $P_L = V_L I_L \times \text{PF} = 230 \times 25 \times 0.8 = 4600$ W.

The supply current during voltage sag compensation is $I_s' = P_L / V_s' = 4600/207 = 22.22$ A.

a. The voltage rating of the DVR of the UPQC-P is computed as follows.

   There is a voltage sag of −10% in the supply system with a base value of 230 V. Therefore, the DVR of the left-hand UPQC-P must inject a voltage out of phase with the load current to provide the required voltage at the load end.

   From the phasor diagram of Figure E6.5b, $V_s' \sin\delta = V_{DVR} \sin\phi$ and $(V_s' \cos\delta + V_{DVR} \cos\phi) = V_L'$.

   Substituting the values $V_s' = 0.9 \times 230$ V, $V_L' = 230$ V, $\cos\phi = 0.8$, and $\sin\phi = 0.6$, and solving the above equations, one gets $V_{DVR} = 29.71$ V, $\delta = 4.94°$, and $\beta = \delta + \phi = 41.83°$.

   The voltage rating of the DVR is $V_{DVR} = 29.71$ V.

b. The current rating of the DVR of the UPQC-P must be same as load current: $I_{DVR} = I_L' = 25$ A.

c. The VA rating of the DVR of the UPQC-P is calculated as follows.
$Q_{DVR}=0.0$ VAR, as the voltage of the DVR is injected in phase with its current.

$$P_{DVR} = V_{DVR}I_{DVR} = 29.71 \times 25\text{ VA} = 742.75\text{ W}.$$

$$S_{DVR} = \sqrt{(P_{DVR}^2 + Q_{DVR}^2)} = P_{DVR} = 742.75\text{ VA}.$$

d. The voltage rating of the DSTATCOM of the UPQC-P is equal to maximum voltage across it: $V_{DST}=V_s=230$ V.
e. The current rating of the DSTATCOM of the UPQC-P is computed as follows.
The DSTATCOM of the UPQC-P needs to correct the power factor of the AC mains to unity. Hence, the required reactive power by the DSTATCOM is now higher than the load reactive power as the angle between the supply voltage and the load current is changed to $\beta=\phi+\delta$.

$$Q_{DST} = V_s' I_L' \sin\beta = 207 \times 25 \sin 41.83° = 3450\text{ VAR}.$$

$$P_{DST} = -P_{DVR} = V_{DVR}I_{DVR} = 29.71 \times 25\text{ VA} = -742.75\text{ W}.$$

f. The VA rating of the DSTATCOM is $S_{DST} = \sqrt{(P_{DST}^2 + Q_{DST}^2)} = 3529.1$ VA.

$$I_{DST} = S_{DST}/V_{DST} = 3529.1/207 = 17.04\text{ A}.$$

Hence, the current rating of the DSTATCOM is 17.04 A.

The VA rating of the UPQC-P during sag compensation is $S_{UPQC\text{-}P(under\ sag)} = S_{DST} + S_{DVR} = 3529 + 742.75 = 4271.75$ VA.

Hence, considering an overall rating (both normal and under voltage sag), ratings of both the compensators are

$$V_{DST} = 230\text{ V},\quad I_{DST} = 17.04\text{ A},\quad S_{DST} = V_{DST}I_{DST} = 230 \times 17.04 = 3921.2\text{ VA}.$$
$$V_{DVR} = 29.71\text{ V},\quad I_{DVR} = 25\text{ A},\quad S_{DVR} = V_{DVR}I_{DVR} = 742.75\text{ VA}.$$

g. The VA rating of the UPQC-P is $S_{UPQC\text{-}P}=S_{DST}+S_{DVR}=3921.2+742.75=4663.95$ VA.

Example 6.6

A single-phase unified power quality compensator (consisting of a DSTATCOM and a DVR using two VSCs with a common DC bus capacitor) is to be designed for a load of 230 V, 50 Hz, 25 A, 0.8 lagging power factor as a left-hand UPQC-P (shown in Figure E6.6). There is a voltage swell of +10% in the supply system with a base value of 230 V. Calculate (a) the voltage rating of the DVR of the UPQC-P, (b) the current rating of the DVR of the UPQC-P, (c) the VA rating of the DVR of the UPQC-P, (d) the voltage rating of the DSTATCOM of the UPQC-P, (e) the current rating of the DSTATCOM of the UPQC-P, (f) the VA rating of the DSTATCOM of the UPQC-P, and (g) total VA rating of the UPQC-P to provide reactive power compensation for unity power factor at the AC mains with a constant regulated voltage of 230 V at 50 Hz across the load.

Solution: Given supply voltage $V_s=230$ V, frequency of the supply $(f)=50$ Hz, and a load of 230 V, 50 Hz, 25 A, 0.8 lagging power factor. There is a voltage swell of +10% in the supply system with a base value of 230 V.

The active power of the load is $P_L=V_LI_L\times \text{PF}=230\times 25\times 0.8=4600$ W.

Grid current under nominal operating condition is $I_s=P_L/V_s=4600/230=20$ A.

During nominal grid condition, injected voltage by the DVR should be zero; hence, $V_{DVRN}=0$ and $I_{DVRN}=I_L=25$ A.

During nominal grid condition, $I_{DSTN}=I_L \sin\phi=15$ A.

Voltage Swell Compensation by a Left-Hand UPQC-P

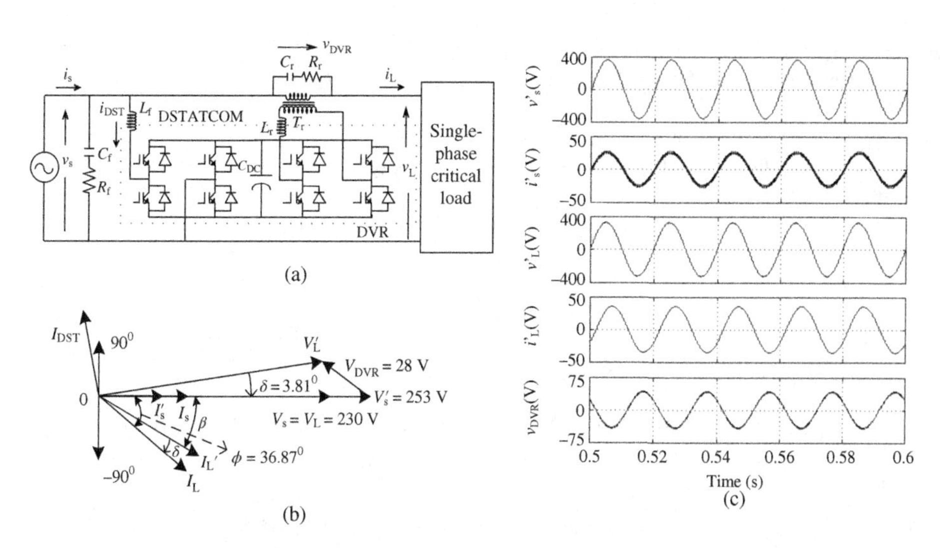

Figure E6.6 (a) A single-phase left-hand UPQC-P. (b) Phasor diagram under voltage swell. (c) Waveforms of a single-phase left-hand UPQC-P under voltage swell

It means that after the voltage swell, the supply voltage increases to $V_s' = 230 \times (1 + 0.10) = 253$ V. The supply current after voltage swell compensation is $I_s' = P_L/V_s' = 4600/253 = 18.18$ A.

a. The voltage rating of the DVR of the UPQC-P is computed as follows.

   There is a voltage swell of +10% in the supply system with a base value of 230 V. Therefore, the DVR of the left-hand UPQC-P must inject a voltage in phase with the load current to provide the required voltage at the load end.

   From the phasor diagram of Figure E6.6b, $V_s' \sin\delta = V_{DVR}\sin\phi$ and $V_s'\cos\delta = V_L' + V_{DVR}\cos\phi$.

   Substituting the values $V_s' = 1.1 \times 230$ V, $V_L' = 230$ V, $\cos\phi = 0.8$, and $\sin\phi = 0.6$, and solving the above equations, one gets $V_{DVR} = 28.05$ V, $\delta = 3.81°$, and $\beta = \phi - \delta = 33°$.

   The voltage rating of the DVR is $V_{DVR} = 28.05$ V.
b. The current rating of the DVR of the UPQC-P must be same as load current: $I_{DVR} = I_L = 25$ A.
c. The VA rating of the DVR of the UPQC-P is calculated as follows.

   $Q_{DVR} = 0.0$ VAR, as the voltage of the DVR is injected in phase with its current.

$$P_{DVR} = V_{DVR} I_{DVR} = 28.05 \times 25 \text{ VA} = 701 \text{ W}.$$

$$S_{DVR} = \sqrt{(P_{DVR}^2 + Q_{DVR}^2)} = P_{DVR} = 701 \text{ VA}.$$

d. The voltage rating of the DSTATCOM of the UPQC-P is equal to AC source voltage ($V_{DST} = V_s' = 253$ V), since it is connected across PCC of 253 V sine waveform.
e. The current rating of the DSTATCOM of the UPQC-P is computed as follows.

   The DSTATCOM of the UPQC-P needs to correct the power factor of the source to unity. Hence, the required reactive power by the DSTATCOM is now lower than the load reactive power as the angle between the supply voltage and the load current is reduced to $\beta = \phi - \delta$.

$$Q_{DST} = V_s' I_L' \sin\beta = 253 \times 25 \sin(33°) = 3450 \text{ VAR}.$$

$$P_{DST} = P_{DVR} = V_{DVR} I_{DVR} = 701 \text{ W}.$$

f. The VA rating of the DSTATCOM is $S_{DST} = \sqrt{(P_{DST}^2 + Q_{DST}^2)} = 3520.5$ VA.

The current through the DSTATCOM during voltage swell is $I_{DST} = S_{DST}/V_s' = 3520.5/253 = 13.91$ A.

Hence, the current rating of the DSTATCOM is 13.91 A.

The VA rating of the UPQC-P during voltage swell is $S_{UPQC\text{-}P(under\ swell)} = S_{DST} + S_{DVR} = 3520.5 + 701 = 4221.5$ VA.

Hence, considering an overall rating (both normal and under voltage swell), ratings of both the compensators are

$$V_{DST} = 253\ \text{V},\quad I_{DST} = 15\ \text{A},\quad S_{DST} = V_{DST}I_{DST} = 253 \times 15 = 3795\ \text{VA}.$$

$$V_{DVR} = 28.05\ \text{V},\quad I_{DVR} = I_L = 25\ \text{A},\quad S_{DVR} = V_{DVR}I_{DVR} = 701\ \text{VA}.$$

g. The overall VA rating of the UPQC-P is $S_{UPQC\text{-}P} = S_{DST} + S_{DVR} = 3795 + 701 = 4496$ VA.

Example 6.7

A single-phase unified power quality compensator (consisting of a DSTATCOM and a DVR using two VSCs with a common DC bus capacitor) is to be designed for a load of 230 V, 50 Hz, 25 A, 0.8 lagging power factor as a right-hand UPQC-S (shown in Figure E6.7). There is a voltage sag of −20% in the supply system with a base value of 230 V. Calculate (a) the voltage rating of the DVR of the UPQC-S, (b) the current rating of the DVR of the UPQC-S, (c) the VA rating of the DVR of the UPQC-S, (d) the voltage rating of the DSTATCOM of the UPQC-S, (e) the current rating of the DSTATCOM of the UPQC-S, (f) the VA rating of the DSTATCOM of the UPQC-S, and (g) total VA rating of the UPQC-S to provide reactive power compensation for unity power factor at PCC with a constant regulated voltage of 230 V at 50 Hz across the load. Consider same rating of both VSCs.

Solution: Given supply voltage $V_s = 230$ V, frequency of the supply $(f) = 50$ Hz, and a load of 230 V, 50 Hz, 25 A, 0.8 lagging power factor. There is a voltage sag of −20% in the supply system with a base value of 230 V. Both VSCs are of same rating.

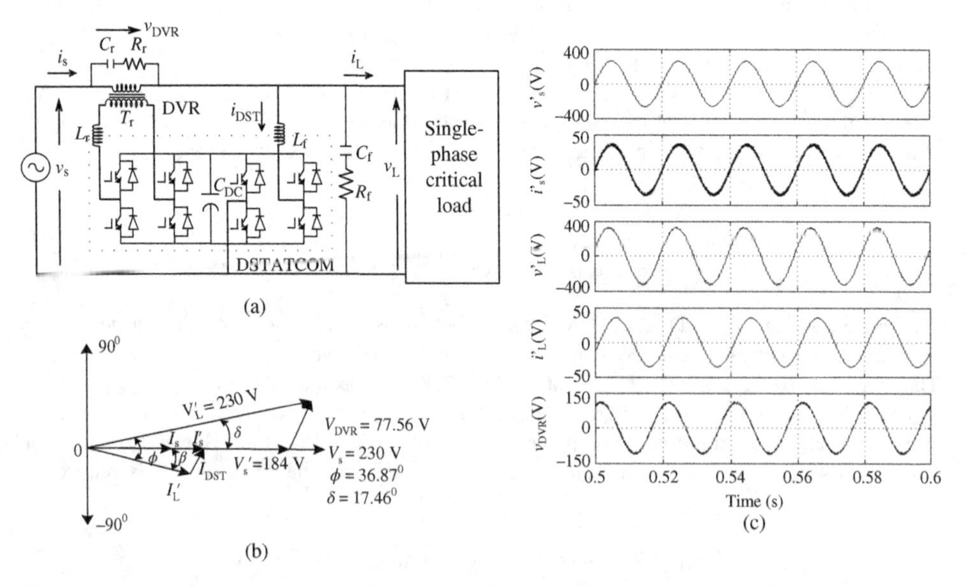

Figure E6.7 (a) A single-phase right-hand UPQC-S. (b) Phasor diagram under voltage sag. (c) Waveforms of a single-phase right-hand UPQC-S under voltage sag

Under steady-state nominal grid condition, both the VSCs operate to share equal reactive power and real power to perform operation of UPF at grid as well as voltage regulation at load terminal.

Under nominal grid operation, voltage and current ratings of both the VSCs are calculated as follows.

Under Steady-State Condition for Unity Power Factor at the AC mains without Voltage Sag

The active power of the load is $P_L = V_L I_L \times PF = 230 \times 25 \times 0.8 = 4600$ W.

The reactive power of the load is $Q_L = V_L I_L \sqrt{(1 - PF^2)} = 230 \times 25 \times 0.6 = 3450$ VAR.

The supply current is $I_s = P_L/V_s = 4600/230 = 20$ A.

The power factor angle is calculated as $\cos^{-1}(PF) = \cos^{-1}(0.8) = 36.86°$.

In the right-hand UPQC-S, for same VA rating of both VSCs, they must have same reactive power rating, as active power rating is same for both VSCs because of the common DC bus capacitor. It means that if one VSC has positive active power (it accepts the active power), another VSC must relieve equal amount of active power (it feeds the active power) as the DC bus capacitor cannot absorb any active power. Moreover, the reactive power rating of both converters is equal to the reactive power of the load. Therefore, each VSC has reactive power rating equal to half of the reactive power of the load: $Q_{DST} = Q_{DVR} = Q_L/2 = 3450/2 = 1725$ VAR.

From the following relation of reactive power of the DVR, the power angle can be computed as follows.

$$Q_{DVR} = V_L I_s \sin\delta = 1725; \text{ therefore, } \delta = 22.02° \text{ and } P_{DVR} = -V_L I_s(1 - \cos\delta) = -339.69 \text{ W}.$$

$$S_{DVR} = \sqrt{(Q_{DVR}^2 + P_{DVR}^2)} = 1758.12 \text{ VA}, \quad I_{DVRN} = I_s = 20 \text{ A}, \quad V_{DVRN} = S_{DVR}/I_{DVRN} = 87.90 \text{ V}.$$

The active power of the DSTATCOM is $P_{DST} = -P_{DVR} = 339.69$ W.

The voltage rating of the DSTATCOM of the UPQC-S is equal to AC load voltage ($V_{DSTN} = 230$ V), since it is connected across the load of 230 V sine waveform.

The current rating of the DSTATCOM of the UPQC-S is computed as follows.

The angle between the supply voltage and the load current is $\beta = \phi - \delta = 36.87° - 22.02° = 14.85°$.

Therefore, DSTATCOM current is $I_{DSTN} = I_L\sqrt{(1 + \cos^2\phi - 2\cos\beta\cos\phi)} = 25\sqrt{(1 + 0.8^2 - 2 \times 0.8 \times 0.967)} = 7.64$ A.

The phase angle between DSTATCOM voltage and current can be calculated as follows.

$$P_{DST} = 339.69 \text{ W} = V_L I_{DSTN}\cos\phi_{DST}, \quad \phi_{DST} = 79.02° \text{ and } Q_{DST} = V_L I_{DST}\sin\phi_{DST} = 1725 \text{ VAR}.$$

Voltage Sag Compensation by a Right-Hand UPQC-S

In the right-hand UPQC-S, for same VA rating of both VSCs, they must have same reactive power rating, as active power rating is same for both VSCs because of the common DC bus capacitor. It means that if one VSC has positive active power (it accepts the active power), another VSC must relieve equal amount of active power (it feeds the active power) as the DC bus capacitor cannot absorb any active power. Moreover, the reactive power rating of both converters is equal to the reactive power of the load. Therefore, each VSC has reactive power rating equal to half of the reactive power of the load: $Q_{DST} = Q_{DVR} = Q_L/2 = 3450/2 = 1725$ VAR.

From the relation of reactive power of the DVR (see the phasor diagram of Figure 6.7b), the power angle is computed as follows.

It means that after the voltage sag, the supply voltage reduces to $V_s' = 230 \times (1 - 0.20) = 184$ V and $X = 0.2$.

$Q_{DVR} = V_L' I_s' \sin\delta = 1725$ VAR; therefore, $\delta = 17.46°$ (as $I_s' = P_L/V_s' = 4600/184 = 25$ A and $V_L' = 230$ V).

a. The voltage rating of the DVR of the UPQC-S is $V_{DVR} = \sqrt{(V_L'^2 + V_s'^2 - 2V_L' V_s' \cos\delta)} = 77.56$ V.
b. The current rating of the DVR of the UPQC-S is $I_{DVR} = I_s' = 25$ A, since it is connected in series with the supply.
c. The VA rating of the DVR of the UPQC-S is $S_{DVR} = V_{DVR} I_{DVR} = 77.56 \times 25 = 1939.02$ VA.

The active power flowing through the DVR is $P_{DVR} = -\sqrt{(S_{DVR}^2 - Q_{DVR}^2)} = -885$ W.

This negative sign denotes that the DVR absorbs the active power that is fed to the DC bus; therefore, the active power of the DSTATCOM is $P_{DST} = -P_{DVR} = 885$ W.

d. The voltage rating of the DSTATCOM of the UPQC-S is equal to AC load voltage ($V_{DST} = 230$ V), since it is connected across the load of 230 V sine waveform.

e. The current rating of the DSTATCOM of the UPQC-S is computed as follows.

The angle between the supply voltage and the load current is $\beta = \phi - \delta = 36.87° - 17.46° = 19.41°$.

Therefore, the DSTATCOM current is $I_{DST} = S_{DST}/V_{DST} = S_{DVR}/V_{DST} = 1939.02/230 = 8.43$ A (as both VSCs have equal VA rating).

The phase angle between DSTATCOM voltage and current can be calculated as follows:

$$P_{DST} = 885 \text{ W} = V_L I_{DST} \cos\phi_{DST},\ \phi_{DST} = 62.86°, \text{ and } Q_{DST} = V'_L I_{DST} \sin\phi_{DST} = 1725 \text{ VAR}.$$

f. The VA rating of the DSTATCOM of the UPQC-S is $S_{DST} = S_{DVR} = \sqrt{(P_{DST}^2 + Q_{DST}^2)} = 1939.02$ VA, which confirms the solution.

The VA rating of the UPQC-S during voltage sag is $S_{UPQC\text{-}S(under\ sag)} = S_{DST} + S_{DVR} = 1939.02 + 1939.02 = 3878.04$ VA.

Hence, considering an overall rating (both normal and under voltage sag), ratings of both the compensators are

$$V_{DVR} = V_{DVRN} = 87.90 \text{ V}, \quad I_{DVR} = 25 \text{ A}, \quad S_{DVR} = V_{DVRN} I_{DVRN} = 87.90 \times 25 = 2197.5 \text{ VA}.$$

$$V_{DST} = 230 \text{ V}, \quad I_{DST} = 8.43 \text{ A}, \quad S_{DST} = V_{DST} I_{DST} = 230 \times 8.43 = 1939 \text{ VA}.$$

g. The VA rating of the UPQC-S is $S_{UPQC\text{-}S} = S_{DST} + S_{DVR} = 1939 + 2197.5 = 4136.5$ VA.

Alternatively, if both the VSCs of same VA rating are considered for same inventory, then $S_{DST} = S_{DVR} = 2197.5$ VA.

The overall VA rating of the UPQC-S is $S_{UPQC\text{-}S} = S_{DST} + S_{DVR} = 2197.5 + 2197.5 = 4395$ VA.

Example 6.8

A single-phase unified power quality compensator (consisting of a DSTATCOM and a DVR using two VSCs with a common DC bus capacitor) is to be designed for a load of 230 V, 50 Hz, 25 A, 0.8 lagging power factor as a left-hand UPQC-S (shown in Figure E6.8). There is a voltage sag of –20% in the supply system with a base value of 230 V. Calculate (a) the voltage rating of the DSTATCOM of the UPQC-S, (b) the current rating of the DSTATCOM of the UPQC-S, (c) the VA rating of the DSTATCOM of the UPQC-S, (d) the voltage rating of the DVR of the UPQC-S, (e) the current rating of the DVR of the UPQC-S, (f) the VA rating of the DVR of the UPQC-S, and (g) total VA rating of the UPQC-S to provide reactive power compensation for unity power factor at PCC with a constant regulated voltage of 230 V at 50 Hz across the load. Consider same rating of both VSCs.

Solution: Given supply voltage $V_s = 230$ V, frequency of the supply $(f) = 50$ Hz, and a load of 230 V, 50 Hz, 25 A, 0.8 lagging power factor. There is a voltage sag of –20% in the supply system with a base value of 230 V. Both VSCs are of same rating.

Under nominal grid condition, voltage and current ratings of both the VSCs are calculated as follows.

The active power of the load is $P_L = V_L I_L \times \text{PF} = 230 \times 25 \times 0.8 = 4600$ W.

The reactive power of the load is $Q_L = V_L I_L \sqrt{(1 - \text{PF}^2)} = 230 \times 25 \times 0.6 = 3450$ VAR.

The supply current is $I_s = P_L/V_s = 4600/230 = 20$ A.

In the left-hand UPQC-S, for same VA rating of both VSCs, they must have same reactive power rating, as active power rating is same for both VSCs because of the common DC bus capacitor. It means that if one VSC has positive active power (it accepts the active power), another VSC must relieve equal amount of active power (it feeds the active power) as the DC bus capacitor cannot absorb any active power.

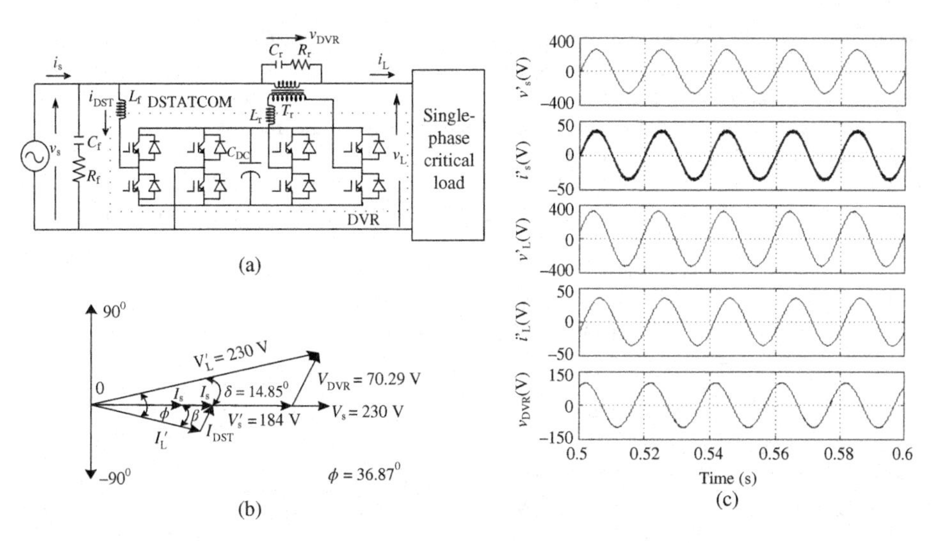

Figure E6.8 (a) A single-phase left-hand UPQC-S. (b) Phasor diagram under voltage sag. (c) Waveforms of a single-phase left-hand UPQC-S under voltage sag

Moreover, the reactive power rating of both converters is equal to the reactive power of the load. Therefore, each VSC has reactive power rating equal to half of the reactive power of the load: $Q_{DST} = Q_{DVR} = Q_L/2 = 3450/2 = 1725$ VAR.

From the following relation of reactive power of the DSTATCOM, the power angle can be computed as follows.

The reactive power of the DSTATCOM is $Q_{DST} = V_s I_L \sin\beta = 1725$ VAR, $\beta = 17.46°$, and $\delta = \phi - \beta = 36.87° - 17.46° = 19.41°$.

The active power of the DSTATCOM is $P_{DST} = V_s I_s - V_s I_L \cos\beta = 4600 - 5485.15 = -885.15$ W.

The current rating of the DSTATCOM is $I_{DSTN} = S_{DST}/V_{DSTN} = \sqrt{(P_{DST}^2 + Q_{DST}^2)}/V_{DSTN} = 8.43$ A.

$$\phi_{DST} = \cos^{-1}(P_{DST}/V_{DSTN} I_{DSTN}) = 117.22°.$$

$$S_{DST} = V_{DSTN} I_{DSTN} = V_s I_{DSTN} = 230 \times 8.43 = 1938.90 \text{ VA}.$$

The active power of the DVR is $P_{DVR} = -P_{DST} = 885.15$ W.

The current rating of the DVR of the UPQC-S is equal to AC load current ($I_{DVR} = I_L = 25$ A), since it is connected in series between the load and the AC mains.

In this case, the VA rating of both VSCs must be the same because of equal sharing of reactive power and same active power due to the common DC bus.

The voltage rating of the DVR of the UPQC-S is $V_{DVRN} = S_{DVR}/I_{DVRN} = 1938.90/25 = 77.56$ V.

$$Q_{DVR} = V_{DVRN} I_{DVRN} \sin\phi_{DVR} = 1725 \text{ VAR}, \ \phi_{DVR} = 62.79°, \text{ and } P_{DVR} = -P_{DST} = 885.15 \text{ W}.$$

$$S_{DVR} = \sqrt{(P_{DVR}^2 + Q_{DVR}^2)} = 1938.95 \text{ VA}.$$

First Method: Voltage Sag Compensation by a Left-Hand UPQC-S

In the left-hand UPQC-S, for same VA rating of both VSCs, they must have same reactive power rating, as active power rating is same for both VSCs because of the common DC bus capacitor. It means that if one VSC has positive active power (it accepts the active power), another VSC must relieve equal amount of active power (it feeds the active power) as the DC bus capacitor cannot absorb any active power. Moreover, the reactive power rating of both converters is equal to the reactive power of the load. Therefore, each VSC has reactive power rating equal to half of the reactive power of the load: $Q_{DST} = Q_{DVR} = Q_L/2 = 3450/2 = 1725$ VAR.

The supply voltage under voltage sag is $V_s' = V_s(1 - X) = V_s(1 - 0.2) = 184$ V.

From the following relation of reactive power of the DSTATCOM, the power angle can be computed as follows.

The reactive power of the DSTATCOM is $Q_{DST} = V_s' I_L' \sin\beta = 1725$ VAR.

On solving the above equation, $\beta = 22.02°$ and $\delta = \phi - \beta = 36.87° - 22.02° = 14.85°$.

The active power of the DSTATCOM is $P_{DST} = V_s' I_s' - V_s' I_L' \cos\beta = V_s I_s - V_s' I_L' \cos\beta = 230 \times 20 - 184 \times 25 \cos 22.02° = 4600 - 4264.31 = 335.69$ W.

a. The voltage rating of the DSTATCOM of the UPQC-S is calculated as follows.

   The DSTATCOM is connected at PCC where the voltage is reduced to 184 V due to a sag of −0.2. Therefore, the voltage rating of the DSTATCOM is $V_{DST} = 184$ V.

b. The current rating of the DSTATCOM is $I_{DST} = S_{DST}/V_{DST} = \sqrt{(P_{DST}^2 + Q_{DST}^2)}/V_{DST} = 1757.36/184 = 9.55$ A.

c. The VA rating of the DSTATCOM of the UPQC-S is $S_{DST} = V_{DST} I_{DST} = V_s' I_{DST} = 184 \times 9.55 = 1757.36$ VA.

   The active power of the DVR is $P_{DVR} = -P_{DST} = -335.69$ W.

   The current rating of the DVR of the UPQC-S is equal to AC load current ($I_{DVR} = I_L = 25$ A), since it is connected in series between the load and the AC mains.

   In this case, the VA rating of both VSCs must be the same because of equal sharing of reactive power and same active power due to the common DC bus.

d. The voltage rating of the DVR of the UPQC-S is $V_{DVR} = S_{DVR}/I_{DVR} = 1757.36/25 = 70.29$ V.

e. The current rating of the DVR is same as load current as it is connected in series with the load: $I_{DVR} = I_L = 25$ A.

$$P_{DVR} = -P_{DST} = -335.69 \text{ W}.$$

$$Q_{DVR} = V_{DVR} I_{DVR} \sin\phi_{DVR} = 1725.11 \text{ VAR}.$$

$$S_{DVR} = \sqrt{(P_{DVR}^2 + Q_{DVR}^2)} = 1757.36 \text{ VA}.$$

f. The VA rating of the UPQC-S is $S_{UPQC\text{-}S} = S_{DST} + S_{DVR} = 1757.36 + 1757.36 = 3514.72$ VA.

Second Method: Voltage Sag Compensation by a Left-Hand UPQC-S

It means after the voltage sag, the supply voltage reduces to $V_s' = V_s(1 - X) = 230 \times (1 - 0.20) = 184$ V, $X = 0.2$, $K_0 = V_s/V_s' = 1.25$, and $n_0 = 1/K_0 = 0.8$.

The supply current after sag compensation is $I_s' = P_L/V_s' = 4600/184 = 25$ A.

In the left-hand UPQC-S, for same VA rating of both VSCs, they must have same reactive power rating, as active power rating is same for both VSCs because of the common DC bus capacitor. It means that if one VSC has positive active power (it accepts the active power), another VSC must relieve equal amount of active power (it feeds the active power) as the DC bus capacitor cannot absorb any active power. Moreover, the reactive power rating of both converters is equal to the reactive power of the load. Therefore, each VSC has reactive power rating equal to half of the reactive power of the load: $Q_{DST} = Q_{DVR} = Q_L/2 = 3450/2 = 1725$ VAR.

a. Since the DSTATCOM is connected across the AC mains, the voltage rating of the DSTATCOM is same as AC mains voltage: $V_{DST} = V_s' = 230 \times (1 - 0.2) = 184$ V.

Under the voltage sag, from the following relation of reactive power of the DSTATCOM, the power angle can be computed as follows.

$$Q_{DST} = V'_s I'_L \sin\beta = n_0 V_s I'_L \sin\beta = 1725\ \text{VAR}, \quad \beta = 22.02°,$$
$$\delta = \phi - \beta = 36.87° - 22.02° = 14.85°.$$

$$P_{DST} = V'_s I'_s - V'_s I'_L \cos\beta = 4600 - 4264.31 = 335.69\ \text{W}.$$

$$\phi_{DST} = \tan^{-1}(Q_{DST}/P_{DST}) = 78.99°.$$

b. The current rating of the DSTATCOM is $I_{DST} = P_{DST}/(V_{DST}\cos\phi_{DST}) = 9.55$ A.
c. The VA rating of the DSTATCOM is $S_{DST} = V_{DST}I_{DST} = V_s I_{DST} = 184 \times 9.55 = 1757.36$ VA.
   The rating of the DVR is computed from the common relations of the DSTATCOM and DVR as follows.
   The active power of the DVR is $P_{DVR} = -P_{DST} = -335.69$ W.
d. The current rating of the DVR of the UPQC-S is equal to AC load current ($I_{DVR} = I_L = 25$A), since it is connected in series between the load and the AC mains.
e. The voltage rating of the DVR of the UPQC-S is $V_{DVR} = S_{DVR}/I_{DVR} = 1757.36/25 = 70.29$ V.

$$P_{DVR} = -P_{DST} = -335.69\ \text{W}.$$

f. The VA rating of the DVR is computed as follows.

$$Q_{DVR} = V_{DVR}I_{DVR}\sin\phi_{DVR} = 1725\ \text{VAR}.$$

$$S_{DVR} = \sqrt{(P^2_{DVR} + Q^2_{DVR})} = 1757.36\ \text{VA}.$$

The VA rating of the UPQC-S during voltage sag is $S_{UPQC\text{-}S(under\ sag)} = S_{DST} + S_{DVR} = 1757.36 + 1757.36 = 3514.72$ VA.

Hence, considering an overall rating (both normal and under voltage sag), ratings of both the compensators are

$$V_{DVR} = V_{DVRN} = 77.56\ \text{V}, \quad I_{DVR} = 25\ \text{A}, \quad S_{DVR} = V_{DVR}I_{DVR} = 77.56 \times 25 = 1939\ \text{VA}.$$

$$V_{DST} = 230\ \text{V}, \quad I_{DST} = 9.55\ \text{A}, \quad S_{DST} = V_{DST}I_{DST} = 230 \times 9.55 = 2196\ \text{VA}.$$

g. The VA rating of the UPQC-S is $S_{UPQC\text{-}S} = S_{DST} + S_{DVR} = 1939 + 2196.5 = 4135.5$ VA.
   Alternatively, if both the VSCs of same VA rating are considered for same inventory, then $S_{DST} = S_{DVR} = 2197.5$ VA.
   The VA rating of the UPQC-S is $S_{UPQC\text{-}S} = S_{DST} + S_{DVR} = 2196.5 + 2196.5 = 4393$ VA.

Example 6.9

A single-phase unified power quality compensator (consisting of a DSTATCOM and a DVR using two VSCs with a common DC bus capacitor) is to be designed for a load of 230 V, 50 Hz, 25 A, 0.8 lagging power factor as a right-hand UPQC-S (shown in Figure E6.9). There is a voltage sag of −20% in the supply system with a base value of 230 V. Calculate (a) the voltage rating of the DVR of the UPQC-S, (b) the current rating of the DVR of the UPQC-S, (c) the VA rating of the DVR of the UPQC-S, (d) the voltage rating of the DSTATCOM of the UPQC-S, (e) the current rating of the DSTATCOM of the UPQC-S, (f) the VA rating of the DSTATCOM of the UPQC-S, and (g) total VA rating of the UPQC-S to provide reactive power compensation for unity power factor at PCC with a constant regulated voltage of

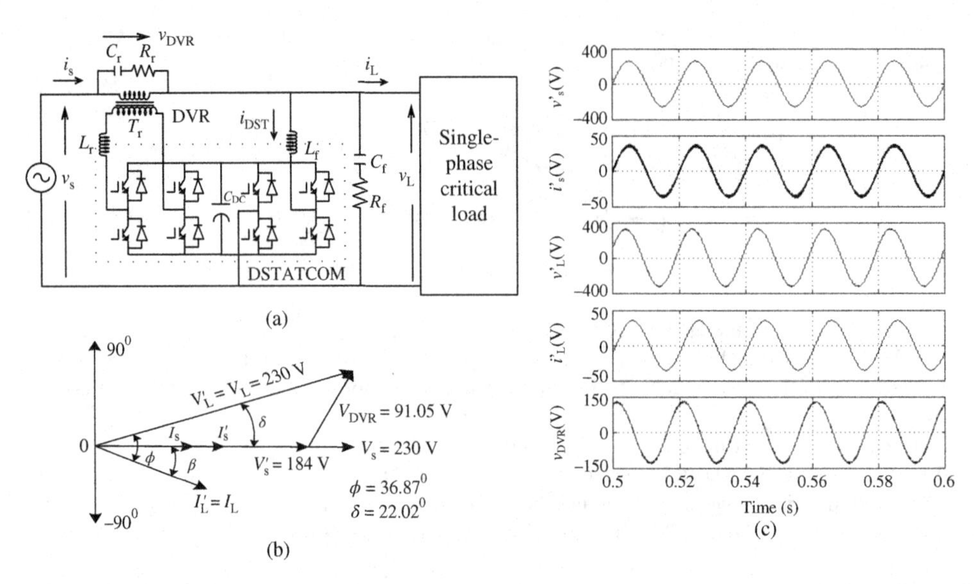

Figure E6.9 (a) A single-phase right-hand UPQC-S. (b) Phasor diagram under voltage sag. (c) Waveforms of a single-phase right-hand UPQC-S under voltage sag

230 V at 50 Hz across the load. Consider same power angle for equal reactive power sharing in both VSCs in steady-state condition for unity power factor at the AC mains without voltage sag condition.

Solution: Given supply voltage $V_s = 230$ V, frequency of the supply $(f) = 50$ Hz, and a load of 230 V, 50 Hz, 25 A, 0.8 lagging power factor. There is a voltage sag of −20% in the supply system with a base value of 230 V. Same power angle is to be maintained for equal reactive power sharing in both VSCs in steady-state condition for unity power factor at the AC mains without voltage sag condition.

Under Steady-State Condition for Unity Power Factor at the AC Mains without Voltage Sag

The active power of the load is $P_L = V_L I_L \times \text{PF} = 230 \times 25 \times 0.8 = 4600$ W.

The reactive power of the load is $Q_L = V_L I_L \sqrt{(1 - \text{PF}^2)} = 230 \times 25 \times 0.6 = 3450$ VAR.

The supply current is $I_s = P_L/V_s = 4600/230 = 20$ A.

The power factor angle can be calculated as $\cos^{-1}(\text{PF}) = \cos^{-1}(0.8) = 36.86°$.

In the right-hand UPQC-S, for same VA rating of both VSCs, they must have same reactive power rating, as active power rating is same for both VSCs because of the common DC bus capacitor. It means that if one VSC has positive active power (it accepts the active power), another VSC must relieve equal amount of active power (it feeds the active power) as the DC bus capacitor cannot absorb any active power. Moreover, the reactive power rating of both converters is equal to the reactive power of the load. Therefore, each VSC has reactive power rating equal to half of the reactive power of the load: $Q_{DST} = Q_{DVR} = Q_L/2 = 3450/2 = 1725$ VAR.

From the following relation of reactive power of the DVR, the power angle is computed as follows.

$$Q_{DVR} = V_L I_s \sin\delta = 1725; \text{ therefore, } \delta = 22.02° \text{ and } P_{DVR} = -V_L I_s (1 - \cos\delta) = -339.69 \text{ W}.$$

$$S_{DVR} = \sqrt{(Q_{DVR}^2 + P_{DVR}^2)} = 1758.12 \text{ VA}, \quad I_{DVRN} = I_s = 20 \text{ A}, \quad V_{DVRN} = S_{DVR}/I_{DVRN} = 87.90 \text{ V}.$$

This negative sign denotes that the DVR absorbs the active power that is fed to the DC bus; therefore, the active power of the DSTATCOM is $P_{DST} = -P_{DVR} = 339.69$ W.

The voltage rating of the DSTATCOM of the UPQC-S is equal to AC load voltage ($V_{DSTN} = 230$ V), since it is connected across the load of 230 V sine waveform.

The current rating of the DSTATCOM of the UPQC-S is computed as follows.

The angle between the supply voltage and the load current is $\beta = \phi - \delta = 36.87° - 22.02° = 14.85°$.

Therefore, DSTATCOM current is $I_{DSTN} = I_L\sqrt{(1 + \cos^2\phi - 2\cos\beta\cos\phi)} = 25\sqrt{(1 + 0.8^2 - 2 \times 0.8 \times 0.967)} = 7.64$ A.

The phase angle between DSTATCOM voltage and current can be calculated as follows.

$$P_{DST} = 339.69 \text{ W} = V_L I_{DSTN}\cos\phi_{DST},\ \phi_{DST} = 79.02°, \text{ and } Q_{DST} = V_L I_{DST}\sin\phi_{DST} = 1725 \text{ VAR}.$$

$$S_{DST} = \sqrt{(P_{DST}^2 + Q_{DST}^2} = 1758.12 \text{ VA}.$$

Voltage Sag Compensation by a Right-Hand UPQC-S

It means after the voltage sag, the supply voltage reduces to $V_s' = V_s(1 - X) = 230 \times (1 - 0.20) = 184$ V, $X = 0.2$, $K_0 = V_s/V_s' = 1.25$, and $n_0 = 1/K_0 = 0.8$.

The supply current after sag compensation is $I_s' = P_L/V_s' = 4600/184 = 25$ A.

In the right-hand UPQC-S, for same power angle of $\delta = 22.02°$, as above, the reactive power rating is to be changed, but the active power rating (however, its value may change) needs to be same for both VSCs because of the common DC bus capacitor. It means that if one VSC has positive active power (it accepts the active power), another VSC must relieve equal amount of active power (it feeds the active power) as the DC bus capacitor cannot absorb any active power.

Under the voltage sag, from the following relation of reactive power of the DVR, the reactive power can be computed as $Q_{DVR} = K_0 V_s' I_s' \sin\delta = (V_s/V_s')V_s' I_s' \sin\delta = 2155.85$ VAR, for same power angle of $\delta = 22.02°$.

$$P_{DVR} = -K_0 V_s' I_s'(n_0 - \cos\delta) = -(V_s/V_s')V_s' I_s'\{(V_s'/V_s) - \cos\delta\} = 730.55 \text{ W}.$$

a. The voltage rating of the DVR is $V_{DVR} = V_L'\sqrt{(1 + n_0^2 - 2n_0\cos\delta)} = 230\sqrt{(1 + 0.8^2 - 2 \times 0.8\cos 22.02°)} = 91.05$ V.

b. The current rating of the DVR is equal to the supply current after sag compensation: $I_{DVR} = I_s = P_L/V_s = 4600/184 = 25$ A.

c. The VA rating of the DVR is $S_{DVR} = V_{DVR} I_{DVR} = 91.05 \times 25 = 2276.27$ VA.

   The active power of the DSTATCOM is $P_{DST} = -P_{DVR} = -730.55$ W.

d. The voltage rating of the DSTATCOM of the UPQC-S is equal to AC load voltage ($V_{DST} = 230$ V), since it is connected across the load of 230 V sine waveform.

e. The current rating of the DSTATCOM of the UPQC-S is computed as follows.

   The angle between the supply voltage and the load current is $\beta = \phi - \delta = 36.87° - 22.02° = 14.85°$.

   Therefore, DSTATCOM current is $I_{DST} = I_L'\sqrt{[(1 + K_0^2\cos^2\phi - 2K_0\cos\beta\cos\phi)} = 25\sqrt{(1 + 1.25^2 \times 0.8^2 - 2 \times 1.25 \times 0.8 \times 0.967)} = 6.46$ A.

   The phase angle between DSTATCOM voltage and current can be calculated as follows.

$$P_{DST} = -730.55 \text{ W} = V_L I_{DST}\cos\phi_{DST},\ \phi_{DST} = 119.5°, \text{ and } Q_{DST} = V_L' I_{DST}\sin\phi_{DST} = 1293.88 \text{ VA}.$$

f. The VA rating of the DSTATCOM of the UPQC-S is $S_{DST} = V_{DST} I_{DST} = 230 \times 6.46$ VA $= 1485.88$ VA.

   The VA rating of the UPQC-S during voltage sag is $S_{UPQC\text{-}S(under\ sag)} = S_{DST} + S_{DVR} = 1485.88 + 2276.27 = 3762.15$ VA.

   Hence, considering an overall rating (both normal and under voltage sag), ratings of both the compensators are

$$V_{DVR} = 91.05 \text{ V},\quad I_{DVR} = 25 \text{ A},\quad S_{DVR} = V_{DVR} I_{DVR} = 91.05 \times 25 = 2276 \text{ VA}.$$

$$V_{DST} = 230 \text{ V},\quad I_{DST} = 7.64 \text{ A},\quad S_{DST} = V_{DST} I_{DST} = 230 \times 7.64 = 1757.2 \text{ VA}.$$

g. The overall VA rating of the UPQC-S is $S_{UPQC\text{-}S} = S_{DST} + S_{DVR} = 1757.2 + 2276.27 = 4033.47$ VA.

Example 6.10

A single-phase unified power quality compensator (consisting of a DSTATCOM and a DVR using two VSCs with a common DC bus capacitor) is to be designed for a load of 230 V, 50 Hz, 25 A, 0.8 lagging power factor as a left-hand UPQC-S (shown in Figure E6.10). There is a voltage sag of –20% in the supply system with a base value of 230 V. Calculate (a) the voltage rating of the DVR of the UPQC-S, (b) the current rating of the DVR of the UPQC-S, (c) the VA rating of the DVR of the UPQC-S, (d) the voltage rating of the DSTATCOM of the UPQC-S, (e) the current rating of the DSTATCOM of the UPQC-S, (f) the VA rating of the DSTATCOM of the UPQC-S, and (g) total VA rating of the UPQC-S to provide reactive power compensation for unity power factor at PCC with a constant regulated voltage of 230 V at 50 Hz across the load. Consider same power angle for equal reactive power sharing in both VSCs in steady-state condition for unity power factor at the AC mains without voltage sag condition.

Solution: Given supply voltage $V_s = 230$ V, frequency of the supply $(f) = 50$ Hz, and a load of 230 V, 50 Hz, 25 A, 0.8 lagging power factor. There is a voltage sag of –20% in the supply system with a base value of 230 V. Both VSCs are of same rating.

It means after the voltage sag, the supply voltage reduces to $V_s = 230 \times (1 - 0.20) = 184$ V and $X = 0.2$.

Under Steady-State Condition for Unity Power Factor at the AC Mains without Voltage Sag Compensation

The active power of the load is $P_L = V_L I_L \times PF = 230 \times 25 \times 0.8 = 4600$ W.

The reactive power of the load is $Q_L = V_L I_L \sqrt{(1 - PF^2)} = 230 \times 25 \times 0.6 = 3450$ VAR.

The supply current is $I_s = P_L / V_s = 4600/230 = 20$ A.

In the left-hand UPQC-S, for same VA rating of both VSCs, they must have same reactive power rating, as active power rating is same for both VSCs because of the common DC bus capacitor. It means that if one VSC has positive active power (it accepts the active power), another VSC must relieve equal amount of active power (it feeds the active power) as the DC bus capacitor cannot absorb any active power. Moreover, the reactive power rating of both converters is equal to the reactive power of the load. Therefore, each VSC has reactive power rating equal to half of the reactive power of the load: $Q_{DST} = Q_{DVR} = Q_L/2 = 3450/2 = 1725$ VAR.

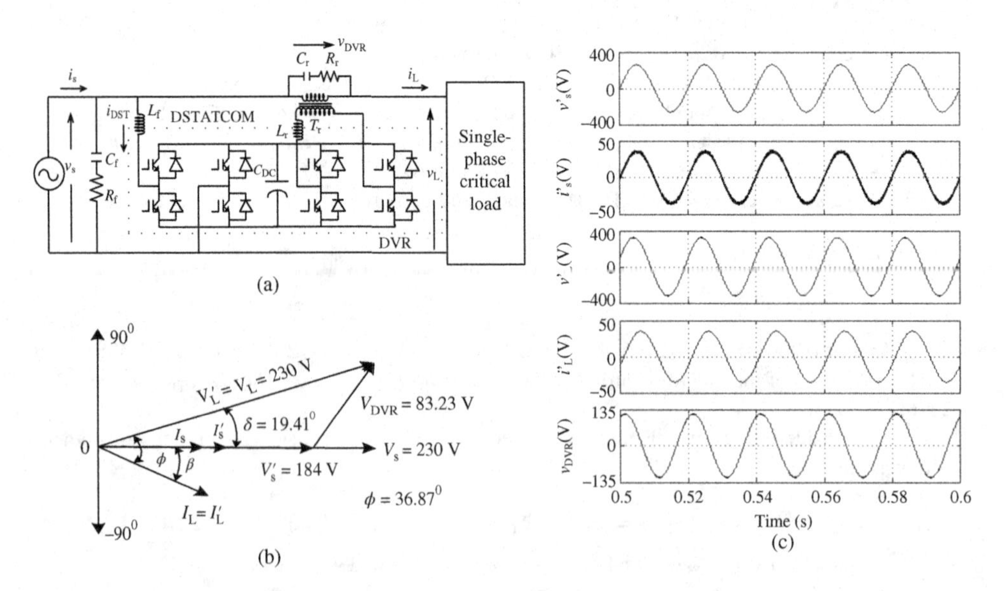

Figure E6.10 (a) A single-phase left-hand UPQC-S. (b) Phasor diagram under voltage sag. (c) Waveforms of a single-phase left-hand UPQC-S under voltage sag

From the following relation of reactive power of the DSTATCOM, the power angle can be computed as follows.

The reactive power of the DSTATCOM is $Q_{DST} = V_s I_L \sin\beta = 1725$ VAR, $\beta = 17.46°$, and $\delta = \phi - \beta = 36.87° - 17.46° = 19.41°$.

The active power of the DSTATCOM is $P_{DST} = V_s I_s - V_s I_L \cos\beta = 4600 - 5485.15 = -885.15$ W.

The current rating of the DSTATCOM is $I_{DSTN} = S_{DST}/V_{DSTN} = \sqrt{(P^2_{DST} + Q^2_{DST})}/V_{DSTN} = 8.43$ A.

$$\phi_{DST} = \cos^{-1}(P_{DST}/V_{DSTN} I_{DSTN}) = 117.22°.$$

$$S_{DST} = V_{DSTN} I_{DSTN} = V_s I_{DSTN} = 230 \times 8.43 = 1938.90 \text{ VA}.$$

This negative sign denotes that the DSTATCOM absorbs the active power that is fed to the DC bus; therefore, the active power of the DVR is $P_{DVR} = -P_{DST} = 885.15$ W.

The current rating of the DVR of the UPQC-S is equal to AC load current ($I_{DVR} = I_L = 25$ A), since it is connected in series between the load and the AC mains.

In this case, the VA rating of both VSCs must be the same because of equal sharing of reactive power and same active power due to the common DC bus.

The voltage rating of the DVR of the UPQC-S is $V_{DVRN} = S_{DVR}/I_{DVRN} = 1938.90/25 = 77.56$ V.

$$Q_{DVR} = V_{DVRN} I_{DVRN} \sin\phi_{DVR} = 1725 \text{ VAR}, \quad \phi_{DVR} = 62.79°.$$

$$P_{DVR} = -P_{DST} = 885.15 \text{ W}.$$

$$S_{DVR} = \sqrt{(P^2_{DVR} + Q^2_{DVR})} = 1938.95 \text{ VA}.$$

Voltage Sag Compensation by a Left-Hand UPQC-S

It means after the voltage sag, the supply voltage reduces to $V'_s = V_s(1 - X) = 230 \times (1 - 0.20) = 184$ V, $X = 0.2$, $K_0 = V_s/V'_s = 1.25$, and $n_0 = 1/K_0 = 0.8$.

The supply current after sag compensation is $I'_s = P_L/V'_s = 4600/184 = 25$ A.

In the left-hand UPQC-S, for same power angle of $\delta = 19.41°$, as above, the reactive power rating is to be changed, but the active power rating (however, its value may change) needs to be same for both VSCs because of the common DC bus capacitor. It means that if one VSC has positive active power (it accepts the active power), another VSC must relieve equal amount of active power (it feeds the active power) as the DC bus capacitor cannot absorb any active power.

a. Since the DSTATCOM is connected across the AC mains, the voltage rating of the DSTATCOM is same as AC mains voltage: $V_{DST} = V'_s = 230 \times (1 - 0.2) = 184$ V.

   Under the voltage sag, the same power angle of $\delta = 19.41°$, from the following relation of reactive power of the DSTATCOM, can be computed as $\delta = \phi - \beta = 36.87° - 17.46° = 19.41°$.

   $Q_{DST} = V'_s I'_L \sin\beta = n_0 V_s I'_L \sin\beta = 1380$ VAR, for same power angle of $\delta = 19.41°$ and $\beta = 17.46°$.

$$P_{DST} = V'_s I'_s - V'_s I'_L \cos\beta = 4600 - 4388.06 = 211.94 \text{ W}.$$

$$\phi_{DST} = \cos^{-1}(P_{DST}/V_{DST} I_{DST}) = 81.31°.$$

b. The current rating of the DSTATCOM is $I_{DST} = P_{DST}/(V_{DST} \cos\phi_{DST}) = 7.59$ A.
c. The VA rating of the DSTATCOM is $S_{DST} = V_{DST} I_{DST} = V_s I_{DST} = 184 \times 7.59 = 1396.37$ VA.

   The rating of the DVR is computed from the common relations of the DSATCOM and the DVR as follows.

   The active power of the DVR is $P_{DVR} = -P_{DST} = -211.94$ W.
d. The current rating of the DVR of the UPQC-S is equal to AC load current ($I_{DVR} = I_L = 25$ A), since it is connected in series between the load and the AC mains.

e. The voltage rating of the DVR of the UPQC-S is computed as follows.
The remaining reactive power of the load (which not supplied by the DSTATCOM) has to be supplied by the DVR, which is computed as follows.

$$Q_{DVR} = 3450 - 1380 = 2070\ \text{VAR}.$$

$$S_{DVR} = \sqrt{(P_{DVR}^2 + Q_{DVR}^2)} = 2080.64\ \text{VA}.$$

$$V_{DVR} = S_{DVR}/I_{DVR} = 2080.64/25 = 83.23\ \text{V}.$$

f. The VA rating of the DVR is $S_{DVR} = \sqrt{(P_{DVR}^2 + Q_{DVR}^2)} = 2080.64$ VA.
The VA rating of the UPQC-S during voltage sag is $S_{UPQC\text{-}S(\text{under sag})} = S_{DST} + S_{DVR} = 1396.37 + 2080.64 = 3477.01$ VA.
Hence, considering an overall rating (both normal and under voltage sag), ratings of both the compensators are

$$V_{DVR} = 83.23\ \text{V}, \quad I_{DVR} = 25\ \text{A}, \quad S_{DVR} = V_{DVR} I_{DVR} = 83.23 \times 25 = 2080.75\ \text{VA}.$$

$$V_{DST} = 230\ \text{V}, \quad I_{DST} = 8.43\ \text{A}, \quad S_{DST} = V_{DST} I_{DST} = 230 \times 8.49 = 1938.9\ \text{VA}.$$

g. The VA rating of the UPQC-S is $S_{UPQC\text{-}S} = S_{DST} + S_{DVR} = 1938.9 + 2080.75 = 4019.55$ VA.
However, in case of same inventory for both the inverters, the VA rating of both the power converters is $S_{DST} = S_{DVR} = 2080.75$ VA.
Hence, overall VA rating of the UPQC-S under same inventory for both the inverters is $S_{UPQC\text{-}S} = S_{DST} + S_{DVR} = 2080.75 + 2080.75 = 4161.5$ VA.

Example 6.11

A single-phase unified power quality compensator (consisting of a DSTATCOM and a DVR using two VSCs with a common DC bus capacitor) is to be designed for a load of 230 V, 50 Hz, 25 A, 0.8 lagging power factor as a left-hand UPQC-S (shown in Figure E6.11). There is a voltage swell of +20% in the supply system with a base value of 230 V. Calculate (a) the voltage rating of the DVR of the UPQC-S, (b) the current rating of the DVR of the UPQC-S, (c) the VA rating of the DVR of the UPQC-S, (d) the voltage rating of the DSTATCOM of the UPQC-S, (e) the current rating of the DSTATCOM of the UPQC-S, (f) the VA rating of the DSTATCOM of the UPQC-S, and (g) total VA rating of the UPQC-S to provide reactive power compensation for unity power factor at PCC with a constant regulated voltage of 230 V at 50 Hz across the load. Consider same power angle for equal reactive power sharing in both VSCs in steady-state condition for unity power factor at the AC mains without voltage swell condition.

Solution: Given supply voltage $V_s = 230$ V, frequency of the supply $(f) = 50$ Hz, and a load of 230 V, 50 Hz, 25 A, 0.8 lagging power factor. There is a voltage swell of +20% in the supply system with a base value of 230 V. Both VSCs are to be designed of same rating.

Under Steady-State Condition without Voltage Swell

The active power of the load is $P_L = V_L I_L \times \text{PF} = 230 \times 25 \times 0.8 = 4600$ W.
The reactive power of the load is $Q_L = V_L I_L \sqrt{(1 - \text{PF}^2)} = 230 \times 25 \times 0.6 = 3450$ VAR.
The supply current is $I_s = P_L/V_s = 4600/230 = 20$ A.
In the left-hand UPQC-S, for same VA rating of both VSCs, they must have same reactive power rating, as active power rating is same for both VSCs because of the common DC bus capacitor. It means that if one VSC has positive active power (it accepts the active power), another VSC must relieve equal amount of active power (it feeds the active power) as the DC bus capacitor cannot absorb any active power. Moreover, the reactive power rating of both converters is equal to the reactive power of the load.

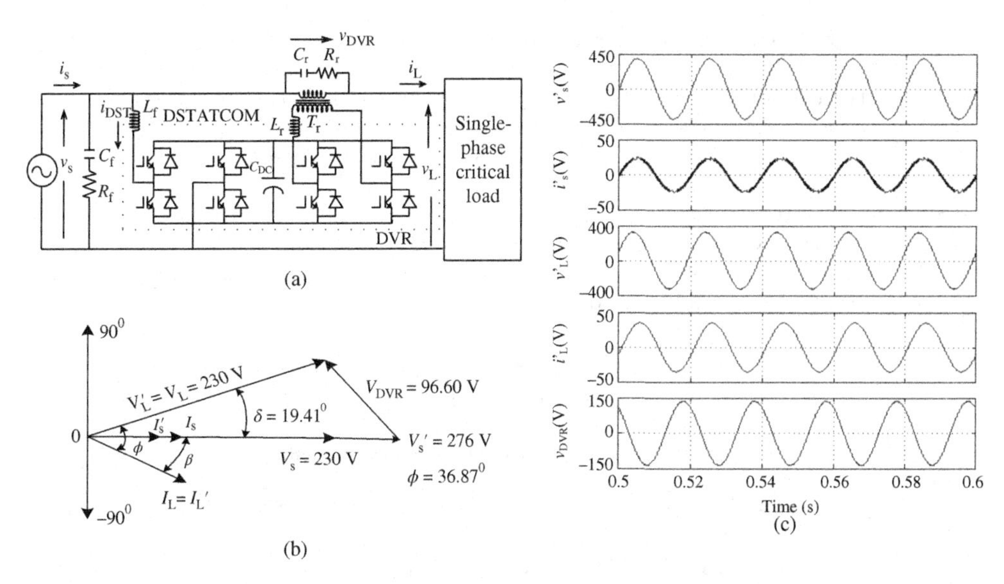

Figure E6.11 (a) A single-phase left-hand UPQC-S. (b) Phasor diagram under voltage swell. (c) Waveforms of a single-phase left-hand UPQC-S under voltage swell

Therefore, each VSC has reactive power rating equal to half of the reactive power of the load: $Q_{DST} = Q_{DVR} = Q_L/2 = 3450/2 = 1725$ VAR.

From the following relation of reactive power of the DSTATCOM, the power angle can be computed as follows.

The reactive power of the DSTATCOM is $Q_{DST} = V_sI_L \sin \beta = 1725$ VAR, $\beta = 17.46°$, and $\delta = \phi - \beta = 36.87° - 17.46° = 19.41°$.

The active power of the DSTATCOM is $P_{DST} = V_sI_s - V_sI_L \cos \beta = 4600 - 5485.15 = -885.15$ W.

The current rating of the DSTATCOM is $I_{DSTN} = S_{DST}/V_{DSTN} = \sqrt{(P_{DST}^2 + Q_{DST}^2)}/V_{DSTN} = 8.43$ A.

$$\phi_{DST} = \cos^{-1}(P_{DST}/V_{DSTN}I_{DSTN}) = 117.22°, \quad S_{DST} = V_{DSTN}I_{DSTN} = V_sI_{DSTN} = 230 \times 8.43 = 1938.90 \text{ VA}.$$

This negative sign denotes that the DSTATCOM absorbs the active power that is fed to the DC bus; therefore, the active power of the DVR is $P_{DVR} = -P_{DST} = 885.15$ W.

The current rating of the DVR of the UPQC-S is equal to AC load current ($I_{DVR} = I_L = 25$ A), since it is connected in series between the load and the AC mains.

In this case, the VA rating of both VSCs must be the same because of equal sharing of reactive power and same active power due to the common DC bus.

The voltage rating of the DVR of the UPQC-S is $V_{DVRN} = S_{DVR}/I_{DVRN} = 1938.90/25 = 77.56$ V.

$$Q_{DVR} = V_{DVRN}I_{DVRN} \sin \phi_{DVR} = 1725 \text{ VAR}, \ \phi_{DVR} = 62.79°, \text{ and } P_{DVR} = -P_{DST} = 885.15 \text{ W}.$$

$S_{DVR} = \sqrt{(P_{DVR}^2 + Q_{DVR}^2)} = 1938.95$ VA, which is same as the DSTATCOM rating.

Voltage Swell Compensation by a Left-Hand UPQC-S

It means that after the voltage swell, the supply voltage increases to $V_s' = V_s(1 + X) = 230 \times (1 + 0.20) = 276$ V, $X = 0.2$, $K_0 = V_s/V_s' = 1/(1 + X) = 1/1.2 = 0.8333$, and $n_0 = 1/K_0 = 1 + X = 1 + 0.2 = 1.2$.

The supply current after swell compensation is $I_s = P_L/V_s' = 4600/276 = 16.67$ A.

In the left-hand UPQC-S, for same power angle of $\delta = 19.41°$, as above, the reactive power rating is to be changed, but the active power rating (however, its value may change) needs to be same for both VSCs because of the common DC bus capacitor. It means that if one VSC has positive active power (it accepts the active power), another VSC must relieve equal amount of active power (it feeds the active power) as the DC bus capacitor cannot absorb any active power.

a. Since the DSTATCOM is connected across the AC mains, the voltage rating of the DSTATCOM is same as AC mains voltage: $V_{\text{DST}} = V'_{\text{s}} = V_{\text{s}}(1 + X) = 230 \times (1 + 0.2) = 276$ V.

   Under the voltage swell, the same power angle of $\delta = 19.41°$, from the following relation of reactive power of the DSTATCOM, can be computed as follows.

$$Q_{\text{DST}} = V'_{\text{s}} I'_{\text{L}} \sin\beta = n_0 V'_{\text{s}} I'_{\text{L}} \sin\beta = 2070.27 \text{ VAR},\ \beta = 17.46°, \text{ and}$$
$$\delta = \phi - \beta = 36.87° - 17.46° = 19.41°.$$

$$P_{\text{DST}} = V'_{\text{s}} I'_{\text{s}} - V'_{\text{s}} I'_{\text{L}} \cos\beta = 4600 - 6582.09 = -1982.09 \text{ W}.$$

$$S_{\text{DST}} = \sqrt{(P_{\text{DST}}^2 + Q_{\text{DST}}^2)} = \sqrt{(1982.09^2 + 2070.27^2)} = 2866.13 \text{ VA}.$$

b. The current rating of the DSTATCOM is $I_{\text{DST}} = S_{\text{DST}}/V_{\text{DST}} = 10.384$ A.

$$\phi_{\text{DST}} = \cos^{-1}(P_{\text{DST}}/V_{\text{DST}} I_{\text{DST}}) = 133.75°.$$

c. The VA rating of the DSTATCOM is $S_{\text{DST}} = V_{\text{DST}} I_{\text{DST}} = V_s I_{\text{DST}} = 276 \times 10.384 = 2866.13$ VA.

   This conforms the earlier calculated rating of the DSTATCOM.

   The rating of the DVR is computed from the common relations of the DSATCOM and DVR as follows.

   The active power of the DVR is $P_{\text{DVR}} = -P_{\text{DST}} = 1982.09$ W.
d. The current rating of the DVR of the UPQC-S is equal to AC load current ($I_{\text{DVR}} = I_{\text{L}} = 25$ A), since it is connected in series between the load and the AC mains.
e. The voltage rating of the DVR of the UPQC-S is computed as follows.

   The remaining reactive power of the load (not supplied by the DSTATCOM) must be supplied by the DVR, which is computed as follows: $Q_{\text{DVR}} = 3450 - 2070.27 = 1379.72$ VAR, $P_{\text{DVR}} = -P_{\text{DST}} = 1982.09$ W, and $S_{\text{DVR}} = \sqrt{(P_{\text{DVR}}^2 + Q_{\text{DVR}}^2)} = 2415$ VA.

$$V_{\text{DVR}} = S_{\text{DVR}}/I_{\text{DVR}} = 2415.02/25 = 96.60 \text{ V}.$$

f. The VA rating of the DVR is computed as follows.

$$Q_{\text{DVR}} = 3450 - 2070.27 = 1379.72 \text{ VAR}.$$

$$P_{\text{DVR}} = -P_{\text{DST}} = 1982.09 \text{ W}, \quad S_{\text{DVR}} = \sqrt{(P_{\text{DVR}}^2 + Q_{\text{DVR}}^2)} = 2415 \text{ VA}.$$

   The VA rating of the UPQC-S during voltage swell is $S_{\text{UPQC-S(under swell)}} = S_{\text{DST}} + S_{\text{DVR}} = 2866.13 + 2415 = 5281.18$ VA.

   Hence, considering an overall rating (both normal and under voltage swell), ratings of both the compensators are

$$V_{\text{DVR}} = 96.60 \text{ V}, \quad I_{\text{DVR}} = 25 \text{ A}, \quad S_{\text{DVR}} = V_{\text{DVR}} I_{\text{DVR}} = 96.6 \times 25 = 2415 \text{ VA}.$$

$$V_{\text{DST}} = 276 \text{ V}, \quad I_{\text{DST}} = 10.38 \text{ A}, \quad S_{\text{DST}} = V_{\text{DST}} I_{\text{DST}} = 276 \times 10.38 = 2866.13 \text{ VA}.$$

g. The VA rating of the left-hand UPQC-S is $S_{\text{UPQC-S}} = S_{\text{DST}} + S_{\text{DVR}} = 2866.13 + 2415 = 5281.18$ VA.

Example 6.12

A single-phase supply has AC mains voltage of 230 V at 50 Hz and a feeder (source) impedance of 1.0 Ω resistance and 3.0 Ω inductive reactance. It is feeding a load having $Z = (16 + j12)$ Ω. Calculate the voltage drop across the source impedance and the voltage across the load. If a single-phase right-hand unified power quality compensator (consisting of a DSTATCOM and a DVR using two VSCs with a common DC bus capacitor shown in Figure E6.12) is used to raise the voltage to the input voltage (230 V) and to maintain unity power factor at the AC mains, calculate (a) the voltage rating of the DVR of the UPQC-Q, (b) the current rating of the DVR of the UPQC-Q, (c) the VA rating of the DVR of the UPQC-Q, (d) the voltage rating of the DSTATCOM of the UPQC-Q, (e) the current rating of the DSTATCOM of the UPQC-Q, (f) the VA rating of the DSTATCOM of the UPQC-Q, and (g) total VA rating of the UPQC-Q to provide reactive power compensation at unity power factor with a constant regulated voltage of 230 V at 50 Hz.

Solution: Given that a single-phase supply has AC mains voltage $V_s = 230$ V, frequency of the supply $(f) = 50$ Hz, and a feeder (source) impedance of 1.0 Ω resistance and 3.0 Ω inductive reactance, $Z_s = (1.0 + j3.0)$ Ω. It is feeding a load having $Z_L = (16 + j12)$ Ω. This load must be compensated to result in unity power factor at the AC mains with a constant regulated load voltage of 230 V at 50 Hz.

The voltage drop across the source impedance and the voltage across the load are computed as follows.

$$Z_T = Z_s + Z_L = (1.0 + j3.0) + (16.0 + j12.0) = (17 + j15)\,\Omega = 22.67\,\Omega, \quad Z_L = 16.0 + j12.0 = 20\angle 36.87^\circ\,\Omega.$$

The supply current without any compensator is $I_s = V_s/Z_T = 230/22.67 = 10.15$ A.

The voltage across the load is $V_L = I_L Z_L = 10.15 \times 20 = 202.89$ V.

The voltage across the source impedance is $V_{Zs} = I_L Z_S = 10.15 \times 3.1\ 6 = 32.08$ V.

Voltage Drop Compensation by a Right-Hand UPQC-Q

The load after compensation must have rated voltage (230 V) across it and the UPQC-Q must result in unity power factor at the AC mains.

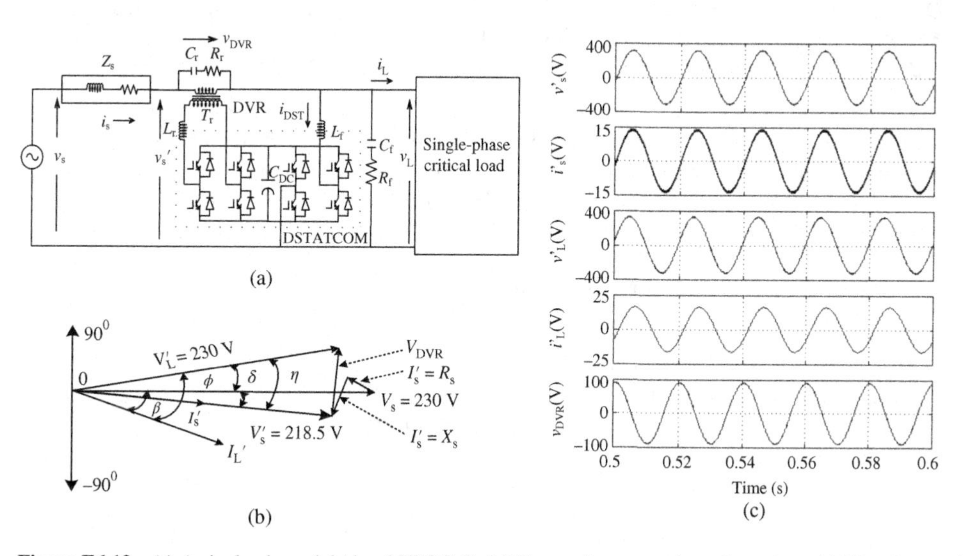

Figure E6.12 (a) A single-phase right-hand UPQC-Q. (b) Phasor diagram under voltage drop. (c) Waveforms of a single-phase right-hand UPQC-Q under voltage drop

The load current is $I'_L = V'_L/Z_L = 230/20 = 11.5$ A.

$$P_L = I_L'^2 R_L = 11.5^2 \times 16.0 = 2116\,\text{W}, \quad Q_L = I_L'^2 X_L = 11.5^2 \times 12.0 = 1587\,\text{VAR}.$$

V'_s and I'_s are the PCC voltage and grid current after compensation, respectively; from real power balance at PCC and supply, the active powers are $V'_s I'_s = P_L$ and $V_s I'_s \cos\gamma = P_L + I_s'^2 R_s$.

For the load, all reactive power is supplied by the UPQC-Q, whereas reactive power needed by the source impedance is supplied by the AC mains (grid). The corresponding equation is $V_s I'_s \sin\gamma = I_s'^2 X_s$.

On solving these equations, (b) $I_{DVR} = I'_s = 9.68$ A and $V'_s = 218.5$ V.

$$\text{(a) } V_{DVR} = \sqrt{(V_L'^2 - V_s'^2)} = 71.9\,\text{V}.$$

$$P_{DVR} = 0, \quad Q_{DVR} = V_{DVR} I_{DVR} = 696.41\,\text{VAR}.$$

$$\text{(c) } S_{DVR} = \sqrt{(P_{DVR}^2 + Q_{DVR}^2)} = 696.41\,\text{VA}.$$

$$P_{DST} = 0, \quad \text{(f) } S_{DST} = Q_L - Q_{DVR} = 1587 - 696.41 = 890.58\,\text{VA}.$$

The DSTATCOM is connected across the load. Hence, its voltage rating is same as load voltage rating. So, (d) $V_{DST} = 230$ V and the current rating of the DSTATCOM is estimated as (e) $I_{DST} = S_{DST}/V_{DST} = 890.58/230 = 3.87$ A.

The overall VA rating of the UPQC-Q is (g) $S_{UPQC\text{-}Q} = S_{DST} + S_{DVR} = 890.58 + 696.41 = 1587$ VA.

Example 6.13

A single-phase supply has AC mains voltage of 230 V at 50 Hz and a feeder (source) impedance of 1.0 Ω resistance and 3.0 Ω inductive reactance. It is feeding a load having $Z = (16 + j12)$ Ω. Calculate the voltage drop across the source impedance and the voltage across the load. If a single-phase right-hand unified power quality compensator (consisting of a DSTATCOM and a DVR using two VSCs with a common DC bus capacitor shown in Figure E6.13) is used to regulate the load voltage to the input voltage (230 V) and to maintain unity power factor at the AC mains, calculate (a) the voltage rating of the DSTATCOM of the UPQC-P, (b) the current rating of the DSTATCOM of the UPQC-P, (c) the VA rating of the DSTATCOM of the UPQC-P, (d) the voltage rating of the DVR of the UPQC-P, (e) the current rating of the DVR of the UPQC-P, (f) the VA rating of the DVR-P of UPQC-P, and (g) total VA rating of the UPQC-P.

Solution: Given that a single-phase supply has AC mains voltage $V_s = 230$ V, frequency of the supply $(f) = 50$ Hz, and a feeder (source) impedance of 1.0 Ω resistance and 3.0 Ω inductive reactance, $Z_s = (1.0 + j3.0)$ Ω. It is feeding a load having $Z_L = (16 + j12)$ Ω. This load must be compensated to result in unity power factor at the AC mains with a constant regulated load voltage of 230 V at 50 Hz.

The voltage drop across the source impedance and the voltage across the load are computed as follows.

$$Z_T = Z_s + Z_L = (1.0 + j3.0) + (16.0 + j12.0) = (17 + j15)\,\Omega = 22.67\,\Omega, \quad Z_L = 16.0 + j12.0 = 20\angle 36.87^\circ\,\Omega.$$

The supply current without any compensator is $I_s = V_s/Z_T = 230/22.67 = 10.15$ A.

The voltage across the load is $V_L = I_L Z_L = 10.15 \times 20 = 202.89$ V.

The voltage across the source impedance is $V_{Zs} = I_L Z_s = 10.15 \times 3.1\ 6 = 32.08$ V.

Voltage Drop Compensation by a Right-Hand UPQC-P

The load after compensation must have rated voltage (230 V) across it and the UPQC-P must result in unity power factor at the AC mains.

The load current is $I'_L = V'_L/Z_L = 230/20 = 11.5$ A.

$$P_L = I_L'^2 R_L = 11.5^2 \times 16.0 = 2116\,\text{W}, \quad Q_L = I_L'^2 X_L = 11.5^2 \times 12.0 = 1587\,\text{VAR}.$$

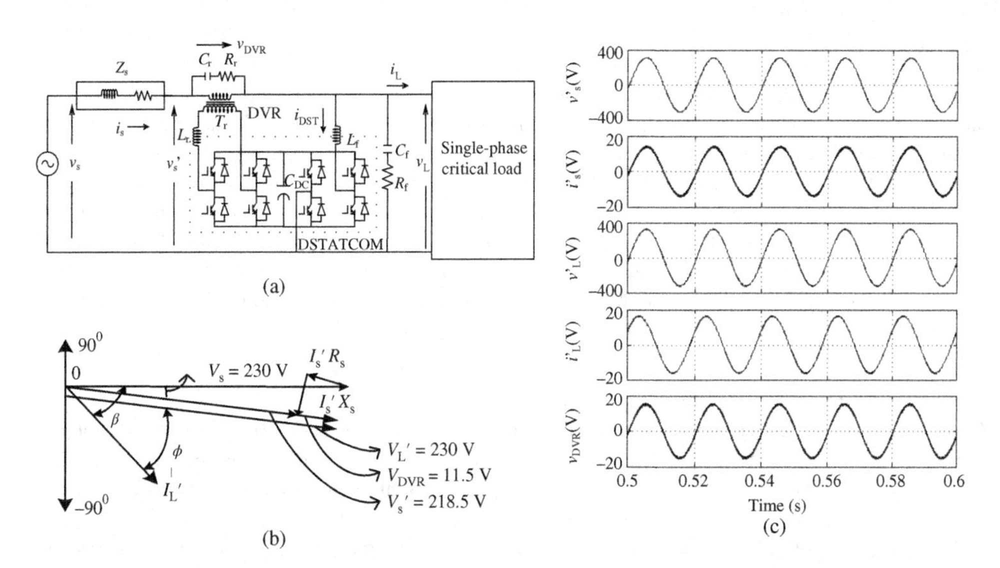

Figure E6.13 (a) A single-phase right-hand UPQC-P. (b) Phasor diagram under voltage drop. (c) Waveforms of a single-phase right-hand UPQC-P under voltage drop

V_s' and I_s' are the PCC voltage and grid current after compensation, respectively; from real power balance at PCC and supply, the active powers are $V_s'I_s' = P_L$ and $V_sI_s'\cos\gamma = P_L + I_s'^2R_s$.

For the load, all reactive power is supplied by the UPQC-P, whereas reactive power needed by the source impedance is supplied by the AC source (grid). The corresponding equation is $V_sI_s'\sin\gamma = I_s'^2X_s$.

On solving these equations, $I_s' = 9.685$ A and $V_s' = 218.5$ V.

The voltage and current ratings of the DVR are estimated as follows.

$$\text{(d) } V_{DVR} = V_L' - V_s' = 11.53 \text{ V}.$$

$$\text{(e) } I_{DVR} = I_s' = 9.685 \text{ A}.$$

$$P_{DVR} = V_{DVR}I_{DVR} = 8 \times 9.685 = 111.66 \text{ W}, \quad Q_{DVR} = 0 \text{ VAR}.$$

$$\text{(f) } S_{DVR} = \sqrt{(P_{DVR}^2 + Q_{DVR}^2)} = 111.66 \text{ VA}.$$

The active power absorbed by the DVR is sent back into the system via the DSTATCOM of the UPQC-P. Hence, $P_{DST} = -P_{DVR} = -111.66$ W.

All load reactive power is supplied by the DSTATCOM of the UPQC-P.

$$Q_{DST} = Q_L = 1587 \text{ VAR}.$$

$$\text{(c) } S_{DST} = \sqrt{(P_{DST}^2 + Q_{DST}^2)} = 1590.9 \text{ VA}.$$

$$\text{(a) } V_{DST} = 230 \text{ V}.$$

$$\text{(b) } I_{DST} = S_{DST}/V_{DST} = 1590.9/230 = 6.91 \text{ A}.$$

The total VA rating of the UPQC-P is (g) $S_{UPQC\text{-}P} = S_{DVR} + S_{DST} = 111.66 + 1590.9 = 1702.6$ VA.

Example 6.14

A single-phase supply has AC mains voltage of 230 V at 50 Hz and a feeder impedance of 1.0 Ω resistance and 3.0 Ω inductive reactance. It is feeding a load having $Z = (16 + j12)$ Ω. Calculate the voltage drop across the source impedance and the voltage across the load. If a single-phase right-hand unified power quality compensator (consisting of a DSTATCOM and a DVR using two VSCs with a common DC bus capacitor shown in Figure E6.14) is used to raise the voltage to the input voltage (230 V) and to maintain unity power factor at the AC mains, calculate (a) the voltage rating of the DVR of the UPQC-S, (b) the current rating of the DVR of the UPQC-S, (c) the VA rating of the DVR of the UPQC-S, (d) the voltage rating of the DSTATCOM of the UPQC-S, (e) the current rating of the DSTATCOM of the UPQC-S, (f) the VA rating of the DSTATCOM of the UPQC-S, and (g) total VA rating of the UPQC-S. Consider same rating of both VSCs.

Solution: Given that a single-phase supply has AC mains voltage $V_s = 230$ V, frequency of the supply $(f) = 50$ Hz, and a feeder (source) impedance of 1.0 Ω resistance and 3.0 Ω inductive reactance, $Z_s = (1.0 + j3.0)$ Ω. It is feeding a load having $Z_L = (16 + j12)$ Ω. This load must be compensated to result in unity power factor at the AC mains with a constant regulated load voltage of 230 V at 50 Hz.

The voltage drop across the source impedance and the voltage across the load are computed as follows.

$$Z_T = Z_s + Z_L = (1.0 + j3.0) + (16.0 + j12.0) = (17 + j15)\,\Omega = 22.67\,\Omega, \quad Z_L = 16.0 + j12.0 = 20\angle 36.87°\,\Omega.$$

The supply current without any compensator is $I_s = V_s/Z_T = 230/22.67 = 10.15$ A.

The voltage across the load is $V_L = I_L Z_L = 10.15 \times 20 = 202.89$ V.

The voltage across the source impedance is $V_{Zs} = I_L Z_s = 10.15 \times 3.1\ 6 = 32.08$ V.

Voltage Drop Compensation by a Right-Hand UPQC-S

The load after compensation must have rated voltage (230 V) across it and the UPQC-S must result in unity power factor at the AC mains.

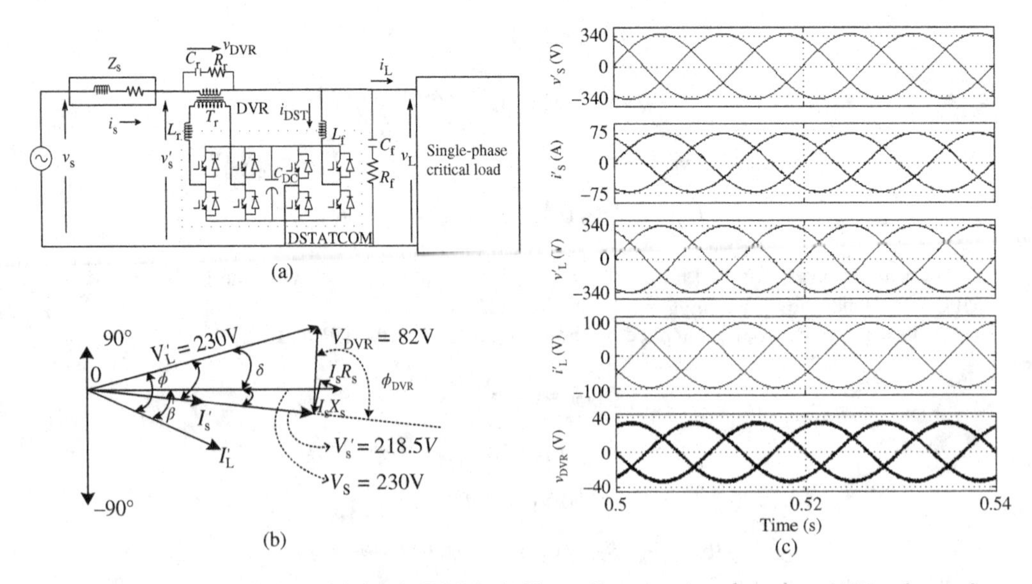

Figure E6.14 (a) A single phase right-hand UPQC-S. (b) Phasor diagram under voltage drop. (c) Waveforms of a single-phase right-hand UPQC-S under voltage drop

The load current is $I'_L = V'_L/Z_L = 230/20 = 11.5$ A.

$$P_L = I'^2_L R_L = 11.5^2 \times 16.0 = 2116 \text{ W}, \quad Q_L = I'^2_L X_L = 11.5^2 \times 12.0 = 1587 \text{ VAR}.$$

V'_s and I'_s are the PCC voltage and grid current after compensation, respectively; from real power balance at PCC and supply, the active powers are $V'_s I'_s = P_L$ and $V_s I'_s \cos\gamma = P_L + I'^2_s R_s$.

For the load, all reactive power is supplied by the UPQC-S, whereas reactive power needed by the source impedance is supplied by the AC source (grid). The corresponding equation is $V_s I'_s \sin\gamma = I'^2_s X_s$.

On solving these equations, $I'_s = 9.686$ A and $V'_s = 218.5$ V.

As both the VSCs have equal rating, both share equal amount of load reactive power.

$$Q_{DVR} = Q_{DST} = Q_L/2 = 1587/2 = 793.5 \text{ VAR}.$$

$$\text{(b) } I_{DVR} = I'_s = 9.686 \text{ A}.$$

The following equations are written from the phasor diagram of Figure E6.14b:

$$Q_{DVR} = V_{DVR} I_{DVR} \sin\phi_{DVR}.$$

$$V'_s + V_{DVR}\cos\phi_{DVR} = V'_L \cos\eta.$$

$$V_{DVR}\sin\phi_{DVR} = V'_L \sin\eta.$$

$$\sin\eta = \frac{V_{DVR}\sin\phi_{DVR}}{V'_L} = \frac{Q_{DVR}}{I_{DVR}V'_L} = 0.36.$$

On solving the above equation, the value of η is 20.88°.

The value of η is back substituted to find value of V_{DVR} and ϕ_{DVR}.

On solving the above equations, the estimated voltage rating of the DVR is (a) $V_{DVR} = 82$ V and $\phi_{DVR} = 87.56°$.

The VA rating of the DVR is estimated as follows.

$$P_{DVR} = V_{DVR} I_{DVR}\cos\phi_{DVR} = 34.46 \text{ W}.$$

$$Q_{DVR} = 793.5 \text{ VAR}.$$

$$\text{(c) } S_{DVR} = \sqrt{(P^2_{DVR} + Q^2_{DVR})} = 794.25 \text{ VA}.$$

$$\text{(f) } S_{DST} = S_{DVR} = 794.25 \text{ VA}.$$

$$\text{(d) } V_{DST} = 230 \text{ V}.$$

$$\text{(e) } I_{DST} = S_{DST}/V_{DST} = 794.25/230 = 3.45 \text{ A}.$$

The overall VA rating of the UPQC-S is (g) $S_{UPQC\text{-}S} = S_{DVR} + S_{DST} = 794.25 + 794.25 = 1588.5$ VA.

Example 6.15

A three-phase supply with AC line voltage output of 415 V (rms) at 50 Hz is feeding a critical linear load of 50 kVA, 415 V, 50 Hz, 0.8 lagging power factor. A voltage sag of −10% occurs in the supply voltage at PCC. Design a three-phase unified power quality compensator (consisting of a DSTATCOM and a DVR using two VSCs with a common DC bus capacitor) to compensate the power factor of the load to unity at the AC mains and to mitigate the voltage sag to avoid the interruption to the load. If two three-phase VSCs are used as a right-hand UPQC-Q (shown in Figure E6.15), calculate (a) the voltage rating of the DVR of

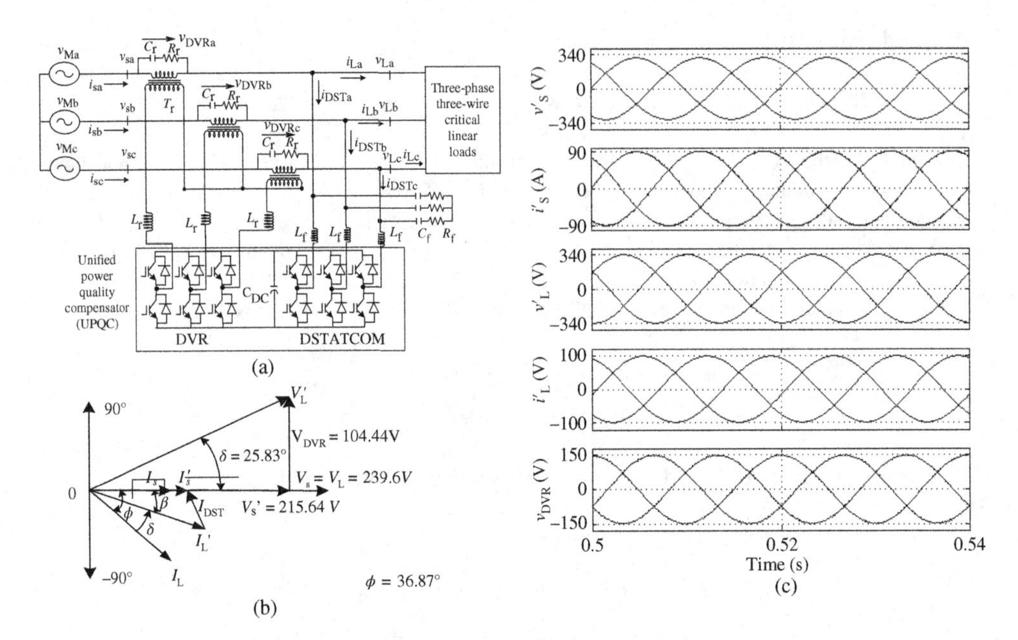

Figure E6.15 (a) A three-phase right-hand UPQC-Q. (b) Phasor diagram under voltage sag. (c) Waveforms of a three-phase right-hand UPQC-Q under voltage sag

the UPQC-Q, (b) the current rating of the DVR of the UPQC-Q, (c) the VA rating of the DVR of the UPQC-Q, (d) the voltage rating of the DSTATCOM of the UPQC-Q, (e) the current rating of the DSTATCOM of the UPQC-Q, (f) the VA rating of the DSTATCOM of the UPQC-Q, and (g) total VA rating of the UPQC-Q to provide reactive power compensation for unity power factor at the AC mains with a constant regulated voltage of 415 V at 50 Hz.

Solution: Given supply voltage $V_{sl} = 415$ V, frequency of the supply $(f) = 50$ Hz, and a load of 415 V, 50 Hz, 50 kVA, 0.8 lagging power factor. There is a voltage sag of −10% in the supply system with a base value of 415 V.

Operation of UPQC-Q under Nominal Grid Condition

Under nominal grid condition, the DVR of the UPQC-Q injects zero voltage similarly to the single-phase case. However, the DSTSTCOM injects current in quadrature with the load voltage for compensation of reactive power of the load.

$$V_{sp} = V_{sl}/\sqrt{3} = 415/\sqrt{3} = 239.6\text{ V},\quad V_{Lp} = V_{Ll}/\sqrt{3} = 415/\sqrt{3} = 239.6\text{ V}.$$

The active power of the load is $P_L = 3V_{Lp}I_{Lp} \times \text{PF} = 50\ 000 \times 0.8 = 40$ kW.
The reactive power of the load is $Q_L = 3V_{Lp}I_{Lp}\sqrt{(1 - \text{PF}^2)} = 50\ 000 \times 0.6 = 30$ kVAR.
The supply current before any voltage sag is $I_s = I_{DVRN} = P_L/3V_{sp} = 40\ 000/(3 \times 239.6) = 55.64$ A.
The current supplied by the DSTATCOM is $I_{DSTN} = Q_L/3V_{Lp} = 41.73$ A.
The reactive power rating of the DSTATCOM is $Q_{DSTN} = Q_L = 30$ kVAR.

Voltage Sag Compensation by a Right-Hand UPQC-Q

a. The voltage rating of the DVR of the UPQC-Q is computed as follows.

There is a voltage sag of −10% in the supply system with a base value of 239.59 V. Therefore, the DVR of the UPQC-Q must inject in quadrature with the supply voltage to provide the required voltage at the load end.

It means that after the voltage sag, the supply voltage reduces to $V_{sl}=415\times(1-0.10)=373.5$ V and $V_{sp}=V_{sl}/\sqrt{3}=215.64$ V.

The phase load voltage is $V_{Lp}=415/\sqrt{3}=239.59$ V.

The phase load current is $I_{Lp}=50\ 000/(3\times 239.59)=69.56$ A.

The supply current during voltage sag compensation is $I'_s = P_L/3V_{sp} = 40\,000/(3\times 215.64) =$ 61.83 A.

The voltage rating of the DVR is $V_{DVR} = \sqrt{(V_L^2 - V_s^2)} = 239.59\sqrt{(1^2 - 0.9^2)} = 0.4359\times$ 239.59 V = 104.44 V.

b. The current rating of the DVR of the UPQC-Q must be same as supply current after shunt compensation: I_{DVR} = the fundamental active component of load current $(I'_s) = P_L/3V_{sp} = 40\,000/(3\times 215.64) = 61.83$ A.

c. The VA rating of the DVR of the UPQC-Q is computed as follows.

$$Q_{DVR} = 3V_{DVR}I_{DVR} = 3\times 104.44\times 61.83\ \text{VA} = 19\,373\ \text{VAR}.$$

$P_{DVR}=0.0$, as the voltage of the DVR is injected in quadrature with its current.

$$S_{DVR} = \sqrt{(P_{DVR}^2 + Q_{DVR}^2)} = Q_{DVR} = 3\times 104.44\times 61.83\ \text{VA} = 19\,373\ \text{VA} = 19.373\ \text{kVA}.$$

d. The voltage rating of the DSTATCOM of the UPQC-Q is equal to AC load voltage ($V_{DST}=239.59$ V), since it is connected across the load of 239.59 V sine waveform.

e. The current rating of the DSTATCOM of the UPQC-Q is computed as follows.

The DSTATCOM of the UPQC-Q needs to correct the power factor of the source to unity. Hence, the required reactive power by the DSTATCOM is now lower than the load reactive power as the angle between the supply voltage and the load current is reduced to $\beta=\phi-\delta$, where $\cos\phi$ is the load power factor and δ is the angle between the load voltage and the PCC voltage after compensation. The angle δ is computed as follows: $\tan\delta = V_{DVR}/V'_s = 104.44/215.64 = 0.4842$, $\delta = 25.83°$, and power factor angle is $\phi=\cos^{-1}(0.8)=36.87°$.

The current rating of the DSTATCOM is $I_{DST} = I_L[\sqrt{\{(1-X)^2+\cos^2\phi - 2\cos\phi\cos(\phi-\delta)(1-X)\}}]/(1-X) = 69.56[\sqrt{\{0.9^2+0.8^2-2\times 0.8\cos(36.87°-25.83°)\times 0.9}]/0.9 = 14.796$ A.

f. The VA rating of the DSTATCOM is computed as follows.

$$Q_{DST} = 3V_{DST}I_{DST} = 3\times 239.59\times 14.79\ \text{VA} = 10\,630\ \text{VAR}.$$

$P_{DST}=0.0$, as the voltage of the DVR is injected in quadrature with its current.

$$S_{DST} = \sqrt{(P_{DST}^2 + Q_{DST}^2)} = Q_{DVR} = 10\,630\ \text{VA}.$$

The VA rating of the UPQC-Q during voltage sag is $S_{UPQC\text{-}Q(under\ sag)} = S_{DST} + S_{DVR} = 10\,630 +$ 19 373 = 30 003 VA = 30.003 kVA.

Hence, considering an overall rating (both normal and under sag), ratings of both the compensators are

$$V_{DST} = 239.6\ \text{V},\quad I_{DST} = I_{DSTN} = 41.73\ \text{A},\quad S_{DST} = 3V_{DST}I_{DSTN} = 3\times 239.6\times 41.73 = 30\,000\ \text{VA}.$$

$$V_{DVR} = 104.44\ \text{V},\quad I_{DVR} = 61.83\ \text{A},\quad S_{DVR} = 3V_{DVR}I_{DVR} = 19\,370\ \text{VA}.$$

g. The VA rating of the UPQC-Q is $S_{UPQC\text{-}Q}=S_{DST}+S_{DVR}=30\ 000+19\ 370=49\ 370$ VA = 49.370 kVA.

Example 6.16

A three-phase supply with AC line voltage of 415 V (rms) at 50 Hz is feeding a critical linear load of 50 kVA, 415 V, 50 Hz, 0.8 lagging power factor. A voltage swell of +10% occurs in the supply at PCC. Design a three-phase unified power quality compensator UPQC-P (consisting of a DSTATCOM and a DVR using two VSCs with a common DC bus capacitor) to compensate the power factor of the load to unity at the AC mains and to mitigate the voltage swell to avoid the interruption to the load. If two three-phase VSCs are used as a right-hand UPQC-P (shown in Figure E6.16), calculate (a) the voltage rating of the DVR of the UPQC-P, (b) the current rating of the DVR of the UPQC-P, (c) the VA rating of the DVR of the UPQC-P, (d) the voltage rating of the DSTATCOM of the UPQC-P, (e) the current rating of the DSTATCOM of the UPQC-P, (f) the VA rating of the DSTATCOM of the UPQC-P, (g) total VA rating of the UPQC-P, (h) turns ratio of the injection transformer for the DVR, (i) DC bus voltage, (j) interfacing inductance, and (k) DC bus capacitance of VSCs. Consider the switching frequency of 20 kHz and DC bus voltage has to be controlled within 8% range of the nominal value and the ripple current in the interfacing inductor is 5%.

Solution: Given supply voltage $V_{sl} = 415$ V, frequency of the supply $(f) = 50$ Hz, and a load of 415 V, 50 Hz, 50 kVA, 0.8 lagging power factor. There is a voltage swell of +10% in the supply system with a base value of 415 V.

The phase load voltage is $V_{Lp} = 415/\sqrt{3} = 239.59$ V.

The phase load current is $I_{Lp} = I_s = 50\ 000/(3 \times 239.59) = 69.56$ A.

The active power of the load is $P_L = 3V_{Lp}I_{Lp} \times PF = 50\ 000 \times 0.8 = 40$ kW.

The supply current before the voltage swell is $I_s = P_L/3V_{sp} = 40\ 000/(3 \times 239.6) = 55.64$ A.

Voltage Swell Compensation by a Right-Hand UPQC-P

After the voltage swell, the supply voltage increases to $V'_{sl} = 415 \times (1 + 0.10) = 456.50$ V and $V'_{sp} = V'_{sl}/\sqrt{3} = 263.56$ V.

a. The voltage rating of the DVR of the UPQC-P is computed as follows.

   For the UPQC-P, the injected voltage is in phase with the source voltage. Hence, the value of injected voltage is the algebraic difference between load and source voltages: $V_{DVR} = V'_{sp} - V_{sp} = 263.56 - 239.6 = 23.96$ V.

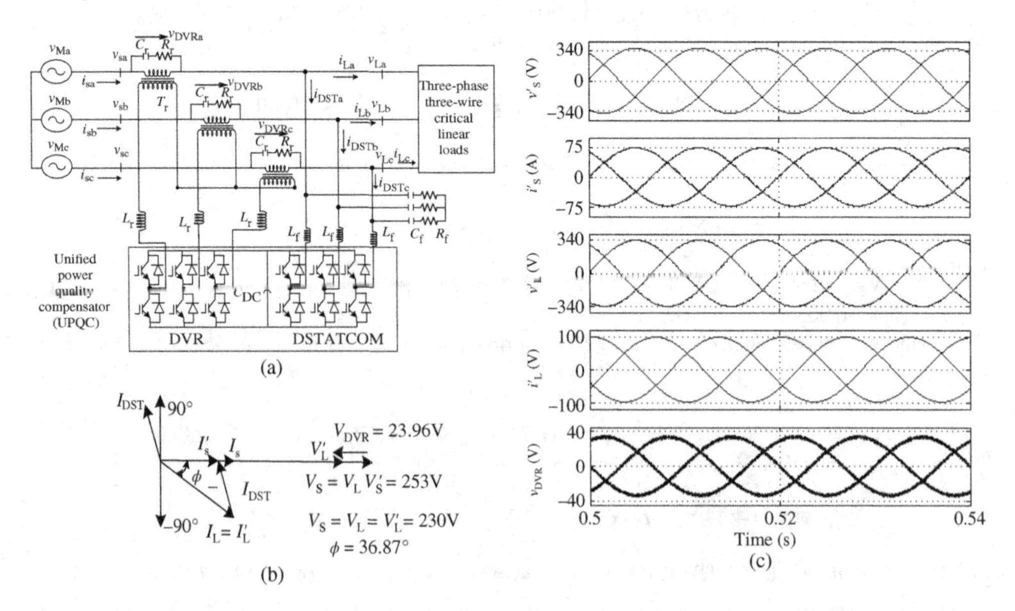

Figure E6.16 (a) A three-phase right-hand UPQC-P. (b) Phasor diagram under voltage swell. (c) Waveforms of a three-phase right-hand UPQC-P under voltage swell

b. The current rating of the DVR of the UPQC-P must be same as supply current during voltage swell compensation. If only voltage swell compensation is considered, then supply current is maximum for nominal operation condition. In case of voltage swell, the supply current is given by $I'_s = P_L/3V'_{sp} = 40\,000/(3 \times 263.56) = 50.589$ A.
c. The VA rating of the DVR of the UPQC-P is calculated as follows.

$$P_{DVR} = 3V_{DVR}I_{DVR} = 3 \times 23.96 \times 50.58 \text{ W} = 3636.4 \text{ W} = 3.6364 \text{ kW}.$$

$Q_{DVR} = 0.0$, as the voltage of the DVR is injected in opposite phase with its current.

$$S_{DVR} = \sqrt{(P_{DVR}^2 + Q_{DVR}^2)} = P_{DVR} = 3636.4 \text{ VA}.$$

d. The voltage rating of the DSTATCOM of the UPQC-P is equal to AC load voltage ($V_{DSTp} = 239.59$ V), since it is connected across the load of 239.59 V sine waveform.
e. The current rating of the DSTATCOM of the UPQC-P is computed as follows.
The DSTATCOM of the UPQC-P needs to correct the power factor of the source to unity. Hence, the required reactive power by the DSTATCOM is equal to the load reactive power: $Q_{DST} = Q_L = 3V'_L I'_L \sin\phi = 50\,000 \times 0.6 = 30\,000$ VAR and $P_{DST} = P_{DVR} = 3636.4$ W.

$$S_{DST} = \sqrt{(P_{DST}^2 + Q_{DST}^2)} = 30220 \text{ VA}.$$

The current rating of the DSTATCOM is $I_{DST} = S_{DST}/3V_{DVST} = 42.04$ A.
f. The VA rating of the DSTATCOM is $S_{DST} = \sqrt{(P_{DST}^2 + Q_{DST}^2)} = 30\,220$ VA.
The VA rating of the UPQC-P during voltage swell is $S_{UPQC\text{-}P} = S_{DST} + S_{DVR} = 30\,220 + 3636.4 = 33\,856.4$ VA.
Hence, considering an overall rating (both normal and under voltage swell), ratings of both the compensators are

$$V_{DST} = 239.6 \text{ V}, \quad I_{DST} = 42.04 \text{ A}, \quad S_{DST} = 3V_{DST}I_{DST} = 30\,220 \text{ VA}.$$

$$V_{DVR} = 23.96 \text{ V}, \quad I_{DVR} = 55.64 \text{ A}, \quad S_{DVR} = 3V_{DVR}I_{DVR} = 3999.4 \text{ VA}.$$

g. The VA rating of the UPQC-P is $S_{UPQC\text{-}P} = S_{DST} + S_{DVR} = 30\,220 + 3999.4 = 34\,219$ VA.
h. The turns ratio of the injection transformer for the DVR is computed as follows.
The DC bus voltage of 750 V (decided for 415 V for the DSTATCOM) can be used to obtain 415 V across the line at the output of the VSC using a PWM controller. However, the DVR requires only 23.96 V per phase. Therefore, the turns ratio of the injection transformer is $K_{DVR} = N_{VSC}/N_{DVR} = 415/(\sqrt{3} \times 23.96) = 10$.
i. The DC bus voltage is decided based on the AC voltages impressed across the VSCs. Here, the DSTATCOM of the UPQC-P is directly used without any transformer; therefore, the AC voltage across the DSTATCOM (load voltage of 415 V) decides the DC bus voltage: $V_{DC} = 2\sqrt{2}(V_L/\sqrt{3})/m_a = 2\sqrt{2}(415/\sqrt{3})/0.9 = 752.99 \approx 750$ V.
j. The interfacing inductances are required in VSCs, which are working as DSTATCOM and DVR.
The interfacing inductance of the DSTATCOM is $L_{DST} = (\sqrt{3}/2)m_a V_{DC}/6af_s\Delta I_{DST} = (\sqrt{3}/2) \times 0.9 \times 750/(6 \times 1.2 \times 20\,000 \times 2.1002) = 1.93$ mH.
The interfacing inductance of the DVR is $L_{DVR} = (\sqrt{3}/2)m_a V_{DC}K_{DVR}/6af_s\Delta I_{DVR} = (\sqrt{3}/2) \times 0.9 \times 750 \times 10/(6 \times 1.2 \times 20\,000 \times 2.53) = 16$ mH.
k. The DC bus capacitance of the UPQC-P is computed from change in stored energy during dynamics as follows.
The change in stored energy during dynamics is $\Delta E = (1/2)C_{DC}(V_{DCapf}^2 - V_{DCminapf}^2) = 3V_{DST}I_{DST}\Delta t$.

Substituting the values, $\Delta E = (1/2)C_{DC}(750^2 - 690^2) = 3 \times 239.6 \times 42.04 \times 10/1000$ (considering $\Delta t = 10\,\text{ms}$).

$$C_{DC} = 7018.28\,\mu\text{F}.$$

Example 6.17

A three-phase supply with AC mains voltage of 440 V at 50 Hz has a voltage sag of −20% due to a rural feeder and a nearby factory. A hospital needs a three-phase 415 V, 100 kVA, 0.8 lagging power factor, 50 Hz linear load in critical equipment. If a right-hand UPQC-S (consisting of a DSTATCOM and a DVR using two VSCs with a common DC bus capacitor) is used as a unified power quality compensator (shown in Figure E6.17), calculate (a) the rating of the shunt and series components of the UPQC-S to provide rated voltage across the load and to realize unity power factor at the AC mains. If an injection transformer is used to inject the voltage in the DVR, calculate (b) turns ratio of the injection transformer, (c) values of interfacing inductors, and (d) the value of the DC bus capacitor. Consider the switching frequency of 10 kHz and DC bus voltage has to be controlled within 8% range and ripple current in the inductor is 10%. Consider same rating of both VSCs.

Solution: Given supply voltage $V_{sl} = 440$ V, frequency of the supply $(f) = 50$ Hz, and a load of 415 V, 50 Hz, 100 kVA, 0.8 lagging power factor. There is a voltage sag of −20% due to a rural feeder and a nearby factory.

It means that after the voltage sag, the supply voltage reduces to $V_{sl} = 440 \times (1 - 0.20) = 352$ V and $V_{sp} = V_{sl}/\sqrt{3} = 203.22$ V.

The phase load voltage is $V_{Lp} = 415/\sqrt{3} = 239.59$ V.

The phase load current is $I_{Lp} = 100\,000/(3 \times 239.59) = 139.12$ A.

The active power of the load is $P_L = 3V_{Lp}I_{Lp} \times \text{PF} = 100\,000 \times 0.8 = 80$ kW.

The supply current during voltage sag compensation is $I_s = P_L/3V_{sp} = 80\,000/(3 \times 203.22) = 131.22$ A.

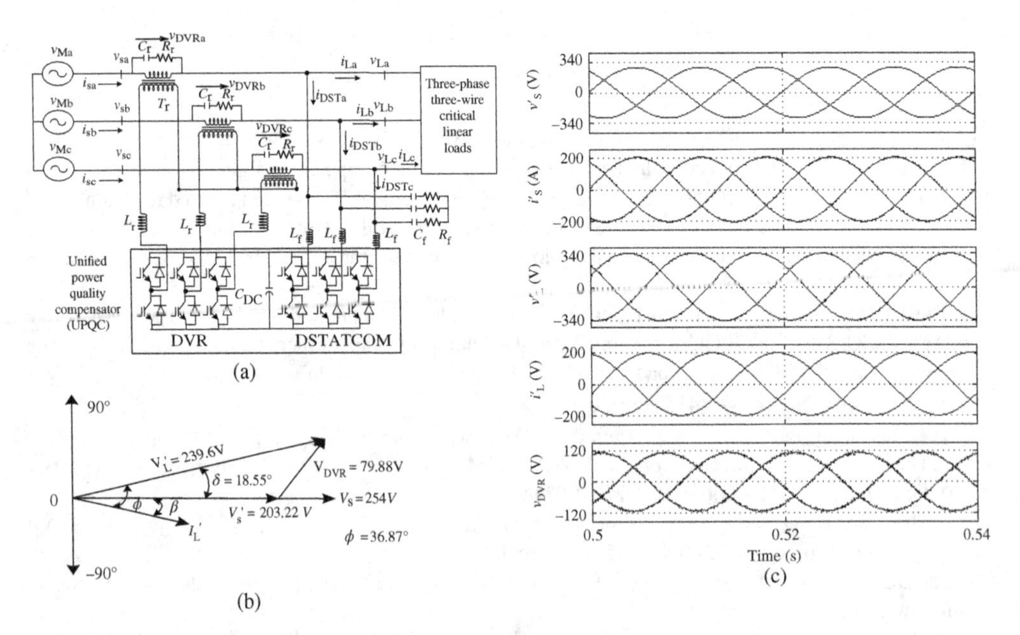

Figure E6.17 (a) A three-phase right-hand UPQC-S. (b) Phasor diagram under voltage sag. (c) Waveforms of a three-phase right-hand UPQC-S under voltage sag

Under Steady-State Condition for Unity Power Factor at the AC Mains and Voltage Compensation across the Load

The active power of the load is $P_L = 3V_{Lp}I_{Lp} \times PF = 100\ 000 \times 0.8 = 80$ kW.

The reactive power of the load is $Q_L = P_L = 3V_{Lp}I_{Lp}\sqrt{(1-PF^2)} = 100\ 000 \times 0.6 = 60$ kVAR.

Voltage Rise Compensation by a Right-Hand UPQC-S under Nominal Operating Condition

Under normally operating condition, the supply voltage is 440 V, but the load voltage is rated for 415 V.

It means that the supply voltage is more than the load voltage even under nominal grid conditions: $V_{sp} = 440/\sqrt{3}$ V = 254 V per phase.

The extra voltage from the supply is compensated in the series compensator of the UPQC-S. The per-unit extra voltage is $X = [\{(V_{sp}/V_{Lp}) - V_{Lp}\}/V_{Lp}] = [\{(254/239.6) - 239.6\}/239.6] = 0.06$ pu, $K_0 = V_{Lp}/V_{sp} = 239.6/254 = 0.943$, and $n_0 = 1/K_0 = 1 + X = 1 + 0.06 = 1.06$.

The supply current during voltage swell compensation is $I_s = P_L/3V_{sp} = 80\,000/(3 \times 254) = 104.99$ A.

In the right-hand UPQC-S, for same VA rating of both VSCs, they must have same reactive power rating, as active power rating is same for both VSCs because of the common DC bus capacitor. It means that if one VSC has positive active power (it accepts the active power), another VSC must relieve equal amount of active power (it feeds the active power) as the DC bus capacitor cannot absorb any active power. Moreover, the reactive power rating of both converters is equal to the reactive power of the load. Therefore, each VSC has reactive power rating equal to half of the reactive power of the load; $Q_{DST} = Q_{DVR} = Q_L/2 = 60/2 = 30$ kVAR.

Under the voltage rise condition, from the following relation of reactive power of the DVR, the power angle is computed as follows.

$$Q_{DVR} = 3K_0V_sI_s \sin\delta = 3(V_L/V_s)V_sI_s \sin\delta = 30\,000 \text{ VAR; therefore, } \delta = 23.42°.$$

$$P_{DVR} = -3K_0V_sI_s(n_0 - \cos\delta) = -3(V_L/V_s)V_sI_s\{(V_s/V_L) - \cos\delta\} = -10\,766.15 \text{ W}.$$

The voltage rating of the DVR is $V_{DVR} = V_L\sqrt{(1 + n_0^2 - 2n_0\cos\delta)} = 239.6\sqrt{(1 + 1.06^2 - 2 \times 1.06 \cos 23.42°)} = 101.12$ V.

The supply current during nominal voltage compensation is equal to current through the DVR. Therefore, the current rating of the DVR is $I_{DVR} = I_s = P_L/3V_{sp} = 80\ 000/(3 \times 254) = 104.99$ A.

The VA rating of the DVR is $S_{DVR} = 3V_{DVR}I_{DVR} = 3 \times 101.2 \times 104.98 = 31\ 873.37$ VA.

The active power of the DSTATCOM is $P_{DST} = -P_{DVR} = 10\ 747.15$ W.

The voltage rating of the DSTATCOM of the UPQC-S is equal to AC load voltage ($V_{DST} = 239.6$ V), since it is connected across the load of 239.6 V sine waveform.

The current rating of the DSTATCOM of the UPQC-S is computed as follows.

The angle between the supply voltage and the load current is $\beta = \phi - \delta = 36.87° - 23.42° = 13.45°$.

Therefore, DSTATCOM current is $I_{DST} = I_L\sqrt{[(1 + K_0^2\cos^2\phi - 2K_0\cos\beta\cos\phi)} = 139.12\sqrt{(1 + 0.943^2 \times 0.8^2 - 2 \times 0.943 \times 0.8 \times 0.973)} = 44.34$ A.

The phase angle between DSTATCOM voltage and current can be calculated as follows.

$$P_{DST} = 10747.15 \text{ W} = 3V_LI_{DST}\cos\phi_{DST}, \quad \phi_{DST} = 70.29°.$$

$$Q_{DST} = V_LI_{DST}\sin\phi_{DST} = 30\,000 \text{ VA}.$$

The VA rating of the DSTATCOM of the UPQC-S is $S_{DST} = 3V_{DST}I_{DST} = 3 \times 239.6 \times 44.34$ VA = 31 873.16 VA.

The VA rating of the UPQC-S is $S_{UPQC\text{-}S} = S_{DST} + S_{DVR} = 31\ 873.16 + 31\ 873.37 = 63\ 746.53$ VA.

Voltage Sag Compensation by a Right-Hand UPQC-S

There is a voltage sag of −20% due to a rural feeder and a nearby factory.

It means that after the voltage sag, the supply voltage reduces to $V_{sl} = 440 \times (1 - 0.20) = 352$ V and $V_{sp} = V_{sl}/\sqrt{3} = 203.22$ V.

The phase load voltage is $V_{Lp} = 415/\sqrt{3} = 239.59$ V.

It means that after the voltage sag, the supply voltage reduces to $V_{sp} = 203.22$ V, $X = 0.152$, $K_0 = 1.179$, and $n_0 = 0.848$.

The supply current after voltage sag compensation is $I_s = P_L/3V_s = 80\ 000/(3 \times 203.22) = 131.212$ A.

In the right-hand UPQC-S, for same VA rating of both VSCs, they must have same reactive power rating, as active power rating is same for both VSCs because of the common DC bus capacitor. It means that if one VSC has positive active power (it accepts the active power), another VSC must relieve equal amount of active power (it feeds the active power) as the DC bus capacitor cannot absorb any active power. Moreover, the reactive power rating of both converters is equal to the reactive power of the load. Therefore, each VSC has reactive power rating equal to half of the reactive power of the load: $Q_{DST} = Q_{DVR} = Q_L/2 = 60\ 000/2 = 30\ 000$ VAR.

Under the sag, from the following relation of reactive power of the DVR, the power angle can be computed as follows.

$$Q_{DVR} = 3K_0V_sI_s \sin\delta = 3(V_L/V_s)V_sI_s \sin\delta = 30\,000 \text{ VAR; therefore, } \delta = 18.55°.$$

$$P_{DVR} = -3K_0V_sI_s(n_0 - \cos\delta) = -3(V_L/V_s)V_sI_s\{(V_s/V_L) - \cos\delta\} = 9419.9 \text{ W}.$$

The voltage rating of the DVR is $V_{DVR} = V_L\sqrt{(1 + n_0^2 - 2n_0\cos\delta)} = 239.6\sqrt{(1 + 0.848^2 - 2 \times 0.848\cos 17.46°)} = 79.88$ V.

The current rating of the DVR is equal to the supply current after sag compensation: $I_{DVR} = I_s = P_L/3V_s = 80\ 000/(3 \times 203.22) = 131.212$ A.

The VA rating of the DVR is $S_{DVR} = 3V_{DVR}I_{DVR} = 3 \times 79.88 \times 131.212 = 31\,444.33$ VA.

The active power of the DSTATCOM is $P_{DST} = -P_{DVR} = -9419.9$ W.

The voltage rating of the DSTATCOM of the UPQC-S is equal to AC load voltage ($V_{DST} = 239.6$ V), since it is connected across the load of 239.6 V sine waveform.

The current rating of the DSTATCOM of the UPQC-S is computed as follows.

The angle between the supply voltage and the load current is $\beta = \phi - \delta = 36.87° - 18.55° = 18.32°$

Therefore, DSTATCOM current is $I_{DST} = I_L\sqrt{[(1 + K_0^2\cos^2\phi - 2K_0\cos\beta\cos\phi)} = 139.12\sqrt{(1 + 1.179^2 \times 0.8^2 - 2 \times 1.179 \times 0.8 \times 0.949)} = 43.74$ A.

The phase angle between DSTATCOM voltage and current is calculated as follows.

$$P_{DST} = -9438.71 \text{ W} = 3V_LI_{DST}\cos\phi_{DST}, \quad \phi_{DST} = 107.47°.$$

$$Q_{DST} = 3V_LI_{DST}\sin\phi_{DST} = 29\,990 = 30\,000 \text{ VA}.$$

The VA rating of the DSTATCOM of the UPQC-S is $S_{DST} = 3V_{DST}I_{DST} = 3 \times 239.6 \times 43.74$ VA = 31 444 VA.

The VA rating of the UPQC-S during voltage sag is $S_{UPQC\text{-}S(under\ sag)} = S_{DST} + S_{DVR} = 31\,444 + 31\,444 = 62\,888$ VA.

Hence, considering an overall rating (both normal and under voltage sag), ratings of both the compensators are

$$V_{DST} = 239.6 \text{ V}, \quad I_{DST} = 44.34 \text{ A}, \quad S_{DST} = 3V_{DST}I_{DST} = 31\,871.59 \text{ VA}.$$

$$V_{DVR} = 101.12 \text{ V}, \quad I_{DVR} = 131.21 \text{ A}, \quad S_{DVR} = 3V_{DVR}I_{DVR} = 39\,803.86 \text{ VA}.$$

a. The total VA rating of the UPQC-S considering both VSCs share equal VA rating all the time is $S_{UPQC\text{-}S} = S_{DST} + S_{DVR} = 31\,871.59 + 39\,803.86 = 71\,674.86$ VA.

   If both the VSCs of same VA rating are considered for same inventory, then $S_{DST} = S_{DVR} = 39\ 803.86$ VA.

   The VA rating of the UPQC-S is $S_{UPQC\text{-}S} = S_{DST} + S_{DVR} = 39\ 803.86 + 39\ 803.86 = 79\ 607.72$ VA.

b. The turns ratio of the injection transformer for the DVR is computed as follows.

The DC bus voltage of 750 V (decided for 415 V for the DSTATCOM) can be used to obtain 415 V across the line at the output of the VSC using a PWM controller. However, the DVR requires only 101.12 V per phase. Therefore, the maximum value of the turns ratio of the injection transformer is $K_{DVR} = N_{VSC}/N_{DVR} = 415/(\sqrt{3} \times 101.12) = 2.37 \approx 2.0$.

The DC bus voltage is decided based on the AC voltages impressed across the VSCs. Here, the DSTATCOM of the UPQC-S is directly used without any transformer; therefore, the AC voltage across the DSTATCOM (load voltage of 415 V) decides the DC bus voltage: $V_{DC} = 2\sqrt{2}(V_L/\sqrt{3})/m_a = 2\sqrt{2}(415/\sqrt{3})/0.9 = 752.99 \approx 750$ V.

c. The interfacing inductances are required in VSCs, which are working as DSTATCOM and DVR.

The interfacing inductance of the DSTATCOM is $L_{DST} = (\sqrt{3}/2)m_a V_{DC}/6af_s\Delta I = (\sqrt{3}/2) \times 0.9 \times 750/(6 \times 1.2 \times 10\ 000 \times 4.374) = 1.86$ mH.

The interfacing inductance of the DVR is $L_{DVR} = (\sqrt{3}/2)m_a V_{DC}/(6af_s\Delta I_{DVR})/K_{DVR} = (\sqrt{3}/2) \times 0.9 \times 750/(6 \times 1.2 \times 10\ 000 \times 13.112/2) = 1.238$ mH.

d. The DC bus capacitance of the UPQC-S is computed from change in stored energy during dynamics as follows.

The change in stored energy during dynamics is $\Delta E = (1/2)C_{DC}(V_{DCapf}^2 - V_{DCminapf}^2) = 3V_{DST}I_{DST}\Delta t$.

Substituting the values, $\Delta E = (1/2)C_{DC}(750^2 - 690^2) = 3 \times 239.6 \times 43.74 \times 10/1000$ (considering $\Delta t = 10$ ms).

$$C_{DC} = 7227.85\,\mu\text{F}.$$

Example 6.18

A three-phase AC supply has a line voltage of 415 V at 50 Hz and a feeder (source) impedance of 1.0 Ω resistance and 3.0 Ω inductive reactance per phase after which a balanced delta connected load having $Z_L = 30\,\Omega$ per phase is connected. Calculate the voltage drop across the source impedance and the voltage across the load. If a three-phase right-hand unified power quality compensator (consisting of a DSTATCOM and a DVR using two VSCs with a common DC bus capacitor shown in Figure E6.18) is used to raise the voltage to the input voltage (415 V) and to maintain unity power factor at the AC mains, calculate (a) the voltage rating of the DVR of the UPQC-S, (b) the current rating of the DVR of the UPQC-S, (c) the VA rating of the DVR of the UPQC-S, (d) the voltage rating of the DSTATCOM of the UPQC-S, (e) the current rating of the DSTATCOM of the UPQC-S, (f) the VA rating of the DSTATCOM of the UPQC-S, and (g) total VA rating of the UPQC-S. Consider same rating of both VSCs.

Solution: Given supply line voltage of 415 V at 50 Hz and a feeder (source) impedance of 1.0 Ω resistance and 3.0 Ω inductive reactance per phase after which a balanced delta connected load having $Z_{LD} = 21\,\Omega$ per phase is connected.

The equivalent load impedance in star connection is $Z_{LY} = Z_{LD}/3 = 30/3 = 10\,\Omega$ per phase.

The total impedance in star connection is $Z_{LT} = Z_s + Z_{LY} = (11 + j3) = 11.40\angle 15.25°\,\Omega$.

The load current before compensation is $I_{sold} = V/|Z_{LT}| = 239.9/11.40 = 21.07$ A.

The voltage drop across the source impedance is $V_{Zs} = I_{old}Z_s = 66.45$ V.

The voltage across the load is $V_{Lold} = V_{ZL} = I_{old}Z_{LY} = 210.14$ V per phase (star) = 363.97 V per phase (for a delta connected load).

Voltage Drop Compensation by a Right-Hand UPQC-S

The load after compensation must have rated voltage ($415/\sqrt{3} = 239.6$ V) across it and the UPQC-S must result in unity power factor at the AC mains.

The load current is $I_L = V_L/Z_{LY} = 239.6/10 = 23.96$ A, $\cos\phi = 1$, and $\sin\phi = 0$.

$$P_{Lp} = I_L^2 R_L = 5740.8\text{ W (load power per phase)}, \quad Q_{Lp} = 0.$$

From real power balance at PCC and supply, the active powers are $V_s' I_s' = P_{Lp}$ and $V_s I_s' \cos\gamma = P_{Lp} + I_s'^2 R_s$.

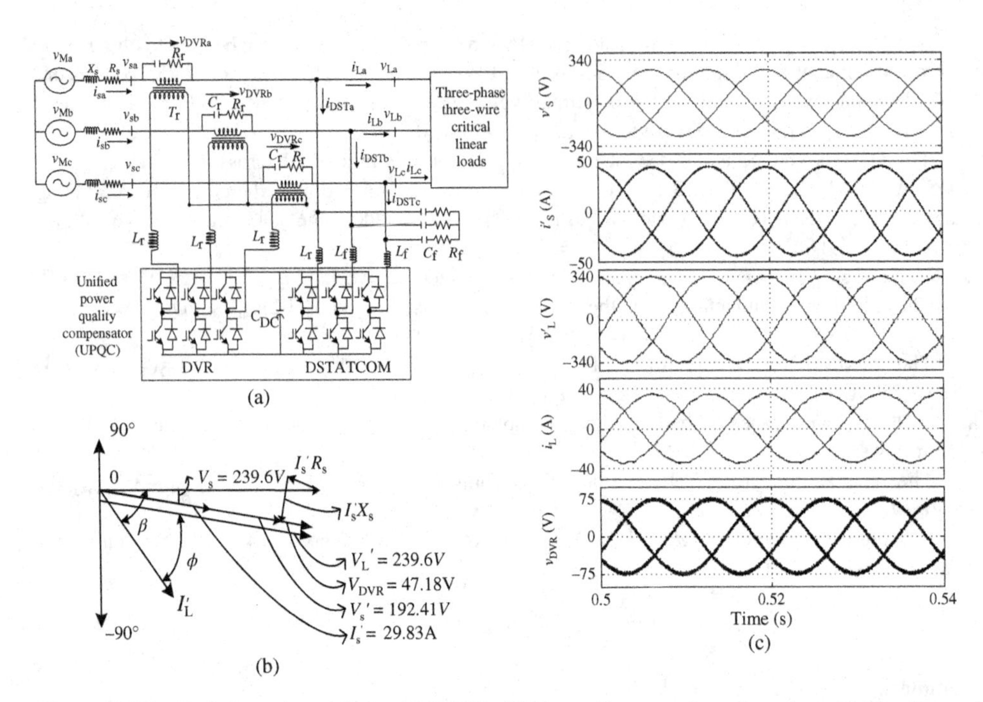

Figure E6.18 (a) A three-phase right-hand UPQC-S. (b) Phasor diagram under voltage drop. (c) Waveforms of a three-phase right-hand UPQC-S under voltage drop

For this load, all reactive power is supplied by the UPQC-S, whereas reactive power needed by the source impedance is supplied by the AC source (grid). The corresponding equation is $V_s I_s' \sin\gamma = I_s'^2 X_s$. On solving these equations, $I_s' = 29.83$ A and $V_s' = 192.41$ V.

As load reactive power is zero, the VSCs circulate only real power for voltage regulation purpose. It can be seen as UPQC-S working as UPQC-P. It means that $Q_{DVR} = Q_{DST} = Q_{Lp}/2 = 0$.

For this case, the voltage and current ratings of the DVR are:

a. $V_{DVR} = V_L' - V_s' = 239.6 - 192.41 = 47.18$ V.
b. $I_{DVR} = I_s'$.
   The power rating of the DVR of the UPQC-S is $P_{DVR} = 3V_{DVR}I_{DVR} = -3 \times 47.18 \times 29.83 = -4222.9$ W, $Q_{DVR} = 0$ VAR.
c. $S_{DVR} = \sqrt{(P_{DVR}^2 + Q_{DVR}^2)} = 4222.9$ VA.
   The DSTATCOM circulates the same real power as the DVR of the UPQC-S. Hence, the real power absorbed by the DVR is $P_{DST} = -P_{DVR} = 4222.9$ W.
   The reactive power required by the load is zero. Hence, the reactive power rating of the DSTATCOM is $Q_{DST} = Q_{Lp}/2 = 0$ VA and (f) $S_{DST} = \sqrt{(P_{DST}^2 + Q_{DST}^2)} = 4222.9$ VAR.
d. $V_{DST} = 239.6$ V per phase.
e. $I_{DST} = S_{DST}/3V_{DST} = 4222.9/(3 \times 239.6) = 5.87$ A.
f. $S_{UPQC\text{-}S} = S_{DVR} + S_{DST} = 4222.9 + 4222.9 = 8445.7$ VA.

Example 6.19

A three-phase right-hand UPQC-P (consisting of a DSTATCOM and a DVR using two VSCs with a common DC bus capacitor shown in Figure E6.19) is designed for a line–line connected single-phase

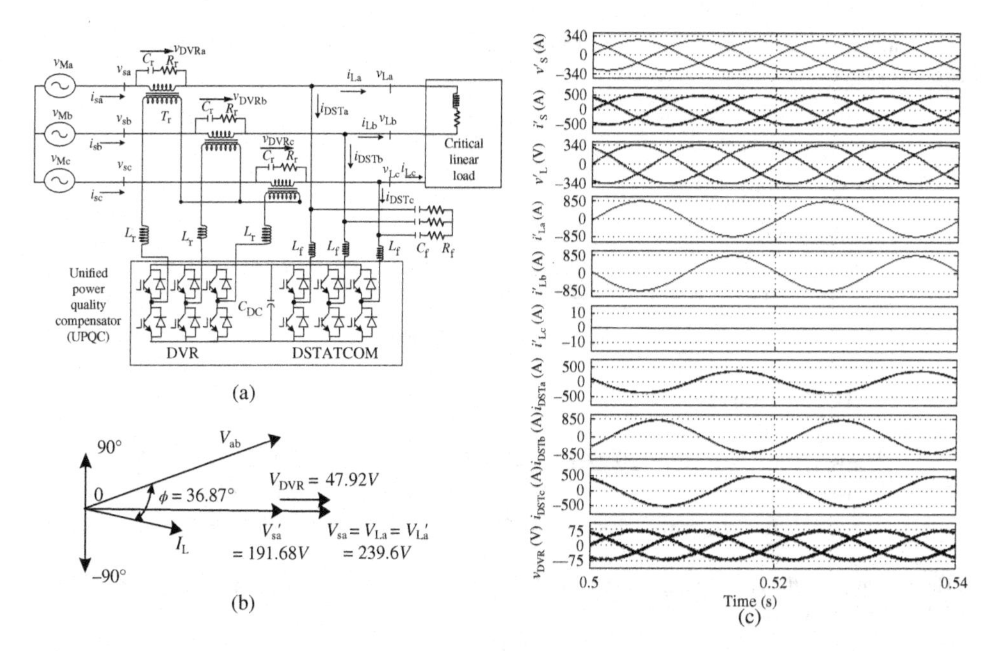

Figure E6.19 (a) A three-phase right-hand UPQC-P. (b) Phasor diagram under voltage sag. (c) Waveforms of a three-phase three-wire right-hand UPQC-P under unbalanced load and voltage sag

415 V, 50 Hz, 250 kVA, 0.8 lagging power factor load. If there is a voltage sag of −20% in the supply system with a base value of 415 V, calculate the VA shared by both DSTATCOM and DVR used in the right-hand UPQC-P to provide a three-phase balanced load at unity power factor with a constant regulated voltage of 415 V at 50 Hz.

Solution: Given supply voltage $V_{sp} = 415/\sqrt{3} = 239.6$ V, frequency of the supply $(f) = 50$ Hz, and a single-phase 239.6 V per phase, 50 Hz, 250 kVA, 0.8 lagging power factor load connected across line to line. There is a voltage sag of −20% in the supply system with a base value of 239.6 V per phase.

It means that under the voltage sag, the supply voltage reduces to $V'_{sp} = V_{sp}(1 - X) = 239.6 \times (1 - 0.20) = 191.68$ V and $X = 0.2$.

The active power of the load is $P_L = 250 \times 0.8 = 200$ kW.

The reactive power of the load is $Q_L = 250 \times 0.6 = 150$ kVAR.

The load impedance is $|Z_L| = V_{LL}^2/S_L = 0.6889\ \Omega$.

Voltage Sag Compensation by a Right-Hand UPQC-P

The supply current under voltage sag compensation, power factor correction, and load balancing is $I'_s = P_L/V'_{sp} = 200\,000/(3 \times 191.68) = 347.80$ A.

$$I_{sa} = 347.8\angle 0^\circ \text{ A}, \quad I_{sb} = 347.8\angle -120^\circ \text{ A}, \quad I_{sc} = 347.8\angle 120^\circ \text{ A}.$$

The voltage rating of the DSTATCOM of the UPQC-P is equal to AC load voltage ($V_{DST} = 239.6$ V), since it is connected across the load of 239.6 V sine waveform.

The current rating of the DSTATCOM of the UPQC-P is computed as follows.

Let the load currents in lines a and b are equal before compensation: $I_{La} = -I_{Lb} = S/V_{ab} = 250\,000/415 = 602.41$ A.

This single-phase load connected across line to line is to be realized as a balanced load on the three-phase supply. A three-phase three-wire DSTATCOM is used for this purpose, which balances this load by circulating the currents on its AC sides. Consider that the load is connected across line ab in a three-phase abc system.

$$I_L = S_L/V_{LL}.$$

$$Z_L = |Z_L|\angle 36.87°\,\Omega.$$

The load currents are

$$I_{La} = V_{ab}/Z_L = 415\angle 30°/0.6889\angle 36.87° = (598 - j72)\text{ A}.$$

$$I_{Lb} = -I_{La} = (-598 + j72)\text{ A}.$$

The DSTATCOM currents are

$$I_{DSTa} = I_{sa} - I_{La} = (-250.29 + j72.057)\text{ A} = |-250.29 + j72.057| = 260.45\text{ A}.$$
$$I_{DSTb} = I_{sb} - I_{Lb} = (424.19 - j373.25)\text{ A} = |424.19 - j373.25| = 565.02\text{ A}.$$
$$I_{DSTc} = I_{sc} - I_{Lc} = (-173.90 + j301.20\,)\text{A} = |-173.90 + j301.20| = 347.79\text{ A}.$$

The VA rating of the DSTATCOM is $S_{DST} = V_{La}|I_{DSTa}| + V_{Lb}|I_{DSTb}| + V_{Lc}|I_{DSTc}| = 239.6 \times (260.45 + 565.02 + 347.79)\text{ VA} = 281.13\text{ kVA}$.

$$Q_{DST} = 150\text{ kVA}.$$

The voltage rating of the DVR of the UPQC-P is computed as follows.

There is a voltage sag of −20% in the supply system with a base value of 239.6 V. Therefore, the DVR of the UPQC-P must inject the difference of these two voltages to provide the required voltage at the load end.

The voltage rating of the DVR is $V_{DVR} = 239.60 \times 0.20 = 47.92$ V.

The current rating of the DVR of the UPQC-P is same as the supply current after shunt compensation: I_{DVR} = the fundamental active component of load current $(I'_s) = P_L/3V'_{sp} = 200\,000/(3 \times 191.68) = 347.80$ A.

The VA rating of the DVR of the UPQC-P is $S_{DVR} = 3V_{DVR}I_{DVR} = 3 \times 47.92 \times 347.80\text{ VA} = 49\,999.73$ VA.

The VA rating of the UPQC-P is $S_{UPQC\text{-}P} = S_{DST} + S_{DVR} = 281\,130 + 49\,999.73 = 331\,130$ VA.

Example 6.20

A three-phase four-wire right-hand UPQC-P (consisting of a four-leg VSC-based DSTATCOM and a DVR using two VSCs with a common DC bus capacitor shown in Figure E6.20) is designed for a line–neutral connected single-phase load of 30 kW with an input AC line voltage of 415 V at 50 Hz in a rural sector for lighting. If there is a voltage sag of −10% in the supply system with a base value of 415 V, calculate the VA ratings shared by both DSTATCOM and DVR used in the right-hand UPQC-P and its total rating to provide a three-phase balanced load at unity power factor with a constant regulated load voltage of 415 V at 50 Hz.

Solution: Given supply voltage $V_{sl} = 415$ V, frequency of the supply $(f) = 50$ Hz, and a line–neutral connected single-phase load of 30 kW with an input AC line voltage of 415 V at 50 Hz. There is a voltage sag of −10% in the supply system with a base value of 415 V.

It means that after the voltage sag, the supply voltage reduces to $V'_{sl} = 415 \times (1 - 0.10) = 373.5$ V and $V'_{sp} = V'_{sl}/\sqrt{3} = 215.64$ V.

The phase load voltage is $V_{Lp} = 415/\sqrt{3} = 239.59$ V and $X = 0.1$.

The load phase current is $I_{Lp} = 30\,000/239.59 = 125.21$ A.

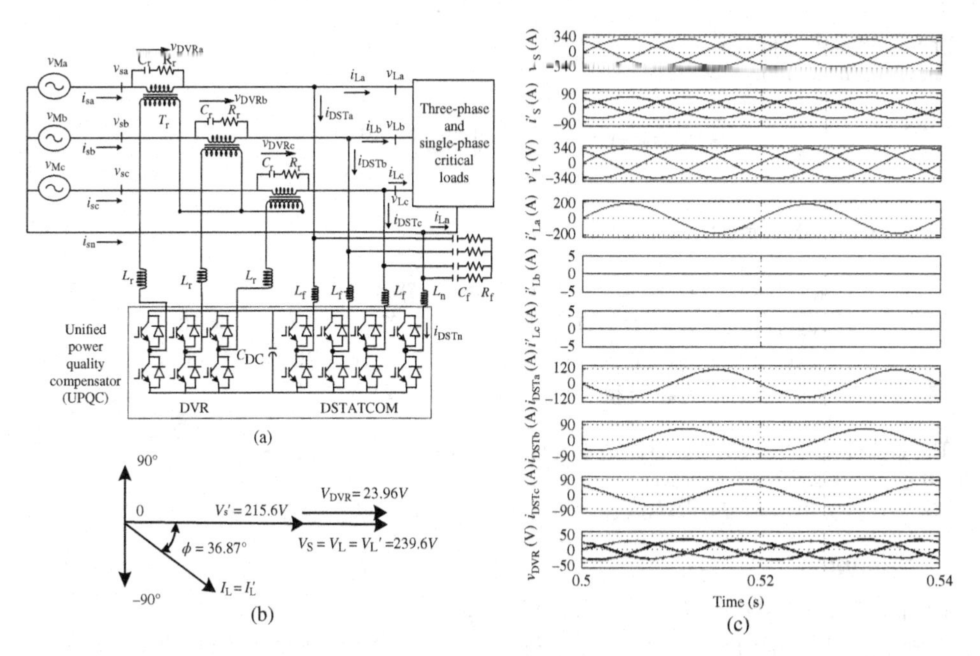

Figure E6.20 (a) A three-phase right-hand UPQC-P. (b) Phasor diagram under voltage sag. (c) Waveforms of a three-phase four-wire right-hand UPQC-P under unbalanced load and voltage sag

The load impedance is $|Z_L| = V_{LP}/I_{LP} = 1.91\ \Omega$.

The supply current during voltage sag compensation is $I_s' = P_L/3V_{sp}' = 30\,000/(3 \times 215.64) = 46.37$ A.

Voltage Sag Compensation by a Right-Hand UPQC-P

The voltage rating of the DVR of the UPQC-P is computed as follows.

There is a voltage sag of −10% in the supply system with a base value of 239.6 V. Therefore, the DVR of the UPQC-P must inject the difference of these two voltages to provide the required voltage at the load end.

The voltage rating of the DVR is $V_{DVR} = 239.60 \times 0.10 = 23.96$ V.

The current rating of the DVR of the UPQC-P is same as supply current after shunt compensation: I_{DVR} = the fundamental active component of load current $(I_s') = P_L/3V_{sp}' = 30\,000/(3 \times 215.64) = 46.37$ A.

The VA rating of the DVR of the UPQC-P is $S_{DVR} = P_{DVR} = 3V_{DVR}I_{DVR} = 3 \times 23.96 \times 46.37$ VA = 3333.08 VA.

$Q_{DVR} = 0.0$, as the voltage of the DVR is injected in phase with its current.

$$S_{DVR} = \sqrt{(P_{DVR}^2 + Q_{DVR}^2)} = P_{DVR} = 3 \times 23.96 \times 46.37 \text{ VA} = 3333.08 \text{ VA}.$$

The voltage rating of the DSTATCOM of the UPQC-P is equal to AC load voltage ($V_{DSTp} = 239.59$ V), since it is connected across the load of 239.59 V sine waveform.

The current rating of the DSTATCOM of the UPQC-P is computed as follows.

Supply line currents after compensation are $I_{sa} = (P_{La}/3V_{sp}')\angle 0°$ A, $I_{sb} = (P_{La}/3V_{sp}')\angle{-120°}$ A, and $I_{sc} = (P_{La}/3V_{sp}')\angle{-240°}$ A.

Substituting the values, $I_{sa} = \{30\,000/(3 \times 215.64)\}\angle 0°\text{ A} = 46.37\angle 0°$ A, $I_{sb} = 46.37\angle{-120°}$ A, $I_{sc} = 46.37\angle{-240°}$ A, and $I_{sn} = 0\angle 0°$ A.

The load phase voltage is $V_{\mathrm{Lpa}} = |V_{\mathrm{Lp}}|\angle 0°$ V.
The load line currents are

$$I_{\mathrm{La}} = V_{\mathrm{Lpa}}/Z_{\mathrm{L}} = 125.2/1.91 = 125.2\angle 0° \text{ A} = -I_{\mathrm{Ln}}.$$

$$I_{\mathrm{Lb}} = I_{\mathrm{Lc}} = 0 \text{ A}.$$

The DSTATCOM currents are

$$\begin{aligned}
I_{\mathrm{DSTa}} &= I_{\mathrm{sa}} - I_{\mathrm{La}} = -78.83 \text{ A} = |-78.83| = 78.83 \text{ A}.\\
I_{\mathrm{DSTb}} &= I_{\mathrm{sb}} - I_{\mathrm{Lb}} = (-23.18 - j40.15) \text{ A} = |-23.18 - j40.15| = 46.30 \text{ A}.\\
I_{\mathrm{DSTc}} &= I_{\mathrm{sc}} - I_{\mathrm{Lc}} = (-23.18 + j40.15) \text{ A} = |-23.18 + j40.15| = 46.30 \text{ A}.\\
I_{\mathrm{DSTn}} &= I_{\mathrm{sn}} - I_{\mathrm{Ln}} = 125.2 \text{ A} = |125.21| = 125.21 \text{ A}.
\end{aligned}$$

The VA rating of the DSTATCOM is $S_{\mathrm{DST}} = V_{\mathrm{Lp}}(|I_{\mathrm{DSTa}}| + |I_{\mathrm{DSTb}}| + |I_{\mathrm{DSTc}}| + |I_{\mathrm{DSTn}}|) = 239.6 \times (78.83 + 46.30 + 46.30 + 125.21)$ VA $= 71.11$ kVA.

The VA rating of the UPQC-P is $S_{\text{UPQC-P}} = S_{\mathrm{DST}} + S_{\mathrm{DVR}} = 71\,110 + 3333.1 = 74\,444.1$ VA $= 74.44$ kVA.

6.8 Summary

A comprehensive study of UPQCs is presented to provide a wide exposure on various aspects of UPQCs to the researchers, designers, and users of these UPQCs for power quality improvement. A classification of UPQCs into three categories with many circuits in each category is expected to help in the selection of an appropriate topology for a particular application. These UPQCs are considered as a better alternative for power quality improvement due to mitigation of voltage- and current-based power quality problems, simple design and control, and high reliability compared with other options of power quality improvement. An analytical study of various performance indices of UPQCs for the compensation of sensitive and critical consumer loads is made in detail with several numerical examples to study the rating of converters and how it is affected by various kinds of consumer loads and supply conditions. UPQCs are observed as one of the best retrofit solutions for mitigating the power quality problems of sensitive loads for reducing the pollution of the AC mains. Moreover, due to a large number of circuits of UPQCs, the user can select the most appropriate topology with required features to suit a specific application. It is expected to be beneficial to the designers, users, manufacturers, and research engineers dealing power quality improvement.

6.9 Review Questions

1. What is a unified power quality compensator used for compensation in an AC distribution system?
2. What are the power quality problems that a UPQC can mitigate?
3. What are the benefits of using a UPQC?
4. What are the economic factors for the justification for using a UPQC in an AC system?
5. Can a current source inverter be used as a UPQC?
6. Why voltage source inverters (VSIs) are preferred in a UPQC?
7. What are the factors that decide the rating of a UPQC?
8. How the value of an interfacing inductor is computed for a UPQC?
9. How the value of a DC bus capacitor is computed for a VSI used in a UPQC?
10. How the self-supporting DC bus of a VSI is achieved in a UPQC?

11. Is it possible to get unity power at the AC mains while regulating the AC voltage across the load by using a UPQC?
12. How the rating of a UPQC for AC load voltage regulation is affected with the power factor of the load?
13. Why the real power exchange is there in a UPQC from the DC side to the AC side of its converters?
14. Which out of DSTATCOM, DVR, and UPQC has the lowest power rating for voltage regulation across the load with power factor correction to unity?
15. What are the variables required for sensing the control of a three-phase three-wire UPQC?
16. In which condition a right shunt UPQC is used?
17. In which condition a left shunt UPQC is used?
18. How the power rating of a UPQC can be reduced for compensation of a lagging power linear load with some definite source impedance (L_s, R_s) at zero voltage regulation?
19. Is it possible to get unity power at the AC mains while regulating the AC voltage across the load with some definite source impedance (L_s, R_s) using a UPQC? If yes, then how?
20. What are the economic factors for the justification for using a UPQC in an AC system?
21. What are the factors that decide the rating of UPQCs?
22. What is the configuration of a UPQC for neutral current compensation with load balancing and zero voltage regulation with some definite source impedance (L_s, R_s)?
23. In which application of a UPQC, both DSTATCOM and DVR need isolation transformers?
24. Which different control techniques are used for the control of UPQCs?
25. What are the benefits of using a lossless UPQC in a three-phase isolated gasoline engine-driven squirrel cage induction generator set (DG set) for economic justification?

6.10 Numerical Problems

1. A single-phase unified power quality compensator (consisting of a DSTATCOM and a DVR using two VSCs with a common DC bus capacitor) is to be designed for a load of 220 V, 50 Hz, 20 A, 0.75 lagging power factor as a right-hand UPQC-Q (shown in Figure E6.1). There is a voltage sag of −15% in the supply system with a base value of 220 V. Calculate (a) the voltage rating of the DVR of the UPQC-Q, (b) the current rating of the DVR of the UPQC-Q, (c) the VA rating of the DVR of the UPQC-Q, (d) the voltage rating of the DSTATCOM of the UPQC-Q, (e) the current rating of the DSTATCOM of the UPQC-Q, (f) the VA rating of the DSTATCOM of the UPQC-Q, and (g) total VA rating of the UPQC-Q to provide reactive power compensation of the load for unity power factor at the AC mains with a constant regulated voltage of 220 V at 50 Hz across the load.

2. A single-phase unified power quality compensator (consisting of a DSTATCOM and a DVR using two VSCs with a common DC bus capacitor) is to be designed for a load of 220 V, 50 Hz, 20 A, 0.75 lagging power factor as a left-hand UPQC-Q (shown in Figure E6.2). There is a voltage sag of −15% in the supply system with a base value of 220 V. Calculate (a) the voltage rating of the DSTATCOM of the UPQC-Q, (b) the current rating of the DSTATCOM of the UPQC-Q, (c) the VA rating of the DSTATCOM of the UPQC-Q, (d) the voltage rating of the DVR of the UPQC-Q, (e) the current rating of the DVR of the UPQC-Q, (f) the VA rating of the DVR of the UPQC-Q, and (g) total VA

rating of the UPQC-Q to provide reactive power compensation at unity power factor with a constant regulated voltage of 220 V at 50 Hz.

3. A single-phase unified power quality compensator (consisting of a DSTATCOM and a DVR using two VSCs with a common DC bus capacitor) is to be designed for a load of 220 V, 50 Hz, 20 A, 0.75 lagging power factor as a left-hand UPQC-Q (shown in Figure E6.3). There is a voltage swell of +15% in the supply system with a base value of 220 V. Calculate (a) the voltage rating of the DSTATCOM of the UPQC-Q, (b) the current rating of the DSTATCOM of the UPQC-Q, (c) the VA rating of the DSTATCOM of the UPQC-Q, (d) the voltage rating of the DVR of the UPQC-Q, (e) the current rating of the DVR of the UPQC-Q, (f) the VA rating of the DVR of the UPQC-Q, and (g) total VA rating of the UPQC-Q to provide reactive power compensation at unity power factor with a constant regulated voltage of 220 V at 50 Hz.
4. A single-phase unified power quality compensator (consisting of a DSTATCOM and a DVR using two VSCs with a common DC bus capacitor) is to be designed for a load of 220 V, 50 Hz, 20 A, 0.75 lagging power factor as a right-hand UPQC-P (shown in Figure E6.4). There is a voltage sag of −15% in the supply system with a base value of 220 V. Calculate (a) the voltage rating of the DSTATCOM of the UPQC-P, (b) the current rating of the DSTATCOM of the UPQC-P, (c) the VA rating of the DSTATCOM of the UPQC-P, (d) the voltage rating of the DVR of the UPQC-P, (e) the current rating of the DVR of the UPQC-P, (f) the VA rating of the DVR of the UPQC-P, and (g) total VA rating of the UPQC-P to provide reactive power compensation at unity power factor with a constant regulated voltage of 220 V at 50 Hz.
5. A single-phase unified power quality compensator (consisting of a DSTATCOM and a DVR using two VSCs with a common DC bus capacitor) is to be designed for a load of 220 V, 50 Hz, 20 A, 0.75 lagging power factor as a left-hand UPQC-P (shown in Figure E6.5). There is a voltage sag of −15% in the supply system with a base value of 220 V. Calculate (a) the voltage rating of the DVR of the UPQC-P, (b) the current rating of the DVR of the UPQC-P, (c) the VA rating of the DVR of the UPQC-P, (d) the voltage rating of the DSTATCOM of the UPQC-P, (e) the current rating of the DSTATCOM of the UPQC-P, (f) the VA rating of the DSTATCOM of the UPQC-P, and (g) total VA rating of the UPQC-P to provide reactive power compensation at unity power factor with a constant regulated voltage of 220 V at 50 Hz.
6. A single-phase unified power quality compensator (consisting of a DSTATCOM and a DVR using two VSCs with a common DC bus capacitor) is to be designed for a load of 220 V, 50 Hz, 20 A, 0.75 lagging power factor as a left-hand UPQC-P (shown in Figure E6.6). There is a voltage swell of +15% in the supply system with a base value of 220 V. Calculate (a) the voltage rating of the DVR of the UPQC-P, (b) the current rating of the DVR of the UPQC-P, (c) the VA rating of the DVR of the UPQC-P, (d) the voltage rating of the DSTATCOM of the UPQC-P, (e) the current rating of the DSTATCOM of the UPQC-P, (f) the VA rating of the DSTATCOM of the UPQC-P, and (g) total VA rating of the UPQC-P to provide reactive power compensation at unity power factor with a constant regulated voltage of 230 V at 50 Hz.
7. A single-phase unified power quality compensator (consisting of a DSTATCOM and a DVR using two VSCs with a common DC bus capacitor) is to be designed for a load of 220 V, 50 Hz, 20 A, 0.75 lagging power factor as a right-hand UPQC-S (shown in Figure E6.7). There is a voltage sag of −15% in the supply system with a base value of 220 V. Calculate (a) the voltage rating of the DVR of the UPQC-S, (b) the current rating of the DVR of the UPQC-S, (c) the VA rating of the DVR of the UPQC-S, (d) the voltage rating of the DSTATCOM of the UPQC-S, (e) the current rating of the DSTATCOM of the UPQC-S, (f) the VA rating of the DSTATCOM of the UPQC-S, and (g) total VA rating of the UPQC-S to provide reactive power compensation for unity power factor at PCC with a constant regulated voltage of 220 V at 50 Hz across the load. Consider same rating of both VSCs.
8. A single-phase unified power quality compensator (consisting of a DSTATCOM and a DVR using two VSCs with a common DC bus capacitor) is to be designed for a load of 220 V, 50 Hz, 20 A,

0.75 lagging power factor as a left-hand UPQC-S (shown in Figure E6.8). There is a voltage sag of −15% in the supply system with a base value of 220 V. Calculate (a) the voltage rating of the DSTATCOM of the UPQC-S, (b) the current rating of the DSTATCOM of the UPQC-S, (c) the VA rating of the DSTATCOM of the UPQC-S, (d) the voltage rating of the DVR of the UPQC-S, (e) the current rating of the DVR of the UPQC-S, (f) the VA rating of the DVR of the UPQC-S, and (g) total VA rating of the UPQC-S to provide reactive power compensation for unity power factor at PCC with a constant regulated voltage of 220 V at 50 Hz across the load. Consider same rating of both VSCs.

9. A single-phase unified power quality compensator (consisting of a DSTATCOM and a DVR using two VSCs with a common DC bus capacitor) is to be designed for a load of 220 V, 50 Hz, 20 A, 0.75 lagging power factor as a right-hand UPQC-S (shown in Figure E6.9). There is a voltage sag of −15% in the supply system with a base value of 220 V. Calculate (a) the voltage rating of the DVR of the UPQC-S, (b) the current rating of the DVR of the UPQC-S, (c) the VA rating of the DVR of the UPQC-S, (d) the voltage rating of the DSTATCOM of the UPQC-S, (e) the current rating of the DSTATCOM of the UPQC-S, (f) the VA rating of the DSTATCOM of the UPQC-S, and (g) total VA rating of the UPQC-S to provide reactive power compensation for unity power factor at PCC with a constant regulated voltage of 220 V at 50 Hz across the load. Consider same power angle for equal reactive power sharing in both VSCs in steady-state condition for unity power factor at the AC mains without voltage sag condition.

10. A single-phase unified power quality compensator (consisting of a DSTATCOM and a DVR using two VSCs with a common DC bus capacitor) is to be designed for a load of 220 V, 50 Hz, 20 A, 0.75 lagging power factor as a left-hand UPQC-S (shown in Figure E6.10). There is a voltage sag of −15% in the supply system with a base value of 220 V. Calculate (a) the voltage rating of the DVR of the UPQC-S, (b) the current rating of the DVR of the UPQC-S, (c) the VA rating of the DVR of the UPQC-S, (d) the voltage rating of the DSTATCOM of the UPQC-S, (e) the current rating of the DSTATCOM of the UPQC-S, (f) the VA rating of the DSTATCOM of the UPQC-S, and (g) total VA rating of the UPQC-S to provide reactive power compensation for unity power factor at PCC with a constant regulated voltage of 220 V at 50 Hz across the load. Consider same power angle for equal reactive power sharing in both VSCs in steady-state condition for unity power factor at the AC mains without voltage sag condition.

11. A single-phase unified power quality compensator (consisting of a DSTATCOM and a DVR using two VSCs with a common DC bus capacitor) is to be designed for a load of 220 V, 50 Hz, 20 A, 0.75 lagging power factor as a left-hand UPQC-S (shown in Figure E6.11). There is a voltage swell of +15% in the supply system with a base value of 220 V. Calculate (a) the voltage rating of the DVR of the UPQC-S, (b) the current rating of the DVR of the UPQC-S, (c) the VA rating of the DVR of the UPQC-S, (d) the voltage rating of the DSTATCOM of the UPQC-S, (e) the current rating of the DSTATCOM of the UPQC-S, (f) the VA rating of the DSTATCOM of the UPQC-S, and (g) total VA rating of the UPQC-S to provide reactive power compensation for unity power factor at PCC with a constant regulated voltage of 220 V at 50 Hz across the load. Consider same power angle for equal reactive power sharing in both VSCs in steady-state condition for unity power factor at the AC mains without voltage swell condition.

12. A single-phase supply has AC mains voltage of 220 V at 50 Hz and a feeder (source) impedance of 1.0 Ω resistance and 3.0 Ω inductive reactance. It is feeding a load having $Z = (20 + j15)\ \Omega$. Calculate the voltage drop across the source impedance and the voltage across the load. If a single-phase right-hand unified power quality compensator (consisting of a DSTATCOM and a DVR using two VSCs with a common DC bus capacitor shown in Figure E6.12) is used to raise the voltage to the input voltage (220 V) and to maintain unity power factor at the AC mains, calculate (a) the voltage rating of the DVR of the UPQC-Q, (b) the current rating of the DVR of the UPQC-Q, (c) the VA rating of the DVR of the UPQC-Q, (d) the voltage rating of the DSTATCOM of the UPQC-Q, (e) the current rating of the DSTATCOM of the UPQC-Q, (f) the VA rating of the DSTATCOM of the UPQC-Q,

and (g) total VA rating of the UPQC-Q to provide reactive power compensation at unity power factor with a constant regulated voltage of 220 V at 50 Hz.

13. A single-phase supply has AC mains voltage of 220 V at 50 Hz and a feeder (source) impedance of 1.0 Ω resistance and 3.0 Ω inductive reactance. It is feeding a load having $Z = (20 - j15)\ \Omega$. Calculate the voltage drop across the source impedance and the voltage across the load. If a single-phase right-hand unified power quality compensator (consisting of a DSTATCOM and a DVR using two VSCs with a common DC bus capacitor shown in Figure E6.13) is used to regulate the load voltage to the input voltage (220 V) and to maintain unity power factor at the AC mains, calculate (a) the voltage rating of the DSTATCOM of the UPQC-P, (b) the current rating of the DSTATCOM of the UPQC-P, (c) the VA rating of the DSTATCOM of the UPQC-P, (d) the voltage rating of the DVR of the UPQC-P, (e) the current rating of the DVR of the UPQC-P, (f) the VA rating of the DVR-P of UPQC-P, and (g) total VA rating of the UPQC-P.

14. A single-phase supply has AC mains voltage of 220 V at 50 Hz and a feeder (source) impedance of 1.0 Ω resistance and 3.0 Ω inductive reactance. It is feeding a load having $Z = (20 + j15)\ \Omega$. Calculate the voltage drop across the source impedance and the voltage across the load. If a single-phase right-hand unified power quality compensator (consisting of a DSTATCOM and a DVR using two VSCs with a common DC bus capacitor shown in Figure E6.14) is used to raise the voltage to the input voltage (220 V) and to maintain unity power factor at the AC mains, calculate (a) the voltage rating of the DVR of the UPQC-S, (b) the current rating of the DVR of the UPQC-S, (c) the VA rating of the DVR of the UPQC-S, (d) the voltage rating of the DSTATCOM of the UPQC-S, (e) the current rating of the DSTATCOM of the UPQC-S, (f) the VA rating of the DSTATCOM of the UPQC-S, and (g) total VA rating of the UPQC-S. Consider same rating of both VSCs.

15. A three-phase supply with AC line voltage output of 400 V (rms) at 50 Hz is feeding a critical linear load of 75 kVA, 400 V, 50 Hz at 0.85 lagging power factor. A voltage sag of −15% occurs in the supply voltage at PCC. Design a three-phase unified power quality compensator (consisting of a DSTATCOM and a DVR using two VSCs with a common DC bus capacitor) to compensate the power factor of the load to unity at the AC mains and to mitigate the voltage sag to avoid the interruption to the load. If two three-phase VSCs are used as a right-hand UPQC-Q (shown in Figure E6.15), calculate (a) the voltage rating of the DVR of the UPQC-Q, (b) the current rating of the DVR of the UPQC-Q, (c) the VA rating of the DVR of the UPQC-Q, (d) the voltage rating of the DSTATCOM of the UPQC-Q, (e) the current rating of the DSTATCOM of the UPQC-Q, (f) the VA rating of the DSTATCOM of the UPQC-Q, and (g) total VA rating of the UPQC-Q to provide reactive power compensation for unity power factor at the AC mains with a constant regulated voltage of 400 V at 50 Hz.

16. A three-phase supply with AC line voltage of 400 V (rms) at 50 Hz is feeding a critical linear load of 75 kVA, 400 V, 50 Hz, 0.85 lagging power factor. A voltage swell of +15% occurs in the supply at PCC. Design a three-phase unified power quality compensator (consisting of a DSTATCOM and a DVR using two VSCs with a common DC bus capacitor) in this system to compensate the power factor of the load to unity at the AC mains and to mitigate the voltage swell to avoid the interruption to the load. If two three-phase VSCs are used as a right-hand UPQC-P (shown in Figure E6.16), calculate (a) the voltage rating of the DVR of the UPQC-P, (b) the current rating of the DVR of the UPQC-P, (c) the VA rating of the DVR of the UPQC-P, (d) the voltage rating of the DSTATCOM of the UPQC-P, (e) the current rating of the DSTATCOM of the UPQC-P, (f) the VA rating of the DSTATCOM of the UPQC-P, (g) total VA rating of the UPQC-P, (h) interfacing inductance, (i) DC bus capacitance of VSCs, (j) DC bus voltage, and (k) turns ratio of the injection transformer for the DVR. Consider the switching frequency of 20 kHz and DC bus voltage has to be controlled within 8% range of the nominal value and the ripple current in the inductor is 5%.

17. A three-phase supply with AC mains voltage of 440 V, 50 Hz has a voltage sag of −15% due to a rural feeder and a nearby factory. A hospital needs a three-phase 400 V, 150 kVA, 0.85 lagging power factor, 50 Hz linear load in critical equipment. If a right-hand UPQC-S (consisting of a DSTATCOM and a DVR using two VSCs with a common DC bus capacitor) is used as a unified power quality compensator (shown in Figure E6.17), then calculate (a) the rating of the shunt and series components of the UPQC-S to provide rated voltage across the load and to realize unity power factor at the AC mains. If an injection transformer is used to inject the voltage in the DVR, calculate (b) turns ratio of the injection transformer, (c) values of interfacing inductors, and (d) the value of the DC bus capacitor. Consider the switching frequency of 10 kHz and DC bus voltage has to be controlled within 8% range and ripple current in the inductor is 10%. Consider same rating of both VSCs.

18. A three-phase AC supply has a line voltage of 400 V at 50 Hz and a feeder (source) impedance of 1.0 Ω resistance and 3.0 Ω inductive reactance per phase after which a balanced delta connected load having $Z_L = 33\ \Omega$ per phase is connected. Calculate the voltage drop across the source impedance and the voltage across the load. If a three-phase right-hand unified power quality compensator (consisting of a DSTATCOM and a DVR using two VSCs with a common DC bus capacitor shown in Figure E6.18) is used to raise the voltage to the input voltage (400 V) and to maintain unity power factor at the AC mains, calculate (a) the voltage rating of the DVR of the UPQC-S, (b) the current rating of the DVR of the UPQC-S, (c) the VA rating of the DVR of the UPQC-S, (d) the voltage rating of the DSTATCOM of the UPQC-S, (e) the current rating of the DSTATCOM of the UPQC-S, (f) the VA rating of the DSTATCOM of the UPQC-S, and (g) total VA rating of the UPQC-S. Consider same rating of both VSCs.

19. A three-phase right-hand UPQC-P (consisting of a DSTATCOM and a DVR using two VSCs with a common DC bus capacitor shown in Figure E6.19) is designed for a line–line connected single-phase 400 V, 50 Hz, 200 kVA, 0.85 lagging power factor load. If there is a voltage sag of −10% in the supply system with a base value of 400 V, calculate the ratings of both DSTATCOM and DVR used in the right-hand UPQC-P to provide a three-phase balanced load at unity power factor with a constant regulated voltage of 400 V at 50 Hz.

20. A three-phase four-wire right-hand UPQC-P (consisting of a four-leg VSC-based DSTATCOM and a DVR using two VSCs with a common DC bus capacitor shown in Figure E6.20) is designed for a line–neutral connected single-phase load of 20 kW with an input AC line voltage of 400 V at 50 Hz in a rural sector for lighting. If there is a voltage sag of −15% in the supply system with a base value of 400 V, calculate the ratings of both DSTATCOM and DVR used in the right-hand UPQC-P and its total rating to provide a three-phase balanced load at unity power factor with a constant regulated load voltage of 400 V at 50 Hz.

6.11 Computer Simulation-Based Problems

1. Design and develop a dynamic model of a single-phase UPQC-Q (a combination of right side DSTATCOM across the load and DVR between PCC and load with common dc bus of both VSCs) and simulate its behavior for regulating PCC voltage and to maintain unity power factor at ac mains under a single-phase load of 220 V, 50 Hz, 20 A, 0.8 lagging power factor (PF). It has source impedance of 0.5 Ω resistive element and 2.5 Ω inductive element. Demonstrate the effect of change of the load from 20 A to 10 A, and then to 20 A, all at 0.8 lagging power factor. Plot the supply voltage and supply current, load ac voltage, load ac current, DVR voltage and DSTATCOM current with time. Compute kVA ratings of DVR, DSTATCOM and UPQC, rms value of ac mains current, power-factor, input active and reactive powers. The UPQC is controlled in UPQC-Q mode for regulating the load voltage and to maintain unity power factor at ac mains.

2. Design and develop a dynamic model of a single-phase UPQC-P (a combination of left side DSTATCOM across the load and DVR between PCC and load with common dc bus of both VSCs) and simulate its behavior for regulating PCC voltage and to maintain unity power factor at ac mains under a single-phase load of 230 V, 50 Hz, 5 kVA, 0.85 lagging power factor. It has source impedance of 0.5 Ω resistive element and 2.5 Ω inductive element. Demonstrate the effect of change of the load from 5 kVA to 1 kVA and then to 5 kVA all at 0.85 lagging PF. Plot the supply voltage and supply current, load ac voltage, load ac current, DVR voltage and DSTATCOM current with time. Compute kVA ratings of DVR, DSTATCOM and UPQC, rms value of ac mains current, power-factor, input active and reactive powers. The UPQC is controlled in UPQC-P mode for regulating the load voltage and to maintain unity power factor at ac mains.

3. Design and develop a dynamic model of a single-phase UPQC-S (a combination of right side DSTATCOM across the load and DVR between PCC and load with common dc bus of both VSCs) and simulate its behavior for regulating PCC voltage and to maintain unity power factor at ac mains under a single-phase load of 220 V, 50 Hz, 25 A, 0.75 lagging power factor. It has source impedance of 0.5 Ω resistive element and 2.5 Ω inductive element. Demonstrate the effect of change of the load from 25 A to 15 A, and then to 25 A, all at 0.75 lagging PF. Plot the supply voltage and supply current, load ac voltage, load ac current, DVR voltage and DSTATCOM current with time. Compute kVA ratings of DVR, DSTATCOM and UPQC, rms value of ac mains current, power-factor, input active and reactive powers. The UPQC is controlled in UPQC-S mode for regulating the load voltage and to maintain unity power factor at ac mains.

4. Design a single-phase, RH-UPQC-P (consisting of VSC based DSTATCOM and DVR using two VSCs with common dc bus capacitor) and simulate its behavior for a single-phase AC supply, which has a voltage of 220 V at 50 Hz and feeder (source) impedance of 0.5 Ω resistance and 3.5 Ω inductive reactance. It feeds a single-phase load of $Z = (24 - j18)\ \Omega$. It has another load of load of $Z = (4 + j3)\ \Omega$ connected at PCC for 20 cycles. Plot the supply voltage and supply current, load ac voltage, load ac current, DVR voltage and DSTATCOM current with time. Compute kVA ratings of DVR, DSTATCOM and UPQC-P, rms value of ac mains current, power-factor, input active and reactive powers. The UPQC is controlled in UPQC-P mode for regulating the load voltage and to maintain unity power factor at ac mains.

5. Design a single-phase, RH-UPQC-P (consisting of VSC based DSTATCOM and DVR using two VSCs with common dc bus capacitor) and simulate its behavior for single-phase AC supply system which has ac voltage of 230 V at 50 Hz and feeder (source) impedance of 0.5 Ω resistance and 3.0 Ω inductive reactance. It is feeding a load of $Z_{L1} = (20 + j10)\ \Omega$. There is another load connected in parallel to it, which a load impedance of $Z_{L2} = (5 - j3)\ \Omega$ for 20 cycles of AC mains. Plot the supply voltage and supply current, load ac voltage, load ac current, DVR voltage and DSTATCOM current with time. Compute kVA ratings of DVR, DSTATCOM and UPQC, rms value of ac mains current, power-factor, input active and reactive powers. The UPQC is controlled in UPQC-P mode for regulating the load voltage and to maintain unity power factor at ac mains.

6. Design and develop a dynamic model of a single-phase UPQC-Q (a combination of left side DSTATCOM across the load and DVR between PCC and load with common dc bus of both VSCs) and simulate its behavior for regulating PCC voltage and to maintain unity power factor at ac mains under a single-phase load of 220 V, 50 Hz, 20 A, 0.8 lagging power factor. It has source impedance of 0.5 Ω resistive element and 2.5 Ω inductive element. Demonstrate the effect of change of the load from 20 A to 10 A, and then to 20 A, all at 0.8 lagging PF. Plot the supply voltage and supply current, load ac voltage, load ac current, DVR voltage and DSTATCOM current with time. Compute kVA ratings of DVR, DSTATCOM and UPQC, rms value of ac mains current, power-factor, input active and reactive powers. The UPQC is controlled in UPQC-Q mode for regulating the load voltage and to maintain unity power factor at ac mains.

7. Design and develop a dynamic model of a single-phase UPQC-S (a combination of right side DSTATCOM across the load and DVR between PCC and load with a common dc bus of both VSCs) and simulate its behavior for a single-phase ac supply system which has ac mains voltage of 230 V at 50 Hz and has a feeder (source) impedance of 0.75 ohms resistance and 3.5 ohms inductive reactance. It is feeding a load of $Z_{L1} = (16 - j12)$ ohms. It has another load of load of $Z_{L2} = (4 + j3)$ ohms connected at PCC for 20 cycles. Plot the supply voltage and supply current, load ac voltage, load ac current, DVR voltage and DSTATCOM current with time. Compute kVA ratings of DVR, DSTATCOM and UPQC, rms value of ac mains current, power-factor, input active, reactive powers. The UPQC is controlled in UPQC-S mode with minimum kVA for regulating the load voltage and to maintain unity power factor at ac mains.

8. Design and develop a dynamic model of a UPQC (a combination of right side DSTATCOM across the load and DVR between PCC and load with a common dc bus of both VSCs) and simulate its behavior for single-phase AC supply system, which has ac voltage of 230 V at 50 Hz and feeder (source) impedance of 0.25 ohms resistance and 2.5 ohms inductive reactance. It is feeding a load of $Z_{L1} = (25 + j15)$ ohms. There is another load connected in parallel to it, which a load impedance of $Z_{L2} = (5 + j3)$ ohms for 20 cycles of ac mains. Plot the supply voltage and supply current, load ac voltage, load ac current, DVR voltage and DSTATCOM current with time. Compute kVA ratings of DVR, DSTATCOM and UPQC, rms value of ac mains current, power-factor, input active, reactive powers. The UPQC is controlled in UPQC-P mode for regulating the load voltage and to maintain unity power factor at ac mains.

9. A single-phase supply has ac mains voltage of 220 V at 50 Hz and feeder (source) impedance of 1.0 ohms resistance and 3.0 ohms inductive reactance. It is feeding a load of $Z_{L1} = (20 + j15)$ ohms. Calculate the voltage drop across the source impedance and voltage across the load. A single-phase Right Hand unified power quality compensator, RH-UPQC-Q (UPQC-Q consisting of DSTATCOM and DVR using two VSCs with a common dc bus capacitor) is used to raise the voltage to same as input voltage (220 V) and a unity power factor at the ac mains. Demonstrate the effect of change of the load from $Z_{L1} = (20 + j15)$ ohms to $Z_{L2} = (10 + j5)$ ohms and then to $Z_{L1} = (20 + j15)$ ohms. Plot the supply voltage and supply current, load ac voltage, load ac current, DVR voltage and DSTATCOM current with time. Compute kVA ratings of DVR, DSTATCOM and UPQC, rms value of ac mains current, power-factor, input active, reactive powers.

10. Design and develop a dynamic model of a UPQC-S (a combination of right side DSTATCOM across the load and DVR between PCC and load with a common dc bus of both VSCs) for an isolated single-phase, 230 V, 7.5 kVA, 50 Hz, 4-pole diesel engine driven permanent magnet synchronous generator (PMSG) system to regulate its voltage across loads and its unity power at DG set. This PMSG system has the winding resistance of 1.5 ohms and the reactance of 4.5 ohms. This PMSG is loaded with a single-phase load of 10 kVA at 0.75 lagging. Plot the PCC voltage and supply current, load ac voltage, load ac current, DVR voltage and DSTATCOM current with time. The UPQC is controlled in UPQC-S mode with minimum kVA for regulating the load voltage and to maintain unity power factor at DG set. Demonstrate the effect of change of the load from 10 kVA to 4 kVA and then to 10 kVA all at 0.75 lagging PF. Compute kVA ratings of DVR, DSTATCOM and UPQC. Compute (a) rms value of PMSG current, (b) rms value of ac load current, (c) PMSG active, reactive powers and load active, reactive powers.

11. Design and develop a dynamic model of a three-phase, three-wire UPQC-S (a combination of right side DSTATCOM across the load and DVR between PCC and load with a common dc bus of both VSCs) and simulate its behavior for providing reactive power compensation for unity power factor of a single-phase load of 25 kVA, 0.8 lagging power factor connected between two lines of ac mains under 10% sag in PCC voltage. Plot the supply voltage and supply current, load ac voltage, load ac current, DVR voltage and DSTATCOM current with time. Demonstrate the effect of change of the load from single-phase, 25 kVA to 10 kVA and then to 25 kVA all at 0.8 lagging PF. Compute kVA ratings of DVR, DSTATCOM and UPQC, rms value of ac mains current, power-factor, input active,

reactive powers. The UPQC is controlled in UPQC-S mode with minimum kVA for regulating the load voltage and to maintain unity power factor at ac mains.

12. A three-phase ac supply has a line voltage of 400 V at 50 Hz and feeder (source) impedance of 1.0 ohms resistance and 3.0 ohms inductive reactance/phase after which a balanced delta connected load $Z_L = 20$ ohms/phase is connected. Compute the voltage drop across the source impedance and voltage across the load. If a three-phase Right Hand unified power quality compensator, RH-UPQC-S (UPQC-S consisting of DSTATCOM and DVR using two VSCs with a common dc bus capacitor) is used to raise the voltage to same as input voltage (400 V) and a unity power factor at the ac mains, Plot the supply voltage and supply current, load ac voltage, load ac current, DVR voltage and DSTATCOM current with time. Compute (a) the voltage rating of DVR of UPQC-S, (b) the current rating DVR of UPQC-S, (c) the VA rating of DVR of UPQC-S, (d) the voltage rating of DSTATCOM of UPQC-S, (e) the current rating of DSTATCOM of UPQC-S, (f) the VA rating of DSTATCOM of UPQC-S, and (g) total VA rating of UPQC-S. Consider same rating of both VSCs.

13. Design and develop a dynamic model of a UPQC (a combination of right side DSTATCOM across the load and DVR between PCC and load with a common dc bus of both VSCs) and simulate its behavior for a three-phase three-wire ac supply, which has a line voltage of 415 V at 50 Hz and feeder (source) impedance of 0.75 ohms resistance and 3.5 ohms inductive reactance/phase. It is feeding an unbalanced delta connected load ($Z_{ab} = 6.0 + j3.0$ pu, $Z_{bc} = 3.0 + j1.5$ pu and $Z_{ca} = 7.5 + j1.5$ pu) with base impedance of 4.15 ohms per phase. It has another load of $Z = (4 + j3)$ ohms connected across two lines at PCC for 20 cycles. Plot the supply voltage and supply current, load ac voltage, load ac current, DVR voltage and DSTATCOM current with time and the harmonic spectra of supply voltage, supply current and load voltage, load current. Compute kVA ratings of DVR, DSTATCOM and UPQC, rms value of ac mains current, power-factor, input active, reactive powers. The UPQC is controlled in UPQC-S mode with minimum kVA for regulating the load voltage and to maintain unity power factor at ac mains.

14. Design and develop a dynamic model of a UPQC (a combination of right side DSTATCOM across the load and DVR between PCC and load with a common dc bus of both VSCs) and simulate its behavior fro a three-phase, three-wire ac supply system which has ac mains voltage of 415 V at 50 Hz and feeder (source) impedance of 0.75 ohms/phase resistance and 3.5 ohms/phase inductive reactance. It feeds a balanced three-phase star connected load $Z_{L1} = (20 + j12)$ ohms/phase. If one another single-phase load of $Z_{L2} = (3 - j5)$ ohms is connected between two lines connected at PCC for 20 cycles. Plot the supply voltage and supply current, load ac voltage, load ac current, DVR voltage and DSTATCOM current with time. Compute kVA ratings of DVR, DSTATCOM and UPQC, rms value of ac mains current, power-factor, input active, reactive powers. The UPQC is controlled in UPQC-S mode with minimum kVA for regulating the load voltage and to maintain unity power factor at ac mains.

15. Design and develop a dynamic model of a UPQC (a combination of right side DSTATCOM across the load and DVR between PCC and load with a common dc bus of both VSCs) and simulate its behavior for a three-phase ac mains of 400 V, 50 Hz has a voltage fluctuation between +10% and −15% and 5% voltage unbalance due to rural feeder and a nearby factory. A hospital needs a three-phase 400 V, 50 kVA, 0.8 lagging pf, 50 Hz supply, for critical equipments. It has source impedance of 0.75 Ω resistive element and 3.5 Ω inductive element. If one another single-phase load of $Z = (10 + j15)$ ohms is connected between two lines connected at PCC for 20 cycles. Plot the supply voltage and supply current, load ac voltage, load ac current, DVR voltage and DSTATCOM current with time. Compute kVA ratings of DVR, DSTATCOM and UPQC, rms value of ac mains current, input active, reactive powers. The UPQC is controlled in UPQC-S mode with minimum kVA for regulating the load voltage and to maintain unity power factor at ac mains.

16. Design and develop a dynamic model of a UPQC (a combination of right side DSTATCOM across the load and DVR between PCC and load with a common dc bus of both VSCs) and simulate its behavior to regulating the PCC voltage and to maintain UPF of a three-phase, 3-wire, 440 V, 50 Hz system. It has a linear load of 30 kVA at 0.8 lagging pf. It has source impedance of 0.25 Ω resistive element and 3.0 Ω inductive element. Plot the supply voltage and supply current, load ac voltage, load ac current, DVR voltage and DSTATCOM current with time. Compute kVA ratings of DVR, DSTATCOM and UPQC, rms value of ac mains current, input active, reactive powers. The UPQC is controlled in UPQC-S mode with minimum kVA for regulating the load voltage and to maintain unity power factor at ac mains.

17. Design and develop a dynamic model of a UPQC (a combination of right side DSTATCOM across the load and DVR between PCC and load with a common dc bus of both VSCs) and simulate its behavior for a three phase 415 V, 50 Hz, $Z_{L1} = (30 + j15)$ ohms/phase load. It has source impedance of 0.5 Ω resistive element and 3.5 Ω inductive element. If one another single-phase load of $Z_{L2} = (10 + j5)$ ohms is connected between two lines connected at PCC for 20 cycles. Plot the supply voltage and supply current, load ac voltage, load ac current, DVR voltage and DSTATCOM current with time. Compute kVA ratings of DVR, DSTATCOM and UPQC, rms value of ac mains current, input active, reactive powers. The UPQC is controlled in UPQC-S mode with minimum kVA for regulating the load voltage and to maintain unity power factor at ac mains.

18. A three-phase ac supply has a line voltage of 400 V at 50 Hz and feeder (source) impedance of 1.0 ohms resistance and 3.0 ohms inductive reactance/phase after which a balanced delta connected load $Z_L = 10$ ohms/phase is connected. Compute the voltage drop across the source impedance and voltage across the load. If a three-phase Right Hand unified power quality compensator, RH-UPQC-S (UPQC-S consisting of DSTATCOM and DVR using two VSCs with a common dc bus capacitor) is used to raise the voltage to same as input voltage (400 V) and a unity power factor at the ac mains. Demonstrate the effect of change of the load from $Z_{L1} = 10$ ohms/phase to $Z_{L2} = 20$ ohms/phase and then to $Z_{L1} = 10$ ohms/phase. Plot the supply voltage and supply current, load ac voltage, load ac current, DVR voltage and DSTATCOM current with time. Compute (a) the voltage rating of DVR of UPQC-S, (b) the current rating DVR of UPQC-S, (c) the VA rating of DVR of UPQC-S, (d) the voltage rating of DSTATCOM of UPQC-S, (e) the current rating of DSTATCOM of UPQC-S, (f) the VA rating of DSTATCOM of UPQC-S, and (g) total VA rating of UPQC-S. Consider same rating of both VSCs.

19. A three-phase supply with ac line voltage of 400 V rms at 50 Hz is feeding a critical linear load of 50 kVA, 400 V, 50 Hz at 0.85 lagging pf. A voltage swell of +15% occurs in the supply at PCC. Design a three-phase unified power quality compensator UPQC-P (UPQC consisting of DSTATCOM and DVR using two VSCs with common dc bus capacitor) in this system to compensate the pf of the load to unity at ac mains and mitigation of the voltage swell to avoid the interruption to the load. Two three-phase VSCs are used as a Right Hand UPQC-P. Demonstrate the effect of change of the load from 50 kVA to 10 kVA and then to 50 kVA all at 0.85 lagging PF. Plot the supply voltage and supply current, load ac voltage, load ac current, DVR voltage and DSTATCOM current with time. Compute (a) the voltage rating of DVR of UPQC-P, (b) the current rating DVR of UPQC-P, (c) the VA rating of DVR of UPQC-P, (d) the voltage rating of DSTATCOM of UPQC-P, (e) the current rating of DSTATCOM of UPQC-P, (f) the VA rating of DSTATCOM of UPQC-P, (g) total VA rating of UPQC-P.

20. A three-phase supply with ac line voltage output of 415 V rms at 50 Hz is feeding a critical linear load of 60 kVA, 415 V, 50 Hz at 0.85 lagging pf. A voltage sag of –15% occurs in the supply voltage at PCC. Design a three-phase unified power quality compensator (UPQC consisting of DSTATCOM and DVR using two VSCs with common dc bus capacitor) in this system to compensate the pf of the load to unity at ac mains and mitigation of the voltage sag to avoid the interruption to the load. Two three-phase VSCs are used as a right hand UPQC-Q. Demonstrate the effect of change of the load from 60 kVA to 20 kVA and then to 60 kVA A all at 0.85 lagging PF. Plot the supply voltage and supply current, load ac voltage, load ac current, DVR voltage and DSTATCOM current with time. Compute (a) the voltage rating of DVR of UPQC-Q, (b) the current rating DVR of UPQC-Q, (c) the

VA rating of DVR of UPQC-Q, (d) the voltage rating of DSTATCOM of UPQC-Q, (e) the current rating of DSTATCOM of UPQC-Q, (f) the VA rating of DSTATCOM of UPQC-Q, (g) total VA rating of UPQC-Q to provide reactive power compensation for unity power factor at ac mains with constant regulated voltage of 415 V at 50 Hz.

21. A three-phase, three-wire, RH-UPQC-P (consisting of DSTATCOM and DVR using two VSCs with common dc bus capacitor) is designed for the compensation of a line-line connected a single-phase load of 400 V, 50 Hz, 100 kVA, 0.85 lagging PF load. If there is a voltage sag −10% in supply system with base value of 400 V. Demonstrate the effect of change of the load from 100 kVA, 0.85 lagging PF load to 20 kVA, 0.85 lagging PF load and then to 100 kVA, 0.85 lagging PF load. Plot the supply voltage and supply current, load ac voltage, load ac current, DVR voltage and DSTATCOM current with time. Compute the rating of both DSTATCOM and DVR used in RH-UPQC-P to provide a three-phase balanced load at unity power factor with constant regulated voltage of 400 V at 50 Hz.

22. A three-phase, four-wire RH-UPQC-P (consisting of 4-leg VSC based DSTATCOM and DVR using two VSCs with common dc bus capacitor) is designed for the compensation of a line-neutral connected a single-phase load of a 20 kW with an input line voltage of 400 V, 50 Hz, ac supply system in a rural sector for lighting. Demonstrate the effect of change of the load from 20 kW to 10 kW and then to 20 kW. If there is a voltage sag −15% in supply system with base value of 400 V. Plot the supply voltage and supply current, load ac voltage, load ac current, DVR voltage and DSTATCOM current with time. Compute the rating of both DSTATCOM and DVR used in RH-UPQC-P and its total rating to provide a three-phase balanced load at unity power factor with constant regulated load voltage of 400 V at 50 Hz.

23. Design a three-phase, four-wire RH-UPQC-S (consisting of 4-leg VSC based DSTATCOM and DVR using two VSCs with common dc bus capacitor) and simulate its behavior for regulating the PCC voltages and maintaining UPF at PCC and load balancing of a single-phase 230 V, 50 Hz load of 125 A at 0.8 lagging PF connected between line and neutral terminals. The supply system has its per-phase impedance of 0.25 Ω resistive element and 1.25 Ω inductive element. Plot the supply voltage and supply current, load ac voltage, load ac current, DVR voltage and DSTATCOM current with time. Compute the rating of both DSTATCOM and DVR used in RH-UPQC-S and its total rating to provide a three-phase balanced load at unity power factor with constant regulated load voltage of 400 V at 50 Hz. Demonstrate the effect of change of the load from 125 A to 25 A and then to 125 A all at 0.8 lagging PF. Compute (a) rms value of supply current, (b) rms value of ac load current, (c) supply active, reactive powers and load active, reactive powers. Compute the rating of both DSTATCOM and DVR used in RH-UPQC-P and its total rating to provide a three-phase balanced load at unity power factor with constant regulated load voltage of 400 V at 50 Hz.

24. A three-phase, four-wire RH-UPQC-P (consisting of 4-leg VSC based DSTATCOM and DVR using two VSCs with common dc bus capacitor) is designed for an isolated three-phase, 4-wire, 415 V, 50 kVA, 50 Hz, 4-pole diesel engine driven permanent magnet synchronous generator (PMSG) system to regulate its voltage, to maintain UPF at PCC and to balance its loads and compensate its neutral current. This PMSG system has the winding resistance of 0.25 ohms/phase and the reactance of 0.75 ohms/phase. This PMSG is loaded with a single-phase line-neutral connected load of 50 kVA at 0.8 lagging. Demonstrate the effect of change of the load from 50 kVA to 20 kVA and then to 50 kVA all at 0.8 lagging PF. Plot the supply voltage and supply current, load ac voltage, load ac current, DVR voltage and DSTATCOM current with time. Compute the rating of both DSTATCOM and DVR used in RH-UPQC-P and its total rating to provide a three-phase balanced load at unity power factor with constant regulated load voltage of 415 V at 50 Hz. Compute (a) rms value of PMSG current, (b) rms value of ac load current, (c) PMSG active, reactive powers and load active, reactive powers.

25. Design and develop a dynamic model of a three-phase, four-wire RH-UPQC-S (consisting of 4-leg VSC based DSTATCOM and DVR using two VSCs with common dc bus capacitor) and simulate its

behavior for a three-phase AC supply has AC voltage of 415 V at 50 Hz and feeder (source) impedance of 0.25 Ω resistance and 2.5 Ω inductive reactance/phase. It is feeding a balanced star connected load of Z = (25 + j15) Ω/phase. Plot the supply voltage and supply current, load ac voltage, load ac current, DVR voltage and DSTATCOM current with time. Compute the rating of both DSTATCOM and DVR used in RH-UPQC-S and its total rating to provide a three-phase balanced load at unity power factor with constant regulated load voltage of 415 V at 50 Hz.

References

1. Moran, S. (1989) A line voltage regulator/conditioner for harmonic-sensitive load isolation. Proceedings of the IEEE IAS Annual Meeting, pp. 945–951.
2. Brennen, M.A. (1994) Active power line conditioner utilizing harmonic frequency injection for improved peak voltage regulation. US Patent 5,349,517.
3. Kamran, F. and Habetler, T.G. (1995) Combined deadbeat control of a series–parallel converter combination used as a universal power filter. Proceedings of IEEE PESC'95, pp. 196–201.
4. Kamran, F. and Habetler, T.G. (1995) A novel on-line UPS with universal filtering capabilities. Proceedings of IEEE PESC'95, pp. 500–506.
5. Muthu, S. and Kim, J.M.S. (1997) Steady-state operating characteristics of unified active power filters. Proceedings of IEEE APEC'97, pp. 199–206.
6. Fujita, H. and Akagi, H. (1998) The unified power quality conditioner: the integration of series- and shunt-active filters. *IEEE Transactions on Power Electronics*, 13(2), 315–322.
7. Vilathgamuwa, M., Zhang, Y.H., and Choi, S.S. (1998) Modelling, analysis and control of unified power quality conditioner. Proceedings of the International Conference on Harmonics and Power Quality, pp. 1035–1040.
8. Singh, B.N., Chandra, A., Al-Haddad, K., and Singh, B. (1998) Fuzzy control algorithm for universal active filter. Proceedings of the IEEE Power Quality Conference, pp. 73–80.
9. Aredes, M., Heumann, K., and Watanabe, E.H. (1998) An universal active power line conditioner. *IEEE Transactions on Power Delivery*, 13(2), 545–551.
10. Tolbert, L.M., Peng, F.Z., and Habetler, T.G. (1999) A multilevel converter-based universal power conditioner. Proceedings of IEEE PESC'99, pp. 393–399.
11. Bester, D.D., le Roux, A.D., du Mouton, T.H., and Enslin, J.H.R. (1999) Evaluation of power-ratings for active power quality compensators. Proceedings of EPE'99.
12. Zhan, C., Wong, M., Han, Y. *et al.* (1999) Universal custom power conditioner (UCPC) in distribution networks. Proceedings of the IEEE PEDS'99, pp. 1067–1072.
13. Chen, Y., Zha, X., Wang, J. *et al.* (2000) Unified power quality conditioner (UPQC): the theory, modeling and application. Proceedings of the IEEE International Conference on Power System Technology, pp. 1329–1333.
14. Li, R., Johns, A.T., and Elkateb, M.M. (2000) Control concept of unified power line conditioner. Proceedings of the IEEE PES Meeting, pp. 2594–2599.
15. Zhan, C., Wong, M., Wang, Z., and Han, Y. (2000) DSP control of power conditioner for improving power quality. Proceedings of the IEEE PES Meeting, pp. 2556–2561.
16. Tolbert, L.M., Peng, F.Z., and Habetler, T.G. (2000) A multilevel converter-based universal power conditioner, *IEEE Transactions on Industry Applications*, 36(2), pp. 596–603.
17. Graovac, D., Katić, V., and Rufer, A. (2000) Universal power quality system – an extension to universal power quality conditioner. Proceedings of EPE-PEMC'00, Košice, pp 4.32–4.38.
18. Graovac, D., Katic, V., and Rufer, A. (2000) Power quality compensation using universal power quality conditioning system. *IEEE Power Engineering Review*, 20(2), 58–60.
19. Graovac, D., Kati, V., and Rufer, A. (2000) Solving supply and load imperfections using universal power quality conditioning system. Proceedings of PCIM'00.
20. Chen, S. and Joos, G. (2001) Rating issues of unified power quality conditioners for load bus voltage control in distribution systems. Proceedings of the IEEE PES Winter Meeting, pp. 944–949.
21. Chen, S. and Joos, G. (2001) A unified series–parallel deadbeat control technique for an active power quality conditioner with full digital implementation. Proceedings of the IEEE IAS Annual Meeting, pp. 172–178.
22. Singh, P., Pacas, J.M., and Bhatia, C.M. (2001) An improved unified power quality conditioner. Proceedings of EPE'01.
23. Graovac, D., Rufer, A., Katic, V., and Knezevic, J. (2001) Unified power quality conditioner based on current source converter topology. Proceedings of EPE'01.

24. Chae, B.-S., Lee, W.-C., Lee, T.-K., and Hyun, D.-S. (2001) A fault protection scheme for unified power quality conditioners. Proceedings of IEEE-PEDS'01, pp. 66–71.
25. Elnady, A. and Salama, M.M.A. (2001) New functionalities of an adaptive unified power quality conditioner. Proceedings of the IEEE PES Winter Meeting, pp. 295–300.
26. Zhan, C., Ramachandaramurthy, V.K., Arulampalam, A. *et al.* (2001) Universal custom power conditioner (UCPC) with integrated control. Proceedings of the IEEE PES Winter Meeting, pp. 1039–1044.
27. Elnady, A. and Salama, M.M.A. (2001) New functionalities of the unified power quality conditioner. Proceedings of the IEEE PES T&D Conference, pp. 415–420.
28. Basu., M., Das, S.P., and Dubey, G.K. (2001) Experimental investigation of performance of a single-phase UPQC for voltage sensitive and non-linear loads. Proceedings of IEEE PEDS'01, pp. 218–222.
29. Elnady, A., Goauda, A., and Salama, M.M.A. (2001) Unified power quality conditioner with a novel control algorithm based on wavelet transform. Proceedings of the IEEE Canadian Conference on Electrical and Computer Engineering, May, pp. 1041–1045.
30. Ghosh, A. and Ledwich, G. (2001) A unified power quality conditioner (UPQC) for simultaneous voltage and current compensation. *Journal of Electric Power Systems Research*, 59, 55–63.
31. Prieto, J., Saimeron, P., Vazquez, J.R., and Alcantara, J. (2002) A series–parallel configuration of active power filters for VAR and harmonic compensation. Proceedings of IEEE IECON'02, pp. 2945–2950.
32. Haque, M.T., Ise, T., and Hosseini, S.H. (2002) A novel control strategy for unified power quality conditioner (UPQC). Proceedings of IEEE PESC'02, pp. 94–98.
33. Elnady, A., El-Khattam, W., and Salama, M.M.A. (2002) Mitigation of AC arc furnace voltage flicker using the unified power quality conditioner. Proceedings of the IEEE PES Winter Meeting, pp. 735–739.
34. Faranda, R. and Valade, I. (2002) UPQC compensation strategy and design aimed at reducing losses. Proceedings of IEEE ISIE'02, pp. 1264–1270.
35. Basu, M., Das, S.P., and Dubey, G.K. (2002) Performance study of UPQC-Q for load compensation and voltage sag mitigation. Proceedings of IEEE IECON'02, pp. 698–703.
36. Newman, M.J. and Holmes, D.G. (2002) A universal custom power conditioner (UCPC) with selective harmonic voltage compensation. Proceedings of IEEE IECON'02, pp. 1261–1266.
37. Jianjun, G., Dianguo, X., Hankui, L., and Maozhong, G. (2002) Unified power quality conditioner (UPQC): the principle, control and application. Proceedings of the IEEE Power Conversion Conference, Osaka, pp. 80–85.
38. Morimoto, H., Ando, M., Mochinaga, Y. *et al.* (2002) Development of railway static power conditioner used at substation for Shinkansen. Proceedings of the IEEE Power Conversion Conference, Osaka, pp. 1108–1111.
39. Tey, L.H., So, P.L., and Chu, Y.C. (2002) Neural network-controlled unified power quality conditioner for system harmonics compensation. Proceedings of the IEEE PES Asia-Pacific Transmission and Distribution Conference and Exhibition, pp. 1038–1043.
40. Haghighat, H., Seifi, H., and Yazdian, A. (2003) An instantaneous power theory based control scheme for unified power flow controller in transient and steady state conditions. *Journal of Electric Power Systems Research*, 64, 175–180.
41. Monteiro, L.F.C., Aredes, M., and Moor Neto, J.A. (2003) A control strategy for unified power quality conditioner. Proceedings of IEEE ISIE'003, June.
42. Yun, M.-S., Lee, W.-C., Suh, I., and Hyun, D.-S. (2004) A new control scheme of unified power quality compensator-Q with minimum power injection. Proceedings of the 30th Annual Conference of the IEEE Industrial Electronics Society, November 2–6, Busan, Korea, pp. 51–56.
43. Kolhatkar, Y.Y. and Das, S.P. (2005) Simulation and experimental investigation of an optimum UPQC with minimum VA loading. Proceedings of IEEE PEDS'05, pp. 526–531.
44. Han, B., Bae, B., Baek, S., and Jang, G. (2006) New configuration of UPQC for medium-voltage application. *IEEE Transactions on Power Delivery*, 21(3), 1438–1444.
45. Khadkikar, V. (2008) Power quality enhancement at distribution level utilizing the unified power quality conditioner (UPQC). Ph.D. thesis, Department of Electrical Engineering, Ecole de Technologie Superieure, Universite du Quebec, Montreal, Canada.
46. Zhili, T., Xun, L., Jian, C. *et al.* (2006) A direct control strategy for UPQC in three-phase four-wire system. Proceedings of the International Power Electronics and Motion Control Conference (IPEMC), August, vol. 2, pp. 1–5.
47. Kolhatkar, Y.Y. and Das, S.P. (2007) Experimental investigation of a single-phase UPQC with minimum VA loading. *IEEE Transactions on Power Delivery*, 22(1), 373–380.
48. Zhili, T., Xun, L., Jian, C. *et al.* (2007) A new control strategy of UPQC in three-phase four-wire system. Proceedings of the IEEE Power Electronics Specialists Conference, June, pp. 1060–1065.

49. Kwan, K.H., Png, Y.S., Chuand, Y.C., and So, P.L. (2007) Model predictive control of unified power quality conditioner for power quality improvement. Proceedings of the 16th IEEE International Conference on Control Applications, Part of IEEE Multi-conference on Systems and Control, Singapore, October 1–3, pp. 910–921.
50. Basu, M., Das, S.P., and Dubey, G.K. (2008) Investigation on the performance of UPQC-Q for voltage sag mitigation and power quality improvement at a critical load point. *IET Generation, Transmission & Distribution*, 2(3), 414–423.

7

Loads That Cause Power Quality Problems

7.1 Introduction

In true sense, most of the electrical loads have nonlinear behavior at the AC mains. As they draw harmonic currents of various types such as characteristic harmonics, noncharacteristic harmonics, interharmonics, subharmonics, reactive power component of current, fluctuating current, unbalanced currents from the AC mains, these loads are known as nonlinear loads. Majority of rotating electric machines and magnetic devices such as transformers, reactors, chokes, magnetic ballasts, and so on behave as nonlinear loads due to saturation in their magnetic circuits, geometry such as presence of teeth and slots, winding distribution, air gap asymmetry, and so on. Many fluctuating loads such as furnaces, electric hammers, and frequently switching devices exhibit highly nonlinear behavior as electrical loads. Even nonsaturating electrical loads such as power capacitors behave as nonlinear loads at the AC mains and they create a number of power quality problems due to switching and resonance with magnetic components in the system and are overloaded due to harmonic currents caused by the presence of harmonic voltages in the supply system.

Moreover, the solid-state control of AC power using diodes, thyristors, and other semiconductor switches is widely used to feed controlled power to electrical loads such as lighting devices with electronic ballasts, controlled heating elements, magnet power supplies, battery chargers, fans, computers, copiers, TVs, switched mode power supplies (SMPS) in computers and other equipments, furnaces, electroplating, electrochemical processes, adjustable speed drives (ASDs) in electric traction, air-conditioning systems, pumps, wastewater treatment plants, elevators, conveyers, cranes, and so on. These AC loads consisting of solid-state converters draw nonsinusoidal currents from the AC mains and behave in a nonlinear manner and therefore they are also known as nonlinear loads. These nonlinear loads consisting of solid-state converters draw harmonic currents and reactive power component of current from the AC mains. In three-phase systems, they could also cause unbalance and sometimes draw excessive neutral current, especially the distributed single-phase nonlinear loads on three-phase four-wire supply system. These solid-state converters may be AC–DC converters, AC voltage controllers, cycloconverters, and so on. The injected harmonic currents, reactive power burden, unbalanced currents, and excessive neutral current caused by these nonlinear loads result in low system efficiency, poor power factor (PF), maloperation of protection systems, AC capacitors overloading and nuisance tripping, noise and vibration in electrical machines, heating of the rotor bars due to negative sequence currents, derating of components of distribution system, user equipment, and so on. They also cause distortion in the supply voltage, disturbance to protective devices and other consumers, and interference in nearby communication networks and digital and analog control systems.

Power Quality Problems and Mitigation Techniques, First Edition.
Bhim Singh, Ambrish Chandra and Kamal Al-Haddad.

These nonlinear loads exhibit different behavior thereby causing different power quality problems, and they are therefore often classified according to their performance. Accordingly, the power quality improvement techniques for mitigating the power quality problems caused by nonlinear loads are also different to reduce the rating and cost of devices used for these purposes. One of the major and broad classifications of these nonlinear loads is based on their behavior either as current fed type or as voltage fed type or a combination of both. The current fed type of nonlinear loads with AC–DC converters having constant DC current used for field winding excitation, magnet power supplies, thyristor converter feeding DC motor drives, converter feeding current source inverter-fed AC motor drives, magnetic devices with saturation, and so on draw the prespecified kind of current pattern. Normally devices used for power quality improvements of such current fed type nonlinear loads are connected in shunt with the loads to supply locally all their current components other than the fundamental active power component of load current. On the contrary, the voltage fed type of nonlinear loads having AC–DC converters feeding almost constant DC voltage loads such as battery chargers, AC–DC converters with large DC filter capacitor as front-end converters in SMPS, AC–DC converters in voltage source inverter feeding AC motor drives, and so on draw highly nonlinear and unpredictable current waveform rich in harmonics with high crest factor (CF). In general, devices used for power quality improvements of such voltage fed type nonlinear loads are connected in series with these loads to block all their harmonic currents with much reduced rating and they do not have reactive power requirement. The mixed nonlinear loads consist of either several current fed type and voltage type of nonlinear loads or typically AC–DC converters with LC DC bus filter. The devices used for power quality improvements of such mixed nonlinear loads are connected in shunt with these loads or consist of hybrid of shunt and reduced rating series devices.

Despite causing power quality problems, the use of nonlinear loads, especially those employing solid-state controllers, is increasing day by day owing to benefits of the low cost and small size, remarkable energy conservation, simplicity in control, reduced wear and tear, and low maintenance requirements in the new and automated electric appliances leading to high productivity. Although these electronically automated energy-efficient electrical loads are most sensitive to power quality problems, they themselves cause additional power quality problems to the supply system. Hence, it is very important to classify and analyze their behavior to identify the proper power quality improvement devices for mitigating the power quality problems or to modify their structure for reducing or eliminating the power quality pollution at the AC mains.

This chapter deals with the classification and analysis of these single-phase and three-phase nonlinear loads and their performance with specific reference to power quality problems. The following sections include the state of the art of these nonlinear loads, their classification, analysis, modeling and simulation of performance, examples, review questions, numerical problems, computer simulation problems, and references.

7.2 State of the Art on Nonlinear Loads

Since the inception of AC power, majority of electrical equipment are developed based on the principle of energy storage, which are used in the process of energy conversion and especially in the magnetic energy storage system. They behave as inductive loads causing burden on the AC mains of the lagging reactive power and thereby poor power factor in the AC network that results in increased losses and poor utilization of components of distribution system such as transformers, feeders, and switchgear due to increased current for a given active power. AC power capacitors and synchronous condensers have been used to supply the reactive power locally and to reduce the burden of reactive power on the AC mains. In addition, because of a number of single-phase loads in the distribution system, especially domestic, residential, and commercial in small power ratings and traction, transportation, rural distribution systems, and so on in medium power ratings, there have been additional problems of load unbalancing and excessive neutral current causing increased losses, voltage imbalance, and derating of the distribution system. Moreover, switching in many electrical loads causes switching transients and inrush currents resulting in various voltage-based power quality problems such as surges, spikes, sags, voltage fluctuations, voltage imbalance, and so on. These power quality problems affect other loads and system components such as protection systems, telecommunication systems, and so on. These power quality problems of voltage imbalance and fluctuations even affect good linear loads such as AC motors, especially induction motors, with negative sequence currents and subsequent rotor heating and increased losses and thus resulting in derating of these motors. Some additional power quality problems are created because of several physical phenomena in electric equipment such as saturation especially in single-phase induction motors, magnetic ballasts, transformers, voltage regulators based on ferroresonant and tap changers, and air gap asymmetry in rotating electric motors. They result in the generation of harmonics and increased neutral current. These harmonics and neutral current result in voltage distortion at the neutral terminal, increased losses, and harmonic voltage at the point of common coupling (PCC).

With the subsequent advancement, the modern automated controlled electrical loads use solid-state converters because of a number of benefits, namely, energy conservation, reduced size, reduced overall cost, and so on. However, even with sinusoidal applied voltage, they draw nonsinusoidal and increased current from the AC mains in addition to the fundamental active power component of current. Some of these nonlinear loads are as follows:

- Fluorescent lighting and other vapor lamps with electronic ballasts
- Switched mode power supplies
- Computers, copiers, and television sets
- Printer, scanners, and fax machines
- High-frequency welding machines
- Fans with electronic regulators
- Microwave ovens and induction heating devices
- Xerox machines and medical equipment
- Variable frequency-based HVAC (heating ventilation and air-conditioning) systems
- Battery chargers and fuel cells
- Electric traction
- Arc furnaces
- Cycloconverters
- Adjustable speed drives
- Static slip energy recovery schemes of wound rotor induction motors
- Wind and solar power generation
- Static VAR compensators (SVCs)
- HVDC transmission systems
- Magnet power supplies
- Plasma power supplies
- Static field excitation systems

These types of nonlinear loads draw harmonic currents and reactive power component of the current from the single-phase AC mains. Some of them have harmonic currents, reactive power component of the current, and unbalanced currents in the three-phase three-wire supply system. The single-phase distributed nonlinear loads also consist of harmonic currents, reactive power component of the current, and unbalanced currents and excessive neutral current in three-phase four-wire system. These increased currents in addition to the fundamental active power component of current cause increased losses, poor power factor, disturbances to other consumers, communication systems, protection systems, and many other electronics appliances, voltage distortion, voltage spikes, voltage notches, surges, dip, sag, swell in voltages, and so on. Owing to the ever-increasing use of such nonlinear loads in present-day distribution system for obvious reasons, the exhaustive study of these nonlinear loads becomes very relevant to find the proper remedy for mitigation of power quality problems caused by them in the supply system.

7.3 Classification of Nonlinear Loads

The nonlinear loads can be classified based on (i) the use of non-solid-state or solid-state devices, for example, the presence or absence of power electronics converter in the circuits of nonlinear loads; (ii) the use of converter types such as AC–DC converter type, AC voltage controller type, and cycloconverter type; (iii) their nature as stiff current fed type or stiff voltage fed type or a combination of both; and (iv) the number of phases such as two-wire single-phase, three-phase three-wire, and four-wire three-phase systems.

7.3.1 Non-Solid-State and Solid-State Device Types of Nonlinear Loads

The nonlinear loads may be classified based on whether they consist of solid-state devices or any other power converters or not. There are a number of electrical loads that are nonlinear in nature, but they do not involve any power converters. Similarly, there are only some nonlinear loads that consist of solid-state converters. These nonlinear loads are further explained with examples.

7.3.1.1 Non-Solid-State Device Type Nonlinear Loads

As already mentioned, there are many electrical loads in nature that do not consist of any solid-state device or power electronics converter. However, they behave as nonlinear loads when they are connected to AC mains. Most of the electrical machines fall in this category of nonlinear loads. A number of physical phenomena in these electrical machines cause their behavior as nonlinear loads. Typically, the saturation in magnetic material of these machines and electromagnetic devices, skin and proximity effects in conductors, nonuniform air gap in rotating machines, effect of teeth and slotting, and so on result in harmonic currents under steady-state and transient conditions in the AC mains when they are connected to the AC supply system. Some practical examples of these types of nonlinear loads are various types of transformers operating at no load or light load conditions, magnetic ballasts of fluorescent lamps, and single-phase induction motors as they are usually designed with high level of no load current (due to high level of saturation) to reduce the cost and size of these motors. They draw harmonic currents and reactive power component of the current, and also cause excessive neutral current in the three-phase four-wire supply system due to such distributed single-phase nonlinear loads.

7.3.1.2 Solid-State Device Type Nonlinear Loads

Many types of electrical equipment consist of different circuits of solid-state devices to process the AC power to suit specific application. They draw nonsinusoidal current from the AC mains and they behave as nonlinear loads. This nonsinusoidal current consists of harmonic currents and the reactive power component of the current along with the fundamental active power component of current. They use various AC–DC converters, AC voltage controllers, cycloconverters, or a combination of all in their front-end converter. In the single-phase configuration, they draw harmonic currents and reactive power from the AC mains. Examples of single-phase nonlinear loads include both domestic and commercial equipment—among the home appliances are microwave oven, induction heaters, television sets, electronic ballasts-based lighting systems, domestic inverter, adjustable speed drive-based air conditioners, and AC voltage regulator-based fans, whereas the commercial and industrial equipment are computers, copiers, fax machines, xerox machines, scanner, printers, small welding sets, and so on. In the three-phase, three-wire supply system, they may also draw unbalanced three-phase currents in addition to harmonic currents and reactive power. Some practical loads are three-phase adjustable speed drives, consisting of converter-fed DC motor drives, synchronous motor drives, induction motor drives, and other electric motors used in HVAC systems, wastewater treatment plants, large industrial fans, pumps, compressors, cranes, elevators, electrochemical process such as electroplating and electromining, and so on. In the three-phase four-wire supply system, there are many single-phase nonlinear loads connected to AC mains causing excessive neutral current. Distributed single-phase loads on all three phases such as electronic ballasts-based lighting systems, computer loads in high storied buildings, and all other single-phase loads cause burden on the AC mains of harmonic currents, reactive power component of currents, unbalanced currents, and excessive neutral current.

7.3.2 Converter-Based Nonlinear Loads

There are various types of converters used in electrical equipments that behave as nonlinear loads. These nonlinear loads mainly consist of AC–DC converters, AC voltage controllers, cycloconverters, or a

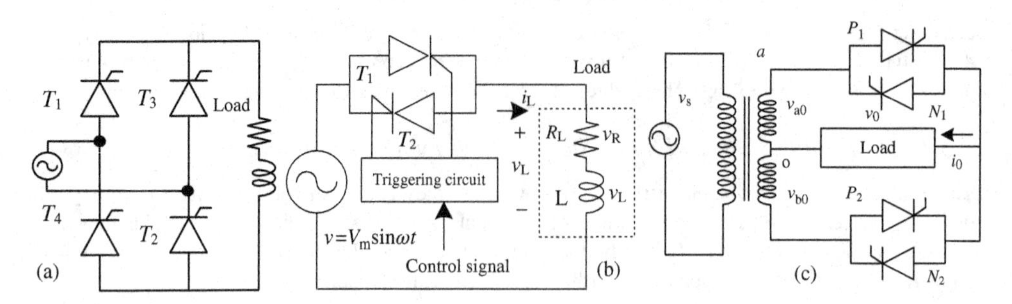

Figure 7.1 Various types of current fed nonlinear loads

combination of all. These are classified on the basis of these converters, but are not confined to them. Figure 7.1 shows some of these types of current fed loads.

7.3.2.1 AC–DC Converter-Based Nonlinear Loads

A large number of loads use AC–DC converters as front-end converters ranging from few watts to megawatt rating. These converters are developed in many circuit configurations such as single-phase and three-phase, uncontrolled, semicontrolled, and fully controlled, and half-wave, full-wave, and bridge converter circuits to suit the requirements of specific application. Depending upon the types of filters used for filtering the rectified DC, their behavior vary in a number of ways at the AC mains. Some of the examples of such nonlinear loads include microwave ovens, SMPS, computers, fax machines, battery chargers, HVDC transmission systems, electric traction, adjustable speed drives, and so on. In some cases they draw current with excessive harmonic contents with high crest factor. However, in many cases they draw current with moderate harmonic contents and reactive power with low crest factor, even less than the sine wave. They exhibit poor power factor at the AC mains generally due to harmonics only, but with reactive power as well.

7.3.2.2 AC Controllers-Based Nonlinear Loads

Some nonlinear loads use AC voltage controllers for the control of AC rms voltage across the electrical loads to control the physical process. They draw the harmonic currents along with the reactive power and cause poor power factor. In single-phase distributed loads on three-phase supply systems, they also cause excessive harmonic currents. Some of the examples of such nonlinear loads include AC voltage regulator in fans, lighting controllers, heating controllers, soft starters, speed controllers, and energy saving controllers of three-phase induction motors operating under light load conditions in a number of applications such as hack saw, electric hammers, wood-cutting machines, and so on. They are also used in static VAR compensators (SVCs) in TCRs (thyristor controlled reactors) and so on.

7.3.2.2.1 Cycloconverter-Based Nonlinear Loads

In many applications, cycloconverters are used to convert AC voltage of a fixed frequency to variable voltage at a variable frequency or vice versa. These cycloconverters-based nonlinear loads draw harmonic currents not only at higher order harmonics but also at subharmonics and reactive power and exhibit a very poor power factor at the AC mains. Some of the examples of such nonlinear loads include cycloconverter-fed large-rating synchronous motor drives in cement mills, ore crushing plants, large-rating squirrel cage induction motors, slip energy recovery scheme of wound rotor induction motor drives, VSCF (variable speed constant frequency) generating systems, and so on.

7.3.3 Nature-Based Classification

Most of the nonlinear loads behave as either stiff current fed type or as stiff voltage fed type, or a combination of both. The stiff current fed loads normally consist of AC–DC converters with constant DC

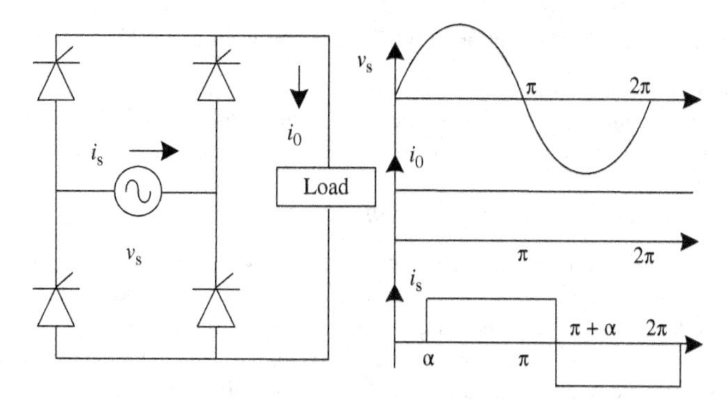

Figure 7.2 A single-phase controlled converter-based current fed type of nonlinear load

current load and a predetermined harmonic pattern in the AC mains with reactive power burden. The voltage stiff loads consist of generally AC–DC converters with a large DC capacitor at the DC bus to provide ideal DC voltage source for the remaining process of solid-state conversion and draw peaky current from the AC mains with high crest factor. Since the analysis of the behavior and remedy for mitigation of power quality problems of these types of loads depend reasonably on this classification, it becomes relevant and important to select a proper compensator.

7.3.3.1 Current Fed Type of Nonlinear Loads

The stiff current fed types of nonlinear loads generally have predetermined pattern of harmonics and sometimes they have reactive power burden on the AC mains. They have flat current waveform drawn from the AC mains with a low value of crest factor. They typically consist of AC–DC converters feeding DC motor drives, magnet power supplies, field excitation system of the alternators, controlled AC–DC converters used to derive DC current source for feeding current source inverter supplying large-rating AC motor drives, HVDC transmission systems, and so on. Figure 7.2 shows such current fed type of nonlinear load.

7.3.3.2 Voltage Fed Type of Nonlinear Loads

The stiff voltage types of nonlinear loads behave as sink of harmonic currents. Typical example of such load is an AC–DC converter with a large DC capacitor at its DC bus to provide an ideal DC voltage source for the remaining process of solid-state conversion and it draws peaky current from the AC mains with high crest factor (as shown in Figure 7.3). They generally do not have reactive power requirement, but they have much greater amount of harmonic currents drawn from the AC mains. Examples of such loads include SMPS, battery chargers, front-end converters of voltage source inverter fed AC motor drives, electronic ballasts, and most of the electronic appliances.

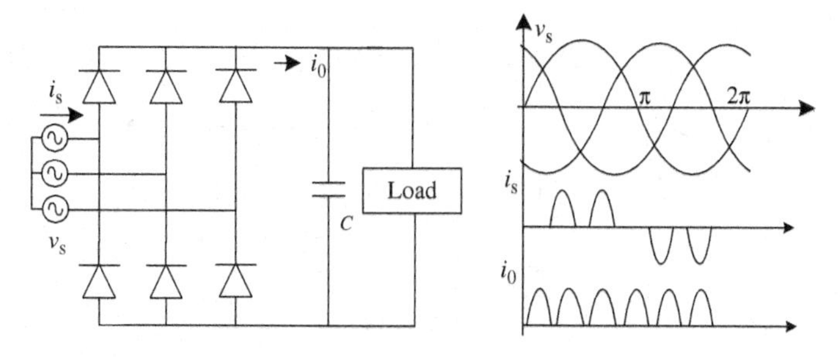

Figure 7.3 A three-phase converter-based voltage fed type of nonlinear load

7.3.3.3 Mix of Current Fed and Voltage Fed Types of Nonlinear Loads

The mixed nonlinear loads are combination of current fed and voltage fed types of loads. A group of nonlinear loads and a combination of linear and nonlinear loads fall under this category. Most of the electrical loads consisting of solid-state converters behave as these types of nonlinear loads.

7.3.4 Supply System-Based Classification

This classification of nonlinear loads is based on the supply system having single-phase (two-wire) and three-phase (three-wire or four-wire) systems. There are many nonlinear loads such as domestic appliances that are fed from single-phase supply systems. Some three-phase nonlinear loads are without neutral conductor, such as ASDs (Adjustable Speed Drives), fed from a three-wire supply system. There are many nonlinear single-phase loads distributed on a four-wire, three-phase supply system, such as computers, commercial lighting, and so on.

7.3.4.1 Two-Wire Nonlinear Loads

There are a very large number of single-phase nonlinear loads supplied by the two-wire single-phase AC mains. All these loads consisting of single-phase diode rectifiers, semiconverters, and thyristor converters behave as nonlinear loads. They draw harmonic currents and sometimes also the reactive power from the AC mains. Typical examples of such loads are power supplies, electronic fan regulators, electronic ballasts, computers, television sets, and traction. Figure 7.4 shows such voltage fed type nonlinear load.

7.3.4.2 Three-Wire Nonlinear Loads

Three-phase, three-wire nonlinear loads inject harmonic currents, and sometimes they draw reactive power from the AC mains and sometimes they also have unbalanced currents. These nonlinear loads are in large numbers and consume major amount of electric power. Typical examples are ASDs using DC and AC motors, HVDC transmission systems, and wind power conversion. Figure 7.5 shows such current fed type nonlinear load.

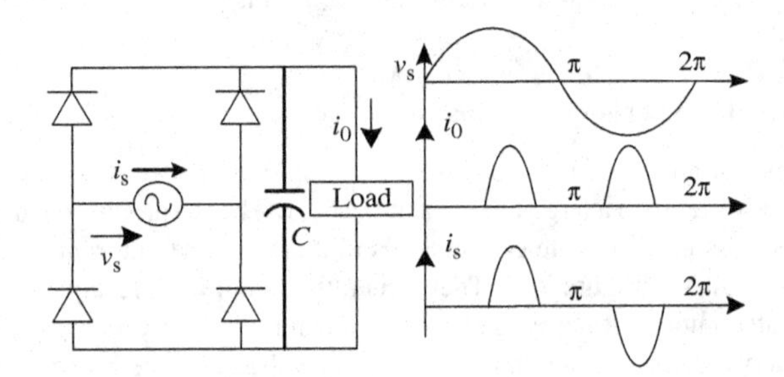

Figure 7.4 A single-phase converter-based voltage fed type of nonlinear load

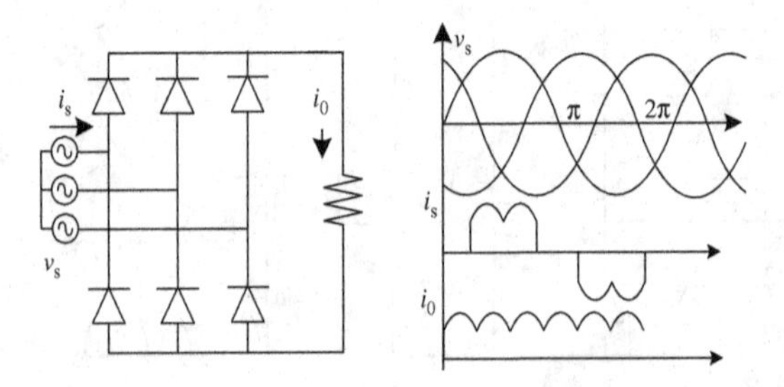

Figure 7.5 A three-phase converter-based current fed type of nonlinear load

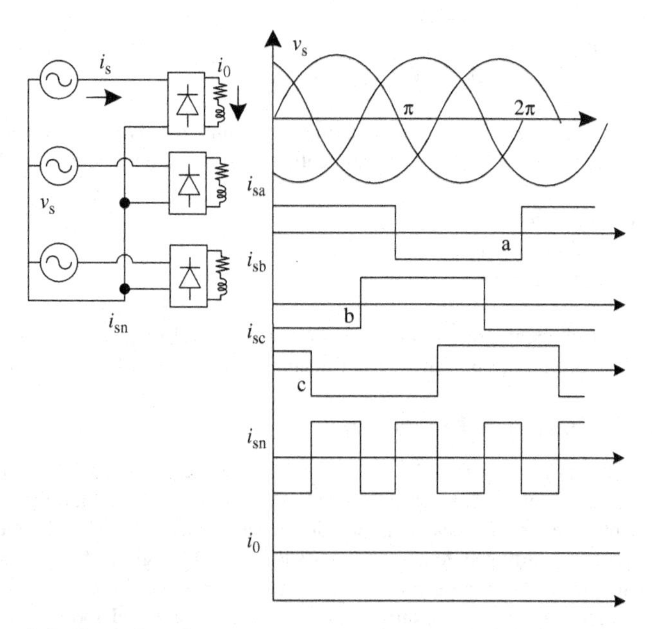

Figure 7.6 Three-phase four wire converter-based current fed type of nonlinear loads

7.3.4.3 Four-Wire Nonlinear Loads

A large number of single-phase nonlinear loads may be supplied from the three-phase AC mains with the neutral conductor. Apart from harmonic currents, reactive power, and unbalanced currents, they also cause excessive neutral current due to harmonic currents and unbalancing of these loads on three phases. Typical examples are computer loads and electronic ballasts-based vapor lighting systems. Besides, they cause voltage distortion and voltage imbalance at the PCC and some potential at the neutral terminal. Figure 7.6 shows such current fed type nonlinear load.

7.4 Power Quality Problems Caused by Nonlinear Loads

The nonlinear loads cause a number of power quality problems in the distribution system. They inject harmonic currents into the AC mains. These harmonic currents increase the rms value of supply current, increase losses, cause poor utilization and heating of components of the distribution system, and also cause distortion and notching in voltage waveforms at the point of common coupling due to voltage drop in the source impedance. Some of the effects are as follows:

- Increased rms value of the supply current
- Increased losses
- Poor power factor
- Poor utilization of distribution system
- Heating of components of distribution system
- Derating of the distribution system
- Distortion in voltage waveform at the point of common coupling, which indirectly affects many types of equipment
- Disturbance to the nearby consumers
- Interference in communication system
- Mal-operation of protection systems such as relays
- Interference in controllers of many other types of equipment
- Capacitor bank failure due to overload, resonance, harmonic amplification, and nuisance fuse operation

- Excessive neutral current
- Harmonic voltage at the neutral point

Some of these nonlinear loads, in addition to harmonics, require reactive power and create unbalancing, which not only increases the severity of the above-mentioned problems but also causes additional problems.

- Voltage regulation and voltage fluctuations
- Imbalance in three-phase voltages
- Derating of cables and feeders

The voltage imbalance creates substantial problems to electrical machines due to negative sequence currents, noise, vibration, torque pulsation, rotor heating, and so on and of course their derating.

7.5 Analysis of Nonlinear Loads

There are varieties of nonlinear loads in the AC network that create power quality problems. Therefore, it has become important and relevant to analyze these loads and thereby select a right technique for power quality improvements. Majority of these nonlinear loads can be analyzed using the measured data at the site and then the power quality problems are identified to select a right technique for their mitigation. However, this technique becomes quite cumbersome, expansive, and sometimes practically difficult as it requires a large manpower, costly measuring equipment, and analytical tools. The other method for analyzing these nonlinear loads is an identification of its input stage with its output requirements and set the circuit parameters for the required performance for particular application reported in the literature. Once the equivalent circuit of the nonlinear load is properly analyzed, it can be used to design, model, and simulate the mitigation technique for power quality improvements.

7.6 Modeling, Simulation, and Performance of Nonlinear Loads

As the quantification and identification of the majority of nonlinear loads may be carried out by using their equivalent circuit and by properly tuning their parameters to match their behavior with practical applications, the modeling of these nonlinear loads is very much essential for this purpose. Moreover, once the model of these nonlinear loads is developed, it can be used for the simulation of its performance. Apart from it, once the performance and identification of the load are done properly, this developed model can be used to select the right mitigation technique for power quality improvements. After the selection of proper mitigation technique, the complete model of the nonlinear load with power quality improvement technique can be validated by simulating its behavior before manufacturing it in practice. The analysis, modeling, simulation, and performance of most popular nonlinear loads are illustrated through several examples in the following sections so that these can be used to study the power quality mitigation techniques in other chapters.

7.7 Numerical Examples

Example 7.1

Consider a single-phase uncontrolled bridge converter (shown in Figure E7.1) with sinusoidal input supply $V_s = 230$ V and constant DC load current of 15 A. Calculate (a) CF, (b) distortion factor (DF), (c) displacement factor (DPF), (d) PF, and (e) total harmonic distortion (THD).

Solution: Given supply voltage $V_s = 230$ V, frequency of the supply $(f) = 50$ Hz, and DC link current $I_0 = 15$ A.

In a single-phase uncontrolled bridge converter, the waveform of the supply current (I_s) is a square wave with the amplitude of the DC link current (I_0). The rms value of the fundamental component of square wave is $(2\sqrt{2}/\pi) = 0.9$ times its amplitude.

Therefore, the rms value of supply current is $I_s = I_0 = 15$ A and the rms value of fundamental current is $I_{s1} = (2\sqrt{2}/\pi)I_0 = 0.9 I_0 = 13.5$ A.

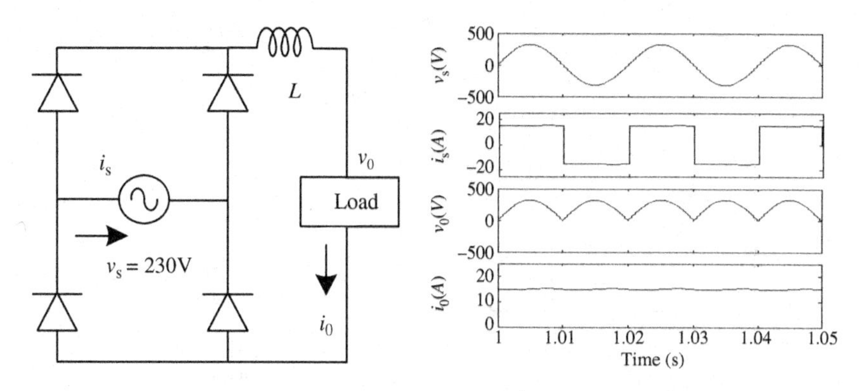

Figure E7.1 A single-phase uncontrolled bridge converter-based current fed type of nonlinear load

a. CF of supply current = supply peak current/rms value of supply current $= I_{peak}/I_{rms} = I_0/I_s = 15/15 = 1$.
b. DF = rms value of fundamental supply current/rms value of supply current $= I_{s1}/I_s = 13.5/15 = 0.9$.
c. DPF $= \cos\theta_1 = 1$ (since fundamental supply current is in phase with supply voltage $\theta_1 = 0°$).
d. PF = DF × DPF $= 0.9 \times 1 = 0.9$.
e. THD of supply current $= \{\sqrt{(I_s^2 - I_{s1}^2)}\}/I_{s1} = \{\sqrt{(15^2 - 13.5^2)}\}/13.5 = 0.4843 = 48.43\%$.

Example 7.2

A single-phase uncontrolled bridge converter (shown in Figure E7.2) has a *RE* load with $R = 5\,\Omega$ and $E = 150$ V. The input AC voltage is $V_s = 230$ V at 50 Hz. Calculate (a) load average current, (b) rms value of supply current, (c) CF, (d) DF, (e) DPF, (f) PF, and (g) THD.

Solution: Given supply voltage $V_s = 230$ V, $V_{sm} = 325.27$ V, frequency of supply $(f) = 50$ Hz, load $R = 5\,\Omega$, and $E = 150$ V.

In a single-phase diode bridge converter, with *RE* load, the current flows from angle α when AC voltage is equal to E to angle β at which AC voltage reduces to E.

$\alpha = \sin^{-1}(E/V_{sm}) = \sin^{-1}(150/325.27) = 27.46°$, $\beta = \pi - \alpha = 152.54°$. The conduction angle $= \beta - \alpha = 125.08°$.

Active power drawn from the AC mains is $P = I_s^2R + EI_0 = 4593.22$ W.

The rms value of fundamental current from the AC mains is $I_{s1} = P/V_s = 19.9705$ A.

Supply AC peak current is $I_{peak} = (V_{sm} - E)/R = 35.054$ A.

a. Load average current is

$$I_0 = \{1/(\pi R)\}(2V_{sm}\cos\alpha + 2E\alpha - \pi E) = 15.886\text{ A}.$$

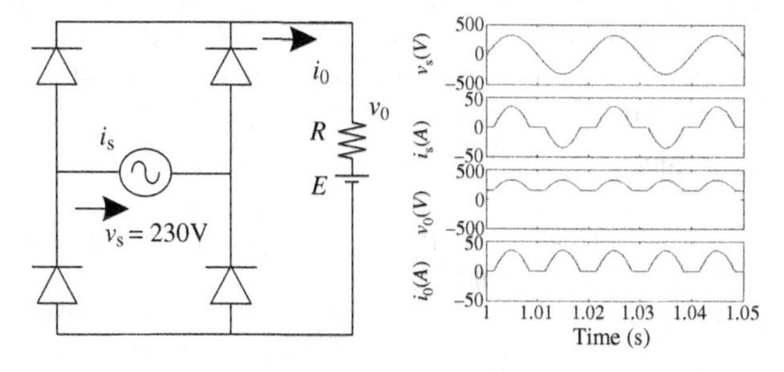

Figure E7.2 A single-phase converter-based voltage fed type of nonlinear load

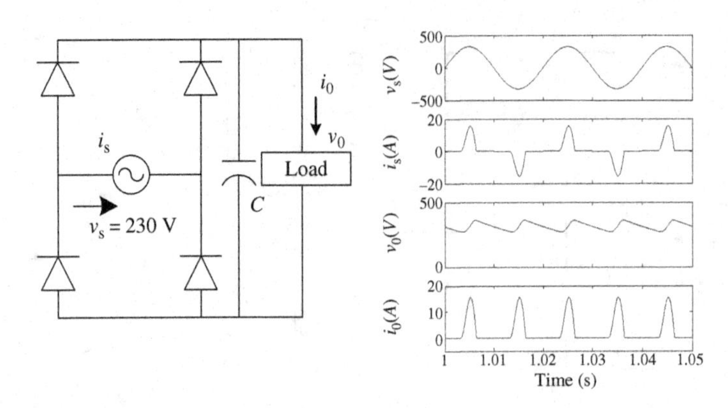

Figure E7.3 Single-phase converter-based voltage fed type of nonlinear load

b. The rms value of supply current I_s is rms value of discontinuous current from the AC mains:

$$I_s = [\{1/(\pi R^2)\}\{(0.5V_{sm}^2 + E^2)(\pi - 2\alpha) + 0.5V_{sm}^2 \sin 2\alpha - 4V_{sm}E\cos\alpha\}]^{1/2} = 21.025 \text{ A}.$$

c. CF of supply current = supply peak current/rms value of supply current = $I_{peak}/I_{rms} = I_{peak}/I_s = 1.66725$.
d. DF = rms value of fundamental supply current/rms value of supply current = $I_{s1}/I_s = 19.97/21.025 = 0.94982$.
e. DPF = $\cos\theta_1 = 1$ (since fundamental supply current is in phase with supply voltage $\theta_1 = 0°$).
f. PF = $P/(V_s I_s) = 0.949846$.
g. THD of AC current = $\{\sqrt{(I_s^2 - I_{s1}^2)}\}/I_{s1} = \{\sqrt{(21.03^2 - 19.97^2)}\}/19.97 = 0.3301 = 33.01\%$.

Example 7.3

A single-phase diode bridge rectifier (shown in Figure E7.3) draws AC current at 0.98 DPF and the THD of AC current is 60%. It draws 1000 W from 230 V, 50 Hz, AC source and CF of AC current is 3. Calculate (a) PF, (b) rms value of current, and (c) peak current of AC mains.

Solution: Given supply voltage $V_s = 230$ V (rms), frequency of the supply $(f) = 50$ Hz, THD of $I_s = 60\%$, $P = 1000$ W, DPF = $\cos\theta_1 = 0.98$, and CF = 3.

DF = $1/\sqrt{(1 + \text{THD}^2)} = 0.857493$.

a. PF = DPF × DF = 0.98 × 0.857493 = 0.84034.
b. The rms value of supply current is $I_s = P/(V_s \times \text{PF}) = 1000/(230 \times 0.84034) = 5.17387$ A.
c. The peak current of AC mains is $I_{peak} = \text{CF} \times I_s = 3 \times 5.17387 = 15.5216$ A.

Example 7.4

Consider single-phase semicontrolled bridge converter (shown in Figure E7.4a) with sinusoidal input supply voltage V_s of 230 V at 50 Hz and constant DC load current of 25 A. (a) Calculate DF, DPF, PF, THD of the supply current for $V_0 = 0.5V_{DC0}$ where V_{DC0} is the DC output at $\alpha = 0°$. (b) Repeat part (a) for a fully controlled bridge converter (shown in Figure E7.4b).

Solution: Given supply voltage $V_s = 230$ V (rms), frequency of the supply $(f) = 50$ Hz, and $I_0 = 25$ A.

a. In single-phase semicontrolled bridge converter, the waveform of the supply current (I_s) is from firing angle α to 180° with the amplitude of the DC link current (I_0).

   If $V_0 = 0.5V_{DC0}$ where V_{DC0} is the DC output at $\alpha = 0°$, then firing angle $\alpha = 90°$ [$V_0 = 0.5V_{DC0}(1 + \cos\alpha)$].

   The rms value of supply current is $I_s = I_0\sqrt{\{(\pi - \alpha)/\pi\}} = I_0/\sqrt{2} = 17.678$ A.

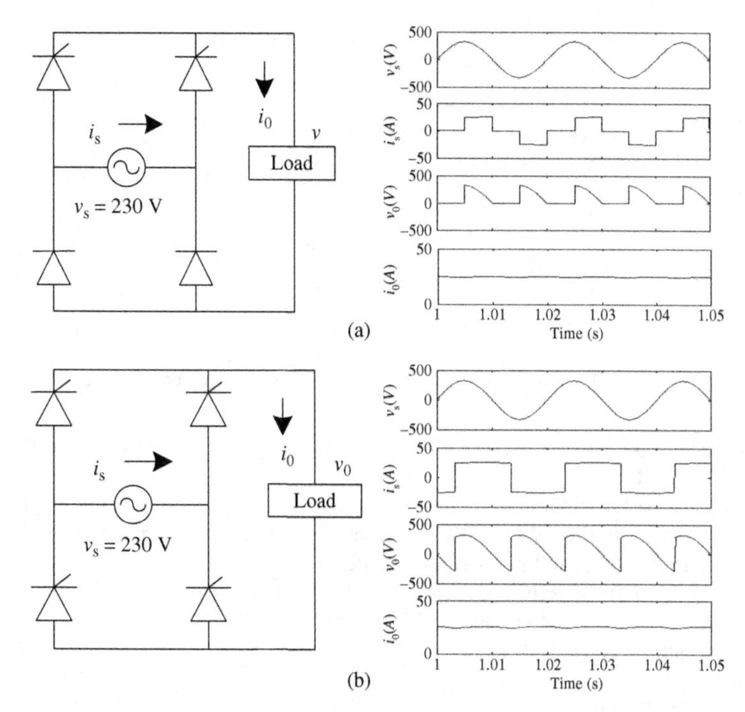

Figure E7.4 (a) A single-phase semi converter-based current fed type of nonlinear load. (b) Single-phase fully control bridge converter-based current fed type of nonlinear load

The rms value of fundamental supply current is $I_{s1} = 0.9I_0 \cos(\alpha/2) = 0.9I_0 \cos(\alpha/2) = 0.9I_0/\sqrt{2} = 15.91$ A.

$\text{DPF} = \cos\theta_1 = \cos(\alpha/2) = 1/\sqrt{2} = 0.7071$.

$\text{DF} = 1/\sqrt{(1 + \text{THD}^2)} = 0.90$.

$\text{PF} = \text{DPF} \times \text{DF} = 0.9 \times 0.7071 = 0.63\,639$.

THD of AC current $= \{\sqrt{(I_s^2 - I_{s1}^2)}\}/I_{s1} = 0.4843 = 48.43\%$.

b. In single-phase fully controlled bridge converter, the waveform of the supply current (I_s) is from firing angle α to $(\pi + \alpha)$ with the amplitude of the DC link current (I_0).

If $V_0 = 0.5V_{\text{DC0}}$ where V_{DC0} is the DC output at $\alpha = 0°$, then firing angle $\alpha = 60°$ [$V_0 = V_{\text{DC0}} \cos\alpha$].

The rms value of supply current is $I_s = I_0 = 25$ A.

The rms value of fundamental supply current is $I_{s1} = 0.9I_0 = 22.5$ A.

$\text{DPF} = \cos\theta_1 = \cos(\alpha) = 1/2 = 0.5$.

$\text{DF} = I_{s1}/I_s = 1/\sqrt{(1 + \text{THD}^2)} = 0.90$.

$\text{PF} = \text{DPF} \times \text{DF} = 0.9 \times 0.5 = 0.45$.

THD of AC current $= \{\sqrt{(I_s^2 - I_{s1}^2)}\}/I_{s1} = 0.4843 = 48.43\%$.

Example 7.5

A single-phase fully controlled bridge converter (shown in Figure E7.5) is used as a line commutated inverter (LCI) to feed power from a battery with 180 V and an internal resistance of 0.1 Ω. The supply voltage is 220 V (rms) and sufficient inductance is included in the output circuit to maintain the current virtually constant at 20 A. Determine the required delay angle α,DF, DPF, total harmonic distortion of AC source current (THD_I), CF of AC source current, PF, and rms value of AC source current (I_s).

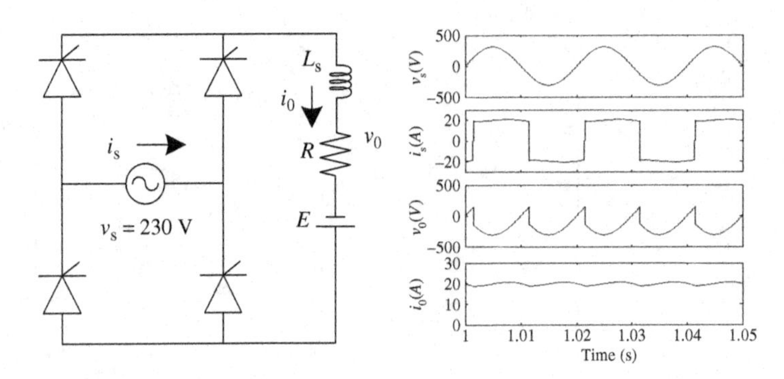

Figure E7.5 Single-phase converter-based current fed type of nonlinear load

Solution: Given supply voltage $V_s = 220$ V (rms), frequency of the supply $(f) = 50$ Hz, $I_0 = 20$ A, $E = 180$ V, and $R = 0.1\ \Omega$.

In single-phase thyristor bridge converter operating as a LCI, the waveform of the supply current (I_s) is a square wave with the amplitude of the DC link current (I_0). Moreover, the rms value of the fundamental component of square wave is 0.9 times its amplitude.

Therefore, the rms value of supply current is $I_s = I_0 = 20$ A.

The rms value of fundamental current is $I_{s1} = 0.9I_0 = 18$ A.

The average output voltage $V_0 = (2\sqrt{2}/\pi)V_s \cos\alpha = 0.9V_s \cos\alpha = -(E - I_0R) = -(180 - 20 \times 0.1) = -178$ V, $\alpha = 154.03°$.

$\mathrm{DF} = 1/\sqrt{(1 + \mathrm{THD}^2)} = 0.90$.

$\mathrm{DPF} = \cos\theta_1 = \cos\alpha = \cos 154.03° = -0.899$.

$\mathrm{PF} = \mathrm{DPF} \times \mathrm{DF} = 0.9 \times 0.899 = 0.8091$.

THD of AC current $= \sqrt{(I_s^2 - I_{s1}^2)}/I_{s1} = 0.4843 = 48.43\%$.

CF of supply current = supply peak current/rms value of supply current $= I_{peak}/I_{rms} = I_{peak}/I_s = 1.0$.

Example 7.6

A single-phase AC voltage controller (shown in Figure E7.6) has a resistive load of 10 Ω. The input voltage is 230 V (rms) at 50 Hz. The thyristor delay angle is $\alpha = 100°$. Calculate (a) rms value of load voltage, (b) power consumed, (c) DPF, (d) DF, (e) total harmonic distortion of AC source current ($\mathrm{THD_I}$), (f) PF, (g) CF of AC source current, and (h) rms value of AC source current (I_s).

Solution: Given supply voltage $V_s = 230$ V (rms), frequency of the supply $(f) = 50$ Hz, $R = 10\ \Omega$, and $\alpha = 100°$.

In a single-phase, phase controlled AC controller, the waveform of the supply current (I_s) has a value of V_s/R from angle α to π. $V_{sm} = 230\sqrt{2} = 325.27$ V.

The rms value of AC mains current is $I_s = V_{sm}[\{1/(2\pi)\}\{(\pi - \alpha) + \sin 2\alpha/2\}]^{1/2}/R = 14.363$ A.

The rms value of fundamental current is $I_{s1} = V_{sm}/(2\pi R\sqrt{2})[(\cos 2\alpha - 1)^2 + \{\sin 2\alpha + 2(\pi - \alpha)\}^2]^{1/2} = 11.44$ A.

$$\theta_1 = \tan^{-1}[(\cos 2\alpha - 1)/\{\sin 2\alpha + 2(\pi - \alpha)\}] = 38.3656°.$$

The fundamental active power drawn by the load is $P_1 = V_sI_{s1}\cos\theta_1 = 2062.957$ W.

The fundamental reactive power drawn by the load is $Q_1 = V_sI_{s1}\sin\theta_1 = 1633.125$ VAR.

a. The rms value of load voltage is $V_{lrms} = V_{sm}[\{1/(2\pi)\}\{(\pi - \alpha) + \sin 2\alpha/2\}]^{1/2} = 143.63$ V.
b. The active power drawn by the load is $P_1 = V_sI_{s1}\cos\theta_1 = 2062.957$ W.
c. $\mathrm{DPF} = \cos\theta_1 = I_{s1a}/I_{s1} = 0.784$.
d. $\mathrm{DF} = I_{s1}/I_s = 0.79\,649$.
e. Total harmonic distortion of AC source current $(\mathrm{THD_I}) = \sqrt{\{(1/\mathrm{DF}^2) - 1\}} = 75.91\%$.
f. $\mathrm{PF} = \mathrm{DPF} \times \mathrm{DF} = 0.62\,445$.

Peak supply current $(I_{peak}) = V_{sm}\sin\alpha/R = 32.03$ A.

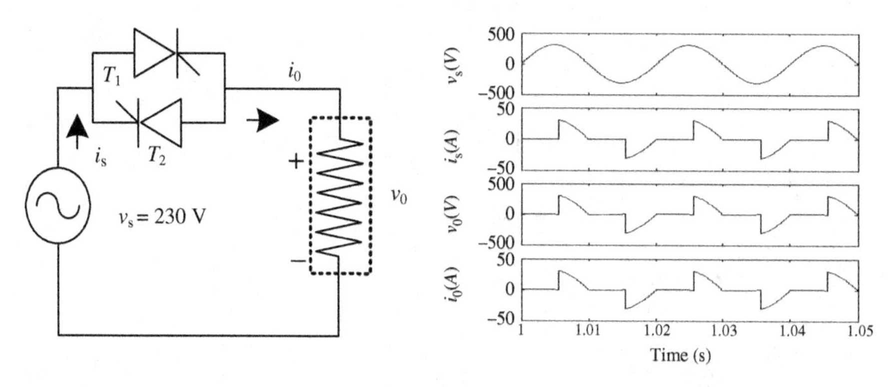

Figure E7.6 A single-phase converter-based current fed type of nonlinear load

g. CF of supply current $= I_{peak}/I_s = 2.23$.
h. The rms value of AC mains current is $I_s = V_{sm}[\{1/(2\pi)\}\{(\pi - \alpha) + \sin 2\alpha/2\}]^{1/2}/R = 14.363$ A.

Example 7.7

An uncontrolled three-phase bridge rectifier (shown in Figure E7.7) is fed by a line voltage of 415 V at 50 Hz. If a continuous constant load current is of 40 A in RL load, calculate (a) mean DC load voltage, (b) load resistance, (c) load power, (d) rms value of supply current, (e) DF, (f) DPF, (g) PF, and (h) THD of supply current.

Solution: Given supply phase voltage $V_s = 415/\sqrt{3} = 239.6$ V (rms), frequency of the supply $(f) = 50$ Hz, and $I_0 = 40$ A.

In three-phase diode bridge converter, the waveform of the supply current (I_s) is a quasi-square wave with the amplitude of the DC link current (I_0).

Therefore, the rms value of the quasi-square wave load current is $I_s = I_0\sqrt{(2/3)} = 32.659$ A.

Moreover, the rms value of the fundamental component of quasi-square wave is $I_{s1} = \{(\sqrt{6})/\pi\}I_0 = 31.188$ A.

The active power drawn by the load is $P = 3V_sI_{s1}\cos\theta_1 = 22.4178$ kW.

a. Average output DC voltage is $V_0 = 3\sqrt{3}\sqrt{2}V_s/\pi = 56.446$ V.
b. $R = V_0/I_0 = 560.446/40 = 14.011\ \Omega$.

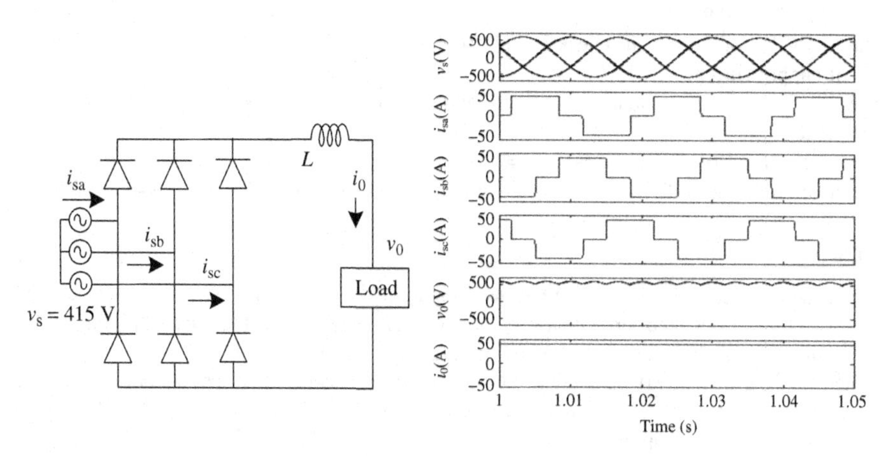

Figure E7.7 A three-phase converter-based current fed type of nonlinear load

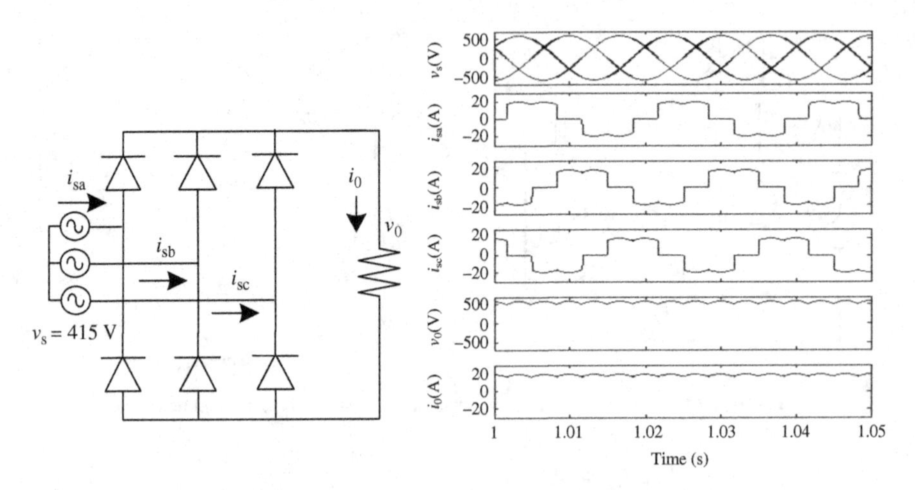

Figure E7.8 A three-phase converter-based current fed type of nonlinear load

c. Load power $= 3V_sI_{s1}\cos\theta_1 = V_0I_0 = 22.41$ kW.
d. The rms value of the quasi-square wave load current is $I_s = I_0\sqrt{(2/3)} = 32.659$ A.
e. $\text{DF} = I_{s1}/I_s = 3/\pi = 0.9549$.
f. $\text{DPF} = \cos\theta_1 = \cos\alpha = \cos 0° = 1.0$.
g. $\text{PF} = \text{DPF} \times \text{DF} = 0.9549 \times 1 = 0.9549$.
h. THD of AC current $= \sqrt{(I_s^2 - I_{s1}^2)}/I_{s1} = 0.3108 = 31.08\%$.

Example 7.8

A three-phase nonlinear load (shown in Figure E7.8) is fed from a three-phase 415 V, 50 Hz supply system consisting of a diode bridge converter that feeds a resistive load of 30 Ω. Calculate (a) fundamental active power drawn by the load, (b) PF, (c) rms value of supply current, (d) DF, (e) rms value of fundamental supply current, (f) peak current of AC mains, and (g) total harmonic distortion of AC source current (THD_I).

Solution: Given supply phase voltage $V_s = 415/\sqrt{3} = 239.6$ V (rms), $V_{sm} = 239.6\sqrt{2}$ V $= 338.85$ V, frequency of the supply $(f) = 50$ Hz, and $R = 30\,\Omega$.

a. The fundamental active power drawn by the load is $P = 3\{V_s^2/(2\pi R)\}[\{(2\pi + 3\sqrt{3})\} =$ 10.48843 kW.
b. $\text{PF} = P/(3V_sI_s) = 0.955\,577$.
c. The rms value of supply current is $I_s = \{V_s/(R)\}[\{(2\pi + 3\sqrt{3})\}/\pi\}^{1/2} = 15.2668$ A.
d. DF of supply current $= I_{s1}/I_s = 0.955\,577$.
e. The rms value of the fundamental component of supply current is $I_{s1} = \{V_s/(R^2\pi)\}(2\pi + 3\sqrt{3}) =$ 14.59 158 A.
f. The peak current of AC mains is $I_{peak} = V_{peak}/R = \sqrt{2} \times 415/30 = 19.56$ A.
g. The total harmonic distortion of AC source current $(\text{THD}_I) = \sqrt{\{(1/\text{DF}^2) - 1\}} = 30.77\%$.

Example 7.9

A three-phase nonlinear load (shown in Figure E7.9) is fed from a 415 V, 50 Hz supply system that has a three-phase diode bridge rectifier, with a very large capacitive filter at its DC bus, supplying DC power to a VSI fed three-phase induction motor variable frequency drive (VFD) at 540 V. The total load circuit resistance is 1 Ω. Calculate (a) the average DC current, (b) rms value of AC supply current, (c) rms value of fundamental supply current, (d) input active power, (e) THD of AC mains current, (f) DPF, (g) DF, (h) PF, and (i) CF of the supply current.

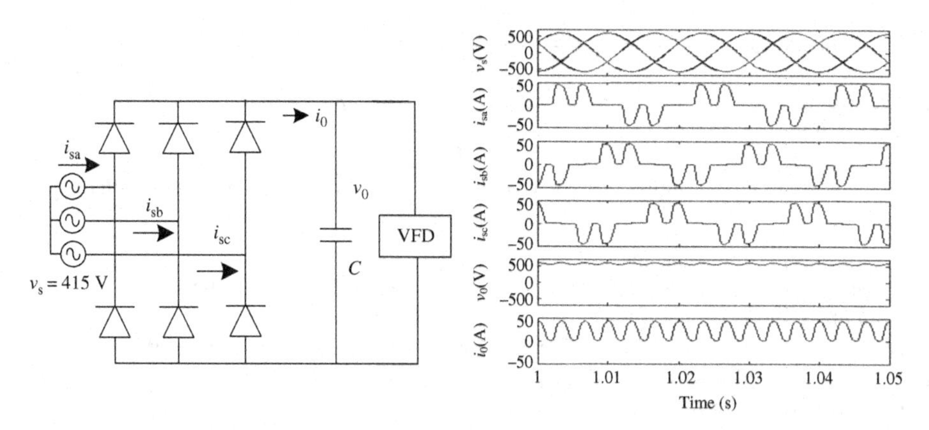

Figure E7.9 A three-phase converter-based voltage fed type of nonlinear load

Solution: Given supply voltage $V_s = 415/\sqrt{3} = 239.6$ V, frequency of the supply $(f) = 50$ Hz, series resistance $R = 1\ \Omega$, and $E = 540$ V.

The three-phase diode bridge rectifier with a very large capacitive filter at its DC bus behaves as a series resistance with back emf load (*RE* load). The AC supply current flows with two pulses in each half cycle from angle θ_1 to θ_2 and θ_3 to θ_4 when segments of AC line voltages are equal to or greater than E (540 V) and it is discontinuous. At angle θ_1, the AC supply current increases and at angle θ_2 it decreases and ceases to zero when AC voltage reduces to E. Similar conduction for second pulse of this AC supply current is from θ_3 to θ_4. These angles are computed as

$$\alpha = \sin^{-1}(E/V_{slm}) = \sin^{-1}\{540/(\sqrt{2} \times 415)\} = 66.94°,$$
$$\beta = \pi - \alpha = 113.06°, \quad \text{the conduction angle } \mu = \beta - \alpha = 46.12°.$$

However, these two angles for first pulse will be on reference of a first line voltage. The second pulse is on the segment of second line voltage: $\gamma = (\pi/3) + \alpha = 126.94°$, $\delta = (4\pi/3) - \gamma = 173.06°$, and the conduction angle $= \delta - \gamma = 46.12°$.

However, these angles are

$$\theta_1 = 66.94° - 30° = 36.94°, \quad \theta_2 = 83.06°, \quad \theta_3 = 96.94°, \quad \theta_4 = 143.06°.$$

Since there are six pulses for charging the DC bus capacitor in one cycle and all six pulses are identical, the DC bus average current is computed as

$$I_0 = \{6/(2\pi R)\}(2V_{slm}\cos\alpha + 2E\alpha - \pi E)$$
$$= 23.9682\text{ A} \quad (R = 1\ \Omega, \quad V_{slm} = \sqrt{2} \times 415\ V, \quad E = 540\text{ V}).$$

Since the AC input current with a couple of pulses in each cycle is an odd and quarter cycle and half cycle symmetry, the rms value of supply current I_s is the rms value of discontinuous current in the AC mains:

$$I_s = \left[\{4//(2\pi R^2)\} \left\{ \int_{\theta_1}^{\theta_2} \{V_{slm}\sin(\theta + 30°) - E\} d\theta \right\}^2 \right]^{1/2} = 24.47\text{ A}.$$

The input power drawn from the AC mains is $P = I_s^2 R + EI_0 = 13541.50$ W.
The rms value of fundamental current from the AC mains is $I_{s1} = P/(3V_s) = 18.839$ A.

The supply AC peak current is $I_{peak} = (V_{slm} - E)/R = 46.89$ A.

a. The average DC current is $I_0 = \{6/(2\pi R)\}(2\ V_{slm}\cos\alpha + 2E\alpha - \pi E) = 23.9682$ A ($R = 1\ \Omega$, $V_{slm} = \sqrt{2}\times 415$ V, $E = 540$ V).
b. The rms value of AC supply current is $I_s = \left[\{4/(2\pi R^2)\}\left\{\int_{\theta_1}^{\theta_2}\{V_{slm}\sin(\theta + 30°) - E\}d\theta\right\}^2\right]^{1/2} =$ 24.47 A.
c. The rms value of fundamental supply current is $I_{s1} = P/(3V_s) = 18.839$ A.
d. The input power drawn from the AC mains is $P = I_s^2 R + EI_0 = 13541.50$ W.
e. THD of AC current $= \{\sqrt{(I_s^2 - I_{s1}^2)}\}/I_{s1} = 82.894\%$.
f. DPF $= \cos\psi_1 = 1$ (since fundamental supply current is in phase with supply voltage $\psi_1 = 0°$).
g. DF = rms value of fundamental supply current/rms value of supply current $= I_{s1}/I_s = 0.76988$.
h. PF $= P/(3V_sI_s) = 0.76988$.
i. CF of supply current = supply peak current/rms value of supply current $= I_{peak}/I_{rms} = I_{peak}/I_s = 1.9166$.

Example 7.10

Consider a three-phase semicontrolled bridge converter (shown in Figure E7.10a) with sinusoidal input line supply voltage V_s of 415 V at 50 Hz and constant DC load current of 50 A. (a) Calculate DF, DPF, PF,

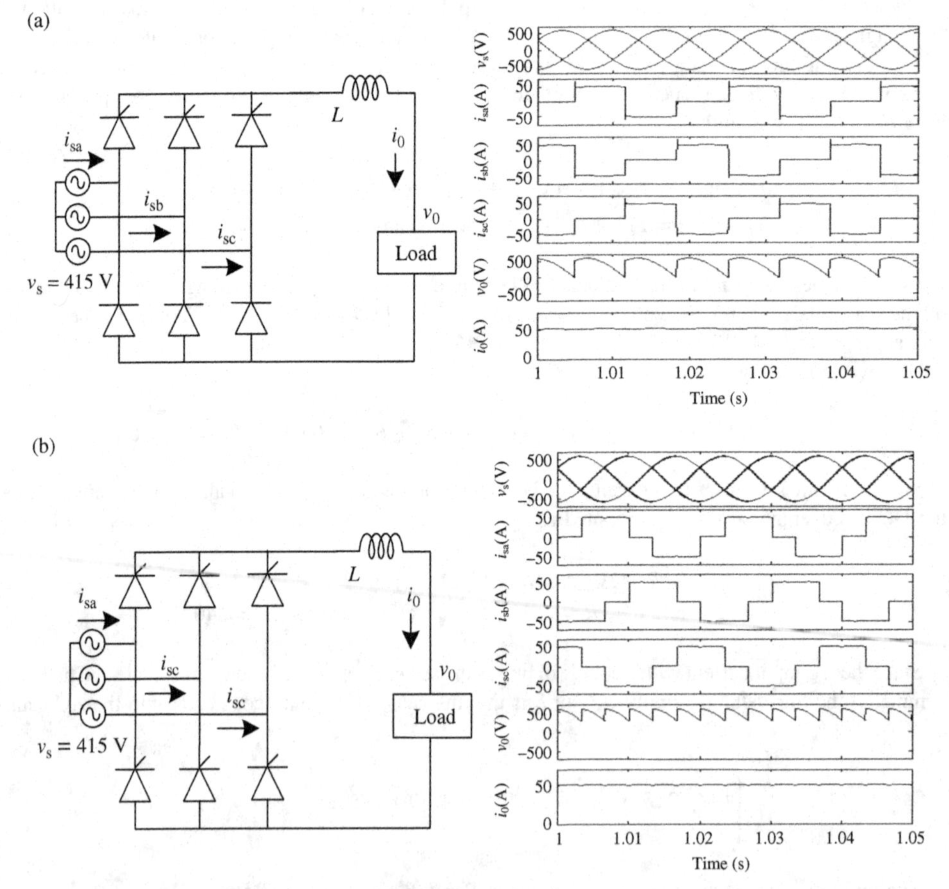

Figure E7.10 (a) A three-phase semicontrolled bridge converter-based current fed type of nonlinear load. (b) Three-phase fully controlled bridge converter-based current fed type of nonlinear load

THD for $V_0 = 0.75V_{d0}$, where V_{d0} is the DC output at $\alpha = 0°$. (b) Repeat part (a) for a fully controlled bridge converter (shown in Figure E7.10b).

Solution: Given supply voltage $V_s = 415/\sqrt{3} = 239.6$ V (rms), supply frequency $(f) = 50$ Hz, and $I_0 = 50$ A.

a. In a three-phase semicontrolled bridge converter, the waveform of the supply current (I_s) is from firing angle α to 180° with the amplitude of the DC link current (I_0).
   If $V_0 = 0.75V_{DC0}$ where V_{DC0} is the DC output at $\alpha = 0°$, then firing angle $\alpha = 60°$.
   The rms value of supply current is $I_s = I_0\sqrt{\{(\pi - \alpha)/\pi\}} = I_0\sqrt{(2/3)} = 40.825$ A.
   The rms value of fundamental supply current is $I_{s1} = I_0\sqrt{6}/\pi \times \cos(\alpha/2) = 33.73$ A.
   $\text{DPF} = \cos\theta_1 = \cos(\alpha/2) = 0.866$.
   $\text{DF} = 1/\sqrt{(1 + \text{THD}^2)} = 0.9549\cos(\alpha/2) = 0.826\,968$.
   $\text{PF} = \text{DPF} \times \text{DF} = 0.71\,615$.
   THD of AC current $= \{(1/\text{DF}^2) - 1\}^{1/2} = 68.006\%$.
b. In a three-phase fully controlled bridge converter (shown in Figure E7.10b), the waveform of the supply current (I_s) is from firing angle α to $(2\pi/3 + \alpha)$ with the amplitude of the DC link current (I_0).
   If $V_0 = 0.75V_{DC0}$ where V_{DC0} is the DC output at $\alpha = 0°$, then firing angle $\alpha = 31.057°$.
   The rms value of supply current is $I_s = I_0\sqrt{(2/3)} = 40.825$ A.
   The rms value of fundamental supply current is $I_{s1} = I_0\sqrt{6}/\pi = 38.985$ A.
   $\text{DPF} = \cos\theta_1 = \cos(\alpha) = 0.85\,665$.
   $\text{DF} = 1/\sqrt{(1 + \text{THD}^2)} = 0.9549$.
   $\text{PF} = \text{DPF} \times \text{DF} = 0.818$.
   THD of AC current $= \{\sqrt{(I_s^2 - I_{s1}^2)}\}/I_{s1} = 0.3108 = 31.08\%$.

Example 7.11

A three-phase nonlinear load (shown in Figure E7.11) is fed from a three-phase 415 V, 50 Hz supply system and has a thyristor bridge converter feeding a resistive load of 15 Ω at a thyristor firing angle of 45°. Calculate (a) fundamental active power drawn by the load, (b) fundamental reactive power drawn by the load, (c) PF, (d) rms value of supply current, (e) DF, (f) rms value of the fundamental component of supply current, (g) peak current of AC mains, and (h) total harmonic distortion of AC source current (THD_I).

Solution: Given supply voltage $V_s = 415/\sqrt{3} = 239.6$ V (rms), frequency of supply $(f) = 50$ Hz, $R = 15$ Ω, and $\alpha = 45°$.

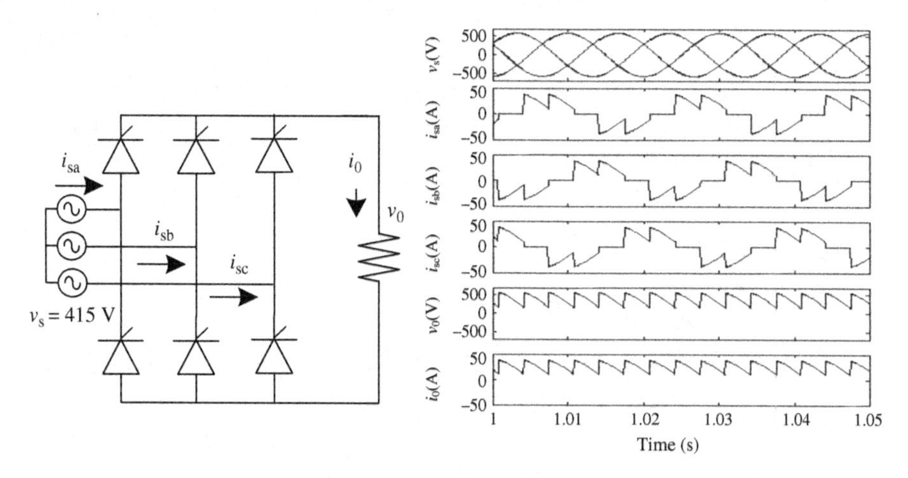

Figure E7.11 Three-phase converter-based current fed type of nonlinear load

In a three-phase thyristor bridge converter, the waveform of the supply current (I_s) is decided by the load resistance (R) and the firing angle (α) as

$I_s = \{V_s/(R)\}[\{(2\pi + 3\sqrt{3}\cos 2\alpha)\}/\pi\}^{1/2} = 22.589$ A.

The rms value of the fundamental component of load current is $I_{s1} = \{(V_s/(\pi R\sqrt{2})\}\{(3\sqrt{3}\sin 2\alpha)^2 + (2\pi + 3\sqrt{3}\cos 2\alpha)^2\}^{1/2} = 20.725$ A.

The active power component of supply current $I_{s1a} = P/(3\ V_s) = 15.97\,338$ A.

$\mathrm{DPF} = \cos\theta_1 = I_{s1a}/I_{s1} = 0.77\,073$; $\sin\theta_1 = \sqrt{(1 - \mathrm{DPF}^2)} = 0.63\,716$.

a. The fundamental active power drawn by the load is $P = 3\{V_S^2/(2\pi R)\}[\{(2\pi + 3\sqrt{3}\cos 2\alpha)\} = 11481.667$ kW.
b. The fundamental reactive power drawn by the load is $Q_1 = 3\ V_s I_{s1}\sin\theta_1 = 9.491\,885$ kVAR.
c. $\mathrm{PF} = \mathrm{DF} \times \mathrm{DPF} = 0.70\,711$.
d. The rms value of supply current is $I_s = \{V_s/(R)\}[\{(2\pi + 3\sqrt{3}\cos 2\alpha)\}/\pi\}^{1/2} = 22.589$ A.
e. $\mathrm{DF} = I_{s1}/I_s = 0.91\,745$.
f. The rms value of fundamental supply current is $I_{s1} = \{(V_s/(\pi\ R\sqrt{2})\}\{(3\sqrt{3}\sin 2\alpha)^2 + (2\pi + 3\sqrt{3}\cos 2\alpha)^2\}^{1/2} = 20.725$ A.
g. The peak current of AC mains is $I_{peak} = V_{peak}/R = \sqrt{2} \times 415/15 = 39.12\,657$ A.
h. Total harmonic distortion of AC source current ($\mathrm{THD_I}$) $= \sqrt{\{(1/\mathrm{DF}^2) - 1\}} = 43.3638\%$

Example 7.12

A three-phase fully controlled bridge converter (shown in Figure E7.12) is used as a LCI to feed power from a battery with 360 V and an internal resistance of 0.2 Ω. The supply line voltage is 415 V (rms) at 50 Hz and sufficient inductance is included in the output circuit to maintain the current virtually constant at 30 A. Determine the required delay angle α, DF, DPF, total harmonic distortion of AC source current ($\mathrm{THD_I}$), CF of AC source current, PF, and rms value of AC source current (I_s).

Solution: Given supply voltage $V_s = 415/\sqrt{3} = 239.6$ V (rms), supply frequency (f) = 50 Hz, $I_0 = 30$ A, $E = 360$ V, and $R_{DC} = 0.2\ \Omega$.

In a three-phase thyristor bridge converter, the waveform of the supply current (I_s) is a quasi-square wave with the amplitude of the DC link current (I_0).

Therefore, $I_s = \sqrt{(2/3)}I_0 = 0.81\,649 I_0 = 24.49$ A.

Moreover, the rms value of the fundamental component of quasi-square wave is $I_{s1} = \{(\sqrt{6})/\pi\}\ I_0 = 23.39$ A.

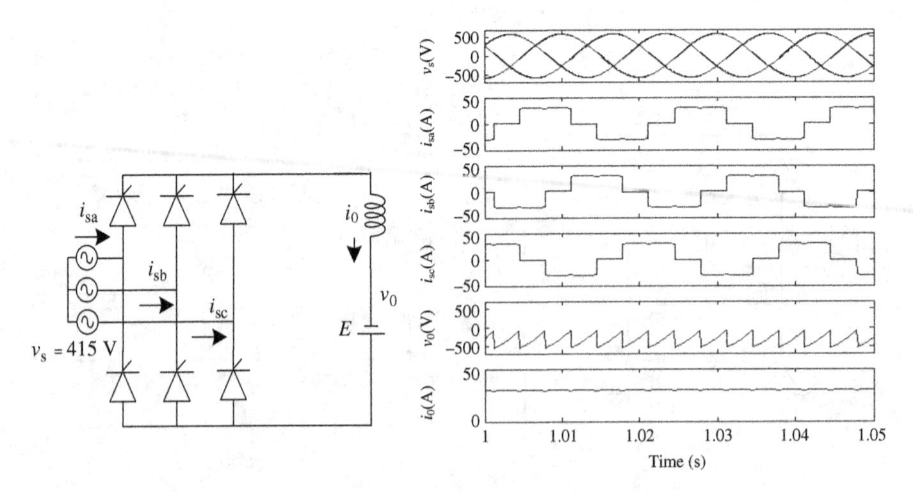

Figure E7.12 Three-phase converter-based current fed type of nonlinear load

The average output voltage $V_0=(3\sqrt{3}\sqrt{2}V_s/\pi)\cos\alpha-(E-I_0R_{DC})=-(360-30\times0.2)=-354$ V, $\alpha=129.1712°$.

$DF=I_{s1}/I_s=3/\pi=0.9549$.

$DPF=\cos\theta_1=\cos\alpha=\cos 129.1712°=0.6\,316\,397$.

THD of AC current $=\{\sqrt{(I_s^2-I_{s1}^2)}\}/I_{s1}=0.3108=31.08\%$.

CF of AC source current $=I_{peak}/I_{rms}=I_0/[\{\sqrt{(2/3)}\}I_0]=\sqrt{(3/2)}=1.22\,474$.

$PF=DPF\times DF=0.9549\times0.6\,316\,397=0.60\,315$.

The rms value of AC source current is $I_s=\{\sqrt{(2/3)}\}I_0=0.81\,649I_0=19.8147$ A.

Example 7.13

A three-phase nonlinear load (shown in Figure E7.13) is fed from a three-phase 415 V, 50 Hz supply system and has a 12-pulse diode bridge converter with 100 A constant DC current. It consists of an ideal transformer with single primary star connected winding and two secondary windings connected in star and delta with same line voltages to provide 30° phase shift between two sets of three-phase output voltages. Two 6-pulse diode bridges are connected in series to provide a 12-pulse AC–DC converter. Calculate (a) fundamental active power drawn by the load, (b) PF, (c) rms value of supply current, (d) DF, (e) rms value of fundamental supply current, (f) peak current of AC mains, and (g) total harmonic distortion of AC source current (THD_I).

Solution: Given supply phase voltage $V_s=415/\sqrt{3}=239.6$ V (rms), frequency of the supply $(f)=50$ Hz, and the amplitude of the DC link current $I_0=100$ A.

In a three-phase 12-pulse diode bridge converter, the waveform of the input AC current (I_s) is a stepped waveform as (i) first step of $\pi/6$ angle (from 0° to $\pi/6$) and input current magnitude of $I_0/\sqrt{3}$, (ii) second step of $\pi/6$ angle (from $\pi/6$ to $\pi/3$) and input current magnitude of $I_0(1+1/\sqrt{3})$, and (iii) third step of $\pi/6$ angle (from $\pi/3$ to $\pi/2$) and input current magnitude of $I_0(1+2/\sqrt{3})$ and it has all four symmetric segments of such steps.

Therefore, the rms value of 12-pulse converter input current is $I_s=I_0[(1/3)+(1+1/\sqrt{3})^2+(1+2/\sqrt{3})^2]^{1/2}=1.57\,735I_0=157.735$ A.

The rms value of the fundamental component of 12-pulse converter input current is $I_{s1}=\{(2\sqrt{6})/\pi\}I_0=1.559\,393I_0=155.939$ A.

a. The active power drawn by the load is $P=3V_sI_{s1}\cos\theta_1=112.08\,895$ kW.
b. $PF=P/(3\,V_sI_s)=0.9\,886\,138$.
c. The rms value of 12-pulse converter input current is $I_s=I_0[(1/3)+(1+1/\sqrt{3})^2+(1+2/\sqrt{3})^2]^{1/2}=1.57\,735\,I_0=157.735$ A.
d. $DF=I_{s1}/I_s=0.9\,886\,138$.

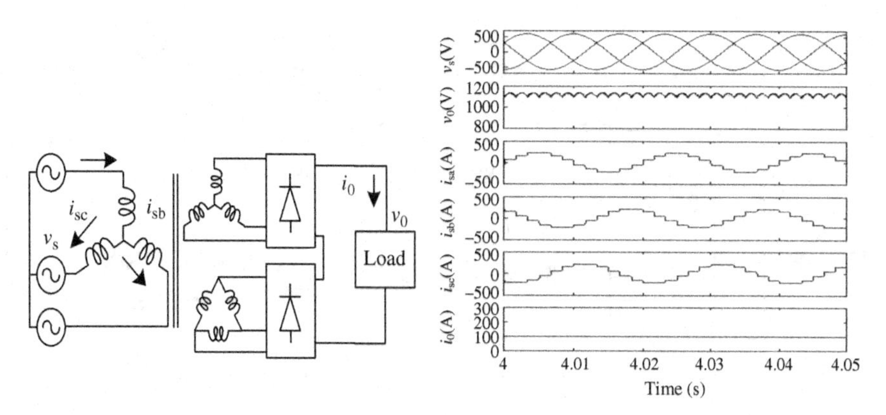

Figure E7.13 Three-phase converter-based current fed type of nonlinear load

e. The rms value of the fundamental component of input AC mains current is $I_{s1} = \{(2\sqrt{6})/\pi\}I_0 = 1.559\,393I_0 = 155.939$ A.
f. The peak current of AC mains is $I_{peak} = \{I_0(1 + 2/\sqrt{3})\} = 2.1547I_0 = 215.45$ A.
g. The total harmonic distortion of AC source current (THD_I) $= \sqrt{\{(1/DF^2) - 1\}} = 15.22\%$.

Example 7.14

A three-phase nonlinear load (shown in Figure E7.14) is fed from a three-phase 415 V, 50 Hz supply system and has a 12-pulse thyristor bridge converter with 200 A constant DC current at a thyristor firing angle of 60°. It consists of an ideal transformer with single primary star connected winding and two secondary windings connected in star and delta with same line voltages to provide 30° phase shift between two sets of three-phase output voltages. Two 6-pulse thyristor bridges are connected in series to provide a 12-pulse AC–DC converter. Calculate (a) fundamental active power drawn by the load, (b) fundamental reactive power drawn by the load, (c) PF, (d) rms value of supply current, (e) DF, (f) rms value of fundamental supply current, (g) peak supply current, and (h) total harmonic distortion of supply current (THD_I).

Solution: Given supply voltage $V_s = 415/\sqrt{3} = 239.6$ V (rms), frequency of the supply $(f) = 50$ Hz, $I_0 = 200$ A, and $\alpha = 60°$.

In a three-phase 12-pulse thyristor bridge converter, the waveform of the input AC current (I_s) is a stepped waveform as (i) first step of $\pi/6$ angle (from α to $(\alpha + \pi/6)$) and input current magnitude of $I_0/\sqrt{3}$, (ii) second step of $\pi/6$ angle (from $(\alpha + \pi/6)$ to $(\alpha + \pi/3)$) and input current magnitude of $I_0(1 + 1/\sqrt{3})$, and (iii) third step of $\pi/6$ angle (from $(\alpha + \pi/3)$ to $(\alpha + \pi/2)$) and input current magnitude of $I_0(1 + 2/\sqrt{3})$ and it has all four symmetric segments of such steps.

Therefore, the rms value of 12-pulse converter input current is $I_s = I_0[(1/3) + (1 + 1/\sqrt{3})^2 + (1 + 2/\sqrt{3})^2]^{1/2} = 1.57\,735I_0 = 315.47$ A.

Moreover, the rms value of 12-pulse converter fundamental AC current is $I_{s1} = \{(2\sqrt{6})/\pi\}I_0 = 1.559\,393I_0 = 311.8786$ A.

The active power component of supply current is $I_{s1a} = I_{s1}\cos\theta_1 = I_{s1}\cos\alpha = 311.8786\cos 60° = 155.9393$ A.

a. The active power drawn by the load is $P = 3V_sI_{s1}\cos\theta_1 = 112.08\,895$ kW.
b. The fundamental reactive power is $Q_1 = 3V_sI_{s1}\sin\theta_1 = 3\ V_sI_{s1}\sin\alpha = 194.1437$ kVAR.
c. $PF = P/(3V_sI_s) = 0.4\,943\,069$.
d. The rms value of 12-pulse converter input current is $I_s = I_0[(1/3) + (1 + 1/\sqrt{3})^2 + (1 + 2/\sqrt{3})^2]^{1/2} = 1.57\,735I_0 = 315.47$ A.
e. $DF = I_{s1}/I_s = 0.9\,886\,138$.

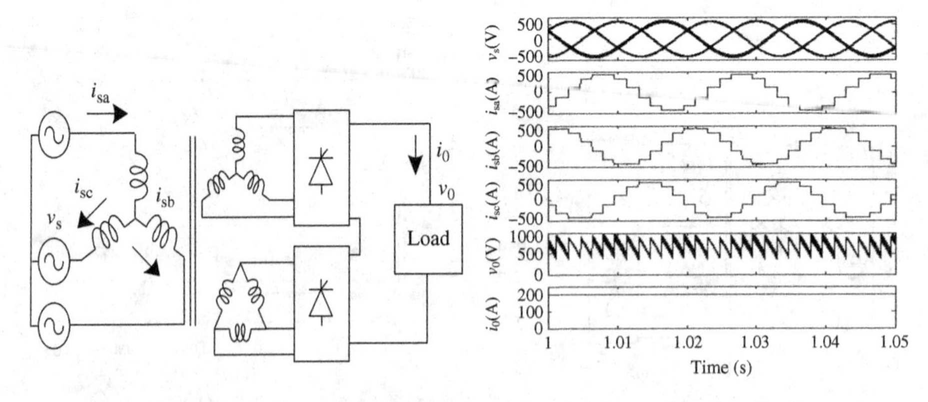

Figure E7.14 Three-phase converter-based current fed type of nonlinear load

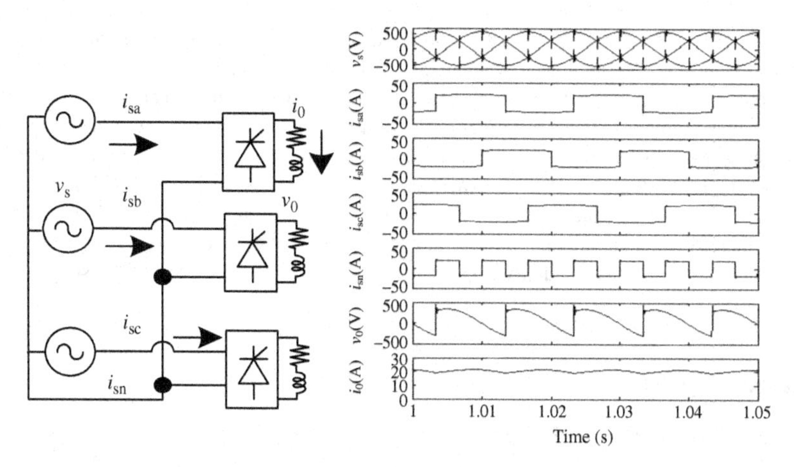

Figure E7.15 Three-phase converter-based current fed type of nonlinear load

f. The rms value of the fundamental component of input AC mains current is $I_{s1}=\{(2\sqrt{6})/\pi\}I_0=1.559\,393I_0=311.8786$ A.
g. The peak current of AC mains is $I_{peak}=\{I_0(1+2/\sqrt{3})\}=2.1547I_0=430.94$ A.
h. The total harmonic distortion of AC source current $(THD_I)=\sqrt{\{(1/DF^2)-1\}}=15.22\%$.

Example 7.15

In a three-phase, 415 V line voltage, 50 Hz four-wire distribution system, three single-phase loads (connected between phases and neutral) have a single-phase thyristor bridge converter drawing equal 20 A constant DC current at a thyristor firing angle of 60° (shown in Figure E7.15). Calculate (a) active power consumed, (b) reactive power drawn, (c) DPF, (d) DF, (e) total harmonic distortion of AC source current (THD_I), (f) PF, (g) CF of AC source current, (h) rms value of AC source current (I_s), and (i) neutral current (I_{sn}).

Solution: Given supply voltage $V_s=239.6$ V, frequency of the supply $(f)=50$ Hz, DC link current $I_0=20$ A, and firing angle $\alpha=60°$.

In a single-phase thyristor bridge converter, the waveform of the supply current (I_s) is a square wave with the amplitude of the DC link current (I_0). The rms value of the fundamental component of square wave is $(2\sqrt{2}/\pi)=0.9$ times its amplitude.

Therefore, $I_s=I_0=20$ A and $I_{s1}=(2\sqrt{2}/\pi)I_0=0.9I_0=18$ A.

The rms value of the fundamental active power component of load current is $I_{s1a}=I_{s1}\cos\alpha=9.0$ A.

The neutral current $I_{sn}=20$ A (since it is a square wave three times the fundamental frequency).

a. The active power consumed is $P_1=3V_sI_{s1a}=6469.21$ W.
b. The reactive power consumed is $Q_1=3V_sI_{s1}\sin\alpha=11\,205.00$ VAR.
c. $DPF=\cos\alpha=0.5$.
d. $DF=I_{s1}/I_s=(2\sqrt{2}/\pi)=0.9$.
e. The total harmonic distortion of AC source current $(THD_I)=\sqrt{\{(1/DF^2)-1\}}=48.43\%$.
f. $PF=(2\sqrt{2}/\pi)\cos\alpha=0.45$.
g. CF of AC source current $=I_{peak}/I_{rms}=I_0/I_s=1$.
h. The rms value of AC source current is $I_s=I_0=20$ A.
i. The neutral current $(I_{sn})=20$ A (since it is a square wave three times the fundamental frequency).

7.8 Summary

Majority of power quality problems are mainly caused by the use of nonlinear loads. The nonlinear loads draw nonsinusoidal current from AC mains, which consists of various harmonic currents such as characteristic harmonics, noncharacteristic harmonics, interharmonics, subharmonics, reactive power component of current, fluctuating current, unbalanced currents, and so on. These nonlinear loads are classified into different categories considering the severity of the created problems. A number of practical examples of these nonlinear loads are given to have a proper exposure of power quality problems. An analytical study of various performance indices of these nonlinear loads is made in detail with several numerical examples to study the level of power quality they may cause in the system. Since these nonlinear loads cannot be dispensed due to many economic advantages, energy conservation, and increase in production; therefore, it is quite important to study the behavior of these nonlinear loads to find out proper mitigation techniques for power quality improvements to reduce the pollution of the supply system.

7.9 Review Questions

1. What are voltage-fed nonlinear loads? Give two examples.
2. What are current-fed nonlinear loads? Give two examples.
3. What are the reasons for which nonlinear loads draw harmonic currents from AC mains?
4. What is the value of THD of the input current of a single-phase diode rectifier with constant DC current?
5. What is the value of THD of the input current of a three-phase diode rectifier with constant DC current?
6. What is the value of CF of the input current of a single-phase diode rectifier with constant DC current?
7. What is the value of CF of the input current of a three-phase diode rectifier with constant DC current?
8. What is the value of DF of the input current of a single-phase diode rectifier with constant DC current?
9. What is the value of DF of the input current of a three-phase diode rectifier with constant DC current?
10. What is the value of PF of the input current of a single-phase diode rectifier with constant DC current?
11. What is the value of PF of the input current of a three-phase diode rectifier with constant DC current?
12. What is the value of PF of a single-phase thyristor bridge converter with constant DC current at a firing angle of 60°?
13. What is the value of PF of a three-phase thyristor bridge converter with constant DC current at a firing angle of 30°?
14. What are the reasons that nonlinear loads cause excessive neutral current?
15. Which nonlinear loads cause excessive neutral current? Give two examples.
16. Which nonlinear loads do not consist of solid-state control and they have the harmonic currents?
17. Which nonlinear loads draw harmonic currents but do not need reactive power? Give two examples.
18. What are the power quality problems due to harmonic currents drawn by nonlinear loads?
19. What are the power quality problems due to reactive power component of currents drawn by nonlinear loads?
20. What is the classification of nonlinear loads based on solid-state converter used in them?
21. What are the reasons that these nonlinear loads are to be used in many types of equipment?
22. What are the reasons that load unbalancing is observed in three-phase supply system?

23. Which solid-state converter used in nonlinear loads has maximum power quality problems and why?
24. What are the reasons that the solid-state controllers are needed in some nonlinear loads?
25. Why magnetic ballasts have harmonic currents in fluorescent lighting system?

7.10 Numerical Problems

1. Consider a single-phase uncontrolled bridge converter (shown in Figure E7.1) with sinusoidal input supply $V_s = 110$ V and constant DC load current of 25 A. Calculate (a) v_O (b) CF, (c) DF, (d) DPF, (e) PF, and (f) THD.
2. A single-phase uncontrolled bridge converter has a *RE* load (shown in Figure E7.2) with $R = 10\,\Omega$, and $E = 60$ V. The input AC voltage is $V_s = 110$ V at 50 Hz. Calculate (a) load average current, (b) rms value of supply current, (c) CF, (d) DF, (e) DPF, (f) PF, and (g) THD.
3. A single-phase diode bridge rectifier (shown in Figure E7.3) is drawing AC current at 0.95 DPF and THD of AC current is 60%. It is drawing 2000 W from 220 V, 50 Hz AC source and CF is 2 of AC current. Calculate (a) PF, (b) rms value of current, and (c) peak current of AC mains.
4. Consider a single-phase semicontrolled bridge converter (shown in Figure E7.4) with sinusoidal input supply V_s of 220 V at 50 Hz and constant DC load current of 20 A. (a) Calculate DF, DPF, PF, THD of the supply current for $V_O = 0.6V_{DCO}$ where V_{DCO} is the DC output at $\alpha = 0°$. (b) Repeat part (a) for a fully controlled bridge converter.
5. A single-phase fully controlled bridge converter is used as a LCI (shown in Figure E7.5) to feed power from a battery with 168 V and an internal resistance of $0.2\,\Omega$. The supply voltage is 230 V (rms) and a sufficient inductance is included in the output circuit to maintain the current virtually constant at 25 A. Determine the (a) required delay angle α, (b) DF, (c) DPF, (d) total harmonic distortion of AC source current (THD_I), (e) CF of AC source current, (f) PF, and (g) rms value of AC source current (I_s).
6. A single-phase AC voltage controller (shown in Figure E7.6) has a resistive load of $5\,\Omega$. The input voltage is 220 V (rms) at 50 Hz. The thyristor delay angle is $\alpha = 130°$. Calculate (a) rms value of load voltage, (b) power consumed, (c) DPF, (d) DF, (e) total harmonic distortion of AC source current (THD_I), (f) PF, (g) CF of AC source current, and (h) rms value of AC source current (I_s).
7. An uncontrolled three-phase bridge rectifier (shown in Figure E7.7) is fed by a line voltage of 440 V at 50 Hz. If a continuous constant load current is of 100 A in *RL* load, calculate (a) the mean DC load voltage, (b) load resistance, (c) load power, (d) rms value of AC mains current, (e) DF, (f) DPF, (g) PF, and (h) THD of the supply current.
8. A three-phase nonlinear load is fed from a three-phase 460 V, 60 Hz supply system and has a diode bridge converter (shown in Figure E7.8) feeding a resistive load of $10\,\Omega$. Calculate (a) fundamental active power drawn by the load, (b) PF, (c) rms value of supply current, (d) DF, (e) rms value of fundamental supply current, (f) peak current of AC mains, and (g) total harmonic distortion of AC source current (THD_I).
9. A three-phase nonlinear load consisting of a 460 V, 60 Hz, three-phase diode bridge rectifier with a very large capacitive filter (shown in Figure E7.9) at its DC bus is supplying DC power to a VSI-fed three-phase induction motor variable frequency drive (VFD) at 580 V with a total circuit resistance of $1.5\,\Omega$. Calculate (a) the average DC current, (b) rms value of AC supply current, (c) rms value of fundamental supply current, (d) input active power, (e) THD of AC mains current, (f) DPF, (g) DF, (h) PF, and (i) CF of the supply current.
10. Consider a three-phase semicontrolled bridge converter (shown in Figure E7.10) with sinusoidal input line supply voltage of 440 V at 50 Hz and constant DC load current of 150 A. (a) Calculate DF,

DPF, PF, and THD for $V_d = 0.6 \times V_{d0}$ where V_{d0} is the DC output at $\alpha = 0°$. (b) Repeat part (a) for a fully controlled bridge converter.

11. A three-phase nonlinear load is fed from a three-phase 460 V, 60 Hz supply system and has a thyristor bridge converter (shown in Figure E7.11) feeding a resistive load of 20 Ω at a thyristor firing angle of 30°. Calculate (a) fundamental active power drawn by the load, (b) fundamental reactive power drawn by the load, (c) PF, (d) rms value of supply current, (e) DF, (f) rms value of fundamental supply current, (g) peak current of AC mains, and (h) total harmonic distortion of AC source current (THD_I).

12. A three-phase fully controlled bridge converter is used as a LCI (shown in Figure E7.12) to feed power from a battery with 400 V and an internal resistance of 0.25 Ω. The supply line voltage is 460 V (rms) at 60 Hz and a sufficient inductance is included in the output circuit to maintain the current virtually constant at 50 A. Determine the (a) required delay angle (α), (b) DF, (c) DPF, (d) total harmonic distortion of AC source current (THD_I), (e) CF of AC source current, (f) PF, and (g) rms value of AC source current (I_s).

13. A three-phase nonlinear load is fed from a three-phase 460 V, 60 Hz supply system and has a 12-pulse diode bridge converter with 200 A constant DC current. It consists of an ideal transformer with single primary star connected winding and two secondary windings connected in star and delta with same line voltages to provide 30° phase shift between two sets of three-phase output voltages. Two 6-pulse diode bridges are connected in series to provide a 12-pulse AC–DC converter (shown in Figure E7.13). Calculate (a) fundamental active power drawn by the load, (b) PF, (c) rms value of supply current, (d) DF, (e) rms value of fundamental supply current, (f) peak current of AC mains, and (g) total harmonic distortion of AC source current (THD_I).

14. A three-phase nonlinear load is fed from a three-phase 440 V, 50 Hz supply system and has a 12-pulse thyristor bridge converter with 500 A constant DC current at a thyristor firing angle of 30°. It consists of an ideal transformer with single primary star connected winding and two secondary windings connected in star and delta with same line voltages to provide 30° phase shift between two sets of three-phase output voltages. Two 6-pulse thyristor bridges are connected in series to provide a 12-pulse AC–DC converter (shown in Figure E7.14). Calculate (a) fundamental active power drawn by the load, (b) fundamental reactive power drawn by the load, (c) PF, (d) rms value of supply current, (e) DF, (f) rms value of fundamental supply current, (g) peak current of AC mains, and (h) total harmonic distortion of AC source current (THD_I).

15. In a three-phase, 400 V line voltage, 50 Hz four-wire distribution system, three single-phase loads (connected between phases and neutral) have a single-phase thyristor bridge converter drawing an equivalent 30 A constant DC current at a thyristor firing angle of 30° (shown in Figure E7.15). Calculate (a) active power consumed, (b) reactive power drawn, (c) DPF, (d) DF, (e) total harmonic distortion of AC source current (THD_I), (f) PF, (g) CF of AC source current, (h) rms value of AC source current (I_s), and (i) neutral current (I_{sn}).

7.11 Computer Simulation-Based Problems

1. Simulate the behavior of a single-phase 230 V, 50 Hz uncontrolled bridge converter with a parallel capacitive DC filter of 1500 μF and an equivalent resistive load of 25 Ω. It has a source impedance of 0.25 Ω resistive element and 1.0 Ω inductive element. Plot the supply voltage and input current, output voltage and current with time, and the harmonic spectra of supply current and voltage at PCC. Compute THD, CF, rms value of AC mains current, DPF, DF, PF, % ripple in load current, output voltage ripple, ripple factor, and input active, reactive, and output powers.

2. Simulate the behavior of a single-phase 230 V, 50 Hz uncontrolled bridge converter used for charging a battery of 240 V. It has a source impedance of 1.5 Ω resistive element and 5.0 Ω inductive element. Plot the supply voltage and input current, charging current with time, and the harmonic spectra of

supply current and voltage at PCC. Compute THD, CF, rms value of AC mains current, DPF, DF, PF, % ripple in load current, and input active, reactive, and output powers.

3. Simulate the behavior of a single-phase 230 V, 50 Hz semicontrolled bridge converter with a series connected inductive load of 10 mH, an equivalent resistive load of 2 Ω, and back emf of 60 V. The thyristor delay angle is $\alpha = 90°$. It has a source impedance of 0.15 Ω resistive element and 1.5 Ω inductive element. Plot the supply voltage and input current, output voltage and current with time, and the harmonic spectra of supply current and voltage at PCC. Compute THD, CF, rms value of AC mains current, DPF, DF, PF, % ripple in load current, and input active, reactive, and output powers.
4. Simulate the behavior of a single-phase 230 V, 50 Hz controlled bridge converter with a series connected inductive load of 100 mH and an equivalent resistive load of 10 Ω. It has a source impedance of 0.15 Ω resistive element and 1.5 Ω inductive element. The thyristor delay angle is $\alpha = 60°$. Plot the supply voltage and input current, output voltage and current with time, and the harmonic spectra of supply current and voltage at PCC. Compute THD, CF, rms value of AC mains current, DPF, DF, PF, % ripple in load current, and input active, reactive, and output powers.
5. Simulate the behavior of a single-phase 230 V, 50 Hz controlled bridge converter with a series connected inductive load of 10 mH, an equivalent resistive load of 2 Ω, and a back emf of 48 V. It has a source impedance of 0.2 Ω resistive element and 2.0 Ω inductive element. The thyristor delay angle is $\alpha = 50°$. Plot the supply voltage and input current, output voltage and current with time, and the harmonic spectra of supply current and voltage at PCC. Compute THD, CF, rms value of AC mains current, DPF, DF, PF, % ripple in load current, output voltage ripple, ripple factor, and input active, reactive, and output powers.
6. Simulate the behavior of a single-phase TRIAC-based light dimmer circuit having *RC* circuit along with the DIAC for firing angle control used to vary the intensity of a 230 V, 50 Hz, 40 W incandescent filament lamp with $R = 3.6$ kΩ (using 4.7 kΩ potentiometer) and $C = 0.22$ μF. The break mover voltage of the DIAC is 50 V. Plot the supply voltage and input current, output voltage and current with time, and the harmonic spectra of supply current and voltage at PCC. Compute THD, CF, rms value of AC mains current, firing angle, DPF, DF, PF, and input active, reactive, and output powers.
7. Simulate the behavior of a single-phase AC voltage controller having a resistive–inductive (*RL*) load of $R = 5$ Ω and $L = 50$ mH. The input voltage is 230 V (rms) at 50 Hz. The thyristor delay angle is $\alpha = 60°$. Plot the supply voltage and input current, output voltage and current with time, and the harmonic spectra of supply current and voltage at PCC. Compute THD, CF, rms value of AC mains current, thyristor conduction angle, DPF, DF, input power factor, and input active, reactive, and output powers.
8. Simulate the behavior of a single-phase, 230 V, 50 Hz supply system feeding a set of nonlinear loads consisting of a thyristor bridge and a diode rectifier connected in parallel. The diode bridge converter is feeding a parallel capacitive DC filter of 470 μF and a resistive load of 30 Ω. The thyristor bridge converter is feeding a resistive–inductive (*RL*) load of $R = 20$ Ω and $L = 50$ mH with a thyristor firing angle of 60°. Plot the supply voltage and input current, output voltages and currents with time, and the harmonic spectra of supply current and voltage at PCC. For this composite nonlinear load, compute (a) active power consumed, (b) reactive power drawn, (c) DPF, (d) DF, (e) total harmonic distortion of AC source current (THD_I), (f) PF, (g) CF of AC source current, and (h) rms value of AC source current (I_s).
9. Simulate the behavior of a three-phase 415 V, 50 Hz uncontrolled bridge rectifier with a source inductance of 20 mH and DC load resistance of 40 Ω. Plot the supply voltage and input current, output voltage and current with time, and the harmonic spectra of supply current and voltage at PCC. Compute (a) average output voltage, (b) overlap angle, (c) DPF, (d) DF, (e) total harmonic distortion, (f) PF, and (g) rms value of AC mains current.
10. Simulate the behavior of a three-phase 415 V, 50 Hz uncontrolled bridge converter with a parallel capacitive DC filter of 5000 μF and a resistive load of 20 Ω. It has a source impedance of 0.5 Ω

resistive element and 2.0 Ω inductive element. Plot the supply voltage and input current, output voltage and current with time, and the harmonic spectra of supply current and voltage at PCC. Compute THD, CF, rms value of AC mains current, DPF, DF, PF, % ripple in load current, output voltage ripple, ripple factor, and input active, reactive, and output powers.

11. Simulate the behavior of a three-phase 415 V, 50 Hz uncontrolled six-pulse bridge converter used for charging a battery of 540 V. It has a source impedance of 0.5 Ω resistive element and 2.0 Ω inductive element. Plot the supply voltage and input current, charging current with time, and the harmonic spectra of supply current and voltage at PCC. Compute THD, CF, rms value of AC mains current, DPF, DF, PF, % ripple in load current, and input active, reactive, and output powers.

12. Simulate the behavior of a three-phase 415 V, 50 Hz controlled bridge converter with a series connected inductive load of 50 mH, an equivalent resistive load of 15 Ω, and a back emf of 240 V. It has a source impedance of 0.25 Ω resistive element and 1.5 Ω inductive element. The thyristor delay angle is $\alpha = 30°$. Plot the supply voltage and input current, output voltage and current with time, and the harmonic spectra of supply current and voltage at PCC. Compute THD, CF, rms value of AC mains current, DPF, DF, PF, % ripple in load current, and input active, reactive, and output powers.

13. Simulate the behavior of a three-phase bidirectional delta connected controller (back-to-back connected thyristors in series with load) having a resistive load of $R = 6\,\Omega$ and fed from 415 V (rms) line voltage at 50 Hz. The thyristor delay angle is $\alpha = 90°$. Plot the supply voltage and input current, output voltage and current with time, and the harmonic spectra of supply current and voltage at PCC. Compute THD, CF, rms value of AC mains current, thyristor conduction angle, rms value of output voltage, DPF, DF, input power factor, and input active, reactive, and output powers.

14. Simulate the behavior of a three-phase bidirectional delta connected controller (back-to-back connected thyristors in series with load) having a resistive load of $R = 5\,\Omega$ and fed from 415 V (rms) line voltage at 50 Hz. The output power of 50 kW is desired. Plot the supply voltage and input current, output voltage and current with time, and the harmonic spectra of supply current and voltage at PCC. Compute THD, CF, rms value of AC mains current, thyristor firing angle, rms value of output voltage, thyristor conduction angle, rms value of phase current, DPF, DF, input power factor, and input active, reactive, and output powers.

15. Simulate the behavior of a three-phase bidirectional star connected controller (back-to-back connected thyristors in series with load) having a resistive load of $R = 10\,\Omega$ and fed from 415 V (rms) line voltage at 50 Hz. The thyristor delay angle is $\alpha = 40°$. Plot the supply voltage and input current, output voltage and current with time, and the harmonic spectra of supply current and voltage at PCC. Compute THD, CF, rms value of AC mains current, thyristor conduction angle, rms value of output voltage, DPF, DF, input power factor, and input active, reactive, and output powers.

16. Simulate the behavior of a three-phase, three-wire, 415 V, 50 Hz supply system feeding a set of nonlinear loads consisting of a thyristor bridge and a diode rectifier connected in parallel. The diode bridge converter is feeding a parallel capacitive DC filter of 4700 μF and a resistive load of 20 Ω. The thyristor bridge converter is feeding a resistive–inductive (*RL*) load of $R = 10\,\Omega$ and $L = 25$ mH at a thyristor firing angle of 30°. Plot the supply voltage and input current, output voltages and currents with time, and the harmonic spectra of supply current and voltage at PCC. For this composite nonlinear load, compute (a) active power consumed, (b) reactive power drawn, (c) DPF, (d) DF, (e) total harmonic distortion of AC source current (THD_I), (f) PF, (g) CF of AC source current, and (h) rms value of AC source current (I_s).

17. Simulate the behavior of a three-phase nonlinear load is fed from a three-phase 415 V, 50 Hz supply system and has a 12-pulse diode bridge converter with a parallel capacitive DC filter of 10 000 μF and a resistive load of 10 Ω. It has a source impedance of 0.15 Ω resistive element and 0.75 Ω inductive element. It consists of an ideal transformer with single primary star connected winding and two secondary windings connected in star and delta with same output line voltages as input voltage at

no load to provide 30° phase shift between two sets of three-phase output equal voltages. Two 6-pulse diode bridges are connected in series to provide a 12-pulse AC–DC converter. Plot the supply voltage and input current, output voltage and current with time, and the harmonic spectra of supply current and voltage at PCC. Compute (a) the fundamental active power drawn by the load, (b) PF, (c) rms value of supply current, (d) DF, (e) rms value of fundamental supply current, (f) peak current of AC mains, and (g) total harmonic distortion of AC source current (THD_I).

18. A three-phase nonlinear load is fed from a three-phase 440 V, 50 Hz supply system and has a 12-pulse thyristor bridge converter with a series connected inductive load of 50 mH and an equivalent resistive load of 5 Ω. It has a source impedance of 0.2 Ω resistive element and 2.0 Ω inductive element and a thyristor firing angle of 30°. It consists of an ideal transformer with single primary star connected winding and two secondary windings connected in star and delta with same line voltages as input supply voltage to provide 30° phase shift between two sets of three-phase output voltages. Two 6-pulse thyristor bridges are connected in series to provide a 12-pulse AC–DC converter. Plot the supply voltage and input current, output voltage and current with time, and the harmonic spectra of supply current and voltage at PCC. Compute (a) fundamental active power drawn by the load, (b) fundamental reactive power drawn by the load, (c) PF, (d) rms value of supply current, (e) DF, (f) rms value of fundamental supply current, (g) peak current of AC mains, and (h) total harmonic distortion of AC source current (THD_I).

19. Simulate the behavior of a three-phase to single-phase cycloconverter (three-pulse type) that is to supply a resistive–inductive (*RL*) series load circuit and sinusoidal modulation of the delay angle α is to be employed. The required output frequency is 5 Hz and the inductive reactance of the load circuit at this frequency is 0.90 Ω. The load circuit resistance is 1.2 Ω. A three-phase 230 V, 60 Hz (line–line rms) supply system is to be used directly without a transformer. The ideal output voltage wave is defined as $V_0 = 156 \sin(10\,\pi t)$ V. Plot the supply voltage and input current, output voltage and current with time, and the harmonic spectra of supply current and voltage at PCC. Compute THD, CF, rms value of AC mains current, thyristor conduction angle, peak thyristor voltage, rms value of thyristor current, rms value of output voltage, DPF, DF, input power factor, and input active, reactive, and output powers.

20. Simulate the behavior of a three-phase to single-phase cycloconverter (dual converter type) that is to supply a resistive–inductive (*RL*) series load circuit and sinusoidal modulation of the delay angle α is to be employed. The required output frequency is 10 Hz and the inductive reactance of the load circuit at this frequency is 1.2 Ω. The load circuit resistance is 1.8 Ω. A three-phase 220 V, 60 Hz (line–line rms) supply system is to be used directly without a transformer. The ideal output voltage wave is defined as $V_0 = 156 \sin(20\,\pi t)$ V. Plot the supply voltage and input current, output voltage and current with time, and the harmonic spectra of supply current and voltage at PCC. Compute THD, CF, rms value of AC mains current, thyristor conduction angle, peak thyristor voltage, rms value of thyristor current, rms value of output voltage, DPF, DF, input power factor, and input active, reactive, and output powers.

21. Simulate the behavior of a three-phase to three-phase cycloconverter (three-pulse type) that is to supply a resistive–inductive (*RL*) series load circuit and sinusoidal modulation of the delay angle α is to be employed. The required output frequency is 15 Hz and the inductive reactance of the load circuit at this frequency is 1.90 Ω. The load circuit resistance is 5.2 Ω. A three-phase 208 V, 60 Hz (line–line rms) supply system is to be used directly without a transformer. The ideal output voltage wave is defined as $V_0 = 156 \sin(30\,\pi t)$ V. Plot the supply voltage and input current, output voltage and current with time, and the harmonic spectra of supply current and voltage at PCC. Compute THD, CF, rms value of AC mains current, thyristor conduction angle, peak thyristor voltage, rms value of thyristor current, rms value of output voltage, DPF, DF, input power factor, and input active, reactive, and output powers.

22. Simulate the behavior of a three-phase to three-phase cycloconverter (dual converter type) that is to supply a resistive–inductive (*RL*) series load circuit and sinusoidal modulation of the delay angle α is to be employed. The required output frequency is 30 Hz and the inductive reactance of the load circuit at this frequency is 1.8 Ω. The load circuit resistance is 4.2 Ω. A three-phase 208 V, 60 Hz (line–line rms) supply system is to be used directly without a transformer. The ideal output phase voltage wave is defined as $V_0 = 156 \sin(60\ \pi t)$ V. Plot the supply voltage and input current, output voltage and current with time, and the harmonic spectra of supply current and voltage at PCC. Compute THD, CF, rms value of AC mains current, thyristor conduction angle, peak thyristor voltage, rms value of thyristor current, rms value of output voltage, DPF, DF, input power factor, and input active, reactive, and output powers.

23. Simulate the behavior of a three-phase, 400 V line voltage, 50 Hz four-wire distribution system, with three single-phase loads (connected between phases and neutral) and a single-phase AC voltage controller with an equivalent series connected inductive load of 10 mH and an equivalent resistive load of 5 Ω at a thyristor firing angle of 90°. Plot the supply voltage and input current, supply neutral current, output voltage and current with time, and the harmonic spectra of supply current and voltage at PCC. Compute (a) active power consumed, (b) reactive power drawn, (c) DPF, (d) DF, (e) total harmonic distortion of AC source current (THD_I), (f) PF, (g) CF of AC source current, (h) rms value of AC source current (I_s), and (i) neutral current (I_{sn}).

24. Simulate the behavior of a three-phase, 400 V line voltage, 50 Hz four-wire distribution system, with three single-phase loads (connected between phases and neutral) and a single-phase thyristor bridge converter with an equivalent series connected inductive load of 50 mH and an equivalent resistive load of 25 Ω at a thyristor firing angle of 30°. Plot the supply voltage and input current, supply neutral current, output voltage and current with time, and the harmonic spectra of supply current and voltage at PCC. Compute (a) active power consumed, (b) reactive power drawn, (c) DPF, (d) DF, (e) total harmonic distortion of AC source current (THD_I), (f) PF, (g) CF of AC source current, (h) rms value of AC source current (I_s), and (i) neutral current (I_{sn}).

25. Simulate the behavior of a three-phase, 380 V line voltage, 50 Hz four-wire distribution system, with three single-phase loads (connected between phases and neutral) and a single-phase uncontrolled bridge converter with a parallel capacitive DC filter of 470 μF and an equivalent resistive load of 20 Ω. It has a source impedance of 0.15 Ω resistive element and 1.5 Ω inductive element. Plot the supply voltage and input current, output voltage and current with time, and the harmonic spectra of supply current and voltage at PCC. Compute the supply voltage and input current supply, neutral current, output voltage and current with time, and the harmonic spectra of supply current and voltage at PCC. Compute (a) active power consumed, (b) reactive power drawn, (c) DPF, (d) DF, (e) total harmonic distortion of AC source current (THD_I), (f) PF, (g) CF of AC source current, (h) rms value of AC source current (I_s), and (i) neutral current (I_{sn}).

References

1. Schaefer, J. (1965) *Rectifier Circuits, Theory and Design*, John Wiley & Sons, Inc., New York.
2. Shepherd, W. and Zand, P. (1979) *Energy Flow and Power Factor in Nonsinusoidal Circuits*, Cambridge University Press, London.
3. IEEE Standard 597 (1983) *IEEE Standard Practices and Requirements for General Purpose Thyristor DC Drives*, IEEE.
4. Arrillage, J., Bradley, D.A., and Bodger, P.S. (1985) *Power System Harmonics*, John Wiley & Sons, Inc., Chichester, UK.
5. Seguier, G. (1986) *Power Electronic Converters: AC/DC Conversion*, McGraw-Hill.
6. Shepherd, W. and Hulley, L.N. (1987) *Power Electronics and Motor Control*, Cambridge University Press, London.
7. IEEE Standard 1030 (1987) *IEEE Guide for Specification of High-Voltage Direct Current Systems Part I: Steady State Performance*, IEEE.
8. Griffith, D.C. (1989) *Uninterruptible Power Supplies*, Marcel Dekker, New York.
9. Clark, J.W. (1990) *AC Power Conditioners: Design Applications*, Academic Press.

10. IEC SC 77 A (1990) *Draft-Revision of Publication IEC 555-2: Harmonics, Equipment for Connection to the Public Low Voltage Supply System*, IEC.
11. IEEE Standard 519 (1992) *IEEE Guide for Harmonic Control and Reactive Compensation of Static Power Converters*, IEEE.
12. Lander, C.G. (1993) *Power Electronics*, 3rd edn, McGraw-Hill Inc., New York.
13. Heydt, G.T. (1994) *Electric Power Quality*, 2nd edn, Stars in a Circle, West Lafayette, IN.
14. Barton, T.H. (1994) *Rectifiers, Cycloconverters, and AC Controllers*, Clarendon Press, New York.
15. IEC 61000-3-2 (1995) *Electromagnetic compatibility (EMC) – Part 3: Limits – Section 2: Limits for Harmonic Current Emissions (Equipment Input Current <16 A Per Phase), 1st Edition*, IEC.
16. Vithayathil, J. (1995) *Power Electronics: Principles and Applications*, McGraw-Hill Inc., New York.
17. Mohan, N., Udeland, T., and Robbins, W. (1995) *Power Electronics: Converters, Applications and Design*, 2nd edn, John Wiley & Sons, Inc., New York.
18. Paice, D.A. (1996) *Power Electronic Converter Harmonics: Multi-Pulse Methods for Clean Power*, IEEE Press, New York.
19. Rashid, M.H. (1996) *Power Electronics Circuit Devices and Applications*, PHI Pvt. Ltd, New Delhi.
20. Duagan, R.C., Mcgranaghan, M.F., and Beaty, H.W. (1996) *Electric Power System Quality*, McGraw-Hill, New York.
21. Porter, G.J. and Sciver, J.A.V. (eds) (1999) *Power Quality Solutions: Case Studies for Troubleshooters*, The Fairmount Press Inc., Lilburn, GA.
22. Arrillaga, J., Watson, N.R., and Chen, S. (2000) *Power System Quality Assessment*, John Wiley & Sons, Inc., New York.
23. Wakileh, M.G.J. (2001) *Power Systems Harmonics*, Springer, New York.
24. Agarwal, J.P. (2001) *Power Electronics Systems: Theory and Design*, Prentice-Hall, New Jersey.
25. Bollen, H.J. (2001) *Understanding Power Quality Problems*, 1st edn, Standard Publishers Distributors, Delhi.
26. Schlabbach, J., Blume, D., and Stephanblome, T. (2001) *Voltage Quality in Electrical Power Systems*, Power Engineering and Energy Series, IEE Press.
27. Wakileh, M.G.J. (2001) *Power Systems Harmonics*, Springer, New York.
28. Ghosh, A. and Ledwich, G. (2002) *Power Quality Enhancement Using Custom Power Devices*, Kluwer Academic Publishers, London.
29. Sankaran, C. (2002) *Power Quality*, CRC Press, New York.
30. Das, J.C. (2002) *Power System Analysis: Short-Circuit Load Flow and Harmonics*, Marcel Dekker, New York.
31. IEEE Standard 1573 (2003) *IEEE Guide for Application and Specification of Harmonic Filters*, IEEE.
32. Emadi, A., Nasiri, A., and Bekiarov, S.B. (2005) *Uninterruptible Power Supplies and Active Filters*, CRC Press, New York.
33. Wu, B. (2006) *High-Power Converters and AC Drives*, IEEE Press, Hoboken, NJ.
34. Bollen, M.H.J. and Yu-Hua Gu, I. (2006) *Signal Processing of Power Quality Disturbances*, IEEE Press, Hoboken, New Jersey.
35. Liew, A.C. (1989) Excessive neutral currents in three-phase fluorescent lighting circuits. *IEEE Transactions on Industry Applications*, **25**(4), 776–782.
36. Gruzs, T.M. (1990) A survey of neutral currents in three-phase computer power systems. *IEEE Transactions on Industry Applications*, **26**(4), 719–725.
37. Subjak, J.S., Jr. and McQuilkin, J.S. (1990) Harmonics – causes, effects, measurements, analysis: an update. *IEEE Transactions on Industry Applications*, **26**(6), 1034–1042.
38. Amoli, M.E. and Florence, T. (1990) Voltage and current harmonic control of a utility system: a summary of 1120 test measurements. *IEEE Transactions on Power Delivery*, **5**(3), 1552–1557.
39. Emanuel, A.E., Orr, J.A., Cyganski, D., and Gulchenski, E.M. (1993) A survey of harmonics voltages and currents at the customer's bus. *IEEE Transactions on Power Delivery*, **8**(1), 411–421.
40. Van Wyk, J.D. (1993) Power quality, power electronics and control. Proceedings of the EPE'93, pp. 17–32.
41. Ling, P.J.A. and Eldridge, C.J. (1994) Designing modern electrical systems with transformers that inherently reduce harmonic distortion in a PC-rich environment. Proceedings of the Power Quality Conference, September, pp. 166–178.
42. Packebush, P. and Enjeti, P. (1994) A survey of neutral current harmonics in campus buildings, suggested remedies. Proceedings of the Power Quality Conference, September, pp. 194–205.
43. Mansoor, A., Grady, W.M., Staats, P.T. *et al.* (1994) Predicting the net harmonic currents produced by large numbers of distributed single-phase computer loads. *IEEE Transactions on Power Delivery*, **10**(4), 2001–2006.
44. IEEE Working Group on Nonsinusoidal Situations (1996) A survey of North American electric utility concerns regarding nonsinusoidal waveforms. *IEEE Transactions on Power Delivery*, **11**(1), 73–78.

45. Domijan, A., Jr., Santander, E.E., Gilani, A. *et al.* (1996) Watthour meter accuracy under controlled unbalanced harmonic voltage and current conditions. *IEEE Transactions on Power Delivery*, **11**(1), 64–72.
46. IEEE Working Group on Nonsinusoidal Situations (1996) Practical definitions for powers in systems with nonsinusoidal waveforms and unbalanced loads: a discussion. *IEEE Transactions on Power Delivery*, **11**(1), 79–101.
47. Peng, F.Z. (1998) Application issues of active power filters. *IEEE Industry Applications Magazine*, **4**(5), 21–30.
48. Peng, F.Z. (2001) Harmonic sources and filtering approaches. *IEEE Industry Applications Magazine*, **7**(4), 18–25.
49. Green, T.C. and Marks, J.H. (2003) Ratings of active power filters. *IEEE Proceedings: Electric Power Applications*, **150**(5), 607–614.

8

Passive Power Filters

8.1 Introduction

Power converters using thyristors and other semiconductor switches are widely used to feed controlled electric power to electrical loads such as adjustable speed drives (ASDs), furnaces, and large power supplies. Such solid-state converters are also used in HVDC transmission systems, AC distribution systems, and renewable electrical power generation. As nonlinear loads, the solid-state converters draw harmonics and reactive power components of current from the AC mains. The injected harmonic currents, and reactive power burden, cause low system efficiency and poor power factor. They also result in disturbance to other consumers and protective devices and interference to nearby communication networks. Traditionally, passive power filters (PPFs) are used to reduce harmonics and capacitors are generally employed to improve the power factor of the AC loads. The passive filters are classified into many categories such as shunt, series, hybrid, single tuned, double tuned, damped, band-pass, and high-pass. In high power rating such as HVDC systems, they are very much in use even nowadays due to simplicity, low cost, robust structure, and benefits of meeting reactive power requirements in most of the applications at fundamental frequency. Moreover, they are also extensively used in hybrid configurations of power filters, where the major portion of filtering is taken care by passive filters.

In medium and low power ratings, especially in distribution systems, the passive filters are used again because of their low cost and simplicity. However, the requirements of passive filters in the distribution systems are much different from those in high power rating applications of transmission and other applications. In many situations, the requirement of reactive power at fundamental frequency is quite low and the design of passive filters becomes very challenging to reduce rms current of the supply where dominance is of the harmonic currents. Typical examples are ASDs and power supplies, among others, consisting of diode rectifiers at the front end of equipment with a capacitive filter at the DC bus of these converters. These applications do not need any amount of reactive power due to the presence of a diode rectifier, but harmonic currents are produced by them in excess. Because of the low value of power capacitors in passive filters, these filters become very sensitive to the parallel resonance between filter capacitors and source impedance (mainly inductive in nature). If the parallel resonance frequency occurs at or near a harmonic produced by the load, a severe voltage distortion and a harmonic current amplification may be produced. It may result in nuisance fuse blowing and/or breaker operation. Therefore, utmost care must be taken in the design of passive filters to avoid such parallel resonance and associated problems. However, if the passive filters are used along with a small active filter that blocks or avoids such parallel resonance, then the objective is confined to reducing rms current of the supply to fully utilize the capabilities of passive filters irrespective of such problems that are taken care by other means. In view of these increasing applications of passive filters, the design and selection of these filters are becoming interesting and challenging. Because of these reasons, in recent years, many texts,

Power Quality Problems and Mitigation Techniques, First Edition.
Bhim Singh, Ambrish Chandra and Kamal Al-Haddad.

standards, and publications have also appeared on the passive power filters. Therefore, it is considered very relevant to present the basic concepts of the design and applications of the passive power filters.

In view of these facts, this chapter deals with an exhaustive analysis and design of PPFs. The PPFs are classified on the basis of (a) topology (e.g., tuned and damped), (b) connection (e.g., series and parallel), and (c) supply system (e.g., single-phase two-wire, three-phase three-wire, and three-phase four-wire). Starting with an introduction, the other sections include state of the art on the PPFs, their classification, principle of operation, analysis and design, modeling and simulation of performance, numerical examples, summary, review questions, numerical problems, computer simulation-based problems, and references.

8.2 State of the Art on Passive Power Filters

The PPF technology is a mature technology for providing compensation for harmonic currents and reactive power in AC networks. It has evolved in the past half century with development in terms of varying configurations. Passive filters are also used to eliminate voltage harmonics, to regulate the terminal voltage, to suppress voltage flicker, and to improve voltage balance in three-phase systems. These objectives are achieved either individually or in combination depending upon the requirements and configuration that needs to be selected appropriately. This section describes the history of development and the current status of the PPF technology.

Because of the widespread use of solid-state control of AC power, the power quality issues have become significant. Therefore, the applications of the passive filters have also increased manifold. In view of these requirements, the passive filters are classified based on (i) topology (e.g., tuned and damped), (ii) connection (e.g., series and parallel/shunt), and (iii) supply system (e.g., single-phase two-wire, three-phase three-wire, and three-phase four-wire) to meet the requirements of various types of nonlinear loads on supply systems. Single-phase loads such as domestic lights, ovens, television sets, computer power supplies, air conditioners, laser printers, and Xerox machines behave as nonlinear loads and cause power quality problems. Single-phase two-wire passive filters are investigated in varying configurations to meet the requirements of single-phase nonlinear loads. Many configurations of PPFs such as passive series filters, passive shunt filters, and a combination of both have been developed and commercialized to meet varying requirements of nonlinear loads.

Major amount of AC power is consumed by three-phase loads such as ASDs with solid-state controllers both with current-fed (line commutated inverter-fed synchronous motor drives, DC motor drives, current source inverter-fed AC motor drives) and with voltage-fed (diode bridge rectifier with capacitive filter in voltage source inverter-fed AC motor drives, power supplies) configurations of converters. Therefore, three-phase three-wire passive filters are used to reduce harmonics and to meet the reactive power requirements of such loads.

In distribution systems, the four-wire configuration of the supply system is very important for balancing the AC network, for taking advantages of three-phase supply systems, and for meeting the requirements of distributed single-phase loads. In such conditions, additional problems not of load balancing but of neutral current are also observed, which have to be taken care by proper design of passive filters.

In majority of the cases, shunt passive filters have been considered more appropriate to mitigate the harmonic currents and partially to meet reactive power requirement of these loads and to relieve the AC network from this problem, especially current-fed types of nonlinear loads (thyristor converters with constant current DC load). However, in voltage-fed types of loads (diode rectifiers with a DC capacitive filter), passive series filters are considered better for blocking of harmonic currents. There are many such situations that need passive power filters but with varying configurations; therefore, an exhaustive study of the passive power filters is considered very much relevant and presented here.

8.3 Classification of Passive Filters

Passive filters can be classified based on the topology, connection, and the number of phases. Figures 8.1 and 8.2 show the classification of the passive power filters based on the topology and the number of phases, respectively. The topology can be shunt, series, and hybrid and further subclassified as tuned and damped to act as low-pass and high-pass for shunt filters or to act as low-block and high-block for series filters. The PPFs may be connected in shunt, series, or a combination of both for compensating different types of nonlinear loads as shown in Figure 8.1. Other major classification is based on the number of phases such as

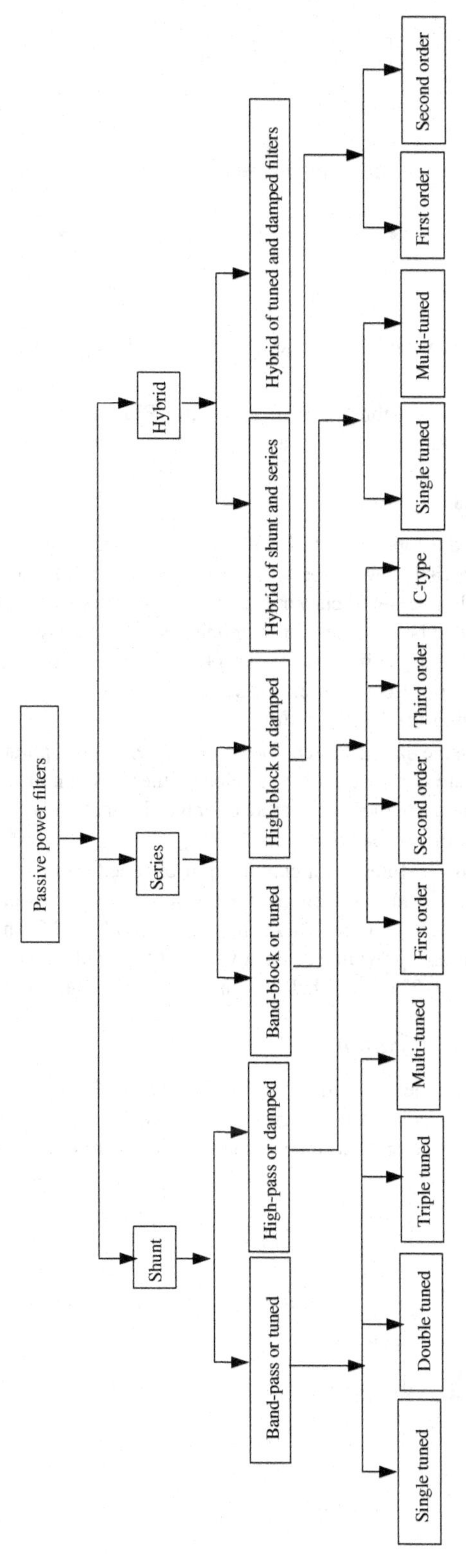

Figure 8.1 Topology-based classification of passive power filters

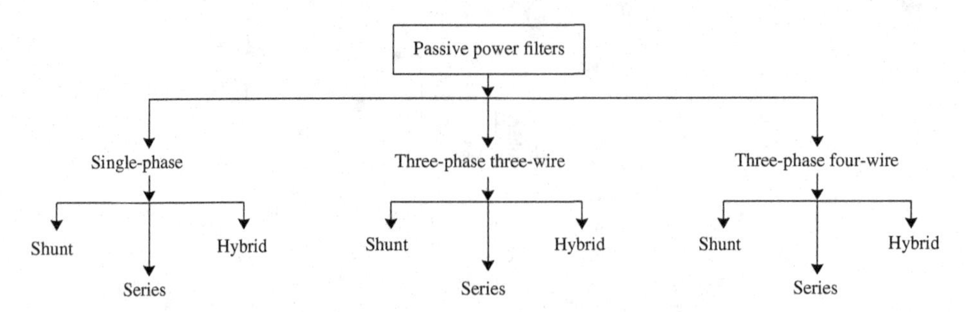

Figure 8.2 Supply-based classification of passive power filters

single-phase (two-wire) and three-phase (three-wire or four-wire) PPFs with these supply systems as shown in Figure 8.2. Various configurations of the passive filters are shown in Figures 8.3–8.18.

8.3.1 Topology-Based Classification

PPFs can be classified based on the topology used, for example, tuned filters, damped filters, or a combination of both. Figures 8.3 and 8.4 show the passive tuned filters for shunt and series configurations that are most widely used for the elimination of current harmonics and for reactive power compensation. These are mainly used at the load end because current harmonics are injected by nonlinear loads. These inject equal compensating currents, opposite in phase, to cancel harmonics and/or reactive components of the nonlinear load current at the point of connection. These can also provide the reactive power in the power system network for improving the voltage profile.

Figures 8.5 and 8.6 show the passive damped filters for shunt and series configurations for eliminating all higher order harmonics. These are connected before the load either in shunt or in series with the AC mains depending upon the requirements of the nonlinear load for the elimination of current harmonics and for regulating the terminal voltage of the load.

Figure 8.7 shows the hybrid passive filters as a combination of tuned and damped filters. Another classification of hybrid passive filters includes a combination of shunt and series filters. These are used in single-phase as well three-phase configurations. These are considered ideal PPFs that eliminate voltage and current harmonics and are capable of providing clean power to critical and harmonic-prone loads such as computers and medical equipment. These can balance and regulate terminal voltages.

8.3.2 Connection-Based Classification

PPFs can also be classified based on the connection used, for example, shunt filters, series filters, or a combination of both. The combinations of passive series and passive shunt filters are known as hybrid filters. These are mainly used at the load end because current harmonics are injected by nonlinear loads.

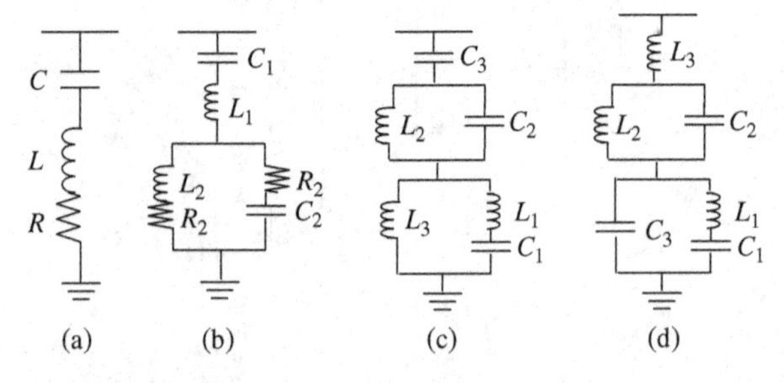

Figure 8.3 Shunt passive tuned or band-pass filters: (a) single tuned; (b) double tuned; (c) triple tuned with a series capacitor; (d) triple tuned with a series inductor

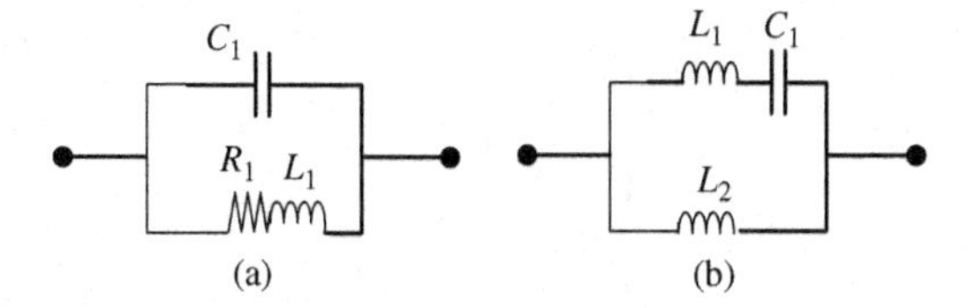

Figure 8.4 Series passive tuned or band-block filters: (a) single tuned; (b) double tuned

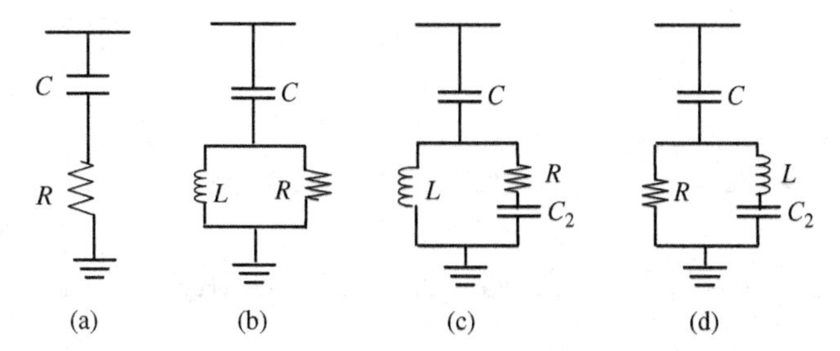

Figure 8.5 Shunt passive damped or high-pass filters: (a) first order; (b) second order; (c) third order; (d) C-type

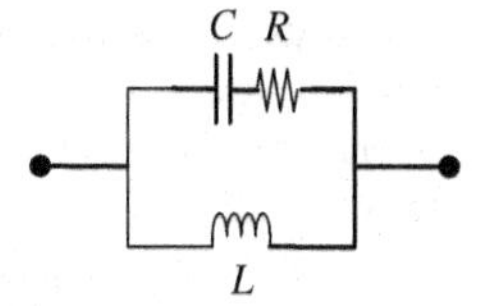

Figure 8.6 A series passive damped or high-block filter

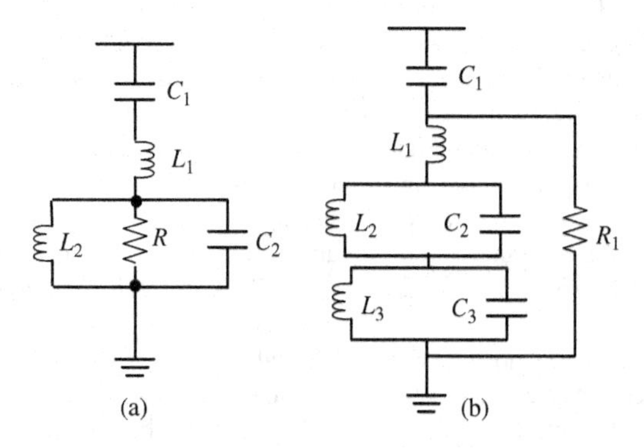

Figure 8.7 Hybrid passive filters: (a) damped double tuned; (b) damped triple tuned

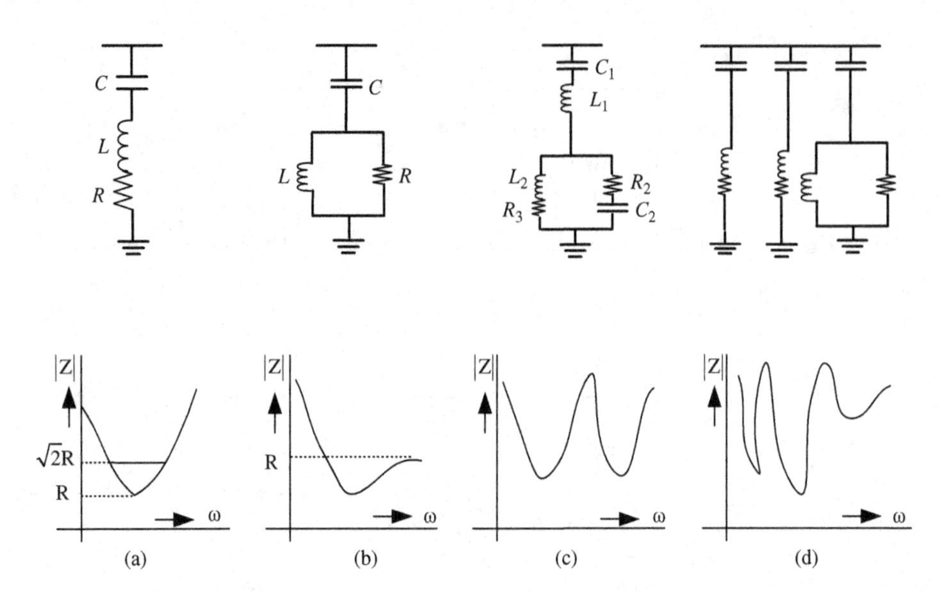

Figure 8.8 Common types of passive shunt filters with impedance–frequency plots: (a) band-pass; (b) high-pass; (c) double band-pass; (d) composite

8.3.2.1 Shunt Filters

Passive shunt filters are connected in parallel to harmonic-producing loads to provide low-impedance paths for harmonic currents so that these harmonic currents do not enter supply systems and are confined to flow in the local passive circuits preferably consisting of lossless passive elements such as inductors (L) and capacitors (C) to reduce losses in the filter system. Practically, capacitors may have very low internal power losses; however, inductors have reasonable resistance and other losses (core loss if the core is made of a ferromagnetic material). Therefore, losses in the inductors cannot be neglected and are considered as an equivalent resistance connected in series with the inductors. It is also represented in terms of quality factor of the inductor. There are various types of passive shunt filters as shown in Figure 8.8. It can be a notch filter sharply tuned at one particular frequency, which is also known as a single tuned filter. It is a simple series RLC circuit, in which R is the resistance of the inductor as shown in Figure 8.8a. The value of the capacitor, also known as the size of the filter, is decided by the reactive power requirements of the loads and its inductor value is decided by the tuned frequency. Therefore, these types of tuned or notch filters provide harmonic current and voltage reduction and power factor correction because of capacitive reactive power at fundamental frequency as this filter circuit behaves as capacitive impedance at fundamental frequency. The resistance of the reactor (inductor) decides the sharpness of tuning and is responsible for limiting the harmonic current to flow in the passive filter. Normally, the notch filters are used at more than one tuned frequency and may have more than one series RLC circuit for multiple harmonics. Sometimes two tuned filters are combined in one circuit. It is known as a double tuned or double band-pass filter, as shown in Figure 8.8c, having minimum impedance at both the tuned frequencies. The main use of the double tuned filter is in high-voltage applications because of reduction in the number of inductors to be subjected to full line impulse voltages. More than two tuned filters (triple and quadruple) can also be combined in one circuit, but no specific advantage is achieved and there is difficulty in adjustment. Moreover, more than two tuned filters are rarely used in practice and only in a few applications.

Other types of passive filters, shown in Figure 8.8b and d, are known as high-pass filters that absorb all higher order harmonics. They are also known as damped filters as they provide damping due to the presence of a resistor in the circuit. These filters have higher losses, but fortunately at high frequencies not much higher currents and power losses are present in the loads. These can be first-order simple series RC

circuits. These help to improve the voltage profile at the point of common coupling (PCC) even for very high frequencies. Normally, a second-order high-pass filter is used as it also reduces the harmonic components in the system. It consists of an external resistor in parallel to the inductor and a capacitor connected in series with the *RL* circuit as shown in Figure 8.8b. Third-order high-pass filters are also used sometimes, as shown in Figure 8.8d, for reducing the losses and for better filtering characteristics.

There are many other types of passive shunt filters, such as automatically tuned filters (by automatically varying capacitor or inductor values), C-type filters, and composite filters used as a combination of these filters.

However, the passive filters have problems of resonance with the source impedance, fixed compensation, and poor power factor at light loads due to excessive leading reactive power injection.

8.3.2.2 Series Filters

Passive series filters are connected in series with harmonic-producing loads to provide high impedance for blocking harmonic currents so that these harmonic currents do not enter supply systems and are confined to flow in the local passive circuits preferably consisting of parallel connected lossless passive elements such as inductors (*L*) and capacitors (*C*) to reduce losses in the filter system. The passive series filter is a simple parallel *LC* circuit, as shown in Figures 8.4 and 8.6. At fundamental frequency, the filter is designed to offer very low impedance, thereby allowing the fundamental current with negligible voltage drop and losses. Series filters are used to block single harmonic current such as third harmonic current. These are used in small power ratings in single-phase systems to block dominant third harmonic current. For blocking multiple harmonic currents, multiple harmonic filters need to be connected in series, as shown in Figures 8.4 and 8.6. These may also have a high-block filter with a parallel *LC* circuit and a resistance in series with the capacitor. Such a configuration of multiple series connected filters has significant series voltage drop and losses at fundamental frequency. In addition, these filters must be designed to carry full rated load current with overcurrent protection. Moreover, at fundamental frequency, these consume lagging reactive power resulting in further voltage drop. Hence, a shunt filter is much cheaper than a series filter for equal effectiveness. Therefore, series filters are much less in use compared with passive shunt filters. However, single-phase series filters at single tuned frequency to block third harmonic current are quite popular in small power rating voltage-fed nonlinear loads.

8.3.2.3 Hybrid Filters

Hybrid filters, consisting of series and shunt passive filters as shown in Figures 8.15–8.18, can be used in many industrial applications. As mentioned earlier, both passive shunt and passive series filters have some drawbacks if they are used individually. However, a passive hybrid filter consisting of a single tuned passive series filter with a single tuned passive shunt filter and a high-pass passive shunt filter offers very good filtering characteristics. A single tuned passive series filter is able to block resonance between the supply and the passive shunt filter and absorbs excess reactive power of the passive shunt filter at light load conditions. This type of hybrid passive filter offers very good filtering characteristics under varying loads. Similarly, other types of passive hybrid filters such as low-pass broadband filters are considered a good option, which consist of leakage reactance of a series transformer for stepping down the voltage for the load and then a capacitor at the load offering good filtering characteristics with a low cutoff frequency and preventing harmonics from penetrating into the high-voltage side above this cutoff frequency.

Moreover, a combination of a passive shunt filter and a small series active filter is considered a low-cost ideal hybrid filter for many large power rating applications. However, such hybrid filters are considered in other chapters.

8.3.3 Supply System-Based Classification

This classification of the PPFs is based on the supply and/or the load system, for example, single-phase (two-wire) and three-phase (three-wire or four-wire) systems. There are many nonlinear loads such as domestic appliances connected to single-phase supply systems. Some three-phase nonlinear loads are

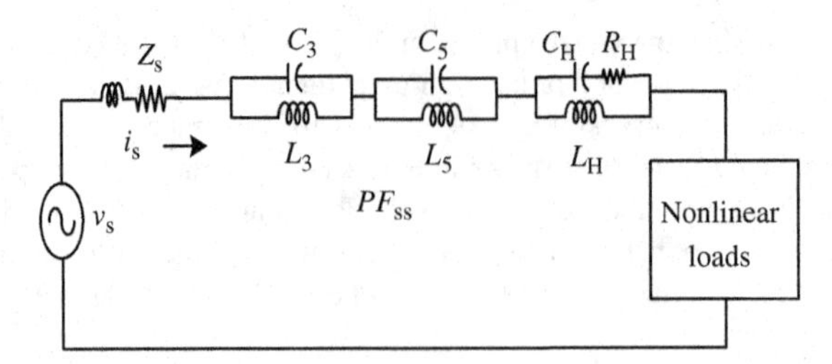

Figure 8.9 A single-phase passive series filter

without neutral, such as ASDs fed from three-wire supply systems. There are many nonlinear single-phase loads distributed on three-phase four-wire supply systems, such as computers and commercial lighting. Hence, PPFs may also be classified accordingly as two-wire, three-wire, and four-wire PPFs.

8.3.3.1 Two-Wire PPFs

Two-wire (single-phase) PPFs are used in all three modes, for example, series, shunt, and a combination of both. Figures 8.9, 8.10, and 8.15–8.18 show the configurations of series, shunt, and hybrid passive filters.

8.3.3.2 Three-Wire PPFs

Three-phase three-wire nonlinear loads such as ASDs are one of the major applications of solid-state power converters and lately many ASDs incorporate passive filters in their front-end design. A large number of publications have appeared on three-wire PPFs with different configurations. All the configurations shown in Figures 8.11 and 8.12 are developed, in three-wire PPFs, with three wires on the AC side and rectifier type nonlinear load.

8.3.3.3 Four-Wire PPFs

A large number of single-phase loads may be supplied from the three-phase AC mains with a neutral conductor. They cause excessive neutral current, harmonic and reactive power burden, and unbalance. To reduce these problems, four-wire PPFs have been developed. Figures 8.13 and 8.14 show typical configurations of series and shunt PPFs. Detailed comparisons of the features of the passive filters are provided for different types of nonlinear loads.

8.4 Principle of Operation of Passive Power Filters

The basic principle of operation of passive power filters may be explained through their objectives, locations, connections, quality, sharpness, rating, size, cost, detuning, applications, and other factors. These are illustrated in the following sections.

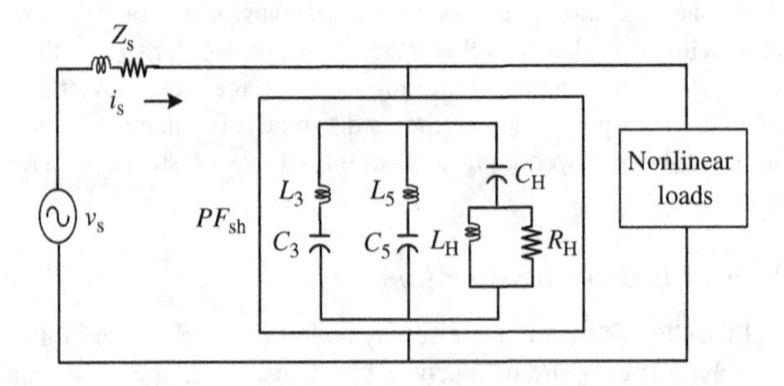

Figure 8.10 A single-phase passive shunt filter

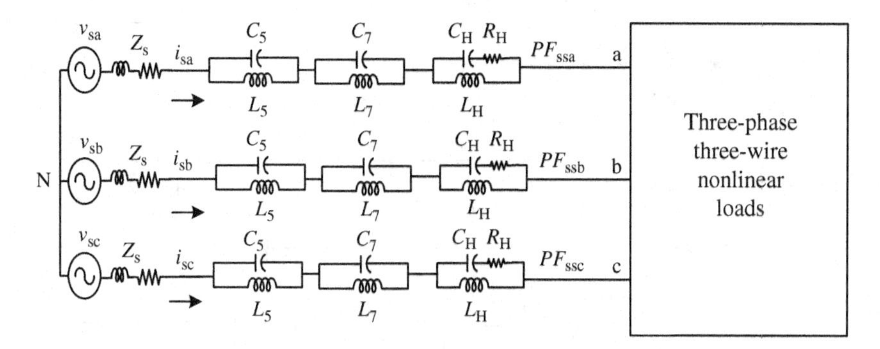

Figure 8.11 A three-phase three-wire passive series filter

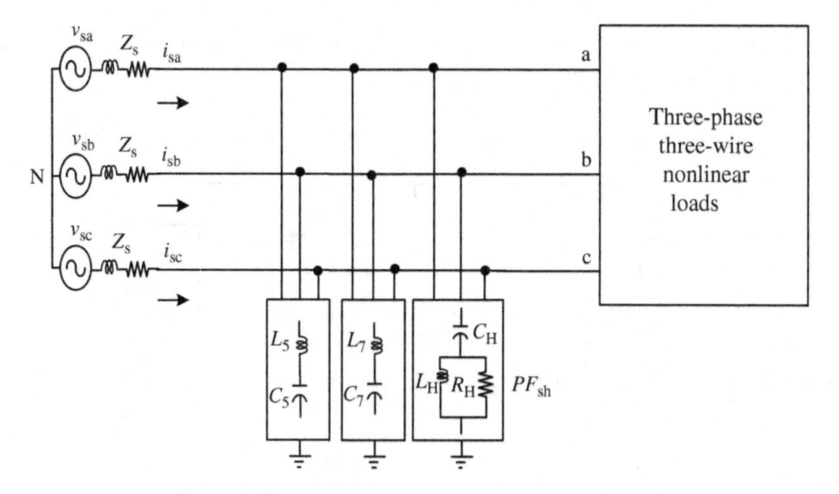

Figure 8.12 A three-phase three-wire passive shunt filter

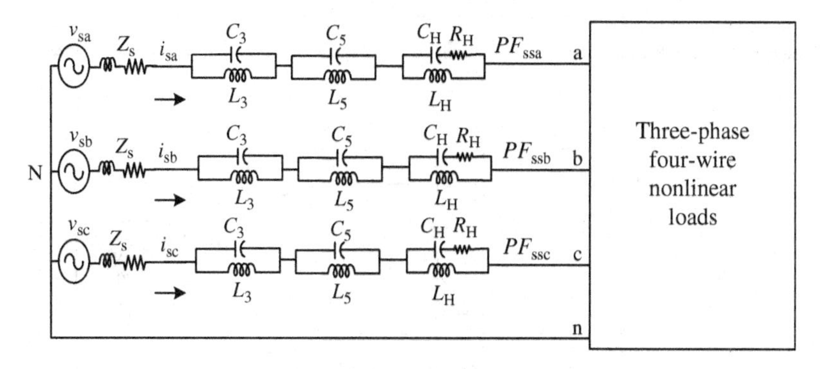

Figure 8.13 A three-phase four-wire passive series filter

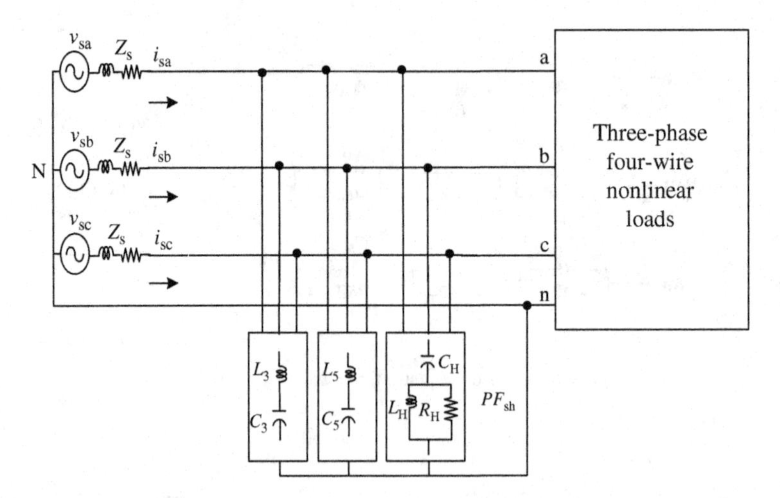

Figure 8.14 A three-phase four-wire passive shunt filter

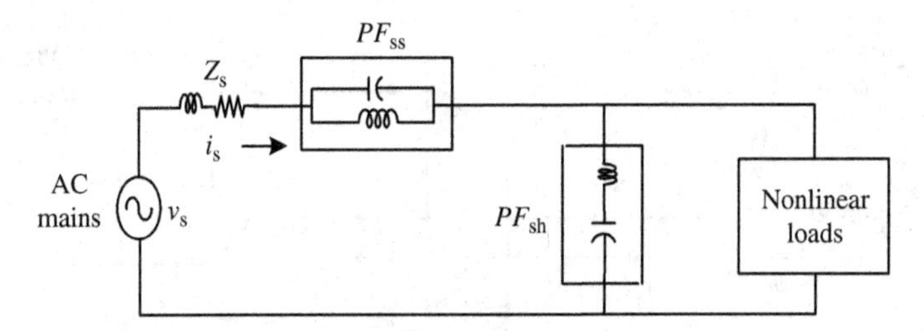

Figure 8.15 A hybrid filter as a combination of passive series (PF_{ss}) and passive shunt (PF_{sh}) filters

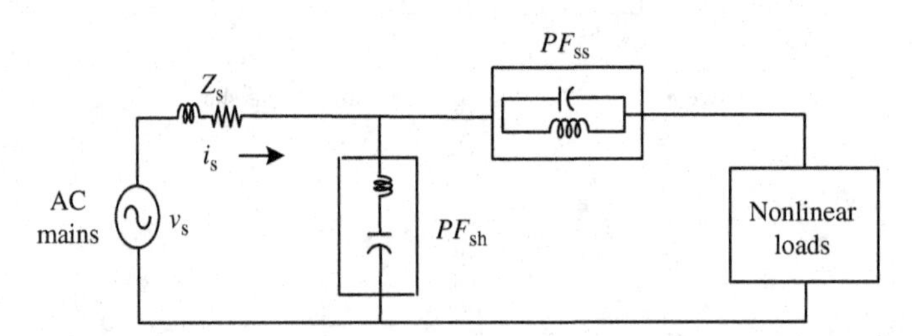

Figure 8.16 A hybrid filter as a combination of passive shunt (PF_{sh}) and passive series (PF_{ss}) filters

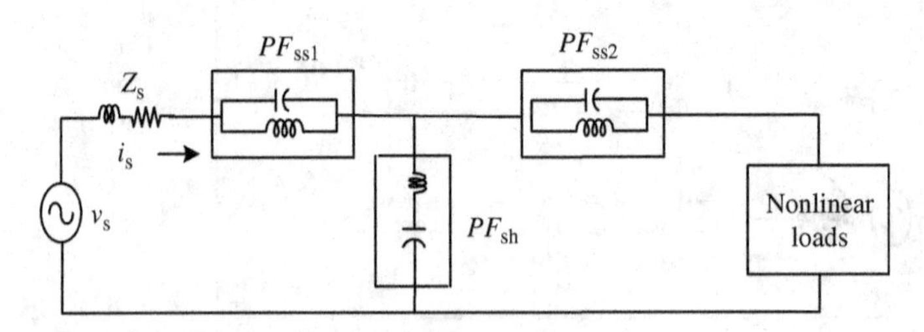

Figure 8.17 A hybrid filter as a combination of passive series (PF_{ss1}), passive shunt (PF_{sh}), and passive series (PF_{ss2}) filters

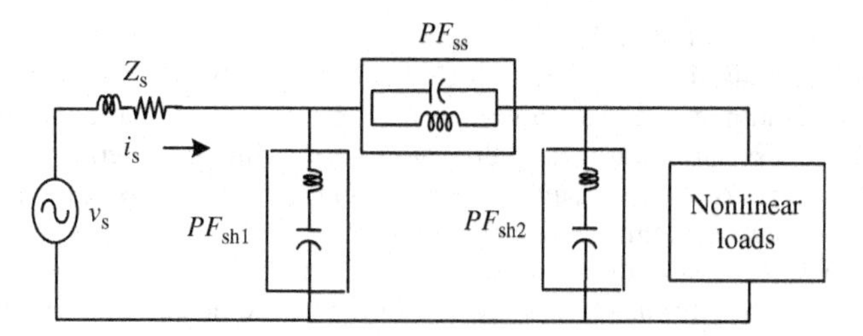

Figure 8.18 A hybrid filter as a combination of passive shunt (PF_{sh1}), passive series (PF_{ss}), and passive shunt (PF_{sh2}) filters

8.4.1 Objectives

The main objective of passive filters is to reduce harmonic voltages and currents in an AC power system to an acceptable level. The AC passive shunt filters also provide the leading reactive power required in most of the nonlinear loads. The DC harmonic filters are used to reduce only harmonics on the DC bus of the load in the system. The basic operating principle of a passive shunt harmonic filter is to absorb harmonic currents in a low-impedance path realized using a tuned series *LC* circuit as shown in Figure 8.19. Similarly, the basic operating principle of a passive series filter is to block harmonic currents entering the AC network by a passive tuned parallel *LC* circuit offering high impedance for harmonic currents as shown in Figure 8.9. Passive shunt filters are connected in parallel to the load and rated for the system voltage at PCC, whereas passive series filters are connected between the AC line and the load and rated for full load current. Passive filters are the engineering solution for harmonic reduction within an acceptable limit and not the elimination of harmonics. In case of passive shunt filters, harmonic voltages are required at PCC to flow the complete harmonic currents in passive series *RLC* circuits of the shunt filter. The passive shunt connected circuit absorbs a part of harmonic currents into it and a fraction of harmonic currents still flows in the network. Therefore, it only reduces harmonic currents and does not completely eliminate them.

8.4.2 Location

The passive filters are located very close to the loads either on AC or DC lines. Most of the time the passive filters are connected at PCC where loads are connected; however, sometimes they are connected to the tertiary winding of the transformer designed for this purpose at optimum voltage to reduce the cost and to increase effectiveness because of the properly designed leakage reactance of this tertiary winding. Moreover, the passive filters may be used for high-voltage harmonic-producing loads that require transformers and tertiary winding of the same transformer is designed to use the passive filters.

8.4.3 Connection and Configuration

The passive filters are used in shunt, series, and hybrid configurations. The passive shunt filters, which are connected in shunt or parallel to the harmonic-producing loads or network, consist of

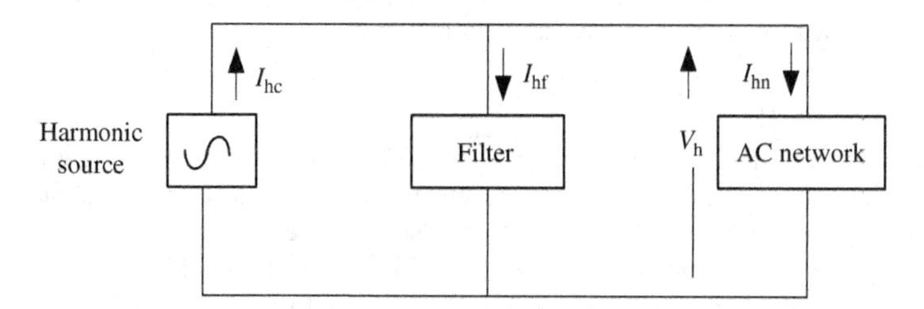

Figure 8.19 A circuit for computation of harmonic currents and voltages on the AC side

series *RLC* circuits tuned at slightly low frequency at which they absorb harmonic currents. For multiple harmonics, multiple tuned series *RLC* circuits are used to absorb the harmonic currents. However, voltage harmonics are also reduced at PCC due to reduced harmonic currents flowing in the AC network and less harmonic voltage drop at PCC resulting in reduction in harmonic voltages. For reducing the number of series circuits of passive elements, a high-pass filter is used in parallel to other branches to absorb all higher order harmonic currents. This high-pass passive filter uses an additional resistor (R_H) parallel to the inductor, which is the offered impedance (R_H) for all higher order harmonics. This resistor has losses because of the harmonic currents. However, higher order harmonic currents are quite low and therefore not much loss is produced in the resistor. Moreover, low-frequency current (fundamental) is mainly confined to the inductor. The passive shunt filters provide leading reactive power at fundamental frequency, which is fortunately an additional requirement of large rating AC–DC converters for which these filters are used. Moreover, size, cost, and weight of the passive shunt filters are decided by AC capacitors that are a part of these filters. Because of these additional advantages, the passive shunt filters are very popular in practice.

The passive series filters, which are connected in series with the AC mains and the harmonic-producing loads or network, consist of parallel *LC* circuits tuned at slightly low frequency at which they have to block harmonic currents. For the mitigation of multiple harmonics, multiple parallel tuned *RLC* circuits are used to block the harmonic currents. For reducing the number of series connected parallel tuned *LC* circuits of passive elements, a high-block filter is used in series with other branches to block all higher order harmonic currents. This high-block passive filter uses an additional resistor (R_H) in series with the capacitor, which is the lower offered impedance (R_H) for all higher order harmonics. This resistor has low losses because of blocking of the harmonic currents; however, higher order harmonic currents are quite low and therefore not much loss is produced in the resistor. Moreover, low-frequency current (fundamental) is mainly confined to the inductor. The passive series filters have lagging reactive power at fundamental frequency, which is an undesired feature, and there is inductive voltage drop at fundamental frequency resulting in poor voltage regulation across the loads. In addition, these passive filters have to be rated for full rated current and their protection becomes an additional requirement. Therefore, the passive series filters are not very popular in practice.

A hybrid passive filter, consisting of one branch of a series passive filter and remaining elements of passive shunt filters, is considered a very popular passive filter in the distribution system. It also avoids the drawbacks of the series filter by using shunt filters and demerits of shunt filters are overcome by a series filter.

8.4.4 Sharpness of Tuning of Passive Filters

This is one of the important properties of the passive filters. It is quantified in terms of a quality factor of the inductor (Q) and is known as sharpness of tuning of the passive filter. A passive shunt filter is considered to be tuned corresponding to a frequency at which its inductive reactance is equal to its capacitive reactance. Passive shunt filters tuned at lower frequencies are sharply tuned and have a high value of quality factor, typically 10–100 and preferably between 30 and 60. Other types of passive shunt filters, known as damped filters and high-pass filters, are tuned for high frequencies and have a low value of quality factor, typically 0.5–5 and preferably between 1 and 2. The high-pass passive filters offer low impedance over a broad band of frequency bandwidth. The admittance and impedance characteristics of these filters along with their circuits are shown in Figure 8.8.

8.4.5 Cost of Passive Filters

The cost of the passive filter is reasonable and sometimes it reaches 15–20% of the equipment for which it is used; therefore, the cost factor should be taken into account while designing the passive filters. Moreover, it has some power losses, which must also be considered in its design. The cost of the passive filter may also be partially supplemented to the reactive power supplied by it. These filters are sometimes designed based on a minimum cost filter. An overall minimum cost filter is a minimum capital cost filter, which adequately reduces harmonics at least cost with some or part of reactive power. A minimum cost filter has to consider the cost of losses along with capital cost of the filter. The major part of the capital cost (about 60%) is the cost of the power capacitors; therefore, a reasonable cost reduction may be achieved by proper selection of its capacitors. Detailed discussions on the cost of these filters are given in the design of these filters.

8.5 Analysis and Design of Passive Power Filters

The analysis and design of passive power filters are normally considered together. It needs the data and nature of the nonlinear load for which a passive filter is to be designed and then a step-by-step procedure is adopted to design the passive filter. It is an iterative procedure because of several issues and constraints.

The design procedure of a passive filter generally involves the following steps:

- Estimate or record the input current frequency spectrum of the nonlinear load and its displacement power factor.
- Obtain the frequency response of the power distribution equivalent impedance at the PCC where a passive filter is to be connected in the system.
- Select the numbers, types, and tuned frequencies of passive filters (out of the tuned – single, double, triple, etc.; damped – first order, second order, and C-type filters; normally a C-type filter is recommended for low frequency and a high-pass filter for high frequency).
- Appropriately assign the reactive power to be generated by each unit of the passive filter.
- Estimate the parameters of each unit of the passive filter.
- Evaluate the attenuation factor of each unit of the passive filter as a function of the frequency.
- Check the existence of resonance frequencies of each unit of the passive filter.
- If these resonance frequencies of passive filter units are close to current harmonics generated by the nonlinear load, then change the tuned frequency of the filter and accordingly calculate new parameters of the passive filter to avoid the parallel resonance with the supply system.
- Validate the performance of the distribution system with filter scheme connected through simulation, and estimate the harmonic distortion of voltage and current and displacement power factor.
- Iterate this design procedure of the passive filter till satisfactory performance is achieved for the distribution system in terms of total harmonic distortion (THD) of the current and voltage and power factor.

The complete design procedure is illustrated through several solved examples in the latter part of the chapter.

8.5.1 Design of Passive Power Filters

The design of the passive power filters includes the detailed analysis for deriving the design equations for calculating the values of different components used in their circuit configurations. As already discussed in the previous section, there are a large number of topologies of the passive power filters; therefore, the design of such large number of circuit configurations is not practically possible to include here due to space constraints. In view of these facts, the design of a selected topology of a shunt passive power filter is given here through a step-by-step design procedure. The design of other topologies such as series passive power filters may be carried out in a similar manner.

8.5.1.1 Design of Passive Shunt Filters

The passive shunt filter basically consists of a series combination of an inductor and a capacitor tuned to a particular frequency and it acts as a low-impedance path for that harmonic. In a single-phase system, the third and fifth harmonic filters are designed using a series tuned filter and a high-pass filter is designed using a second-order damped filter. In a three-phase system, the fifth and seventh harmonic filters are designed using a series tuned filter and a high-pass filter is designed using a second-order damped filter. Initially, the size of the capacitors is calculated from the reactive power requirement (Q_C) of the load.

The absolute value of capacitance (C_n) is calculated as

$$C_n = \frac{Q_C}{m\omega V_s^2}, \tag{8.1}$$

where m is the number of branches in the passive filter for all the phases, V_s is per-phase fundamental voltage across it, and ω is the fundamental frequency of the supply.

To trap the nth harmonic current, the inductance for the nth-order filter is calculated as

$$L_n = \frac{1}{n^2\omega^2 C_n}. \tag{8.2}$$

The series resistance for the inductor of the nth-order filter is calculated as

$$R_n = \frac{n\omega L_n}{Q_n}, \tag{8.3}$$

where Q_n is the quality factor of the inductor of the nth-order filter, which is normally considered as $10 < Q < 100$.

For the design of a second-order damped filter, the filter parameters C_H, L_H, and R_H are calculated using Equations 8.1–8.3. However, the next dominant harmonic is considered as the next value of n for the high-pass filter and the quality factor (Q_H) of the inductor of the high-pass filter is considered as $0.5 < Q < 5$.

This design procedure is used for the design of shunt passive power filters. Similarly, this procedure may easily be modified for the design of series passive power filters. This design procedure is illustrated for different types of passive power filters in the solved examples.

8.6 Modeling, Simulation, and Performance of Passive Power Filters

Modeling and simulation of passive shunt and series filters are carried out to demonstrate their performance for their effectiveness and presence of various phenomena such as resonance through voltage and current waveforms. After the design of the passive filters, these are connected in the system configuration and waveform analysis is done through simulation to study their effect on the system and to observe their interactions with the system and occurrence of any phenomena such as parallel resonance considering all the practical conditions, which are not considered in the design of the passive filters. Earlier, the simulation study of these filters with the system has been quite cumbersome. However, with various available simulation tools such as MATLAB, PSCAD, EMTP, PSPICE, SABER, PSIM, ETAP, and desilent, the simulation of the performance of these filters has become quite simple and straightforward. Nowadays, for a particular application, after the design of these filters, their performance is studied in simulation before these are implemented in practice. In view of these requirements, a few examples of the simulation studies are presented here to give an exposure to the designers and application engineers.

8.6.1 Modeling, Simulation, and Performance of Passive Series Power Filters

Figure 8.20 shows the schematic circuit of a series passive power filter configuration for the compensation of the voltage-fed load. For the simulation study, the supply voltage is considered as 415 V at 50 Hz. A voltage-fed nonlinear load with load power of 25 kW is considered for the simulation study. Various designed and selected parameters of the passive series filter are given in Table 8.1. This series passive filter is modeled in MATLAB using the SimPowerSystems toolbox for simulation of its performance. The response of the series passive power filter configuration for this load is shown in Figure 8.21 and the

Table 8.1 Parameters of a passive series power filter

C_5	70 μF	C_7	60 μF	R_{HSF}	3 Ω
L_5	5.8 mH	L_7	3.4 mH	C_{HSF}	48 μF
				L_{HSF}	1.8 mH

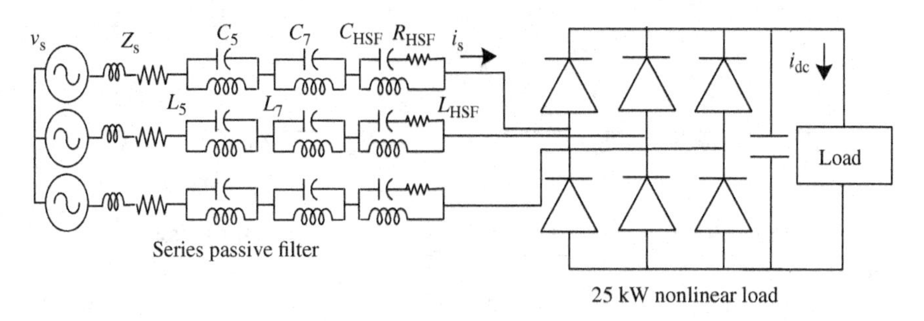

Figure 8.20 A schematic circuit of a three-phase diode rectifier bridge with a passive series power filter

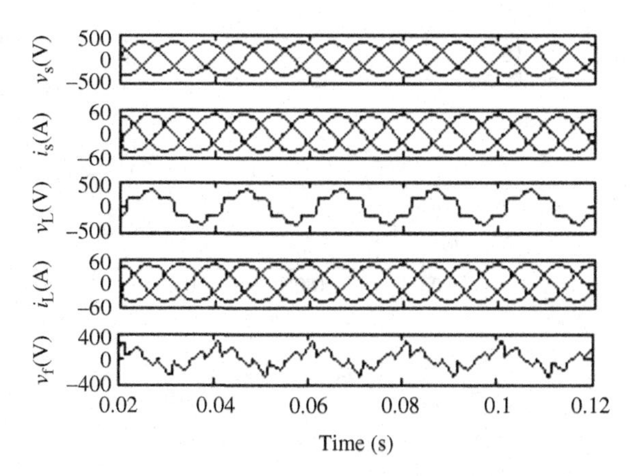

Figure 8.21 Response of a passive power filter configuration for the compensation of a voltage-fed nonlinear load of 25 kW

harmonic spectra for (a) supply current, (b) source voltage, and (c) load voltage are shown in Figure 8.22 with THDs of 4.89, 0.05, and 25.29%, respectively.

8.6.2 Modeling, Simulation, and Performance of Passive Shunt Power Filters for the Compensation of a Six-Pulse Thyristor Converter Load

Figure 8.23 shows the schematic circuit of a shunt passive power filter configuration for the compensation of a six-pulse thyristor converter load. For the simulation study, the supply voltage is considered as 415 V at 50 Hz. A current-fed nonlinear load (three-phase thyristor bridge with $\alpha = 30°$) with load power of 25 kW is considered for the simulation study. The passive shunt filter is designed using the same procedure as described earlier. Various designed and selected parameters of the passive shunt filter are given in Table 8.2. This passive power filter is modeled in MATLAB using the SimPowerSystems toolbox for simulation of its performance. The response of the passive shunt power filter configuration for this load is shown in Figure 8.24 and the harmonic spectra for (a) supply current, (b) PCC voltage, and (c) load current are shown in Figure 8.25 with THDs of 8.55, 3.47, and 27.81%, respectively.

8.6.3 Modeling, Simulation, and Performance of Passive Shunt Power Filters for the Compensation of a 12-Pulse Thyristor Converter Load

Figure 8.26 shows the schematic circuit of a shunt passive power filter configuration for the compensation of a 12-pulse thyristor converter load. For the simulation study, the three-phase supply voltage is considered as 415 V at 50 Hz. A nonlinear load consisting of a 12-pulse thyristor bridge converter with 100 A constant DC current at 30° firing angle of its thyristors is considered for the simulation study. This

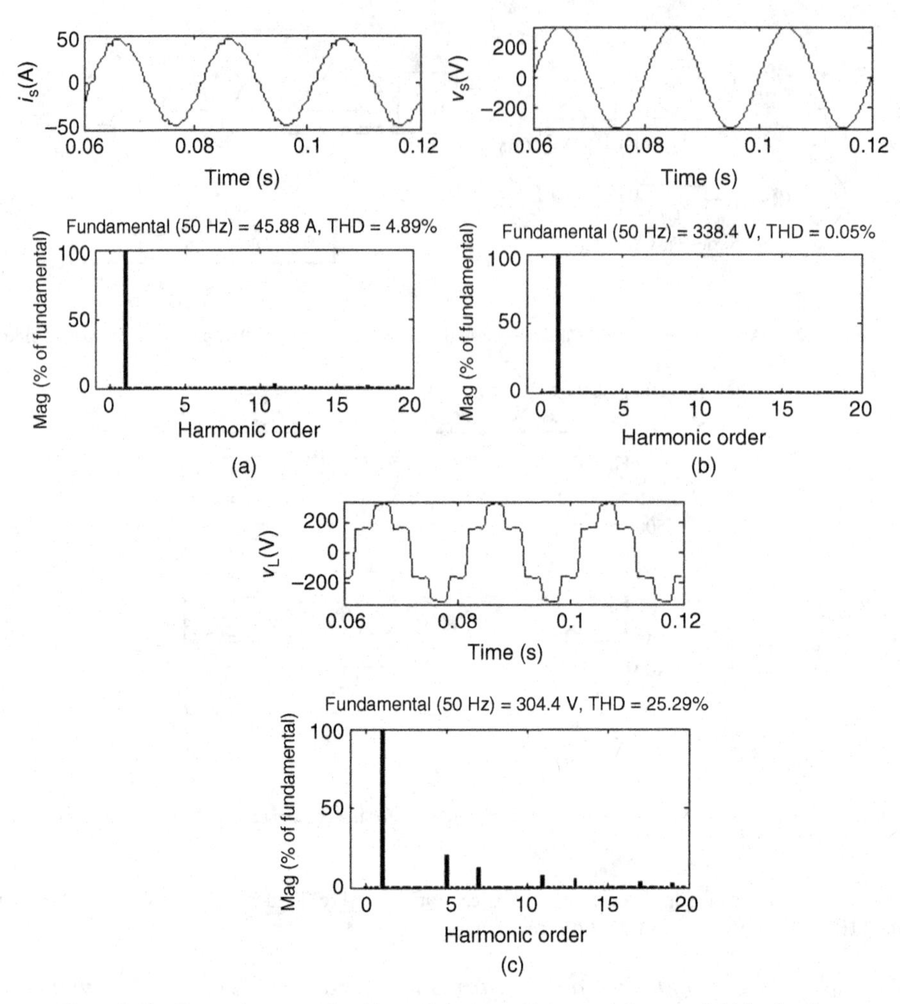

Figure 8.22 Harmonic spectra for (a) supply current, (b) source voltage, and (c) load voltage

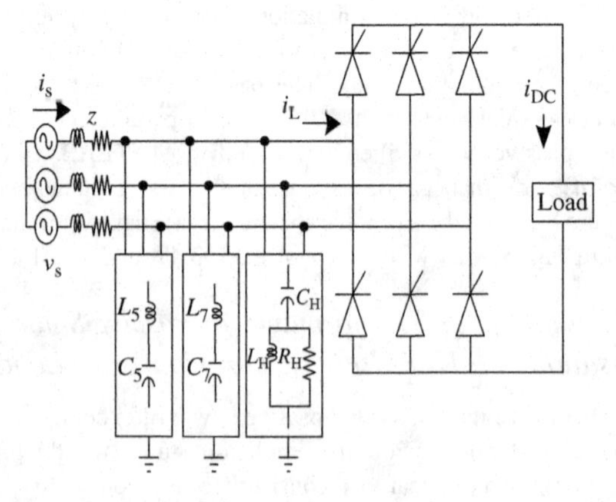

Figure 8.23 A schematic circuit of a three-phase thyristor bridge converter with a passive shunt power filter

Table 8.2 Parameters of a passive shunt power filter

Order n	C (μF)	L (mH)	R (Ω)
5th	89.77	4.4	0.2364
7th	89.77	2.3	0.1688
11th high-pass	89.77	0.932	1.6117

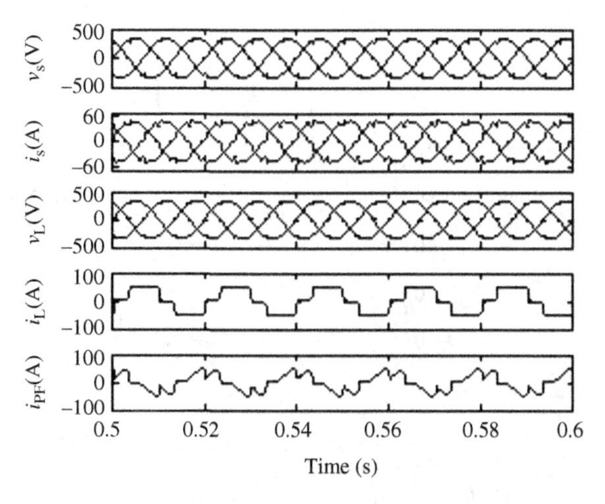

Figure 8.24 Response of a passive power filter configuration for a load of 25 kW

converter consists of an ideal transformer with single primary star connected winding and two secondary windings connected in star and delta with same line voltages to provide 30° phase shift between two sets of three-phase output voltages. Two 6-pulse thyristor bridges are connected in series to form this 12-pulse AC–DC converter with load power of 115 kW for the simulation study. The passive shunt filter is designed using the same procedure as described earlier. Various designed and selected parameters of the passive shunt filter are given in Table 8.3. This passive power shunt filter is modeled in MATLAB using the SimPowerSystems toolbox for simulation of its performance. The response of the passive shunt power filter configuration for this load is shown in Figure 8.27 and the harmonic spectra for (a) supply current, (b) PCC voltage, and (c) load current are shown in Figure 8.28 with THDs of 5.34, 2.91, and 15.23%, respectively.

8.7 Limitations of Passive Filters

Passive filters are extensively used in many applications. However, the passive filters have the following limitations.

- The passive filters are not adaptable to varying system conditions and remain rigid once they are installed in an application. The size and tuned frequency cannot be altered easily.
- The change in operating conditions of the system may result in detuning of the filter and it may cause increased distortion. Such a change may happen undetected provided there is online detection or monitoring in the system.
- The design of the passive filter is reasonably affected by the source impedance. For an effective filter design, its impedance must be less than the source impedance. It may result in large size of the filter in a stiff system with low source impedance, which may result in overcompensation of the reactive power. This overcompensation may cause overvoltage on switching in and undervoltage on switching out the passive filter.

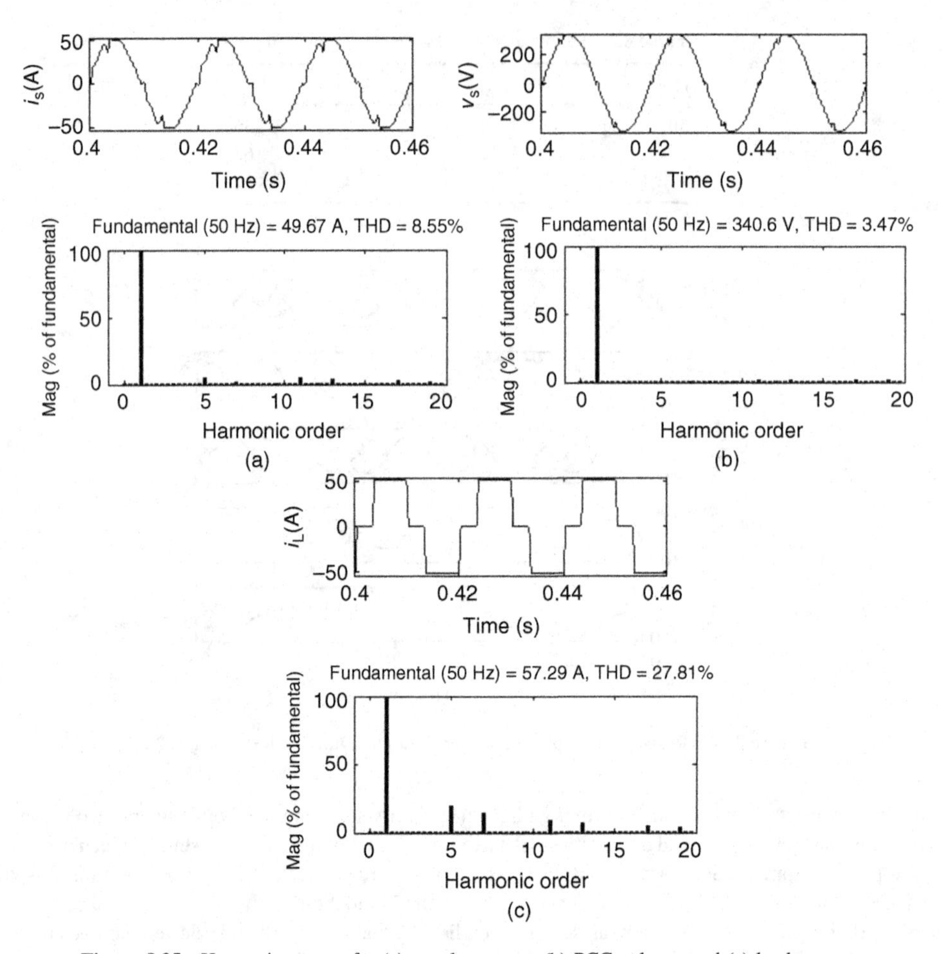

Figure 8.25 Harmonic spectra for (a) supply current, (b) PCC voltage, and (c) load current

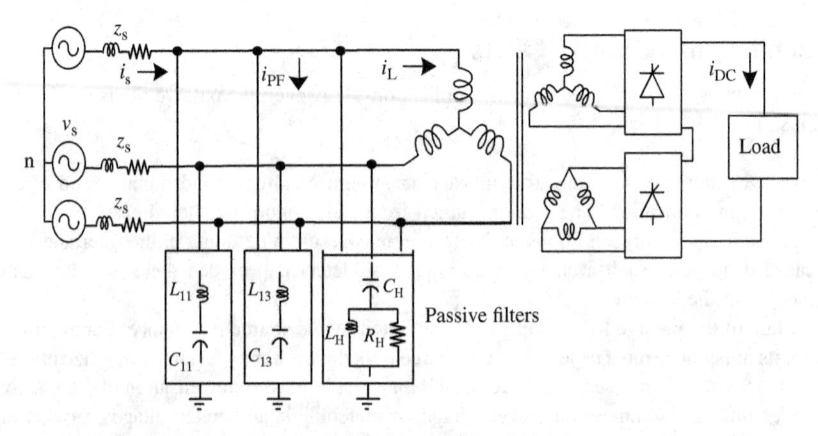

Figure 8.26 A schematic circuit diagram of a 12-pulse thyristor bridge converter with a passive power filter

Table 8.3 Parameters of a passive shunt power filter

Order n	C (μF)	L (mH)	R (Ω)
11th	345.8	0.242	0.0419
13th	345.8	0.1736	0.0355
23rd high-pass	345.8	0.055 47	0.204

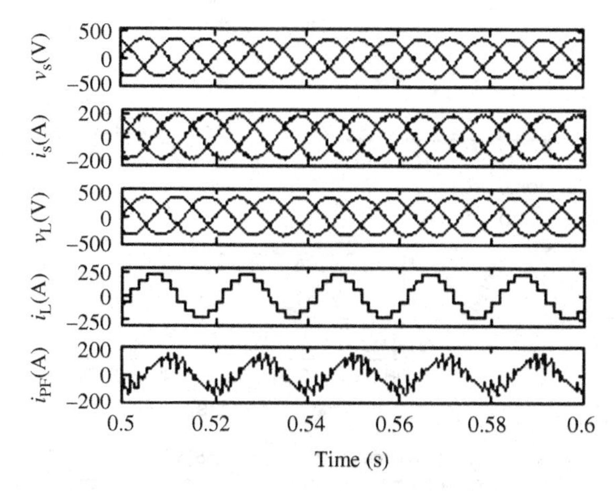

Figure 8.27 Response of a passive power filter configuration for the compensation of a load of 115 kW

- The passive filters are designed with a large number of elements and loss/damage of some of the elements may change its resonance frequencies. This may result in increased distortion in the distribution above the permissible limits.
- In case of large filters, the power losses may be substantial because of resistive elements.
- The parallel resonance due to interaction between the source and the filter can cause amplification of some characteristic and noncharacteristic harmonics. Such problems enforce constraints on the designer in selecting tuned frequency for avoiding such resonances.
- The size of the damped filter becomes large in handling the fundamental and harmonic frequencies.
- The environmental effects such as aging, deterioration, and temperature change and detune the filters in a random manner.
- In some cases, even the presence of a small DC component and harmonic current may cause saturation of the reactors of the filter.
- A special switching is required for switching in and switching out passive filters to avoid the switching transients.
- The grounded neutral of star connected capacitor banks may cause amplification of third harmonic currents in some cases.
- Special protective and monitoring devices are required in passive filters.

These are general limitations; other constraints depend on the application and nature of the loads.

8.8 Parallel Resonance of Passive Filters with the Supply System and Its Mitigation

The parallel resonance with the supply system is a common problem in a passive shunt filter or even in a power factor correction capacitor at PCC where nonlinear harmonic-producing loads are connected. It is

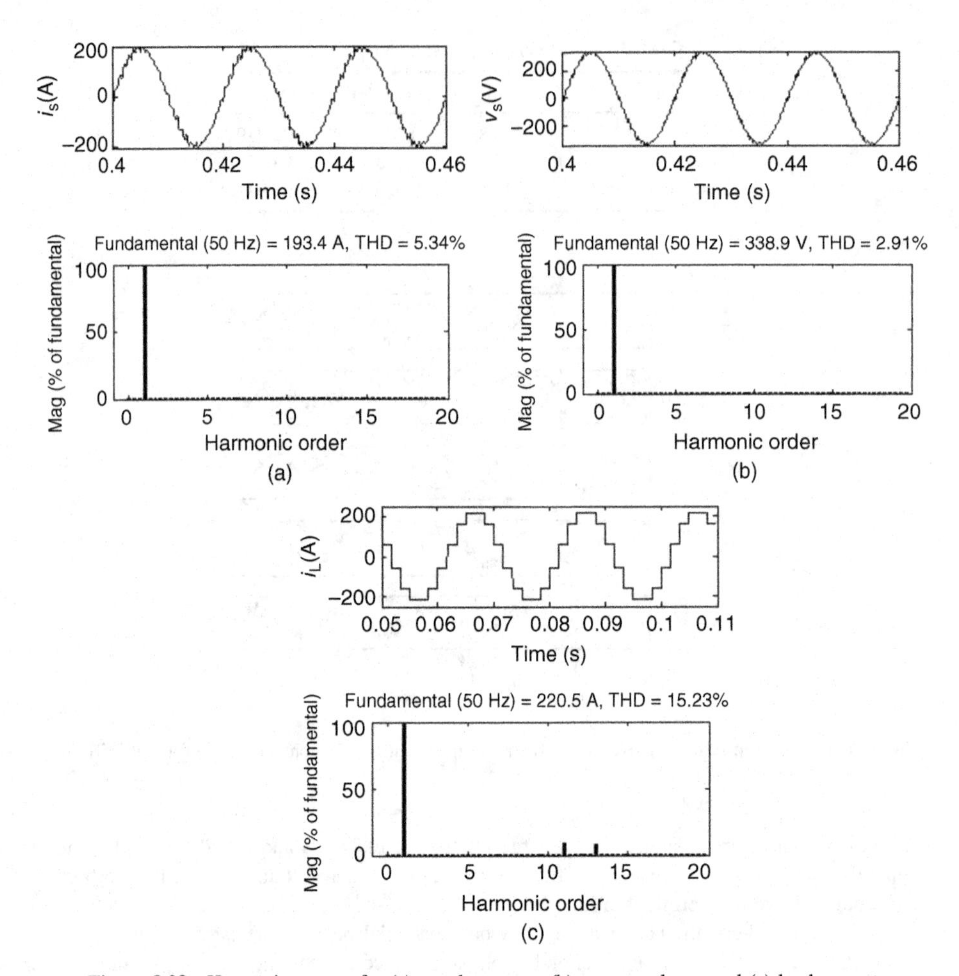

Figure 8.28 Harmonic spectra for (a) supply current, (b) source voltage, and (c) load current

because of unknown source impedance required in the design of passive shunt filters. This phenomenon is illustrated in this section along with its mitigation.

8.8.1 Resonance with the Supply System

Sometimes, when a passive shunt filter or even a power factor correction capacitor is connected at PCC, it may have resonance with the supply impedance if some harmonic-producing loads are connected at the same PCC. This resonance may happen between the inductive reactance of the supply impedance and the capacitive reactance of the passive shunt filter or the capacitor used for power factor correction. As a passive shunt filter is connected in parallel to the supply, a parallel resonance occurs between the inductive reactance of the supply impedance and the capacitive reactance of the passive shunt filter. This parallel combination of the inductive reactance of the supply and the capacitive reactance of the passive filter offers infinite parallel impedance at a frequency (f_r) when these two reactances become equal. This frequency (f_r) is known as a parallel resonance frequency. If any frequency of harmonic-producing loads connected at the same PCC is at or close to this resonance frequency (f_r), a severe voltage distortion and a harmonic current amplification may result in the circuit. This amplification of harmonic currents may be many times and it may be sufficient enough to blow the fuse or cause undesired breaker tripping. A passive shunt filter or a power factor correction capacitor does not produce such harmonic currents, but it

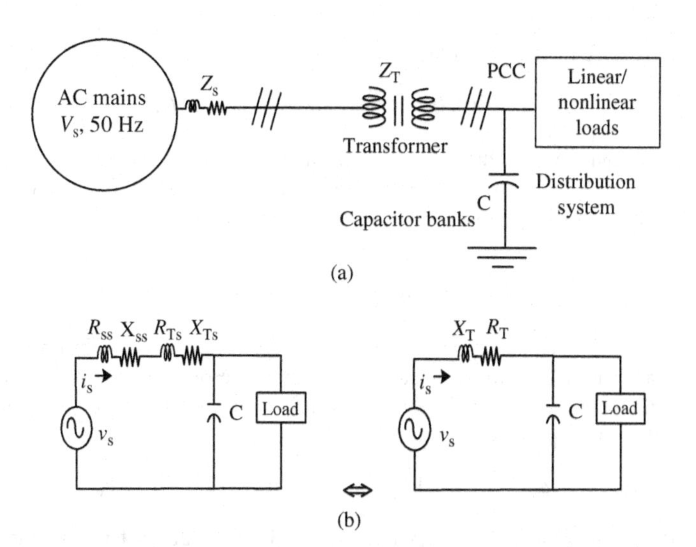

Figure 8.29 (a) Single line diagram of a distribution system and (b) equivalent circuit (impedance diagram)

causes an amplification of the harmonic currents, which is known as a resonance problem. The parallel resonance frequency of any system can be determined by using data of the system, which can be explained by an analytical method as follows.

Figure 8.29a shows a single line diagram of a distribution system and Figure 8.29b shows its equivalent circuit to find the parallel resonance frequency of this system. It has a supply consisting of feeder impedance (R_s, X_s), a step-down transformer (short-circuit test parameters at the LV side are R_{Tr} and X_{Tr}), linear loads, and harmonic-producing loads. A power factor correction capacitor (Q_C kVAR) is also connected at PCC for improving the power factor of the total load.

All these parameters are referred to PCC at which loads are connected to find equivalent circuit parameters shown in Figure 8.29b.

Normally, the feeder impedance is given in terms of short-circuit MVA (MVA_{SC}), *X*/*R* ratio, and voltage level. Similarly, data of the transformer are given in terms of pu or percent, voltages, and kVA rating. The rating of the power factor correction capacitor is also given in terms of kVAR and voltage. These data may be given in other forms as well, which can easily be calculated in the required form.

The values of the feeder resistance and reactance (in ohms) are calculated as

$$R_s = (kV_{LL}^2/MVA_{SC})\cos\{\tan^{-1}(X_s/R_s)\}, \quad X_s = (kV_{LL}^2/MVA_{SC})\sin\{\tan^{-1}(X_s/R_s)\}. \tag{8.4}$$

These data of the feeder (in ohms) are referred to the LV side of the transformer as follows:

$$R_{ss} = R_s/n^2, \quad X_{ss} = X_s/n^2. \tag{8.5}$$

The equivalent circuit parameters of the transformer referred to the secondary winding are calculated as

$$R_{Ts} = R_{Trpu}(1000\,kV^2/kVA_{Tr}), \quad X_{Ts} = X_{Trpu}(1000\,kV^2/kVA_{Tr}), \tag{8.6}$$

where R_{Trpu} and X_{Trpu} are pu resistance and reactance parameters of the transformer, respectively, kVA_{Tr} is the three-phase kVA rating, and kV is the line voltage of the secondary winding of the transformer.

The total equivalent circuit parameters of the system shown in Figure 8.29b are

$$R_T = R_{ss} + R_{Ts}, \quad X_T = X_{ss} + X_{Ts}, \quad L_T = X_T/2\pi f. \tag{8.7}$$

The equivalent reactance of the capacitor bank is

$$X_C = 1000\,\text{kV}_C^2/\text{kVAR}_C, \quad C = 1/2\pi f X_C, \tag{8.8}$$

where kVAR_C, kV_C, and C are the three-phase kVAR rating, line voltage rating, and capacitance value (F) of the power factor correction capacitor, respectively, and f is the supply frequency in Hz.

The total impedance of the equivalent circuit shown in Figure 8.29b at any frequency ω_a can be calculated as

$$Z_i = [(R_T + j\omega_a L_T)(-j/\omega_a C)]/(R_T + j\omega_a L_T - j/\omega_a C). \tag{8.9}$$

The impedance Z_i can be calculated for a wide range of frequency ω_a to locate the resonance frequency at which this impedance has the maximum value. Under parallel resonance, the denominator of impedance Z_i must be the minimum value. It will be the minimum value at a frequency ω_r, when the inductive reactance is equal to the capacitive reactance:

$$\omega_r L_T = 1/\omega_r C, \quad \omega_r = 1/\sqrt{(L_T C)}, \quad f_r = \omega_r/2\pi = 1/[2\pi\sqrt{(L_T C)}]. \tag{8.10}$$

From these relations, a resonance frequency may be calculated for the system for which data are given. Similarly, a capacitance value may be calculated to check the resonance at known frequency or to avoid/mitigate the resonance.

8.8.2 Mitigation of Parallel Resonance with the Supply System

The parallel resonance phenomenon may also occur with a passive shunt tuned filter. Figure 8.30a shows the circuit of a passive shunt tuned filter. Figure 8.30b shows an equivalent circuit of the system using a passive shunt filter.

The total impedance of the equivalent circuit shown in Figure 8.30b at any frequency ω_a can be calculated as

$$Z_i = [(R_T + j\omega_a L_T)(j\omega_a L_f - j/\omega_a C)]/(R_T + j\omega_a L_T + j\omega_a L_f - j/\omega_a C), \tag{8.11}$$

where L_f is the value of the filter inductor.

The impedance Z_i can be calculated for a wide range of frequency ω_a to locate the parallel resonance frequency at which this impedance has the maximum value. Under parallel resonance, the denominator of

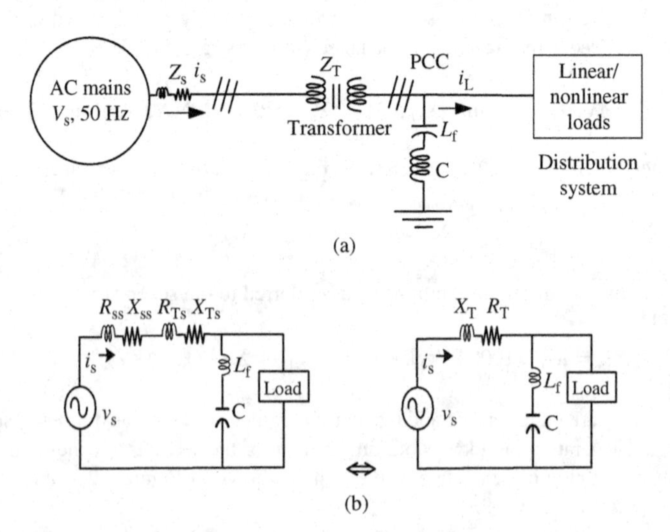

Figure 8.30 (a) Single line diagram of a distribution system with a passive tuned filter and (b) equivalent circuit (impedance diagram)

impedance Z_i must be the minimum value or the imaginary part of its denominator is equal to zero. It will be the minimum value at a frequency ω_r, when the inductive reactance is equal to the capacitive reactance:

$$\omega_r(L_T + L_f) = 1/\omega_r C, \quad \omega_r = 1/\sqrt{\{(L_T + L_f)C\}}, \quad f_r = \omega_r/2\pi = 1/[2\pi\sqrt{\{(L_T + L_f)C\}}]. \tag{8.12}$$

The parallel resonance phenomenon can be avoided by shifting the parallel resonance frequency (f_r) to a value less than the lowest order harmonic current present in the harmonic-producing loads. Therefore, this shifting of parallel resonance frequency (f_r) to much lower frequency than harmonics present in the loads avoids the amplification of harmonic currents and PCC voltage distortion can be minimized in the system. This is also one of the reasons why tuned filters are tuned at a frequency 5–10% lower than the actual harmonic to be reduced in the system.

The voltage distortion and harmonic currents flowing in the supply system can be computed as follows. The currents flowing in the harmonic filter and the supply system at any frequency ω_a are

$$I_{fh} = I_h(R_T + j\omega_a L_T)/(R_T + j\omega_a L_T + j\omega_a L_f - j/\omega_a C),$$
$$I_{sh} = I_h(j\omega_a L_f - j/\omega_a C)/(R_T + j\omega_a L_T + j\omega_a L_f - j/\omega_a C). \tag{8.13}$$

The harmonic voltage at each harmonic present at PCC may be calculated as

$$V_h = I_{sh}(R_T + j\omega_a L_T). \tag{8.14}$$

These equations may be used to calculate all harmonic currents and voltages to find the duty of the harmonic filter and THDs of the PCC voltage and supply current.

However, fundamental currents of the supply and filter can be calculated as

$$I_{s1} = \text{total kVA of the load}/(\sqrt{3} \times \text{nominal PCC voltage}),$$
$$I_{f1} = \text{nominal phase voltage}/(j\omega_s L_f - j/\omega_s C), \tag{8.15}$$

where ω_s is the fundamental frequency of the system.

The filter capacitor and inductor elements have currents and voltages at all harmonics, including fundamental frequency. Because of series circuit and resonance, these elements have overvoltage and increased current compared with their fundamental voltage and current ratings. Therefore, the voltage and current of these elements may be checked for their selection. These harmonic voltages across the capacitor and an inductor may be calculated as

$$V_{Ch} = I_{fh}/\omega_h C, \quad V_{Lh} = I_{fh}\omega_h L_f. \tag{8.16}$$

As the voltage across the filter capacitor is higher by 5–10%, these capacitors are selected of high voltage rating. However, fundamental reactive power rating has to be derated as

$$\text{kVAR}_A = \text{kVAR}_N(f_A/f_N)(V_A/V_N)^2, \tag{8.17}$$

where A represents actual and N represents nominal or rated quantities.

This method of avoiding the phenomenon of parallel resonance with the supply system is used by designing the passive filter, which shifts it at a different frequency from the harmonics of the nonlinear loads.

However, other methods of avoiding the phenomenon of parallel resonance with the supply system are available, including the use of a small active power filter either in series with the AC mains or in series with the shunt passive filter, and the use of a series passive filter in series with the AC mains before a shunt passive filter. These methods of avoiding the phenomenon of parallel resonance with the supply system are demonstrated in Chapter 11.

8.9 Numerical Examples

Example 8.1

A single-phase diode bridge rectifier is supplied from a 230 V, 50 Hz AC mains as shown in Figure E8.1. The load resistance is $R = 100\ \Omega$. (a) Design a capacitive filter so that the ripple factor (RF) of the output voltage is less than 5%. (b) With this value of the capacitor, calculate the average load voltage V_{DC}.

Solution: Given supply rms voltage $V_s = 230$ V, frequency of the supply (f) = 50 Hz, $R = 100\ \Omega$, and RF = 5%.

The amplitude of AC mains voltage is $V_{sm} = 230 \times \sqrt{2} = 325.269$ V.

In the diode bridge rectifier shown in Figure E8.1, the diodes conduct only when instantaneous AC input voltage is higher than the voltage across the DC bus capacitor. As the DC bus capacitor is discharged through the load when diodes are not conducting, the voltage across the DC capacitor varies from the minimum value V_{DCmin} (at the instant when diodes start conduction) to the maximum value V_{DCmax} (when diodes cease conduction). Considering T_c and T_{nc} as conducting and nonconducting periods within the half cycle of the AC mains, respectively, diodes cease conduction at peak voltage of the AC mains (V_{sm}). However, the capacitor discharges exponentially through load resistance (R) and it can be represented mathematically as $(1C_{DC})\int i_{DC}\,dt + v_{DC}(t=0) + Ri_{DC} = 0$.

It has an initial condition of $v_{DC}(t = 0) = V_{sm}$, which is solved for the DC load current as $i_{DC} = (V_{sm}/R)e^{-T/RC_{DC}}$.

The output DC bus voltage across the capacitor during the discharging period may be solved as $v_{DC} = Ri_{DC} = V_{sm}\,e^{-T/RC_{DC}}$.

The peak-to-peak ripple voltage is solved as $V_{DCrpp} = v_{DCTc} - v_{DCTnc} = V_{sm} - V_{sm}\,e^{-T/RC_{DC}} = V_{sm}(1 - e^{-T/RC_{DC}})$.

Substituting $e^{-y} = 1 - y$, peak ripple voltage is estimated as $V_{DCrpp} = V_{sm}(1 - 1 + T_{nc}/RC_{DC}) = V_{sm}T_{nc}/RC_{DC} = V_{sm}/2fRC_{DC}$ (considering $T_{nc} = 1/2f$).

The mean DC bus voltage is calculated as $V_{DC} = V_{sm} - V_{DCrpp}/2 = V_{sm} - V_s/4fRC_{DC}$.

The rms DC bus voltage ripple is calculated as $V_{AC} = V_{DCrpp}/2\sqrt{2} = V_{sm}/(4\sqrt{2}fRC_{DC})$.

The ripple factor is defined as $\mathrm{RF} = V_{AC}/V_{DC} = 1/\{(4fRC_{DC} - 1)\sqrt{2}\}$.

a. The value of the DC bus capacitor is $C_{DC} = \{1 + 1/(\mathrm{RF}\sqrt{2})\}/4fR = 757.11\ \mu$F.
b. The mean DC bus voltage is $V_{DC} = V_{sm} - V_{sm}/4fRC_{DC} = 303.78$ V.

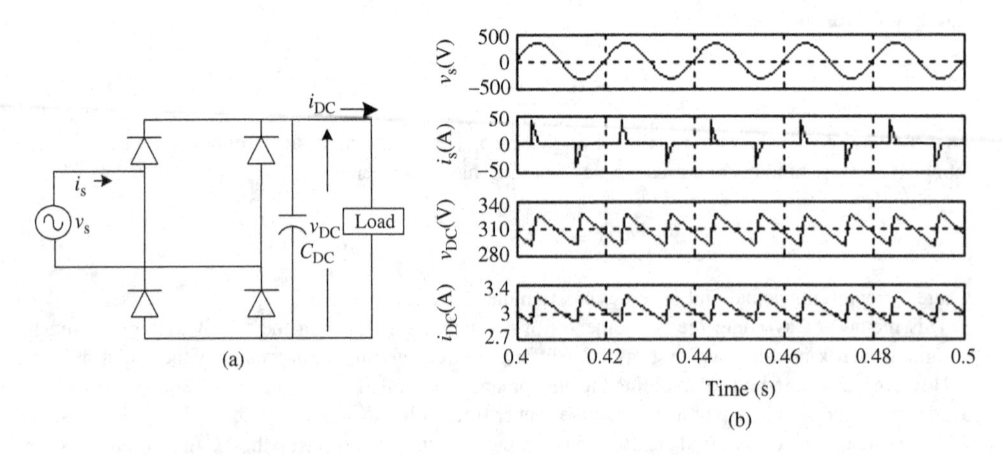

Figure E8.1 (a) A rectifier with a DC side C-type filter and (b) performance waveforms

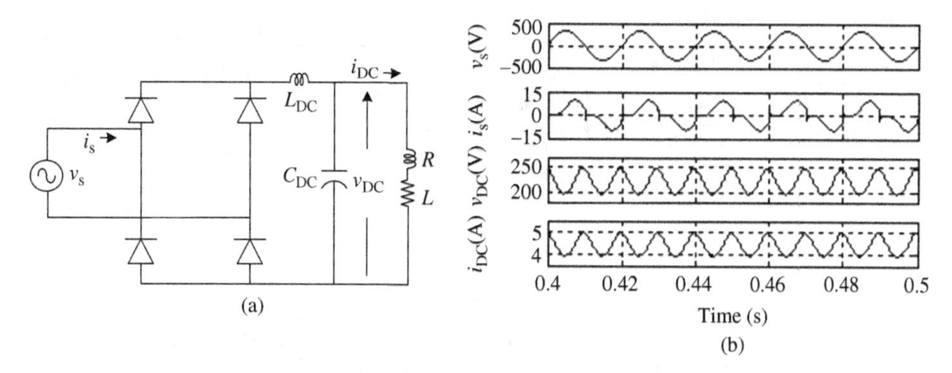

Figure E8.2 (a) A rectifier with a DC side *LC* filter and (b) performance waveforms

Example 8.2

A single-phase diode bridge rectifier is supplied from a 230 V, 50 Hz AC mains as shown in Figure E8.2. The DC load resistance is $R = 50\ \Omega$ and load inductance is $L = 10$ mH. Design a DC side LC filter so that the ripple factor of the output voltage is less than 10%.

Solution: Given supply rms voltage $V_s = 230$ V, frequency of the supply (f) = 50 Hz, $R = 50\ \Omega$, $L = 10$ mH, and RF = 10%.

The amplitude of AC mains voltage is $V_{sm} = 230 \times \sqrt{2} = 325.269$ V.

In the filter equivalent circuit of the converter system, the diode bridge generates harmonics in the output voltage that flows through the output filter (L_{DC}, C_{DC}) and the load. To reduce the ripple in the output voltage, major amount of ripple (harmonics) must flow in the filter capacitor. For this, the load impedance for these harmonics must be very high. It means analytically as $\{R^2 + (n\omega L)^2\}^{1/2} \gg 1/n\omega C_{DC}$.

Considering the case that load impedance is 10 times the filter impedance, $\{R^2 + (n\omega L)^2\}^{1/2} = 10/n\omega C_{DC}$.

The nth harmonic voltage appearing across the load is computed as

$$V_{DCn} = [-1/(n\omega C_{DC})/\{(n\omega L_{DC}) - 1/(n\omega C_{DC})\}]V_{nh} = [-1/\{(n\omega)^2 L_{DC}C_{DC} - 1\}]V_{nh},$$

where V_{nh} is the harmonic voltage present in the diode rectifier output voltage.

The total ripple voltage caused by all harmonics is estimated as $V_{AC} = (\sum V_{DCn}^2)^{1/2}$ for $n = 2, 4, 6, \ldots$.

In this case, only the dominant harmonic may be considered to limit the total harmonics as all other are lower than this dominant harmonic. Considering the second harmonic voltage, which is the dominant one, its rms value is $V_{2h} = 4V_{sm}/3\pi\sqrt{2}$. Its mean voltage is $V_{DC} = 2V_{sm}/\pi$.

For the second harmonic, $n = 2$, it results in $V_{AC} = V_{DC2} = [-1/(2\omega L_{DC}C_{DC} - 1)]V_{2h}$.

The value of the filter capacitor is computed as $\{R^2 + (2\omega L)^2\}^{1/2} = 10/2\omega C_{DC}$ or $C_{DC} = 10/[4\pi f\{R^2 + (2\omega L)^2\}^{1/2}] = 315.83\ \mu\text{F}$.

The ripple factor is defined as $\text{RF} = V_{AC}/V_{DC} = V_{2h}/[V_{DC}\{(4\pi f)^2 L_{DC}C_{DC} - 1\}] = \sqrt{2}/[3\{(4\pi f)^2 L_{DC}C_{DC} - 1\}] = 0.1$.

It results in $(4\pi f)^2 L_{DC}C_{DC} - 1 = 4.714$ or $L_{DC} = 45.8$ mH.

Example 8.3

A single-phase three-branch shunt passive filter (third, fifth, and high-pass) is employed to reduce the THD of the supply current and to improve the displacement factor to unity of a single-phase 230 V, 50 Hz fed diode bridge converter with an overlap angle of 30° drawing 20 A constant DC current as shown in Figure E8.3. Calculate (a) fundamental active power drawn by the load, (b) fundamental reactive power drawn by the load, (c) element values of the passive filter, and (d) current and VA ratings of the passive filter.

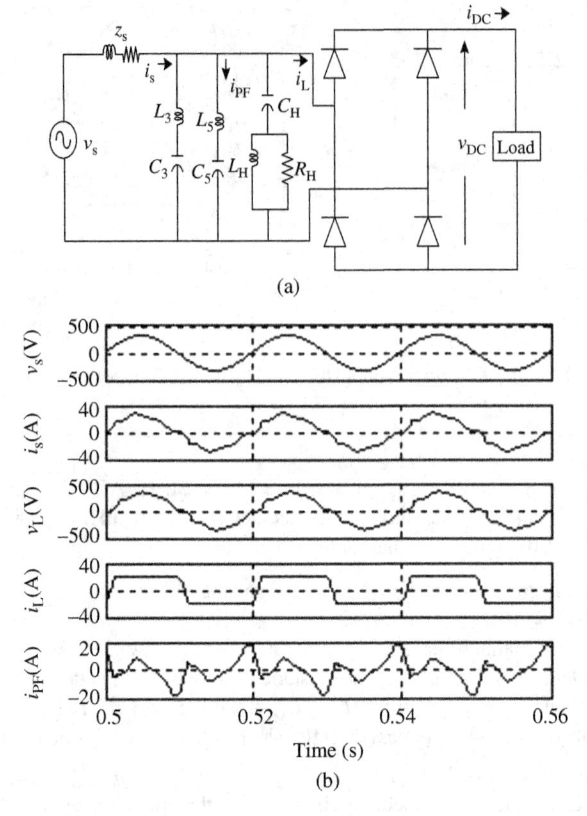

Figure E8.3 (a) A rectifier with an AC side filter and (b) performance waveforms

Solution: Given supply voltage $V_s = 230$ V, frequency of the supply (f) = 50 Hz, a load of a single-phase uncontrolled bridge converter fed from 230 V, 50 Hz AC supply with an overlap angle $\mu = 30°$, and DC link current $I_{DC} = 20$ A.

In a single-phase uncontrolled bridge converter with an overlap angle, the waveform of the supply current (I_s) is a trapezoidal wave with the amplitude of DC link current (I_{DC}). Since this trapezoidal wave of input current is odd and has quarter-cycle and half-cycle symmetry, its rms value is

$$I_s = I_L = \sqrt{[(1/2\pi)(\text{area of the total waveform})]} = \sqrt{\left[(1/2\pi)\left(4\int_0^{\mu/2}(2I_{DC}\theta/\mu)^2\,d\theta + 2\int_{\mu/2}^{\pi-\mu/2} I_{DC}^2\,d\theta\right)\right]}$$

$$= I_{DC}\sqrt{[(1/2\pi)\{(2\mu/3) + 2(\pi-\mu)\}]} = 0.94281 \times 20 = 18.856\text{ A.}$$

The rms value of the nth harmonic current I_{sn} is

$$I_{snrms} = B_n/\sqrt{2} = \left[(8/2\pi)\left(\int_0^{\pi/2} f(\theta)\sin n\theta\,d\theta\right)\right]$$

$$= \left[(8/2\pi)\left(\int_0^{\mu/2}(2I_{DC}\theta/\mu)\sin n\theta\,d\theta + \int_{\mu/2}^{\pi-\mu/2} I_{DC}\sin n\theta\,d\theta\right)\right]$$

$$= I_{DC}[(8/\pi\mu n^2)\sin(n\mu/2)]/\sqrt{2}\quad \text{(since the trapezoidal wave will not have } A_n\text{).}$$

Therefore, the rms value of the fundamental current is $I_{L1} = I_{s1} = I_{DC}[\{8/(\pi\mu\sqrt{2})\}\sin(\mu/2)] = 0.89 \times I_{DC} = 17.80$ A.

The active power component of the supply current is $I_{s1a} = I_{L1a} = I_{s1}\cos(\mu/2) = 17.19$ A.

a. The input active power is $P = V_s I_{s1}\cos(\mu/2) = 3954.50$ W.
b. The input reactive power is $Q_1 = V_s I_{s1}\sin(\mu/2) = 1059.61$ VAR.
   The mean output voltage is $V_{DC} = (\sqrt{2}/\pi)V_s(1 + \cos\mu) = 0.9V_s(1 + \cos\mu)/2 = 193.13$ V.
   The mean voltage is $V_{DC} = (2\sqrt{2}V_s - 2\omega L_s I_{DC})/\pi$.
   Therefore, the source inductance is $L_s = 3.51$ mH.
   Considering the source inductance as $L_s = 5$ mH, the source impedance is $Z_s = jX_s = j\omega L_s = 314 \times 0.005 = j1.57\ \Omega$.
   The voltage drop in the source impedance is $V_{Zs} = j17.19 \times 1.57 = j26.988$ V.
   The fundamental voltage across the load is $V_L = \sqrt{(V_s^2 - V_{Zs}^2)} = 228.412 \approx 230$ V.
c. The passive shunt filter has a three-branch shunt passive filter (third, fifth, and high-pass damped). The reactive power of 1059.61 VAR required by the single-phase diode rectifier has to be provided by all three branches of the passive shunt filter. Considering that all branches of the passive filter have equal capacitors, the value of this capacitor is $C = Q/3V_s^2\omega = 1059.61/(3 \times 230^2 \times 314) = 21.26\ \mu F$.
   Therefore, $C_3 = C_5 = C_H = C = 21.26\ \mu F$.
   The value of the inductor for the third harmonic tuned filter is $L_3 = 1/\omega_3^2 C_3 = 52.99$ mH.
   The resistance of the inductor of the third harmonic tuned filter is $R_3 = X_3/Q_3 = 49.924/50 = 0.9985\ \Omega$ (considering $Q_3 = 50$ as it may be in the range of 10–100 depending upon the design of the inductor).
   The third harmonic current in the load is $I_{L3} = I_{DC}[\{8/(9\pi\mu\sqrt{2})\}\sin(3\mu/2)] = 5.4$ A.
   The third harmonic current in the supply is $I_{s3} = I_{L3}Z_{PF3}/(Z_{PF3} + Z_{s3}) = I_{L3}R_3/(R_3 + jX_{s3}) = 5.40 \times 0.9985/\sqrt{\{0.9985^2 + (3 \times 1.57)^2\}} = 1.12$ A.
   The value of the inductor for the fifth harmonic filter is $L_5 = 1/\omega_5^2 C_5 = 19.1$ mH.
   The resistance in parallel of the inductor of the fifth harmonic tuned filter is $R_5 = X_5/Q_5 = 29.96/50 = 0.599\ \Omega$ (considering $Q_3 = 50$ as it may be in the range of 10–100 depending upon the design of the inductor).
   The fifth harmonic current in the load is $I_{L5} = I_{DC}[\{8/(25\pi\mu\sqrt{2})\}\sin(5\mu/2)] = 2.665$ A.
   The fifth harmonic current in the supply is $I_{s5} = I_{L5}Z_{PF5}/(Z_{PF5} + Z_{s5}) = I_{L5}R_5/(R_5 + jX_{s5}) = 2.665 \times 0.599/\sqrt{\{0.599^2 + (5 \times 1.57)^2\}} = 0.2022$ A.
   The value of the inductor for the high-pass damped harmonic filter (tuned at seventh harmonic) is $L_H = 1/\omega_7^2 C_H = 9.734$ mH.
   The resistance in parallel of the inductor of the high-pass damped harmonic tuned filter is $R_H = X_H/Q_H = 21.39/1 = 21.39\ \Omega$ (considering $Q_H = 1$ as it may be in the range of 0.5–5 depending upon the attenuation required).
   All other harmonic load currents to flow in the high-pass damped harmonic filter are

$$I_{LH} = \sqrt{[I_L^2 - I_{L1}^2 - I_{L3}^2 - I_{L5}^2]} = \sqrt{(18.856^2 - 17.80^2 - 5.4^2 - 2.665^2)} = 1.575\text{ A}.$$

$$I_{sH} = I_{LH}Z_{PFH}/(Z_{PFH} + Z_{sH}) = I_{LH}R_H/(R_H + jX_{sH}) = 1.564 \times 21.39/\sqrt{\{21.39^2 + (7 \times 1.57^2\}} = 1.4011\text{ A}.$$

$$THD_{Is} = \sqrt{(I_{s3}^2 + I_{s5}^2 + I_{sH}^2)}/I_{s1} = 0.1014 = 10.14\%.$$

$$THD_{VL} = \sqrt{\{(X_{s3}I_{s3}^2) + (X_{s5}I_{s5})^2 + (X_{sH}I_{sH})^2\}}/V_{s1} = 0.0711 = 7.11\%.$$

d. $I_{PF} = \sqrt{(I_s^2 - I_{s1a}^2 - I_{s3}^2 - I_{s5}^2 - I_{sH}^2)} = 7.53$ A.
   The VA rating of the passive shunt filter is $VA_{PF} = V_s I_{PF} = 230 \times 7.53 = 1732.1$ VA.

Example 8.4

A single-phase four-branch shunt passive filter (third, fifth, seventh, and high-pass) is used in a single-phase 220 V, 50 Hz system to reduce the THD of the supply current and to improve the displacement factor to unity. It has a load of a thyristor bridge converter operating at 60° firing angle of its thyristors drawing constant 25 A DC current as shown in Figure E8.4. Calculate (a) element values of the passive filter, (b) total harmonic distortion of the supply current, (c) total harmonic distortion of the terminal voltage at the load end, (d) the current rating of the passive filter, and (e) its kVA rating to provide harmonic and reactive power compensation. Let the supply has 5% source impedance mainly inductive.

Solution: Given supply rms voltage $V_s = 220$ V, frequency of the supply $(f) = 50$ Hz, and a source impedance of 5% mainly inductive feeding a nonlinear load of a 220 V, 50 Hz single-phase thyristor bridge converter with constant DC current of 25 A at 60° firing angle of its thyristors.

In this system, the load current harmonic and reactive power compensation is provided by a single-phase four-branch shunt passive filter (third, fifth, seventh, and high-pass damped) to reduce the THD of the supply current and to improve the displacement factor close to unity.

The AC load rms current is $I_L = I_{DC} = 25$ A.

The fundamental rms input current of the thyristor bridge converter is $I_{L1} = (2\sqrt{2}/\pi)I_L = 0.9 \times 25 = 22.5$ A.

The fundamental active component of load current is $I_{L1a} = I_{L1} \cos\alpha = 0.9 I_{DC} \cos 60° = 11.25$ A.

The fundamental active power of the load is $P_1 = V_{s1} I_{L1} \cos\theta_1 = V_{s1} I_{L1} \cos\alpha = V_{s1} I_{L1a} = 220 \times 11.25 = 2475$ W.

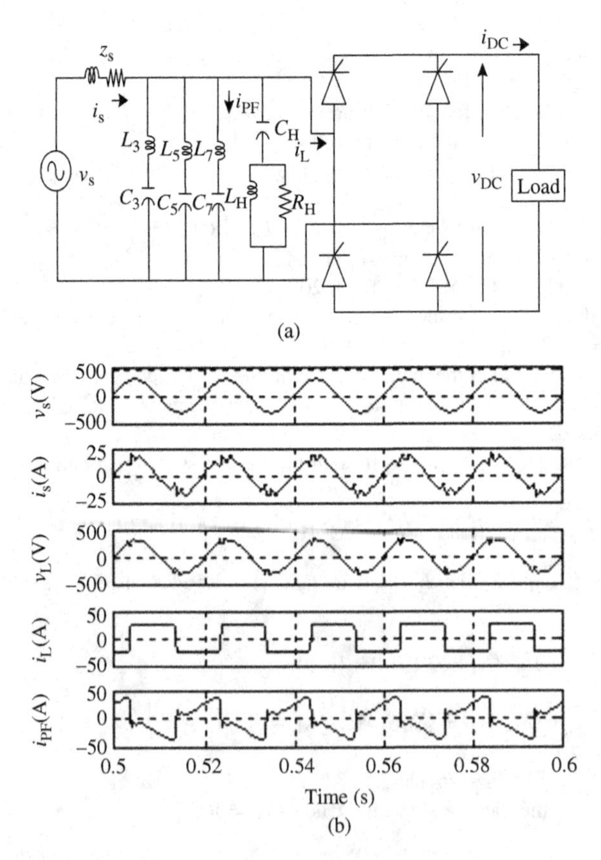

Figure E8.4 (a) A passive power filter and (b) performance waveforms

The fundamental reactive power of the load is $Q_1 = V_{s1}I_{L1}\sin\theta_1 = V_{s1}I_{L1}\sin\alpha = 220\times22.5\times0.866 = 4286.8$ VAR.

The source impedance is $Z_s = jX_s = j0.05\times220/25 = j0.44\ \Omega$.

The voltage drop in the source impedance is $V_{Zs} = j11.25\times0.44 = j4.95$ V.

The fundamental voltage across the load is $V_L = \sqrt{(V_s^2 - V_{Zs}^2)} = 219.94 \approx 220$ V.

a. The passive shunt filter has a four-branch shunt passive filter (third, fifth, seventh, and high-pass damped). The reactive power of 4286.8 VAR required by the single-phase thyristor rectifier has to be provided by all four branches of the passive shunt filter. Considering that all branches of the passive filter have equal capacitors, the value of this capacitor is $C = Q_1/4V_s^2\omega = 4286.8/(4\times220^2\times314) = 70.48\ \mu\text{F}$.

Therefore, $C_3 = C_5 = C_7 = C_H = C = 70.48\ \mu\text{F}$.

The value of the inductor for the third harmonic tuned filter is $L_3 = 1/\omega_3^2C_3 = 16$ mH.

The resistance of the inductor of the third harmonic tuned filter is $R_3 = X_3/Q_3 = 15.072/50 = 0.301\ \Omega$ (considering $Q_3 = 50$ as it may be in the range of 10–100 depending upon the design of the inductor).

The third harmonic current in the supply is $I_{s3} = I_{L3}Z_{PF3}/(Z_{PF3} + Z_{s3}) = (I_{L1}/3)R_3/(R_3 + jX_{s3}) = (18/3)\times0.301/\sqrt{\{0.301^2 + (3\times0.44)^2\}} = 1.667$ A.

The third harmonic voltage at PCC is $V_{s3} = I_{s3}Z_{s3} = 1.667\times3\times0.44 = 2.2016$ V.

The value of the inductor for the fifth harmonic tuned filter is $L_5 = 1/\omega_5^2C_5 = 5.8$ mH.

The resistance of the inductor of the fifth harmonic tuned filter is $R_5 = X_5/Q_5 = 8.949/50 = 0.1806\ \Omega$ (considering $Q_5 = 50$ as it may be in the range of 10–100 depending upon the design of the inductor).

The fifth harmonic current in the supply is $I_{s5} = I_{L5}Z_{PF5}/(Z_{PF5} + Z_{s5}) = (I_{L1}/5)R_5/(R_5 + jX_{s5}) = (18/5)\times0.1806/\sqrt{\{0.1806^2 + (5\times0.44)^2\}} = 0.3683$ A.

The fifth harmonic voltage at PCC is $V_{s5} = I_{s5}Z_{s5} = 0.3683\times5\times0.44 = 0.81$ V.

The value of the inductor for the seventh harmonic tuned filter is $L_7 = 1/\omega_7^2C_7 = 2.9$ mH.

The resistance of the inductor of the seventh harmonic tuned filter is $R_7 = X_7/Q_7 = 6.37/50 = 0.129\ \Omega$ (considering $Q_7 = 50$ as it may be in the range of 10–100 depending upon the design of the inductor).

The seventh harmonic current in the supply is $I_{s7} = I_{L7}Z_{PF7}/(Z_{PF7} + Z_{s7}) = (I_{L1}/7)R_7/(R_7 + jX_{s7}) = (18/7)\times0.129/\sqrt{\{0.129^2 + (7\times0.44)^2\}} = 0.134$ A.

The seventh harmonic voltage at PCC is $V_{s7} = I_{s7}Z_{s7} = 0.134\times7\times0.44 = 0.4144$ V.

The value of the inductor for the high-pass damped harmonic filter (tuned at ninth harmonic) is $L_H = 1/\omega_9^2C_H = 1.8$ mH.

The resistance in parallel of the inductor of the high-pass damped harmonic tuned filter is $R_H = X_H/Q_H = 5.08/2 = 2.508\ \Omega$ (considering $Q_H = 2$ as it may be in the range of 0.5–5 depending upon the attenuation required).

All other harmonic load currents to flow in the high-pass damped harmonic filter are

$$I_{LH} = \sqrt{[I_L^2 - I_{L1}^2 - I_3^2 - I_5^2 - I_7^2]} = \sqrt{[25^2 - 22.5^2 - (22.5/3)^2 - (22.5/5)^2 - (22.5/7)^2]} = 5.649\ \text{A}.$$

$$\begin{aligned} I_{sH} &= I_{LH}Z_{PFH}/(Z_{PFH} + Z_{sH}) = I_{LH}R_H/(R_H + jX_{sH}) \\ &= 5.6126\times2.508/\sqrt{\{2.508^2 + (9\times0.44)^2\}} = 3.0237\ \text{A}. \end{aligned}$$

The high-pass harmonic voltage at PCC is $V_{sH} = I_{sH}Z_{sH} = 3.0237\times9\times0.44 = 11.97$ V.

b. $\text{THD}_{Is} = \{\sqrt{(I_{s3}^2 + I_{s5}^2 + I_{sH}^2)}/I_{s1}\} = 0.1545 = 15.45\%$.

c. $\text{THD}_{VL} = [\sqrt{\{(X_{s3}I_{s3}^2 + (X_{s5}I_{s5})^2 + (X_{sH}I_{SH})^2\}}/V_{s1}] = 0.0437 = 4.3\%$.

d. The current rating of the passive shunt filter is $I_{PF} = \sqrt{[I_L^2 - I_{L1a}^2 - I_{s3}^2 - I_{s5}^2 - I_{s7}^2 - I_{sH}^2]} = 22.054$ A.

e. The VA rating of the passive shunt filter is $\text{VA}_{PF} = V_{PF}I_{PF} = 220\times22.0548 = 4851.9$ VA.

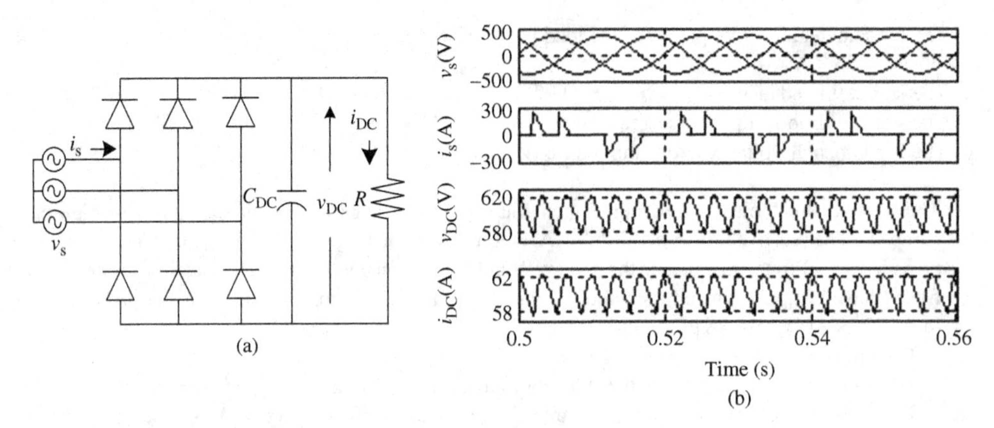

Figure E8.5 (a) A rectifier with a DC side filter and (b) performance waveforms

Example 8.5

A three-phase diode bridge rectifier is supplied from a 440 V, 50 Hz AC mains as shown in Figure E8.5. The load resistance is $R = 10\ \Omega$. (a) Design a DC bus parallel capacitive filter so that the ripple factor of the output voltage is less than 5%. (b) With this value of the capacitor, calculate the average load voltage V_{DC}.

Solution: Given supply rms voltage $V_s = 440$ V, frequency of the supply (f) = 50 Hz, $R = 10\ \Omega$, and RF = 5%.

The amplitude of AC mains voltage is $V_{sm} = 440 \times \sqrt{2} = 622.25$ V.

a. The DC bus capacitance for a given ripple factor in a three-phase diode bridge rectifier is $C_{DC} = (1/12fR)\{1 + 1/(RF\sqrt{2})\} = 2500\ \mu F$.
b. The average load voltage V_{DC} is estimated as $V_{DC} = V_{sm} - V_{sm}/12fRC_{DC} = 581.16$ V.

Example 8.6

A three-phase diode bridge rectifier is supplied from a 415 V, 50 Hz AC mains as shown in Figure E8.6. The DC load resistance is $R = 10\ \Omega$ and the load inductance is 10 mH. Design a DC side LC filter so that the ripple factor of the output voltage is less than 10%.

Solution: Given supply rms voltage $V_s = 415$ V, frequency of the supply (f) = 50 Hz, $R = 10\ \Omega$, L = 10 mH, and RF = 10%.

In the filter equivalent circuit of the converter system, the diode bridge generates harmonics in the output voltage that flows through the output filter (L_{DC}, C_{DC}) and the load. To reduce the ripple in the output voltage, major amount of ripple (harmonics) must flow in the filter capacitor. For this, the load impedance for these harmonics must be very high. It means analytically as $\{R^2 + (n\omega L)^2\}^{1/2} \gg 1/n\omega C_{DC}$.

Considering the case that load impedance is 10 times the filter impedance, $\{R^2 + (n\omega L)^2\}^{1/2} = 10/n\omega C_{DC}$.

The nth harmonic voltage appearing across the load is computed as

$$V_{DCn} = [-1/(n\omega C_{DC})/\{(n\omega L_{DC}) - 1/(n\omega C_{DC})\}]V_{nh} = [-1/\{(n\omega)^2 L_{DC} C_{DC} - 1\}]V_{nh},$$

where V_{nh} is the harmonics present in the diode rectifier output voltage.

The total ripple voltage caused by all harmonics is $V_{AC} = (\sum V_{DCn}^2)^{1/2}$ for $n = 6, 12, 18, \ldots$.

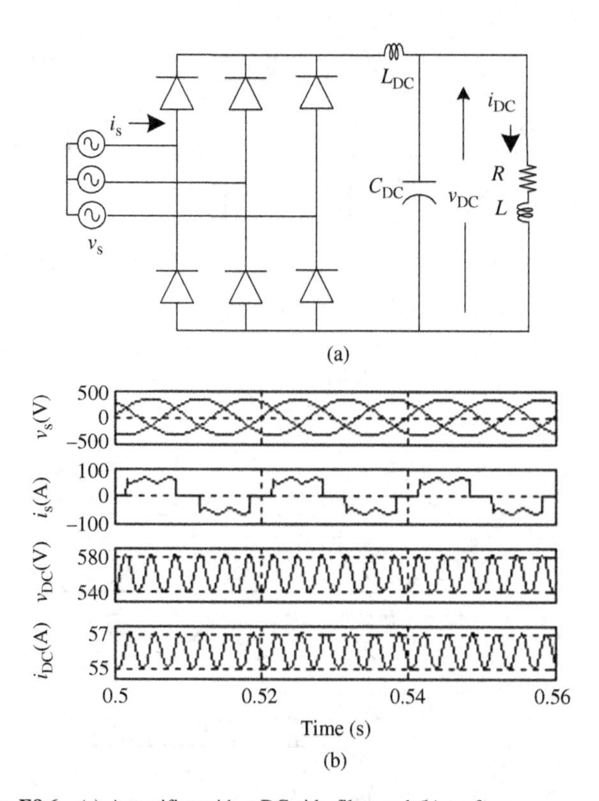

Figure E8.6 (a) A rectifier with a DC side filter and (b) performance waveforms

In this case, only the dominant harmonic may be considered to limit the total harmonics as all other are lower than this dominant harmonic. Considering the sixth harmonic, which is the dominant one, its rms value is $V_{6h} = 6V_{sm}/35\pi$. Its mean voltage is $V_{DC} = 3V_{sm}/\pi$.

For the sixth harmonic, $n = 6$, it results in $V_{AC} = V_{DC6} = [-1/\{(6\omega)^2 L_{DC}C_{DC} - 1\}]V_{6\,h}$.

The value of the filter capacitor is computed as $\{R^2 + (6\omega L)^2\}^{1/2} = 10/6\omega C_{DC}$ or $C_{DC} = 10/[12\pi f\{R^2 + (6\omega L)^2\}1/2] = 248.63\ \mu F$.

The ripple factor is estimated as $RF = V_{AC}/V_{DC} = V_{6h}/[V_{DC}\{(12\pi f)^2 L_{DC}C_{DC} - 1\}] = 2/[560\{(12\pi f)^2 L_{DC}C_{DC} - 1\}] = 0.1$.

It results in $(12\pi f)^2 L_{DC}C_{DC} - 1 = 1.75$ or $L_{DC} = 1.8$ mH.

Example 8.7

A three-phase three-branch shunt passive filter (tuned fifth, seventh, and high-pass) is employed to reduce the THD of the supply current and to improve the displacement factor to unity for a three-phase 415 V, 50 Hz fed six-pulse thyristor bridge converter drawing 100 A constant DC current at 30° firing angle of its thyristors as shown in Figure E8.7. Calculate (a) fundamental active power drawn by the load, (b) fundamental reactive power drawn by the load, (c) values of filter elements, (d) THD of the supply current, and (e) THD of the load voltage, and (f) the voltage, current, and VA ratings of the passive filter. Let the supply has 5% source impedance mainly inductive.

Solution: Given supply rms voltage $V_s = 415/\sqrt{3} = 239.6$ V, frequency of the supply (f) = 50 Hz, and a source impedance of 5% mainly inductive feeding a nonlinear load of a 415 V, 50 Hz three-phase thyristor bridge converter with constant DC current of 100 A at 30° firing angle of its thyristors.

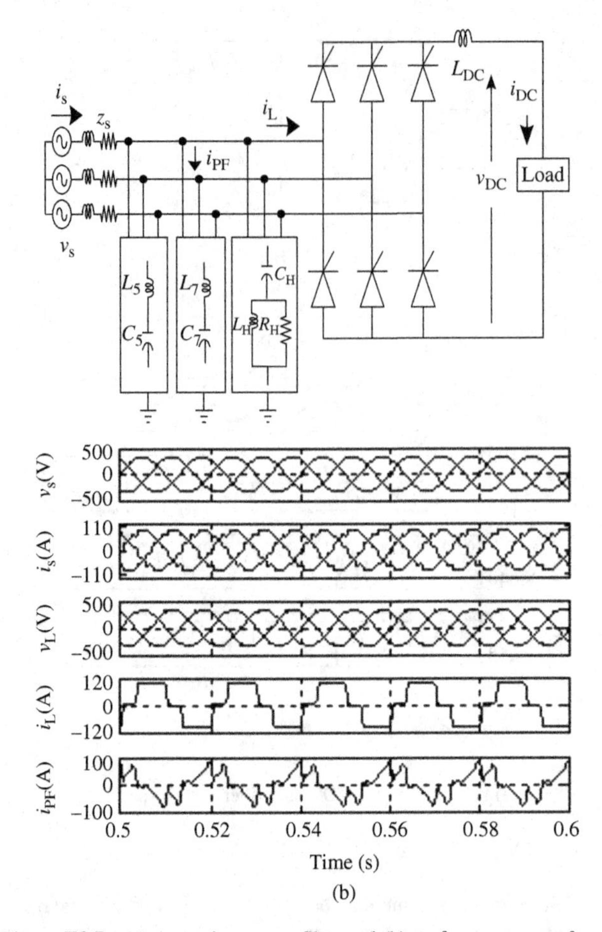

Figure E8.7 (a) A passive power filter and (b) performance waveforms

In this system, the load current harmonic and reactive power compensation is provided by a three-phase three-branch shunt passive filter (fifth, seventh, and high-pass damped) to reduce the THD of the supply current and to improve the displacement factor close to unity.

The AC load rms current is $I_L = I_{DC}\sqrt{2}/\sqrt{3} = 81.65$ A.

The fundamental rms input current of the thyristor bridge converter is $I_{L1} = (\sqrt{6}/\pi)I_L = 0.7797 \times 100 = 77.98$ A.

The fundamental active component of load current is $I_{L1a} = I_{1T} \cos \alpha = 77.98 \cos 30° = 67.52$ A.

a. The fundamental active power of the load is $P_1 = 3V_{s1}I_{L1} \cos \theta_1 = 3V_{s1}I_{L1} \cos \alpha = 3 \times 239.6 \times 67.52 = 48\ 536.05$ W.
b. The fundamental reactive power of the load is $Q_1 = 3V_{s1}I_{L1} \sin \theta_1 = 3V_{s1}I_{L1} \sin \alpha = 3 \times 239.6 \times 77.98 \times 0.5 = 28\ 026.01$ VAR.
   The source impedance is $Z_s = jX_s = j0.05 \times 239.6/81.6 = j0.1468\ \Omega$.
   The voltage drop in the source impedance is $V_{Zs} = j67.52 \times 0.1468 = j9.91$ V.
   The fundamental voltage across the load is $V_L = \sqrt{(V_s^2 - V_{Zs}^2)} = 239.17 \approx 239.2$ V.
c. The passive shunt filter has a three-branch shunt passive filter (fifth, seventh, and high-pass damped). The reactive power of 28 026.01 VAR required by the three-phase thyristor rectifier has to be provided by all three branches of the passive shunt filter. Considering that all branches of the passive filter have equal capacitors, the value of this capacitor is $C = Q_1/9V_s^2\omega = 28026.01/(9 \times 239.6^2 \times 314) = 172.75\ \mu$F.

Therefore, $C_5 = C_7 = C_H = C = 172.75$ μF.

The value of the inductor for the fifth harmonic tuned filter is $L_5 = 1/\omega_5^2 C_5 = 2.35$ mH.

The resistance of the inductor of the fifth harmonic tuned filter is $R_5 = X_5/Q_5 = 3.687/30 = 0.1229\ \Omega$ (considering $Q_5 = 30$ as it may be in the range of 10–100 depending upon the design of the inductor).

The fifth harmonic current in the supply is $I_{s5} = I_{L5}Z_{PF5}/(Z_{PF5} + Z_{s5}) = (I_{L1}/5)R_5/(R_5 + jX_{s5}) = (77.98/5) \times 0.1229/\sqrt{\{0.1229^2 + (5 \times 0.1536)^2\}} = 2.5765$ A.

The fifth harmonic voltage at PCC is $V_{s5} = I_{s5}Z_{s5} = 2.5765 \times 5 \times 0.1468 = 1.89$ V.

The value of the inductor for the seventh harmonic tuned filter is $L_7 = 1/\omega_7^2 C_7 = 1.198$ mH.

The resistance of the inductor of the seventh harmonic tuned filter is $R_7 = X_7/Q_7 = 2.634/30 = 0.087\ 79\ \Omega$ (considering $Q_7 = 30$ as it may be in the range of 10–100 depending upon the design of the inductor).

The seventh harmonic current in the supply is $I_{s7} = I_{L7}Z_{PF7}/(Z_{PF7} + Z_{s7}) = (I_{L1}/7)R_7/(R_7 + jX_{s7}) = (77.98/7) \times 0.08779/\sqrt{\{0.08779^2 + (7 \times 0.1468)^2\}} = 0.948$ A.

The seventh harmonic voltage at PCC is $V_{s7} = I_{s7}Z_{s7} = 0.948 \times 7 \times 0.1468 = 0.9744$ V.

The value of the inductor for the high-pass damped harmonic filter (tuned at 11th harmonic) is $L_H = 1/\omega_{11}^2 C_H = 0.4852$ mH.

The resistance in parallel of the inductor of the high-pass damped harmonic tuned filter is $R_H = X_H/Q_H = 1.676/2 = 0.835\ \Omega$ (considering $Q_H = 5$ as it may be in the range of 0.5–5 depending upon the attenuation required).

All other harmonic load currents to flow in the high pass damped harmonic filter are

$$I_{LH} = \sqrt{[I_L^2 - I_{L1}^2 - I_5^2 - I_7^2]} = \sqrt{[81.65^2 - 77.98^2 - (77.98/5)^2 - (77.98/7)^2]} = \sqrt{218.51} = 14.78 \text{ A.}$$

$$I_{sH} = I_{LH}Z_{PFH}/(Z_{PFH} + Z_{sH}) = I_{LH}R_H/(R_H + jX_{sH}) = 14.78 \times 0.835/\sqrt{\{0.835^2 + (11 \times 0.1468)^2\}} = 6.837 \text{ A.}$$

The high-pass harmonic voltage at PCC is $V_{sH} = I_{sH}Z_{sH} = 6.837 \times 11 \times 0.1468 = 11.035$ V.

d. $\text{THD}_{Is} = \{\sqrt{(I_{s5}^2 + I_{s7}^2 + I_{sH}^2)}/I_{s1}\} = 0.1091 = 10.91\%$.

e. $\text{THD}_{VL} = [\sqrt{\{(X_{s5}I_{s5}^2 + (X_{s7}I_{s7})^2 + (X_{sH}I_{SH})^2\}}/V_{s1}] = 4.69\%$.

f. The current rating of the passive shunt filter is $I_{PF} = \sqrt{[I_L^2 - I_{L1a}^2 - I_{s5}^2 - I_{s7}^2 - I_{sH}^2]} = 45.309$ A.

The VA rating of the passive shunt filter is $\text{VA}_{PF} = 3V_{PF}I_{PF} = 3 \times 239.6 \times 45.309 = 32\ 568$ VA.

Example 8.8

A three-phase one-branch shunt passive filter (11th-order high-pass damped) is employed to reduce the THD of the supply current and to improve the displacement factor to unity of a three-phase 415 V, 50 Hz, nonlinear load consisting of a 12-pulse thyristor bridge converter with 100 A constant DC current at 30° firing angle of its thyristors as shown in Figure E8.8. This converter consists of an ideal transformer with single primary star connected winding and two secondary windings connected in star and delta with same line voltages to provide 30° phase shift between two sets of three-phase output voltages. Two 6-pulse thyristor bridges are connected in series to form this 12-pulse AC–DC converter. Calculate (a) fundamental active power drawn by the load, (b) fundamental reactive power drawn by the load, (c) values of filter elements, (d) THD of the supply current with the passive filter, and (e) THD of the load voltage with the passive filter, and (f) the voltage, current, and VA ratings of the passive filter. Let the supply has 5% source impedance mainly inductive.

Solution: Given supply rms voltage $V_s = 415/\sqrt{3} = 239.6$ V, frequency of the supply (f) = 50 Hz, $I_{DC} = 100$ A, and $\alpha = 30°$.

In a three-phase 12-pulse thyristor bridge converter, the waveform of the input AC current (I_s) is a stepped waveform with (i) the first step of $\pi/6$ angle (from α to $\alpha + \pi/6$) and input current magnitude of $I_{DC}/\sqrt{3}$, (ii) the second step of $\pi/6$ angle (from $\alpha + \pi/6$ to $\alpha + \pi/3$) and input current magnitude of $I_{DC}(1 + 1/\sqrt{3})$, and (iii) the third step of $\pi/6$ angle (from $\alpha + \pi/3$ to $\alpha + \pi/2$) and input current magnitude of $I_{DC}(1 + 2/\sqrt{3})$, and it has all four symmetric segments of such steps.

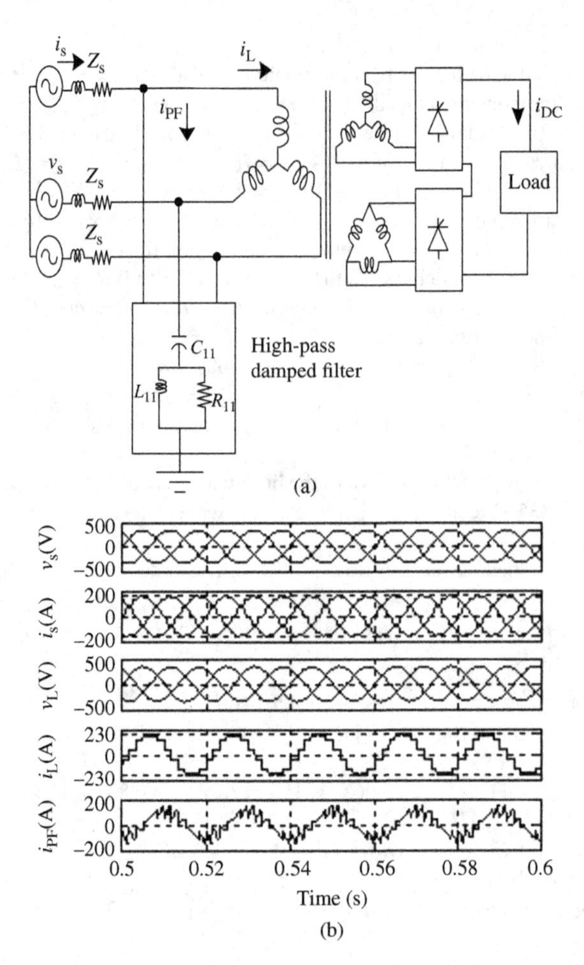

Figure E8.8 (a) A 12-pluse controlled rectifier with an AC side high-pass damped filter and (b) performance waveforms

The rms value of the 12-pulse converter input current is $I_s = 1.577\ 35I_{DC} = 157.77$ A.

The rms value of the 12-pulse converter fundamental AC current is $I_{s1} = \{(2\sqrt{6})/\pi\}I_{DC} = 1.559\ 393I_{DC} = 155.94$ A.

The active power component of supply current is $I_{s1a} = I_{s1}\cos\theta_1 = I_{s1}\cos\alpha = 155.94\cos 30° = 135.05$ A.

a. The active power drawn by the load is $P = 3V_sI_{s1}\cos\theta_1 = 97.073$ kW.
b. The fundamental reactive power is $Q_1 = 3V_sI_{s1}\sin\theta_1 = 3V_sI_{s1}\sin\alpha = 56.045$ kVAR.
   The source impedance is $Z_s = jX_s = j0.05 \times 239.6/157.77 = j0.075\ 933\ \Omega$.
   The voltage drop in the source impedance is $V_{Zs} = j135.05 \times 0.075\ 933 = j10.254$ V.
   The fundamental voltage across the load is $V_L = \sqrt{(V_s^2 - V_{Zs}^2)} = 239.38 \approx 239.6$ V.
c. The passive shunt filter has one-branch shunt passive filter (11th-order high-pass damped). The reactive power of 56.045 kVAR required by the three-phase thyristor rectifier has to be provided by one branch of the passive shunt filter. Therefore, the value of this capacitor is $C = Q_1/3V_s^2\omega = 56045/(3 \times 239.6^2 \times 314) = 1036.36$ μF.
   Therefore, $C_H = C = 1036.36$ μF.

The value of the inductor for the high-pass damped harmonic filter (tuned at 11th harmonic) is $L_H = 1/\omega_{11}^2 C_H = 0.080\,88$ mH.

The resistance in parallel of the inductor of the high-pass damped harmonic tuned filter is $R_H = X_H/Q_H = 0.2794/2 = 0.1397\ \Omega$ (considering $Q_H = 2$ as it may be in the range of 0.5–5 depending upon the attenuation required).

All other harmonic load currents to flow in the high-pass damped harmonic filter are

$$I_{LH} = \sqrt{(I_s^2 - I_{s1}^2)} = \sqrt{(157.77^2 - 155.94^2)} = \sqrt{574.089} = 23.49\text{ A}.$$

$$I_{sH} = I_{LH} Z_{PFH}/(Z_{PFH} + Z_{sH}) = I_{LH} R_H/(R_H + jX_{sH})$$
$$= 23.49 \times 0.1397/\sqrt{\{0.1397^2 + (11 \times 0.075\,933)^2\}} = 3.87\text{ A}.$$

The high-pass harmonic voltage at PCC is $V_{sH} = I_{sH} Z_{sH} = 3.87 \times 11 \times 0.1397 = 3.2575$ V

d. $\text{THD}_{Is} = \sqrt{I_{sH}^2}/I_{s1} = 0.0248 = 2.48\%$.

e. $\text{THD}_{VL} = \sqrt{(X_{sH} I_{sH})^2}/V_{s1} = 0.0135 = 1.35\%$.

f. The current rating of the passive shunt filter is $I_{PF} = \sqrt{(I_s^2 - I_{s1a}^2 - I_{sH}^2)} = 81.3417$ A.

The VA rating of the passive shunt filter is $VA_{PF} = 3V_{PF} I_{PF} = 3V_s I_{PF} = 3 \times 239.6 \times 81.3417 =$ 58 468 VA.

Example 8.9

A three-phase three-branch shunt passive filter (tuned 11th, 13th, and high–pass) is employed to reduce the THD of the supply current and to improve the displacement factor to unity of a three-phase 415 V, 50 Hz, nonlinear load consisting of a 12-pulse thyristor bridge converter with 200 A constant DC current at 60° firing angle of its thyristors as shown in Figure E8.9. This converter consists of an ideal transformer with single primary star connected winding and two secondary windings connected in star and delta with same line voltages to provide 30° phase shift between two sets of three-phase output voltages. Two 6-pulse thyristor bridges are connected in series to form this 12-pulse AC–DC converter. Calculate (a) fundamental active power drawn by the load, (b) fundamental reactive power drawn by the load, (c) values of filter elements, (d) THD of the supply current with the passive filter, (e) THD of the load voltage with the passive filter, and (f) the voltage, current, and VA ratings of the passive filter. Let the supply has 5% source impedance mainly inductive.

Solution: Given supply rms voltage $V_s = 415/\sqrt{3} = 239.6$ V, frequency of the supply (f) = 50 Hz, $I_{DC} = 200$ A, and $\alpha = 60°$.

In a three-phase 12-pulse thyristor bridge converter, the waveform of the input AC current (I_s) is a stepped waveform with (i) the first step of $\pi/6$ angle (from α to $\alpha + \pi/6$) and input current magnitude of $I_{DC}/\sqrt{3}$, (ii) the second step of $\pi/6$ angle (from $\alpha + \pi/6$ to $\alpha + \pi/3$) and input current magnitude of $I_{DC}(1 + 1/\sqrt{3})$, and (iii) the third step of $\pi/6$ angle (from $\alpha + \pi/3$ to $\alpha + \pi/2$) and input current magnitude of $I_{DC}(1 + 2/\sqrt{3})$, and it has all four symmetric segments of such steps.

The rms value of the 12-pulse converter input current is $I_s = I_L = 1.577\,35 I_{DC} = 315.47$ A.

The rms value of the 12-pulse converter fundamental AC current is $I_{s1} = I_{L1} = \{(2\sqrt{6})/\pi\} I_{DC} = 1.559\,393 I_{DC} = 311.8786$ A.

The active power component of supply current is $I_{s1a} = I_{s1} \cos\theta_1 = I_{s1} \cos\alpha = 311.8786 \cos 60° = 155.9393$ A.

a. The active power drawn by the load is $P = 3V_s I_{s1} \cos\theta_1 = 112.088\,95$ kW.

b. The fundamental reactive power is $Q_1 = 3V_s I_{s1} \sin\theta_1 = 3V_s I_{s1} \sin\alpha = 194.1437$ kVAR.

The source impedance is $Z_s = jX_s = j0.05 \times 239.6/315.47 = j0.037\,975\ \Omega$.

The voltage drop in the source impedance is $V_{Zs} = j155.9393 \times 0.037\,975 = j5.922$ V.

The fundamental voltage across the load is $V_L = \sqrt{(V_s^2 - V_{Zs}^2)} = 239.53 \approx 239.6$ V.

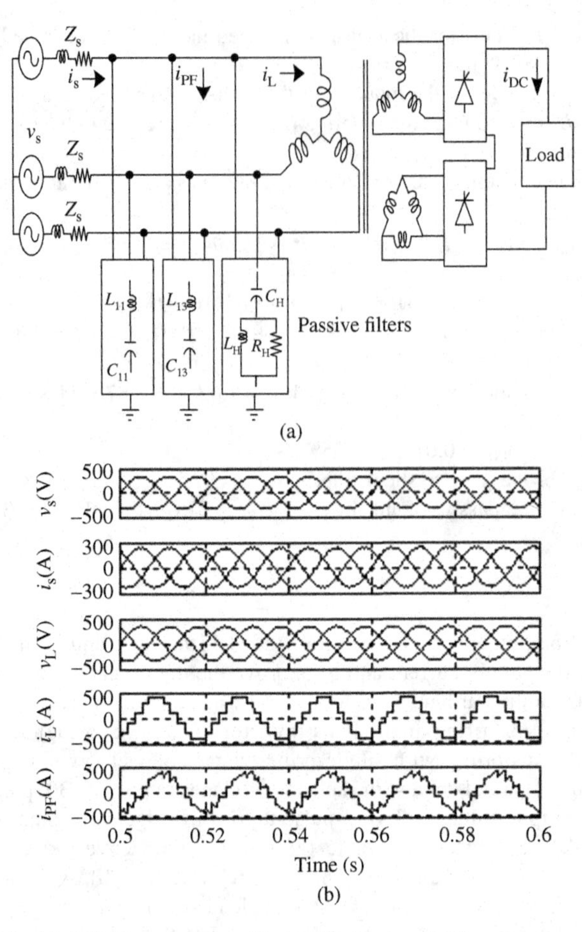

Figure E8.9 (a) A 12-pluse rectifier with AC side filters and (b) performance waveforms

c. The passive shunt filter has a three-branch shunt passive filter (11th, 13th, and high-pass damped). The reactive power of 194 143.7 VAR required by the three-phase thyristor rectifier has to be provided by all three branches of the passive shunt filter. Considering that all branches of the passive filter have equal capacitors, the value of this capacitor is $C = Q_1/9V_s^2\omega = 194\,143.7/(9 \times 239.6^2 \times 314) = 1196.68\ \mu\text{F}$.

Therefore, $C_{11} = C_{13} = C_H = C = 1196.68\ \mu\text{F}$.

The value of the inductor for the 11th harmonic tuned filter is $L_{11} = 1/\omega_{11}^2 C_{11} = 0.07$ mH.

The resistance of the inductor of the 11th harmonic tuned filter is $R_{11} = X_{11}/Q_{11} = 0.2419/20 = 0.0121\ \Omega$ (considering $Q_{11} = 20$ as it may be in the range of 10–100 depending upon the design of the inductor).

The 11th harmonic current in the supply is $I_{s11} = I_{L11}Z_{PF11}/(Z_{PF11} + Z_{s11}) = (I_{L1}/11)R_{11}/(R_{11} + jX_{s11}) = (311.88/11) \times 0.0121/\sqrt{\{0.0121^2 + (11 \times 0.037\,975)^2\}} = 0.821$ A.

The 11th harmonic voltage at PCC is $V_{s11} = I_{s11}Z_{s11} = 0.821 \times 11 \times 0.037\,975 = 0.343$ V.

The value of the inductor for the 13th harmonic tuned filter is $L_{13} = 1/\omega_{13}^2 C_{13} = 0.0502$ mH.

The resistance of the inductor of the 13th harmonic tuned filter is $R_{13} = X_{13}/Q_{13} = 0.2047/20 = 0.010\,24\ \Omega$ (considering $Q_{13} = 20$ as it may be in the range of 10–100 depending upon the design of the inductor).

The 13th harmonic current in the supply is $I_{s13} = I_{L13}Z_{PF13}/(Z_{PF13} + Z_{s13}) = (I_{L1}/13)R_{13}/(R_{13} + jX_{s13}) = (311.88/13) \times 0.0102/\sqrt{\{0.0102^2 + (13 \times 0.0379)^2\}} = 0.497$ A.

The 13th harmonic voltage at PCC is $V_{s13} = I_{s13}Z_{s13} = 0.497 \times 13 \times 0.0379 = 0.2455$ V.

The value of the inductor for the high-pass damped harmonic filter (tuned at 23rd harmonic) is $L_H = 1/\omega_{23}^2 C_H = 0.016$ mH.

The resistance in parallel of the inductor of the high-pass damped harmonic tuned filter is $R_H = X_H/Q_H = 0.1157/2 = 0.0579\ \Omega$ (considering $Q_H = 2$ as it may be in the range of 0.5–5 depending upon the attenuation required).

All other harmonic load currents to flow in the high-pass damped harmonic filter are

$$I_{LH} = \sqrt{(I_L^2 - I_{L1}^2 - I_{11}^2 - I_{13}^2)} = \sqrt{[315.47^2 - 311.88^2 - (311.88/11)^2 - (311.88/13)^2]} = 28.799 \text{ A.}$$

$$\begin{aligned} I_{sH} &= I_{LH}Z_{PFH}/(Z_{PFH} + Z_{sH}) = I_{LH}R_H/(R_H + jX_{sH}) \\ &= (28.799) \times 0.0579/\sqrt{\{0.0579^2 + (23 \times 0.0379)^2\}} = 1.903 \text{ A.} \end{aligned}$$

The high-pass harmonic voltage at PCC is $V_{sH} = I_{sH}Z_{sH} = 1.903 \times 23 \times 0.0379 = 1.66$ V.

d. $\text{THD}_{Is} = \sqrt{(I_{s11}^2 + I_{s13}^2 + I_{sH}^2)}/I_{s1} = 0.0068 = 0.68\%$.
e. $\text{THD}_{VL} = \sqrt{\{(X_{s11}I_{s11})^2 + (X_{s13}I_{s13})^2 + (X_{sH}I_{sH})^2\}}/V_{s1} = 0.0072 = 0.72\%$.
f. The current rating of the passive shunt filter is $I_{PF} = \sqrt{(I_s^2 - I_{s1a}^2 - I_{s11}^2 - I_{s13}^2 - I_{sH}^2)} = 274.14$ A.
   The VA rating of the passive shunt filter is $\text{VA}_{PF} = 3V_{PF}I_{PF} = 3 \times 239.6 \times 274.14 = 197060$ VA.

Example 8.10

An industry is fed electric power from a three-phase 33 kV, 50 Hz AC mains. The data of the supply feeder are as follows: short-circuit level of 150 MVA and an *X/R* ratio of 3. A step-down transformer is placed between the AC mains and this industry with a rating of 1500 kVA, 33 kV/440 V, $R = 1\%$, and $X = 5\%$. It is supplying linear and nonlinear loads as shown in Figure E8.10. If an AC capacitor bank is used at the PCC for power factor correction of the industry, calculate rating of this capacitor bank, which may excite parallel resonance at fifth harmonic with the supply system.

Solution: Given that a three-phase 33 kV, 50 Hz AC mains has a feeder with the following data: short-circuit level of 150 MVA and an *X/R* ratio of 3. A step-down transformer is placed between the AC mains and this industry with a rating of 1500 kVA, 33 kV/440 V, $R = 1\%$, and $X = 5\%$. An AC capacitor bank is used at the PCC for power factor correction of the industry. The total system and its equivalent circuit are shown in Figure E8.10.

Here feeder impedance is given in terms of short-circuit MVA (MVA_{SC}), *X/R* ratio, and voltage level. The values of the feeder resistance and reactance (in ohms) are calculated as

$$R_s = (\text{kV}_{LL}^2/\text{MVA}_{SC})\cos\{\tan^{-1}(X_s/R_s)\} = (33^2/150)\cos\{\tan^{-1}(3)\} = 2.2958\ \Omega.$$

$$X_s = (\text{kV}_{LL}^2/\text{MVA}_{SC})\sin\{\tan^{-1}(X_s/R_s)\} = (33^2/150)\sin\{\tan^{-1}(3)\} = 6.8874\ \Omega.$$

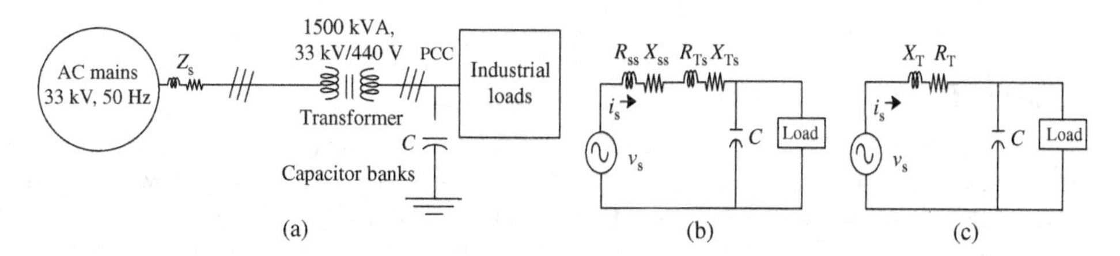

Figure E8.10 (a) An industrial load with a three-phase grid supply and (b and c) impedance diagrams

These data of the feeder (in ohms) are referred to the LV side of the transformer as follows:

$$R_{ss} = R_s/n^2 = 2.2958/(33\,000/440)^2 = 0.000\,408\,14\,\Omega.$$

$$X_{ss} = X_s/n^2 = 6.8874/(33\,000/440)^2 = 0.001\,2244\,\Omega.$$

Similarly, data of the transformer are given in terms of pu or percent, voltages, and kVA rating. The rating of the power factor correction capacitor is also given in terms of kVAR and voltage. These data may be given in other forms as well, which can easily be calculated in the required form.

The equivalent circuit parameters of the transformer referred to the secondary winding are calculated as

$$R_{Ts} = R_{Trpu}(1000\,\text{kV}^2/\text{kVA}_{Tr}) = 0.01 \times 1000 \times 0.440^2/1500 = 0.001\,291\,\Omega,$$

$$X_{Ts} = X_{Trpu}(1000\,\text{kV}^2/\text{kVA}_{Tr}) = 0.05 \times 1000 \times 0.440^2/1500 = 0.006\,453\,\Omega,$$

where R_{Trpu} and X_{Trpu} are pu resistance and reactance parameters of the transformer, respectively, kVA_{Tr} is the three-phase kVA rating, and kV is the line voltage of the secondary winding of the transformer.

The total equivalent circuit parameters of the system shown in Figure E8.10b are as follows:

$$R_T = R_{ss} + R_{Ts} = 0.000\,408\,14 + 0.001\,291 = 0.001\,699\,\Omega.$$

$$X_T = X_{ss} + X_{Ts} = 0.001\,2244 + 0.006\,453 = 0.007\,6777\,\Omega.$$

$$L_T = X_T/2\pi f = 24.451\,38\,\mu\text{H}.$$

The rating of the capacitor bank to excite parallel resonance at fifth harmonic with the supply system is calculated as follows.

Since parallel resonance frequency is given fifth harmonic, its frequency is $f_r = 250$ Hz.

The total impedance of the equivalent circuit shown in Figure E8.10c at any frequency ω_a can be calculated as $Z_i = [(R_T + j\omega_a L_T)(-j/\omega_a C)]/(R_T + j\omega_a L_T - j/\omega_a C)$.

Under parallel resonance, the denominator of impedance Z_i must be the minimum value. It will be the minimum value at a frequency ω_r, when the inductive reactance is equal to the capacitive reactance:

$$\omega_r L_T = 1/\omega_r C \quad \text{or} \quad \omega_r = 1/\sqrt{(L_T C)}.$$

$$f_r = \omega_r/2\pi = 1/\{2\pi\sqrt{(L_T C)}\}.$$

From this relation, the equivalent capacitance of the capacitor bank is

$$C = 1/[\{5(2\pi f)\}^2 L_T] = 1/[\{5(2\pi \times 50)\}^2 \times 24.451\,38 \times 10^{-6}] = 16\,591.95\,\mu\text{F} \quad \text{or}$$

$$X_C = 0.191\,9433\,\Omega.$$

$$\text{kVAR}_C = (1000\,\text{kV}_C^2)\omega C = 1000 \times 0.44^2 \times 314 \times 16\,591.95 \times 10^{-6} = 1008.631\,\text{kVAR}.$$

Example 8.11

An industry is fed electric power from a three-phase 33 kV, 50 Hz AC mains. The data of the supply feeder are as follows: short-circuit level of 150 MVA and an *X/R* ratio of 3. A step-down transformer is placed between the AC mains and this industry with a rating of 1500 kVA, 33 kV/440 V, $R = 1\%$, and $X = 5\%$. A 300 kVA, 440 V, three-phase thyristor bridge converter-fed DC motor drive is used in this industry along with other linear loads as shown in Figure E8.11. If a 1000 kVAR AC capacitor bank is

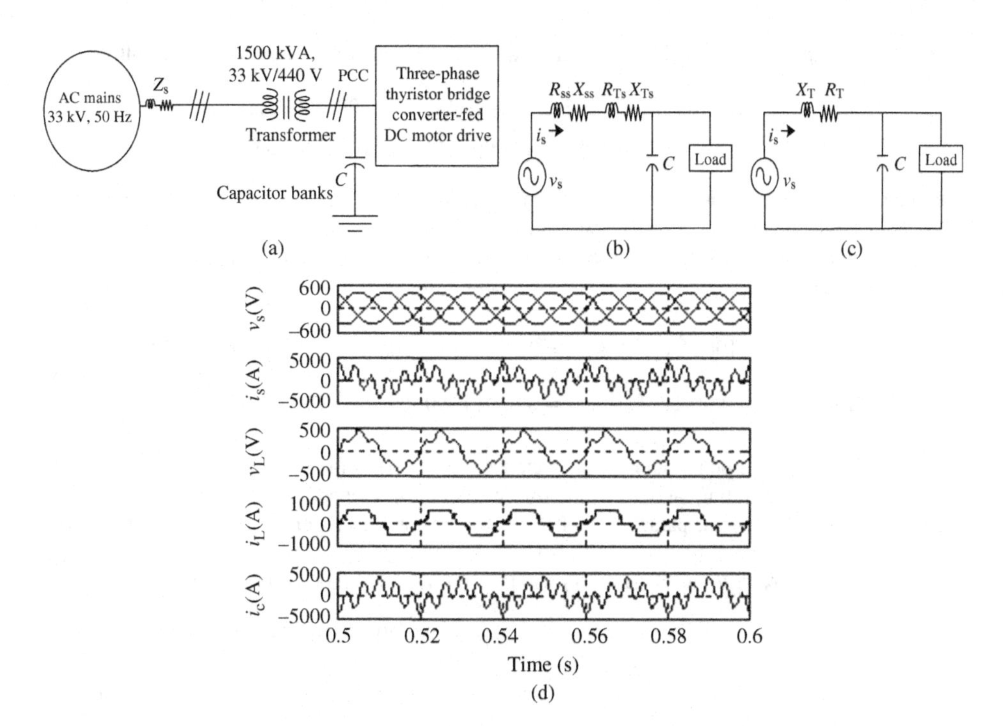

Figure E8.11 (a) A loaded three-phase control rectifier with a three-phase grid supply, (b and c) impedance diagrams, and (d) performance waveforms

Table E8.11-1 Harmonic spectrum of a 300 kVA, 440 V, three-phase thyristor converter-fed DC motor drive

Harmonic order	Frequency (Hz)	I_h (A)
5th	250	70
7th	350	40
11th	550	20
13th	650	10
17th	850	3

used at the PCC for power factor correction of the industry, calculate (a) PCC voltage and its THD without the capacitor bank, (b) PCC voltage and its THD with the capacitor bank, and (c) current in the capacitor bank. The harmonic spectrum of AC current of this 300 kVA, 440 V, three-phase thyristor bridge converter-fed DC motor drive is given in Table E8.11-1.

Solution: Given that a three-phase 33 kV, 50 Hz AC mains has a feeder with the following data: short-circuit level of 150 MVA and an *X*/*R* ratio of 3. A step-down transformer is placed between the AC mains and this industry with a rating of 1500 kVA, 33 kV/440 V, $R = 1\%$, and $X = 5\%$. A 300 kVA, 440 V, three-phase thyristor bridge converter-fed DC motor drive is used in this industry along with other linear loads. A 1000 kVAR AC capacitor bank is used at the PCC for power factor correction of the industry. The total system and its equivalent circuit are shown in Figure E8.11.

Here feeder impedance is given in terms of short-circuit MVA (MVA_{SC}), *X*/*R* ratio, and voltage level. The values of the feeder resistance and reactance (in ohms) are calculated as

$$R_s = (\text{kV}_{\text{LL}}^2/\text{MVA}_{\text{SC}})\cos\{\tan^{-1}(X_s/R_s)\} = (33^2/150)\cos\{\tan^{-1}(3)\} = 2.2958\,\Omega.$$

$$X_s = (\text{kV}_{\text{LL}}^2/\text{MVA}_{\text{SC}})\sin\{\tan^{-1}(X_s/R_s)\} = (33^2/150)\sin\{\tan^{-1}(3)\} = 6.8874\,\Omega.$$

These data of the feeder (in ohms) are referred to the LV side of the transformer as follows:

$$R_{ss} = R_s/n^2 = 2.2958/(33\,000/440)^2 = 0.000\,408\,14\,\Omega.$$

$$X_{ss} = X_s/n^2 = 6.8874/(33\,000/440)^2 = 0.001\,2244\,\Omega.$$

Similarly, data of the transformer are given in terms of pu or percent, voltages, and kVA rating. The rating of the power factor correction capacitor is also given in terms of kVAR and voltage. These data may be given in other forms as well, which can easily be calculated in the required form.

The equivalent circuit parameters of the transformer referred to the secondary winding are calculated as

$$R_{\mathrm{Ts}} = R_{\mathrm{Trpu}}(1000\,\mathrm{kV}^2/\mathrm{kVA_{Tr}}) = 0.01 \times 1000 \times 0.440^2/1500 = 0.001\,291\,\Omega,$$

$$X_{\mathrm{Ts}} = X_{\mathrm{Trpu}}(1000\,\mathrm{kV}^2/\mathrm{kVA_{Tr}}) = 0.05 \times 1000 \times 0.440^2/1500 = 0.006\,453\,\Omega,$$

where R_{Trpu} and X_{Trpu} are pu resistance and reactance parameters of the transformer, respectively, $\mathrm{kVA_{Tr}}$ is the three-phase kVA rating, and kV is the line voltage of the secondary winding of the transformer.

The total equivalent circuit parameters of the system (shown in Figure E8.11c) are as follows:

$$R_{\mathrm{T}} = R_{ss} + R_{\mathrm{Ts}} = 0.000\,408\,14 + 0.001\,291 = 0.001\,699\,14\,\Omega.$$

$$X_{\mathrm{T}} = X_{ss} + X_{\mathrm{Ts}} = 0.001\,2244 + 0.006\,453 = 0.007\,6777\,\Omega.$$

$$L_{\mathrm{T}} = X_{\mathrm{T}}/2\pi f = 24.451\,38\,\mu\mathrm{H}.$$

The equivalent reactance of the capacitor bank is

$$X_{\mathrm{C}} = 1000\,\mathrm{kV_C^2}/\mathrm{kVAR_C} = 1000 \times 0.44^2/1000 = 0.1936\,\Omega.$$

$$C = 1/2\pi f X_{\mathrm{C}} = 16\,449.97\,\mu\mathrm{F}.$$

The magnitude of fundamental current of the converter supplying a DC motor is $I_{s1} = 300 \times 1000/(\sqrt{3} \times 440) = 393.65$ A.

a. PCC voltage and its THD without the capacitor bank are computed as follows.

Without the capacitor bank, the supply impedance is computed at harmonics. The voltage drop because of each harmonic current of the load is computed by multiplying the supply impedance at that harmonic frequency and harmonic current. The computed values are given in Table E8.11-2.

The rms value of the supply current is $I_{srms} = \sqrt{(393.65^2 + 70^2 + 40^2 + 20^2 + 10^2 + 3^2)} = 402.45$ A.

The THD of the supply current is $\mathrm{THD_{Is}} = \sqrt{(I_{srms}^2 - I_{s1}^2)}/I_{s1} = \sqrt{(402.45^2 - 393.65^2)}/393.65 = 0.2127 = 21.27\%$.

The rms value of the PCC voltage is $V_{srms} = \sqrt{(254^2 + 2.69^2 + 2.15^2 + 1.69^2 + 0.99^2 + 0.39^2)} = 254.0633$ V.

Table E8.11-2 Harmonic voltages at PCC without a capacitor bank

Harmonic order	Frequency (Hz)	R_{T} (Ω)	X_{T} (Ω)	Z_{T} (Ω)	I_h (A)	V_h (V)
5th	250	0.001 699 14	0.038 3885	0.038 426 085	70	2.69
7th	350	0.001 699 14	0.053 7439	0.053 770 753	40	2.15
11th	550	0.001 699 14	0.084 4547	0.084 471 791	20	1.69
13th	650	0.001 699 14	0.099 8101	0.099 824 562	10	0.99
17th	850	0.001 699 14	0.130 5209	0.130 531 959	3	0.39

Table E8.11-3 Harmonic voltages at PCC with a capacitor bank

h	f (Hz)	R_T (Ω)	$X_T = \omega L_T$ (Ω)	Z_T (Ω)	$X_C = 1/\omega C$ (Ω)	$Z_i = (R + jX_T)(-jX_C)/(R + jX_T - jX_C)$ (Ω)	I_h (A)	V_h (V)	I_C (A)
5	250	0.001 699 14	0.038 3885	0.038 426 085	0.038 72	0.859 449 499 73	70	60.16	1553.76
7	350	0.001 699 14	0.053 7439	0.053 770 753	0.027 66	0.056 864 597 37	40	2.27	82.07
11	550	0.001 699 14	0.084 4547	0.084 471 791	0.017 60	0.022 230 652 31	20	0.445	25.26
13	650	0.001 699 14	0.099 8101	0.099 824 562	0.014 89	0.017 503 052 58	10	0.175	11.75
17	850	0.001 699 14	0.130 5209	0.130 531 959	0.011 39	0.012 476 657 79	3	0.037	3.29

The THD of the PCC voltage is $\text{THD}_{Vs} = \sqrt{(V_{\text{srms}}^2 - V_{s1}^2)}/V_{s1} = \sqrt{(254.0633^2 - 254^2)}/254 = 0.0157 = 1.57\%$.

b. PCC voltage and its THD with the capacitor bank are computed as follows.

The parallel impedance Z_i at each frequency is calculated from capacitor value and then harmonic voltage across the parallel impedance and then capacitor current from that harmonic voltage. The calculated values are given in Table E8.11-3.

The rms value of the PCC voltage is $V_{\text{srms}} = \sqrt{(254^2 + 60.16^2 + 2.27^2 + 0.445^2 + 0.175^2 + 0.037^2)} = 261.04$ V.

The THD of the PCC voltage is $\text{THD}_{Vs} = \sqrt{(V_{\text{srms}}^2 - V_{s1}^2)}/V_{s1} = \sqrt{(261.04^2 - 254^2)}/254 = 0.2370 = 23.70\%$.

This large voltage distortion of 23.70% is because of parallel resonance of this power factor correction capacitor with supply impedance.

c. The current in the capacitor bank is calculated as follows.

The fundamental value of the capacitor current is $I_{C1} = 1000\,000/(3 \times 254) = 1332.34$ A.

The rms value of the capacitor current is $I_{\text{Crms}} = \sqrt{(1332.34^2 + 1553.76^2 + 82.07^2 + 25.26^2 + 11.75^2 + 3.29^2)} = 2035.66$ A.

This capacitor current is 153% of its rated current because of parallel resonance of this power factor correction capacitor with supply impedance. This magnitude of capacitor current is likely to blow the fuse of capacitor bank or cause its breaker to trip. It may also damage the capacitor bank.

Example 8.12

An industry is fed electric power from a three-phase 33 kV, 50 Hz AC mains. The data of the supply feeder are as follows: short-circuit level of 150 MVA and an *X/R* ratio of 3. A step-down transformer is placed between the AC mains and this industry with a rating of 1500 kVA, 33 kV/440 V, $R = 1\%$, and $X = 5\%$. A 300 kVA, 440 V, three-phase thyristor bridge converter-fed DC motor drive is used in this industry along with other linear loads as shown in Figure E8.12. If a 1000 kVAR AC capacitor bank used at the PCC for power factor correction of the industry is tuned to the lowest harmonic (fifth) passive shunt filter, calculate (a) parallel resonance frequency, (b) the filter harmonic current spectrum and rms current of the filter, (c) PCC voltage and its THD with the passive filter, and (d) rms capacitor voltage. Harmonic spectrum of input AC current of this 300 kVA, 440 V, three-phase thyristor bridge converter-fed DC motor drive is given in Table E8.12-1.

Solution: Given that a three-phase 33 kV, 50 Hz AC mains has a feeder with the following data: short-circuit level of 150 MVA and an *X/R* ratio of 3. A step-down transformer is placed between the AC mains and this industry with a rating of 1500 kVA, 33 kV/440 V, $R = 1\%$, and $X = 5\%$. A 300 kVA, 440 V, three-phase thyristor bridge converter-fed DC motor drive is used in this industry along with other linear loads. The harmonic spectrum of AC current of this 300 kVA, 440 V, three-phase thyristor bridge converter-fed DC motor drive is given in Table E8.12-1. A 1000 kVAR AC capacitor bank used at the PCC for power factor correction of the industry is tuned to the lowest harmonic (fifth) passive shunt filter. The total system and its equivalent circuit are shown in Figure E8.12.

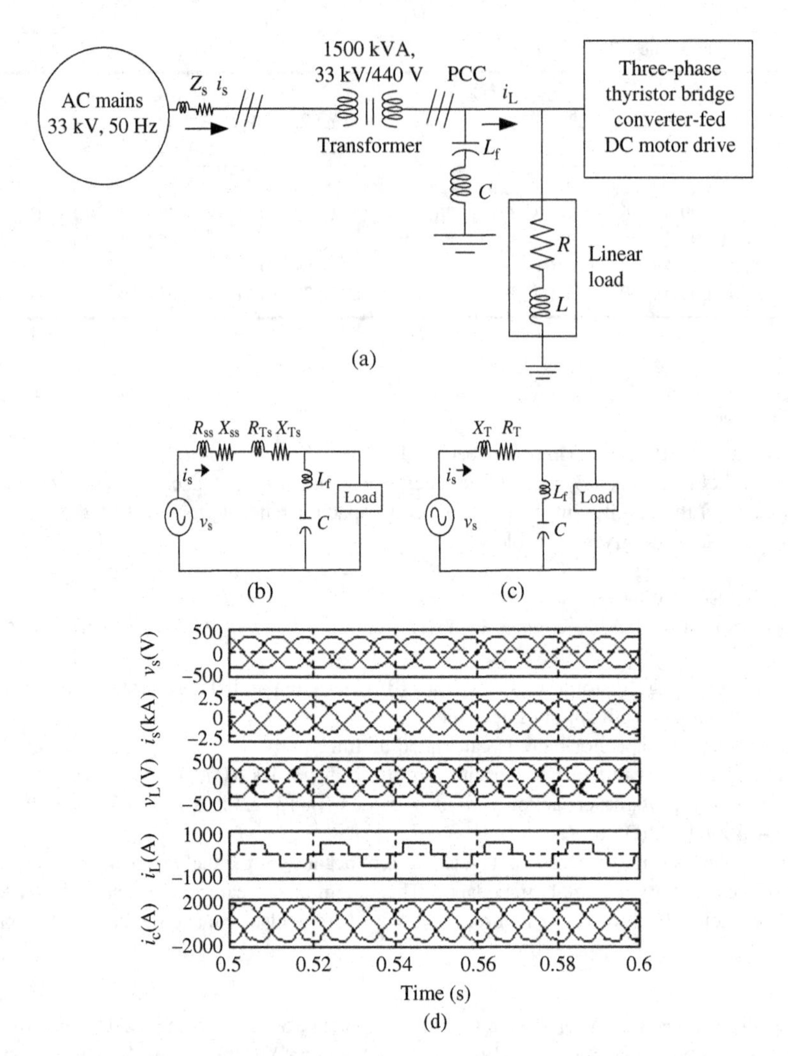

Figure E8.12 (a) A loaded three-phase controlled rectifier along with linear loads connected to a three-phase grid supply, (b and c) impedance diagrams, and (d) performance waveforms

Table E8.12-1 Harmonic spectrum of a 300 kVA, 440 V, three-phase thyristor converter-fed DC motor drive

Harmonic order	Frequency (Hz)	I_h (A)
5th	250	70
7th	350	40
11th	550	20
13th	650	10
17th	850	3

Here feeder impedance is given in terms of short-circuit MVA (MVA_{SC}), X/R ratio, and voltage level. The values of the feeder resistance and reactance (in ohms) are calculated as

$$R_s = (kV_{LL}^2/MVA_{SC})\cos\{\tan^{-1}(X_s/R_s)\} = (33^2/150)\cos\{\tan^{-1}(3)\} = 2.2958\ \Omega.$$

$$X_s = (kV_{LL}^2/MVA_{SC})\sin\{\tan^{-1}(X_s/R_s)\} = (33^2/150)\sin\{\tan^{-1}(3)\} = 6.8874\ \Omega.$$

These data of the feeder (in ohms) are referred to the LV side of the transformer as follows:

$$R_{ss} = R_s/n^2 = 2.2958/(33\,000/440)^2 = 0.000\,408\,14\ \Omega.$$

$$X_{ss} = X_s/n^2 = 6.8874/(33\,000/440)^2 = 0.001\,2244\ \Omega.$$

Similarly, data of the transformer are given in terms of pu or percent, voltages, and kVA rating. The rating of the power factor correction capacitor is also given in terms of kVAR and voltage. These data may be given in other forms as well, which can easily be calculated in the required form.

The equivalent circuit parameters of the transformer referred to the secondary winding are calculated as

$$R_{Ts} = R_{Trpu}(1000\ kV^2/kVA_{Tr}) = 0.01 \times 1000 \times 0.440^2/1500 = 0.001\,291\ \Omega,$$

$$X_{Ts} = X_{Trpu}(1000\ kV^2/kVA_{Tr}) = 0.05 \times 1000 \times 0.440^2/1500 = 0.006\,453\ \Omega,$$

where R_{Trpu} and X_{Trpu} are pu resistance and reactance parameters of the transformer, respectively, kVA_{Tr} is the three-phase kVA rating, and kV is the line voltage of the secondary winding of the transformer.

The total equivalent circuit parameters of the system (shown in Figure E8.12c) are as follows:

$$R_T = R_{ss} + R_{Ts} = 0.000\,408\,14 + 0.001\,291 = 0.001\,699\,14\ \Omega.$$

$$X_T = X_{ss} + X_{Ts} = 0.001\,2244 + 0.006\,453 = 0.007\,6777\ \Omega.$$

$$L_T = X_T/2\pi f = 24.451\,38\ \mu H.$$

The equivalent reactance of the capacitor bank is

$$X_C = 1000 \times kV_C^2/kVAR_C = 1000 \times 0.44^2/1000 = 0.1936\ \Omega.$$

$$C = 1/2\pi f X_C = 16\,449.97\ \mu F.$$

The passive shunt tuned filters are tuned at a frequency 5–10% lower than the actual harmonic to be reduced in the system. Normally, the fifth harmonic filter is tuned at 4.7th-order frequency of the supply system. Therefore, tuning the fifth harmonic filter at 4.7th-order frequency of the supply system, the inductor of the filter is calculated as

$$L_f = 1/\{(4.7 \times 2 \times \pi f)^2 C\} = 1/\{(4.7 \times 2 \times \pi \times 50)^2 \times 16\,449.97 \times 10^{-6}\} = 27.883\ \mu H.$$

$$X_{Lf} = 2\pi f L_f = 0.008\,7597\ \Omega.$$

The magnitude of fundamental current of the converter supplying a DC motor is $I_{s1} = 300 \times 1000/(\sqrt{3} \times 440) = 393.65$ A.

a. The parallel resonance frequency with the passive tuned filter can be calculated as follows.

   The parallel resonance phenomenon may occur with a passive shunt tuned filter. Figure E8.12a shows the circuit of a passive shunt tuned filter. Figure E8.12b shows an equivalent circuit of the system using a passive shunt filter. Total impedance of the equivalent circuit shown in Figure E8.12c at any frequency ω_a can be calculated as

$$Z_i = [(R_T + j\omega_a L_T)(j\omega_a L_f - j/\omega_a C)]/(R_T + j\omega_a L_T + j\omega_a L_f - j/\omega_a C),$$

where L_f is the value of the filter inductor.

Under parallel resonance, the denominator of impedance Z_i must be the minimum value or the imaginary part of its denominator is equal to zero. It will be the minimum value at a frequency ω_r, when the inductive reactance is equal to the capacitive reactance:

$$\omega_r(L_T + L_f) = 1/\omega_r C, \quad \omega_r = 1/\sqrt{\{(L_T + L_f)C\}},$$

$$f_r = \omega_r/2\pi = 1/[2\pi\sqrt{\{(L_T + L_f)C\}}]$$
$$= 1/[2\pi\sqrt{\{(24.451\,38 \times 10^{-6} + 27.883 \times 10^{-6}) \times 16\,449.97 \times 10^{-6}\}}] = 171.51 \text{ Hz}.$$

The parallel resonance phenomenon is now avoided from fifth harmonic and the parallel resonance frequency is shifted (f_r) to a value less than the lowest order harmonic current present in the harmonic-producing loads at 3.43rd-order harmonic. Therefore, this shifting of parallel resonance frequency (f_r) to much lower frequency (3.43rd-order harmonic) than harmonics present in the loads avoids the amplification of harmonic currents and PCC voltage distortion is minimized in the system.

b. The filter harmonic current spectrum and rms current of the filter can be calculated as follows.

The fundamental current of the filter can be calculated as I_{f1} = nominal phase voltage/$(j\omega_s L_f - j/\omega_s C)$ = $(440/\sqrt{3})/[j\{2 \times \pi \times 50 \times 27.883 \times 10^{-6} - 1/(2 \times \pi \times 50 \times 16\,449.97 \times 10^{-6})\}]$ = $254/\{j(0.008\ 755\ 262 - 0.193\ 60)\}$ = 1374.13 A.

The currents flowing in the harmonic filter at any frequency ω_a are calculated using the following relation:

$$I_{fh} = I_h(R_T + j\omega_a L_T)/(R_T + j\omega_a L_T + j\omega_a L_f - j/\omega_a C).$$

Using this equation, harmonic currents in the passive filter are calculated and the calculated values are given in Table E8.12-2.

The rms filter current is $I_{frms} = \sqrt{(1374.13^2 + 61.58^2 + 24.60^2 + 10.35^2 + 5.02^2 + 1.46^2)}$ = 1375.78 A.

The rated current of the capacitor is $I_{Crms} = 1000\,000/(3 \times 254) = 1312.34$ A.

The rms filter current as percentage of rated current of the capacitor is I_{frms}/I_{Crms} = $1375.78/1312.34 = 1.048\,34 = 104.834\%$

This filter current flowing through the capacitor is only 105% of its rated current, which is well within its considered rating.

c. The PCC voltage and its THD with the passive filter are calculated as follows.

For calculation of the voltage distortion, harmonic currents flowing in the supply system at a frequency ω_a are computed and shown in Table E8.12-2 using the following relation:

$$I_{sh} = I_h(j\omega_a L_f - j/\omega_a C)/(R_T + j\omega_a L_T + j\omega_a L_f - j/\omega_a C).$$

Table E8.12-2 Harmonic voltages at PCC with a passive filter

h	f (Hz)	R_T (Ω)	$X_T = \omega L_T$ (Ω)	Z_T (Ω)	$X_C = 1/\omega C$ (Ω)	$X_f = \omega L_f$ (Ω)	I_h (A)	I_{fh} (A)	I_{sh} (A)	V_h (V)	V_{Ch} (V)	V_{Lh} (V)
5	250	0.001 699 14	0.038 3885	0.038 426 085	0.038 72	0.043 7985	70	61.58	8.14	0.313	2.384	2.697
7	350	0.001 699 14	0.053 7439	0.053 770 753	0.027 66	0.061 3179	40	24.60	15.40	0.828	0.681	1.509
11	550	0.001 699 14	0.084 4547	0.084 471 791	0.017 60	0.096 3567	20	10.35	9.65	0.815	0.182	0.997
13	650	0.001 699 14	0.099 8101	0.099 824 562	0.014 89	0.113 8761	10	5.02	4.98	0.497	0.075	0.572
17	850	0.001 699 14	0.130 5209	0.130 531 959	0.011 39	0.148 9149	3	1.46	1.54	0.201	0.017	0.217

The fundamental PCC voltage is $V_{s1} = 440/\sqrt{3} = 254$ V.

The harmonic voltage at each harmonic present at PCC may be calculated as $V_h = I_{sh}(R_T + j\omega_a L_T)$.

These equations are used to calculate all harmonic voltages to find the THD of the PCC voltage and the calculated values are given Table E8.12-2.

The rms value of the supply current from Table E8.12-2 is $I_{srms} = \sqrt{(393.65^2 + 8.14^2 + 15.4^2 + 9.65^2 + 4.98^2 + 1.54^2)} = 394.187$ A.

The THD of the supply current is $THD_{Is} = \sqrt{(393.187^2 - 393.65^2)}/393.65 = 0.0522 = 5.22\%$.

The rms value of the PCC voltage from Table E8.12-2 is $V_{srms} = \sqrt{(254^2 + 0.313^2 + 0.828^2 + 0.815^2 + 0.497^2 + 0.201^2)} = 254.003\,415$ V.

The THD of the PCC voltage is $THD_{Vs} = \sqrt{(254.003\,415^2 - 254^2)}/254 = 0.005\,186 = 0.518\%$.

This distortion in the PCC voltage of 0.5186% is reduced from 2.234% without the filter. This distortion in the PCC voltage of 0.5186% is also reduced from 23.70% with the power factor correction capacitor.

d. The rms capacitor voltage is calculated as follows.

The fundamental current filter can be calculated as I_{f1} = nominal phase voltage/$(j\omega_s L_f - j/\omega_s C) = (440/\sqrt{3})/[j\{2 \times \pi \times 50 \times 27.883 \times 10^{-6} - 1/(2 \times \pi \times 50 \times 16\,449.97 \times 10^{-6})\}] = 254/\{j(0.008\,755\,262 - 0.193\,60\} = 1374.13$ A.

The fundamental capacitor voltage is $V_{C1} = I_{f1}X_{C1} = 1374.13/(2 \times \pi \times 50 \times 16\,449.97 \times 10^{-6}) = 1374.13 \times 0.193\,60 = 266.032$ V.

The filter capacitor and inductor elements have currents and voltages at all harmonics, including fundamental frequency. Because of series circuit and resonance, these elements have overvoltage and increased current compared with their fundamental voltage and current ratings. Therefore, the voltage and current of these elements may be checked for their selection. These harmonic voltages across the capacitor and an inductor are calculated as $V_{Ch} = I_{fh}/\omega_h C$ and $V_{Lh} = I_{fh}\omega_h L_f$.

These relations are used to compute harmonic voltages of the filter capacitor and inductor elements and the computed values are given in Table E8.12-2.

The rms voltage across the capacitor is $V_{Crms} = \sqrt{(266.032^2 + 2.384^2 + 0.681^2 + 0.182^2 + 0.075^2 + 0.017^2)} = 266.043\,19$ V.

The rms capacitor voltage as percentage of rated voltage is $V_{Crms}/V_s = 266.043\,19/254 = 1.047\,41 = 104.741\%$.

This rms capacitor voltage is only 104.741% of the rated supply voltage, which is within acceptable range.

The voltage across the filter capacitor is higher by 4.741%; therefore, this capacitor is selected of high voltage rating. However, fundamental reactive power rating has to be derated as

$$kVAR_A = kVAR_N(f_A/f_N)(V_A/V_N)^2 = 1000(50/50)(266.043\,19/254)^2 = 1097.076\,\text{kVAR},$$

where A represents actual and N represents nominal or rated quantities.

Therefore, a capacitor bank of 1097.076 kVAR has to be used in place of 1000 kVAR; it means 9.7076% derating has to be there for the capacitor bank.

Example 8.13

A three-phase 415 V, 50 Hz, six-pulse diode bridge rectifier with a DC bus capacitor filter is feeding a 20 kW variable-frequency VSI-fed induction motor drive. A three-phase three-branch passive series filter (tuned fifth, seventh, and high-block) is used in series with this rectifier–inverter system to reduce the harmonics in AC mains current as shown in Figure E8.13. Calculate (a) element values of the passive series filter, (b) rms line voltage at the input of the diode rectifier, (c) the line current, (d) the rms current of the passive series filter, (e) rms voltage across the passive series filter, and (f) VA rating of the passive series filter.

Solution: Given supply voltage $V_s = 415/\sqrt{3} = 239.6$ V, frequency of the supply (f) = 50 Hz, and active power of the load (P) = 20 000 W. A three-phase three-branch passive series power filter (tuned

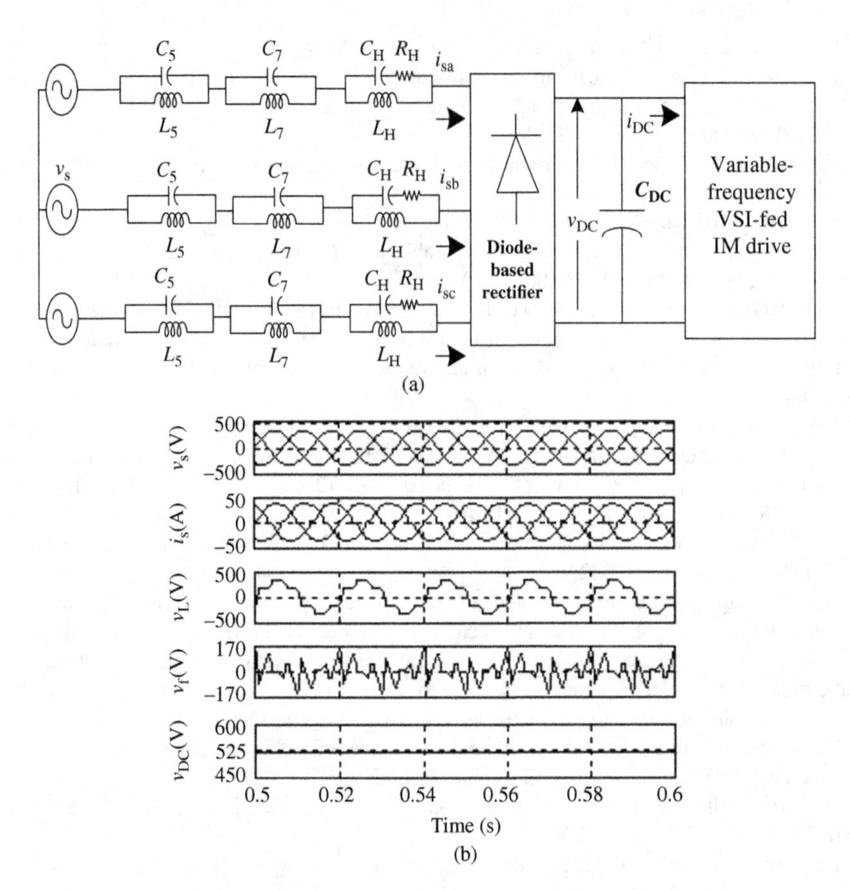

Figure E8.13 (a) A VSI-based induction motor drive with a six-pulse diode bridge rectifier and filter arrangement and (b) performance waveforms

fifth, seventh, and high-block) is designed to be used in series with this rectifier–inverter system to reduce the harmonics in AC mains current.

The passive series filters are more effective for such voltage-fed loads as their rating is lower than the passive shunt filters. Moreover, in place of the drop in line, an inductive effect of the passive series filter results in little boosting of the output DC bus voltage of the converter. Therefore, output DC bus voltage does not change much due to the use of a passive series filter.

Considering around 4% voltage drop of each series filter (totaling 3 × 4 = 12%) for the inductive effect of the passive series filter in series with the line between the AC mains and the diode rectifier, the corresponding reactance is $X_{PSF} = 0.12 \times Z_B = 0.12 \times V^2/P = 0.12 \times 415^2/20\,000 = 1.033\,35\ \Omega$.

The supply current after compensation by the series passive filter is $I_{s1} = 20\ 000/(3 \times 239.6) = 27.82$ A.

The approximate fundamental voltage at the input of the diode rectifier is calculated as $V_{CI} = \sqrt{\{V_s^2 - (I_{s1}X_{PSF})^2\}} = \sqrt{\{239.6^2 - (27.82 \times 1.033\,35)^2\}} = 237.86$ V.

The input voltage of the diode rectifier is a stepped wave as the segments of the DC bus voltage. From this phase fundamental voltage V_{C1}, the DC bus voltage of the diode rectifier is computed as follows:

$$V_{CI} = 4V_{DC} \sin 60°/(\sqrt{2}\sqrt{3}\pi) = 0.450\,158\,V_{DC}.$$

$$V_{DC} = 2.221\,44\,V_{CI} = 528.28\ \text{V}.$$

The sine wave supply current after compensation results in continuous conduction of diodes of the three-phase diode rectifier (each diode conducting for 180°). It also results in a stepped waveform of the phase voltage at the input of the diode (V_{PCCph}) with (i) the first step of $\pi/3$ angle (from 0° to $\pi/3$) and a magnitude of $V_{DCload}/3$, (ii) the second step of $\pi/3$ angle (from $\pi/3$ to $2\pi/3$) and a magnitude of $2V_{DCload}/3$, and (iii) the third step of $\pi/3$ angle (from $2\pi/3$ to π) and a magnitude of $V_{DCload}/3$, and it has both half cycles of symmetric segments of such steps.

a. The element values of the passive series filter are calculated as follows.
   Figure E8.13 shows the circuit of a three-element three-phase passive filter. It consists of fifth harmonic, seventh harmonic, and high-block passive filters. The inductances of these three series connected passive filter elements are considered equal. Therefore, the value of the inductance is $L_5 = L_7 = L_{HB} = X_{PSF}/3\omega = 1.033\,35/(3\times2\times\pi\times50) = 1.097$ mH.
   The tuned capacitor value for the fifth harmonic filter, which is connected in parallel to the inductor in the passive series filter, is $C_5 = 1/\{(5\omega)^2L_5\} = 1/\{(5\times2\times\pi\times50)^2\times1.097\times0.001\} =$ 369.45 μF.
   For a quality factor Q_5 of 20, the value of the resistance R_5 is $R_5 = X_5/20 = 5\times2\times\pi\times50\times1.097/(1000\times20) = 0.086\,\Omega$.
   The tuned capacitor value for the seventh harmonic filter, which is connected in parallel to the inductor in the passive series filter, is $C_7 = 1/\{(7\omega)^2L_7\} = 1/\{(7\times2\times\pi\times50)^2\times1.097\times0.001\} =$ 188.49 μF.
   For a quality factor Q_7 of 20, the value of the resistance R_7 is $R_7 = X_7/20 = 7\times2\times\pi\times50\times1.097/(1000\times20) = 0.1206\,\Omega$.
   The tuned capacitor value for the high-block filter (tuned at 11th harmonic), which is connected in parallel to the inductor in the passive series filter, is $C_{HP} = C_{11} = 1/\{(11\omega)^2L_7\} = 1/\{(11\times2\times\pi\times50)^2\times1.097\times0.001\} = 76.33$ μF.
   For a quality factor Q_{HB} of 5, the value of the resistance R_{HB} is $R_{HB} = X_{11}/5 = 11\times2\times\pi\times50\times1.097/(1000\times5) = 0.7578\,\Omega$.
b. The rms line voltage at the input of the diode rectifier is computed as follows.
   The sine wave supply current after compensation will result in continuous conduction of diodes of the rectifier (180°) and it results in quasi-square wave AC voltage at PCC with the amplitude of the DC bus voltage $V_{PCC} = V_{DC}(\sqrt{2}/\sqrt{3}) = 431.43$ V. Therefore, $V_{PCC} = 431.43$ V.
c. The supply current after compensation by the series passive filter is $I_{s1} = 20\,000/(3\times239.6) = 27.82$ A.
d. The current rating of the series passive filter is $I_f = I_{s1} = 20\,000/(3\times239.6) = 27.82$ A (since the passive filter is connected in series with the supply).
e. The rms voltage rating of the series passive filter (V_f) is computed by taking the difference of the supply phase voltage and phase voltage at the input of the diode rectifier:

$$V_f = \sqrt{\left[(1/\pi)\left\{\int_0^{\pi/3}(239.6\sqrt{2}\sin\theta - 176.09)^2\,d\theta + \int_{\pi/3}^{2\pi/3}(239.6\sqrt{2}\sin\theta - 352.18)^2\,d\theta + \int_{2\pi/3}^{\pi}(239.6\sqrt{2}\sin\theta - 176.09)^2\,d\theta\right\}\right]} = 73.96\text{ V}.$$

f. The kVA rating of the series passive filter is $S = 3V_fI_s = 3\times73.96\times27.82 = 6.173$ kVA.
   This filter results in a THD_{Vs} of 0.12% and a THD_{Is} of 7.5% through simulation as shown in Figure E8.13b.

Example 8.14

A passive filter system is to be designed for a 12-pulse converter HVDC station rated at ±110 kV, 100 MW DC, operating at $\alpha = 15°$, connected to a 275 kV, 50 Hz AC system via a converter transformer with a reactance of 3.75% as shown in Figure E8.14. If rated voltage is obtained with $\alpha = 15°$, then the transformer ratio must be 275/83 kV. A passive shunt filter system (consisting of a C-type filter tuned at 12th harmonic and a shunt tuned filter arm at 19th harmonic) is connected on the primary winding side rated at 50 MVAR total. Consider 90% of this reactive power is to be provided by the C-type filter. Calculate the element values of a passive shunt filter system.

Solution: Given AC mains rms voltage V_s = 275 kV, frequency of the supply (f) = 50 Hz, and a 12-pulse converter HVDC station rated at 110 kV, 100 MW DC, operating at $\alpha = 15°$, connected to a 275 kV, 50 Hz AC system via a converter transformer with a reactance of 3.75%. A passive shunt filter system (consisting of a C-type filter tuned at 12th harmonic and a shunt tuned filter arm at 19th harmonic) is connected on the primary winding side rated at 50 MVAR total. Consider 90% of this reactive power is to be provided by the C-type filter.

The base impedance is $Z_B = \sqrt{3V_L^2/P} = \sqrt{3 \times (275 \times 10^3)^2/(100 \times 10^6)} = 1309.86\,\Omega$.

$$X_s = 0.0375 Z_B = 0.375 \times 1309.86 = 49.12\,\Omega.$$

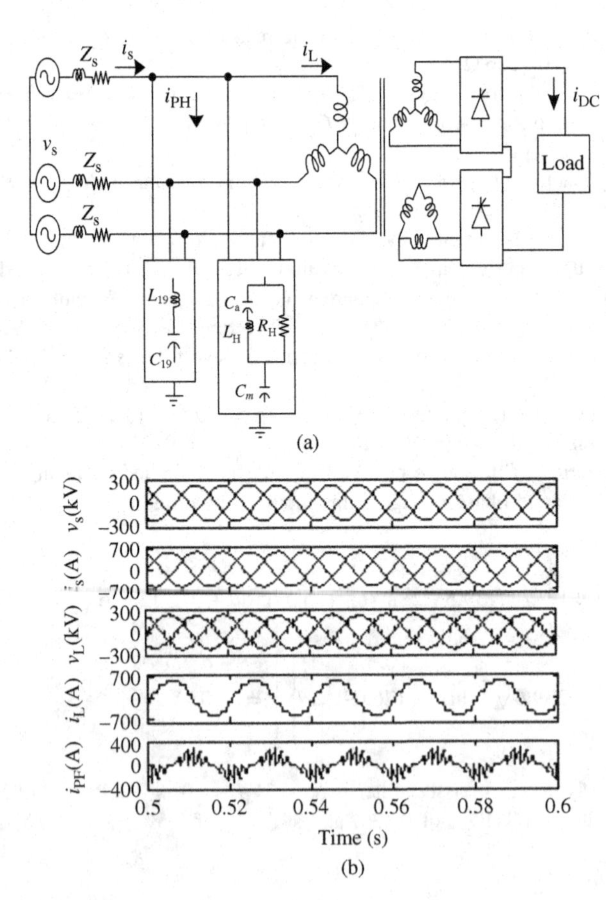

Figure E8.14 (a) A 12-pulse HVDC converter station with a single-phase tuned filter and a C-type filter and (b) performance waveforms

The C-type filter is a high-pass filter with the capacitor C_a connected in series with the reactor X. Its capacitive reactance needs to be equal to the inductive reactance of X at the fundamental frequency. The series connection of C_a and X avoids the fundamental frequency current to flow through R and it reduces power losses in the filter. The C-type filter may also be tuned to a lower frequency than a conventional high-pass filter and it may provide harmonic attenuation at its tuned frequency. As the C-type filter provides adequate attenuation at higher frequencies, the harmonic level can be brought within the permissible limits by addition of a single tuned filter of 19th harmonic. Here, a single tuned filter is tuned at $h_T = 19$ to reduce a sufficient amount of 23rd and 25th harmonic currents without shifting the parallel resonance to lower frequencies. The C-type filter also has low initial cost, low losses, superior noncharacteristic harmonics and interharmonic damping, better damping of energizing transient, and higher robustness against tuning uncertainties compared with a conventional tuned and high-pass filter system. The parameters of C-type filter can be obtained as follows.

From the reactive power supplied by the C-type filter, the value of C_m is calculated as follows:

$$X_{Cm} = V_L^2/Q = (275 \times 10^3)^2/(0.9 \times 50 \times 10^6) = 1680.55\,\Omega.$$

$$C_m = 1/X_{Cm}\omega = 1/(1680.55 \times 2 \times \pi \times 50) = 1.894\,\mu\text{F}.$$

$$L = X/\omega = V_L^2/\{Q\omega(h_T^2 - 1)\} = (275 \times 10^3)^2/\{(0.9 \times 50 \times 10^6) \times (2 \times \pi \times 50) \times (12^2 - 1)\}$$

$$= 0.037\,408 = 37.41\,\text{mH}.$$

$$X = \omega L = 11.746\,\Omega.$$

$$C_a = 1/X\omega = 1/L\omega^2 = 271.1276\,\mu\text{F}.$$

$$R_{12} = qV_L^2/Qh_T = 2 \times (275 \times 10^3)^2/\{(0.9 \times 50 \times 10^6) \times 12\}$$

$$= 280.093\,\Omega \quad \text{(considering a quality factor } q \text{ of 2)}.$$

A passive shunt single tuned filter is tuned at $h_T = 19$ and its design parameters are calculated as follows.

It needs to provide only 10% of the total reactive power of 50 MVAR.

$$L_{19} = V_L^2/\{Q\omega(h_T^2 - 1)\} = (275 \times 10^3)^2/\{(0.1 \times 50 \times 10^6) \times (2 \times \pi \times 50) \times (19^2 - 1)\}$$

$$= 0.1338\,\text{H} = 133.8\,\text{mH}.$$

$$C_{19} = (h_T^2 - 1)Q/V_L^2\omega h_T^2 = (19^2 - 1) \times (0.9 \times 50 \times 10^6)/\{(275 \times 10^3)^2 \times (2 \times \pi \times 50) \times (19^2)\}$$

$$= 0.2098\,\mu\text{F}.$$

The resistance of the 19th-order passive shunt tuned filter is $R_{19} = h_T\omega L/q = 19 \times (2 \times \pi \times 50) \times 0.13378/50 = 15.964\,\Omega$.

This filter results in a THD_{Is} of 1.8% through simulation as shown in Figure E8.14b.

Example 8.15

A three-phase double tuned passive shunt filter (double tuned fifth and seventh and high-pass) is employed to reduce the THD of the supply current and to improve the displacement factor to unity for a three-phase 415 V, 50 Hz fed 300 kVA, 415 V, three-phase thyristor bridge converter feeding a line commutated inverter synchronous motor drive used in an industry at 30° firing angle of its thyristors as shown in Figure E8.15. Calculate (a) fundamental active power drawn by the load, (b) fundamental reactive power drawn by the load, and (c) values of filter elements. It has per-phase source impedance of $R_T = 0.01\ \Omega$ and $X_T = 0.078$ at 50 Hz.

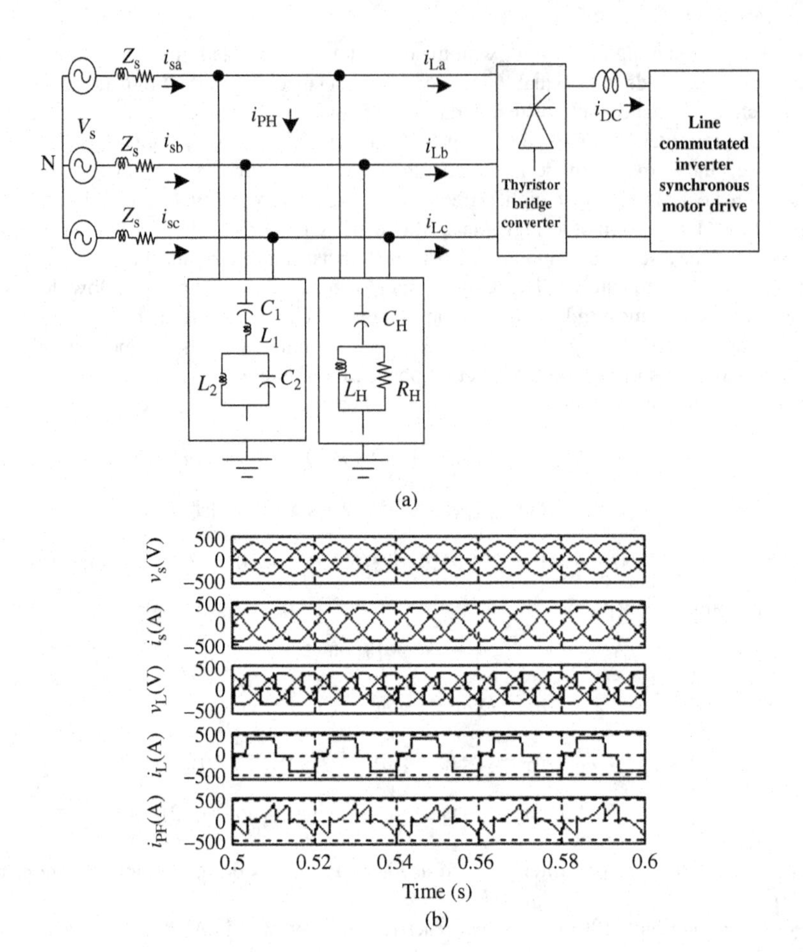

Figure E8.15 (a) A three-phase thyristor bridge converter feeding a LCI synchronous motor drive with a passive filter and (b) performance waveforms

Solution: Given supply rms voltage $V_s = 415/\sqrt{3} = 239.6$ V, frequency of the supply (f) = 50 Hz, and a 415 V, 50 Hz fed 300 kVA, 415 V, three-phase thyristor bridge converter feeding a line commutated inverter synchronous motor drive used in an industry at 30° firing angle of its thyristors.

In this system, the load current harmonic and reactive power compensation is provided by a three-phase double tuned passive shunt filter (double tuned fifth and seventh and high-pass damped) to reduce the THD of the supply current and to improve the displacement factor close to unity.

The AC load rms current is I_L = kVA × 1000/3V_s = 300 × 1000/(3 × 239.6) = 417.36 A.

$$I_{DC} = (\sqrt{3}/\sqrt{2})I_L = 511.16\,\text{A}.$$

The fundamental rms input current of the thyristor bridge converter is $I_{L1} = (\sqrt{6}/\pi)I_{load} = 0.7797 \times 511.16 = 398.5$ A.

The fundamental active component of load current is $I_{L1a} = I_{L1}\cos\alpha = 398.5\cos 30° = 345.15$ A.

a. The fundamental active power of the load is $P_1 = 3V_{s1}I_{L1}\cos\theta_1 = 3V_{s1}I_{L1}\cos\alpha = 3 \times 239.6 \times 345.15 = 248.1$ kW.
b. The fundamental reactive power of the load is $Q_1 = 3V_{s1}I_{L1}\sin\theta_1 = 3V_{s1}I_{L1}\sin\alpha = 3 \times 239.6 \times 398.5 \times 0.5 = 143.2$ kVAR.

The source impedance is $Z_s = R_T + jX_T = 0.01 + j0.078 = 0.0786\ \Omega$.

The voltage drop in the source impedance is $V_{Zs} = 345.15 \times 0.0786 = 27.128$ V.

The fundamental voltage across the load is $V_L = \sqrt{(V_s^2 - V_{Zs}^2)} = 238.06$ V.

c. A three-phase double tuned passive shunt filter (double tuned fifth and seventh and high-pass damped) is used to reduce harmonics and reactive power. The reactive power of 143.2 kVAR required by the three-phase thyristor rectifier has to be provided by two branches of the passive shunt filter.

A double tuned passive shunt filter for fifth and seventh harmonics and high-pass damped filter is designed as follows.

Considering that all branches of the double tuned filter and the high-pass damped filter have equal capacitors, the values of these capacitors are $C_a = C_b = C_H = Q_1/9V_s^2\omega = 882.4\ \mu F$, where C_a and C_b are the capacitances of two equivalent single tuned filters of the double tuned filter.

$$L_a = 1/\omega_5^2 C_a = 0.459\ \text{mH},$$

$$L_b = 1/\omega_7^2 C_b = 0.234\ \text{mH},$$

where L_a and L_b are the inductances of two equivalent single tuned filters of the double tuned filter.

$$C_1 = C_a + C_b = 1800\ \mu F.$$

$$C_2 = C_a C_b (C_a + C_b)(L_a + L_b)^2/(L_a C_a - L_b C_b)^2 = 1680\ \mu F.$$

$$L_1 = L_a L_b/(L_a + L_b) = 0.155\ \text{mH}.$$

$$L_2 = (L_a C_a - L_b C_b)^2/\{(C_a + C_b)^2 (L_a + L_b)\} = 0.018\ \text{mH}.$$

$$L_H = 1/\omega_{11}^2 C_H = 0.0948\ \text{mH}.$$

$R_H = X_H/Q_H = 0.65\ \Omega$ (considering $Q_H = 0.5$ as it may be in the range of 0.5–5 depending upon the attenuation required).

This filter results in a THD_{Is} of 3% through simulation as shown in Figure E8.15b.

8.10 Summary

Passive filters are widely used to limit harmonic propagation, to improve power quality, to reduce harmonic distortion, and to provide reactive power compensation. These are designed for high-current and high-voltage applications. Many such filters are in operation for HVDC transmission systems, large industrial drives, static VAR compensators, and so on. The passive filters are also used along with small rating active filters as hybrid filters where the major portion of filtering is taken care by passive filters.

8.11 Review Questions

1. What is a tuned passive shunt power filter?
2. What are the limitations and demerits of a tuned passive shunt power filter?
3. What are the factors on which the performance of a tuned passive shunt power filter depends?
4. What is the value of quality factor of a tuned passive shunt power filter and on what factors it depends?
5. What are the factors on which the rating, cost, and size of a tuned passive shunt power filter depend?
6. Why the resonance is caused due to source impedance and it appears in a tuned passive shunt power filter?
7. What are the problems that result in the system due to the resonance caused by a tuned passive shunt power filter?

8. What is a damped passive shunt power filter?
9. What are the limitations and demerits of a damped passive shunt power filter?
10. What are the factors on which the performance of a damped passive shunt power filter depends?
11. What is the value of quality factor of a damped passive shunt power filter and on what factors it depends?
12. What are the factors on which the rating, cost, and size of a damped passive shunt power filter depend?
13. What is a tuned passive series power filter?
14. What are the limitations and demerits of a tuned passive series power filter?
15. What are the factors on which the performance of a tuned passive series power filter depends?
16. What is the value of quality factor of a tuned passive series power filter and on what factors it depends?
17. What are the factors on which the rating, cost, and size of a tuned passive series power filter depend?
18. What is a damped passive series power filter?
19. What are the limitations and demerits of a damped passive series power filter?
20. What are the factors on which the performance of a damped passive series power filter depends?
21. What is the value of quality factor of a damped passive series power filter and on what factors it depends?
22. What are the factors on which the rating, cost, and size of a damped passive series power filter depend?
23. What are the passive ripple filters?
24. What are the factors on which the rating, cost, and size of passive ripple filters depend?
25. What are the factors that cause detuning of passive power filters?

8.12 Numerical Problems

1. A single-phase diode bridge rectifier is supplied from a 220 V, 50 Hz AC mains as shown in Figure E8.1. The load resistance is $R = 50\ \Omega$. (a) Design a capacitive filter so that the ripple factor of the output voltage is less than 10%. (b) With this value of the capacitor, calculate the average load voltage V_{DC}.

2. A single-phase diode bridge rectifier is supplied from a 220 V, 50 Hz AC mains as shown in Figure E8.2. The DC load resistance is $R = 100\ \Omega$ and load inductance is $L = 20$ mH. Design a DC side LC filter so that the ripple factor of the output voltage is less than 10%.

3. A single-phase three-branch shunt passive filter (third, fifth, and high-pass) is employed to reduce the THD of the supply current and to improve the displacement factor to unity of a single-phase 220 V, 50 Hz fed diode bridge converter with an overlap angle of 20° drawing 15 A constant DC current as shown in Figure E8.3. Calculate (a) fundamental active power drawn by the load, (b) fundamental reactive power drawn by the load, (c) element values of the passive filter, and (d) current and VA ratings of the passive filter. Let the supply has 7% source impedance mainly inductive.

4. A single-phase four-branch shunt passive filter (third, fifth, seventh, and high-pass) is used in a single-phase 220 V, 50 Hz system to reduce the THD of the supply current and to improve the displacement factor to unity. It has a load of a thyristor bridge converter operating at 30° firing angle

of its thyristors drawing constant 15 A DC current as shown in Figure E8.4. Calculate (a) element values of the passive filter, (b) total harmonic distortion of the supply current, (c) total harmonic distortion of the terminal voltage at the load end, (d) the current rating of the passive filter, and (e) its kVA rating to provide harmonic and reactive power compensation. Let the supply has 5% source impedance mainly inductive.

5. A three-phase diode bridge rectifier is supplied from a 415 V, 50 Hz AC mains as shown in Figure E8.5. The load resistance is $R = 20\ \Omega$. (a) Design a DC bus parallel capacitive filter so that the ripple factor of the output voltage is less than 6%. (b) With this value of the capacitor, calculate the average load voltage V_{DC}.

6. A three-phase diode bridge rectifier is supplied from a 440 V, 50 Hz AC mains as shown in Figure E8.6. The DC load resistance is $R = 25\ \Omega$ and load inductance is 10 mH. Design a DC side LC filter so that the ripple factor of the output voltage is less than 5%.

7. A three-phase three-branch shunt passive filter (tuned fifth, seventh, and high-pass) is employed to reduce the THD of the supply current and to improve the displacement factor to unity for a three-phase 400 V, 50 Hz fed six-pulse thyristor bridge converter drawing 200 A constant DC current at 45° firing angle of its thyristors as shown in Figure E8.7. Calculate (a) fundamental active power drawn by the load, (b) fundamental reactive power drawn by the load, (c) values of filter elements, (d) THD of the supply current, (e) THD of the load voltage, and (f) the voltage, current, and VA ratings of the passive filter. Let the supply has 5% source impedance mainly inductive.

8. A three-phase one-branch shunt passive filter (11th harmonic high-pass damped) is employed to reduce the THD of the supply current and to improve the displacement factor to unity of a three-phase 440 V, 50 Hz, nonlinear load consisting of a 12-pulse thyristor bridge converter with 300 A constant DC current at 20° firing angle of its thyristors as shown in Figure E8.8. This converter consists of an ideal transformer with single primary star connected winding and two secondary windings connected in star and delta with same line voltages to provide 30° phase shift between two sets of three-phase output voltages. Two 6-pulse thyristor bridges are connected in series to form this 12-pulse AC–DC converter. Calculate (a) fundamental active power drawn by the load, (b) fundamental reactive power drawn by the load, (c) values of filter elements, (d) THD of the supply current with the passive filter, (e) THD of the load voltage with the passive filter, and (f) the voltage, current, and VA ratings of the passive filter. Let the supply has 6% source impedance mainly inductive.

9. A three-phase three-branch shunt passive filter (tuned 11th, 13th, and high-pass) is employed to reduce the THD of the supply current and to improve the displacement factor to unity of a three-phase 400 V, 50 Hz, nonlinear load consisting of a 12-pulse thyristor bridge converter with 200 A constant DC current at 45° firing angle of its thyristors as shown in Figure E8.9. This converter consists of an ideal transformer with single primary star connected winding and two secondary windings connected in star and delta with same line voltages to provide 30° phase shift between two sets of three-phase output voltages. Two 6-pulse thyristor bridges are connected in series to form this 12-pulse AC–DC converter. Calculate (a) fundamental active power drawn by the load, (b) fundamental reactive power drawn by the load, (c) values of filter elements, (d) THD of the supply current with the passive filter, (e) THD of the load voltage with the passive filter, and (f) the voltage, current, and VA ratings of the passive filter. Let the supply has 6% source impedance mainly inductive.

10. An industry is fed electric power from a three-phase 11 kV, 50 Hz AC mains as shown in Figure E8.10. The data of the supply feeder are as follows: short-circuit level of 100 MVA and an X/R ratio of 4. A step-down transformer is placed between the AC mains and this industry with a rating of 1000 kVA, 11 kV/440 V, $R = 1\%$, and $X = 6\%$. It is supplying linear and nonlinear loads. If an AC capacitor bank is used at the PCC for power factor correction of the industry, calculate rating of this capacitor bank, which may excite parallel resonance at fifth harmonic with the supply system.

11. An industry is fed electric power from a three-phase 11 kV, 50 Hz AC mains as shown in Figure E8.11. The data of the supply feeder are as follows: short-circuit level of 100 MVA and an

Table P8.11-1 Harmonic spectrum of a 200 kVA, 440 V, three-phase thyristor converter-fed DC motor drive

Harmonic order	Frequency (Hz)	I_h (A)
5th	250	50
7th	350	30
11th	550	20
13th	650	5
17th	850	3

X/R ratio of 4. A step-down transformer is placed between the AC mains and this industry with a rating of 1000 kVA, 11 kV/440 V, $R = 1\%$, and $X = 6\%$. A 200 kVA, 440 V, a three-phase thyristor bridge converter-fed DC motor drive is used in this industry along with other linear loads. If a 600 kVAR AC capacitor bank is used at the PCC for power factor correction of the industry, calculate (a) PCC voltage and its THD without the capacitor bank, (b) PCC voltage and its THD with the capacitor bank, and (c) current in the capacitor bank. The harmonic spectrum of AC current of this 200 kVA, 440 V, three-phase thyristor bridge converter-fed DC motor drive is given in Table P8.11-1.

12. An industry is fed electric power from a three-phase 11 kV, 50 Hz AC mains as shown in Figure E8.12. The data of the supply feeder are as follows: short-circuit level of 100 MVA and an X/R ratio of 3. A step-down transformer is placed between the AC mains and this industry with a rating of 1000 kVA, 11 kV/440 V, $R = 1\%$, and $X = 6\%$. A 200 kVA, 440 V, a three-phase thyristor bridge converter-fed DC motor drive is used in this industry along with other linear loads. If a 600 kVAR AC capacitor bank used at the PCC for power factor correction of the industry is tuned to the lowest harmonic (fifth) passive shunt filter, calculate (a) parallel resonance frequency, (b) the filter harmonic current spectrum and rms current of the filter, (c) PCC voltage and its THD with the passive filter, and (d) rms capacitor voltage. Harmonic spectrum of input AC current of this 200 kVA, 440 V, three-phase thyristor bridge converter-fed DC motor drive is given in Table P8.12-1.

13. A three-phase 415 V, 50 Hz, six-pulse diode bridge rectifier with a DC bus capacitor filter is feeding a 10 kW variable-frequency VSI-fed induction motor drive. A three-phase three-branch passive series filter (tuned fifth, seventh, and high-block) is used in series with this rectifier–inverter system to reduce the harmonics in AC mains current as shown in Figure E8.13. Calculate (a) element values of the passive series filter, (b) rms line voltage at the input of the diode rectifier, (c) the line current, (d) the rms current of the passive series filter, (e) rms voltage across the passive series filter, and (f) VA rating of the passive series filter.

14. A passive filter system is to be designed for a 12-pulse converter HVDC station rated at ±220 kV, 200 MW DC, operating at $\alpha = 15°$, connected to a 275 kV, 50 Hz AC system via a converter transformer with a reactance of 4% as shown in Figure E8.14. If rated voltage is obtained with $\alpha = 15°$, then the transformer ratio must be 275/83 kV. A passive shunt filter system (consisting of a C-type filter tuned at 12th harmonic and a shunt tuned filter arm at 19th harmonic) is connected on the

Table P8.12-1 Harmonic spectrum of a 200 kVA, 440 V, three-phase thyristor converter-fed DC motor drive

Harmonic order	Frequency (Hz)	I_h (A)
5th	250	50
7th	350	30
11th	550	20
13th	650	5
17th	850	3

primary winding side rated at 100 MVAR total. Consider 90% of this reactive power is to be provided by the C-type filter. Calculate the element values of a passive filter system.

15. A three-phase double tuned passive shunt filter (double tuned fifth and seventh and high-pass) is employed to reduce the THD of the supply current and to improve the displacement factor to unity for a three-phase 415 V, 50 Hz fed 200 kVA, 415 V, three-phase thyristor bridge converter feeding a line commutated inverter synchronous motor drive used in an industry at 30° firing angle of its thyristors as shown in Figure E8.15. Calculate (a) fundamental active power drawn by the load, (b) fundamental reactive power drawn by the load, and (c) values of filter elements. It has per-phase source impedance of $R_T = 0.01\ \Omega$ and $X_T = 0.076\ \Omega$ at 50 Hz.

8.13 Computer Simulation-Based Problems

1. Design an AC shunt passive filter (consisting of two tuned branches and a high-pass branch) and simulate its behavior for reducing harmonic currents and minimizing AC mains current of a single-phase 230 V, 50 Hz, uncontrolled bridge converter with a parallel capacitive DC filter of 470 μF and an equivalent resistive load of 20 Ω. It has a source impedance of 0.25 Ω resistive element and 1.0 Ω inductive element. Plot supply voltage and current, AC load current and voltage at PCC, passive filter currents with time, and harmonic spectra of AC mains current, AC load current, and voltage at PCC. Calculate THD, crest factor, rms value of AC mains current, displacement factor, distortion factor, power factor, input active, reactive, and output powers, and rating of passive filters.
2. Design an AC shunt passive filter (consisting of two tuned branches and a high-pass branch) and simulate its behavior for reducing harmonic currents and minimizing AC mains current of a single-phase AC voltage controller having a RL load of $R = 10\ \Omega$ and $L = 20$ mH. The input voltage is 230 V (rms) at 50 Hz. The delay angle of thyristors is $\alpha = 60°$. It has a source impedance of 0.15 Ω resistive element and 0.75 Ω inductive element. Plot supply voltage and current, AC load current and voltage at PCC, passive filter currents with time, and harmonic spectra of AC mains current, AC load current, and voltage at PCC. Calculate THD, crest factor, rms value of AC mains current, displacement factor, distortion factor, power factor, input active, reactive, and output powers, and rating of passive filters.
3. Design an AC series passive filter (consisting of two tuned branches and a high-block branch) and simulate its behavior for reducing harmonic currents and minimizing AC mains current of a single-phase 230 V, 50 Hz, uncontrolled bridge converter with a parallel capacitive DC filter of 1000 μF and an equivalent resistive load of 10 Ω. It has a source impedance of 0.15 Ω resistive element and 0.75 Ω inductive element. Plot supply voltage and current, AC load current and voltage at PCC, voltage across passive filter with time, and harmonic spectra of AC mains current and voltage at PCC. Calculate THD, crest factor, rms value of AC mains current, displacement factor, distortion factor, power factor, input active, reactive, and output powers, and rating of passive filters.
4. Design an AC passive filter (consisting of two tuned series branches and a high-pass parallel branch) and simulate its behavior for reducing harmonic currents and minimizing AC mains current of a single-phase 220 V, 50 Hz, uncontrolled bridge converter with a parallel capacitive DC filter of 1500 μF and an equivalent resistive load of 15 Ω. It has a source impedance of 0.1 Ω resistive element and 0.5 Ω inductive element. Plot supply voltage and current, AC load current and voltage at PCC, voltage across and current through passive filters with time, and harmonic spectra of AC mains current, AC load current, and voltage at PCC. Calculate THD, crest factor, rms value of AC mains current, displacement factor, distortion factor, power factor, input active, reactive, and output powers, and rating of passive filters.
5. Design an AC passive filter (consisting of a single tuned series branch and a compensating capacitor parallel branch) and simulate its behavior for reducing harmonic currents and minimizing AC mains current of a single-phase 230 V, 50 Hz, uncontrolled bridge converter with a parallel capacitive DC

filter of 680 μF and an equivalent resistive load of 10 Ω. It has a source impedance of 0.15 Ω resistive element and 0.75 Ω inductive element. Plot supply voltage and current, AC load current and voltage at PCC, voltage across and current through passive filters with time, and harmonic spectra of AC mains current, AC load current, and voltage at PCC. Calculate THD, crest factor, rms value of AC mains current, displacement factor, distortion factor, power factor, input active, reactive, and output powers, and rating of passive filters.

6. Design an AC shunt passive filter (consisting of two tuned branches and a high-pass branch) and simulate its behavior for reducing harmonic currents and minimizing AC mains current of a single-phase 230 V, 50 Hz, controlled bridge converter with a series connected inductive load of 20 mH, an equivalent resistive load of 2 Ω, and a back emf of 60 V. It has a source impedance of 0.2 Ω resistive element and 1.5 Ω inductive element. The delay angle of its thyristors is $\alpha = 60°$. Plot supply voltage and current, AC load current and voltage at PCC, current through passive filters with time, and harmonic spectra of AC mains current, AC load current, and voltage at PCC. Calculate THD, crest factor, rms value of AC mains current, displacement factor, distortion factor, power factor, input active, reactive, and output powers, and rating of passive filters.

7. Design an AC shunt passive filter (consisting of two tuned branches and a high-pass branch) and simulate its behavior for reducing harmonic currents and minimizing AC mains current of a three-phase 415 V, 50 Hz, uncontrolled bridge converter with a parallel capacitive DC filter of 4700 μF and a resistive load of 10 Ω. It has a source impedance of 0.25 Ω resistive element and 1.5 Ω inductive element. Plot supply voltage and current, AC load current and voltage at PCC, current through passive filters with time, and harmonic spectra of AC mains current, AC load current, and voltage at PCC. Calculate THD, crest factor, rms value of AC mains current, displacement factor, distortion factor, power factor, input active, reactive, and output powers, and rating of passive filters.

8. Design an AC shunt passive filter (consisting of four tuned branches and a high-pass branch) and simulate its behavior for reducing harmonic currents and minimizing AC mains current of a three-phase 440 V, 50 Hz, uncontrolled bridge converter with a parallel capacitive DC filter of 3000 μF and a resistive load of 20 Ω. It has a source impedance of 0.15 Ω resistive element and 0.75 Ω inductive element. Plot supply voltage and current, AC load current and voltage at PCC, current through passive filters with time, and harmonic spectra of AC mains current, AC load current, and voltage at PCC. Calculate THD, crest factor, rms value of AC mains current, displacement factor, distortion factor, power factor, input active, reactive, and output powers, and rating of passive filters.

9. Design an AC shunt 11th harmonic high-pass passive filter and simulate its behavior for reducing harmonic currents and minimizing AC mains current of a three-phase 415 V, 50 Hz system, having a 12-pulse diode bridge converter with a parallel capacitive DC filter of 4700 μF and a resistive load of 10 Ω. It consists of an ideal transformer with single primary star connected winding and two secondary windings connected in star and delta with same output line voltages as the input voltage at no load to provide 30° phase shift between two sets of three-phase output voltages. Two 6-pulse diode bridges are connected in series to provide this 12-pulse AC–DC converter. It has a source impedance of 0.15 Ω resistive element and 0.75 Ω inductive element. Plot supply voltage and current, AC load current and voltage at PCC, current through passive filters with time, and harmonic spectra of AC mains current, AC load current, and voltage at PCC. Calculate THD, crest factor, rms value of AC mains current, displacement factor, distortion factor, power factor, input active, reactive, and output powers, and rating of passive filters.

10. Design an AC shunt passive filter (consisting of two tuned branches and a high-pass branch) and simulate its behavior for reducing harmonic currents and minimizing AC mains current of a three-phase 440 V, 50 Hz system, having a 12-pulse diode bridge converter with a parallel capacitive DC filter of 3000 μF and a resistive load of 20 Ω. It consists of an ideal transformer with single primary star connected winding and two secondary windings connected in star and delta with same output line voltages as the input voltage at no load to provide 30° phase shift between two sets of three-phase

output voltages. Two 6-pulse diode bridges are connected in series to provide this 12-pulse AC–DC converter. It has a source impedance of 0.15 Ω resistive element and 0.75 Ω inductive element. Plot supply voltage and current, AC load current and voltage at PCC, current through passive filters with time, and harmonic spectra of AC mains current, AC load current, and voltage at PCC. Calculate THD, crest factor, rms value of AC mains current, displacement factor, distortion factor, power factor, input active, reactive, and output powers, and rating of passive filters.

11. Design an AC shunt passive filter (consisting of two tuned branches and a high-pass branch) and simulate its behavior for a three-phase uncontrolled six-pulse isolated bridge converter with a parallel capacitive DC filter and a resistive load. Plot supply voltage and current, load current and voltage, device voltage and current with time, and its harmonic spectrum. Calculate THD, crest factor, rms value of AC mains current, displacement factor, distortion factor, power factor, % ripple in load currents, output voltage ripple, ripple factor, and powers at two loading conditions.

12. Design an AC shunt passive filter (consisting of four tuned branches and a high-pass branch) and simulate its behavior for a three-phase uncontrolled six-pulse isolated bridge converter with a parallel capacitive DC filter and a resistive load. Plot supply voltage and current, load current and voltage, device voltage and current with time, and its harmonic spectrum. Calculate THD, crest factor, rms value of AC mains current, displacement factor, distortion factor, power factor, % ripple in load currents, output voltage ripple, ripple factor, and powers at two loading conditions.

13. Design an AC shunt 11th harmonic high-pass passive filter and simulate its behavior for a three-phase uncontrolled 12-pulse isolated bridge converter with a parallel capacitive DC filter and a resistive load. Plot supply voltage and current, load current and voltage, device voltage and current with time, and its harmonic spectrum. Calculate THD, crest factor, rms value of AC mains current, displacement factor, distortion factor, power factor, % ripple in load currents, output voltage ripple, ripple factor, and powers at two loading conditions.

14. Design an AC shunt passive filter (consisting of two tuned branches and a high-pass branch) and simulate its behavior for a three-phase uncontrolled 12-pulse isolated bridge converter with a parallel capacitive DC filter and a resistive load. Plot supply voltage and current, load current and voltage, device voltage and current with time, and its harmonic spectrum. Calculate THD, crest factor, rms value of AC mains current, displacement factor, distortion factor, power factor, % ripple in load currents, output voltage ripple, ripple factor, and powers at two loading conditions.

15. Design an AC shunt passive filter (consisting of two tuned branches and a high-pass branch) and simulate its behavior for a three-phase uncontrolled six-pulse bridge converter with *RL* and *RLE* loads. Plot supply voltage and current, load current and voltage, device voltage and current with time, and its harmonic spectrum. Calculate THD, crest factor, rms value of AC mains current, displacement factor, distortion factor, power factor, % ripple in load currents, output voltage ripple, ripple factor, and powers at two loading conditions.

16. Design an AC shunt passive filter (consisting of two tuned branches and a high-pass branch) and simulate its behavior for a three-phase uncontrolled 12-pulse bridge converter with *RL* and *RLE* loads. Plot supply voltage and current, load current and voltage, device voltage and current with time, and its harmonic spectrum. Calculate THD, crest factor, rms value of AC mains current, displacement factor, distortion factor, power factor, % ripple in load currents, output voltage ripple, ripple factor, and powers at two loading conditions.

17. Design an AC shunt passive filter (consisting of two tuned branches and a high-pass branch) and simulate its behavior for a three-phase uncontrolled six-pulse isolated bridge converter with *RL* and *RLE* loads. Plot supply voltage and current, load current and voltage, device voltage and current with time, and its harmonic spectrum. Calculate THD, crest factor, rms value of AC mains current, displacement factor, distortion factor, power factor, % ripple in load currents, output voltage ripple, ripple factor, and powers at two loading conditions.

18. Design an AC shunt passive filter (consisting of two tuned branches and a high-pass branch) and simulate its behavior for a three-phase uncontrolled 12-pulse isolated bridge converter with *RL* and *RLE* loads. Plot supply voltage and current, load current and voltage, device voltage and current with time, and its harmonic spectrum. Calculate THD, crest factor, rms value of AC mains current, displacement factor, distortion factor, power factor, % ripple in load currents, output voltage ripple, ripple factor, and powers at two loading conditions.

19. Design an AC shunt passive filter (consisting of two tuned branches and a high-pass branch) and simulate its behavior for a three-phase controlled six-pulse bridge converter with *RL* and *RLE* loads. Plot supply voltage and current, load current and voltage, device voltage and current with time, and its harmonic spectrum. Calculate THD, crest factor, rms value of AC mains current, displacement factor, distortion factor, power factor, % ripple in load currents, output voltage ripple, ripple factor, and powers at two loading conditions.

20. Design an AC shunt passive filter (consisting of two tuned branches and a high-pass branch) and simulate its behavior for a three-phase controlled 12-pulse bridge converter with *RL* and *RLE* loads. Plot supply voltage and current, load current and voltage, device voltage and current with time, and its harmonic spectrum. Calculate THD, crest factor, rms value of AC mains current, displacement factor, distortion factor, power factor, % ripple in load currents, output voltage ripple, ripple factor, and powers at two loading conditions.

21. Design an AC shunt passive filter (consisting of two tuned branches and a high-pass branch) and simulate its behavior for a three-phase controlled six-pulse isolated bridge converter with *RL* and *RLE* loads. Plot supply voltage and current, load current and voltage, device voltage and current with time, and its harmonic spectrum. Calculate THD, crest factor, rms value of AC mains current, displacement factor, distortion factor, power factor, % ripple in load currents, output voltage ripple, ripple factor, and powers at two loading conditions.

22. Design an AC shunt passive filter (consisting of two tuned branches and a high-pass branch) and simulate its behavior for a three-phase controlled 12-pulse isolated bridge converter with *RL* and *RLE* loads. Plot supply voltage and current, load current and voltage, device voltage and current with time, and its harmonic spectrum. Calculate THD, crest factor, rms value of AC mains current, displacement factor, distortion factor, power factor, % ripple in load currents, output voltage ripple, ripple factor, and powers at two loading conditions.

23. Design an AC series passive filter (consisting of two tuned branches and a high-block branch) and simulate its behavior for a three-phase uncontrolled six-pulse bridge converter with a parallel capacitive DC filter and a resistive load. Plot supply voltage and current, load current and voltage, device voltage and current with time, and its harmonic spectrum. Calculate THD, crest factor, rms value of AC mains current, displacement factor, distortion factor, power factor, % ripple in load currents, output voltage ripple, ripple factor, and powers at two loading conditions.

24. Design an AC passive filter (consisting of two tuned series branches and a high-pass parallel branch) and simulate its behavior for a three-phase uncontrolled six-pulse bridge converter with a parallel capacitive DC filter and a resistive load. Plot supply voltage and current, load current and voltage, device voltage and current with time, and its harmonic spectrum. Calculate THD, crest factor, rms value of AC mains current, displacement factor, distortion factor, power factor, % ripple in load currents, output voltage ripple, ripple factor, and powers at two loading conditions.

25. Design an AC shunt passive filter (consisting of two tuned branches and a high-pass branch) and simulate its behavior for a three-phase AC voltage controller with a resistive load. Plot supply voltage and current, load current and voltage, device voltage and current with time, and its harmonic spectrum. Calculate THD, crest factor, rms value of AC mains current, displacement factor, distortion factor, power factor, % ripple in load currents, output voltage, and powers at two loading conditions.

References

1. Kimbark, E.W. (1971) *Direct Current Transmission*, vol. 1, John Wiley & Sons, Inc., New York.
2. Bosela, T.R. (1997) *Introduction to Electrical Power System Technology*, Prentice-Hall, Upper Saddle River, NJ.
3. Das, J.C. (2002) *Power System Analysis: Short-Circuit Load Flow and Harmonics*, Marcel Dekker, New York.
4. IEEE Standard 1573 (2003) *IEEE Guide for Application and Specification of Harmonic Filters*, IEEE.
5. Vedam, R.S. and Sarma, M.S. (2008) *Power Quality: VAR Compensation in Power Systems*, CRC Press, New York.
6. Steeper, D.E. and Stratford, R.P. (1976) Reactive compensation and harmonic suppression for industrial power systems using thyristor converters. *IEEE Transactions on Industry Applications*, 12(3), 232–254.
7. Shipp, D.D. (1979) Harmonic analysis and suppression for electrical systems supplying static power converters and other non-linear loads. *IEEE Transactions on Industry Applications*, 15(5), 453–458.
8. Dewan, S.B. and Shahrodi, E.B. (1985) Design of an input filter for the six-pulse bridge rectifier. *IEEE Transactions on Industry Applications*, 21(5), 1168–1175.
9. Gonalez, D.A. and Maccall, J.C. (1987) Design of filters to reduce harmonic distortion in industrial power systems. *IEEE Transactions on Industry Applications*, 23(3), 504–511.
10. Hammond, P.W. (1988) A harmonic filter installation to reduce voltage distortion from static power converters. *IEEE Transactions on Industry Applications*, 24(1), 53–58.
11. Ludbrook, A. (1988) Harmonic filters for notch reduction. *IEEE Transactions on Industry Applications*, 24, 947–954.
12. Sueker, K.H., Hummel, S.D., and Argent, R.D. (1989) Power factor correction and harmonic mitigation in a thyristor controlled glass melter. *IEEE Transactions on Industry Applications*, 25(6), 972–975.
13. Cameron, M.M. (1993) Trends in power factor correction with harmonic filtering. *IEEE Transactions on Industry Applications*, 29(1), 60–65.
14. Lowenstein, M.Z. and Hibbard, J.F. (1993) Modeling and application of passive-harmonic trap filters for harmonic reduction and power factor improvement. Proceedings of the IEEE IAS Meeting, vol. 2, pp. 1570–1578.
15. Lowenstein, M.Z. (1993) Improving power factor in the presence of harmonics using low-voltage tuned filters. *IEEE Transactions on Industry Applications*, 29(3), 528–535.
16. Makran, E.B., Subramaniam, E.V., Girgis, A.A., and Catoe, R. (1993) Harmonic filter design using actual recorded data. *IEEE Transactions on Industry Applications*, 29(6), 1176–1183.
17. Sharaf, A.M. and Fisher, M.E. (1994) An optimization based technique for power system harmonic filter design. *Journal of Electric Power Systems Research*, 30, 63–67.
18. Swamy, M.M., Rossiter, S.L., Spencer, M.C., and Richardson, M. (1994) Case studies on mitigating harmonics in ASD systems to meet IEEE 519-1992 standards. Proceedings of the IEEE IAS Annual Meeting, pp. 685–692.
19. Lawrance, W.B., Michalik, G., Mielczarski, W., and Szczepanik, J. (1995) Reduction of harmonic pollution in distribution networks. Proceedings of the IEEE International Conference on Energy Management and Power Delivery (EMPD'95), vol. 1, pp. 198–202.
20. Sharaf, A.M. and Huang, H. (1995) Flicker control using rule based modulated passive power filters. *Journal of Electric Power Systems Research*, 33, 49–52.
21. Sharaf, A.M., Huang, H., and Chang, L. (1995) Power quality and nonlinear load voltage stabilization using error-driven switched passive power filter. Proceedings of IEEE-ISIE'95, pp. 616–621.
22. Peeran, S.M. and Cascadden, C.W.P. (1995) Application, design and specification of harmonic filters for variable frequency drives. *IEEE Transactions on Industry Applications*, 31(4), 841–847.
23. Maset, E., Sanchis, E., Sebastian, J., and de la Cruz, E. (1996) Improved passive solutions to meet IEC 1000-3-2 regulation in low-cost power supplies. Proceedings of IEEE INTELEC'96, pp. 99–106.
24. Kawann, C. and Emanuel, A.E. (1996) Passive shunt harmonic filters for low and medium voltage: a cost comparison study. *IEEE Transactions on Power Systems*, 11, 1825–1831.
25. Phipps, J.K. (1997) A transfer function approach to harmonic filter design. *IEEE Industry Applications Magazine*, 3, 68–82.
26. Ji, Y. and Wang, F. (1998) Single-phase diode rectifier with novel passive filter. *IEE Proceedings – Circuits, Devices and Systems*, 145(4), 493–413.
27. Chang, T.T. and Chang, H.C. (1998) Application of differential evolution to passive shunt harmonic filter planning. Proceedings of IEEE-ICHPQ'98, vol. 1, pp. 149–153.
28. Pretorius, J.H.C., Van Wyk, J.D., and Swart, P.H. (1998) Evaluation of the effectiveness of passive and dynamic filters for non active power in a large industrial plant. Proceedings of IEEE-ICHPQ'98, vol. 1, pp. 331–336.

29. Upadhyaya, S.D. and Atre, Y.R. (1998) Determination of the design parameters of passive harmonic filters using nonlinear optimization. Proceedings of the IEEE Industrial and Commercial Power Systems Technical Conference, pp. 155–164.
30. Wu, C.J., Chiang, J.C., Yen, S.S. *et al.* (1998) Investigation and mitigation of harmonic amplification problems caused by single-tuned filters. *IEEE Transactions on Power Delivery*, 13, 800–806.
31. Lin, K.P., Lin, M.H., and Lin, T.P. (1998) An advanced computer code for single-tuned harmonic filter design. *IEEE Transactions on Industry Applications*, 34, 640–648.
32. McGranaghan, M.F. and Mueller, D.R. (1999) Designing harmonic filters for adjustable-speed drives to comply with IEEE-519 harmonic limits. *IEEE Transactions on Industry Applications*, 35, 312–318.
33. Peng, F.Z., Su, G.-J., and Farquharson, G. (1999) A series LC filter for harmonic compensation of AC drives. Proceedings of IEEE-PESC'99, pp. 213–218.
34. Cavalloni, D., Gagliardi, L., Pinato, P. *et al.* (2000) Design criteria of passive filters for power electronic converters used in electric traction. Proceedings of the IEEE International Conference on Harmonics and Quality of Power, vol. 1, pp. 134–141.
35. Medora, N.K. and Kusko, A. (2000) Computer-aided design of power harmonic filters. *IEEE Transactions on Industry Applications*, 36(2), 604–613.
36. El-Saadany, E.F., Salama, M.M.A., and Chikhani, A.Y. (2000) Passive filter design for harmonic reactive power compensation in single-phase circuits supplying nonlinear loads. *IEE Proceedings – Generation, Transmission and Distribution*, 147, 373–380.
37. Berizzi, A. and Bovo, C. (2000) The use of genetic algorithms for the localization and the sizing of passive filters. Proceedings of IEEE-ICHPQ'00, vol. 1, pp. 19–25.
38. Luor, T.S. (2000) Influence of load characteristics on the applications of passive and active harmonic filters. Proceedings of IEEE-ICHPQ'00, vol. 1, pp. 128–133.
39. Zaninelli, D. (2000) Nonlinear passive filters in power systems. Proceedings of the IEEE PES Meeting, vol. 2, pp. 773–777.
40. Mattavelli, P. (2000) Design aspects of harmonic filters for high-power AC/DC converters. Proceedings of the IEEE PES Meeting, vol. 2, pp. 795–799.
41. Abdel-Galil, T.K., El-Saadany, E.F., and Salama, M.M.A. (2001) Implementation of different mitigation techniques for reducing harmonic distortion in medium voltage industrial distribution system. Proceedings of the IEEE PES Conference and Exposition on Transmission and Distribution, vol. 1, pp. 561–566.
42. Turkay, B. (2001) Harmonic filter design and power factor correction in a cement factory. Proceedings of the IEEE Porto Power Tech Conference.
43. Chang, G.W., Chu, S.Y., and Wang, H.L. (2002) A new approach for placement of single-tuned passive harmonic filters in a power system. Proceedings of the IEEE PES Meeting, vol. 2, pp. 814–817.
44. Neugebauer, T.C., Phinney, J.W., and Perreault, D.J. (2002) Filters and components with inductance cancellation. Proceedings of the IEEE IAS Annual Meeting, pp. 939–947.
45. Chen, Y.-M. (2003) Passive filter design using genetic algorithms. *IEEE Transactions on Industrial Electronics*, 50, 202–207.
46. Das, J.C. (2004) Passive filters – potentialities and limitations. *IEEE Transactions on Industry Applications*, 40(1), 232–241.
47. Chang, G.W., Wang, H.-L., and Chu, S.-Y. (2004) Strategic placement and sizing of passive filters in a power system for controlling voltage distortion. *IEEE Transactions on Power Delivery*, 19(3), 1204–1211.
48. Chang, G.W., Chu, S.-Y., and Wang, H.-L. (2006) A new method of passive harmonic filter planning for controlling voltage distortion in a power system. *IEEE Transactions on Power Delivery*, 21(1), 305–312.
49. Chang, Y.-P. and Low, C. (2008) Optimization of a passive harmonic filter based on the neural-genetic algorithm with fuzzy logic for a steel manufacturing plant. *Expert Systems with Applications*, 34, 2059–2070.
50. Verma, V. and Singh, B. (2010) Genetic algorithm based design of passive filters for offshore applications. *IEEE Transactions on Industry Applications*, 46, 1295–1303.

9

Shunt Active Power Filters

9.1 Introduction

Solid-state control of AC power using diodes, thyristors, triacs, and other semiconductor switches is widely employed to feed controlled power to electrical loads such as computers, printers, fax machines, copiers, TV power supplies, lighting devices especially vapor lamps consisting of magnetic or electronic ballasts, solid-state AC voltage controllers feeding fans, furnaces, adjustable speed drives (ASDs) consisting of solid-state controllers for both DC and AC motors, uninterruptible power supplies (UPS), high-frequency transformer-isolated welding machines, magnet power supplies, electrochemical industries such as electroplating, electromining, and so on. Such solid-state controllers are also used in electric traction, high-voltage direct current (HVDC) systems, flexible alternating current transmission system (FACTS), and renewable electrical power generation. As nonlinear loads (NLLs), these solid-state converters draw harmonic currents and the reactive power component of the current from the AC mains. In three-phase systems, they could also cause unbalanced currents and draw excessive neutral current. The injected harmonic currents, reactive power burden, and unbalanced and excessive neutral current cause low efficiency of distribution system, poor power factor, mal-operation of protection systems, power capacitor banks overloading and their nuisance tripping, noise and vibration in electrical machines, derating of distribution and user equipment, and so on. They also cause disturbance to other consumers and interference in nearby communication networks. Traditionally, passive L–C filters have been used to reduce harmonics and power capacitors have been employed to improve the power factor of the AC loads. However, passive filters have the demerits of fixed compensation, large size, resonance, and so on. The increased severity of harmonic pollution in power networks has attracted attention of power electronics and power system engineers to develop dynamic and adjustable solutions to the power quality problems. Such equipment generally known as active power filter (APF) is also called active power line conditioner (APLC), instantaneous reactive power compensator (IRPC), and active power quality conditioner (APQC). In recent years, many studies have also appeared on harmonics, reactive power, load balancing, and neutral current compensation associated with linear and nonlinear loads. It is thus relevant to present the analysis, design, and control of SAPF considering their increasing applications for the compensation of nonlinear loads.

Shunt active power filters are classified on the basis of converter used such as current source converter (CSC), voltage source converter (VSC), topology used such as half-bridge converters, full-bridge converters, and so on, and supply systems. On the basis of supply system, SAPFs are classified into three major categories: single-phase two-wire, three-phase three-wire, and three-phase four-wire SAPFs. This chapter deals with the state of the art on these SAPFs, classification, principle of operation and control, analysis and design, modeling and simulation of performance, numerical examples, summary, review questions, numerical problems, computer simulation-based problems, and references.

Power Quality Problems and Mitigation Techniques, First Edition.
Bhim Singh, Ambrish Chandra and Kamal Al-Haddad.

9.2 State of the Art on Shunt Active Power Filters

The SAPF technology is now a mature technology for providing harmonic current compensation, reactive power compensation, and neutral current compensation in AC distribution networks. It has evolved in the past quarter century with development in terms of varying configurations, control strategies, and solid-state devices. Shunt active power filters are also used to regulate the terminal voltage and suppress voltage flicker in three-phase systems. These objectives are achieved either individually or in combination depending upon the requirements, control strategy, and configuration that need to be selected appropriately.

With the widespread use of solid-state control of AC power, the power quality issues have also become significant. A large number of publications are reported on the power quality survey, measurements, analysis, cause and effects of harmonics, and reactive power in the electric networks. Shunt active power filter is considered as an ideal device for mitigating power quality problems. The shunt active power filters are basically categorized into three types, namely, single-phase two-wire, three-phase three-wire, and three-phase four-wire configurations, to meet the requirements of the three types of nonlinear loads on supply systems. Some single-phase loads such as domestic lights, ovens, TVs, computer power supplies, air conditioners, laser printers, and Xerox machines behave as nonlinear loads and cause power quality problems. Single-phase two-wire active power filters of varying configurations and control strategies have been investigated to meet the needs of single-phase nonlinear loads. Starting from 1971, many configurations of shunt active power filters have been developed and commercialized in UPS applications. Both current source converter with inductive energy storage and voltage source converter with capacitive energy storage are used to develop single-phase SAPFs.

A major amount of AC power is consumed by three-phase loads such as ASDs with solid-state control and lately in the front-end design of many other electrical loads the active power filters have been incorporated. A substantial work has been reported on three-phase, three-wire APFs. Starting from 1976, many control strategies such as instantaneous reactive power theory (IRPT), synchronous frame d–q theory, synchronous detection method, notch filter method are used in the development of three-phase APFs.

The problem of excessive neutral current is observed in three-phase four-wire systems mainly due to nonlinear unbalanced loads such as computer power supplies, fluorescent lighting, and so on. These problems of neutral current and unbalanced load currents in four-wire systems have been attempted to resolve through elimination/ reduction of neutral current, harmonic compensation, load balancing, and reactive power compensation.

One of the major factors in advancing the APF technology is the advent of fast, self-commutating solid-state devices. In the initial stages, thyristors, bipolar junction transistors (BJTs), and power MOSFETs (metal–oxide–semiconductor field-effect transistors) have been used to develop APFs; later, SITs and GTOs have been employed to develop APFs. With the introduction of insulated gate bipolar transistors (IGBTs), the APF technology has got a real boost and at present it is considered as an ideal solid-state device for APFs. The improved sensor technology has also contributed to the enhanced performance of the APF. The availability of Hall effect sensors and isolation amplifiers at reasonable cost and with adequate ratings has improved the APF performance.

The next breakthrough in the APF development has resulted from the microelectronics revolution. Starting from the use of discrete analog and digital electronics components, the progression has been to microprocessors, microcontrollers, and DSPs. Now it is possible to implement complex algorithms online for the control of APF at a reasonable cost. This development in APF has made it possible to use different control algorithms such as PI (proportional–integral) control, variable structure control, fuzzy logic control, and neural network-based control algorithms for improving the dynamic and steady-state performance of APFs. With these improvements, APFs are capable of providing fast corrective action even with dynamically changing nonlinear loads. Moreover, these APFs are found to compensate higher order harmonics (typically up to 25th harmonics).

9.3 Classification of Shunt Active Power Filters

Shunt active power filters can be classified based on the type of converter used, topology, and the number of phases. The converter used in the SAPF can be either a current source converter or a voltage source converter. Different topologies of SAPF can be realized by using various circuits of VSCs. The third

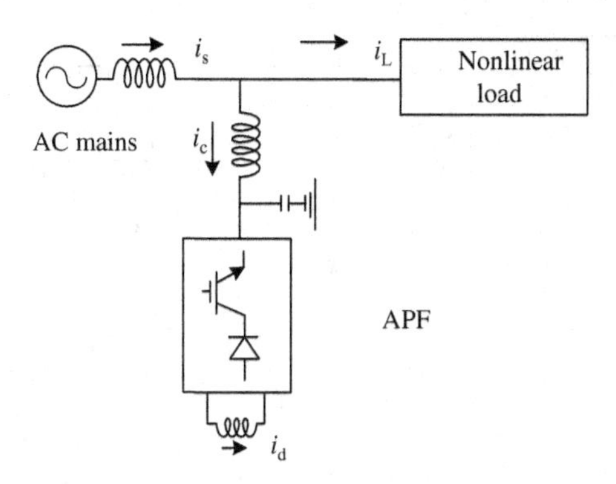

Figure 9.1 A CSC-based SAPF

classification is based on the number of phases: single-phase two-wire, three-phase three-wire, and three-phase four-wire APF systems.

9.3.1 Converter-Based Classification

Two types of converters are used to develop APFs. Figure 9.1 shows a SAPF using a current fed PWM (pulse-width modulation) converter or a CSC bridge. It behaves as a nonsinusoidal current source to meet the harmonic current requirement of the nonlinear loads. A diode is used in series with the self-commutating device (IGBT) for reverse voltage blocking. However, GTO-based circuit configurations do not need the series diode, but they have restricted frequency of switching. These CSC-based SAPFs are considered sufficiently reliable, but have higher losses and require higher values of parallel AC power capacitors. Moreover, they cannot be used in multilevel or multistep modes to improve the performance of SAPFs in higher ratings.

The other converter used in APF is a voltage-fed PWM converter or voltage source converter shown in Figure 9.2. It has a self-supporting DC voltage bus with a large DC capacitor. It is more widely used because it is lighter, cheaper, and expandable to multilevel and multistep versions to enhance the performance with lower switching frequencies. It is more popular in UPS-based applications because in the presence of AC mains, the same converter bridge can be used in SAPF to eliminate harmonics of critical nonlinear loads.

9.3.2 Topology-Based Classification

SAPFs can also be classified based on topology, namely, half-bridge topology, full-bridge topology, and H-bridge topology. Figures 9.3–9.6 show these topologies of SAPFs. The VSC-based half-bridge

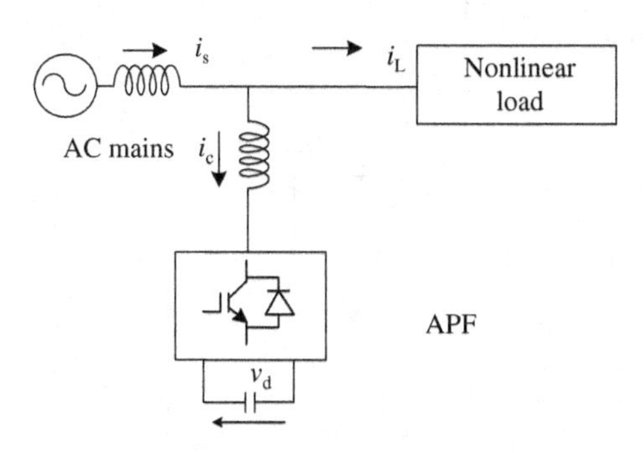

Figure 9.2 A VSC-based SAPF

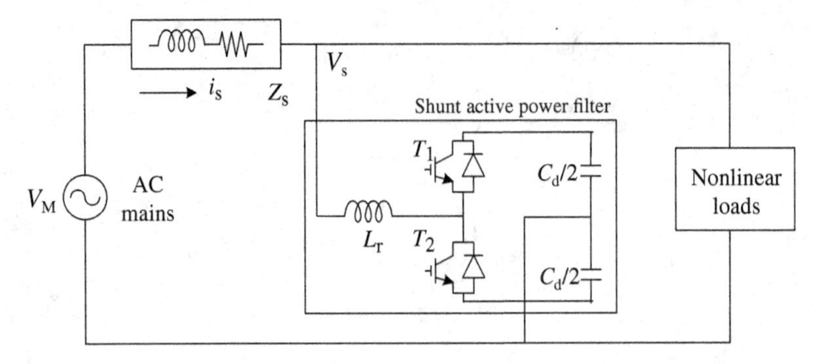

Figure 9.3 Half-bridge topology of the VSC-based single-phase shunt active power filter

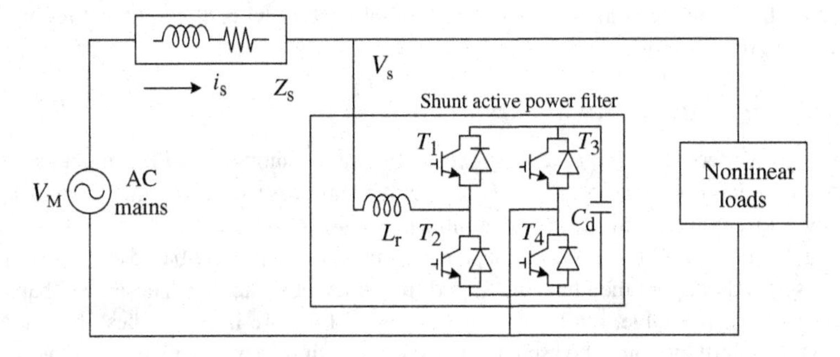

Figure 9.4 Full-bridge topology of the VSC-based single-phase shunt active power filter

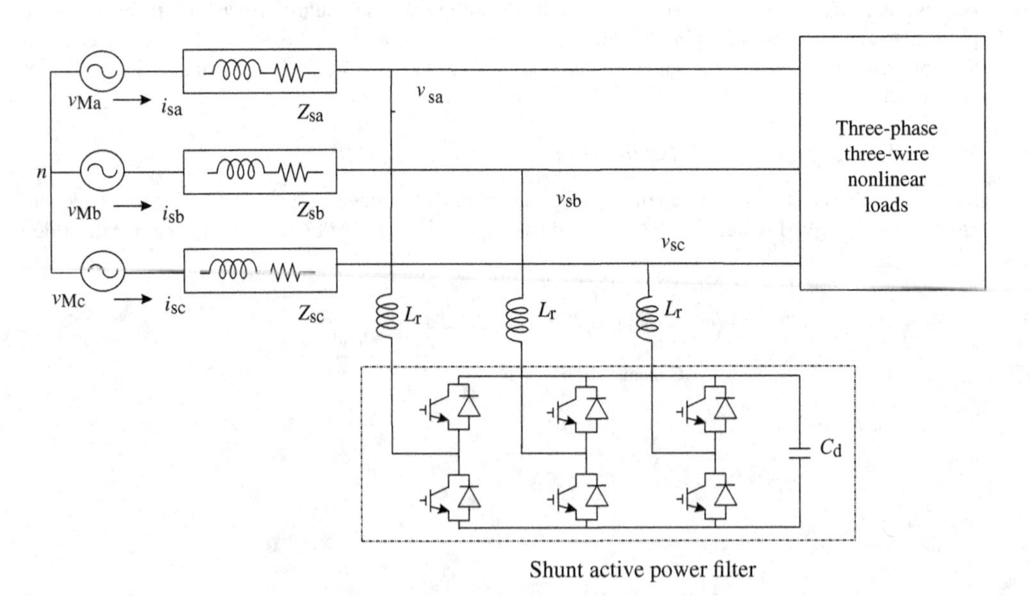

Figure 9.5 A three-phase three-wire shunt active power filter

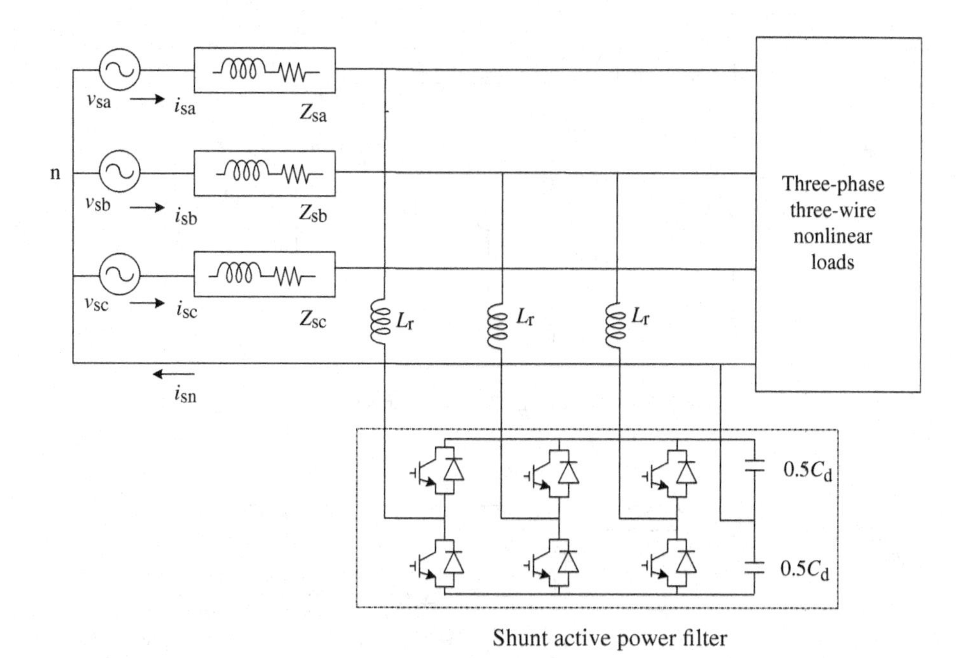

Figure 9.6 A three-phase four-wire shunt active power filter with capacitor midpoint topology

topology of SAPFs involves less number of solid-state devices and their control and hence is cheap and cost-effective. The VSC-based full-bridge topology of SAPFs is considered ideal for three-phase three-wire and three-phase four-wire AC systems and it does not require transformers for isolation. The VSC-based H-bridge topology of SAPFs consists of single-phase full H-bridges with two legs and four switching devices with independent control of each phase with unipolar switching to reduce the switching frequency and losses. Each H-bridge for each phase of VSC-based SAPFs needs a separate transformer for isolation, voltage matching, and reliability; from safety point of view, this is the most preferred configuration in SAPFs by the industries.

9.3.3 Supply System-Based Classification

This classification of SAPFs is based on the supply and/or the load system, namely, single-phase two-wire, three-phase three-wire, and three-phase four-wire systems. There are many nonlinear loads such as domestic appliances connected to single-phase supply systems. Some three-phase nonlinear loads are without neutral terminal, such as ASDs, fed from three-phase three-wire supply systems. There are many single-phase nonlinear loads distributed on three-phase four-wire supply systems, such as computers and commercial lighting Hence, these SAPFs may also be classified accordingly as two-wire, three-wire, and four-wire SAPFs.

9.3.3.1 Two-Wire SAPFs

Single-phase two-wire SAPFs are used in both converter configurations, namely, current source converter PWM bridge with inductive energy storage element and voltage source converter PWM bridge with capacitive DC bus energy storage element to form two-wire SAPF circuits. In some cases, active filtering is included in the power conversion stage to improve input characteristics at the supply end.

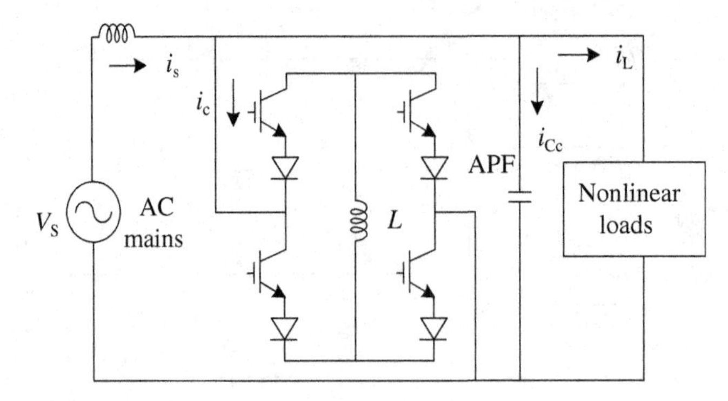

Figure 9.7 A two-wire SAPF with a current source converter

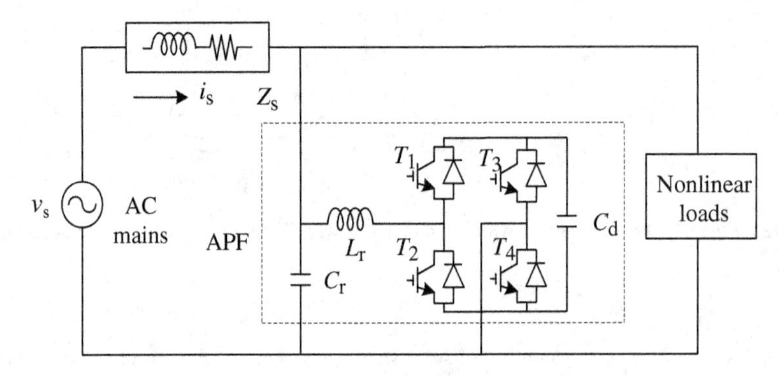

Figure 9.8 A two-wire SAPF with a voltage source converter

Figures 9.7 and 9.8 show two detailed configurations of shunt active power filter with a current source converter using inductive storage element and a voltage source converter with capacitive DC bus energy storage element.

9.3.3.2 Three-Wire SAPFs

Solid-state power converters have been widely used in three-phase three-wire nonlinear loads such as ASDs and lately many other electrical loads have also incorporated active power filters in their front design. A large number of publications have appeared on three-wire APFs with different configurations. All the configurations shown in Figures 9.1–9.8 are developed in three wire SAPFs, with three wires on the AC side and two wires on the DC side. SAPFs are developed in current fed type (Figure 9.1) or voltage fed type with single-stage (Figure 9.2) or multistep/multilevel and multiseries configurations. Figure 9.9 shows a typical VSC-based three-wire SAPF. SAPFs are also designed with three single-phase APFs with isolation transformers for proper voltage matching, independent phase control, and reliable compensation with unbalanced systems.

9.3.3.3 Four-Wire SAPFs

A large number of single-phase loads may be supplied from a three-phase AC mains with a neutral conductor. They cause excessive neutral current, harmonics, and reactive power burden and unbalanced currents. To reduce these problems, four-wire SAPFs have been used in four-wire distribution systems. They have been developed as active shunt mode with current fed converter and voltage fed converter.

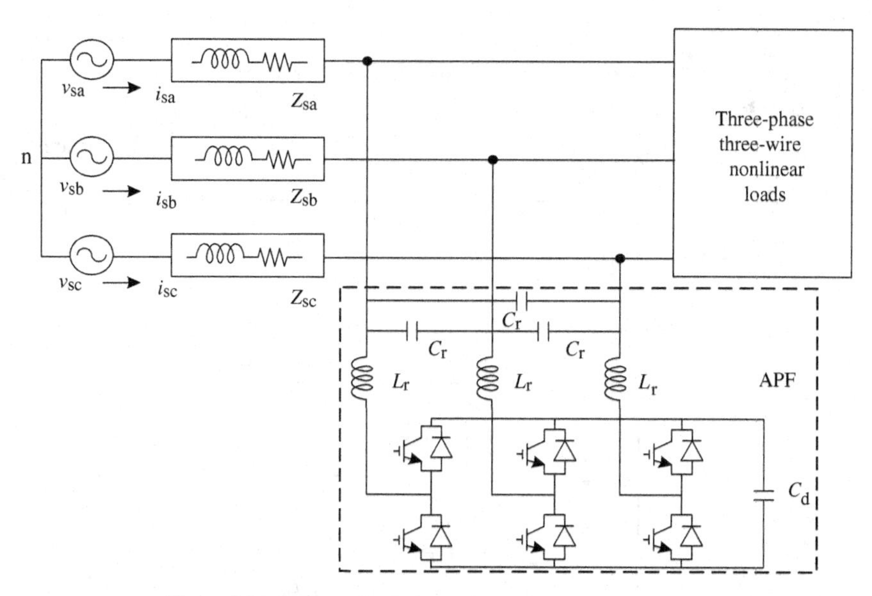

Figure 9.9 A three-wire SAPF with a voltage source converter

Figures 9.10–9.12 show three typical configurations of three-phase four-wire SAPFs. The first configuration of four-wire SAPFs is known as capacitor midpoint type used in smaller ratings. Here, the total neutral current flows through DC bus capacitors that are of large value. Figure 9.11 shows another configuration known as four-pole type, in which the fourth pole is used to stabilize the neutral terminal of the APF. A three single-phase H-bridge configuration, shown in Figure 9.12, is quite common and

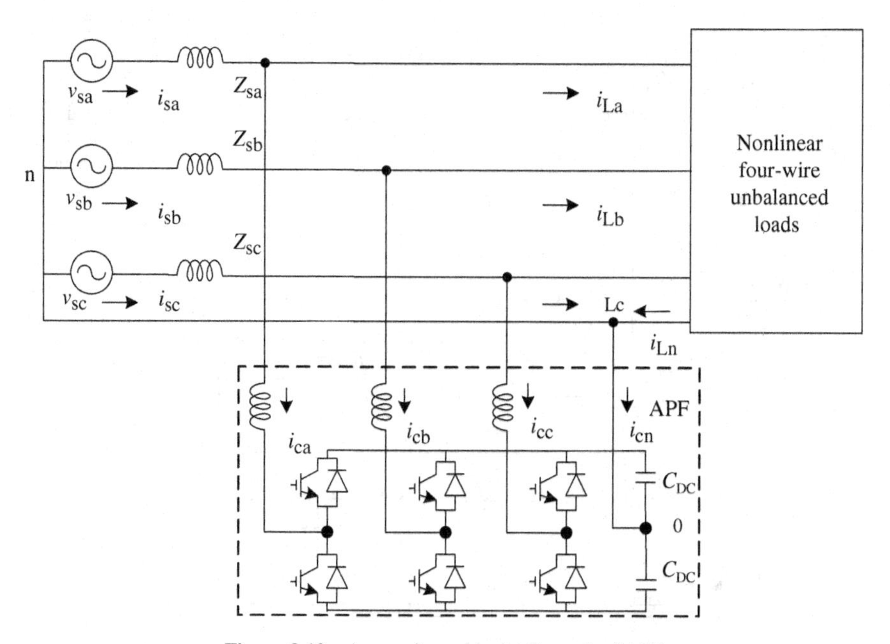

Figure 9.10 A capacitor midpoint four-wire SAPF

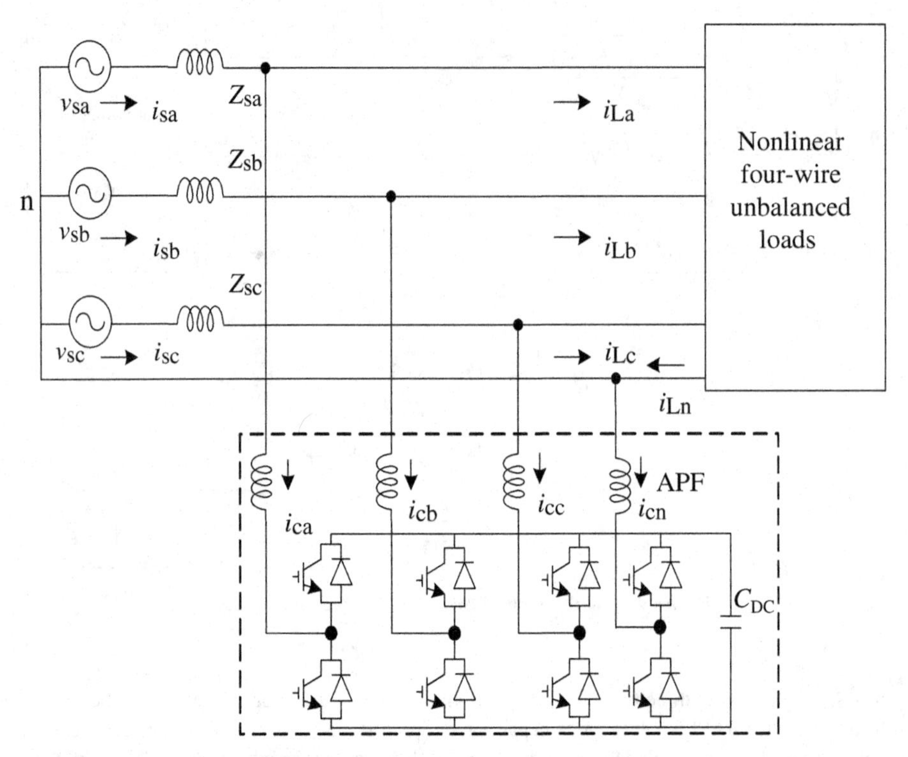

Figure 9.11 A four-pole, four-wire SAPF

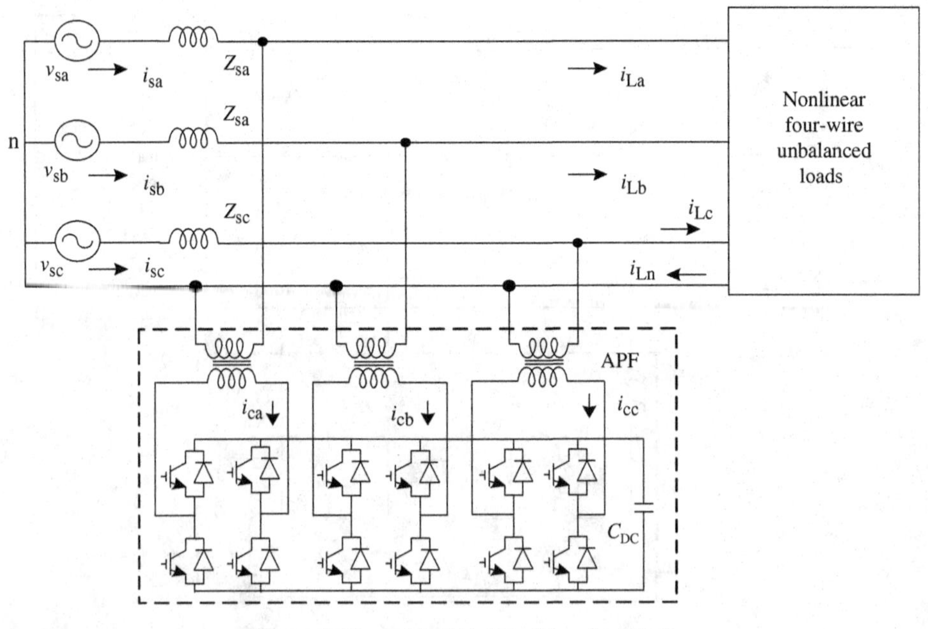

Figure 9.12 A three H-bridge, four-wire SAPF

this version uses a proper voltage matching for solid-state devices and enhances the reliability of the APF system.

9.4 Principle of Operation and Control of Shunt Active Power Filters

A fundamental circuit of SAPF for a three-phase, three-wire AC system with balanced/unbalanced NLL is shown in Figure 9.9. An IGBT-based current-controlled voltage source converter (CC-VSC) with a DC bus capacitor is used as SAPF.

Using a control algorithm, the reference APF currents are directly controlled by estimating the reference APF currents. However, in place of APF currents, the reference currents may be estimated for an indirect current control of the VSC. The gating pulses to the APF are generated by employing hysteresis (carrierless PWM) or PWM (fixed frequency) current control over reference and sensed supply currents resulting in an indirect current control. Using SAPF, the supply current harmonics compensation, reactive power compensation, and unbalanced currents compensation are achieved in all the control algorithms. In addition, zero voltage regulation (ZVR) at the point of common coupling (PCC) is also achieved by modifying the control algorithm suitably.

9.4.1 Principle of Operation of Shunt Active Power Filters

The main objective of shunt active power filters is to mitigate multiple power quality problems in a distribution system. SAPF mitigates most of the current quality problems, such as reactive power, unbalanced currents, neutral current, harmonics, and fluctuations, present in the consumer loads or otherwise in the system and provides sinusoidal balanced currents in the supply along with its DC bus voltage control.

In general, a SAPF has a VSC connected to a DC bus and its AC side is connected in shunt normally across the consumer loads or across the PCC, as shown in Figure 9.9. The VSC uses PWM current control; therefore, it requires small ripple filters to mitigate switching ripples. It requires Hall effect voltage and current sensors for feedback signals and normally a digital signal processor (DSP) is used to implement the required control algorithm to generate gating signals for the solid-state devices of the VSC of the SAPF. The VSC used as SAPF is normally controlled in PWM current control mode to inject appropriate currents into the system. The SAPF also needs many passive elements such as a DC bus capacitor, AC interacting inductors, and small passive filters.

9.4.2 Control of Shunt Active Power Filters

Reference current signals for the control of SAPF have to be derived accordingly and these signals may be estimated using a number of control algorithms. There are many control algorithms reported in the literature for the control of SAPFs, which are classified as time-domain and frequency-domain control algorithms. There are more than a dozen of time-domain control algorithms that are used for the control of SAPFs. A few of these time-domain control algorithms are as follows:

- Synchronous reference frame (SRF) theory, also known as d–q theory
- Unit template technique or proportional–integral controller-based theory
- Instantaneous reactive power theory, also known as PQ theory or α–β theory
- Instantaneous symmetrical component (ISC) theory
- Power balance theory (BPT)
- Neural network theory (Widrow's LMS-based Adaline algorithm)
- Current synchronous detection (CSD) method
- I-cos φ algorithm
- Singe-phase PQ theory
- Singe-phase DQ theory
- Enhanced phase locked loop (EPLL)-based control algorithm
- Conductance-based control algorithm
- Adaptive detecting algorithm, also known as adaptive interference canceling theory

Similarly, there are around the same number of frequency-domain control algorithms. Some of them are as follows:

- Fourier series theory
- Discrete Fourier transform theory
- Fast Fourier transform theory
- Recursive discrete Fourier transform theory
- Kalman filter-based control algorithm
- Wavelet transformation theory
- Stockwell transformation (S-transform) theory
- Empirical decomposition (EMD) transformation theory
- Hilbert–Huang transformation theory

Most of these frequency-domain control algorithms are used for power quality monitoring for a number of purposes in the power analyzers, PQ instruments, and so on. Some of them have been used for the control of SAPFs. However, these algorithms are sluggish and slow, requiring heavy computation burden; therefore, these control methods are not much preferred for real-time control of SAPFs compared with time-domain control algorithms.

All these control algorithms may be used in the control of SAPFs. However, because of space limitation and to give just a basic understanding, only SRF theory—also known as d–q theory, unit template technique or PI controller-based theory, and IRPT—also known as PQ theory or α–β theory—are explained here.

9.4.2.1 SRF Theory-Based Control Algorithm of APFs

The synchronous reference frame theory is employed for the control of three-phase three-leg VSC of the APF. A block diagram of the control scheme is shown in Figure 9.13. The load currents (i_{La}, i_{Lb}, i_{Lc}), the PCC voltages (v_{sa}, v_{sb}, v_{sc}), and the DC bus voltage (v_{DC}) of the APF are sensed as feedback signals. The load currents in the three phases are converted into the dq0 frame using the Park's transformation as follows:

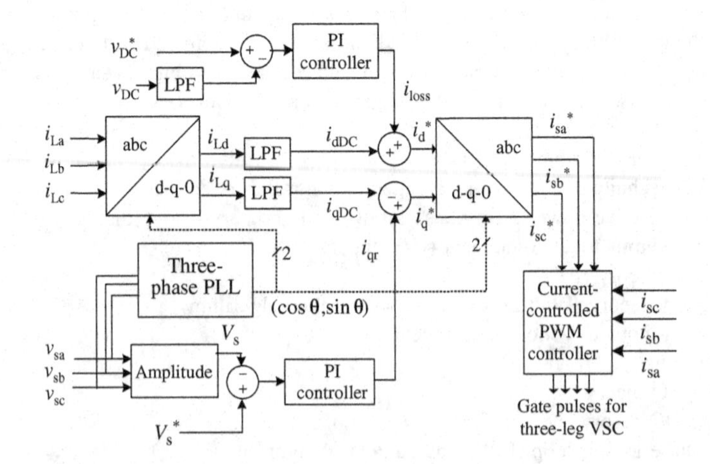

Figure 9.13 Block diagram of SRF theory-based control algorithm

$$\begin{bmatrix} i_{\mathrm{Ld}} \\ i_{\mathrm{Lq}} \\ i_{\mathrm{L0}} \end{bmatrix} = \frac{2}{3} \begin{bmatrix} \cos\theta & -\sin\theta & \frac{1}{2} \\ \cos\left(\theta - \frac{2\pi}{3}\right) & -\sin\left(\theta - \frac{2\pi}{3}\right) & \frac{1}{2} \\ \cos\left(\theta + \frac{2\pi}{3}\right) & \sin\left(\theta + \frac{2\pi}{3}\right) & \frac{1}{2} \end{bmatrix} \begin{bmatrix} i_{\mathrm{La}} \\ i_{\mathrm{Lb}} \\ i_{\mathrm{Lc}} \end{bmatrix}. \tag{9.1}$$

A three-phase PLL (phase locked loop) is used to synchronize these signals with the PCC voltages. These d–q current components are then passed through low-pass filters (LPFs) to extract the DC components of i_{Ld} and i_{Lq}. The d-axis and q-axis currents consist of fundamental and harmonic components as

$$i_{\mathrm{Ld}} = i_{\mathrm{dDC}} + i_{\mathrm{dAC}}, \tag{9.2}$$

$$i_{\mathrm{Lq}} = i_{\mathrm{qDC}} + i_{\mathrm{qAC}}. \tag{9.3}$$

A SRF controller extracts DC quantities by a LPF and hence the non-DC quantities (harmonics) are separated from the reference signals. It can be operated in unity power factor (UPF) and zero voltage regulation modes as described in the following sections.

9.4.2.1.1 UPF Operation of APFs

The control strategy for reactive power compensation for UPF operation considers that the source must deliver the DC component of the direct-axis component of the load current (i_{dDC}) along with the active power component of the load current for maintaining the DC bus and meeting the losses (i_{loss}) in the APF. The output of the PI controller at the DC bus voltage of the APF is considered as the current (i_{loss}) for meeting its losses:

$$i_{\mathrm{loss}(n)} = i_{\mathrm{loss}(n-1)} + K_{\mathrm{pd}}\left(v_{\mathrm{de}(n)} - v_{\mathrm{de}(n-1)}\right) + K_{\mathrm{id}} v_{\mathrm{de}(n)}, \tag{9.4}$$

where $v_{\mathrm{de}(n)} = v_{\mathrm{DC}}^{*} - v_{\mathrm{DC}(n)}$ is the error between the reference (v_{DC}^{*}) and the sensed (v_{DC}) DC voltage at the nth sampling instant and K_{pd} and K_{id} are the proportional and integral gain constants of the DC bus voltage PI controller, respectively.

Therefore, the reference direct-axis supply current is

$$i_{\mathrm{d}}^{*} = i_{\mathrm{dDC}} + i_{\mathrm{loss}}. \tag{9.5}$$

The reference source current must be in phase with the voltage at PCC but with no zero-sequence component. It is therefore obtained by the following reverse Park's transformation with i_{d}^{*} as in Equation 9.5 and i_{q}^{*} and i_{0}^{*} as zero:

$$\begin{bmatrix} i_{\mathrm{sa}}^{*} \\ i_{\mathrm{sb}}^{*} \\ i_{\mathrm{sc}}^{*} \end{bmatrix} = \begin{bmatrix} \cos\theta & \sin\theta & 1 \\ \cos\left(\theta - \frac{2\pi}{3}\right) & \sin\left(\theta - \frac{2\pi}{3}\right) & 1 \\ \cos\left(\theta + \frac{2\pi}{3}\right) & \sin\left(\theta + \frac{2\pi}{3}\right) & 1 \end{bmatrix} \begin{bmatrix} i_{\mathrm{d}}^{*} \\ i_{\mathrm{q}}^{*} \\ i_{0}^{*} \end{bmatrix}. \tag{9.6}$$

9.4.2.1.2 ZVR Operation of APFs

The control strategy for the ZVR operation of APF considers that the supply must deliver the same direct-axis component i_{d}^{*} as mentioned in Equation 9.5 along with the difference of quadrature-axis current of the load (i_{qDC}) and the component obtained from the PI voltage controller (i_{qr}) used for regulating the voltage at PCC. The amplitude of the AC terminal voltage (V_S) at PCC is controlled to its reference voltage (V_s^*) using the PI voltage controller. The output of the PI voltage controller is considered as the reactive component of the current (i_{qr}) for zero voltage regulation of the AC voltage at PCC. The amplitude of the AC voltage (V_S) at PCC is calculated from the AC voltages (v_{sa}, v_{sb}, v_{sc}) as

$$V_{sp} = (2/3)^{1/2}(v_{sa}^2 + v_{sb}^2 + v_{sc}^2)^{1/2}. \tag{9.7}$$

Then, a PI voltage controller is used to regulate this voltage to a reference value as

$$i_{qr(n)} = i_{qr(n-1)} + K_{pq}(v_{te(n)} - v_{te(n-1)}) + K_{iq}v_{te(n)}, \tag{9.8}$$

where $V_{te}(n) = V_{sp}^*(n) - V_{sp}(n)$ denotes the error voltage between reference $V_{sp}^*(n)$ and actual $V_{sp}(n)$ terminal voltage amplitudes at the nth sampling instant and K_{pq} and K_{iq} are the proportional and integral gain constants of the AC voltage PI controller, respectively.

The reference supply quadrature-axis current is

$$i_q^* = i_{qr} - i_{qDC}. \tag{9.9}$$

Three-phase reference supply currents are obtained by reverse Park's transformation using Equation 9.6 with i_d^* as in Equation 9.5, i_q^* as in Equation 9.9, and i_0^* as zero.

9.4.2.2 Unit Template-Based Control Algorithm of APFs

A simple control algorithm of a SAPF for the AC voltage regulation at load terminals (at PCC), harmonics elimination, and load balancing of nonlinear loads is explained using the unit template algorithm. This control algorithm of the SAPF is made flexible and can be modified for power factor correction (unity), harmonics elimination, and load balancing of nonlinear loads. The proposed control algorithm inherently provides a self-supporting DC bus of the APF. An indirect current control technique is employed to obtain PWM switching signals for the devices used in the CC-VSC working as an APF. Three-phase reference supply currents are derived using sensed AC voltages (at PCC) and the DC bus voltage of the APF as feedback signals. Two PI voltage controllers are used to estimate the amplitudes of in-phase and quadrature components of reference supply currents.

9.4.2.2.1 Control of APFs in UPF Mode of Operation

Figure 9.14 shows the basic control algorithm of the APF using unit template control. Three-phase voltages at PCC along with the DC bus voltage of the APF are used for implementing this control algorithm. In real-time implementation of the APF, a band-pass filter (BPF) plays an important role.

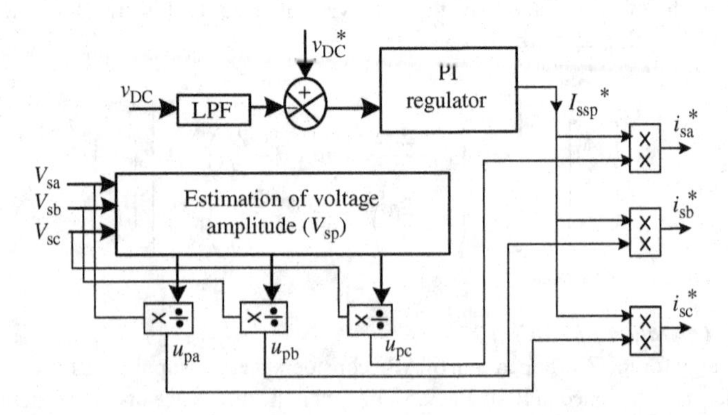

Figure 9.14 Unit template-based control algorithm of APF for power factor correction mode of operation

Three-phase voltages are sensed at PCC and are conditioned in a band-pass filter to filter out any distortion. The three-phase load voltages are inputs and three-phase filtered voltages (v_{sa}, v_{sb}, v_{sc}) are outputs of band-pass filters.

For the control of the active power filter, the self-supporting DC bus is realized using a PI voltage controller over the sensed (v_{DC}) and reference (v_{DC}^*) values of the DC bus voltage of the APF. The PI voltage controller on the DC bus voltage of the APF provides the amplitude (I_{spp}^*) of in-phase components ($i_{sa}^*, i_{sb}^*, i_{sc}^*$) of reference supply currents. The three-phase unit current vectors (u_{sa}, u_{sb}, u_{sc}) are derived in phase with the filtered supply voltages (v_{sa}, v_{sb}, v_{sc}). The multiplication of the in-phase amplitude with in-phase unit current vectors results in the in-phase components of three-phase reference supply currents ($i_{sa}^*, i_{sb}^*, i_{sc}^*$). Hence, for fundamental unity power factor supply currents, the in-phase reference supply currents, which are estimated in the above-described procedure, become the reference supply currents.

The amplitude of reference supply currents is computed using a PI voltage controller over the average value of the DC bus voltage (v_{DCa}) of the APF and its reference value (v_{DC}^*). A comparison of average and reference values of the DC bus voltage of the APF results in a voltage error, which is fed to a PI voltage controller:

$$v_{DCerror} = v_{DC}^* - v_{DCa}. \tag{9.10}$$

Here, proportional (K_p) and integral (K_i) gain constants are chosen such that a desired DC bus voltage response is achieved. The output of the PI controller is taken as the amplitude (I_{spp}^*) of the reference supply currents. Now, three-phase in-phase components of the reference supply currents are computed using their amplitude and in-phase unit current vectors derived in phase with the supply voltages, and are given as

$$i_{sa}^* = I_{spp}^* u_{sa}, \quad i_{sb}^* = I_{spp}^* u_{sb}, \quad i_{sc}^* = I_{spp}^* u_{sc}, \tag{9.11}$$

where u_{sa}, u_{sb}, and u_{sc} are in-phase unit current vectors and are derived as

$$u_{sa} = v_{sa}/V_{sp}, \quad u_{sb} = v_{sb}/V_{sp}, \quad u_{sc} = v_{sc}/V_{sp}, \tag{9.12}$$

where V_{sp} is the amplitude of the supply voltage and is computed as

$$V_{sp} = \{2/3(v_{sa}^2 + v_{sb}^2 + v_{sc}^2)\}^{1/2}. \tag{9.13}$$

9.4.2.2.2 Control of APFs in ZVR Mode of Operation

Shunt active power filters can compensate the reactive power, negative-sequence, and harmonic currents of the loads. However, because of finite (nonzero) internal impedance of the utility, which is represented by Z_s (L_s, R_s), the voltage waveforms at PCC to other loads are distorted and result in a voltage drop. The APF systems should operate and meet IEEE-519 Standard under all supply and load conditions, for example, in the presence of maximum allowable supply voltage harmonics, supply voltage swells/sags, supply voltage unbalance, and supply current subharmonics. In such cases, a voltage drop is also of great concern in addition to harmonic currents. The voltage drops are caused by inrush currents and by the direct start of motors, and in this case the proportion of the linear load is higher than that of the nonlinear load. As a result, harmonic currents are no longer a serious problem. Thus, it is necessary to switch the operation mode of the APF to a voltage regulator.

In addition to harmonics elimination, the APF can also be operated to maintain constant voltage at PCC. For this purpose, the APF takes a leading current component (in general) due to lagging power factor loads and is explained using phasor diagrams shown in Figure 9.15. When the system is operating without the APF, the voltage at PCC (V_s) is less than the supply voltage (V_M) due to the drop in the supply impedance Z_s (L_s, R_s), as shown in Figure 9.15a. Now with an APF connected in the system and drawing a leading current component, the supply current and hence the drop across the supply impedance can be controlled

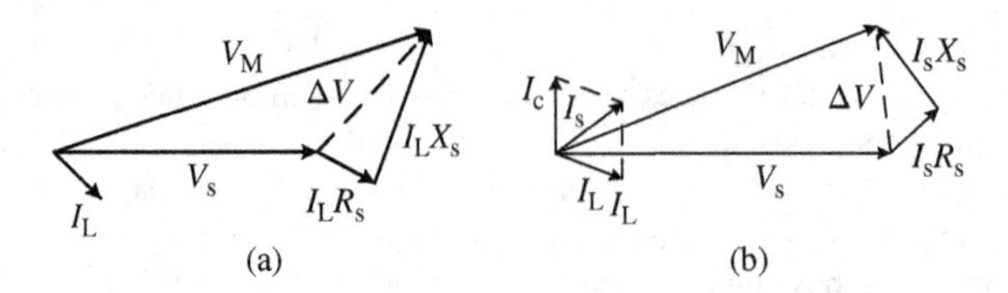

Figure 9.15 Phasor diagrams (a) without an APF and (b) with an APF in ZVR mode of operation

so that the magnitudes of the PCC voltage and supply voltage become equal ($|V_s| = |V_M|$), as shown in Figure 9.15b. By controlling the APF current, the amplitude and phase of the supply current may be changed to maintain the desired load voltage. Hence, at the same time, both UPF and ZVR functions cannot be achieved.

The control algorithm to maintain the desired PCC voltage, the APF for ZVR operation at PCC, is shown in Figure 9.16. Using this algorithm, one can achieve AC voltage regulation at load terminals (at PCC), harmonic current elimination, and load balancing of nonlinear loads. For regulation of voltage at

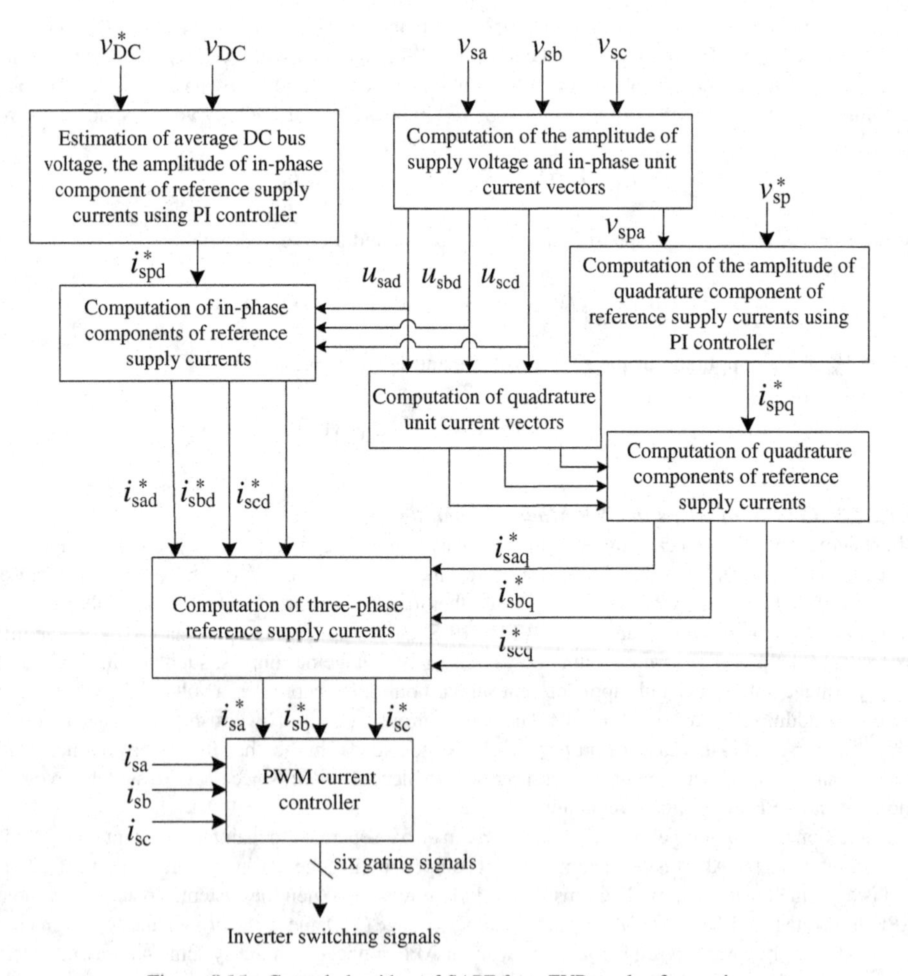

Figure 9.16 Control algorithm of SAPF for a ZVR mode of operation

PCC, the three-phase reference supply currents (i^*_{sa}, i^*_{sb}, i^*_{sc}) have two components. The first component (i^*_{sad}, i^*_{sbd}, i^*_{scd}) is in-phase with the voltage at PCC to feed active power to the loads and the losses of the APF. The second component (i^*_{saq}, i^*_{sbq}, i^*_{scq}) is in quadrature with the voltage at PCC to feed reactive power to the loads and to compensate the line voltage drop by reactive power injection at PCC. For power factor correction to unity, harmonic elimination, and balancing of nonlinear loads, the quadrature component of reference supply currents is set to zero. For voltage regulation at PCC, the supply currents should lead the supply voltages for lagging PF loads, while for the power factor control to unity, the supply currents should be in phase with the supply voltages. These two conditions, namely, voltage regulation at PCC and power factor control to unity, cannot be achieved simultaneously. Therefore, the control algorithm of the APF is made flexible to achieve either voltage regulation or power factor correction to unity, harmonics compensation, and load balancing. The operation of APFs in UPF mode is already explained in the previous section. Therefore, the three in-phase components are the quantities computed from Equation 9.11.

The amplitude (I^*_{spq}) of the quadrature component of reference supply currents is estimated using another PI controller over the amplitude (V_{spa}) of the supply voltage and its reference value (V^*_{spa}). A comparison of the reference value with the amplitude of the supply voltage results in a voltage error (V_{spe}). This voltage error signal is processed in a PI controller. The output signal of the PI voltage controller $I^*_{spq}(n)$ for maintaining the PCC terminal voltage at a constant value at the nth sampling instant is expressed as

$$I^*_{spq}(n) = I^*_{spq}(n-1) + K_{pt}\{v_{spe}(n) - v_{spe}(n-1)\} + K_{it}v_{spe}(n), \tag{9.14}$$

where K_{pt} and K_{it} are the proportional and integral gain constants of the AC bus PI voltage controller, respectively, $v_{spe}(n)$ and $v_{spe}(n-1)$ are the voltage errors at the nth and $(n-1)$th instants, respectively, and $I^*_{spq}(n-1)$ is the required reactive power component at the $(n-1)$th instant. The term $I^*_{spq}(n)$ is considered as the amplitude (I^*_{spq}) of the quadrature component of reference supply currents. Three-phase quadrature components of the reference supply currents are computed using their amplitude and quadrature unit current vectors as

$$i^*_{saq} = I_{spq}{}^*u_{saq}, \quad i^*_{sbq} = I_{spq}{}^*u_{sbq}, \quad i^*_{scq} = I_{spq}{}^*u_{scq}, \tag{9.15}$$

where u_{saq}, u_{sbq}, and u_{scq} are quadrature unit current vectors and are derived as

$$\begin{aligned} u_{saq} &= (-u_{sbd} + u_{scd})/\sqrt{3}, \\ u_{sbq} &= (3u_{sad} + u_{sbd} - u_{scd})/(2\sqrt{3}), \\ u_{scq} &= (-3u_{sad} + u_{sbd} - u_{scd})/(2\sqrt{3}). \end{aligned} \tag{9.16}$$

Three-phase instantaneous reference supply currents are computed by adding in-phase and quadrature components expressed in Equations 9.11 and 9.15. For power factor correction along with harmonic elimination and load balancing, the amplitude of quadrature components is set to zero and in this condition the in-phase components of reference supply currents become the total reference supply currents. The computed three-phase reference supply currents and sensed supply currents are given to a hysteresis/PWM current controller to generate the switching signals for switches of the VSC of the APF.

9.4.2.3 IRPT-Based Control Algorithm of APFs

The control algorithm of the APF using IRPT is shown in Figure 9.17. Three-phase load currents and the PCC voltages are sensed and used to calculate the active and reactive powers due to harmonic components. Three-phase load (PCC) voltages are sensed and processed through BPF before their transformation to eliminate their ripple contents and are denoted as (v_{sa}, v_{sb}, v_{sc}). A first-order Butterworth filter is used as a BPF.

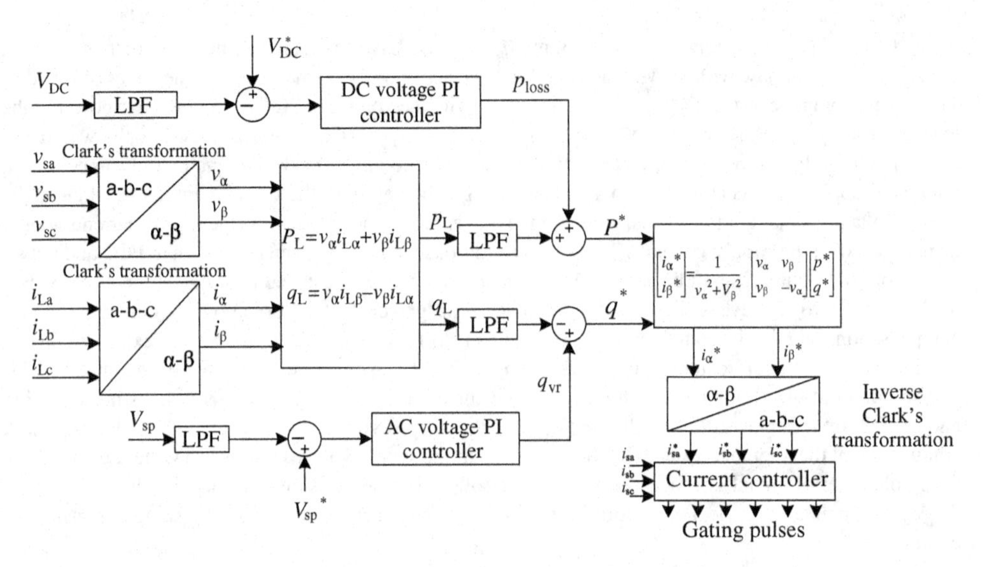

Figure 9.17 Control algorithm of APF using instantaneous reactive power theory

These three-phase filtered load voltages are transformed into two-phase α–β orthogonal coordinates (v_α, v_β):

$$\begin{pmatrix} v_\alpha \\ v_\beta \end{pmatrix} = \sqrt{\frac{2}{3}} \begin{pmatrix} 1 & -\frac{1}{2} & -\frac{1}{2} \\ 0 & \frac{\sqrt{3}}{2} & -\frac{\sqrt{3}}{2} \end{pmatrix} \begin{pmatrix} v_{sa} \\ v_{sb} \\ v_{sc} \end{pmatrix}. \tag{9.17}$$

Similarly, the three-phase load currents (i_{La}, i_{Lb}, i_{Lc}) are transformed into two-phase α–β orthogonal coordinates $(i_{L\alpha}, i_{L\beta})$ as

$$\begin{pmatrix} i_{L\alpha} \\ i_{L\beta} \end{pmatrix} = \sqrt{\frac{2}{3}} \begin{pmatrix} 1 & -\frac{1}{2} & -\frac{1}{2} \\ 0 & \frac{\sqrt{3}}{2} & -\frac{\sqrt{3}}{2} \end{pmatrix} \begin{pmatrix} i_{La} \\ i_{Lb} \\ i_{Lc} \end{pmatrix}. \tag{9.18}$$

From these two expressions, the instantaneous active power p_L and the instantaneous reactive power q_L flowing into the load side are computed as

$$\begin{pmatrix} p_L \\ q_L \end{pmatrix} = \begin{pmatrix} v_\alpha & v_\beta \\ v_\beta & -v_\alpha \end{pmatrix} \begin{pmatrix} i_{L\alpha} \\ i_{L\beta} \end{pmatrix}. \tag{9.19}$$

Let $\bar{p}_L$ and $\tilde{p}_L$ are the DC component and the harmonic component of p_L, respectively, and $\bar{q}_L$ and $\tilde{q}_L$ are the DC component and the harmonic component of q_L, respectively, Therefore, these may be expressed as

$$\begin{aligned} p_L &= \bar{p}_L + \tilde{p}_L, \\ q_L &= \bar{q}_L + \tilde{q}_L. \end{aligned} \tag{9.20}$$

In these expressions, the fundamental component of the load power is transformed to DC components $\bar{p}_L$ and $\bar{q}_L$, and the harmonics are transformed to AC components $\tilde{p}_L$ and $\tilde{q}_L$. Now, the AC components of active and reactive powers are extracted by using two low-pass filters and the reference three-phase supply currents $i_{sa}^*, i_{sb}^*, i_{sc}^*$ are obtained as

$$\begin{pmatrix} i_{sa}^* \\ i_{sb}^* \\ i_{sc}^* \end{pmatrix} = \sqrt{\frac{2}{3}} \begin{pmatrix} 1 & 0 \\ -\frac{1}{2} & \frac{\sqrt{3}}{2} \\ -\frac{1}{2} & -\frac{\sqrt{3}}{2} \end{pmatrix} \begin{pmatrix} v_\alpha & v_\beta \\ -v_\beta & v_\alpha \end{pmatrix}^{-1} \begin{pmatrix} p^* \\ q^* \end{pmatrix}, \tag{9.21}$$

where p_{loss} and q_{vr} are, respectively, the instantaneous active power necessary to adjust the voltage of the DC capacitor to its reference value and the instantaneous reactive power necessary to adjust the voltage of the AC bus to its reference value (it may be achieved using a PI controller similar to above algorithms), and $\overline{p}_L$ and $\overline{q}_L$ are the extracted load fundamental active and reactive power components.

This IRPT-based control algorithm may easily be modified for the control on source currents for indirect current control. In this case, for power factor correction mode of operation of the SAPF, $p^* = \overline{p}_L + p_{loss}$ and $q^* = q_L - q_{vr} = 0$ in Equation 9.21 and after the transformation from the α–β frame to the a–b–c frame, three-phase transformed currents are reference source currents and these must be compared with sensed source currents as shown in Figure 9.17 for indirect current control of the APF.

In the case of voltage regulation at PCC (voltage regulation mode of operation of the APF), a PI voltage controller is used similar to the above algorithms and its output (q_{vq}) is subtracted from or added to q_L to estimate p^* and q^* as $p^* = \overline{p}_L + p_{loss}$ and $q^* = q_{vp} - q_L$ as shown in Equation 9.21 and after the transformation, three-phase transformed currents are reference source currents and these are compared with sensed source currents as shown in Figure 9.17 for indirect current control of the APF.

9.5 Analysis and Design of Shunt Active Power Filters

The design of three-phase three-wire shunt active power filters includes the design of the VSC and its other passive components. The shunt active power filter includes a VSC, interfacing inductors, and a ripple filter. The design of the VSC includes the DC bus voltage level, the DC capacitance, and the rating of IGBTs.

A three-phase three-wire shunt active power filter topology is considered for detailed analysis. Figure 9.9 shows a schematic diagram of one of the shunt active power filters for a three-phase three-wire distribution system. It uses a three-leg VSC-based shunt active power filter. The design of the shunt active power filter is discussed in the following sections through the example of a 50 kVA shunt active power filter.

9.5.1 Design Example of a Three-Phase Three-Wire Shunt Active Power Filter for Industrial Loads

An APF is a device used in the AC distribution system where harmonic current mitigation, reactive current compensation, and load balancing are necessary. The building block of the APF is a voltage source converter consisting of semiconductor valves and a capacitor on the DC bus. The device is shunt connected to the power distribution network through a coupling inductance that is usually represented by the transformer leakage reactance. In general, an APF can provide power factor correction, harmonics compensation, and load balancing. Design of different parts of APF is given as follows:

- **Three Phase three-wire distribution system:** The design procedure of a three-leg, PWM-controlled IGBT- and VSC-based shunt active power filter consists of estimation and selection of its components.
- **Supply:** A three-phase, 415 V power supply at 50 Hz, with a source resistance (R_s) = 0.04 Ω and a source inductance (L_s) = 1 mH, is considered here.
- **Load:** A three-phase, three-wire rectifier is used as a nonlinear load, with a rectifier output current I_d = 224.17 A.

The rms value of rectifier input current is $I_{Lrms} = 0.816 \times 224.17 = 182.92$ A.
The value of the fundamental component of rectifier input current is $I_{L1} = 0.779 \times 224.17 = 174.78$ A.

$$\text{Harmonic current } I_h = \sqrt{(I_{Lrms}^2 - I_{L1}^2)} = 55.86 \text{ A}. \tag{9.22}$$

The rating of the SAPF is $S = 3V_{ph}^{*} I_h = 3 \times 240 \times 55.86 = 40.21 \times 1.25$ (25% extra for dynamics) = 50.27 kVA (considering 50 kVA).

DC Bus Voltage

The value of the DC bus voltage (V_{DC}) of the APF depends on the PCC line voltage. For a three-phase VSC, the DC bus voltage is defined as

$$V_{DC} = 2\sqrt{2}V_{LL}/(\sqrt{3}m), \tag{9.23}$$

where m is the modulation index and is considered as 1 and V_{LL} is the AC line output voltage of the APF. Thus, V_{DC} is obtained as 677.69 V for a V_{LL} of 415 V and is selected as 700 V.

DC Bus Capacitor

The design of the DC capacitor is governed by the reduction in the DC bus voltage upon the application of the load and rise in the DC bus voltage on removal of the load. Using the principle of energy conservation, the equation governing C_{DC} is

$$0.5C_{DC}[(V_{DC}^2) - (V_{DC1}^2)] = 3V_{ph}(aI)t, \tag{9.24}$$

where V_{DC} is the reference DC voltage and V_{DC1} is the minimum voltage level of the DC bus, a is the overloading factor, V_{ph} is the phase voltage, I is the phase current of the VSC, and t is the time by which the DC bus voltage is to be recovered.

Considering $V_{DC} = 700$ V, $V_{DC1} = 678$ V, $a = 1.2$, $V_{ph} = 240$ V, $I = 69.82$ A (25% more than phase current of the VSC), $t = 0.05$ and 5–10% of $3V_{ph}(aI)t$, the calculated value of C_{DC} is 19 897.084 μF and it is selected as 20 000 μF.

An alternative method for calculating the DC link voltage is based on the second harmonic ripple voltage at DC bus of the VSC because of load unbalancing as

$$C_{DC} = I_d/(2\omega V_{DCripple}), \tag{9.25}$$

where I_d is the active filter supply current, $V_{DCripple}$ = maximum DC link ripple voltage, and ω equal to $2\pi f$ (where $f = 50$ Hz):

$$I_d = \text{active filter kVA/DC link voltage} = 50\,000/700 = 71.42 \text{ A},$$

$$V_{DCripple} = 7 \text{ V (considering } 1-3\% \text{ of } V_{dc}).$$

After calculation the value of C_{DC} is 16 238.83 μF and it is selected as 20 000 μF.

AC Inductor

The selection of the AC inductance depends on the current ripple I_{crpp}, which is a part of the VSC current (69.82 A). Considering 15% (approximately) current ripple, at switching frequency f_s, the AC inductance is given as

$$L_f = \sqrt{3}mV_{DC}/(12af_s I_{crpp}). \tag{9.26}$$

Considering the switching frequency (f_s) = 10 kHz, modulation index (m) = 1, DC bus voltage (V_{DC}) of 700 V, and overload factor $a = 1.2$, the value of L_f is calculated to be 0.8 mH. The selected value of interfacing AC inductor (L_f) is considered as 0.75 mH.

Voltage Rating of IGBTs

The voltage rating (v_{sw}) of the IGBT under dynamic conditions is

$$V_{sw} = V_{DC} + V_d, \tag{9.27}$$

where V_d is the 10% overshoot in the DC link voltage under dynamic conditions. The voltage rating of the switch is calculated as 770 V. With an appropriate safety factor, IGBTs of 1200 V are selected to design the VSC used in the APF.

Current Rating of IGBTs

The current rating (I_{sw}) of the IGBT under dynamic conditions is

$$I_w = 1.25(I_{cr} + I_{sp}), \tag{9.28}$$

where I_{sp} and I_{cr} are the peak value of active filter current and the allowable value of ripple current, respectively. The minimum current rating of switch is 136.51 A. With an appropriate safety factor, IGBTs of 300 A are selected to design the VSC used in the APF.

Ripple Filter

A low-pass first-order filter tuned at half the switching frequency is used to filter the high-frequency noise from the compensating current and the voltage at the point of common coupling. Considering a low impedance for the harmonic voltage at a frequency of 10 kHz, the ripple filter capacitor is designed as $C_f = 25$ μF. A series resistance (R_f) of 4 Ω is included in series with the capacitor (C_f).

9.5.2 *Design Example of a Three-Phase, Four-Wire Shunt Active Power Filter for Industrial Loads*

Figure 9.18 shows a schematic diagram of a four-leg active filter connected to a three-phase AC mains feeding a three-phase four-wire load. The three-phase load may be a lagging power factor load or an unbalanced load or a nonlinear load or a combination of all these loads. For reducing ripple in

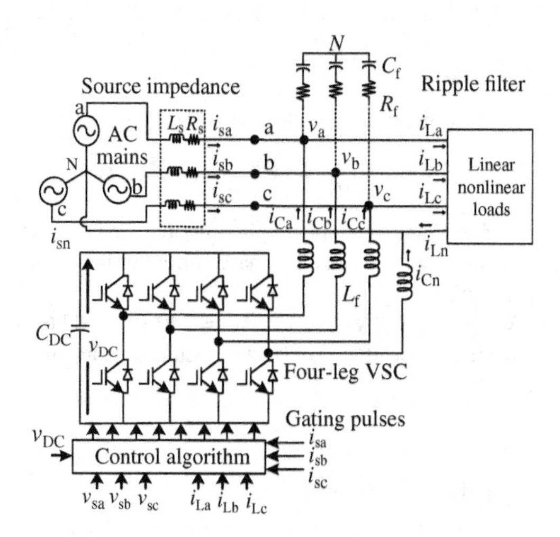

Figure 9.18 Design of components of a four-leg VSC-based SAPF

compensating currents, interfacing inductors (L_f) are used at AC side of the voltage source converter. A small series-connected capacitor (C_f) and resistor (R_f) represent the ripple filter installed at PCC in parallel with the load and the compensator to filter the high-frequency switching noise of the voltage at PCC. The harmonics/reactive currents (i_{Cabc}) are injected by the active filter to cancel the harmonic/reactive power component of load currents so that the supply currents are harmonic free (reduction in harmonic) and the load reactive power is also compensated. The rating of the switches is based on the voltage and current rating of the required compensation. For the considered load, $I_d = 77.65$ A and the rating of the VSC for the reactive power compensation/harmonics elimination is found to be 45 kVA (approximately 25% higher than the reactive current of rated value). Design and selected values of the DC bus voltage, DC bus capacitor, AC inductors, and the ripple filter of the active filter are given as follows:

- **Supply:** A three-phase, 415 V power supply at 50 Hz, with a source resistance (R_s) = 0.04 Ω and a source inductance (L_s) = 1 mH, is considered here.
- **Load:** A set of three single-phase rectifiers with $R = 2.5$ Ω and $L = 25$ mH is considered as a nonlinear load and its rectifier output current is $I_d = 86.28$ A.

The rms value of rectifier input current is $I_{Lrms} = I_d = 86.28$ A.
The rms value of rectifier input fundamental current is $I_{L1} = 0.9I_d = 77.65$ A.

$$\text{The rms value of harmonics current is } (I_h) = \sqrt{(I^2_{Lrms} - I^2_{L1})} = 37.61 \text{ A.} \tag{9.29}$$

The APF neutral current $I_{fn} = -I_{Ln} = 37.61$ A (since it has to cancel total load neutral current).

$$\text{The VA rating of the APF is } S = 3V_f I_f + V_f I_{fn} = 36.096 \text{ kVA (since } V_f = V_s = 239.6 \text{ V).} \tag{9.30}$$

The rating of the SAPF = 36.096 × 1.25 (25% extra for dynamics) = 45.12 kVA (consider 45 kVA).

Selection of the DC Voltage

The value of the DC bus voltage (V_{DC}) of the APF depends on the PCC line voltage. For a three-phase VSC, the DC bus voltage is defined as

$$\text{DC link voltage } (V_{DC}) = 2\sqrt{2}V_{LL}/(\sqrt{3}m), \tag{9.31}$$

where m is the modulation index considered as 1 and V_{LL} is the AC line output voltage of the VSC used in the active filter. Calculated value of the DC link voltage is 677.69 V and its selected value is 700 V.

Selection of a DC Link Capacitor

The value of the DC bus capacitor of the VSC of the APF is computed from energy relation at DC bus as

$$0.5C_{DC}[(V^2_{DC}) - (V^2_{DC1})] = 3V_{ph}(aI)t, \tag{9.32}$$

where V_{DC} is the reference DC voltage and V_{DC1} is the minimum voltage level of the DC bus, a is the overloading factor, V_{ph} is the phase voltage, I is the phase current of the VSC, and t is the time for which DC bus voltage is to be recovered.

Considering $V_{DC} = 700$ V, $V_{DC1} = 678$ V, $a = 1.2$, $V_{ph} = 240$ V, $I = 62.60$ A (25% more than phase current of the VSC), $t = 0.06$ and 5–10% of $3V_{ph}(aI)t$, the calculated value of C_{DC} is 12 845.36 μF and it is selected as 15 000 μF.

Selection of an AC Inductor

The value of the interfacing inductor is estimated as

$$\text{AC inductance } (L_f) = (\sqrt{3}mV_{DC})/(12 \times a \times f_s \times I_{crpp}), \tag{9.33}$$

where v_{DC} is the DC link voltage, $a = 1.2$, $m = 1$, f_s is the switching frequency, and I_{crpp} is the current ripple. Considering 5–10% current ripple in active current, the value of AC inductor is calculated to be 1.68 mH and its selected value is 2 mH.

Voltage Rating of IGBTs

The voltage rating (V_{sw}) of the IGBT under dynamic conditions is

$$V_{sw} = V_{DC} + V_d, \tag{9.34}$$

where V_d is the 10% overshoot in the DC link voltage under dynamic conditions. The voltage rating of the switch is calculated as 770 V. With an appropriate safety factor, IGBTs of 1200 V are selected to design the VSC used in the APF.

Current Rating of IGBTs

The current rating (I_{sw}) of the IGBT under dynamic conditions is

$$I_{sw} = 1.25(I_{cr} + I_{sp}), \tag{9.35}$$

where I_{sp} and I_{cr} are the peak value of active filter current and the allowable value of ripple current, respectively. The minimum current rating of the switch is 130.21 A. With an appropriate safety factor, IGBTs of 300 A are selected to design the VSC used in the APF.

Ripple Filter

A first-order high-pass filter tuned at half of the switching frequency, as shown in Figure 9.18, is used to filter the high-frequency noise from the voltage at the point of common coupling. A capacitor with series resistance is selected as a ripple filter. The value of the ripple filter capacitor and the resistance are considered as 5 μF and 5 Ω, respectively. This filter offers a high impedance of 636.64 Ω at fundamental frequency and a low impendence of 8.09 Ω at half of the switching frequency (here 5 kHz), which prevents the flow of the fundamental supply components in ripple filter branch and allows the flow of high-frequency noises through the ripple filter branch at higher than the fundamental frequency.

9.6 Modeling, Simulation, and Performance of Shunt Active Power Filters

The model of the SAPF is developed along with a non-stiff source, nonlinear loads, and the VSC with other passive components. The nonlinear load is modeled using a three-phase uncontrolled rectifier with constant DC current. The VSC of the APF is modeled using IGBT switches with a DC capacitor connected at the DC bus. The active filter is connected at PCC with a ripple filter and an interfacing inductor to eliminate high-frequency switching components. The switch-in response of the APF and load dynamics are implemented by incorporating circuit breaker models in the system. The simulation of three APFs with a three-wire and a four-wire system using MATLAB along with Simulink and Sim Power System toolboxes are described for the application of harmonics elimination, load balancing, and power factor correction.

9.6.1 Modeling of a Three-Phase Three-Wire Shunt Active Power Filter System

The model of a SRF-based control algorithm is developed and simulation is carried out using MATLAB/Simulink. To extract the DC transformed components, a first-order Butterworth LPF is used. A rectifier as a harmonics-producing nonlinear load is connected to the AC mains.

9.6.1.1 Simulation Parameters of a Three-Phase Three-Wire Shunt Active Power Filter System

The MATLAB platform is used for modeling and simulation of a three-phase three-wire active filter. The considered data for analysis of performance are given as follows:

AC mains	
Line voltage and frequency	Three phase, 415 V, 50 Hz (grid)
Source impedance	$R_s = 0.04\,\Omega$ and $L_s = 1$ mH
Nonlinear loads	Three-phase rectifier with $R = 2.5\,\Omega$ and $L = 25$ mH
Cutoff frequency of DC link low-pass filter	10 Hz
Switching frequency	10 kHz
Sampling time	10 μs

9.6.1.2 Performance of a Three-Phase Three-Wire Shunt Active Power Filter System

The dynamic performance of APF with performance indices as phase voltages at PCC (v_s), balanced supply currents (i_s), load currents (i_{La}, i_{Lb}, i_{Lc}), APF currents (i_{Ca}, i_{Cb}, i_{Cc}), and the DC bus voltage (v_{DC}) are shown under time-varying nonlinear load (at $t = 1.15$–1.25 s) conditions and waveforms of phase "a" voltage at PCC (v_{sa}), supply current (i_{sa}), and load current (i_{La}) with their harmonic spectra are shown in Figures 9.19 and 9.20a–c, respectively. Total harmonic distortion (THD) of the phase "a" at the PCC voltage, supply current, and load current are 5.96%, 1.61%, and 23.65% respectively. These results show satisfactory performance of the APF for harmonics elimination, load balancing, and power factor correction in nonlinear loads.

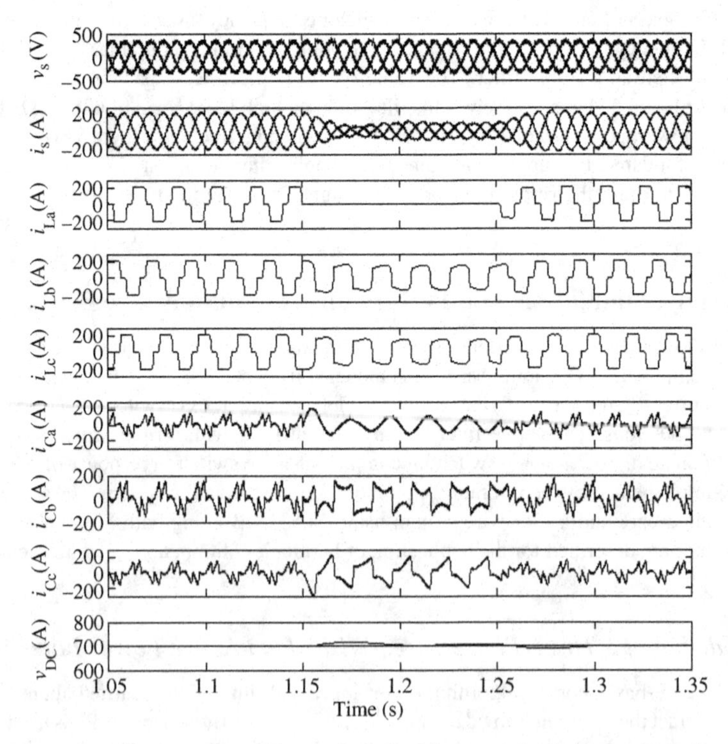

Figure 9.19 Performance of a three-wire APF at nonlinear load for harmonic elimination (load change 1.15–1.25 s)

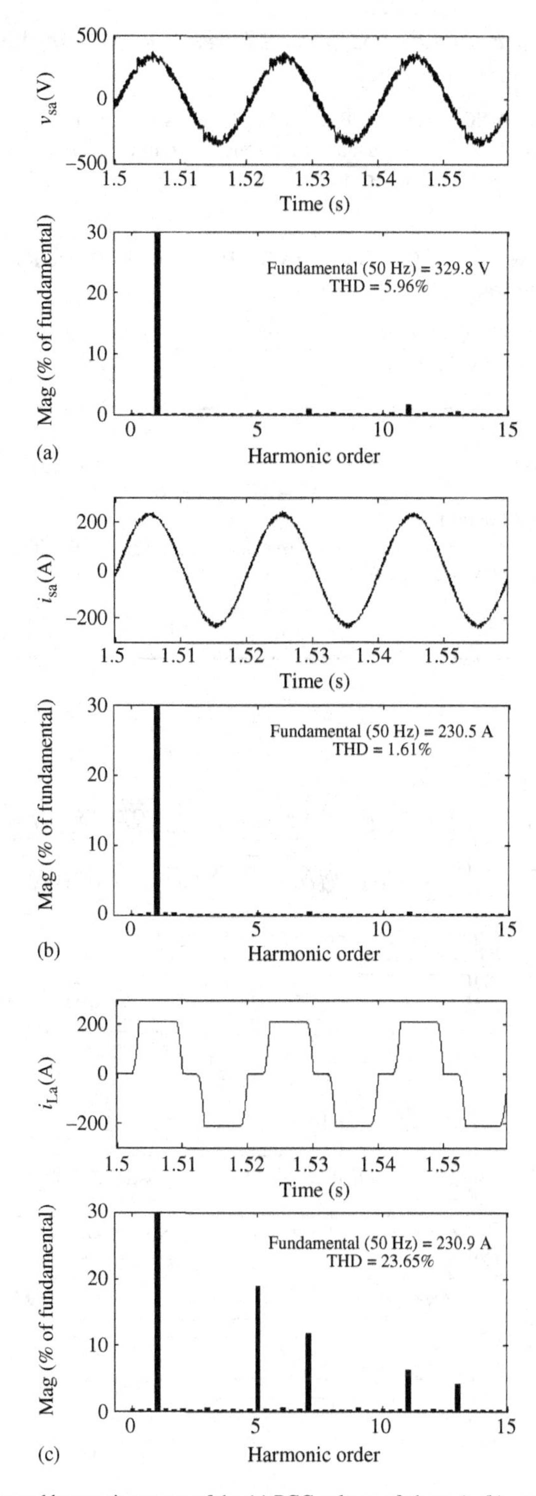

Figure 9.20 Waveforms and harmonic spectra of the (a) PCC voltage of phase A, (b) source current of phase A, and (c) load current of phase A

9.6.2 *Modeling of a Four-Leg VSC-Based Three-Phase Four-Wire Shunt Active Filter System*

The model of the four-leg VSC-based SAPF is developed and simulation is carried out using MATLAB/Simulink. The SRF control algorithm is used for reference current estimation. A rectifier as a harmonics-producing nonlinear load is connected to the AC mains.

9.6.2.1 Simulation Parameters of a Four-Leg VSC-Based Three-Phase Four-Wire Shunt Active Power Filter System

The MATLAB platform is used for modeling and simulation of performance of a three-phase three-wire active filter. The considered data for analysis of performance are given as follows.

AC mains	
Line voltage and frequency	Three phase, 415 V, 50 Hz (grid source)
Source impedance	$R_s = 0.04\,\Omega$ and $L_s = 1$ mH
Active filter control circuit parameter	
Nonlinear loads	Three single-phase rectifier with $R = 2.5\,\Omega$ and $L = 25$ mH
Cutoff frequency of DC link low-pass filter	10 Hz
Switching frequency	10 kHz
Sampling time	10 μs

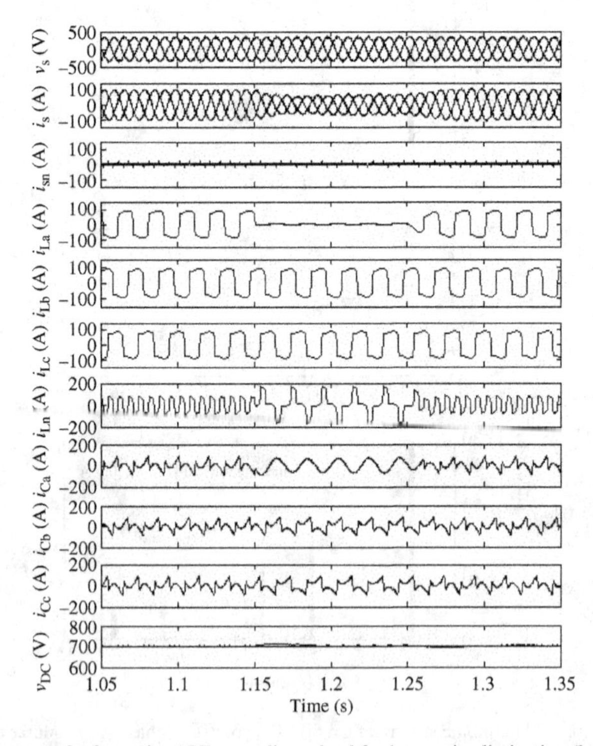

Figure 9.21 Performance of a four-wire APF at nonlinear load for harmonic elimination (load change 1.15–1.25 s)

9.6.2.2 Performance of a Four-Leg VSC-Based Three-Phase Four-Wire Shunt Active Power Filter System

The dynamic performance of the APF with performance indices as phase voltages at PCC (v_s), balanced supply currents (i_s), supply neutral current (i_{sn}), load currents (i_{La}, i_{Lb}, i_{Lc}), load neutral current (i_{Ln}), APF currents (i_{Ca}, i_{Cb}, i_{Cc}), and DC bus voltage (v_{DC}) is shown in Figure 9.21 under nonlinear load (at $t = 1.15$–1.25 s) conditions. Waveforms of phase "a" voltage at PCC (v_{sa}), supply current (i_{sa}), and load current (i_{La}) with their harmonic spectra are shown in Figure 9.22a–c, respectively. THD of phase "a" at the PCC voltage, supply current, and load current are 3.24% and 36.46%, respectively. These results show satisfactory performance of the APF for harmonics elimination, load balancing, neutral current compensation, and power factor correction in nonlinear loads.

Similarly, other configurations of shunt active power filters may be modeled and their performance may be simulated in detail to demonstrate their effectiveness and used in control algorithms. Some of them are illustrated in solved numerical examples in the next section.

9.7 Numerical Examples

Example 9.1

A single-phase shunt active power filter (shown in Figure E9.1) is employed for harmonic current compensation for a single-phase 230 V, 50 Hz, fed diode bridge converter drawing 20 A constant DC current. Calculate current, voltage, and the VA rating of the APF to provide harmonic compensation at unity power factor. Let the supply be stiff enough so that the distortion in voltage at the point of common coupling is negligible.

Solution: Given supply voltage, $V_s = 230$ V, frequency of the supply (f) = 50 Hz, and DC link current $I_{DC} = 20$ A.

In a single-phase diode bridge converter, the waveform of the supply current (I_s) is a square wave with the amplitude of the DC link current (I_{DC}). Moreover, the rms value of the fundamental component of the square wave is $(2\sqrt{2}/\pi) = 0.9$ times its amplitude.

Therefore, $I_s = I_{DC} = 20$ A and $I_{s1} = (2\sqrt{2}/\pi)I_{DC} = 0.9I_{DC} = 18$ A.

The current rating of the APF = current flowing through the filter $= I_f = I_h = \sqrt{(I_s^2 - I_{s1}^2)} = \sqrt{(20^2 - 18^2)} = 8.718$ A.

The voltage rating of the APF = voltage across the filter $= V_f = V_s = 230$ V.

The VA rating of the APF is $S = V_f I_f = 230 \times 8.718 = 2.008$ kVA.

Example 9.2

A shunt active power filter (voltage source inverter (VSI) with an AC series inductor and a DC bus capacitor) (shown in Figure E9.2) is used in parallel with a single-phase AC supply of 230 V at 50 Hz feeding a diode rectifier charging a battery of 180 V at 15 A average current with a small current limiting resistor to reduce the harmonics in AC mains current and to almost maintain UPF. Calculate (a) the value of the current limiting resistor, (b) AC input current to the diode rectifier, (c) the voltage rating of the APF, (d) the current rating of the APF, (e) the VA rating of the APF, (f) the value of the DC bus voltage of the APF, (g) the value of the AC inductor of the APF, and (h) the value of the DC bus capacitor of the APF. Consider the switching frequency is 20 kHz and the DC bus voltage has to be controlled within 8% range and the ripple current in inductor is 5%.

Solution: Given supply voltage $V_s = 230$ V, $V_{sm} = 325.27$ V, frequency of the supply (f) = 50 Hz, load $I_{DC} = 15$ A, $E = 180$ V, $f_s = 20$ kHz, $\Delta I_f = 5\%$, and $\Delta V_{DC\ APF} = 8\%$.

In a single-phase diode bridge converter, with RE load, the current will flow from angle α when the AC voltage is equal to E to angle β at which AC voltage reduces to E:

$$\alpha = \sin^{-1}(E/V_{sm}) = \sin^{-1}(180/325.27) = 33.59°, \quad \beta = \pi - \alpha = 146.40°, \quad \text{conduction angle} = \beta - \alpha = 112.81°.$$

The current limiting resistor (R) is $R = \{1/(\pi I_{DC})\}(2V_{sm}\cos\alpha + 2E\alpha - \pi E) = 3.97828\ \Omega$.

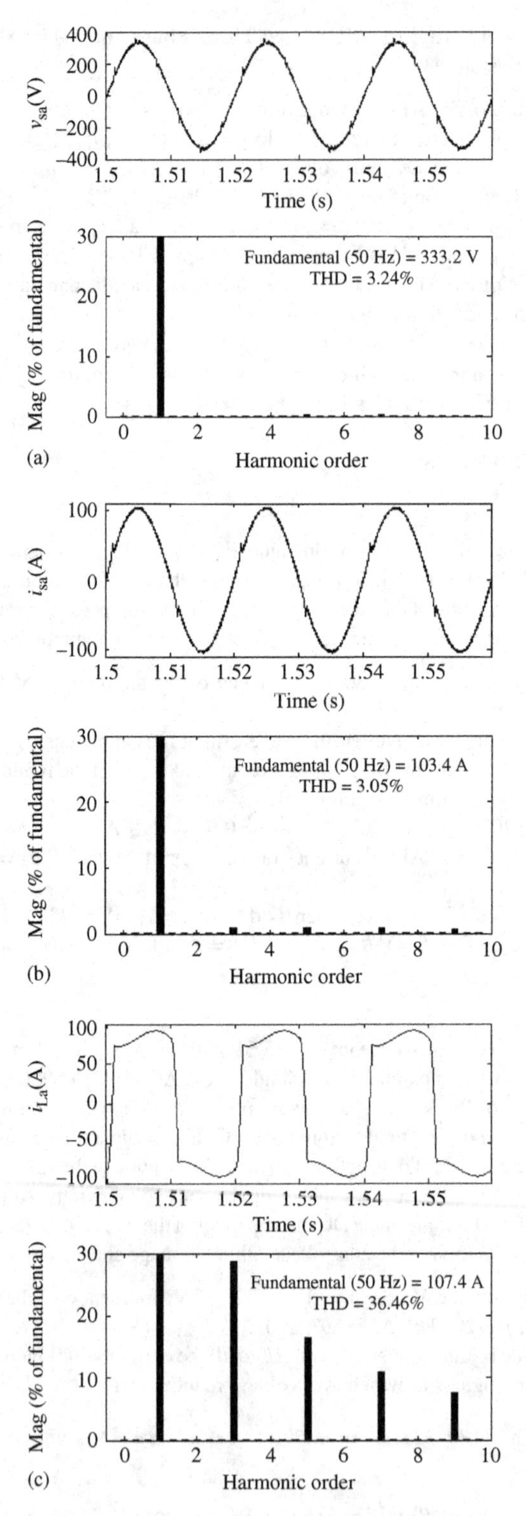

Figure 9.22 Waveforms and THD spectra of the (a) PCC voltage of phase A, (b) source current of phase A, and (c) load current of phase A

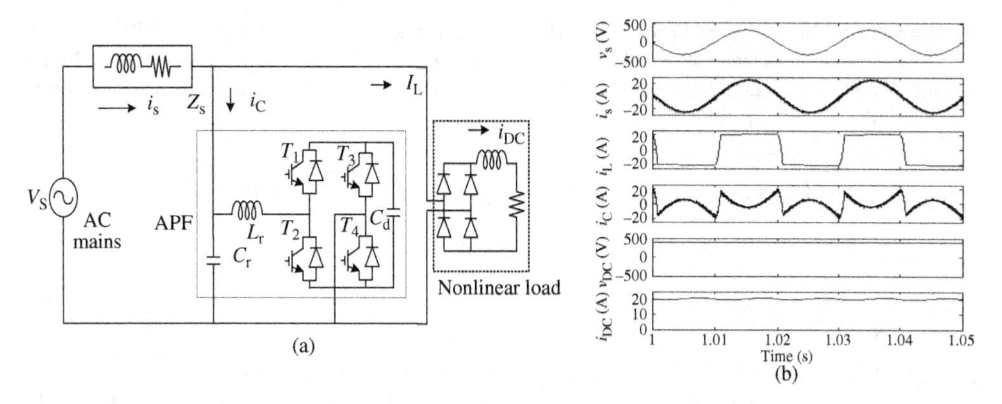

Figure E9.1 A single-phase APF for compensation of a current fed type of nonlinear load and waveforms

The rms value of AC input current to diode rectifier (I_{dAC}) is the rms value of discontinuous current in the AC mains:

$$I_{dAC} = [\{1/(\pi R^2)\}\{(0.5V_{sm}^2 + E^2)(\pi - 2\alpha) + 0.5V_{sm}^2 \sin 2\alpha - 4\,V_{sm}E\cos\alpha\}]^{1/2} = 20.8522\ \text{A}.$$

Active power drawn from the AC mains is $P = I_s^2 R + EI_{DC} = 4429.83$ W.
The fundamental rms value of current from the AC mains is $I_{s1} = P/V_s = 4429.833/230 = 19.26$ A.

a. The current limiting resistor $R = \{1/(\pi I_{DC})\}(2V_{sm}\cos\alpha + 2E\alpha - \pi E) = 3.97\ 828\ \Omega$.
b. The rms value of AC input current to a diode rectifier (I_{dAC}) is the rms value of discontinuous current in the AC mains:

$$I_{dAC} = [\{1/(\pi R^2)\}\{(0.5V_{sm}^2 + E^2)(\pi - 2\alpha) + 0.5V_{sm}^2 \sin 2\alpha - 4V_{sm}E\cos\alpha\}]^{1/2} = 20.8522\ \text{A}.$$

c. The voltage rating of the APF = voltage across the filter = $V_f = V_s = 230$ V since it connected at PCC.
d. The current rating of the APF = current flowing through the filter = $I_f = I_h = \sqrt{(I_{dAC}^2 - I_{s1}^2)} =$ 7.991 A.
e. The VA rating of the APF is $S = V_f I_f = 230 \times 7.99\ 166 = 1.8380$ kVA.

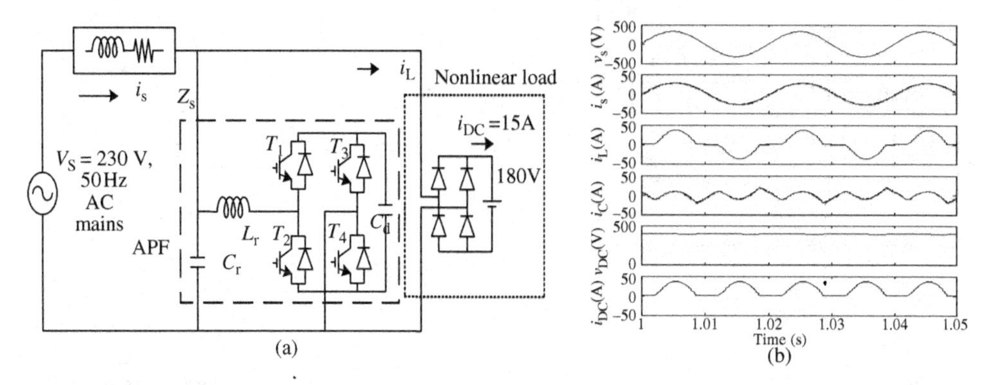

Figure E9.2 A single-phase APF for compensation of a voltage fed type of nonlinear load and waveforms

f. The value of the DC bus voltage of the APF is $V_{DC\ APF} = \sqrt{2}V_f/m_a = \sqrt{2}V_f/0.8 = 406.59\ V \approx 400\ V$.
g. The interfacing inductance of the APF is $L_f = V_{DC\ APF}/(4f_s\Delta I_f) = 400/(4 \times 20000 \times 0.39995) = 1.25\ mH$.
h. The DC bus capacitance of the APF is computed from change in the stored energy during dynamics: $\Delta E = 1/2C_{DC}(V^2_{DCAPF} - V^2_{DC\ min\ APF}) = V_f \times I_f \times \Delta t$,

$$\Delta E = 1/2C_{DC}(400^2 - 368^2) = 230 \times 7.99166 \times 10/1000\ (\text{considering } \Delta t = 10\ \text{ms}),$$

$$C_{dc} = 1495.83\ \mu F.$$

Example 9.3

A single-phase shunt active power filter (shown in Figure E9.3) is employed for harmonic current and reactive power compensation for a single-phase 220 V, 50 Hz, fed thyristor with fully controlled bridge converter drawing 20 A constant DC current operating at a thyristor firing angle of 60°. Calculate current, voltage, and the VA rating of the SAPF to provide (a) only harmonic compensation, (b) only reactive power compensation, and (c) harmonic and reactive power compensation at unity power factor. Let the supply be stiff enough so that the distortion in voltage at the point of common coupling is negligible.

Solution: Given supply rms voltage $V_s = 220$ V, frequency of the supply $(f) = 50$ Hz, $I_{DC} = 20$ A, and $\alpha = 60°$.

In a single-phase thyristor bridge converter, the waveform of the supply current (I_s) is a square wave with the amplitude of the DC link current (I_{DC}). Moreover, the rms value of the fundamental component of the square wave is 0.9 times its amplitude. Therefore, $I_s = I_{DC} = 20$ A and $I_{s1} = 0.9I_{DC} = 18$ A.

The total rms value of harmonic current is $I_h = \sqrt{(I_s^2 - I_{s1}^2)} = \sqrt{(20^2 - 18^2)} = 8.718$ A.

The active power component of the supply current is $I_{s1a} = I_{s1} \cos\theta_1 = I_{s1} \cos\alpha = 18 \cos 60° = 9$ A.

Total harmonic and reactive current is $I_f = \sqrt{(I_s^2 - I_{s1a}^2)} = \sqrt{(20^2 - 9^2)} = 17.86$ A.

a. Only harmonic compensation:
   Current rating = current flowing through the filter = $I_f = I_h = \sqrt{(I_s^2 - I_{s1}^2)} = \sqrt{(20^2 - 18^2)} =$ 8.718 A.

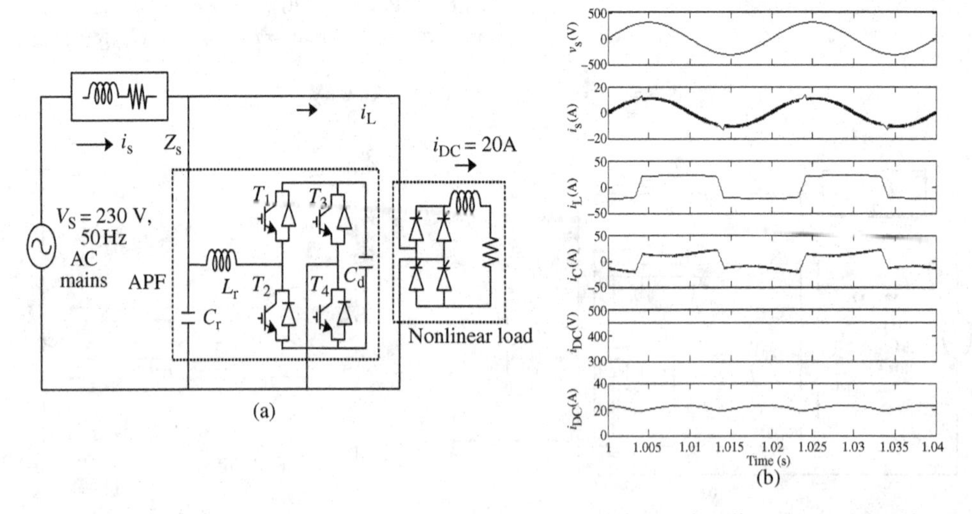

Figure E9.3 A single-phase APF for compensation of a current fed type of nonlinear load and waveforms

Voltage rating = voltage across the filter = $V_f = 220$ V.
The VA rating of the APF is $S = V_f\, I_f = 220 \times 8.718 = 1.91\ 796$ kVA.

b. Only reactive power compensation:
Current rating = filter current I_f = reactive current $I_R = I_{s1} \sin\theta_1 = I_{s1} \sin\alpha = 18 \sin 60° = 15.588$ A.
Voltage rating of the filter = voltage across the filter = $V_f = V_s = 220$ V.
The VA rating of the APF is $S = V_f I_f = 220 \times 15.588 = 3.429$ kVA.

c. Harmonic current and reactive power compensation:

$$I_f = \sqrt{(I_s^2 - I_{s1a}^2)} = \sqrt{(20^2 - 9^2)} = 17.86 \text{ A}.$$

The voltage rating of the filter = voltage across the filter = $V_f = V_s = 220$ V.
The VA rating of the APF is $S = V_f\, I_f = 220 \times 17.86 = 3.929$ kVA.

Example 9.4

A resistive heating load of 10 Ω (shown in Figure E9.4) is fed from a single-phase 230 V (rms), 50 Hz, AC source through a phase-controlled AC controller at a firing angle of 120°. Calculate (a) fundamental active power drawn by the load, (b) fundamental reactive power drawn by the load, and (c) the VA rating of the APF to provide (i) only harmonic compensation, (ii) only reactive power compensation, and (iii) harmonic and reactive power compensation at unity power factor. Let the supply be stiff enough so that the distortion in voltage at the point of common coupling is negligible.

Solution: Given supply rms voltage $V_s = 230$ V, frequency of the supply $(f) = 50$ Hz, $R = 10\ \Omega$, and $\alpha = 120°$.

In a single-phase phase-controlled AC controller, the waveform of the supply current (I_s) has a value of v_s/R from angle α to π. $V_{sm} = 230\sqrt{2} = 325.27$ V.

The rms value of the supply current is $I_s = V_{sm}[\{1/(2\pi)\}\{(\pi - \alpha) + \sin 2\alpha/2\}]^{1/2}/R = 10.168\ 699$ A.

The fundamental rms value of current is $I_{s1} = V_{sm}/(2\pi R\sqrt{2})[(\cos 2\alpha - 1)^2 + \{\sin 2\alpha + 2(\pi - \alpha)\}^2]^{1/2} =$ 7.097 A.

$$\theta_1 = \tan^{-1}[(\cos 2\alpha - 1)/\{\sin 2\alpha + 2(\pi - \alpha)\}] = 50.69°.$$

The rms value of the active power fundamental component of the supply current is $I_{s1a} = I_{s1} \cos\theta_1 = 4.49\ 647$ A.

Hence, the total rms harmonic current is $I_h = \sqrt{(I_s^2 - I_{s1}^2)} = \sqrt{(10.168699^2 - 7.097^2)} = 7.2829$ A.

The rms value of the reactive power fundamental component of the supply current is $I_{s1r} = I_{s1} \sin\theta_1 = I_{s1}\sqrt{(1 - \cos\theta_1^2)} = 5.49$ A.

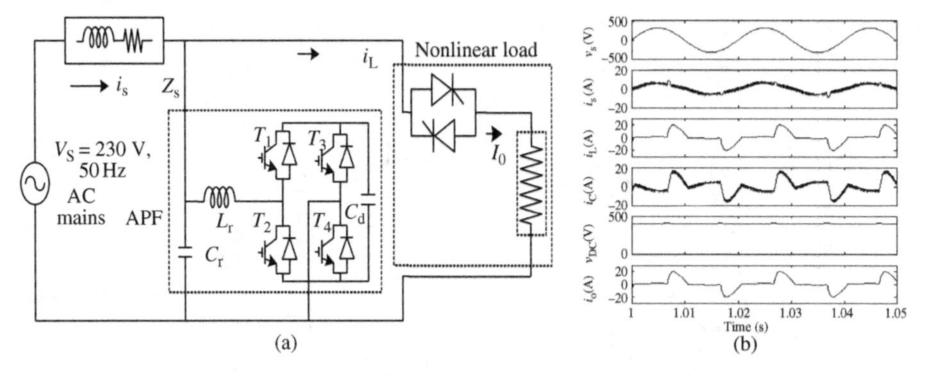

Figure E9.4 A single-phase APF for compensation of a current fed type of nonlinear load and waveforms

Total harmonic and reactive current is $I_f = \sqrt{(I_s^2 - I_{s1a}^2)} = \sqrt{(10.168699^2 - 4.49647^2)} = 9.11974$ A.

a. The fundamental active power drawn by the load is $P_1 = V_s I_{s1} \cos\theta_1 = 1034.19$ W.
b. The fundamental reactive power drawn by the load is $Q_1 = V_s I_{s1} \sin\theta_1 = 1262.786$ VAR.
c. The VA rating of the APF to provide the following:
   i. Only harmonic compensation:
      Current rating = current flowing through the filter = $I_f = I_h = \sqrt{(I_s^2 - I_{s1}^2)} = \sqrt{(10.168699^2 - 7.097^2)} =$ 7.2829 A.
      Voltage rating = voltage across the filter = $V_f = 230$ V.
      The VA rating of the APF is $S = V_f I_f = 230 \times 7.2829 = 1.675$ kVA.
   ii. Only reactive power compensation:
      Reactive power $Q = P \sin\theta_1/\cos\theta_1 = 1262.786$ VAR.
      Current rating = filter current I_f = reactive current $I_R = I_{s1}\sin\theta_1 = I_{s1}\sin\theta_1 = I_{s1}\sqrt{(1 - \cos\theta_1^2)} = 5.49$ A.
      The voltage rating of the filter = voltage across the filter = $V_f = V_s = 230$ V.
      The VA rating of the APF is $S = V_f I_f = Q_1 = V_s I_{s1} \sin\theta_1 = 1262.786$ VAR.
   iii. Harmonic current and reactive power compensation:

$$I_f = I_f = \sqrt{(I_s^2 - I_{s1a}^2)} = \sqrt{(10.168699^2 - 4.49647^2)} = 9.11974 \text{ A.}$$

The voltage rating of the filter = voltage across the filter = $V_f = V_s = 230$ V.
The VA rating of the APF is $S = V_f I_f = 230 \times 9.11\ 974 = 2097.54$ VA.

Example 9.5

A three-phase shunt active power filter (shown in Figure E9.5) is employed for harmonic current compensation for a three-phase 415 V, 50 Hz, fed diode bridge converter drawing 200 A constant DC current. Calculate (a) active power drawn by the load, (b) the current rating of the APF, (c) the voltage rating of the APF, (d) the VA rating of the APF to provide harmonic current compensation at unity power factor, (e) the value of the DC bus voltage of the APF, (f) the value of the AC inductor of the APF, and (g) the value of the DC bus capacitor of the APF. Consider the switching frequency is 20 kHz and the DC bus voltage has to be controlled within 8% range and the ripple current in inductor is 5%. Let the supply be stiff enough so that the distortion in voltage at the point of common coupling is negligible.

Solution: Given supply rms phase voltage $V_s = 415/\sqrt{3} = 239.6$ V, frequency of the supply $(f) = 50$ Hz, and $I_{DC} = 200$ A.

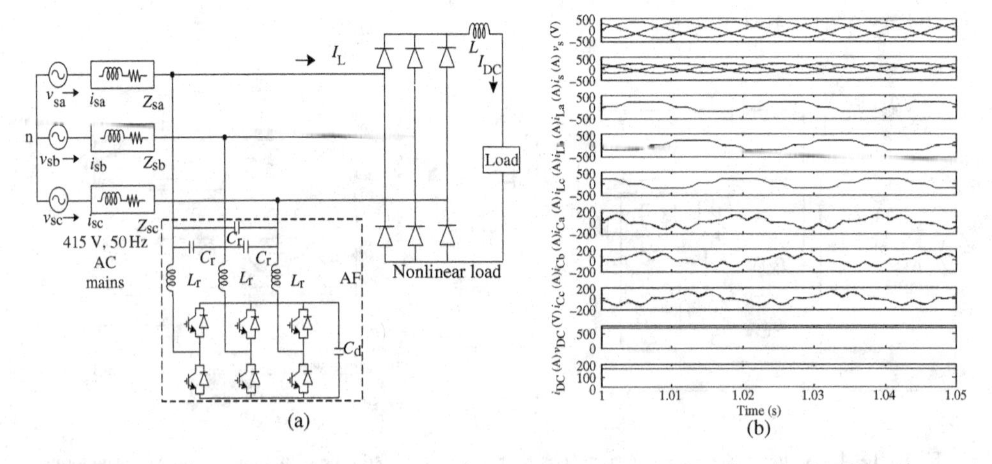

Figure E9.5 A three-phase APF for compensation of a current fed type of nonlinear load and waveforms

In a three-phase diode bridge converter, the waveform of the supply current (I_s) is a quasi-square wave with the amplitude of the DC link current (I_{DC}).

Therefore, the rms value of the quasi-square wave load current is $I_s = I_{DC}\sqrt{(2/3)} = 163.29$ A.

Moreover, the rms value of the fundamental component of the quasi-square wave is $I_{s1} = \{(\sqrt{6})/\pi\}I_{DC} = 155.939$ A.

a. The active power drawn by the load is $P = 3V_sI_{s1}\cos\theta_1 = 112.08$ kW.
b. The current rating of the APF = the total rms harmonic current $= I_{APF} = I_f = I_h = \sqrt{(I_s^2 - I_{s1}^2)} =$ 48.439 A.
c. The voltage rating of the filter = voltage across the filter $= V_f = V_s = 239.60$ V.
d. The VA rating of the APF is $S = 3V_fI_f = 3 \times 239.60 \times 48.439 = 34.818$ kVA.
e. The value of the DC bus voltage of the APF is $V_{DC\ APF} = 2\sqrt{2}V_f/m_a = 2\sqrt{2}V_f/0.9 = 752.99 \approx 750$ V.
f. The interfacing inductance of the APF is $L_f = \{(\sqrt{3})/2\}m_a V_{DC\ APF}/(6af_s\Delta I_f) = \{(\sqrt{3})/2\} \times 0.9 \times 750/(6 \times 1.2 \times 20000 \times 2.42195) = 1.676$ mH.
g. The DC bus capacitance of the APF is computed from change in the stored energy during dynamics as follows:

$$\Delta E = 1/2C_{DC}(V_{DCAPF}^2 - V_{DC\ min\ APF}^2) = 3 \times V_f \times I_f \times \Delta t.$$

$\Delta E = 1/2C_{DC}\ (750^2 - 690^2) = 3 \times 239.6 \times 48.439 \times 10/1000$ (considering $\Delta t = 10$ ms).
$C_{DC} = 8059.711\ \mu$F.

Example 9.6

A three-phase shunt active power filter (shown in Figure E9.6) is employed for reactive power and harmonics compensation for a three-phase, three-wire, 415 V, 50 Hz system. A nonlinear load consisting of a three-phase diode bridge rectifier is drawing AC current at 0.96 displacement factor (DPF) and THD of its AC current is 65%. It is drawing 25 kW from AC source and crest factor (CF) is 2.5 of AC input current. Calculate current, voltage, and the VA rating of the APF to provide (a) only harmonic current compensation, (b) only reactive power compensation, and (c) harmonic current and reactive power compensation at unity power factor. Let the supply be stiff enough so that the distortion in voltage at the point of common coupling is negligible.

Solution: Given supply rms voltage $V_s = 415/\sqrt{3} = 239.6$ V, frequency of the supply (f) = 50 Hz, THD of $I_s = 65\%$, DPF = 0.96, crest factor CF of $I_s = 2.5$, active power, $P = 25$ kW.

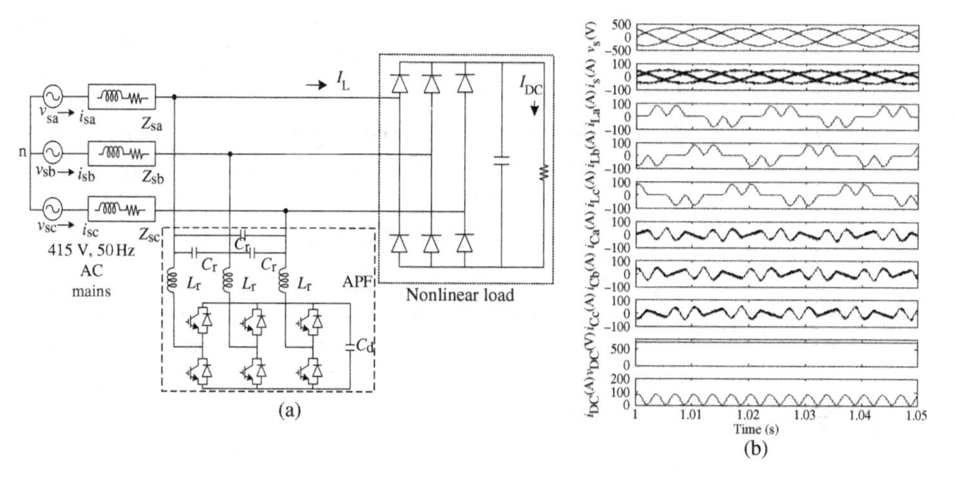

Figure E9.6 A three-phase APF for compensation of a voltage fed type of nonlinear load and waveforms

In a three-phase diode bridge converter, the fundamental active power component of the supply current is $I_{s1a} = P/(3V_s) = 34.78$ A.

The rms fundamental supply current is $I_{s1} = I_{s1a}/\text{DPF} = 34.78/0.96 = 36.28$ A.

The distortion factor is $\text{DF} = 1/\sqrt{(1 + \text{THD}^2)} = 0.83\ 844$.

The power factor is $\text{PF} = \text{DF} \times \text{DPF} = 0.8049$.

Moreover, the rms value of supply current is $I_s = P/(3V_s \times \text{PF}) = 43.21$ A.

The VA rating of the APF to provide the following:

a. Only harmonic compensation:

   Current rating = current flowing through the filter $= I_f = I_h = \sqrt{(I_s^2 - I_{s1}^2)} = \sqrt{(43.21^2 - 36.28^2)} =$ 23.55 A.

   Voltage rating = voltage across the filter $= V_f = 239.60$ V.

   The VA rating of the APF is $S = 3V_f I_f = 3 \times 239.6 \times 23.55 = 16.926$ kVA.

b. Only reactive power compensation:

   Current rating = filter current I_f = reactive current $I_R = I_{s1} \sin\theta_1 = I_{s1}\sqrt{(1 - \text{DPF}^2)} = 10.144$ A.

   The voltage rating of the filter = voltage across the filter $= V_f = V_s = 239.60$ V.

   The VA rating of the APF is $S = 3V_f I_f = 3V_s I_{s1} \sin\theta_1 = 7.29\ 166$ kVA.

c. Harmonic current and reactive power compensation:

$$I_f = \sqrt{(I_s^2 - I_{s1a}^2)} = \sqrt{(43.21^2 - 34.78^2)} = 25.64 \text{ A}.$$

   The voltage rating of the filter = voltage across the filter $= V_f = V_s = 239.60$ V.

   The VA rating of the APF is $S = 3V_f I_f = 18.431$ kVA.

Example 9.7

A three-phase shunt active power filter (shown in Figure E9.7) is employed for harmonic current and reactive power compensation for a three-phase 415 V, 50 Hz, fed thyristor bridge converter feeding a resistive load of 10 Ω at a thyristor firing angle of 30°. Calculate (a) fundamental active power drawn by the load, (b) fundamental reactive power drawn by the load, (c) the VA rating of the APF to provide (i) only harmonic current compensation, (ii) only reactive power compensation, and (iii) full harmonic current and reactive power compensation at unity power factor. Let the supply be stiff enough so that the distortion in voltage at the point of common coupling is negligible.

Solution: Given supply rms voltage $V_s = 415/\sqrt{3} = 239.6$ V, frequency of supply $(f) = 50$ Hz, $R = 10\ \Omega$, and $\alpha = 30°$.

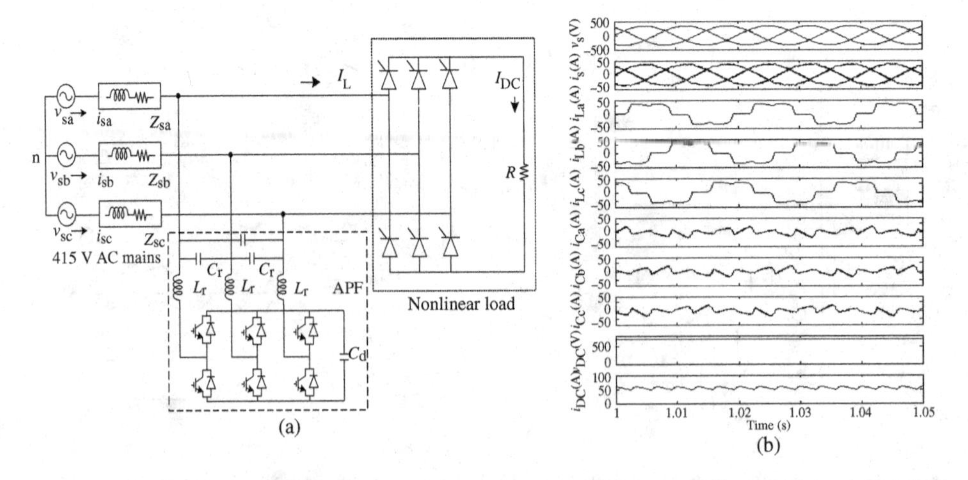

Figure E9.7 A three-phase APF for compensation of a current fed type of nonlinear load and waveforms

In a three-phase thyristor bridge converter, the waveform of the supply current (I_s) is decided by the load resistance and firing angle. Therefore, $I_s = \{V_s/(R)\}[\{(2\pi + 3\sqrt{3}\ \cos\ 2\alpha)\}/\pi\}^{1/2} = 40.286$ A.

The rms value of the fundamental component of the load current is $I_{s1} = \{(V_s/(2\pi R)\}\{(3\sqrt{3}\ \sin\ 2\alpha)^2 + (2\pi + 3\sqrt{3}\ \cos\ 2\alpha)^2\}^{1/2} = 37.966$ A.

Hence, the total rms value of the harmonic current is $I_h = \sqrt{(I_s^2 - I_{s1}^2)} = \sqrt{(40.28^2 - 37.98^2)} =$ 13.472 A.

The fundamental active power drawn by the load is $P = 3\{V_s^2/(2\pi R)\}[\{(2\pi + 3\sqrt{3}\cos 2\alpha)\} =$ 24.34387 kW.

The active power component of the supply current is $I_{s1a} = P/(3V_s) = 33.8667$ A.

The displacement factor is DPF $= \cos\ \theta_1 = I_{s1a}/I_{s1} = 0.892$; $\sin\ \theta_1 = \sqrt{(1 - \text{DPF}^2)} = 0.45\ 197$.

Total harmonics and reactive current $= I_f = \sqrt{(I_s^2 - I_{s1a}^2)} = \sqrt{(40.28^2 - 33.86^2)} = 21.8176$ A.

a. The fundamental active power drawn by the load is $P_1 = 3V_sI_{s1}\ \cos\ \theta_1 = 24.34\ 387$ kW.
b. The fundamental reactive power drawn by the load is $Q_1 = 3V_sI_{s1}\ \sin\ \theta_1 = 12.334$ kVAR.
c. The VA rating of the APF to provide the following:
   i. Only harmonic compensation:
      Current rating = current flowing through the filter $= I_f = I_h = \sqrt{(I_s^2 - I_{s1}^2)} = 13.472$ A.
      Voltage rating = voltage across the filter $= V_f = 239.60$ V.
      The VA rating of the APF is $S = 3V_f\ I_f = 3 \times 239.6 \times 13.472 = 9.68\ 367$ kVA.
   ii. Only reactive power compensation:
      Current rating = filter current I_f = reactive current $I_R = I_{s1}\ \sin\ \theta_1 = 17.1597$ A.
      The voltage rating of the filter = voltage across the filter $= V_f = V_s = 239.60$ V.
      The VA rating of the APF is $S = 3V_fI_f = 3V_sI_{s1}\ \sin\ \theta_1 = 12.33\ 424$ kVA.
   iii. Harmonic current and reactive power compensation:

$$I_f = \sqrt{(I_s^2 - I_{s1a}^2)} = \sqrt{(40.28^2 - 33.86^2)} = 21.8176 \text{ A.}$$

      The voltage rating of the filter = voltage across the filter $= V_f = V_s = 239.60$ V.
      The VA rating of the APF is $S = 3V_fI_f = 15.68\ 261$ kVA.

Example 9.8

A three-phase shunt active power filter (shown in Figure E9.8) is employed for reactive power and harmonics compensation for a three-phase, three-wire, 415 V, 50 Hz system feeding a set of nonlinear loads consisting of a thyristor bridge and a diode rectifier connected in parallel. The diode bridge

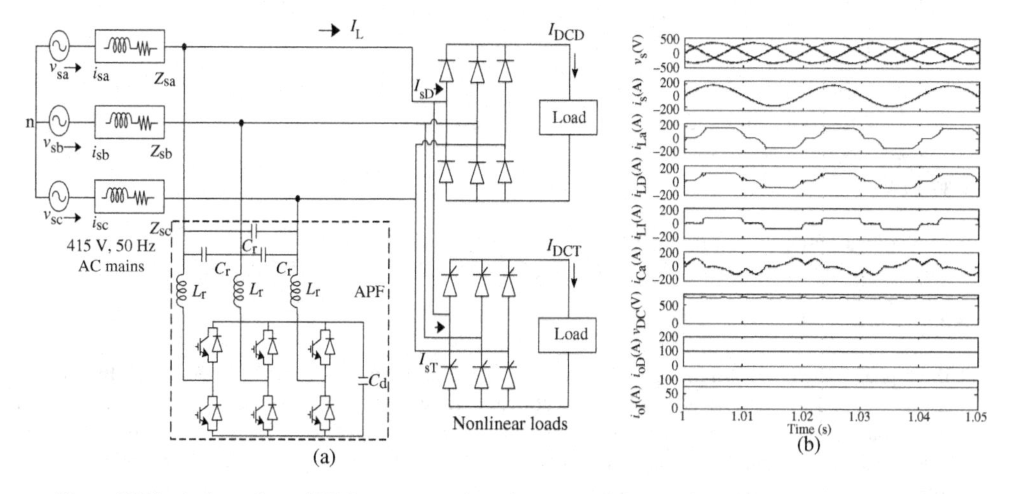

Figure E9.8 A three-phase APF for compensation of a current fed type of nonlinear load and waveforms

converter is drawing 100 A constant DC current. The thyristor bridge converter is drawing 75 A constant DC current at a thyristor firing angle of 30°. Calculate the current rating of the APF to provide harmonic current and reactive power compensation at unity power factor. Let the supply be stiff enough so that the distortion in voltage at the point of common coupling is negligible.

Solution: Given supply rms voltage $V_s = 415/\sqrt{3} = 239.6$ V and frequency of the supply $(f) = 50$ Hz feeding a set of nonlinear loads consisting of a thyristor bridge and a diode bridge rectifier connected in parallel. The diode bridge converter is drawing 100 A constant DC current. The AC–DC converter with a thyristor bridge is drawing 75 A constant DC current at a thyristor firing angle of 30°.

In a three-phase diode bridge converter, the waveform of the input AC current (I_{sD}) is a quasi-square wave with the amplitude of the DC link current (I_{DCD}).

Therefore, the rms value of the quasi-square wave load current is $I_{sD} = I_{DCD}\sqrt{(2/3)} = 81.65$ A.

Moreover, the rms value of the fundamental component of the quasi-square wave is $I_{sD1} = \{(\sqrt{6})/\pi\}$ $I_{DCD} = 77.97$ A.

Moreover, the active power component of the AC current of the diode converter is $I_{sD1a} = I_{sD1} = 77.97$ A.

In a three-phase thyristor bridge converter, the waveform of the input AC current (I_{sT}) is a quasi-square wave with the amplitude of the DC link current (I_{DCT}).

Therefore, $I_{sT} = \sqrt{(2/3)}I_{DCT} = 0.81\,649 I_{DCT} = 61.24$ A.

Moreover, the rms value of the fundamental component of the quasi-square wave is $I_{sT1} = \{(\sqrt{6})/\pi\}$ $I_{DCT} = 58.48$ A.

The active power component of the supply current is $I_{sT1a} = I_{sT1}\cos\theta_1 = I_{sT1}\cos\alpha = 58.48$ $\cos 30° = 50.64$ A.

Therefore, the total rms value of the active power component of the supply current of both loads is $I_{s1a} = I_{sD1a} + I_{sT1a} = 128.61$ A.

Therefore, the ideal supply current will be $i_s = I_{s1a}\sin\theta = 128.61\sqrt{2}\sin\theta$ A.

The AC current of the diode rectifier is a quasi-square wave with unity at displacement factor. However, the AC current of the thyristor bridge converter is also a quasi-square wave, but is phase shifted by a firing angle $\alpha = 30°$ with respect to AC supply voltage. The supply current after harmonic current and reactive power compensation will be at unity power factor in phase with supply voltage. The shunt filter must supply harmonic and reactive currents; therefore, it will be the difference of the required supply current and the combined load current that consists of currents of the diode and thyristor bridge converters.

Therefore, the rms value of the current of the shunt filter is computed with the half cycle integration as follows:

$$I_f = \sqrt{\left[1/\pi\left\{\int_0^{\pi/6}(128.61\sqrt{2}\sin\theta)^2 d\theta + \int_{\pi/6}^{\pi/6+\alpha}(128.61\sqrt{2}\sin\theta - 100)^2 d\theta + \int_{\pi/6+\alpha}^{5\pi/6}(128.61\sqrt{2}\sin\theta - 100 - 75)^2 d\theta \right.\right.}$$
$$\overline{\left.\left. + \int_{5\pi/6}^{5\pi/6+\alpha}(128.61\sqrt{2}\sin\theta - 75)^2 d\theta \int_{(5\pi/6+\alpha)}^{\pi}(128.61\sqrt{2}\sin\theta)^2 d\theta\right\}\right]}$$
$$= 37.24\text{A}.$$

The voltage rating of the filter = voltage across the filter = $V_f = V_s = 239.6$ V.

The VA rating of the APF is $S = 3V_f I_f = 3 \times 239.6 \times 37.24 = 26\,771.25$ VA.

Example 9.9

A three-phase SAPF (shown in Figure E9.9) is employed for harmonic current compensation for a three-phase 415 V, 50 Hz, fed 12-pulse diode bridge converter drawing 500 A constant DC current. It consists of an ideal transformer with a single primary star connected winding and two secondary windings connected in star and delta with same line voltages and unity turn ratios to provide 30° phase shift between two sets of three-phase output voltages. Two 6-pulse diode bridges are connected in series to form this 12-pulse

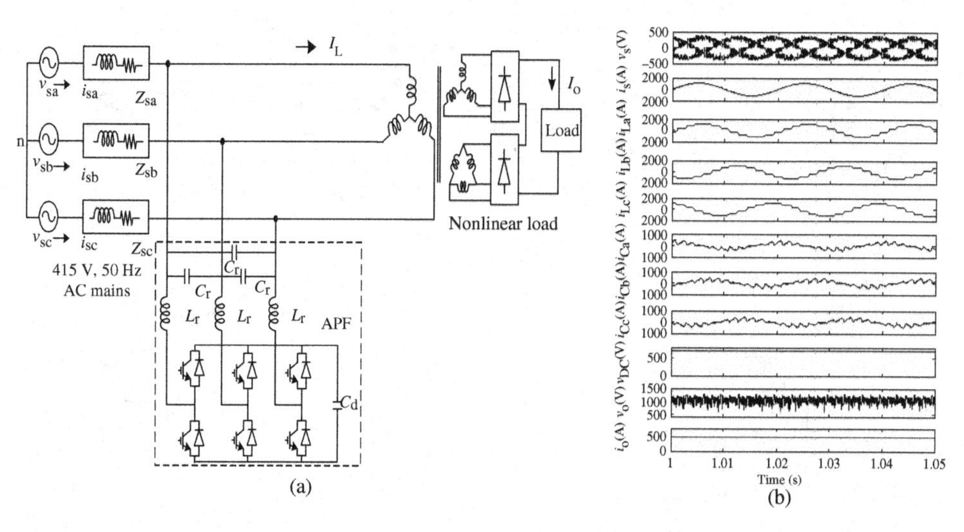

Figure E9.9 A three-phase APF for compensation of a current fed type of nonlinear load and waveforms

AC–DC converter. Calculate active power drawn by the load, current, voltage, and the VA rating of the APF to provide harmonic current compensation at unity power factor. Let the supply be stiff enough so that the distortion in voltage at the point of common coupling is negligible.

Solution: Given supply rms phase voltage $V_s = 415/\sqrt{3} = 239.6$ V, frequency of the supply $(f) = 50$ Hz, the amplitude of the DC link current $I_o = I_{DC} = 500$ A.

In a three-phase 12-pulse diode bridge converter, the waveform of the input AC current (I_s) is a stepped waveform with (i) the first step of $\pi/6$ angle (from 0° to $\pi/6$) and input current magnitude of $(I_{DC}/\sqrt{3})$, (ii) the second step of $\pi/6$ angle (from $\pi/6$ to $\pi/3$) and input current magnitude of $\{I_{DC}(1 + 1/\sqrt{3})\}$, and (iii) the third step of $\pi/6$ angle (from $\pi/3$ to $\pi/2$) and input current magnitude of $\{I_{DC}(1 + 2/\sqrt{3})\}$ and it has all four symmetric segments of such steps.

Therefore, the rms value of the 12-pulse converter input current is $I_s = I_{DC}[(1/3) + (1 + 1/\sqrt{3})^2 + (1 + 2/\sqrt{3})^2]^{1/2} = 1.57\ 735 I_{DC} = 788.675$ A.

The rms value of the fundamental component of the 12-pulse converter input current is $I_{s1} = \{(2\sqrt{6})/\pi\} I_{DC} = 1.559\ 393 I_{DC} = 779.69$ A.

The active power drawn by the load is $P = 3V_sI_{s1}\cos\theta_1 = 560.441$ kW.

Hence, the total rms harmonic current is $I_{APF} = I_f = I_h = \sqrt{(I_s^2 - I_{s1}^2)} = 118.74$ A.

Voltage rating of the filter = voltage across the filter = $V_f = V_s = 239.60$ V.

The VA rating of the APF is $S = 3V_fI_f = 3 \times 239.60 \times 118.74 = 85.3517$ kVA.

Example 9.10

A three-phase SAPF (shown in Figure E9.10) is employed for harmonic current and reactive power compensation for a three-phase 415 V, 50 Hz, fed 12-pulse thyristor bridge converter drawing 400 A constant DC current at a thyristor firing angle of 30°. It consists of an ideal transformer with single primary star connected winding and two secondary windings connected in star and delta with same line voltages and unity turn ratios to provide 30° phase shift between two sets of three-phase output voltages. Two 6-pulse thyristors bridges are connected in series to form this 12-pulse AC–DC converter. Calculate (a) fundamental active power drawn by the load, (b) fundamental reactive power drawn by the load, (c) the VA rating of the APF to provide (i) only harmonic current compensation, (ii) only reactive power compensation, and (iii) full harmonic current and reactive power compensation at unity power factor. Let the supply be stiff enough so that the distortion in voltage at the point of common coupling is negligible.

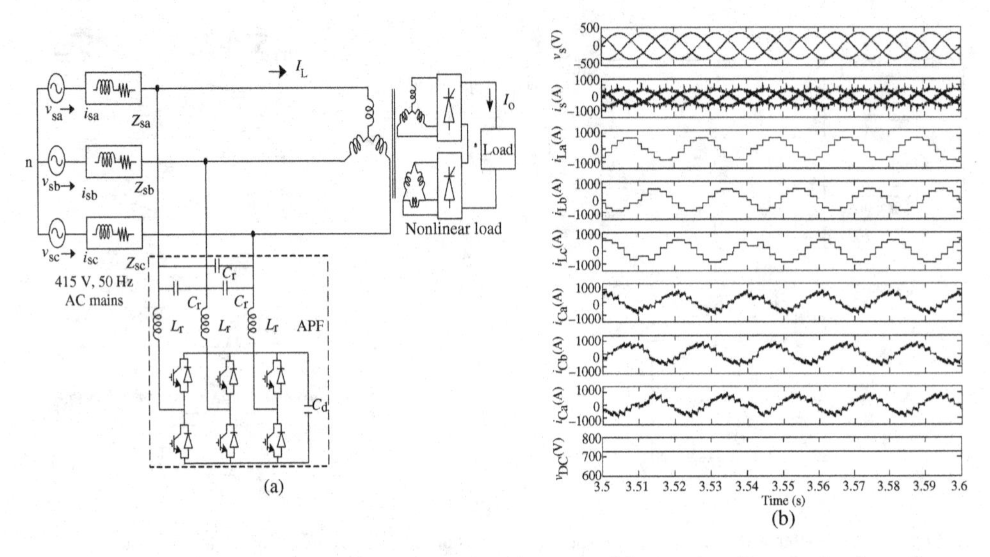

Figure E9.10 A three-phase APF for compensation of a current fed type of nonlinear load and waveforms

Solution: Given supply rms voltage $V_s = 415/\sqrt{3} = 239.6$ V, frequency of supply $(f) = 50$ Hz, $I_o = I_{DC} =$ 400 A, and $\alpha = 30°$.

In a three-phase 12-pulse thyristor bridge converter, the waveform of the input AC current (I_s) is a stepped waveforms with (i) the first step of $\pi/6$ angle (from α to $(\alpha + \pi/6)$) and input current magnitude of $(I_{DC}/\sqrt{3})$, (ii) the second step of $\pi/6$ angle (from $(\alpha + \pi/6)$ to $(\alpha + \pi/3)$) and input current magnitude of $(I_{DC}(1 + 1/\sqrt{3}))$, and (iii) the third step of $\pi/6$ angle (from $(\alpha + \pi/3)$ to $(\alpha + \pi/2)$) and input current magnitude of $\{I_{DC}(1 + 2/\sqrt{3})\}$ and it has all four symmetric segments of such steps.

Therefore, the rms value of the 12-pulse converter input current is $I_s = I_{DC}[(1/3) + (1 + 1/\sqrt{3})^2 + (1 + 2/\sqrt{3})^2]^{1/2} = 1.57\ 735 I_{DC} = 630.94$ A.

Moreover, the rms value of the 12-pulse converter fundamental AC current is $I_{s1} = \{(2\sqrt{6})/\pi\}$ $I_{DC} = 1.559\ 393 I_{DC}$ 623.757 A.

Hence, the total rms value of harmonic current is $I_h = \sqrt{(I_s^2 - I_{s1}^2)} = \sqrt{(630.94^2 - 623.757^2)} =$ 94.934 A.

The active power component of the supply current is $I_{s1a} = I_{s1} \cos\theta_1 = I_{s1} \cos\alpha = 623.757 \cos 30° =$ 540.19 A.

Total harmonics and reactive current $= I_f = \sqrt{(I_s^2 - I_{s1a}^2)} = \sqrt{(630.94^2 - 540.19^2)} = 326.01$ A.

a. The fundamental active power drawn by the load is $P_1 = 3V_sI_{s1} \cos\theta_1 = 3V_sI_{s1} \cos\alpha = 388.289$ kW.
b. The fundamental reactive power drawn by the load is $Q_1 = 3V_sI_{s1} \sin\theta_1 = 194.144$ kVAR.
c. The VA rating of the APF to provide the following:
   i. Only harmonic compensation:
   Current rating = current flowing through the filter $= I_f = I_h = \sqrt{(I_s^2 - I_{s1}^2)} = \sqrt{(630.94^2 - 623.757^2)} = 94.934$ A.
   Voltage rating = voltage across the filter $= V_f = 239.60$ V.
   The VA rating of the APF is $S = 3V_f\,I_f = 3 \times 239.6 \times 94.934 = 68.2386$ kVA.
   ii. Only reactive power compensation:
   Current rating = filter current I_f = reactive current $I_R = I_{s1} \sin\theta_1 = I_{s1} \sin\alpha = 623.757 \sin 30° =$ 311.87 A.
   The voltage rating of the filter = voltage across the filter $= V_f = V_s = 239.60$ V.
   The VA rating of the APF is $S = 3V_fI_f = 3V_sI_{s1} \sin\theta_1 = 224.172$ kVA.

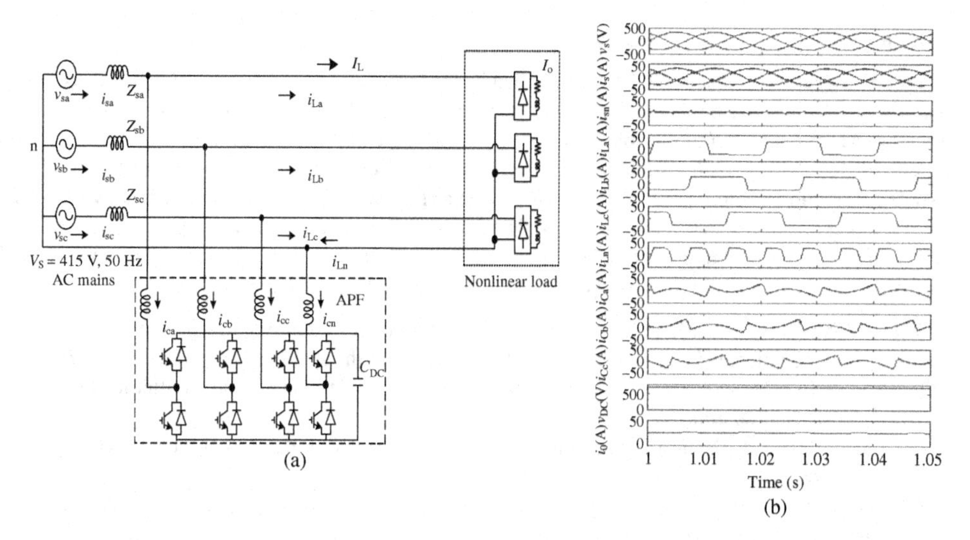

Figure E9.11 A three-phase APF for compensation of a current fed type of nonlinear load and waveforms

iii. Harmonics current and reactive power compensation:

$$I_f = \sqrt{(I_s^2 - I_{s1a}^2)} = \sqrt{(630.94^2 - 540.19^2)} = 326.01 \text{ A}.$$

The voltage rating of the filter = voltage across the filter = $V_f = V_s = 239.60$ V.
The VA rating of the APF is $S = 3V_f I_f = 234.336$ kVA.

Example 9.11

A three-phase four-wire distribution system, with a line voltage of 415 V, at 50 Hz, is feeding three single-phase loads (connected between phases and neutral) having a diode bridge converter drawing equivalent constant 25 A DC current (shown in Figure E9.11). A four-leg VSI with DC bus capacitor is used as APF. Calculate the (a) load neutral current, (b) APF phase current, (c) APF neutral current, (d) kVA rating of the APF to provide harmonic current compensation at unity power factor. Let the supply be stiff enough so that the distortion in voltage at the point of common coupling is negligible.

Solution: Given supply voltage $V_s = 239.6$ V, frequency of supply $(f) = 50$ Hz, and the DC link current $I_o = I_{DC} = 25$ A.

In a single-phase diode bridge converter, the waveform of the supply current (I_s) is a square wave with the amplitude of the DC link current (I_{DC}). Moreover, the rms value of the fundamental component of the square wave is $(2\sqrt{2}/\pi) = 0.9$ times its amplitude.

Therefore, $I_s = I_{DC} = 25$ A and $I_{s1} = (2\sqrt{2}/\pi)I_{DC} = 0.9 I_{DC} = 22.5$ A.

The current rating of the APF = current flowing through the filter = $I_f = I_h = \sqrt{(I_s^2 - I_{s1}^2)} = \sqrt{(25^2 - 22.5^2)} =$ 10.897 A.

The voltage rating of the APF = voltage across the filter = $V_f = V_s = 239.6$ V.

a. The load neutral current $I_{Ln} = 25$ A (since it will also be square wave three times the fundamental frequency).
b. The phase current rating of the APF = $I_f = I_h = \sqrt{(I_s^2 - I_{s1}^2)} = \sqrt{(25^2 - 22.5^2)} = 10.897$ A.
c. The APF neutral current $I_{fn} = -I_{Ln} = 25$ A (since it has to cancel total load neutral current).
d. The VA rating of the APF is $S = 3V_f I_f + V_f I_{fn} = 13.822$ kVA (since $V_f = V_s = 239.6$ V).

Example 9.12

A three-phase four-wire distribution system with a line voltage of 415 V, at 50 Hz, is feeding three single-phase loads (connected between phases and neutral terminal) having a set of diode bridge rectifiers drawing AC current at 0.96 displacement factor and THD of its AC current is 60% (shown in Figure E9.12). It is drawing 1500 W from AC source and crest factor is 2 of AC input current. A four-leg VSI with DC bus capacitors is used to realize as four-wire SAPF. Calculate current, voltage, and the VA rating of the APF to provide (a) harmonic current compensation, (b) reactive power compensation, and (c) harmonic current and reactive power compensation at unity power factor. Let the supply be stiff enough so that the distortion in voltage at the point of common coupling is negligible.

Solution: Given a three-phase four-wire distribution system with a line voltage of 415 V at 50 Hz, the supply rms phase voltage $V_s = 415/\sqrt{3} = 239.6$ V, frequency of the supply (f) = 50 Hz, THD of $I_s = 60\%$, DPF = 0.96, CF of $I_s = 2.0$, $P = 1500$ W on each phase, and a single-phase diode rectifier load connected between phases and neutral.

In a single-phase diode bridge converter, the fundamental active power component of the supply current is $I_{s1a} = P/(V_s) = 6.26\ 043$ A.

The rms value of fundamental supply current is $I_{s1} = I_{s1a}/\text{DPF} = 6.26\ 043/0.96 = 6.52$ A.

The distortion factor is $\text{DF} = 1/\sqrt{(1 + \text{THD}^2)} = 0.8575$.

The power factor is $\text{PF} = \text{DF} \times \text{DPF} = 0.8232$.

Moreover, the rms value of supply current is $I_s = P/(V_s \times \text{PF}) = 7.6036$ A.

The VA rating of the APF to provide the following:

a. Only harmonic compensation:

Current rating = current flowing through the filter = $I_f = I_h = \sqrt{(I_s^2 - I_{s1}^2)} = \sqrt{(7.6036^2 - 6.52^2)} = 3.912$ A.

Voltage rating = voltage across the filter = $V_f = 239.60$ V.

The VA rating of the APF is $S = 3V_f\, I_f = 3 \times 239.6 \times 3.912 = 2.812$ kVA.

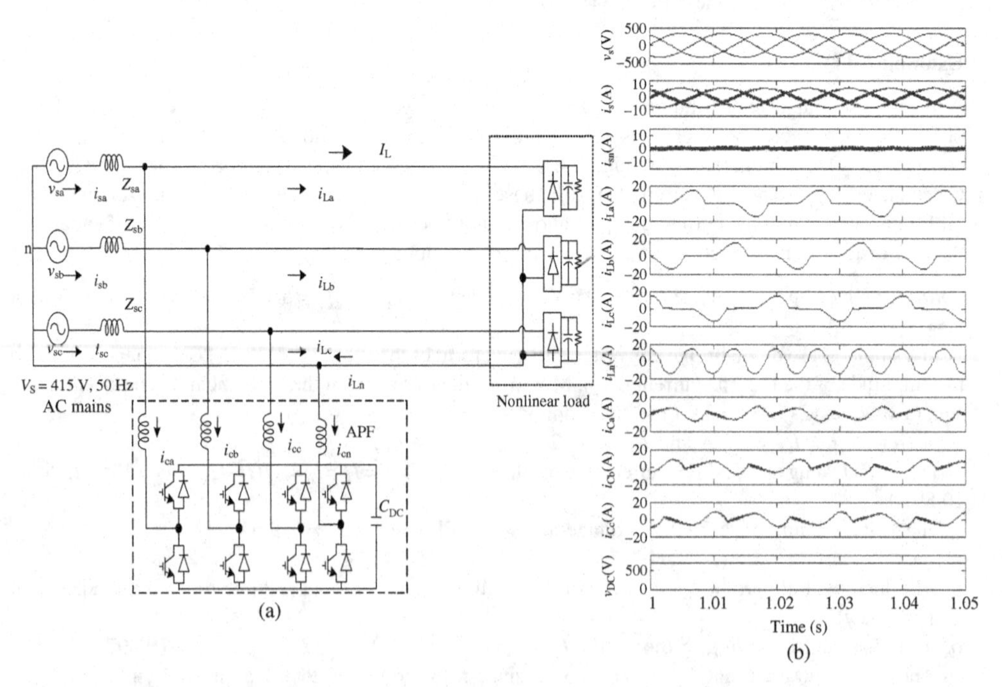

Figure E9.12 A three-phase APF for compensation of a voltage fed type of nonlinear load and waveforms

b. Only reactive power compensation:
   Current rating = filter current I_f = reactive current $I_R = I_{s1} \sin\theta_1 = I_{s1}\sqrt{(1-\text{DPF}^2)} = 1.8256$ A.
   The voltage rating of the filter = voltage across the filter = $V_f = V_s = 239.60$ V.
   The VA rating of the APF is $S = 3V_fI_f = 3V_sI_{s1}\sin\theta_1 = 1.312$ kVA.
c. Harmonic current and reactive power compensation:

$$I_f = \sqrt{(I_s^2 - I_{s1a}^2)} = \sqrt{(7.6036^2 - 6.26043^2)} = 4.315 \text{ A}.$$

   The voltage rating of the filter = voltage across the filter = $V_f = V_s = 239.60$ V.
   The VA rating of the APF is $S = 3V_fI_f = 3.1018$ kVA.

Example 9.13

A three-phase four-wire distribution system, with a line voltage of 415 V at 50 Hz, is feeding three single-phase loads (connected between phases and neutral) having a set of single-phase uncontrolled diode bridge converters, with a *RE* load of $R = 2\,\Omega$ and $E = 264$ V (shown in Figure E9.13). A three-leg VSI with midpoint DC bus capacitors for neutral connection is used to realize as four-wire SAPF. Calculate (a) fundamental active power drawn by the load and (b) the voltage, current, and VA rating of the APF to provide unity power factor. Let the supply be stiff enough so that the distortion in voltage at the point of common coupling is negligible.

Solution: Given supply phase voltage $V_s = 239.6$ V, $V_{sm} = 239.6 \times \sqrt{2}$ V $= 338.85$ V, frequency of the supply (f) = 50 Hz, load $R = 2\,\Omega$, and $E = 264$ V.

In a single-phase diode bridge converter, with *RE* load, the current will flow from angle α when the AC voltage is equal to E to angle β at which the AC voltage reduces to E:

$\alpha = \sin^{-1}(E/V_{sm}) = \sin^{-1}(264/338.87) = 51.179°$, $\beta = \pi - \alpha = 128.82°$, and the conduction angle = $\beta - \alpha = 77.641°$.

The rms value of supply current (I_s) is the rms value of the discontinuous current in the AC mains:

$I_s = [\{1/(\pi R^2)\}\{(0.5V_{sm}^2 + E^2)(\pi - 2\alpha) + 0.5V_{sm}^2 \sin 2\alpha - 4V_{sm}E\cos\alpha\}]^{1/2} = 17.84$ A.

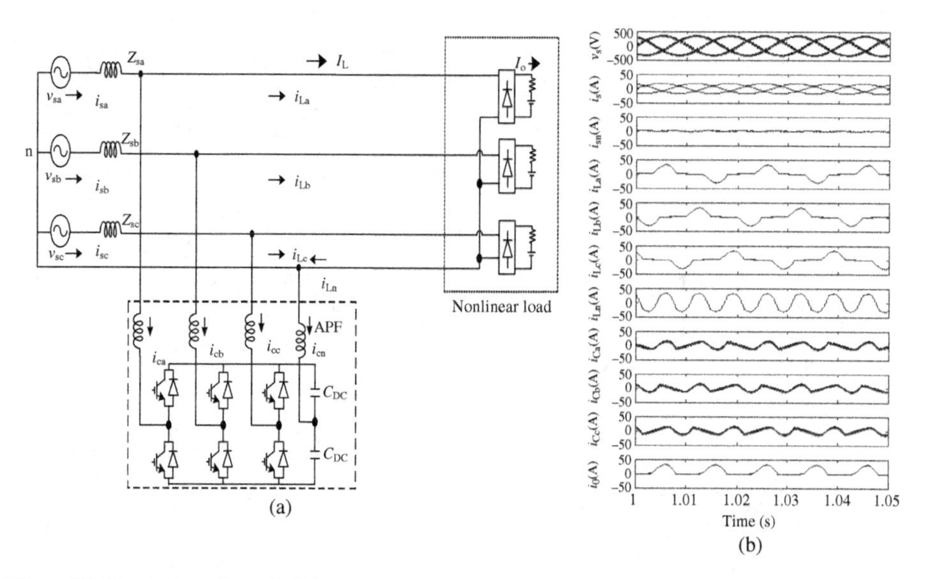

Figure E9.13 A three-phase APF for compensation of a voltage fed type of nonlinear load and waveforms

The average current ($I_o = I_{DC}$) in the battery is
$I_{DC} = \{1/(\pi R)\}(2V_{sm} \cos \alpha + 2E\alpha - \pi E) = 10.67$ A.
The active power drawn from the AC mains is$P = I_s^2 R + EI_{DC} = 3453.94$ W.
The fundamental rms value of current from the AC mains is $I_{s1} = P/V_s = 14.415$ A.
The current rating of the APF = current flowing through the filter = $I_f = I_h = \sqrt{(I_s^2 - I_{s1}^2)} = 10.522$ A.
The voltage rating of the APF = voltage across the filter = $V_f = V_s = 239.6$ V.
The VA rating of the APF is $S = 3V_f I_f = 3 \times 239.6 \times 10.522 = 7.563$ kVA.

Example 9.14

A three-phase four-wire distribution system, with a line voltage of 415 V at 50 Hz, is feeding three single-phase loads (connected between phases and neutral) having a single-phase thyristor bridge converter drawing equivalent 40 A constant DC current at a thyristor firing angle of 30° (shown in Figure E9.14). A four-leg VSI with DC bus capacitor is used as APF. Calculate current, voltage, and the VA rating of the APF to provide (a) harmonic current compensation and (b) harmonic current and reactive power compensation at unity power factor. Let the supply be stiff enough so that the distortion in voltage at the point of common coupling is negligible.

Solution: Given supply voltage $V_s = 239.6$ V, frequency of the supply (f) = 50 Hz, DC link current $I_o = I_{DC} = 40$ A, and firing angle $\alpha = 30°$.

In a single-phase thyristor bridge converter, the waveform of the supply current (I_s) is a square wave with the amplitude of the DC link current (I_{DC}). Moreover, the rms value of the fundamental component of the square wave is $(2\sqrt{2}/\pi) = 0.9$ times its amplitude.

Therefore, $I_s = I_{DC} = 40$ A and $I_{s1} = (2\sqrt{2}/\pi)I_{DC} = 0.9I_{DC} = 36$ A.

The rms value of fundamental active power component of the load current is $I_{s1a} = I_{s1} \cos \alpha = 31.177$ A.

The load neutral current $I_{Ln} = 40$ A (since it will also be a square wave three times the fundamental frequency).

The APF fourth leg neutral current is $I_{fn} = I_{Ln} = 40$ A (since it must be opposite to load neutral to cancel it).

The VA rating of the APF to provide the following:

a. For harmonic compensation:
   Current rating = current flowing through the filter = $I_f = I_h = \sqrt{(I_s^2 - I_{s1}^2)} = \sqrt{(40^2 - 36^2)} =$ 17.44 A.

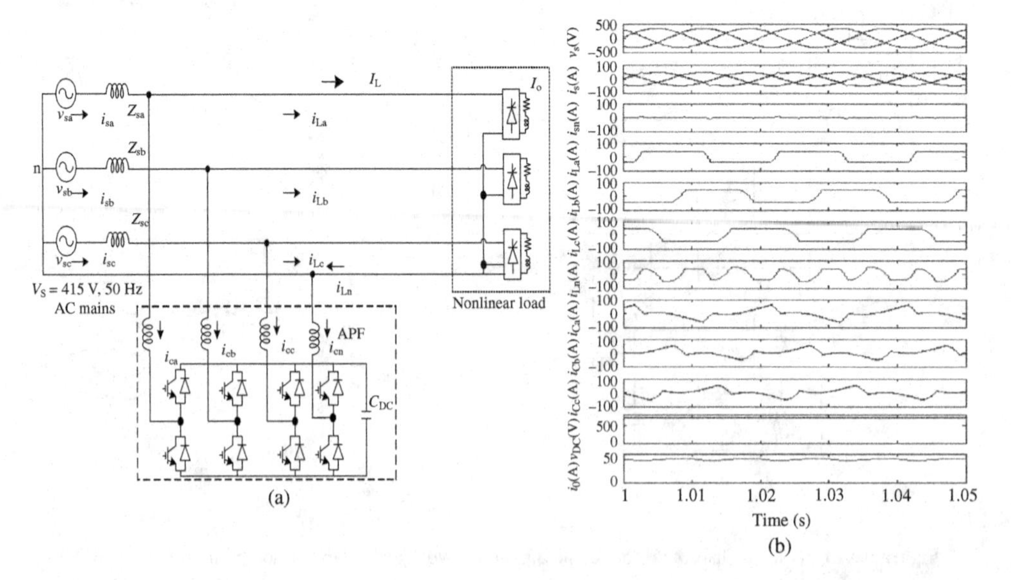

Figure E9.14 A three-phase APF for compensation of a current fed type of nonlinear load and waveforms

Voltage rating = voltage across the filter = $V_f = 239.60$ V.
The VA rating of the APF is $S = 3V_fI_f + V_fI_{fn} = 22.11$ kVA.

b. For harmonics current and reactive power compensation:

$$I_f = \sqrt{(I_s^2 - I_{s1a}^2)} = \sqrt{(40^2 - 31.177^2)} = 25.0598\text{ A}.$$

Voltage rating of the filter = voltage across the filter = $V_f = V_s = 239.60$ V.
The VA rating of the APF is $S = 3V_f\, I_f + V_fI_{fn} = 27.59$ kVA.

Example 9.15

A three-phase four-wire distribution system, with a line voltage of 415 V at 50 Hz, is feeding three single-phase loads (connected between phases and neutral) having a resistive heating load of 15 Ω through a phase-controlled AC controller at a firing angle of 60° (shown in Figure E9.15). A four-leg VSI with DC bus capacitors is used to realize as four-wire SAPF. Calculate (a) fundamental active power drawn by the load, (b) fundamental reactive power drawn by the load, and (c) the VA rating of the APF to provide (i) only harmonic compensation, (ii) only reactive power compensation, and (iii) harmonic and reactive power compensation at unity power factor. Let the supply be stiff enough so that the distortion in voltage at the point of common coupling is negligible.

Solution: Given supply rms voltage $V_s = 415/\sqrt{3} = 239.60$ V, frequency of the supply $(f) = 50$ Hz, $R = 15\ \Omega$, and $\alpha = 60°$.

In a single-phase, phase-controlled AC controller, the waveform of the supply current (I_s) has a value of v_s/R from angle α to π. $V_{sm} = 239.60\sqrt{2} = 338.8456$ V.

The rms value of supply current is $I_s = V_{sm}[\{1/(2\pi)\}\{(\pi - \alpha) + \sin 2\alpha/2\}]^{1/2}/R = 14.3267$ A.

The fundamental rms value of current is $I_{s1} = V_{sm}/(2\pi R\sqrt{2})[(\cos 2\alpha - 1)^2 + \{\sin 2\alpha + 2(\pi - \alpha)\}^2]^{1/2} = 13.4044$ A.

$\theta_1 = \tan^{-1}[(\cos 2\alpha - 1)/\{\sin 2\alpha + 2(\pi - \alpha)\}] = -16.53°$.

The rms value of active power fundamental component of the supply current is $I_{s1a} = I_{s1} \cos\theta_1 = 12.85$ A.

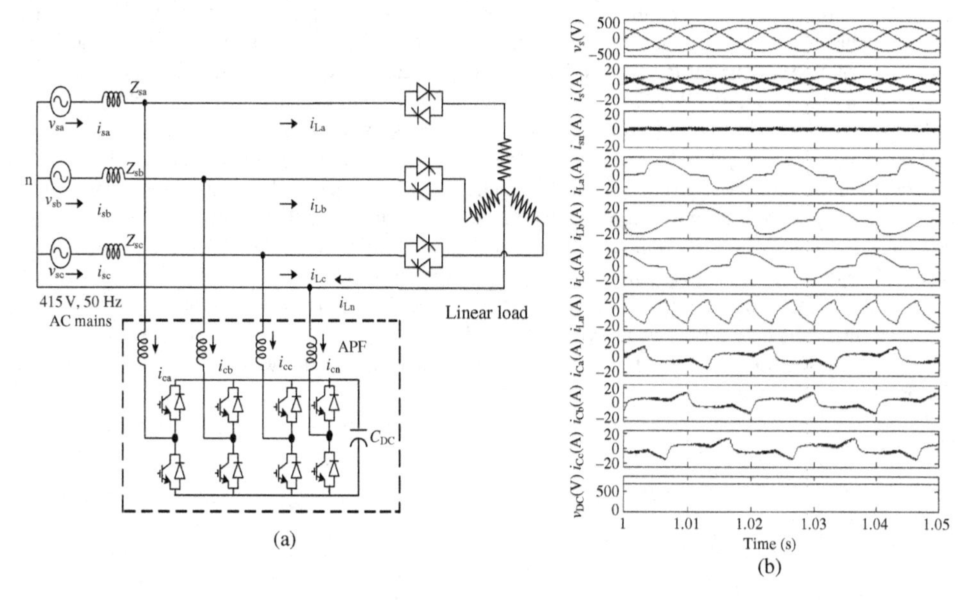

Figure E9.15 A three-phase APF for compensation of a current fed type of nonlinear load and waveforms

Hence, the total rms value of harmonic current is $I_h = \sqrt{(I_s^2 - I_{s1}^2)} = \sqrt{(14.3267^2 - 13.40400^2)} = 5.0573$ A.

The rms value of reactive power fundamental component of the supply current is $I_{s1r} = I_{s1} \sin\theta_1 = I_{s1}\sqrt{(1 - \cos\theta_1^2)} = -3.81378$ A.

Total harmonics and reactive current $= I_f = \sqrt{(I_s^2 - I_{s1a}^2)} = \sqrt{(14.3267^2 - 12.85^2)} = 6.335$ A.

a. The fundamental active power drawn by the load is $P_1 = 3V_sI_{s1} \cos\theta_1 = 9236.59$ W.
b. The fundamental reactive power drawn by the load is $Q_1 = 3V_sI_{s1} \sin\theta_1 = 2741.35$ VAR.
c. The VA rating of the APF to provide the following:
   i. Only harmonic compensation:
      Current rating = current flowing through the filter $= I_f = I_h = \sqrt{(I_s^2 - I_{s1}^2)} = \sqrt{(14.3267^2 - 13.4044^2)} = 5.0573$ A.
      Voltage rating = voltage across the filter $= V_f = 239.60$ V.
      The VA rating of the APF is $S = 3V_fI_f = 3 \times 239.60 \times 5.0573 = 3.6351$ kVA.
   ii. Only reactive power compensation:
      Reactive power $Q = 3P \sin\theta_1/\cos\theta_1 = 2741.81$ VAR.
      Current rating = filter current I_f = reactive current $I_R = I_{s1} \sin\theta_1 = I_{s1} \sin\theta_1 = I_{s1}\sqrt{(1 - \cos\theta_1^2)} = 3.814428$ A.
      The voltage rating of the filter = voltage across the filter $= V_f = V_s = 239.60$ V.
      The VA rating of the APF is $S = 3V_fI_f = Q_1 = 3V_sI_{s1} \sin\theta_1 = 2741.81$ kVA.
   iii. Harmonic current and reactive power compensation:

$$I_f = I_f = \sqrt{(I_s^2 - I_{s1a}^2)} = \sqrt{(14.3267^2 - 12.85^2)} = 6.334 \text{ A.}$$

      The voltage rating of the filter = voltage across the filter $= V_f = V_s = 239.60$ V.
      The VA rating of the APF is $S = 3V_fI_f = 3 \times 239.60 \times 6.334 = 4.553$ kVA.

Moreover, the current rating of neutral leg of VSC is decided by load neutral current, which in this case mainly consists of triplen harmonics current. For this particular example the current rating for neutral leg is approximately 12 A.

9.8 Summary

SAPFs are used to compensate the power quality problems of harmonic currents such as characteristic harmonics, noncharacteristic harmonics, interharmonics, subharmonics, reactive current, fluctuating current, unbalanced currents, and excessive neutral current produced by nonlinear loads. These SAPFs offer best solution for current fed type and combination of current fed and voltage fed type of nonlinear loads with moderate rating. PWM-based voltage source inverters are preferred to develop SAPFs because of low cost, reduced size, light weight, and reduced losses. A passive ripple filter is used at PCC where SAPF is connected to improve the voltage profile and to eliminate higher order harmonics, which a SAPF cannot eliminate. This passive ripple filter does not have much losses even being sometimes damped type because energy associated with higher order harmonics is quite small. However, these higher order harmonics produced by either nonlinear loads or switching of SAPF cause disturbance to communication systems and other electronic appliances. An analytical study of various performance indices of SAPFs for the compensation of nonlinear loads is made in detail with several numerical examples to study the rating of power filters and how it is affected by various kinds of nonlinear loads. SAPFs are observed as one of the best retrofit solutions for mitigating power quality problems caused due to nonlinear loads for reducing the pollution of AC mains.

9.9 Review Questions

1. What is a shunt active power filter?
2. What are the power quality problems that a SAPF can mitigate?
3. Can a current source inverter be used as a SAPF?

4. Why voltage source inverters are preferred in a SAPF?
5. What are the factors that decide the rating of a SAPF?
6. How the value of an interfacing inductor is computed for a VSI used in a SAPF?
7. How the value of a DC bus capacitor is computed for a VSI used in a SAPF?
8. How the self-supporting DC bus of a VSI is achieved in a SAPF?
9. How does the crest factor of the load current affect the rating of a SAPF?
10. Why an indirect current control is considered superior than a direct current control scheme of a SAPF?
11. Which of the two has higher current rating of SAPF when used only for harmonics compensation: current fed or voltage fed type of nonlinear load?
12. In which kind of nonlinear load the SAPF is more effective?
13. Why an H-bridge VSI-based SAPFs are preferred by the industries?
14. Among nonlinear loads using diode bridge, thyristor bridge, or AC voltage controller and cyclo-converter, which one needs highest rating of SAPF?
15. Among nonlinear loads using diode bridge, thyristor bridge, or AC voltage controller and cyclo-converter, which one needs lowest rating of SAPF?
16. Can a SAPF be used for regulating the AC voltage at PCC?
17. Which configuration of a SAPF is most effective to eliminate the neutral current?
18. How the profile of the voltage at PCC can be improved in a SAPF?
19. Up to what highest order of harmonic can be eliminated by a SAPF if the switching frequency is 5 kHz?
20. How the harmonics of higher order than the switching frequency can be eliminated in a SAPF?
21. What is the power rating of a SAPF in terms of power rating of a nonlinear load consisting of a single-phase diode rectifier with constant DC current?
22. What is the power rating of a SAPF in terms of power rating of a nonlinear load consisting of a three-phase diode rectifier with constant DC current?
23. What is the power rating of a SAPF in terms of power rating of a nonlinear load consisting of a single-phase thyristor bridge rectifier with constant DC current at a firing angle of 30° for unity power of the AC mains?
24. What is the power rating of a SAPF in terms of power rating of a nonlinear load consisting of a three-phase thyristor bridge rectifier with constant DC current at a firing angle of 60° for unity power of the AC mains?
25. What is the power rating of a SAPF in terms of power rating of a nonlinear load consisting of a single-phase thyristor AC voltage controller with a resistive load at a firing angle of 60° for unity power of the AC mains?

9.10 Numerical Problems

1. A single-phase shunt active power filter is employed for harmonic current compensation for a single-phase 220 V, 50 Hz, fed diode bridge converter drawing 25 A constant DC current. Calculate current, voltage, and the VA rating of the APF to provide harmonic compensation at unity power factor. Let the supply be stiff enough so that the distortion in voltage at the point of common coupling is negligible.

2. A shunt active power filter (VSI with an AC series inductor and a DC bus capacitor) is used in parallel with a single-phase AC supply of 220 V at 50 Hz feeding a diode rectifier charging a battery of 160 V at 20 A average current with a small current limiting resistor to reduce the harmonics in AC mains current and to almost maintain UPF. Calculate (a) the value of the current limiting resistor, (b) AC input current to the diode rectifier, (c) the voltage rating of the APF, (d) the current rating of the APF, (e) the VA rating of the APF, (f) the value of the DC bus voltage of the APF, (g) the value of the AC inductor of the APF, and (h) the value of the DC bus capacitor of the APF. Consider the switching frequency is 20 kHz and the DC bus voltage has to be controlled within 5% range and the ripple current in inductor is 3%.

3. A single-phase SAPF is employed for harmonic current and reactive power compensation for a single-phase 230 V, 50 Hz, fed thyristor with fully controlled bridge converter drawing 30 A constant DC current operating at a thyristor firing angle of 30°. Calculate current, voltage, and the VA rating of the SAPF to provide (a) only harmonic compensation, (b) only reactive power compensation, and (c) harmonic and reactive power compensation at unity power factor. Let the supply be stiff enough so that the distortion in voltage at the point of common coupling is negligible.

4. A resistive heating load of 20 Ω is fed from a single-phase 220 V (rms), 50 Hz, AC source through a phase-controlled AC controller at a firing angle of 90°. Calculate (a) fundamental active power drawn by the load, (b) fundamental reactive power drawn by the load, (c) the VA rating of the APF to provide (i) only harmonic compensation, (ii) only reactive power compensation, and (iii) harmonic and reactive power compensation at unity power factor. Let the supply be stiff enough so that the distortion in voltage at the point of common coupling is negligible.

5. A three-phase shunt active power filter is employed for harmonic current compensation for a three-phase 440 V, 50 Hz, fed diode bridge converter drawing 100 A constant DC current. Calculate (a) active power drawn by the load, (b) the current rating of the APF, (c) the voltage of the APF, (d) the VA rating of the APF to provide harmonic current compensation at unity power factor, (e) the value of the DC bus voltage of the APF, (f) the value of the AC inductor of the APF, and (g) the value of the DC bus capacitor of the APF. Consider the switching frequency is 20 kHz and the DC bus voltage has to be controlled within 10% range and the ripple current in inductor is 2.5%. Let the supply be stiff enough so that the distortion in voltage at the point of common coupling is negligible.

6. A three-phase shunt active power filter is employed for reactive power and harmonics compensation for a three-phase, three-wire, 440 V, 50 Hz system. A nonlinear load consisting of a three-phase diode bridge rectifier is drawing AC current at 0.95 displacement factor and THD of its AC current is 45%. It is drawing 45 kW from the AC source and crest factor is 2.0 of AC input current. Calculate current, voltage, and the VA rating of the APF to provide (a) only harmonic current compensation, (b) only reactive power and harmonics compensation, and (c) harmonic current and reactive power and harmonics compensation at unity power factor. Let the supply be stiff enough so that the distortion in voltage at the point of common coupling is negligible.

7. A three-phase shunt active power filter is employed for harmonic current and reactive power compensation for a three-phase 440 V, 50 Hz, fed thyristor bridge converter feeding a resistive load of 20 Ω at a thyristor firing angle of 45°. Calculate (a) fundamental active power drawn by the load, (b) fundamental reactive power drawn by the load, (c) the VA rating of the APF to provide (i) only harmonic current compensation, (ii) only reactive power compensation, and (iii) full harmonic current and reactive power compensation at unity power factor. Let the supply be stiff enough so that the distortion in voltage at the point of common coupling is negligible.

8. A three-phase shunt active power filter is employed for reactive power and harmonics compensation for a three-phase, three-wire, 440 V, 50 Hz system feeding a set of nonlinear loads consisting of a thyristor bridge and a diode rectifier connected in parallel. The diode bridge converter is drawing 200 A constant DC current. The thyristor bridge converter is drawing 150 A constant DC current at a thyristor firing angle of 60°. Calculate the current rating of the APF to provide harmonic current and

reactive power and harmonics compensation at unity power factor. Let the supply be stiff enough so that the distortion in voltage at the point of common coupling is negligible.

9. A three-phase shunt active power filter is employed for harmonic current compensation for a three-phase 440 V, 50 Hz, fed 12-pulse diode bridge converter drawing 400 A constant DC current. It consists of an ideal transformer with single primary star connected winding and two secondary windings connected in star and delta with same line voltages and unity turn ratios to provide 30° phase shift between two sets of three-phase output voltages. Two 6-pulse diode bridges are connected in series to form this 12-pulse AC–DC converter. Calculate active power drawn by the load, current, voltage, and the VA rating of the APF to provide harmonic current compensation at unity power factor. Let the supply be stiff enough so that the distortion in voltage at the point of common coupling is negligible.

10. A three-phase shunt active power filter is employed for harmonic current and reactive power compensation for a three-phase 440 V, 50 Hz, fed 12-pulse thyristor bridge converter drawing 500 A constant DC current at a thyristor firing angle of 60°. It consists of an ideal transformer with single primary star connected winding and two secondary windings connected in star and delta with same line voltages and unity turn ratios to provide 30° phase shift between two sets of three-phase output voltages. Two 6-pulse thyristors bridges are connected in series to form this 12-pulse AC–DC converter. Calculate (a) fundamental active power drawn by the load, (b) fundamental reactive power drawn by the load, (c) the VA rating of the APF to provide (i) only harmonic current compensation, (ii) only reactive power compensation, and (iii) full harmonic current and reactive power compensation at unity power factor. Let the supply be stiff enough so that the distortion in voltage at the point of common coupling be negligible.

11. A three-phase four-wire distribution system, with a line voltage of 400 V, at 50 Hz, is feeding three single-phase loads (connected between phases and neutral) having diode bridge converter drawing equivalent constant 15 A DC current. A four-leg VSI with DC bus capacitor is used as APF. Calculate (a) load neutral current, (b) APF phase current, (c) APF neutral current, (d) the kVA rating of the APF to provide harmonic current compensation at unity power factor. Let the supply be stiff enough so that the distortion in voltage at the point of common coupling is negligible.

12. A three-phase four-wire distribution system, with a line voltage of 380 V, at 50 Hz, is feeding three single-phase loads (connected between phases and neutral) having a set of diode bridge rectifiers drawing AC current at 0.95 displacement factor and THD of its AC current is 80%. It is drawing 4000 W from AC source and crest factor is 2.5 of AC input current. A three-leg VSI with midpoint DC bus capacitors for neutral connection is used to realize as four-wire SAPF. Calculate current, voltage, and the VA rating of the APF to provide (a) harmonic current compensation, (b) reactive power compensation, and (c) harmonic current and reactive power compensation at unity power factor. Let the supply be stiff enough so that the distortion in voltage at the point of common coupling is negligible.

13. A three-phase four-wire distribution system, with a line voltage of 380 V, at 50 Hz, is feeding three single-phase loads (connected between phases and neutral) having a set of single-phase uncontrolled diode bridge converters, which has a *RE* load with $R = 5\,\Omega$, and $E = 240$ V. A three-leg VSI with midpoint DC bus capacitors for neutral connection is used to realize as four-wire SAPF. Calculate (a) fundamental active power drawn by the load, (b) voltage, current, and the VA rating of the APF to provide unity power factor. Let the supply be stiff enough so that the distortion in voltage at the point of common coupling is negligible.

14. A three-phase four-wire distribution system, with a line voltage of 400 V, at 50 Hz, is feeding three single-phase loads (connected between phases and neutral) having a single-phase thyristor bridge converter drawing equivalent 30 A constant DC current at a thyristor firing angle of 60°. A four-leg VSI with a DC bus capacitor is used as the APF. Calculate current, voltage, and the VA rating of the APF to provide (a) harmonic current compensation and (b) harmonic current and reactive power compensation at unity power factor. Let the supply be stiff enough so that the distortion in voltage at the point of common coupling is negligible.

15. A three-phase four-wire distribution system, with a line voltage of 380 V, at 50 Hz, is feeding three single-phase loads (connected between phases and neutral) having a resistive heating load of 10 Ω through a phase-controlled AC controller at a firing angle of 90°. A three-leg VSI with midpoint DC bus capacitors for neutral connection is used to realize as four-wire SAPF. Calculate (a) fundamental active power drawn by the load, (b) fundamental reactive power drawn by the load, (c) the VA rating of the APF to provide (i) only harmonic compensation, (ii) only reactive power compensation, and (iii) harmonic and reactive power compensation at unity power factor. Let the supply be stiff enough so that the distortion in voltage at the point of common coupling is negligible.

9.11 Computer Simulation-Based Problems

1. Design an AC shunt active filter and simulate its behavior to reduce harmonic currents and to maintain almost UPF of a single-phase 230 V, 50 Hz, uncontrolled bridge converter with a parallel capacitive DC filter of 470 μF and equivalent resistive load of 20 Ω. It has a source impedance of 0.25 Ω resistive element and 1.5 Ω inductive element. Plot the supply voltage and input current, active filter current, DC bus voltage of the filter, output voltage and current with time, and the harmonic spectra of supply current, load current, and voltage at PCC. Compute (a) THD, crest factor, and rms value of AC mains current, (b) THD, crest factor, and rms value of AC load current, (c) displacement factor, (d) distortion factor, (e) power factor, (f) output voltage ripple, (g) ripple factor, (h) input active, reactive, and output powers, (i) the voltage rating of the APF, (j) the current rating of the APF, and (k) the VA rating of the APF.

2. Design an AC shunt active filter and simulate its behavior to reduce the harmonic currents and to maintain almost UPF of a single-phase AC supply of 230 V, at 50 Hz, feeding a diode rectifier charging a battery of 180 V with a large inductor in DC link to result in constant current. It has a source impedance of 0.5 Ω resistive element and 2.5 Ω inductive element. Plot the supply voltage and input current, active filter current, DC bus voltage of the filter, output voltage and current with time, and the harmonic spectra of supply current, AC load current, and voltage at PCC. Compute (a) THD, crest factor, and rms value of AC mains current, (b) THD, crest factor, rms value of AC load current, (c) displacement factor, (d) distortion factor, (e) power factor, (f) output voltage ripple, (g) ripple factor, (h) input active, reactive, and output powers, (i) the voltage rating of the APF, (j) the current rating of the APF, and (k) the VA rating of the APF.

3. Design an AC shunt active filter and simulate its behavior to reduce the harmonic currents and to maintain almost UPF of a single-phase AC voltage controller with a resistive–inductive load and a single-phase AC voltage controller has a resistive–inductive (*RL*) load of $R = 5\ \Omega$ and $L = 20$ mH. The input voltage is 220 V (rms) at 50 Hz. The delay angle of thyristors is $\alpha = 60°$. Plot the supply voltage and input current, active filter current, DC bus voltage of the filter, output voltage and current with time, and the harmonic spectra of supply current, load current, and voltage at PCC. Compute (a) THD, crest factor, rms value of AC mains current, (b) THD, crest factor, rms value of AC load current, (c) displacement factor, (d) distortion factor, (e) power factor, (f) input active, reactive, and output powers, (g) the voltage rating of the APF, (h) the current rating of the APF, and (i) the VA rating of the APF.

4. Design an AC shunt active filter and simulate its behavior to reduce the harmonic currents and to maintain almost UPF of a single-phase 220 V, 50 Hz, controlled bridge converter at *RLE* (resistive–inductive–back emf) load with inductance of 20 mH, an equivalent resistance of 1 Ω, and a back emf of 60 V. It has a source impedance of 0.15 Ω resistive element and 1.25 Ω inductive element. The delay angle of thyristors is $\alpha = 40°$. Plot the supply voltage and input current, active filter current, DC bus voltage of the filter, output voltage and current with time, and the harmonic spectra of supply current, load current, and voltage at PCC. Compute (a) THD, crest factor, and rms value of AC mains current, (b) THD, crest factor, and rms value of AC load current, (c) displacement factor, (d) distortion factor, (e) power factor, (f) input active, reactive, and output powers, (g) the voltage rating of the APF, (h) the current rating of the APF, and (i) the VA rating of the APF.

5. Design an AC shunt active filter and simulate its behavior to reduce the harmonic currents and to maintain almost UPF of a single-phase, 230 V, 50 Hz supply system, which is feeding a set of nonlinear loads consisting of a thyristor bridge and a diode rectifier connected in parallel. The diode bridge converter is feeding a parallel capacitive DC filter of 1000 μF and a resistive load of 10 Ω. The thyristor bridge converter is feeding a *RL* load of $R = 5\ \Omega$ and $L = 50$ mH with a thyristor firing angle of 60°. Plot the supply voltage and input current, active filter current, DC bus voltage of the filter, load current with time, and the harmonic spectra of supply current, load current, and voltage at PCC. Compute (a) THD, crest factor, and rms value of AC mains current, (b) THD, crest factor, and rms value of AC load current, (c) displacement factor, (d) distortion factor, (e) power factor, (f) input active, reactive, and output powers, (g) the voltage rating of the APF, (h) the current rating of the APF, and (i) the VA rating of the APF.

6. Design an AC shunt active filter and simulate its behavior to reduce the harmonic currents and to maintain almost UPF of a three-phase 415 V, 50 Hz, uncontrolled bridge converter with a parallel capacitive DC filter of 3300 μF and a resistive load of 10 Ω. It has a source impedance of 0.15 Ω resistive element and 1.25 Ω inductive element. Plot the supply voltage and input current, active filter current, DC bus voltage of the filter, output voltage and current with time, and the harmonic spectra of supply current, load current, and voltage at PCC. Compute (a) THD, crest factor, and rms value of AC mains current, (b) THD, crest factor, and rms value of AC load current, (c) displacement factor, (d) distortion factor, (e) power factor, (f) input active, reactive, and output powers, (g) the voltage rating of the APF, (h) the current rating of the APF, and (i) the VA rating of the APF.

7. Design an AC shunt active filter and simulate its behavior to reduce the harmonic currents and to maintain almost UPF of a three-phase AC supply of 415 V, at 50 Hz, feeding a three-phase 12-pulse diode rectifier (consisting of two parallel connected 6-pulse diode bridge rectifiers with star/delta and delta/delta transformers to output equivalent line voltage of 415 V) having a parallel DC capacitor filter of 4700 μF with a 5 Ω equivalent resistive load. It has a source impedance of 0.15 Ω resistive element and 0.75 Ω inductive element. Plot the supply voltage and input current, active filter current, DC bus voltage of the filter, output voltage and current with time, and the harmonic spectra of supply current, load current, and voltage at PCC. Compute (a) THD, crest factor, and rms value of AC mains current, (b) THD, crest factor, and rms value of AC load current, (c) displacement factor, (d) distortion factor, (e) power factor, (f) input active, reactive, and output powers, (g) the voltage rating of the APF, (h) the current rating of the APF, and (i) the VA rating of the APF.

8. Design an AC shunt active filter and simulate its behavior to reduce harmonic currents and to maintain almost UPF of a three-phase 415 V, 50 Hz uncontrolled bridge rectifier, which has a source inductance of 5 mH and an equivalent DC load resistance of 5 Ω. Plot the supply voltage and input current, active filter current, DC bus voltage of the filter, output voltage and current with time, and the harmonic spectra of supply current, load current, and voltage at PCC. Compute (a) THD, crest factor, and rms value of AC mains current, (b) THD, crest factor, and rms value of AC load current, (c) displacement factor, (d) distortion factor, (e) power factor, (f) input active, reactive, and output powers, (g) the voltage rating of the APF, (h) the current rating of the APF, and (i) the VA rating of the APF.

9. Design an AC shunt active filter and simulate its behavior to reduce the harmonic currents and to maintain almost UPF of a three-phase AC supply of 415 V, at 50 Hz, feeding a three-phase 12-pulse diode rectifier (consisting of two parallel connected 6-pulse diode bridge rectifiers with star/delta and delta/delta transformers to output equivalent line voltage of 415 V) having a *RL* load of $R = 5\ \Omega$ and $L = 50$ mH. It has a source impedance of 0.15 Ω resistive element and 1.5 Ω inductive element. Plot the supply voltage and input current, active filter current, DC bus voltage of the filter, output voltage and current with time, and the harmonic spectra of supply current, load current, and voltage at PCC. Compute (a) THD, crest factor, and rms value of AC mains current, (b) THD, crest factor, and rms value of AC load current, (c) displacement factor, (d) distortion factor, (e) power factor, (f) input active, reactive, and output powers, (g) the voltage rating of the APF, (h) the current rating of the APF, and (i) the VA rating of the APF.

10. Design an AC shunt active filter and simulate its behavior to reduce the harmonic currents and to maintain almost UPF of a three-phase 415 V, 50 Hz, fed thyristor bridge converter feeding a resistive load of 10 Ω at a thyristor firing angle of 60°. Plot the supply voltage and input current, active filter current, DC bus voltage of the filter, output voltage and current with time, and the harmonic spectra of supply current, load current, and voltage at PCC. Compute (a) THD, crest factor, and rms value of AC mains current, (b) THD, crest factor, and rms value of AC load current, (c) displacement factor, (d) distortion factor, (e) power factor, (f) input active, reactive, and output powers, (g) the voltage rating of the APF, (h) the current rating of the APF, (i) and the VA rating of the APF.

11. Design an AC shunt active filter and simulate its behavior to reduce the harmonic currents and to maintain almost UPF of a three-phase 415 V, 50 Hz, fed 12-pulse thyristor bridge converter with series connected inductive load of 50 mH and an equivalent resistive load of 10 Ω. It has a source impedance of 0.1 Ω resistive element and 1.5 Ω inductive element and a thyristor firing angle of 30°. It consists of an ideal transformer with single primary star connected winding and two secondary windings connected in star and delta with same line voltages as input supply voltage to provide 30° phase shift between two sets of three-phase output voltages. Two 6-pulse thyristor bridges are connected in series to form this 12-pulse AC–DC converter. Plot the supply voltage and input current, active filter current, DC bus voltage of the filter, output voltage and current with time, and the harmonic spectra of supply current, load current, and voltage at PCC. Compute (a) THD, crest factor, and rms value of AC mains current, (b) THD, crest factor, and rms value of AC load current, (c) displacement factor, (d) distortion factor, (e) power factor, (f) input active, reactive, and output powers, (g) the voltage rating of the APF, (h) the current rating of the APF, and (i) the VA rating of the APF.

12. Design an AC shunt active filter and simulate its behavior to reduce the harmonic currents and to maintain almost UPF of a three-phase bidirectional delta connected controller (back-to-back connected thyristors in series with load) having resistive load of $R = 10\,\Omega$ and fed from 415 V (rms) line voltage at 50 Hz. The delay angle of thyristors is $\alpha = 90°$. Plot the supply voltage and input current, active filter current, DC bus voltage of the filter, output voltage and current with time, and the harmonic spectra of supply current, load current, and voltage at PCC. Compute (a) THD, crest factor, and rms value of AC mains current, (b) THD, crest factor, and rms value of AC load current, (c) displacement factor, (d) distortion factor, (e) power factor, (f) input active, reactive, and output powers, (g) the voltage rating of the APF, (h) the current rating of the APF, and (i) the VA rating of the APF.

13. Design an AC shunt active filter and simulate its behavior to reduce the harmonic currents and to maintain almost UPF of a three-phase diode bridge converter used to convert 415 V AC to DC voltage. The load on its DC side is a parallel capacitive DC filter of 2200 μF and resistive load of 10 Ω. Along with this rectifier, an AC inductive load of 20 kVAR is connected to the AC same bus. Plot the supply voltage and input current, active filter current, DC bus voltage of the filter, output voltage and current with time, and the harmonic spectra of supply current, load current, and voltage at PCC. Compute (a) THD, crest factor, and rms value of AC mains current, (b) THD, crest factor, and rms value of AC load current, (c) displacement factor, (d) distortion factor, (e) power factor, (f) input active, reactive, and output powers, (g) the voltage rating of the APF, (h) the current rating of the APF, and (i) the VA rating of the APF.

14. Design an AC shunt active filter and simulate its behavior to reduce the harmonic currents and to maintain almost UPF of a three-phase, three-wire, 415 V, 50 Hz supply system, which is feeding a set of nonlinear loads consisting of a thyristor bridge and a diode rectifier connected in parallel. The diode bridge converter is feeding a parallel capacitive DC filter of 2200 μF and resistive load of 30 Ω. The thyristor bridge converter is feeding a *RL* load of $R = 5\,\Omega$ and $L = 25$ mH at a thyristor firing angle of 60°. Plot the supply voltage and input current, active filter current, DC bus voltage of the filter, output voltage and current with time, and the harmonic spectra of supply current, load current, and voltage at PCC. Compute (a) THD, crest factor, and rms value of AC mains current, (b) THD, crest factor, and rms value of AC load current, (c) displacement factor, (d) distortion factor, (e) power

factor, (f) input active, reactive, and output powers, (g) the voltage rating of the APF, (h) the current rating of the APF, and (i) the VA rating of the APF.

15. Design a three-phase four-wire three-leg midpoint capacitor with a VSI-based AC shunt active filter and simulate its behavior to reduce the harmonic currents and to maintain almost UPF of a three single-phase AC voltage controller with a resistive load of $R = 5\,\Omega$ connected between line and neutral and create unbalanced load conditions. It is fed from 415 V (rms) line voltage at 50 Hz. The delay angle of thyristors is $\alpha = 120°$. Plot the supply voltage and input current, active filter current, DC bus voltage of the filter, supply and load neutral currents, output voltage and current with time, and the harmonic spectra of supply current, load current, and voltage at PCC. Compute (a) THD, crest factor, and rms value of AC mains current, (b) THD, crest factor, and rms value of AC load current, (c) displacement factor, (d) distortion factor, (e) power factor, (f) input active, reactive, and output powers, (g) the voltage rating of the APF, (h) the current rating of the APF, and (i) the VA rating of the APF.

16. Design a three-phase four-wire three-leg midpoint capacitor with a VSI-based AC shunt active filter and simulate its behavior to reduce the harmonic currents and to maintain almost UPF of a three single-phase 230 V, 50 Hz, uncontrolled bridge converter used for charging a battery of 240 V connected between line and neutral and create unbalanced load conditions. It has a source impedance of 1.5 Ω resistive element and 5.0 Ω inductive element. Plot the supply voltage and input current, active filter current, DC bus voltage of the filter, supply and load neutral currents, output voltage and current with time, and the harmonic spectra of supply current, load current, and voltage at PCC. Compute (a) THD, crest factor, and rms value of AC mains current, (b) THD, crest factor, and rms value of AC load current, (c) displacement factor, (d) distortion factor, (e) power factor, (f) input active, reactive, and output powers, (g) the voltage rating of the APF, (h) the current rating of the APF, and (i) the VA rating of the APF.

17. Design a three-phase four-wire three-leg midpoint capacitor with a VSI-based AC shunt active filter and simulate its behavior to reduce the harmonic currents and to maintain almost UPF of a three single-phase 230 V, 50 Hz, uncontrolled bridge converter with a parallel capacitive DC filter of 470 μF and an equivalent resistive load of 20 Ω connected between line and neutral and create unbalanced load conditions. It has a source impedance of 0.25 Ω resistive element and 1.5 Ω inductive element. Plot the supply voltage and input current, active filter current, DC bus voltage of the filter, supply and load neutral currents, output voltage and current with time, and the harmonic spectra of supply current, load current, and voltage at PCC. Compute (a) THD, crest factor, and rms value of AC mains current, (b) THD, crest factor, and rms value of AC load current, (c) displacement factor, (d) distortion factor, (e) power factor, (f) input active, reactive, and output powers, (g) the voltage rating of the APF, (h) the current rating of the APF, and (i) the VA rating of the APF.

18. Design a three-phase four-wire three-leg midpoint capacitor with a VSI-based AC shunt active filter and simulate its behavior to reduce the harmonic currents and to maintain almost UPF of a three single-phase 230 V, 50 Hz, controlled bridge converter with series connected inductive load of 100 mH and an equivalent resistive load of 5 Ω connected between line and neutral and create unbalanced load conditions. It has a source impedance of 0.15 Ω resistive element and 1.25 Ω inductive element. The delay angle of thyristors is $\alpha = 30°$. Plot the supply voltage and input current, active filter current, DC bus voltage of the filter, supply and load neutral currents, output voltage and current with time, and the harmonic spectra of supply current, load current, and voltage at PCC. Compute (a) THD, crest factor, and rms value of AC mains current, (b) THD, crest factor, and rms value of AC load current, (c) displacement factor, (d) distortion factor, (e) power factor, (f) input active, reactive, and output powers, (g) the voltage rating of the APF, (h) the current rating of the APF, and (i) the VA rating of the APF.

19. Design a three-phase four-wire three-leg midpoint capacitor with a VSI-based AC shunt active filter and simulate its behavior to reduce the harmonic currents and to maintain almost UPF of a three single-phase 230 V, 50 Hz, controlled bridge converter with series connected inductive load of

10 mH, an equivalent resistive load of 2 Ω, and a back emf of 60 V connected between line and neutral and create unbalanced load conditions. It has a source impedance of 0.2 Ω resistive element and 2.0 Ω inductive element. The delay angle of thyristors is $\alpha = 60°$. Plot the supply voltage and input current, active filter current, DC bus voltage of the filter, supply and load neutral currents, output voltage and current with time, and the harmonic spectra of supply current, load current, and voltage at PCC. Compute (a) THD, crest factor, and rms value of AC mains current, (b) THD, crest factor, and rms value of AC load current, (c) displacement factor, (d) distortion factor, (e) power factor, (f) input active, reactive, and output powers, (g) the voltage rating of the APF, (h) the current rating of the APF, and (i) the VA rating of the APF.

20. Design a three-phase four-wire four-leg VSI-based AC shunt active filter and simulate its behavior to reduce the harmonic currents and to maintain almost UPF of a three single-phase AC voltage controllers having a *RL* load of $R = 3\ \Omega$ and $L = 20$ mH connected between line and neutral and create unbalanced load conditions. The input voltage is 230 V (rms) at 50 Hz. The delay angle of thyristors is $\alpha = 90°$. Plot the supply voltage and input current, active filter current, DC bus voltage of the filter, supply and load neutral currents, output voltage and current with time, and the harmonic spectra of supply current, load current, and voltage at PCC. Compute (a) THD, crest factor, and rms value of AC mains current, (b) THD, crest factor, and rms value of AC load current, (c) displacement factor, (d) distortion factor, (e) power factor, (f) input active, reactive, and output powers, (g) the voltage rating of the APF, (h) the current rating of the APF, and (i) the VA rating of the APF.

21. Design a three-phase four-wire four-leg VSI-based AC shunt active filter and simulate its behavior to reduce the harmonic currents and to maintain almost UPF of a three single-phase 230 V, 50 Hz, uncontrolled bridge converter with a parallel capacitive DC filter of 1500 μF and an equivalent resistive load of 25 Ω connected between line and neutral and create unbalanced load conditions. It has a source impedance of 0.25 Ω resistive element and 1.0 Ω inductive element. Plot the supply voltage and input current, active filter current, DC bus voltage of the filter, supply and load neutral currents, output voltage and current with time, and the harmonic spectra of supply current, load current, and voltage at PCC. Compute (a) THD, crest factor, and rms value of AC mains current, (b) THD, crest factor, and rms value of AC load current, (c) displacement factor, (d) distortion factor, (e) power factor, (f) input active, reactive, and output powers, (g) the voltage rating of the APF, (h) the current rating of the APF, and (i) the VA rating of the APF.

22. Design a three-phase four-wire four-leg VSI-based AC shunt active filter and simulate its behavior to reduce the harmonic currents and to maintain almost UPF of a three single-phase 230 V, 50 Hz, controlled bridge converter with series connected inductive load of 100 mH and an equivalent resistive load of 5 Ω connected between line and neutral and create unbalanced load conditions. It has a source impedance of 0.15 Ω resistive element and 1.5 Ω inductive element. The delay angle of thyristors is $\alpha = 30°$. Plot the supply voltage and input current, active filter current, DC bus voltage of the filter, supply and load neutral currents, output voltage and current with time, and the harmonic spectra of supply current, load current, and voltage at PCC. Compute (a) THD, crest factor, and rms value of AC mains current, (b) THD, crest factor, and rms value of AC load current, (c) displacement factor, (d) distortion factor, (e) power factor, (f) input active, reactive, and output powers, (g) the voltage rating of the APF, (h) the current rating of the APF, and (i) the VA rating of the APF.

23. Design a three-phase four-wire three single-phase VSI-based AC shunt active filter and simulate its behavior to reduce the harmonic currents and to maintain almost UPF of a three single-phase 230 V, 50 Hz, uncontrolled bridge converter with a parallel capacitive DC filter of 470 μF and an equivalent resistive load of 20 Ω connected between line and neutral and create unbalanced load conditions. It has a source impedance of 0.25 Ω resistive element and 1.25 Ω inductive element. Plot the supply voltage and input current, active filter current, DC bus voltage of the filter, supply and load neutral currents, output voltage and current with time, and the harmonic spectra of supply current, load current, and voltage at PCC. Compute (a) THD, crest factor, and rms value of AC mains current, (b) THD, crest factor, and rms value of AC load current, (c) displacement factor, (d) distortion factor,

(e) power factor, (f) input active, reactive, and output powers, (g) the voltage rating of the APF, (h) the current rating of the APF, and (i) the VA rating of the APF.

24. Design a three-phase four-wire three single-phase VSI-based AC shunt active filter and simulate its behavior to reduce the harmonic currents and to maintain almost UPF of a three single-phase 230 V, 50 Hz, controlled bridge converter with series connected inductive load of 50 mH, an equivalent resistive load of 1 Ω, and a back emf of 60 V connected between line and neutral and create unbalanced load conditions. It has a source impedance of 0.2 Ω resistive element and 1.0 Ω inductive element. The delay angle of thyristors is $\alpha = 30°$. Plot the supply voltage and input current, active filter current, DC bus voltage of the filter, supply and load neutral currents, output voltage and current with time, and the harmonic spectra of supply current, load current, and voltage at PCC. Compute (a) THD, crest factor, and rms value of AC mains current, (b) THD, crest factor, and rms value of AC load current, (c) displacement factor, (d) distortion factor, (e) power factor, (f) input active, reactive, and output powers, (g) the voltage rating of the APF, (h) the current rating of the APF, and (i) the VA rating of the APF.

25. Design a three-phase VSI-based AC shunt active filter and simulate its behavior to reduce the harmonic currents and to maintain almost UPF of a three-phase cycloconverter (three-pulse type) supplying a *RL* series load circuit and a sinusoidal modulation of the delay angle α is to be employed. The required output frequency is 15 Hz and the inductive reactance of the load circuit at this frequency is 2.5 Ω. The load circuit resistance is 7.5 Ω. A three-phase supply of 415 V, at 50 Hz (line–line rms), is to be used directly without transformer. The ideal output voltage wave is defined as $V_0 = 156 \sin(30\pi t)$ V. Plot the supply voltage and input current, filter current, output voltage and current with time, and the harmonic spectra of supply current and voltage at PCC. Compute (a) fundamental active power drawn by the load, (b) fundamental reactive power drawn by the load, (c) power factor, (d) rms supply current, (e) distortion factor, (f) fundamental rms supply current, (g) peak current of AC mains, (h) total harmonic distortion of AC source current (THD_I), (i) voltage, current, and the VA rating of the filter.

References

1. Mitsubishi Electric Corporation (1989) Active Filters: Technical Document, 2100/1100 Series, pp. 1–36.
2. ABB Power Systems (1988) Harmonic Currents and Static VAR Systems, Information NR500-015E, pp. 1–13.
3. Kikuchi, A.H. (1992) Active Power Filters, Toshiba GTR Module (IGBT) Application Notes, pp. 44–45.
4. Sasaki, H. and Machida, T. (1971) A new method to eliminate AC harmonic currents by magnetic flux compensation-considerations on basic design. *IEEE Transactions on Power Apparatus and Systems*, **90**(5), 2009–2019.
5. Gyugyi, L. and Strycula, E. (1976) Active AC power filters, IEEE-IAS Annual Meeting Record, pp. 529–535.
6. Uceda, J., Aldana, F., and Martinez, P. (1983) Active filters for static power converters. *Proceedings of the IEEE*, **130**(5), 347–354.
7. Drouin, P.L.A., Sévigny, A., Jacob, A., and Rajagopalan, V. (1983) Studies on line current harmonics compensation scheme suitable for an electric distribution system. *Canadian Journal of Electrical and Computer Engineering*, **8**(4), 123–129.
8. Choe, G.H. and Park, M.H. (1986) Analysis and control of active power filter with optimized injection. IEEE Transactions on Power Electronics, pp. 401–409.
9. Enjeti, P., Shireen, W., and Pitel, I. (1992) Analysis and design of an active power filter to cancel harmonic currents in low voltage electric power distribution systems. Proceedings of the 1992 International Conference on Industrial Electronics, Control, Instrumentation, and Automation, pp. 368–373.
10. Tuttas, C. (1992) Compensation of capacitive loads by a voltage-source active filter. *European Transactions on Electrical Power Engineering*, **2**(1), 15–19.
11. Duke, R.M. and Round, S.D. (1993) The steady state performance of a controlled current active filter. *IEEE Transactions on Power Electronics*, **8**(3), 140–146.
12. Round, S.D. and Mohan, N. (1993) Comparison of frequency and time domain neural network controllers for an active power filter. Proceedings of the IECON'93, pp. 1099–1104.

13. Rim, G.H., Kang, Y., Kim, W.H., and Kim, J.S. (1995) Performance improvement of a voltage source active filter. Proceedings of the Tenth Annual Applied Power Electronics Conference and Exposition, pp. 613–619.
14. Choi, J.H., Park, G.W., and Dewan, S.B. (1995) Standby power supply with active power filter ability using digital controller. Proceedings of the Tenth Annual Applied Power Electronics Conference and Exposition, pp. 783–789.
15. Torrey, D.A. and Al-Zamel, A.M.A.M. (1995) Single-phase active power filters for multiple non-linear loads. *IEEE Transactions on Power Electronics*, **10**(3), 263–272.
16. Qin, Y. and Du, S. (1995) A DSP-based active power filter for line interactive UPS. Proceedings of the IEEE-IECON 21st International Conference on Industrial Electronics, Control, and Instrumentation, pp. 884–888.
17. Wu, J.C. and Jou, H.L. (1995) A new UPS scheme provides harmonic suppression, input power factor correction. *IEEE Transactions on Industrial Electronics*, **42**(6), 629–635.
18. Hsu, C.Y. and Wu, H.Y. (1996) A new single-phase active power filter with reduced energy storage capacity. *IEEE Proceedings: Electric Power Applications*, **143**(1), 25–30.
19. Mohan, N., Peterson, H.A., Long, W.F. *et al.* (1977) Active filters for AC harmonic suppression. IEEE PES Winter Meeting, pp. 168–174.
20. Takahashi, I. and Nabae, A. (1980) Universal power distortion compensator of line commutated thyristor converter. IEEE-IAS Annual Meeting Record, pp. 858–864.
21. Kawahira, H., Nakamura, T., Nakazawa, S., and Nomura, M. (1983) Active power filter. *IPEC-Tokyo*, Institute of Electrical Engineers of Japan, pp. 981–992.
22. Akagi, H., Kanazawa, Y., and Nabae, A. (1984) Instantaneous reactive power compensators comprising switching devices without energy storage components. *IEEE Transactions on Industry Applications*, **IA-20** (3), 625–630.
23. Hayafune, K., Ueshiba, T., Masada, E., and Ogiwara, Y. (1984) Microcomputer controlled active power filter. IEEE-IECON Record, pp. 1221–1226.
24. Akagi, H., Nabae, A., and Atoh, S. (1986) Control strategy of active power filters using multiple voltage-source PWM converters. *IEEE Transactions on Industry Applications*, **22**(3), 460–465.
25. Akagi, H., Atoh, S., and Nabae, A. (1986) Compensation characteristics of active power filter using multiseries voltage-source PWM converters. *Electrical Engineering in Japan*, **106**(5), 28–36.
26. Komatsugi, K. and Imura, T. (1986) Harmonic current compensator composed of static power converter. Proceedings of the17th annual IEEE Power Electronics Specialists Conference, pp. 283–290.
27. Malesani, L., Rossetto, L., and Tenti, P. (1986) Active filter for reactive power, harmonic compensation. Proceedings of the17th annual IEEE Power Electronics Specialists Conference, pp. 321–330.
28. Nakajima, T., Masada, E., and Ogihara, Y. (1987) Compensation of the cycloconverter input current harmonics using active power filters. IInd EPE Record, pp. 1227–1232.
29. Fisher, R. and Hoft, R. (1987) Three-phase power line conditioner for harmonic compensation and power-factor correction. IEEE-IAS Annual Meeting Record, pp. 803–807.
30. Takeda, M., Ikeda, K., and Tominaga, Y. (1987) Harmonic current compensation with active filter. IEEE-IAS Annual Meeting Record, pp. 808–815.
31. Hammond, P.W. (1988) A harmonic filter installation to reduce voltage distortion from static power converter. *IEEE Transactions on Industry Applications*, **24**(1), 53–58.
32. Choe, J.H. and Park, M.H. (1988) A new injection method for AC harmonic elimination by active power filter. *IEEE Transactions on Industrial Electronics*, **35**(1), 141–147.
33. Peng, F.Z., Akagi, H., and Nabae, A. (1988) A novel harmonic power filter. Proceedings of the 19th Annual IEEE-Power Electronics Specialists Conference, pp. 1151–1159.
34. Nakajima, T., Tamura, M., and Masada, E. (1988) Compensation of non-stationary harmonics using active power filter with Prony's spectral estimation. Proceedings of the 19th Annual IEEE-Power Electronics Specialists Conference, pp. 1160–1167.
35. Nakajima, A., Oku, K., Nishidai, J. *et al.* (1988) Development of active filter with series resonant circuit. Proceedings of the 19th Annual IEEE-Power Electronics Specialists Conference, pp. 1168–1173.
36. Takeda, M., Ikeda, K., Teramoto, A., and Aritsuka, T. (1988) Harmonic current, reactive power compensation with an active filter. Proceedings of the 19th Annual IEEE-Power Electronics Specialists Conference, pp. 1174–1179.
37. Hayashi, Y., Sato, N., and Takahashi, K. (1988) A novel control of a current source active filter for AC power system harmonic compensation. IEEE-IAS Annual Meeting Record, pp. 837–842.
38. Kohata, M., Shiota, T., Watanabe, Y. *et al.* (1988) A novel compensation using static induction thyristors for reactive power, harmonics. IEEE-IAS Annual Meeting Record, pp. 843–849.

39. Choe, G.H., Wallace, A.K., and Park, M.H. (1988) Control technique of active power filter for harmonic elimination, reactive power control. IEEE-IAS Annual Meeting Record, pp. 859–866.
40. Peng, F.Z., Akagi, H., and Nabae, A. (1988) A new approach to harmonic compensation in power systems. IEEE-IAS Annual Meeting Record, pp. 874–880.
41. Wong, C., Mohan, N., Wright, S.E., and Mortensen, K.N. (1989) Feasibility study of AC, DC-side active filters for HVDC converter terminals. *IEEE Transactions on Power Delivery*, **4**(4), 2067–2075.
42. Moran, L.T., Ziogas, P.D., and Joos, G. (1989) Analysis and design of a novel 3-phase solid-state power factor compensator, harmonic suppressor system. *IEEE Transactions on Industry Applications*, **25**(4), 609–619.
43. Peng, F.Z., Akagi, H., and Nabae, A. (1990) A study of active power filters using quad-series voltage-source PWM converters for harmonic compensation. *IEEE Transactions on Power Electronics*, **5**(1), 9–15.
44. Akagi, H., Tsukamoto, Y., and Nabae, A. (1990) Analysis and design of an active power filter using quad-series voltage source PWM converters. *IEEE Transactions on Industrial Electronics*, **26**(1), 93–98.
45. Grady, W.M., Samotyj, M.J., and Noyola, A.H. (1990) Survey of active power line conditioning methodologies. *IEEE Transactions on Power Delivery*, **5**(3), 1536–1542.
46. Rossetto, L. and Tenti, P. (1990) Using AC-fed PWM converters as instantaneous reactive power compensators. Proceedings of the 21st Annual IEEE-Power Electronics Specialists Conference, pp. 855–861.
47. Fukuda, S. and Yamaji, M. (1990) Design, characteristics of active power filter using current source converter. IEEE-IAS Annual Meeting Record, pp. 965–970.
48. Grady, W.M., Samotyj, M.J., and Noyola, A.H. (1991) Minimizing network harmonic voltage distortion with an active power line conditioner. *IEEE Transactions on Power Delivery*, **6**, 1690–1697.
49. Malasani, L., Rossetto, L., and Tenti, P. (1991) Active power filter with hybrid energy storage. *IEEE Transactions on Power Electronics*, **6**(3), 392–397.
50. Williams, S.M. and Hoft, R.G. (1991) Adaptive frequency domain control of PWM switched power line conditioner. *IEEE Transactions on Power Electronics*, **6**(4), 665–670.
51. Acharya, B., Divan, D.M., and Gascoigne, R.W. (1992) Active power filters using resonant pole inverters. *IEEE Transactions on Industry Applications*, **28**(6), 1269–1276.
52. Akagi, H. (1992) Trends in active power line conditioners. IEEE-IECON Record, pp. 19–24.
53. Moran, L., Diaz, M., Higuera, V. *et al.* (1992) A three-phase active power filter operating with fixed switching frequency for reactive power and current harmonic compensation, IEEE-IECON Record, pp. 362–367.
54. Song, E.H. and Kwon, B.H. (1992) A novel digital control for active power filter. IEEE-IECON Record, pp. 1168–1173.
55. Wojciak, P.F. and Torrey, D.A. (1992) The design and implementation of active power filters based on variable structure system concepts. IEEE-IAS Annual Meeting Record, pp. 850–857.
56. Emanuel, A.E. and Yang, M. (1993) On the harmonic compensation in nonsinusoidal systems. *IEEE Transactions on Power Delivery*, **8**(1), 393–399.
57. Bhavaraju, V.B. and Enjeti, P.N. (1993) Analysis and design of an active power filter for balancing unbalanced loads. *IEEE Transactions on Power Electronics*, **8**(4), 640–647.
58. Chicharo, J.F., Dejsakulrit, D., and Perera, B.S.P. (1993) A centroid based switching strategy for active power filters. *IEEE Transactions on Power Electronics*, **8**(4), 648–653.
59. Moran, L., Godoy, P., Wallace, R., and Dixon, J. (1993) A new current control strategy for active power filters using three PWM voltage source inverters. Proceedings of the 24th Annual IEEE-Power Electronics Specialists Conference, pp. 3–9.
60. Fukuda, S. and Endoh, J. (1993) Control method and characteristics of active power filters. EPE Conference Record, pp. 139–144.
61. Tuttas, C. (1993) Sliding mode control of a voltage-source active filter. EPE Conference Record, pp. 156–161.
62. Wang, M.X., Pouliquen, H., and Grandpierre, M. (1993) Performance of an active filter using PWM current source inverter. EPE Conference Record, pp. 218–223.
63. Xu, J.H., Lott, C., Saadate, S., and Davat, B. (1993) Compensation of AC–DC converter input current harmonics using a voltage-source active power filter. EPE Conference Record, pp. 233–238.
64. Bhavaraju, V.B. and Enjeti, P. (1993) A novel active line conditioner for a three-phase system. IEEE-IAS Annual Meeting Record, pp. 979–985.
65. Dixon, J.W., Garcia, J.C., and Moran, L.T. (1993) A control system for a three-phase active power filter which simultaneously compensates power factor, unbalanced loads. IEEE-IECON Record, pp. 1083–1087.
66. Humberto, P. and Albenes, Z.C. (1993) A simple control strategy for shunt power line conditioner with inductive energy storage. IEEE-IECON Record, pp. 1093–1098.
67. Hoffman, K. and Ledwich, G. (1993) Fast compensation by a pulsed resonant current source active power filter. IEEE-IECON Record, pp. 1297–1302.

68. Le, J.N., Pereira, M., Renz, K., and Vaupel, G. (1994) Active damping of resonances in power systems. *IEEE Transactions on Power Delivery*, **9**(2), 1001–1008.
69. Akagi, H. (1994) Trends in active power line conditioners. *IEEE Transactions on Power Electronics*, **9**(3), 263–268.
70. Campos, A., Joos, G., Ziogas, P.D., and Lindsay, J.F. (1994) Analysis and design of a series voltage unbalance compensator based on a three-phase VSI operating with unbalanced switching functions. *IEEE Transactions on Power Electronics*, **9**(3), 269–274.
71. Cavallini, A. and Montanari, G.C. (1994) Compensation strategies for shunt active-filter control. *IEEE Transactions on Power Electronics*, **9**(6), 587–593.
72. Dejsakulrit, D., Perera, B.S.P., and Chicharo, J.F. (1994) A novel equal sampling switching strategy for active power filters. *Electric Machines & Power Systems*, **22**, 405–421.
73. Xu, J.H., Lott, C., Saadate, S., and Davat, B. (1994) Simulation and experimentation of a voltage source active filter compensating current harmonics and power factor. IEEE-IECON, pp. 411–415.
74. Krah, J.O. and Holtz, J. (1994) Total compensation of line side switching harmonics in converter fed AC locomotives. IEEE-IAS Annual Meeting Record, pp. 913–920.
75. Radulovic, Z. and Sabanovic, A. (1994) Active filter control using a sliding mode approach. Proceedings of the 25th Annual IEEE-Power Electronics Specialists Conference, pp. 177–182.
76. Pahmer, C., Capolino, G.A., and Henao, H. (1994) Computer-aided design for control of shunt active filter. IEEE-IECON Record, pp. 669–674.
77. Le Magoarou, F. and Monteil, F. (1994) Influence of the load on the design process of an active power filter. IEEE-IECON Record, pp. 416–421.
78. Verdelho, P. and Marques, G.D. (1994) Design and performance of an active power filter and unbalanced current compensator. IEEE-IECON Record, pp. 422–427.
79. Ledwich, G. and Doulai, P. (1995) Multiple converter performance and active filtering. *IEEE Transactions on Power Electronics*, **10**(3), 273–279.
80. Bhattacharya, S., Divan, D.M., and Banerjee, B.B. (1995) Active filter solutions for utility interface. IEEE-ISIE Conference Record, pp. 1–11.
81. Yao, Z., Lahaie, S., and Rajagopolan, V. (1995) Robust compensator of harmonics and reactive power. Proceedings of the 26th Annual IEEE-Power Electronics Specialists Conference, pp. 215–221.
82. Akagi, H. and Fujita, H. (1995) A new power line conditioner for harmonic compensation in power systems. *IEEE Transactions on Power Delivery*, **10**(3), 1570–1575.
83. Saetieo, S., Devaraj, R., and Torrey, D.A. (1995) The design and implementation of a three-phase active power filter based on sliding mode control. *IEEE Transactions on Industry Applications*, **31**(5), 993–1000.
84. Bhattacharya, S., Veltman, A., Divan, D.M., and Lorenz, R.D. (1995) Flux based active filter controller. IEEE-IAS Annual Meeting Record, pp. 2483–2491.
85. Jeong, S.G. and Woo, M.H. (1995) DSP based active power filter with predictive current control. IEEE-IECON Record, pp. 645–650.
86. Li, Z., Jin, H., and Joos, G. (1995) Control of active filters using digital signal processors. IEEE-IECON Record, pp. 651–655.
87. Jou, H.L. (1995) Performance comparison of the three-phase active power-filter algorithms. *IEE Proceedings of the Generation, Transmission and Distribution*, **142**(6), 646–652.
88. Dixon, J.W., Garcia, J.J., and Moran, L. (1995) Control system for three-phase active power filter which simultaneously compensates power factor and unbalanced loads. *IEEE Transactions on Industrial Electronics*, **42**(6), 636–641.
89. Taleb, M., Kamal, A., Sowaied, A.J., and Khan, M.R. (1996) An alternative active power filter. IEEE-PEDES Conference Record, pp. 410–416.
90. Akagi, H. (1996) New trends in active filters for improving power quality. IEEE-PEDES Conference Record, pp. 417–425.
91. Malesani, L., Mattavelli, P., and Tamasin, P. (1996) High-performance hysteresis modulation technique for active filters. IEEE-APEC Record, pp. 939–946.
92. Van Schoor, G. and Wyk, J.D.V. (1987) A study of a system of current fed converters as an active three-phase filter. IEEE-PESC Record, pp. 482–490.
93. Lin, C.E., Chen, C.L., and Huang, C.H. (1991) Reactive, harmonic current compensation for unbalanced three-phase system. International Conference on High Technology in the Power Industry, pp. 317–321.
94. Quinn, C.A. and Mohan, N. (1992) Active filtering of harmonic currents in three-phase, four-wire systems with three-phase, single-phase non-linear loads. IEEE-APEC Record, pp. 829–836.

95. Lin, C.E., Chen, C.L., and Huang, C.L. (1992) Calculating approach and implementation for active filters in unbalanced three-phase system using synchronous detection method. IEEE-IECON Record, pp. 374–380.
96. Quinn, C.A., Mohan, H., and Mehta, H. (1993) A four-wire, current controlled converter provides harmonic neutralization in three-phase, four-wire systems. IEEE-APEC Record, pp. 841–846.
97. Enjeti, P., Shireen, W., Packebush, P., and Pitel, I. (1993) Analysis and design of a new active power filter to cancel neutral current harmonics in three-phase four-wire electric distribution systems. IEEE-IAS Annual Meeting Record, pp. 939–946.
98. Chen, C.L., Lin, C.E., and Huang, C.L. (1994) An active filter for unbalanced three-phase system using synchronous detection method. Proceedings of the 25th Annual IEEE-Power Electronics Specialists Conference, pp. 1451–1455.
99. Kamath, G., Mohan, N., and Albertson, D. (1995) Hardware implementation of a novel reduced rating active filter for 3-phase, 4-wire loads. IEEE-APEC Record, pp. 984–989.
100. Aredes, M. and Watanabe, E.H. (1995) New control algorithms for series, shunt three-phase four-wire active power filter. *IEEE Transactions on Power Delivery*, **10**(3), 1649–1656.
101. Singh, B., Jayaprakash, P., Kothari, D.P. *et al.* (2014) Comprehensive study of DSTATCOM configurations. *IEEE Transactions on Industrial Informatics*, **10**(2), 854–870.

10

Series Active Power Filters

10.1 Introduction

There are a number of voltage quality problems in the AC mains nowadays, such as harmonics, sag, dip, flicker, swell, fluctuations, and imbalance, and these problems increase losses in many loads and sometimes trip the sensitive loads causing loss of production. The DVRs (dynamic voltage restorers), which are mainly used for dynamic compensation of voltage quality problems such as sag, dip, flicker, swell, fluctuations, and imbalance, have already been explained in Chapter 5. However, the series active power filter (APF) protects the sensitive loads from these distortions (especially harmonics) in the voltage of the AC mains. As its name represents, a series active filter is expected to filter voltage harmonics appearing in the supply systems so that the loads are supplied with clean sinusoidal supply voltage. Moreover, the solid-state control of AC power employing diodes, thyristors, and other semiconductor switches is extensively used to feed controlled power to electrical loads such as adjustable speed drives (ASDs), furnaces, computer power supplies, fax machines, copier, and printers. As nonlinear loads, the solid-state converters draw harmonics and reactive power components of current in addition to fundamental active power component of the current from the AC mains. Moreover, voltage-fed nonlinear loads (such as a diode rectifier with a large DC bus capacitor filter) used to realize the DC voltage source with the DC capacitor are increasingly used nowadays for feeding the voltage source inverter (VSI) in many applications. Such voltage-fed nonlinear loads draw peaky and discontinuous current and inject a large amount of harmonic currents into the AC mains. In such situations, series APFs are quite effective for harmonic current compensation with moderate rating. In current-fed nonlinear loads, a small-rating series active filter (approximately 3–5% of the load rating) is used along with a large-rating passive shunt filter to improve the filtering characteristics of the passive filter and the hybrid filter as a combination of these two filters is an adjustable solution for the harmonic compensation of varying loads. This type of filter is considered as a cost-effective filter especially in large rating. Therefore, series active power filters are having substantial applications for eliminating voltage harmonics present in the supply system or created by the nonlinear loads at the point of common coupling (PCC) and current harmonics in both voltage-fed and current-fed loads depending upon the applications. Therefore, it is essential and relevant to study the series active filters. This chapter deals with an exhaustive analysis and design of series APFs. The series APFs are classified into three major categories: single-phase two-wire, three-phase three-wire, and three-phase four-wire series APFs. Starting with the introduction, the other sections include state of the art on the series APFs, their classification, principle of operation and control, analysis and design, modeling and simulation of performance, numerical examples, summary, review questions, numerical and computer simulation-based problems, and references.

Power Quality Problems and Mitigation Techniques, First Edition.
Bhim Singh, Ambrish Chandra and Kamal Al-Haddad.

10.2 State of the Art on Series Active Power Filters

The series APF technology is now a mature technology for providing compensation for harmonics present in the voltages and currents in AC networks. It has evolved in the past quarter century with development in terms of varying configurations, control strategies, and solid-state devices. The series active power filters are mainly used to eliminate voltage harmonics. In addition, they can be used to regulate the terminal voltage, to suppress voltage flicker, and to improve voltage balance in three-phase systems, of course, with additional cost and rating. The series active power filters are also used to eliminate harmonic currents in voltage-fed nonlinear loads. Moreover, they are also used in current-fed nonlinear loads along with passive filters. These objectives of the series active power filters are achieved either individually or in combination depending upon the requirements and control strategy and configuration that need to be selected appropriately. This section describes the history of development and the current status of the series APF technology.

Following the widespread use of solid-state control of AC power, the power quality problems have become significant. The series active power filters are basically categorized into three types, namely, single-phase two-wire, three-phase three-wire, and three-phase four-wire configurations, to meet the requirements of the three types of nonlinear loads on supply systems. Single-phase loads such as domestic lights and ovens, television sets, computer power supplies, air conditioners, laser printers, and Xerox machines behave as voltage-fed nonlinear loads and cause substantial power quality problems. Single-phase two-wire series active power filters are investigated in varying configurations and control strategies to meet the needs of single-phase nonlinear loads. Starting from 1976, many configurations of the series active power filter with current source converters (CSCs), voltage source converters (VSCs), and so on have been evolved for the compensation of voltage- and current-based power quality problems. Both current source converters with inductive energy storage and voltage source converters with capacitive energy storage are used to develop the single-phase series APFs.

A large amount of AC power is consumed by three-phase loads such as ASDs with solid-state control and lately many ASDs incorporate active power filters in their front-end design. A substantial work has been reported on three-phase three-wire series APFs, since 1976. Many control strategies such as instantaneous reactive power theory, synchronous frame d–q theory, synchronous detection method, and notch filter method are used in the development of three-phase series APFs.

One of the major factors in advancing the series APF technology is the advent of fast, self-commutating solid-state devices. In the initial stages, BJTs (bipolar junction transistors) and power MOSFETs (metal-oxide semiconductor field-effect transistors) have been used to develop series APFs; later, SITs (static induction thyristors) and GTOs (gate turn-off thyristors) have been employed to develop series APFs. With the introduction of IGBTs (insulated gate bipolar transistors), the series APF technology has got a real boost and at present it is considered as an ideal solid-state device for series APFs. The improved sensor technology has also contributed to the enhanced performance of the series APFs. The availability of Hall effect voltage and current sensors and isolation amplifiers at reasonable cost with adequate ratings has improved the performance of the series APFs.

10.3 Classification of Series Active Power Filters

Series active power filters can be classified based on the type of converter used, topology, and the number of phases. The converter can be either a current source converter or a voltage source converter. The topology can be circuit configurations used to develop the series APF, such as half bridge and full bridge. The third classification is based on the number of phases, for example, single-phase two-wire, three-phase three-wire, and three-phase four-wire series APF systems.

10.3.1 Converter-Based Classification of Series APFs

Two types of converters are used in the development of the series APFs. Figure 10.1 shows a single-phase series APF using a current source converter. It behaves as a nonsinusoidal voltage source to meet the harmonic voltage requirement to feed clean sinusoidal voltage to the consumer loads. A diode is used in

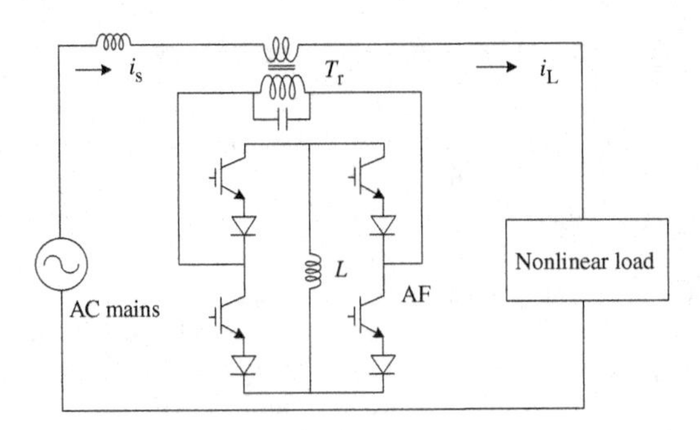

Figure 10.1 A two-wire series APF with a current source converter

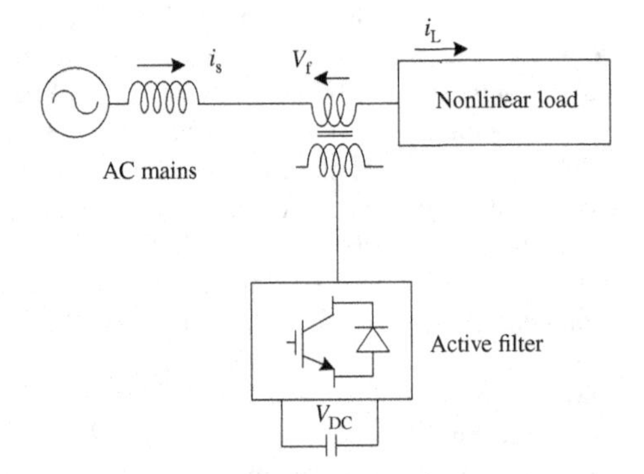

Figure 10.2 A series APF with a voltage source converter

series with the self-commutating device (IGBT) for reverse voltage blocking. However, GTO-based configurations do not need the series diode, but they have restricted frequency of switching. They are considered sufficiently reliable, but have high losses and require high values of parallel AC power capacitors. Moreover, they cannot be used in multilevel or multistep modes to improve performance in higher ratings.

The other power converter used as a series APF is a voltage source converter shown in Figure 10.2. It has a self-supporting DC voltage bus with a large DC capacitor. It is more widely used because it is light, cheap, and expandable to multilevel and multistep versions, to enhance the performance with lower switching frequencies.

10.3.2 Topology-Based Classification of Series APFs

Series APFs can also be classified based on the topology used. Combinations of an active series filter and a passive shunt filter are known as hybrid filters. Figures 10.3 and 10.4 show half-bridge and full-bridge topologies of VSC-based series APFs. The active series filters are most widely used to eliminate voltage harmonics at the load end to provide clean power to consumer loads. They are also used to block harmonic currents of voltage-fed nonlinear loads, which effectively reduces voltage harmonics at PCC.

Figure 10.5 shows a basic block of a series active power filter. It is connected before the load in series with the AC mains, using a matching transformer, to eliminate voltage harmonics and to balance and regulate the terminal voltage across the load. It can also be used to reduce negative-sequence voltage and to regulate the voltage on three-phase systems but at the cost of additional rating. It can be installed by electric utilities to compensate voltage harmonics and to damp out harmonic propagation caused by resonance with line impedances and/or passive shunt compensators.

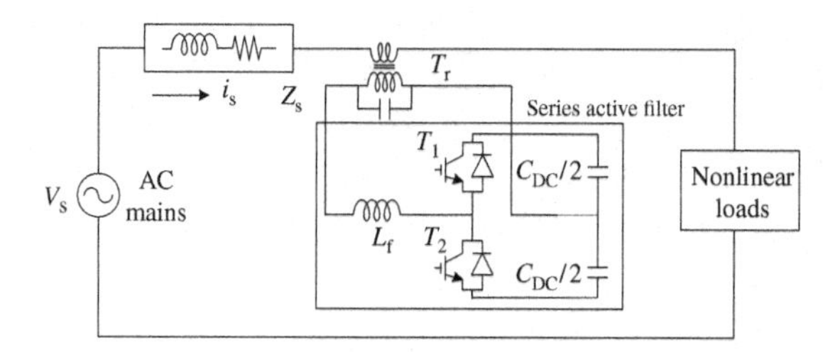

Figure 10.3 Half-bridge topology of a VSC-based single-phase series filter

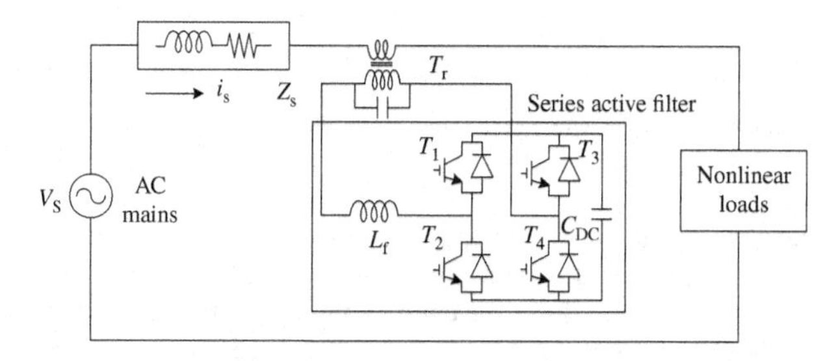

Figure 10.4 Full-bridge topology of a VSC-based single-phase series filter

10.3.3 Supply System-Based Classification of Series APFs

This classification of series APFs is based on the supply and/or the load system having single-phase (two-wire) and three-phase (three-wire or four-wire) series APF systems. There are many nonlinear loads such as domestic appliances connected to single-phase supply systems. Some three-phase nonlinear loads are without neutral terminal, such as ASDs fed from three-wire supply systems. There are many nonlinear single-phase loads distributed on three-phase four-wire supply systems, such as computers and commercial lighting. Hence, the series APFs may also be classified accordingly as two-wire, three-wire, and four-wire series APFs.

10.3.3.1 Two-Wire Series APFs

Single-phase VSCs and CSCs are used as two-wire series active power filters. Both power converter configurations, current source converters with inductive energy storage elements and voltage source

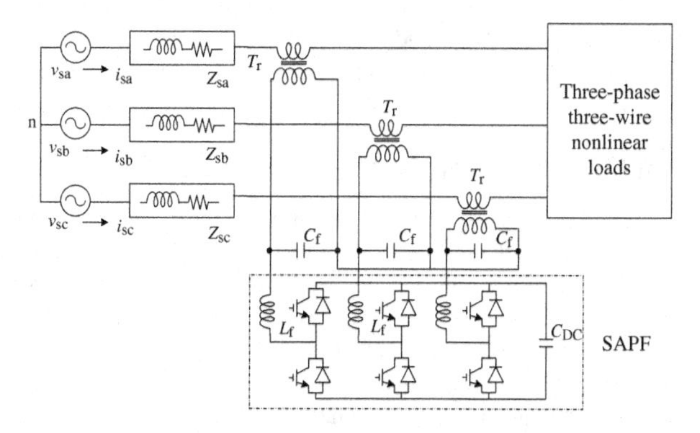

Figure 10.5 A three-phase three-wire active series filter

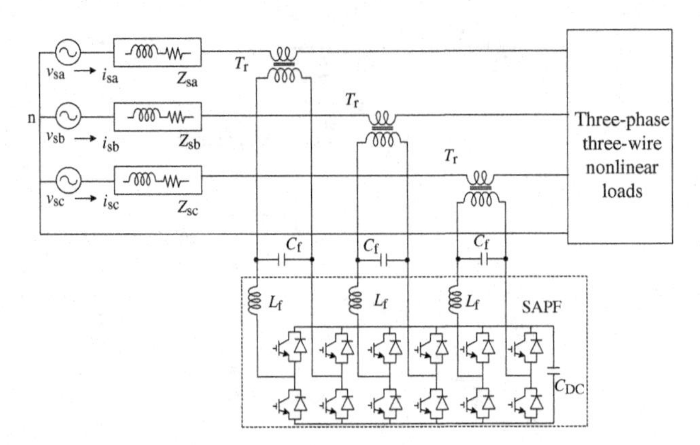

Figure 10.6 A three-phase four-wire series active filter with three single-phase VSC bridge topology

converters with capacitive DC bus energy storage elements, are used to form two-wire series AF circuits. In some cases, active filtering is included in the power conversion stage to improve input characteristics at the supply end. Figures 10.3 and 10.4 show the configurations of series active power filters with voltage source converters. In the case of a series APF with a voltage source converter, sometimes the transformer is removed and the load is shunted with passive *LC* components. The series APF is normally used to eliminate voltage harmonics, spikes, sags, notches, and so on.

10.3.3.2 Three-Wire Series APFs

Three-phase three-wire nonlinear loads such as ASDs are one of the major applications of solid-state power converters and lately many ASDs incorporate active power filters in their front-end design. Figure 10.5 shows a three-wire series APF, with three wires on the AC side and two wires on the DC bus of the VSC used as the series APF. Series APFs are developed using CSCs (Figure 10.1), VSCs (Figure 10.2), or multistep/multilevel and multiseries configurations. Series APFs are also designed with three single-phase APFs with isolation transformers for proper voltage matching, independent phase control, and reliable compensation with unbalanced systems.

10.3.3.3 Four-Wire Series APFs

A large number of single-phase loads may be supplied from three-phase AC mains with a neutral conductor. They cause excessive neutral current, injection of current harmonics and subsequently voltage harmonics, and unbalance. To reduce these problems, four-wire series APFs are used in four-wire distribution systems. They are developed as series active power filters in the typical configuration shown in Figure 10.6. A three single-phase VSC bridge configuration, shown in Figure 10.6, is quite common and this version allows proper voltage matching for solid-state devices and enhances the reliability of the APF system.

10.4 Principle of Operation and Control of Series Active Power Filters

The basic function of series active power filters is to mitigate most of the voltage-based power quality problems, mainly voltage harmonics present at PCC, and to provide sinusoidal balanced voltages even across linear loads with its self-supporting DC bus by injecting suitable voltages in series between the PCC and the load. The series active power filters are also found quite effective in eliminating harmonics in supply currents in voltage-fed nonlinear loads (such as a diode rectifier with a large DC bus capacitor filter) with quite small rating by injecting suitable voltages. In addition, the series APFs are also used in current-fed nonlinear loads along with passive filters to eliminate supply current harmonics. These objectives of the series active power filters are achieved either individually or in combination depending

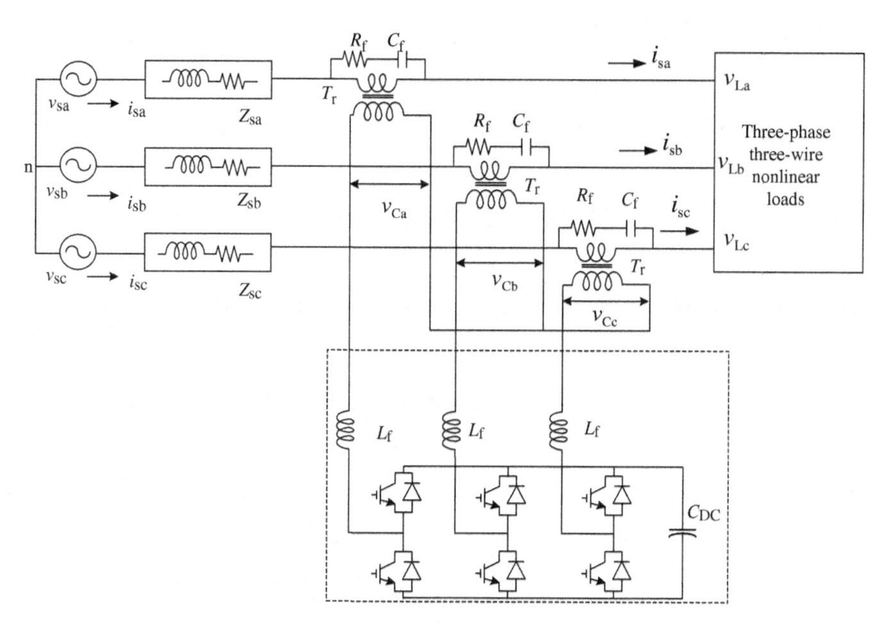

Figure 10.7 System configuration of a three-phase series active filter

upon the requirements and control strategy and configuration that need to be selected appropriately. A fundamental circuit of the series APF for a three-phase three-wire AC system is shown in Figure 10.7. An IGBT-based voltage source converter (CC-VSC) with a DC bus capacitor is used as a series APF. Using a control algorithm, the reference voltages or currents are estimated and the sensed voltages or currents are directly controlled close to reference voltages or currents by the voltage source converter used as the series APF.

10.4.1 Principle of Operation of Series Active Power Filters

Figure 10.7 shows the circuit diagram of a series active power filter system, which consists of a three-phase VSC connected in series with three-phase supply through three single-phase coupling transformers. A three-phase VSC with a DC bus capacitor is used as a series active power filter. A small-rating *RC* filter is connected across secondary of each series transformer to eliminate high switching ripple content in the series active power filter injected voltage. The loads may include linear loads requiring elimination of voltage harmonics across them or voltage-fed nonlinear loads, such variable-frequency AC motor drives, as balanced harmonic-producing loads requiring elimination of supply current harmonics. A single series filter may be installed at the PCC for multiple diverse types of loads for the elimination of voltage harmonics across them. However, such a configuration is susceptible to danger under short-circuit condition in utility line and thus requires an adequate protection. The series APF is controlled to eliminate harmonics in the three-phase supply currents or distortion and unbalance in the PCC voltages by injecting suitable voltage in series with the supply.

For the voltage-fed nonlinear loads, which consist of a capacitive filter and an equivalent load at the DC link of a three-phase diode rectifier, a series APF alone can effectively maintain sinusoidal supply currents. However, for the current-fed nonlinear loads, which consist of the series connection of a resistor and an inductor at the DC link of a three-phase diode rectifier or a three-phase thyristor bridge converter, a combined system of the shunt passive filters and a series active power filter needs to be employed to effectively maintain sinusoidal supply currents. The control algorithm of the series active filter to eliminate current harmonics is suitable for both the series active filter and hybrid configurations of a series active power filter with a shunt passive filter. Moreover, for voltage-sensitive loads, to eliminate the

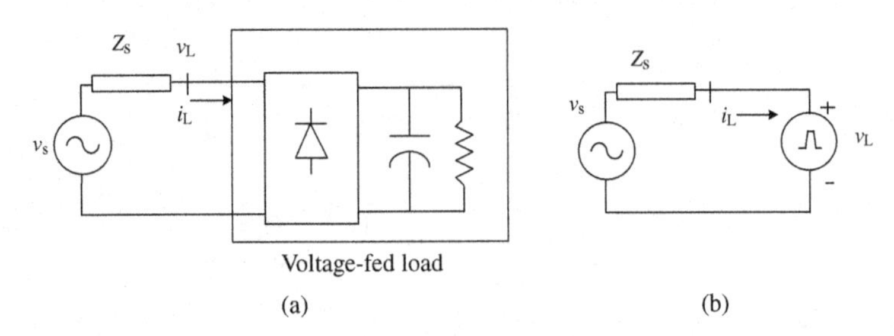

Figure 10.8 (a) Single line diagram of a voltage-fed load using a diode rectifier with a capacitor filter and (b) its equivalent circuit

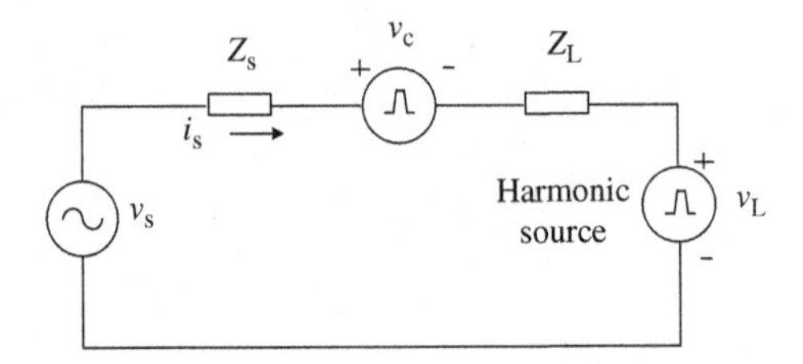

Figure 10.9 A series active filter and the harmonic voltage source load

voltage harmonics and unbalance and to maintain zero voltage regulation at PCC, the series APF is directly controlled to inject sufficient voltage in series with the supply; that is, the sum of supply voltage and injected voltage becomes sinusoidal with desired amplitude across the loads.

The single line diagram of a voltage-fed harmonic-producing load is shown in Figure 10.8a and its equivalent circuit is shown in Figure 10.8b. The load is represented as a voltage source (v_L) with fundamental as well as harmonic voltage. The harmonic current (i_L) originates from this rectifier voltage (v_L) and the source impedance (Z_s) as shown in Figure 10.8b. Figure 10.9 shows the operation of a series AF for a voltage-fed nonlinear load. The series AF is controlled as a current-controlled harmonic voltage source (V_C) to offer low impedance at fundamental frequency and acts as a high-valued resistor for harmonic currents in the AC mains. This satisfies the need of harmonic currents required by the load and prevents the flow of harmonic currents into the AC source. Figure 10.10 shows the waveforms of PCC voltage (v_s), load voltage (v_L), source current (i_s), and SAF voltage (v_f) of a capacitor-supported series AF for the compensation of harmonic currents in voltage-fed nonlinear loads. Along with this, the series AF requires a small fundamental voltage drop across the coupling transformer to draw active power for maintaining its DC bus.

10.4.2 Control of Series Active Power Filters

The main objective of a control algorithm of series active power filters is to estimate the reference voltages or currents using feedback signals depending upon their applications. The reference voltages or currents along with corresponding sensed voltages or currents are used in PWM (pulse-width modulated) voltage or current controllers to derive PWM gating signals for switching devices (IGBTs) of the VSC used as the series active power filter. Reference voltages or currents for the control of series active power filters have to be derived accordingly and these signals may be estimated using a number of control algorithms. There are many control algorithms reported in the literature for the control of series active power filters, which are classified as time-domain and frequency-domain control algorithms. There are more than a dozen of time-domain and frequency-domain control algorithms that are used for the control of series

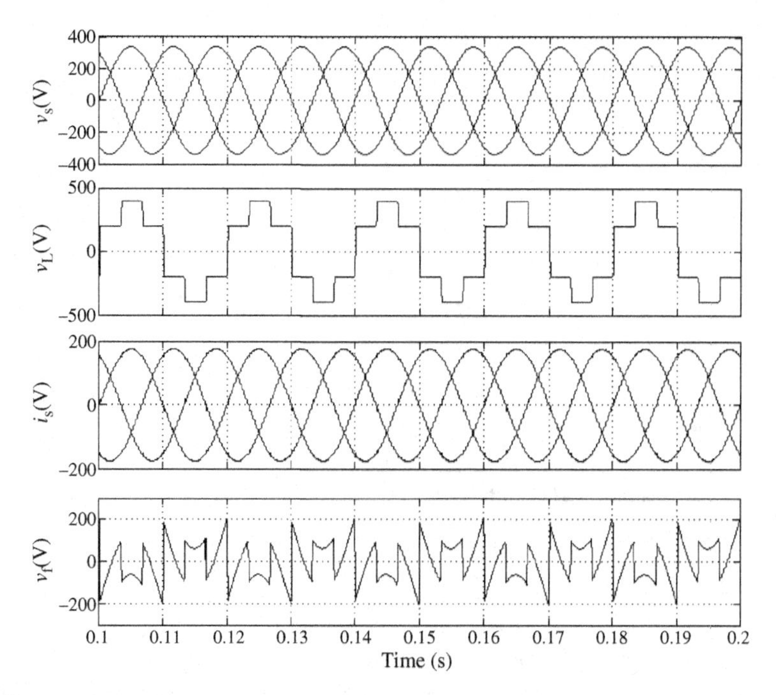

Figure 10.10 Waveforms of a series AF showing PCC voltages (v_s), load voltage (v_L), source currents (i_s), and SAF voltage (v_f)

active power filters. These control algorithms are explained in Chapter 4, which may easily be modified for the control of series active power filters. However, because of space limitation and to give basic understanding, only some of them are explained here for different functions.

10.4.2.1 Control Algorithm for Elimination of Voltage Harmonics

Figure 10.11 shows the control algorithm of the series AF in which the synchronous reference frame (SRF) theory is used for the control of a self-supported series AF. The voltages at PCC (v_s) are converted to the rotating reference frame using the Park's transformation. The harmonics and the oscillatory component of the voltages are eliminated using low-pass filters (LPFs). The components of voltages in the d-axis and q-axis are

$$v_{sd} = v_{dDC} + v_{dAC}, \tag{10.1}$$

$$v_{sq} = v_{qDC} + v_{qAC}. \tag{10.2}$$

The compensation strategy for the compensation of voltage quality problems considers that the load terminal voltage should be of rated magnitude and undistorted in nature. In order to maintain the DC bus voltage of the self-supported capacitor of the VSC used as a series active filter, a PI (proportional–integral) controller is used at the DC bus voltage of the series AF and the output is considered as the voltage (v_{loss}) for meeting its losses.

Therefore, the reference direct-axis load voltage is

$$v_d^* = v_{dDC} - v_{loss}. \tag{10.3}$$

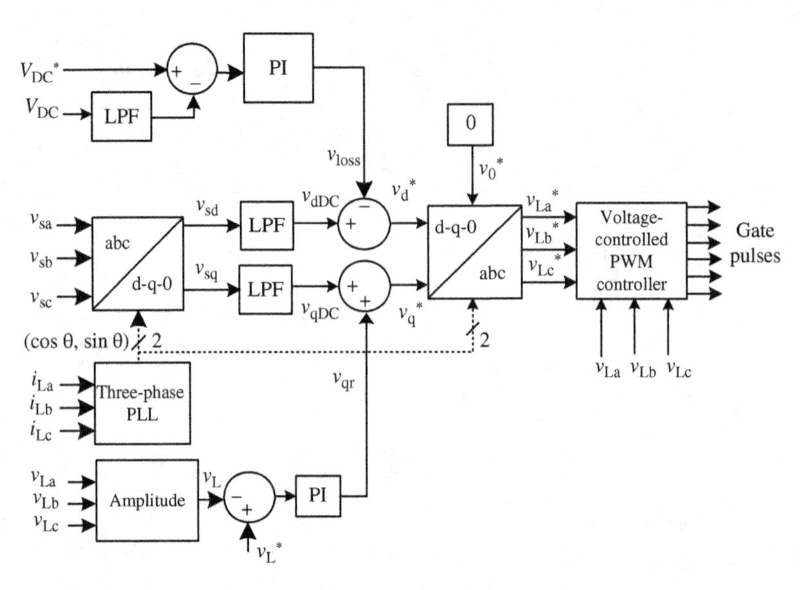

Figure 10.11 Control scheme of a series APF for voltage harmonic elimination

The amplitude of load terminal voltage (V_L) is controlled to its reference voltage (V_L^*) using another PI controller. The output of the PI controller is considered as the reactive component of voltage (v_{qr}) for voltage control across the load terminal.

The reference quadrature-axis load voltage is

$$v_q^* = v_{qDC} + v_{qr}. \tag{10.4}$$

The reference load voltages ($v_{La}^*, v_{Lb}^*, v_{Lc}^*$) in the abc frame are obtained from the reverse Park's transformation. The error between the sensed load voltages (v_{La}, v_{Lb}, v_{Lc}) and the reference load voltages ($v_{La}^*, v_{Lb}^*, v_{Lc}^*$) is used over a PWM controller to generate gating pulses to the VSC of the series AF.

10.4.2.2 Control Algorithm for Elimination of Current Harmonics

This control algorithm is based on the estimation of reference currents. The control algorithm for the control of the series AF is depicted in Figure 10.12. The series AF is used to inject a voltage in series with the terminal voltage to block harmonic currents. The harmonics in the supply currents are compensated by controlling the series AF and the algorithm inherently provides a self-supporting DC bus for the series AF. Three-phase reference supply currents ($i_{sa}^*, i_{sb}^*, i_{sc}^*$) are derived using the sensed load voltages (v_{La}, v_{Lb}, v_{Lc}) and DC bus voltage (v_{DC}) of the VSC as feedback signals.

The synchronous reference frame theory-based method is used to obtain the direct-axis (i_{Ld}) and quadrature-axis (i_{Lq}) components of the supply/load currents. The load currents in the three phases are converted into the dq0 frame using the Park's transformation as follows:

$$\begin{bmatrix} i_{Ld} \\ i_{Lq} \\ i_{L0} \end{bmatrix} = \frac{2}{3} \begin{bmatrix} \cos\theta & -\sin\theta & \frac{1}{2} \\ \cos\left(\theta - \frac{2\pi}{3}\right) & -\sin\left(\theta - \frac{2\pi}{3}\right) & \frac{1}{2} \\ \cos\left(\theta + \frac{2\pi}{3}\right) & \sin\left(\theta + \frac{2\pi}{3}\right) & \frac{1}{2} \end{bmatrix} \begin{bmatrix} i_{La} \\ i_{Lb} \\ i_{Lc} \end{bmatrix}. \tag{10.5}$$

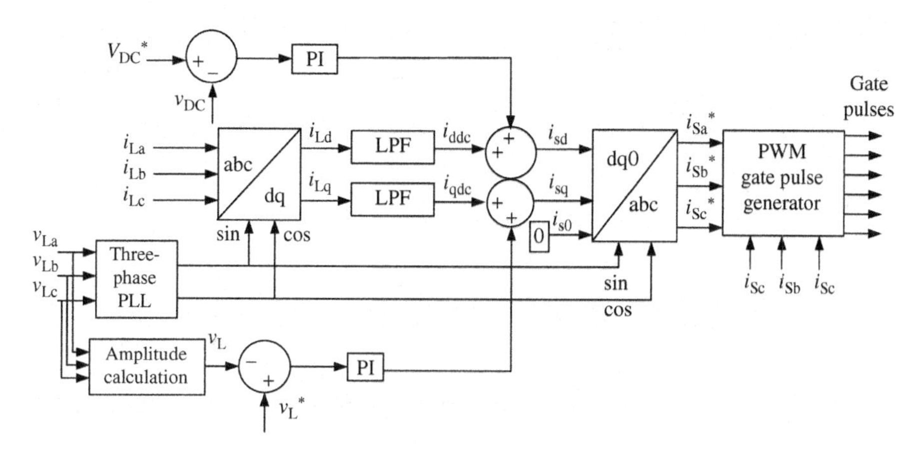

Figure 10.12 Control scheme of a series APF for current harmonic elimination

A three-phase PLL (phase locked loop) is used to synchronize these signals with the PCC/load voltages (v_{La}, v_{Lb}, v_{Lc}). The d–q components are then passed through low-pass filters to extract the DC components of i_{Ld} and i_{Lq}. The error between the reference DC capacitor voltage and the sensed DC bus voltage of the VSC is given to a PI controller, the output of which is considered as the loss component of current and is added to the DC component of i_{Ld}. The amplitude of the load voltage (V_L) is estimated as

$$V_L = (2/3)^{1/2}(v_{La}^2 + v_{Lb}^2 + v_{Lc}^2)^{1/2}. \tag{10.6}$$

Another PI controller is used to regulate the amplitude of the load voltage (V_L). The amplitude of the load terminal voltage is employed over the reference amplitude and the output of the PI controller is added to the DC component of i_{Lq}. The resultant currents are converted into the reference supply currents using the reverse Park's transformation as follows:

$$\begin{bmatrix} i_{sa}^* \\ i_{sb}^* \\ i_{sc}^* \end{bmatrix} = \begin{bmatrix} \cos\theta & \sin\theta & 1 \\ \cos\left(\theta - \dfrac{2\pi}{3}\right) & \sin\left(\theta - \dfrac{2\pi}{3}\right) & 1 \\ \cos\left(\theta + \dfrac{2\pi}{3}\right) & \sin\left(\theta + \dfrac{2\pi}{3}\right) & 1 \end{bmatrix} \begin{bmatrix} i_{sd}^* \\ i_{sq}^* \\ i_{s0}^* \end{bmatrix}. \tag{10.7}$$

The reference supply currents (i_{sa}^*, i_{sb}^*, i_{sc}^*) and the sensed supply currents (i_{sa}, i_{sb}, i_{sc}) are used in a PWM current controller to generate gating pulses for the switches of the VSC used as the series APF.

10.4.2.3 Control Algorithm Based on Unit Vector Generation for Elimination of Current Harmonics

A simple control algorithm of series active filters for current harmonic elimination with a self-supporting DC bus is shown in Figure 10.13. It is implemented to control the series APF for eliminating supply current harmonics. The series active filter is controlled such that it injects a set of voltages (v_{Ca}, v_{Cb}, v_{Cc}) that cancels out the distortions present in the supply currents (i_{sa}, i_{sb}, i_{sc}) and provides perfectly balanced and sinusoidal currents with the desired amplitude. Since the PCC voltages (v_{sa}, v_{sb}, v_{sc}) are unbalanced and/or distorted, a PLL is used to get the synchronization with the PCC voltages. Three-phase distorted/unbalanced PCC voltages are sensed and given to a PLL that generates two quadrature unit vectors (sin θ,

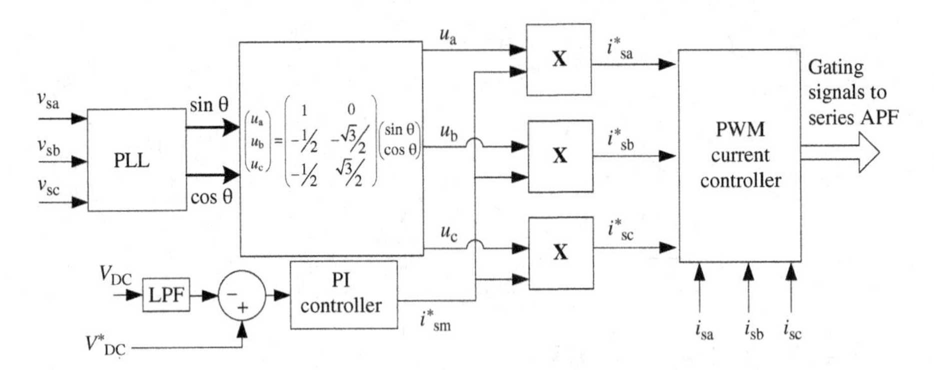

Figure 10.13 Control algorithm of a series APF for current harmonic elimination with a self-supporting DC bus of a VSC of a series AF

cos θ). The in-phase sine and cosine outputs from the PLL are used to compute the PCC in-phase, 120° displaced unit vectors (u_a, u_b, u_c) as

$$\begin{pmatrix} u_a \\ u_b \\ u_c \end{pmatrix} = \begin{pmatrix} 1 & 0 \\ -\frac{1}{2} & -\frac{\sqrt{3}}{2} \\ -\frac{1}{2} & \frac{\sqrt{3}}{2} \end{pmatrix} \begin{pmatrix} \sin\theta \\ \cos\theta \end{pmatrix}. \tag{10.8}$$

The reference supply currents are generated by controlling the DC bus voltage of the VSC. The DC link voltage V_{DC} is sensed and compared with the reference value of the DC link voltage V^*_{DC}. The voltage error is passed through a PI controller to generate the peak value of reference supply currents ($i^*_{sa}, i^*_{sb}, i^*_{sc}$). This peak value is multiplied by the unit vectors to generate the reference supply/load currents as

$$\begin{pmatrix} i^*_{sa} \\ i^*_{sb} \\ i^*_{sc} \end{pmatrix} = I^*_{sm} \begin{pmatrix} u_a \\ u_b \\ u_c \end{pmatrix}. \tag{10.9}$$

Switching signals for the IGBTs of the VSC are generated by a PWM current controller that takes sensed supply currents (i_{sa}, i_{sb}, i_{sc}) and reference supply currents ($i^*_{sa}, i^*_{sb}, i^*_{sc}$) as inputs.

Similarly, a control algorithm of series APFs based on unit vector generation for voltage harmonic elimination with a self-supporting DC bus can also be developed.

10.5 Analysis and Design of Series Active Power Filters

The analysis and design of the series active power filters include the detailed analysis for deriving the design equations for calculating the values of different components used in their circuit configurations. As already discussed in the previous section, there are a large number of topologies of the series active power filters; therefore, the design of a large number of circuit configurations is not practically possible to include here due to space constraints. In view of these facts, the step-by-step design procedure of a selected topology of a series active power filter is given here. The design of a three-phase three-wire series active power filter includes the design of a VSC, interfacing inductors, and a ripple filter. The design of the VSC includes the selection of the DC bus voltage level, the DC capacitance, and the rating of IGBTs.

10.5.1 *Design of a Series Active Power Filter for a Voltage-Fed Nonlinear Load*

A three-phase series active filter consists of a VSC with an AC series inductor, a coupling transformer, and a DC bus capacitor used in series with a three-phase AC supply feeding a nonlinear load having a three-phase diode rectifier with a capacitive filter. The series AF is used to reduce the harmonics in AC mains currents and to maintain almost UPF (unity power factor) at the AC mains. The design of the series APF includes calculation of the voltage, current, and VA ratings of the VSC of the series AF, value of the AC inductor, value of the DC bus capacitor, and its DC voltage. Figure 10.7 shows a schematic diagram of the series active filter connected to a system with a nonlinear rectifier load. A three-phase 415 V (phase voltage $V_{ph} = 239.6$ V) supply system with a 25 kW load is considered to illustrate the design of the series AF. The series AF is connected in series with the supply system through an injection transformer (T_r). The capacitor (C_{DC}) is connected at the DC bus of the VSC to act as a DC voltage source. The DC bus capacitor is selected based on the transient energy requirement and the DC bus voltage is selected based on the injection voltage level. The series AF acts as a harmonic isolator between the supply and the load. It has a ripple filter to absorb current ripples generated by the VSC.

Design of a VSC of a Series Active Power Filter

The design of the series active power filter is based on the DC bus voltage of the three-phase rectifier load. The series active power filter is operated for eliminating harmonics in the supply currents and hence it injects only harmonic voltages. It is clear that the fundamental component of load voltage is the PCC voltage as the series AF injects the harmonic component of the load voltage. For the series AF, the switching frequency (f_s) of 20 kHz is used. The DC link voltage of the VSC (V_{DC}) is selected as 700 V and it has to be controlled within 5% range and the ripple current in the inductor is to be constrained to 5% of current flowing through the series APF. Hence, the fundamental component of load AC voltage is

$$V_{LL} = (\sqrt{6}/\pi)V_d = 0.779V_d. \tag{10.10}$$

For a given line voltage of 415 V and a DC load voltage (V_d) of 540 V, the voltage rating of the series AF is obtained from the difference of PCC and load voltages and hence the SAF voltage is calculated as

$$V_f = \sqrt{\left[(1/\pi)\left\{\int_0^{\pi/3}(V_{ph}\sqrt{2}\sin\theta - V_d/3)^2\,d\theta + \int_{\pi/3}^{2\pi/3}(V_{ph}\sqrt{2}\sin\theta - 2V_d/3)^2\,d\theta + \int_{2\pi/3}^{\pi}(V_{ph}\sqrt{2}\sin\theta - V_d/3)^2\,d\theta\right\}\right]}$$

$$= \sqrt{\left[(1/\pi)\left\{\int_0^{\pi/3}(239.6\sqrt{2}\sin\theta - 180)^2\,d\theta + \int_{\pi/3}^{2\pi/3}(239.6\sqrt{2}\sin\theta - 360)^2\,d\theta + \int_{2\pi/3}^{\pi}(239.6\sqrt{2}\sin\theta - 180)^2\,d\theta\right\}\right]}$$

$$= 75.6415\ \text{V}. \tag{10.11}$$

(Since the sine wave supply current after compensation results in continuous conduction of diodes of the three-phase diode rectifier load (each diode conducting for 180°) and it results in a stepped waveform of the phase voltage at the input of the diode bridge with (i) the first step of $\pi/3$ angle (from 0° to $\pi/3$) and a magnitude of $V_d/3$, (ii) the second step of $\pi/3$ angle (from $\pi/3$ to $2\pi/3$) and a magnitude of $2V_d/3$, and (iii) the third step of $\pi/3$ angle (from $2\pi/3$ to π) and a magnitude of $V_d/3$, and it has both half cycles of symmetric segments of such steps.)

The voltage rating of the VSC is obtained as 75.6415 V.

Design of Current Rating of a VSC of a Series Active Power Filter

The current rating of the series APF depends on the fundamental component of load current and it is obtained as follows.

The load power is calculated as

$$P_{DC} = V_d^2/R, \tag{10.12}$$

where R is the equivalent resistance of the DC load at the output of the diode bridge rectifier. For a given load of 25 kW at 540 V DC bus voltage of the load, the equivalent resistance is $R = 19.6\,\Omega$.

Considering a UPF supply current and a lossless series AF, the rms supply current is calculated as

$$I_{sa} = P/\sqrt{3}V_{LL}, \tag{10.13}$$

where P is the input power equal to 25 kW. Considering $V_{LL} = 415$ V, the supply current is 34.78 A. The current rating of the VSC is obtained as $I_f = 34.78$ A.

Design of kVA Rating of a VSC of a Series Active Power Filter

The kVA rating of the VSC of the SAF is calculated as

$$\text{kVA} = 3V_f I_f/1000 = 3 \times 75.6415 \times 34.78 = 7.892\,\text{kVA}. \tag{10.14}$$

Design of Rating of an Injection Transformer of a Series Active Power Filter

The injection transformer is designed considering the optimum voltage level of the VSC. The maximum AC voltage on the AC side of the VSC of the series APF may be $m_a V_{DC}/(2\sqrt{2}) = 0.8 \times 700/(2 \times \sqrt{2}) = 197.99$ V (considering the modulation index $m_a = 0.8$) and on the supply side it must be $V_{supply} = V_f$. The turns ratio of the coupling transformer is

$$N_{VSC}/N_{supply} = V_{VSC}/V_f = 197.99\,\text{V}/75.6415\,\text{V} = 2.62. \tag{10.15}$$

The kVA rating of the injection transformer is same as that of the VSC and is calculated as

$$\text{kVA} = 3V_f I_f/1000 = 3 \times 75.6415 \times 34.78 = 7.892\,\text{kVA}. \tag{10.16}$$

Hence, the rating of the injection transformer is 7.892 kVA, 197.99 V/75.6415 V.

Design of DC Capacitance of a VSC of a Series Active Power Filter

The DC bus capacitance is selected based on the transient energy required during change in the loads. Considering that the energy stored in the capacitor is for meeting the energy demand of the load for a fraction of power cycle, the relation can be expressed as

$$(1/2)C_{DC}(V_{DC}^2 - V_{DC1}^2) = 3V_f I_f \Delta t, \tag{10.17}$$

where V_{DC} is the rated voltage, V_{DC1} is the drop in DC bus voltage allowed during transients, Δt is the time for which support is required, and C_{DC} is the DC bus capacitance.

Considering $\Delta t = 0.1$ ms, $V_{DC} = 700$ V, and $V_{DC1} = 700 - (5\%$ of $700) = 665$ V, Equation 10.17 results in

$$(1/2)C_{DC}(700^2 - 665^2) = 3 \times 75.6415 \times 34.78 \times 0.1 \times 10^{-3}.$$

It gives $C_{DC} = 3304\,\mu$F. Hence, a DC bus capacitor of 4000 μF, 700 V is selected for the series AF.

Design of an Interfacing Inductor for a VSC of a Series Active Power Filter

The value of the interfacing inductor is selected based on the current ripple in the series AF. Considering that ripple current in the inductor is 5% and overloading factor $a = 1.2$, the inductor is calculated as

$$L_f = (N_{VSI}/N_{supply})(\sqrt{3}/2)m_a V_{DC}/(6af_s\Delta I_f). \quad (10.18)$$

Substituting the values in Equation 10.18, the value of the interfacing inductor is estimated as

$$L_f = 3.27(\sqrt{3}/2) \times 0.8 \times 700/(6 \times 1.2 \times 20\,000 \times 1.739) = 6.3 \text{ mH}.$$

Hence, an interfacing inductor of 6.3 mH and 40 A current-carrying capacity is selected for the series AF.

Design of a Ripple Filter

The ripple filter is designed for eliminating the switching frequency ripples from the injected voltage of the series APF. The ripple filter, a combination of a capacitor (C_f) and a resistor (R_f) connected in series, is generally tuned at half of the switching frequency (f_r), which is calculated as

$$f_r = 1/2\pi R_f C_f, \quad (10.19)$$

$$f_r = f_s/2 = 10\,000 = 1/2\pi R_f C_f. \quad (10.20)$$

Considering $R_f = 5\,\Omega$, $C_f = 3.2\,\mu$F. Hence, $R_f = 5\,\Omega$ and $C_f = 5\,\mu$F are selected for designing a ripple filter.

Similarly, the design of series active power filters may be carried out for other applications and other configurations.

10.6 Modeling, Simulation, and Performance of Series Active Power Filters

The MATLAB-based models of different topologies of series APF systems are developed using Simulink and SimPowerSystems (SPS) toolboxes to simulate the performance of the series active power filters in single-phase and three-phase distribution systems. A large number of cases of the topologies of series active power filters are given in solved examples. Here, the performance of typical developed models of series APFs is illustrated with stiff supply under (a) linear load, (b) nonlinear load, and (c) DC link of the series APF connected to the DC bus of the load.

10.6.1 Series APF for Elimination of Voltage Harmonics

Its supply block consists of a three-phase balanced and distorted voltage source. Supply/load currents also contain harmonics due to the distorted supply voltages. The linear load is modeled as a 25 kW, 0.8 lagging power factor star connected real and reactive load. The VSC of the series APF is modeled using IGBT switches with a DC capacitor connected at the DC link. The series APF is connected in series between the PCC and the load using three single-phase transformers. Figure 10.14 shows the configuration and performance of a series APF compared with the system without a series APF. Figure 10.14a shows a series

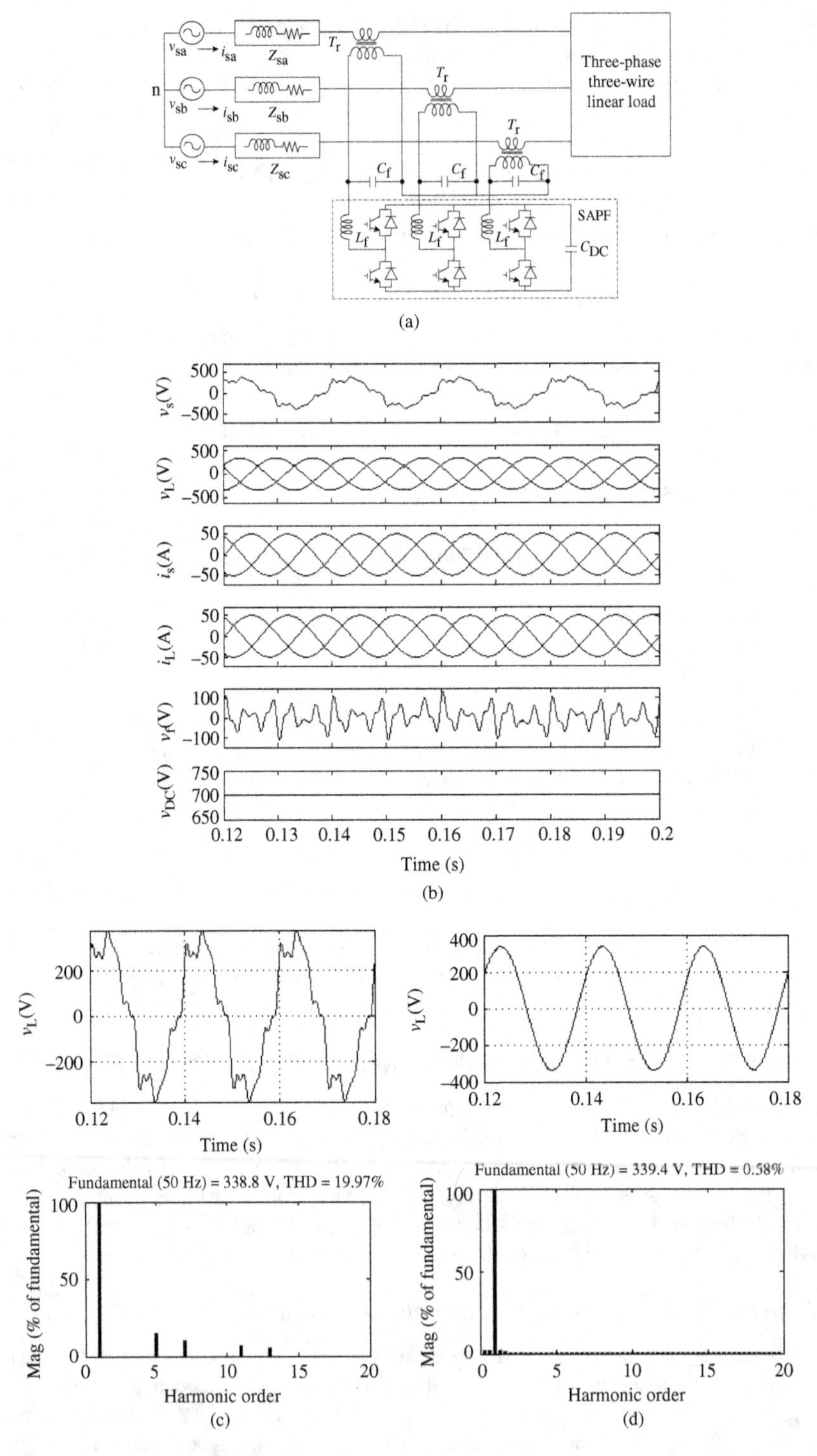

Figure 10.14 Performance of a series APF for voltage harmonic elimination. (a) Configuration, (b) waveforms after compensation, (c) load voltage and its harmonic spectrum without a series AF, and (d) load voltage and its harmonic spectrum with a series AF

APF used to inject the voltages such that the harmonics in load voltages are eliminated across the loads. Figure 10.14b shows the performance of a series active filter for eliminating the voltage harmonics across the load. Figure 10.14c shows the load voltage and its harmonic spectrum without a series AF and Figure 10.14d shows the load voltage and its harmonic spectrum with a series AF when a linear load is connected to the system.

10.6.2 Series APF for Elimination of Current Harmonics

Its supply block consists of three-phase balanced and sinusoidal voltages. A voltage-fed nonlinear load is modeled as a three-phase diode bridge rectifier feeding power to a resistive load with a capacitor filter at its DC bus. The VSC of the series APF is modeled using IGBT switches with a DC capacitor connected at the DC link. The series APF is connected in series between the PCC and the load using three single-phase transformers. Figure 10.15 shows the configuration and performance of a series APF compared with the system without a series APF. Figure 10.15a shows a series APF used for injecting the voltage such that the harmonics in the supply/load current are eliminated under a voltage-fed nonlinear load. Figure 10.15b shows the performance of a series APF for injecting the voltages such that the harmonics in the supply/load currents are eliminated under a voltage-fed nonlinear load. Figure 10.15c shows supply current and its harmonic spectrum without a series AF and Figure 10.15d shows supply current and its harmonic spectrum with a series AF when a voltage-fed load is connected to the system.

10.6.3 Series APF for Elimination of Current Harmonics with a Common DC Bus of the Load and the Series APF

The series active filter is also used to eliminate supply current harmonics under a voltage-fed nonlinear load when the DC bus of the voltage-fed nonlinear load can be shared with the DC bus of the series active filter. It saves a DC bus capacitor and it can also be used to regulate the DC bus voltage of the load. It may also avoid the derating of the loads due to fluctuations in the supply voltages and makes the load immune to voltage-based power quality problems. It may also provide override facility to the loads. However, all these functions of the series APF are achieved at the cost of increased rating of the series active filter. Here, the supply, transformers, and the load are modeled as mentioned in Section 10.6.2. The VSC of the series APF is modeled using IGBT switches with a common DC capacitor connected at the combined DC bus of the load and the VSC. Figure 10.16 shows the configuration and performance of a series APF system compared with the system without a series APF. Figure 10.16a shows a series APF used for injecting the voltage such that the harmonics in the supply/load current are eliminated under a voltage-fed nonlinear load. Figure 10.16b shows the performance of a series APF for injecting the voltages such that the harmonics in the supply/load currents are eliminated under a voltage-fed nonlinear load. Figure 10.16c shows supply current and its harmonic spectrum without a series AF and Figure 10.16d shows supply current and its harmonic spectrum with a series AF when a voltage-fed load is connected to the system.

10.7 Numerical Examples

Example 10.1

Design a single-phase series active power filter (shown in Figure E10.1) for filtering the voltage harmonics in a 220 V, 50 Hz AC mains (15% third, 10% fifth, and 5% seventh harmonics) present due to other loads and source impedance, before it is connected to a critical linear load of 5 kVA, 220 V, 50 Hz at 0.8 lagging power factor. If a single-phase VSC is used as a series APF, calculate (a) voltage and current ratings of the APF, (b) VA rating of the VSC of the series APF, and (c) interfacing inductance of the APF. Consider the switching frequency of 20 kHz, DC bus voltage of 200 V, and ripple current in the inductor is 10%.

Solution: Given supply voltage $V_s = 220$ V, frequency of the supply $(f) = 50$ Hz, $V_{s3} = 15\%$, $V_{s5} = 10\%$, $V_{s7} = 5\%$, a critical load of 5 kVA at 0.8 lagging power factor, $V_{DC} = 200$ V, $f_s = 20$ kHz, $\Delta I_f = 10\%$, and $\Delta V_{DC} = 5\%$.

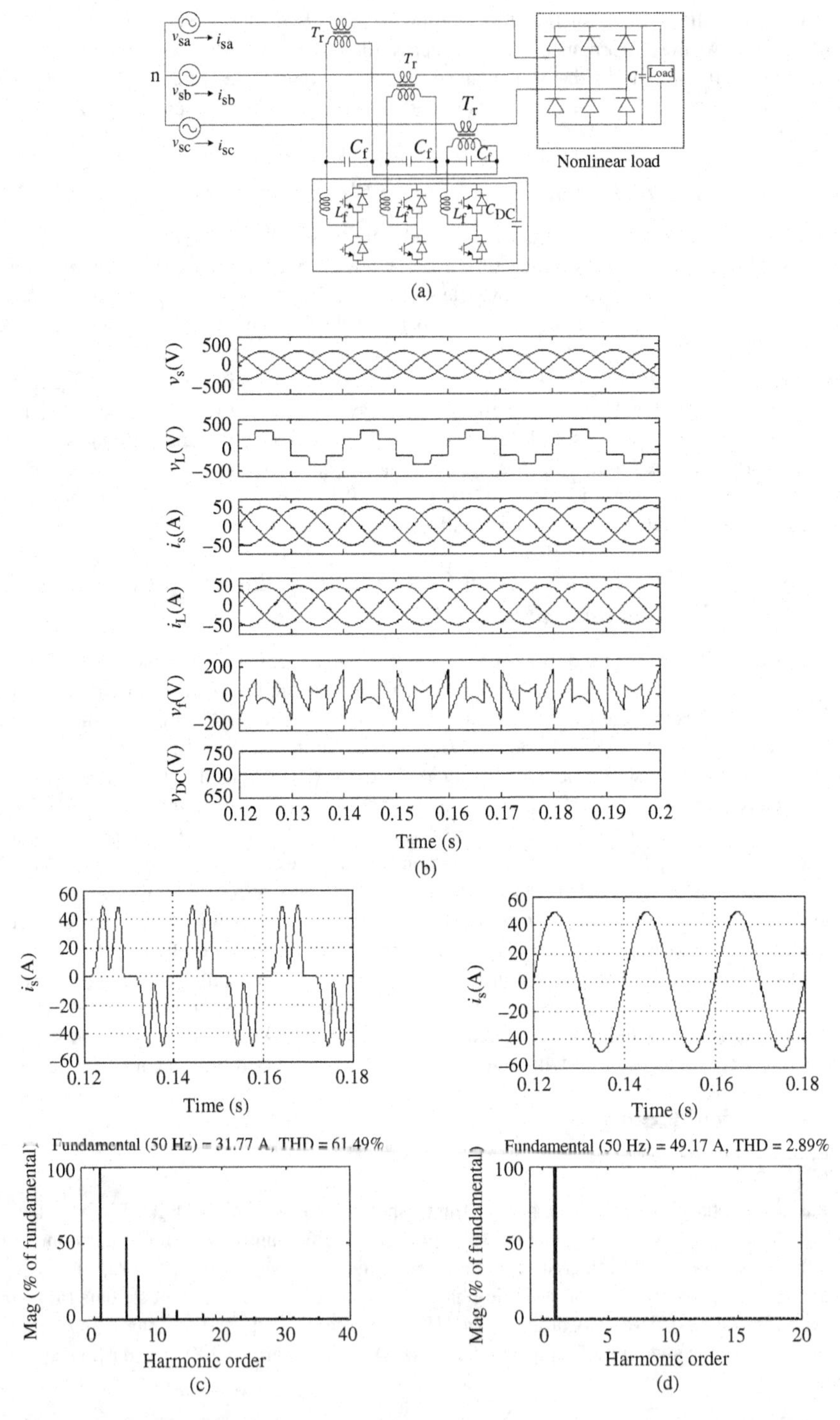

Figure 10.15 Performance of a series APF for current harmonic elimination. (a) Configuration, (b) waveforms after compensation, (c) supply current and its harmonic spectrum without a series AF, and (d) supply current and its harmonic spectrum with a series AF

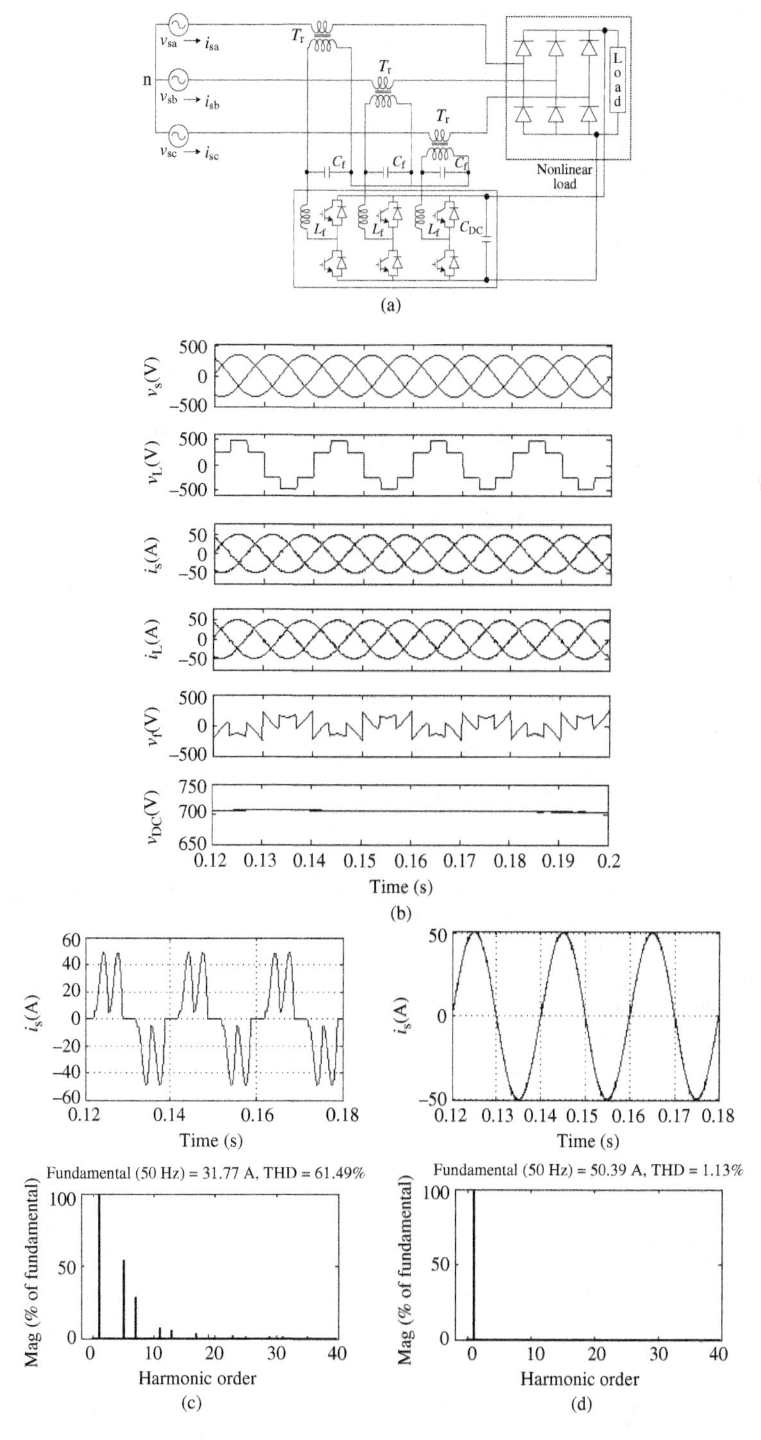

Figure 10.16 Performance of a series APF for current harmonic elimination when the DC bus of the load is connected to the DC bus of the VSC. (a) Configuration, (b) waveforms after compensation, (c) supply current and its harmonic spectrum without a series AF, and (d) supply current and its harmonic spectrum with a series AF

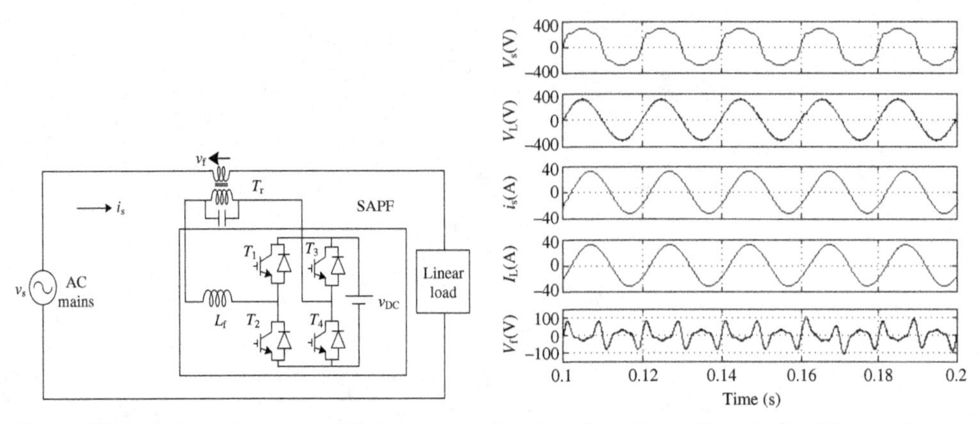

Figure E10.1 A single-phase series APF for compensation of a voltage-fed nonlinear load and its waveforms

The load current is $I_s = 5000/220 = 22.73$ A.

$$\Delta I_f = 10\% \text{ of } I_s = 2.273 \text{ A}, \quad \Delta V_{DC} = 5\% \text{ of } V_{DC} = 10 \text{ V}, \quad V_{DCmin} = 200 - \Delta V_{DC} = 190 \text{ V}.$$

$$V_{s3} = 15\% \text{ of } V_s = 33 \text{ V}, \quad V_{s5} = 10\% \text{ of } V_s = 22 \text{ V}, \quad V_{s7} = 5\% \text{ of } V_s = 11 \text{ V}.$$

a. The voltage and current ratings of the series APF are $V_f = \sqrt{(V_{s3}^2 + V_{s5}^2 + V_{s7}^2)} = 41.158$ V and $I_f = 22.73$ A.
b. The VA rating of the VSC of the APF is $S = V_f I_s = 41.158 \times 22.73$ VA $= 935.41$ VA.
c. The interfacing inductance of the APF is $L_f = V_{DC}/(4f_s\Delta I_f) = 200/(4 \times 20\,000 \times 2.273) = 1.1$ mH.

Example 10.2

A single-phase VSI with a square-wave AC output of 220 V (rms) at 50 Hz (shown in Figure E10.2) is feeding a critical linear load of 3 kVA, 220 V, 50 Hz at 0.8 lagging power factor. Design a single-phase series active power filter for filtering the voltage harmonics in this system to eliminate voltage harmonics and to regulate fundamental 220 V (rms) across the load. If a single-phase VSC is used as an APF, calculate (a) voltage rating of the APF, (b) current rating of the APF (c) VA rating of the VSC of the APF,

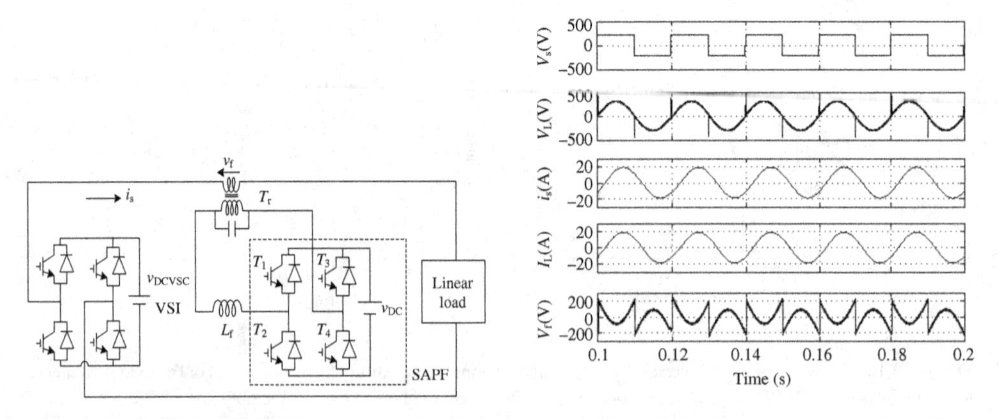

Figure E10.2 A single-phase series APF for compensation of a voltage-fed nonlinear load and its waveforms

and (d) interfacing inductance. Consider the switching frequency of 20 kHz, DC bus voltage of 200 V, and ripple current in the inductor is 5%.

Solution: Given supply voltage $V_s = 220$ V (rms square wave), frequency of the supply $(f) = 50$ Hz, a critical load of 3 kVA at 0.8 lagging power factor, $V_{DC} = 200$ V, $f_s = 20$ kHz, $\Delta I_f = 5\%$, and $\Delta V_{DC} = 10\%$.

The load current is $I_s = 3000/220 = 13.636$ A.

$$\Delta I_f = 5\% \text{ of } I_s = 0.6818 \text{ A}, \quad \Delta V_{DC} = 10\% \text{ of } V_{DC} = 20 \text{ V}, \quad V_{DCmin} = 200 - \Delta V_{DC} = 180 \text{ V}.$$

$$V_s = 220 \text{ V (square wave)}, \quad V_{PCC} = 220 \text{ V (sine wave)}.$$

a. The voltage rating of the series APF is

$$V_f = \sqrt{\left\{(1/\pi)\int_0^{\pi}(220 - 220\sqrt{2}\sin\theta)^2\, d\theta\right\}} = 98.23 \text{ V}.$$

b. The current rating of the series APF is $I_f = I_s = 3000/220 = 13.636$ A.
c. The VA rating of the VSC of the APF is $S = V_f I_s = 98.23 \times 13.636$ VA $= 1339.46$ VA.
d. The interfacing inductance of the APF is $L_f = V_{DC}/(4 f_s \Delta I_f) = 200/(4 \times 20\,000 \times 0.6818) = 3.667$ mH.

Example 10.3

A series active filter (consisting of a VSC with an AC series inductor and a DC bus capacitor) (shown in Figure E10.3) is used in series with single-phase AC supply of 220 V at 50 Hz feeding a diode rectifier used for charging a battery of 252 V at 15 A average current to reduce the harmonics in AC mains current and to maintain almost UPF. Calculate (a) rms voltage at the input of the diode rectifier, (b) line current, (c) voltage rating of the APF, (d) current rating of the APF, (e) VA rating of the APF, (f) value of the DC bus voltage of the APF, and (g) value of the AC inductor of the APF. Consider the switching frequency of 20 kHz and ripple current in the inductor is 5%.

Solution: Given supply voltage $V_s = 220$ V, frequency of the supply $(f) = 50$ Hz, a load of $V_{DCload} = 252$ V and $I_{DC} = 15$ A, $f_s = 20$ kHz, $\Delta I_f = 5\%$, and $\Delta V_{DCAPF} = 8\%$.

The active power is $P = V_{DCload} I_{DC} = 252 \times 15 = 3780$ W.

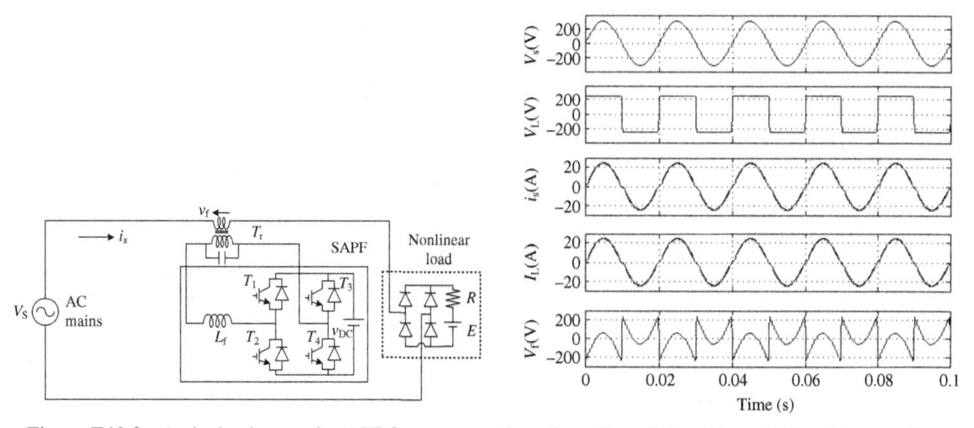

Figure E10.3 A single-phase series APF for compensation of a voltage-fed nonlinear load and its waveforms

The supply current is $I_s = 3780/220 = 17.1818$ A.

$$\Delta I_f = 5\% \text{ of } I_s = 0.859\,09 \text{ A}, \quad \Delta V_{DC} = 8\% \text{ of } V_{DC} \text{ of the APF.}$$

a. The rms voltage at the input of the diode rectifier is computed as follows.
   The sine wave supply current after compensation results in continuous conduction of diodes of the rectifier (180°) and it results in square wave AC voltage at PCC with an amplitude of DC bus voltage. Therefore, $V_{PCC} = 252$ V.
b. The supply line current at unity power factor is $I_s = 3780/220 = 17.1818$ A.
c. The voltage rating of the series APF is

$$V_f = V_{frms} = \sqrt{\left\{(1/\pi)\int_0^{\pi}(220\sqrt{2}\sin\theta - 252)^2\, d\theta\right\}} = 109.8954 \text{ V}.$$

d. The current rating of the series APF is $I_f = I_s = 17.1818$ A (since the APF is connected in series with the supply).
e. The VA rating of the VSI of the APF is $S = V_f I_s = 109.8954 \times 17.1818$ VA $= 1888.20$ VA.
f. The value of the DC bus voltage of the APF is $V_{DCAPF} = \sqrt{2}V_f/m_a = \sqrt{2}V_f/0.8 = 194.269\text{ V} \approx 200$ V.
g. The interfacing inductance of the APF is $L_f = V_{DCAPF}/(4f_s\Delta I_f) = 200/(4 \times 20\,000 \times 0.859\,09) = 2.91$ mH.

Example 10.4

A series active filter (consisting of a VSC with an AC series inductor, a coupling transformer, and its DC bus connected to a battery functioning as the load) (shown in Figure E10.4) is used in series with single-phase AC supply of 220 V at 50 Hz feeding a diode rectifier used for charging a battery of 192 V at 25 A average current to reduce the harmonics in AC mains current and to maintain almost UPF. Calculate (a) rms voltage at the input of the diode rectifier, (b) line current, (c) voltage rating of the APF, (d) current rating of the APF, (e) VA rating of the APF, (f) value of the AC inductor, and (g) turns ratio of the coupling transformer. Consider the switching frequency of 20 kHz and ripple current in the inductor is 5%.

Solution: Given supply voltage $V_s = 220$ V (rms), frequency of the supply $(f) = 50$ Hz, and a load consisting of a diode rectifier having current $I_{DC} = 25$ A and a battery of $V_{DC} = 192$ V.
The active power is $P = V_{DC}I_{DC} = 192 \times 25$ W $= 4800$ W.
$\Delta I_f = 5\%$ of $I_s = 1.090\,905$ A.

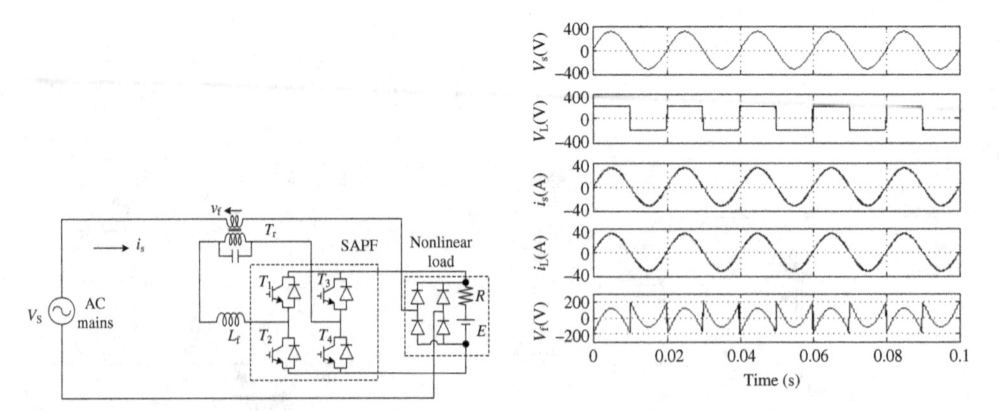

Figure E10.4 A single-phase series APF for compensation of a voltage-fed nonlinear load and its waveforms

a. The sine wave supply current after compensation results in continuous conduction of diodes of the rectifier (180°) and it results in square wave AC voltage at PCC with an amplitude of DC bus voltage. Therefore, $V_{PCC} = 192$ V.
b. The line current at unity power factor is $I_s = P/V_s = 4800/220 = 21.8181$ A.
c. The voltage rating of the series APF is

$$V_f = V_{frms} = \sqrt{\left\{(1/\pi)\int_0^{\pi}(220\sqrt{2}\sin\theta - 192)^2\,d\theta\right\}} = 95.94\text{ V}.$$

d. The current rating of the series APF is $I_f = I_s = 21.8181$ A.
e. The VA rating of the VSI of the APF is $S = V_f I_s = 95.94 \times 21.8181$ VA $= 2093.3$ VA.
f. The interfacing inductance of the APF is $L_f = V_{DCAPF}/(4f_s\Delta I_f) = 192/(4 \times 20\,000 \times 1.090\,905) = 2.2$ mH.
g. The turns ratio of the coupling transformer is computed as follows.
   The maximum AC voltage on the AC side of the VSC is $V_{DC}/\sqrt{2} = 192/\sqrt{2} = 135.76$ V and on the supply side it must be $V_{supply} = V_f$. The turns ratio of the coupling transformer is $N_{VSI}/N_{supply} = 135.76/95.94 = 1.415\,05$.

Example 10.5

A series active power filter (consisting of a VSC with an AC series inductor, a coupling transformer, and its DC bus connected to the DC bus of the load) (shown in Figure E10.5) is used to reduce the harmonics in AC mains current and to maintain almost UPF in the series of single-phase AC supply of 230 V at 50 Hz feeding a diode rectifier with a capacitive filter of 2000 μF and a resistive load of 20 Ω. The DC bus voltage of the load is decided to result in minimum injected voltage of the APF. Calculate (a) rms voltage at the input of the diode rectifier, (b) line current, (c) voltage rating of the APF, (d) current rating of the APF, (e) VA rating of the APF, (f) DC bus voltage, and (g) turns ratio of the coupling transformer.

Solution: Given supply voltage $V_s = 230$ V (rms), frequency of the supply $(f) = 50$ Hz, and a load consisting of a diode rectifier having $R_{DC} = 20\ \Omega$.

If x volts is the DC bus voltage of the rectifier load, then the injected voltage of the APF is

$$V_f = V_{frms} = \sqrt{\left\{(1/\pi)\int_0^{\pi}(230\sqrt{2}\sin\theta - x)^2\,d\theta\right\}}.$$

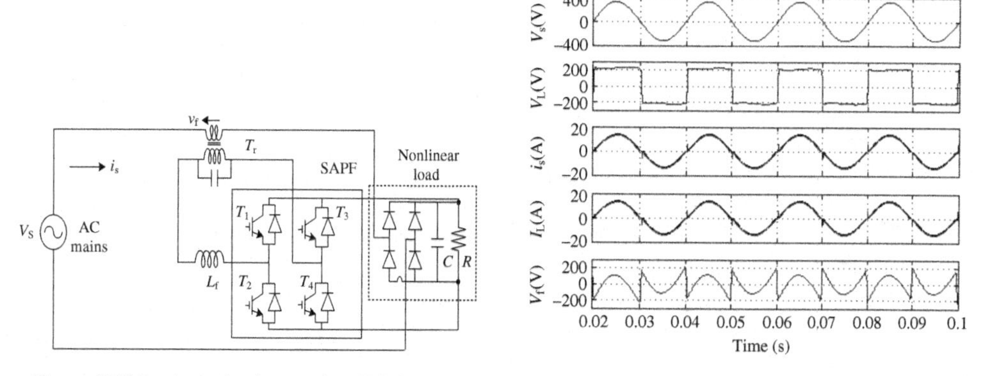

Figure E10.5 A single-phase series APF for compensation of a voltage-fed nonlinear load and its waveforms

After integrating and equating the derivative with respect to x to zero, it gives $x = 0.9V_s = V_{DC} = 207$ V.
The DC load current is $I_{DC} = V_{DC}/R_{DC} = 207/20 = 10.35$ A.
The active power is $P = V_{DC}I_{DC} = 207 \times 10.35$ W $= 2142.45$ W.

a. The sine wave supply current after compensation results in continuous conduction of diodes of the rectifier (180°) and it results in square wave AC voltage at PCC with an amplitude of DC bus voltage. Therefore, $V_{PCC} = 207$ V.
b. The line current at unity power factor is $I_s = P/V_s = 2142.45/230 = 9.315$ A.
c. The voltage rating of the series APF is

$$V_f = V_{frms} = \sqrt{\left\{(1/\pi)\int_0^{\pi}(230\sqrt{2}\sin\theta - 207)^2\,d\theta\right\}} = 100.01 \text{ V}.$$

d. The current rating of the series APF is $I_f = I_s = 9.315$ A.
e. The VA rating of the VSI of the APF is $S = V_f I_s = 100.01 \times 9.315$ VA $= 931.597$ VA.
f. The DC bus voltage of the load is $V_{DC} = x = 207$ V.
g. The turns ratio of the coupling transformer is computed as follows.
   The maximum AC voltage on the AC side of the VSC of the APF may be $V_{DC}/\sqrt{2} = 207/\sqrt{2} = 146.37$ V and on the supply side it must be $V_{supply} = V_f$. The turns ratio of the coupling transformer is $N_{VSI}/N_{supply} = 146.37/100.01 = 1.463\,55$.

Example 10.6

A series active power filter (consisting of a VSC with an AC series inductor, a coupling transformer, and its DC bus connected to the DC bus of the load) (shown in Figure E10.6) is used to reduce the harmonics in AC mains current and to maintain almost UPF in the series of single-phase AC supply of 220 V at 50 Hz feeding a diode rectifier with a capacitive filter of 1000 μF and a resistive load of 10 Ω. If DC bus voltage of the load is to be maintained at constant ripple-free 320 V, then calculate (a) rms voltage at the input of the diode rectifier, (b) line current, (c) voltage rating of the APF, (d) current rating of the APF, (e) VA rating of the APF, and (f) turns ratio of the coupling transformer.

Solution: Given supply voltage $V_s = 220$ V (rms), frequency of the supply (f) = 50 Hz, and a load consisting of a diode rectifier having $R_{DC} = 10\,\Omega$ and $V_{DC} = 320$ V.
The load current is $I_{DC} = 320/10 = 32$ A.
The active power is $P = V_{DC}I_{DC} = 320 \times 32$ W $= 10\,240$ W.

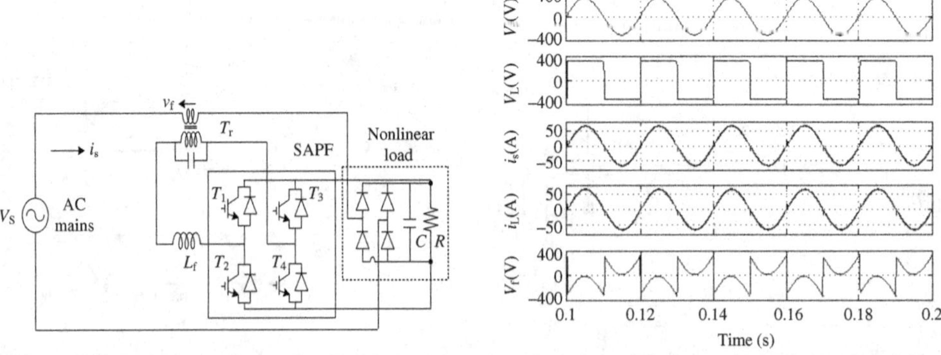

Figure E10.6 A single-phase series APF for compensation of a voltage-fed nonlinear load and its waveforms

a. The sine wave supply current after compensation results in continuous conduction of diodes of the rectifier (180°) and it results in square wave AC voltage at PCC with an amplitude of DC bus voltage. Therefore, $V_{PCC} = 320$ V.
b. The line current at unity power factor is $I_s = P/V_s = 10\,240/220 = 46.545$ A.
c. The voltage rating of the series APF is

$$V_f = V_{frms} = \sqrt{\left\{(1/\pi)\int_0^{\pi}(220\sqrt{2}\sin\theta - 320)^2\,d\theta\right\}} = 155.032\text{ V}.$$

d. The current rating of the series APF is $I_f = I_s = 46.545$ A.
e. The VA rating of the VSI of the APF is $S = V_f I_s = 155.032 \times 46.545$ VA $= 7215.98$ VA.
f. The turns ratio of the coupling transformer is computed as follows.
The maximum AC voltage on the AC side of the VSC of the APF may be $V_{DC}/\sqrt{2} = 320/\sqrt{2} = 226.27$ V and on the supply side it must be $V_{supply} = V_f$. The turns ratio of the coupling transformer is $N_{VSI}/N_{supply} = 226.27/155.032 = 1.4595$.

Example 10.7

A single-phase 230 V, 50 Hz diode bridge rectifier with a DC capacitor filter is feeding a DC of 340 V at 10 A average current to a variable-frequency three-phase VSI-fed induction motor drive in an air conditioner. A single-phase series active power filter (consisting of a VSC with an AC series inductor, a coupling transformer, and a DC bus capacitor) (shown in Figure E10.7) is used in series with this rectifier–inverter system to reduce the harmonics in AC mains current, to maintain almost UPF, and to regulate the DC bus voltage of the rectifier to 340 V. Calculate (a) rms voltage at the input of the single-phase diode rectifier, (b) line current, (c) rms current of the APF, (d) rms voltage across the APF, and (e) VA rating of the APF.

Solution: Given supply voltage $V_s = 230$ V (rms), frequency of the supply (f) = 50 Hz, and a load consisting of a diode rectifier having $I_{DC} = 10$ A and $V_{DC} = 340$ V.
The active power is $P = V_{DC} I_{DC} = 340 \times 10$ W $= 3400$ W.

a. The sine wave supply current after compensation results in continuous conduction of diodes of the rectifier (180°) and it results in square wave AC voltage at PCC with an amplitude of DC bus voltage. Therefore, $V_{PCC} = 340$ V.
b. The line current at unity power factor is $I_s = P/V_s = 3400/230 = 14.783$ A.
c. The current rating of the series APF is $I_f = I_s = 14.783$ A.

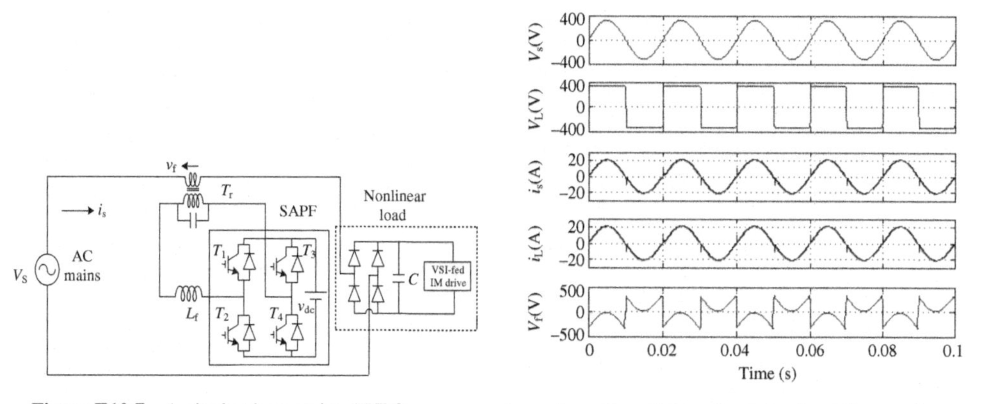

Figure E10.7 A single-phase series APF for compensation of a voltage-fed nonlinear load and its waveforms

d. The voltage rating of the series APF is

$$V_f = V_{frms} = \sqrt{\left\{(1/\pi)\int_0^{\pi}(230\sqrt{2}\sin\theta - 340)^2\, d\theta\right\}} = 166.4047\ \text{V}.$$

e. The VA rating of the VSI of the APF is $S = V_f I_s = 166.4047 \times 14.783\ \text{VA} = 2459.96\ \text{VA}$.

Example 10.8

Design a three-phase series active power filter (shown in Figure E10.8) for filtering the voltage harmonics in a 415 V, 50 Hz AC mains (having 15% 5th, 10% 7th, 7% 11th, and 5% 13th harmonics) present due to other loads and source impedance, before it is connected to a critical linear load of 50 kVA, 415 V, 50 Hz at 0.8 lagging power factor. If a three-phase VSC is used as a series APF, calculate (a) voltage and current ratings of the APF, (b) VA rating of the VSC of the APF, (c) interfacing inductance, and (d) turns ratio of the coupling transformer. Consider the switching frequency of 20 kHz, DC bus voltage of 200 V, and ripple current in the inductor is 2.5%.

Solution: Given supply voltage $V_s = 415/\sqrt{3} = 239.6$ V, frequency of the supply $(f) = 50$ Hz, $V_{s5} = 15\%$, $V_{s7} = 10\%$, $V_{s11} = 7\%$, $V_{s13} = 5\%$, a critical load of 50 kVA at 0.8 lagging power factor, $V_{DC} = 200$ V, $f_s = 20$ kHz, $\Delta I_f = 2.5\%$, and $\Delta V_{DC} = 8\%$.

The load current is $I_s = 50\,000/(3 \times 239.6) = 69.56$ A.

$$\Delta I_f = 2.5\%\ \text{of}\ I_s = 1.739\ \text{A},\quad \Delta V_{DC} = 8\%\ \text{of}\ V_{DC} = 16\ \text{V},\quad V_{DCmin} = 200 - \Delta V_{DC} = 184\ \text{V}.$$

$$V_{s5} = 15\%\ \text{of}\ V_s = 35.94\ \text{V},\quad V_{s7} = 10\%\ \text{of}\ V_s = 23.96\ \text{V},\quad V_{s11} = 7\%\ \text{of}\ V_s = 16.77\ \text{V},$$

$$V_{s13} = 5\%\ \text{of}\ V_s = 11.98\ \text{V}.$$

a. The voltage and current ratings of the series APF are $V_f = \sqrt{(V_{s5}^2 + V_{s7}^2 + V_{s11}^2 + V_{s13}^2)} = 47.86$ V and $I_f = 69.56$ A.
b. The VA rating of the VSC of the APF is $S = 3V_f I_s = 3 \times 47.86 \times 69.56\ \text{VA} = 9.9875$ kVA.
c. The interfacing inductance of the APF is $L_f = (N_{VSI}/N_{supply})(\sqrt{3}/2)m_a V_{DCAPF}/(6af_s\Delta I_f) = 1.477 \times (\sqrt{3}/2) \times 0.8 \times 200/(6 \times 1.2 \times 20\,000 \times 1.739) = 0.817$ mH.
d. The turns ratio of the coupling transformer is computed as follows.

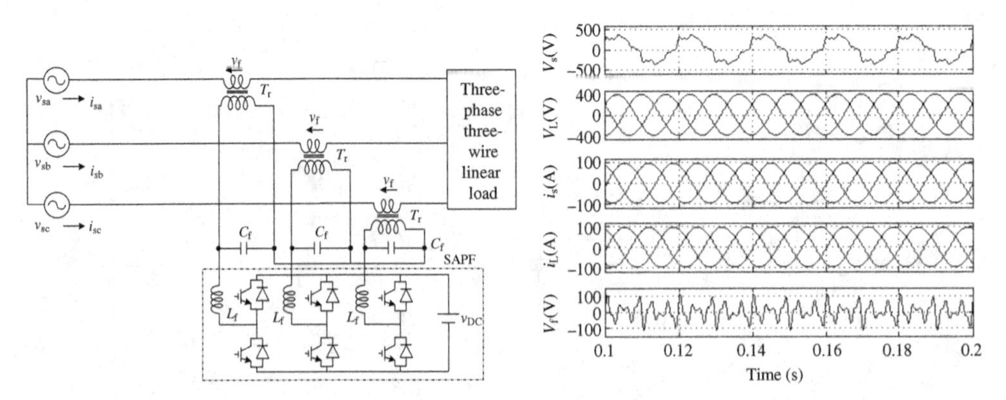

Figure E10.8 A three-phase APF for compensation of a voltage-fed nonlinear load and its waveforms

The maximum AC voltage on the AC side of the VSC of the APF may be $V_{DC}/(2\sqrt{2}) = 200/(2\times\sqrt{2}) = 70.71$ V and on the supply side it must be $V_{supply} = V_f$. The turns ratio of the coupling transformer is $N_{VSI}/N_{supply} = 70.71/47.86 = 1.477$.

Example 10.9

A three-phase series active power filter (consisting of a VSC with an AC series inductor, a coupling transformer, and a DC bus capacitor) (shown in Figure E10.9) is used in series with three-phase AC supply of 415 V at 50 Hz feeding a three-phase diode rectifier used for charging a battery of 540 V at 50 A average current to reduce the harmonics in AC mains current and to maintain almost UPF. Calculate (a) rms line voltage at the input of the diode rectifier (almost quasi-square wave), (b) AC line current, (c) rms current of the APF, (d) rms voltage across the APF (e) VA rating of the APF, (f) AC inductor value, and (g) turns ratio of the coupling transformer if the DC bus voltage of the APF is 400 V. Consider the switching frequency of 20 kHz and ripple current in the inductor is 2%.

Solution: Given supply voltage $V_s = 415/\sqrt{3} = 239.6$ V, frequency of the supply $(f) = 50$ Hz, $V_{DCload} = E = 540$ V, $I_{DC} = 50$ A, $f_s = 20$ kHz, $\Delta I_f = 2\%$, $\Delta V_{DC} = 8\%$, and $V_{DCAPF} = 400$ V.

The active power is $P = V_{DC}I_{DC} = 540 \times 50$ W $= 27\,000$ W.

The supply current is $I_s = 27\,000/(3 \times 239.6) = 37.563$ A.

$$\Delta I_f = 2\% \text{ of } I_s = 0.75125 \text{ A}, \quad \Delta V_{DC} = 8\% \text{ of } V_{DC} = 32 \text{ V}, \quad V_{DCmin} = 400 - \Delta V_{DC} = 368 \text{ V}.$$

The sine wave supply current after compensation results in continuous conduction of diodes of the three-phase diode rectifier (each diode conducting for 180°) and it results in a stepped waveform of the phase voltage at the input of the diode (V_{PCCph}) with (i) the first step of $\pi/3$ angle (from 0° to $\pi/3$) and a magnitude of $V_{DCload}/3$, (ii) the second step of $\pi/3$ angle (from $\pi/3$ to $2\pi/3$) and a magnitude of $2V_{DCload}/3$, and (iii) the third step of $\pi/3$ angle (from $2\pi/3$ to π) and a magnitude of $V_{DCload}/3$, and it has both half cycles of symmetric segments of such steps.

a. The rms line voltage at the input of the diode rectifier is computed as follows.

   The sine wave supply current after compensation results in continuous conduction of diodes of the rectifier (180°) and it results in quasi-square wave AC voltage at PCC with an amplitude of DC bus voltage. Therefore, $V_{PCC} = V_{DCload}(\sqrt{2}/\sqrt{3}) = 440.91$ V.
b. The supply line current at unity power factor is $I_s = 27\,000/(3 \times 239.6) = 37.563$ A.
c. The current rating of the series APF is $I_f = I_s = 37.563$ A (since the APF is connected in series with the supply).

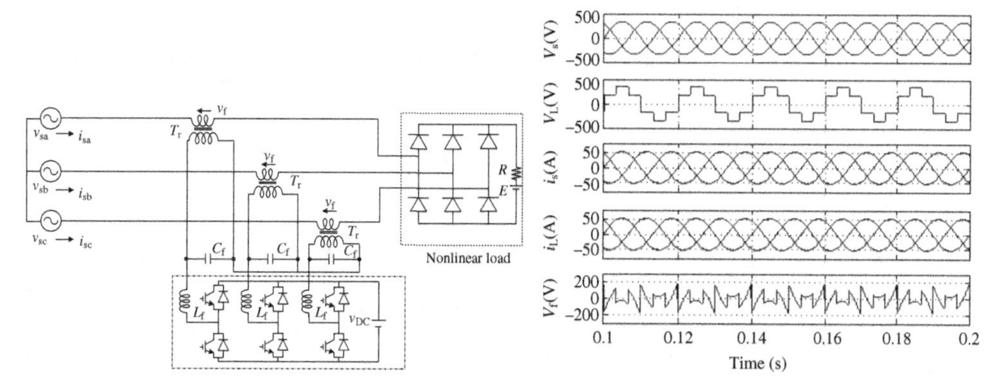

Figure E10.9 A three-phase APF for compensation of a voltage-fed nonlinear load and its waveforms

d. The rms voltage rating of the series APF is computed by taking the difference of the supply phase voltage and phase voltage at the input of the diode rectifier as follows:

$$V_f = \sqrt{\left[(1/\pi)\left\{\int_0^{\pi/3}(239.6\sqrt{2}\sin\theta - 180)^2\, d\theta + \int_{\pi/3}^{2\pi/3}(239.6\sqrt{2}\sin\theta - 360)^2\, d\theta + \int_{2\pi/3}^{\pi}(239.6\sqrt{2}\sin\theta - 180)^2\, d\theta\right\}\right]}$$

$$= 75.6415\ \text{V}.$$

e. The VA rating of the VSC of the APF is $S = 3V_fI_s = 3 \times 75.6415 \times 37.563$ VA $= 8524$ VA.
f. The interfacing inductance of the APF is $L_f = (N_{VSI}/N_{supply})(\sqrt{3}/2)m_aV_{DCAPF}/(6af_s\Delta I_f) = 1.4957 \times (\sqrt{3}/2) \times 0.8 \times 400/(6 \times 1.2 \times 20\,000 \times 0.751\,25) = 3.83$ mH.
g. The turns ratio of the coupling transformer is computed as follows.
The maximum AC voltage on the AC side of the VSC of the APF may be $m_aV_{DC}/(2\sqrt{2}) = 0.8 \times 400/(2 \times \sqrt{2}) = 113.137$ V and on the supply side it must be $V_{supply} = V_f$. The turns ratio of the coupling transformer is $N_{VSI}/N_{supply} = 113.137/75.6415 = 1.4957$.

Example 10.10

A three-phase 415 V, 50 Hz, six-pulse diode bridge rectifier with a DC capacitor filter is feeding a DC of 600 V at 150 A average current to a variable-frequency VSI-fed induction motor drive. A three-phase series active power filter (consisting of a VSC with an AC series inductor, a coupling transformer, and a DC bus capacitor) (shown in Figure E10.10) is used in series with this rectifier–inverter system to reduce the harmonics in AC mains current, to maintain almost UPF, and to regulate the DC bus voltage of the rectifier to 600 V. Calculate (a) rms line voltage at the input of the diode rectifier, (b) line current, (c) rms current of the APF, (d) rms voltage across the APF, and (e) VA rating of the APF.

Solution: Given supply voltage $V_s = 415/\sqrt{3} = 239.6$ V, frequency of the supply $(f) = 50$ Hz, $V_{DCload} = 600$ V, and $I_{DC} = 150$ A.
The active power is $P = V_{DC}I_{DC} = 600 \times 150$ W $= 90\,000$ W.
The supply current is $I_s = 90\,000/(3 \times 239.6) = 125.21$ A.

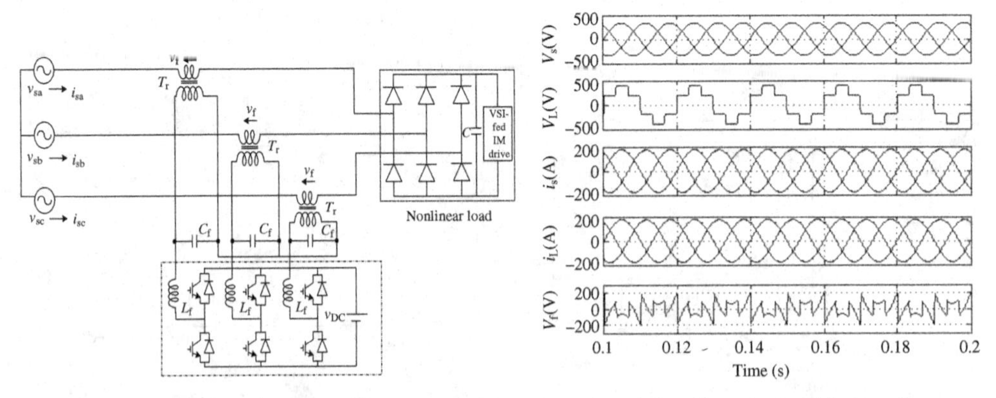

Figure E10.10 A three-phase APF for compensation of a voltage-fed nonlinear load and its waveforms

The sine wave supply current after compensation results in continuous conduction of diodes of the three-phase diode rectifier (each diode conducting for 180°) and it results in a stepped waveform of the phase voltage at the input of the diode (V_{PCCph}) with (i) the first step of $\pi/3$ angle (from 0° to $\pi/3$) and a magnitude of $V_{DCload}/3$, (ii) the second step of $\pi/3$ angle (from $\pi/3$ to $2\pi/3$) and a magnitude of $2V_{DCload}/3$, and (iii) the third step of $\pi/3$ angle (from $2\pi/3$ to π) and a magnitude of $V_{DCload}/3$, and it has both half cycles of symmetric segments of such steps.

a. The rms line voltage at the input of the diode rectifier is computed as follows.
   The sine wave supply current after compensation results in continuous conduction of diodes of the rectifier (180°) and it results in quasi-square wave AC voltage at PCC with an amplitude of DC bus voltage. Therefore, $V_{PCC} = V_{DCload}(\sqrt{2}/\sqrt{3}) = 489.89$ V.
b. The supply line current at unity power factor is $I_s = 90\ 000/(3 \times 239.6) = 125.21$ A.
c. The current rating of the series APF is $I_f = I_s = 125.21$ A (since the APF is connected in series with the supply).
d. The rms voltage rating of the series APF is computed by taking the difference of the supply phase voltage and phase voltage at the input of the diode rectifier as follows:

$$V_f = \sqrt{\left[(1/\pi)\left\{\int_0^{\pi/3} (239.6\sqrt{2}\sin\theta - 200)^2\, d\theta + \int_{\pi/3}^{2\pi/3} (239.6\sqrt{2}\sin\theta - 400)^2\, d\theta + \int_{2\pi/3}^{\pi} (239.6\sqrt{2}\sin\theta - 200)^2\, d\theta\right\}\right]}$$

$$= 89.323 \text{ V.}$$

e. The VA rating of the VSC of the APF is $S = 3V_f I_s = 3 \times 89.323 \times 125.21$ VA $= 33.552$ kVA.

Example 10.11

A three-phase series active power filter (consisting of a VSC with an AC series inductor, a coupling transformer, and its DC bus connected to a battery functioning as the load) is used in series with three-phase AC supply of 440 V at 50 Hz feeding a three-phase diode bridge rectifier used for charging a battery of 580 V at 60 A average current to reduce the harmonics in AC mains current and to maintain almost UPF as shown in Figure E10.11. Calculate (a) rms line voltage at the input of the diode rectifier,

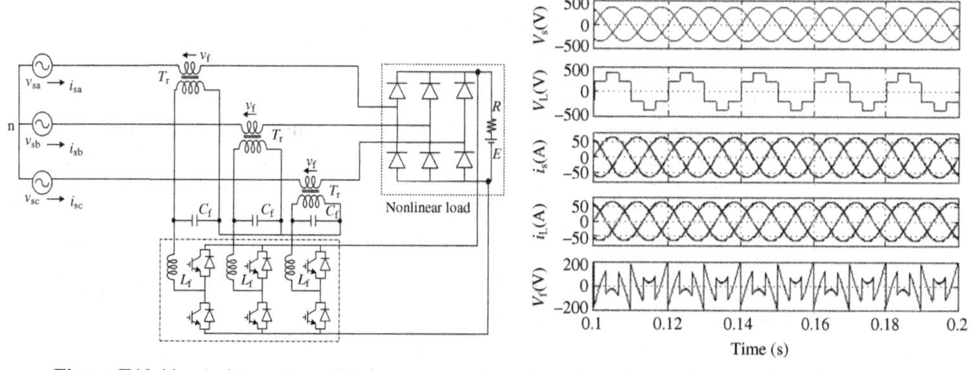

Figure E10.11 A three-phase APF for compensation of a voltage-fed nonlinear load and its waveforms

(b) AC line current, (c) rms current of the APF, (d) rms voltage across the APF, and (e) VA rating of the APF.

Solution: Given supply voltage $V_s = 440/\sqrt{3} = 254.03$ V, frequency of the supply (f) = 50 Hz, V_{DCload} = 580 V, and $I_{DC} = 60$ A.

The active power is $P = V_{DC}I_{DC} = 580 \times 60$ W = 34 800 W.

The supply current is $I_s = 34\ 800/(3 \times 254.03) = 45.664$ A.

The sine wave supply current after compensation results in continuous conduction of diodes of the three-phase diode rectifier (each diode conducting for 180°) and it results in a stepped waveform of the phase voltage at the input of the diode (V_{PCCph}) with (i) the first step of $\pi/3$ angle (from 0° to $\pi/3$) and a magnitude of $V_{DCload}/3$, (ii) the second step of $\pi/3$ angle (from $\pi/3$ to $2\pi/3$) and a magnitude of $2V_{DCload}/3$, and (iii) the third step of $\pi/3$ angle (from $2\pi/3$ to π) and a magnitude of $V_{DCload}/3$, and it has both half cycles of symmetric segments of such steps.

a. The rms line voltage at the input of the diode rectifier is computed as follows.

   The sine wave supply current after compensation results in continuous conduction of diodes of the rectifier (180°) and it results in quasi-square wave AC voltage at PCC with an amplitude of DC bus voltage. Therefore, $V_{PCC} = V_{DCload}(\sqrt{2}/\sqrt{3}) = 473.57$ V.
b. The supply line current at unity power factor is $I_s = 34\ 800/(3 \times 254.03) = 45.664$ A
c. The current rating of the series APF is $I_f = I_s = 45.664$ A (since the APF is connected in series with the supply).
d. The rms voltage rating of the series APF is computed by taking the difference of the supply phase voltage and phase voltage at the input of the diode rectifier as follows:

$$V_f = \sqrt{\left[(1/\pi)\left\{\int_0^{\pi/3}(254\sqrt{2}\sin\theta - 193.3)^2\,d\theta + \int_{\pi/3}^{2\pi/3}(254\sqrt{2}\sin\theta - 386.67)^2\,d\theta \right.\right.}$$

$$\left.\left. + \int_{2\pi/3}^{\pi}(254\sqrt{2}\sin\theta - 193.3)^2\,d\theta\right\}\right]$$

$$= 81.4622\ \text{V}.$$

e. The VA rating of the VSC of the APF is $S = 3V_fI_s = 3 \times 81.4622 \times 45.664$ VA = 11.16 kVA.

Example 10.12

A three-phase series active power filter (consisting of a VSC with an AC series inductor, a coupling transformer, and its DC bus connected to the DC bus of the load) is used in series with the three-phase AC supply of 415 V at 50 Hz feeding a three-phase diode rectifier with a parallel DC capacitor at a load of 50 A average current to reduce the harmonics in AC mains current, to maintain UPF of the supply current, and to regulate the DC bus voltage of the load at almost constant 650 V as shown in Figure E10.12. Calculate (a) rms line voltage at the input of the diode rectifier, (b) AC line current, (c) rms current of the APF, (d) rms voltage across the APF, (e) VA rating of the APF, and (f) turns ratio of the coupling transformer.

Solution: Given supply voltage $V_s = 415/\sqrt{3} = 239.6$ V, frequency of the supply (f) = 50 Hz, V_{DC} = 650 V, and $I_{DC} = 50$ A.

The active power is $P = V_{DC}I_{DC} = 650 \times 50$ W = 32 500 W.

The supply current is $I_s = 32\ 500/(3 \times 239.6) = 45.21$ A.

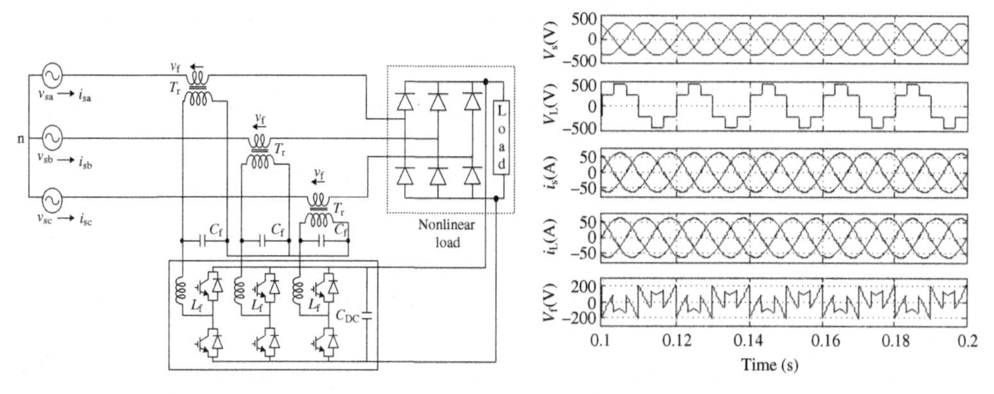

Figure E10.12 A three-phase APF for compensation of a voltage-fed nonlinear load and its waveforms

The sine wave supply current after compensation results in continuous conduction of diodes of the three-phase diode rectifier (each diode conducting for 180°) and it results in a stepped waveform of the phase voltage at the input of the diode (V_{PCCph}) with (i) the first step of $\pi/3$ angle (from 0° to $\pi/3$) and a magnitude of $V_{DC}/3$, (ii) the second step of $\pi/3$ angle (from $\pi/3$ to $2\pi/3$) and a magnitude of $2V_{DC}/3$, and (iii) the third step of $\pi/3$ angle (from $2\pi/3$ to π) and a magnitude of $V_{DC}/3$, and it has both half cycles of symmetric segments of such steps.

a. The rms line voltage at the input of the diode rectifier is computed as follows.
   The sine wave supply current after compensation results in continuous conduction of diodes of the rectifier (180°) and it results in quasi-square wave AC voltage at PCC with an amplitude of DC bus voltage. Therefore, $V_{PCC} = V_{DC}(\sqrt{2}/\sqrt{3}) = 530.72$ V.
b. The supply line current at unity power factor is $I_s = 32\,500/(3 \times 239.6) = 45.21$ A
c. The current rating of the series APF is $I_f = I_s = 45.21$ A (since the APF is connected in series with the supply).
d. The rms voltage rating of the series APF is computed by taking the difference of the supply phase voltage and phase voltage at the input of the diode rectifier as follows:

$$V_f = \sqrt{\left[(1/\pi)\left\{\int_0^{\pi/3}(239.6\sqrt{2}\sin\theta - 216.7)^2\,d\theta + \int_{\pi/3}^{2\pi/3}(239.6\sqrt{2}\sin\theta - 433.33)^2\,d\theta + \int_{2\pi/3}^{\pi}(239.6\sqrt{2}\sin\theta - 216.7)^2\,d\theta\right\}\right]}$$

$$= 105.27\text{ V}.$$

e. The VA rating of the VSC of the APF is $S = 3V_f I_s = 3 \times 105.27 \times 45.21$ VA $= 14.278$ kVA.
f. The turns ratio of the coupling transformer is computed as follows.
   The maximum AC voltage on the AC side of the VSC of the APF may be $m_a V_{DC}/(2\sqrt{2}) = 0.8 \times 650/(2 \times \sqrt{2}) = 183.85$ V and on the supply side it must be $V_{supply} = V_f$. The turns ratio of the coupling transformer is $N_{VSI}/N_{supply} = 183.85/105.27 = 1.746$.

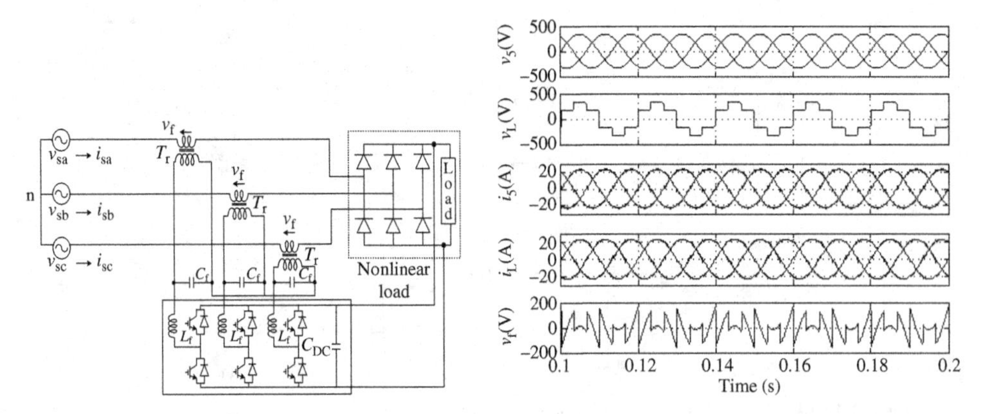

Figure E10.13 A three-phase APF for compensation of a voltage-fed nonlinear load and its waveforms

Example 10.13

A series active power filter (consisting of a VSC with an AC series inductor, a coupling transformer, and its DC bus connected to the DC bus of the load) is used to reduce the harmonics in AC mains current and to maintain almost UPF in the series of three-phase AC supply of 415 V at 50 Hz feeding a three-phase diode rectifier with a capacitive filter of 10 000 μF and a resistive load of 20 Ω as shown in Figure E10.13. The DC bus voltage of the load is decided to result in minimum injected voltage of the APF. Calculate (a) rms voltage at the input of the diode rectifier, (b) line current, (c) VA rating of the APF, (d) DC bus voltage of the load, and (e) turns ratio of the coupling transformer.

Solution: Given supply voltage $V_s = 415/\sqrt{3} = 239.6$ V, frequency of the supply $(f) = 50$ Hz, and $R_{DC} = 20\,\Omega$.

If x volts is the DC bus voltage of the rectifier load, then the injected voltage of the APF is

$$V_f = \sqrt{\left[(1/\pi)\left\{\int_0^{\pi/3}(V_s\sqrt{2}\sin\theta - x/3)^2\,d\theta + \int_{\pi/3}^{2\pi/3}(V_s\sqrt{2}\sin\theta - 2x/3)^2\,d\theta + \int_{2\pi/3}^{\pi}(V_s\sqrt{2}\sin\theta - x/3)^2\,d\theta\right\}\right]}.$$

After integrating and equating the derivative with respect to x to zero, it gives $x = (\sqrt{2}/\pi)(27/6)V_s = V_{DC} = 485.36$ V.

The DC load current is $I_{DC} = V_{DC}/R_{DC} = 485.36/20 = 24.268$ A.

The active power is $P = V_{DC}I_{DC} = 485.36 \times 24.268$ W $= 11\ 778.747$ W.

The supply current is $I_s = 11\ 778.747/(3 \times 239.6) = 16.39$ A.

The sine wave supply current after compensation results in continuous conduction of diodes of the three-phase diode rectifier (each diode conducting for 180°) and it results in a stepped waveform of the phase voltage at the input of the diode (V_{PCCph}) with (i) the first step of $\pi/3$ angle (from 0° to $\pi/3$) and a magnitude of $V_{DC}/3$, (ii) the second step of $\pi/3$ angle (from $\pi/3$ to $2\pi/3$) and a magnitude of $2V_{DC}/3$, and (iii) the third step of $\pi/3$ angle (from $2\pi/3$ to π) and a magnitude of $V_{DC}/3$, and it has both half cycles of symmetric segments of such steps.

a. The rms line voltage at the input of the diode rectifier is computed as follows.

   The sine wave supply current after compensation results in continuous conduction of diodes of the rectifier (180°) and it results in quasi-square wave AC voltage at PCC with an amplitude of DC bus voltage. Therefore, $V_{PCC} = V_{DC}(\sqrt{2}/\sqrt{3}) = 396.29$ V.

b. The supply line current at unity power factor is $I_s = 16.39$ A.

The current rating of the series APF is $I_f = I_s = 16.39$ A (since the APF is connected in series with the supply).

The rms voltage rating of the series APF is computed by taking the difference of the supply phase voltage and phase voltage at the input of the diode rectifier as follows:

$$V_f = \sqrt{\left[(1/\pi)\left\{\int_0^{\pi/3}(239.6\sqrt{2}\sin\theta - 161.79)^2\, d\theta + \int_{\pi/3}^{2\pi/3}(239.6\sqrt{2}\sin\theta - 323.57)^2\, d\theta + \int_{2\pi/3}^{\pi}(239.6\sqrt{2}\sin\theta - 161.79)^2\, d\theta\right\}\right]}$$

$$= 71.12\ \text{V}.$$

c. The VA rating of the VSC of the APF is $S = 3V_f I_s = 3 \times 71.12 \times 16.39$ VA $= 3.497$ kVA.

d. The DC bus voltage of the load is $V_{DC} = x = 485.36$ V.

e. The turns ratio of the coupling transformer is computed as follows.

The maximum AC voltage on the AC side of the VSC of the APF may be $m_a V_{DC}/(2\sqrt{2}) = 0.8 \times 485.36/(2 \times \sqrt{2}) = 137.28$ V and on the supply side it must be $V_{supply} = V_f$. The turns ratio of the coupling transformer is $N_{VSI}/N_{supply} = 137.28/71.12 = 1.93$.

Example 10.14

A series active power filter (consisting of a VSC with an AC series inductor, a coupling transformer, and its DC bus connected to the DC bus of the load) is used to reduce the harmonics in AC mains current and to maintain almost UPF in the series of three-phase AC supply of 400 V at 50 Hz feeding a three-phase diode rectifier with a capacitive filter of 40 000 μF and a resistive load of 10 Ω as shown in Figure E10.14. If DC bus voltage of the load is to be maintained at constant ripple-free 600 V, then calculate (a) rms line voltage at the input of the diode rectifier, (b) AC line current, (c) rms current of the APF, (d) rms voltage across the APF, (e) VA rating of the APF, and (f) turns ratio of the coupling transformer.

Solution: Given supply voltage $V_s = 400/\sqrt{3} = 230.94$ V, frequency of the supply $(f) = 50$ Hz, $V_{DC} = 600$ V, and $R_{DC} = 10\ \Omega$.

The DC load current is $I_{DC} = V_{DC}/R_{DC} = 600/10 = 60$ A.

The active power is $P = V_{DC} I_{DC} = 600 \times 60$ W $= 36\ 000$ W.

The supply current is $I_s = 36\ 000/(3 \times 230.94) = 51.96$ A.

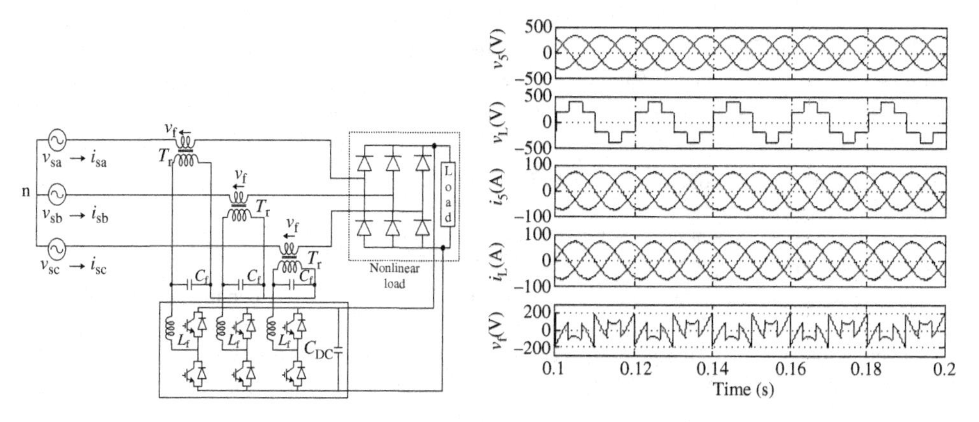

Figure E10.14 A three-phase APF for compensation of a voltage-fed nonlinear load and its waveforms

The sine wave supply current after compensation results in continuous conduction of diodes of the three-phase diode rectifier (each diode conducting for 180°) and it results in a stepped waveform of the phase voltage at the input of the diode (V_{PCCph}) with (i) the first step of $\pi/3$ angle (from 0° to $\pi/3$) and a magnitude of $V_{DC}/3$, (ii) the second step of $\pi/3$ angle (from $\pi/3$ to $2\pi/3$) and a magnitude of $2V_{DC}/3$, and (iii) the third step of $\pi/3$ angle (from $2\pi/3$ to π) and a magnitude of $V_{DC}/3$, and it has both half cycles of symmetric segments of such steps.

a. The rms line voltage at the input of the diode rectifier is computed as follows.
   The sine wave supply current after compensation results in continuous conduction of diodes of the rectifier (180°) and it results in quasi-square wave AC voltage at PCC with an amplitude of DC bus voltage. Therefore, $V_{PCC} = V_{DC}(\sqrt{2}/\sqrt{3})$= 489.89 V.
b. The supply line current at unity power factor is $I_s = 36\ 000/(3 \times 230.94) = 51.96$ A.
c. The current rating of the series APF is $I_f = I_s = 51.96$ A (since the APF is connected in series with the supply).
d. The rms voltage rating of the series APF is computed by taking the difference of the supply phase voltage and phase voltage at the input of the diode rectifier as follows:

$$V_f = \sqrt{\left[(1/\pi)\left\{\int_0^{\pi/3} (230.94\sqrt{2}\sin\theta - 200)^2\, d\theta + \int_{\pi/3}^{2\pi/3} (230.94\sqrt{2}\sin\theta - 400)^2\, d\theta + \int_{2\pi/3}^{\pi} (230.94\sqrt{2}\sin\theta - 200)^2\, d\theta\right\}\right]}$$

$$= 92.6382\ \text{V}.$$

e. The VA rating of the VSI of the APF is $S = 3V_f I_s = 3 \times 92.6382 \times 51.96$ VA $= 14.44$ kVA.
f. The turns ratio of the coupling transformer is computed as follows.
   The maximum AC voltage on the AC side of the VSC of the APF may be $m_a V_{DC}/(2\sqrt{2}) = 0.8 \times 600/(2 \times \sqrt{2}) = 169.7$ V and on the supply side it must be $V_{supply} = V_f$. The turns ratio of the coupling transformer is N_{VSI}/N_{supply} = 169.7/92.6382 = 1.83.

Example 10.15

A three-phase series active power filter (consisting of a VSC with AC series inductors, coupling transformers, and a DC bus capacitor) is used in series with the three-phase AC supply of 415 V at 50 Hz feeding a three-phase 12-pulse diode rectifier (consisting of two parallel connected 6-pulse diode bridge rectifiers with star/delta and delta/delta transformers to output equal line voltages of 415 V) having a parallel DC capacitor filter with 400 V DC at a load of 50 A average current to reduce the harmonics in AC mains current and to maintain UPF of the supply current as shown in Figure E10.15. Calculate (a) the rms phase voltage at the input of the 12-pulse diode rectifier, (b) the AC line current, (c) the rms current of the APF, (d) the rms voltage across the APF, and (e) the VA rating of the APF.

Solution: Given supply voltage $V_s = 415/\sqrt{3} = 239.6$ V, frequency of the supply (f) = 50 Hz, V_{DC} = 400 V, and I_{DC} = 50 A.

The active power is $P = V_{DC} I_{DC} = 400 \times 50$ W = 20 000 W.

The supply current is $I_s = 20\ 000/(3 \times 239.6) = 27.82$ A.

The sine wave supply current after compensation results in continuous conduction of diodes of the three-phase diode rectifier (each diode conducting for 180°) and it results in a stepped waveform of the phase voltage at the input of the 12-pulse diode rectifier (V_{PCCph}) with (i) the first step of $\pi/6$ angle (from 0° to $\pi/6$) and a magnitude of $V_{DC}(2/\sqrt{3} - 1)$, (ii) the second step of $\pi/6$ angle (from $\pi/6$ to $\pi/3$) and a

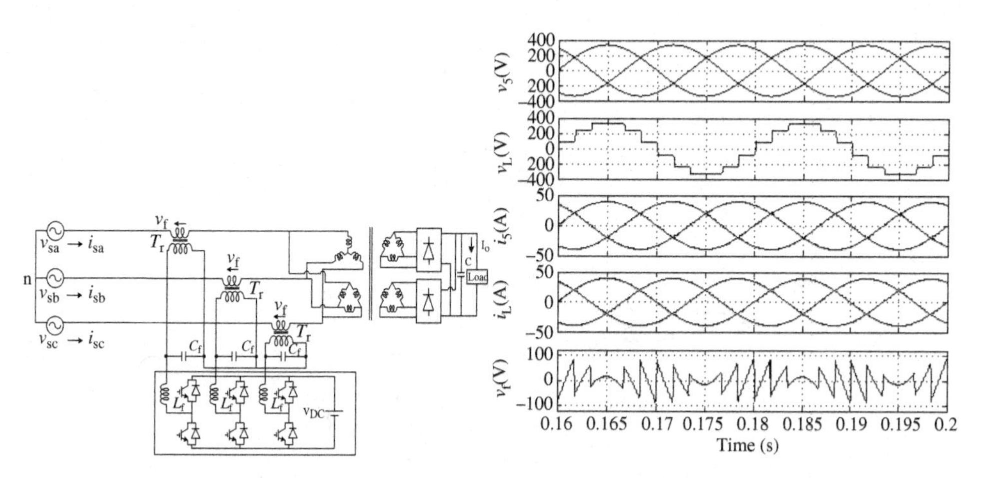

Figure E10.15 A three-phase APF for compensation of a voltage-fed nonlinear load and its waveforms

magnitude of $V_{DC}(1 - 1/\sqrt{3})$, and (iii) the third step of $\pi/6$ angle (from $\pi/3$ to $\pi/2$) and a magnitude of $V_{DC}(1/\sqrt{3})$, and it has all four symmetric segments of such steps.

The turns ratio of the transformer is calculated as follows.

$$3\sqrt{2}V_{line,sec}/\pi = 400\text{ V}, \quad V_{line,sec} = 296.19\text{ V}, \quad V_{line,pri} = 415\text{ V}.$$

Primary to secondary turns ratio of the star/delta transformer is $N_{ps}/N_{sd} = (415/\sqrt{3})/296.19 = 0.81$.
Primary to secondary turns ratio of the delta/delta transformer is $N_{pd}/N_{sd} = 415/296.19 = 1.4$.
Hence, the turns ratio of 1.4 will be multiplied with each step.

a. The rms phase voltage at the input of the 12-pulse diode rectifier is computed as follows:

$$V_{PCCph} = 1.4V_{DC}\sqrt{\frac{1}{\pi}\left[\int_0^{\pi/6}\left(\frac{2}{\sqrt{3}}-1\right)^2 d\theta + \int_{\pi/6}^{\pi/3}\left(1-\frac{1}{\sqrt{3}}\right)^2 d\theta + \int_{\pi/3}^{2\pi/3}\left(\frac{1}{\sqrt{3}}\right)^2 d\theta + \int_{2\pi/3}^{5\pi/6}\left(1-\frac{1}{\sqrt{3}}\right)^2 d\theta + \int_{5\pi/6}^{\pi}\left(\frac{2}{\sqrt{3}}-1\right)^2 d\theta\right]}$$

$$= 236.8\text{ V}.$$

b. The supply line current at unity power factor is $I_s = 20\,000/(3 \times 239.6) = 27.82$ A.
c. The current rating of the series APF is $I_f = I_s = 27.82$ A (since the APF is connected in series with the supply).
d. The rms voltage rating of the series APF is computed by taking the difference of the supply phase voltage and phase voltage at the input of the diode rectifier as follows:

$$V_f = \sqrt{\frac{1}{\pi}\left[\begin{array}{l}\int_0^{\pi/6}(239.6\sqrt{2}\sin\theta - 86.7)^2 d\theta + \int_{\pi/6}^{\pi/3}(239.6\sqrt{2}\sin\theta - 236.8)^2 d\theta + \int_{\pi/3}^{2\pi/3}(239.6\sqrt{2}\sin\theta - 323.6)^2 d\theta \\ + \int_{2\pi/3}^{5\pi/6}(239.6\sqrt{2}\sin\theta - 236.8)^2 d\theta + \int_{5\pi/6}^{\pi}(239.6\sqrt{2}\sin\theta - 86.7)^2 d\theta\end{array}\right]}$$

$$= 36\text{ V}.$$

e. The VA rating of the VSI of the APF is $S = 3V_fI_s = 3 \times 36 \times 27.82\text{ VA} = 3\text{ kVA}$.

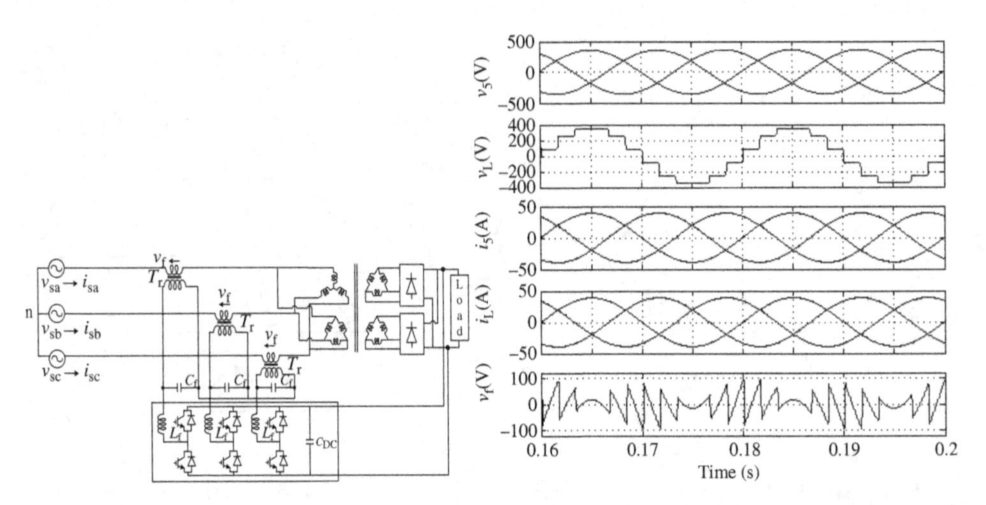

Figure E10.16 A three-phase APF for compensation of a voltage-fed nonlinear load and its waveforms

Example 10.16

A three-phase series active power filter (consisting of a VSC with AC series inductors, coupling transformers, and a DC bus connected to the DC bus of the load) is used in series with the three-phase AC supply of 440 V at 50 Hz feeding a three-phase 12-pulse diode rectifier (consisting of two parallel connected 6-pulse diode bridge rectifiers with star/delta and delta/delta transformers to output equal line voltages of 440 V) having a parallel DC capacitor filter of 10 000 μF with a 20 Ω resistive load to reduce the harmonics in AC mains current and to maintain UPF of the supply current as shown in Figure E10.16. If the DC bus of the load is to be maintained at 650 V, then calculate (a) the rms phase voltage at the input of the 12-pulse diode rectifier, (b) the AC line current, (c) the rms current of the APF, (d) the rms voltage across the APF, (e) the VA rating of the APF, and (f) turns ratio of the coupling transformer.

Solution: Given supply voltage $V_s = 440/\sqrt{3} = 254$ V, frequency of the supply (f) = 50 Hz, $V_{DC} = 650$ V, and $R_{DC} = 20\ \Omega$.

The DC load current is $I_{DC} = 650/20 = 32.5$ A.

The active power is $P = V_{DC}I_{DC} = 650 \times 32.5$ W = 21 125 W.

The supply current is $I_s = 21\,125/(3 \times 254) = 27.72$ A.

The sine wave supply current after compensation results in continuous conduction of diodes of the three-phase diode rectifier (each diode conducting for 180°) and it results in a stepped waveform of the phase voltage at the input of the 12-pulse diode rectifier (V_{PCCph}) with (i) the first step of $\pi/6$ angle (from 0° to $\pi/6$) and a magnitude of $V_{DC}(2/\sqrt{3} - 1)$, (ii) the second step of $\pi/6$ angle (from $\pi/6$ to $\pi/3$) and a magnitude of $V_{DC}(1 - 1/\sqrt{3})$, and (iii) the third step of $\pi/6$ angle (from $\pi/3$ to $\pi/2$) and a magnitude of $V_{DC}(1/\sqrt{3})$, and it has all four symmetric segments of such steps.

The turns ratio of the transformers is calculated as follows.

$$3\sqrt{2}V_{\text{line,sec}}/\pi = 650\ \text{V}, \quad V_{\text{line,sec}} = 481.312\ \text{V}, \quad V_{\text{line,pri}} = 440\ \text{V}.$$

Primary to secondary turns ratio of the star/delta transformer is $N_{ps}/N_{sd} = (440/\sqrt{3})/481.312 = 0.528$.
Primary to secondary turns ratio of the delta/delta transformer is $N_{pd}/N_{sd} = 440/481.312 = 0.914$.
Hence, the turns ratio of 0.914 will be multiplied with each step.

a. The rms phase voltage at the input of the 12-pulse diode rectifier is computed as follows:

$$V_{PCCph} = 0.914V_{DC}\sqrt{\frac{1}{\pi}\left[\int_0^{\pi/6}\left(\frac{2}{\sqrt{3}}-1\right)^2 d\theta + \int_{\pi/6}^{\pi/3}\left(1-\frac{1}{\sqrt{3}}\right)^2 d\theta + \int_{\pi/3}^{2\pi/3}\left(\frac{1}{\sqrt{3}}\right)^2 d\theta + \int_{2\pi/3}^{5\pi/6}\left(1-\frac{1}{\sqrt{3}}\right)^2 d\theta + \int_{5\pi/6}^{\pi}\left(\frac{2}{\sqrt{3}}-1\right)^2 d\theta\right]}$$

$$= 251.1\text{ V}.$$

b. The supply line current at unity power factor is $I_s = 21\,125/(3\times 254) = 27.72$ A.
c. The current rating of the series APF is $I_f = I_s = 27.72$ A (since the APF is connected in series with the supply).
d. The rms voltage rating of the series APF is computed by taking the difference of the supply phase voltage and phase voltage at the input of the diode rectifier as follows:

$$V_f = \sqrt{\frac{1}{\pi}\left[\begin{array}{l}\int_0^{\pi/6}(254\sqrt{2}\sin\theta - 91.91)^2 d\theta + \int_{\pi/6}^{\pi/3}(254\sqrt{2}\sin\theta - 251.1)^2 d\theta + \int_{\pi/3}^{2\pi/3}(254\sqrt{2}\sin\theta - 343)^2 d\theta \\ + \int_{2\pi/3}^{5\pi/6}(254\sqrt{2}\sin\theta - 251.1)^2 d\theta + \int_{5\pi/6}^{\pi}(254\sqrt{2}\sin\theta - 91.91)^2 d\theta\end{array}\right]}$$

$$= 38.22\text{ V}.$$

e. The VA rating of the VSC of the APF is $S = 3V_f I_s = 3\times 38.22\times 27.72$ VA $= 3.18$ kVA.
f. The turns ratio of the coupling transformer is computed as follows.
   The maximum AC voltage on the AC side of the VSC of the APF may be $m_a V_{DC}/(2\sqrt{2}) = 0.8\times 650/(2\times\sqrt{2}) = 183.85$ V and on the supply side it must be $V_{supply} = V_f$. The turns ratio of the coupling transformer is $N_{VSI}/N_{supply} = 183.85/38.22 = 4.81$.

Example 10.17

A three-phase series active power filter (consisting of a VSC with AC series inductors and its DC bus connected to a battery behaving as the load) is connected in series with the three-phase AC supply of 415 V at 50 Hz feeding a three-phase 12-pulse diode rectifier (consisting of two parallel connected 6-pulse diode bridge rectifiers with star/delta and delta/delta transformers to output equal line voltages of 415 V) charging a battery of 500 V at 100 A average current as shown in Figure E10.17. The series APF is employed to reduce the harmonics in AC mains current and to maintain almost UPF of the supply current. Calculate (a) the rms phase voltage at the input of the 12-pulse diode rectifier, (b) the AC line current, (c) the rms current of the APF, (d) the rms voltage across the APF, (e) the VA rating of the APF, and (f) turns ratio of the coupling transformer.

Solution: Given supply voltage $V_s = 415/\sqrt{3} = 239.6$ V, frequency of the supply (f) = 50 Hz, $V_{DC} = 500$ V, and $I_{DC} = 100$ A.

The active power is $P = V_{DC}I_{DC} = 500\times 100$ W $= 50\,000$ W.

The supply current is $I_s = 50\,000/(3\times 239.6) = 69.56$ A

The sine wave supply current after compensation results in continuous conduction of diodes of the three-phase diode rectifier (each diode conducting for 180°) and it results in a stepped waveform of the phase voltage at the input of the 12-pulse diode rectifier (V_{PCCph}) with (i) the first step of $\pi/6$ angle (from 0° to $\pi/6$) and a magnitude of $V_{DC}(2/\sqrt{3}-1)$, (ii) the second step of $\pi/6$ angle (from $\pi/6$ to $\pi/3$) and a

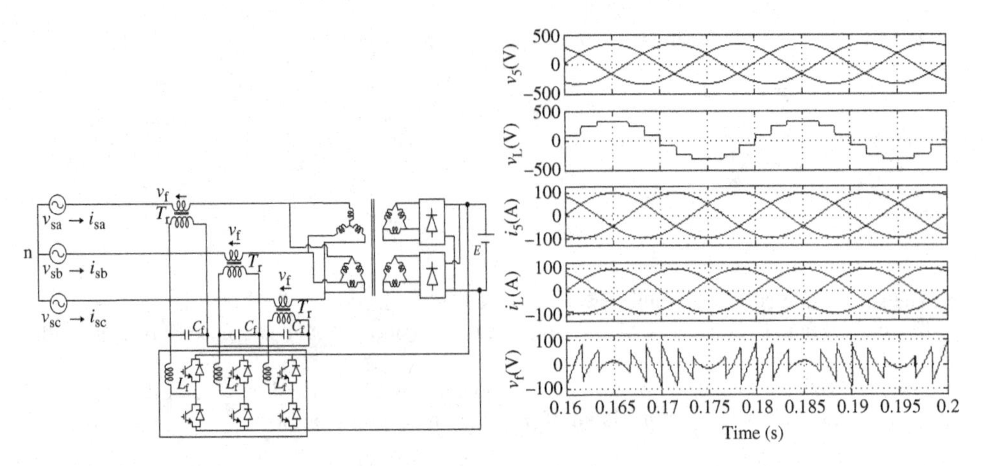

Figure E10.17 A three-phase APF for compensation of a voltage-fed nonlinear load and its waveforms

magnitude of $V_{DC}(1 - 1/\sqrt{3})$, and (iii) the third step of $\pi/6$ angle (from $\pi/3$ to $\pi/2$) and a magnitude of $V_{DC}(1/\sqrt{3})$, and it has all four symmetric segments of such steps.

The turns ratio of the transformers is calculated as follows.

$$3\sqrt{2}V_{\text{line,sec}}/\pi = 500\text{ V},\quad V_{\text{line,sec}} = 370.24\text{ V},\quad V_{\text{line,pri}} = 415\text{ V}.$$

Primary to secondary turns ratio of the star/delta transformer is $N_{ps}/N_{sd} = (415/\sqrt{3})/370.24 = 0.647$.
Primary to secondary turns ratio of the delta/delta transformer is $N_{pd}/N_{sd} = 415/370.24 = 1.121$.
Hence, the turns ratio of 1.121 will be multiplied with each step.

a. The rms phase voltage at the input of the 12-pulse diode rectifier is computed as follows:

$$V_{PCCph} = 1.121V_{DC}\sqrt{\frac{1}{\pi}\left[\int_0^{\pi/6}\left(\frac{2}{\sqrt{3}}-1\right)^2 d\theta + \int_{\pi/6}^{\pi/3}\left(1-\frac{1}{\sqrt{3}}\right)^2 d\theta + \int_{\pi/3}^{2\pi/3}\left(\frac{1}{\sqrt{3}}\right)^2 d\theta + \int_{2\pi/3}^{5\pi/6}\left(1-\frac{1}{\sqrt{3}}\right)^2 d\theta + \int_{5\pi/6}^{\pi}\left(\frac{2}{\sqrt{3}}-1\right)^2 d\theta\right]}$$

$$= 236.87\text{ V}.$$

b. The supply line current at unity power factor is $I_s = 50\,000/(3 \times 239.6) = 69.56$ A.
c. The current rating of the series APF is $I_f = I_s = 69.56$ A (since the APF is connected in series with the supply).
d. The rms voltage rating of the series APF is computed by taking the difference of the supply phase voltage and phase voltage at the input of the diode rectifier as follows:

$$V_f = \sqrt{\frac{1}{\pi}\left[\begin{aligned}&\int_0^{\pi/6}(239.6\sqrt{2}\sin\theta - 86.7)^2 d\theta + \int_{\pi/6}^{\pi/3}(239.6\sqrt{2}\sin\theta - 236.8)^2 d\theta + \int_{\pi/3}^{2\pi/3}(239.6\sqrt{2}\sin\theta - 323.6)^2 d\theta\\ &+ \int_{2\pi/3}^{5\pi/6}(239.6\sqrt{2}\sin\theta - 236.8)^2 d\theta + \int_{5\pi/6}^{\pi}(239.6\sqrt{2}\sin\theta - 86.7)^2 d\theta\end{aligned}\right]}$$

$$= 36\text{ V}.$$

e. The VA rating of the VSI of the APF is $S = 3V_f I_s = 3 \times 36 \times 69.56\,\text{VA} = 7.5\,\text{kVA}$.
f. The turns ratio of the coupling transformer is computed as follows.
   The maximum AC voltage on the AC side of the VSC of the APF may be $m_a V_{DC}/(2\sqrt{2}) = 0.8 \times 500/(2 \times \sqrt{2}) = 141.42\,\text{V}$. The turns ratio of the coupling transformer is $N_{VSI}/N_{supply} = 141.42/36 = 3.92$.

Example 10.18

In a three-phase, line voltage of 380 V, 50 Hz, four-wire distribution system, three single-phase loads (connected between phases and neutral) have a set of single-phase uncontrolled diode bridge converter, which has a *RE* load with $R = 10\,\Omega$ and $E = 204$ V as shown in Figure E10.18. It is desired to charge the battery *E* with an average current of 10 A. If a three-phase series active power filter (consisting of a VSC with AC series inductors, coupling transformers, and its DC bus connected to the DC bus capacitor) is connected in series with the three-phase supply lines to maintain UPF at the AC mains, calculate (a) supply current, (b) the rms current of the APF, (c) the rms voltage at the input of the rectifier, (d) the rms voltage across the APF, (e) the VA rating of the APF, and (f) neutral current.

Solution: Given supply phase voltage $V_s = 380/\sqrt{3} = 219.39$ V, frequency of the supply $(f) = 50$ Hz, a load of $R = 10\,\Omega$, $E = 204$ V, and $I_{DC} = 10$ A.

A series active filter is used to make the supply current sinusoidal. The rms supply current corresponding to average current of I_{DC} is $I_s = \pi I_{DC}/(2\sqrt{2}) = 11.11$ A.

The active power is $P = 3(I_s^2 R + E I_{DC}) = 3(11.11^2 \times 10 + 10 \times 204) = 3 \times 3273.70 = 9823$ W.

The supply current is $I_s = 11.11$ A.

a. The supply line current at unity power factor is $I_s = 11.11$ A.
b. The current rating of the series APF is $I_f = I_s = 11.11$ A (since the APF is connected in series with the supply).
c. The rms voltage at the input of the diode rectifier is computed as follows.
   The sine wave supply current after compensation results in continuous conduction of diodes of the rectifier (180°) and it results in a sum of square wave AC voltage of *E* and drop in *R* as follows: $V_{PCC} = \sqrt{\{E^2 + (I_s R)^2\}} = 232.29$ V.

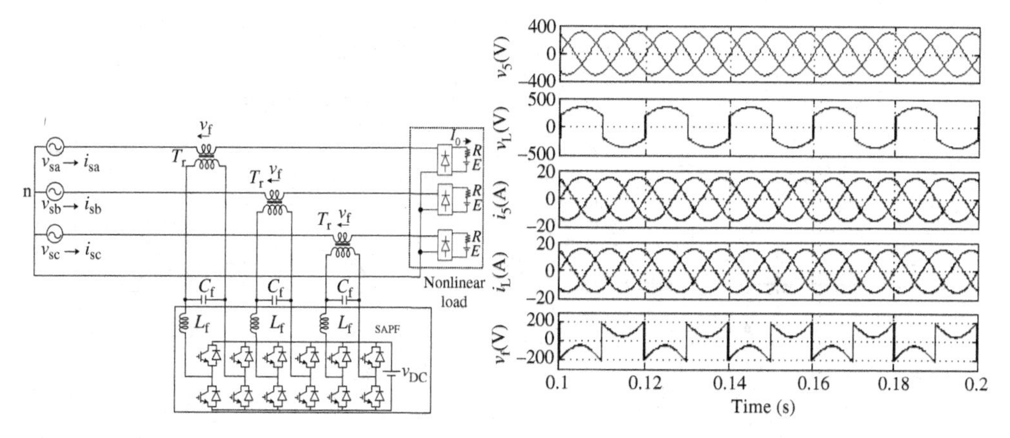

Figure E10.18 A three-phase series APF for compensation of a voltage-fed nonlinear load and its waveforms

d. The voltage rating of the series APF is

$$V_f = V_{frms} = \sqrt{\left\{(1/\pi)\int_0^{\pi}(219.39\sqrt{2}\sin\theta - 204 - 111.1\sqrt{2}\sin\theta)^2\,d\theta\right\}}$$

$$= \sqrt{\left\{(1/\pi)\int_0^{\pi}(108.29\sqrt{2}\sin\theta - 204)^2\,d\theta\right\}} = 116.46\text{ V}.$$

e. The VA rating of the VSC of the APF is $S = 3V_f I_s = 3 \times 116.46 \times 11.11$ VA $= 3.88$ kVA.
f. Since all three-phase currents are sinusoidal and balanced, the neutral current of the supply is zero.

Example 10.19

In a three-phase, line voltage of 415 V, 50 Hz, four-wire distribution system, three single-phase loads (connected between phases and neutral) have a set of single-phase uncontrolled diode bridge converter, which has a *RE* load with $R = 2\,\Omega$ and a battery voltage $E = 264$ V as shown in Figure E10.19. It is desired to charge the battery with average current of 15 A. If a three-phase series active power filter (a three single-phase VSC with AC series inductors, coupling transformers, and its DC bus connected to the battery) is connected in series with the three-phase supply lines to maintain UPF at the AC mains, calculate (a) the AC line current, (b) the rms current of the APF, (c) the rms phase voltage at the input of the diode rectifier, (d) the rms voltage across the APF, (e) the VA rating of the APF, and (f) turns ratio of the coupling transformer.

Solution: Given supply phase voltage $V_s = 415/\sqrt{3} = 239.6$ V, frequency of the supply $(f) = 50$ Hz, a load of $R = 2\,\Omega$, $E = 264$ V, and $I_{DC} = 15$ A.

A series active filter is used to make the supply current sinusoidal in each phase. The rms supply current corresponding to average current of I_{DC} is $I_s = \pi I_{DC}/(2\sqrt{2}) = 16.66$ A.

The active power is $P = 3(I_s^2 R + E I_{DC}) = 3(16.66^2 \times 2 + 15 \times 264) = 3 \times 4515 = 13\,545$ W.

a. The rms supply line current at unity power factor is $I_s = 16.66$ A.
b. The rms current rating of the series APF is $I_f = I_s = 16.66$ A (since the AF is connected in series with the supply).
c. The rms voltage at the input of the diode rectifier is computed as follows.

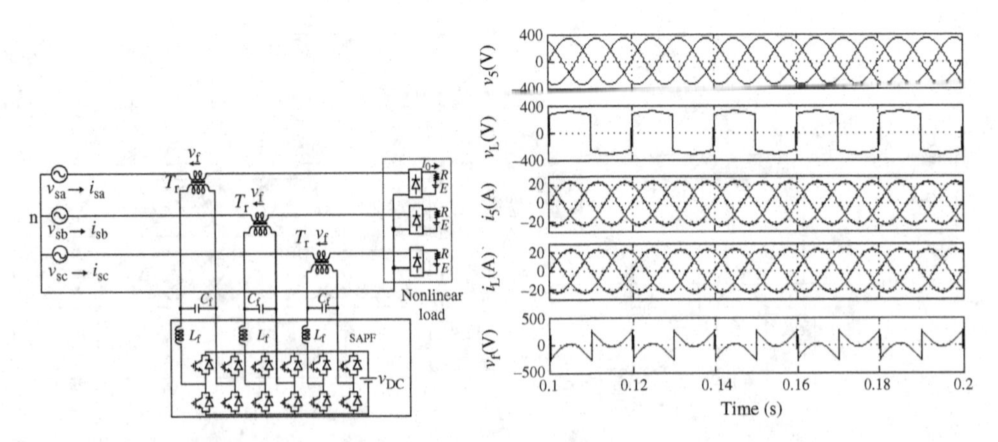

Figure E10.19 A three-phase APF for compensation of a voltage-fed nonlinear load and its waveforms

The sine wave supply current after compensation results in continuous conduction of diodes of the rectifier (180°) and it results in a sum of square wave AC voltage of E and drop in R as follows: $V_{PCC} = \sqrt{\{E^2 + (I_s R)^2\}} = 266.09$ V.

d. The voltage rating of the series APF is

$$V_f = V_{frms} = \sqrt{\left\{(1/\pi)\int_0^{\pi}(239.6\sqrt{2}\sin\theta - 264 - 33.32\sqrt{2}\sin\theta)^2\, d\theta\right\}}$$

$$= \sqrt{\left\{(1/\pi)\int_0^{\pi}(206.28\sqrt{2}\sin\theta - 264)^2\, d\theta\right\}} = 119.1164 \text{ V}.$$

e. The VA rating of the VSC of the APF is $S = 3V_f I_s = 3 \times 119.1164 \times 16.66$ VA $= 5953.4$ VA.

f. The turns ratio of the coupling transformer is computed as follows.

The maximum AC voltage on the AC side of the VSC of the APF may be $m_a E/\sqrt{2} = 0.8 \times 264/(2 \times \sqrt{2}) = 74.67$ V and on the supply side it must be $V_{supply} = V_f$. The turns ratio of the coupling transformer is $N_{VSI}/N_{supply} = 74.67/119.1164 = 0.6268$.

Example 10.20

In a three-phase, line voltage of 380 V, 50 Hz, four-wire distribution system, three single-phase loads (connected between phases and neutral) have a set of single-phase uncontrolled diode bridge converter, which has a resistive load with $R = 10\,\Omega$ and a parallel capacitive filter of 1000 μF as shown in Figure E10.20. It is desired to maintain the DC bus voltage at 200 V. If a three-phase series active power filter (a three singe-phase VSC with AC series inductors, coupling transformers, and its DC bus connected to a battery with a voltage of 200 V) is connected in series with the three-phase supply lines to eliminate neutral current and to maintain UPF at the AC mains, calculate (a) the AC line current, (b) the rms current of the APF, (c) the rms phase voltage at the input of the diode rectifier, (d) the rms voltage across the APF, (e) the VA rating of the APF, and (f) turns ratio of the coupling transformer.

Solution: Given supply phase voltage $V_s = 380/\sqrt{3} = 219.39$ V, frequency of the supply (f) = 50 Hz, a load of $R = 10\,\Omega$, and $V_{DC} = 200$ V.

The DC load current is $I_{DC} = 200/10 = 20$ A.

A series active filter is used to make the supply current sinusoidal in each phase. The rms supply current corresponding to it is $I_s = 200 \times 20/219.39 = 18.23$ A.

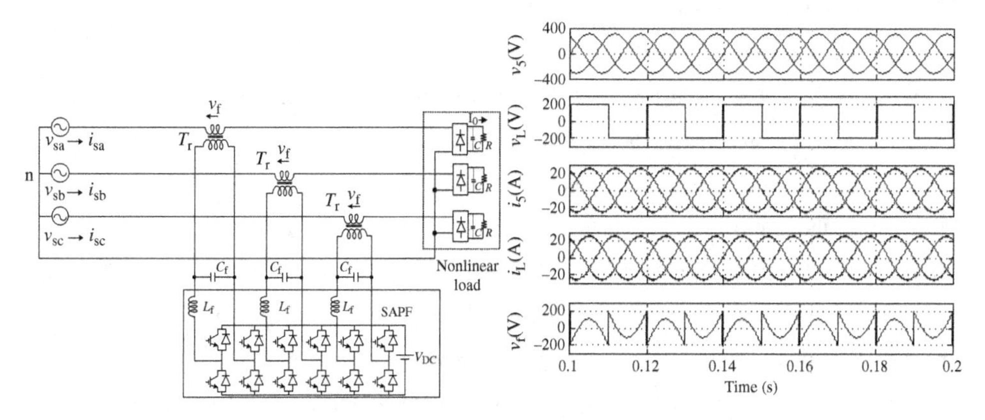

Figure E10.20 A three-phase APF for compensation of a voltage-fed nonlinear load and its waveforms

The active power is $P = 3(V_{DC}I_{DC}) = 3(200 \times 20) = 3 \times 4000 = 12\,000$ W.

a. The rms supply line current at unity power factor is $I_s = 18.23$ A.
b. The rms current rating of the series AF is $I_f = I_s = 18.23$ A (since the AF is connected in series with the supply).
c. The sine wave supply current after compensation results in continuous conduction of diodes of the rectifier (180°) and it results in square wave AC voltage at PCC with an amplitude of DC bus voltage. Therefore, $V_{PCC} = 200$ V.
d. The voltage rating of the series APF is

$$V_f = V_{frms} = \sqrt{\left\{(1/\pi)\int_0^{\pi}(219.39\sqrt{2}\sin\theta - 200)^2\,d\theta\right\}} = 95.52\text{ V}.$$

e. The VA rating of the VSI of the APF is $S = 3V_fI_s = 3 \times 95.52 \times 18.23$ VA $= 3 \times 1740.14 = 5.22$ kVA.
f. The turns ratio of the coupling transformer is computed as follows.

   The maximum AC voltage on the AC side of the VSC of the APF may be $m_aV_{DC}/(2\sqrt{2}) = 0.8 \times 200/(2 \times \sqrt{2}) = 56.568$ V and on the supply side it must be $V_{supply} = V_f$. The turns ratio of the coupling transformer is $N_{VSI}/N_{supply} = 56.568/95.52 = 0.592$.

10.8 Summary

Series APFs are used to compensate the voltage quality problems of the supply system such as sag, dip, flicker, swell, fluctuations, imbalance, and harmonics, and they protect sensitive loads from interruptions, which cause loss of production and mal-operation of other critical equipment such as medical and healthcare systems. The series APFs are also used to compensate various harmonic currents (e.g., characteristic harmonics, noncharacteristic harmonics, interharmonics, and subharmonics) of voltage-fed nonlinear loads by injecting the appropriate voltage to block the harmonic currents. The series APFs offer the best solution for voltage-fed nonlinear loads with moderate rating. The PWM-based VSCs are preferred to realize series APFs because of low cost, reduced size, light weight, and reduced losses. An analytical study of various performance indices of series APFs for the compensation of voltage-based power quality problems of different types of loads and harmonic current compensation of voltage-fed nonlinear loads is made in detail with several numerical examples to study the rating of power filters and how it is affected by various kinds of nonlinear loads. Series APFs are observed as one of the best retrofit solutions for mitigating the voltage quality problems across various types of loads and harmonic currents of voltage-fed nonlinear loads for reducing the pollution of the AC mains.

10.9 Review Questions

1. What is a series active power filter?
2. What are the power quality problems that a series APF can mitigate?
3. Can a current source inverter be used as a series APF?
4. Why voltage source inverters are preferred in a series APF?
5. What are the factors that decide the rating of a series APF?
6. How the value of an interfacing inductor is computed for a VSI used as a series APF?
7. How the value of a DC bus capacitor is computed for a VSI used as a series APF?
8. How the self-supporting DC bus of a VSI is achieved in a series APF?
9. Why the indirect current control is considered superior to the direct current control scheme of a series APF?

10. For the compensation of which kind of nonlinear load the series APF is more effective?
11. How a series APF can be used to compensate the harmonic currents of current-fed loads?
12. Can a series APF be used for regulating AC voltage at PCC?
13. How a series APF can avoid the resonance between the AC mains and a load?
14. Up to what highest order of harmonic can be eliminated by a series APF if the switching frequency is 3 kHz?
15. How the harmonics of higher order than switching frequency can be eliminated in a series APF?
16. What is the power rating of a series APF in terms of power rating of a nonlinear load consisting of a single-phase diode rectifier with constant DC voltage?
17. What is the power rating of a series APF in terms of power rating of a nonlinear load consisting of a three-phase diode rectifier with constant DC voltage?
18. How series filters protect sensitive loads from distortion in supply voltages?
19. Why is it difficult to achieve a self-supporting DC bus of a series APF?
20. What are the factors that must be considered in designing the injection transformers for a series APF?
21. Under which conditions series APFs do not need injection transformers?
22. Can a series APF in three-phase four-wire systems eliminate the neutral current produced by harmonic currents?
23. Under what conditions series APFs may operate in current control mode when a VSI is used as a series APF?
24. Under what conditions series APFs may operate in voltage control mode when a VSI is used as a series APF?
25. How the power rating of a series APF may be reduced when it is used for harmonic current compensation?

10.10 Numerical Problems

1. Design a single-phase series active power filter for filtering the voltage harmonics in a 110 V, 60 Hz AC mains (having 15% third, 10% fifth, and 5% seventh harmonics) present due to other loads and source impedance, before it is connected to a critical linear load of 3 kVA, 110 V, 60 Hz at 0.8 lagging power factor. If a single-phase VSI is used as a series APF, calculate (a) voltage and current ratings of the APF, (b) VA rating of the VSI of the series APF, and (c) interfacing inductance of the APF. Consider the switching frequency of 10 kHz, DC bus voltage of 200 V, and ripple current in the inductor is 5%.

2. A single-phase VSI with a square wave AC output of 230 V (rms) at 50 Hz is feeding a critical linear load of 5 kVA, 230 V, 50 Hz at 0.8 lagging power factor. Design a single-phase series active power filter for filtering the voltage harmonics in this system to eliminate voltage harmonics and to regulate fundamental 230 V (rms) across the load. If a single-phase VSI is used as an APF, calculate (a) voltage rating of the APF, (b) current rating of the APF (c) VA rating of the VSI of the APF, and (d) interfacing inductance. Consider the switching frequency of 20 kHz, DC bus voltage of 400 V, and ripple current in the inductor is 10%.

3. A series active power filter (a VSC with an AC series inductor and a DC bus capacitor) is used in series with single-phase AC supply of 230 V at 50 Hz feeding a diode rectifier charging a battery of 228 V at 25 A average current to reduce the harmonics in AC mains current and to maintain almost

UPF. Calculate (a) rms voltage at the input of the diode rectifier, (b) line current, (c) voltage rating of the APF, (d) current rating of the APF, (e) VA rating of the APF, (f) value of the DC bus voltage of the APF, and (g) value of the AC inductor of the APF. Consider the switching frequency of 10 kHz and ripple current in the inductor is 10%.

4. A series active power filter (a VSI with an AC series inductor, a coupling transformer, and its DC bus connected to a battery functioning as the load) is used in series with single-phase AC supply of 110 V at 60 Hz feeding a diode rectifier charging a battery of 72 V at 15 A average current to reduce the harmonics in AC mains current and to maintain almost UPF. Calculate (a) rms voltage at the input of the diode rectifier, (b) line current, (c) voltage rating of the APF, (d) current rating of the APF, (e) VA rating of the APF, (f) the value of the AC inductor, and (g) turns ratio of the coupling transformer. Consider the switching frequency of 10 kHz and ripple current in the inductor is 2%.
5. A series active power filter (a VSC with an AC series inductor, a coupling transformer, and its DC bus connected to the DC bus of the load) is used to reduce the harmonics in AC mains current and to maintain almost UPF in the series of single-phase AC supply of 110 V at 60 Hz feeding a diode rectifier with a capacitive filter of 2000 μF and a resistive load of 10 Ω. The DC bus voltage of the load is decided to result in minimum injected voltage of the APF. Calculate (a) rms voltage at the input of the diode rectifier, (b) line current, (c) voltage rating of the APF, (d) current rating of the APF, (e) VA rating of the APF, (f) DC bus voltage, and (g) turns ratio of the coupling transformer.
6. A series active power filter (a VSC with an AC series inductor, a coupling transformer, and its DC bus connected to the DC bus of the load) is used to reduce the harmonics in AC mains current and to maintain almost UPF in the series of single-phase AC supply of 110 V at 60 Hz feeding a diode rectifier with a capacitive filter of 2000 μF and a resistive load of 5 Ω. If the DC bus voltage of the load is to be maintained at constant ripple-free 120 V, then calculate (a) rms voltage at the input of the diode rectifier, (b) line current, (c) voltage rating of the APF, (d) current rating of the APF, (e) VA rating of the APF, and (f) turns ratio of the coupling transformer.
7. A single-phase 220 V, 50 Hz diode bridge rectifier with a DC capacitor filter is feeding a DC of 380 V at 20 A average current to a variable-frequency three-phase VSI-fed induction motor drive in an air conditioner. A single-phase series active power filter (a VSC with an AC series inductor, a coupling transformer, and a DC bus capacitor) is used in series with this rectifier–inverter system to reduce the harmonics in AC mains current, to maintain almost UPF, and to regulate the DC bus voltage of the rectifier to 380 V. Calculate (a) rms voltage at the input of the single-phase diode rectifier, (b) line current, (c) rms current of the APF, (d) rms voltage across the APF, and (e) VA rating of the APF.
8. Design a three-phase series active power filter for filtering the voltage harmonics in a 440 V, 50 Hz AC mains (15% 5th, 10% 7th, 7% 11th, and 5% 13th harmonics) present due to other loads and source impedance, before it is connected to a critical linear load of 25 kVA, 440 V, 50 Hz at 0.8 lagging power factor. If a three-phase VSI is used as an APF, calculate (a) voltage and current ratings of APF, (b) VA rating of the VSI of the APF, (c) interfacing inductance, and (d) turns ratio of the coupling transformer. Consider the switching frequency of 10 kHz, DC bus voltage of 400 V DC, and ripple current in the inductor is 5%.
9. A three-phase series active power filter (a VSI with an AC series inductor, a coupling transformer, and a DC bus capacitor) is used in series with three-phase AC supply of 440 V at 50 Hz feeding a three-phase diode rectifier charging a battery of 516 V at 60 A average current to reduce the harmonics in AC mains current and to maintain almost UPF. Calculate (a) rms line voltage at the input of the diode rectifier (almost quasi-square wave), (b) AC line current, (c) rms current of the APF, (d) rms voltage across the APF, (e) VA rating of the APF, (f) AC inductor value, and (g) turns ratio of the coupling transformer if the DC bus voltage of the APF is 400 V. Consider the switching frequency of 5 kHz and ripple current in the inductor is 5%.

10. A three-phase 440 V, 50 Hz six-pulse diode bridge rectifier with a DC capacitor filter is feeding a DC of 650 V at 250 A average current to a variable-frequency VSI-fed induction motor drive. A three-phase series active power filter (a VSI with an AC series inductor, a coupling transformer, and a DC bus capacitor) is used in series with this rectifier–inverter system to reduce the harmonics in AC mains current, to maintain almost UPF, and to regulate the DC bus voltage of the rectifier to 650 V. Calculate (a) rms line voltage at the input of the diode rectifier, (b) line current, (c) rms current of the APF, (d) rms voltage across the APF, and (e) VA rating of the APF.

11. A three-phase series active power filter (a VSC with an AC series inductor, a coupling transformer, and its DC bus connected to a battery functioning as the load) is used in series with three-phase AC supply of 400 V at 50 Hz feeding a three-phase diode bridge rectifier charging a battery of 564 V at 50 A average current to reduce the harmonics in AC mains current and to maintain almost UPF. Calculate (a) rms line voltage at the input of the diode rectifier, (b) AC line current, (c) rms current of the APF, (d) rms voltage across the APF, and (e) VA rating of the APF.

12. A three-phase series active power filter (a VSC with an AC series inductor, a coupling transformer, and its DC bus connected to the DC bus of the load) is used in series with the three-phase AC supply of 440 V at 50 Hz feeding a three-phase diode rectifier with a parallel DC capacitor at a load of 100 A average current to reduce the harmonics in AC mains current, to maintain UPF of the supply current, and to regulate the DC bus voltage of the load at almost constant 700 V. Calculate (a) rms line voltage at the input of the diode rectifier, (b) AC line current, (c) rms current of the APF, (d) rms voltage across the APF, (e) VA rating of the APF, and (f) turns ratio of the coupling transformer.

13. A series active power filter (a VSI with an AC series inductor, a coupling transformer, and its DC bus connected to the DC bus of the load) is used to reduce the harmonics in AC mains current and to maintain almost UPF in the series of three-phase AC supply of 400 V at 50 Hz feeding a three-phase diode rectifier with a capacitive filter of 5000 μF and a resistive load of 10 Ω. The DC bus voltage of the load is decided to result in minimum injected voltage of the APF. Calculate (a) rms voltage at the input of the diode rectifier, (b) line current, (c) VA rating of the APF, (d) DC bus voltage of the load, and (e) turns ratio of the coupling transformer.

14. A series active power filter (a VSI with an AC series inductor, a coupling transformer, and its DC bus connected to the DC bus of the load) is used to reduce the harmonics in AC mains current and to maintain almost UPF in the series of three-phase AC supply of 440 V at 50 Hz feeding a three-phase diode rectifier with a capacitive filter of 10 000 μF and a resistive load of 15 Ω. If DC bus voltage of the load is to be maintained at constant ripple-free 650 V, then calculate (a) rms line voltage at the input of the diode rectifier, (b) AC line current, (c) rms current of the APF, (d) rms voltage across the APF, (e) VA rating of the APF, and (f) turns ratio of the coupling transformer.

15. A three-phase series active power filter (a VSC with AC series inductors, coupling transformers, and a DC bus capacitor) is used in series with the three-phase AC supply of 440 V at 50 Hz feeding a three-phase 12-pulse diode rectifier (consisting of two parallel connected 6-pulse diode bridge rectifiers with star/delta and delta/delta transformers to output equal line voltages of 440 V) having a parallel DC capacitor filter with 400 V DC at a load of 100 A average current to reduce the harmonics in AC mains current and to maintain UPF of the supply current. Calculate (a) the rms phase voltage at the input of the 12-pulse diode rectifier, (b) the AC line current, (c) the rms current of the APF, (d) the rms voltage across the APF, and (e) the VA rating of the APF.

16. A three-phase series active power filter (a VSI with AC series inductors, coupling transformers, and a DC bus connected to the DC bus of the load) is used in series with the three-phase AC supply of 415 V at 50 Hz feeding a three-phase 12-pulse diode rectifier (consisting of two parallel connected 6-pulse diode bridge rectifiers with star/delta and delta/delta transformers to output equal line voltages of 415 V) having a parallel DC capacitor filter of 10 000 μF with a 10 Ω resistive load to reduce the harmonics in AC mains current and to maintain UPF of the supply current. If the DC bus of the load is to be maintained at 600 V, then calculate (a) the rms phase voltage at the input of the 12-pulse diode

rectifier, (b) the AC line current, (c) the rms current of the APF, (d) the rms voltage across the APF, (e) the VA rating of the APF, and (f) turns ratio of the coupling transformer.

17. A three-phase series active power filter (a VSC with AC series inductors and its DC bus connected to a battery behaving as the load) is connected in series with the three-phase AC supply of 440 V at 50 Hz feeding a three-phase 12-pulse diode rectifier (consisting of two parallel connected 6-pulse diode bridge rectifiers with star/delta and delta/delta transformers to output equal line voltages of 440 V) charging a battery of 600 V at 200 A average current. The series APF is employed to reduce the harmonics in AC mains current and to maintain almost UPF of the supply current. Calculate (a) the rms phase voltage at the input of the 12-pulse diode rectifier, (b) the AC line current, (c) the rms current of the APF, (d) the rms voltage across the APF, (e) the VA rating of the APF, and (f) check the need of the coupling transformer.

18. In a three-phase, line voltage of 415 V, 50 Hz, four-wire distribution system, three single-phase loads (connected between phases and neutral) have a set of single-phase uncontrolled diode bridge converter, which has a *RE* load with $R = 5\,\Omega$ and $E = 240$ V. It is desired to charge the battery E with an average current of 20 A. If a three-phase series active power filter (a VSC with AC series inductors, coupling transformers, and its DC bus connected to the DC bus capacitor) is connected in series with the three-phase supply lines to maintain UPF at the AC mains, calculate (a) supply current, (b) the rms current of the APF, (c) the rms voltage at the input of the rectifier, (d) the rms voltage across the APF, (e) the VA rating of the APF, and (f) neutral current.

19. In a three-phase, line voltage of 400 V, 50 Hz, four-wire distribution system, three single-phase loads (connected between phases and neutral) have a set of single-phase uncontrolled diode bridge converter, which has a *RE* load with $R = 1\,\Omega$ and battery voltage $E = 216$ V. It is desired to charge the battery with average current of 25 A. If a three-phase series active power filter (a three single-phase VSC with AC series inductors, coupling transformers, and its DC bus connected to the battery) is connected in series with the three-phase supply lines to maintain UPF at the AC mains, calculate (a) the AC line current, (b) the rms current of the APF, (c) the rms phase voltage at the input of the diode rectifier, (d) the rms voltage across the APF, (e) the VA rating of the APF, and (f) turns ratio of the coupling transformer.

20. In a three-phase, line voltage of 415 V, 50 Hz, four-wire distribution system, three single-phase loads (connected between phases and neutral) have a set of single-phase uncontrolled diode bridge converter, which has a resistive load with $R = 5\,\Omega$ and a parallel capacitive filter of 2000 μF. It is desired to maintain the DC bus voltage at 240 V. If a three-phase series active power filter (a three singe-phase VSI with AC series inductors, coupling transformers, and its DC bus connected to a DC capacitor) is connected in series with the three-phase supply lines to eliminate neutral current and to maintain UPF at the AC mains, calculate (a) the AC line current, (b) the rms current of the APF, (c) the rms phase voltage at the input of the diode rectifier, (d) the rms voltage across the APF, (e) the VA rating of the APF, and (f) turns ratio of the coupling transformer.

10.11 Computer Simulation-Based Problems

1. Design an AC series active filter and simulate its behavior for a single-phase 220 V, 50 Hz, uncontrolled bridge converter with a parallel capacitive DC filter of 470 μF and an equivalent resistive load of 10 Ω. It has a source impedance of 0.15 Ω resistive element and 1.0 Ω inductive element. Plot the supply voltage and input current, output voltage and current with time, and the harmonic spectra of supply current and voltage at PCC. Compute THD, crest factor, rms value of AC mains current, displacement factor, distortion factor, power factor, % ripple in load current, output voltage ripple, ripple factor, and input active, reactive, and output powers.

2. Design an AC series active filter and simulate its behavior for a single-phase VSI with a quasi-square wave AC output of 230 V (rms) at 50 Hz feeding a critical linear load of 5 kVA, 220 V, 50 Hz at

0.8 lagging power factor to eliminate voltage harmonics, to regulate fundamental 220 V (rms) across this load, and to maintain UPF at the source. Plot the supply voltage and input current, load voltage and DC link voltage of the APF with time, and the harmonic spectra of supply current and voltage at PCC. Compute THD, crest factor, rms value of AC mains current, displacement factor, distortion factor, power factor, input active, reactive, and output powers, voltage rating of the APF, current rating of the APF, and VA rating of the APF.

3. Design an AC series active filter and simulate its behavior for a single-phase 230 V, 50 Hz, uncontrolled bridge converter used for charging a battery of 180 V. It has a source impedance of 0.5 Ω resistive element and 5.0 Ω inductive element. Plot the supply voltage and input current, charging current with time, and the harmonic spectra of supply current and voltage at PCC. Compute THD, crest factor, rms value of AC mains current, displacement factor, distortion factor, power factor, and input active, reactive, and output powers.

4. Design a single-phase series active power filter and simulate its behavior for filtering the voltage harmonics and to regulate the sinusoidal voltage of 220 V and to maintain UPF at a source of a 230 V, 50 Hz AC mains (10% third, 8% fifth, and 5% seventh harmonics) present due to other loads and source impedance, before it is connected to a critical linear load of 5 kVA, 220 V, 50 Hz at 0.8 lagging power factor. Plot the supply voltage and input current, load voltage with time, and the harmonic spectra of supply current and voltage at PCC. Compute voltage and current ratings of the APF and VA rating of the VSI of the series APF.

5. Design an AC series active power filter (a VSI with an AC series inductor, a coupling transformer, and its DC bus connected to a battery functioning as the load) and simulate its behavior for a single-phase AC supply of 230 V at 50 Hz feeding a diode rectifier charging a battery of 220 V and average load current of 25 A to reduce the harmonics in AC mains current and to maintain almost UPF. It has a source impedance of 0.25 Ω resistive element and 2.5 Ω inductive element. Plot the supply voltage and input current, charging current and diode rectifier input voltage with time, and the harmonic spectra of supply current and voltage at PCC. Compute rms voltage at the input of the diode rectifier, line current, voltage rating of the APF, current rating of the APF, and VA rating of the APF.

6. Design an AC series active power filter (a VSI with an AC series inductor, a coupling transformer, and its DC bus connected to the DC bus of the load) and simulate its behavior for a single-phase 220 V, 50 Hz diode bridge rectifier with a DC capacitor filter feeding a DC current of 20 A to a variable-frequency three-phase VSI-fed induction motor drive in an air conditioner to reduce the harmonics in AC mains current, to maintain almost UPF, and to regulate the DC bus voltage of the rectifier to 400 V. It has a source impedance of 0.15 Ω resistive element and 1.5 Ω inductive element. Plot the supply voltage and input current, rectifier output current and input voltage with time, and the harmonic spectra of supply current and voltage at PCC. Compute THD, crest factor, rms value of AC mains current, displacement factor, distortion factor, power factor, input active, reactive, and output powers, rms voltage at the input of the diode rectifier, voltage rating of the APF, current rating of the APF, and VA rating of the APF.

7. Design an AC series active filter and simulate its behavior for filtering the single-phase square wave supply of 230 V (rms) to be converted to sine wave of 220 V (rms) to feed a critical linear lagging power load with $R = 20\,\Omega$ and $X = 15\,\Omega$. Plot supply voltage and current, load current and voltage, device voltage and current with time, and harmonic spectra of supply current and voltage at PCC. Compute THD, crest factor, rms value of AC mains current, displacement factor, distortion factor, power factor, output voltage, and input and output powers.

8. Design an AC series active filter and simulate its behavior for filtering the single-phase quasi-square wave supply of 240 V (rms) to be converted to sine wave of 230 V (rms) to feed a critical linear lagging power load with $R = 16\,\Omega$ and $X = 12\,\Omega$. Plot the supply voltage and input current, output voltage and current with time, and the harmonic spectra of supply current and voltage at PCC.

Compute THD, crest factor, rms value of AC mains current, displacement factor, distortion factor, power factor, and input active, reactive, and output powers.

9. Design an AC series active power filter (a VSI with an AC series inductor, a coupling transformer, and its DC bus connected to the DC bus of the load) and simulate its behavior to reduce the harmonics in AC mains current and to maintain almost UPF in the series of single-phase AC supply of 220 V at 50 Hz feeding a diode rectifier with a capacitive filter of 1000 μF and a resistive load of 10 Ω. The DC bus voltage of the load is to be maintained at constant ripple-free 240 V. Plot supply voltage and current, rectifier input voltage and output current, device voltage and current with time, and harmonic spectra of supply current and voltage at PCC. Compute rms voltage at the input of the diode rectifier, line current, voltage rating of the APF, current rating of the APF, VA rating of the APF, and turns ratio of the coupling transformer.

10. Design an AC series active power filter (a VSI with an AC series inductor, a coupling transformer, and its DC bus connected to the DC bus of the load) and simulate its behavior to reduce the harmonics in AC mains current and to maintain almost UPF in the series of single-phase AC supply of 230 V at 50 Hz feeding a three-phase diode rectifier with a capacitive filter of 470 μF and a resistive load of 10 Ω. The DC bus voltage of the load is decided to result in minimum injected voltage of the APF. It has a source impedance of 0.15 Ω resistive element and 0.75 Ω inductive element. Plot supply voltage and current, DC load current and voltage, injected voltage and rectifier input voltage with time, and harmonic spectra of supply current and voltage at PCC. Compute THD, crest factor, rms value of AC mains current, displacement factor, distortion factor, power factor, input active, reactive, and output powers, rms line voltage at the input of the diode rectifier, rms voltage across the APF, VA rating of the APF, and turns ratio of the coupling transformer.

11. Design an AC series active filter and simulate its behavior for a three-phase 415 V, 50 Hz, uncontrolled bridge converter with a parallel capacitive DC filter of 4700 μF and a resistive load of 10 Ω. It has a source impedance of 0.25 Ω resistive element and 1.0 Ω inductive element. Plot the supply voltage and input current, output voltage and current with time, and the harmonic spectra of supply current and voltage at PCC. Compute THD, crest factor, rms value of AC mains current, displacement factor, distortion factor, power factor, % ripple in load current, output voltage ripple, ripple factor, and input active, reactive, and output powers.

12. Design an AC series active filter and simulate its behavior for a three-phase 415 V, 50 Hz, uncontrolled six-pulse bridge converter used for charging a battery of 552 V. It has a source impedance of 0.25 Ω resistive element and 1.5 Ω inductive element. Plot the supply voltage and input current, charging current with time, and the harmonic spectra of supply current and voltage at PCC. Compute THD, crest factor, rms value of AC mains current, displacement factor, distortion factor, power factor, % ripple in load current, and input active, reactive, and output powers.

13. Design an AC series active filter and simulate its behavior for filtering the three-phase square wave supply of 440 V (rms) at 50 Hz to be converted to sine wave of 415 V (rms) to feed a critical linear load of 25 kVA at 0.8 lagging power factor. Plot supply voltage and current, load current and voltage, device voltage and current with time, and harmonic spectra of supply current and voltage at PCC. Calculate THD, crest factor, rms value of AC mains current, displacement factor, distortion factor, power factor, output voltage, and input active, reactive, and output active powers.

14. Design an AC series active filter and simulate its behavior for filtering the three-phase quasi-square wave supply of 415 V (rms) at 50 Hz to be converted to sine wave of 400 V (rms) to feed a critical linear load of 50 kVA at 0.8 lagging power factor. Plot supply voltage and current, load current and voltage, device voltage and current with time, and harmonic spectra of supply current and voltage at PCC. Calculate THD, crest factor, rms value of AC mains current, displacement factor, distortion factor, power factor, output voltage, and input active, reactive, and output active powers.

15. Design an AC series active filter and simulate its behavior for a three-phase 415 V, 50 Hz six-pulse diode bridge rectifier with a DC capacitor filter feeding a DC of 600 V at 150 A average current to a variable-frequency VSI-fed induction motor drive. It has a source impedance of 0.25 Ω resistive element and 1.25 Ω inductive element. Plot the supply voltage and input current, rectifier output current with time, and the harmonic spectra of supply current and voltage at PCC. Compute THD, crest factor, rms value of AC mains current, displacement factor, distortion factor, power factor, and input active, reactive, and output powers.

16. Design an AC series active power filter (a VSI with an AC series inductor, a coupling transformer, and its DC bus connected to the DC bus of the load) and simulate its behavior to reduce the harmonics in AC mains current and to maintain almost UPF in the series of three-phase AC supply of 460 V at 60 Hz feeding a three-phase diode rectifier with a capacitive filter of 10 000 μF and a resistive load of 10 Ω. The DC bus voltage of the load is to be maintained at constant ripple-free 650 V. It has a source impedance of 0.15 Ω resistive element and 0.75 Ω inductive element. Plot the supply voltage and input current, rectifier output current with time, and the harmonic spectra of supply current and voltage at PCC. Compute THD, crest factor, rms value of AC mains current, displacement factor, distortion factor, power factor, input active, reactive, and output powers, rms line voltage at the input of the diode rectifier, rms voltage across the APF, VA rating of the APF, and turns ratio of the coupling transformer.

17. Design an AC series active power filter (a VSI with an AC series inductor, a coupling transformer, and its DC bus connected to the DC bus of the load) and simulate its behavior to reduce the harmonics in AC mains current and to maintain almost UPF in the series of three-phase AC supply of 400 V at 50 Hz feeding a three-phase diode rectifier with a capacitive filter of 5000 μF and a resistive load of 20 Ω. The DC bus voltage of the load is decided to result in minimum injected voltage of the APF. It has a source impedance of 0.10 Ω resistive element and 0.5 Ω inductive element. Plot supply voltage and current, DC load current and voltage, injected voltage and rectifier input voltage with time, and harmonic spectra of supply current and voltage at PCC. Compute THD, crest factor, rms value of AC mains current, displacement factor, distortion factor, power factor, input active, reactive, and output powers, rms line voltage at the input of the diode rectifier, rms voltage across the APF, VA rating of the APF, and turns ratio of the coupling transformer.

18. Design an AC series active filter and simulate its behavior for a three-phase 415 V, 50 Hz supply having a 12-pulse diode bridge converter with a parallel capacitive DC filter of 4700 μF and a resistive load of 20 Ω. It has a source impedance of 0.10 Ω resistive element and 0.50 Ω inductive element. It consists of an ideal transformer with single primary star connected winding and two secondary windings connected in star and delta with same output line voltages as the input voltage at no load to provide 30° phase shift between two sets of three-phase output voltages. Two 6-pulse diode bridges are connected in series to provide this 12-pulse AC–DC converter. Plot the supply voltage and input current, output voltage and current with time, and the harmonic spectra of supply current and voltage at PCC. Compute the fundamental active power drawn by the load, power factor, rms supply current, distortion factor, fundamental rms supply current, peak current of the AC mains, and total harmonic distortion of AC source current.

19. Design an AC series active filter and simulate its behavior for a three-phase 380 V, 50 Hz supply having a 12-pulse diode bridge converter used for charging a battery of 840 V. It has a source impedance of 0.15 Ω resistive element and 0.75 Ω inductive element. It consists of an ideal transformer with single primary star connected winding and two secondary windings connected in star and delta with same output line voltages as the input voltage at no load to provide 30° phase shift between two sets of three-phase output voltages. Two 6-pulse diode bridges are connected in series to provide this 12-pulse AC–DC converter. Plot the supply voltage and input current, output voltage and current with time, and the harmonic spectra of supply current and voltage at PCC. Compute the fundamental active power drawn by the load, power factor, rms supply current,

distortion factor, fundamental rms supply current, peak current of the AC mains, and total harmonic distortion of AC source current.

20. Design an AC three-phase series active power filter (a VSI with AC series inductors, coupling transformers, and a DC bus connected to the DC bus of the load) and simulate its behavior for a three-phase AC supply of 415 V at 50 Hz feeding a three-phase 12-pulse diode rectifier (consisting of two parallel connected 6-pulse diode bridge rectifiers with star/delta and delta/delta transformers to output equal line voltages of 415 V) having a parallel DC capacitor filter of 20 000 μF with a 5 Ω resistive load to reduce the harmonics in AC mains current and to maintain UPF of the supply current. The DC bus of the load is to be maintained at 600 V. It has a source impedance of 0.05 Ω resistive element and 0.25 Ω inductive element. Plot the supply voltage and input current, injected voltage and input voltage of the rectifier transformer, output DC voltage and current with time, and the harmonic spectra of supply current and voltage at PCC. Compute THD, crest factor, rms value of AC mains current, displacement factor, distortion factor, power factor, input active, reactive, and output powers, rms line voltage at the input of the diode rectifiers, rms voltage across the APF, VA rating of the APF, and turns ratio of the coupling transformer.

21. Design an AC three-phase series active power filter (a VSI with AC series inductors, coupling transformers, and a DC bus connected to the DC bus of the load) used in series with the three-phase AC supply of 415 V at 50 Hz feeding a three-phase 12-pulse diode rectifier (consisting of two series connected 6-pulse diode bridge rectifiers with star/delta and delta/delta transformers to output equal line voltages of 415 V) having a parallel DC capacitor filter of 5000 μF with a 10 Ω resistive load to reduce the harmonics in AC mains current and to maintain UPF of the supply current. The DC bus of the load is to be maintained at 1000 V. It has a source impedance of 0.01 Ω resistive element and 0.5 Ω inductive element. Plot the supply voltage and input current, injected voltage and input voltage of the rectifier transformer, output DC voltage and current with time, and the harmonic spectra of supply current and voltage at PCC. Compute THD, crest factor, rms value of AC mains current, displacement factor, distortion factor, power factor, input active, reactive, and output powers, rms line voltage at the input of the diode rectifiers, rms voltage across the APF, VA rating of the APF, and turns ratio of the coupling transformer.

22. Design a three-phase four-wire three-leg VSI with midpoint capacitor-based AC series active filter and simulate its behavior for a three-phase, line voltage of 380 V, 50 Hz, four-wire distribution system with the three single-phase loads (connected between phases and neutral) of a single-phase uncontrolled bridge converter with a parallel capacitive DC filter of 470 μF and an equivalent resistive load of 10 Ω. It has a source impedance of 0.15 Ω resistive element and 1.25 Ω inductive element. Plot the supply voltage and input current, neutral current of supply and load, output voltage and current with time, and the harmonic spectra of supply current, load current, and voltage at PCC. Compute active power consumed, reactive power drawn, displacement factor, distortion factor, total harmonic distortion of AC source current, power factor, crest factor of AC source current, AC source rms current, and neutral current.

23. Design a three-phase four-wire four-leg VSI-based AC series active filter and simulate its behavior for a three-phase, line voltage of 400 V, 50 Hz, four-wire distribution system with the three single-phase loads (connected between phases and neutral) of a single-phase uncontrolled bridge converter with a parallel capacitive DC filter of 220 μF and an equivalent resistive load of 20 Ω. It has a source impedance of 0.25 Ω resistive element and 2.25 Ω inductive element. Plot the supply voltage and input current, neutral current of supply and load, output voltage and current with time, and the harmonic spectra of supply current, load current, and voltage at PCC. Compute active power consumed, reactive power drawn, displacement factor, distortion factor, total harmonic distortion of AC source current, power factor, crest factor of AC source current, AC source rms current, and neutral current.

24. Design a three-phase four-wire three single-phase VSI-based AC series active filter and simulate its behavior for a three-phase, line voltage of 415 V, 50 Hz, four-wire distribution system with the three single-phase loads (connected between phases and neutral) of a single-phase uncontrolled bridge converter with a parallel capacitive DC filter of 680 μF and an equivalent resistive load of 15 Ω. It has a source impedance of 0.5 Ω resistive element and 2.5 Ω inductive element. Plot the supply voltage and input current, neutral current of supply and load, output voltage and current with time, and the harmonic spectra of supply current, load current, and voltage at PCC. Compute active power consumed, reactive power drawn, displacement factor, distortion factor, total harmonic distortion of AC source current, power factor, crest factor of AC source current, AC source rms current, and neutral current.

25. Design a three-phase four-wire three single-phase VSI-based AC series active filter and simulate its behavior for a three-phase, line voltage of 380 V, 50 Hz, four-wire distribution system with a three single-phase uncontrolled bridge converter used for charging a battery of 172 V. It has a source impedance of 0.5 Ω resistive element and 3.5 Ω inductive element. Plot the supply voltage and input current, charging current, neutral current of the supply and load with time, and the harmonic spectra of supply current and voltage at PCC. Compute THD, crest factor, rms value of AC mains current, displacement factor, distortion factor, power factor, and input active, reactive, and output powers.

References

1. Gyugyi, L. and Strycula, E. (1976) Active AC power filters. IEEE IAS Annual Meeting Record, pp. 529–535.
2. Moran, S. (1989) A line voltage regulator/conditioner for harmonic-sensitive load isolation. IEEE IAS Annual Meeting Record, pp. 945–951.
3. Nastran, J., Cajhen, R., Seliger, M., and Jereb, P. (1994) Active power filter for non-linear AC loads. *IEEE Transactions on Power Electronics*, 9(1), 92–96.
4. Bhattacharya, S., Divan, D.M., and Banerjee, B. (1991) Synchronous frame harmonic isolator using active series filter. Fourth EPE Conference Record.
5. Fujita, H. and Akagi, H. (1991) A practical approach to harmonic compensation in power systems – series connection of passive and active filters. *IEEE Transactions on Industry Applications*, 27(6), 1020–1025.
6. Bhattacharya, S., Divan, D.M., and Banerjee, B.B. (1993) Control and reduction of terminal voltage total harmonic distortion (THD) in a hybrid series active and parallel passive filter system. IEEE PESC Record, pp. 779–786.
7. Bhavaraju, V.B. and Enjeti, P. (1993) A novel active line conditioner for a three-phase system. IEEE IAS Annual Meeting Record, pp. 979–985.
8. Dixon, J.W., Garcia, J.C., and Moran, L.T. (1993) A control system for a three-phase active power filter which simultaneously compensates power factor and unbalanced loads. IEEE IECON Record, pp. 1083–1087.
9. Campos, A., Joos, G., Ziogas, P.D., and Lindsay, J.F. (1994) Analysis and design of a series voltage unbalance compensator based on a three-phase VSI operating with unbalanced switching functions. *IEEE Transactions on Power Electronics*, 9(3), 269–274.
10. Vincenti, D., Jin, H., and Ziogas, P. (1994) Design and implementation of a 25 kVA three-phase PWM AC line conditioner. *IEEE Transactions on Power Electronics*, 9(4), 384–389.
11. Aredes, M. and Watanabe, E.H. (1995) New control algorithms for series and shunt three-phase four-wire active power filter. *IEEE Transactions on Power Delivery*, 10(3), 1649–1656.
12. Nelson, R.J. and Ramey, D.G. (1997) Dynamic power and voltage regulator for an ac transmission line. US Patent 5,610,501.
13. Hochgraf, C.G. (1999) Power inverter apparatus using a transformer with its primary winding connected to the source end and a secondary winding connected to the load end of an ac power line to insert series compensation. US Patent 5,905,367.
14. Cheng, P.T., Lasseter, R., and Divan, D. (1999) Dynamic series voltage restoration for sensitive loads in unbalanced power systems. US Patent 5,883,796.
15. Dahler, P. and Knap, G. (2003) Protection of a dynamic voltage restorer. US Patent 6,633,092.
16. Bhavraju, V.B. and Enjeti, P.N. (1994) A fast active power filter to correct voltage sags. *IEEE Transactions on Industry Electronics*, 41(3), 333–338.
17. Bhavraju, V.B. and Enjeti, P.N. (1996) An active line conditioner to balance voltages in a three-phase system. *IEEE Transactions on Industry Applications*, 32(2), 287–292.
18. Campos, A., Joos, G., Ziogas, P.D., and Lindsay, J.F. (1994) Analysis and design of a series unbalanced compensator based on a three-phase VSI operating with unbalanced switching functions. *IEEE Transactions on Power Electronics*, 9, 269–274.

19. Kara, A., Dahler, P., Amhof, D., and Gruning, H. (1998) Power supply quality improvement with a dynamic voltage restorer (DVR). Proceedings of IEEE APEC'98, vol. 2, pp. 986–993.
20. Woodley, N.H., Morgan, L., and Sundaram, A. (1999) Experience with an inverter-based dynamic voltage restorer. *IEEE Transactions on Power Delivery*, 14(3), 1181–1186.
21. Vilathgamuwa, M., Perera, R., Choi, S., and Tseng, K. (1999) Control of energy optimized dynamic voltage restorer. Proceedings of IEEE IECON'99, vol. 2, pp. 873–878.
22. Choi, S.S., Li, B.H., and Vilathgamuwa, D.M. (2000) Dynamic voltage restoration with minimum energy injection. Proceedings of the IEEE Power Engineering Society Winter Meeting, vol. 2, pp. 1156–1161.
23. Daehler, P. and Affolter, R. (2000) Requirements and solutions for dynamic voltage restorer: a case study. Proceedings of the IEEE Power Engineering Society Winter Meeting, vol. 4, pp. 2881–2885.
24. Nielsen, J.G., Blaabjerg, F., and Mohan, N. (2001) Control strategies for dynamic voltage restorer compensating voltage sags with phase jump. Proceedings of IEEE APEC'01, vol. 2, pp. 1267–1273.
25. Brumsickle, W.E., Schneider, R.S., Luckjiff, G.A. *et al.* (2001) Dynamic sag correctors: cost-effective industrial power line conditioning. *IEEE Transactions on Industry Applications*, 37, 212–217.
26. Ghosh, A. and Ledwich, G. (2002) Compensation of distribution system voltage using DVR. *IEEE Transactions on Power Delivery*, 17(4), 1030–1036.
27. Ding, H., Shuangyan, S., Xianzhong, D., and Jun, G. (2002) A novel dynamic voltage restorer and its unbalanced control strategy based on space vector PWM. *International Journal of Electrical Power & Energy Systems*, 24(9), 693–699.
28. Ghosh, A. and Joshi, A. (2002) A new algorithm for the generation of reference voltages of a DVR using the method of instantaneous symmetrical components. *IEEE Power Engineering Review*, 22(1), 63–65.
29. Awad, H., Svensson, J., and Bollen, M.H.J. (2003) Static series compensator for voltage dips mitigation. Proceedings of the IEEE Power Tech Conference, June, Bologna, vol. 3, pp. 23–26.
30. Aeloíza, E.C., Enjeti, P.N., Morán, L.A. *et al.* (2003) Analysis and design of a new voltage sag compensator for critical loads in electrical power distribution systems. *IEEE Transactions on Industry Applications*, 39(4), 1143–1150.
31. Liu, J.W., Choi, S.S., and Chen, S. (2003) Design of step dynamic voltage regulator for power quality enhancement. *IEEE Transactions on Power Delivery*, 18(4), 1403–1409.
32. Affolter, R. and Connell, B. (2003) Experience with a dynamic voltage restorer for a critical manufacturing facility. Proceedings of the IEEE Transmission and Distribution Conference and Exposition, September, vol. 3, pp. 937–939.
33. Chung, I.-Y., Won, D.-J., Park, S.-Y. *et al.* (2003) The DC link energy control method in dynamic voltage restorer system. *International Journal of Electrical Power & Energy Systems*, 25(7), 525–531.
34. Ghosh, A., Jindal, A.K., and Joshi, A. (2004) Design of a capacitor-supported dynamic voltage restorer (DVR) for unbalanced and distorted loads. *IEEE Transactions on Power Delivery*, 19(1), 405–413.
35. Singh, B.N. and Simina, M. (2004) Intelligent solid-state voltage restorer for voltage swell/sag and harmonics. *IEE Proceedings – Electric Power Applications*, 151(1), 98–106.
36. Newman, M.J., Holmes, D.G., Nielsen, J.G., and Blaabjerg, F. (2005) A dynamic voltage restorer (DVR) with selective harmonic compensation at medium voltage level. *IEEE Transactions on Industrial Electronics*, 51(6), 1744–1753.
37. Nielsen, J.G. and Blaabjerg, F. (2005) A detailed comparison of system topologies for dynamic voltage restorers. *IEEE Transactions on Industry Applications*, 41(5), 1272–1280.
38. Banaei, M.R., Hosseini, S.H., Khanmohamadi, S., and Gharehpetian, G.B. (2006) Verification of a new energy control strategy for dynamic voltage restorer by simulation. *Simulation Modelling Practice and Theory*, 14(2), 112–125.
39. Banaei, M.R., Hosseini, S.H., and Gharehpetian, G.B. (2006) Inter-line dynamic voltage restorer control using a novel optimum energy consumption strategy. *Simulation Modelling Practice and Theory*, 14(7), 989–999.
40. Vilathgamuwa, D.M., Wijekoon, H.M., and Choi, S.S. (2007) A novel technique to compensate voltage sags in multiline distribution system—the interline dynamic voltage restorer. *IEEE Transactions on Industrial Electronics*, 54(4), 1603–1611.
41. Jindal, A.K., Ghosh, A., and Joshi, A. (2008) Critical load bus voltage control using DVR under system frequency variation. *Electric Power Systems Research*, 78(2), 255–263.
42. Nabae, A., Yamaguchi, M., and Peng, F.Z. (1992) Suppression of resonance and flicker in an arc-furnace supply system. Proceedings of the Third AFRICON Conference, September 22–24, pp. 382–385.
43. Qun, W., Weizheng, Y., Jinjun, L., and Zhaoan, W. (1999) A control approach for detecting source current and series active power filter. Proceedings of the IEEE International Conference on Power Electronics and Drive Systems, vol. 2, pp. 910–914.

44. Qun, W., Weizheng, Y., Jinjun, L., and Zhaoan, W. (1999) Voltage type harmonic source and series active power filter adopting new control approach. Proceedings of the IEEE Industrial Electronics Conference, vol. 2, pp. 843–848.
45. Zhaoan, W., Qun, W., Weizheng, Y., and Jinjun, L. (2001) A series active power filter adopting hybrid control approach. *IEEE Transactions on Power Electronics*, 16, 301–310.
46. Xu, L., Acha, E., and Agelidis, V.G. (2001) A new synchronous frame based control strategy for a series voltage and harmonic compensator. Proceedings of the IEEE Applied Power Electronics Conference, vol. 2, pp. 1274–1280.
47. Singh, B. and Verma, V. (2003) A new control scheme of series active filter for varying rectifier loads. Proceedings of the Fifth International Conference on Power Electronics and Drive Systems (PEDS), November, vol. 1, pp. 554–559.
48. Pan, Z., Peng, F.Z., and Wang, S. (2005) Power factor correction using a series active filter. *IEEE Transactions on Power Electronics*, 20(1), 148–154.
49. Libano, F.B., Muller, S.L., Braga, R.A.M. *et al.* (2006) Simplified control of the series active power filter for voltage conditioning. Proceedings of IEEE ISIE, July 9–12, pp. 1706–1712.
50. Singh, B. and Verma, V. (2006) An indirect current control of hybrid power filter for varying loads. *IEEE Transactions on Power Delivery*, 21(1), 178–184.
51. Nastran, J., Cajhen, R., Seliger, M., and Jereb, P. (1994) Active power filter for non-linear AC loads. *IEEE Transactions on Power Electronics*, 9(1), 92–96.
52. Bhattacharya, S., Divan, D.M., and Banerjee, B. (1991) Synchronous frame harmonic isolator using active series filter. Fourth EPE Conference Record.
53. Aredes, M. and Watanabe, E.H. (1995) New control algorithms for series and shunt three-phase four-wire active power filter. *IEEE Transactions on Power Delivery*, 10(3), 1649–1656.
54. Qun, W., Weizheng, Y., Jinjun, L., and Zhaoan, W. (1999) A control approach for detecting source current and series active power filter. Proceedings of the IEEE International Conference on Power Electronics and Drive Systems, vol. 2, pp. 910–914.
55. Qun, W., Weizheng, Y., Jinjun, L., and Zhaoan, W. (1999) Voltage type harmonic source and series active power filter adopting new control approach. Proceedings of the IEEE Industrial Electronics Conference, vol. 2, pp. 843–848.
56. Zhaoan, W., Qun, W., Weizheng, Y., and Jinjun, L. (2001) A series active power filter adopting hybrid control approach. *IEEE Transactions on Power Electronics*, 16, 301–310.
57. Ribeiro, E.R. and Barbi, I. (2001) A series active power filter for harmonic voltage suppression. Proceedings of IEEE INTELEC'01, pp. 514–519.
58. Singh, B. and Verma, V. (2003) A new control scheme of series active filter for varying rectifier loads. Proceedings of the Fifth International Conference on Power Electronics and Drive Systems (PEDS), November, vol. 1, pp. 554–559.
59. Singh, B., Jayaprakash, P., Kothari, D.P. *et al.* (2008) Neural network based control of series active filter for compensation of voltage fed nonlinear loads. Proceedings of the IEEE International Symposium on Industrial Electronics (ISIE08), June 30–July 2, Cambridge, UK.
60. Jayaprakash, P., Singh, B., and Kothari, D.P. (2007) Control strategies of series active filters for harmonic current compensation in voltage fed nonlinear loads. NPEC 2007, December 17–19, Indian Institute of Science, Bangalore.

11

Hybrid Power Filters

11.1 Introduction

Solid-state conversion of AC power using diodes and thyristors is widely adopted to control a number of processes such as adjustable speed drives (ASDs), furnaces, chemical processes such as electroplating, power supplies, welding, and heating. The solid-state converters are also used in power industries such as HVDC transmission systems, battery energy storage systems, and interfacing renewable energy electricity generating systems. Some solid-state controllers draw harmonic currents and reactive power from the AC mains and behave as nonlinear loads. Moreover, in three-phase AC mains, they cause unbalance and excessive neutral current resulting in low power factor and poor efficiency of the system. In addition, they cause poor utilization of the distribution system, RFI and EMI noise, interference to the communication system, voltage distortion, disturbance to neighboring consumers, and poor power quality at the AC source due to notch, sag, swell, noise, spikes, surge, flicker, unbalance, low-frequency oscillations, and malfunction of protection systems. Because of the severity of power quality problems, several standards are developed and are being enforced on the consumers, manufacturers, and utilities. Moreover, the power community has become more conscious about these problems and a number of technology options are reported in the texts and research publications. Initially, lossless passive filters (*LC*) have been used to reduce harmonics and capacitors have been chosen for power factor correction of the nonlinear loads. But passive filters have the demerits of fixed compensation, large size, and resonance with the supply system. Active power filters (APFs) have been explored in shunt and series configurations to compensate different types of nonlinear loads. However, they have drawbacks that their rating sometimes is very close to the load (up to 80%) in some typical applications and thus they become a costly option for power quality improvement in a number of situations. Moreover, a single active power filter does not provide a complete solution for compensation in many cases of nonlinear loads due to the presence of both voltage- and current-based power quality problems. However, many researchers have classified different types of nonlinear loads and have suggested various filter options for their compensation. Because of higher rating of APFs and cost considerations, the acceptability of the APFs by the users has faced a hindrance in practical situations. In response to these factors, a series of hybrid power filters (HPFs) is evolved and extensively used in practice as a cost-effective solution for the compensation of nonlinear loads. Moreover, the HPFs are found to be more effective in providing complete compensation of various types of nonlinear loads. The rating of active filters is reduced by adding passive filters to form hybrid filters, which reduces the overall cost, and in many instances they provide better compensation than either passive or active filters. Therefore, it is considered a timely attempt to present a broad perspective on the hybrid filter technology for the power community dealing with power quality issues.

This chapter deals with state of the art, classification, principle of operation, control, analysis, design, modeling, simulation, and performance of HPFs. The HPFs are classified based on the number of

---

Power Quality Problems and Mitigation Techniques, First Edition.
Bhim Singh, Ambrish Chandra and Kamal Al-Haddad.

elements in the topology, supply system, and type of converter used in their circuits. The supply system can be a single-phase (two-wire) or three-phase (three-wire and four-wire) system to feed a variety of nonlinear loads. The converter can be a voltage source converter (VSC) or a current source converter (CSC) to realize the APF part of the hybrid filter with appropriate control. The number of elements in the topology can be two, three, or more, which may be either active filters (AFs) or passive filters (PFs). Here, the main classification is made on the basis of the supply system with further subclassification on the basis of filter elements. Starting with the introduction, the other sections include state of the art on the HPFs, their classification, principle of operation and control, analysis and design, modeling and simulation of performance, numerical examples, summary, review questions, numerical problems, computer simulation-based problems, and references.

11.2 State of the Art on Hybrid Power Filters

The technology of power filters is now a mature technology for compensating different types of nonlinear loads through current-based compensation and for improving the power quality of AC supply through voltage-based compensation techniques that eliminate voltage harmonics, sags, swell, notches, glitches, spikes, flickers, and voltage unbalance and provide voltage regulation. Moreover, these filters are also identified according to the nature of nonlinear loads such as voltage-fed loads (voltage stiff or voltage source on the DC side of the rectifier through the capacitive filter), current-fed loads (current stiff or current source on the DC side of the DC motor drive or current source for the CSI-fed AC motor drive), and a combination of both. Various topologies such as passive, active, and hybrid filters in shunt, series, and a combination of both configurations for single-phase two-wire, three-phase three-wire, and three-phase four-wire systems have been proposed using current source and voltage source converters to improve the power quality at the AC mains. As mentioned earlier, hybrid filters are a cost-effective and perfect solution for the compensation of nonlinear loads and for providing clean and ideal AC supply to a variety of loads. This section describes the chronological development and the current status of the HF technology.

Because of the extensive use of solid-state converters, the pollution level in the AC supply system is increasing rapidly and power quality has become an important area of research. A number of standards, surveys, and texts have been published for improving the power quality and maintaining it to the prescribed level through different approaches in single-phase two-wire, three-phase three-wire, and three-phase four-wire systems. Moreover, hybrid filters are developed using one, two, or three passive and active filters either to improve their performance or to reduce the cost of the system compared with single active or passive filters. Lossless passive filters (*LC*) have been used for a long time as a combination of single tuned, double tuned, and damped high-pass filters either to absorb current harmonics by creating a harmonic valley in shunt with current-fed nonlinear loads (thyristor-based DC motor drive, HVDC, DC current source for CSI, etc.) or to block harmonic currents by creating a harmonic dam in series with voltage-fed nonlinear loads. However, the passive filters have the limitations of fixed compensation and resonance with the supply system, which are normally overcome by using AFs). A single unit of AF normally has a high rating resulting in high cost and even does not provide complete compensation. The rating of active filters is reduced by adding passive filters to form hybrid filters, which reduces the overall cost, and in many instances they provide better compensation than either passive or active filters. However, if one can afford the cost, then a hybrid of two active filters provides the perfect and best solution and thus it is known as a universal power quality conditioner or universal active power conditioner (UAPC). Therefore, the development in hybrid filter technology has been from a hybrid of passive filters to a hybrid of active filters, that is, from a cost-effective solution to a perfect solution.

In a single-phase system, there are a large number of nonlinear loads such as fluorescent lamps, ovens, TVs, computers, air conditioners, power supplies, printers, copiers, and high-rating furnaces and traction systems. These loads are compensated using a hybrid of passive filters as a low-cost solution and a hybrid of active filters in the traction. A major amount of power is processed in a three-phase three-wire system, either in ASDs in small rating to reasonable power level or in the HVDC transmission system in high power rating and they behave as nonlinear loads. These loads are also compensated by using either a group of passive filters or a combination of active and passive filters in different configurations depending upon their nature to the AC system, such as current-fed loads, voltage-fed loads, or a combination of both. Vastly distributed single-phase nonlinear loads cause power quality problems in a three-phase four-wire AC system and are compensated by using a number of passive filters, active filters, or hybrid filters.

One of the major reasons for the advancement in hybrid filter technology using active filter elements is the development of fast self-commutating solid-state devices such as MOSFETs (metal-oxide semiconductor field-effect transistors) and IGBTs (insulated gate bipolar transistors). An improved and low-cost sensor technology is also responsible for reducing the cost and improving the response of HPFs. Fast Half effect sensors and compact isolation amplifiers have resulted in HPFs with affordable cost. Another major factor that has contributed to the HPF technology is the evolution of microelectronics. The development of low-cost, high-accuracy, and fast digital signal processors (DSPs), microcontrollers, and application-specific integrated circuits (ASICs) has made it possible to implement complex control algorithms for online control at an affordable price. A number of control theories of HPFs such as instantaneous reactive power theory (IRPT), synchronously rotating frame (SRF) theory, and many more with several low-pass, high-pass, and band-pass digital filters along with several closed-loop controllers such as proportional–integral (PI) controller and sliding mode controller (SMC) have been employed to implement hybrid filters. Moreover, many manufacturers are developing hybrid filters even in quite large power rating to improve the power quality of a variety of nonlinear loads.

11.3 Classification of Hybrid Power Filters

HPFs can be classified based on the number of elements in the topology, supply system, and type of converter used in their circuits. The supply system can be a single-phase two-wire, three-phase three-wire, or three-phase four-wire system to feed a variety of nonlinear loads. The converter can be a VSC or a CSC to realize the APF part of the hybrid power filter with appropriate control. The number of elements in the topology can be either two, three, or more, which may be either APFs or passive power filters (PPFs). Here, the main classification is made on the basis of the supply system with further subclassification on the basis of filter elements. Figure 11.1 shows the proposed classification of hybrid power filters based on the supply system with topology as further subclassification. However, there is common subclassification in each case of the supply system. Therefore, major classification is made on the basis of number (two and three) and types of elements (passive and active filters) in different topologies in each case of the supply system. The hybrid filters consisting of two passive elements have two circuit configurations, as shown in Figures 11.2 and 11.3, and those consisting of three passive elements also have two circuit configurations, as shown in Figures 11.4 and 11.5. The hybrid filters consisting of two elements, one active and one passive filter, have eight valid circuit configurations, as shown in Figures 11.6–11.13. Similarly, the hybrid filters consisting of three elements, two passive with one active and one passive with two active filter elements, have 18 valid circuit configurations each, resulting in 36 circuit configurations, as shown

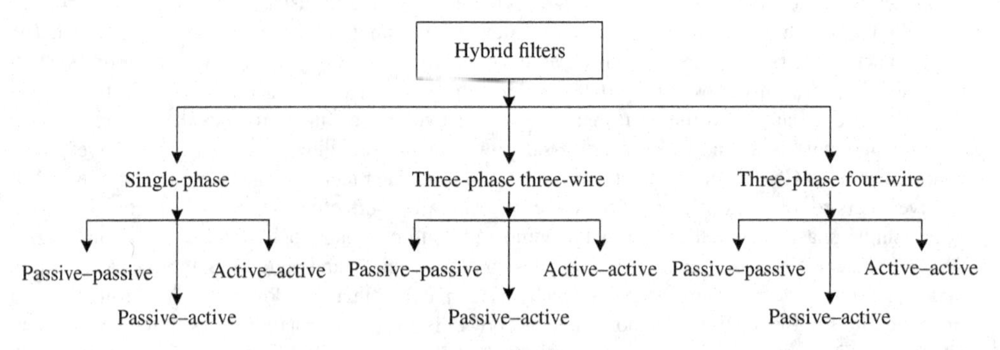

Figure 11.1 Classification of hybrid filters for power quality improvement

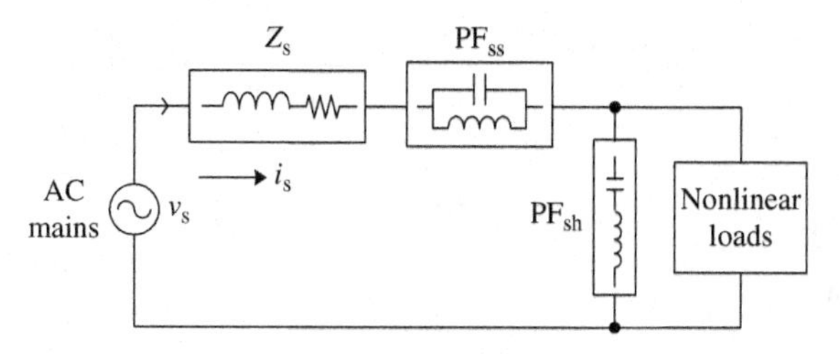

Figure 11.2 A hybrid filter as a combination of passive series (PF_{ss}) and passive shunt (PF_{sh}) filters

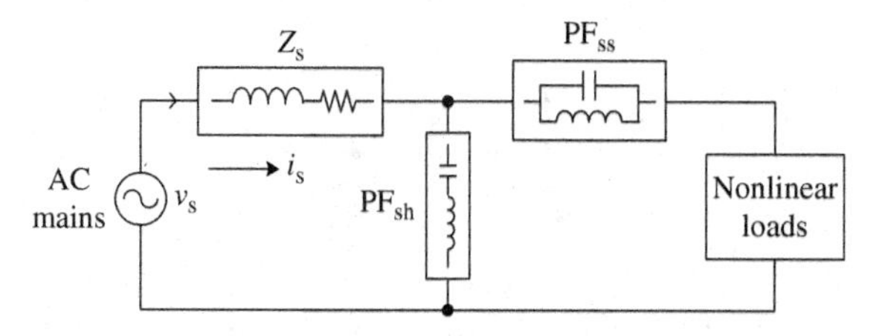

Figure 11.3 A hybrid filter as a combination of passive shunt (PF_{sh}) and passive series (PF_{ss}) filters

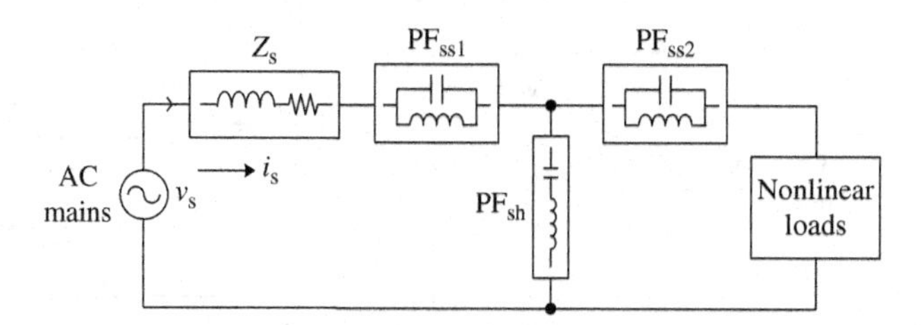

Figure 11.4 A hybrid filter as a combination of passive series (PF_{ss1}), passive shunt (PF_{sh}), and passive series (PF_{ss2}) filters

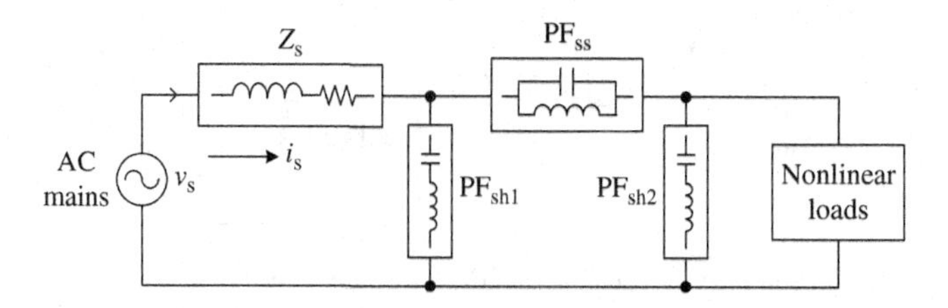

Figure 11.5 A hybrid filter as a combination of passive shunt (PF_{sh1}), passive series (PF_{ss}), and passive shunt (PF_{sh2}) filters

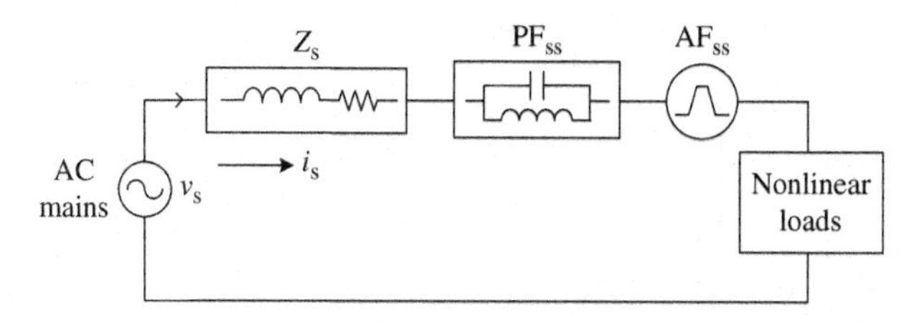

Figure 11.6 A hybrid filter as a combination of series connected passive series (PF_{ss}) and active series (AF_{ss}) filters

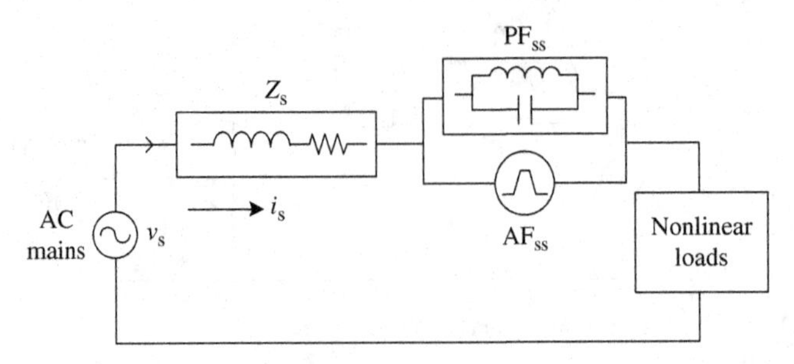

Figure 11.7 A hybrid filter as a combination of parallel connected passive series (PF_{ss}) and active series (AF_{ss}) filters

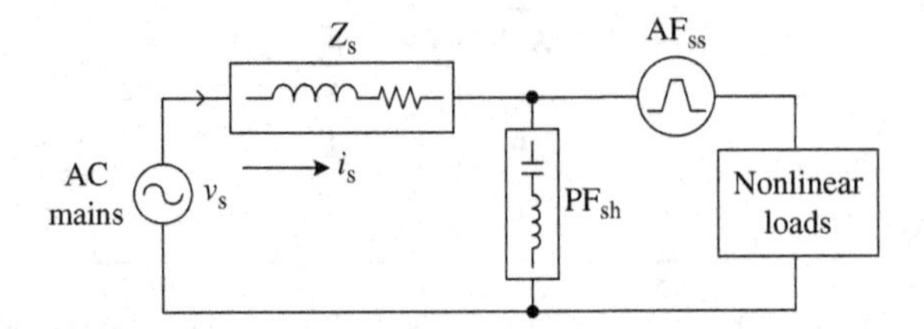

Figure 11.8 A hybrid filter as a combination of passive shunt (PF_{sh}) and active series (AF_{ss}) filters

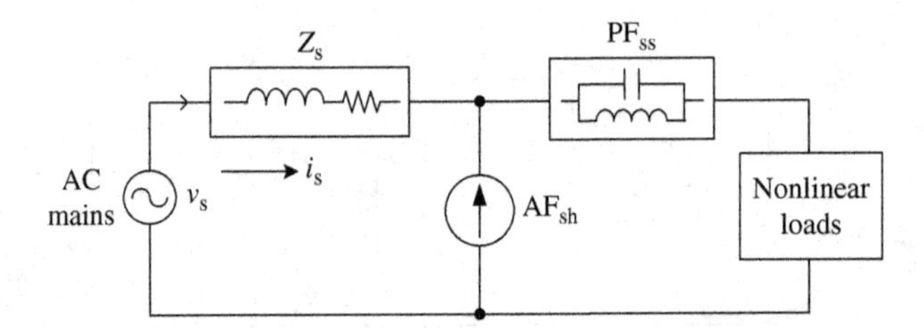

Figure 11.9 A hybrid filter as a combination of active shunt (AF_{sh}) and passive series (PF_{ss}) filters

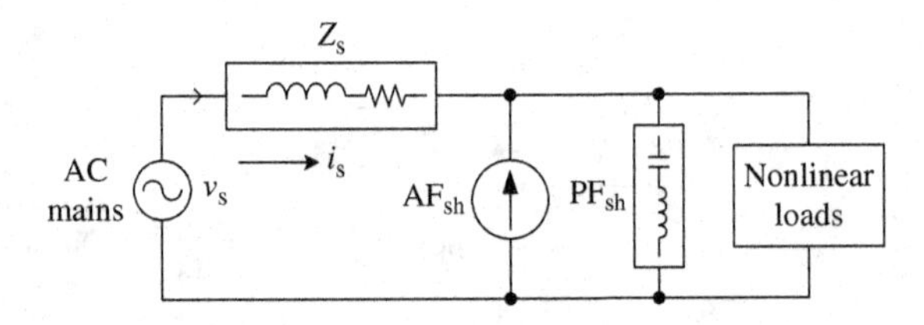

Figure 11.10 A hybrid filter as a combination of active shunt (AF_{sh}) and passive shunt (PF_{sh}) filters

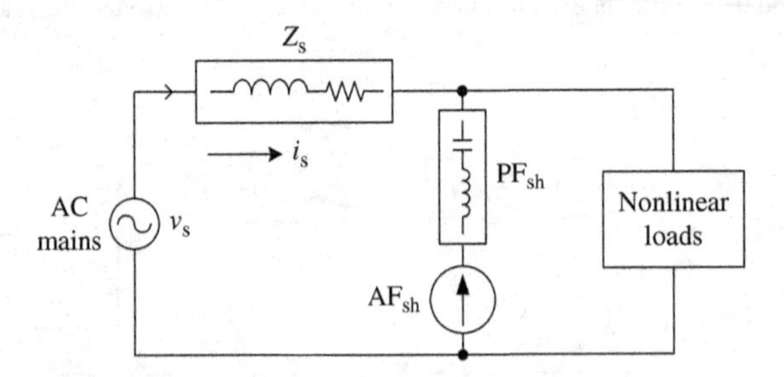

Figure 11.11 A hybrid filter as a combination of series connected passive shunt (PF_{sh}) and active shunt (AF_{sh}) filters

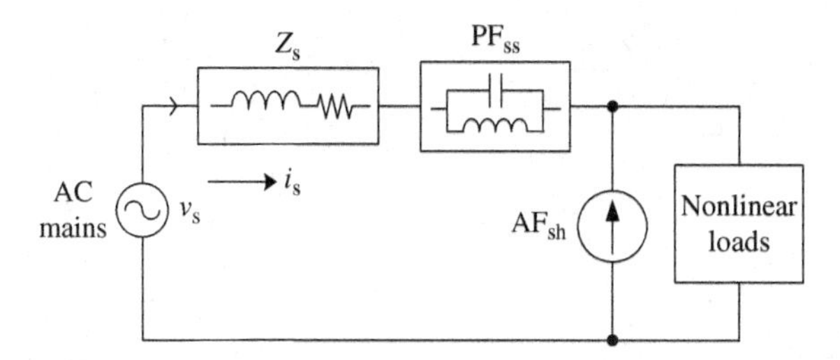

Figure 11.12 A hybrid filter as a combination of passive series (PF_{ss}) and active shunt (AF_{sh}) filters

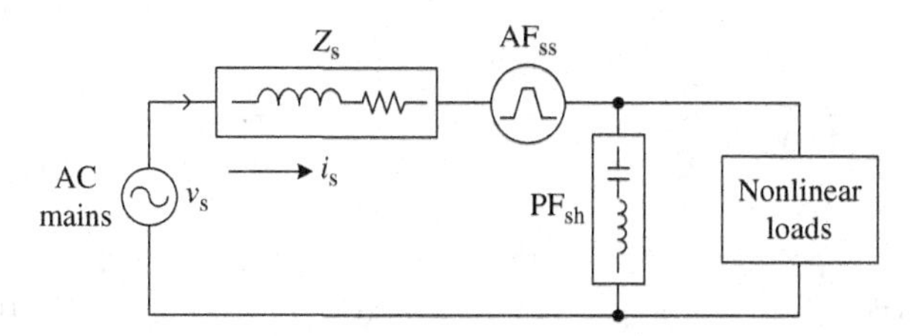

Figure 11.13 A hybrid filter as a combination of active series (AF_{ss}) and passive shunt (PF_{sh}) filters

in Figures 11.14–11.49. The hybrid filters consisting of two and three active filter elements have two circuit configurations each, as shown in Figures 11.50–11.53. The hybrid filters consisting of more than three elements are rarely used due to cost and complexity considerations and hence are not included here. The hybrid filters consisting of two and three active and passive elements result in 52 practically valid circuit configurations.

These 52 circuit configurations of hybrid filters are valid for each case of the supply system, for example, single-phase two-wire, three-phase three-wire, and three-phase four-wire AC systems. In each case of the supply system, four basic elements of the filter circuit, such as passive series (PF_{ss}), passive shunt (PF_{sh}), active series (AF_{ss}), and active shunt (AF_{sh}), are required to develop complete hybrid filter circuit configurations. However, there may be many more combinations such as active filter elements using current source converters or voltage source converters.

Normally, each passive filter element employs three tuned filters, the first two being of lowest dominant harmonics followed by a high-pass filter element. However, in some high-power applications such as HVDC systems, five tuned filter elements are used, the first four tuned for four lower dominant harmonics and fifth one as a high-pass filter element. In a passive series filter element (PF_{ss}), two lossless *LC* components are connected in parallel for creating a harmonic dam to block harmonic currents. All the three or five components of the passive series filter are connected in a series configuration. However, in a passive shunt filter element (PF_{sh}), two lossless *LC* components are connected in series for creating a

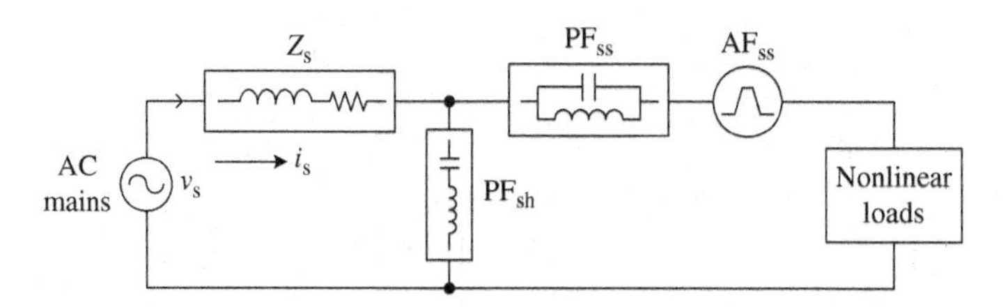

Figure 11.14 A hybrid filter as a combination of passive shunt (PF_{sh}), passive series (PF_{ss}), and active series (AF_{ss}) filters

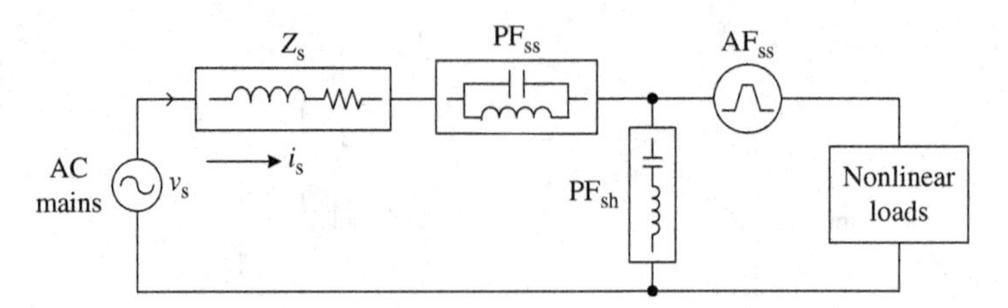

Figure 11.15 A hybrid filter as a combination of passive series (PF_{ss}), passive shunt (PF_{sh}), and active series (AF_{ss}) filters

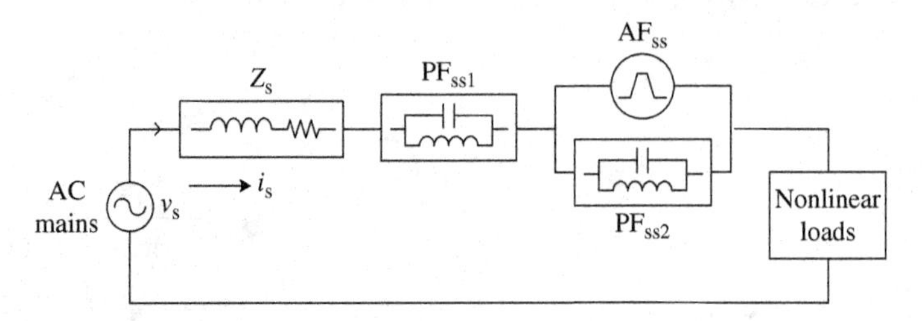

Figure 11.16 A hybrid filter as a combination of passive series (PF_{ss1}) in series with a parallel connected active series (AF_{ss}) and passive series (PF_{ss2}) filters

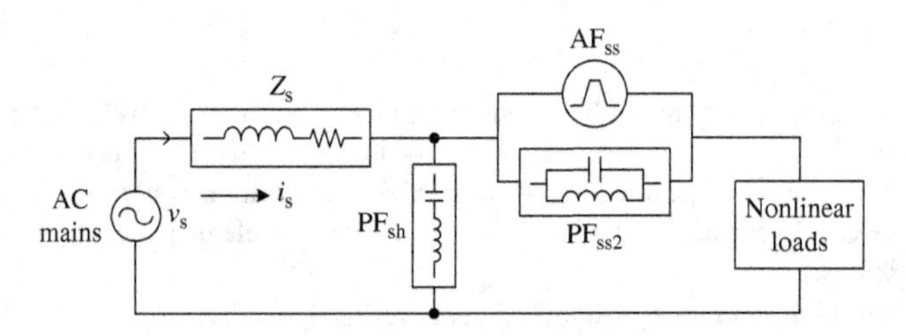

Figure 11.17 A hybrid filter as a combination of passive shunt (PF_{sh}) and parallel connected active series (AF_{ss}) and passive series (PF_{ss}) filters

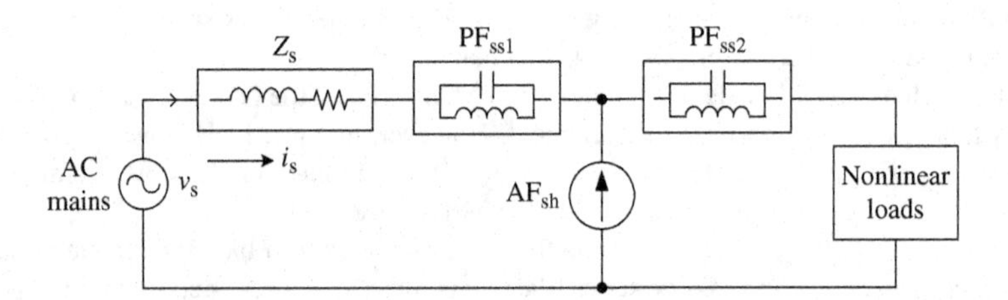

Figure 11.18 A hybrid filter as a combination of passive series (PF_{ss1}), active shunt (AF_{sh}), and passive series (PF_{ss2}) filters

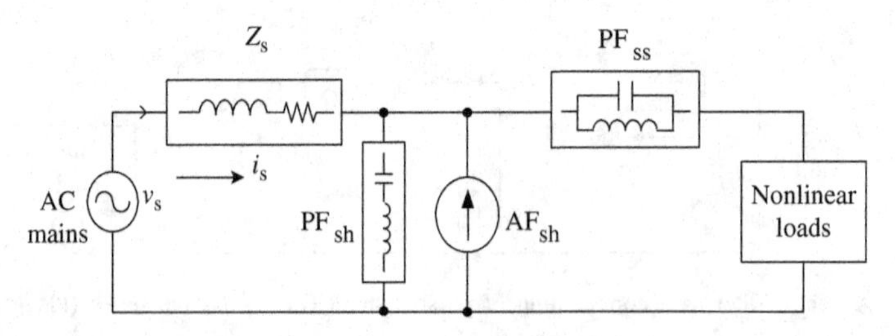

Figure 11.19 Hybrid filter as a combination of parallel connected passive shunt (PF_{sh}) with active shunt (AF_{sh}) and passive series (PF_{ss}) Filters

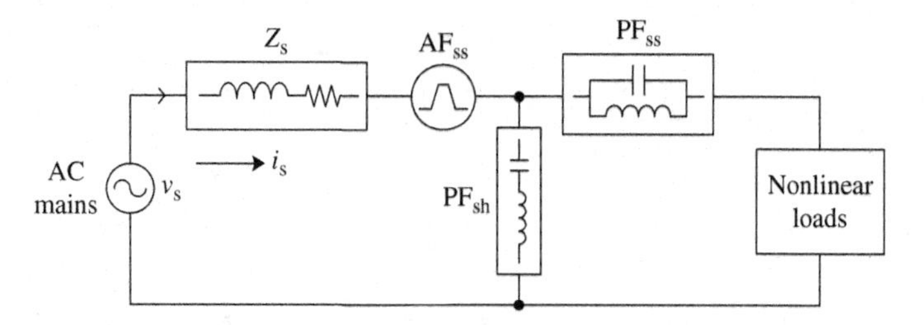

Figure 11.20 A hybrid filter as a combination of active series (AF_{ss}), passive shunt (PF_{sh}), and passive series (PF_{ss}) filters

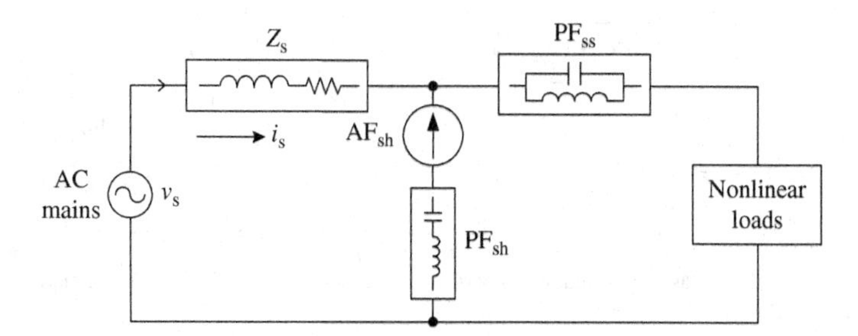

Figure 11.21 A hybrid filter as a combination of series connected passive shunt (PF_{sh}) with active shunt (AF_{sh}) and passive series (PF_{ss}) filters

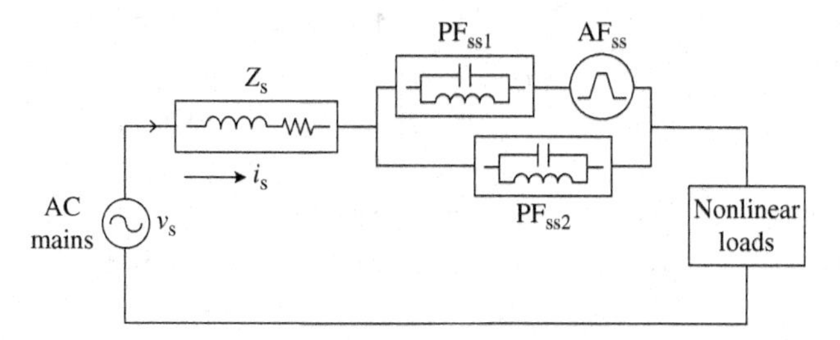

Figure 11.22 A hybrid filter as a combination of series connected passive series (PF_{ss1}) with active series (AF_{ss}) in parallel with passive series (PF_{ss2}) filters

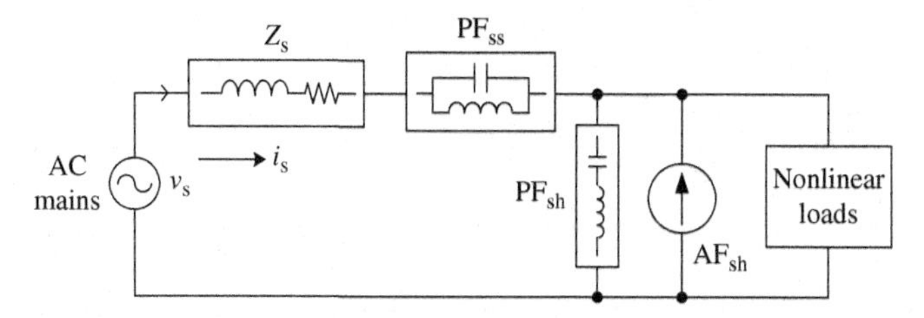

Figure 11.23 A hybrid filter as a combination of passive series (PF_{ss}) and parallel connected passive shunt (PF_{sh}) with active shunt (AF_{sh}) filters

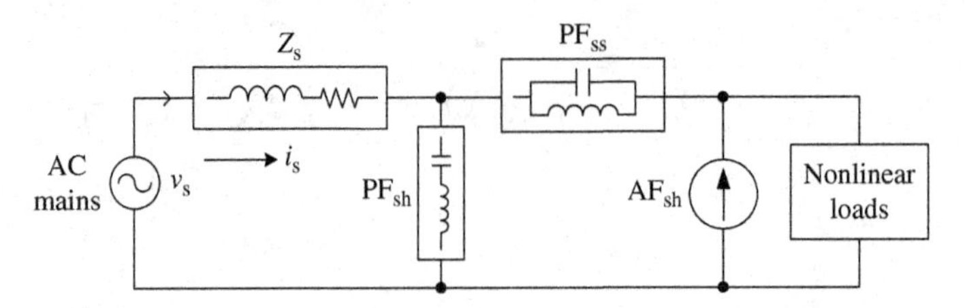

Figure 11.24 A hybrid filter as a combination of passive shunt (PF_{sh}), passive series (PF_{ss}), and active shunt (AF_{sh}) filters

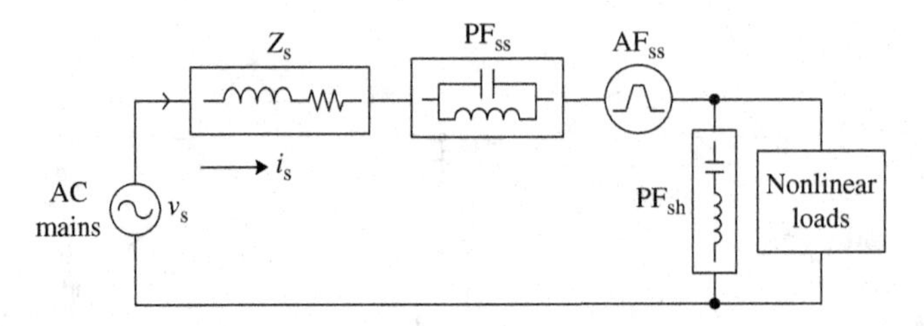

Figure 11.25 A hybrid filter as a combination of series connected passive series (PF_{ss}) with active series (AF_{ss}) and passive shunt (PF_{sh}) filters

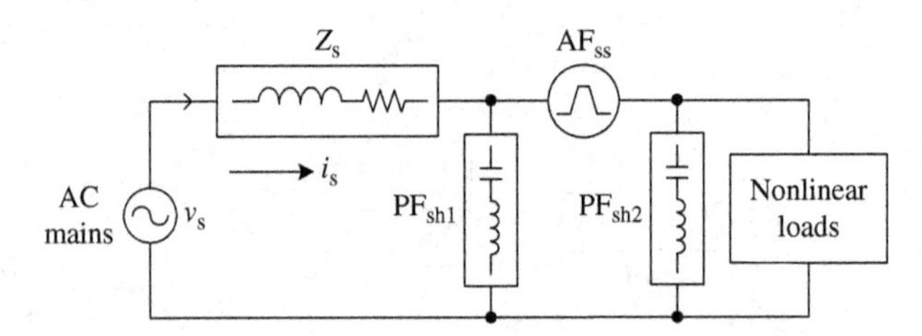

Figure 11.26 A hybrid filter as a combination of passive shunt (PF_{sh1}), active series (AF_{ss}), and passive shunt (PF_{sh2}) filters

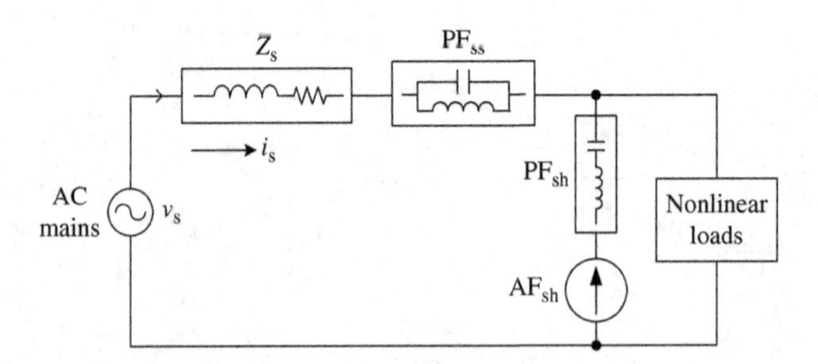

Figure 11.27 A hybrid filter as a combination of passive series (PF_{ss}) and series connected passive shunt (PF_{sh}) with active shunt (AF_{sh}) filters

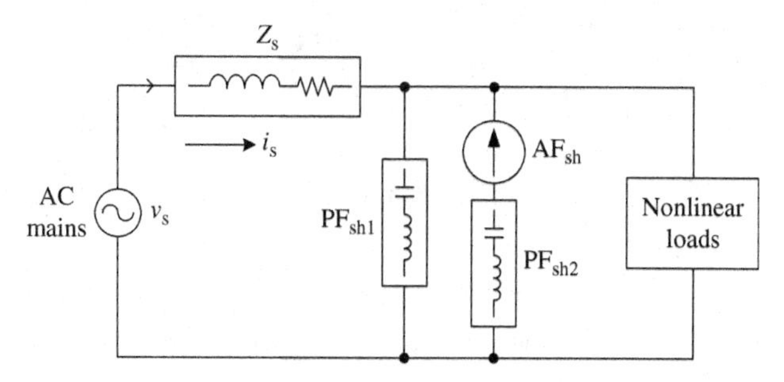

Figure 11.28 A hybrid filter as a combination of passive shunt (PF_{sh1}) and series connected active shunt (AF_{sh}) with passive shunt (PF_{sh2}) filters

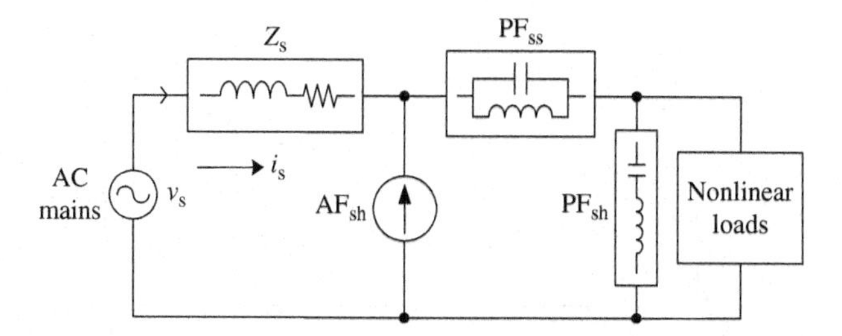

Figure 11.29 A hybrid filter as a combination of active shunt (AF_{sh}), passive series (PF_{ss}), and passive shunt (PF_{sh}) filters

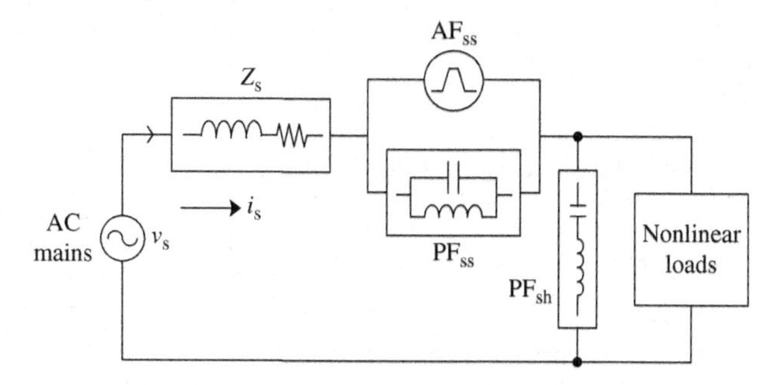

Figure 11.30 A hybrid filter as a combination of parallel connected active series (AF_{ss}) with passive series (PF_{ss}) and passive shunt (PF_{sh}) filters

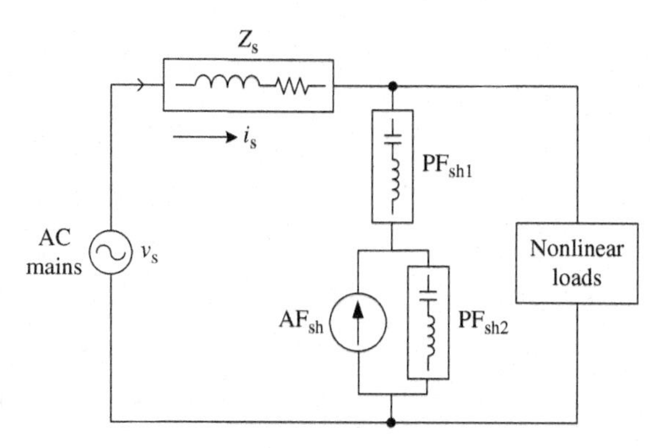

Figure 11.31 A hybrid filter as a combination of passive shunt (PF_{sh1}) and parallel connected passive shunt (PF_{sh2}) with active shunt (AF_{sh}) filters

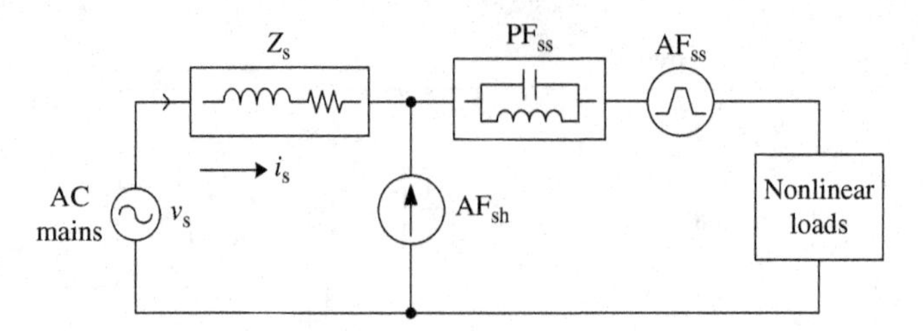

Figure 11.32 A hybrid filter as a combination of active shunt (AF_{sh}), passive series (PF_{ss}), and active series (AF_{ss}) filters

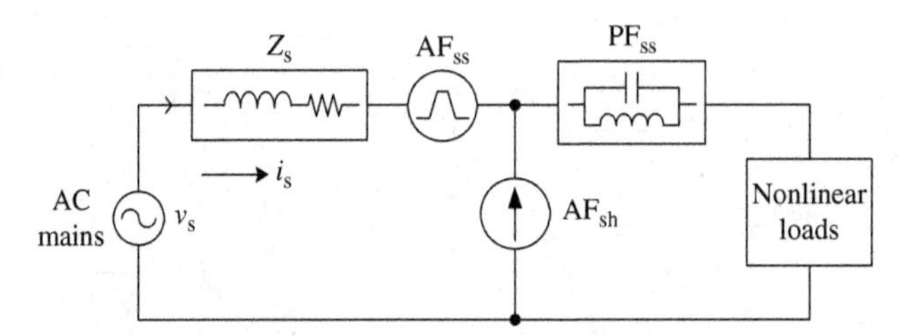

Figure 11.33 A hybrid filter as a combination of active series (AF_{ss}), active shunt (AF_{sh}), and passive series (PF_{ss}) filters

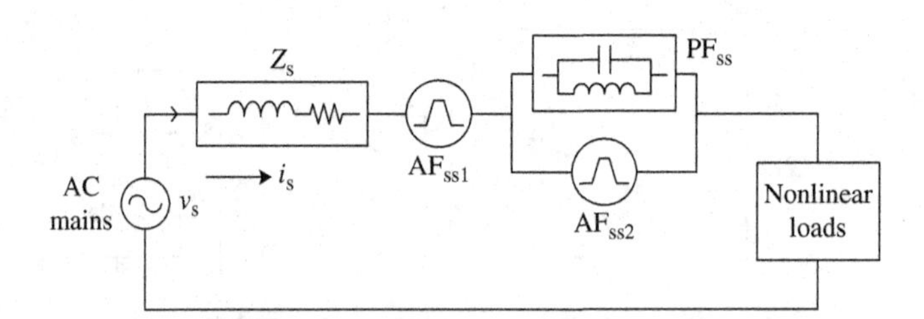

Figure 11.34 A hybrid filter as a combination of active series (AF_{ss1}) and parallel connected passive series (PF_{ss}) with active series (AF_{ss2}) filters

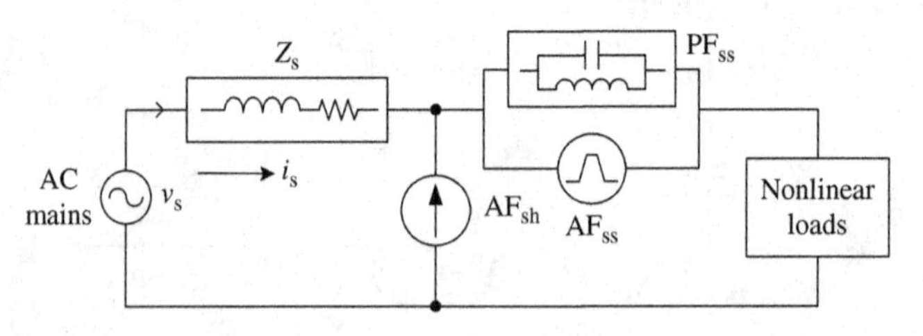

Figure 11.35 A hybrid filter as a combination of active shunt (AF_{sh}) and parallel connected passive series (PF_{ss}) with active series (AF_{ss}) filters

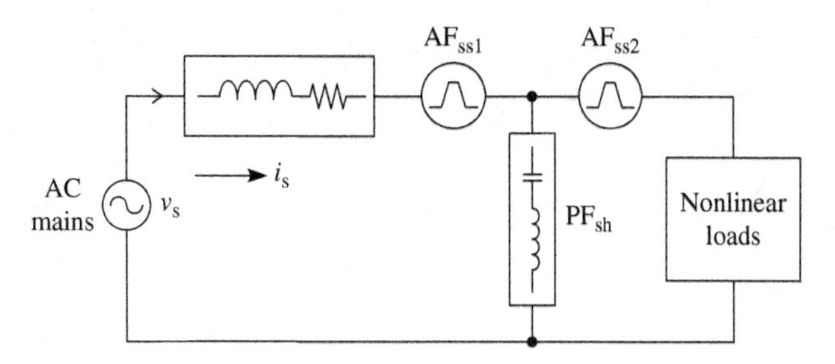

Figure 11.36 A hybrid filter as a combination of active series (AF$_{ss1}$), passive shunt (PF$_{sh}$), and active series (AF$_{ss2}$) filters

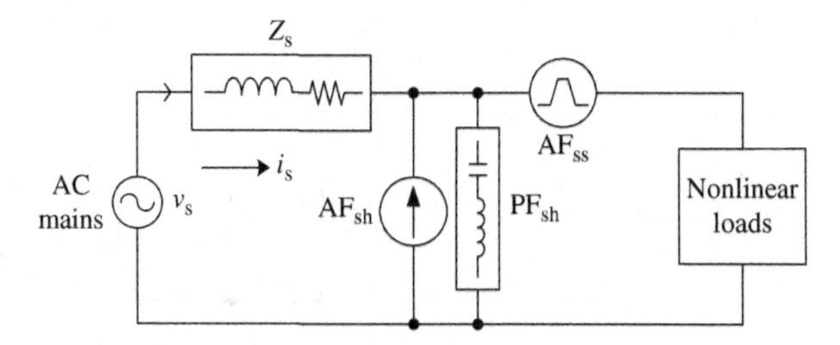

Figure 11.37 A hybrid filter as a combination of active shunt (AF$_{sh}$), passive shunt (PF$_{sh}$), and active series (AF$_{ss}$) filters

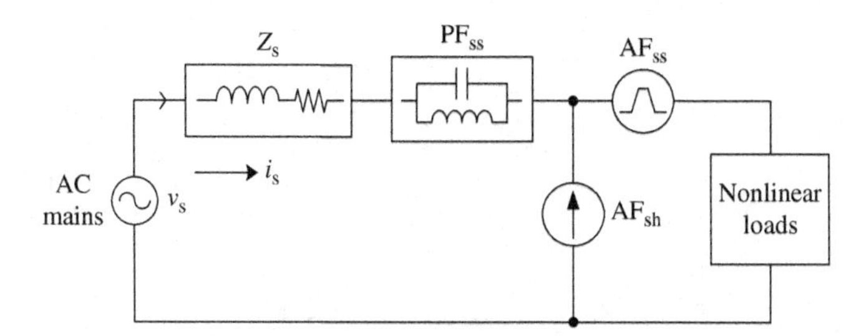

Figure 11.38 A hybrid filter as a combination of passive series (PF$_{ss}$), active shunt (AF$_{sh}$), and active series (AF$_{ss}$) filters

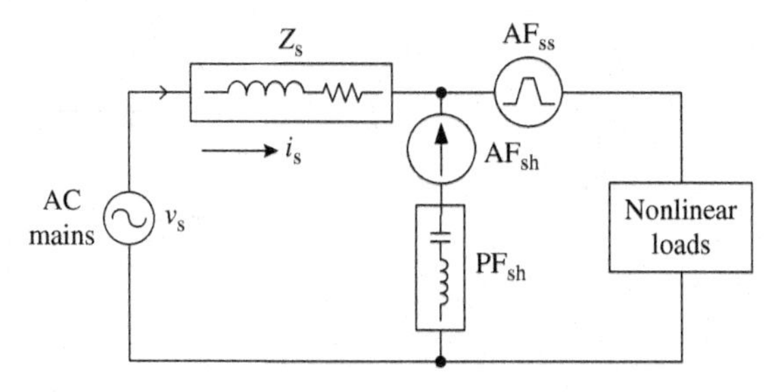

Figure 11.39 A hybrid filter as a combination of series connected active shunt (AF$_{sh}$) with passive shunt (PF$_{sh}$) and active series (AF$_{ss}$) filters

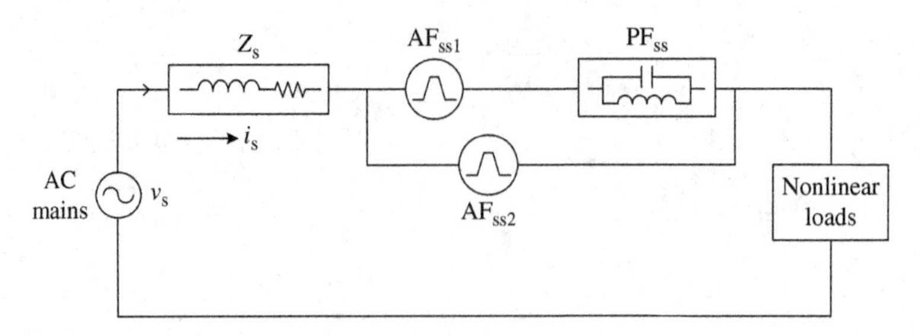

Figure 11.40 A hybrid filter as a combination of active series (AF_{ss1}), passive series (PF_{ss}), and active series (AF_{ss2}) filters

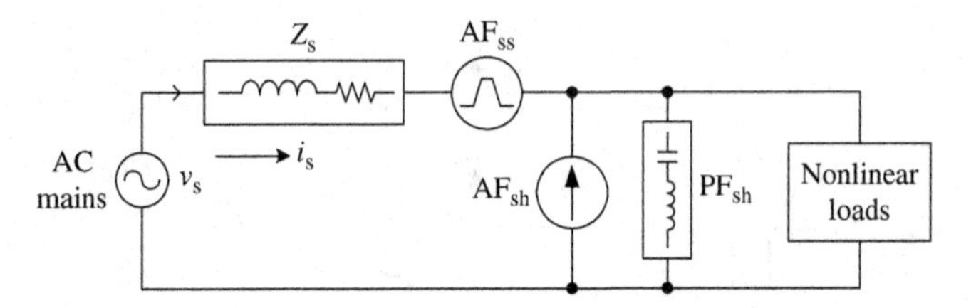

Figure 11.41 A hybrid filter as a combination of active series (AF_{ss}), active shunt (AF_{sh}), and passive shunt (PF_{sh}) filters

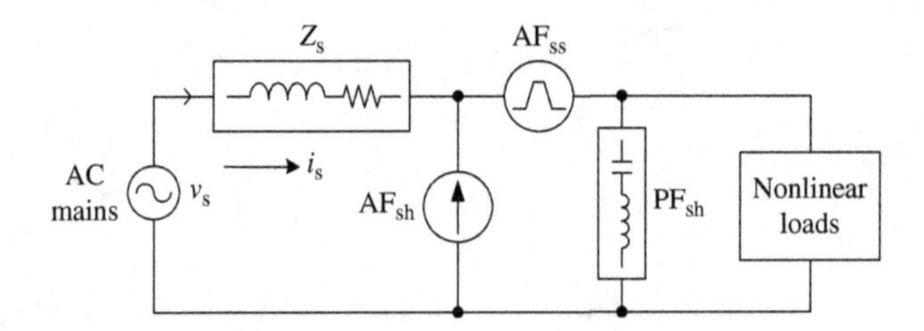

Figure 11.42 A hybrid filter as a combination of active shunt (AF_{sh}), active series (AF_{ss}), and passive shunt (PF_{sh}) filters

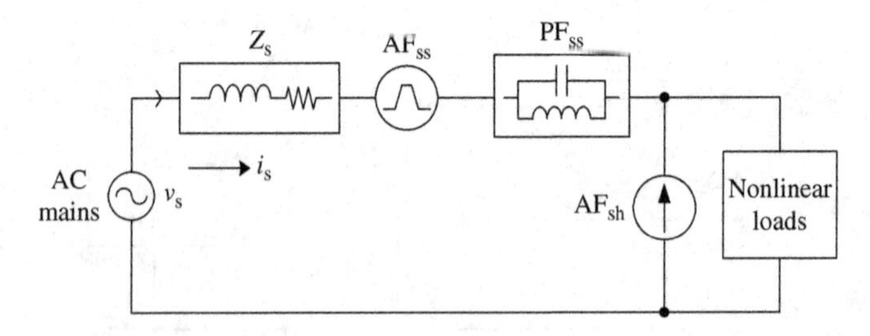

Figure 11.43 A hybrid filter as a combination of active series (AF_{ss}), passive series (PF_{ss}), and active shunt (AF_{sh}) filters

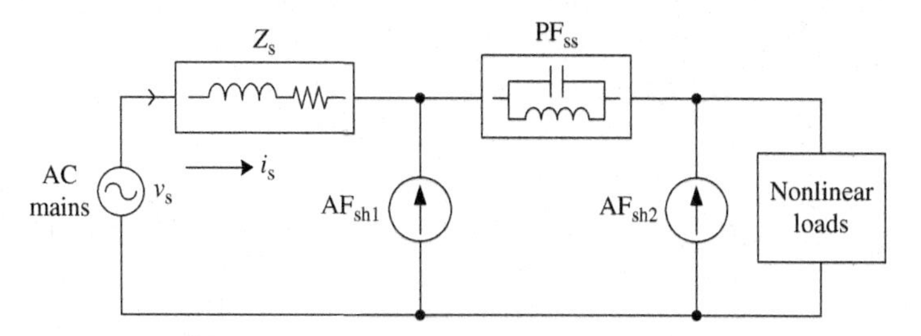

Figure 11.44 A hybrid filter as a combination of active shunt (AF_{sh1}), passive series (PF_{ss}), and active shunt (AF_{sh2}) filters

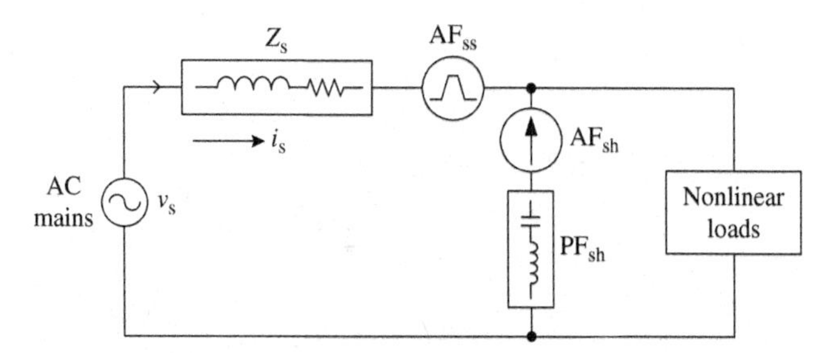

Figure 11.45 A hybrid filter as a combination of active series (AF_{ss}) and series connected active shunt (AF_{sh}) and passive shunt (PF_{sh}) filters

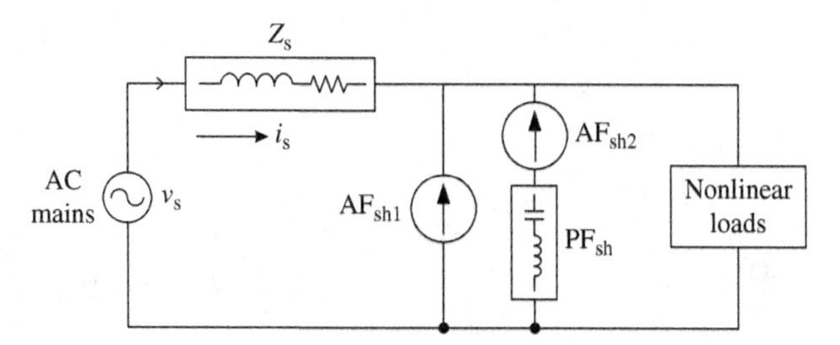

Figure 11.46 A hybrid filter as a combination of active shunt (AF_{sh1}) and series connected active shunt (AF_{sh2}) and passive shunt (PF_{sh}) filters

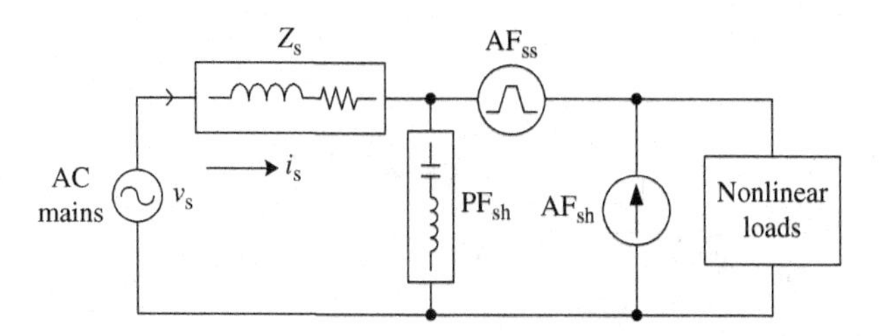

Figure 11.47 A hybrid filter as a combination of passive shunt (PF_{sh}), active series (AF_{ss}), and active shunt (AF_{sh}) filters

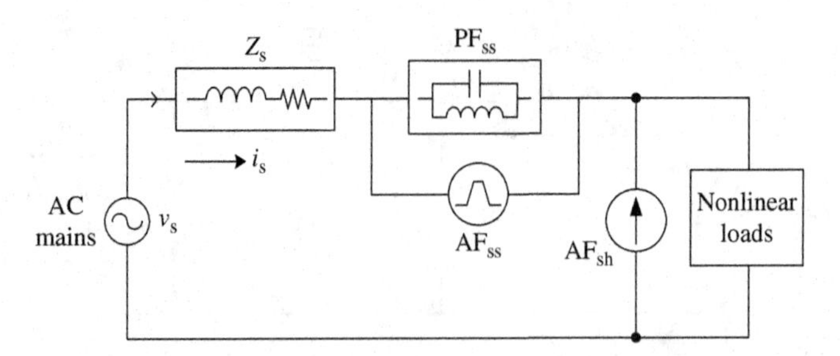

Figure 11.48 A hybrid filter as a combination of parallel connected passive series (PF_{ss}) with active series (AF_{ss}) and active shunt (AF_{sh}) filters

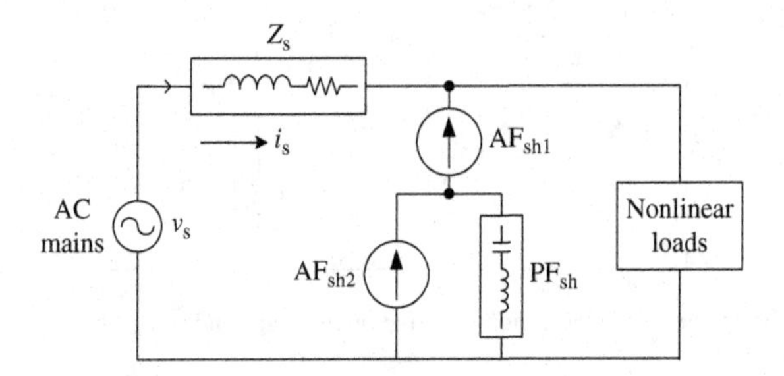

Figure 11.49 A hybrid filter as a combination of active shunt (AF_{sh1}) in series with parallel connected active shunt (AF_{sh2}) and passive shunt (PF_{sh}) filters

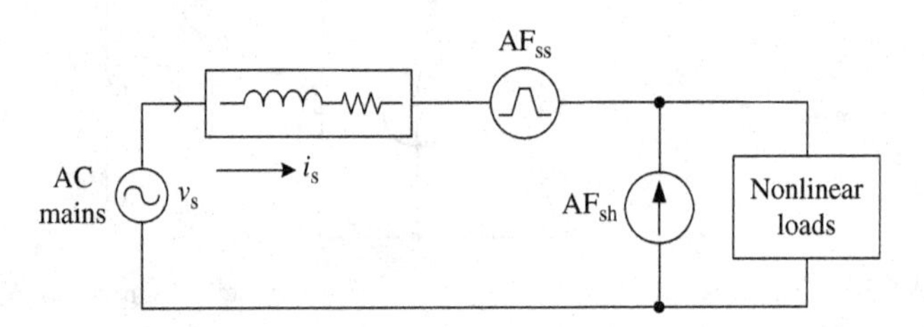

Figure 11.50 A hybrid filter as a combination of active series (AF_{ss}) and active shunt (AF_{sh}) filters

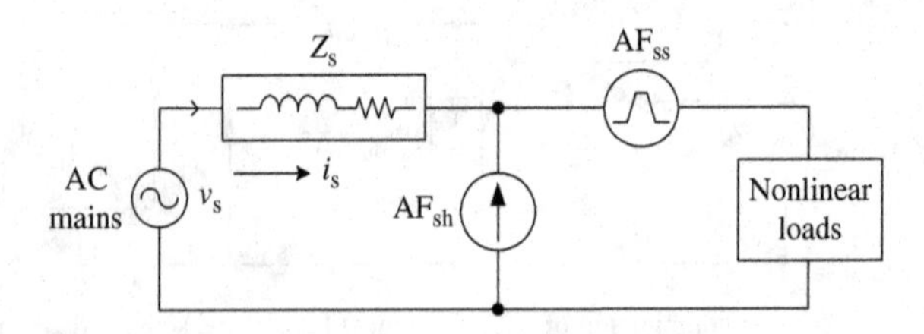

Figure 11.51 A hybrid filter as a combination of active shunt (AF_{sh}) and active series (AF_{ss}) filters

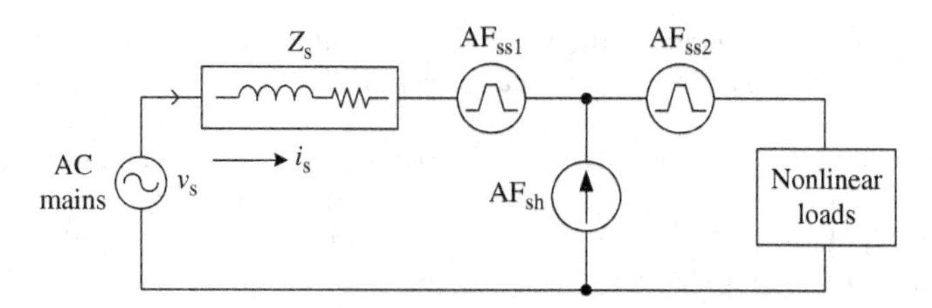

Figure 11.52 A hybrid filter as a combination of active series (AF_{ss1}), active shunt (AF_{sh}), and active series (AF_{ss2}) filters

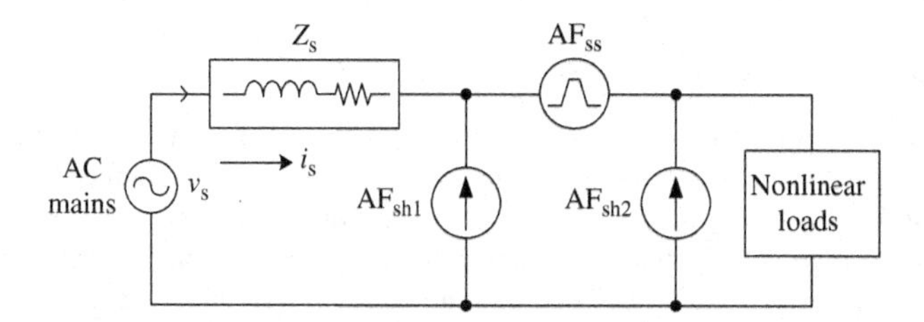

Figure 11.53 A hybrid filter as a combination of active shunt (AF_{sh1}), active series (AF_{ss}), and active shunt (AF_{sh2}) filters

harmonic valley to absorb harmonic currents. All the three or five components of the passive shunt filter (PF_{sh}) are connected in a parallel configuration.

Similarly, each active filter element employs a VSC preferably with a self-supporting DC bus having an electrolytic capacitor (C_d) and an AC inductor (L_r) along with an optional small AC capacitor (C_r) to form a ripple filter to eliminate the switching ripple. It may also use a CSC with inductive energy storage at DC link using current control along with shunt AC capacitors to form an active filter element. However, a VSC is normally preferred due to various advantages such low losses, small size, and low noise. Depending upon the supply system, the VSC-based active filter element may be single-phase two-arm H-bridge, three-phase three-arm bridge, and three-phase four-arm, midpoint, or three single-phase VSC. These units can be connected in series directly in single phase to reduce the cost or through injunction transformers usually with higher turns on the VSC side to form the active series filter element (AF_{ss}) for two-wire, three-wire, and four-wire systems to act as a high active impedance to block harmonic currents and a low impedance for fundamental frequency current. In the same manner, the active shunt filter element (AF_{sh}) may be connected either directly or through step-down transformers to connect the VSC at optimum voltage to act as an adjustable sink for harmonic currents for three cases of the AC supply system.

There are 156 valid basic circuit configurations of HFs for all three cases of the supply system to suit majority of applications for improving the power quality of the system having either nonlinear loads or polluted AC supply. Moreover, there may be many more variations in the active or passive filter element, but the basic concept of HFs remains out of these circuit configurations.

11.4 Principle of Operation and Control of Hybrid Power Filters

Many configurations of hybrid power filters have been discussed in the previous section for mitigating various power quality problems in addition to eliminating voltage and current harmonics. A large number of these configurations of hybrid power filters are reported in the literature for power quality improvement by compensation of various types of nonlinear loads. Here mainly four configurations of hybrid power filters are discussed, which are most prominently used in practice as a combination of passive filters, a combination of active filters, and a combination of an active filter and a passive filter to provide a cost-effective universal filter for mitigating multiple power quality problems caused by nonlinear loads and supply systems. Conceptually, these HPFs consist of (a) a combination of passive series (PF_{ss}) and passive shunt (PF_{sh}) filters (Figure 11.2), (b) a combination of series connected passive shunt (PF_{sh}) and active shunt (AF_{sh}) filters (Figure 11.11), (c) a combination of active series (AF_{ss}) and passive shunt (PF_{sh}) filters (Figure 11.13), and (d) a combination of active series (AF_{ss}) and active shunt (AF_{sh}) filters (Figure 11.50). Out of the 52 configurations of HPFs, these 4 configurations have been preferred due to a number of benefits and to meet the requirements of various types of nonlinear loads. Therefore, the principle of operation and control of HPFs are limited to these four hybrid power filters. However, a large number of configurations of HPFs are illustrated in numerical examples. Here, most of the concepts are given for three-phase HPFs, which can also be extended to single-phase hybrid power filters.

11.4.1 Principle of Operation of Hybrid Power Filters

The main objective of HPFs is to mitigate multiple power quality problems of nonlinear loads and/or the supply system depending upon the requirements. A hybrid power filter mitigates most of the power quality problems for satisfactory operation of the consumer equipment even in the case of polluted supply system. At the same time, it mitigates most of the voltage-based power quality problems such as voltage harmonics, unbalance, sag, and swell and current-based power quality problems such as current harmonics, reactive power, unbalanced currents, neutral current, and current fluctuations present in the consumer loads and/or the supply system and provides sinusoidal balanced currents in the supply, with the DC bus voltage control in proper coordination with the series filter in a universal active filter as a combination of two VSCs. Here, the basic principle of the most commonly used four topologies of HPFs is given in detail.

11.4.1.1 Principle of Operation of a Passive Series and Passive Shunt Based Hybrid Power Filter

The passive series and passive shunt based hybrid power filter shown in Figure 11.54 offers an excellent performance for harmonic compensation of voltage-fed nonlinear loads. In voltage-fed nonlinear loads, a passive shunt filter suffers from inherent problems of series and parallel resonances with distribution network components and fixed compensation in addition to its inability to operate for such voltage-fed nonlinear loads consisting of a diode rectifier-fed load with a large DC smoothing capacitor. Moreover, its application is considered inadequate for voltage-fed type of harmonic-producing loads, such as a diode rectifier-fed variable-frequency drive (VFD), and its compensation characteristics are dominated by an unknown source impedance, which poses a serious problem in its design. On the other hand, a series passive filter has been found suitable for voltage-fed type of harmonic-producing loads, but suffers

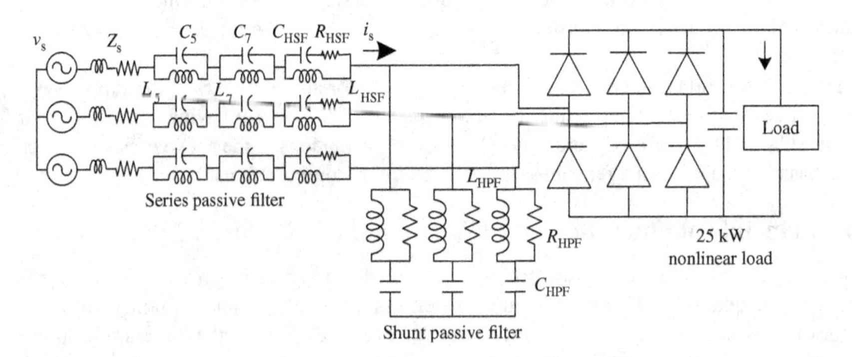

Figure 11.54 A schematic diagram of a passive series and passive shunt based hybrid filter configuration

heavily due to poor voltage regulation at the point of common coupling (PCC) and poor power factor throughout its range of operation. Therefore, a hybrid passive filter shown in Figure 11.54 provides harmonic compensation at par with active filters, whose design is insensitive to the source impedance, eliminates the chances of resonance over wide spectra, and reduces large variation of power factor and terminal voltage with varying rectifier-type nonlinear loads. In addition, this topology also helps fast settling of transients and blocking of harmonics. This passive hybrid filter provides an effective compensation of current harmonics for such voltage-fed loads with slightly detuned resonant points (corresponding to available component values) for effective and efficient operation under varying load conditions. Moreover, drawbacks of a series passive filter having an inductive impedance drop at the fundamental frequency are eliminated by using a passive shunt filter. In addition, drawbacks of a shunt passive filter having resonance with the source impedance and fixed compensation are eliminated by using a passive series filter.

11.4.1.2 Principle of Operation of a Passive Shunt and Active Series Based Hybrid Power Filter

The passive shunt and active series based hybrid power filter shown in Figure 11.55 provides harmonic compensation of voltage-fed nonlinear loads and is a cost-effective and adjustable solution for eliminating current harmonics. This hybrid filter is formed by series connection of a passive filter and a small-capacity active filter. The passive filter suppresses harmonic currents produced by the load, whereas the active filter improves the filtering characteristics of the passive filter. As a result, the hybrid filter system can solve the problems involved in using the passive filter only. The series connected active filter is controlled to act as a harmonic compensator for the load by constraining all the harmonic currents to sink into passive filters. This eliminates the possibility of series and parallel resonances. By actively improving the compensation characteristics of the tuned passive filters, the need for precise tuning of the passive filters is avoided and the design of the passive filter becomes insensitive to the supply impedance up to some extent. This topology is also suited for the harmonic compensation of the load connected to stiff supply.

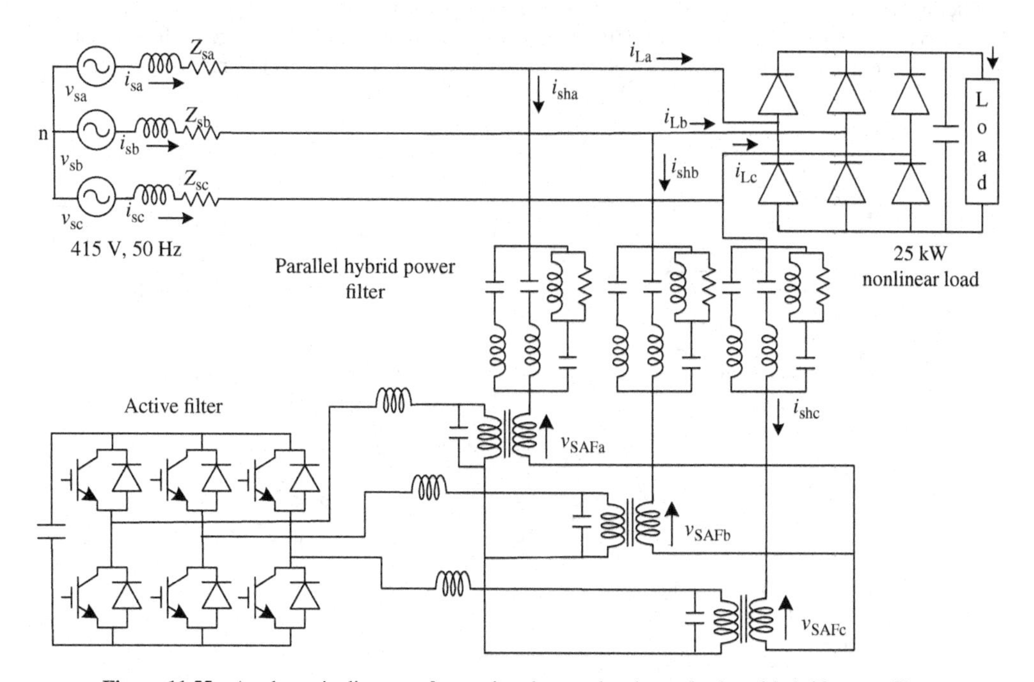

Figure 11.55 A schematic diagram of a passive shunt and active series based hybrid power filter

This topology uses an active filter to actively tune the passive filters connected in series with it. This ensures the compensation of harmonics in the supply current by enhancing the compensation characteristics of the passive filters in addition to eliminating the chances of resonance. The active filter element acts as a harmonic voltage source, compensating voltage drop in passive filters at harmonic frequencies at PCC, thereby producing a short circuit across passive filters at harmonic frequencies. This HPF eliminates current harmonics and limits supply voltage distortions, since it acts as a harmonic voltage source, compensating voltage drop in passive filters at harmonic frequencies at PCC. However, the distortions in the utility voltage add to the required voltage injected, and hence the required rating of the active filter may increase. Unlike hybrid filters having an APF in series with the AC mains, these are less susceptible to danger under short-circuit condition in utility line and do not increase voltage harmonics any further.

11.4.1.3 Principle of Operation of an Active Series and Passive Shunt Based Hybrid Power Filter

The active series and passive shunt based hybrid power filter shown in Figure 11.56 provides a cost-effective and practical harmonic compensation solution by providing harmonic isolation between the supply and the load in addition to effectively mitigating the problems of both passive filters and the pure active filter. The combination of low cost of passive filters and control capability of a small-rating active filter effectively improves the compensation characteristics of passive filters and hence reduces the rating of the active filters (<5%), compared with pure shunt or series active filters.

However, series active filtering is identified as an appropriate solution for voltage-fed type of harmonic-producing loads. However, in this type of hybrid power filter shown in Figure 11.56, for effective application and for reducing series active filter rating, a set of shunt connected passive filters are required to suitably compensate harmonics and retain the characteristics of multiple diverse connected loads. It is capable of eliminating harmonics generated by the nonlinear loads and meeting the variable limited reactive power requirements based on the available rating of the VSC employed as an AF.

This hybrid filter is formed by series connection of a passive filter and a small-capacity active filter. The passive filter suppresses harmonic currents produced by the load, whereas the active filter improves the filtering characteristics of the passive filter. As a result, the hybrid filter system can solve the problems involved in using the passive filter only. The series connected active filter is controlled to act as a harmonic compensator for the load by constraining all the harmonic currents to sink into passive filters.

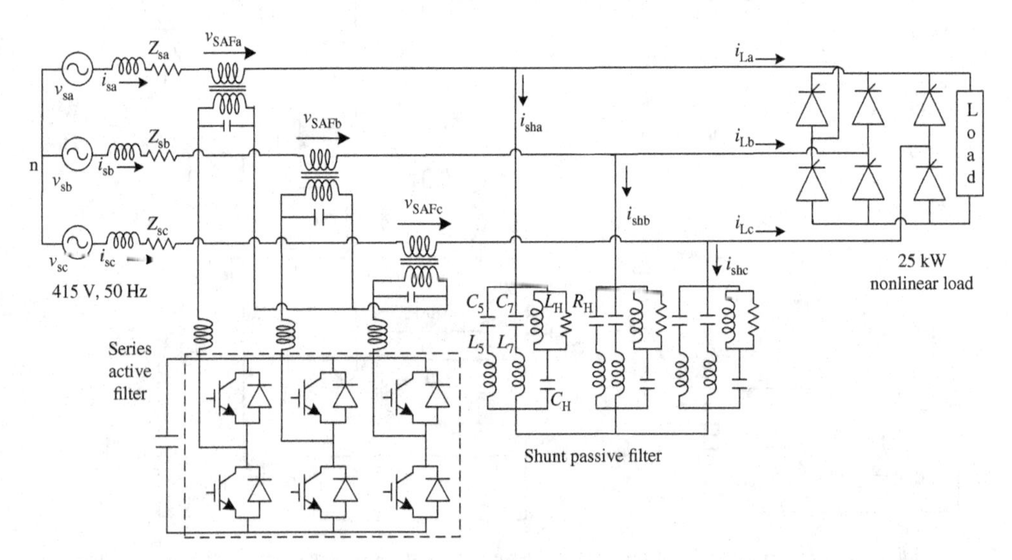

Figure 11.56 A schematic diagram of an active series and passive shunt based hybrid filter configuration

This eliminates the possibility of series and parallel resonances. By actively improving the compensation characteristics of the tuned passive filters, the need for precise tuning of the passive filters is avoided and the design of the passive filter becomes insensitive to the supply impedance up to some extent. This topology is also suited for the harmonic compensation of the load connected to stiff supply.

This topology of HPF also uses an active filter to actively force the flow of harmonic currents of the nonlinear load into the passive filters. This ensures the compensation of harmonics in the supply current by enhancing the compensation characteristics of the passive filters in addition to eliminating the chances of resonance. The active filter element acts as a harmonic voltage source at PCC to effectively force the flow of harmonic currents into the passive filter. This HPF eliminates current harmonics, limits supply voltage distortions at PCC, and provides an adjustable solution for varying nonlinear loads.

11.4.1.4 Principle of Operation of an Active Series and Active Shunt Based Hybrid Power Filter

In general, this HPF (known as a universal APF) has two VSCs connected to a common DC bus, one VSC is connected in series (known as a series APF) with AC lines through an injection transformer and another VSC is connected in shunt (known as a shunt APF) normally connected across the consumer loads or across the PCC, as shown in Figure 11.57 as a combination of active series (AF_{ss}) and active shunt (AF_{sh}) filters. Both the VSCs use PWM control; therefore, they require small ripple filters to mitigate switching ripples. It requires Hall effect voltage and current sensors for feedback signals and normally a DSP is used to implement the required control algorithm to generate gating signals for the solid-state devices of both VSCs of the universal APF. The series VSC used as an active series filter (AF_{ss}) is normally controlled in PWM voltage control mode to inject appropriate voltages in series with the AC mains and the shunt VSC used as an active shunt (AF_{sh}) filter is normally controlled in PWM current control mode to inject appropriate currents in parallel to the load in the system. This universal APF is considered as a perfect solution for eliminating voltage harmonics at PCC and current harmonics produced by nonlinear loads. It is also capable of providing the compensation of other voltage- and current-based power quality problems but at the cost of additional rating and at an increased cost. This universal APF also needs many passive elements such as a DC bus capacitor, AC interacting inductors, injection and isolation transformers, and small passive filters.

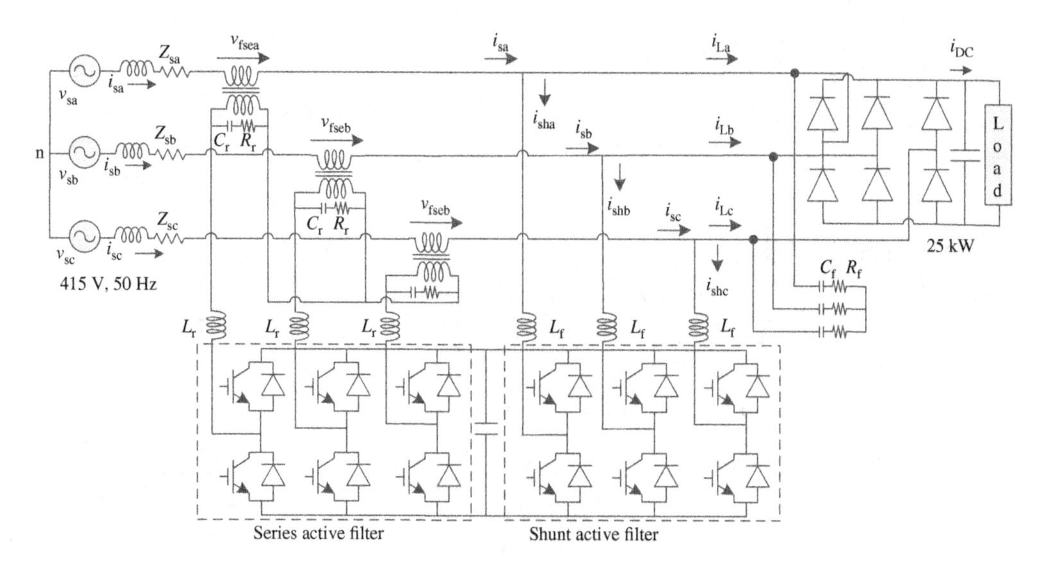

Figure 11.57 A schematic diagram of an active series and active shunt based hybrid filter configuration

11.4.2 Control of Hybrid Power Filters

Out of the 52 configurations of HPFs, the 4 configurations shown in Figures 11.54–11.57 are considered here in detail for improving the power quality at the AC mains under various types of nonlinear loads and distortion at PCC. The first topology of HPF shown in Figure 11.54 does not need any control as it consists of passive filters only. Moreover, for the control of the universal active filter type of HPF shown in Figure 11.57, either the control algorithms of UPQCs discussed in Chapter 6 may be used directly or these may be slightly modified to provide required compensation. Therefore, the control algorithm of only one typology of HPF shown in Figure 11.55 is illustrated here. However, the control of other topologies of HPFs may easily be derived in a similar manner.

11.4.2.1 Control of a Passive Shunt and Active Series Based Hybrid Power Filter

The main objective of a control algorithm of hybrid power filters is to estimate the reference voltages or reference currents using feedback signals depending upon their applications. These reference voltages or reference currents along with corresponding sensed voltages or currents are used in PWM voltage or current controllers to derive PWM gating signals for switching devices (IGBTs) of the VSC used as the series active power filter. Reference voltages or currents for the control of the series active power filter have to be derived accordingly and these signals may be estimated using a number of control algorithms. There are many control algorithms reported in the literature for the control of the series active power filter, which are classified as time-domain and frequency-domain control algorithms. There are more than a dozen of time-domain and frequency-domain control algorithms that are used for the control of the series active power filter. These control algorithms are explained in Chapter 4, which may easily be modified for the control of the series active power filter. However, because of space limitation and to give basic understanding, only a couple of them are explained here for different functions.

11.4.2.1.1 Control of a Hybrid Filter Using Instantaneous Reactive Power Theory

In the hybrid filter configuration shown in Figure 11.55, three-phase supply currents are sensed and their harmonic contents are estimated using the IRPT algorithm adopted for the conventional active filter as shown in Figure 11.58. But, instead of load currents, supply currents are sensed and processed using the IRPT-based control algorithm.

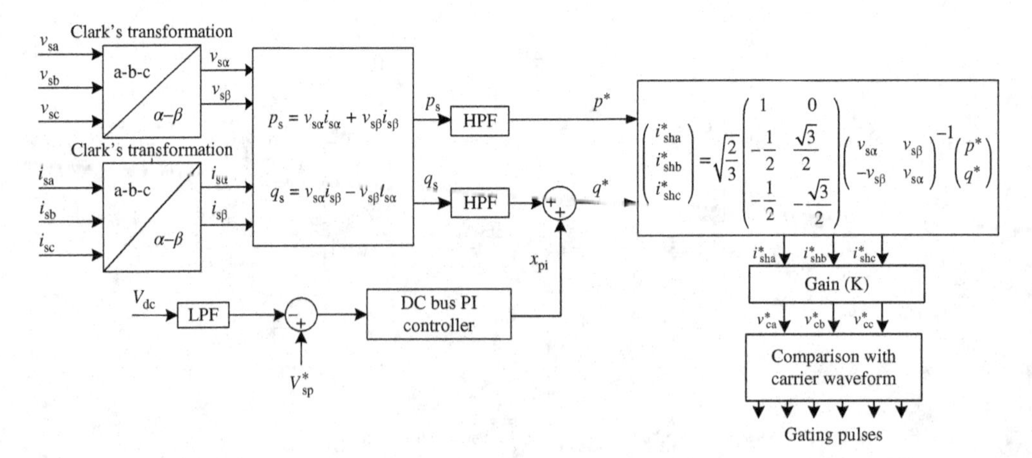

Figure 11.58 Instantaneous reactive power theory control algorithm for the active filter element of the hybrid filter

The three-phase filtered supply voltages are transformed into two-phase α–β orthogonal coordinates $(v_{s\alpha}, v_{s\beta})$ as

$$\begin{pmatrix} v_{s\alpha} \\ v_{s\beta} \end{pmatrix} = \sqrt{\frac{2}{3}} \begin{pmatrix} 1 & -\frac{1}{2} & -\frac{1}{2} \\ 0 & \frac{\sqrt{3}}{2} & -\frac{\sqrt{3}}{2} \end{pmatrix} \begin{pmatrix} v_{sa} \\ v_{sb} \\ v_{sc} \end{pmatrix}. \tag{11.1}$$

Similarly, three-phase supply currents (i_{sa}, i_{sb}, i_{sc}) are transformed into two-phase α–β orthogonal coordinates $(i_{s\alpha}, i_{s\beta})$ as

$$\begin{pmatrix} i_{s\alpha} \\ i_{s\beta} \end{pmatrix} = \sqrt{\frac{2}{3}} \begin{pmatrix} 1 & -\frac{1}{2} & -\frac{1}{2} \\ 0 & \frac{\sqrt{3}}{2} & -\frac{\sqrt{3}}{2} \end{pmatrix} \begin{pmatrix} i_{sa} \\ i_{sb} \\ i_{sc} \end{pmatrix}. \tag{11.2}$$

From these two sets of expressions, the instantaneous active power p_s and the instantaneous reactive power q_s flowing from the supply side are calculated as

$$\begin{pmatrix} p_s \\ q_s \end{pmatrix} = \begin{pmatrix} v_{s\alpha} & v_{s\beta} \\ v_{s\beta} & -v_{s\alpha} \end{pmatrix} \begin{pmatrix} i_{s\alpha} \\ i_{s\beta} \end{pmatrix}. \tag{11.3}$$

Let $\overline{p}_s$ and $\tilde{p}_s$ are the DC component and the AC component of p_s, respectively, and $\overline{q}_s$ and $\tilde{q}_s$ are the DC component and the AC component of q_s, respectively. Therefore, these may be expressed as

$$p_s = \overline{p}_s + \tilde{p}_s, \quad q_s = \overline{q}_s + \tilde{q}_s. \tag{11.4}$$

In these expressions, the fundamental load power is transformed to DC components $\overline{p}_s$ and $\overline{q}_s$, and the distortion or negative sequence is transformed to AC components $\tilde{p}_s$ and $\tilde{q}_s$. The DC components of active and reactive powers are extracted by using two high-pass filters.

In addition, the active filter regulates its DC capacitor voltage without any external power supply. If the active filter outputs a fundamental voltage that is in phase with the fundamental leading current of the passive filter, then the active power formed by the leading current and the fundamental voltage is supplied to the DC capacitor. Therefore, the electrical quantity to be controlled in the DC voltage of the AF feedback loop is added to the reactive power $(\tilde{q}_s)$. For the buildup of the DC bus voltage of the AF, the error between the sensed DC bus voltage and the reference DC bus voltage is given to a PI controller. The output of the PI controller is added to the AC component of instantaneous reactive power. Hence, the reference powers for the harmonic currents are estimated as

$$\begin{aligned} p^* &= \tilde{p}_L, \\ q^* &= \tilde{q}_L + x_{pi}, \end{aligned} \tag{11.5}$$

where x_{pi} is the output signal from the PI controller. Here, a low-pass filter is also used in the DC bus voltage feedback loop to eliminate high-frequency ripples in DC bus voltages.

The reference three-phase harmonic currents $(i^*_{sha}, i^*_{shb}, i^*_{shc})$ are estimated as

$$\begin{pmatrix} i^*_{sha} \\ i^*_{shb} \\ i^*_{shc} \end{pmatrix} = \sqrt{\frac{2}{3}} \begin{pmatrix} 1 & 0 \\ -\frac{1}{2} & \frac{\sqrt{3}}{2} \\ -\frac{1}{2} & -\frac{\sqrt{3}}{2} \end{pmatrix} \begin{pmatrix} v_{s\alpha} & v_{s\beta} \\ -v_{s\beta} & v_{s\alpha} \end{pmatrix}^{-1} \begin{pmatrix} p^* \\ q^* \end{pmatrix}. \tag{11.6}$$

The estimated harmonic current in each phase is amplified by a gain K and given as input to a PWM controller as a reference voltage $V_c^* = KI_{sh}$. The gating signals are generated by comparing the reference voltage V_c^* with a triangle wave carrier frequency. This induces a voltage in series with the three-phase passive filter, thus improving the performance of the passive filter alone, as explained in the compensation principle of a hybrid filter.

11.4.2.1.2 Control of a Hybrid Filter Using Indirect Current Control

In this control algorithm, the computational delay and number of sensors are reduced compared with the instantaneous reactive power theory, by indirectly controlling the three-phase supply currents. In this hybrid filter configuration shown in Figure 11.55, three-phase supply currents are sensed and their fundamental components are estimated with a similar algorithm adopted for the active filter controlled using SRF theory. However, instead of load currents, supply currents are sensed and processed in this algorithm. This control algorithm of the hybrid filter is shown in Figure 11.59. The SRF isolator extracts the fundamental positive-sequence component of the supply currents by transformation of (i_{sa}, i_{sb}, i_{sc}) into the d–q reference frame as

$$\begin{pmatrix} i_{s\alpha} \\ i_{s\beta} \end{pmatrix} = \sqrt{\frac{2}{3}} \begin{pmatrix} 1 & -\frac{1}{2} & -\frac{1}{2} \\ 0 & \frac{\sqrt{3}}{2} & -\frac{\sqrt{3}}{2} \end{pmatrix} \begin{pmatrix} i_{sa} \\ i_{sb} \\ i_{sc} \end{pmatrix}. \tag{11.7}$$

$$\begin{pmatrix} i_{sd} \\ i_{sq} \end{pmatrix} = \begin{pmatrix} \cos\theta & -\sin\theta \\ \sin\theta & \cos\theta \end{pmatrix} \begin{pmatrix} i_{s\alpha} \\ i_{s\beta} \end{pmatrix}. \tag{11.8}$$

The DC (stationary) transformed fundamental supply current components i_{sDCd} and i_{sDCq} are extracted using two first-order low-pass filters of low cutoff frequency (typically 9–25 Hz). The fundamental component of the supply current is extracted using the inverse Park's transformation as

$$\begin{pmatrix} i_{s1\alpha} \\ i_{s1\beta} \end{pmatrix} = \sqrt{\frac{2}{3}} \begin{pmatrix} \cos\theta & \sin\theta \\ -\sin\theta & \cos\theta \end{pmatrix} \begin{pmatrix} i_{sDCd} \\ i_{sDCq} \end{pmatrix}. \tag{11.9}$$

$$\begin{pmatrix} i_{s1a} \\ i_{s1b} \\ i_{s1c} \end{pmatrix} = \sqrt{\frac{2}{3}} \begin{pmatrix} 1 & 0 \\ -1/2 & -\sqrt{3}/2 \\ -1/2 & \sqrt{3}/2 \end{pmatrix} \begin{pmatrix} i_{s1\alpha} \\ i_{s1\beta} \end{pmatrix}. \tag{11.10}$$

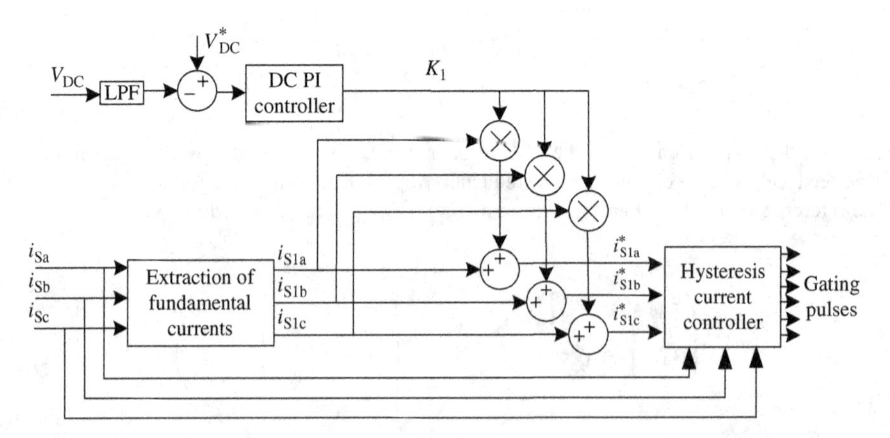

Figure 11.59 Indirect current control algorithm for the active filter element of the hybrid filter configuration

The DC bus voltage control of the AF is achieved using a PI controller. A low-pass filter is required to filter ripples in the feedback path of the DC bus voltage. The output of the DC bus filter is represented as a gain K_1. The reference signal for the current controller is given by

$$\begin{pmatrix} i_{sla}^* \\ i_{slb}^* \\ i_{slc}^* \end{pmatrix} = \begin{pmatrix} i_{sla} \\ i_{slb} \\ i_{slc} \end{pmatrix} + K_1 \begin{pmatrix} i_{sla} \\ i_{slb} \\ i_{slc} \end{pmatrix}. \tag{11.11}$$

Three-phase supply currents are controlled to follow the reference fundamental positive-sequence currents by switching the VSC, through a hysteresis current controller, which results in an indirect current control. This induces a voltage in series with the three-phase passive filter, thus improving the performance of the passive filter alone, as explained in the compensation principle of a hybrid filter.

11.5 Analysis and Design of Hybrid Power Filters

Since the considered configuration of the hybrid filter shown in Figure 11.55 consists of a passive filter along with a small active filter, its design consists of both the components. This design procedure involves the design of a passive filter for a voltage-fed load consisting of a diode rectifier with a filter capacitor and an equivalent resistive load of 25 kW fed from a 415 V, 50 Hz three-phase supply system.

11.5.1 Design of a Passive Filter

The passive filter basically consists of a series combination of an inductor and a capacitor tuned to a particular frequency and acts as a low-impedance path for that harmonic. The fifth and seventh harmonic filters are designed using a series tuned filter and the high-pass filter is designed using a second-order damped filter.

The design procedure is explained step by step for the design of a passive filter for a voltage-fed load consisting of a diode rectifier with a filter capacitor and an equivalent resistive load of 25 kW fed from a 415 V, 50 Hz three-phase supply system.

Initially, the capacitor size is calculated from the reactive power requirement (Q_c) of the load. The absolute value of capacitance is calculated as

$$C_n = \frac{Q_c}{m\omega V_s^2}. \tag{11.12}$$

To trap the nth harmonic current, the inductance for the nth-order filter is calculated as

$$L_n = \frac{1}{n^2\omega^2 C_n}. \tag{11.13}$$

The series resistance for the inductor of the nth-order filter is calculated as

$$R_n = \frac{n\omega L_n}{Q_n}, \tag{11.14}$$

where Q_n is the quality factor of the inductor of the nth-order filter, which is normally considered as $30 < Q < 100$.

In the design of a second-order damped filter, the filter parameters C_H, L_H, and R_H are calculated using Equations 11.12–11.14. However, the next dominant harmonic is considered as the next value of n for the high-pass filter and the quality factor (Q_H) for the high-pass filter inductor is considered as $0.5 < Q < 5$.

11.5.2 Design of an Active Series Filter

Since the considered configuration of the hybrid filter shown in Figure 11.55 consists of a passive filter along with a small-rating active filter, its design consists of both the components. The active filter has a small rating of the order of 5% of the load rating. Moreover, this small-rating active series filter is designed using the same procedure as given in Chapter 10. In addition, several numerical examples are illustrated for different topologies of HPFs consisting of such small-rating series active filters.

11.5.3 Design of an Active Shunt Filter

The active shunt filter is used in the UAPC as shown in Figure 11.57. The shunt active filter is designed using the same procedure as given in Chapter 9. In addition, several numerical examples are illustrated for different topologies of UAPCs consisting of such active shunt filters.

11.6 Modeling, Simulation, and Performance of Hybrid Power Filters

The MATLAB-based models of different topologies of HPF systems are developed using Simulink and SimPowerSystems (SPS) toolboxes to simulate the performance of hybrid power filters in single-phase and three-phase distribution systems. A large number of cases of the topologies of hybrid power filters are given in solved examples. Here, the performance of typical MATLAB-based models of HPFs is illustrated for four configurations of three-phase HPFs: (a) a combination of passive series (PF_{ss}) and passive shunt (PF_{sh}) filters (Figure 11.54), (b) a combination of series connected passive shunt (PF_{sh}) and active series (AF_{ss}) filters (Figure 11.55), (c) a combination of active series (AF_{ss}) and passive shunt (PF_{sh}) filters (Figure 11.56), and (d) a combination of active series (AF_{ss}) and active shunt (AF_{sh}) filters (Figure 11.57).

11.6.1 Modeling, Simulation, and Performance of a Passive Series and Passive Shunt Based Hybrid Power Filter

Figure 11.54 shows the circuit schematic of a passive series and passive shunt based hybrid filter configuration. For the simulation studies, the supply voltage is considered as 415 V at 50 Hz. A voltage-fed nonlinear load with load power of 25 kW is considered for the simulation study. The reactive power for the passive shunt filter design is considered as approximately 15% of active power of the load. The passive series and passive shunt filters can be designed using the same procedure as given in Chapter 8. The various filter parameters are given in Table 11.1. This hybrid filter is modeled in MATLAB using the SPS toolbox for simulation. The response of the passive series and passive shunt based hybrid filter configuration for this load is shown in Figure 11.60, in which from $t_s = 0.2$ to 0.06 s, only the passive series filter compensates, and after that both passive series and passive shunt filters compensate the harmonics. The harmonic spectra are shown in Figure 11.61 for supply current (a) with a hybrid filter and (b) with a passive series filter having THDs of 2.97 and 5.33%, respectively. The harmonic spectra for (a) supply voltage, (b) load voltage, and (c) load current are shown in Figure 11.62 with THDs of 0.03, 19.69, and 12.60%, respectively.

Table 11.1 Parameters of a passive series and passive shunt based hybrid filter

Passive series filter						Passive shunt filter	
				R_{HSF}	3 Ω	R_{HPF}	16 Ω
C_5	70 μF	C_7	60 μF	C_{HSF}	48 μF	C_{HPF}	40 μF
L_5	5.8 mH	L_7	3.4 mH	L_{HSF}	1.8 mH	L_{HPF}	10 mH

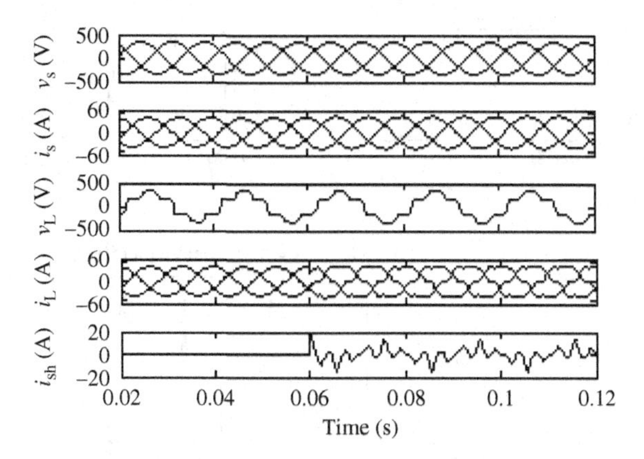

Figure 11.60 Response of a passive series and passive shunt based hybrid filter configuration for a load of 25 kW

11.6.2 Modeling, Simulation, and Performance of a Passive Shunt and Active Series Based Hybrid Power Filter

Figure 11.55 shows the circuit schematic of a passive shunt and active series based hybrid filter configuration. For the simulation studies, the supply voltage is considered as 415 V at 50 Hz. A voltage-fed nonlinear load with load power of 25 kW is considered for the simulation study. The reactive power for the passive shunt filter design is considered as approximately 15% of active power of the load. The active series filter is designed using the same procedure as given in Chapter 10. However, the passive shunt filters can be designed using the same procedure as given in Chapter 8. The various parameters for this type of filter are given in Table 11.2. This parallel hybrid filter is modeled in MATLAB using the SPS toolbox for simulation. It consists of AC mains system, where the AC source contains fundamental and harmonic components of voltages. This block is constructed with the help of a controlled voltage source. The three-phase VSC with a DC link capacitor uses IGBTs as switches. The converter block contains a three-phase bridge and a DC bus measurement block. A ripple filter is also incorporated in the system to crunch current ripples generated by the VSC. The size of the inductor is determined on the basis of

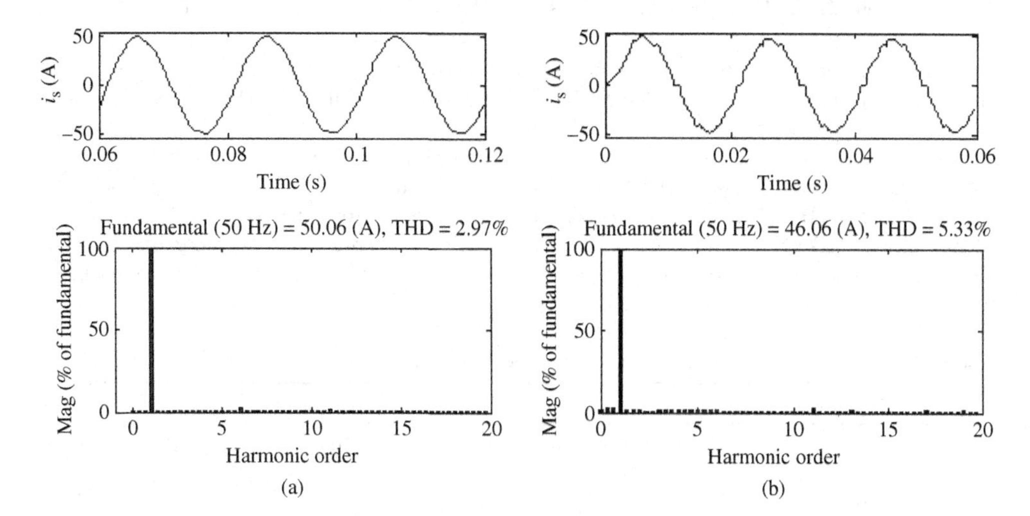

Figure 11.61 Harmonic spectra for supply current (a) with a hybrid filter and (b) with a passive series filter

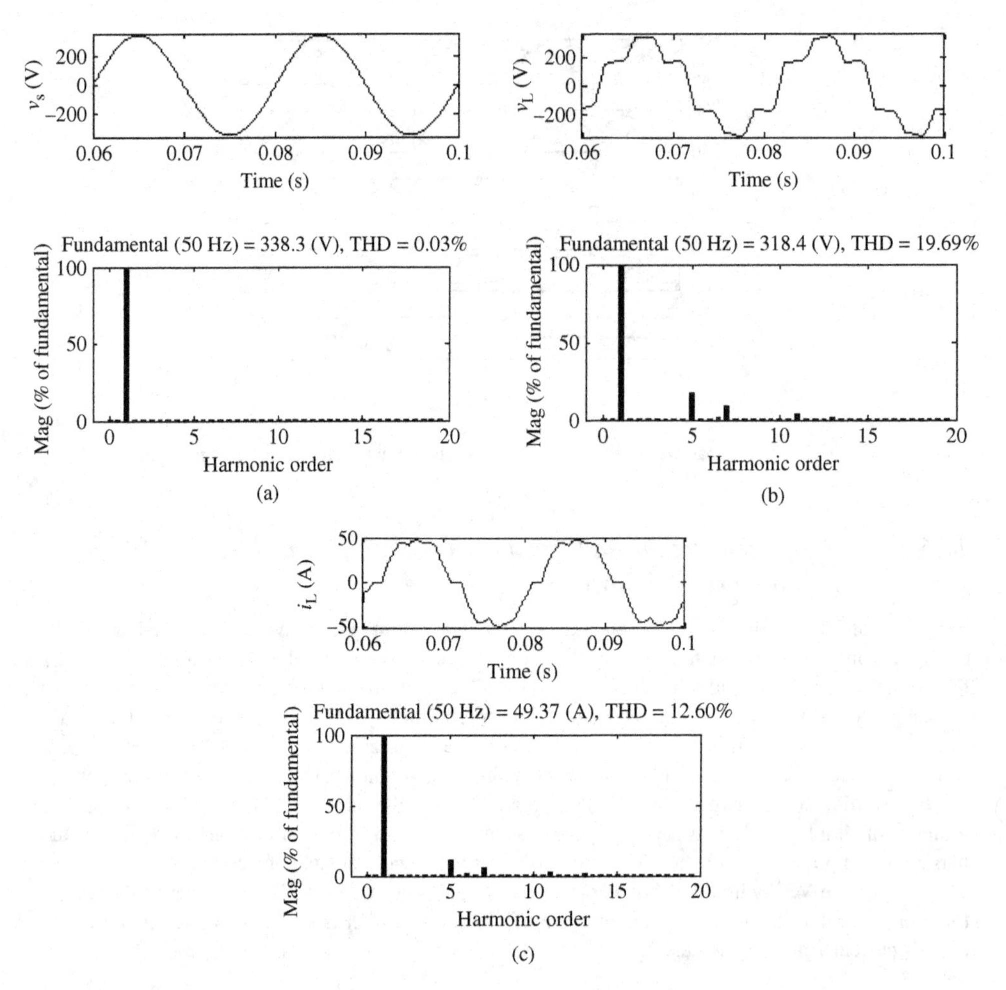

Figure 11.62 Harmonic spectra for (a) supply voltage, (b) load voltage, and (c) load current

required bandwidth of desired current compensation. The coupling transformers have the ratio 1 : 1. The passive filter block is constructed with the help of *RLC* branch given in the SPS. The system is considered as supplying a nonlinear voltage-fed or current-fed load.

The simulated results demonstrate the response of the hybrid power filter for a nonlinear load of 25 kW, with indirect current control of the AF. Figure 11.63 shows the steady-state response of the parallel hybrid power filter. The THD of the supply current is reduced to 1.52% as shown in Figure 11.64, while the load current has a THD of 27.87%. From the harmonic spectra, it is observed that there exists significant resonance between the source impedance and the passive filter at 11th harmonic frequency, leading to

Table 11.2 Parameters of a passive shunt and active series based hybrid filter

Active series filter				Passive shunt filter			
V_{DC}	450 V	R_f	5 Ω	Order *n*	*C* (μF)	*L* (mH)	*R* (Ω)
C_{DC}	3000 μF			5 th	25	16.4	1.29
L_f	3 mH	C_f	10 μF	7 th	25	8.4	0.9226
				11 th	25	3.4	4.67

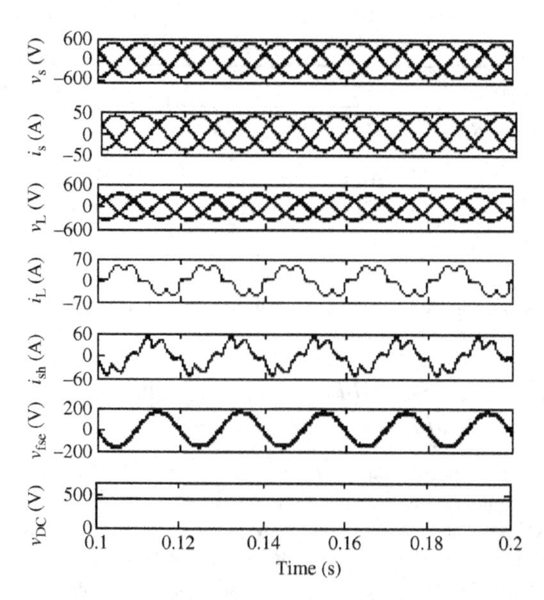

Figure 11.63 Response of a parallel hybrid power filter configuration for a load of 25 kW

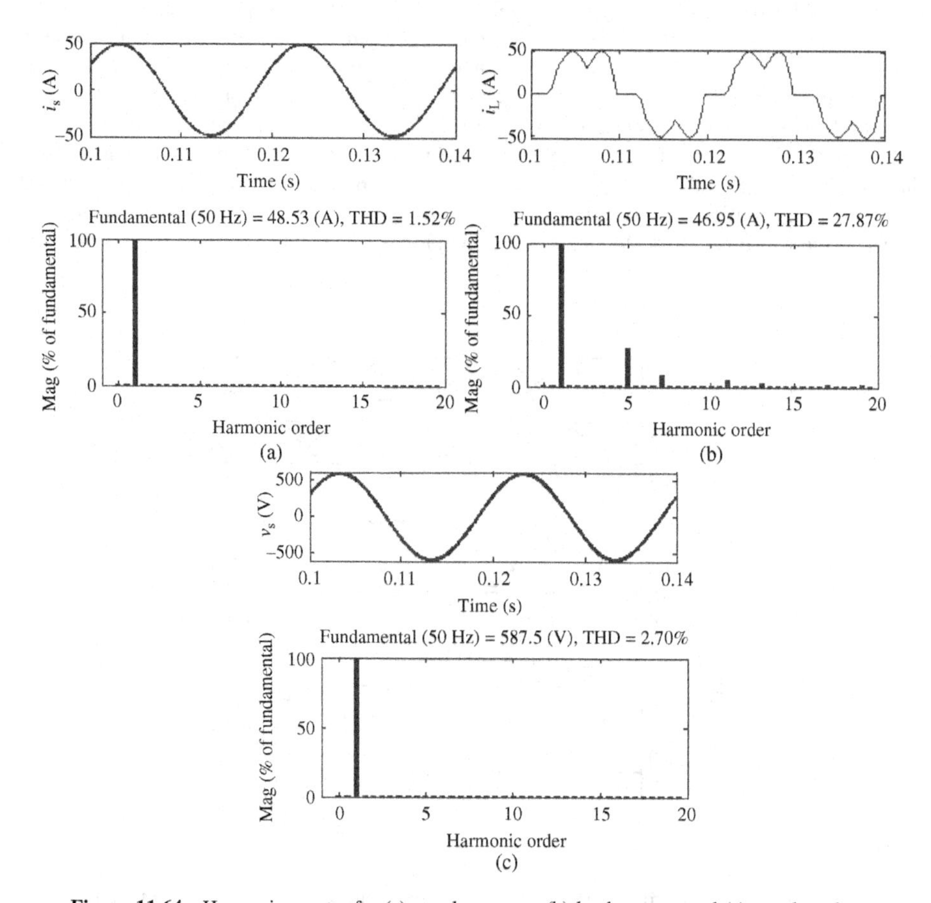

Figure 11.64 Harmonic spectra for (a) supply current, (b) load current, and (c) supply voltage

Table 11.3 Parameters of an active series and passive shunt based hybrid filter

Active series filter				Passive shunt filter			
V_{DC}	100 V	R_f	5 Ω	Order *n*	*C* (μF)	*L* (mH)	*R* (Ω)
C_{DC}	3000 μF			5 th	25	16.4	1.29
L_f	3 mH	C_f	10 μF	7 th	25	8.4	0.9226
				11 th	25	3.4	4.67

amplification of the 11th harmonic component in supply current. It is compensated by using the hybrid filter configuration shown in Figure 11.64.

11.6.3 Modeling, Simulation, and Performance of an Active Series and Passive Shunt Based Hybrid Power Filter

Figure 11.56 shows the circuit schematic of an active series and passive shunt based hybrid filter configuration. For the simulation studies, the supply voltage is considered as 415 V at 50 Hz. A diode bridge rectifier-based current-fed load with load power of 25 kW is considered for the simulation study. The reactive power for the passive shunt filter design is considered as approximately 15% of active power of the load. The active series filter is designed using the same procedure as given in Chapter 10. However, the passive shunt filters can be designed using the same procedure as given in Chapter 8. The various parameters for this type of filter are given in Table 11.3. This hybrid filter is modeled in MATLAB using the SPS toolbox for simulation. This hybrid filter consists of a shunt passive filter along with a series filter. A diode bridge with *RL* load is considered as a nonlinear load. An indirect current-controlled voltage source converter is considered as an active series filter. The size of the inductor is determined on the basis of required bandwidth of desired current compensation. The coupling transformers have the ratio 1: 1. The simulated results demonstrate the response of the hybrid power filter for a nonlinear load of 25 kW, with indirect current control of the AF. Figure 11.65 shows the response of the series active and shunt passive based hybrid filter configuration for a load of 25 kW. The DC bus voltage is effectively maintained at its reference value with this control algorithm as shown in Figure 11.65. The harmonic spectra are shown in

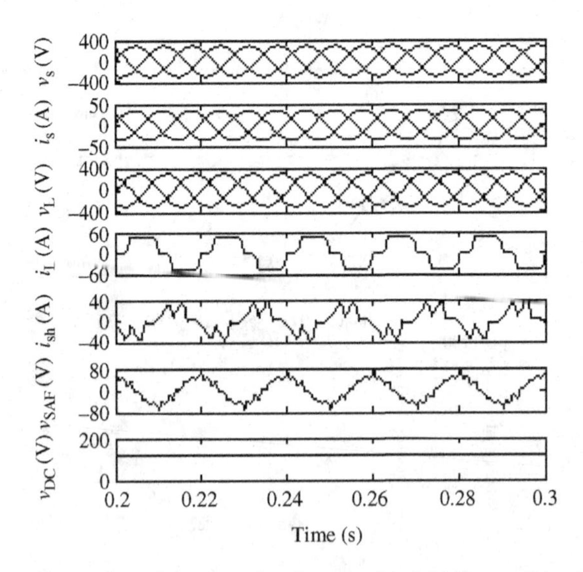

Figure 11.65 Response of an active series and passive shunt based hybrid filter configuration for a load of 25 kW

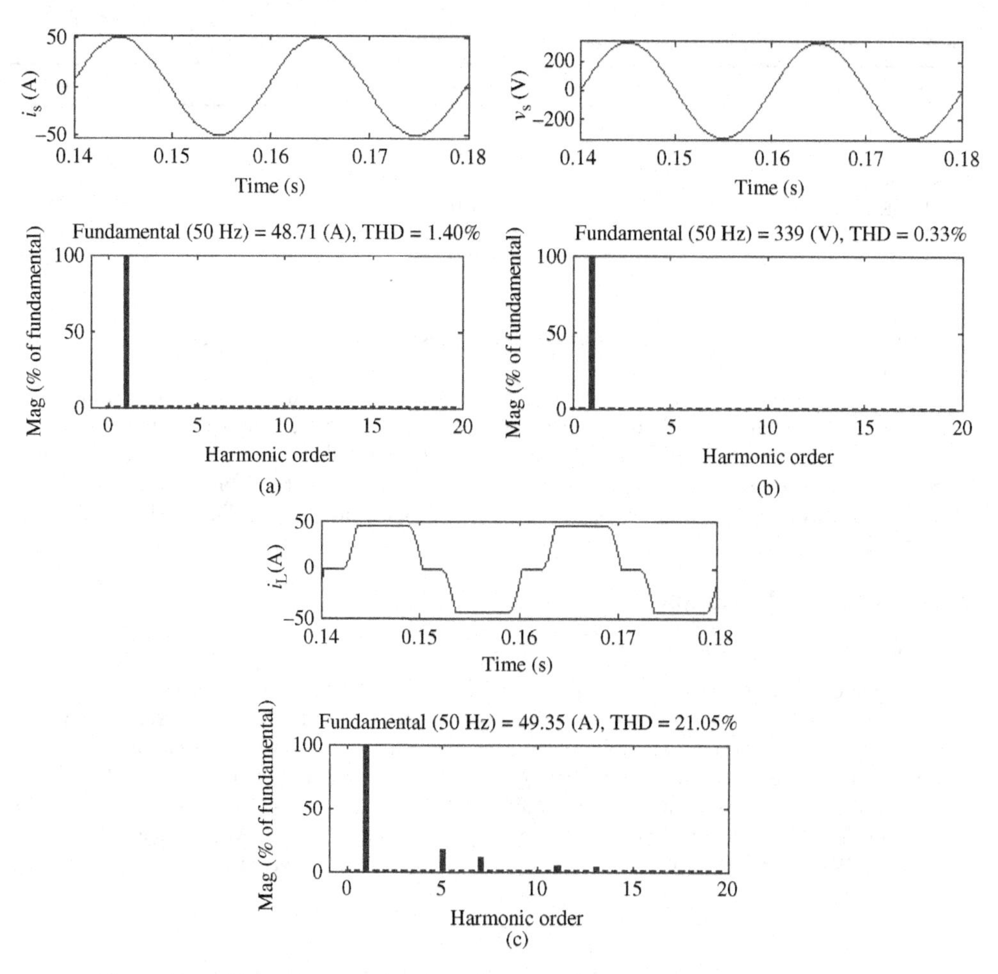

Figure 11.66 Harmonic spectra for (a) supply current, (b) supply voltage, and (c) load current

Figure 11.66 for (a) supply current, (b) supply voltage, and (c) load current with THDs of 1.4, 0.33, and 21.05%, respectively. The THD of the supply current is reduced to 1.4%, which is in limit and in accordance with the IEEE 519 Standard.

11.6.4 Modeling, Simulation, and Performance of an Active Series and Active Shunt Based Hybrid Power Filter

Figure 11.57 shows the circuit schematic of an active series and active shunt based hybrid power filter configuration. This system configuration is often referred to as the UAPC. For the simulation studies, the supply voltage is considered as 415 V at 50 Hz along with harmonics in supply voltage (0.2 pu (fifth harmonic) and 0.1 pu (seventh harmonic)). A voltage-fed nonlinear load with load power of 25 kW is considered for the simulation study. The active series filter is designed using the same procedure as given in Chapter 10. However, the active shunt filter can be designed using the same procedure as given in Chapter 9. The various parameters for this type of filter are given in Table 11.4. This hybrid filter is modeled in MATLAB using the SPS toolbox for simulation. This hybrid filter consists of an active shunt filter along with an active series filter. An indirect current-controlled voltage source converter is considered as an active series filter. The size of the inductor is

Table 11.4 Simulation parameters of an active series and active shunt based hybrid filter

Parameters	Active series filter	Active shunt filter
V_{DC}	700 V	
C_{DC}	5000 μF	
Interfacing inductor	$L_r = 4$ mH	$L_f = 2$ mH
Ripple filter capacitor	$C_r = 10$ μF	$C_f = 5$ μF
Ripple filter resistor	$R_r = 5\,\Omega$	$R_f = 5\,\Omega$

determined on the basis of required bandwidth of desired current compensation. The coupling transformers have the ratio 1 : 1.

Figure 11.67 shows the steady-state response of the hybrid filter under a nonlinear load of 25 kW connected to the nonideal distribution system. A diode bridge rectifier with a DC filter to maintain constant DC voltage is considered as a nonlinear load. The DC bus voltage is effectively maintained at its reference value with this control algorithm as shown in Figure 11.67. Harmonic spectra for (a) load current, (b) supply current, (c) supply voltage, and (d) load voltage are shown in Figure 11.68 with THDs of 80.53, 2.18, 22.53, and 1.47%, respectively.

11.7 Numerical Examples

Example 11.1

A single-phase VSI with a square waveform AC output of 230 V (rms) at 50 Hz is feeding a critical load of 230 V, 50 Hz single-phase 4 kVA at 0.8 lagging power factor. A single-phase universal active power conditioner (consisting of shunt and series APFs using two VSCs with a common bus capacitor as shown in Figure E11.1) is designed for this critical linear load. Calculate (a) the voltage rating of the shunt element of the UAPC, (b) the current rating of the shunt element of the UAPC, (c) the VA rating of the shunt element of the UAPC, (d) the voltage rating of the series element of the UAPC, (e) the current rating of the series element of the UAPC, and (f) the VA rating of the series element of the UAPC to provide load

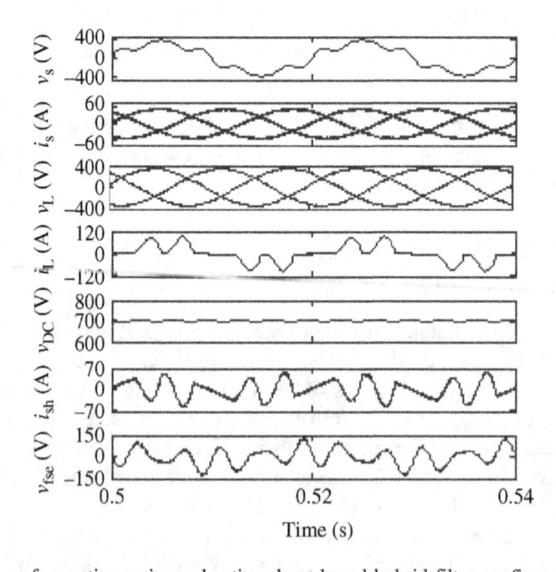

Figure 11.67 Response of an active series and active shunt based hybrid filter configuration for a load of 25 kW

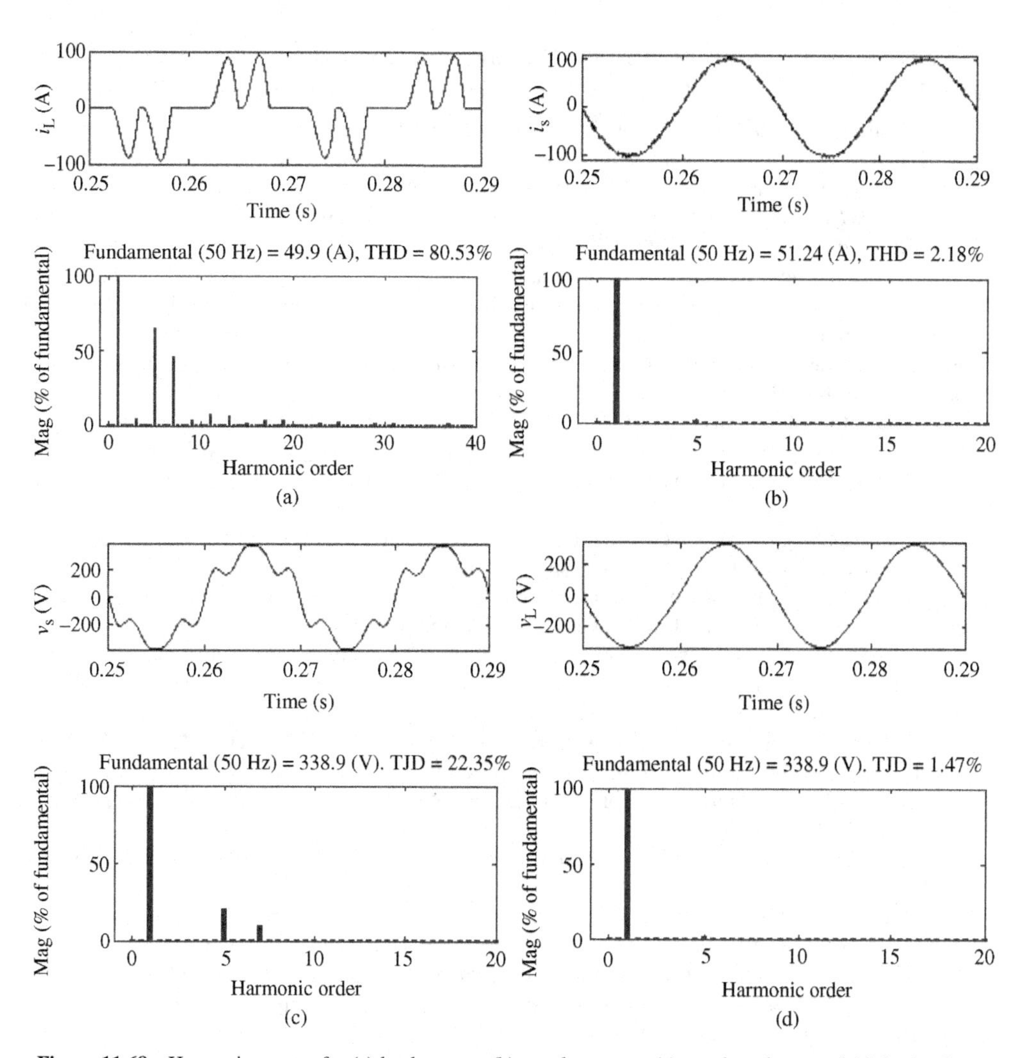

Figure 11.68 Harmonic spectra for (a) load current, (b) supply current, (c) supply voltage, and (d) load voltage

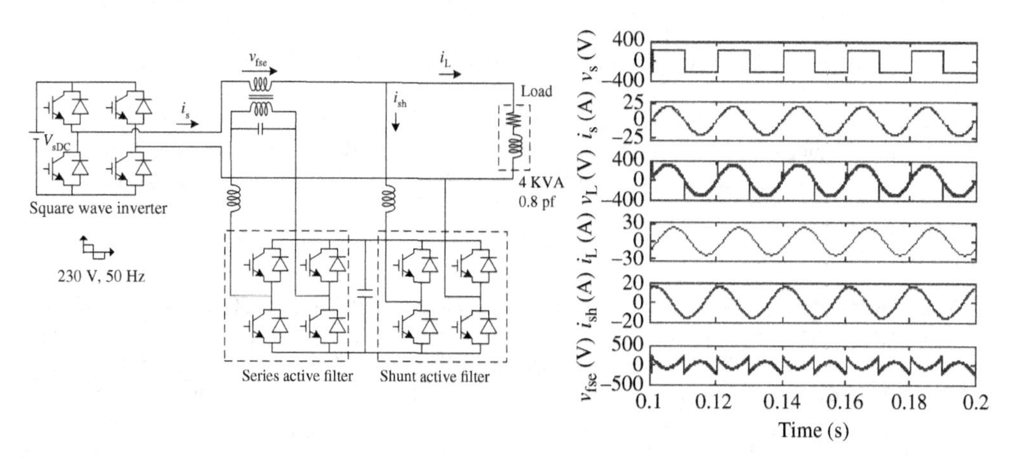

Figure E11.1 A hybrid power filter and its waveforms

compensation for unity power factor at the AC mains by the shunt element of the UAPC and constant regulated sine wave voltage of 230 V (rms) at 50 Hz across the load by the series element of the UAPC.

Solution: Given supply voltage $V_s = 230$ V (rms square wave), frequency of the supply (f) = 50 Hz, and a critical load of 230 V, 50 Hz single-phase 4 kVA at 0.8 lagging power factor.

The fundamental component of supply voltage for square wave voltage is estimated as $V_{s1} = 0.9V_s = 0.9 \times 230 = 207$ V. The load voltage is to be regulated at nominal sine wave voltage; hence, $V_L = 230$ V.

The AC load current is $I_L = S_L/V_L = 4000/230 = 17.39$ A.

In this system, load reactive power compensation is provided by the shunt active filter of the UAPC. The voltage compensation is provided by the series filter of the UAPC. However, there is a difference in magnitude of fundamental voltage in the supply and load terminals, and to compensate that an active power is circulated between series and shunt active filters as explained in Chapter 6. The rating calculations for both the VSCs of the UAPC are as follows.

The load active power is calculated as $P_L = S_L \times \text{PF} = 4000 \times 0.8 = 3200$ W.

The active power component of load current is estimated as $I_{L1a} = P_L/V_L = 3200/230 = 13.91$ A.

The supply current after compensation is estimated as $I_s = P_L/V_{s1} = 3200/207 = 15.46$ A.

a. The voltage rating of the shunt element of the UAPC is equal to AC load voltage of $V_{fsh} = 230$ V, since it is connected across the load of 230 V sine waveform.
b. The current rating of the shunt element of the UAPC is computed as follows.

   The shunt element of the UAPC needs to correct the power factor of the load to unity; hence, it supplies the reactive power required for the load. Therefore, the reactive power component of the shunt APF current is estimated as I_{shr} = reactive power component of the load current = $I_L\sqrt{(1 - \text{PF}^2)} = 17.39 \times 0.6 = 10.43$ A.

   The supply fundamental voltage is lower than the required load voltage. Hence, the shunt APF absorbs active power and that active power is delivered back into the system via the series APF. The active power component of the shunt APF current is estimated as $I_{sha} = I_s - I_{L1a} = (15.46 - 13.91)\text{ A} = 1.55$ A.

   The net current rating of the shunt APF is estimated as $I_{sh} = \sqrt{(I_{sha}^2 + I_{shr}^2)} = \sqrt{(1.55^2 + 10.43^2)} = 10.54$ A.
c. The VA rating of the VSC of the shunt APF is $S_{sh} = V_{fsh}I_{sh} = 230 \times 10.54 = 2424.2$ VA.
d. The voltage rating of the series active filter of the UAPC is computed as follows.

   The supply voltage is a square wave of $V_s = 230$ V (rms) and the load voltage at PCC must be a sine wave of $V_L = 230$ V. Therefore, the series APF must inject the difference of these two voltages to provide the required voltage at the load end.

   The voltage rating of the series APF is

$$V_{fse} = \sqrt{\left\{(1/\pi)\int_0^{\pi}(230 - 230\sqrt{2}\sin\theta)^2\, d\theta\right\}} = 102.69\text{ V}.$$

e. The current rating of the series active filter of the UAPC is same as supply current: $I_{se} = I_s = 15.46$ A.
f. The VA rating of the series active filter of the UAPC is $S_{se} = V_{fse}I_{se} = 102.86 \times 15.46 = 1590.2$ VA.

Example 11.2

A single-phase VSI with a quasi-square waveform AC output of 220 V (rms) at 50 Hz is feeding a critical load of 220 V, 50 Hz single-phase 3 kVA at 0.8 lagging power factor. A single-phase universal active power conditioner (consisting of shunt and series APFs using two VSCs with a common bus capacitor as shown in Figure E11.2) is designed for this critical linear load. Calculate (a) the voltage rating of the shunt element of the UAPC, (b) the current rating of the shunt element of the UAPC, (c) the VA rating of the shunt element of the UAPC, (d) the voltage rating of the series element of the UAPC, (e) the current rating

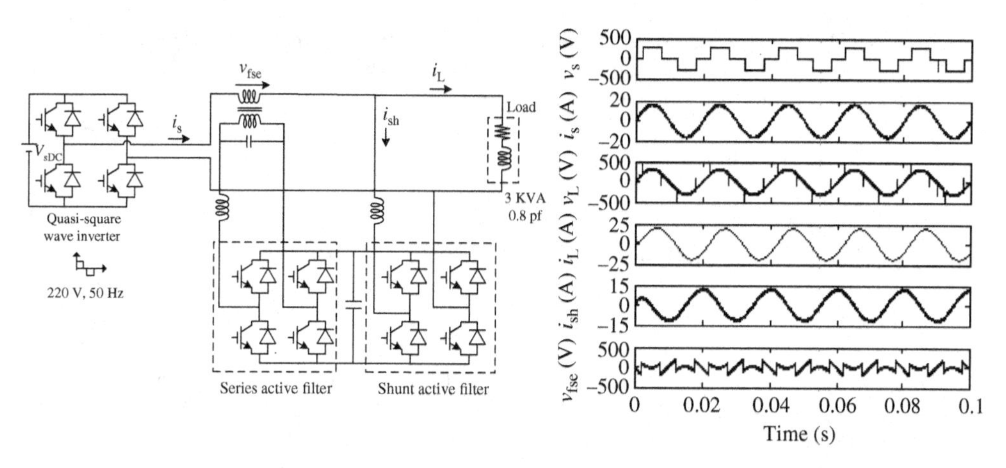

Figure E11.2 A hybrid power filter and its waveforms

of the series element of the UAPC, and (f) the VA rating of the series element of the UAPC to provide load compensation for unity power factor at the AC mains by the shunt element of the UAPC and constant regulated sine wave voltage of 220 V (rms) at 50 Hz across the load by the series element of the UAPC.

Solution: Given supply voltage $V_s = 220$ V (rms quasi-square wave), frequency of the supply (f) = 50 Hz, and a critical load of 220 V, 50 Hz single-phase 3 kVA at 0.8 lagging power factor.

The AC load current is $I_L = 3000/220 = 13.64$ A.

In this system, load reactive power compensation is provided by the shunt active filter of the UAPC. The voltage compensation is provided by the series filter of the UAPC. However, there is a difference in magnitude of fundamental voltage in the supply and load terminals, and to compensate that an active power is circulated between series and shunt active filters as explained in Chapter 6. The rating calculations for both the VSCs of the UAPC are as follows.

The load active power is calculated as $P_L = S_L \times PF = 3000 \times 0.8 = 2400$ W.

The active power component of load current is estimated as $I_{L1a} = P_L/V_L = 2400/220 = 10.91$ A.

The amplitude of the quasi-square wave is estimated as $V_{sDC} = V_s/\sqrt{(2/3)} = 269.44$ V.

The fundamental supply voltage for the quasi-square wave is estimated as $V_{s1} = (\sqrt{6}/\pi)V_{sDC} = 210.08$ V.

The supply current after compensation is estimated as $I_s = P_L/V_{s1} = 2400/210.08 = 11.42$ A.

a. The voltage rating of the shunt element of the UAPC is equal to AC load voltage of $V_{fsh} = 220$ V, since it is connected across the load of 220 V sine waveform.
b. The current rating of the shunt element of the UAPC is computed as follows.

   The shunt element of the UAPC needs to correct the power factor of the load to unity; hence, it supplies the reactive power required for the load. Therefore, the reactive power component of the shunt APF current is estimated as I_{shr} = reactive power component of the load current $= I_L\sqrt{(1 - PF^2)} = 13.64 \times 0.6 = 8.18$ A.

   The supply fundamental voltage is lower than the required load voltage. Hence, the shunt APF absorbs active power and that active power is delivered back into the system via the series APF. The active power component of the shunt APF current is estimated as $I_{sha} = I_s - I_{L1a} = (11.42 - 10.91)$ A $= 0.51$ A.

   The net current rating of the shunt APF is estimated as $I_{sh} = \sqrt{(I_{sha}^2 + I_{shr}^2)} = \sqrt{(0.51^2 + 8.18^2)} = 8.19$ A.
c. The VA rating of the VSC of the shunt APF is $S_{sh} = V_{fsh}I_{sh} = 220 \times 8.19 = 1802$ VA.
d. The voltage rating of the series element of the UAPC is computed as follows.

The supply voltage is a quasi-square wave of $V_s = 220$ V (rms) and the voltage at the load terminal is a sine wave of $V_L = 220$ V. The amplitude of the quasi-square wave is $V_{sDC} = 269.44$ V. Therefore, the series APF must inject the difference of these two voltages to provide the required voltage at the load end.

The voltage rating of the series APF is

$$V_{fse} = \sqrt{\left[(1/\pi)\left\{\int_0^{\pi/6}(-220\sqrt{2}\sin\theta)^2\,d\theta + \int_{\pi/6}^{5\pi/6}(269.44 - 220\sqrt{2}\sin\theta)^2\,d\theta + \int_{5\pi/6}^{\pi}(-220\sqrt{2}\sin\theta)^2\,d\theta\right\}\right]}$$

$= 66.05$ V.

e. The current rating of the series element of the UAPC is same as supply current: $I_{se} = I_s = 3000 \times 0.8/210.08 = 11.42$ A.
f. The VA rating of the series element of the UAPC is $S_{se} = V_{fse}I_{se} = 66.05 \times 11.42 = 754.29$ VA.

Example 11.3

A single-phase active hybrid filter (consisting of a three-branch passive shunt filter and a series APF using a VSC with a bus capacitor as shown in Figure E11.3) is designed for a critical load of a 230 V, 50 Hz single-phase thyristor bridge with constant DC current of 25 A at 60° firing angle of its thyristors. Calculate (a) element values of the passive shunt filter and (b) voltage, (c) current, and (d) VA ratings of both (i) the series APF and (ii) the passive shunt filter used in the hybrid filter to provide harmonic and reactive power compensation for unity power factor at the AC mains.

Solution: Given supply voltage $V_s = 230$ V (rms), frequency of the supply (f) = 50 Hz, and a nonlinear load of a 230 V, 50 Hz single-phase thyristor bridge converter with constant DC current of 25 A at 60° firing angle of its thyristors.

In this system, load current harmonic and reactive power compensation is provided by the shunt passive filter of the hybrid filter and small harmonic voltages are provided by the series active filter of the hybrid filter to force all harmonic currents into the passive shunt filter.

The AC load rms current is $I_L = I_{DC} = 25$ A.

The fundamental rms input current of the thyristor bridge converter is $I_{1T} = (2\sqrt{2}/\pi)I_L = 0.9 \times 25 = 22.5$ A.

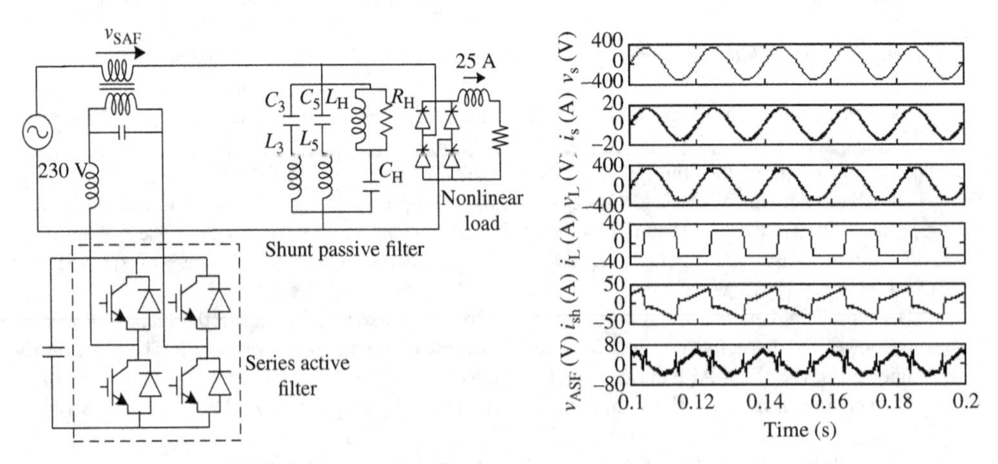

Figure E11.3 A hybrid power filter and its waveforms

The fundamental active power component of load current is $I_{L1a} = I_{L1} \cos \alpha = 0.9 I_{DC} \cos 60° = 11.25$ A.

The fundamental active power of the load is $P_1 = V_{s1} I_{L1} \cos \theta_1 = V_{s1} I_{L1a} = 230 \times 11.25 = 2587.5$ W.

The fundamental reactive power of the load is $Q_1 = V_{s1} I_{L1} \sin \theta_1 = V_{s1} I_{L1} \sin \alpha = 230 \times 22.5 \times 0.866 =$ 4481.7 VAR.

a. The passive shunt filter has three branches (third- and fifth-order tuned filters and a high-pass damped filter). The reactive power of 4481.7 VAR required by the single-phase thyristor rectifier has to be provided by all three branches of the passive shunt filter. Considering that all branches of the passive filter have equal capacitors, the value of the capacitor is $C = Q/3V_s^2\omega = 4481.7/(3 \times 230^2 \times 314) =$ 89.89 μF. Thus, $C_3 = C_5 = C_H = C = 89.89$ μF.

   The value of the inductor for the third harmonic tuned filter is $L_3 = 1/\omega_3^2 C_3 = 12.5$ mH.

   The resistance of the inductor of the third harmonic tuned filter is $R_3 = X_3/Q_3 = 11.775/20 =$ 0.5902 Ω (considering $Q_3 = 20$ as it may be in the range of 10–100 depending upon the design of the inductor).

   The value of the inductor of the fifth harmonic tuned filter is $L_5 = 1/\omega_5^2 C_5 = 4.5$ mH.

   The resistance of the inductor of the fifth harmonic tuned filter is $R_5 = X_5/Q_5 = 7.065/20 = 0.3541$ Ω (considering $Q_5 = 20$ as it may be in the range of 10–100 depending upon the design of the inductor).

   The value of the inductor for the high-pass damped harmonic filter (tuned at seventh harmonic) is $L_H = 1/\omega_7^2 C_H = 2.3$ mH.

   The resistance in parallel of the inductor of the high-pass damped harmonic tuned filter is $R_H = X_H/Q_H = 5.0554/1 = 5.0554$ Ω (considering $Q_H = 1$ as it may be in the range of 0.5–5 depending upon the attenuation required.).

   The third harmonic load current to flow in the third harmonic tuned filter is $I_3 = I_{1T}/3 = 22.5/3 =$ 7.5 A.

   The third harmonic voltage at the load end and across the passive filter is $V_3 = I_3 R_3 =$ $7.5 \times 0.5901 = 4.4264$ V.

   The fifth harmonic load current to flow in the fifth harmonic tuned filter is $I_5 = I_{1T}/5 = 22.5/5 =$ 4.5 A.

   The fifth harmonic voltage at the load end and across the passive filter is $V_5 = I_5 R_5 = 4.5 \times 0.3541 =$ 1.5935 V.

   All other harmonic load currents to flow in the high-pass damped harmonic filter are $I_H = \sqrt{(I_L^2 - I_{1T}^2 - I_3^2 - I_3^2)} = \sqrt{(25^2 - 22.5^2 - 7.5^2 - 4.5^2)} = \sqrt{42.25} = 6.5$ A.

   All higher order harmonic voltages at the load end and across the passive filter are $V_H = I_H R_H =$ $6.5 \times 5.0554 = 32.8815$ V.

   All harmonic voltages other than the fundamental voltage at the load end and across the passive filter are $V_{LH} = \sqrt{(V_3^2 + V_5^2 + V_H^2)} = 33.2163$ V.

   (i) The various ratings of the series APF are as follows.

b. The series active filter must inject all harmonic voltages to force all harmonic currents into the passive filter. Therefore, the voltage rating of the series active filter is $V_{SAF} = V_{LH} = \sqrt{(V_3^2 + V_5^2 + V_H^2)} =$ 33.2163 V.

c. The current rating of the series active filter is $I_{SAF} = I_s =$ the fundamental active power component of load current $(I_{L1a}) = I_{L1} \cos \alpha = 0.9 I_{DC} \cos 60° = 11.25$ A (since the series APF is connected in series with the AC mains before the passive shunt filter that is providing harmonic and reactive power compensation resulting in UPF at the AC mains).

d. The VA rating of the series APF is $S_{APF} = V_{SAF} I_{SAF} = 33.2163 \times 11.25 = 373.6 = 0.37$ kVA.

   The pu rating of the series APF is $S_{APFpu} = S_{APF}/S_L = 373.6/(230 \times 25) = 0.0656 = 6.5\%$ of the load rating.

(ii) The various ratings of the passive filter are as follows.

b. The voltage rating of shunt passive filters of the hybrid filter is equal to AC load voltage, which is rms of supply voltage of 230 V (rms) at 50 Hz and small harmonic voltages to be generated by the series active filter (decided by the shunt passive filter to force harmonic currents into it): $V_{PF} = \sqrt{(230^2 + 33.2163^2)} = 232.3862$ V.

c. The current rating of the shunt passive filter is computed as follows.
   The shunt passive filter needs to compensate the load current harmonics and the reactive power; hence, it supplies the required harmonic currents and reactive power of the load. Therefore, total harmonic and reactive power component of load current is $I_{sh} = \sqrt{(I_L^2 - I_{L1a}^2)} = \sqrt{(25^2 - 11.25^2)} = 22.32$ A.

d. The VA rating of the shunt passive filter is $S_{sh} = V_{fsh} I_{sh} = 232.38 \times 22.32$ VA $= 5188.3$ VA $= 5.188$ kVA.
   The pu rating of the shunt passive filter is $S_{PFpu} = S_{PF}/S_L = 5188/(230 \times 25) = 0.9023 = 90.23\%$ of the load rating.

Example 11.4

A single-phase universal active power conditioner (consisting of shunt and series APFs using two VSCs with a common bus capacitor as shown in Figure E11.4) is designed for a typical load of a 230 V, 50 Hz single-phase thyristor bridge with constant DC current of 20 A at 60° firing angle of its thyristors. If there is a voltage fluctuation of +10 and −20% in the supply system with a base value of 230 V, calculate (a) the voltage rating of the shunt element of the UAPC, (b) the current rating of the shunt element of the UAPC, (c) the VA rating of the shunt element of the UAPC, (d) the voltage rating of the series element of the UAPC, (e) the current rating of the series element of the UAPC, and (f) the VA rating of the series element of the UAPC to provide harmonic and reactive power compensation for unity power factor at the AC mains with constant regulated sine wave voltage of 230 V at 50 Hz across the load.

Solution: Given supply voltage $V_s = 230$ V (rms), frequency of the supply $(f) = 50$ Hz, and a nonlinear load of a 230 V, 50 Hz single-phase thyristor bridge converter with constant DC current of 20 A at 60° firing angle of its thyristors. There is a voltage fluctuation of +10 and −20% in the supply system with a base value of 230 V. Let X be the pu voltage variation and V_s' be the PCC voltage under voltage variation.

In this system, load current harmonic and reactive power compensation is provided by the shunt active filter of the UAPC. The voltage sag/swell compensation is provided by the series filter of the UAPC. While compensating for sag/swell, an active power is circulated between series and shunt active filters as explained in Chapter 6. Under maximum voltage dip, the maximum rating for both the VSCs is realized. The various rating calculations are as follows.

The AC load rms current is $I_L = 20$ A.

The fundamental active power component of load current is $I_{L1a} = I_{L1} \cos\alpha = 0.9 \times 20 \cos 60° = 9$ A.

The active power consumed by the load is $P_L = V_s I_{L1a} = 230 \times 9 = 2070$ W.

The supply voltage under maximum voltage sag is $V_s' = V_s(1 - X) = 230 \times (1 - 0.2) = 184$ V.

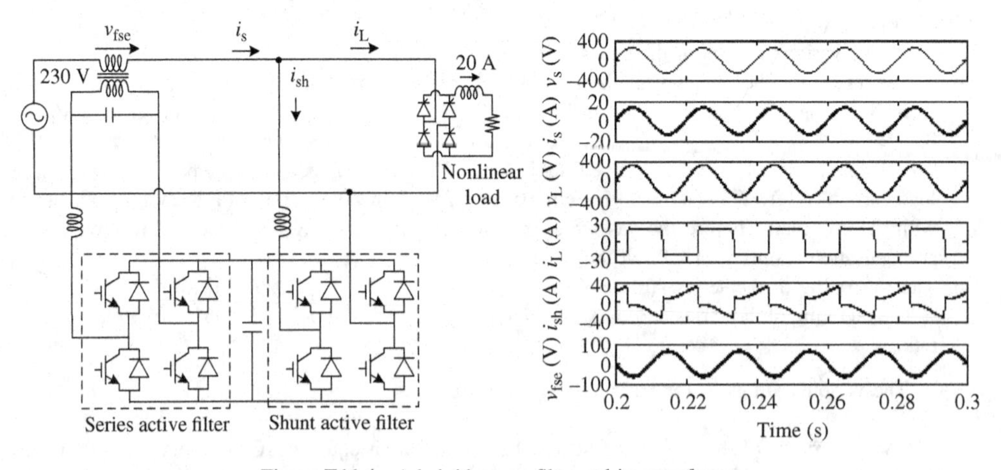

Figure E11.4 A hybrid power filter and its waveforms

The supply current under maximum voltage variation (−20% sag) is $I'_s = P_L/V'_s = 2070/184 = 11.25$ A.

a. The voltage rating of the shunt element of the UAPC is equal to AC load voltage of $V_{fsh} = 230$ V, since it is connected across the load of 230 V sine waveform.
b. The current rating of the shunt element of the UAPC is computed as follows.
   The shunt element of UAPC needs to provide load current harmonic and reactive power compensation; hence, it supplies the required harmonic currents and reactive power of the load. Therefore, total harmonic and reactive power component of current of the shunt filter is $I_{shr} = \sqrt{(I_L^2 - I_{L1a}^2)} = \sqrt{(20^2 - 9^2)} = 17.86$ A.
   The supply fundamental voltage is lower than the required load voltage. Hence, the shunt APF absorbs active power and that active power is delivered back into the system via the series APF. Under voltage sag, the active power component of the shunt APF current is calculated as $I_{sha} = I'_s - I_{L1a} = (11.25 - 9)$ A $= 2.25$ A.
   The net current rating of the shunt active filter is calculated as $I_{sh} = \sqrt{(I_{sha}^2 + I_{shr}^2)} = \sqrt{(2.25^2 + 17.86^2)} = 18.00$ A.
c. The VA rating of the VSC of the shunt APF is $S_{sh} = V_{fsh}I_{sh} = 230 \times 18$ VA $= 4140$ VA.
d. The voltage rating of the series element of the UAPC is computed as follows.
   There is a voltage fluctuation of +10 and −20% in the supply system with a base value of 230 V. Therefore, the series APF must inject the difference of the maxima of these two voltages to provide the required voltage at the load end.
   The voltage rating of the series APF is $V_{fse} = 230 \times 0.20 = 46$ V.
e. The current rating of the series element of the UAPC is same as supply current under supply voltage sag: $I_{se} = I'_s = 11.25$ A.
f. The VA rating of the series element is estimated as $S_{se} = V_{fse}I_{se} = 46 \times 11.25$ VA $= 517.5$ VA.

Example 11.5

A single-phase universal active power conditioner (consisting of shunt and series APFs using two VSCs with a common bus capacitor as shown in Figure E11.5) is designed for a 230 V, 50 Hz single-phase resistive load of 10 Ω through a phase-controlled AC voltage controller at 45° firing angle of its thyristors. If there is a voltage fluctuation of +10 and −20% in the supply system with a base value of 230 V, calculate (a) the voltage rating of the shunt element of the UAPC, (b) the current rating of the shunt

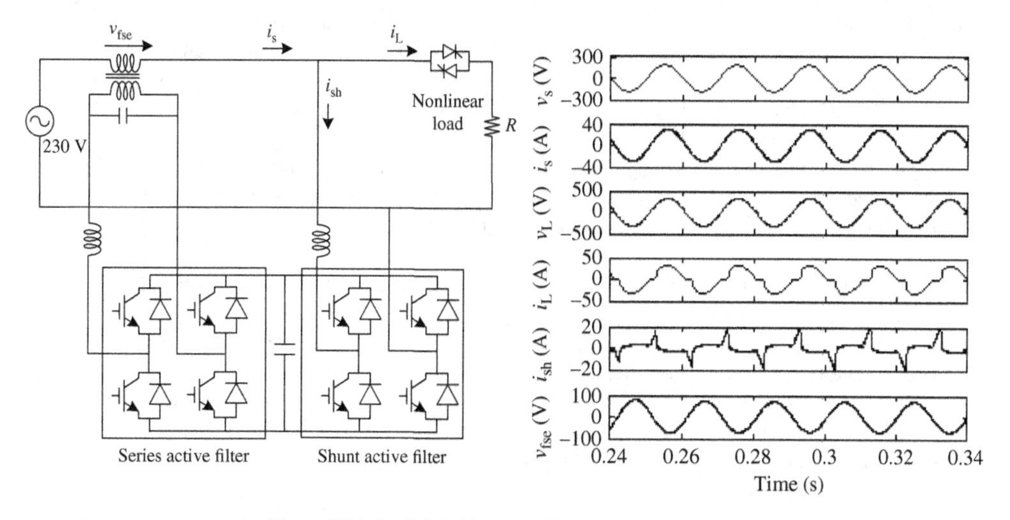

Figure E11.5 A hybrid power filter and its waveforms

element of the UAPC, (c) the VA rating of the shunt element of the UAPC, (d) the voltage rating of the series element of the UAPC, (e) the current rating of the series element of the UAPC, and (f) the VA rating of the series element of the UAPC to provide harmonic and reactive power compensation for unity power factor at the AC mains with constant sine wave voltage of 230 V at 50 Hz across the load.

Solution: Given supply voltage $V_s = 230$ V (rms), frequency of the supply $(f) = 50$ Hz, and a nonlinear load of a phase-controlled AC voltage controller consisting of 230 V, 50 Hz, $R = 10\,\Omega$, and $\alpha = 45°$ ($\pi/4$ rad). There is a voltage fluctuation of +10 and −20% in the supply system with a base value of 230 V. A single-phase universal active power conditioner needs to provide unity power factor at the AC mains with constant sine wave voltage of 230 V at 50 Hz across the load. Let X be the pu voltage variation and V_s' be the PCC voltage under voltage variation.

In this system, load current harmonic and reactive power compensation is provided by the shunt active filter of the UAPC. The voltage sag/swell compensation is provided by the series filter of the UAPC. While compensating for sag/swell, an active power is circulated between series and shunt active filters as explained in Chapter 6. Under maximum voltage dip, the maximum rating for both the VSCs is realized. The various rating calculations are as follows.

The supply voltage under maximum voltage sag is $V_s' = V_s(1 - X) = 230 \times (1 - 0.2) = 184$ V.

In a single-phase, phase-controlled AC controller, the waveform of the supply current (I_s) has a value of V_s/R from angle α to π. The peak load voltage is $V_{LA} = 230\sqrt{2} = 325.27$ V.

The load rms current is $I_L = V_{LA}[(1/2\pi)\{(\pi - \alpha) + \sin 2\alpha/2\}]^{1/2}/R = 21.72$ A.

The fundamental rms load current is $I_{L1} = V_{LA}/(2\pi R\sqrt{2})[(\cos 2\alpha - 1)^2 + \{\sin 2\alpha + 2(\pi - \alpha)\}^2]^{1/2} = 20.58$ A.

The angle between fundamental voltage and current is $\theta_1 = \tan^{-1}[(\cos 2\alpha - 1)/\{\sin 2\alpha + 2(\pi - \alpha)\}] = 9.929°$.

The rms active power fundamental component of load current is $I_{L1a} = I_{L1} \cos \theta_1 = 20.58 \cos 45° = 20.27$ A.

The active power consumed by the load is $P_L = V_s I_{L1a} = 230 \times 20.27 = 4662.1$ W.

The supply current under maximum voltage variation (−20% sag) is $I_s' = P_L/V_s' = 4662.1/184$ A = 25.33 A.

Total harmonic and reactive power component of load current is calculated as $I_{Lr} = \sqrt{(I_L^2 - I_{L1a}^2)} = \sqrt{(21.72^2 - 20.27^2)} = 7.7955$ A.

a. The voltage rating of the shunt element of the UAPC is equal to AC load voltage of $V_{sh} = 230$ V, since it is connected across the load of 230 V sine waveform.
b. The current rating of the shunt element of the UAPC is computed as follows.

   The shunt element of UAPC needs to provide load current harmonic and reactive power compensation; hence, it supplies the required harmonic currents and reactive power of the load. Therefore, total harmonic and reactive power component of load current to be supplied by the shunt active filter is $I_{shr} = I_{Lr} = 7.7955$ A.

   The supply fundamental voltage is lower than the required load voltage. Hence, the shunt APF absorbs active power and that active power is delivered back into the system via the series APF. The active power component of the shunt APF current is estimated as $I_{sha} = I_s' - I_{L1a} = 25.33 - 20.27 = 5.06$ A.

   The overall current rating of the shunt active filter is estimated as $I_{sh} = \sqrt{(I_{sha}^2 + I_{shr}^2)} = \sqrt{(5.06^2 + 7.79^2)} = 9.29$ A.
c. The VA rating of the VSC of the shunt APF is estimated as $S_{sh} = V_{sh} I_{sh} = 230 \times 9.29$ VA = 2137.24 VA.
d. The voltage rating of the series element of the UAPC is computed as follows.

   There is a voltage fluctuation of +10 and −20% in the supply system with a base value of 230 V. Therefore, the series APF must inject the difference of the maxima of these two voltages to regulate the voltage at the load terminal. The voltage rating of the series APF under maximum sag condition is $V_{fse} = 230 \times 0.20 = 46$ V.

e. The current rating of the series element of the UAPC is same as supply current under voltage sag: $I_{se} = I'_s = 25.33$ A.
f. The VA rating of the series element of the UAPC is $S_{se} = V_{fse}I_{se} = 46 \times 25.33$ VA = 1165.18 VA.

Example 11.6

A single-phase VSI with a quasi-square wave AC output of 230 V (rms) at 50 Hz is feeding a critical load of a 230 V, 50 Hz single-phase thyristor bridge with constant DC current of 20 A at 30° firing angle of its thyristors. A single-phase universal active power conditioner (consisting of shunt and series APFs using two VSCs with a common bus capacitor as shown in Figure E11.6) is designed for this critical nonlinear load. Calculate (a) the voltage rating of the shunt element of the UAPC, (b) the current rating of the shunt element of the UAPC, (c) the VA rating of the shunt element of the UAPC, (d) the voltage rating of the series element of the UAPC, (e) the current rating of the series element of the UAPC, and (f) the VA rating of the series element of the UAPC to provide harmonic and reactive power compensation for unity power factor at the AC mains by the shunt element of the UAPC and constant regulated sine wave voltage of 230 V (rms) at 50 Hz across the load by the series element of the UAPC.

Solution: Given supply voltage $V_s = 230$ V (rms quasi-square wave), frequency of the supply $(f) = 50$ Hz, and a critical load of a 230 V, 50 Hz single-phase thyristor bridge with constant DC current of 20 A at 30° ($\pi/6$ rad) firing angle of its thyristors.

In this system, load reactive power and harmonic current compensation is provided by the shunt active filter of the UAPC. The voltage compensation is provided by the series filter of the UAPC. However, there is a difference in magnitude of fundamental voltage in the supply and load terminals, and to compensate that an active power is circulated between series and shunt active filters as explained in Chapter 6. The rating calculations for both the VSCs of the UAPC are as follows.

The load current is a square wave current with amplitude I_{DC}. The AC load rms current is $I_L = 20$ A.

The fundamental component of load current is estimated as $I_{L1} = 0.9I_{DC}$.

The fundamental active power component of load current is $I_{L1a} = I_{L1} \cos\alpha = 0.9 \times 20 \cos 30° =$ 15.59 A.

The harmonic and reactive power component of load current is estimated as $I_{Lr} = \sqrt{(I_L^2 - I_{L1a}^2)} =$ 12.52 A.

The amplitude of the quasi-square wave is estimated as $V_{sDC} = 230/\sqrt{(2/3)} = 281.69$ V.

The fundamental component of supply voltage is estimated as $V_{s1} = (\sqrt{6}/\pi)V_{sDC} = 0.779 \times 281.69 =$ 219.63 V.

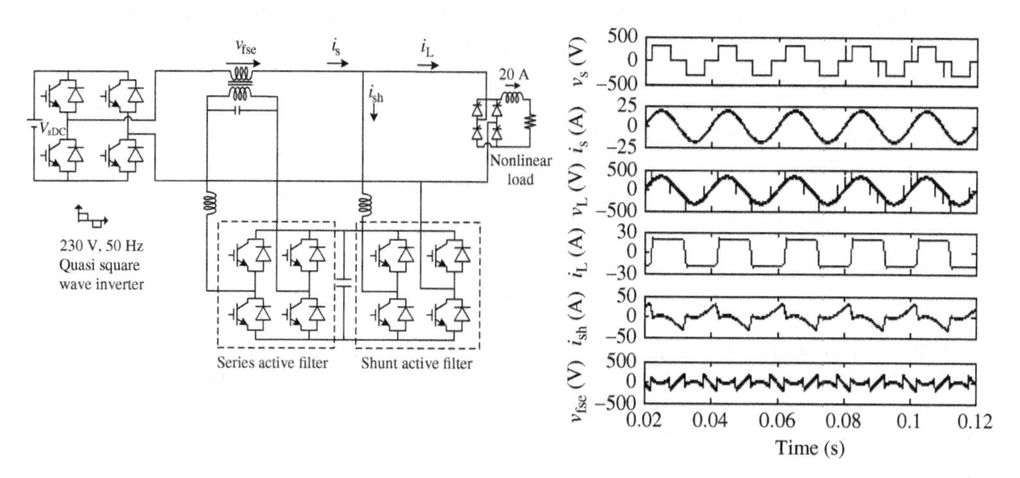

Figure E11.6 A hybrid power filter and its waveforms

The active power consumed by the load is $P_L = V_L I_{L1a} = 230 \times 15.59 = 3585.7$ W.
The supply current is estimated as $I_s = P_L/V_{s1} = 3585.7/219.63 = 16.33$ A.

a. The voltage rating of the shunt element of the UAPC is equal to AC load voltage of $V_{fsh} = 230$ V, since it is connected across the load of 230 V sine waveform.
b. The current rating of the shunt element of the UAPC is computed as follows.
   The shunt element of the UAPC needs to supply load current harmonic and reactive power compensation; hence, total harmonic and reactive power component of load current through the shunt active filter is $I_{shr} = \sqrt{(I_L^2 - I_{L1a}^2)} = \sqrt{(20^2 - 15.59^2)} = 12.52$ A.
   The supply fundamental voltage is lower than the required load voltage. Hence, the shunt APF absorbs active power and that active power is delivered back into the system via the series APF. The active power component of the shunt APF current is estimated as $I_{sha} = I_s - I_{L1a} = (16.33 - 15.59)$ A $= 0.74$ A.
   The net current rating of the shunt APF is estimated as $I_{sh} = \sqrt{(I_{sha}^2 + I_{shr}^2)} = \sqrt{(0.74^2 + 12.52^2)} = 12.56$ A.
c. The VA rating of the VSC of the shunt APF is $S_{sh} = V_{fsh} I_{sh} = 230 \times 12.56 = 2888.8$ VA.
d. The voltage rating of the series element of the UAPC is computed as follows.
   The supply voltage is a quasi-square wave of $V_s = 230$ V (rms) and the voltage at the load terminal must be a sine wave of $V_L = 230$ V. Therefore, the series APF must inject the difference of these two voltages to provide the required voltage at the load end.
   The voltage rating of the series APF is

$$V_{fse} = \sqrt{\left[(1/\pi)\left\{\int_0^{\pi/6} (-230\sqrt{2}\sin\theta)^2 d\theta + \int_{\pi/6}^{5\pi/6} (281.69 - 230\sqrt{2}\sin\theta)^2 d\theta + \int_{5\pi/6}^{\pi} (-230\sqrt{2}\sin\theta)^2 d\theta\right\}\right]}$$
$$= 69.05\,\text{V}.$$

e. The current rating of the series element of the UAPC is same as supply current: $I_{se} = I_s = 16.33$ A.
f. The VA rating of the series element of the UAPC is $S_{se} = V_{fse} I_{se} = 69.29 \times 16.33$ VA $= 1131.5$ VA.

Example 11.7

A single-phase VSI with a square wave AC output of 230 V (rms) at 50 Hz is feeding a critical load of a 220 V, 50 Hz single-phase uncontrolled diode bridge converter having a *RE* load with $R = 4\,\Omega$ and $E = 216$ V. A single-phase universal active power conditioner (consisting of shunt and series APFs using two VSCs with a common bus capacitor as shown in Figure E11.7) is designed for this critical nonlinear load. Calculate (a) the voltage rating of the shunt element of the UAPC, (b) the current rating of the shunt element of the UAPC, (c) the VA rating of the shunt element of the UAPC, (d) the voltage rating of the series element of the UAPC, (e) the current rating of the series element of the UAPC, and (f) the VA rating of the series element of the UAPC to provide harmonic and reactive power compensation for unity power factor at the AC mains by the shunt element of the UAPC and constant regulated sine wave voltage of 220 V (rms) at 50 Hz across the load by the series element of the UAPC.

Solution: Given supply voltage $V_s = 230$ V (rms square wave), frequency of the supply $(f) = 50$ Hz, and a critical load of a 220 V, 50 Hz single-phase uncontrolled diode bridge converter having a *RE* load with $R = 4\,\Omega$ and $E = 216$ V.

In this system, load reactive power and harmonic current compensation is provided by the shunt active filter of the UAPC. The voltage compensation is provided by the series filter of the UAPC. However, there is a difference in magnitude of fundamental voltage in the supply and load terminals, and to compensate that an active power is circulated between series and shunt active filters as explained in Chapter 6. The rating calculations for both the VSCs of the UAPC are as follows. The amplitude of the square wave is $V_{sDC} = 230$ V.

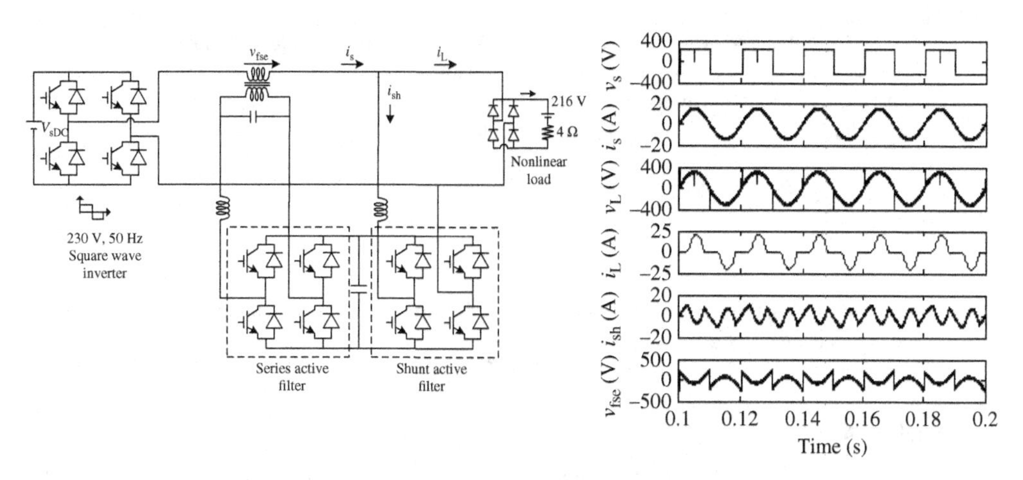

Figure E11.7 A hybrid power filter and its waveforms

The fundamental component of supply voltage is estimated as $V_{s1} = 0.9V_{sDC} = 0.9 \times 230 = 207$ V. However, the load terminal voltage needs to be regulated at $V_L = 220$ V.

For this reduced fundamental voltage, the rating calculations for both the filters are as follows.

In a single-phase diode bridge converter with *RE* load, the current flows from angle α when AC voltage is equal to E to angle β at which AC voltage reduces to E.

The peak load voltage is $V_{Lm} = 220 \times \sqrt{2}\text{ V} = 311.13$ V.

$\alpha = \sin^{-1}(E/V_{Lm}) = \sin^{-1}(216/311.13) = 43.96°$, $\beta = \pi - \alpha = 136.03°$, and the conduction angle $= \beta - \alpha = 92.06°$.

The rms load current (I_L) is rms of discontinuous current in the AC mains, which is estimated as $I_L = [(1/\pi R^2)\{(0.5V_{Lm}^2 + E^2)(\pi - 2\alpha) + 0.5V_{Lm}^2 \sin 2\alpha - 4V_{Lm}E\cos\alpha\}]^{1/2} = 12.32$ A.

The average current (I_{DC}) flowing into the battery is $I_{DC} = (1/\pi R)(2V_{Lm}\cos\alpha + 2E\alpha - \pi E) = 8.02$ A.

The active power drawn from the AC mains is $P_L = I_L^2 R + EI_{DC} = 12.32^2 \times 4 + 216 \times 8.02 = 2339.8$ W.

The active power component of fundamental load current from the AC mains is $I_{L1a} = P_L/V_L = 2339.8/220 = 10.63$ A.

The supply current is estimated as $I_s = P_L/V_{s1} = 2339.8/207 = 11.30$ A.

The harmonic and reactive power component of load current is estimated as $I_{Lr} = \sqrt{(I_L^2 - I_{L1a}^2)} = \sqrt{(12.32^2 - 10.36^2)} = 6.22$ A.

a. The voltage rating of the shunt element of the UAPC is equal to AC load voltage of $V_{fsh} = 220$ V, since it is connected across the load of 220 V sine waveform.
b. The current rating of the shunt element of the UAPC is computed as follows.

   The shunt element of the UAPC supplies all current harmonics and reactive power component of load current; hence, the harmonic and reactive power component of current from the shunt active filter is $I_{shr} = \sqrt{(I_L^2 - I_{L1a}^2)} = \sqrt{(12.32^2 - 10.36^2)} = 6.22$ A.

   The supply fundamental voltage is lower than the required load voltage. Hence, the shunt APF absorbs active power and that active power is delivered back into the system via the series APF. The active power component of current from the shunt APF is estimated as $I_{sha} = I_s - I_{L1a} = (11.30 - 10.63)\text{ A} = 0.67$ A.

   The net current rating of the shunt APF is estimated as $I_{sh} = \sqrt{(I_{sha}^2 + I_{shr}^2)} = \sqrt{(0.67^2 + 6.22^2)} = 6.25$ A.
c. The VA rating of the VSC of the shunt APF is $S_{sh} = V_{fsh}I_{sh} = 220 \times 6.25 = 1434.4$ VA.
d. The voltage rating of the series element of the UAPC is computed as follows.

The supply voltage is a square wave of $V_s = 230$ V (rms) and the voltage at the load terminal must be a sine wave of $V_L = 220$ V. Therefore, the series APF must inject the difference of these two voltages to provide the required voltage at the load end.

The voltage rating of the series APF is estimated as

$$V_{fse} = \sqrt{\left[(1/\pi)\int_0^{\pi}(230 - 220\sqrt{2}\sin\theta)^2\, d\theta\right]} = 100.93\ \text{V}.$$

e. The current through the series element of the UAPC is same as supply current; hence, the current rating of the series element is estimated as $I_{se} = I_s = 11.30$ A.
f. The VA rating of the series element of the UAPC is $S_{se} = V_{fse}I_{se} = 100.93 \times 11.30$ VA $= 1140.5$ VA.

Example 11.8

A single-phase active hybrid filter (consisting of a one-branch passive shunt filter as a high-pass damped filter and a series APF connected in series with the AC mains using a VSC with a bus capacitor as shown in Figure E11.8) is designed for a load of a 230 V, 50 Hz thyristor AC voltage controller with a resistive load of 10.0 Ω at 100° firing angle of its thyristors. Calculate element values of the passive shunt filter and the ratings of both the passive shunt filter and the series APF used in the hybrid filter to provide harmonic compensation and reactive power compensation for unity power factor at the AC mains.

Solution: Given supply rms voltage $V_s = 230$ V, frequency of the supply (f) = 50 Hz, $R = 10\ \Omega$, and $\alpha = 100°$. A single-phase active hybrid filter (consisting of a shunt passive filter and a series APF using a VSC with a DC bus capacitor) connected in series with the AC mains is designed for this load compensation.

In a single-phase, phase-controlled AC controller, the waveform of the supply current (I_s) has a value of V_s/R from angle α to π. The peak load voltage is $V_{LA} = 230\sqrt{2} = 325.27$ V.

The AC load rms current is $I_L = V_{LA}[(1/2\pi)\{(\pi - \alpha) + \sin 2\alpha/2\}]^{1/2}/R = 14.363$ A.

The fundamental rms load current is $I_{L1} = V_{LA}/(2\pi R\sqrt{2})[(\cos 2\alpha - 1)^2 + \{\sin 2\alpha + 2(\pi - \alpha)\}^2]^{1/2} =$ 11.44 A.

The displacement angle is $\theta_1 = \tan^{-1}[(\cos 2\alpha - 1)/\{\sin 2\alpha + 2(\pi - \alpha)\}] = 38.36°$.

The fundamental active power drawn by the load is $P_{L1} = V_sI_{L1}\cos\theta_1 = 2063.2$ W.

The fundamental reactive power drawn by the load is $Q_{L1} = V_sI_{L1}\sin\theta_1 = 1633.1$ VAR.

The rms load voltage is $V_L = V_{LA}[(1/2\pi)\{(\pi - \alpha) + \sin 2\alpha/2\}]^{1/2} = 143.63$ V.

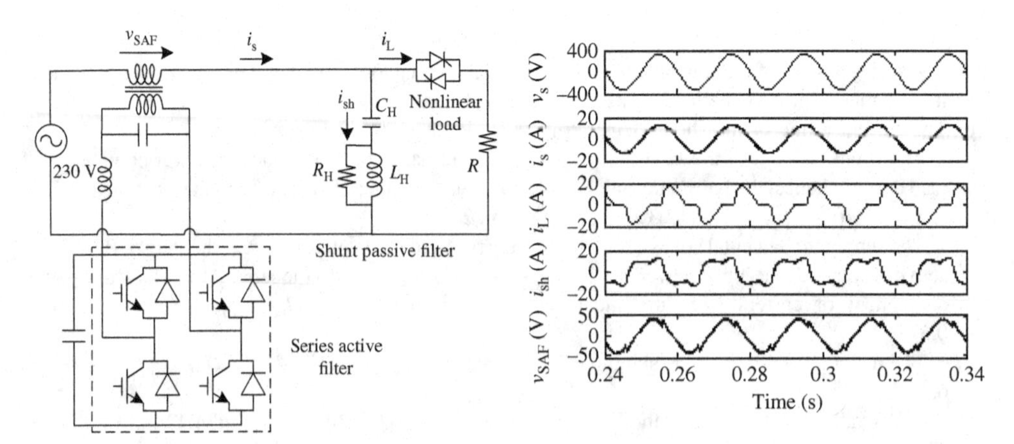

Figure E11.8 A hybrid power filter and its waveforms

The load displacement factor is $DPF = \cos\theta_1 = \cos 38.36° = 0.784$.

The load distortion factor is $DF = I_{L1}/I_L = 11.44/14.36 = 0.79649$.

The load power factor is $PF = DPF \times DF = 0.62445$.

The AC mains current after compensation is $I_s = I_{L1}\cos\theta_1 = P_{L1}/V_s = 2062.957/230 = 9$ A.

The shunt passive filter has to meet the reactive power requirement of the load for UPF at the AC mains. Therefore, the capacitor of the passive shunt filter must be selected to provide this required reactive power.

The value of the capacitor is $C = Q_1/V_s^2\omega = 1633.1/(230^2 \times 314) = 98.26\,\mu F$.

Since there is only one branch in the passive shunt filter, which must be tuned for the lowest harmonic (third-order harmonic in this case) present in the load current as a damped filter to take care of all the harmonic currents, this value of the capacitor is $C_H = C = 98.26\,\mu F$.

The value of the inductor for the high-pass damped harmonic filter (tuned at third harmonic) is $L_H = 1/\omega_3^2 C_H = 11.5$ mH.

The resistance in parallel of the inductor of the high-pass damped harmonic tuned filter is $R_H = X_H/Q_H = 5 \times 314 \times 11.5/5 = 2.15\,\Omega$ (considering $Q_H = 5$ for lower losses in it as it may be in the range of 0.5–5 depending upon the attenuation required).

All other harmonic load currents to flow in the high-pass damped harmonic filter are $I_H = \sqrt{(I_L^2 - I_{L1}^2)} = \sqrt{(14^2 - 11.44^2)} = 8.68$ A.

All higher order harmonic voltages at the load end and across the passive filter are $V_H = I_H R_H = 8.68 \times 2.15 = 18.75$ V.

All harmonic voltages other than the fundamental voltage at the load end and across the passive filter are $V_{LH} = V_H = I_H R_H = 18.75$ V.

The voltage rating of the passive shunt filter is $V_{PF} = \sqrt{(V_s^2 + V_{LH}^2)} = \sqrt{(230^2 + 18.75^2)} = 230.76$ V.

The current rating of the passive shunt filter is $I_{PF} = \sqrt{(I_L^2 - I_s^2)} = \sqrt{(14.363^2 - 9^2)} = 11.18$ A.

The VA rating of the passive shunt filter is $S_{PF} = V_{PF} I_{PF} = 230.76 \times 11.18 = 2581.7 = 2.5817$ kVA.

The pu rating of the passive shunt filter is $S_{PFpu} = S_{PF}/S_L = 2581.7/(230 \times 14.363) = 0.7815 = 78.15\%$ of the load rating.

The series active filter must inject all harmonic voltages to force all harmonic currents into the passive filter. Therefore, the voltage rating of the series active filter is $V_{SAF} = V_{LH} = V_H = 18.76$ V.

The current rating of the series active filter is $I_{SAF} = I_s$ = fundamental active power component of load current $(I_s) = I_{L1}\cos\theta_1 = P_1/V_s = 2062.957/230 = 8.969$ A (since the series APF is connected in series with the AC mains before the passive shunt filter that is providing harmonic and reactive power compensation resulting in UPF at the AC mains).

The VA rating of the series APF is $S_{APF} = V_{SAF} I_{SAF} = 18.76 \times 8.969 = 168.24 = 0.1682$ kVA.

The pu rating of the series APF is $S_{APFpu} = S_{APF}/S_L = 0.16824/(230 \times 14.363) = 0.0509 = 5.09\%$ of the load rating.

Example 11.9

A three-phase VSI with a quasi-square wave AC output line voltage of 400 V (rms) at 50 Hz is feeding a critical load of 415 V, 50 Hz, three-phase 30 kVA at 0.8 lagging power factor. A three-phase universal active power conditioner (consisting of series and shunt APFs using two VSCs with a common bus capacitor as shown in Figure E11.9) is designed for this critical linear load. Calculate (a) the voltage rating of the shunt element of the UAPC, (b) the current rating of the shunt element of the UAPC, (c) the VA rating of the shunt element of the UAPC, (d) the voltage rating of the series element of the UAPC, (e) the current rating of the series element of the UAPC, and (f) the VA rating of the series element of the UAPC to provide harmonic and reactive power compensation for unity power factor at the AC mains by the shunt element of the UAPC and constant regulated sine wave voltage of 415 V (rms) at 50 Hz across the load by the series element of the UAPC.

Solution: Given supply line voltage $V_{sL} = 400$ V (rms quasi-square wave), frequency of the supply $(f) = 50$ Hz, and a critical linear load of 415 V, 50 Hz three-phase 30 kVA at 0.8 lagging power factor.

The desired load phase voltage is $V_{Lp} = 415/\sqrt{3} = 239.6$ V of sine wave.

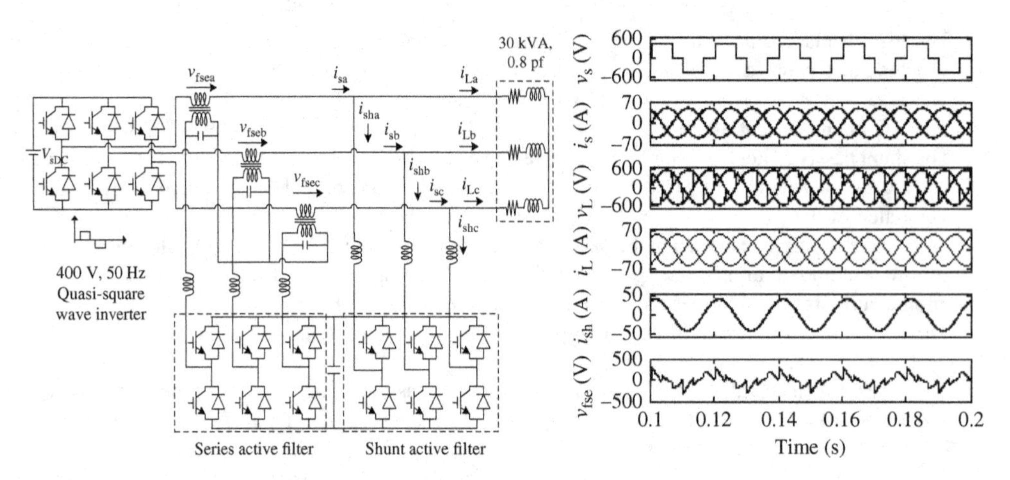

Figure E11.9 A hybrid power filter and its waveforms

The magnitude of DC side voltage of the supply side VSI is estimated as $V_{sDC} = V_{sL}/\sqrt{(2/3)} = 400/0.8165 = 489.89$ V.

The load line current is $I_L = I_{L1} = 30\,000/(3 \times 239.6) = 41.73$ A.

The active power component of load line current is $I_{L1a} = S_L \times PF/3V_{Lp} = 30\,000 \times 0.8/(3 \times 239.6) = 33.39$ A.

The fundamental component of supply side line voltage is $V_{sL1} = \sqrt{6}V_{sDC}/\pi = 0.779 \times 489.89 = 381.62$ V.

The fundamental component of supply phase voltage is $V_{sp1} = V_{sL1}/\sqrt{3} = 220.32$ V.

The supply current after compensation is $I_s = S_L \times PF/3V_{sp1} = 36.31$ A.

In this system, load current harmonic and reactive power compensation is provided by the shunt active filter of the UAPC. The voltage compensation is provided by the series filter of the UAPC. However, there is a difference in magnitude of fundamental voltage in the supply and load terminals, and to compensate that an active power is circulated between series and shunt active filters as explained in Chapter 6. The rating calculations for both the VSCs of the UAPC are as follows.

a. The voltage rating of the shunt element of the UAPC is equal to AC load phase voltage of $V_{sh} = 239.6$ V, since it is connected across the load of 239.6 V sine waveform.
b. The current rating of the shunt element of the UAPC is computed as follows.

   The shunt element of the UAPC compensates for the load reactive power. Hence, the required reactive power component of the shunt active filter current is estimated as $I_{shr} = \sqrt{(I_L^2 - I_{L1a}^2)} = \sqrt{(41.73^2 - 33.39^2)} = 25.03$ A.

   The supply fundamental voltage is lower than the required load voltage. Hence, the shunt APF absorbs active power and that active power is delivered back into the system via the series APF. The active power component of the shunt APF current is estimated as $I_{sha} = I_s - I_{L1a} = (36.31 - 33.39)$ A $= 2.92$ A.

   The net current rating of the shunt APF is estimated as $I_{sh} = \sqrt{(I_{sha}^2 + I_{shr}^2)} = \sqrt{(2.92^2 + 25.03^2)} = 25.2$ A.
c. The VA rating of the VSC of the shunt APF is $S_{sh} = 3V_{sh}I_{sh} = 3 \times 239.6 \times 25.2$ VA $= 18\,113.57$ VA.
d. The supply of quasi-square line voltage results in a stepped phase voltage. The waveform of the phase voltage at the input of the series active filter is a stepped waveform with (i) the first step of $\pi/3$ angle (from 0° to $\pi/3$) and a magnitude of $V_{sDC}/3$, (ii) the second step of $\pi/3$ angle (from $\pi/3$ to $2\pi/3$) and a magnitude of $2V_{sDC}/3$, and (iii) the third step of $\pi/3$ angle (from $2\pi/3$ to π) and a magnitude of $V_{sDC}/3$, and it has both half cycles of symmetric segments of such steps.

The voltage rating of the series APF (V_{fse}) is computed by taking the difference of the supply phase voltage and required load sine phase voltage at the input of the linear load as

$$V_{fse} = \sqrt{\left[(1/\pi)\left\{\int_0^{\pi/3}(163.29 - 239.6\sqrt{2}\sin\theta)^2\,d\theta + \int_{\pi/3}^{2\pi/3}(326.59 - 239.6\sqrt{2}\sin\theta)^2\,d\theta + \int_{2\pi/3}^{\pi}(163.29 - 239.6\sqrt{2}\sin\theta)^2\,d\theta\right\}\right]}$$

$$= 71.15\text{ V}.$$

e. The current rating of the series APF is $I_{se} = I_s = 36.31$ A (since the series filter is connected in series with the supply VSI).
f. The VA rating of the series APF is $S_{se} = 3V_{fse}I_{se} = 3 \times 71.15 \times 36.31$ VA $= 7750.36$ VA.

Example 11.10

A three-phase active filter connected in series with the AC mains and a three-branch shunt passive filter (fifth, seventh, and high-pass filters as shown in Figure E11.10) is used for harmonic current and reactive power compensation in a three-phase 415 V, 50 Hz system to reduce the THD of the supply current and to improve the displacement factor to unity. It has a load of a three-phase thyristor bridge converter operating at 30° firing angle drawing constant 150 A DC current. Calculate (a) the designed values of passive filter components, (b) line current, and (c) the VA rating of the series APF.

Solution: Given that a three-phase AC line voltage of 415 V (rms) at 50 Hz is feeding a load of a three-phase thyristor bridge converter operating at 30° firing angle drawing constant 150 A DC current. A set of shunt connected AC tuned passive filters are used to compensate the reactive power and fifth, seventh, and high-pass damped filters are used to compensate all higher order harmonics at the input AC mains.

In this system, load current harmonic and reactive power compensation is provided by the shunt passive filter and a series active filter is used to force all harmonic currents through this shunt passive filter.

In a three-phase thyristor bridge converter, the waveform of the load current (I_L) is a quasi-square wave with the amplitude of DC link current (I_{DC}).

Therefore, $I_L = \sqrt{(2/3)}I_{DC} = 0.81649 \times 150 = 122.47$ A.

The rms value of the fundamental component of quasi-square wave load current is $I_{L1} = (\sqrt{6}/\pi)I_{DC} = 0.779 \times 150 = 117.01$ A.

The active power component of supply current is $I_{L1a} = I_{L1}\cos\theta_1 = I_{L1}\cos\alpha = 117.01\cos 30° = 101.33$ A.

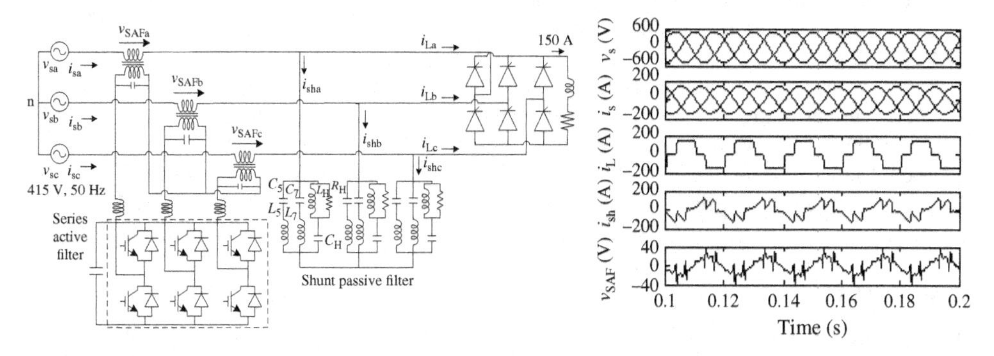

Figure E11.10 A hybrid power filter and its waveforms

The fundamental active power of the load is $P_L = 3V_L I_{L1a} = 3 \times 239.6 \times 101.33 = 72.836$ kW.

The fundamental reactive power of the load is $Q_L = 3V_L I_{L1}$ sin $\alpha = 3 \times 239.6 \times 117.01 \times 0.5 = 42.053$ kVAR.

a. The passive shunt filter has nine branches (three for each phase and nine for all three phases). The total reactive power of 42.053 kVAR required by the thyristor rectifier has to be provided by all nine branches of the passive shunt filter. Considering that all branches of the passive filter have equal capacitors, the value of this capacitor is $C = Q/9V_s^2\omega = 42\,053/(9 \times 239.6^2 \times 314) = 259.21$ μF. Therefore, $C_5 = C_7 = C_H = C = 259.21$ μF.

   The value of the inductor for the fifth harmonic tuned filter is $L_5 = 1/\omega_5^2 C_5 = 1.565$ mH.

   The resistance of the inductor of the fifth harmonic tuned filter is $R_5 = X_5/Q_5 = 2.547/20 = 0.123\,\Omega$ (considering $Q_5 = 20$ as it may be in the range of 10–100 depending upon the design of the inductor).

   The value of the inductor of the seventh harmonic tuned filter is $L_7 = 1/\omega_7^2 C_7 = 0.798$ mH.

   The resistance of the inductor of the seventh harmonic tuned filter is $R_7 = X_7/Q_7 = 1.7655/20 = 0.088\,\Omega$ (considering $Q_7 = 20$ as it may be in the range of 10–100 depending upon the design of the inductor).

   The value of the inductor for the high-pass damped harmonic filter (tuned at 11th harmonic) is $L_H = 1/\omega_{11}^2 C_H = 0.323\,37$ mH.

   The resistance in parallel of the inductor of the high-pass damped harmonic tuned filter is $R_H = X_H/Q_H = 1.1169/2 = 0.558\,\Omega$ (considering $Q_H = 2$ as it may be in the range of 0.5–5 depending upon the attenuation required.).

   The fundamental rms current of the three-phase thyristor rectifier with quasi-square AC mains current is $I_{1TS} = I_{L1} = 117.01$ A.

   The fifth harmonic load current to flow in the fifth harmonic tuned filter is $I_5 = I_{1TS}/5 = 117.01/5 = 23.402$ A.

   The fifth harmonic voltage at the load end and across the passive filter is $V_5 = I_5 R_5 = 23.402 \times 0.122\,862 = 2.875$ V.

   The seventh harmonic load current to flow in the seventh harmonic tuned filter is $I_7 = I_{1TS}/7 = 117.01/7 = 16.72$ A.

   The seventh harmonic voltage at the load end and across the passive filter is $V_7 = I_7 R_7 = 16.72 \times 0.088\,275 = 1.48$ V.

   All other harmonic load currents to flow in the high-pass damped harmonic filter are $I_H = \sqrt{(I_s^2 - I_{1TS}^2 - I_5^2 - I_7^2)} = \sqrt{(122.47^2 - 117.01^2 - 23.402^2 - 16.72^2)} = 21.92$ A.

   All higher order harmonic voltages at the load end and across the passive filter are $V_H = I_H R_H = 21.92 \times 0.558\,45 = 12.24$ V.

   All harmonic voltages other than the fundamental voltage at the load end and across the passive filter are $V_{LH} = \sqrt{(V_5^2 + V_7^2 + V_H^2)} = 12.66$ V.

   The series active filter must inject all harmonic voltages to force all harmonic currents into the passive filter. Therefore, the voltage rating of the series active filter is $V_{SAF} = V_{LH} = \sqrt{(V_5^2 + V_7^2 + V_H^2)} = 12.66$ V.

b. The line current after compensation is $I_{L1a} = I_{L1} \cos\theta_1 = I_{L1} \cos\alpha = 117.01 \cos 30° = 101.33$ A.

   The current rating of the series active filter is $I_{SAF} = I_{L1a} = 101.33$ A (since the series APF is connected in series with the AC mains).

c. The VA rating of the series APF is $S_{APF} = 3V_{SAF} I_{SAF} = 3 \times 12.66 \times 101.33 = 3848.5 = 3.8485$ kVA.

   The pu rating of the series APF is $S_{pu} = S_{APF}/S_L = 3848.5/(3 \times 239.6 \times 122.47) = 0.0437 = 4.37\%$ of the load rating.

Example 11.11

A three-phase series active filter connected in series with a three-branch shunt passive filter (fifth, seventh, and high-pass filters as shown in Figure E11.11) is used for harmonic current and reactive power

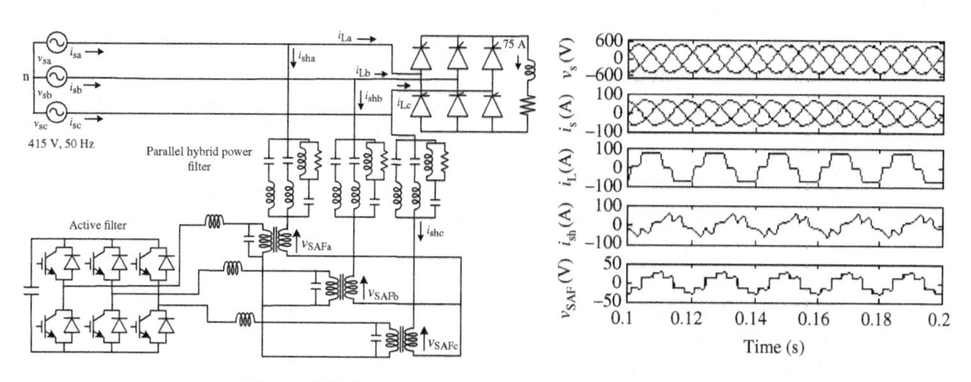

Figure E11.11 A hybrid power filter and its waveforms

compensation in a three-phase 415 V, 50 Hz system to reduce the THD of the supply current and to improve the displacement factor to unity. It has a load of a three-phase thyristor bridge converter operating at 30° firing angle drawing constant 75 A DC current. Calculate (a) the designed values of passive filter components, (b) line current, (c) the VA rating of the APF, (d) the AC inductor value of the APF, and (e) its DC bus voltage, and (f) the DC bus capacitor value of the APF. Consider the switching frequency of 20 kHz and the DC bus voltage has to be controlled within 8% range and ripple current in the inductor is 15%. The turns ratio of the injection transformer is 1:10.

Solution: Given that a three-phase AC line voltage of 415 V (rms) at 50 Hz is feeding a load of a three-phase thyristor bridge converter operating at 30° firing angle drawing constant 75 A DC current. A three-phase series active filter connected in series with a three-branch shunt passive filter (fifth, seventh, and high-pass filters) is used for harmonic current and reactive power compensation in a three-phase 415 V, 50 Hz system to reduce the THD of the supply current and to improve the displacement factor to unity. Consider the switching frequency of 20 kHz and the DC bus voltage has to be controlled within 8% range and ripple current in the inductor is 15%. The turns ratio of the injection transformer is 1:10.

In this system, load current harmonic and reactive power compensation is provided by the shunt passive filter and a series filter is used to force all harmonic currents through this shunt passive filter.

In a three-phase thyristor bridge converter, the waveform of the load current (I_L) is a quasi-square wave with the amplitude of DC link current (I_{DC}).

Therefore, $I_L = \sqrt{(2/3)}I_{DC} = 0.816 \times 75 = 61.24$ A.

Moreover, the rms value of the fundamental component of the quasi-square wave is $I_{L1} = (\sqrt{6}/\pi) I_{DC} = 58.48$ A.

The active power component of supply current is $I_{L1a} = I_{L1} \cos\theta_1 = I_{L1} \cos\alpha = 58.48 \cos 30° = 50.64$ A.

Total harmonic and reactive power component of load current is $I_f = \sqrt{(I_L^2 - I_{L1a}^2)} = \sqrt{(61.24^2 - 50.64^2)} = 34.43$ A.

The fundamental reactive power of the load is $Q_L = 3V_L I_{L1} \sin\alpha = 3 \times 239.6 \times 58.48 \times 0.5 = 21.015$ kVAR.

a. The passive shunt filter has nine branches (three for each phase and nine for all three phases). The total reactive power of 21.0156 kVAR required for the thyristor converter has to be provided by all nine branches of the passive shunt filter. Considering that all branches of the passive filter have equal capacitors, the value of this capacitor is $C = Q_L/9V_L^2\omega = 21\,015.6/(9 \times 239.6^2 \times 314) = 129.47$ μF. So, $C_5 = C_7 = C_H = C = 129.47$ μF.

   The value of the inductor for the fifth harmonic tuned filter is $L_5 = 1/\omega_5^2 C_5 = 3.131$ mH.

The resistance of the inductor of the fifth harmonic tuned filter is $R_5 = X_5/Q_5 = 4.917/20 = 0.245\,\Omega$ (considering $Q_5 = 20$ as it may be in the range of 10–100 depending upon the design of the inductor).

The value of the inductor of the seventh harmonic tuned filter is $L_7 = 1/\omega_7^2 C_7 = 1.597$ mH.

The resistance of the inductor of the seventh harmonic tuned filter is $R_7 = X_7/Q_7 = 3.5122/20 = 0.1756\,\Omega$ (considering $Q_5 = 20$ as it may be in the range of 10–100 depending upon the design of the inductor).

The value of the inductor for the high-pass damped harmonic filter (tuned at 11th harmonic) is $L_H = 1/\omega_{11}^2 C_H = 0.6468$ mH.

The resistance in parallel of the inductor of the high-pass damped harmonic tuned filter is $R_H = X_H/Q_H = 2.235/2.5 = 0.8945\,\Omega$ (considering $Q_H = 2.5$ as it may be in the range of 0.5–5 depending upon the attenuation required).

The fundamental rms current of the three-phase thyristor rectifier with quasi-square AC input current is $I_{1TS} = I_{L1} = (\sqrt{6}/\pi)I_{DC} = 58.48$ A.

The fifth harmonic load current to flow in the fifth harmonic tuned filter is $I_5 = I_{1TS}/5 = 58.48/5 = 11.696$ A.

The fifth harmonic voltage at the load end and across the passive filter is $V_5 = I_5 R_5 = 11.696 \times 0.2459 = 2.876$ V.

The seventh harmonic load current to flow in the seventh harmonic tuned filter is $I_7 = I_{1TS}/7 = 58.48/7 = 8.35$ A.

The seventh harmonic voltage at the load end and across the passive filter is $V_7 = I_7 R_7 = 8.35 \times 0.1756 = 1.467$ V.

All other harmonic load currents to flow in the high-pass damped harmonic filter are $I_H = \sqrt{(I_L^2 - I_{1TS}^2 - I_5^2 - I_7^2)} = \sqrt{(61.24^2 - 58.48^2 - 11.696^2 - 8.35^2)} = 10.99$ A.

All higher order harmonic voltages at the load end and across the passive filter are $V_H = I_H R_H = 11.13 \times 0.8945 = 9.828$ V.

All harmonic voltages other than the fundamental voltage at the load end and across the passive filter are $V_{LH} = \sqrt{(V_5^2 + V_7^2 + V_H^2)} = 10.3445$ V.

The series active filter must inject all harmonic voltages to force all harmonic currents into the passive filter. Therefore, the voltage rating of the series active filter is $V_{SAF} = V_{LH} = \sqrt{(V_5^2 + V_7^2 + V_H^2)} = 10.3445$ V.

b. The line current after compensation is $I_{s1a} = I_{s1} \cos\theta_1 = I_{s1} \cos\alpha = 58.48 \cos 30° = 50.64$ A.

   The current rating of the series active filter is $I_{SAF} = \sqrt{(I_L^2 - I_{L1a}^2)} = 34.44$ A. (since the series APF is connected in series with the passive shunt filter).

c. The VA rating of the series APF is $S_{APF} = 3V_{SAF}I_{SAF} = 3 \times 10.3445 \times 34.44 = 1067.5$ VA.

   The pu rating of the series APF is $S_{pu} = S_{APF}/S_L = 1080.934/(3 \times 239.6 \times 61.24) = 2.43\%$ of the load rating.

d. The AC inductor value of the series APF is $L_f = N_2/N_1(\sqrt{3}/2)m_a V_{DCAPF}/(6af_s\Delta I_f) = (\sqrt{3}/2) \times 0.8 \times 365.73/(6 \times 1.2 \times 20\,000 \times 0.15 \times 34.44/10) = 3.831$ mH.

e. The DC bus voltage of the series APF connected in series with the passive shunt filter can be calculated from its injected voltage.

   The VSC side AC injected voltage is $V_{inj} = V_{SAF}N_2/N_1 = 10.3445 \times 10 = 103.4452$ V.

   The DC bus voltage to the VSC to inject this phase rms voltage is $V_{DCAPF} = 2\sqrt{2}V_{inj}/m_a = 2\sqrt{2} \times 103.4452/0.8 = 365.73$ V, where m_a is the modulation index.

f. The DC bus capacitance of the APF is computed from change in stored energy during dynamics as follows.

   The change in stored energy during dynamics is $\Delta E = (1/2)C_{DC}(V_{DCAPF}^2 - V_{DCminAPF}^2) = 3V_{SAF}I_{SAF}\Delta t$.

   Substituting the values, $\Delta E = (1/2)C_{DC}(365.73^2 - 336.4753^2) = 3 \times 10.3445 \times 34.44 \times 10/1000$ (considering $\Delta t = 10$ ms).

$$C_{DC} = 1000\,\mu\text{F}.$$

Example 11.12

A three-phase universal active power conditioner (consisting of shunt and series APFs using two VSCs with a common bus capacitor as shown in Figure E11.12) is designed for a load of a 415 V, 50 Hz thyristor bridge with constant DC current of 150 A at 60° firing angle of its thyristors. If there is a voltage fluctuation of +10 and −20% in the supply system with a base value of 415 V, calculate (a) the voltage rating of the shunt element of the UAPC, (b) the current rating of the shunt element of the UAPC, (c) the VA rating of the shunt element of the UAPC, (d) the voltage rating of the series element of the UAPC, (e) the current rating of the series element of the UAPC, and (f) the VA rating of the series element of the UAPC to provide harmonic and reactive power compensation for unity power factor at the AC mains with constant regulated sine wave voltage of 415 V at 50 Hz across the load.

Solution: Given supply rms voltage per phase $V_{sp} = 415/\sqrt{3} = 239.6\,\text{V}$, frequency of the supply $(f) = 50\,\text{Hz}$, and a nonlinear load of a 415 V, 50 Hz thyristor bridge with constant DC current of 150 A at 60° firing angle of its thyristors. There is a voltage fluctuation of +10 and −20% in the supply system with a base value of 415 V.

Let X be the pu voltage variation and V'_{sp} be the PCC voltage under voltage variation.

The per-phase load voltage is $V_{Lp} = V_{LL}/\sqrt{3} = 415/\sqrt{3} = 239.6\,\text{V}$.

In this system, load current harmonic and reactive power compensation is provided by the shunt active filter of the UAPC. The voltage sag/swell compensation is provided by the series filter of the UAPC. While compensating for sag/swell, an active power is circulated between series and shunt active filters as explained in Chapter 6. Under maximum voltage sag, the maximum rating for both the VSCs is realized. The various rating calculations are as follows.

The supply voltage under maximum voltage sag is $V'_{sp} = V_s(1 - X) = 239.6 \times (1 - 0.2) = 191.68\,\text{V}$.

In a three-phase thyristor bridge converter, the waveform of the load line current (I_L) is a quasi-square wave with the amplitude of DC link current (I_{DC}).

Therefore, $I_L = \sqrt{(2/3)}I_{DC} = 0.81649 \times 150 = 122.47\,\text{A}$.

Moreover, the rms value of the fundamental component of the quasi-square wave is $I_{L1} = (\sqrt{6}/\pi)I_{DC} = 0.779 \times 150 = 116.95\,\text{A}$.

The active power component of supply current is $I_{L1a} = I_{L1}\cos\theta_1 = I_{L1}\cos\alpha = 116.95\cos 60° = 58.48\,\text{A}$.

The active power rating of the load is estimated as $P_L = 3V_{Lp}I_{L1a} = 3 \times 239.6 \times 58.48 = 42\,035.42\,\text{W}$.

The supply current under voltage sag is estimated as $I'_s = P_L/3V'_{sp} = 42\,035.42/191.68 = 73.1\,\text{A}$.

Total harmonic and reactive power component of load current is estimated as $I_{Lr} = \sqrt{(I_L^2 - I_{L1a}^2)} = \sqrt{(122.47^2 - 58.48^2)} = 107.61\,\text{A}$.

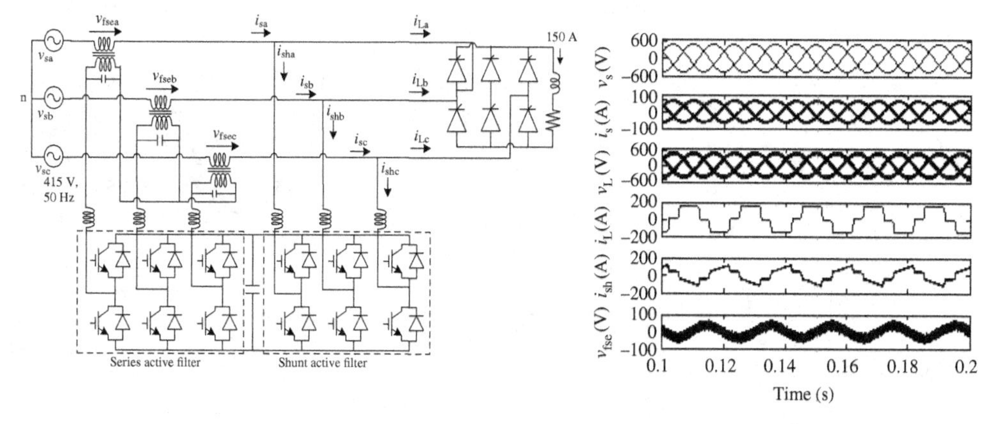

Figure E11.12 A hybrid power filter and its waveforms

a. The voltage rating of the shunt element of the UAPC is equal to AC load phase voltage of $V_{sh} = 239.6$ V per phase, since it is connected across the load of 239.6 V sine waveform. The line voltage rating for the UAPC is $\sqrt{3} \times 239.6 = 415$ V.

b. The current rating of the shunt element of the UAPC is computed as follows.

The shunt element of the UAPC needs to provide load current harmonic and reactive power compensation; hence, it supplies the required harmonic currents and reactive power of the load. Therefore, total harmonic and reactive power component of current through the shunt active filter is estimated as $I_{shr} = \sqrt{(I_L^2 - I_{L1a}^2)} = \sqrt{(122.47^2 - 58.48^2)} = 107.51$ A.

The supply fundamental voltage is lower than the required load voltage. Hence, the shunt APF absorbs active power and that active power is delivered back into the system via the series APF. The active power component of the shunt active filter is estimated as $I_{sha} = I'_s - I_{L1a} = (73.1 - 58.48)$ A = 14.62 A.

The overall current rating of the shunt active filter is estimated as $I_{sh} = \sqrt{(I_{sha}^2 + I_{shr}^2)} = \sqrt{(14.62^2 + 107.51^2)} = 108.49$ A.

c. The VA rating of the VSC of the shunt APF is $S_{sh} = 3V_{sh}I_{sh} = 3 \times 239.6 \times 108.49$ VA = 77 989.45 VA.

d. The voltage rating of the series element of the UAPC is computed as follows.

There is a voltage fluctuation of +10 and −20% in the supply system with a base value of 239.60 V. Therefore, the series APF must inject the difference of the maxima of these two voltages to provide the required voltage at the load end.

The voltage rating of the series APF is $V_{fse} = 239.6 \times 0.20 = 47.92$ V.

e. The current rating of the series element of the UAPC is same as supply current under sag compensation: $I_{se} = I'_s = 73.1$ A.

f. The VA rating of the series element of the UAPC is $S_{se} = 3V_{fse}I_{se} = 3 \times 47.92 \times 73.1$ VA = 10 508.85 VA.

Example 11.13

A three-phase universal active power conditioner (consisting of shunt and series APFs using two VSCs with a common bus capacitor as shown in Figure E11.13) is designed for feeding a critical load of a 415 V, 50 Hz three-phase diode bridge rectifier drawing AC current at 0.96 displacement factor and THD of its AC current is 60%. It is drawing 50 kW from the AC source and the crest factor is 2.5 of AC input current. If there is a voltage fluctuation of +10 and −20% in the supply system with a base value of 415 V, calculate (a) the voltage rating of the shunt element of the UAPC, (b) the current rating of the shunt element of the UAPC, (c) the VA rating of the shunt element of the UAPC, (d) the voltage rating of the series element of the UAPC, (e) the current rating of the series element of the UAPC, and (f) the VA rating

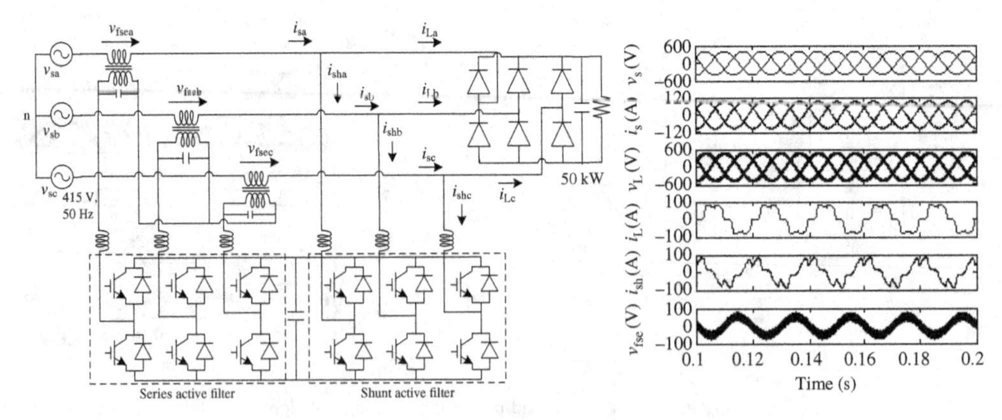

Figure E11.13 A hybrid power filter and its waveforms

of the series element of the UAPC to provide harmonic and reactive power compensation for unity power factor at the AC mains with constant regulated sine wave voltage of 415 V at 50 Hz across the load.

Solution: Given supply rms voltage $V_{sp} = 415/\sqrt{3} = 239.6$ V, frequency of the supply (f) = 50 Hz, and a critical load of a 415 V, 50 Hz three-phase diode bridge rectifier drawing AC current at 0.96 displacement factor and THD of its AC current is 60%. It is drawing 50 kW from the AC source and the crest factor is 2.5 of AC input current. There is a voltage fluctuation of +10 and −20% in the supply system with a base value of 415 V. The load phase voltage is to be regulated to nominal supply voltage; hence, $V_{Lp} = V_{sp} = 239.6$ V.

Let X be the pu voltage variation and V'_{sp} be the PCC voltage under voltage variation.

In this system, load current harmonic and reactive power compensation is provided by the shunt active filter of the UAPC. The voltage sag/swell compensation is provided by the series filter of the UAPC. While compensating for sag/swell, an active power is circulated between series and shunt active filters as explained in Chapter 6. Under maximum voltage sag, the maximum rating for both the VSCs is realized. The various rating calculations are as follows.

The supply voltage under maximum voltage sag is $V'_{sp} = V_s(1 - X) = 239.6 \times (1 - 0.2) = 191.68$ V.

The active power component of the load current is estimated as $I_{L1a} = P_L/3V_{Lp} = 50\,000/(3 \times 239.6) = 69.56$ A.

The fundamental current of the load is $I_{L1} = I_{L1a}/\cos\theta_1 = 69.56/0.96 = 72.46$ A.

The load rms current is $I_L = I_{L1}\{(1 + \mathrm{THD}^2)^{1/2}\} = 72.46 \times \sqrt{(1 + 0.6^2)} = 84.50$ A.

Total harmonic and reactive power component of current is $I_{Lr} = \sqrt{(I_L^2 - I_{L1a}^2)} = \sqrt{(84.50^2 - 69.56^2)} = 47.97$ A.

The supply current under voltage sag is estimated as $I'_s = P_L/3V'_{sp} = 50\,000/(3 \times 191.68) = 86.95$ A.

a. The voltage rating of the shunt element of the UAPC is equal to AC load phase voltage of $V_{sh} = 239.6$ V, since it is connected across the load of 239.6 V sine waveform.
b. The current rating of the shunt element of the UAPC is computed as follows.

   The shunt element of the UAPC needs to provide load current harmonic and reactive power compensation; hence, it supplies the required harmonic currents and reactive power of the load. Therefore, total harmonic and reactive power component of current through the shunt active filter is estimated as $I_{shr} = \sqrt{(I_L^2 - I_{L1a}^2)} = \sqrt{(84.50^2 - 69.56^2)} = 47.97$ A.

   The supply fundamental voltage is lower than the required load voltage. Hence, the shunt APF absorbs active power and that active power is delivered back into the system via the series APF. The active power component of the shunt active filter is estimated as $I_{sha} = I'_s - I_{L1a} = (86.95 - 69.56)$ A = 17.39 A.

   The overall current rating of the shunt active filter is estimated as $I_{sh} = \sqrt{(I_{sha}^2 + I_{shr}^2)} = \sqrt{(17.39^2 + 47.97^2)} = 51.02$ A.
c. The VA rating of the VSC of the shunt APF is $S_{sh} = 3V_{sh}I_{sh} = 3 \times 239.6 \times 51.02$ VA = 36 676.76 VA.
d. The voltage rating of the series element of the UAPC is computed as follows.

   There is a voltage fluctuation of +10 and −20% in the supply system with a base value of 239.60 V. Therefore, the series APF must inject the difference of the maxima of these two voltages to provide the required voltage at the load end.

   The voltage rating of the series APF is $V_{fse} = 239.6 \times 0.20 = 47.92$ V.
e. The current rating of the series element of the UAPC is same as supply current under sag compensation: $I_{se} = I'_s = 86.95$ A.
f. The VA rating of the series element of the UAPC is $S_{se} = 3V_{fse}I_{se} = 3 \times 47.92 \times 86.95 = 12\,500$ VA.

Example 11.14

A three-phase VSI with a quasi-square wave AC output line voltage of 440 V (rms) at 50 Hz is feeding a critical 415 V (line), 50 Hz three-phase thyristor bridge with constant DC current of 50 A at 30° firing angle of its thyristors. A three-phase universal active power conditioner (consisting of shunt and series APFs using two VSCs with a common bus capacitor as shown in Figure E11.14) is designed for this

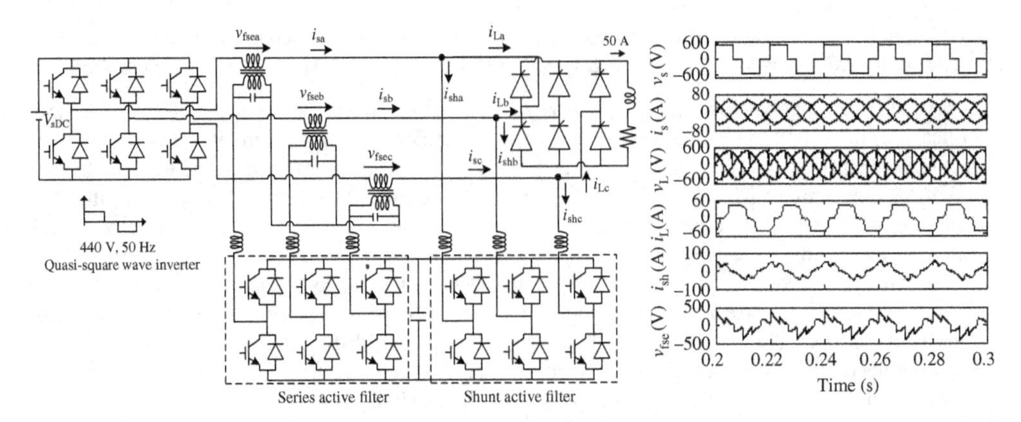

Figure E11.14 A hybrid power filter and its waveforms

critical nonlinear load. Calculate (a) the voltage rating of the shunt element of the UAPC, (b) the current rating of the shunt element of the UAPC, (c) the VA rating of the shunt element of the UAPC, (d) the voltage rating of the series element of the UAPC, (e) the current rating of the series element of the UAPC, and (f) the VA rating of the series element of the UAPC to provide harmonic and reactive power compensation for unity power factor at the AC mains by the shunt element of the UAPC and constant regulated sine wave voltage of 415 V (rms) at 50 Hz across the load by the series element of the UAPC.

Solution: Given that a three-phase VSI with a quasi-square wave AC output line voltage $V_{sL} = 440$ V (rms) at 50 Hz is feeding a critical 415 V (line), 50 Hz three-phase thyristor bridge with constant DC current of 50 A at 30° firing angle of its thyristors.

In this system, load current harmonic and reactive power compensation is provided by the shunt active filter of the UAPC and voltage regulation and its harmonic compensation are provided by the series filter of the UAPC. However, for voltage regulation an active power is circulated between the series and shunt active filters of the UAPC.

The supply is a quasi-square line voltage of 440 V (rms). The amplitude of the quasi-square waveform is estimated as $V_{sDC} = V_{sL}(\sqrt{3}/\sqrt{2}) = 440 \times \sqrt{3}/\sqrt{2} = 538.89$ V.

Moreover, the rms value of the fundamental component of quasi-square supply voltage is $V_{sL1} = (\sqrt{6}/\pi)V_{sDC} = 420.17$ V.

The fundament component of supply phase voltage is $V_{sp1} = V_{sL1}/\sqrt{3} = 242.58$ V.

However, the desired load phase voltage is $V_{Lp} = 415/\sqrt{3} = 239.6$ V sine wave. This voltage is applied across a critical nonlinear load of a 415 V (line), 50 Hz three-phase thyristor bridge with constant DC current of 50 A at 30° firing angle of its thyristors. In a three-phase thyristor bridge converter, the waveform of the load current (I_L) is a quasi-square wave with the amplitude of DC link current (I_{DC}).

Therefore, $I_L = \sqrt{(2/3)}I_{DC} = 0.816\,49 \times 50 = 40.82$ A.

Moreover, the rms value of the fundamental component of the quasi-square wave is $I_{L1} = (\sqrt{6}/\pi)$ $I_{DC} = 0.779 \times 50 = 38.98$ A.

The active power component of supply current is $I_{L1a} = I_{L1} \cos\theta_1 = I_{L1} \cos\alpha = 38.98 \cos 30° =$ 33.76 A.

Total harmonic and reactive power component of load current is estimated as $I_{Lr} = \sqrt{(I_L^2 - I_{L1a}^2)} =$ $\sqrt{(40.82^2 - 33.76^2)} = 22.95$ A

The active power consumed by the load is $P_L = 3V_{Lp}I_{L1a} = 3 \times 239.6 \times 33.76 = 24\,266.69$ W.

The supply current is estimated as $I_s = P_L/3V_{sp1} = 24\,266.69/(3 \times 242.58) = 33.34$ A.

a. The voltage rating of the shunt element of the UAPC is equal to AC load voltage of $V_{sh} = 239.6$ V, since it is connected across the load of 239.6 V sine waveform.

b. The current rating of the shunt element of the UAPC is computed as follows.

The shunt active filter of the UAPC facilitates harmonic and reactive power compensation. Therefore, total harmonic and reactive power component of current through the shunt active filter is estimated as $I_{shr} = \sqrt{(I_L^2 - I_{L1a}^2)} = \sqrt{(40.83^2 - 33.76^2)} = 22.94$ A.

The supply fundamental voltage is higher than the required load voltage. Hence, the series APF absorbs active power and that active power is delivered back into the system via the shunt APF. The active power component of the shunt active filter is estimated as $I_{sha} = I_s - I_{L1a} = (33.34 - 33.76)$ A $= -0.42$ A.

The negative sign for the active power component denotes that the shunt active filter supplies active power (absorbed by the series filter) into the system.

The overall current rating of the shunt active filter is estimated as $I_{sh} = \sqrt{(I_{sha}^2 + I_{shr}^2)} = \sqrt{(0.42^2 + 22.94^2)} = 22.94$ A.

c. The VA rating of the VSC of the shunt APF is $S_{sh} = 3V_{sh}I_{sh} = 3 \times 239.6 \times 22.94$ VA $= 16\,489.27$ VA.

d. The voltage rating of the series APF (V_{fse}) is computed as follows.

The supply of quasi-square line voltage results in stepped phase voltage. The series filter injects a voltage that is equal to the difference between the supply voltage and the load terminal voltage.

The waveform of the phase voltage at the input of the series active filter is a stepped waveform with (i) the first step of $\pi/3$ angle (from 0° to $\pi/3$) and a magnitude of $V_{sDC}/3$, (ii) the second step of $\pi/3$ angle (from $\pi/3$ to $2\pi/3$) and a magnitude of $2V_{sDC}/3$, and (iii) the third step of $\pi/3$ angle (from $2\pi/3$ to π) and a magnitude of $V_{sDC}/3$, and it has both half cycles of symmetric segments of such steps.

The voltage rating of the series APF (V_{fse}) is computed by taking the difference of the supply phase voltage and required load sine phase voltage at the input of the linear load as

$$V_{fse} = \sqrt{\left[(1/\pi)\int_0^{\pi/3} (179.63 - 239.6\sqrt{2}\sin\theta)^2\, d\theta + \int_{\pi/3}^{2\pi/3} (359.26 - 239.6\sqrt{2}\sin\theta)^2\, d\theta \right.}$$

$$\left. + \int_{2\pi/3}^{\pi} (179.63 - 239.6\sqrt{2}\sin\theta)^2\, d\theta\right]$$

$$= 75.46 \text{ V}.$$

e. The current rating of the series filter is $I_{se} = I_s = 33.34$ A (since the series filter is connected in series with the supply).

f. The VA rating of the series filter is $S = 3V_{fse}I_{se} = 3 \times 75.46 \times 33.34$ VA $= 7549$ VA.

Example 11.15

In a three-phase four-wire distribution system with a three-phase VSI with a stepped wave AC line voltage output of 415 V (rms) at 50 Hz (quasi-square wave phase voltage), three single-phase loads (connected between phases and neutral) have a single-phase 230 V, 50 Hz thyristor bridge converter drawing equal 30 A constant DC current at 30° firing angle of its thyristors. A three-phase universal active power conditioner (consisting of a four-leg VSC as a shunt APF and a three-leg VSC as a series APF with a common bus capacitor as shown in Figure E11.15) is designed for this critical nonlinear load. Calculate (a) the voltage rating of the shunt element of the UAPC, (b) the current rating of the shunt element of the UAPC, (c) the VA rating of the shunt element of the UAPC, (d) the voltage rating of the series element of the UAPC, (e) the current rating of the series element of the UAPC, and (f) the VA rating of the series element of the UAPC to provide harmonic and reactive power compensation for unity power factor at the AC mains and zero neutral current by the shunt element of the UAPC and constant regulated sine wave voltage of 415 V (line rms) at 50 Hz across the load by the series element of the UAPC.

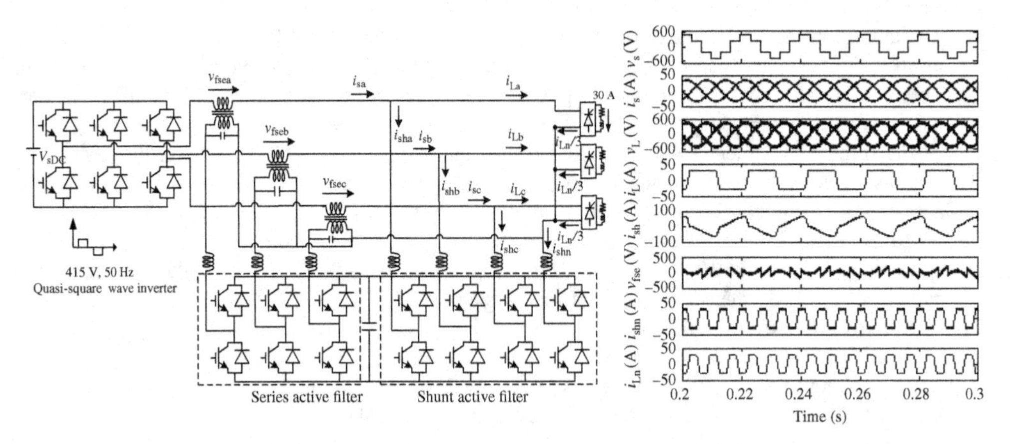

Figure E11.15 A hybrid power filter and its waveforms

Solution: Given that a three-phase VSI with a stepped wave AC line voltage output of 415 V (rms) at 50 Hz (quasi-square wave phase voltage) is feeding a single-phase 230 V, 50 Hz thyristor bridge converter drawing equal 30 A constant DC current at 30° firing angle of its thyristors in each of the three phases.

In this system, load current harmonic and reactive power compensation is provided by the shunt active filter of the UAPC and voltage regulation and its harmonic compensation are provided by the series filter of the UAPC. However, for voltage regulation, an active power is circulated between the series and shunt active filters of the UAPC.

The supply is a stepped wave line voltage of 415 V (rms). The amplitude of DC side voltage of the supply side VSI is estimated as $V_{sDC} = V_{sL}\sqrt{2} = 415 \times \sqrt{2} = 586.89$ V.

The fundamental component of supply side line voltage is $V_{sL1} = 3V_{sdc}/\sqrt{2}\pi = 0.675 \times V_{sDC} = 396.15$ V.

The fundamental component of supply phase voltage is $V_{sp1} = V_{sL1}/\sqrt{3} = 228.71$ V.

The desired load phase voltage is $V_{Lp} = 230$ V (sine wave). This voltage is applied across a critical linear single-phase load of a 230 V, 50 Hz thyristor bridge converter drawing equal 30 A constant DC current at 30° firing angle of its thyristors in each of the three phases.

In a three-phase thyristor bridge converter, the waveform of the load line current (I_L) is a square wave with the amplitude of DC link current (I_{DC}).

The AC load line current is $I_L = 30$ A.

The fundamental active power component of load current is estimated as $I_{L1a} = I_{L1}\cos\alpha = 0.9I_{DC}\cos\alpha = 0.9 \times 30\cos\ 30° = 23.38$ A.

Total harmonic and reactive power component of load current is estimated as $I_{Lr} = \sqrt{(I_L^2 - I_{L1a}^2)} = \sqrt{(30^2 - 23.38^2)} = 18.79$ A.

The active power consumed by the load is $P_L = 3V_{Lp}I_{L1a} = 3 \times 230 \times 23.38 = 16\,134$ W.

The supply current is estimated as $I_s = P_L/3V_{sp1} = 16\,134/(3 \times 228.71) = 23.5$ A.

a. The voltage rating of the shunt element of the UAPC is equal to AC load voltage of $V_{sh} = 230$ V, since it is connected across the load of 230 V sine waveform.
b. The current rating of the shunt element of the UAPC is computed as follows.

   The shunt active filter of the UAPC facilitates harmonic and reactive power compensation. Therefore, total harmonic and reactive power component of current through the shunt active filter is estimated as $I_{shr} = \sqrt{(I_L^2 - I_{L1a}^2)} = \sqrt{(30^2 - 23.38^2)} = 18.79$ A.

   The supply fundamental voltage is lower than the required load voltage. Hence, the shunt APF absorbs active power and that active power is delivered back into the system via the series APF.

The active power component of the shunt active filter is estimated as $I_{sha} = I_s - I_{L1a} = (23.5 - 23.38)$ A = 0.12 A.

The phase current rating of the shunt active filter is estimated as $I_{shp} = \sqrt{(I_{sha}^2 + I_{shr}^2)} = \sqrt{(0.12^2 + 18.79^2)} = 18.79$ A.

However, the fourth leg supplies the neutral current of the load. The neural current for this type of load is a third harmonic square wave with amplitude equal to DC side load current. Hence, the rms current of the fourth leg of the shunt active filter is $I_{shn} = I_{Ln} = 30$ A.

c. The VA rating of the VSC of the shunt active filter is $S_{sh} = 3V_{sh}I_{shp} + V_{sh}I_{shn} = 3 \times 230 \times 18.79 + 230 \times 30$ VA = 19 865 VA.

d. The voltage rating of the series APF (V_{fse}) is computed as follows.

The three-phase VSI is operated in 120° conduction mode to get stepped wave AC line voltage. The supply stepped wave AC line voltage of $V_s = 415$ V (rms) (the stepped waveform has (i) the first step of $\pi/3$ angle (from 0° to $\pi/3$) and a magnitude of V_{sDC}, (ii) the second step of $\pi/3$ angle (from $\pi/3$ to $2\pi/3$) and a magnitude of $V_{sDC}/2$, and (iii) the third step of $\pi/3$ angle (from $2\pi/3$ to π) and a magnitude of $-V_{sDC}/2$, and it has both half cycles of symmetric segments of such steps) has an amplitude of $V_{sDC} = \sqrt{2} \times 415 = 586.89$ V. Therefore, the supply stepped wave AC line voltage of amplitude $V_{sDC} = 586.89$ V results in a quasi-square waveform of the phase voltage with the amplitude of $V_{sDC}/2 = 293.45$ V.

However, a phase shift may be considered for phase voltage to coincide with the sine waveform of the load phase voltage $V_L = 230$ V.

The voltage rating of series filter is computed as,

$$V_{fse} = \sqrt{\left[1/\pi \int_0^{\pi/6} (-230\sqrt{2}\sin\theta)^2\,d\theta + \int_{\pi/6}^{5\pi/6} (293.45 - 230\sqrt{2}\sin\theta)^2\,d\theta + \int_{5\pi/6}^{\pi} (-230\sqrt{2}\sin\theta)^2\,d\theta\right]}$$

$= 71.33$ V.

e. The current rating of series filter, $I_{se} = I_s = 23.5$ A.

f. The VA rating of the VSC of the series filter is $S = 3 \times V_{fse} \times I_{se} = 3 \times 71.33 \times 23.5$ VA = 5.028 kVA.

11.8 Summary

A comprehensive study of HFs is presented to provide a wide exposure on various aspects of the HFs to the researchers, designers, and users of these filters for power quality improvement. A classification of HFs into nine categories with many circuits in each category is expected to help in the selection of an appropriate topology for a particular application. The hybrid filters are considered as a better alternative for power quality improvement due to reduced cost, simple design and control, and high reliability compared with other options of power quality improvement. Some of the circuit configurations of HFs avoid the problems involved in passive and active filters, and therefore provide a cost-effective and better solution for harmonic elimination of nonlinear loads. An analytical study of various performance indices of HFs for the compensation of sensitive nonlinear loads is made in detail with several numerical examples to study the rating of power filters and how it is affected by various kinds of nonlinear loads and supply conditions. HFs are observed as one of the best retrofit solutions for mitigating the power quality problems due to nonlinear loads for reducing the pollution of the AC mains. Moreover, due to a large number of circuits of HFs, the user can select the most appropriate topology with required features to suit a specific application. It is expected to be beneficial to the designers, users, manufacturers, and research engineers dealing with power quality improvement.

11.9 Review Questions

1. What is a hybrid power filter?
2. What are the power quality problems that a HPF can mitigate?

3. Can current source converters be used in the AF element of a HPF?
4. Why voltage source converters are preferred in the AF element of a HPF?
5. What are the factors that decide the rating of a HPF?
6. How the overall rating of a HPF (with one AF and one PF) may be reduced for the compensation of voltage-fed loads?
7. How the overall rating of a HPF (with one AF and one PF) may be reduced for the compensation of current-fed loads?
8. How the overall rating of a HPF (with both AF elements) may be reduced for the compensation of current-fed loads?
9. How the overall rating of a HPF (with both AF elements) may be reduced for the compensation of voltage-fed loads?
10. How the value of an interfacing inductor is computed for VSCs used in the AF element of a HPF?
11. How the value of a DC bus capacitor is computed for VSCs used in the AF element of a HPF?
12. How the self-supporting DC bus of a VSC is achieved in the AF element of a HPF?
13. How does the crest factor of the load current affect the rating of a HPF?
14. Why the indirect current controller is considered superior to the direct current controller in the AF element of a HPF?
15. Which out of current-fed and voltage-fed types of nonlinear loads has higher power rating of a HPF when used only for harmonic compensation?
16. Which topology of a HPF (with one AF and one PF) is more effective for the compensation of voltage-fed loads?
17. Which topology of a HPF (with one AF and one PF) is more effective for the compensation of current-fed loads?
18. How many combinations of HPFs are possible using two elements where one element is PF and other element is AF in a three-phase three-wire supply system?
19. How many combinations of HPFs are possible using three elements where one element (AF or PF) is different from other two elements (AF or PF) in a three-phase three-wire supply system?
20. How many combinations of HPFs are possible using four elements where one element (AF or PF) is different from other three elements (AF or PF) in a three-phase three-wire supply system?
21. Which configuration of HPFs is most effective in eliminating the neutral current?
22. Up to what highest order of harmonic can be eliminated by a HPF if the switching frequency of the AF element of a HF is 2.5 kHz?
23. How the harmonics of higher order than switching frequency can be eliminated in a HPF?
24. What is the power rating of a HPF (with one AF and one PF) in terms of power rating of a nonlinear load consisting of a three-phase diode rectifier with constant DC current?
25. What is the power rating of a HPF (with one AF and one PF) in terms of power rating of a nonlinear load consisting of a single-phase thyristor bridge rectifier with constant DC current at a firing angle of 30° for unity power factor at the AC mains?

11.10 Numerical Problems

1. A single-phase VSI with a square waveform AC output of 220 V (rms) at 50 Hz is feeding a critical load of 220 V, 50 Hz single-phase 5 kVA at 0.85 lagging power factor. A single-phase universal active power conditioner (consisting of shunt and series APFs using two VSCs with a common bus capacitor as shown in Figure E11.1) is designed for this critical linear load. Calculate (a) the voltage rating of the shunt element of the UAPC, (b) the current rating of the shunt element of the UAPC, (c) the VA rating of the shunt element of the UAPC, (d) the voltage rating of the series element of the UAPC, (e) the current rating of the series element of the UAPC, and (f) the VA rating of the series element of the UAPC to provide load compensation for unity power factor at the AC mains by the shunt element of the UAPC and constant regulated sine wave voltage of 220 V (rms sine wave) at 50 Hz across the load by the series element of the UAPC.

2. A single-phase VSI with a quasi-square waveform AC output of 220 V (rms) at 50 Hz is feeding a critical load of 230 V, 50 Hz single-phase 4 kVA at 0.85 lagging power factor. A single-phase universal active power conditioner (consisting of shunt and series APFs using two VSCs with a common bus capacitor as shown in Figure E11.2) is designed for this critical linear load. Calculate (a) the voltage rating of the shunt element of the UAPC, (b) the current rating of the shunt element of the UAPC, (c) the VA rating of the shunt element of the UAPC, (d) the voltage rating of the series element of the UAPC, (e) the current rating of the series element of the UAPC, and (f) the VA rating of the series element of the UAPC to provide load compensation for unity power factor at the AC mains by the shunt element of the UAPC and constant regulated sine wave voltage of 230 V (rms sine wave) at 50 Hz across the load by the series element of the UAPC.

3. A single-phase active hybrid filter (consisting of a three-branch passive shunt filter and a series APF using a VSC with a bus capacitor as shown in Figure E11.3) is designed for a critical load of a 220 V, 50 Hz single-phase thyristor bridge with constant DC current of 20 A at 30° firing angle of its thyristors. Calculate the voltage, current, and VA ratings of both the series APF and the passive shunt filter used in the hybrid filter to provide harmonic and reactive power compensation for unity power factor at the AC mains.

4. A single-phase universal active power conditioner (consisting of shunt and series APFs using two VSCs with a common bus capacitor as shown in Figure E11.4) is designed for a typical load of a 220 V, 50 Hz single-phase thyristor bridge with constant DC current of 25 A at 30° firing angle of its thyristors. If there is a voltage fluctuation of +10 and −20% in the supply system with a base value of 220 V, calculate (a) the voltage rating of the shunt element of the UAPC, (b) the current rating of the shunt element of the UAPC, (c) the VA rating of the shunt element of the UAPC, (d) the voltage rating of the series element of the UAPC, (e) the current rating of the series element of the UAPC, and (f) the VA rating of the series element of the UAPC to provide harmonic and reactive power compensation for unity power factor at the AC mains with constant regulated sine wave voltage of 220 V at 50 Hz.

5. A single-phase universal active power conditioner (consisting of shunt and series APFs using two VSCs with a common bus capacitor as shown in Figure E11.5) is designed for a 220 V, 50 Hz single-phase resistive load of 8 Ω through a phase-controlled AC voltage controller at a firing angle of 60° of its thyristors. If there is a voltage fluctuation of +10 and −20% in the supply system with a base value of 220 V, calculate (a) the voltage rating of the shunt element of the UAPC, (b) the current rating of the shunt element of the UAPC, (c) the VA rating of the shunt element of the UAPC, (d) the voltage rating of the series element of the UAPC, (e) the current rating of the series element of the UAPC, and (f) the VA rating of the series element of the UAPC to provide harmonic and reactive power compensation for unity power factor at the AC mains with constant voltage of 220 V at 50 Hz.

6. A single-phase VSI with a quasi-square wave AC output of 220 V (rms) at 50 Hz is feeding a critical load of a 230 V, 50 Hz single-phase thyristor bridge with constant DC current of 25 A at 60° firing angle of its thyristors. A single-phase universal active power conditioner (consisting of shunt

and series APFs using two VSCs with a common bus capacitor as shown in Figure E11.6) is designed for this critical nonlinear load. Calculate (a) the voltage rating of the shunt element of the UAPC, (b) the current rating of the shunt element of the UAPC, (c) the VA rating of the shunt element of the UAPC, (d) the voltage rating of the series element of the UAPC, (e) the current rating of the series element of the UAPC, and (f) the VA rating of the series element of the UAPC to provide harmonic and reactive power compensation for unity power factor at the AC mains by the shunt element of the UAPC and constant regulated sine wave voltage of 230 V (rms sine wave) at 50 Hz across the load by the series element of the UAPC.

7. A single-phase VSI with a square wave AC output of 210 V (rms) at 50 Hz is feeding a critical load of a 220 V, 50 Hz single-phase uncontrolled diode bridge converter having a *RE* load with $R = 2\,\Omega$ and $E = 200$ V. A single-phase universal active power conditioner (consisting of shunt and series APFs using two VSCs with a common bus capacitor as shown in Figure E11.7) is designed for this critical nonlinear load. Calculate (a) the voltage rating of the shunt element of the UAPC, (b) the current rating of the shunt element of the UAPC, (c) the VA rating of the shunt element of the UAPC, (d) the voltage rating of the series element of the UAPC, (e) the current rating of the series element of the UAPC, and (f) the VA rating of the series element of the UAPC to provide harmonic and reactive power compensation for unity power factor at the AC mains by the shunt element of the UAPC and constant regulated sine wave voltage of 220 V (rms sine wave) at 50 Hz across the load by the series element of the UAPC.

8. A single-phase active hybrid filter (consisting of a one-branch passive shunt filter as a high-pass damped filter and a series APF connected in series with the AC mains using a VSC with a bus capacitor as shown in Figure E11.8) is designed for a load of a 220 V, 50 Hz thyristor AC voltage controller with a resistive load of $12.0\,\Omega$ at 45° firing angle of its thyristors. Calculate the ratings of both the series APF and the passive shunt filter used in the hybrid filter to provide harmonic and reactive power compensation with unity power factor at full load and 60° firing angle [224].

9. A three-phase VSI with a quasi-square wave AC output line voltage of 420 V (rms) at 50 Hz is feeding a critical load of 400 V, 50 Hz three-phase 25 kVA at 0.85 lagging power factor. A three-phase universal active power conditioner (consisting of shunt and series APFs using two VSCs with a common bus capacitor as shown in Figure E11.9) is designed for this critical linear load. Calculate (a) the voltage rating of the shunt element of the UAPC, (b) the current rating of the shunt element of the UAPC, (c) the VA rating of the shunt element of the UAPC, (d) the voltage rating of the series element of the UAPC, (e) the current rating of the series element of the UAPC, and (f) the VA rating of the series element of the UAPC to provide harmonic and reactive power compensation for unity power factor at the AC mains by the shunt element of the UAPC and constant regulated sine wave voltage of 400 V (rms sine wave) at 50 Hz across the load by the series element of the UAPC.

10. A three-phase active filter connected in series with a three-branch shunt passive filter (fifth, seventh, and high-pass filters as shown in Figure E11.10) is used for harmonic current and reactive power compensation in a three-phase 400 V, 50 Hz system to reduce the THD of the supply current and to improve the displacement factor to unity. It has a load of a thyristor bridge converter operating at 60° firing angle drawing constant 100 A DC current. Calculate (a) the designed values of passive filter components, (b) line current, and (c) the VA rating of the APF.

11. A three-phase series active filter along with a three-branch shunt passive filter (fifth, seventh, and high-pass filters as shown in Figure E11.11) is used for harmonic current and reactive power compensation in a three-phase 440 V, 50 Hz system to reduce the THD of the supply current and to improve the displacement factor to unity. It has a load of a thyristor bridge converter operating at 60° firing angle drawing constant 125 A DC current. Calculate (a) the designed values of passive filter components, (b) line current, (c) the VA rating of the APF, (d) the AC inductor value of the APF, and (e) the bus capacitor value of the APF and its DC voltage. Consider the switching frequency of 10 kHz and the DC bus voltage has to be controlled within 5% range and ripple current in the inductor is 10%.

12. A three-phase universal active power conditioner (consisting of shunt and series APFs using two VSCs with a common DC bus capacitor as shown in Figure E11.12) is designed for a load of a 440 V, 50 Hz thyristor bridge with constant DC current of 250 A at 30° firing angle of its thyristors. If there is a voltage fluctuation of +10 and −20% in the supply system with a base value of 440 V, calculate (a) the voltage rating of the shunt element of the UAPC, (b) the current rating of the shunt element of the UAPC, (c) the VA rating of the shunt element of the UAPC, (d) the voltage rating of the series element of the UAPC, (e) the current rating of the series element of the UAPC, and (f) the VA rating of the series element of the UAPC to provide harmonic and reactive power compensation for unity power factor at the AC mains with constant regulated sine wave voltage of 440 V at 50 Hz.
13. A three-phase universal active power conditioner (consisting of shunt and series APFs using two VSCs with a common bus capacitor as shown in Figure E11.13) is designed for feeding a critical load of a 440 V, 50 Hz three-phase diode bridge rectifier drawing AC current at 0.94 displacement factor and THD of its AC current is 65%. It is drawing 80 kW from the AC source and the crest factor is 2.5 of AC input current. If there is a voltage fluctuation of +10 and −10% in the supply system with a base value of 415 V, calculate (a) the voltage rating of the shunt element of the UAPC, (b) the current rating of the shunt element of the UAPC, (c) the VA rating of the shunt element of the UAPC, (d) the voltage rating of the series element of the UAPC, (e) the current rating of the series element of the UAPC, and (f) the VA rating of the series element of the UAPC to provide harmonic and reactive power compensation for unity power factor at the AC mains with constant regulated sine wave voltage of 440 V at 50 Hz.
14. A three-phase VSI with a quasi-square wave AC output line voltage of 440 V (rms) at 50 Hz is feeding a critical 400 V (line), 50 Hz three-phase thyristor bridge with constant DC current of 150 A at 30° firing angle of its thyristors. A three-phase universal active power conditioner (consisting of shunt and series APFs using two VSCs with a common bus capacitor as shown in Figure E11.14) is designed for this critical nonlinear load. Calculate (a) the voltage rating of the shunt element of the UAPC, (b) the current rating of the shunt element of the UAPC, (c) the VA rating of the shunt element of the UAPC, (d) the voltage rating of the series element of the UAPC, (e) the current rating of the series element of the UAPC, and (f) the VA rating of the series element of the UAPC to provide harmonic and reactive power compensation for unity power factor at the AC mains by the shunt element of the UAPC and constant regulated sine wave voltage of 400 V (rms sine wave) at 50 Hz across the load by the series element of the UAPC.
15. In a three-phase four-wire distribution system with a three-phase VSI with a stepped wave AC line voltage output of 400 V (rms) at 50 Hz (quasi-square wave phase voltage), three single-phase loads (connected between phases and neutral) have a single-phase 220 V, 50 Hz thyristor bridge converter drawing equal 25 A constant DC current at 45° firing angle of its thyristors. A three-phase universal active power conditioner (consisting of a four-leg VSC as a shunt APF and a three-leg VSC as a series APF with a common bus capacitor as shown in Figure E11.15) is designed for this critical nonlinear load. Calculate (a) the voltage rating of the shunt element of the UAPC, (b) the current rating of the shunt element of the UAPC, (c) the VA rating of the shunt element of the UAPC, (d) the voltage rating of the series element of the UAPC, (e) the current rating of the series element of the UAPC, and (f) the VA rating of the series element of the UAPC to provide harmonic and reactive power compensation for unity power factor at the AC mains and zero neutral current by the shunt element of the UAPC and constant regulated sine wave voltage of 400 V (line rms sine wave) at 50 Hz across the load by the series element of the UAPC.

11.11 Computer Simulation-Based Problems

1. Design and develop a dynamic model of a hybrid filter (an active filter in series with a three-branch passive shunt filter) and simulate its behavior for a single-phase 230 V, 50 Hz, uncontrolled bridge converter with a parallel capacitive DC filter of 1000 μF and an equivalent resistive load of 10 Ω.

It has a source impedance of 0.25 Ω resistive element and 1.25 Ω inductive element. Plot the supply voltage and input current, passive filter currents, output voltage and current with time, and the harmonic spectra of supply current and voltage at PCC. Compute THD, crest factor, rms value of AC mains current, displacement factor, distortion factor, power factor, % ripple in load current, output voltage ripple, ripple factor, and input active, reactive, and output powers.

2. Design and develop a dynamic model of a hybrid filter (an active filter in series with a three-branch passive shunt filter) and simulate its behavior for a single-phase 220 V, 50 Hz, uncontrolled bridge converter used for charging a battery of 200 V. It has a source impedance of 0.25 Ω resistive element and 2.0 Ω inductive element. Plot the supply voltage and input current, passive filter currents, charging current with time, and the harmonic spectra of supply current and voltage at PCC. Compute THD, crest factor, rms value of AC mains current, displacement factor, distortion factor, power factor, and input active, reactive, and output powers.

3. Design and develop a dynamic model of a hybrid filter (an active filter (a VSI with an AC series inductor, a coupling transformer, and its DC bus connected to a battery functioning as the load) in series with a three-branch passive shunt filter) and simulate its behavior for a single-phase AC supply of 230 V at 50 Hz feeding a diode rectifier charging a battery of 196 V and an average load current of 20 A to reduce the harmonics in AC mains current and to maintain almost UPF. It has a source impedance of 0.25 Ω resistive element and 2.5 Ω inductive element. Plot the supply voltage and input current, passive filter currents, charging current and diode rectifier input voltage with time, and the harmonic spectra of supply current and voltage at PCC. Compute THD, crest factor, rms value of AC mains current, displacement factor, distortion factor, power factor, input active, reactive, and output powers, rms voltage at the input of the diode rectifier, voltage rating of the APF, current rating of the APF, and VA rating of the APF.

4. Design and develop a dynamic model of a hybrid filter (an active filter (a VSI with an AC series inductor, a coupling transformer, and its DC bus connected to the DC bus of the load) in series with a three-branch passive shunt filter) and simulate its behavior for a single-phase 230 V, 50 Hz diode bridge rectifier with a DC capacitor filter feeding a DC current of 16 A to a variable-frequency three-phase VSI-fed induction motor drive in an air conditioner to reduce the harmonics in AC mains current, to maintain almost UPF, and to regulate the DC bus voltage of the rectifier to 400 V. It has a source impedance of 0.15 Ω resistive element and 1.5 Ω inductive element. Plot the supply voltage and input current, passive filter currents, rectifier output current and input voltage with time, and the harmonic spectra of supply current and voltage at PCC. Compute THD, crest factor, rms value of AC mains current, displacement factor, distortion factor, power factor, input active, reactive, and output powers, rms voltage at the input of the diode rectifier, voltage rating of the APF, current rating of the APF, and VA rating of the APF.

5. Design and develop a dynamic model of a hybrid filter (an active filter (a VSI with an AC series inductor, a coupling transformer, and its DC bus connected to the DC bus of the load) in series with a three-branch passive shunt filter) and simulate its behavior to reduce the harmonics in AC mains current and to maintain almost UPF in the series of a single-phase AC supply of 230 V at 50 Hz feeding a diode rectifier with a capacitive filter of 1500 μF and a resistive load of 5 Ω. The DC bus voltage of the load needs to be maintained at constant ripple-free 220 V. Plot supply voltage and current, passive filter currents, rectifier input voltage and output current, device voltage and current with time, and the harmonic spectra of supply current and voltage at PCC. Compute THD, crest factor, rms value of AC mains current, displacement factor, distortion factor, power factor, input active, reactive, and output powers, rms voltage at the input of the diode rectifier, voltage rating of the APF, current rating of the APF, VA rating of the APF, and turns ratio of the coupling transformer.

6. Design and develop a dynamic model of a hybrid filter (an active filter (a VSI with an AC series inductor, a coupling transformer, and its DC bus connected to the DC bus of the load) in series with

a three-branch passive shunt filter) and simulate its behavior to reduce the harmonics in AC mains current and to maintain almost UPF in the series of a single-phase AC supply of 230 V at 50 Hz feeding a three-phase diode rectifier with a capacitive filter of 470 μF and a resistive load of 10 Ω. The DC bus voltage of the load is maintained to result in minimum injected voltage of the APF. It has a source impedance of 0.25 Ω resistive element and 1.25 Ω inductive element. Plot supply voltage and current, DC load current and voltage, passive filter currents, injected voltage and rectifier input voltage with time, and the harmonic spectra of supply current and voltage at PCC. Compute THD, crest factor, rms value of AC mains current, displacement factor, distortion factor, power factor, input active, reactive, and output powers, rms line voltage at the input of the diode rectifier, rms voltage across the APF, VA rating of the APF, and turns ratio of the coupling transformer.

7. Design and develop a dynamic model of a hybrid filter (an active filter in series with a three-branch passive shunt filter) and simulate its behavior for a single-phase 230 V, 50 Hz, controlled bridge converter with a series connected inductive load of 100 mH and an equivalent resistive load of 5 Ω. It has a source impedance of 0.25 Ω resistive element and 1.25 Ω inductive element. The delay angle of its thyristors is $\alpha = 60°$. Plot the supply voltage and input current, passive filter currents, output voltage and current with time, and the harmonic spectra of supply current, AC load current, and voltage at PCC. Compute THD, crest factor, rms value of AC mains current, displacement factor, distortion factor, power factor, % ripple in load current, and input active, reactive, and output powers.
8. Design and develop a dynamic model of a hybrid filter (an active filter in series with a three-branch passive shunt filter) and simulate its behavior for a single-phase 230 V, 50 Hz, controlled bridge converter having a *RLE* load with an inductance of 10 mH, an equivalent resistance of 2 Ω, and a back emf of 48 V. It has a source impedance of 0.25 Ω resistive element and 1.5 Ω inductive element. The delay angle of its thyristors is $\alpha = 30°$. Plot the supply voltage and input current, passive filter currents, output voltage and current with time, and the harmonic spectra of supply current, AC load current, and voltage at PCC. Compute THD, crest factor, rms value of AC mains current, displacement factor, distortion factor, power factor, % ripple in load current, and input active, reactive, and output powers.
9. Design and develop a dynamic model of a hybrid filter (an active series filter + a one-branch passive shunt filter) and simulate its behavior for a single-phase AC voltage controller having a *RL* load of $R = 2\,\Omega$ and $L = 10$ mH. The input voltage is 230 V (rms) at 50 Hz. The delay angle of its thyristors is $\alpha = 90°$. Plot supply voltage and current, passive filter current, load current and voltage, device voltage and current with time, and the harmonic spectra of supply current and load current. Calculate THD, crest factor, rms value of AC mains current, displacement factor, distortion factor, power factor, and input active, reactive, and output powers.
10. Design and develop a dynamic model of a hybrid filter (an active series filter + a three-branch passive shunt filter) and simulate its behavior for a single-phase 230 V, 50 Hz supply system, which is feeding a set of nonlinear loads consisting of a thyristor bridge and a diode rectifier connected in parallel. The diode bridge converter is feeding a parallel capacitive DC filter of 470 μF and a resistive load of 20 Ω. The thyristor bridge converter is feeding a *RL* load of $R = 10\,\Omega$ and $L = 50$ mH with 30° firing angle of its thyristors. Plot supply voltage and current, passive filter currents, load currents and voltages with time, and the harmonic spectra of supply current and load currents. Calculate THD, crest factor, rms value of AC mains current, displacement factor, distortion factor, power factor, and input active, reactive, and output powers.
11. Design and develop a dynamic model of a hybrid filter (an active series filter + a three-branch passive shunt filter) and simulate its behavior for a three-phase 415 V, 50 Hz, uncontrolled bridge converter with a parallel capacitive DC filter of 4700 μF and a resistive load of 20 Ω. It has a source impedance of 0.25 Ω resistive element and 1.5 Ω inductive element. Plot the supply voltage and input current, passive filter currents, output voltage and current with time, and the harmonic spectra of supply current and voltage at PCC. Compute THD, crest factor, rms value of AC mains current, displacement

factor, distortion factor, power factor, % ripple in load current, output voltage ripple, ripple factor, and input active, reactive, and output powers.

12. Design and develop a dynamic model of a hybrid filter (an active series filter + a three-branch passive shunt filter) and simulate its behavior to reduce harmonic currents and to maintain almost UPF of a three-phase 415 V, 50 Hz uncontrolled bridge rectifier, which has a source inductance of 5 mH and a DC load resistance of 10 Ω. Plot the supply voltage and input current, passive filter currents, output voltage and current with time, and the harmonic spectra of supply current and voltage at PCC. Compute THD, crest factor, rms value of AC mains current, displacement factor, distortion factor, power factor, % ripple in load current, output voltage ripple, ripple factor, and input active, reactive, and output powers.

13. Design and develop a dynamic model of a hybrid filter (an active filter (a VSI with an AC series inductor, a coupling transformer, and its DC bus connected to the DC bus of the load) in series with a three-branch passive shunt filter) and simulate its behavior to reduce harmonic currents and to maintain almost UPF of a three-phase 415 V, 50 Hz, uncontrolled six-pulse bridge converter used for charging a battery of 560 V. It has a source impedance of 0.5 Ω resistive element and 1.5 Ω inductive element. Plot the supply voltage and input current, passive filter currents, charging current with time, and the harmonic spectra of supply current and voltage at PCC. Compute THD, crest factor, rms value of AC mains current, displacement factor, distortion factor, power factor, input active, reactive, and output powers, rms voltage across the APF, VA rating of the APF, and turns ratio of the coupling transformer.

14. Design and develop a dynamic model of a hybrid filter (an active series filter connected in series with the AC mains + a three-branch passive shunt filter) and simulate its behavior to reduce harmonic currents and to maintain almost UPF of a three-phase 415 V, 50 Hz fed thyristor bridge converter feeding a resistive load of 10 Ω at 50° firing angle of its thyristors. Plot the supply voltage and input current, passive filter currents, injected voltage of the APF, output voltage and current with time, and the harmonic spectra of supply current and voltage at PCC. Compute fundamental active power drawn by the load, fundamental reactive power drawn by the load, power factor, rms supply current, distortion factor, fundamental rms supply current, peak current of the AC mains, total harmonic distortion of AC supply current, and (i) voltage, current, and VA ratings of the APF and the passive filter.

15. Design and develop a dynamic model of a hybrid filter (an active filter connected in series with a three-branch passive shunt filter) and simulate its behavior to reduce harmonic currents and to maintain almost UPF of a three-phase 415 V, 50 Hz fed thyristor bridge converter feeding a resistive load of 20 Ω at 30° firing angle of its thyristors. Plot the supply voltage and input current, passive filter currents, injected voltage of the APF, output voltage and current with time, and the harmonic spectra of supply current and voltage at PCC. Compute fundamental active power drawn by the load, fundamental reactive power drawn by the load, power factor, rms supply current, distortion factor, fundamental rms supply current, peak current of the AC mains, total harmonic distortion of AC supply current, and voltage, current, and VA ratings of the APF and the passive filter.

16. Design and develop a dynamic model of a hybrid filter (an active series filter + a three-branch passive shunt filter) and simulate its behavior to reduce harmonic currents and to maintain almost UPF of a three-phase 415 V, 50 Hz, controlled bridge converter with a series connected inductive load of 50 mH, an equivalent resistive load of 5 Ω, and a back emf of 240 V. It has a source impedance of 0.15 Ω resistive element and 1.5 Ω inductive element. The delay angle of its thyristors is $\alpha = 60°$. Plot the supply voltage and input current, passive filter currents, injected voltage by the APF, output voltage and current with time, and the harmonic spectra of supply current and voltage at PCC. Compute THD, crest factor, rms value of AC mains current, displacement factor, distortion factor, power factor, % ripple in load current, and input active, reactive, and output powers.

17. Design and develop a dynamic model of a hybrid filter (an active series filter + a three-branch passive shunt filter) and simulate its behavior to reduce harmonic currents and to maintain almost UPF of a three-phase bidirectional delta connected controller (back-to-back connected thyristors in series with the load) having a resistive load of $R = 5\,\Omega$ and fed from a line voltage of 415 V (rms) at 50 Hz. The delay angle of its thyristors is $\alpha = 80°$. Plot the supply voltage and input current, passive filter currents, injected voltage of the APF, output voltage and current with time, and the harmonic spectra of supply current and voltage at PCC. Compute THD, crest factor, rms value of AC mains current, conduction angle of thyristors, the rms output voltage, displacement factor, distortion factor, input power factor, and input active, reactive, and output powers.

18. Design and develop a dynamic model of a hybrid filter (an active series filter + a three-branch passive shunt filter) and simulate its behavior to reduce harmonic currents and to maintain almost UPF of a three-phase bidirectional star connected controller (back-to-back connected thyristors in series with the load) having a resistive load of $R = 20\,\Omega$ and fed from a line voltage of 415 V (rms) at 50 Hz. The delay angle of its thyristors is $\alpha = 50°$. Plot the supply voltage and input current, passive filter currents, injected voltage of the APF, output voltage and current with time, and the harmonic spectra of supply current and voltage at PCC. Compute THD, crest factor, rms value of AC mains current, conduction angle of thyristors, the rms output voltage, displacement factor, distortion factor, input power factor, and input active, reactive, and output powers.

19. Design and develop a dynamic model of a hybrid filter (an active series filter + a three-branch passive shunt filter) and simulate its behavior to reduce harmonic currents and to maintain almost UPF of a three-phase 415 V, 50 Hz supply feeding a 12-pulse diode bridge converter with a parallel capacitive DC filter of 4700 μF and a resistive load of 10 Ω. It has a source impedance of 0.15 Ω resistive element and 0.75 Ω inductive element. It consists of an ideal transformer with single primary star connected winding and two secondary windings connected in star and delta with same output line voltages as the input voltage at no load to provide 30° phase shift between two sets of three-phase output voltages. Two 6-pulse diode bridges are connected in series to provide this 12-pulse AC–DC converter. Plot the supply voltage and input current, passive filter currents, injected voltage of the APF, output voltage and current with time, and the harmonic spectra of supply current and voltage at PCC. Compute the fundamental active power drawn by the load, power factor, rms supply current, distortion factor, fundamental rms supply current, peak current of the AC mains, and total harmonic distortion of AC supply current.

20. Design and develop a dynamic model of a hybrid filter (an active filter (a VSI with an AC series inductor, a coupling transformer, and its DC bus connected to the DC bus of the load) in series with a three-branch passive shunt filter) and simulate its behavior to reduce harmonic currents and to maintain almost UPF of a three-phase AC supply of 415 V at 50 Hz feeding a three-phase 12-pulse diode rectifier (consisting of two parallel connected 6-pulse diode bridge rectifiers with star/delta and delta/delta transformers to output equal line voltages of 415 V) having a parallel DC capacitor filter of 2000 μF with a 10 Ω resistive load to reduce the harmonics in AC mains current and to maintain UPF of supply current. The DC bus of the load needs to be maintained at 600 V. It has a source impedance of 0.15 Ω resistive element and 0.75 Ω inductive element. Plot the supply voltage and input current, passive filter currents, injected voltage, input voltage of the rectifier transformer, output DC voltage and current with time, and the harmonic spectra of supply current and voltage at PCC. Compute THD, crest factor, rms value of AC mains current, displacement factor, distortion factor, power factor, input active, reactive, and output powers, rms line voltage at the input of diode rectifiers, rms voltage across the APF, VA rating of the APF, and turns ratio of the coupling transformer.

21. Design and develop a dynamic model of a hybrid filter (an active series filter + a three-branch passive shunt filter) and simulate its behavior to reduce harmonic currents and to maintain almost UPF of a three-phase 415 V, 50 Hz supply feeding a 12-pulse thyristor bridge converter with a series connected inductive load of 50 mH and an equivalent resistive load of 5 Ω. It has a source impedance of 0.1 Ω resistive element and 1.0 Ω inductive element and a 60° firing angle of its thyristors. It consists of an ideal transformer with single primary star connected winding and two secondary windings connected in star and delta with same line voltages as the input supply voltage to provide 30° phase shift between two sets of three-phase output voltages. Two 6-pulse thyristors bridges are connected in series to provide this 12-pulse AC–DC converter. Plot the supply voltage and input current, passive filter currents, injected voltage of the APF, output voltage and current with time, and the harmonic spectra of supply current and voltage at PCC. Compute fundamental active power drawn by the load, fundamental reactive power drawn by the load, power factor, rms supply current, distortion factor, fundamental rms supply current, peak current of the AC mains, total harmonic distortion of AC supply current, and voltage, current, and VA ratings of the APF and the passive filter.

22. Design and develop a dynamic model of a hybrid filter (an active series filter + a three-branch passive shunt filter) and simulate its behavior to reduce harmonic currents and to maintain almost UPF of a three-phase diode bridge used to convert 415 V AC to DC voltage. The load on its DC side is a parallel capacitive DC filter of 3300 μF and a resistive load of 5 Ω. Along with this rectifier, an AC inductive load of 25 kVAR is connected to the same AC bus. Plot the supply voltage and input current, passive filter currents, injected voltage of the APF, output voltage and current with time, and the harmonic spectra of supply current and voltage at PCC. Compute fundamental active power drawn by the load, fundamental reactive power drawn by the load, power factor, rms supply current, distortion factor, fundamental rms supply current, peak current of the AC mains, total harmonic distortion of AC supply current, and voltage, current, and VA ratings of the APF and the passive filter.

23. Design and develop a dynamic model of a hybrid filter (an active series filter + a three-branch passive shunt filter) and simulate its behavior to reduce harmonic currents and to maintain almost UPF of a three-phase three-wire 415 V, 50 Hz supply system, which is feeding a set of nonlinear loads consisting of a thyristor bridge and a diode rectifier connected in parallel. The diode bridge converter is feeding a parallel capacitive DC filter of 4700 μF and a resistive load of 10 Ω. The thyristor bridge converter is feeding a *RL* load of $R = 5\,\Omega$ and $L = 25\,\text{mH}$ at 60° firing angle of its thyristors. Plot the supply voltage and input current, passive filter currents, injected voltage of the APF, output voltage and current with time, and the harmonic spectra of supply current and voltage at PCC. Compute fundamental active power drawn by the load, fundamental reactive power drawn by the load, power factor, rms supply current, distortion factor, fundamental rms supply current, peak current of the AC mains, total harmonic distortion of AC supply current, and voltage, current, and VA ratings of the APF and the passive filter.

24. Design and develop a dynamic model of a hybrid filter (an active series filter + a three-branch passive shunt filter) and simulate its behavior to reduce harmonic currents and to maintain almost UPF of a three-phase three-wire 440 V, 50 Hz system. It has a linear load of 30 kVA at 0.8 lagging power factor and a thyristor bridge converter drawing 100 A constant DC current at 30° firing angle of its thyristors. Plot the supply voltage and input current, passive filter currents, injected voltage of the APF, output voltage and current with time, and the harmonic spectra of supply current and voltage at PCC. Compute fundamental active power drawn by the load, fundamental reactive power drawn by the load, power factor, rms supply current, distortion factor, fundamental rms supply current, peak current of the AC mains, total harmonic distortion of AC supply current, and voltage, current, and VA ratings of the APF and the passive filter.

25. Design and develop a dynamic model of a hybrid filter (an active series filter + a three-branch passive shunt filter) and simulate its behavior to reduce harmonic currents and to maintain almost UPF of a three-phase to three-phase cycloconverter (three-pulse type) supplying a *RL* series load circuit and sinusoidal modulation of the delay angle α is to be employed. The required output frequency is 15 Hz and the inductive reactance of the load circuit at this frequency is 2.5 Ω. The load circuit resistance is 7.5 Ω. A three-phase supply of 415 V at 50 Hz (line–line rms) is to be used directly without a transformer. If the ideal output voltage wave is defined as $V_0 = 156\,\sin(30\pi t)$ V, plot the supply voltage and input current, passive filter currents, injected voltage of the APF, output voltage and current with time, and the harmonic spectra of supply current and voltage at PCC. Compute

fundamental active power drawn by the load, fundamental reactive power drawn by the load, power factor, rms supply current, distortion factor, fundamental rms supply current, peak current of the AC mains, total harmonic distortion of AC supply current, and voltage, current, and VA ratings of the APF and the passive filter.

References

1. Stacy, E.J. and Strycula, E.C. (1974) Hybrid power filter employing both active and passive elements. US Patent 3,849,677.
2. Gyugyi, L. and Strycula, E. (1976) Active AC power filters. Proceedings of the IEEE IAS Annual Meeting, pp. 529–535.
3. Harashima, F., Inaba, H., and Tsuboi, K. (1976) A closed-loop control system for the reduction of reactive power required by electronic converters. *IEEE Transactions on Industrial Electronics and Control Instrumentation*, **23**(2), 162–166.
4. Epstein, E., Yair, A., and Alexandrovitz, A. (1979) Analysis of a reactive current source used to improve current drawn by static inverters. *IEEE Transactions on Industrial Electronics and Control Instrumentation*, **26**(3), 172–177.
5. Uceda, J., Aldana, F., and Martinez, P. (1983) Active filters for static power converters. *IEE Proceedings – Part B*, **130**(5), 347–354.
6. Rajagopalan, V., Jacob, A., Sévigny, A. *et al.* (1983) Harmonic currents compensation-scheme for electrical distribution systems. Proceedings of the IFAC on Control in Power Electronics and Electrical Drives, Lausanne, Switzerland, pp. 683–690.
7. Drouin, P.L.A., Sévigny, A., Jacob, A., and Rajagopalan, V. (1983) Studies on line current harmonics' compensation scheme suitable for an electric distribution system. *Canadian Electrical Engineering Journal*, **8**(4), 123–129.
8. Stacy, E.J. and Brennen, M.A. (1987) Active power conditioner system. US Patent 4,651,265.
9. Takeda, M., Ikeda, K., and Tominaga, Y. (1987) Harmonic current compensation with active filter. Proceedings of the IEEE IAS Annual Meeting, pp. 808–815.
10. Peng, F.Z., Akagi, H., and Nabae, A. (1988) A novel harmonic power filter. Proceedings of IEEE PESC'88, pp. 1151–1159.
11. Nakajima, A., Oku, K., Nishidai, J. *et al.* (1988) Development of active filter with series resonant circuit. Proceedings of IEEE PESC'88, pp. 1168–1173.
12. Kohata, M., Shiota, T., Watanabe, Y. *et al.* (1988) A novel compensation using static induction thyristors for reactive power and harmonics. Proceedings of the IEEE IAS Annual Meeting, pp. 843–849.
13. Moran, S. (1989) A line voltage regulator/conditioner for harmonic-sensitive load isolation. Proceedings of the IEEE IAS Annual Meeting, pp. 945–951.
14. Shimamura, T., Kurosawa, R., Hirano, M., and Uchino, H. (1989) Parallel operation of active and passive filters for variable speed cycloconverter drive systems. Proceedings of IEEE IECON'89, pp. 186–191.
15. Wong, C., Mohan, N., Wright, S.E., and Mortensen, K.N. (1989) Feasibility study of AC- and DC-side active filters for HVDC converter terminals. *IEEE Transactions on Power Delivery*, **4**(4), 2067–2075.
16. Peng, F.Z., Akagi, H., and Nabae, A. (1990) A new approach to harmonic compensation in power systems – a combined system of shunt passive and series active filters. *IEEE Transactions on Industry Applications*, **26**(6), 983–990.
17. Moran, S. (1990) Line voltage regulator. US Patent 4,950,916.
18. Bou-rabee, M., Chang, C.S., Sutanto, D., and Tam, K.S. (1991) Passive and active harmonic filters for industrial power systems. Proceedings of IEEE TENCON'91, pp. 222–226.
19. Fujita, H. and Akagi, H. (1991) Design strategy for the combined system of shunt passive and series active filters. Proceedings of IEEE PESC'91, pp. 898–903.
20. Williams, S.M. and Hoft, R.G. (1991) Adaptive frequency domain control of PWM switched power line conditioner. *IEEE Transactions on Power Electronics*, **6**(4), 665–670.
21. Tanaka, T. and Akagi, H. (1991) A new combined system of series active and shunt passive filters aiming at harmonic compensation for large capacity thyristor converters. Proceedings of IEEE IECON'91, pp. 723–728.
22. Fujita, H. and Akagi, H. (1991) A practical approach to harmonic compensation in power systems – series connection of passive and active filters. *IEEE Transactions on Industry Applications*, **27**(6), 1020–1025.
23. Banerjee, B.B., Pileggi, D., Atwood, D. *et al.* (1992) Design of an active series/passive parallel harmonic filter for ASD loads at a wastewater treatment plant. Proceedings of the IEEE PQA Conference, pp. 1–7.

24. Bocchetti, G., Carpita, M., Giannini, G., and Tenconi, S. (1993) Line filter for high power inverter locomotive using active circuit for harmonic reduction. Proceedings of the Fifth EPE, pp. 267–271.
25. Mochinaga, Y., Hisamizu, Y., Takeda, M. *et al.* (1993) Static power conditioner using GTO converters for AC electric railway. Proceedings of the IEEE Power Conversion Conference, Yokohama, pp. 641–646.
26. Peng, F.Z., Akagi, H., and Nabae, A. (1993) Compensation characteristics of the combined system of shunt passive and series active filters. *IEEE Transactions on Industry Applications*, **29**(1), 144–152.
27. Bhattacharya, S., Divan, D.M., and Banerjee, B.B. (1993) Control and reduction of terminal voltage total harmonic distortion (THD) in a hybrid series active and parallel passive filter system. Proceedings of IEEE PESC'93, pp. 779–786.
28. Balbo, N., Sella, D., Penzo, R. *et al.* (1993) Hybrid active filter for parallel harmonic compensation. Proceedings of EPE'93, pp. 133–138.
29. Le, J.N., Pereira, M., Renz, K., and Vaupel, G. (1994) Active damping of resonances in power systems. *IEEE Transactions on Power Delivery*, **9**(2), 1001–1008.
30. Czarnecki, L.S. (1994) Combined time-domain and frequency-domain approach to hybrid compensation in unbalanced nonsinusoidal systems. *European Transactions on Electrical Power Engineering*, **4**(6), 477–484.
31. Aliouane, K., Saadate, S., and Davat, B. (1994) Analytical study and numerical simulation of the static and dynamic performances of combined shunt passive and series active filters. Proceedings of the Fifth International Conference on Power Electronics and Variable-Speed Drives, pp. 147–151.
32. Brennen, M.A. and Moran, S.A. (1994) Active power line conditioner with low cost surge protection and fast overload recovery. US Patent 5,287,288.
33. Schauder, C.D. and Moran, S.A. (1994) Multiple reference frame controller for active filters and power line conditioners. US Patent 5,309,353.
34. Brennen, M.A. (1994) Series–parallel active power line conditioner utilizing reduced-turns-ratio transformer for enhanced peak voltage regulation capability. US Patent 5,319,534.
35. Brennen, M.A. (1994) Active power line conditioner having capability for rejection of common-mode disturbances. US Patent 5,319,535.
36. Brennen, M.A. (1994) Active power line conditioner utilizing harmonic frequency injection for improved peak voltage regulation. US Patent 5,349,517.
37. Brennen, M.A. and Moran, S.A. (1994) Active power line conditioner with a derived load current fundamental signal for fast dynamic response. US Patent 5,351,178.
38. Brennen, M.A. and Moran, S.A. (1994) Highly fault tolerant active power line conditioner. US Patent 5,351,180.
39. Brennen, M.A., Moran, S.A., and Gyugyi, L. (1994) Low cost active power line conditioner. US Patent 5,351,181.
40. Brennen, M.A. and Moran, S.A. (1994) Active power line conditioner with synchronous transformation control. US Patent 5,355,025.
41. Brennen, M. (1994) Series–parallel active power line conditioner utilizing temporary link energy boosting for enhanced peak voltage regulation capability. US Patent 5,355,295.
42. Krah, J.O. and Holtz, J. (1994) Total compensation of line side switching harmonics in converter fed AC locomotives. Proceedings of the IEEE IAS Annual Meeting, pp. 913–920.
43. Rastogi, M., Mohan, N., and Edris, A.A. (1995) Hybrid-active filtering of harmonic currents in power systems. *IEEE Transactions on Power Delivery*, **10**(4), 1994–2000.
44. Chang, C.S., Lock, K.S., Wang, F., and Liew, A.C. (1995) Harmonic level control schemes and evaluation methods in power system. Proceedings of the International Conference on Energy Management and Power Delivery (EMPD'95), vol. 1, pp. 146–151.
45. Ledwich, G. and Doulai, P. (1995) Multiple converter performance and active filtering. *IEEE Transactions on Power Electronics*, **10**(3), 273–279.
46. Kamran, F. and Habetler, T.G. (1995) Combined deadbeat control of a series–parallel converter combination used as a universal power filter. Proceedings of IEEE PESC'95, pp. 196–201.
47. Raju, N.R., Venkata, S.S., Kagalwala, R.A., and Sastry, V.V. (1995) An active power quality conditioner for reactive power and harmonics compensation. Proceedings of IEEE PESC'95, pp. 209–214.
48. Moran, L., Werlinger, P., Dixon, J., and Wallace, R. (1995) A series active power filter which compensates current harmonics and voltage unbalance simultaneously. Proceedings of IEEE PESC'95, pp. 222–227.
49. Kamran, F. and Habetler, T.G. (1995) A novel on-line UPS with universal filtering capabilities. Proceedings of IEEE PESC'95, pp. 500–506.
50. Lim, Y.C., Park, J.K., Jung, Y.G. *et al.* (1995) Development of a simulator for compensation performance evaluation of hybrid active power filter using three-dimensional current co-ordinate. Proceedings of the IEEE International Conference on Power Electronics and Drive Systems, pp. 427–432.

51. Moran, S.A. and Brennen, M.B. (1995) Active power line conditioner with fundamental negative sequence compensation. US Patent 5,384,696.
52. Bhattacharya, S. and Divan, D.M. (1995) Hybrid series active/parallel passive power line conditioner with controlled harmonic injection. US Patent 5,465,203.
53. Bhattacharya, S., Divan, D.M., and Banerjee, B. (1995) Active filter solutions for utility interface. Proceedings of IEEE ISIE'95, pp. 53–63.
54. Bhattacharya, S. and Divan, D. (1995) Design and implementation of a hybrid series active filter system. Proceedings of IEEE PESC'95, pp. 189–195.
55. Rastogi, M., Mohan, N., and Edris, A.A. (1995) Filtering of harmonic currents and damping of resonances in power systems with a hybrid-active filter. Proceedings of IEEE APEC'95, pp. 607–612.
56. Akagi, H. and Fujita, H. (1995) A new power line conditioner for harmonic compensation in power systems. *IEEE Transactions on Power Delivery*, **10**(3), 1570–1575.
57. Lin, C.E., Su, W.F., Lu, S.L. *et al.* (1995) Operation strategy of hybrid harmonic filter in demand-side system. Proceedings of the IEEE IAS Annual Meeting, pp. 1862–1866.
58. Bhattacharya, S. and Divan, D. (1995) Synchronous frame based controller implementation for a hybrid series active filter system. Proceedings of the IEEE IAS Annual Meeting, pp. 2531–2540.
59. Dixon, J., Venegas, G., and Moran, L. (1995) A series active power filter based on a sinusoidal current controlled voltage source inverter. Proceedings of IEEE IECON'95, pp. 639–644.
60. Kamath, G., Mohan, N., and Albertson, D. (1995) Hardware implementation of a novel reduced rating active filter for 3-phase, 4-wire loads. Proceedings of IEEE APEC'95, pp. 984–989.
61. Tolbert, L.M., Hollis, H.D., and Hale, P.S., Jr. (1996) Evaluation of harmonic suppression devices. Proceedings of the IEEE IAS Annual Meeting, pp. 2340–2347.
62. Rim, G.H., Kang, I., Kim, W.H., and Kim, J.S. (1996) A shunt hybrid active filter with two passive filters in tandem. Proceedings of IEEE APEC'96, pp. 361–366.
63. Bhattachaya, S. and Divan, D.M. (1996) Hybrid series active, parallel passive, power line conditioner for harmonic isolation between a supply and load. US Patent 5,513,090.
64. Mohan, N. and Rastogi, M. (1996) Hybrid filter for reducing distortion in power system. US Patent 5,548,165.
65. Zyl, A.V., Enslin, J.H.R., Steyn, W.H., and Spee, R. (1996) A new unified approach to power quality management. *IEEE Transactions on Power Electronics*, **11**(5), 691–697.
66. Libano, F.B., Simonetti, D.S.L., and Uceda, J. (1996) Frequency characteristics of hybrid filter systems. Proceedings of IEEE PESC'96, pp. 1142–1148.
67. Kurowski, T., Strzelecki, R., and Supronowicz, H. (1996) A new method of alternating current harmonic compensation in parallel systems of hybrid filters. Proceedings of IEEE ISIE'96, vol. 2, pp. 596–601.
68. Strzelecki, R., Benysek, G., and Frackowiak, L. (1996) Dynamic properties of hybrid filters in regenerative braking thyristor systems. Proceedings of IEEE ISIE'96, vol. 2, pp. 612–617.
69. Yoshioka, Y., Konishi, S., Eguchi, N., and Hino, K. (1996) Self-commutated static flicker compensator for arc furnaces. Proceedings of IEEE APEC'96, pp. 891–897.
70. Raju, N.R., Venkata, S.S., and Sastry, V.V. (1996) A decoupled series compensator for voltage regulation and harmonic compensation. Proceedings of IEEE PESC'96, pp. 527–531.
71. Akagi, H. (1996) New trends in active filters for power conditioning. *IEEE Transactions on Industry Applications*, **32**(6), 1312–1322.
72. Muthu, S. and Kim, J.M.S. (1997) Steady-state operating characteristics of unified active power filters. Proceedings of IEEE APEC'97, pp. 199–206.
73. Hafner, J., Aredes, M., and Huemann, K. (1997) A shunt active power filter applied to high voltage distribution lines. *IEEE Transactions on Power Delivery*, **12**(1), 266–272.
74. Mariscotti, A. (1997) Low frequency conducted disturbances compensation using a hybrid filter system. Proceedings of IEEE ISIE'97, vol. 2, pp. 405–410.
75. Segura, N., Iglcsias, R.S.J., Sanchez, P.B., and Perez, J.N. (1997) Experimental performance of passive and hybrid filters applied to AC/DC converters fed by a weak AC system. Proceedings of IEEE ISIE'97, vol. 2, pp. 600–605.
76. Bhattacharya, S., Cheng, P.T., and Divan, D.M. (1997) Hybrid solutions for improving passive filter performance in high power applications. *IEEE Transactions on Industry Applications*, **33**(3), 732–747.
77. Rujula, A.A.B., Arasanz, J.S., Badia, M.S., and Estopinan, A.L. (1997) Searching for the better topology and control strategy in hybrid power filters. Proceedings of EPE'97, pp. 4.825–4.830.
78. Libano, F., Cobos, J., and Uceda, J. (1997) Simplified control strategy for hybrid active filters. Proceedings of IEEE PESC'97, vol. 2, pp. 1102–1108.
79. Khositkasame, S. and Sangwongwanich, S. (1997) Design of harmonic current detector and stability analysis of a hybrid parallel active filter. Proceedings of the IEEE Power Conversion Conference, Nagaoka, pp. 181–186.

80. Pittorino, L.A., du Toit, J.A., and Enslin, J.H.R. (1997) Evaluation of converter topologies and controllers for power quality compensators under unbalanced conditions. Proceedings of IEEE PESC'97, pp. 1127–1133.
81. Jeon, S.-J. and Cho, G.-H. (1997) A series–parallel compensated uninterruptible power supply with sinusoidal input current and sinusoidal output voltage. Proceedings of IEEE PESC'97, pp. 297–303.
82. Maeda, T., Watanabe, T., Mechi, A. *et al.* (1997) A hybrid single-phase power active filter for high order harmonics compensation in converter-fed high speed trains. Proceedings of the Power Conversion Conference, Nagaoka, pp. 711–717.
83. Fujita, H. and Akagi, H. (1997) An approach to harmonic current-free AC/DC power conversion for large industrial loads: the integration of a series active filter with a double series diode rectifier. *IEEE Transactions on Industry Applications*, **33**(5), 1233–1240.
84. Su, W.F., Lin, C.E., and Huang, C.L. (1998) Hybrid filter application for power quality improvement. *Electric Power Systems Research*, **47**, 165–171.
85. Cheng, P.T., Bhattacharya, S., and Divan, D.M. (1998) Power line harmonic reduction by hybrid parallel active/passive filter system with square wave inverter and DC bus control. US Patent 5,731,965.
86. Aredes, M., Heumann, K., and Watanabe, E.H. (1998) An universal active power line conditioner. *IEEE Transactions on Power Delivery*, **13**(2), 545–551.
87. Cheng, P.T., Bhattacharya, S., and Divan, D.M. (1998) Hybrid parallel active/passive filter system with dynamically variable inductance. US Patent 5,757,099.
88. Cheng, P.T., Bhattacharya, S., and Diwan, D.M. (1998) Control of square-wave inverters in high-power hybrid active filter systems. *IEEE Transactions on Industry Applications*, **34**(3), 458–472.
89. Sitaram, I.M., Padiyar, K.R., and Ramanarayanan, V. (1998) Digital simulation of a hybrid active filter – an active filter in series with shunt passive filter. Proceedings of the IEEE PES Meeting on Power Quality, pp. 65–71.
90. Jung, G.H. and Cho, G.H. (1998) New power active filter with simple low cost structure without tuned filters. Proceedings of IEEE PESC'98, pp. 217–222.
91. Basic, D., Ramsden, V.S., and Muttick, P.K. (1998) Performance of combined power filters in harmonic compensation of high-power cycloconverter drives. Proceedings of the IEEE Conference on Power Electronics and Variable Speed Drives, Publication No. 456, pp. 674–679.
92. Venkta, S.S., Raju, N.R., Kagalwala, R.A., and Sastry, V.V. (1998) Active power conditioner for reactive and harmonic compensation having PWM and stepped wave inverters. US Patent 5,751,138.
93. Singh, B.N., Chandra, A., Al-Haddad, K., and Singh, B. (1998) Fuzzy control algorithm for universal active filter. Proceedings of the IEEE Power Quality Conference, pp. 73–80.
94. Huang, S.J., Wu, J.C., and Jou, H.L. (1998) Electric-power-quality improvement using parallel active-power conditioners. *IEE Proceedings – Generation, Transmission and Distribution*, **145**(5), 597–603.
95. Barbosa, P.G., Santisteban, J.A., and Watanabe, E.H. (1998) Shunt–series active power filter for rectifiers AC and DC sides. *IEE Proceedings – Electric Power Applications*, **145**(6), 577–584.
96. Su, W.F., Lin, C.E., and Huang, C.L. (1998) Hybrid filter application for power quality improvement. *Electric Power Systems Research*, **47**, 165–171.
97. Peng, F.Z., McKeever, J.W., and Adams, D.J. (1998) A power line conditioner using cascade multilevel inverters for distribution systems. *IEEE Transactions on Industry Applications*, **34**(6), 1293–1298.
98. Chen, G., Li, M., Zhongming, Y., and Zhaoming, Q. (1998) A hybrid solution to active power filter for the purpose of harmonic suppression and resonance damping. Proceedings of IEEE POWERCON'98, pp. 1542–1546.
99. Peng, F.Z. (1998) Application issues of active power filters. *IEEE Industry Applications Magazine*, **4**(5), 21–30.
100. Peng, F.Z. and Adams, D.J. (1999) Harmonic sources and filtering approaches – series/parallel, active/passive, and their combined power filters. Proceedings of the IEEE IAS Annual Meeting, pp. 448–455.
101. Koczara, W. and Dakyo, B. (1999) AC voltage hybrid filter. Proceedings of IEEE INTELEC'99, pp. 9 2.
102. Liu, J., Yang, J., and Wang, Z. (1999) A new approach for single-phase harmonic current detecting and its application in a hybrid active power filter. Proceedings of IEEE IECON'99, pp. 849–854.
103. Huang, S.J. and Wu, J.C. (1999) Design and operation of cascaded active power filters for the reduction of harmonic distortions in power system. *IEE Proceedings – Generation, Transmission and Distribution*, **146**(2), 193–199.
104. Tolbert, L.M., Peng, F.Z., and Habetler, T.G. (1999) A multilevel converter-based universal power conditioner. Proceedings of IEEE PESC'99, pp. 393–399.
105. Valouch, V. and Dolejskova, S. (1999) Evaluation of performance criteria of hybrid power filters. Proceedings of EPE'99, p. 9.
106. Bester, D.D., le Roux, A.D., du Mouton, T.H., and Enslin, J.H.R. (1999) Evaluation of power-ratings for active power quality compensators. Proceedings of EPE'99.

107. Alexa, D. (1999) Combined filtering system consisting of passive filter with capacitors in parallel with diodes and low-power inverter. *IEE Proceedings – Electric Power Applications*, **146**(1), 88–94.
108. Li, R., Johns, A.R., Elkateb, M.M., and Robinson, F.V.P. (1999) Comparative study of parallel hybrid filters in resonance damping. Proceedings of the IEEE International Conference on Electrical Power Engineering, Power Tech Budapest'99, p. 230.
109. Yonghai, X., Xiangning, X., and Lianguang, L. (1999) New approaches of low cost hybrid active filter. Proceedings of the IEEE International Conference on Electric Power Engineering, Power Tech Budapest'99, pp. 291–295.
110. Singh, B.N., Singh, B., Chandra, A., and Al-Haddad, K. (1999) Digital implementation of a new type of hybrid filter with simplified control strategy. Proceedings of IEEE APEC'99, vol. 1, pp. 642–648.
111. Al-Zamel, A.M. and Torrey, D.A. (1999) A three-phase hybrid series passive/shunt active filter system. Proceedings of IEEE APEC'99, pp. 875–881.
112. Duro, B., Ramsden, V.S., and Muttik, P. (1999) Minimization of active filter rating in high power hybrid filter systems. Proceedings of IEEE PEDS'99, July, vol. 2, pp. 1043–1048.
113. Singh, B.N., Singh, B., Chandra, A., and Al-Haddad, K. (1999) A new control scheme of series hybrid active filter. Proceedings of IEEE PESC'99, vol. 1, pp. 249–154.
114. Park, S., Sung, J.H., and Nam, K. (1999) A new parallel hybrid filter configuration minimizing active filter size. Proceedings of IEEE PESC'99, vol. 1, pp. 400–405.
115. Laren, E. and Delmerico, R. (1999) Hybrid active power filter with programmed impedance characteristics. US Patent 5,910,889.
116. Shahalami, S.H., Benchaita, L., and Saadate, S. (1999) A comparative study between two structures of hybrid active filter for harmonic compensation of a 18-thyristors cycloconverter fed induction motor drive. Proceedings of EPE'99.
117. Li, R., Elkateb, M.M., Jones, A.T., and Robinson, F.V. (1999) Performance of parallel hybrid filters in damping harmonic resonance. Proceedings of EPE'99, p. 10.
118. Ramirez, S., Visairo, N., Oliver, M. *et al.* (2000) Harmonic compensation in the AC mains by the use of current and voltage active filters controlled by a passivity-based law. Proceedings of the IEEE International Power Electronics Congress (CIEP'02), pp. 87–92.
119. Valouch, V. (2000) Compensation of harmonic voltage source by using parallel and series active filters. Proceedings of EPE-PEMC'00, Košice, pp. 4.20–4.25.
120. Fujita, H., Yamasaki, T., and Akagi, H. (2000) A hybrid active filter for damping of harmonic resonance in industrial power systems. *IEEE Transactions on Power Electronics*, **15**(2), 215–222.
121. Sung, J.H., Park, S., and Nam, K. (2000) New hybrid parallel active filter configuration minimising active filter size. *IEE Proceedings – Electric Power Applications*, **147**, 93–98.
122. Basic, D., Ramsden, V.S., and Muttick, P.K. (2000) Hybrid filter control system with adaptive filters for selective elimination of harmonics and interharmonics. *IEE Proceedings – Electric Power Applications*, **147**(3), 295–303.
123. Park, S., Han, S.B., Jung, B.M. *et al.* (2000) A current control scheme based on multiple synchronous reference frames for parallel hybrid active filter. Proceedings of the Third International Conference on Power Electronics and Motion Control (PIEMC'00), pp. 218–223.
124. Yuwen, B., Jiang, X., and Zhu, D. (2000) Study on the performance of the combined power filter with unsymmetrical condition. Proceedings of the Third International Conference on Power Electronics and Motion Control (PIEMC'00), pp. 365–370.
125. Guozhu, C., Lu, Z., and Zhaoming, Q. (2000) The design and implement of series hybrid active power filter for variable nonlinear loads. Proceedings of the Third International Conference on Power Electronics and Motion Control (PIEMC'00), pp. 1041–1044.
126. Karthik, K. and Quaicoe, J.E. (2000) Voltage compensation and harmonic suppression using series active and shunt passive filters. Proceedings of the IEEE Canadian Conference on Electrical and Computer Engineering, pp. 582–586.
127. Rivas, D., Moran, L., Dixon, J., and Espinoza, J. (2000) A simple control scheme for hybrid active power filter. Proceedings of IEEE PESC'00, pp. 991–996.
128. Basic, D., Ramsden, V.S., and Muttik, P.K. (2000) Selective compensation of cycloconverter harmonics and interharmonics by using a hybrid power filter system. Proceedings of IEEE PESC'00, vol. 3, pp. 1137–1142.
129. Akagi, H. (2000) Active and hybrid filters for power conditioning. Proceedings of IEEE ISIE'00, vol. 1, pp. TU26–TU36.
130. Tong, M., Feng, P., and Jiang, J. (2000) The design of a hybrid power filter based on variable structure control. Proceedings of the Third World Congress on Intelligent Control and Automation, pp. 2982–2986.
131. Senini, S. and Wolfs, P.J. (2000) Hybrid active filter for harmonically unbalanced three-phase three-wire railway traction loads. *IEEE Transactions on Power Electronics*, **15**(4), 702–710.

132. Cheng, P.T., Bhattacharya, S., and Divan, D.M. (2000) Operations of the dominant harmonic active filter (DHAF) under realistic utility conditions. Proceedings of IEEE IAS'00, pp. 2135–2142.
133. van Schoor, G., van Wyk, J.D., and Shaw, I.S. (2000) Modelling of a power network compensated by hybrid power filters with different distorted loads. Proceedings of IEEE-IHQPC'00, October, pp. 301–306.
134. Alexa, D., Rosu, E., Lazar, A. *et al.* (2000) Analysis of operation for the combined filtering system consisting of passive filter with capacitors in parallel with diode. Proceedings of PCIM'00.
135. Zhan, C., Wong, M., Wang, Z., and Han, Y. (2000) DSP control of power conditioner for improving power quality. Proceedings of the IEEE PES Meeting, pp. 2556–2561.
136. Tolbert, L.M., Peng, F.Z., and Habetler, T.G. (2000) A multilevel converter-based universal power conditioner. *IEEE Transactions on Industry Applications*, **36**(6), 596–603.
137. Singh, P., Pacas, J.M., and Bhatia, C.M. (2000) A novel 3-phase hybrid harmonic and reactive power compensator. Proceedings of EPE-PEMC'00, Košice, pp. 4.26–4.31.
138. Singh, B. and Verma, V. (2000) Modeling and control of series hybrid power filter with self supporting DC bus. Eleventh National Power System Conference, December 20–22, IISc, Bangalore, vol. 2, pp. 620–626.
139. Graovac, D., Katić, V., and Rufer, A. (2000) Universal power quality system – an extension to universal power quality conditioner. Proceedings of EPE-PEMC'00, Košice, pp. 4.32–4.38.
140. Graovac, D., Katic, V., and Rufer, A. (2000) Power quality compensation using universal power quality conditioning system. *IEEE Power Engineering Review*, **20**(12), 58–60.
141. Graovac, D., Kati, V., and Rufer, A. (2000) Solving supply and load imperfections using universal power quality conditioning system. Proceedings of PCIM'00.
142. Moran, L., Pastorini, I., Dixon, J., and Wallace, R. (2000) Series active power filter compensates current harmonics and voltage unbalance simultaneously. *IEE Proceedings – Generation, Transmission and Distribution*, **147**(1), 31–36.
143. da Silva, S.A.O., Donoso-Garcia, P.F., Cortizo, P.C., and Seixas, P.F. (2000) A comparative analysis of control algorithms for three-phase line-interactive UPS systems with series–parallel active power-line conditioning using SRF method. Proceedings of IEEE PESC'00, pp. 1023–1028.
144. Peng, F.Z. (2000) Power line conditioner using cascade multilevel inverters for voltage regulation, reactive power correction, and harmonic filtering. US Patent 6,075,350.
145. Sérgio, A., da Silva, O., Donso-Garcia, P. *et al.* (2000) Three-phase line-interactive UPS systems with series–parallel active power line conditioning for high power quality. Proceedings of EPE-PEMC'00, Košice, pp. 3.130–3.135.
146. Suzuki, S., Baba, J., Shutoh, K., and Masada, E. (2000) Application of unified power flow controller for power quality control. Proceedings of EPE-PEMC'00, Košice, pp. 4.165–4.170.
147. Prieto, J. and Salmerón, P.R. (2000) Control design of an active conditioner for three-phase load compensation. Proceedings of PCIM'00.
148. Hojo, M., Matsui, N., and Ohnishi, T. (2000) Instantaneous line voltage controlled harmonics compensator. Proceedings of IEEE IECON'00, pp. 754–759.
149. El Shatshat, R., Kazerani, M., and Salama, M.M.A. (2001) Multi converter approach to active power filtering using current source converters. *IEEE Transactions on Power Delivery*, **16**(1), 38–45.
150. le Roux, A.D. and Mouton, H.D.T. (2001) A series–shunt compensator with combined UPS operation. Proceedings of the IEEE International Symposium on Industrial Electronics (ISIE'01), pp. 2038–2043.
151. Basu., M., Das, S.P., and Dubey, G.K. (2001) Experimental investigation of performance of a single-phase UPQC for voltage sensitive and non-linear loads. Proceedings of IEEE PEDS'01, pp. 218–222.
152. Lin, B.R. and Yang, B.R. (2001) Current harmonics elimination with a series hybrid active filter. Proceedings of the IEEE International Symposium (ISIE'01), pp. 566–570.
153. Zhao, C., Li, G., Chen, Z., and Li, G. (2001) Design and realization of a new hybrid power filter system used in single-phase circuit. Proceedings of IEEE IECON'01, pp. 1067–1071.
154. Chae, B.-S., Lee, W.-C., Lee, T.-K., and Hyun, D.-S. (2001) A fault protection scheme for unified power quality conditioners. Proceedings of IEEE-PEDS'01, pp. 66–71.
155. Kandil, T.A. and Quaicoe, J.E. (2001) A new approach to voltage and harmonic compensation. Proceedings of the IEEE Canadian Conference on Electrical and Computer Engineering, pp. 747–752.
156. Cao, R., Zhao, J., Shi, W. *et al.* (2001) Series power quality compensator for voltage sags, swells, harmonics and unbalance. Proceedings of the IEEE PES Conference and Exposition on Transmission and Distribution, pp. 543–547.
157. Chen, S. and Joos, G. (2001) A unified series–parallel deadbeat control technique for an active power quality conditioner with full digital implementation. Proceedings of the IEEE IAS Annual Meeting, pp. 172–178.
158. Graovac, D., Rufer, A., Katic, V., and Knezevic, J. (2001) Unified power quality conditioner based on current source converter topology. Proceedings of EPE'01.

159. Singh, P., Pacas, J.M., and Bhatia, C.M. (2001) A novel 3-phase hybrid harmonic and reactive power compensator. *EPE Journal*, **11**(4), 14–19.
160. Jou, H.-L., Wu, J.C., and Wu, K.D. (2001) Parallel operation of passive power filter and hybrid power filter for harmonic suppression. *IEE Proceedings –Generation, Transmission and Distribution*, **148**(1), 8–14.
161. Al-Zamil, A.M. and Torrey, D.A. (2001) A passive series, active shunt filter for high power applications. *IEEE Transactions on Power Electronics*, **16**(1), 101–109.
162. Guozhu, C., Zhengyu, L., and Zhaoming, Q. (2001) The special design considerations for series hybrid active power filter. Proceedings of IEEE PESC'01, pp. 560–564.
163. Chen, L. and von Jouanne, A. (2001) A comparison and assessment of hybrid filter topologies and control algorithms. Proceedings of IEEE PESC'01, vol. 2, pp. 565–570.
164. Guozhu, C., Zhengyu, L., and Zhaoming, Q. (2001) A novel hybrid active power filter with two passive channels for high power application. Proceedings of IEEE PESC'01, pp. 1889–1892.
165. Tnani, S., Bosche, J., Gaubert, J.P., and Champenois, G. (2001) Sliding mode control of parallel hybrid filters. Proceedings of EPE'01, p. 9.
166. Zhuo, F., Li, H., Li, H., and Wang, Z. (2001) Main circuit consists of multiplex use for active power filter. Proceedings of the International Conference on Electrical Machines and Systems (ICEMS'01), pp. 504–507.
167. Jacobs, J., Fischer, A., Detjen, D., and De Doncker, R.W. (2001) An optimized hybrid power filter and VAR compensator. Proceedings of the IEEE IAS Meeting, vol. 4, pp. 2412–2418.
168. Braga, R.A.M., Libano, F.B., and Lemos, F.A.B. (2001) Development environment for control strategies of hybrid active power filters using Matlab® and dSpace® DSP. Proceedings of the IEEE Power Tech Conference, pp. 6–12.
169. Detjen, D., Jacobs, J., De Doncker, R.W., and Mall, H.G. (2001) A new hybrid filter to dampen resonances and compensate harmonic currents in industrial power systems with power factor correction equipment. *IEEE Transactions on Power Electronics*, **16**(6), 821–827.
170. Chae, B.-S., Lee, W.-C., Hyun, D.-S., and Lee, T.-K. (2001) An overcurrent protection scheme for series active compensators. Proceedings of IEEE-IECON'01, pp. 1509–1514.
171. Basic, D., Ramsden, V.S., and Muttik, P.K. (2001) Harmonic filtering of high-power 12-pulse rectifier loads with a selective hybrid filter system. *IEEE Transactions on Industrial Electronics*, **48**(6), 1118–1127.
172. Joep, J., Dirk, D., and Rik, D.D. (2001) A new hybrid filter versus a shunt active power filter. Proceedings of EPE'01, p. 11.
173. Prieto, J., Salmerón, P., and Vázquez, J.R. (2001) Control implementation of a three-phase load compensation active conditioner. Proceedings of EPE'01, p. 10.
174. Alexa, D., Lazar, A., Rosu, E. *et al.* (2001) A new efficient filtering system having passive filters with capacitors in parallel with diodes for large rated harmonic currents. Proceedings of PCIM'01.
175. Peng, F.Z. (2001) Harmonic sources and filtering approaches. *IEEE Industry Applications Magazine*, **7**(4), 18–25.
176. Wang, Y., Yang, J., Wang, Z. *et al.* (2002) Rating analysis and design of coupling transformer for single-phase parallel hybrid active filter. Proceedings of IEEE PESC'02, pp. 602–606.
177. Lin, B.R., Huang, C.H., and Yang, B.R. (2002) Control scheme of hybrid active filter for power quality improvement. Proceedings of the IEEE International Conference on Industrial Technology (ICIT'02), pp. 317–322.
178. Lin, B.-R., Yang, B.-R., and Tsai, H.-R. (2002) Analysis and operation of hybrid active filter for harmonic elimination. *Electric Power Systems Research*, **62**(3), 191–200.
179. Senini, S.T. and Wolfs, P.J. (2002) Systematic identification and review of hybrid active filter topologies. Proceedings of IEEE PESC'02, pp. 394–399.
180. Kim, S. and Enjeti, P. (2002) A new hybrid active power filter (APF) topology. *IEEE Transactions on Power Electronics*, **17**(1), 48–54.
181. Morimoto, H., Ando, M., Mochinaga, Y. *et al.* (2002) Development of railway static power conditioner used at substation for Shinkansen. Proceedings of the IEEE Power Conversion Conference, Osaka, pp. 1108–1111.
182. Rodriguez, P., Pindado, R., and Bergas, J. (2002) Alternative topology for three-phase four-wire PWM converters applied to a shunt active power filter. Proceedings of IEEE IECON'02, pp. 2939–2944.
183. Deng, Z.F., Jiang, X.J., and Zhu, D.Q. (2002) A novel hybrid filter to cancel the neutral harmonic current. Proceedings of the IEEE IAS Meeting, vol. 1, pp. 59–63.
184. Prieto, J., Saimeron, P., Vazquez, J.R., and Alcantara, J. (2002) A series–parallel configuration of active power filters for VAR and harmonic compensation. Proceedings of IEEE IECON'02, pp. 2945–2950.
185. da Silva, S.A.O., Donoso-Garcia, P.F., Cortizo, P.C., and Seixas, P.F. (2002) A three-phase line-interactive UPS system implementation with series–parallel active power-line conditioning capabilities. *IEEE Transactions on Industry Applications*, **38**(6), 1581–1590.

186. Chiang, S.J., Ai, W.J., and Lin, F.J. (2002) Parallel operation of capacity-limited three-phase four-wire active power filters. *IEE Proceedings – Electric Power Applications*, **149**(5), 329–336.
187. Lin, B.R., Yang, B.R., and Hung, T.L. (2002) Implementation of a hybrid series active filter for harmonic current and voltage compensations. Proceedings of the IEEE International Conference on Power Electronics, Machines and Drives, pp. 598–603.
188. Escobar, G., Stankovic, A.M., Cardenas, V., and Mattavelli, P. (2002) A controller based on resonant filters for a series active filter used to compensate current harmonics and voltage unbalance. Proceedings of the IEEE International Conference on Control Applications, pp. 7–12.
189. Sadek, K. and Pereira, M. (2002) Hybrid filter for an alternating current network. US Patent 6,385,063B1.
190. Rechka, S., Ngandui, T., Jianhong, X., and Sicard, P. (2002) A comparative study of harmonic detection algorithms for active filters and hybrid active filters. Proceedings of IEEE PESC'02, pp. 357–363.
191. Guozhu, C., Zhengyu, L., Zhaoming, Q., and Peng, F.Z. (2002) A new serial hybrid active power filter using controllable current source. Proceedings of IEEE PESC'02, pp. 364–368.
192. Rivas, D., Moran, L., Dixon, J., and Espinoza, J. (2002) A simple control scheme for hybrid active power filter. *IEE Proceedings – Generation, Transmission and Distribution*, **149**(4), 485–490.
193. Darwish, M.K., El-Habrouk, M., and Kasikci, I. (2002) EMC compliant harmonic and reactive power compensation using passive filter cascaded with shunt active filter. *EPE Journal*, **12**(3), 43–50.
194. Senini, S. and Wolfs, P.J. (2002) Analysis and design of a multiple-loop control system for a hybrid active filter. *IEEE Transactions on Industrial Electronics*, **49**(6), 1283–1292.
195. Sun, Z., Jiang, X., and Zhu, D. (2002) Study of novel traction substation hybrid power quality compensator. Proceedings of IEEE POWERCON'02, pp. 480–484.
196. Lin, B.R., Yang, B.R., and Tsai, H.R. (2002) Analysis and operation of hybrid active filter for harmonic elimination. *Electric Power Systems Research*, **62**, 191–200.
197. Jacobs, J., Detjen, D., and De Doncker, R. (2002) A new hybrid filter versus a shunt active power filter. Proceedings of EPE'02, Aachen.
198. Rahmani, S., Al-Haddad, K., and Fnaiech, F. (2002) A series hybrid power filter to compensate harmonic currents and voltages. Proceedings of IEEE IECON'02, pp. 644–649.
199. Rahmani, S., Al-Haddad, K., and Fnaiech, F. (2002) A new PWM control technique applied to three-phase shunt hybrid power filter. Proceedings of IEEE IECON'02, pp. 727–732.
200. Rahmani, S., Al-Haddad, K., and Fnaiech, F. (2002) A hybrid structure of series active and passive filters to achieving power quality criteria. Proceedings of the IEEE International Conference on Systems, Man and Cybernetics, pp. 1–6.
201. Singh, B.N. (2002) Implementation of a hybrid filter with a potential application to adjustable speed compressor drives for air quality control. *Journal of Electric Power Components & Systems (EPCS)*, **30**(11), 1091–1126.
202. Rivas, D., Moran, L., Dixon, J., and Espinoza, J. (2003) Improving passive filter compensation performance with active techniques. *IEEE Transactions on Industrial Electronics*, **50**(1), 161–170.
203. Ba, A.O. and Barry, A.O. (2003) Active filter analysis by the harmonic impedance compensation method. Part II. Proceedings of the IEEE Canadian Conference on Electrical and Computer Engineering.
204. Mendalek, N., Al-Haddad, K., Dessaint, L.-A., and Casoria, S. (2003) A new regulation algorithm applied to a hybrid power filter. Proceedings of the IEEE Canadian Conference on Electrical and Computer Engineering.
205. Srianthumrong, S. and Akagi, H. (2003) A medium-voltage transformerless AC/DC power conversion system consisting of a diode rectifier and a shunt hybrid filter. *IEEE Transactions on Industry Applications*, **39**(3), 874–882.
206. van Schoor, G., van Wyk, J.D., and Shaw, I.S. (2003) Training and optimization of an artificial neural network controlling a hybrid power filter. *IEEE Transactions on Industrial Electronics*, **50**(3), 546–553.
207. Barrero, F., Martinez, S., Yeves, F. *et al.* (2003) Universal and reconfigurable to UPS active power filter for line conditioning. *IEEE Transactions on Power Delivery*, **18**(1), 283–290.
208. Chiang, S.J. (2003) A three-phase four-wire power conditioner with load-dependent voltage regulation for energy saving. Proceedings of IEEE APEC'03, pp. 159–164.
209. Jintakosonwit, P., Fujita, H., Akagi, H., and Ogasawara, S. (2003) Implementation and performance of cooperative control of shunt active filters for harmonic damping throughout a power distribution system. *IEEE Transactions on Industry Applications*, **39**(2), 556–564.
210. Monteiro, L.F.C., Aredes, M., and Moor Neto, J.A. (2003) A control strategy for unified power quality conditioner. Proceedings of IEEE ISIE'003.
211. Sannino, A., Stevenson, J., and Larsson, T. (2003) Power-electronic solutions to power quality problems. *Electric Power Systems Research*, **66**, 71–82.

212. Bakhshai, A.R., Karimi, H., and Saeedifard, M. (2003) A new adaptive harmonic extraction scheme for single-phase active power filters. Proceedings of IEEE International Symposium on Circuits and Systems (ISCAS'03), pp. 268–271.
213. Singh, B. and Verma, V. (2003) Control of hybrid filter with self-supporting DC bus. *Journal of the Institution of Engineers (India)*, **83**, 307–312.
214. Srianthumrong, S. and Akagi, H. (2003) A medium-voltage transformerless AC/DC power conversion system consisting of a diode rectifier and a shunt hybrid filter. *IEEE Transactions on Industry Applications*, **39**(3), 874–882.
215. Akagi, H., Srianthumrong, S., and Tamai, Y. (2003) Comparisons in circuit configuration and filtering performance between hybrid and pure shunt active filters. Proceedings of the IEEE IAS Annual Meeting, pp. 1195–1202.
216. Singh, B. and Verma, V. (2004) Hybrid of tandem connected series active and series passive filters for varying rectifier loads. Proceedings of the 13th National Power Systems Conference (NPSC'04), December 27–30, IIT Madras, vol. II, pp. 929–935.
217. Singh, B., Verma, V., Chandra, A., and Al-Haddad, K. (2005) Hybrid filters for power quality improvement. *IEE Proceedings – Generation, Transmission and Distribution*, **152**(3), 365–378.
218. Singh, B., Verma, V., and Garg, V. (2005) Passive hybrid filter for varying rectifier loads. Proceedings of the IEEE Conference on Power Electronics and Drive Systems (PEDS'05), November 28–December 1, Kuala Lumpur, Malaysia, vol. 2, pp. 1306–1311.
219. Akagi, H. (2005) Active harmonic filters. *Proceedings of the IEEE*, **93**(12), 2128–2141.
220. Singh, B. and Verma, V. (2006) An indirect current control of hybrid power filter for varying loads. *IEEE Transactions on Power Delivery*, **21**(1), 178–183.
221. Singh, B. and Verma, V. (2006) Indirect current control of series hybrid filter: an experimental study. Proceedings of the IEEE International Symposium on Industrial Electronics (ISIE-2006), July 9–12, Montréal, Québec, Canada, pp. 1364–1369.
222. Singh, B. and Verma, V. (2007) An improved hybrid filter for compensation of current and voltage harmonics for varying rectifier loads. *International Journal of Electrical Power & Energy Systems*, **29**(4), 312–321.
223. Verma, V. and Singh, B. (2009) Design and implementation of a current controlled parallel hybrid power filter. *IEEE Transactions on Industry Applications*, **45**(5), 1910–1917.
224. Nastran, J., Cajhen, R., Seliger, M., and Jereb, P. (1994) Active power filter for non-linear AC loads. *IEEE Transactions on Power Electronics*, **9**(1), 92–96.

Index

Power Quality Problems and Mitigation Techniques, First Edition.
Bhim Singh, Ambrish Chandra and Kamal Al-Haddad.

Printed in the United States
By Bookmasters